SOLUTIONS MANUAL

Lucio Gelmini • Robert W. Hilts

Grant MacEwan College

Robert K. Wismer

Millersville University

GENERAL CHEMISTRY

Principles and Modern Applications

Petrucci • Harwood • Herring

Eighth Edition

Prentice
Hall

Upper Saddle River, NJ 07458

Editor in Chief: John Challice
Project Manager: Kristen Kaiser
Executive Managing Editor: Kathleen Schiaparelli
Assistant Managing Editor: Dinah Thong
Production Editor: Veronica Malone
Supplement Cover Manager: Paul Gourhan
Supplement Cover Designer: PM Workshop Inc.
Manufacturing Buyer: Ilene Kahn
*Photo Credits: Martin Bond/Science Photo Library/Photo Researchers - Richard
Megna/Fundamental Photographs - Ed Degginger/Color Pic, Inc - Chemical Design/Science
Photo Library/Photo Researchers - Oxford Molecular Biophysics Laboratory/Science Photo
Library/Photo Reseachers - David Parker/IMI/Univ of Birmingham/Science Photo
Library/Photo Researchers*

© 2002 by Prentice-Hall, Inc.
Upper Saddle River, NJ 07458

Printed in the United States of America

10 9 8 7 6 5 4 3 2 1

ISBN   0-13-017683-4

Prentice-Hall International (UK) Limited, London
Prentice-Hall of Australia Pty. Limited, Sydney
Prentice-Hall Canada, Inc., Toronto
Prentice-Hall Hispanoamericana, S.A., Mexico City
Prentice-Hall of India Private Limited, New Delhi
Pearson Education Asia Pte. Ltd., Singapore
Prentice-Hall of Japan, Inc., Tokyo
Editora Prentice-Hall do Brazil, Ltda., Rio de Janeiro

CONTENTS

DEDICATION

This manual is dedicated to the memory of Dr. Grant MacEwan (1902-2000), our college namesake, whose outstanding lifetime achievements and selfless service to the people of Western Canada, continue to be an inspiration to young and old alike.

FOREWORD

In this manual you will find the solutions to all of the questions at the end of each chapter in the companion textbook, General Chemistry: Principles and Modern Applications, by Ralph H. Petrucci, William S. Harwood and F. Geoffrey Herring, 8th Edition, Prentice-Hall, Upper Saddle River, NJ., with the exception of the Integrative and Advanced Exercises. The answers to all of the Integrative and Advanced Exercises can be found in the Instructors Resource Manual.

July 2001

Robert K. Wismer
Millersville University
Millersville, Pennsylvannia

Lucio Gelmini
Grant MacEwan College
Edmonton, Alberta, Canada

Robert W. Hilts
Grant MacEwan College
Edmonton, Alberta, Canada

ACKNOWLEDGEMENTS

In the process of writing this manual, we have incurred many debts of gratitude. The deepest, without question, are to our wives, Sanny Chan and Cornelia Bica, not only for tolerating with good cheer our extended absences from the home during the twelve-month period over which the manual was prepared, but also for carefully poring over the chapters and finding errors that had eluded the authors and all of the other accuracy checkers. Our wives, above all others, are glad to see the ordeal end and normal family life return.

We must commend Dr. David Shinn for the tremendous effort he put into thoroughly checking each and every solution in the manual. Put simply , David Shinn's comments and corrections were invaluable to us.

Also lending enormous assistance was Dr. Ralph Petrucci. During the course of the project, we asked Dr. Petrucci innumerable questions, some of a highly complex nature, and, each and every time, he expeditiously fired us back a comprehensive answer. Hats off to you Ralph for a job well done!

Some help with proofreading and keying in corrections was provided by Dr. Dusan Ristic-Petrovic. We thank Dr. Geoff Herring for providing us with a solution to Question 87 in Chapter 12. We are grateful to Kristen Kaiser and John Chalice at Prentice Hall for their help and encouragement throughout the enterprise.

Any errors or misinterpretations contained in this manual are solely our responsibility. We have done our best to make this manual error free, but because of its great size and complexity, some errors undoubtedly have gone undetected. Any errors that come to light will, of course, be dealt with during the next round of corrections.

Finally, we would like to extend a very special thanks to the administration of Grant MacEwan College for allowing us to use the college's in-house computing and printing facilities to prepare this manual.

Robert W. Hilts   July 2001
Lucio Gelmini   Edmonton, Alberta, Canada

CHAPTER 1
MATTER—ITS PROPERTIES AND MEASUREMENT
PRACTICE EXAMPLES

1A   Convert the Fahrenheit temperature to Celsius and compare.

$^\circ C = \left(^\circ F - 32\right)\tfrac{5}{9} = \left(103^\circ F - 32\right)\tfrac{5}{9} = 39.4^\circ C$. The temperature of $41^\circ C$ in New Delhi is higher than the predicted high of $103^\circ F = 39.4^\circ C$ in Phoenix.

1B   We convert the Fahrenheit temperature to Celsius. $^\circ C = \left(^\circ F - 32\right)\tfrac{5}{9} = \left(-15^\circ F - 32\right)\tfrac{5}{9} = -26.1^\circ C$. The antifreeze only protects to $-22^\circ C$ and will not offer protection to temperatures as low as $-15^\circ F = -26.1^\circ C$.

2A   The mass is the difference between the mass of the full and empty flask.

$$\text{density} = \frac{291.4 \text{ g} - 108.6 \text{ g}}{125 \text{ mL}} = 1.46 \text{ g / mL}$$

2B   The volume of the stone is the difference between the level in the graduated cylinder with the stone present and with it absent.

$$\text{density} = \frac{\text{mass}}{\text{volume}} = \frac{28.4 \text{ g rock}}{44.1 \text{ mL rock \& water} - 33.8 \text{ mL water}} = 2.76 \text{ g / mL}$$

3A   Use density as a conversion factor.

$$\text{solution mass} = 125 \text{ mL soln} \times \frac{1.081 \text{ g soln}}{1 \text{ mL soln}} = 135 \text{ g soln.}$$

3B $\text{ethanol volume} = 50.0 \text{ kg ethanol} \times \dfrac{1000 \text{ g}}{1 \text{ kg}} \times \dfrac{1 \text{ mL ethanol}}{0.789 \text{ g ethanol}} \times \dfrac{1 \text{L}}{1000 \text{ mL}} = 63.4 \text{ L ethanol}$

4A   We first use the density to determine the mass of gasohol.

$\text{ethanol mass} = 25 \text{ L gasohol} \times \dfrac{1000 \text{ mL}}{1 \text{ L}} \times \dfrac{0.71 \text{ g gasohol}}{1 \text{ mL gasohol}} \times \dfrac{10 \text{ g ethanol}}{100 \text{ g gasohol}} \times \dfrac{1 \text{ kg ethanol}}{1000 \text{ g ethanol}}$

$= 1.8 \text{ kg ethanol}$

4B   We use the mass percent to determine the mass of the 25.0-mL sample.

$\text{rubbing alcohol mass} = 15.0 \text{ g (2-propanol)} \times \dfrac{100.0 \text{ g rubbing alcohol}}{70.0 \text{ g (2-propanol)}} = 21.4 \text{ g rubbing alcohol}$

$\text{rubbing alcohol density} = \dfrac{21.4 \text{ g}}{25.0 \text{ mL}} = 0.856 \text{ g/mL}$

5A   The factor 0.00456 has three significant figures. $\dfrac{62.356}{0.000456 \times 6.422 \times 10^3} = 21.3$

5B   The factor 1.3×10^{-3} determines the number of significant figures.

$$\frac{8.21 \times 10^4 \times 1.3 \times 10^{-3}}{0.00236 \times 4.071 \times 10^{-2}} = 1.1 \times 10^6$$

6A   The last term has one digit to the right of the decimal. $0.236 + 128.55 - 102.1 = 26.7$

6B   This is easier to visualize if the numbers are not in scientific notation.

$$\frac{(1.302 \times 10^3) + 952.7}{(1.57 \times 10^2) - 12.22} = \frac{1302 + 952.7}{157 - 12.22} = \frac{2255}{145} = 15.6$$

REVIEW QUESTIONS

1. **(a)** A $\mathbf{m}^3$ is a cubic meter. As a regular solid, it is a cube one meter on a side, in volume very crudely equal to one cubic yard.

   (b) **% by mass** is read "percent by mass." It is the mass in grams of a substance present in precisely 100 grams of the sample it is found in.

   (c) $°\mathbf{C}$ is the temperature of a substance expressed on a scale (the Celsius scale) where the freezing point of water has a value of "zero" and the boiling point of water has a value of "one hundred."

   (d) **Density** is the concentration of the mass of a material. It is calculated as the mass of the material (in grams) divided by its volume (in mL or cm^3).

   (e) An **element** is a substance that cannot be altered or decomposed chemically. Each element has a definite name and a specific position on the periodic table.

2. **(a)** The seven **SI base units** are those from which all other units are derived. Among them are the meter for length, the kilogram for mass, the kelvin for temperature, the second for time, and the mole for amount of substance.

   (b) **Significant figures** are those digits in a number that are the result of experimental measurement, or are derived from such a measurement.

   (c) A **natural law** is a summary of experimental results or observations, often expressed in mathematical terms.

3. **(a)** The **mass** of an object is a measure of the amount of material in that object. Its **weight**, on the other hand, is the force that the object exerts due to gravitational attraction.

   (b) An **extensive property** is one that depends on the quantity of material present; an **intensive** property is like a property; it does not depend on the quantity of material present.

(c)   A **substance** is a pure form of matter; it is either an element or a compound. A **mixture** is a blend of two or more substances, in no particular proportion.

(d)   **Precision** refers to the reproducibility of an experimental measurement; **accuracy** describes the agreement between the measurement and the accepted value of the same property.

(e)   A **hypothesis** is a tentative explanation of a natural law. A **theory** is a hypothesis that has survived the test of repeated experiments.

4.   **(a)** $1.55 \text{ kg} \times \dfrac{1000 \text{ g}}{1 \text{ kg}} = 1.55 \times 10^3 \text{ g}$ **(b)** $642 \text{ g} \times \dfrac{1 \text{ kg}}{1000 \text{ g}} = 0.642 \text{ kg}$

   (c) $2896 \text{ mm} \times \dfrac{1 \text{ cm}}{10 \text{ mm}} = 289.6 \text{ cm}$ **(d)** $0.086 \text{ cm} \times \dfrac{10 \text{ mm}}{1 \text{ cm}} = 0.86 \text{ mm}$

5.   **(a)** $0.127 \text{ L} \times \dfrac{1000 \text{ mL}}{1 \text{ L}} = 127 \text{ mL}$ **(b)** $15.8 \text{ mL} \times \dfrac{1 \text{ L}}{1000 \text{ mL}} = 0.0158 \text{ L}$

   (c) $981 \text{ cm}^3 \times \dfrac{1 \text{ L}}{1000 \text{ cm}^3} = 0.981 \text{ L}$ **(d)** $2.65 \text{ m}^3 \times \left(\dfrac{100 \text{ cm}}{1 \text{ m}}\right)^3 = 2.65 \times 10^6 \text{ cm}^3$

6.   **(a)** $68.4 \text{ in.} \times \dfrac{2.54 \text{ cm}}{1 \text{ in.}} = 174 \text{ cm}$ **(b)** $94 \text{ ft} \times \dfrac{12 \text{ in.}}{1 \text{ ft}} \times \dfrac{2.54 \text{ cm}}{1 \text{ in.}} \times \dfrac{1 \text{ m}}{100 \text{ cm}} = 29 \text{ m}$

   (c) $1.42 \text{ lb} \times \dfrac{453.6 \text{ g}}{1 \text{ lb}} = 644 \text{ g}$ **(d)** $248 \text{ lb} \times \dfrac{0.4536 \text{ kg}}{1 \text{ lb}} = 112 \text{ kg}$

   (e) $1.85 \text{ gal} \times \dfrac{4 \text{ qt}}{1 \text{ gal}} \times \dfrac{0.9464 \text{ L}}{1 \text{ qt}} = 7.00 \text{ L}$ **(f)** $3.72 \text{ qt} \times \dfrac{0.9464 \text{ L}}{1 \text{ qt}} \times \dfrac{1000 \text{ mL}}{1 \text{ L}} = 3.52 \times 10^3 \text{ mL}$

7.   **(a)** $1.00 \text{ km}^2 \times \left(\dfrac{1000 \text{ m}}{1 \text{ km}}\right)^2 = 1.00 \times 10^6 \text{ m}^2$

   (b) $1.00 \text{ m}^2 \times \left(\dfrac{100 \text{ cm}}{1 \text{ m}}\right)^2 = 1.00 \times 10^4 \text{ cm}^2$

   (c) $1.00 \text{ mi}^2 \times \left(\dfrac{5280 \text{ ft}}{1 \text{ mi}} \times \dfrac{12 \text{ in.}}{1 \text{ ft}} \times \dfrac{2.54 \text{ cm}}{1 \text{ in.}} \times \dfrac{1 \text{ m}}{100 \text{ cm}}\right)^2 = 2.59 \times 10^6 \text{ m}^2$

8.   The boiling point of water can serve as our reference. $204°\text{F}$ is below the $212°\text{F}$ boiling point of water, while $102°\text{C}$ is above the $100°\text{C}$ boiling point of water. Thus, $102°\text{C}$ is the higher temperature.

9. The 80.0 g ethanol seems least massive. The 100.0 mL of benzene, with a density less than 1 g/mL, must have a mass less than 100.0 g (it is actually 87 g). On the other hand, 90.0 mL of carbon disulfide, with a density of 1.26 g/mL, should have a mass somewhat in excess of 100.0 g (it is actually 113 g). Thus, 90.0 mL of carbon disulfide is the most massive.

10. $\text{Butyric acid density} = \dfrac{\text{mass}}{\text{volume}} = \dfrac{2088 \text{ g}}{2.18 \text{ L}} \times \dfrac{1 \text{ L}}{1000 \text{ mL}} = 0.958 \text{ g / mL}$

11. $\text{Mercury density} = \dfrac{\text{mass}}{\text{volume}} = \dfrac{5.23 \text{ kg}}{385 \text{ mL}} \times \dfrac{1000 \text{ g}}{1 \text{ kg}} = 13.6 \text{ g / mL}$

12. **(a)** $\text{mass} = 452 \text{ mL} \times \dfrac{1.11 \text{ g}}{1 \text{ mL}} = 502 \text{ g ethylene glycol}$

(b) $\text{mass} = 18.6 \text{ L} \times \dfrac{1000 \text{ mL}}{1 \text{ L}} \times \dfrac{1.11 \text{ g}}{1 \text{ mL}} \times \dfrac{1 \text{ kg}}{1000 \text{ g}} = 20.6 \text{ kg ethylene glycol}$

(c) $\text{volume} = 65.0 \text{ g} \times \dfrac{1 \text{ mL}}{1.11 \text{ g}} = 58.6 \text{ mL ethylene glycol}$

(d) $\text{volume} = 23.9 \text{ kg} \times \dfrac{1000 \text{ g}}{1 \text{ kg}} \times \dfrac{1 \text{ mL}}{1.11 \text{ g}} \times \dfrac{1 \text{ L}}{1000 \text{ mL}} = 21.5 \text{ L ethylene glycol}$

13. $\text{Acetone mass} = 7.50 \text{ L antifreeze} \times \dfrac{1000 \text{ mL}}{1 \text{ L}} \times \dfrac{0.9867 \text{ g antifreeze}}{1 \text{ mL antifreeze}} \times \dfrac{8.50 \text{ g acetone}}{100.0 \text{ g antifreeze}}$

$\times \dfrac{1 \text{ kg}}{1000 \text{ g}} = 0.629 \text{ kg acetone}$

14. $\text{Acetic acid mass} = 1.00 \text{ lb vinegar} \times \dfrac{453.6 \text{ g}}{1 \text{ lb}} \times \dfrac{5.4 \text{ g acetic acid}}{100.0 \text{ g vinegar}} = 24 \text{ g acetic acid}$

15. $\text{Solution mass} = 1.00 \text{ kg sucrose} \times \dfrac{1000 \text{ g}}{1 \text{ kg}} \times \dfrac{100.00 \text{ g solution}}{12.62 \text{ g sucrose}} = 7.92 \times 10^3 \text{ g solution}$

16. $\text{Fertilizer mass} = 775 \text{ g nitrogen} \times \dfrac{1 \text{ kg N}}{1000 \text{ g N}} \times \dfrac{100 \text{ kg fertilizer}}{21 \text{ kg N}} = 3.69 \text{ kg fertilizer}$

17. **(a)** $8950. = 8.950 \times 10^3$ **(b)** $10{,}700. = 1.0700 \times 10^4$ **(c)** $0.0240 = 2.40 \times 10^{-2}$

(d) $0.0047 = 4.7 \times 10^{-3}$ **(e)** $938.3 = 9.383 \times 10^2$ **(f)** $275{,}482 = 2.75482 \times 10^5$

18. **(a)** $3.21 \times 10^{-2} = 0.0321$ **(b)** $5.08 \times 10^{-4} = 0.000508$

(c) $121.9 \times 10^{-5} = 0.001219$ **(d)** $16.2 \times 10^{-2} = 0.162$

19. **(a)** 450 has two or three significant figures; trailing zeros left of the decimal are indeterminate, if no decimal point is shown.

(b) 98.6 has three significant figures; non-zero digits are significant.

(c) 0.0033 has two significant digits; leading zeros are not significant.

(d) 902.10 has five significant digits; trailing zeros to the right of the decimal point are significant, as are zeros surrounded by non-zero digits.

(e) 0.02173 has four significant digits; leading zeros are not significant.

(f) 7000 has one to four significant figures; trailing zeros left of the decimal are indeterminate, if no decimal point is shown.

(g) 7.02 has three significant figures; zeros surrounded by non-zero digits are significant.

(h) 67,000,000 may have from two to eight significant figures; there is no way to determine which, if any, of the zeros are significant, without the presence of a decimal point.

20. Each of the following is expressed with four significant figures.

(a) $3984.6 \approx 3985$ **(b)** $422.04 \approx 422.0$ **(c)** $186{,}000 = 1.860 \times 10^5$

(d) $33{,}900 \approx 3.390 \times 10^4$ **(e)** 6.321×10^4 is correct   **(f)** $5.0472 \times 10^{-4} \approx 5.047 \times 10^{-4}$

21. **(a)** $0.406 \times 0.0023 = 9.3 \times 10^{-4}$ **(b)** $0.1357 \times 16.80 \times 0.096 = 2.2 \times 10^{-1}$

(c) $0.458 + 0.12 - 0.037 = 5.4 \times 10^{-1}$ **(d)** $32.18 + 0.055 - 1.652 = 3.058 \times 10^1$

22. **(a)** $\dfrac{320 \times 24.9}{0.080} = \dfrac{3.2 \times 10^2 \times 2.49 \times 10^1}{8.0 \times 10^{-2}} = 1.0 \times 10^5$

(b) $\dfrac{432.7 \times 6.5 \times 0.002300}{62 \times 0.103} = \dfrac{4.327 \times 10^2 \times 6.5 \times 2.300 \times 10^{-3}}{6.2 \times 10^1 \times 1.03 \times 10^{-1}} = 1.0$

(c) $\dfrac{32.44 + 4.9 - 0.304}{82.94} = \dfrac{3.244 \times 10^1 + 4.9 - 3.04 \times 10^{-1}}{8.294 \times 10^1} = 4.47 \times 10^{-1}$

(d) $\dfrac{8.002 + 0.3040}{13.4 - 0.066 + 1.02} = \dfrac{8.002 + 3.040 \times 10^{-1}}{1.34 \times 10^1 - 6.6 \times 10^{-2} + 1.02} = 5.79 \times 10^{-1}$

23. The calculated volume of the block is converted to its mass with the density of iron.

$$\text{Mass} = 52.8 \text{ cm} \times 6.74 \text{ cm} \times 3.73 \text{ cm} \times 7.86 \frac{\text{g}}{\text{cm}^3} = 1.04 \times 10^4 \text{ g iron}$$

24. The calculated volume of the cylinder is converted to its mass with the density of steel.

$$\text{Mass} = V(\text{density}) = \pi r^2 h(d) = 3.14159 (1.88 \text{ cm})^2 \, 18.35 \text{ cm} \times 7.75 \frac{\text{g}}{\text{cm}^3} = 1.58 \times 10^3 \text{ g steel}$$

EXERCISES

Scientific Method

25. No. The greater the number of experiments that conform to the predictions of the law, the more confidence we have in the law. There is no point at which the law is ever verified with absolute certainty.

26. One theory is preferred over another if it can predict a wider range of phenomena and if it has fewer assumptions.

27. A given set of conditions, a cause, is expected to produce a certain result, an effect. Although these cause-and-effect relationships may be difficult to establish at times ("God is subtle"), they nevertheless do exist ("he is not malicious").

28. The scientific method requires that *all* observations, the results of *all* experiments, be consistent with the predictions of a theory ("the rule"). Even one exception is sufficient reason to challenge a theory and to search for a modification to explain the exception.

29. The experiments should be carefully set up so as to create a controlled situation in which one can make careful observations after altering the experimental parameters, preferably one at a time. The results must be reproducible (to within experimental error) and, as more and more experiments are conducted, a pattern should begin to emerge, from which a comparison to the current theory can be made.

30. For a theory to be considered as plausible, it must, first and foremost, agree with and/or predict the results from controlled experiments. It should also involve the fewest number of assumptions (i.e. follow Occam's Razor). The best theories predict new phenomena that are subsequently observed after the appropriate experiments have been performed.

Properties and Classification of Matter

31. An object displaying a physical property retains its basic chemical identity. Display of a chemical property is accompanied by a change in composition.

(a) Physical: The iron nail is not changed in any significant way when it is attracted to a magnet. Its basic chemical identity is unchanged.

(b) Chemical: The liquid lighter fluid is converted into a gas (carbon dioxide) and water vapor, along with the evolution of considerable energy.

(c) Chemical: The green patina is the result of the combination of water, oxygen, and carbon dioxide with the copper in the bronze to produce basic copper carbonate.

(d) Physical: Neither the block of wood nor the water has changed its identity.

32.   An object displaying a physical property retains its basic chemical identity. The display of a chemical property is accompanied by a change in composition.

   (a)   Chemical: The change in the color of the apple indicates that a new substance (oxidized apple) has formed by reaction with air.

   (b)   Physical: The marble slab is not changed into another substance by feeling it.

   (c)   Physical: The sapphire retains its identity as it displays its color.

   (d)   Chemical: After firing, the properties of the clay have changed from soft and pliable to rigid and brittle. New substances have formed. (Many of the changes involve driving off water and slightly melting the silicates that remain. These molten substances cool and harden when removed from the kiln.)

33.   **(a)**   Heterogeneous mixture: We can clearly see air pockets within the solid matrix. On close examination, we can distinguish different kinds of solids by their colors.

   (b)   Homogeneous mixture: Modern inks are solutions of dyes in water. Older inks often were heterogeneous mixtures: suspensions of particles of carbon black (soot) in water.

   (c)   Substance: Assuming that no gases or organic chemicals are dissolved in the water.

   (d)   Heterogeneous mixture: The pieces of orange pulp can be seen through a microscope. Most "cloudy" liquids are heterogeneous mixtures; the small particles impede the transmission of light.

34.   **(a)**   Homogeneous mixture: Air is a mixture of nitrogen, oxygen, argon, and traces of other gases. By "fresh," we mean no particles of smoke, pollen, etc. are present. They would produce a heterogeneous mixture.

   (b)   Homogeneous mixture: Most brass is a solid solution of copper and zinc. (Older brass contains zones of slightly different compositions. These show up once the surface is etched, which occurs with repeated handling on items such as doorknobs.)

   (c)   Heterogeneous mixture: Pieces of garlic can be distinguished from those of salt by careful examination.

   (d)   Substance: Ice is simply solid water (assuming no air bubbles).

35.   **(a)**   Physical: This is simply a mixture of sand and sugar (i.e. not chemically bonded).

   (b)   Chemical: Oxygen needs to be removed from the iron oxide.

   (c)   Physical: Seawater is a solution of various substances dissolved in water.

   (d)   Physical: The water sand is simply a heterogeneous mixture.

36.   **(a)**   If a magnet is drawn through the mixture, the iron filings will be attracted to the magnet and the wood will be left behind.

   (b)   When the glass-sucrose mixture is mixed with water, the sucrose will dissolve, whereas the glass will not. The water can then be boiled off to produce pure sucrose.

(c) By heating the ice-salt mixture, the ice will first melt to form liquid water and then the liquid water will be converted into water vapor. Once all of the water has been driven off, a pure sample of anhydrous salt will be left behind. A pure sample of liquid water can be obtained by cooling the water vapor.

(d) The gold flakes will settle to the bottom if the mixture is left undisturbed. The water then can be decanted, that is, carefully poured off.

Exponential Arithmetic

37. (a) $34,000$ centimeters / second $= 3.4 \times 10^4$ cm/s

(b) six thousand three hundred seventy eight kilometers $= 6378$ km $= 6.378 \times 10^3$ km

(c) (trillionth $= 1 \times 10^{-12}$) hence, 74×10^{-12} m or 7.4×10^{-11} m

(d) $\dfrac{(2.2\times10^3)+(4.7\times10^2)}{5.8\times10^{-3}} = \dfrac{2.7\times10^3}{5.8\times10^{-3}} = 4.6\times10^5$

38. (a) 173 thousand trillion watts $= 173,000,000,000,000,000$ W $= 1.73 \times 10^{17}$ W

(b) ten millionths of a meter $= 10 \times 0.000\,001$ m $= 1\times10^{-5}$ m

(c) (trillionth $= 1 \times 10^{-12}$) hence, 142×10^{-12} m or 1.42×10^{-10} m

(d) $\dfrac{5.07\times10^4 \times \left(1.8\times10^{-3}\right)^2}{0.065+\left(3.3\times10^{-2}\right)} = \dfrac{0.16}{0.098} = 1.6$

Significant figures

39. (a) An exact number—24 soda cans in a case.

(b) Pouring the milk into the jug is a process that is subject to error; there can be slightly more or slightly less than one gallon of milk in the jug. This is a measured quantity.

(c) The distance between any pair of planetary bodies can only be determined through certain measurements, which are subject to error.

(d) Measured quantity: the internuclear separation quoted for H_2 is an estimated value derived from experimental data, which contains some inherent error.

40. (a) The number of pages in the text is determined by counting; the result is an exact number.

(b) An exact number. Although the number of days can vary from one month to another (say from January to February), the month of January always has 31 days.

(c) The area is determined by calculations based on measurements. These measurements are subject to error.

(d) Measured quantity: Average internuclear distance for adjacent atoms in silver metal is an estimated value derived from X-ray diffraction data, which contain some inherent error.

41. (a) 2.44×10^4 (b) 1.5×10^3 (c) 40.0

    (d) 2.131×10^3 (e) 4.8×10^{-3}

42. (a) 7.5×10^{-2} (b) 6.3×10^{12} (c) 4.6×10^3

    (d) 1.058×10^{-1} (e) 1.0×10^{-2}

43. (a) The average speed is obtained by dividing the distance traveled (in miles) by the elapsed time (in hours). First, we need to obtain the elapsed time, in hours.

$$9 \text{ days} \times \frac{24 \text{ h}}{1 \text{ d}} = 216.000 \text{ h} \quad 3 \text{ min} \times \frac{1 \text{ h}}{60 \text{ min}} = 0.050 \text{ h} \quad 44 \text{ s} \times \frac{1 \text{ h}}{3600 \text{ s}} = 0.012 \text{ h}$$

Total time $= 216.000 \text{ h} + 0.050 \text{ h} + 0.012 \text{ h} = 216.062 \text{ h}$

$$\text{average speed} = \frac{25,012 \text{ mi}}{216.062 \text{ h}} = 115.76 \text{ mi} / \text{h}$$

(b) First compute the mass of fuel remaining

$$\text{mass} = 14 \text{ gal} \times \frac{4 \text{ qt}}{1 \text{ gal}} \times \frac{0.9464 \text{ L}}{1 \text{ qt}} \times \frac{1000 \text{ mL}}{1 \text{ L}} \times \frac{0.70 \text{ g}}{1 \text{ mL}} \times \frac{1 \text{ lb}}{453.6 \text{ g}} = 82 \text{ lb}$$

Then determine the mass of fuel used, and finally, the fuel consumption. Notice that we know the initial quantity of fuel quite imprecisely, perhaps at best to the nearest 10 lb, certainly ("nearly 9000 lb") not to the nearest pound.

mass of fuel used $= 9000 \text{ lb} - 82 \text{ lb} \cong 8920 \text{ lb}$

$$\text{fuel consumption} = \frac{25,012 \text{ mi}}{8920 \text{ lb}} = 2.80 \text{ mi/lb}$$

44. If the proved reserve truly was an estimate, rather than an actual measurement, it would have been difficult to estimate it to the nearest trillion cubic feet. A statement such as 2,911,000 trillion cubic feet (or even 3×10^{18} ft^3) would have more accurately reflected the precision with which the proved reserve was known.

Units of Measurement

45. Express both masses in the same units for comparison. $2172 \mu g \left(\frac{1 \text{ g}}{10^6 \mu g} \right) \left(\frac{10^3 \text{ mg}}{1 \text{ g}} \right) = 2.172$ mg, which is larger than 0.00515 mg.

46. Express both masses in the same units for comparison. $0.00475 \text{ kg} \times \frac{1000 \text{ g}}{1 \text{ kg}} = 4.75$ g, which is larger than $3257 \text{ mg} \times \frac{1 \text{ g}}{10^3 \text{ mg}} = 3.257$ g.

47. $\text{height} = 15\,\text{hands} \times \dfrac{4\,\text{in.}}{1\,\text{hand}} \times \dfrac{2.54\,\text{cm}}{1\,\text{in.}} \times \dfrac{1\,\text{m}}{100\,\text{cm}} = 1.5\ \text{m}$

48. The length of a mile in feet $(5280\,\text{ft} = 1\,\text{mi})$ and can use that as a conversion factor.

$1.00\,\text{link} \times \dfrac{1\,\text{chain}}{100\,\text{links}} \times \dfrac{1\,\text{furlong}}{10\,\text{chains}} \times \dfrac{1\,\text{mile}}{8\,\text{furlongs}} \times \dfrac{5280\,\text{ft}}{1\,\text{mi}} \times \dfrac{12\,\text{in.}}{1\,\text{ft}} = 7.92\,\text{in.}$

49. **(a)** We use the speed as a conversion factor, but need to convert yards into meters.

$\text{time} = 100.0\,\text{m} \times \dfrac{9.3\,\text{s}}{100\,\text{yd}} \times \dfrac{1\,\text{yd}}{36\,\text{in.}} \times \dfrac{39.37\,\text{in.}}{1\,\text{m}} = 10.\ \text{s}$ Keep two significant figures.

 (b) We need to convert yards to meters.

$\text{speed} = \dfrac{100\,\text{yd}}{9.3\,\text{s}} \times \dfrac{36\,\text{in.}}{1\,\text{yd}} \times \dfrac{2.54\,\text{cm}}{1\,\text{in.}} \times \dfrac{1\,\text{m}}{100\,\text{cm}} = 9.8\underline{3}\ \text{m/s}$

 (c) The speed is used as a conversion factor.

$\text{time} = 1.45\,\text{km} \times \dfrac{1000\,\text{m}}{1\,\text{km}} \times \dfrac{1\,\text{s}}{9.8\,\text{m}} \times \dfrac{1\,\text{min}}{60\,\text{s}} = 2.5\,\text{min}$

50. **(a)** $\text{mass}\,(\text{mg}) = 2\,\text{tablets} \times \dfrac{5.0\,\text{gr}}{1\,\text{tablet}} \times \dfrac{1.0\,\text{g}}{15\,\text{gr}} \times \dfrac{1000\,\text{mg}}{1\,\text{g}} = 6.7 \times 10^{2}\ \text{mg}$

 (b) $\text{dosage rate} = \dfrac{6.7 \times 10^{2}\,\text{mg}}{155\,\text{lb}} \times \dfrac{1\,\text{lb}}{453.6\,\text{g}} \times \dfrac{1000\,\text{g}}{1\,\text{kg}} = 9.5\ \text{mg aspirin/kg body weight}$

 (c) $\text{time} = 1.0\,\text{lb} \times \dfrac{453.6\,\text{g}}{1\,\text{lb}} \times \dfrac{2\,\text{tablets}}{0.67\,\text{g}} \times \dfrac{1\,\text{day}}{2\,\text{tablets}} = 6.8 \times 10^{2}\ \text{days}$

51. $1\,\text{hectare} = 1\,\text{hm}^{2} \times \left(\dfrac{100\,\text{m}}{1\,\text{hm}} \times \dfrac{100\,\text{cm}}{1\,\text{m}} \times \dfrac{1\,\text{in.}}{2.54\,\text{cm}} \times \dfrac{1\,\text{ft}}{12\,\text{in.}} \times \dfrac{1\,\text{mi}}{5280\,\text{ft}} \right)^{2} \times \dfrac{640\,\text{acres}}{1\,\text{mi}^{2}}$

$1\,\text{hectare} = 2.47\,\text{acres}$

52. Here we must convert pounds per cubic inch into grams per cubic centimeter:

$\text{density for metallic iron in } \dfrac{\text{g}}{\text{cm}^{3}} = \dfrac{0.284\,\text{lb}}{\text{in}^{3}} \times \dfrac{454\,\text{g}}{1\,\text{lb}} \times \dfrac{(1\,\text{in})^{3}}{(2.54\,\text{cm})^{3}} = 7.87\ \dfrac{\text{g}}{\text{cm}^{3}}$

53. $\text{pressure} = \dfrac{32\,\text{lb}}{1\,\text{in.}^{2}} \times \dfrac{453.6\,\text{g}}{1\,\text{lb}} \times \left(\dfrac{1\,\text{in.}}{2.54\,\text{cm}} \right)^{2} = 2.2 \times 10^{3}\ \text{g/cm}^{2}$

$\text{pressure} = \dfrac{2.2 \times 10^{3}\,\text{g}}{1\,\text{cm}^{2}} \times \dfrac{1\,\text{kg}}{1000\,\text{g}} \times \left(\dfrac{100\,\text{cm}}{1\,\text{m}} \right)^{2} = 2.2 \times 10^{4}\ \text{kg/m}^{2}$

54.   First we will calculate the radius for a typical red blood cell using the equation for the volume of a sphere. $V = 4/3\pi r^3 = 9.00 \times 10^{-11}$ cm^3

$r^3 = 2.15 \times 10^{-11}$ cm^3 and r $= 2.78 \times 10^{-4}$ cm

Thus, the diameter is $2 \times r = 2 \times 2.78 \times 10^{-4}$ cm $= 5.56 \times 10^{-4}$ cm

Expressed in inches: 5.56×10^{-4} cm $\times \dfrac{1 \text{ in}}{2.54 \text{ cm}} = 2.19 \times 10^{-4}$ inches in diameter

Temperature Scales

<u>55.</u>   high: $^\circ C = \frac{5}{9}\left(^\circ F - 32\right) = \frac{5}{9}\left(118^\circ F - 32\right) = 47.8^\circ C \approx 48 \ ^\circ C$

low: $^\circ C = \frac{5}{9}\left(^\circ F - 32\right) = \frac{5}{9}\left(17^\circ F - 32\right) = -8.3^\circ C$

56.   low: $^\circ F = \frac{9}{5}^\circ C + 32 = \frac{9}{5}\left(-15^\circ C\right) + 32 = 5^\circ F$

high: $^\circ F = \frac{9}{5}^\circ C + 32 = \frac{9}{5}\left(60^\circ C\right) + 32 = 140^\circ F$

<u>57.</u>   Determine the Celsius temperature that corresponds to the highest Fahrenheit temperature, $240^\circ F$. $^\circ C = \frac{5}{9}\left(^\circ F - 32\right) = \frac{5}{9}\left(240^\circ F - 32\right) = 116^\circ C$ Because $116^\circ C$ is above the range of the thermometer, this thermometer cannot be used in this candy making assignment.

58.   Let us determine the Fahrenheit equivalent of absolute zero.

$^\circ F = \left(\frac{9}{5}\right)^\circ C + 32 = \left(\frac{9}{5} \times (-273.15)\right) + 32 = -459.67^\circ F$

A temperature of $-465^\circ F$ cannot be achieved because it is below absolute zero.

<u>59.</u> **(a)**   From the data provided we can write down the following relationship: -38.9°C = 0 °M and 356.9 °C = 100 °M. To find the mathematical relationship between these two scales, we can treat each relationship as a point on a two-dimensional Cartesian graph (see next page)

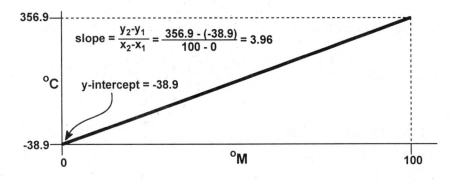

Therefore, the equation for the line is y = 3.96x - 38.9   The algebraic relationship between the two temperature scales is

$$t(°C) = 3.96(°M) - 38.9 \text{ or rearranging, } t(°M) = \frac{t(°C) + 38.9}{3.96}$$

Alternatively, note that the change in temperature in °C corresponding to a change of 100 °M is [356.9 – (-38.9)] = 395.8 °C, hence, (100 °M/395.8 °C) = 1 °M/3.96 °C. This factor must be multiplied by the number of degrees Celsius above zero on the M scale. This number of degrees is t(°C) + 38.9, which leads to the general equation t(°M) = [t(°C) + 38.9]/3.96.

The boiling point of water is 100 °C, corresponding to $t(°M) = \dfrac{100 + 38.9}{3.96} = 35.1°M$

(b) $t(°M) = \dfrac{-273.15 + 38.9}{3.96} = -59.2 °M$

60. (a) From the data provided we can write down the following relationship: -77.75 = 0 °A and -33.35 °C = 100 °A. To find the mathematical relationship between these two scales, we can treat each relationship as a point on a two-dimensional Cartesian graph

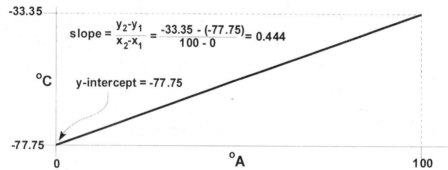

Therefore, the equation for the line is y =0.444x - 77.75
The algebraic relationship between the two temperature scales is

$$t(°C) = 0.444(°A) - 77.75 \text{ or rearranging } t(°A) = \frac{t(°C) + 77.75}{0.444}$$

The boiling point of water(100 °C) corresponds to $t(°A) = \dfrac{100 + 77.75}{0.444} = 400. °A$

(b) $t(°A) = \dfrac{-273.15 + 77.75}{0.444} = -440. °A$

Density

61. The mass of acetone is the difference in masses between empty and filled masses.

$$\text{Density} = \frac{437.5 \text{ lb} - 75.0 \text{ lb}}{55.0 \text{ gal}} \times \frac{453.6 \text{ g}}{1 \text{ lb}} \times \frac{1 \text{ gal}}{3.785 \text{ L}} \times \frac{1 \text{ L}}{1000 \text{ mL}} = 0.790 \text{ g / mL}$$

62. Density is a conversion factor.

$$\text{volume} = (283.2 \text{ g filled} - 121.3 \text{ g empty}) \times \frac{1 \text{ mL}}{1.59 \text{ g}} = 102 \text{ mL}$$

63. Determine the mass of each item.

(1) mass of iron $= (81.5 \text{ cm} \times 2.1 \text{ cm} \times 1.6 \text{ cm}) \times 7.86 \text{ g} / \text{cm}^3 = 2.2 \times 10^3 \text{ g iron}$

(2) mass of aluminum

$$= (12.12 \text{ m} \times 3.62 \text{ m} \times 0.003 \text{ cm}) \times \left(\frac{100 \text{ cm}}{1 \text{ m}}\right)^2 \times 2.70 \text{ g/cm}^3 = 4 \times 10^3 \text{ g aluminum}$$

(3) mass of water $= 4.051 \text{ L} \times \dfrac{1000 \text{ cm}^3}{1 \text{ L}} \times 0.998 \text{ g} / \text{cm}^3 = 4.04 \times 10^3 \text{ g water}$

In order of increasing mass, the items are: iron bar < aluminum foil < water. Realize, however, that the rules for significant figures do not allow us to distinguish between the masses of aluminum and water.

64. First determine the volume of the aluminum foil, then its area, and finally its thickness.

$$\text{volume} = 2.568 \text{ g} \times \frac{1 \text{ cm}^3}{2.70 \text{ g}} = 0.951 \text{ cm}^3 \quad \text{area} = \left(9.0 \text{ in.} \times \frac{2.54 \text{ cm}}{\text{in.}}\right)^2 = 5.2 \times 10^2 \text{ cm}^2$$

$$\text{thickness} = \frac{\text{volume}}{\text{area}} = \frac{0.951 \text{ cm}^3}{5.2 \times 10^2 \text{ cm}^2} \times \frac{10 \text{ mm}}{1 \text{ cm}} = 1.8 \times 10^{-2} \text{ mm}$$

65. Total volume of 125 pieces of shot

$$V = 8.9 \text{ mL} - 8.4 \text{ mL} = 0.5 \text{ mL} \quad \frac{\text{mass}}{\text{shot}} = \frac{0.5 \text{ mL}}{125 \text{ shot}} \times \frac{1 \text{cm}^3}{1 \text{mL}} \times \frac{8.92 \text{ g}}{1 \text{ cm}^3} = 0.04 \text{ g/shot}$$

66. The vertical piece of steel has a volume $= 12.78 \text{ cm} \times 1.35 \text{ cm} \times 2.75 \text{ cm} = 47.4 \text{ cm}^3$
The horizontal piece of steel has a volume $= 10.26 \text{ cm} \times 1.35 \text{ cm} \times 2.75 \text{ cm} = 38.1 \text{ cm}^3$
$V_{\text{total}} = 47.4 \text{ cm}^3 + 38.1 \text{ cm}^3 = 85.5 \text{ cm}^3$. Mass $= 85.5 \text{ cm}^3 \times 7.78 \text{ g/cm}^3 = 665 \text{ g of steel}$

67. Here we are asked to calculate the number of liters of whole blood that have to be collected in order to end up with 0.5 kg of red blood cells. Each red blood cell has a mass of

$$9.00 \times 10^{-11} \text{ cm}^3 \times 1.096 \text{ g cm}^{-3} = 9.864 \times 10^{-11} \text{ g}$$

$$\text{red blood cells (mass per mL)} = \frac{9.864 \times 10^{-11} \text{ g}}{1 \text{ cell}} \times \frac{5.4 \times 10^9 \text{ cells}}{1 \text{ mL}} = \frac{0.533 \text{ g red blood cells}}{1 \text{ mL of blood}}$$

For 0.5 kg or 5×10^2 g of red blood cells, we require

$$= 5 \times 10^2 \text{ g red blood cells} \times \frac{1 \text{ mL of blood}}{0.533 \text{ g red blood cells}} = 9 \times 10^2 \text{ mL of blood or 0.9 L blood}$$

68. The mass of the liquid mixture used for density measurements can be found by subtracting the mass of the full bottle from the mass of the empty bottle = 15.4448 g − 12.4631 g = 2.9817 g liquid. Using a similar approach, the total mass of the water that can be accommodated in the bottle is 13.5441 g − 12.4631 g = 1.081 g H_2O. The volume of the water and hence the internal volume for the bottle is equal to

$$1.081 \text{ g } H_2O \times \frac{1 \text{ mL } H_2O}{0.9970 \text{ g } H_2O} = 1.084 \text{ mL } H_2O \ (25 \text{ °C})$$

Thus, the density of the liquid mixture $= \dfrac{3.8122 \text{ g liquid}}{1.084 \text{ mL}} = 3.516 \text{ g mL}^{-1}$

Since the calcite just floats in this mixture of liquids, it must have the same density as the mixture. Consequently, the solid calcite sample must have a density of 3.517 g mL^{-1} as well.

Percent Composition

69. The percent of students with each grade is obtained by dividing the number of students with that grade by the total number of students. $\%A = \dfrac{9 \text{ A's}}{76 \text{ students}} \times 100\% = 12\% \text{ A}$

$\%B = \dfrac{21 \text{ B's}}{76 \text{ students}} \times 100\% = 28\% \text{ B}$ $\%C = \dfrac{36 \text{ C's}}{76 \text{ students}} \times 100\% = 47\% \text{ C}$

$\%D = \dfrac{8 \text{ D's}}{76 \text{ students}} \times 100\% = 11\% \text{ D}$ $\%F = \dfrac{2 \text{ F's}}{76 \text{ students}} \times 100\% = 3\% \text{ F}$

Note that the percentages add to 101% due to rounding effects.

70. The number of students with a certain grade is determined by multiplying the total number of students by the fraction of students who earned that grade.

$\text{no. A} = 84 \text{ students} \times \dfrac{18 \text{ A's}}{100 \text{ students}} = 15 \text{ A's}$

$\text{no. B} = 84 \text{ students} \times \dfrac{25 \text{ B's}}{100 \text{ students}} = 21 \text{ B's}$ $\text{no. C} = 84 \text{ students} \times \dfrac{32 \text{ C's}}{100 \text{ students}} = 27 \text{ C's}$

$\text{no. D} = 84 \text{ students} \times \dfrac{13 \text{ D's}}{100 \text{ students}} = 11 \text{ D's}$ $\text{no. F} = 84 \text{ students} \times \dfrac{12 \text{ F's}}{100 \text{ students}} = 10 \text{ F's}$

71. Use the percent composition as a conversion factor.

$$\text{mass of sucrose} = 2.75 \text{ L} \times \frac{1000 \text{ mL}}{1 \text{ L}} \times \frac{1.118 \text{ g soln}}{1 \text{ mL}} \times \frac{28.0 \text{ g sucrose}}{100 \text{ g soln}} = 8.61 \times 10^2 \text{ g sucrose}$$

72. Again, percent composition is used as a conversion factor. We are careful to label both the numerator and denominator for each factor.

$$V_{solution} = 3.50 \text{ kg sodium hydroxide} \times \frac{1000 \text{ g}}{1 \text{ kg}} \times \frac{100.0 \text{ g soln}}{12.0 \text{ g sodium hydroxide}} \times \frac{1 \text{ mL}}{1.131 \text{ g soln}}$$

$$V_{solution} = 2.58 \times 10^4 \text{ mL soln} \times \frac{1 \text{ L}}{1000 \text{ mL}} = 25.8 \text{ L soln}$$

FEATURE PROBLEMS

89. All of the pennies minted before 1982 weigh more than 3.00 g, while all of those minted after 1982 weigh less than 2.60 g. One might infer that the composition of a penny changed in 1982. In fact, pennies minted prior to 1982 are composed of almost pure copper (about 96% pure). Those minted after 1982 are composed of zinc with a thin copper cladding. Some pennies of each type were minted in 1982.

90. After sitting in a bathtub that was nearly full and observing the water splashing over the side, Archimedes realized that the crown—when submerged in water—would displace a volume of water equal to its volume. Once Archimedes determined the volume in this way and determined the mass of the crown with a balance, he was able to calculate the crown's density. Since the gold-silver alloy has a different density (it is lower) than pure gold, Archimedes could tell that the crown was not pure gold.

91. Notice that the liquid does not fill each of the floating glass balls. The quantity of liquid in each glass ball is sufficient to give each ball a slightly different density. Note that the density of the glass ball is determined by the density of the liquid, the density of the glass (greater than the liquid's density), and the density of the air. Since the density of the liquid in the cylinder varies slightly with temperature—the liquid's volume increases as temperature goes up, but its mass does not change—different balls will be buoyant at different temperatures.

92. The density of the canoe is determined by the density of the concrete and the density of the empty space inside the canoe, where the passengers sit. It is the empty space, (filled with air), that makes the density of the canoe less than that of water (1.0 g/cm^3). If the concrete canoe fills with water, it will sink to the bottom, unlike a wooden canoe.

93. One needs to convert (lb of force) into (Newtons) 1 lb of force = 1 slug $\times$ 1 ft s^{-2}

(1 slug = 14.59 kg). Therefore, 1 lb of force = $\dfrac{14.59 \text{ kg x 1 ft}}{s^2}$

$$= \left(\frac{14.59 \text{ kg x 1 ft}}{s^2} \right) \left(\frac{12 \text{ in}}{1 \text{ ft}} \right) \left(\frac{2.54 \text{ cm}}{1 \text{ in}} \right) \left(\frac{1 \text{ m}}{100 \text{ cm}} \right) = \frac{4.45 \text{ kg m}}{s^2} = 4.45 \text{ Newtons}$$

From this result it is clear that 1 lb of force = 4.45 Newtons.

CHAPTER 2
ATOMS AND THE ATOMIC THEORY

PRACTICE EXAMPLES

1A   The total mass must be the same before and after reaction.

mass before reaction = 0.382g magnesium + 2.652g nitrogen = 3.034g

mass after reaction = magnesium nitride mass + 2.505g nitrogen = 3.034g

magnesium nitride mass = 3.034g − 2.505g = 0.529g magnesium nitride

1B   Again, the total mass is the same before and after the reaction.

mass before reaction = 7.12g magnesium + 1.80g bromine = 8.92g

mass after reaction = 2.07g magnesium bromide + magnesium mass = 8.92g

magnesium mass = 8.92g − 2.07g = 6.85g magnesium

2A   In Example 2-2 we are told that 0.100 g magnesium produces 0.166 g magnesium oxide.

$$\text{mass of Mg} = 0.500\text{g MgO} \times \frac{0.100\text{g Mg}}{0.166\text{g MgO}} = 0.301\text{g Mg}$$

2B   In Example 2-2 we are told that 0.100 g Mg forms 0.166 g MgO. With this information, we can determine the mass of magnesium needed to form 2.00 g magnesium oxide.

$$\text{mass of Mg} = 2.00\text{g MgO} \times \frac{0.100\text{g Mg}}{0.166\text{g MgO}} = 1.20\text{g Mg}$$

The remainder of the 2.00 g of magnesium oxide is the mass of oxygen

mass of oxygen = 2.00g magnesium oxide − 1.20g magnesium = 0.80 g oxygen

3A   The number of protons and electrons are equal, and thus the species has no charge. The mass number is the sum of the atomic number and the number of neutrons:

$47\text{p} + 61\text{n} = A = 108$.

The atomic number 47 is that of the element silver. Thus the symbol is $^{108}_{47}\text{Ag}$.

3B   The element sulfur has an atomic number of 16 and thus has 16 protons. A charge of 2− indicates two more electrons than protons; there are $16 + 2 = 18$ electrons. The number of neutrons is the mass number minus the number of protons; there are $35 - 16 = 19$ neutrons.

4A   We know that the mass of $^{16}\text{O} = 15.9949$ u and that mass of $^{16}\text{O} = 1.06632 \times$ mass of ^{15}N. We combine these two equations and solve the resulting expression.

$$15.9949\text{u} = 1.06632 \times \text{mass of }^{15}\text{N} \qquad \therefore \text{mass of }^{15}\text{N} = \frac{15.9949\text{u}}{1.06632} = 15.0001\text{u}$$

4B   We know the isotopic mass of ^{12}C is 12 u. Thus, the mass ratio is found by substitution.

$$\frac{^{202}Hg}{^{12}C} = \frac{201.970617}{12u} = 16.8308848$$

Note that the number of significant figures in the result is determined by the precision of the mass of ^{202}Hg, because the mass of ^{12}C is established by definition as an exact number.

5A   The average atomic mass of boron is 10.811, which is closer to 11.009305 than to 10.012937. Thus, boron-11 is the isotope present in greater abundance.

5B   We let x be the fractional abundance of lithium-6.

$$6.941u = [x \times 6.01513u] + [(1-x) \times 7.01601u] = 6.01513xu + 7.01601u - 7.01601xu$$

$$6.941u - 7.01601u = 6.01513xu - 7.01601xu = -1.00088xu$$

$$x = \frac{6.941u - 7.01601u}{-1.00088u} = 0.075 \quad \text{Percent abundances}: 7.5\% \text{ lithium - 6, } 92.5\% \text{ lithium - 7}$$

6A   We assume that atoms lose or gain relatively few electrons to become ions. Thus, elements that will form cations will be on the left-hand side of the periodic table, while elements that will form anions will be on the right-hand side. The number of electrons "lost" when a cation forms is the periodic group number; the number of electrons added when an anion forms is eight minus the group number.

Li is in group 1(1A); it should form a cation by losing one electron: Li^+.

S is in group 6(6A); it should form an anion by adding two electrons: S^{2-}.

Ra is in group 2(2A); it should form a cation by losing two electrons: Ra^{2+}.

F and I are both group 17(7A); they should form anions by gaining an electron: F^- and I^-.

Al is in group 13(3A); it should form a cation by losing three electrons: Al^{3+}.

6B   Main group elements are in the "A" families, while transition elements are in the "B" families. Metals, nonmetals, metalloids, and noble gases are color coded in the periodic table inside the front cover.

Na is a main-group metal in group 1(1A).	Re is a transition metal in group 7
S is a main-group nonmetal in group 16(6A).	I is a main-group nonmetal in group 17.
Kr is a noble gas in group 18(8A).	Mg is a main-group metal in group 2.
U is an inner transition metal, an actinide.	Si is a main-group metalloid in group 14.
B is a main-group metalloid in group 13(3A).	Al is a main-group metal in group 13.
As is a main-group metalloid in group 15(5A).	H is a main-group nonmetal in group 1.

7A Avogadro's number serves as a conversion factor.

$$\text{no. Au atoms} = 5.07 \times 10^{-3} \text{mol Au} \times \frac{6.022 \times 10^{23} \text{Au atoms}}{1 \text{mol Au}} = 3.05 \times 10^{21} \text{Au atoms}$$

7B Of all lead atoms, 24.1% are lead-206, or 241 ^{206}Pb atoms in every 1000 lead atoms

$$^{206}\text{Pb atoms} = 8.27 \times 10^{-3} \text{mol Pb} \times \frac{6.022 \times 10^{23} \text{Pb atoms}}{1 \text{mol Pb}} \times \frac{241 \, ^{206}\text{Pb atoms}}{1000 \text{Pb atoms}} = 1.20 \times 10^{21} \, ^{206}\text{Pb atoms}$$

8A This is similar to Practice Examples 2-8A and 2-8B.

$$\text{Cu mass} = 2.35 \times 10^{24} \text{Cu atoms} \times \frac{1 \text{mol Cu}}{6.022 \times 10^{23} \text{atoms}} \times \frac{63.546 \text{g Cu}}{1 \text{mol Cu}} = 248 \text{ g Cu}$$

8B Atoms of He $= 22.6$ g He $\times \dfrac{1 \text{mol He}}{4.0026 \text{g He}} \times \dfrac{6.022 \times 10^{23} \text{He atoms}}{1 \text{mol He}} = 3.40 \times 10^{24}$ He atoms

9A Both the density and the molar mass of Pb serve as conversion factors.

$$\text{atoms of Pb} = 0.105 \text{cm}^3 \text{Pb} \times \frac{11.34 \text{g}}{1 \text{cm}^3} \times \frac{1 \text{mol Pb}}{207.2 \text{g}} \times \frac{6.022 \times 10^{23} \text{Pb atoms}}{1 \text{mol Pb}} = 3.46 \times 10^{21} \text{Pb atoms}$$

9B First we find the number of rhenium atoms in 0.100 mg of the element.

$$0.100 \text{mg} \times \frac{1 \text{g}}{1000 \text{mg}} \times \frac{1 \text{mol Re}}{186.207 \text{ g Re}} \times \frac{6.022 \times 10^{23} \text{Re atoms}}{1 \text{ mol Re}} = 3.23 \times 10^{17} \text{ Re atoms}$$

$$\% \, ^{187}\text{Re} = \frac{2.02 \times 10^{17} \text{atoms } ^{187}\text{Re}}{3.23 \times 10^{17} \text{ Re atoms}} \times 100\% = 62.5\%$$

REVIEW QUESTIONS

1. **(a)** ^A_ZE is the symbol for a nuclide. E represents the symbol of the element; Z is the atomic number, the number of protons in the nucleus; and A is the mass number, the number of protons plus neutrons.

    (b) A β particle refers to an electron ejected by the nucleus, and is one of the three forms of natural radioactivity.

    (c) An isotope is one of at least two forms of an atom of an element which have the same number of protons in the nucleus, but different numbers of neutrons.

    (d) ^{16}O is the symbol for the isotope of oxygen that has 16 nucleons in its nucleus: 8 protons (characteristic of the element oxygen) and 8 neutrons.

    (e) Molar mass is the mass of a quantity of an element (or a compound) that contains Avogadro's number 6.022×10^{23} of atoms (or formula units).

2.   **(a)**   The law of conservation of mass states that there is no gain or loss of mass during a chemical reaction.

  (b)   The atom as described by Rutherford consists of a very small (approximately 10^{-13} cm diameter), positively charged, and massive (more than 99.5% of the mass) nucleus surrounded by a relatively large (approximately 10^{-8} cm diameter), tenuous (less than 0.5% of the mass), and negatively charged cloud of electrons.

  (c)   The atomic mass that appears in the periodic table for each element is a weighted average, with contributions from each naturally-occurring isotope of the element, each weighted by the relative abundance of that isotope.

  (d)   Radioactivity refers to the spontaneous emission from the nucleus of a photon (γ radiation) or particles (α or β particles).

3.   **(a)**   Cathode rays are beams of electrons that are generated when a large potential difference is applied across two metal plates in a sealed evacuated tube. X-rays are high-energy photons emitted when these cathode rays strike the anode within the glass tube.

  (b)   Protons and neutrons are both particles in the nucleus of the atom, and both have a mass of approximately 1 u. However, protons are positively charged, while neutrons have no electric charge.

  (c)   The nuclear charge of an atom is a positive charge equal to the number of protons in the nucleus. The ionic charge equals the nuclear charge minus the number of electrons; as a consequence, the ionic charge may be negative.

  (d)   A period is a horizontal row in the periodic table. A group is a vertical column, containing elements of similar chemical and physical properties.

  (e)   A metal is an element that has a luster, is malleable, and conducts electricity and heat well. Also, metal atoms tend to form cations in chemical reactions. A nonmetal does not conduct heat or electricity well, and solid nonmetals typically are dull and brittle. Nonmetal atoms tend to form anions in chemical reactions.

  (f)   Avogadro's constant is equal to the number of particles of any type that are present in a mole.

4.   By the law of conservation of mass, all of the magnesium initially present and all of the oxygen that reacted are present in the product. Thus, the mass of oxygen that has reacted is obtained by difference.

   mass of oxygen = 0.674g magnesium oxide − 0.406g magnesium = 0.268g oxygen

<u>**5.**</u>   Again we use the law of conservation of mass. The mass of the starting materials equals the mass of substances present after the reaction is complete.
   mass of potassium + mass of chlorine = mass of potassium chloride + mass of unreacted chlorine
   1.205g potassium + 6.815g chlorine = mass of potassium chloride + 3.300g unreacted chlorine
   mass of potassium chloride = (1.205g + 6.815g) - 3.300g = 4.720 g potassium chloride

6.   No solid residue is produced when (solid) sulfur completely burns because the product of combustion is sulfur dioxide gas. The law of conservation of mass is satisfied because the mass of the sulfur dioxide equals the sum of the masses of sulfur and oxygen that react.

7.   If the two elements combine in the ratio 1:1, there will be one atom of sodium present for each atom of chlorine. To determine the mass percent sodium, we simply convert these atomic quantities to masses (in u) and convert the resulting ratio to a percent.

$$\text{percent Na} = \frac{1\,\text{Na atom} \times \dfrac{22.99\,\text{u Na}}{1\,\text{Na atom}}}{\left(1\,\text{Na atom} \times \dfrac{22.99\,\text{u Na}}{1\,\text{Na atom}}\right) + \left(1\,\text{Cl atom} \times \dfrac{35.45\,\text{u Cl}}{1\,\text{Cl atom}}\right)} \times 100\% = 39.34\%\,\text{Na}$$

8.   **(a)**   The mass of oxygen present in 0.166 g magnesium oxide is the remainder when the 0.100 g magnesium is deducted, or 0.066 g oxygen. Hence, there is 0.066 g oxygen/0.166 g magnesium oxide.

(b)   From the numbers we have already obtained, we see that there is 0.066 g oxygen/0.100 g magnesium, or 0.66 g oxygen/1.00 g magnesium.

(c) % Mg, by $\text{mass} = \dfrac{0.100\text{g Mg}}{0.166\text{g MgO}} \times 100\% = 60.2\%$ Mg

9.   **(a)**   We can determine that carbon dioxide has a fixed composition by finding the % C in each sample. (In the calculations below, abbreviation "cmpd" is short for compound.)

$$\%C = \frac{3.62\text{g C}}{13.26\text{g cmpd}} \times 100\% = 27.3\%\,C \qquad \%C = \frac{5.91\text{g C}}{21.66\text{g cmpd}} \times 100\% = 27.3\%\,C$$

$$\%C = \frac{7.07\text{g C}}{25.91\text{g cmpd}} \times 100\% = 27.3\%\,C$$

Since all three samples have the same percent of carbon, these data do establish that carbon dioxide has a fixed composition.

(b)   Carbon dioxide contains only carbon and oxygen. The percent of oxygen in carbon dioxide is obtained by difference. $\%O = 100.0\% - 27.3\%\,C = 72.7\%O$

10.   By dividing the mass of the oxygen per gram of sulfur in the second sulfur-oxygen compound (compound 2) by the mass of oxygen per gram of sulfur in the first sulfur-oxygen compound (compound 1), we obtain the ratio (right):

$$\frac{\dfrac{1.497\text{ g of O}}{1.000\text{ g of S}}\,(\text{comp 2})}{\dfrac{0.998\text{ g of O}}{1.000\text{ g of S}}\,(\text{comp 1})} = \frac{1.500}{1}$$

To get the simplest whole number ratio we need to multiply both the numerator and the denominator by 2. This gives the simple whole number ratio 3/2. In other words, for a given mass of sulfur, the mass of oxygen in the second compound (SO_3) relative to the mass of oxygen in the first compound (SO_2) is in a ratio of 3:2. These results are entirely consistent with the Law of Multiple Proportions because the same two elements, sulfur and oxygen in this case, have reacted together to give two different compounds that have masses of oxygen that are in the ratio of small positive integers for a fixed amount of sulfur.

11. This question is similar to question 10 in that two elements, phosphorus and chlorine in this case, have combined to give two different compounds. This time, however, different masses have been used for both of the elements in the second reaction. To see if the Law of Multiple Proportions is being followed, the mass of one of the two elements must be set to the same value in both reactions. This can be achieved by dividing the masses of both phosphorus and chlorine in reaction 2 by 2.500:

$$\text{"normalized" mass of phosphorus} = \frac{2.500 \text{ g phosphorus}}{2.500} = 1.000 \text{ g of phosphorus}$$

$$\text{"normalized" mass of chlorine} = \frac{14.308 \text{ g chlorine}}{2.500} = 5.723 \text{ g of chlorine}$$

Now the mass of phosphorus for both reactions is fixed at 1.000 g. Next, we will divide each amount of chlorine by the fixed mass of phosphorus with which they are combined. This gives

$$\frac{\dfrac{3.433 \text{ g of Cl}}{1.000 \text{ g P}} \text{(reaction 1)}}{\dfrac{5.723 \text{ g of Cl}}{1.000 \text{ g P}} \text{(reaction 2)}} = 0.600 = 6{:}10 \text{ or } 3{:}5$$

12. By knowing that all of the 4.15 g of magnesium reacts, producing only magnesium bromide and leaving excess bromine unreacted, we are unable at this point to calculate the mass of magnesium bromide produced. In order to perform this calculation, we need to know how many moles of bromine are combined with each mole of magnesium in the compound.

13.

Name	Symbol	# of protons	# of electrons	# of neutrons	Mass number
sodium	^{23}Na	11	11	12	23
silicon	^{28}Si	14	14[a]	14	28
rubidium	^{85}Rb	37	37[a]	48	85
potassium	^{40}K	19	19	21	40
arsenic[a]	^{75}As	33[a]	33	42	75
neon	^{20}Ne^{2+}	10	8	10	20
bromine[b]	^{80}Br	35	35	45	80
lead[b]	^{208}Pb	82	82	126	208

[a]This result assumes that a neutral atom is involved.

[b]Insuficient data. Does not characterize a specific nuclide; several possibilities exist.

The minimum information needed is the atomic number (or some way to obtain it: the name or the symbol of the element involved), the number of electrons (or some way to obtain it, such as the charge on the species), and the mass number (or the number of neutrons).

14. **(a)** Since all of these species are neutral atoms, the number of electrons are the atomic numbers, the subscript numbers. The symbols must be arranged in order of increasing value of these subscripts.

$$^{40}_{18}Ar < ^{39}_{19}K < ^{58}_{27}Co < ^{59}_{29}Cu < ^{120}_{48}Cd < ^{112}_{50}Sn < ^{122}_{52}Te$$

(b) The number of neutrons is given by the difference between the mass number and the atomic number, $A - Z$. This is the difference between superscripted and subscripted values and are provided (in parentheses) after each element in the following list.

$$^{39}_{19}K(20) < ^{40}_{18}Ar(22) < ^{59}_{29}Cu(30) < ^{58}_{27}Co(31) < ^{112}_{50}Sn(62) < ^{122}_{52}Te(70) < ^{120}_{48}Cd(72)$$

(c) Here the nuclides are arranged by increasing mass number, given by the superscripts.

$$^{39}_{19}K < ^{40}_{18}Ar < ^{58}_{27}Co < ^{59}_{29}Cu < ^{112}_{50}Sn < ^{120}_{48}Cd < ^{122}_{52}Te$$

15. **(a)** cobalt-60 $^{60}_{27}Co$ **(b)** phosphorus-32 $^{32}_{15}P$ **(c)** iodine-131 $^{131}_{53}I$ **(d)** sulfur-35 $^{35}_{16}S$

16. The nucleus of $^{138}_{56}Ba$ contains 56 protons and $(138 - 56) = 82$ neutrons. Thus, the percent of nucleons that are neutrons is given by

$$\% \text{ neutrons} = \frac{82 \text{ neutrons}}{138 \text{ nucleons}} \times 100 = 59\% \text{ neutrons}$$

17. The weighted-average atomic mass of the element iridium is just slightly more than 192 u. The mass of the first isotope is a bit less than 191 u. Hence, the mass of the second isotope must more than 192 u; that isotope must be ^{193}Ir.

18. If we let x represent the number of protons, then $x + 2$ is the number of neutrons. The mass number is the sum of the number of protons and the number of neutrons: $38 = x + (x + 2) = 2x + 2$. We solve this expression for x, and obtain $x = 18$. This is the number of protons of the nuclide and equals the atomic number. Reference to the periodic table indicates that 18 is the atomic number of the element argon.

19. Each isotopic mass must be divided by the isotopic mass of ^{12}C, 12 u, an exact number.

(a) $^{35}Cl \div ^{12}C = 34.96885u \div 12u = 2.914071$

(b) $^{26}Mg \div ^{12}C = 25.98259u \div 12u = 2.165216$

(c) $^{222}Rn \div ^{12}C = 222.0175u \div 12u = 18.50146$

20. We need to work through the mass ratios in sequence to determine the mass of ^{81}Br.

mass of ^{19}F = mass of $^{12}C \times 1.5832 = 12.00000\,u \times 1.5832 = 18.998\,u$

mass of ^{35}Cl = mass of $^{19}F \times 1.8406 = 18.998\,u \times 1.8406 = 34.968\,u$

mass of ^{81}Br = mass of $^{35}Cl \times 2.3140 = 34.968\,u \times 2.3140 = 80.917\,u$

21. Each of the isotopic masses is multiplied by its fractional abundance. The resulting products are summed to obtain the average atomic mass.

contribution from ^{40}Ar = $39.9624\,u \times 0.99600 = 39.803u$

contribution from ^{36}Ar = $35.96755\,u \times 0.00337 = 0.121u$

contribution from ^{38}Ar = $37.96272\,u \times 0.00063 = 0.024u$

average atomic mass of argon = $39.803u + 0.121u + 0.024u = 39.948u$

Of course, this calculation can be performed in one step:

$(39.9624u \times 0.99600)+(35.96755u \times 0.00337)+(37.96272u \times 0.00063) = 39.948u$

22. **(a)** In is in group 13(3A) and in the fifth period.

(b) Other elements in group 16(6A) are similar to S: O, Se, Te. Most of the elements in the periodic table are unlike S, but particularly metals such as Na, K, Rb.

(c) The alkali metal in the sixth period is in group 1(1A), Cs.

(d) The halogen (group 17(7A)) in the fifth period is I.

(e) The element with atomic number 18 is Ar, a noble gas. Xe is a noble gas with atomic number (54) greater than 50.

(f) If an element forms an anion with charge 3-, it is in group 15(5A).

(g) If an element forms a cation with charge 2+, it is in group 2(2A)

23. If the seventh period of the periodic table is 32 members long, it will be the same length as the sixth period. Elements in the same family will have atomic numbers 32 units higher. The noble gas following radon will have atomic number = $86+32 =118$. The alkali metal following francium will have atomic number = $87+32 =119$.

24. One mole of any element contains 6.022×10^{23} atoms, the Avogadro constant.

(a) $12.7\,mol\,Ca \times \dfrac{6.022 \times 10^{23}\,Ca\,atoms}{1\,mol\,Ca} = 7.65 \times 10^{24}\,Ca\,atoms$

(b) $0.00361\,mol\,Ne \times \dfrac{6.022 \times 10^{23}\,Ne\,atoms}{1\,mol\,Ne} = 2.17 \times 10^{21}\,Ne\,atoms$

(c) $1.8 \times 10^{-12}\,mol\,Pu \times \dfrac{6.022 \times 10^{23}\,Pu\,atoms}{1\,mol\,Pu} = 1.1 \times 10^{12}\,Pu\,atoms$

25. In these problems we use the Avogadro constant and the fact that one mole of atoms of an element has a weight in grams equal to its atomic mass.

(a) $\text{no. moles Fe} = 2.18 \times 10^{26} \text{ Fe atoms} \times \dfrac{1 \text{ mol Fe}}{6.022 \times 10^{23} \text{ Fe atoms}} = 362 \text{ mol Fe}$

(b) $\text{mass of Kr, g} = 7.71 \text{ mol Kr} \times \dfrac{83.80 \text{ g Kr}}{1 \text{ mol Kr}} = 646 \text{ g Kr}$

(c) $\text{Au mass, mg} = 6.15 \times 10^{19} \text{ Au atoms} \times \dfrac{1 \text{ mol Au}}{6.022 \times 10^{23} \text{ atoms}} \times \dfrac{196.97 \text{ g Au}}{1 \text{ mol Au}} \times \dfrac{1000 \text{ mg}}{1 \text{ g}}$

$\text{Au atoms} = 20.1 \text{ mg Au}$

(d) $\text{Fe atoms} = 112 \text{ cm}^3 \text{ Fe} \times \dfrac{7.86 \text{ g Fe}}{1 \text{ cm}^3 \text{ Fe}} \times \dfrac{1 \text{ mol Fe}}{55.85 \text{ g Fe}} \times \dfrac{6.022 \times 10^{23} \text{ atoms}}{1 \text{ mol Fe}}$

$\text{Fe atoms} = 9.49 \times 10^{24} \text{ Fe atoms}$

26. Since the molar mass of nitrogen is 14.0 g/mol, 25.0 g N is almost two moles (1.79 mol N), while 6.02×10^{23} Ni atoms is about one mole, and 52.0 g Cr (52.00 g/mol Cr) is also almost one mole. Finally, 10.0 cm^3 Fe (55.85 g/mol Fe) has a mass of about 79 g, and contains about 1.4 moles of atoms. Thus, 25.0 g N contains the greatest number of atoms.

27. We first determine the number of Pb atoms of all types in 1.57 g of Pb, and then use the percent abundance to determine the number of ^{204}Pb atoms present.

$\text{atoms of } ^{204}\text{Pb} = 215 \text{ mg Pb} \times \dfrac{1 \text{ g}}{1000 \text{ mg}} \times \dfrac{1 \text{ mol Pb}}{207.2 \text{ g Pb}} \times \dfrac{6.022 \times 10^{23} \text{ atoms}}{1 \text{ mol Pb}} \times \dfrac{14 \ ^{204}\text{Pb atoms}}{1000 \text{ Pb atoms}}$

$= 8.7 \times 10^{18} \text{ atoms } ^{204}\text{Pb}$

28. $\text{mass of alloy} = 6.50 \times 10^{23} \text{ Cd atoms} \times \dfrac{1 \text{ mol Cd}}{6.022 \times 10^{23} \text{ Cd atoms}} \times \dfrac{112.4 \text{ g Cd}}{1 \text{ mol Cd}} \times \dfrac{100.0 \text{ g alloy}}{8.0 \text{ g Cd}}$

$= 1.5 \times 10^3 \text{ g alloy}$

EXERCISES

Law of Conservation of Mass

29. The observations cited do not necessarily violate the law of conservation of mass. The oxide formed when iron rusts is a solid and remains with the solid iron, increasing the mass of the solid by an amount equal to the mass of the oxygen that has combined. The oxide formed when a match burns is a gas and will not remain with the solid product (the ash); the mass of the ash thus is less than that of the match. We would have to collect all reactants and all products and weigh them to determine if the law of conservation of mass is obeyed or violated.

30. The magnesium that is burned in air combines with some of the oxygen in the air and this oxygen (which, of course, was not weighed when the magnesium metal was weighed) adds its mass to the mass of the magnesium, making the magnesium oxide product weigh more than did the original magnesium. When this same reaction is carried out in a photoflash bulb, the oxygen (in fact, some excess oxygen) that will combine with the magnesium is already present in the bulb before the reaction. Consequently, the product contains no unweighed oxygen.

31. We compare the mass before reaction (initial) with that after reaction (final).

initial mass = 10.500 g calcium hydroxide + 11.125 g ammonium chloride = 21.625 g

final mass = 14.336 g solid residue + (69.605 –62.316) g of gases = 21.625 g

These data support the law of conservation of mass. Note that the gain in the mass of water is due to the gases that it absorbs.

32. We compute the mass of the reactants and compare that with the mass of the products.

reactant mass = mass of calcium carbonate + mass of hydrochloric acid solution

$$= 10.00\text{g calcium carbonate} + 100.0\,\text{mL soln} \times \frac{1.148\,\text{g}}{1\text{mL soln}}$$

$$= 10.00\text{g calcium carbonate} + 114.8\,\text{g solution} = 124.8\text{g reactants}$$

product mass = mass of solution + mass of carbon dioxide

$$= 120.40\,\text{g soln} + 2.22\text{L gas} \times \frac{1.9769\,\text{g}}{1\text{L gas}} = 120.40\,\text{g soln} + 4.39\,\text{g carbon dioxide}$$

$$= 124.79\,\text{g products} \begin{pmatrix} \text{Same mass within experimental error,} \\ \text{Law of conservation of mass obeyed} \end{pmatrix}$$

Law of Constant Composition

33. In the first experiment, 2.18 g of sodium produces 5.54 g of sodium chloride. In the second experiment, 2.10 g of chlorine produces 3.46 g of sodium chloride. The amount of sodium contained in this second sample of sodium chloride is given by
mass of sodium = 3.46 g sodium chloride –2.10 g chlorine = 1.36 g sodium
We now have sufficient information to determine the % Na in each of the samples of sodium chloride.

$$\%\text{Na} = \frac{2.18\,\text{g Na}}{5.54\,\text{g cmpd}} \times 100\% = 39.4\% \text{ Na} \qquad \%\text{Na} = \frac{1.36\,\text{g Na}}{3.46\,\text{g cmpd}} \times 100\% = 39.3\% \text{ Na}$$

Thus, the two samples of sodium chloride have the same composition. Recognize that, according to the interpretation of numbers based on significant figures, each percent has an uncertainty of ±0.1%.

34. If the two samples of water have the same % H, the law of constant composition is demonstrated. Notice that, in the second experiment, the mass of the compound is equal to the sum of the masses of the elements produced from it.

$$\%H = \frac{3.06\,g}{27.35\,g\ H_2O} \times 100\% = 11.2\%\ H \qquad \%H = \frac{1.45g\ H}{(1.45+11.51)g\ H_2O} \times 100\% = 11.2\%\ H$$

The results are consistent with the law of constant composition.

35. The mass of sulfur (0.312 g) needed to produce 0.623 g sulfur dioxide provides the information for the conversion factor.

$$\text{sulfur mass} = 0.842\,g\ \text{sulfur dioxide} \times \frac{0.312\,g\ \text{sulfur}}{0.623\,g\ \text{sulfur dioxide}} = 0.422\,g\ \text{sulfur}$$

36. (a) From the first experiment we see that 1.16 g of compound is produced per gram of Hg. These masses enable us to determine the mass of compound produced from 1.50 g Hg.

$$\text{mass of cmpd} = 1.50\,g\ Hg \times \frac{1.16\,g\ \text{cmpd}}{1.00\,g\ Hg} = 1.74\,g\ \text{cmpd}$$

(b) Since the compound weighs 0.24 g more than the mass of mercury (1.50 g) that was used, 0.24 g of sulfur must have reacted. Thus, the unreacted sulfur has a mass of 0.76 g (=1.00 g initially present −0.24 g reacted).

Law of Multiple Proportions

37. (a) First of all we need to fix the mass of nitrogen in all three compounds at 1.000 g. This can be accomplished by multiplying the masses of hydrogen and nitrogen in compound A by 2 and the amount of hydrogen and nitrogen in compound C by 4/3 (1.333):

Comp. A: "normalized" mass of nitrogen   = 0.500 g N × 2 = 1.000 g N

"normalized" mass of hydrogen   = 0.108 g H × 2 = 0.216 g H

Comp. C: "normalized" mass of nitrogen   = 0.750 g N × 1.333 = 1.000 g N

"normalized" mass of hydrogen   = 0.108 g H × 1.333 = 0.144 g H

Next, we divide the mass of hydrogen in each compound by the smallest mass of hydrogen, namely, 0.0720 g. This gives 3.000 for compound A, 1.000 for compound B and 2.00 for compound C. The ratio of the amounts of hydrogen in the three compounds is 3 (comp A) : 1 (comp B) : 2 (comp C)

These results are consistent with the Law of Multiple Proportions because the masses of hydrogen in the three compounds end up in a ratio of small whole numbers when the mass of nitrogen in all three compounds is normalized to a simple value (1.000 g here).

(b)   The text states that compound B is N_2H_2. This means that, based on the relative amounts of hydrogen calculated in part (a), compound A might be N_2H_6 and compound C, N_2H_4. Actually, compound A is NH_3, but we have no way of knowing this from the data. Note that the H:N ratio in NH_3 and N_2H_6 are the same, 3H:1N.

38. **(a)**   As with the previous problem, one of the two elements must have the same mass in all of the compounds. This can be most readily achieved by setting the mass of iodine in all four compounds to 1.000 g. With this approach we only need to manipulate the data for compounds B and C. To normalize the amount of iodine in compound B to 1.000 g, we need to multiply the masses of both iodine and fluorine by 2. To accomplish the analogous normalization of compound C, we must multiply by 4/3 (1.333).

Comp. B: "normalized" mass of iodine = 0. 500 g I × 2   = 1.000 g I
          "normalized" mass of fluorine = 0.2246 g F × 2   = 0.4492 g F

Comp. C: "normalized" mass of iodine   = 0.750 g I   × 1.333 = 1.000 g I
          "normalized" mass of fluorine = 0.5614 g F × 1.333 = 0.7485 g F

Next we divide the mass of fluorine in each compound by the smallest mass of fluorine, namely, 0.1497 g. This gives 1.000 for compound A, 3.001 for compound B, 5.000 for compound C and 7.001 for compound D. The ratios of the amounts of fluorine in the four compounds A : B : C : D is 1 : 3 : 5 : 7. These results are consistent with the law of multiple proportions because for a fixed amount of iodine (1.000 g), the masses of fluorine in the four compounds are in the ratio of small whole numbers.

(b)   As with the preceding problem, we can figure out the empirical formulas for the four iodine-fluorine containing compounds from the ratios of the amounts of fluorine that were determined in 38(a):   Comp A: IF   Comp B: IF_3 Comp C: IF_5 Comp C: IF_7

39.   One oxide of copper has about 20% oxygen by mass. If we assume a 100 gram sample, then ~ 20 grams of the sample is oxygen (~1.25 moles) and 80 grams is copper (~1.26 moles). This would give an empirical formula of CuO (copper (II) oxide). The second oxide has less oxygen by mass, hence the empirical formula must have less oxygen or more copper (Cu:O ratio greater than 1). If we keep whole number ratios of atoms, a plausible formula would be Cu_2O (copper (I) oxide), where the mass percent oxygen is ≈11%.

40.   Assuming the intermediate is "half-way" between CO (oxygen: carbon mass ratio = 16:12 or 1.333) and CO_2 (oxygen: carbon mass ratio = 32:12 or 2.6667), then the oxygen: carbon ratio would be 2:1, or O:C = 24:12. This mass ratio gives a mole ratio of O:C = 1.5:1. Empirical formulas are simple whole number ratios of elements; hence, a formula of C_3O_2 must be the correct empirical formula for this carbon oxide.

Fundamental Particles

41. A fundamental particle would be expected to be found in all samples of matter. For instance, cathode rays, which are beams of "free" electrons, have the same properties no matter how they are generated. These properties are independent of the material that was used to construct the cathode ray tube, of the gas that filled the tube when it was constructed (and was subsequently pumped out), and of the method used to generate electricity.

42. The detection and characterization of electrons was based on the fact that they are charged particles. For instance, in Thompson's experiment, a beam of electrons is made to curve by the force between a magnetic field and a beam of charged particles. A neutron, however, does not have a charge and thus cannot be detected or characterized by the methods used for electrons.

Fundamental Charges and Mass-to-Charge Ratios

43. We can calculate the charge on each drop, express each in terms of 10^{-19} C, and finally express each in terms of $e = 1.6 \times 10^{-19}$ C.

drop 1: $\quad 1.28 \times 10^{-18}$ $\qquad\qquad\qquad = 12.8 \times 10^{-19}$ C $\quad = 8e$

drops 2 & 3: $1.28 \times 10^{-18} \div 2 = 0.640 \times 10^{-18}$ C $= 6.40 \times 10^{-19}$ C $\quad = 4e$

drop 4: $\quad 1.28 \times 10^{-18} \div 8 = 0.160 \times 10^{-18}$ C $= 1.60 \times 10^{-19}$ C $\quad = 1e$

drop 5: $\quad 1.28 \times 10^{-18} \times 4 = 5.12 \times 10^{-18}$ C $\quad = 51.2 \times 10^{-19}$ C $\quad = 32e$

We see that these values are consistent with the charge that Millikan found for that of the electron, and he could have inferred the correct charge from these data, since they are all multiples of e.

44. We calculate each drop's charge; express each in terms of 10^{-19} C, and then in terms of $e = 1.6 \times 10^{-19}$ C.

drop 1: $\quad 6.41 \times 10^{-19}$ C $\qquad\qquad\qquad = 6.41 \times 10^{-19}$ C $\quad = 4e$

drop 2: $\quad 6.41 \times 10^{-19} \div 2$ $\quad = 3.21 \times 10^{-19}$ C $\quad = 3.21 \times 10^{-19}$ C $\quad = 2e$

drop 3: $\quad 6.41 \times 10^{-19} \times 2$ $\quad = 1.28 \times 10^{-18}$ C $\quad = 12.8 \times 10^{-19}$ C $\quad = 8e$

drop 4: $\quad 1.44 \times 10^{-18}$ $\qquad\qquad\qquad = 14.4 \times 10^{-19}$ C $\quad = 9e$

drop 3: $\quad 1.44 \times 10^{-18} \div 3$ $\quad = 4.8 \times 10^{-19}$ C $\quad = 4.8 \times 10^{-19}$ C $\quad = 3e$

We see that these values are consistent with the charge that Millikan found for that of the electron. He could have inferred the correct charge from these values, since they are all multiples of e, and have no other common factor.

45. **(a)** Determine the ratio of the mass of a hydrogen atom to that of an electron. We use the mass of a proton plus that of an electron for the mass of a hydrogen atom.

$$\frac{\text{mass of proton} + \text{mass of electron}}{\text{mass of electron}} = \frac{1.0073\,u + 0.00055\,u}{0.00055\,u} = 1.8 \times 10^3$$

$$or \quad \frac{\text{mass of electron}}{\text{mass of proton} + \text{mass of electron}} = \frac{1}{1.8 \times 10^3} = 5.6 \times 10^{-4}$$

(b) The only two mass-to-charge ratios that we can determine from the data in Table 2-1 are those for the proton, a hydrogen ion, H^+; and that for the electron.

For the proton : $\quad \dfrac{\text{mass}}{\text{charge}} = \dfrac{1.673 \times 10^{-24}\,g}{1.602 \times 10^{-19}\,C} = 1.044 \times 10^{-5}\,g/C$

For the electron : $\quad \dfrac{\text{mass}}{\text{charge}} = \dfrac{9.109 \times 10^{-28}\,g}{1.602 \times 10^{-19}\,C} = 5.686 \times 10^{-9}\,g/C$

The hydrogen ion is the lightest positive ion available. We see that the mass-to-charge ratio for a positive particle is considerably larger than that for an electron.

46. We do not have the precise isotopic masses for the two ions. The values of the mass-to-charge ratios are only approximate. Consequently, we have used a three-significant figure mass for a nucleon, rather than the more precisely known proton and neutron masses. (Recall that the term "nucleon" refers to a nuclear particle— either a proton or a neutron.)

$^{127}I^-$ $\quad \dfrac{m}{e} = \dfrac{127\,\text{nucleons}}{1\,\text{electron}} \times \dfrac{1e}{1.602 \times 10^{-19}\,C} \times \dfrac{1.67 \times 10^{-24}\,g}{1\,\text{nucleon}} = 1.32 \times 10^{-3}\,g/C \;\; (7.55 \times 10^2\,C/g)$

$^{32}S^{2-}$ $\quad \dfrac{m}{e} = \dfrac{32\,\text{nucleons}}{2\,\text{electrons}} \times \dfrac{1e}{1.602 \times 10^{-19}\,C} \times \dfrac{1.67 \times 10^{-24}\,g}{1\,\text{nucleon}} = 1.67 \times 10^{-4}\,g/C \;\; (6.00 \times 10^3\,C/g)$

Atomic Number, Mass Number, and Isotopes

47. **(a)** A ^{108}Pd atom has 46 protons, and 46 electrons. The atom described is neutral, hence, the number of electrons must equal the nunber of protons. Since there are 108 nucleons in the nucleus, the number of neutrons is 62 ($= 108$ nucleons $- 46$ protons).

(b) The ratio of the two masses is determined as follows:
$$\frac{^{108}Pd}{^{12}C} = \frac{107.90389\,u}{12\,u} = 8.9919908$$

48. **(a)** The atomic number of Ra is 88 and equals the number of protons in the nucleus. The ion's charge is 2+ and, thus, there are two more protons than electrons: no. protons = no. electrons $+ 2 = 88$; no. electrons $= 88 - 2 = 86$. The mass number (228) is the sum of the atomic number and the number of neutrons: $228 = 88 +$ no. neutrons; Hence, the number of neutrons $= 228 - 88 = 140$ neutrons.

(b) The mass of ^{16}O is 15.9949 u.

$$\text{ratio} = \frac{\text{mass of isotope}}{\text{mass of}^{16}O} = \frac{228.030\,u}{15.9949\,u} = 14.2564$$

49. The mass of ^{16}O is 15.9949 u. Isotopic mass $= 15.9949\,u \times 6.68374 = 106.906\,u$

50. The mass of ^{16}O is 15.9949 u.

mass of heavier isotope $= 15.9949\,u \times 7.1838 = 114.90\,u = $ mass of ^{115}In

mass of lighter isotope $= \dfrac{114.90\,u}{1.0177} = 112.90\,u = $ mass of ^{113}In

51. **(a)** Atoms with equal numbers of protons and neutrons will have mass numbers that are approximately twice the size of their atomic numbers. The following species are approximately suitable (with numbers of protons and neutrons in parentheses).

$^{24}_{12}Mg^{2+}$ (12 p, 12 n), $^{47}_{24}Cr$ (24 p, 23 n), $^{60}_{27}Co^{3+}$ (27 p, 32 n), and $^{35}_{17}Cl^-$ (17 p, 18 n).

Of these four nuclides, only $^{24}_{12}Mg^{2+}$ has just as many protons as neutrons.

(b) A species in which protons have more than 50% of the mass must have a mass number smaller than twice the atomic number. Of these species, only in $^{47}_{24}Cr$ is more than 50% of the mass contributed by the protons.

(c) A species with 50% more neutrons than protons will have a mass number equal to 2.5 times the atomic number. $^{226}_{90}Th$ has more than 50% more neutrons than protons.

52. To answer these questions, we determine the number of protons, neutrons, and electrons in each species.

species:	$^{24}_{12}Mg^{2+}$	$^{47}_{24}Cr$	$^{60}_{27}Co^{3+}$	$^{35}_{17}Cl^-$	$^{120}_{50}Sn^{2+}$	$^{226}_{90}Th$	$^{90}_{38}Sr$
no. protons	12	24	27	17	50	90	38
no. neutrons	12	23	33	18	70	136	52
no. electrons	10	24	24	18	48	90	38

(a) The number of neutrons and electrons is equal for $^{35}_{17}Cl^-$.

(b) $^{60}_{27}Co^{3+}$ has protons (27), neutrons (33), and electrons (24) in the ratio 9:11:8.

(c) The species $^{124}_{50}Sn^{2+}$ has a number of neutrons (74) equal to its number of protons (50) plus one-half its number of electrons $(48 \div 2 = 24)$.

Atomic Mass Units, Atomic Masses

53. It is exceedingly unlikely that another nuclide would have an exact integral mass. The mass of carbon-12 is *defined* as precisely 12 u. Each nuclidic mass is close to integral, but none that we have encountered in this chapter are precisely integral. The reason is that each nuclide is composed of protons, neutrons, and electrons, none of which have integral masses, and there is a small quantity of the mass of each nucleon (nuclear particle) lost in the binding energy holding the nuclides together. It would be highly unlikely that all of these contributions would add up to a precisely integral mass.

54. There are no copper atoms that have a mass of 63.546 u. The masses of individual atoms are close to integers and this mass (63.546 u) is about midway between two integers. It is an average atomic mass, the result of averaging two (or more) isotopic masses, each weighted by its natural abundance.

55. To determine the average atomic mass, we use the following expression:
$$\text{average atomic mass} = \sum (\text{isotopic mass} \times \text{fractional natural abundance})$$
Each of the three percents given is converted to a fractional abundance by dividing it by 100.

$$\text{Mg atomic mass} = (23.985042\,u \times 0.7899) + (24.985837\,u \times 0.1000) + (25.982593\,u \times 0.1101)$$
$$= 18.95\,u + 2.499\,u + 2.861\,u = 24.31\,u$$

56. To determine the average atomic mass, we use the following expression
$$\text{average atomic mass} = \sum (\text{isotopic mass} \times \text{fractional natural abundance})$$
Each of the three percents given is converted to a fractional abundance by dividing it by 100.

$$\text{Cr atomic mass} = (49.9461 \times 0.0435) + (51.9405 \times 0.8379) + (52.9407 \times 0.0950) + (53.9389 \times 0.0236)$$
$$= 2.17\,u + 43.52\,u + 5.03\,u + 1.27\,u = 51.99\,u$$

57. We use the expression for determining the weighted-average atomic mass.

$$107.868\,u = (106.905092\,u \times 0.5184) + ({}^{109}Ag \times 0.4816) = 55.42\,u + 0.4816\,{}^{109}Ag$$

$$107.868\,u - 55.42\,u = 0.4816\,{}^{109}Ag = 52.45\,u \qquad {}^{109}Ag = \frac{52.45\,u}{0.4816} = 108.9\,u$$

58. The percent abundances of the two isotopes must add to 100.00%, since there are only two naturally occurring isotopes of bromine. Thus, we can determine the percent natural abundance of the second isotope by difference.

$$\% \text{ second isotope} = 100.00\% - 50.69\% = 49.31\%$$

From the periodic table, we see that the weighted-average atomic mass of bromine is 79.904 u. We use this value in the expression for determining the weighted-average atomic mass, along with the isotopic mass of ^{79}Br and the fractional abundances of the two isotopes (the percent abundances divided by 100).

$$79.904\,u = (0.5069 \times 78.918336\,u) + (0.4931 \times \text{other isotope}) = 40.00\,u + (0.4931 \times \text{other isotope})$$

$$\text{other isotope} = \frac{79.904\,u - 40.00\,u}{0.4931} = 80.92\,u = \text{mass of } {}^{81}\text{Br, the other isotope}$$

59. Since the three percent abundances total 100%, the percent abundance of ^{40}K is found by difference.

$$\%\ {}^{40}\text{K} = 100.0000\% - 93.2581\% - 6.7302\% = 0.0117\%$$

Then the expression for the weighted-average atomic mass is used, with the percent abundances converted to fractional abundances by dividing by 100. The average atomic mass of potassium is 39.0983 u.

$$39.0983\,u = (0.932581 \times 38.963707\,u) + (0.000117 \times 39.963999\,u) + (0.067302 \times {}^{41}\text{K})$$

$$= 36.3368\,u + 0.00468\,u + (0.067302 \times {}^{41}\text{K})$$

$$\text{mass of } {}^{41}\text{K} = \frac{39.0983\,u - (36.3368\,u + 0.00468\,u)}{0.067302} = 40.962\ u$$

60. We use the expression for determining the weighted-average atomic mass, where x represents the fractional abundance of ^{10}B and $(1-x)$ the fractional abundance of ^{11}B

$$10.81\,u = (10.012937\,u \times x) + [11.009305 \times (1-x)] = 10.012937x + 11.009305 - 11.009305x$$

$$10.81 - 11.009305 = -0.20 = 10.012937x - 11.009305x = -0.996368x$$

$$x = \frac{0.20}{0.996368} = 0.20 \qquad \therefore 20.\%\ {}^{10}\text{B} \quad \text{and} \quad (100.0 - 20.) = 80.\%\ {}^{11}\text{B}$$

Mass spectrometry

61. (a)

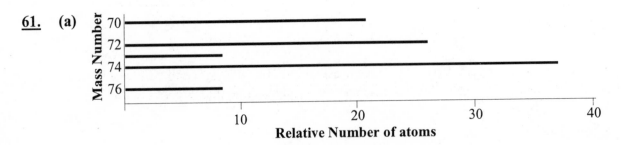

(b) As before, we multiply each isotopic mass by its fractional abundance; after which, we sum these products to obtain the (average) atomic mass for the element.

$$(0.205 \times 70) + (0.274 \times 72) + (0.078 \times 73) + (0.365 \times 74) + (0.078 \times 76)$$

$$14 + 20. + 5.7 + 27 + 5.9 = 72._6 = \text{average atomic mass of germanium}$$

The result is only approximately correct because the isotopic masses are given to only two significant figures. Thus, only a two-significant-figure result can be quoted.

62. (a) Six unique HCl molecules are possible (called isotopomers):

$$^{1}H^{35}Cl, \quad ^{2}H^{35}Cl, \quad ^{3}H^{35}Cl, \quad ^{1}H^{37}Cl, \quad ^{2}H^{37}Cl, \quad \text{and } ^{3}H^{37}Cl$$

The mass numbers of the six different possible types of molecules are obtained by summing the mass numbers of the two atoms in each molecule:

$^{1}H^{35}Cl$ has $A = 36$ $\qquad$ $^{2}H^{35}Cl$ has $A = 37$ $\qquad$ $^{3}H^{35}Cl$ has $A = 38$

$^{1}H^{37}Cl$ has $A = 38$ $\qquad$ $^{2}H^{37}Cl$ has $A = 39$ $\qquad$ $^{3}H^{37}Cl$ has $A = 40$

(b) The most abundant molecule contains the most abundant of each element's isotope. It is $^{1}H^{35}Cl$. The second most abundant molecule is $^{1}H^{37}Cl$. The relative abundance of each type of molecule is determined by multiplying together the fractional abundances of the two isotopes present. Relative abundances of the molecules are as follows.

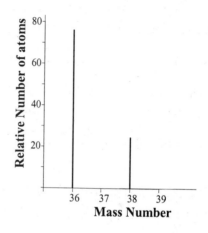

$^{1}H^{35}Cl: 75.76\%$ $\qquad$ $^{1}H^{37}Cl: 24.23\%$

$^{2}H^{35}Cl: < 0.011\%$ $\qquad$ $^{2}H^{37}Cl: < 0.0036\%$

$^{3}H^{35}Cl: < 0.0008\%$ $\qquad$ $^{3}H^{37}Cl: < 0.0002\%$

The Avogadro Constant and the Mole

63. Each of these calculations employs the average atomic mass as a conversion factor.

(a) amount of $Rb = 167.0 \text{ g Rb} \times \dfrac{1 \text{ mol Rb}}{85.468 \text{ g Rb}} = 1.954 \text{ mol Rb}$

(b) number of Fe atoms $= 363.2 \text{ kg Fe} \times \dfrac{1000 \text{ g}}{1 \text{ kg}} \times \dfrac{1 \text{ mol Fe}}{55.847 \text{ g}} \times \dfrac{6.022 \times 10^{23} \text{ Fe atoms}}{1 \text{ mol Fe}}$

$$= 3.916 \times 10^{27} \text{ Fe atoms}$$

(c) Ag mass $= 1.0 \times 10^{12} \text{ Ag atoms} \times \dfrac{1 \text{ mol Ag}}{6.022 \times 10^{23} \text{ Ag atoms}} \times \dfrac{107.87 \text{ g Ag}}{1 \text{ mol Ag}}$

$$= 1.8 \times 10^{-10} \text{ g Ag}$$

(d) Mass of one fluorine atom: 1 mole F = 18.9984 g. Divide both sides by 6.022142×10^{23} atoms (1 mol F). We find the mass of 1 F atom is 3.15476×10^{-23} grams.

64. Each of these calculations employs the average atomic mass as a conversion factor.

(a) $\text{number of Ar atoms} = 5.25 \text{ mg Ar} \times \dfrac{1 \text{ g}}{1000 \text{ mg}} \times \dfrac{1 \text{ mol Ar}}{39.948 \text{ g}} \times \dfrac{6.022 \times 10^{23} \text{ Ar atoms}}{1 \text{ mol Ar}}$

$= 7.91 \times 10^{19} \text{ Ar atoms}$

(b) $\text{molar mass} = \dfrac{4.24 \text{ g}}{2.80 \times 10^{22} \text{ atoms}} \times \dfrac{6.022 \times 10^{23} \text{ atoms}}{1 \text{ mole}} = 91.2 \text{ g/mol}$

(c) $\text{Al mass} = 35.55 \text{ g Zn} \times \dfrac{1 \text{ mol Zn}}{65.39 \text{ g Zn}} \times \dfrac{1 \text{ mol Al}}{1 \text{ mol Zn}} \times \dfrac{26.982 \text{ g Al}}{1 \text{ mol Al}} = 14.67 \text{ g Al}$

Quantities with the same number of moles have the same number of atoms.

65. We determine the mass of Ag in the piece of jewelry and then the number of Ag atoms.

$\text{no. Ag atoms} = 38.7 \text{ g sterling} \times \dfrac{92.5 \text{ g Ag}}{100.0 \text{ g sterling}} \times \dfrac{1 \text{ mol Ag}}{107.9 \text{ g Ag}} \times \dfrac{6.022 \times 10^{23} \text{ atoms}}{1 \text{ mol Ag}}$

$\text{no. Ag atoms} = 2.00 \times 10^{23} \text{ Ag atoms}$

66. We first determine the amount in moles of each metal.

$\text{amount of Pb} = 75.0 \text{ cm}^3 \text{ solder} \times \dfrac{9.4 \text{ g solder}}{1 \text{ cm}^3} \times \dfrac{67 \text{ g Pb}}{100 \text{ g solder}} \times \dfrac{1 \text{ mol Pb}}{207.2 \text{ g Pb}} = 2.3 \text{ mol Pb}$

$\text{amount of Sn} = 75.0 \text{ cm}^3 \text{ solder} \times \dfrac{9.4 \text{ g solder}}{1 \text{ cm}^3} \times \dfrac{33 \text{ g Sn}}{100 \text{ g solder}} \times \dfrac{1 \text{ mol Sn}}{118.7 \text{ g Sn}} = 2.0 \text{ mol Sn}$

$\text{total atoms} = (2.3 \text{ mol Pb} + 2.0 \text{ mol Sn}) \times \dfrac{6.022 \times 10^{23} \text{ atoms}}{1 \text{ mol}} = 2.6 \times 10^{24} \text{ atoms}$

67. We use the average atomic mass of lead, 207.2 g/mol.

(a) $\dfrac{30 \text{ μg Pb}}{1 \text{ dL}} \times \dfrac{1 \text{ dL}}{0.1 \text{ L}} \times \dfrac{1 \text{ g Pb}}{10^6 \text{ μg Pb}} \times \dfrac{1 \text{ mol Pb}}{207.2 \text{ g}} = 1.4 \times 10^{-6} \text{ mol Pb}/\text{L}$

(b) $\dfrac{1.4 \times 10^{-6} \text{ mol Pb}}{\text{L}} \times \dfrac{1 \text{ L}}{1000 \text{ ml}} \times \dfrac{6.022 \times 10^{23} \text{ atoms}}{1 \text{ mol}} = 8.4 \times 10^{14} \text{ Pb atoms}/\text{mL}$

68. The concentration of Pb in air provides the principal conversion factor. Other conversion factors are needed to convert to and from its units, beginning with the 0.500-L volume, and ending with the number of atoms.

$\text{no. Pb atoms} = 0.500 \text{ L} \times \dfrac{1 \text{ m}^3}{1000 \text{ L}} \times \dfrac{3.01 \text{ μg Pb}}{1 \text{ m}^3} \times \dfrac{1 \text{ g Pb}}{10^6 \text{ μg Pb}} \times \dfrac{1 \text{ mol Pb}}{207.2 \text{ g Pb}} \times \dfrac{6.022 \times 10^{23} \text{ Pb atoms}}{1 \text{ mol Pb}}$

$= 4.37 \times 10^{12} \text{ Pb atoms}$

69. Let's begin by finding the volume of copper metal.

Wire diameter (cm) = 0.03196 in $\times \dfrac{2.54 \text{ cm}}{1 \text{ in}}$ = 0.08118 cm

The radius in cm is 0.08118 cm $\times$ 1/2 = 0.04059 cm

The volume of Cu(cm^3) = (0.04059 cm)$^2 \times$ (π) $\times$ (1.00 m $\times \dfrac{100 \text{ cm}}{1 \text{ m}}$) = 0.5176 cm^3

So, the mass of Cu = 0.5176 cm$^3 \times \dfrac{8.92 \text{ g Cu}}{1 \text{ cm}^3}$ = 4.62 g Cu

The number of moles of Cu = 4.62 g Cu $\times \dfrac{1 \text{ mol Cu}}{63.546 \text{ g Cu}}$ = 0.0727 mol Cu

Cu atoms in the wire = 0.0727 mol Cu $\times \dfrac{6.022 \times 10^{23} \text{ atoms Cu}}{1 \text{ mol Cu}}$ = 4.38 $\times$ 10^{22} atoms

70. To answer this question we simply need to calculate the ratio of the mass (in grams) of each sample to its molar mass. Whichever elemental sample gives the largest ratio will be the one that has the greatest number of atoms. Hydrogen sample:

$\dfrac{1.00 \times 10^3 \text{g H}_2}{2 \times (1.00794 \text{ g H})} \times$ 1 mol H = 496 mol of H$_2$ molecules = 992 mol of H atoms

Mercury sample: 76 lb Hg $\times \dfrac{454 \text{ g Hg}}{1 \text{ lb Hg}} \times \dfrac{1 \text{ mol Hg}}{200.6 \text{ g Hg}}$ = 172 mol of Hg atoms

Sulfur sample: $\dfrac{2.00 \times 10^4 \text{ g S}}{32.066 \text{ g S}} \times$ 1 mol S = 624 moles of S atoms

Iron sample: 10 cm $\times$ 10 cm $\times$ 10 cm $\times$ 7.86 g cm^{-3} = 7860 g Fe

$\dfrac{7860 \text{ g Fe}}{55.847 \text{ g Fe}} \times$ 1 mol Fe = 141 moles of Fe atoms

Clearly then, it is the 1.00 kg sample of hydrogen that contains the greatest number of atoms.

FEATURE PROBLEMS

90. The product mass differs from that of the reactants by $(5.62 - 2.50 =)$ 3.12 grains. In order to determine the percent gain in mass, we need to convert the reactant mass to grains.

$$13 \text{ onces} \times \frac{8 \text{ gros}}{1 \text{ once}} = 104 \text{ gros} \times (104 + 2) \text{ gros} \times \frac{72 \text{ grains}}{1 \text{ gros}} = 7632 \text{ grains}$$

$$\% \text{ mass increase} = \frac{3.12 \text{ grains increase}}{(7632 + 2.50) \text{ grains original}} \times 100\% = 0.0409\% \text{ mass increase}$$

The sensitivity of Lavoisier's balance can be as little as 0.01 grain, which seems to be the limit of the readability of the balance; or it can be as large as 3.12 grains, which assumes that all of the error in the experiment is due to the (in)sensitivity of the balance. Let us convert 0.01 grains to a mass in grams.

$$\text{minimum error} = 0.01 \text{ gr} \times \frac{1 \text{ gros}}{72 \text{ gr}} \times \frac{1 \text{ once}}{8 \text{ gros}} \times \frac{1 \text{ livre}}{16 \text{ once}} \times \frac{30.59 \text{ g}}{1 \text{ livre}} = 3 \times 10^{-5} \text{g} = 0.03 \text{ mg}$$

$$\text{maximum error} = 3.12 \text{ gr} \times \frac{3 \times 10^{-5} \text{ g}}{0.01 \text{ gr}} = 9 \times 10^{-3} \text{ g} = 9 \text{ mg}$$

The maximum error is close to that of a common modern laboratory balance, which has a sensitivity of 1 mg. The minimum error is approximated by a good quality analytical balance.

91. One way to determine the common factor of which all 13 numbers are multiples is to first divide all of them by the smallest. The ratios thus obtained may either be integers or they may be rational numbers whose decimal equivalents are easy to recognize.

Obs.	1	2	3	4	5	6	7	8	9	10	11	12	13
Quan.	19.66	24.60	29.62	34.47	39.38	44.42	49.41	53.91	59.12	63.68	68.65	78.34	83.22
Ratio	1.000	1.251	1.507	1.753	2.003	2.259	2.513	2.742	3.007	3.239	3.492	3.984	4.233
Mult.	4.000	5.005	6.026	7.013	8.012	9.038	10.05	10.97	12.03	12.96	13.97	15.94	16.93
Int.	4	5	6	7	8	9	10	11	12	13	14	16	17

The row labeled "Mult." is obtained by multiplying the row "ratio" by 4.000. In the row labeled "Int." we give the integer closest to each of these multipliers. It is obvious that each of the 13 measurements is exceedingly close to a common quantity multiplied by an integer.

92. In a 50-year-old chemistry textbook the atomic mass for oxygen would be 16.000 because chemists assigned precisely 16 as the atomic mass of the naturally occurring mixture of oxygen isotopes. This value is slightly higher than the value of 15.9994 in modern chemistry textbooks. Thus, we would expect all other atomic masses to be slightly higher as well in the older textbooks.

93. We begin with the amount of reparations and obtain the volume in cubic kilometers with a series of conversion factors.

$$\text{volume} = \$28.8 \times 10^9 \times \frac{1 \text{ troy oz Au}}{\$21.25} \times \frac{31.103 \text{ g Au}}{1 \text{ troy oz Au}} \times \frac{1 \text{ mol Au}}{196.97 \text{ g Au}} \times \frac{6.022 \times 10^{23} \text{ atoms Au}}{1 \text{ mol Au}}$$

$$\times \frac{1 \text{ ton sea water}}{4.67 \times 10^{17} \text{ Au atoms}} \times \frac{2000 \text{ lb}}{1 \text{ ton}} \times \frac{453.6 \text{ g}}{1 \text{ lb sea water}} \times \frac{1 \text{ cm}^3}{1.03 \text{ g}} \times \left(\frac{1 \text{ m}}{100 \text{ cm}} \times \frac{1 \text{ km}}{1000 \text{ m}} \right)^3$$

$$\text{volume} = 2.43 \times 10^5 \text{ km}^3$$

94. We start by using the percent natural abundances for ^{87}Rb and ^{85}Rb along with the data in the "spiked" mass spectrum to find the total mass of Rb in the sample. Then, we calculate the Rb content in the rock sample in ppm by mass by dividing the mass of Rb by the total mass of the rock sample, and then multiplying the result by 10^6 to convert to ppm.

^{87}Rb = 27.83 % natural abundance ^{85}Rb = 72.17 % natural abundance

Therefore, $\dfrac{^{87}\text{Rb(natural)}}{^{85}\text{Rb(natural)}} = \dfrac{27.83 \text{ \%}}{72.17 \text{ \%}} = 0.3856$

For the ^{87}Rb(spiked) sample, the ^{87}Rb peak in the mass spectrum is 1.12 times as tall as the ^{85}Rb peak. Thus, for this sample $\dfrac{^{87}\text{Rb(natural)}+^{87}\text{Rb(spiked)}}{^{85}\text{Rb(natural)}} = 1.12$

Using this relationship, we can now find the masses of both ^{85}Rb and ^{87}Rb in the sample.

So, $\dfrac{^{87}\text{Rb(natural)}}{^{85}\text{Rb(natural)}} = 0.3856;$ $^{85}\text{Rb(natural)} = \dfrac{^{87}\text{Rb(natural)}}{0.3856}$

$^{87}\text{Rb(natural)} + ^{87}\text{Rb(spiked)} = \dfrac{1.12 \times ^{87}\text{Rb (natural)}}{0.3856} = 2.905 \ ^{87}\text{Rb(natural)}$

$^{87}\text{Rb(spiked)} = 1.905 \ ^{87}\text{Rb(natural)}$

and $\dfrac{^{87}\text{Rb(natural)}+^{87}\text{Rb(spiked)}}{^{85}\text{Rb(natural)}} = \dfrac{^{87}\text{Rb(natural)}+^{87}\text{Rb(spiked)}}{\dfrac{^{87}\text{Rb(natural)}}{0.3856}} = 1.12$

Since the mass of ^{87}Rb(spike) is equal to 29.45 μg, the mass of ^{87}Rb(natural) must be

$\dfrac{29.45 \text{ μg}}{1.905} = 15.46$ μg of ^{87}Rb(natural)

So, the mass of ^{85}Rb(natural) = $\dfrac{15.46 \text{ μg of } ^{87}\text{Rb(natural)}}{0.3856} = 40.09$ μg of ^{85}Rb(natural)

Therefore, the total mass of Rb in the sample = 15.46 μg of ^{87}Rb(natural) + 40.09 μg of ^{85}Rb(natural) = 55.55 μg of Rb convert to grams:

$= 55.55$ μg of Rb $\times \dfrac{1 \text{ g Rb}}{1 \times 10^6 \text{ μg Rb}} = 5.555 \times 10^{-5}$ g Rb

Rb content (ppm) = $\dfrac{5.555 \times 10^{-5} \text{ g Rb}}{0.350 \text{ g of rock}} \times 10^6 = 159$ ppm Rb

CHAPTER 3
CHEMICAL COMPOUNDS
PRACTICE EXAMPLES

1A For one conversion factor we need the molar mass of ZnO.

$$M = \left(\frac{1\,\text{mol Zn}}{1\,\text{mol ZnO}} \times \frac{65.39\,\text{g Zn}}{1\,\text{mol Zn}} \right) + \left(\frac{1\,\text{mol O}}{1\,\text{mol ZnO}} \times \frac{16.00\,\text{g O}}{1\,\text{mol O}} \right) = \frac{81.39\,\text{g ZnO}}{1\,\text{mol ZnO}}$$

Then determine the number of ions in 1.0 g of ZnO. Note that each mole of ZnO contains two moles of ions: 1 mole of Zn^{2+} ions, and 1 mole of O^{2-} ions.

$$? \text{ ions} = 1.0\,\text{g ZnO} \times \frac{1\,\text{mol ZnO}}{81.39\,\text{g ZnO}} \times \frac{2\,\text{mol ions}}{1\,\text{mol ZnO}} \times \frac{6.022 \times 10^{23}\,\text{ions}}{1\,\text{mol ions}} = 1.5 \times 10^{22}\,\text{ions}$$

1B For one conversion factor we need the molar mass of $MgCl_2$.

$$M = \left(\frac{1\,\text{mol Mg}}{1\,\text{mol MgCl}_2} \times \frac{24.305\,\text{g Mg}}{1\,\text{mol Mg}} \right) + \left(\frac{2\,\text{mol Cl}}{1\,\text{mol MgCl}_2} \times \frac{35.453\,\text{g Cl}}{1\,\text{mol Cl}} \right) = \frac{95.211\,\text{g MgCl}_2}{1\,\text{mol MgCl}_2}$$

Then convert the number of chloride ions to the mass of $MgCl_2$.

$$MgCl_2 \text{ mass} = 5.0 \times 10^{23}\,Cl^- \text{ ions} \times \frac{1\,\text{f.u. MgCl}_2}{2\,Cl^- \text{ ions}} \times \frac{1\,\text{mol MgCl}_2}{6.022 \times 10^{23}\,\text{f.u.}} \times \frac{95.211\,\text{g MgCl}_2}{1\,\text{mol MgCl}_2}$$

$$= 4.0 \times 10^1\,\text{g MgCl}_2$$

2A The volume of gold is converted to its mass and then to the amount in moles.

$$? \text{Au atoms} = (2.50\,\text{cm})^2 \times \left(0.100\,\text{mm} \times \frac{1\,\text{cm}}{10\,\text{mm}} \right) \times \frac{19.32\,\text{g}}{1\,\text{cm}^3} \times \frac{1\,\text{mol Au}}{196.97\,\text{g Au}} \times \frac{6.022 \times 10^{23}\,\text{atoms}}{1\,\text{mol Au}}$$

$$= 3.69 \times 10^{21}\,\text{Au atoms}$$

2B We need the molar mass of ethyl mercaptan for one conversion factor.

$$M = (2 \times 12.011\,\text{g C}) + (6 \times 1.008\,\text{g H}) + (1 \times 32.066\,\text{g S}) = 62.136\,\text{g/mol C}_2\text{H}_6\text{S}$$

$$C_2H_6S \text{ conc.} = \frac{1.0\,\mu\text{L C}_2\text{H}_6\text{S}}{1500\,\text{m}^3} \times \frac{1\,\text{L}}{1 \times 10^6\,\mu\text{L}} \times \frac{1000\,\text{mL}}{1\,\text{L}} \times \frac{0.84\,\text{g}}{1\,\text{mL}} \times \frac{1\,\text{mol C}_2\text{H}_6\text{S}}{62.136\,\text{g}} \times \frac{10^6\,\mu\text{mol}}{1\,\text{mol}}$$

$$= 9 \times 10^{-3}\,\mu\text{mol/m}^3 > 0.9 \times 10^{-3}\,\mu\text{mol/m}^3 = \text{the detectable limit}$$

Thus, the vapor will be detectable.

3A The molar mass of halothane is given in Example 3-3 *in the text* as 197.4 g/mol. The rest of the solution uses conversion factors to change units.

$$C \text{ mass} = 75.0 \, \text{mL } C_2HBrClF_3 \times \frac{1.871 \, \text{g}}{1 \, \text{mL}} \times \frac{1 \, \text{mol halothane}}{197.4 \, \text{g}} \times \frac{2 \, \text{mol C}}{1 \, \text{mol } C_2HBrClF_3} \times \frac{12.01 \, \text{g C}}{1 \, \text{mol C}}$$

$$= 17.1 \text{g C}$$

3B Again, the molar mass of halothane is given in Example 3-3 *in the text* as 197.4 g/mol.

$$V_{halothane} = 100.0 \, \text{g Br} \times \frac{1 \, \text{mol Br}}{79.904 \, \text{g Br}} \times \frac{1 \, \text{mol } C_2HBrClF_3}{1 \, \text{mol Br}} \times \frac{197.4 \, \text{g } C_2HBrClF_3}{1 \, \text{mol } C_2HBrClF_3} \times \frac{1 \, \text{mL}}{1.871 \, \text{g}}$$

$$= 132.0 \, \text{mL } C_2HBrClF_3$$

4A The molecular formula of acetic acid is $C_2H_4O_2$. Determine the molar mass of acetic acid.

$$M = \left(\frac{2 \, \text{mol C}}{1 \, \text{mol acid}} \times \frac{12.011 \, \text{g C}}{1 \, \text{mol C}} \right) + \left(\frac{4 \, \text{mol H}}{1 \, \text{mol acid}} \times \frac{1.008 \, \text{g H}}{1 \, \text{mol H}} \right) + \left(\frac{2 \, \text{mol O}}{1 \, \text{mol acid}} \times \frac{15.9994 \, \text{g O}}{1 \, \text{mol O}} \right)$$

$$M = \frac{24.022 \, \text{g C}}{1 \, \text{mol } C_2H_4O_2} + \frac{4.032 \, \text{g H}}{1 \, \text{mol } C_2H_4O_2} + \frac{31.9988 \, \text{g O}}{1 \, \text{mol } C_2H_4O_2} = \frac{60.053 \, \text{g } C_2H_4O_2}{1 \, \text{mol } C_2H_4O_2}$$

The mass percent for each element is determined by dividing the mass of the element present in a mole of the compound by the molar mass for the compound as a whole, and then multiplying the result by 100%.

$$\%C = \frac{24.022 \, \text{g C}/1 \, \text{mol acid}}{60.053 \, \text{g acetic acid}/1 \, \text{mol acid}} \times 100\% = 40.001\% \, C$$

$$\%H = \frac{4.032 \, \text{g H}}{60.053 \, \text{g } C_2H_4O_2} \times 100\% = 6.714\% \, H \qquad \%O = \frac{31.9988 \, \text{g O}}{60.053 \, \text{g } C_2H_4O_2} \times 100\% = 53.284\% \, O$$

Note: As expected, the percentages sum to 100%, within the limits of significant figures.

4B We use the same technique as before: determine the mass of each element in a mole of the compound. Their sum is the molar mass of the compound. The percent composition is determined by comparing the mass of each element with the molar mass of the compound.

$$M = (10 \times 12.011 \, \text{g C}) + (16 \times 1.008 \, \text{g H}) + (5 \times 14.01 \, \text{g N}) + (3 \times 30.97 \, \text{g P}) + (13 \times 15.999 \, \text{g O})$$

$$= 120.11 \text{g C} + 16.13 \text{g H} + 70.05 \text{g N} + 92.91 \text{g P} + 207.99 \text{g O} = 507.19 \text{g ATP/mol}$$

$$\%C = \frac{120.11 \, \text{g C}}{507.19 \, \text{g ATP}} \times 100\% = 23.681\% \, C \quad \%H = \frac{16.13 \, \text{g H}}{507.19 \, \text{g ATP}} \times 100\% = 3.180\% \, H$$

$$\%N = \frac{70.05 \, \text{g N}}{507.19 \, \text{g ATP}} \times 100\% = 13.81\% \, N \quad \%P = \frac{92.91 \, \text{g P}}{507.19 \, \text{g ATP}} \times 100\% = 18.32\% \, P$$

$$\%O = \frac{207.99 \, \text{g O}}{507.19 \, \text{g ATP}} \times 100\% = 41.008\% \, O \text{ (NOTE: sums to 99.999\%)}$$

5A To answer this question, we start with a 100.00 g sample of the compound. In this way, each elemental mass in grams is numerically equal to its percent. We convert each mass to an amount in moles, and then determine the simplest integer set of molar amounts. This determination begins by dividing all three molar amounts by the smallest.

$$55.37 \text{g C} \times \frac{1 \text{mol C}}{12.011 \text{g C}} = 4.610 \text{mol C} \div 2.305 \rightarrow 2.000 \text{mol C} \times 3.000 = 6.000 \text{mol C}$$

$$7.75 \text{g H} \times \frac{1 \text{mol H}}{1.008 \text{g H}} = 7.69 \text{mol H} \div 2.305 \rightarrow 3.34 \text{mol H} \times 3.000 = 10.02 \text{mol H}$$

$$36.88 \text{g O} \times \frac{1 \text{mol O}}{15.9994 \text{g O}} = 2.305 \text{mol O} \div 2.305 \rightarrow 1.000 \text{mol O} \times 3.000 = 3.000 \text{mol O}$$

Thus, the empirical formula of the compound is $C_6H_{10}O_3$. The empirical molar mass of this compound is:

$$(6 \times 12.01 \text{g C}) + (10 \times 1.008 \text{g H}) + (3 \times 16.00 \text{g O}) = 72.06 \text{g} + 10.08 \text{g} + 48.00 \text{g} = 130.14 \text{ g/mol}$$

The empirical mass is almost precisely one half the reported molar mass, leading to the conclusion that the molecular formula must be twice the empirical formula in order to double the molar mass. Thus, the molecular formula is $C_{12}H_{20}O_6$.

5B Once again, we begin with a 100.00 g sample of the compound. In this way, each elemental mass in grams is numerically equal to its percent. We convert each mass to an amount in moles, and then determine the simplest integer set of molar amounts. This determination begins by dividing all three molar amounts by the smallest.

$$39.56 \text{g C} \times \frac{1 \text{mol C}}{12.011 \text{g C}} = 3.294 \text{mol C} \div 3.294 \rightarrow 1.000 \text{mol C} \times 3.000 = 3.000 \text{mol C}$$

$$7.74 \text{g H} \times \frac{1 \text{mol H}}{1.008 \text{g H}} = 7.68 \text{mol H} \div 3.294 \rightarrow 2.33 \text{mol H} \times 3.000 = 6.99 \text{mol H}$$

$$52.70 \text{g O} \times \frac{1 \text{mol O}}{15.9994 \text{g O}} = 3.294 \text{mol O} \div 3.294 \rightarrow 1.000 \text{mol O} \times 3.000 = 3.000 \text{mol O}$$

Thus, the empirical formula of the compound is $C_3H_7O_3$. The empirical molar mass of this compound is:

$$(3 \times 12.01 \text{g C}) + (7 \times 1.008 \text{g H}) + (3 \times 16.00 \text{g O}) = 36.03 \text{ g} + 7.056 \text{ g} + 48.00 \text{ g} = 91.09 \text{g/mol}$$

The empirical mass is almost precisely one half the reported molar mass, leading to the conclusion that the molecular formula must be twice the empirical formula in order to double the molar mass. Thus, the molecular formula is $C_6H_{14}O_6$.

6A We calculate the amount in moles of each element in the sample (determining the mass of oxygen by difference) and transform these molar amounts to the simplest integral amounts, by first dividing all three by the smallest.

$$2.726\,g\,CO_2 \times \frac{1\,mol\,CO_2}{44.010\,g\,CO_2} \times \frac{1\,mol\,C}{1\,mol\,CO_2} = 0.06194\,mol\,C \times \frac{12.011\,g\,C}{1\,mol\,C} = 0.7440\,g\,C$$

$$1.116\,g\,H_2O \times \frac{1\,mol\,H_2O}{18.015\,g\,H_2O} \times \frac{2\,mol\,H}{1\,mol\,H_2O} = 0.1239\,mol\,H \times \frac{1.008\,g\,H}{1\,mol\,H} = 0.1249\,g\,H$$

$$(1.152\,g\,cmpd - 0.7440\,g\,C - 0.1249\,g\,H) = 0.283\,g\,O \times \frac{1\,mol\,O}{16.00\,g\,O} = 0.0177\,mol\,O$$

$$\left.\begin{array}{l} 0.06194\,mol\,C \div 0.0177 \quad \rightarrow 3.50 \\ 0.1239\,mol\,H \div 0.0177 \quad \rightarrow 7.00 \\ 0.0177\,mol\,O \div 0.0177 \quad \rightarrow 1.00 \end{array}\right\}$$ All of these amounts in moles are multiplied by 2 to make them integral. Thus, the empirical formula of isobutyl propionate is $C_7H_{14}O_2$.

6B Notice that we do not have to obtain the mass of any element in this compound by difference; there is no oxygen present in the compound. We calculate the amount in mole of each element in the sample and transform these molar amounts to the simplest integral amounts, by first dividing all three by the smallest.

$$3.149\,g\,CO_2 \times \frac{1\,mol\,CO_2}{44.010\,g\,CO_2} \times \frac{1\,mol\,C}{1\,mol\,CO_2} = 0.07155\,mol\,C \quad \div 0.01789 = 3.999\,mol\,C$$

$$0.645\,g\,H_2O \times \frac{1\,mol\,H_2O}{18.015\,g\,H_2O} \times \frac{2\,mol\,H}{1\,mol\,H_2O} = 0.0716\,mol\,H \quad \div 0.01789 = 4.00\,mol\,H$$

$$1.146\,g\,SO_2 \times \frac{1\,mol\,SO_2}{64.065\,g\,SO_2} \times \frac{1\,mol\,S}{1\,mol\,SO_2} = 0.01789\,mol\,S \quad \div 0.01789 = 1.000\,mol\,S$$

Thus, the empirical formula of thiophene is C_4H_4S.

7A $\underline{S_8}$ For an atom of a free element, the oxidation state is 0 (rule 1).

$\underline{Cr_2O_7}^{2-}$ The sum of all the oxidation numbers in the ion is -2 (rule 2). The O.S. of each oxygen is -2 (rule 6). Thus, the total for all seven oxygens is -14. The total for both chromiums must be $+12$. Thus, each Cr has an O.S. $= +6$.

$\underline{Cl_2O}$ The sum of all oxidation numbers in the compound is 0 (rule 2). The O.S. of oxygen is -2 (rule 6). The total for the two chlorines must be $+2$. Thus, each chlorine must have O.S. $= +1$.

$K\underline{O_2}$ The sum for all the oxidation numbers in the compound is 0 (rule 2). The O.S. of potassium is $+1$ (rule 3). The sum of the oxidation numbers of the two oxygens must be -1. Thus, each oxygen must have O.S. $= -1/2$.

7B $S_2O_3^{2-}$ The sum of all the oxidation numbers in the ion is -2 (rule 2). The O.S. of oxygen is -2 (rule 6). Thus, the total for three oxygens must be -6. The total for both sulfurs must be $+4$. Thus, each S has an O.S. $= +2$.

  $\underline{Hg}_2Cl_2$ The O.S. of each Cl is -1 (rule 7). The sum of all O.S. is 0 (rule 2). Thus, the total for two Hg is $+2$ and each Hg has O.S. $= +1$.

  $K\underline{Mn}O_4$ The O.S. of each O is -2 (rule 6). Thus, the total for 4 oxygens must be -8. The K has O.S. $= +1$ (rule 3). The total of all O.S. is 0 (rule 2). Thus, the O.S. of Mn is $+7$.

  $H_2\underline{C}O$ The O.S. of each H is $+1$ (rule 5), producing a total for both hydrogens of $+2$. The O.S. of O is -2 (rule 6). Thus, the O.S. of C is 0, because total of all O.S. is 0 (rule 2).

8A In each case, we determine the formula *with its accompanying charge* of each ion in the compound. We then produce a formula for the compound in which the total positive charge equals the total negative charge.

lithium oxide	Li^+ and O^{2-}	*two* Li^+ and *one* O^{2-}	Li_2O
tin(II) fluoride	Sn^{2+} and F^-	*one* Sn^{2+} and *two* F^-	SnF_2
lithium nitride	Li^+ and N^{3-}	*three* Li^+ and *one* N^{3-}	Li_3N

8B Using a similar procedure as that provided in **8A**

aluminum sulfide	Al^{3+} and S^{2-}	*two* Al^{3+} and *three* S^{2-}	Al_2S_3
magnesium nitride	Mg^{2+} and N^{3-}	*three* Mg^{2+} and *two* N^{3-}	Mg_3N_2
vanadium(III) oxide	V^{3+} and O^{2-}	*two* V^{3+} and *three* O^{2-}	V_2O_3

9A The name of each of these ionic compounds is the name of the cation followed by that of the anion. Each anion name is a modified (with the ending "ide") version of the name of the element. Each cation name is the name of the metal, with the oxidation state appended in Roman numerals in parentheses if there is more than one type of cation for that metal.

  CsI   cesium iodide

  CaF_2 calcium fluoride

  FeO   The O.S. of O $= -2$ (rule 6). Thus, the O.S. of Fe $= +2$ (rule 2). The cation is iron(II). The name of the compound is iron(II) oxide.

  $CrCl_3$ The O.S. of Cl $= -1$ (rule 7). Thus, the O.S. of Cr $= +3$ (rule 2). The cation is chromium (III). The compound is chromium (III) chloride.

9B The name of each of these ionic compounds is the name of the cation followed by that of the anion. Each anion name is a modified (with the ending "ide") version of the name of the element. Each cation name is the name of the metal, with the oxidation state appended in Roman numerals in parentheses if there is more than one type of cation for that metal.

CaH_2 calcium hydride Ag_2S silver sulfide

In the next two compounds, the oxidation state of chlorine is −1 (rule 7) and thus the oxidation state of the metal in each cation must be +1 (rule 2).

$CuCl$ copper(I) chloride Hg_2Cl_2 mercury(I) chloride

10A SF_6 — Both S and F are nonmetals. This is a binary molecular compound: sulfur hexafluoride.

HNO_2 — The NO_2^- ion is the nitrite ion. Its acid is nitrous acid.

$Ca(HCO_3)_2$ — HCO_3^- is the bicarbonate ion or the hydrogen carbonate ion. This compound is calcium bicarbonate or calcium hydrogen carbonate.

$FeSO_4$ — The SO_4^{2-} ion is the sulfate ion. The cation is Fe^{2+}, iron(II). This compound is iron(II) sulfate.

10B NH_4NO_3 — The cation is NH_4^+, ammonium ion. The anion is NO_3^-, nitrate ion. This compound is ammonium nitrate.

PCl_3 — Both P and Cl are nonmetals. This is a binary molecular compound: phosphorus trichloride.

$HBrO$ — BrO^- is hypobromite, this is hypobromous acid.

$AgClO_4$ — The anion is perchlorate ion, ClO_4^-. The compound is silver perchlorate.

$Fe_2(SO_4)_3$ — The SO_4^{2-} ion is the sulfate ion. The cation is Fe^{3+}, iron(III). This compound is iron(III) sulfate.

11A boron trifluoride — Both elements are nonmetals. This is a binary molecular compound: BF_3

potassium dichromate — Potassium ion is K^+, and dichromate ion is $Cr_2O_7^{2-}$. This is $K_2Cr_2O_7$.

sulfuric acid — The anion is sulphate, SO_4^{2-}. There must be two H^+s. This is H_2SO_4.

calcium chloride — The ions are Ca^{2+} and Cl^-. There must be one Ca^{2+} and two Cl^-s: $CaCl_2$.

11B aluminum nitrate — Aluminum is Al^{3+}; the nitrate ion is NO_3^-. This is $Al(NO_3)_3$.

tetraphosphorus decoxide — Both elements are nonmetals. This is a binary molecular compound; P_4O_{10}

chromium(III) hydroxide — Chromium(III) ion is Cr^{3+}; the hydroxide ion is OH^-. This is $Cr(OH)_3$.

iodic acid — The halogen "ic" acid has the halogen in a +5 oxidation state. This is HIO_3.

12A **(a)**   Not isomers-molecular formulas are different (C_8H_{18} vs C_9H_{20}).

    (b)   Molecules are isomers (same formula C_7H_{16})

12B **(a)**   Molecules are isomers (same formula C_7H_{14})

    (b)   Not isomers-molecular formulas are different (C_4H_8 vs C_5H_{10}).

13A **(a)**   The carbon to carbon bonds are all single bonds in this hydrocarbon. This compound is an alkane.

    (b)   In this compound, there are only single bonds, and a Cl atom has replaced one H atom. This compound is a chloroalkane.

    (c)   The presence of the carboxyl group (—CO_2H) in this molecule means that the compound is a carboxylic acid.

    (d)   There is a carbon to carbon double bond in this hydrocarbon. This is an alkene.

13B **(a)**   The presence of the hydroxyl group (—OH) in this molecule means that this compound is an alcohol.

    (b)   The presence of the carboxyl group (—CO_2H) in this molecule means that the compound is a carboxylic acid. This molecule also contains the hydroxyl group(—OH).

    (c)   The presence of the carboxyl group (—CO_2H) in this molecule means that the compound is a carboxylic acid. As well, a Cl atom has replaced one H atom. This compound is a chloroalkane. The compound is a chloro carboxylic acid.

    (d)   There is a carbon to carbon double bond in this compound; hence, it is an alkene. There is also one H atom that has been replaced by a Br atom. This compound is also a bromoalkene.

14A **(a)**   The structure is that of an alcohol with the hydroxyl group on the second carbon atom of a three carbon chain. The compound is 2-propanol (commonly isopropyl alcohol).

    (b)   The structure is that of an iodoalkane molecule with the I atom on the first carbon of a three-carbon chain. The compound is called 1-iodopropane.

    (c)   The carbon chain in this structure is four carbon atoms long with the end C atom in a carboxyl group. There is also a methyl group on the third carbon in the chain. The compound is 3-methylbutanoic acid.

    (d)   The structure is that of a three carbon chain that contains a carbon to carbon double bond. This compound is propene.

14B **(a)**   2-chloropropane   **(b)**   1,4-dichlorobutane   **(c)**   2-methyl propanoic acid

15A **(a)**   pentane: $CH_3(CH_2)_3CH_3$ **(b)**   ethanoic acid: CH_3CO_2H

    (c)   1-iodooctane: $ICH_2(CH_2)_6CH_3$ **(d)**   1-pentanol: $CH_2(OH)(CH_2)_3CH_3$

15B **(a)**   propene: CH_3CHCH_2 **(b)**   1-heptanol: $CH_2(OH)(CH_2)_5CH_3$

(c)   chloroacetic acid: CH_2ClCO_2H (d)   hexanoic acid: $CH_3(CH_2)_4CO_2H$

REVIEW QUESTIONS

1.   (a)   The formula unit of a compound is a group of atoms that has atoms of the same type
            and number as are present in the formula of that compound. For example, if " Na_2 "
            appears in the formula of the compound, then there will be two sodium atoms in the
            formula unit of that compound.

      (b) S_8 is a molecule of elemental sulfur. Like several other elements (i.e. H_2, F_2, Cl_2, Br_2,
            I_2, N_2, O_2, and P_4), elemental sulfur exists as molecules rather than as isolated atoms.

      (c)   An ionic compound is one that is composed of (positively charged) cations and
            (negatively charged) anions. Most binary ionic compounds are composed of a metal
            (which becomes the cation) and a nonmetal (which becomes the anion).

      (d)   An oxoacid is an acid that contains the element oxygen, in addition to some other
            element and the element hydrogen. H_2SO_4, HNO_3, and $HClO_4$ all are oxoacids, in
            which the "other element" is S, N, and Cl, respectively.

      (e)   A hydrate is a compound that contains water, rather loosely bound. Usually mild
            heating of the compound can drive off this water of hydration. Addition of water to
            the anhydrous salt reforms the hydrate (reversible loss/gain of water)

2.   (a)   A molecule of an element refers to a small independent particle of that element.
            Usually this is an atom, but in some cases (notably H_2, F_2, Cl_2, Br_2, I_2, N_2, O_2,
            S_8, and P_4) it is a grouping of two or more atoms.

      (b)   The structural formula of a compound not only indicates which atoms are present in the
            formula unit, but also how they are joined together (represents bonding in the molecule)

      (c)   The oxidation state of an element in a compound is an indication of how many
            electrons each atom of that element has lost (positive oxidation state) or gained
            (negative). Since oxidation state is determined by a set of rules, rather than by
            experiment, its connection to the number of electrons actually transferred is rather
            tenuous. It is used in naming compounds and balancing some chemical equations.

      (d)   The determination of the carbon–hydrogen–oxygen content of a compound by
            combustion analysis involves realizing that all of the carbon has formed carbon
            dioxide, all of the hydrogen has formed water, and the amount of oxygen present in
            the original compound must be determined by difference.

3.   (a)   Ionic compounds are made up of positively charged ions (usually metal ions) and
            negatively charged ions (usually non-metal ions or polyatomic anions) held together
            by electrostatic forces of attraction. Molecular compounds are made up of discrete
            units called molecules. Generally they consist of a small number of nonmetal atoms
            held together by covalent bonds (sharing of electrons).

      (b)   An empirical formula indicates the simplest grouping of atoms that has the same ratio
            of elements as are present in the compound. The molecular formula indicates the

actual number of atoms of each type present in the molecule. The molecular formula is an integral multiple of the empirical formula.

(c)   A systematic name is based on the elements present in a compound, indicating its composition. The trivial or common name is simply a label for the substance.

(d)   A binary acid consists of hydrogen and one other element. A ternary acid consists of hydrogen, the other element, and the element oxygen: three elements in all.

4.   **(a)**   The atomic mass of oxygen is the mass of one (average) atom, 15.9994 u.

(b)   Molecular mass of oxygen is the mass of one (average) molecule of O_2, 31.9988 u.

(c)   The molar mass of molecular oxygen is the mass of one mole of oxygen molecules, 31.9988 g. The molar mass of atomic oxygen is the mass of one mole of oxygen atoms, 15.9994 g.

5.   **(a)**   A nitroglycerine molecule, $C_3H_5(NO_3)_3$, contains 3 C atoms, 5 H atoms, 3 N atoms, and $3 \times 3 = 9$ O atoms, for a total of $(3+5+3+9) = 20$ atoms.

(b)   Each molecule of C_2H_6 contains 6 H atoms and 2 C atoms, 8 atoms total.

$$\text{Number of atoms} = 0.00102\,\text{mol C}_2\text{H}_6 \times \frac{6.022 \times 10^{23}\,\text{C}_2\text{H}_6\,\text{molecules}}{1\,\text{mol C}_2\text{H}_6} \times \frac{8\,\text{atoms}}{\text{C}_2\text{H}_6\,\text{molecule}}$$

$$= 4.91 \times 10^{21}\,\text{atoms}$$

(c) $$\text{number of F atoms} = 12.15\,\text{mol C}_2\text{HBrClF}_3 \times \frac{3\,\text{mol F}}{1\,\text{mol C}_2\text{HBrClF}_3} \times \frac{6.022 \times 10^{23}\,\text{F atoms}}{1\,\text{mol F atoms}}$$

$$= 2.195 \times 10^{25}\,\text{F atoms}$$

6.   **(a)**   To convert the amount in moles to mass, we need the molar mass of N_2O_4.

$$\text{molar mass N}_2\text{O}_4 = \left(2\,\text{mol N} \times \frac{14.01\,\text{g N}}{1\,\text{mol N}}\right) + \left(4\,\text{mol O} \times \frac{16.00\,\text{g O}}{1\,\text{mol O}}\right) = 92.02\,\text{g/mol N}_2\text{O}_4$$

$$\text{mass N}_2\text{O}_4 = 7.34\,\text{mol N}_2\text{O}_4 \times \frac{92.02\,\text{g N}_2\text{O}_4}{1\,\text{mol}} = 675\,\text{g N}_2\text{O}_4$$

(b) $$\text{mass of O}_2 = 3.16 \times 10^{24}\,\text{O}_2\,\text{molecules} \times \frac{1\,\text{mol O}_2}{6.022 \times 10^{23}\,\text{molecules}} \times \frac{32.00\,\text{g O}_2}{1\,\text{mol O}_2} = 168\,\text{g O}_2$$

(c) $$\text{molar mass CuSO}_4 \cdot 5\text{H}_2\text{O} = 63.5\,\text{g Cu} + 32.1\,\text{g S} + (9 \times 16.0\,\text{g O}) + (10 \times 1.01\,\text{g H})$$

$$= 249.7\,\text{g/mol CuSO}_4 \cdot 5\text{H}_2\text{O}$$

$$\text{mass of CuSO}_4 \cdot 5\text{H}_2\text{O} = 18.6\,\text{mol} \times \frac{249.7\,\text{g CuSO}_4 \cdot 5\text{H}_2\text{O}}{1\,\text{mol}} = 4.64 \times 10^3\,\text{g CuSO}_4 \cdot 5\text{H}_2\text{O}$$

(d) $$\text{molar mass C}_2\text{H}_4(\text{OH})_2 = (2 \times 12.01\,\text{g C}) + (6 \times 1.01\,\text{g H}) + (2 \times 16.00\,\text{g O}) = 62.08\,\text{g/mol}$$

$$\text{mass of } C_2H_4(OH)_2 = 4.18\times10^{24}\text{molecules}\times\frac{1\,\text{mol}}{6.022\times10^{23}\text{molecules}}\times\frac{62.08\ \text{g}}{1\,\text{mol } C_2H_4(OH)_2}$$

$$= 431\text{g } C_2H_4(OH)_2$$

7. **(a)** $\text{amount of } Br_2 = 8.08\times10^{22}\,Br_2\ \text{molecules}\times\dfrac{1\,\text{mole } Br_2}{6.022\times10^{23}\,Br_2\ \text{molecules}}$

$$= 0.134\,\text{mol } Br_2$$

(b) $\text{amount of } Br_2 = 2.17\times10^{24}\,Br\ \text{atoms}\times\dfrac{1\,Br_2\ \text{molecule}}{2\,Br\ \text{atoms}}\times\dfrac{1\,\text{mole } Br_2}{6.022\times10^{23}\,Br_2\ \text{molecules}}$

$$= 1.80\,\text{mol } Br_2$$

(c) $\text{amount of } Br_2 = 11.3\ \text{kg } Br_2\times\dfrac{1000\,\text{g}}{1\,\text{kg}}\times\dfrac{1\,\text{mol } Br_2}{159.8\,\text{g } Br_2}=70.7\,\text{mol } Br_2$

(d) $\text{amount of } Br_2 = 2.65\,\text{L } Br_2\times\dfrac{1000\ \text{mL}}{1\,\text{L}}\times\dfrac{3.10\,\text{g } Br_2}{1\,\text{mL } Br_2}\times\dfrac{1\,\text{mol } Br_2}{159.8\,\text{g } Br_2}=51.4\,\text{mol } Br_2$

8. **(a)** molecular mass (mass of one molecule).

$$C_5H_{11}NO_2S = (5\times12.0\,\text{u C})+(11\times1.01\,\text{u H})+14.0\,\text{u N}+(2\times16.0\,\text{u O})+32.1\,\text{u S}$$
$$= 149.2\,\text{u/}C_5H_{11}NO_2S\ \text{molecule}$$

(b) Since there are 11 H atoms in each $C_5H_{11}NO_2S$ molecule, there are 11 moles of H atoms in each mole of $C_5H_{11}NO_2S$ molecules

(c) $\text{mass C} = 1\,\text{mol } C_5H_{11}NO_2S\times\dfrac{5\ \text{mol C}}{1\,\text{mol } C_5H_{11}NO_2S}\times\dfrac{12.011\,\text{g C}}{1\,\text{mol C}}=60.055\,\text{g C}$

(d) $\text{no. C atoms} = 9.07\ \text{mol } C_5H_{11}NO_2S\times\dfrac{5\ \text{mol C}}{1\,\text{mol } C_5H_{11}NO_2S}\times\dfrac{6.022\times10^{23}\,\text{atoms}}{1\,\text{mol C}}$

$$= 2.73\times10^{25}\,\text{C atoms}$$

9. The information obtained in the course of calculating the molar mass is used to determine the mass percent of H in decane.

$$\text{molar mass } C_{10}H_{22} = \left(\frac{10\,\text{mol C}}{1\,\text{mol } C_{10}H_{22}}\times\frac{12.011\,\text{g C}}{1\,\text{mol C}}\right)+\left(\frac{22\,\text{mol H}}{1\,\text{mol } C_{10}H_{22}}\times\frac{1.00794\,\text{g H}}{1\,\text{mol H}}\right)$$

$$= \frac{120.11\,\text{g C}}{1\,\text{mol } C_{10}H_{22}}+\frac{22.1747\,\text{g H}}{1\,\text{mol } C_{10}H_{22}}=\frac{142.28\,\text{g}}{1\,\text{mol } C_{10}H_{22}}$$

$$\%H=\frac{22.1747\,\text{g H/mol decane}}{142.28\,\text{g } C_{10}H_{22}/\text{mol decane}}\times100\%=15.585\%\,H$$

10. Determine the mass of O in a mol of $Cu_2(OH)_2CO_3$ and the molar mass of $Cu_2(OH)_2CO_3$.

$$\text{mass O/mol Cu}_2(OH)_2CO_3 = \frac{5\,\text{mol O}}{1\,\text{mol Cu}_2(OH)_2CO_3} \times \frac{16.00\,\text{g O}}{1\,\text{mol O}} = 80.00\,\text{g O/mol Cu}_2(OH)_2CO_3$$

$$\text{molar mass Cu}_2(OH)_2CO_3 = (2\times63.55\,\text{g Cu})+(5\times16.00\,\text{g O})+(2\times1.01\,\text{g H})+12.01\,\text{g C}$$
$$= 221.13\,\text{g/mol Cu}_2(OH)_2CO_3$$

$$\text{Percent oxygen in sample} = \frac{80.00\,\text{g}}{221.13\,\text{g}} \times 100\% = 36.18\%\,\text{O}$$

11. Determine the mass of a mole of $Cr(NO_3)_3 \cdot 9H_2O$, and then the mass of water in a mole.

$$\text{molar mass Cr}(NO_3)_3 \cdot 9H_2O = 52.00\,\text{g Cr} + (3\times14.01\,\text{g N}) + (18\times16.00\,\text{g O}) + (18\times1.01\,\text{g H})$$
$$= 400.2\,\text{g/mol Cr}(NO_3)_3 \cdot 9H_2O$$

$$\text{mass H}_2O = \frac{9\,\text{mol H}_2O}{1\,\text{mol Cr}(NO_3)_3 \cdot 9H_2O} \times \frac{18.02\,\text{g H}_2O}{1\,\text{mol H}_2O} = 162.2\,\text{g H}_2O\big/\text{mol Cr}(NO_3)_3 \cdot 9H_2O$$

$$\frac{162.2\,\text{g H}_2O\big/\text{mol Cr}(NO_3)_3 \cdot 9H_2O}{400.2\,\text{g}\big/\text{mol Cr}(NO_3)_3 \cdot 9H_2O} \times 100\% = 40.53\%\,\text{H}_2O$$

12. In each case, we first determine the molar mass of the compound, and then the mass of the indicated element in one mole of the compound. Finally, we determine the percent by mass of the indicated element to four significant figures.

(a)
$$\text{molar mass Pb}(C_2H_5)_4 = 207.2\,\text{g Pb} + (8\times12.01\,\text{g C}) + (20\times1.008\,\text{g H})$$
$$= 323.4\,\text{g/mol Pb}(C_2H_5)_4$$

$$\text{mass Pb/mol Pb}(C_2H_5)_4 = \frac{1\,\text{mol Pb}}{1\,\text{mol Pb}(C_2H_5)_4} \times \frac{207.2\,\text{g Pb}}{1\,\text{mol Pb}} = 207.2\,\text{g Pb/mol Pb}(C_2H_5)_4$$

$$\%\,\text{Pb} = \frac{207.2\,\text{g Pb}}{323.4\,\text{g Pb}(C_2H_5)_4} \times 100\% = 64.07\%\,\text{Pb}$$

(b)
$$\text{molar mass Fe}_4[Fe(CN)_6]_3 = (7\times55.85\,\text{g Fe}) + (18\times12.01\,\text{g C}) + (18\times14.01\,\text{g N})$$
$$= 859.3\,\text{g/mol Fe}_4[Fe(CN)_6]_3$$

$$\frac{\text{mass Fe}}{\text{mol Fe}_4[Fe(CN)_6]_3} = \frac{7\,\text{mol Fe}}{1\,\text{mol Fe}_4[Fe(CN)_6]_3} \times \frac{55.85\,\text{g Fe}}{1\,\text{mol Fe}} = 391.0\,\text{g Fe/mol Fe}_4[Fe(CN)_6]_3$$

$$\%\,\text{Fe} = \frac{391.0\,\text{g Fe}}{859.3\,\text{g Fe}_4[Fe(CN)_6]_3} \times 100\% = 45.50\%\,\text{Fe}$$

(c)   molar mass $C_{55}H_{72}MgN_4O_5$

$$=(55 \times 12.011\,g\,C)+(72 \times 1.008\,g\,H)+(1 \times 24.305\,g\,Mg)+(4 \times 14.01\,g\,N)+(5 \times 15.999\,g\,O)$$

$$= 893.521\,g/mol\ C_{55}H_{72}MgN_4O_5$$

$$\frac{mass\ Mg}{mol\ C_{55}H_{72}MgN_4O_5}=\frac{1\,mol\ Mg}{1\,mol\ C_{55}H_{72}MgN_4O_5}\times\frac{24.305\,g\,Mg}{1\,mol\ Mg}=\frac{24.305\,g\,Mg}{mol\ C_{55}H_{72}MgN_4O_5}$$

$$\%Mg=\frac{24.305\,g\,Mg}{893.521g\ C_{55}H_{72}MgN_4O_5}\times 100\%=2.7201\%\ Mg$$

13.   For SO_2 and Na_2S, a mole of each contains a mole of S and two moles of another element; in the case of SO_2 the other element (oxygen) has a smaller atomic mass than the other element in Na_2S (Na), causing SO_2 to have a higher mass percent sulfur. For S_2Cl_2 and $Na_2S_2O_3$, a mole of each contains two moles of S; for S_2Cl_2 the rest of the mole has a mass of 71.0 g, while for $Na_2S_2O_3$ it would be $(2 \times 23) + (3 \times 16) = 94$ g. Sulfur makes up the greater proportion of the mass in S_2Cl_2, giving it the larger percent of S. Now we compare SO_2 and S_2Cl_2: S_2O_4 has the same mass proportions as does SO_2 but also has the two moles of S, as does S_2Cl_2. In S_2Cl_2 the remainder of a mole has a mass of 71.0 g, while in S_2O_4 the remainder of a mole would be $4 \times 16.0 = 64.0$ g. Thus, SO_2 has the highest percent of S so far. For CH_3CH_2SH compared to SO_2, we see that both compounds have one S-atom, SO_2 has two O-atoms (each with a molar mass of ~16 g mol^{-1}) and CH_3CH_2SH effectively has two CH_3 groups (each CH_3 group with a mass of ~15 g mol^{-1}). Thus, CH_3CH_2SH has the highest percentage sulfur by mass of the compounds listed.

14.   Determine the % oxygen by difference. $\%O=100.00\%-45.27\%\,C-9.50\%\,H=45.23\%\,O$

$$no.\,mol\ O= 45.23\,g \times \frac{1\,mol\ O}{16.00\,g\ O}=2.827\,mol\ O \quad \div 2.827 \rightarrow 1.000\,mol\ O$$

$$no\ mol\ C= 45.27\,g\ C \times \frac{1\,mol\ C}{12.01\,g\ C}=3.769\,mol\ C \quad \div 2.827 \rightarrow 1.333\,mol\ C$$

$$no.\,mol\ H= 9.50\,g\ H \times \frac{1\,mol\ H}{1.008\,g\ H}=9.42\,mol\ H \quad \div 2.827 \rightarrow 3.33\,mol\ H$$

Multiply all amounts by 3 to obtain integers. Empirical formula is $C_4H_{10}O_3$.

15. We base our calculation on 100.0 g of monosodium glutamate.

$$13.6 \,g\, Na \times \frac{1 \,mol\, Na}{22.99 \,g\, Na} = 0.592 \,mol\, Na \quad \div 0.592 \rightarrow 1.00 \,mol\, Na$$

$$35.5 \,g\, C \times \frac{1 \,mol\, C}{12.01 \,g\, C} = 2.96 \,mol\, C \quad \div 0.592 \rightarrow 5.00 \,mol\, C$$

$$4.8 \,g\, H \times \frac{1 \,mol\, H}{1.01 \,g\, H} = 4.8 \,mol\, H \quad \div 0.592 \rightarrow 8.1 \,mol\, H$$

$$8.3 \,g\, N \times \frac{1 \,mol\, N}{14.01 \,g\, N} = 0.59 \,mol\, N \quad \div 0.592 \rightarrow 1.0 \,mol\, N$$

$$37.8 \,g\, O \times \frac{1 \,mol\, O}{16.00 \,g\, O} = 2.36 \,mol\, O \quad \div 0.592 \rightarrow 3.99 \,mol\, O$$

Empirical formula : $NaC_5H_8NO_4$

16. First determine the empirical formula. Begin by determining the percent oxygen by difference. $\%O = 100\% - 57.83\% C - 3.64\% H = 38.53\% O$

$$no. \,mol\, O = 38.53 \,g \times \frac{1 \,mol\, O}{16.00 \,g\, O} = 2.408 \,mol\, O \quad \div 2.408 \rightarrow 1.000 \,mol\, O$$

$$no. \,mol\, C = 57.83 \,g\, C \times \frac{1 \,mol\, C}{12.01 \,g\, C} = 4.815 \,mol\, C \quad \div 2.408 \rightarrow 2.000 \,mol\, C$$

$$no. \,mol\, H = 3.64 \,g\, H \times \frac{1 \,mol\, H}{1.008 \,g\, H} = 3.61 \,mol\, H \quad \div 2.408 \rightarrow 1.50 \,mol\, H$$

Empirical formula is $C_4H_3O_2$.

The empirical molecular mass $= (4 \times 12.0 \,u\, C) + (3 \times 1.0 \,u\, H) + (2 \times 16.0 \,u\, O) = 83.0 \,u$

This empirical molecular mass is one-half of the measured molecular mass. Thus, the molecular formula of terephthalic acid is twice the empirical formula: $C_8H_6O_4$

17. (a) First determine the masses of carbon and hydrogen in the original sample.

$$mass \,C = 6.029 \,g\, CO_2 \times \frac{1 \,mol\, CO_2}{44.010 \,g} \times \frac{1 \,mol\, C}{1 \,mol\, CO_2} = 0.1370 \,mol\, C \times \frac{12.011 \,g\, C}{1 \,mol\, C} = 1.646 \,g\, C$$

$$mass \,H = 1.709 \,g\, H_2O \times \frac{1 \,mol\, H_2O}{18.02 \,g\, H_2O} \times \frac{2 \,mol\, H}{1 \,mol\, H_2O} = 0.1897 \,mol\, H \times \frac{1.008 \,g\, H}{1 \,mol\, H} = 0.1912 \,g\, H$$

Then the percents of the two elements in the compound are computed.

$$\%\,C = \frac{1.646\,g\,C}{2.174\,g\,cmpd} \times 100\% = 75.71\%\ C \quad \%\ H = \frac{0.1912\,g\,H}{2.174\,g\,cmpd} \times 100\% = 8.795\%\,H$$

The % O is determined by difference.

$$\%\,O = 100\% - 75.71\%\,C - 8.795\%\,H = 15.50\%\,O$$

(b)   In part (a), we determined the number of moles of C and H in the original sample of the compound. We can determine the mass of oxygen in that sample by difference, and then the number of moles of oxygen in that sample. We divide each of these numbers of moles by the smallest number to determine the empirical formula.

$$mass\ O = 2.174\,g\ cmpd - 1.646\,g\ C - 0.1912\,g\ H = 0.337\,g\ O$$

$$mol\ O = 0.337\,g\ O \times \frac{1\,mol\ O}{16.00\,g\ O} = 0.0211\,mol\ O \qquad \div 0.02111 \rightarrow 1.00\,mol\ O$$

$$0.1897\,mol\ H \qquad \div 0.02111 \rightarrow 8.99\,mol\ H$$

$$0.1370\,mol\ C \qquad \div 0.02111 \rightarrow 6.49\,mol\ C$$

Multiply all amounts by 2 to obtain integers; the empirical formula of ibuprofen is $C_{13}H_{18}O_2$.

18.   The element chromium has an atomic mass of 52.0 u. Thus, there can only be one chromium atom per formula unit of the compound. (Two atoms of chromium have a mass of 104 u, more than the formula mass of the compound.) Three of the four remaining atoms in the formula unit must be oxygen. Thus, the oxide is CrO_3, chromium(VI) oxide.

19. SO_3 (40.05% S) and S_2O (80.0 % S) (2 O atoms ~ 1 S atom in terms of atomic masses)

20.   **(a)** Pb^{2+} lead(II) ion   **(b)** Co^{3+} cobalt(III) ion
   (c) Ba^{2+} barium ion   **(d)** Cr^{2+} chromium(II) ion
   (e) IO_4^- periodate ion   **(f)** ClO_2^- chlorite ion
   (g) Au^{3+} gold(III) ion   **(h)** HSO_3^- hydrogen sulfite ion
   (i) HCO_3^- hydrogen carbonate ion   **(j)** CN^- cyanide ion

21.   **(a)** KBr   potassium bromide   **(b)** $SrCl_2$ strontium chloride
   (c) ClF_3 chlorine trifluoride   **(d)** N_2O_4 dinitrogen tetroxide
   (e) PCl_5 phosphorus pentachloride

22.   **(a)** KCN   potassium cyanide   **(b)** HClO   hypochlorous acid
   (c) $(NH_4)_2SO_4$ ammonium sulfate   **(d)** KIO_3 potassium iodate

23. The desired oxidation state is given first, followed by the method used to assign the oxidation state.

   (a) $Zn = 0$ Oxidation state (O.S.) of an uncombined, neutral element is 0.

   (b) $S = -2$ in BaS The O.S. of Ba in its compounds is $+2$.

                                 Oxidation states in a compound must sum to zero.

   (c) $N = +4$ in NO_2 The O.S. of O in its compounds is -2 (in most cases).

   (d) $N = +3$ in HNO_2 The O.S. of H in molecular compounds is $+1$; that of O is -2.

   (e) $V = +4$ in VO^{2+} The O.S. of O in its compounds is -2. O.S. in a polyatomic ion must sum to the charge on that ion.

   (f) $P = +5$ in $H_2PO_4^-$ The O.S. of H in its compounds is $+1$; that of O is -2. O.S. in a polyatomic ion must sum to the charge on that ion.

24. **(a)** $MgBr_2$ magnesium bromide   **(b)** BaO barium oxide

     (c) $Hg(C_2H_3O_2)_2$ mercury(II) acetate   **(d)** $Fe_2(C_2O_4)_3$ iron(III) oxalate

     (e) $Sr(ClO_4)_2$ strontium perchlorate   **(f)** $KHSO_4$ potassium hydrogen sulfate

     (g) NCl_3 nitrogen trichloride   **(h)** BrF_5 bromine pentafluoride

25. **(a)** $HClO_2$ chlorous acid   **(b)** H_2SO_3 sulfurous acid

     (c) H_2Se hydroselenic acid   **(d)** HNO_2 nitrous acid

26. **(a)** HI (aq)   hydroiodic acid   **(b)** HNO_3 nitric acid

     (c) H_3PO_4 phosphoric acid   **(d)** H_2SO_4 sulfuric acid

27. Answer is (b), 2-butanol is the most appropriate name for this molecule. It has a four carbon atom chain with a hydroxyl group on the carbon second from the end.

28. Answer (c), butanoic acid is the most appropriate name for this molecule. It has a four carbon atom chain with an acid group on the 1st carbon (terminal carbon atom)

EXERCISES

Representing Molecules

29. **(a)** H_2O_2 **(b)** CH_3CH_2Cl **(c)** P_4O_{10}

     (d) $CH_3CH(OH)CH_3$ **(e)** HCO_2H

30. **(a)** N_2H_4 **(b)** CH_3CH_2OH **(c)** P_4O_6

     (d) $CH_3(CH_2)_3CO_2H$ **(e)** $CH_2(OH)CH_2Cl$

31. **(b)** CH_3CH_2Cl **(d)** $CH_3CH(OH)CH_3$ **(e)** HCO_2H

32. **(b)** CH_3CH_2OH **(d)** $CH_3(CH_2)_3CO_2H$ **(e)** $CH_2(OH)CH_2Cl$

The Avogadro Constant and the Mole

33. The greatest number of S atoms is contained in the compound with the greatest number of moles of S. The solid sulfur contains $8 \times 0.12\,mol = 0.96$ mol S atoms. There are 0.50×2 mol S atoms in 0.50 mol S_2O. There is slightly greater than 1 mole (64.1 g) of SO_2 in 65 g, and thus a bit more than 1 mole of S atoms. The molar mass of thiophene is: $(4\,mol\,C \times 12.0\,g\,C) + (4\,mol\,H \times 1.0\,g\,H) + (1\,mol\,S \times 32.1\,g\,S) = 84.1\,g$; 75 mL has a mass of 79.8 g and thus contains less than 1 mole of S. So, 65 g SO_2 has the greatest number of S atoms.

34. The greatest number of N atoms is found in the compound with the greatest number of moles of N. The molar mass of $N_2O = (2\,mol\,N \times 14.0\,g\,N) + (1\,mol\,O \times 16.0\,g\,O) = 44.0\,g/mol\,N_2O$. Thus, 50.0 g N_2O is slightly more than 1 mole of N_2O, and contains slightly more than 2 moles of N. Each mole of N_2 contains 2 moles of N. The molar mass of NH_3 is 17.0 g. Thus, there is 1 mole of NH_3 present, which contains 1 mole of N. The molar mass of pyridine is $(5\,mol\,C \times 12.0\,g\,C) + (5\,mol\,H \times 1.01\,g\,H) + 14.0\,g\,N = 79.1\,g/mol$. Because each mole of pyridine contains 1 mole of N, we need slightly more than 2 moles of pyridine to have more N than is present in the N_2O. But that would be a mass of about 158 g pyridine, and 150 mL has a mass of less than 150 g. Thus, the greatest number of N atoms is present in 50.0 g N_2O.

35. **(a)** $P \text{ mass} = 6.25 \times 10^{-2}\,mol\,P_4 \times \dfrac{4\,mol\,P}{1\,mol\,P_4} \times \dfrac{30.97\,g\,P}{1\,mol\,P} = 7.74\,g\,P$

(b) First we need the molar mass of $C_{18}H_{36}O_2$, stearic acid:

$$\text{molar mass} = (18\,mol\,C \times 12.01\,g\,C) + (36\,mol\,H \times 1.01\,g\,H) + (2\,mol\,O \times 16.00\,g\,O)$$
$$= 284.5\,g/mol$$

$$\text{Stearic acid mass} = 4.03 \times 10^{24}\,\text{molecules} \times \dfrac{1\,mole}{6.022 \times 10^{23}\,\text{molecules}} \times \dfrac{284.5\,g}{1\,mole\,C_{18}H_{36}O_2}$$
$$= 1.90 \times 10^3\,g\,\text{stearic acid.}$$

(c) $\text{molar mass} = (6\,mol\,C \times 12.01\,g\,C) + (14\,mol\,H \times 1.01\,g\,H) + (2\,mol\,N \times 14.00\,g\,N)$
$$+ (2\,mol\,O \times 16.00\,g\,O) = 146.2\,g/mol$$

$$\text{lysine mass} = 1.15\,mol\,N \times \dfrac{1\,mol\,C_6H_{14}N_2O_2}{2\,mol\,N} \times \dfrac{146.2\,g\,\text{lysine}}{1\,mol\,\text{lysine}} = 84.1\,g\,\text{lysine}$$

36. **(a)** $\text{molar mass} = \left(2\,\text{mol N} \times 14.01\,\text{g N}\right) + \left(4\,\text{mol O} \times 16.00\,\text{g O}\right) = 92.02\,\text{g/mol}$

$$\text{moles N}_2\text{O}_4 = 82.5\,\text{g N}_2\text{O}_4 \times \frac{1\,\text{mol N}_2\text{O}_4}{92.02\,\text{g N}_2\text{O}_4} = 0.897\,\text{mol N}_2\text{O}_4$$

(b) $\text{molar mass} = \left(1\,\text{mol Mg} \times 24.31\,\text{g Mg}\right) + \left(2\,\text{mol N} \times 14.01\,\text{g N}\right) + \left(6\,\text{mol O} \times 16.00\,\text{g O}\right)$
$\qquad = 148.33\,\text{g/mol}$

$$\text{moles N} = 106\,\text{g Mg}(\text{NO}_3)_2 \times \frac{1\,\text{mol Mg}(\text{NO}_3)_2}{148.33\,\text{g}} \times \frac{2\,\text{mol N}}{1\,\text{mol Mg}(\text{NO}_3)_2} = 1.43\,\text{mol N atoms}$$

(c) $\text{molar mass} = \left(6\,\text{mol C} \times 12.01\,\text{g C}\right) + \left(12\,\text{mol H} \times 1.008\,\text{g H}\right) + \left(6\,\text{mol O} \times 16.00\,\text{g O}\right)$
$\qquad = 180.16\,\text{g/mol C}_6\text{H}_{12}\text{O}_6$

$$\text{moles N} = 56.5\,\text{g C}_6\text{H}_{12}\text{O}_6 \times \frac{1\,\text{mol C}_6\text{H}_{12}\text{O}_6}{180.16\,\text{g}} \times \frac{6\,\text{mol O}}{1\,\text{mol C}_6\text{H}_{12}\text{O}_6} \times \frac{1\,\text{mol C}_7\text{H}_5(\text{NO}_2)_3}{6\,\text{mol O}}$$

$$\times \frac{3\,\text{mol N}}{1\,\text{mol C}_7\text{H}_5(\text{NO}_2)_3} = 0.941\,\text{mol N}$$

37. **(a)** $\text{moles S}_8 = 0.568\,\text{mm}^3 \times \dfrac{1\,\text{cm}^3}{1000\,\text{mm}^3} \times \dfrac{2.07\,\text{g}}{1\,\text{cm}^3} \times \dfrac{1\,\text{mol S}}{32.07\,\text{g}} \times \dfrac{1\,\text{mol S}_8}{8\,\text{mol S}} = 4.58 \times 10^{-6}\,\text{mol S}_8$

(b) $\text{no. S atoms} = 4.58 \times 10^{-6}\,\text{mol S}_8 \times \dfrac{8\,\text{mol S}}{1\,\text{mol S}_8} \times \dfrac{6.022 \times 10^{23}\,\text{atoms}}{1\,\text{mol S}} = 2.21 \times 10^{19}\,\text{S atoms}$

38.   Number of Fe atoms

$$= 6\,\text{L blood} \times \frac{1000\,\text{mL}}{1\,\text{L}} \times \frac{15.5\,\text{g Hb}}{100\,\text{mL blood}} \times \frac{1\,\text{mol Hb}}{64{,}500\,\text{g}} \times \frac{4\,\text{mol Fe}}{1\,\text{mol Hb}} \times \frac{6.022 \times 10^{23}\,\text{atoms}}{1\,\text{mol Fe}}$$

$$= 3 \times 10^{22}\,\text{Fe atoms}$$

Chemical Formulas

39.   For glucose (blood sugar), $\text{C}_6\text{H}_{12}\text{O}_6$,

(a)   FALSE   The percentages by mass of C and O are *different* than in CO. For one thing, CO contains no hydrogen.

(b)   TRUE   In dihydroxyacetone, $(\text{CH}_2\text{OH})_2\text{CO}$ or $\text{C}_3\text{H}_6\text{O}_3$, the ratio of C : H : O = 3 : 6 : 3 = 1 : 2 : 1. In glucose, this ratio is C : H : O = 6 : 12 : 6 = 1 : 2 : 1. Thus, the ratios are the *same*.

(c) FALSE The proportions, by number of atoms, of C and O are the same in glucose. Since, however, C and O have different molar masses, their proportions by mass must be *different*.

(d) FALSE Each mole of glucose contains $(12 \times 1.01 =)12.1$ g H. But each mole also contains 72.0 g C and 96.0 g O. Thus, the highest percentage, by mass, is that of O. The highest percentage, by number of atoms, is that of H.

40. For sorbic acid, $C_6H_8O_2$,

(a) FALSE The C:H:O mole ratio is 3:4:1, but the mass ratio differs because moles of different elements have different molar masses.

(b) TRUE Since the two compounds have the same empirical formula, they have the same mass percent composition.

(c) TRUE Aspidinol, $C_{12}H_{16}O_4$, and sorbic acid have the same empirical formula, C_3H_4O.

(d) TRUE The ratio of H atoms to O atoms is $8:2 = 4:1$. Thus, the mass ratio is $(4\,\text{mol H} \times 1\text{g H}):(1\,\text{mol O} \times 16.0\text{g O}) = 4\text{g H}:16\text{g O} = 1\text{g H}:4\text{g O}$.

41. **(a)** A formula unit of $C_2HBrClF_3$ contains:

2 C atoms   1 H atom   1 Br atom   1 Cl atom   3 F atoms

For a total of $2+1+1+1+3 = 8$ atoms

(b) $\dfrac{\text{no. F atoms}}{\text{no. C atoms}} = \dfrac{3 \text{ F atoms}}{2 \text{ C atoms}}$

(c) $\dfrac{\text{mass Br}}{\text{mass F}} = \dfrac{1\,\text{mol Br}}{3\,\text{mol F}} \times \dfrac{79.90\,\text{g Br}}{1\,\text{mol Br}} \times \dfrac{1\,\text{mol F}}{19.00\,\text{g F}} = 1.402\,\text{g Br/g F}$

(d) The element present in greatest mass percent is the one with the greatest mass in one mole. To determine that, we determine molar mass.

$$\text{molar mass} = (2\,\text{mol C} \times 12.01\,\text{g C}) + (1\,\text{mol H} \times 1.008\,\text{g H}) + (1\,\text{mol Br} \times 79.90\,\text{g Br})$$
$$+ (1\,\text{mol Cl} \times 35.45\,\text{g Cl}) + (3\,\text{mol F} \times 19.00\,\text{g F})$$
$$= 24.02\,\text{g C} + 1.008\,\text{g H} + 79.09\,\text{g Br} + 35.45\,\text{g Cl} + 57.00\,\text{g F} = 196.38\,\text{g/mol}$$

Bromine is present in greatest mass percent.

(e) $\text{compound mass} = 1.00\,\text{g F} \times \dfrac{1\,\text{mol F}}{19.00\,\text{g F}} \times \dfrac{1\,\text{mol cmpd}}{3\,\text{mol F}} \times \dfrac{196.38\,\text{g cmpd}}{1\,\text{mol cmpd}} = 3.45\,\text{g cmpd}$

42.   **(a)**   A formula unit of $Ge\left[S(CH_2)_4 CH_3\right]_4$ contains:

1 Ge atom   4 S atoms $4(4+1) = 20$ C atoms $4[4(2)+3] = 44$ H atoms

For a total of $1+4+20+44 = 69$ atoms per formula unit

(b) $\dfrac{no.\,C\,atoms}{no.\,H\,atoms} = \dfrac{20\,C\,atoms}{44\,H\,atoms} = \dfrac{5\,C\,atoms}{11\,H\,atoms} = 0.455\,C\,atom/H\,atom$

(c) $\dfrac{mass\,Ge}{mass\,S} = \dfrac{1\,mol\,Ge \times \dfrac{72.6\,g\,Ge}{1\,mol\,Ge}}{4\,mol\,S \times \dfrac{32.07\,g\,S}{1\,mol\,S}} = \dfrac{72.6\,g\,Ge}{128.3\,g\,S} = 0.566\,g\,Ge/g\,S$

(d)   molar mass $= 72.6\,g\,Ge + (4 \times 32.1\,g\,S) + (20 \times 12.01\,g\,C) + (44 \times 1.01\,g\,H)$

molar mass $= 485.6\,g/mol$

mass of $S = 1\,mol\,Ge\left[S(CH_2)_4 CH_3\right]_4 \times \dfrac{4\,mol\,S}{1\,mol\,Ge\left[S(CH_2)_4 CH_3\right]_4} \times \dfrac{32.07\,g\,S}{1\,mol\,S}$

mass of $S = 128.3\,g\,S$

(e)   no. C atoms $= 33.10\,g\,cmpd \times \dfrac{1\,mol\,cmpd}{485.6\,g\,cmpd} \times \dfrac{20\,mol\,C}{1\,mol\,cmpd} \times \dfrac{6.022 \times 10^{23}\,C\,atoms}{1\,mol\,C}$

$= 8.210 \times 10^{23}\,C\,atoms$

Percent Composition of Compounds

43.   The information obtained in the course of calculating the molar mass is used to determine the mass percent of each element in stearic acid, $C_{18}H_{36}O_2$, abbreviated as SA below. Molar mass $C_8H_{18} = M$

$M = \left(\dfrac{18\,mol\,C}{1\,mol\,SA} \times \dfrac{12.011\,g\,C}{1\,mol\,C}\right) + \left(\dfrac{36\,mol\,H}{1\,mol\,SA} \times \dfrac{1.00794\,g\,H}{1\,mol\,H}\right) + \left(\dfrac{2\,mol\,O}{1\,mol\,SA} \times \dfrac{15.9994\,g\,O}{1\,mol\,O}\right)$

$M = \dfrac{216.20\,g\,C}{1\,mol\,SA} + \dfrac{36.2858\,g\,H}{1\,mol\,SA} + \dfrac{31.9988\,g\,O}{1\,mol\,SA} = \dfrac{284.48\,g}{1\,mol\,SA}$

$\%C = \dfrac{216.20\,g\,C/mol\,SA}{284.48\,g\,C/mol\,SA} \times 100\% = 75.998\%\,C;$

$\%H = \dfrac{36.2858\,g\,H/mol\,SA}{284.48\,g\,SA/mol\,SA} \times 100\% = 12.755\%\,H$

$\%O = \dfrac{31.9988\,g\,O/mol\,SA}{284.48\,g\,SA/mol\,SA} \times 100\% = 11.248\%\,O$

44. The method of solution of Exercise 43 is abbreviated in the following problem solution.

$$\text{molar mass} = (20\,\text{mol C} \times 12.011\,\text{g C}) + (24\,\text{mol H} \times 1.00794\,\text{g H}) + (2\,\text{mol N} \times 14.0067\,\text{g N})$$
$$+ (2\,\text{mol O} \times 15.9994\,\text{g O}) = 324.42\,\text{g/mol}$$

$$\%C = \frac{240.22}{324.42} \times 100\% = 74.046\,\%C \qquad \%H = \frac{24.1906}{324.42} \times 100\% = 7.4566\,\%H$$

$$\%N = \frac{28.0134}{324.42} \times 100\% = 8.6349\,\%N \qquad \%O = \frac{31.9988}{324.42} \times 100\% = 9.8634\,\%O$$

45. **(a)** $\%Zr = \dfrac{1\,\text{mol Zr}}{1\,\text{mol ZrSiO}_4} \times \dfrac{1\,\text{mol ZrSiO}_4}{183.31\,\text{g ZrSiO}_4} \times \dfrac{91.224\,\text{g Zr}}{1\,\text{mol Zr}} \times 100\% = 49.765\%\,\text{Zr}$

(b) $\%Be = \dfrac{3\,\text{mol Fe}}{1\,\text{mol Be}_3\text{Al}_2\text{Si}_6\text{O}_{18}} \times \dfrac{1\,\text{mol Be}_3\text{Al}_2\text{Si}_6\text{O}_{18}}{537.502\,\text{g Be}_3\text{Al}_2\text{Si}_6\text{O}_{18}} \times \dfrac{9.01218\,\text{g Be}}{1\,\text{mol Be}} \times 100\%$

$\%\,\text{Be} = 5.03004\,\%\,\text{Be}$

(c) $\%Fe = \dfrac{3\,\text{mol Fe}}{1\,\text{mol Fe}_3\text{Al}_2\text{Si}_3\text{O}_{12}} \times \dfrac{1\,\text{mol Fe}_3\text{Al}_2\text{Si}_3\text{O}_{12}}{497.753\,\text{g Fe}_3\text{Al}_2\text{Si}_3\text{O}_{12}} \times \dfrac{55.847\,\text{g Fe}}{1\,\text{mol Fe}} \times 100\%$

$\%Fe = 33.659\%\,\text{Fe}$

(d) $\%S = \dfrac{1\,\text{mol S}}{1\,\text{mol Na}_4\text{SSi}_3\text{Al}_3\text{O}_{12}} \times \dfrac{1\,\text{mol Na}_4\text{SSi}_3\text{Al}_3\text{O}_{12}}{481.219\,\text{g Na}_4\text{SSi}_3\text{Al}_3\text{O}_{12}} \times \dfrac{32.066\,\text{g S}}{1\,\text{mol S}} \times 100\%$

$\%S = 6.6635\%\,\text{S}$

46. **(a)** $\%K = \dfrac{1\,\text{mol K}}{1\,\text{mol C}_{16}\text{H}_{17}\text{KN}_2\text{O}_4\text{S}} \times \dfrac{1\,\text{mol C}_{16}\text{H}_{17}\text{KN}_2\text{O}_4\text{S}}{372.49\,\text{g C}_{16}\text{H}_{17}\text{KN}_2\text{O}_4\text{S}} \times \dfrac{39.0983\,\text{g K}}{1\,\text{mol K}} \times 100\%$

$\%K = 10.497\%\,\text{K}$

(b) $\%N = \dfrac{3\,\text{mol}}{1\,\text{mol C}_{14}\text{H}_{21}\text{N}_3\text{O}_6\text{S}} \times \dfrac{1\,\text{mol C}_{14}\text{H}_{21}\text{N}_3\text{O}_6\text{S}}{359.403\,\text{g C}_{14}\text{H}_{21}\text{N}_3\text{O}_6\text{S}} \times \dfrac{14.0067\,\text{g N}}{1\,\text{mol N}} \times 100\%$

$\%N = 11.6916\%\,\text{N}$

(c) $\%S = \dfrac{2\,\text{mol S}}{1\,\text{mol C}_{14}\text{H}_{18}\text{ClKN}_2\text{O}_4\text{S}_2} \times \dfrac{1\,\text{mol C}_{14}\text{H}_{18}\text{ClKN}_2\text{O}_4\text{S}_2}{416.99\,\text{g C}_{14}\text{H}_{18}\text{ClKN}_2\text{O}_4\text{S}_2} \times \dfrac{32.066\,\text{g S}}{1\,\text{mol S}} \times 100\%$

$\%S = 15.380\%\,\text{S}$

(d) $\%C = \dfrac{32\,\text{mol C}}{1\,\text{mol Ca}(\text{C}_{16}\text{H}_{17}\text{N}_2\text{O}_4\text{S})_2} \times \dfrac{1\,\text{mol Ca}(\text{C}_{16}\text{H}_{17}\text{N}_2\text{O}_4\text{S})_2}{706.85\,\text{g Ca}(\text{C}_{16}\text{H}_{17}\text{N}_2\text{O}_4\text{S})_2} \times \dfrac{12.011\,\text{g C}}{1\,\text{mol C}} \times 100\%$

$\%C = 54.375\%\,\text{C}$

47. Oxide with the largest %Cr will have the largest number of moles of Cr per mole of oxygen.

$$\text{CrO:} \frac{1\,\text{mol Cr}}{1\,\text{mol O}} = 1\,\text{mol Cr/mol O} \qquad\qquad \text{Cr}_2\text{O}_3: \frac{2\,\text{mol Cr}}{3\,\text{mol O}} = 0.667\,\text{mol Cr/mol O}$$

$$\text{CrO}_2: \frac{1\,\text{mol Cr}}{2\,\text{mol O}} = 0.500\,\text{mol Cr/mol O} \qquad\qquad \text{CrO}_3: \frac{1\,\text{mol Cr}}{3\,\text{mol O}} = 0.333\,\text{mol Cr/mol O}$$

Arranged in order of increasing %Cr: $\text{CrO}_3 < \text{CrO}_2 < \text{Cr}_2\text{O}_3 < \text{CrO}$

48. In each case, a P atom is associated with four O atoms. The species with the largest percent phosphorus will have the smallest mass (per P atom) for the atoms other than P and O.

In H_3PO_4, the mass of those other atoms is (3 H atoms ×1 u/H atom) = 3 u

In Na_3PO_4, their mass is (3 Na atoms ×23 u/Na atom) = 69 u

In $(\text{NH}_3)_2\text{HPO}_4$, the mass is (2 N atoms ×14 u/N atom) + (7 H atoms ×1 u/H atom) = 35 u

In $\text{Ca}(\text{H}_2\text{PO}_4)_2$, their mass is (1/2 Ca atom ×40 u/Ca atom) + (2 H atoms×1 u/H atom) = 22 u

Thus, arranged in order of increasing %P: $\text{Na}_3\text{PO}_4 < (\text{NH}_4)_2\text{HPO}_4 < \text{Ca}(\text{H}_2\text{PO}_4)_2 < \text{H}_3\text{PO}_4$

Chemical Formulas from Percent Composition

49. Convert each percentage into the mass in 100.00 g, and then to the moles of that element.

$$93.71\,\text{g C} \times \frac{1\,\text{mol C}}{12.01\,\text{g C}} = 7.803\,\text{mol C} \quad \div 6.23 \rightarrow 1.25\,\text{mol C}$$

$$6.29\,\text{g H} \times \frac{1\,\text{mol H}}{1.01\,\text{g H}} = 6.23\,\text{mol H} \qquad \div 6.23 \rightarrow 1.00\,\text{mol H}$$

Empirical formula is C_5H_4, and molar mass [(5 × 12.0 g C)+(4 × 1.0 g H)] = 64.0 g/mol. Since this empirical molar mass is one-half of the 128 g/mol correct molar mass, the molecular formula must be twice the empirical formula. Molecular formula: C_{10}H_8

50. The percent of selenium in each oxide is found by difference.

First oxide: %Se = 100.0% − 28.8% O = 71.2% Se

A 100 gram sample would contain 28.8 g O and 71.2 g Se

$$28.8\,\text{g O} \times \frac{1\,\text{mol O}}{16.0\,\text{g O}} = 1.80\,\text{mol O} \qquad \div 0.901 \qquad \rightarrow 2.00\,\text{mol O}$$

$$71.2\,\text{g Se} \times \frac{1\,\text{mol Se}}{79.0\,\text{g Se}} = 0.901\,\text{mol Se} \quad \div 0.901 \qquad \rightarrow 1.00\,\text{mol Se}$$

The empirical formula is SeO_2. An appropriate name is selenium dioxide.

Second oxide: $\%\,Se = 100.0\% - 37.8\%\,O = 62.2\%\,Se\,\%$

A 100 gram sample would contain 37.8 g O and 62.2 g Se

$$37.8\,g\ O\times\frac{1\,mol\ O}{16.0\,g\ O}=2.36\,mol\ O \qquad \div 0.787 \to 3.00\,mol\ O$$

$$62.2\,g\ Se\times\frac{1\,mol\ Se}{79.0\,g\ Se}=0.787\,mol\ Se \div 0.787 \to 1.00\,mol\,Se$$

The empirical formula is SeO_3. An appropriate name is selenium trioxide.

51. **(a)** $74.01g\ C\times\dfrac{1\,mol\ C}{12.01g\ C}=6.162\,mol\ C \div 1.298 \to 4.747\,mol\ C$

$5.23g\ H\times\dfrac{1\,mol\ H}{1.01g\ H}=5.18\,mol\ H \div 1.298 \to 3.99\,mol\ H$

$20.76g\ O\times\dfrac{1\,mol\ O}{16.00g\ O}=1.298\,mol\ O \div 1.298 \to 1.00\,mol\ O$

Multiply each of the mole numbers by 4 to obtain an empirical formula of $C_{19}H_{16}O_4$.

(b) $30.20g\ C\times\dfrac{1\,mol\ C}{12.01g\ C}=2.515\,mol\ C \div 0.6288 \to 4.000\,mol\ C$

$5.07g\ H\times\dfrac{1\,mol\ H}{1.01g\ H}=5.02\,mol\ H \div 0.6288 \to 7.98\,mol\ H$

$44.58g\ Cl\times\dfrac{1\,mol\ Cl}{35.45g\ Cl}=1.258\,mol\ Cl \div 0.6288 \to 2.001\,mol\ Cl$

$20.16g\ S\times\dfrac{1\,mol\ S}{32.06g\ S}=0.6288\,mol\ S \div 0.6288 \to 1.000\,mol\ S$

The empirical formula is $C_4H_8Cl_2S$.

52. **(a)** $95.21g\ C\times\dfrac{1\,mol\ C}{12.01g\ C}=7.928\,mol\ C \div 4.74 \to 1.67\,mol\ C$

$4.79g\ H\times\dfrac{1\,mol\ H}{1.01g\ H}=4.74\,mol\ H \div 4.74 \to 1.00\,mol\ H$

Multiply each of the mole numbers by 3 to obtain an empirical formula of C_5H_3.

(b)   Each percent is numerically equal to the mass of that element present in 100.00 g of the compound. These masses then are converted to amounts of the elements, in moles.

$$\text{amount C} = 38.37 \text{g C} \times \frac{1 \text{ mol C}}{12.01 \text{g C}} = 3.195 \text{ mol C} \qquad \div 0.491 \rightarrow 6.51 \text{ mol C}$$

$$\text{amount H} = 1.49 \text{g H} \times \frac{1 \text{ mol H}}{1.01 \text{g H}} = 1.48 \text{ mol H} \qquad \div 0.491 \rightarrow 3.01 \text{ mol H}$$

$$\text{amount Cl} = 52.28 \text{g Cl} \times \frac{1 \text{ mol Cl}}{35.45 \text{g Cl}} = 1.475 \text{ mol Cl} \quad \div 0.491 \rightarrow 3.004 \text{ mol Cl}$$

$$\text{amount O} = 7.86 \text{g O} \times \frac{1 \text{mol O}}{16.0 \text{g O}} = 0.491 \text{ mol O} \qquad \div 0.491 \rightarrow 1.00 \text{ mol O}$$

Multiply each number of moles by 2 to obtain the empirical formula: $C_{13}H_6Cl_6O_2$.

53. Determine the mass of oxygen by difference. Then convert all masses to amounts in moles.
oxygen mass $= 100.00 \text{g} - 73.27 \text{g C} - 3.84 \text{g H} - 10.68 \text{g N} = 12.21 \text{g O}$

$$\text{amount C} = 73.27 \text{g C} \times \frac{1 \text{mol C}}{12.011 \text{g C}} = 6.100 \text{ mol C} \qquad \div 0.7625 \qquad \rightarrow 8.000 \text{ mol C}$$

$$\text{amount H} = 3.84 \text{g H} \times \frac{1 \text{mol H}}{1.008 \text{g H}} = 3.81 \text{ mol H} \qquad \div 0.7625 \qquad \rightarrow 5.00 \text{ mol H}$$

$$\text{amount N} = 10.68 \text{g N} \times \frac{1 \text{mol N}}{14.007 \text{g N}} = 0.7625 \text{ mol N} \qquad \div 0.7625 \qquad \rightarrow 1.000 \text{ mol N}$$

$$\text{amount O} = 12.21 \text{g O} \times \frac{1 \text{mol O}}{15.999 \text{g O}} = 0.7632 \text{ mol O} \qquad \div 0.7625 \qquad \rightarrow 1.001 \text{ mol O}$$

The empirical formula is C_8H_5NO, which has an empirical mass of

$(8 \text{ mol C} \times 12.0 \text{u C}) + (5 \text{ mol H} \times 1.0 \text{u H}) + (1 \text{ mol N} \times 14.0 \text{u N}) + (1 \text{ mol O} \times 16.0 \text{u O}) = 131$ u.

This is almost exactly half the molecular mass of 262.3 u.

Thus, the molecular formula is twice the empirical formula and is $C_{16}H_{10}N_2O_2$.

54. β-Carotene contains only carbon and hydrogen.

$$\text{amount C} = 89.49 \text{g C} \times \frac{1 \text{mol C}}{12.011 \text{g C}} = 7.451 \text{ mol C} \qquad \div 7.451 \rightarrow 1.000 \text{ mol C}$$

$$\text{amount H} = 10.51 \text{g H} \times \frac{1 \text{mol H}}{1.0079 \text{g H}} = 10.43 \text{ mol H} \qquad \div 7.451 \rightarrow 1.400 \text{ mol H}$$

The empirical formula, C_5H_7, has an empirical mass of:

$(5 \text{ mol C} \times 12.01 \text{u C}) + (7 \text{ mol H} \times 1.01 \text{u H}) = 67.12$ u.

The molecular mass is $536.9/67.12 = 7.999 \sim 8$ times the empirical mass.

Thus, the molecular formula is eight times the empirical formula, *viz.* $C_{40}H_{56}$.

55. In each 100 g of the compound there are 65 g of F and 35 g of X. The number of moles of X is given by no. mol $X = 65\,\text{g F} \times \dfrac{1\,\text{mol F}}{19.0\,\text{g F}} \times \dfrac{1\,\text{mol X}}{3\,\text{mol F}} = 1.14\,\text{mol X}$

Thus, the molar mass of $X = \dfrac{35\,\text{g X}}{1.14\,\text{mol X}} = 31\,\text{g/mol X}$, and the atomic mass is 31 u.

The element is most likely P.

56. The molar mass of element X has the units of grams per mole. We can determine the amount, in moles of Cl, and convert that to the amount of X, equivalent to 25.0 g of X.

$$\text{molar mass} = \dfrac{25.0\,\text{g X}}{75.0\,\text{g Cl}} \times \dfrac{35.453\,\text{g Cl}}{1\,\text{mol Cl}} \times \dfrac{4\,\text{mol Cl}}{1\,\text{mol X}} = \dfrac{47.3\,\text{g X}}{1\,\text{mol X}}$$

The atomic mass is 47.3 u. This atomic mass is close to that of the element titanium, which therefore is identified as element X.

57. Each chlorophyll molecule contains one Mg atom, which makes up 2.72 % of the total mass for the molecule. Thus, 2.72 % of the molecular mass is Mg (24.305 g mol^{-1}).

$$\text{Consequently, the molar mass for chlorophyll} = \dfrac{100 \ \%(\text{total mass})}{2.72\ \%(\text{by mass Mg})} \times 24.305 \text{ g mol}^{-1}$$
$$= 894 \text{ g mol}^{-1}$$

Therefore, the molecular mass of chlorophyll is 894 u.

58. Compound I has a molecular mass of 137 u. We are told that chlorine constitutes 77.5 % of the mass, so the mass of chlorine in each molecule is $137 \text{ u} \times \dfrac{77.5}{100} = 106 \text{ u}$.

This corresponds to three chlorine atoms (106 u ÷ 35.453 u/Cl atom = 2.99 or 3 Cl atoms). The remaining 31 u, (137 u - 106 u), is the mass for element X in one molecule of Compound I. Compound II has 85.1 % chlorine by mass, so the mass of chlorine in each molecule of Compound II is $208 \text{ u} \times \dfrac{85.1}{100} = 177 \text{ u}$.

This corresponds to five Cl atoms (177 u ÷ 35.453 u/Cl atom = 4.99 ~5 chlorine atoms). The remaining mass is 31 u (208 u - 177 u), which is very close to the mass of X found in each molecule of compound I. Thus, we have two compounds: X_nCl_3, which has a molecular mass of 137 u, and X_nCl_5, which has a molecular mass of 208 u

(We also know that the mass of X in both molecular species is ~31 u). If we assume that n = 1 in the formulas above, then element X must be phosphorus (30.974 u) and the formulas for the compounds are PCl_3 (Compound I) and PCl_5 (Compound II).

Combustion Analysis

59. **(a)** Determine the mass of carbon and of hydrogen present in the sample. A hydrocarbon only contains the two elements hydrogen and carbon.

$$0.8661\,g\,CO_2 \times \frac{1\,mol\,CO_2}{44.010\,g\,CO_2} \times \frac{1\,mol\,C}{1\,mol\,CO_2} = 0.01968\,mol\,C \times \frac{12.011\,g\,C}{1\,mol\,C} = 0.2364\,g\,C$$

$$0.2216\,g\,H_2O \times \frac{1\,mol\,H_2O}{18.015\,g\,H_2O} \times \frac{2\,mol\,H}{1\,mol\,H_2O} = 0.02460\,mol\,H \times \frac{1.0079\,g\,H}{1\,mol\,H} = 0.02479\,g\,H$$

Then the % C and % H are found.

$$\%\,C = \frac{0.2364\,g\,C}{0.2612\,g\,cmpd} \times 100\% = 90.51\%\,C \quad \%\,H = \frac{0.02479\,g\,H}{0.2612\,g\,cmpd} \times 100\% = 9.491\%\,H$$

(b) Use the moles of C and H from part (a), and divide both by the smallest. 0.01968 mol C ÷ 0.01968 = 1.000 mol C; 0.02460 mol H ÷ 0.01968 = 1.250 mol H. The empirical formula is obtained by multiplying these mole numbers by 4. It is C_4H_5.

(c) The empirical formula C_4H_5 has an empirical molar mass of

$$\left[(4 \times 12.0\,g\,C) + (5 \times 1.0\,g\,H) \right] = 53.0 \text{ g/mol. This value is 1/2 of the actual molar}$$

mass. The molecular formula is twice the empirical formula.

Molecular formula: C_8H_{10}.

60. Determine the mass of carbon and of hydrogen present in the sample.

$$1.0420g\,CO_2 \times \frac{1\,mol\,CO_2}{44.01\,g\,CO_2} \times \frac{1\,mol\,C}{1\,mol\,CO_2} = 0.02368\,mol\,C \times \frac{12.01\,g\,C}{1\,mol\,C} = 0.2844g\,C$$

$$0.2437g\,H_2O \times \frac{1\,mol\,H_2O}{18.02\,g\,H_2O} \times \frac{2\,mol\,H}{1\,mol\,H_2O} = 0.02705\,mol\,H \times \frac{1.008g\,H}{1\,mol\,H} = 0.02726g\,H$$

(a) The percent composition can be determined using the masses of C and H.

$$\%\,C = \frac{0.2844\,g\,C}{0.3654\,g\,cmpd} \times 100\% = 77.83\%\,C \quad \%\,H = \frac{0.02726\,g\,H}{0.3654\,g\,cmpd} \times 100\% = 7.460\%\,H$$

%O=100.00% − 77.83%C − 7.460%H=14.71%O

These percents can be used in determining the empirical formula if one wishes.

(b) To find the empirical formula determine the mass of oxygen by difference, and its amount in moles. Mass O = 0.3654g cmpd − 0.2844g C − 0.02726g H = 0.0537g O

$$0.0537 \, g \, O \, \times \frac{1 \, mol \, O}{16.0 \, g \, O} = 0.00336 \, mol \, O \div 0.00336 \rightarrow 1.00 \, mol \, O$$

$$0.02368 \, mol \, C \div 0.00336 \rightarrow 7.05 \, mol \, C$$

$$0.02705 \, mol \, H \div 0.00336 \rightarrow 8.05 \, mol \, H$$

This gives an empirical formula of C_7H_8O

(c) The molecular formula is found by realizing that a mole of empirical units has a mass of $(7 \times 12.0 \, g \, C + 8 \times 1.0 \, g \, H + 16.0 \, g \, O) = 108.0$ g. Since this agrees with the molecular mass, the molecular formula is the same as the empirical formula: C_7H_8O.

61. First determine the mass of carbon and hydrogen present in the sample.

$$0.741 \, g \, CO_2 \times \frac{1 \, mol \, CO_2}{44.01 \, g \, CO_2} \times \frac{1 \, mol \, C}{1 \, mol \, CO_2} = 0.0168 \, mol \, C \times \frac{12.01 \, g \, C}{1 \, mol \, C} = 0.202 \, g \, C$$

$$0.605 \, g \, H_2O \times \frac{1 \, mol \, H_2O}{18.02 \, g \, H_2O} \times \frac{2 \, mol \, H}{1 \, mol \, H_2O} = 0.0671 \, mol \, H \times \frac{1.008 \, g \, H}{1 \, mol \, H} = 0.0677 \, g \, H$$

Then, the mass of N that this sample would have produced is determined. (Note that this is also the mass of N_2 produced in the reaction.)

$$0.226 \, g \, N_2 \times \frac{0.505 \, g \, 1st \, sample}{0.486 \, g \, 2nd \, sample} = 0.235 \, g \, N_2$$

We determine the mass of N by difference: $505 \, g$ cmpd $- 0.202 \, g \, C - 0.0677 \, g \, H = 0.235$ g N

Then, we can calculate the relative number of moles of each element.

$$0.235 \, g \, N \times \frac{1 \, mol \, N}{14.01 \, g \, N} = 0.0168 \, mol \, N \quad \div 0.0168 \rightarrow 1.00 \, mol \, N$$

$$0.0168 \, mol \, C \quad \div 0.0168 \rightarrow 1.00 \, mol \, C$$

$$0.0671 \, mol \, H \quad \div 0.0168 \rightarrow 3.99 \, mol \, H$$

Thus, the empirical formula is CH_4N

62. Thiophene contains only carbon, hydrogen, and sulfur, so there is no need to determine the mass of oxygen by difference. We simply determine the amount of each element from the mass of its combustion product.

$$2.272 \, g \, CO_2 \times \frac{1 \, mol \, CO_2}{44.010 \, g \, CO_2} \times \frac{1 \, mol \, C}{1 \, mol \, CO_2} = 0.05162 \, mol \, C \div 0.0129 \rightarrow 4.00 \, mol \, C$$

$$0.465 \, g \, H_2O \times \frac{1 \, mol \, H_2O}{18.02 \, g \, H_2O} \times \frac{2 \, mol \, H}{1 \, mol \, H_2O} = 0.0516 \, mol \, H \div 0.0129 \rightarrow 4.00 \, mol \, H$$

$$0.827 \, g \, SO_2 \times \frac{1 \, mol \, SO_2}{64.06 \, g \, SO_2} \times \frac{1 \, mol \, S}{1 \, mol \, SO_2} = 0.0129 \, mol \, S \div 0.0129 \rightarrow 1.00 \, mol \, S$$

The empirical formula of thiophene is C_4H_4S.

63. Each mole of CO_2 is produced from a mole of C. Therefore, the compound with the largest number of moles of C per mole of the compound will produce the largest amount of CO_2 and, thus, also the largest mass of CO_2. Of the compounds listed — $CH_4, C_2H_5OH, C_{10}H_8$, and C_6H_5OH, $C_{10}H_8$ has the largest number of moles of C per mole of the compound and will produce the greatest mass of CO_2 per mole on complete combustion.

64. The compound that produces the largest mass of water per gram of the compound will have the largest amount of hydrogen per gram of the compound. Thus, we need to compare the ratios of amount of hydrogen per mole to the molar mass for each compound.
Note that C_2H_5OH has as much H per mole as does C_6H_5OH, but C_6H_5OH has a higher molar mass. Thus, C_2H_5OH produces more H_2O per gram than does C_6H_5OH.
Notice also that CH_4 has 4 H's per C, while $C_{10}H_8$ has 8 H's per 10 C's or 0.8 H per C. Thus CH_4 will produce more H_2O than will $C_{10}H_8$. Thus, we are left with comparing CH_4 to C_2H_5OH. The O in the second compound has about the same mass (16 u) as does C (12 u). Thus, in CH_4 there are 4 H's per C, while in C_2H_5OH there are about 2 H's per C. CH_4 will produce the most water per gram on combustion, of all four compounds.

65. mass CO_2
$$= 1.562 \text{ g } C_7H_{16} \times \frac{1 \text{ mol } C_7H_{16}}{100.20 \text{ g } C_7H_{16}} \times \frac{7 \text{ mol C}}{1 \text{ mol } C_7H_{16}} \times \frac{1 \text{ mol } CO_2}{1 \text{ mol C}} \times \frac{44.011 \text{ g } CO_2}{1 \text{ mol } CO_2}$$
$$= 4.803 \text{ g } CO_2$$

mass H_2O
$$= 1.562 \text{ g } C_7H_{16} \times \frac{1 \text{ mol } C_7H_{16}}{100.20 \text{ g } C_7H_{16}} \times \frac{16 \text{ mol H}}{1 \text{ mol } C_7H_{16}} \times \frac{1 \text{ mol } H_2O}{2 \text{ mol H}} \times \frac{18.015 \text{ g } H_2O}{1 \text{ mol } H_2O}$$
$$= 2.247 \text{ g } H_2O$$

66. moles of $C_2H_6S = 1.50 \text{ mL} \times \frac{0.84 \text{ g } C_2H_6S}{1 \text{ mL } C_2H_6S} \times \frac{1 \text{ mol } C_2H_6S}{62.134 \text{ g } C_2H_6S} = 2.0\underline{3} \times 10^{-2} \text{ mol } C_2H_6S$

Thus, the mass of CO_2 expected is

$$= 2.0\underline{3} \times 10^{-2} \text{ mol } C_2H_6S \times \frac{2 \text{ mol } CO_2}{1 \text{ mol } C_2H_6S} \times \frac{44.011 \text{ g } CO_2}{1 \text{ mol } CO_2} = 1.8 \text{ g of } CO_2(g)$$

The mass of $SO_2(g)$ expected from the complete combustion is

$$= 2.03 \times 10^{-2} \text{ mol } C_2H_6S \times \frac{1 \text{ mol } SO_2}{1 \text{ mol } C_2H_6S} \times \frac{64.064 \text{ g } SO_2}{1 \text{ mol } H_2O} = 1.3 \text{ g of } SO_2(g)$$

The mass of $H_2O(l)$ expected from the complete combustion is

$$= 2.03 \times 10^{-2} \text{ mol } C_2H_6S \times \frac{3 \text{ mol } H_2O}{1 \text{ mol } C_2H_6S} \times \frac{18.015 \text{ g } H_2O}{1 \text{ mol } H_2O} = 1.1 \text{ g of } H_2O(l)$$

Oxidation States

67. The oxidation state (O.S.) is given first, followed by the explanation for its assignment.

    (a) $C = -4$ in CH_4 H has an oxidation state of $+1$ in its non-metal compounds (Remember that the sum of the oxidation states in a compound equals 0.)

    (b) $S = +4$ in SF_4 F has $O.S. = -1$ in its compounds.

    (c) $O = -1$ in Na_2O_2 Na has $O.S. = +1$ in its compounds.

    (d) $C = 0$ in $C_2H_3O_2^-$ H has $O.S. = +1$ in its non-metal compounds; that of $O = -2$ (usually). (Remember that the sum of the oxidation states in a polyatomic ion equals the charge on that ion.)

    (e) $Fe = +6$ in FeO_4^{2-} O has $O.S. = -2$ in most of its compounds (especially metal containing compounds).

68. The oxidation state of sulfur in each species is determined below. Remember that the oxidation state of O is -2 in its compounds. And the sum of the oxidation states in an ion equals the charge on that ion.

    (a) $S = +4$ in SO_3^{2-} **(b)** $S = +2$ in $S_2O_3^{2-}$ **(c)** $S = +7$ in $S_2O_8^{2-}$

    (d) $S = +6$ in HSO_4^- **(e)** $S = -2.5$ in $S_4O_6^{2-}$

69. Remember that the oxidation state of oxygen is -2 in its compounds. Cr^{3+} and O^{2-} form Cr_2O_3, chromium(III) oxide. Cr^{4+} and O^{2-} form CrO_2, chromium (IV) oxide. Cr^{6+} and O^{2-} form CrO_3, chromium(VI) oxide.

70. Remember that oxygen has an oxidation state of -2 in its compounds.

    $N = +1$ in N_2O, dinitrogen monoxide $N = +2$ in NO, nitric oxide or nitrogen monoxide

    $N = +3$ in N_2O_3, dinitrogen trioxide $N = +4$ in NO_2, nitrogen dioxide

    $N = +5$ in N_2O_5, dinitrogen pentoxide

Nomenclature

71. **(a)** SrO   strontium oxide   **(b)** ZnS   zinc sulfide

    (c) K_2CrO_4 potassium chromate   **(d)** Cs_2SO_4 cesium sulfate

    (e) Cr_2O_3 chromium(III) oxide   **(f)** $Fe_2(SO_4)_3$ iron(III) sulfate

    (g) $Mg(HCO_3)_2$ magnesium hydrogen carbonate   **(h)** $(NH_4)_2HPO_4$ ammonium hydrogen phosphate

(i) $Ca(HSO_3)_2$ calcium hydrogen sulfite   **(j)** $Cu(OH)_2$ copper(II) hydroxide

(k) HNO_3 nitric acid   **(l)** $KClO_4$ potassium perchlorate

(m) $HBrO_3$ bromic acid   **(n)** H_3PO_3 phosphorous acid

72. **(a)** $Ba(NO_3)_2$ barium nitrate   **(b)** HNO_2 nitrous acid

(c) CrO_2 chromium(IV) oxide   **(d)** KIO_3 potassium iodate

(e) $LiCN$ lithium cyanide   **(f)** KIO potassium hypoiodite

(g) $Fe(OH)_2$ iron(II) hydroxide   **(h)** $Ca(H_2PO_4)_2$ calcium dihydrogen phosphate

(i) H_3PO_4 phosphoric acid   **(j)** $NaHSO_4$ sodium hydrogen sulfate

(k) $Na_2Cr_2O_7$ sodium dichromate   **(l)** $NH_4C_2H_3O_2$ ammonium acetate

(m) MgC_2O_4 magnesium oxalate   **(n)** $Na_2C_2O_4$ sodium oxalate

73. **(a)** CS_2 carbon disulfide   **(b)** SiF_4 silicon tetrafluoride

(c) ClF_5 chlorine pentafluoride   **(d)** N_2O_5 dinitrogen pentoxide

(e) SF_6 sulfur hexafluoride   **(f)** I_2Cl_6 diiodine hexachloride

74. **(a)** ICl iodine monochloride   **(b)** ClF_3 chlorine trifluoride

(c) SF_4 sulfur tetrafluoride   **(d)** BrF_5 bromine pentafluoride

(e) N_2O_4 dinitrogen tetroxide   **(g)** S_4N_4 tetrasulfur tetranitride

75. **(a)** $Al_2(SO_4)_3$ aluminum sulfate   **(b)** $(NH_4)_2Cr_2O_7$ ammonium dichromate

(c) SiF_4 silicon tetrafluoride   **(d)** Fe_2O_3 iron(III) oxide

(e) C_3S_2 tricarbon disulfide   **(f)** $Co(NO_3)_2$ cobalt(II) nitrate

(g) $Sr(NO_2)_2$ strontium nitrite   **(h)** $HBr(aq)$ hydrobromic acid

(i) HIO_3 iodic acid   **(j)** PCl_2F_3 phosphorus dichloride trifluoride

76. **(a)** $Mg(ClO_4)_2$ magnesium perchlorate   **(b)** $Pb(C_2H_3O_2)_2$ lead(II) acetate

(c) SnO_2 tin(IV) oxide   **(d)** $HI(aq)$ hydroiodic acid

(e) $HClO_2$ chlorous acid   **(f)** $NaHSO_3$ sodium hydrogen sulfite

(g) $Ca(H_2PO_4)_2$ calcium dihydrogen phosphate   **(h)** $AlPO_4$ aluminum phosphate

(i) N_2O_4 dinitrogen tetroxide   **(j)** S_2Cl_2 disulfur dichloride

77. **(a)** Ti^{4+} and Cl^- produce $TiCl_4$ **(b)** Fe^{3+} and SO_4^{2-} produce $Fe_2(SO_4)_3$

    (c) Cl^{7+} and O^{2-} produce Cl_2O_7 **(d)** S^{7+} and O^{2-} produce $S_2O_8^{2-}$

78. **(a)** N^{5+} and O^{2-} produce N_2O_5 **(b)** N^{3+} and O^{2-} produce N_2O_3

    (c) $C^{+4/3}$ and O^{2-} produce C_3O_2 **(d)** $S^{+2.5}$ and O^{2-} produce $S_4O_6^{2-}$

Hydrates

79. The hydrate with the greatest mass percent H_2O is the one that gives the largest result for the number of moles of water in the hydrate's empirical formula, divided by the mass of one mole of the anhydrous salt for the hydrate.

$$\frac{5\,H_2O}{CuSO_4} = \frac{5\,mol\,H_2O}{159.6\,g} = 0.03133 \qquad \frac{6\,H_2O}{MgCl_2} = \frac{6\,mol\,H_2O}{95.2\,g} = 0.0630$$

$$\frac{18\,H_2O}{Cr_2(SO_4)_3} = \frac{18\,mol\,H_2O}{392.3\,g} = 0.04588 \qquad \frac{2\,H_2O}{LiC_2H_3O_2} = \frac{2\,mol\,H_2O}{66.0\,g} = 0.0303$$

The hydrate with the greatest % H_2O therefore is $MgCl_2 \cdot 6H_2O$

80. A mole of this hydrate will contain about the same mass of H_2O and of Na_2SO_3.

Molar mass $Na_2SO_3 = (2 \times 23.0\,g\,Na) + 32.1\,g\,S + (3 \times 16.0\,g\,O) = 126.1\,g/mol$

$$\text{no. mol } H_2O = 126.1\,g \times \frac{1\,mol\,H_2O}{18.0\,g\,H_2O} = 7.01\,mol\,H_2O$$

Thus, the formula of the hydrate is $Na_2SO_3 \cdot 7H_2O$.

81. Molar mass $CuSO_4 = 63.5\,g\,Cu + 32.1\,g\,S + (4 \times 16.0\,g\,O) = 159.6\,g\,CuSO_4/mol$.

The mass of solid needed to combine with 8.5 g of water depends on the absorption of 5 moles of water by one mole of the solid, based on the formula $CuSO_4 \cdot 5H_2O$.

$$\text{mass } CuSO_4 = 8.5\,g\,H_2O \times \frac{1\,mol\,H_2O}{18.0\,g\,H_2O} \times \frac{1\,mol\,CuSO_4}{5\,mol\,H_2O} \times \frac{159.6\,g\,CuSO_4}{1\,mol\,CuSO_4}$$

$$= 15\,g\,CuSO_4 \quad \text{required to remove all the water}$$

82. The increase in mass of the solid is related to each mole of the solid absorbing ten moles of water.

$$\text{increase in mass} = 3.50\,g\,Na_2SO_4 \times \frac{1\,mol\,Na_2SO_4}{142.1\,g\,Na_2SO_4} \times \frac{10\,mol\,H_2O\,added}{1\,mol\,Na_2SO_4} \times \frac{18.015\,g\,H_2O}{1\,mol\,H_2O}$$

$$= 4.44\,g\,H_2O\,added$$

83. Converting to molar amounts based on 100.0g:

$$20.3 \text{ g Cu} \times \frac{1 \text{ mol Cu}}{63.55 \text{ g Cu}} = 0.319 \text{ mol Cu} \quad \div 0.319 \quad \rightarrow \quad 1.00 \text{ mol Cu}$$

$$8.95 \text{ g Si} \times \frac{1 \text{ mol Si}}{28.09 \text{ g Si}} = 0.319 \text{ mol Si} \quad \div 0.319 \quad \rightarrow \quad 1.00 \text{ mol Si}$$

$$36.3 \text{ g F} \times \frac{1 \text{ mol F}}{19.00 \text{ g F}} = 1.91 \text{ mol F} \quad \div 0.319 \quad \rightarrow \quad 6.00 \text{ mol F}$$

$$34.5 \text{ g H}_2\text{O} \times \frac{1 \text{ mol H}_2\text{O}}{18.02\text{g}} = 1.91 \text{ mol H}_2\text{O} \div 0.319 \quad \rightarrow \quad 6.00 \text{ mol H}_2\text{O}$$

Thus the empirical formula for the hydrate is $CuSiF_6 \cdot 6H_2O$.

84. mass of anhydrous compound = 3.967 g

mass of water = 8.129 g – 3.967 g = 4.162 g

$$\text{moles of anhydrous compound} = 3.967 \text{ g MgSO}_4 \times \frac{1 \text{ mol MgSO}_4}{120.37 \text{ g}} = 0.03296 \text{ mol}$$

$$\text{moles of H}_2\text{O} = 4.162 \text{ g} \times \frac{1 \text{ mol H}_2\text{O}}{18.02 \text{ g H}_2\text{O}} = 0.2310 \text{ mol H}_2\text{O}$$

$$\text{setting up proportions} \quad \frac{0.2310 \text{ mol H}_2\text{O}}{0.03296 \text{ mol anhydrous compound}} = \frac{x \text{ mol H}_2\text{O}}{1.00 \text{ mol anhydrous compound}}$$

$x = 7.01$ Thus, the formula of the hydrate is $MgSO_4 \cdot 7H_2O$.

Organic Compounds and Organic Nomenclature

85. Molecules (a), (b), (c) and (d) are structural isomers. They share a common formula, namely $C_5H_{12}O$, but have different molecular structures. Molecule (e) has a different chemical formula ($C_6H_{14}O$) and hence cannot be classified as an isomer.

86. Molecules (a), (b) and (c) are structural isomers. They share a common formula, namely $C_5H_{11}Cl$, but have different molecular structures. Molecule (d) has a different chemical formula ($C_6H_{13}Cl$) and hence cannot be classified as an isomer.

87. (a) $CH_3(CH_2)_4CH_3$ (b) HCO_2H

      (c) $CH_3CH_2CH(CH_3)CH_2OH$ (d) $ClCH_2CH_3$

88. (a) $CH_3(CH_2)_6CH_3$ (b) $CH_3(CH_2)_5CO_2H$

      (c) $CH_3(CH_2)_3CH_2OH$ (d) CH_3Cl

89. **(a)** Methanol; CH_3OH; Molecular mass = 32.04 u

(b) 2-chlorohexane; $CH_3(CH_2)_3CHClCH_3$ Molecular mass = 120.6 u

(c) pentanoic acid; $CH_3(CH_2)_3CO_2H$ Molecular mass = 102.1 u

(d) 2-methyl-1-propanol; $CH_3CH(CH_3)CH_2OH$ Molecular mass = 74.12 u

90. **(a)** 2-pentanol; $CH_3CH_2CH_2CH(OH)CH_3$ Molecular mass = 88.15 u

(b) Propanoic acid; $CH_3CH_2CO_2H$ Molecular mass = 74.08 u

(c) 1-bromobutane; $CH_3(CH_2)_2CH_2Br$ Molecular mass = 137.0 u

(d) 3-chlorobutanoic acid; $CH_3CHClCH_2CO_2H$ Molecular mass = 122.6 u

FEATURE PROBLEMS

108. (a) "5-10-5" fertilizer contains 5 g N (that is, 5% N), 10 g P_2O_5, and 5 g K_2O in 100 g fertilizer. We convert the last two numbers into masses of the two elements.

(1) $\%P = 10\%P_2O_5 \times \dfrac{1\,mol\,P_2O_5}{141.9\,g\,P_2O_5} \times \dfrac{2\,mol\,P}{1\,mol\,P_2O_5} \times \dfrac{30.97\,g\,P}{1\,mol\,P} = 4.4\%\,P$

(2) $\%K = 5\%K_2O \times \dfrac{1\,mol\,K_2O}{94.20\,g\,K_2O} \times \dfrac{2\,mol\,K}{1\,mol\,K_2O} \times \dfrac{39.10\,g\,K}{1\,mol\,K} = 4.2\%\,K$

(b) First, we determine %P and then convert it to $\%P_2O_5$, given that 10.0% P_2O_5 is equivalent to 4.37% P.

(1) $\%\,P_2O_5 = \dfrac{2\,mol\,P}{1\,mol\,Ca(H_2PO_4)_2} \times \dfrac{30.97\,g\,P}{1\,mol\,P} \times \dfrac{1\,mol\,Ca(H_2PO_4)_2}{234.05\,g\,Ca(H_2PO_4)_2} \times 100\%$

$\times \dfrac{10.0\%P_2O_5}{4.37\%P} = 60.6\%\,P_2O_5$

(2) $\%\,P_2O_5 = \dfrac{1\,mol\,P}{1\,mol(NH_4)_2\,HPO_4} \times \dfrac{30.97\,g\,P}{1\,mol\,P} \times \dfrac{1\,mol(NH_4)_2\,HPO_4}{132.06\,g(NH_4)_2\,HPO_4} \times 100\%$

$\times \dfrac{10.0\%P_2O_5}{4.37\%P} = 53.7\%\,P_2O_5$

109. (a) First calculate the mass of water that was present in the hydrate prior to heating.
Mass of H_2O = 2.574 g $CuSO_4 \cdot x\,H_2O$ - 1.647 g $CuSO_4$ = 0.927 g H_2O
Next we need to find the number of moles of anhydrous copper(II) sulfate and water that were initially present together in the original hydrate sample.

Moles of H_2O = 0.927 g $H_2O \times \dfrac{1\,mol\,H_2O}{18.015\,g\,H_2O}$ = 0.05146 moles of water

The empirical formula is obtained by dividing the number of moles of water by the number of moles of $CuSO_4$ (x = ratio of moles of water to moles of $CuSO_4$)

$$x = \frac{0.05146 \ \text{moles} \ H_2O}{0.01032 \ \text{moles} \ CuSO_4} = 4.99 \sim 5 \quad \text{The empirical formula is } CuSO_4 \cdot 5 \ H_2O.$$

(b) Mass of water present in hydrate = 2.574 g - 1.833 g = 0.741 g H_2O

$$\text{moles of water} = 0.741 \ g \ H_2O \times \frac{1 \text{mol} \ H_2O}{18.015 \ g \ H_2O} = 0.0411 \text{ moles of water}$$

Mass of $CuSO_4$ present in hydrate = 1.833 g $CuSO_4$

$$\text{moles of } CuSO_4 = 1.833 \ g \ CuSO_4 \times \frac{1 \text{mol} \ CuSO_4}{159.61 \ g \ CuSO_4} = 0.0115 \text{ mol } CuSO_4$$

The empirical formula is obtained by dividing the number of moles of water by the number of moles of $CuSO_4$ (x = ratio of moles of water to moles of $CuSO_4$)

$$x = \frac{0.0411 \ \text{moles} \ H_2O}{0.0115 \ \text{moles} \ CuSO_4} = 3.58 \sim 4.$$

Since the hydrate has not been completely dehydrated, there is no problem with obtaining non-integer "garbage" values.

So, the empirical formula is $CuSO_4 \bullet 4H_2O$.

(c) When copper(II) sulfate is strongly heated, it decomposes to give $SO_3(g)$ and $CuO(s)$. The black residue formed at 1000 °C in this experiment is probably CuO. The empirical formula for copper(II) oxide is CuO. Let's calculate the percentages of Cu and O by mass for CuO:

$$\text{Mass percent copper} = \frac{63.546 \ g \ Cu}{79.545 \ g \ CuO} \times 100 \ \% = 79.89 \ \% \text{ by mass Cu}$$

$$\text{Mass percent oxygen} = \frac{15.9994 \ g \ O}{79.545 \ g \ CuO} \times 100 \ \% = 20.11 \ \% \text{ by mass O}$$

The number of moles of CuO formed (by reheating to 1000 °C)

$$= 0.812 \ g \ CuO \times \frac{1 \ \text{mol} \ CuO}{79.545 \ g \ CuO} = 0.0102 \text{ moles of CuO}$$

This is very close to the number of moles of anhydrous $CuSO_4$ formed at 400. °C. Thus, it would appear that upon heating to 1000 °C, the sample of $CuSO_4$ was essentially completely converted to CuO.

110. **(a)** The formula for stearic acid, obtained from the molecular model, is $CH_3(CH_2)_{16}CO_2H$. The number of moles of stearic acid in 10.0 grams is

$$= 10.0 \text{ g stearic acid} \times \frac{1 \text{ mol stearic acid}}{284.48 \text{ g stearic acid}} = 3.51\underline{5} \times 10^{-2} \text{ mol of stearic acid.}$$

The layer of stearic acid is one molecule thick. According to the figure provided with the question, each stearic acid molecule has a cross-sectional area of ~0.22 nm^2. In order to find the stearic acid coverage in square meters, we must multiply the total number of stearic acid molecules by the cross-sectional area for an individual stearic acid molecule. The number of stearic acid molecules is:

$$= 3.51\underline{5} \times 10^{-2} \text{ mol of stearic acid} \times \frac{6.022 \times 10^{23} \text{ molecules}}{1 \text{ mol of stearic acid}} = 2.11\underline{7} \times 10^{22} \text{ molecules}$$

$$\text{Area in } m^2 = 2.11\underline{7} \times 10^{22} \text{ molecules of stearic acid} \times \frac{0.22 \text{ nm}^2}{\text{molecule}} \times \frac{(1 \text{ m})^2}{(1 \times 10^9 \text{ nm})^2}$$

The area in m^2 = 4657 m^2 or 4.7 $\times$ 10^3 m^2 (with correct number of sig. fig.)

(b) The density for stearic acid is 0.85 g cm^{-3}. Thus, 0.85 grams of stearic acid occupies 1 cm^3. Find the number of moles of stearic acid in 0.85 g of stearic acid

$$= 0.85 \text{ grams of stearic acid} \times \frac{1 \text{ mol stearic acid}}{284.48 \text{ g stearic acid}} = 3.0 \times 10^{-3} \text{ mol of stearic}$$

acid. This number of moles of acid occupies 1 cm^3 of space. So, the number of stearic acid molecules in 1 cm^3

$$= 3.0 \times 10^{-3} \text{ mol of stearic acid} \times \frac{6.022 \times 10^{23} \text{ molecules}}{1 \text{ mol of stearic acid}}$$

$$= 1.8 \times 10^{21} \text{ stearic acid molecules.}$$

Thus, the volume for a single stearic acid molecule in nm^3

$$= 1 \text{ cm}^3 \times \frac{1}{1.8 \times 10^{21} \text{ molecules stearic acid}} \times \frac{(1.0 \times 10^7 \text{ nm})^3}{(1 \text{ cm})^3} = 0.55\underline{6} \text{ nm}^3$$

The volume of a rectangular column is simply its area of the base multiplied by its height (i.e. V = area of base (in nm^2) × height (in nm)).

$$\text{So, the average height of a stearic acid molecule} = \frac{0.556 \text{ nm}^3}{0.22 \text{ nm}^2} = 2.5 \text{ nm}$$

(c) The density for oleic acid = 0.895 g mL^{-1}. So, the concentration for oleic acid is

$$= \frac{0.895 \text{ g acid}}{10.00 \text{ mL}} = 0.0895 \text{ g mL}^{-1} \text{ (solution 1)}$$

This solution is then divided by ten, three more times to give a final concentration of $8.9\underline{5} \times 10^{-5}$ g mL^{-1}. A 0.10 mL sample of this solution contains:

$$= \frac{8.95 \times 10^{-5} \text{ g acid}}{1.00\text{mL}} \times 0.10 \text{ mL} = 8.9\underline{5} \times 10^{-6} \text{ g of acid.}$$

The number of acid molecules $= 85 \text{ cm}^2 \times \dfrac{1}{4.6 \times 10^{-15}\text{cm}^2 \text{ per molecule}}$

$$= 1.8\underline{5} \times 10^{16} \text{ oleic acid molecules.}$$

So, 8.95×10^{-6} g of oleic acid corresponds to 1.85×10^{16} oleic acid molecules.

The molar mass for oleic acid, $C_{18}H_{34}O_2$, is 282.47 g mol^{-1}.

The number of moles of oleic acid is

$$= 8.9\underline{5} \times 10^{-6} \text{ g} \times \frac{1 \text{ mol oleic acid}}{282.47 \text{ g}} = 3.1\underline{7} \times 10^{-8} \text{ mol}$$

So, Avogadro's number here would be equal to:

$$= \frac{1.85 \times 10^{16} \text{ oleic acid molecules}}{3.1\underline{7} \times 10^{-8} \text{ oleic acid moles}} = 5.8 \times 10^{23} \text{ molecules per mole of oleic acid.}$$

CHAPTER 4
CHEMICAL REACTIONS
PRACTICE EXAMPLES

1A **(a)**

Unbalanced reaction:	$H_3PO_4(aq) + CaO(s)$	$\rightarrow$	$Ca_3(PO_4)_2(aq) + H_2O(l)$
Balance Ca & PO_4^{3-}:	$2\ H_3PO_4(aq) + 3\ CaO(s)$	$\rightarrow$	$Ca_3(PO_4)_2(aq) + H_2O(l)$
Balance H atoms:	$2\ H_3PO_4(aq) + 3\ CaO(s)$	$\rightarrow$	$Ca_3(PO_4)_2(aq) + 3\ H_2O(l)$
Self Check:	$6\ H + 2\ P + 11\ O + 3\ Ca$	$\rightarrow$	$6\ H + 2\ P + 11\ O + 3\ Ca$

(b)

Unbalanced reaction:	$C_3H_8(g) + O_2(g)$	$\rightarrow$	$CO_2(g) + H_2O(g)$
Balance C & H:	$C_3H_8(g) + O_2(g)$	$\rightarrow$	$3\ CO_2(g) + 4\ H_2O(g)$
Balance O atoms:	$C_3H_8(g) + 5\ O_2(g)$	$\rightarrow$	$3\ CO_2(g) + 4\ H_2O(g)$
Self Check:	$3\ C + 8\ H + 10\ O$	$\rightarrow$	$3\ C + 8\ H + 10\ O$

1B **(a)**

Unbalanced reaction:	$NH_3(g) + O_2(g)$	$\rightarrow$	$NO_2(g) + H_2O(g)$
Balance N and H:	$NH_3(g) + O_2(g)$	$\rightarrow$	$NO_2(g) + 3/2\ H_2O(g)$
Balance O atoms:	$NH_3(g) + 7/4\ O_2(g)$	$\rightarrow$	$NO_2(g) + 3/2\ H_2O(g)$
Multiply by 4 (whole #):	$4\ NH_3(g) + 7\ O_2(g)$	$\rightarrow$	$4\ NO_2(g) + 6\ H_2O(g)$
Self Check:	$4\ N + 12\ H + 14\ O$	$\rightarrow$	$4\ N + 12\ H + 14\ O$

(b)

Unbalanced reaction:	$NO_2(g) + NH_3(g)$	$\rightarrow$	$N_2(g) + H_2O(g)$
Balance H atoms:	$NO_2(g) + 2\ NH_3(g)$	$\rightarrow$	$N_2(g) + 3\ H_2O(g)$
Balance O atoms:	$3/2\ NO_2(g) + 2\ NH_3(g)$	$\rightarrow$	$N_2(g) + 3\ H_2O(g)$
Balance N atoms:	$3/2\ NO_2(g) + 2\ NH_3(g)$	$\rightarrow$	$7/4\ N_2(g) + 3\ H_2O(g)$
Multiply by 4 (whole #)	$6\ NO_2(g) + 8\ NH_3(g)$	$\rightarrow$	$7\ N_2(g) + 12\ H_2O(g)$
Self Check:	$14\ N + 24\ H + 12\ O$	$\rightarrow$	$14\ N + 24\ H + 12\ O$

2A

Unbalanced reaction:	$HgS(s) + CaO(s)$	$\rightarrow CaS(s) + CaSO_4(s) + Hg(l)$
Balance 0 atoms:	$HgS(s) + 4\ CaO(s)$	$\rightarrow CaS(s) + CaSO_4(s) + Hg(l)$
Balance Ca atoms:	$HgS(s) + 4\ CaO(s)$	$\rightarrow 3\ CaS(s) + CaSO_4(s) + Hg(l)$
Balance S atoms:	$4\ HgS(s) + 4\ CaO(s)$	$\rightarrow 3\ CaS(s) + CaSO_4(s) + Hg(l)$
Balance Hg atoms:	$4\ HgS(s) + 4\ CaO(s)$	$\rightarrow 3\ CaS(s) + CaSO_4(s) + 4\ Hg(l)$
Self Check:	$4\ Hg + 4\ S + 4\ O + 4\ Ca$	$\rightarrow 4\ Hg + 4\ S + 4\ O + 4\ Ca$

2B Unbalanced reaction: $C_7H_6O_2S(l) + O_2(g) \rightarrow CO_2(g) + H_2O(l) + SO_2(g)$

Balance C atoms: $C_7H_6O_2S(l) + O_2(g) \rightarrow 7\,CO_2(g) + H_2O(l) + SO_2(g)$

Balance S atoms: $C_7H_6O_2S(l) + O_2(g) \rightarrow 7\,CO_2(g) + H_2O(l) + SO_2(g)$

Balance H atoms: $C_7H_6O_2S(l) + O_2(g) \rightarrow 7\,CO_2(g) + 3\,H_2O(l) + SO_2(g)$

Balance O atoms: $C_7H_6O_2S(l) + 8.5\,O_2(g) \rightarrow 7\,CO_2(g) + 3\,H_2O(l) + SO_2(g)$

Multiply by 2 (whole #): $2\,C_7H_6O_2S(l) + 17\,O_2(g) \rightarrow 14\,CO_2(g) + 6\,H_2O(l) + 2\,SO_2(g)$

Self Check: $14\,C + 12\,H + 2\,S + 38\,O \rightarrow 14\,C + 12\,H + 2\,S + 38\,O$

3A The balanced chemical equation provides the factor needed to convert from moles $KClO_3$ to moles O_2. Amount $O_2 = 1.76\,\text{mol KClO}_3 \times \dfrac{3\,\text{mol O}_2}{2\,\text{mol KClO}_3} = 2.64\,\text{mol O}_2$

3B First, find the molar mass of Ag_2O.

$(2\,\text{mol Ag} \times 107.87\,\text{g Ag}) + 16.00\,\text{g O} = 231.74\,\text{g Ag}_2O\,/\,\text{mol}$

amount $Ag = 1.00\,\text{kg Ag}_2O \times \dfrac{1000\,\text{g}}{1.00\,\text{kg}} \times \dfrac{1\,\text{mol Ag}_2O}{231.74\,\text{g Ag}_2O} \times \dfrac{2\,\text{mol Ag}}{1\,\text{mol Ag}_2O} = 8.63\,\text{mol Ag}$

4A The balanced chemical equation provides the factor to convert from amount of Mg to amount of Mg_3N_2. First, we determine the molar mass of Mg_3N_2.

molar mass $= (3\,\text{mol Mg} \times 24.305\,\text{g Mg}) + (2\,\text{mol N} \times 14.007\,\text{g N}) = 100.93\,\text{g Mg}_3N_2$

mass $Mg_3N_2 = 3.82\,\text{g Mg} \times \dfrac{1\,\text{mol Mg}}{24.31\,\text{g Mg}} \times \dfrac{1\,\text{mol Mg}_3N_2}{3\,\text{mol Mg}} \times \dfrac{100.93\,\text{g Mg}_3N_2}{1\,\text{mol Mg}_3N_2} = 5.29\,\text{g Mg}_3N_2$

4B The pivotal conversion is from $H_2(g)$ to $CH_3OH(l)$. For this we use the balanced equation, which requires that we use the amounts in moles of both substances. The solution involves converting to and from amounts, using molar masses.

mass $H_2(g) = 1.00\,\text{kg CH}_3OH(l) \times \dfrac{1000\,\text{g}}{1\,\text{kg}} \times \dfrac{1\,\text{mol CH}_3OH}{32.04\,\text{g CH}_3OH} \times \dfrac{2\,\text{mol H}_2}{1\,\text{mol CH}_3OH} \times \dfrac{2.016\,\text{g H}_2}{1\,\text{mol H}_2}$

mass $H_2(g) = 126\,\text{g H}_2$

5A The equation for the cited reaction is: $2\,H_2(g) + O_2(g) \rightarrow 2\,H_2O(l)$.

The pivotal conversion is from one substance to another, in moles with the balanced chemical equation providing the conversion factor.

mass $H_2(g) = 1.00\,\text{g O}_2(g) \times \dfrac{1\,\text{mol O}_2}{32.00\,\text{g O}_2} \times \dfrac{2\,\text{mol H}_2}{1\,\text{mol O}_2} \times \dfrac{2.016\,\text{g H}_2}{1\,\text{mol H}_2} = 0.126\,\text{g H}_2$

5B The equation for the combustion reaction is: $C_8H_{18}(l) + \dfrac{25}{2}O_2(g) \rightarrow 8\,CO_2(g) + 9\,H_2O(l)$

$$\text{mass } O_2 = 1.00\,g\,C_8H_{18} \times \frac{1\,mol\,C_8H_{18}}{114.23\,g\,C_8H_{18}} \times \frac{12.5\,mol\,O_2}{1\,mol\,C_8H_{18}} \times \frac{32.00\,g\,O_2}{1\,mol\,O_2} = 3.50\,g\,O_2(g)$$

6A We must convert mass $H_2 \rightarrow$ amount of $H_2 \rightarrow$ amount of $Al \rightarrow$ mass of $Al \rightarrow$ mass of alloy $\rightarrow$ volume of alloy. The calculation is performed as follows: each arrow in the preceding sentence requires a conversion factor.

$$V_{alloy} = 1.000\,g\,H_2 \times \frac{1\,mol\,H_2}{2.016\,g\,H_2} \times \frac{2\,mol\,Al}{3\,mol\,H_2} \times \frac{26.98\,g\,Al}{1\,mol\,Al} \times \frac{100.0\,g\,alloy}{93.7\,g\,Al} \times \frac{1\,cm^3\,alloy}{2.85\,g\,alloy}$$

Volume of alloy $= 3.34\,cm^3$ alloy

6B In the example, $0.207\,g\,H_2$ is collected from $1.97\,g$ alloy; the alloy is 6.3% Cu by mass. This information provides the conversion factors we need.

$$\text{mass Cu} = 1.31\,g\,H_2 \times \frac{1.97\,g\,alloy}{0.207\,g\,H_2} \times \frac{6.3\,g\,Cu}{100.0\,g\,alloy} = 0.79\,g\,Cu$$

Notice that we do not have to consider each step separately. We can simply use values produced in the course of the calculation as conversion factors.

7A The cited reaction is $2\,Al(s) + 6\,HCl(aq) \rightarrow 2\,AlCl_3(aq) + 3\,H_2(g)$. The HCl(aq) solution has a density of 1.14 g/mL and contains 28.0% HCl. We need to convert between the substances HCl and H_2; the important conversion factor comes from the balanced chemical equation. The sequence of conversions is: volume of HCl(aq) $\rightarrow$ mass of HCl(aq) $\rightarrow$ mass of pure HCl $\rightarrow$ amount of HCl $\rightarrow$ amount of $H_2 \rightarrow$ mass of H_2. In the calculation below, each arrow in the sequence is replaced by a conversion factor.

$$\text{mass } H_2 = 0.05\,mL\,HCl(aq) \times \frac{1.14\,g\,sol}{1\,mL\,soln} \times \frac{28.0\,g\,HCl}{100.0\,g\,soln} \times \frac{1\,mol\,HCl}{36.46\,g\,HCl} \times \frac{3\,mol\,H_2}{6\,mol\,HCl} \times \frac{2.016\,g\,H_2}{1\,mol\,H_2}$$

$$\text{mass } H_2 = 4 \times 10^{-4}\,g\,H_2(g) = 0.4\,mg\,H_2(g)$$

7B Density is necessary to determine the mass of the vinegar, and then the mass of acetic acid.

$$\text{mass } CO_2(g) = 5.00\,mL\,\text{vinegar} \times \frac{1.01\,g}{1\,mL} \times \frac{0.040\,g\,acid}{1\,g\,\text{vinegar}} \times \frac{1\,mol\,HC_2H_3O_2}{60.05\,g\,HC_2H_3O_2} \times \frac{1\,mol\,CO_2}{1\,mol\,HC_2H_3O_2} \times \frac{44.01\,g\,CO_2}{1\,mol\,CO_2}$$

$$= 0.15\,g\,CO_2$$

8A Determine the amount in moles of acetone and the volume in liters of the solution.

$$\text{molarity of acetone} = \frac{22.3\,g\,(CH_3)_2\,CO \times \dfrac{1\,mol\,(CH_3)_2\,CO}{58.08\,g\,(CH_3)_2\,CO}}{1.25\,L\,soln} = 0.307\,M$$

8B The molar mass of acetic acid, $HC_2H_3O_2$, is 60.05 g/mol. We begin with the quantity of acetic acid in the numerator and that of the solution in the denominator, and transform to the appropriate units for each.

$$\text{molarity} = \frac{15.0\,\text{mL }HC_2H_3O_2}{500.0\,\text{mL soln}} \times \frac{1000\,\text{mL}}{1\,\text{L soln}} \times \frac{1.048\,\text{g }HC_2H_3O_2}{1\,\text{mL }HC_2H_3O_2} \times \frac{1\,\text{mol }HC_2H_3O_2}{60.05\,\text{g }HC_2H_3O_2} = 0.524\,\text{M}$$

9A The molar mass of $NaNO_3$ is 84.99 g/mol. We recall that "M" stands for "mol /L soln."

$$\text{mass }NaNO_3 = 125\,\text{mL soln} \times \frac{1\,\text{L}}{1000\,\text{mL}} \times \frac{10.8\,\text{mol }NaNO_3}{1\,\text{L soln}} \times \frac{84.99\,\text{g }NaNO_3}{1\,\text{mol NaCl}} = 115\,\text{g }NaNO_3$$

9B We begin by determining the molar mass of $Na_2SO_4 \cdot 10H_2O$. The amount of solute needed is computed from the concentration and volume of the solution.

$$\text{mass }Na_2SO_4 \cdot 10H_2O = 355\,\text{mL soln} \times \frac{1\,\text{L}}{1000\,\text{mL}} \times \frac{0.445\,\text{mol }Na_2SO_4}{1\,\text{L soln}} \times \frac{1\,\text{mol }Na_2SO_4 \cdot 10H_2O}{1\,\text{mol }Na_2SO_4}$$

$$\times \frac{322.21\,\text{g }Na_2SO_4 \cdot 10H_2O}{1\,\text{mol }Na_2SO_4 \cdot 10H_2O} = 50.9\,\text{g }Na_2SO_4 \cdot 10H_2O$$

10A The amount of solute in the concentrated solution doesn't change when the solution is diluted. We take advantage of an alternate definition of molarity to answer the question: millimoles of solute/milliliter of solution.

$$\text{amount }K_2CrO_4 = 15.00\,\text{mL} \times \frac{0.450\,\text{mmol }K_2CrO_4}{1\,\text{mL soln}} = 6.75\,\text{mmol }K_2CrO_4$$

$$K_2CrO_4 \text{ molarity, dilute solution} = \frac{6.75\,\text{mmol }K_2CrO_4}{100.00\,\text{mL soln}} = 0.0675\,\text{M}$$

10B We know the initial concentration (0.105 M) and volume (275 mL) of the solution, along with its final volume (237 mL). The final concentration equals the initial concentration times a ratio of the two volumes.

$$c_f = c_i \times \frac{V_i}{V_f} = 0.105\,\text{M} \times \frac{275\,\text{mL}}{237\,\text{mL}} = 0.122\,\text{M}$$

11A The balanced equation is $K_2CrO_4(aq) + 2\,AgNO_3(aq) \rightarrow Ag_2CrO_4(s) + 2\,KNO_3(aq)$. The molar mass of Ag_2CrO_4 is 331.73 g/mol. The conversions needed are mass

$Ag_2CrO_4 \rightarrow$ amount Ag_2CrO_4 (moles) $\rightarrow$ amount K_2CrO_4 (moles) $\rightarrow$ volume K_2CrO_4 (aq).

$$V_{K_2CrO_4} = 1.50\,\text{g }Ag_2CrO_4 \times \frac{1\,\text{mol }Ag_2CrO_4}{331.73\,\text{g }Ag_2CrO_4} \times \frac{1\,\text{mol }K_2CrO_4}{1\,\text{mol }Ag_2CrO_4} \times \frac{1\,\text{L soln}}{0.250\,\text{mol }K_2CrO_4}$$

$$\times \frac{1000\,\text{mL solution}}{1\,\text{L solution}} = 18.1\,\text{mL}$$

11B　Balanced reaction:　$2\ AgNO_3(aq) + K_2CrO_4(aq)\ \rightarrow\ Ag_2CrO_4(s) + 2\ KNO_3(aq)$

moles of $K_2CrO_4 = C \times V = 0.0855\ M \times 0.175\ L\ sol = 0.0149\underline{6}$ moles K_2CrO_4

moles of $AgNO_3 = 0.0149\underline{6}$ mol $K_2CrO_4 \times \dfrac{2\ mol\ AgNO_3}{1\ mol\ K_2CrO_4} = 0.0299$ mol $AgNO_3$

$V_{AgNO_3} = \dfrac{n}{C} = \dfrac{0.0299\ mol\ AgNO_3}{0.150\ \dfrac{mol}{L}\ AgNO_3} = 0.199\underline{5}\ L$ or $2.00 \times 10^2\ mL$ (0.200 L) of $AgNO_3$

Mass of Ag_2CrO_4 formed $= 0.0149\underline{6}$ moles $K_2CrO_4 \times \dfrac{1mol\ Ag_2CrO_4}{1\ mol\ K_2CrO_4} \times \dfrac{331.73\ g\ Ag_2CrO_4}{1mol\ Ag_2CrO_4}$

Mass of Ag_2CrO_4 formed $= 4.96\ g\ Ag_2CrO_4$

12A　Reaction: $P_4\,(s) + 6\,Cl_2\,(g) \rightarrow 4\,PCl_3\,(l)$. Determine mass of PCl_3 formed by each reactant.

mass $PCl_3 = 215\ g\ P_4 \times \dfrac{1\,mol\ P_4}{123.90\,g\ P_4} \times \dfrac{4\,mol\ PCl_3}{1\,mol\ P_4} \times \dfrac{137.33\,g\ PCl_3}{1\ mol\ PCl_3} = 953\,g\ PCl_3$

mass $PCl_3 = 725\ g\ Cl_2 \times \dfrac{1\,mol\ Cl_2}{70.91\,g\ Cl_2} \times \dfrac{4\,mol\ PCl_3}{6\,mol\ Cl_2} \times \dfrac{137.33\,g\ PCl_3}{1\ mol\ PCl_3} = 936\,g\ PCl_3$

Thus, $936\,g\ PCl_3$ are produced; there is not enough Cl_2 to produce any more.

12B　Since data are supplied and the answer is requested in kilograms (thousands of grams), we can use kilomoles (thousands of moles) to solve the problem. We calculate the amount in kilomoles of $POCl_3$ that would be produced if each of the reactants were completely converted to product. The smallest of these amounts is the one that is actually produced. (This is a limiting reactant question).

amount $POCl_3 = 1.00\,kg\ PCl_3 \times \dfrac{1\,kmol\ PCl_3}{137.33\,kg\ PCl_3} \times \dfrac{10\,kmol\ POCl_3}{6\,kmol\ PCl_3} = 0.0121\,kmol\ POCl_3$

amount $POCl_3 = 1.00\,kg\ Cl_2 \times \dfrac{1\,kmol\ Cl_2}{70.905\,kg\ Cl_2} \times \dfrac{10\,kmol\ POCl_3}{6\,kmol\ Cl_2} = 0.0235\,kmol\ POCl_3$

amount $POCl_3 = 1.00\,kg\ P_4O_{10} \times \dfrac{1\,kmol\ P_4O_{10}}{283.89\,kg\ P_4O_{10}} \times \dfrac{10\,kmol\ POCl_3}{1\,kmol\ P_4O_{10}} = 0.0352\,kmol\ POCl_3$

Thus, $0.0121\,kmol\ POCl_3$ is produced. We determine the mass of the product.

mass $POCl_3 = 0.0121\,kmol\ POCl_3 \times \dfrac{153.33\,kg\ POCl_3}{1\ kmol\ POCl_3} = 1.86\,kg\ POCl_3$

13A The 725 g Cl_2 limits the mass of product formed. The $P_4(s)$ therefore is the reactant in excess. The quantity of excess reactant is sufficient to form the excess product: 953 g PCl_3 $-$ 936 g PCl_3 $=$ 17 g PCl_3. We calculate how much P_4 this is, both in the traditional way and by using the initial $(215$ g $P_4)$ and final $(953$ g $PCl_3)$ values of the previous calculation.

$$\text{mass } P_4 = 17 \text{ g } PCl_3 \times \frac{1 \text{ mol } PCl_3}{137.33 \text{ g } PCl_3} \times \frac{1 \text{ mol } P_4}{4 \text{ mol } PCl_3} \times \frac{123.90 \text{ g } P_4}{1 \text{ mol } P_4} = 3.8 \text{ g } P_4$$

13B Find the amount of $H_2O(l)$ formed by each reactant, to determine the limiting reactant.

$$\text{amount } H_2O = 12.2 \text{ g } H_2 \times \frac{1 \text{ mol } H_2}{2.016 \text{ g } H_2} \times \frac{2 \text{ mol } H_2O}{2 \text{ mol } H_2} = 6.05 \text{ mol } H_2O$$

$$\text{amount } H_2O = 154 \text{ g } O_2 \times \frac{1 \text{ mol } O_2}{32.00 \text{ g } O_2} \times \frac{2 \text{ mol } H_2O}{1 \text{ mol } O_2} = 9.63 \text{ mol } H_2O$$

Since H_2 is limiting, we compute the mass of O_2 needed to react with all of the H_2

$$\text{mass } O_2 \text{ reacting} = 6.05 \text{ mol } H_2O \text{ produced} \times \frac{1 \text{ mol } O_2}{2 \text{ mol } H_2O} \times \frac{32.00 \text{ g } O_2}{1 \text{ mol } O_2} = 96.8 \text{ g } O_2 \text{ reacting}$$

$$\text{mass } O_2 \text{ remaining} = 154 \text{ g originally present } - 96.8 \text{ g } O_2 \text{ reacting} = 57 \text{ g } O_2 \text{ remaining}$$

14A (a) The theoretical yield is the product mass we predict by calculation.

$$\text{mass } CH_2O(g) = 1.00 \text{ mol } CH_3OH \times \frac{1 \text{ mol } CH_2O}{1 \text{ mol } CH_3OH} \times \frac{30.03 \text{ g } CH_2O}{1 \text{ mol } CH_2O} = 30.0 \text{ g } CH_2O$$

(b) The actual yield is what is obtained experimentally: 25.7 g CH_2O (g).

(c) The percent yield is the ratio of actual to theoretical yields, multiplied by 100%:

$$\% \text{ yield} = \frac{25.7 \text{ g } CH_2O \text{ produced}}{30.0 \text{ g } CH_2O \text{ calculated}} \times 100 \% = 85.6 \% \text{ yield}$$

14B Determine the mass of product formed by each reactant.

$$\text{mass } PCl_3 = 25.0 \text{ g } P_4 \times \frac{1 \text{ mol } P_4}{123.90 \text{ g } P_4} \times \frac{4 \text{ mol } PCl_3}{1 \text{ mol } P_4} \times \frac{137.33 \text{ g } PCl_3}{1 \text{ mol } PCl_3} = 111 \text{ g } PCl_3$$

$$\text{mass } PCl_3 = 91.5 \text{ g } Cl_2 \times \frac{1 \text{ mol } Cl_2}{70.91 \text{ g } Cl_2} \times \frac{4 \text{ mol } PCl_3}{6 \text{ mol } Cl_2} \times \frac{137.33 \text{ g } PCl_3}{1 \text{ mol } PCl_3} = 118 \text{ g } PCl_3$$

The limiting reactant is P_4, and 111 g PCl_3 should be produced. This is the theoretical yield. The actual yield is 104 g PCl_3. Thus, the percent yield of the reaction is

$$\frac{104 \text{ g } PCl_3 \text{ produced}}{111 \text{ g } PCl_3 \text{ calculated}} \times 100 \% = 93.7\% \text{ yield.}$$

15A The reaction is $2\,NH_3(g)+CO_2(g)\rightarrow CO(NH_2)_2(s)+H_2O(l)$. We need to distinguish between mass of urea produced (actual yield) and mass of urea predicted (theoretical yield).

$$\text{mass } CO_2 = 50.0\,g\,CO(NH_2)_2 \text{ produced} \times \frac{100.0\,g\,\text{predicted}}{87.5\,g\,\text{produced}} \times \frac{1\,mol\,CO(NH_2)_2}{60.1\,g\,CO(NH_2)_2} \times \frac{1\,mol\,CO_2}{1\,mol\,CO(NH_2)_2}$$

$$\times \frac{44.01\,g\,CO_2}{1\,mol\,CO_2} = 41.8\,g\,CO_2 \text{ needed}$$

15B Care must be taken to use the proper units/label in each conversion factor. Note, you cannot calculate the molar mass of an impure material or mixture.

$$\text{mass } C_6H_{11}OH = 45.0\,g\,C_6H_{10} \text{ produced} \times \frac{100.0\,g\,C_6H_{10}\,\text{cal'd}}{86.2\,g\,C_6H_{10}\,\text{produc'd}} \times \frac{1\,mol\,C_6H_{10}}{82.1\,g\,C_6H_{10}} \times \frac{1\,mol\,C_6H_{11}OH}{1\,mol\,C_6H_{10}}$$

$$\times \frac{100.2\,g\,\text{pure } C_6H_{11}OH}{1\,mol\,C_6H_{11}OH} \times \frac{100.0\,g\,\text{impure } C_6H_{11}OH}{92.3\,g\,\text{pure } C_6H_{11}OH} = 69.0\,g\,\text{impure } C_6H_{11}OH$$

16A We can trace the nitrogen through the sequence of reactions. We notice that 4 moles of N (as 4 mol NH_3) are consumed in the first reaction, and 4 moles of N (as 4 mole NO) are produced. In the second reaction, 2 moles of N (as 2 mol NO) is consumed and 2 moles of N (as 2 mol NO_2) are produced. In the last reaction, 3 moles of N (as 3 mol NO_2) are consumed and just 2 moles of N (as 2 mol HNO_3) are produced.

$$\text{mass } HNO_3 = 1.00\,kg\,NH_3 \times \frac{1000\,g\,NH_3}{1\,kg\,NH_3} \times \frac{1\,mol\,NH_3}{17.03\,g\,NH_3} \times \frac{4\,mol\,NO}{4\,mol\,NH_3} \times \frac{2\,mol\,NO_2}{2\,mol\,NO}$$

$$\times \frac{2\,mol\,HNO_3}{3\,mol\,NO_2} \times \frac{63.01\,g\,HNO_3}{1\,mol\,HNO_3} = 2.47 \times 10^3\,g\,HNO_3$$

16B $$\text{mass } H_2\,(Al) = 0.710\,g\,\text{alloy} \times \frac{0.700\,g\,Al}{1.000\,g\,\text{alloy}} \times \frac{1\,mol\,Al}{26.98\,g\,Al} \times \frac{3\,mol\,H_2}{2\,mol\,Al} \times \frac{2.016\,g\,H_2}{1\,mol\,H_2} = 0.0557\,g$$

$$\text{mass } H_2\,(Mg) = 0.710\,g\,\text{alloy} \times \frac{0.300\,g\,Mg}{1.000\,g\,\text{alloy}} \times \frac{1\,mol\,Mg}{24.31\,g\,Mg} \times \frac{1\,mol\,H_2}{1\,mol\,Mg} \times \frac{2.016\,g\,H_2}{1\,mol\,H_2} = 0.0177\,g$$

total mass of $H_2 = 0.0557\,g\,H_2$ from $Al + 0.0177\,g\,H_2$ from $Mg = 0.0734\,g\,H_2$

REVIEW QUESTIONS

1. **(a)** The symbol " $\xrightarrow{\Delta}$ " indicates that the mixture is heated to produce the reaction.

    (b) The symbol "(aq)" indicates that the species preceding this symbol is dissolved in aqueous solution, that is, it indicates a solution with water as the solvent.

    (c) The stoichiometric coefficient is the number that appears in a chemical equation immediately before the chemical formula of a species.

    (d) The net "overall" equation is the chemical equation that remains after species that appear on both sides of an equation are "cancelled." The term also is used to describe an equation that summarizes the overall result of a process consisting of several reactions.

2. **(a)** One "balances a chemical equation" by inserting stoichiometric coefficients into the formula expression, so that the resulting equation has the same number and type of atoms on each side.

    (b) When one "prepares a solution by dilution" one begins with a more concentrated solution (a homogeneous mixture with a larger concentration of solute) and adds solvent, thus producing a less concentrated (or more dilute) solution.

    (c) One "determines the limiting reactant in a reaction" by discovering which reactant will produce the smallest quantity of product. That reactant will limit the quantity of product that can be formed from the other reactants, and also limit the quantity that will be consumed of the other reactants.

3. **(a)** A chemical formula is a short-hand representation of a chemical species: atom, ion, or molecule. A chemical equation is a written representation of a chemical reaction; it typically involves two or more species. Whereas a chemical formula is rather analogous to a "word," chemical equations parallel "sentences."

    (b) A decomposition reaction is one in which a compound is broken down into simpler substances. In a synthesis reaction two or more substances combine to form a third.

    (c) The solute is the substance that is dispersed in a solution. The solvent is the substance that does the dispersing. Usually, a solution is of the same physical state (solid, liquid, or gas) as the solvent, and the solvent is the component present in the larger amount.

    (d) The actual yield of a chemical reaction is the quantity of product that actually was formed. The percent yield relates the actual yield to the quantity of product that was calculated to be produced, assuming that all reactants produced only one set of products and the reaction continued until one reactant was exhausted.

4. **(a)** $Na_2SO_4(s) + 4C(s) \rightarrow Na_2S(s) + 4CO(g)$

    (b) $4HCl(g) + O_2(g) \rightarrow 2H_2O(l) + 2Cl_2(g)$

    (c) $PCl_5(l) + 4H_2O(l) \rightarrow H_3PO_4(aq) + 5HCl(aq)$

    (d) $3PbO(s) + 2NH_3(g) \rightarrow 3Pb(s) + N_2(g) + 3H_2O(l)$

    (e) $Mg_3N_2(s) + 6H_2O(l) \rightarrow 3Mg(OH)_2(s) + 2NH_3(g)$

5.   **(a)** $2\,Mg(s) + O_2(g) \rightarrow 2\,MgO(s)$

    (b) $2\,NO(g) + O_2(g) \rightarrow 2\,NO_2(g)$

    (c) $2\,C_2H_6(g) + 7\,O_2(g) \rightarrow 4\,CO_2(g) + 6\,H_2O(l)$

    (d) $Ag_2SO_4(aq) + BaI_2(aq) \rightarrow BaSO_4(s) + 2\,AgI(s)$

6.   **(a)** $C_7H_{16}(l) + 11\,O_2(g) \rightarrow 7\,CO_2(g) + 8\,H_2O(l)$

    (b) $C_4H_9OH(l) + 6\,O_2(g) \rightarrow 4\,CO_2(g) + 5\,H_2O(l)$

    (c) $2\,HI(aq) + Na_2CO_3(aq) \rightarrow 2\,NaI(aq) + H_2O(l) + CO_2(g)$

    (d) $3\,NaOH(aq) + FeCl_3(aq) \rightarrow Fe(OH)_3(s) + 3\,NaCl(aq)$

7.   Expression (c) is incorrect because KClO is potassium hypochlorite, but the stated product is potassium chloride, KCl. Expression (a) and (b) are incorrect because O(g) is not normally produced in chemical reactions; $O_2(g)$ is more thermodynamically stable

    The correct equation is $2\,KClO_3(s) \rightarrow 2\,KCl(s) + 3\,O_2(g)$.

8.   For the reaction $2\,H_2S(g) + SO_2(g) \rightarrow 3S(s) + 2\,H_2O(l)$

    (1)   FALSE   3 moles of S are produced per *two* moles of H_2S.

    (2)   FALSE   3 *moles* of S are produced for every *mole* of SO_2 consumed.

    (3)   TRUE   1 mole of H_2O *is* produced per mole of H_2S consumed.

    (4)   TRUE   Two-thirds of the S produced *does* come from the H_2S.

    (5)   FALSE   There are *five* moles of products and *three* moles of reactants.

9.   The conversion factor is obtained from the balanced chemical equation.

$$\text{moles } FeCl_3 = 7.26\,\text{mol } Cl_2 \times \frac{2\,\text{mol } FeCl_3}{3\,\text{mol } Cl_2} = 4.84\,\text{mol } FeCl_3$$

10.   The pivotal conversion factor, from the balanced equation, enables one to related the amounts of O_2 and $KClO_3$.

$$\text{mass } O_2 = 43.4\,\text{g } KClO_3 \times \frac{1\,\text{mol } KClO_3}{122.5\,\text{g } KClO_3} \times \frac{3\,\text{mol } O_2}{2\,\text{mol } KClO_3} \times \frac{32.00\,\text{g } O_2}{1\,\text{mol } O_2} = 17.0\,\text{g } O_2$$

11.   Each calculation uses the stoichiometric coefficients from the balanced chemical equation and the molar mass of the reactant.

$$\text{mass } Cl_2 = 0.337\,\text{mol } PCl_3 \times \frac{6\,\text{mol } Cl_2}{4\,\text{mol } PCl_3} \times \frac{70.91\,\text{g } Cl_2}{1\,\text{mol } Cl_2} = 35.8\,\text{g } Cl_2$$

$$\text{mass } P_4 = 0.337\,\text{mol } PCl_3 \times \frac{1\,\text{mol } P_4}{4\,\text{mol } PCl_3} \times \frac{123.9\,\text{g } P_4}{1\,\text{mol } P_4} = 10.4\,\text{g } P_4$$

12. **(a)** $\text{amount O}_2 = 156\,\text{g CO}_2 \times \dfrac{1\,\text{mol CO}_2}{44.01\,\text{g CO}_2} \times \dfrac{3\,\text{mol O}_2}{2\,\text{mol CO}_2} = 5.32\,\text{mol O}_2$

(b) $\text{mass KO}_2 = 100.0\,\text{g CO}_2 \times \dfrac{1\,\text{mol CO}_2}{44.01\,\text{g CO}_2} \times \dfrac{4\,\text{mol KO}_2}{2\,\text{mol CO}_2} \times \dfrac{71.10\,\text{g KO}_2}{1\,\text{mol KO}_2} = 323.1\,\text{g KO}_2$

(c) $\text{no. O}_2\text{ molecules} = 1.00\,\text{mg KO}_2 \times \dfrac{1\,\text{g KO}_2}{1000\,\text{mg}} \times \dfrac{1\,\text{mol KO}_2}{71.10\,\text{g KO}_2} \times \dfrac{3\,\text{mol O}_2}{4\,\text{mol KO}_2}$

$\times \dfrac{6.022 \times 10^{23}\,\text{molecules}}{1\,\text{mol O}_2} = 6.35 \times 10^{18}\,\text{O}_2\text{ molecules}$

13. **(a)** $\text{CH}_3\text{OH molarity} = \dfrac{2.92\,\text{mol CH}_3\text{OH}}{7.16\,\text{L}} = 0.408\,\text{M}$

(b) $\text{C}_2\text{H}_5\text{OH molarity} = \dfrac{7.69\,\text{mmol C}_2\text{H}_5\text{OH}}{50.00\,\text{mL}} = 0.154\,\text{M}$

(c) $\text{CO}(\text{NH}_2)_2\text{ molarity} = \dfrac{25.2\,\text{g CO}(\text{NH}_2)_2}{275\,\text{mL}} \times \dfrac{1\,\text{mol CO}(\text{NH}_2)_2}{60.06\,\text{g CO}(\text{NH}_2)_2} \times \dfrac{1000\,\text{mL}}{1\,\text{L}} = 1.53\,\text{M}$

(d) $\text{molarity} = \dfrac{18.5\,\text{mL C}_3\text{H}_5(\text{OH})_3}{375\,\text{mL soln}} \times \dfrac{1.26\,\text{g}}{1\,\text{mL}} \times \dfrac{1\,\text{mol C}_3\text{H}_5(\text{OH})_3}{92.09\,\text{g C}_3\text{H}_5(\text{OH})_3} \times \dfrac{1000\,\text{mL}}{1\,\text{L}} = 0.675\,\text{M}$

14. **(a)** $\text{mol NaI} = 2.55 \times 10^3\,\text{L} \times \dfrac{0.125\,\text{mol NaI}}{1\,\text{L soln}} = 319\,\text{mol NaI}$

(b) $\text{mass Na}_2\text{CO}_3 = 475\,\text{mL} \times \dfrac{1\,\text{L}}{1000\,\text{mL}} \times \dfrac{0.398\,\text{mol Na}_2\text{CO}_3}{1\,\text{L}} \times \dfrac{106.0\,\text{g Na}_2\text{CO}_3}{1\,\text{mol Na}_2\text{CO}_3}$

$= 20.0\,\text{g Na}_2\text{CO}_3$

(c) $\text{mg CaCl}_2 = 1.00\,\text{mL} \times \dfrac{0.148\,\text{mmol CaCl}_2}{1\,\text{mL}} \times \dfrac{111.0\,\text{mg CaCl}_2}{1\,\text{mmol CaCl}_2} = 16.4\,\text{mg CaCl}_2$

15. A 1.00 M KCl solution contains 1 mol KCl per liter of solution. The molar mass of KCl is 74.6 g. Thus, a 1.00 M KCl solution contains 74.6 g KCl per liter of solution. "1.00 L containing 100 g" is incorrect; 1.00 L should contain 74.6 g. "500 mL containing 74.6 g" also is incorrect; 74.6 g should be contained in 1000 mL. "a solution containing 7.46 mg KCl/mL" is incorrect; there should be 74.6 mg. 5.00 L of 1.00 M KCl contains five times the mass of solute as does 1.00 L of this solution, 373 g. The last description is correct.

16. Volume of concentrated AgNO_3 solution

$V_{\text{AgNO}_3} = 250.0\,\text{mL dilute soln} \times \dfrac{0.423\,\text{mmol AgNO}_3}{1\,\text{mL dilute soln}} \times \dfrac{1\,\text{mL conc. soln.}}{0.625\,\text{mmol AgNO}_3}$

$V_{\text{AgNO}_3} = 169\,\text{mL AgNO}_3$

17. The same amount in moles of K_2SO_4 is present in both solutions. That amount is given in the numerator of the following expression.

$$K_2SO_4 \text{ molarity} = \frac{135\,\text{mL} \times \dfrac{0.188\,\text{mmol}\,K_2SO_4}{1\,\text{mL}}}{105\,\text{mL}} = 0.242\,\text{M}$$

18. The balanced chemical equation provides a conversion factor between the two compounds.

$$\text{mass } CuCO_3 = 415\,\text{mL} \times \frac{1\,\text{L}}{1000\,\text{mL}} \times \frac{0.275\,\text{mol}\,Cu(NO_3)_2}{1\,\text{L}} \times \frac{1\,\text{mol}\,CuCO_3}{1\,\text{mol}\,Cu(NO_3)_2} \times \frac{123.6\,\text{g}\,CuCO_3}{1\,\text{mol}\,CuCO_3}$$

$$\text{mass } CuCO_3 = 14.1\,\text{g}\,CuCO_3$$

19. After determining the amount of $CaCO_3$ (100.09 g/mol), we find the volume of HCl.

$$\text{volume HCl(aq)} = 1.75\,\text{g}\,CaCO_3 \times \frac{1\,\text{mol}\,CaCO_3}{100.09\,\text{g}} \times \frac{2\,\text{mol HCl}}{1\,\text{mol}\,CaCO_3} \times \frac{1\,\text{L soln}}{2.35\,\text{mol HCl}} \times \frac{1000\,\text{mL}}{1\,\text{L}}$$

$$= 14.9 \text{ mL HCl(aq) solution}$$

20. In this situation, since 2 mol H_2O are required per mol CaH_2, 1.52 mol H_2O is the limiting reactant. Thus, the amount in moles of H_2 can be computed as follows.

$$\text{amount } H_2 = 1.52\,\text{mol}\,H_2O \times \frac{2\,\text{mol}\,H_2}{2\,\text{mol}\,H_2O} = 1.52\,\text{mol}\,H_2$$

21. Determine the number of moles of NO produced by each reactant. The one producing the smaller amount of NO is the limiting reactant.

$$\text{mol NO} = 0.696\,\text{mol}\,Cu \times \frac{2\,\text{mol NO}}{3\,\text{mol Cu}} = 0.464\,\text{mol NO}$$

$$\text{mol NO} = 136\,\text{mL}\,HNO_3\,(aq) \times \frac{1\,\text{L}}{1000\,\text{mL}} \times \frac{6.0\,\text{mol}\,HNO_3}{1\,\text{L}} \times \frac{2\,\text{mol NO}}{8\,\text{mol}\,HNO_3} = 0.204\,\text{mol NO}$$

Since HNO_3 (aq) is the limiting reactant, it will be completely consumed, leaving some Cu unreacted.

22. **(a)** Since the stoichiometry indicates that 1 mole CCl_2F_2 is produced per mole CCl_4, the use of 1.80 mole CCl_4 will produce 1.80 mole CCl_2F_2. This is the theoretical yield of the reaction.

(b) The actual yield of the reaction is the amount actually produced, 1.55 mol CCl_2F_2.

(c) $$\% \text{ yield} = \frac{1.55\,\text{mol}\,CCl_2F_2 \text{ obtained}}{1.80\,\text{mol}\,CCl_2F_2 \text{ calculated}} \times 100\% = 86.1\% \text{ yield}$$

23. (a) $\text{mass } C_6H_{10} = 100.0\,\text{g } C_6H_{11}OH \times \dfrac{1\,\text{mol } C_6H_{11}OH}{100.16\,\text{g } C_6H_{11}OH} \times \dfrac{1\,\text{mol } C_6H_{10}}{1\,\text{mol } C_6H_{11}OH}$

$\times \dfrac{82.146\,\text{g } C_6H_{10}}{1\,\text{mol } C_6H_{10}} = 82.01\,\text{g } C_6H_{10} = \text{theoretical yield}$

(b) $\text{percent yield} = \dfrac{64.0\,\text{g } C_6H_{10}\text{ produced}}{82.01\,\text{g } C_6H_{10}\text{ calculated}} \times 100\% = 78.0\%\text{ yield}$

(c) $\text{mass } C_6H_{11}OH = 100.0\,\text{g } C_6H_{10}\text{ produced} \times \dfrac{1.000\,\text{g calculated}}{0.780\,\text{g produced}} \times \dfrac{1\,\text{mol } C_6H_{10}}{82.15\,\text{g } C_6H_{10}}$

$\times \dfrac{1\,\text{mol } C_6H_{11}OH}{1\,\text{mol } C_6H_{10}} \times \dfrac{100.2\,\text{g } C_6H_{11}OH}{1\,\text{mol } C_6H_{11}OH} = 156\,\text{g } C_6H_{11}OH\text{ needed}$

24. $\text{mass CaCO}_3 = 0.981\,\text{g CO}_2 \times \dfrac{1\,\text{mol CO}_2}{44.01\,\text{g CO}_2} \times \dfrac{1\,\text{mol CaCO}_3}{1\,\text{mol CO}_2} \times \dfrac{100.1\,\text{g CaCO}_3}{1\,\text{mol CaCO}_3} = 2.23\,\text{g CaCO}_3$

$\%\,\text{CaCO}_3 = \dfrac{2.23\,\text{g CaCO}_3}{3.28\,\text{g sample}} \times 100\% = 68.0\%\ \text{CaCO}_3\,(\text{by mass})$

EXERCISES

Writing and Balancing Chemical Equations

25. (a) $Cr_2O_3\,(s) + 2\,Al(s) \xrightarrow{\Delta} Al_2O_3\,(s) + 2\,Cr\,(l)$

(b) $CaC_2\,(s) + 2\,H_2O(l) \rightarrow Ca(OH)_2\,(s) + C_2H_2\,(g)$

(c) $3\,H_2\,(g) + Fe_2O_3\,(s) \xrightarrow{\Delta} 2\,Fe(l) + 3\,H_2O(g)$

(d) $NCl_3\,(g) + 3\,H_2O(l) \rightarrow NH_3\,(g) + 3\,HOCl\,(aq)$

26. (a) $(NH_4)_2\,Cr_2O_7\,(s) \xrightarrow{\Delta} Cr_2O_3\,(s) + N_2\,(g) + 4\,H_2O(g)$

(b) $3\,NO_2\,(g) + H_2O(l) \rightarrow 2\,HNO_3\,(aq) + NO(g)$

(c) $2\,H_2S(g) + SO_2\,(g) \rightarrow 3\,S(s) + 2\,H_2O(g)$

(d) $SO_2Cl_2\,(l) + 8\,HI(aq) \rightarrow H_2S(g) + 2\,H_2O(l) + 2\,HCl(aq) + 4\,I_2\,(aq)$

27. (a) $2\,C_4H_{10}(l) + 13\,O_2\,(g) \rightarrow 8\,CO_2\,(g) + 10\,H_2O(l)$

(b) $2\,C_3H_7OH(l) + 9\,O_2\,(g) \rightarrow 6\,CO_2\,(g) + 8\,H_2O(l)$

(c) $HC_3H_5O_3\,(s) + 3\,O_2\,(g) \rightarrow 3\,CO_2\,(g) + 3\,H_2O(l)$

28. **(a)** $2 C_3H_6(g) + 9 O_2(g) \rightarrow 6 CO_2(g) + 6 H_2O(l)$

(b) $2 CH_2OHCHOHCH_2OH(l) + 7 O_2(g) \rightarrow 6 CO_2(g) + 8 H_2O(l)$

(c) $C_6H_5COSH(s) + 9 O_2(g) \rightarrow 7 CO_2(g) + 3 H_2O(l) + SO_2(g)$

29. **(a)** $NH_4NO_3(s) \xrightarrow{\Delta} N_2O(g) + 2 H_2O(g)$

(b) $Na_2CO_3(aq) + 2 HCl(aq) \rightarrow 2 NaCl(aq) + H_2O(l) + CO_2(g)$

(c) $2 CH_4(g) + 2 NH_3(g) + 3 O_2(g) \rightarrow 2 HCN(g) + 6 H_2O(g)$

30. **(a)** $2 SO_2(g) + O_2(g) \rightarrow 2 SO_3(g)$

(b) $CaCO_3(s) + H_2O(l) + CO_2(aq) \rightarrow Ca(HCO_3)_2(aq)$

(c) $4 NH_3(g) + 6 NO(g) \rightarrow 5 N_2(g) + 6 H_2O(l)$

31.

Unbalanced reaction:	$N_2H_4(g) + N_2O_4(g)$	$\rightarrow H_2O(g) + N_2(g)$
Balance H atoms:	$N_2H_4(g) + N_2O_4(g)$	$\rightarrow 2 H_2O(g) + N_2(g)$
Balance O atoms:	$N_2H_4(g) + 1/2 N_2O_4(g)$	$\rightarrow 2 H_2O(g) + N_2(g)$
Balance N atoms:	$N_2H_4(g) + 1/2 N_2O_4(g)$	$\rightarrow 2 H_2O(g) + 3/2 N_2(g)$
Multiply by 2 (whole #)	$2 N_2H_4(g) + N_2O_4(g)$	$\rightarrow 4 H_2O(g) + 3 N_2(g)$
Self Check:	$6 N + 8 H + 4 O$	$\rightarrow 6 N + 8 H + 4 O$

32.

Unbalanced reaction:	$NH_3(g) + O_2(g)$	$\rightarrow H_2O(g) + NO(g)$
Balance H atoms:	$2 NH_3(g) + O_2(g)$	$\rightarrow 3 H_2O(g) + NO(g)$
Balance N atoms:	$2 NH_3(g) + O_2(g)$	$\rightarrow 3 H_2O(g) + 2 NO(g)$
Balance O atoms:	$2 NH_3(g) + 5/2 O_2(g)$	$\rightarrow 3 H_2O(g) + 2 NO(g)$
Multiply by 2 (whole #)	$4 NH_3(g) + 5 O_2(g)$	$\rightarrow 6 H_2O(g) + 4 NO(g)$
Self Check:	$4 N + 12 H + 10 O$	$\rightarrow 4 N + 12 H + 10 O$

Stoichiometry of Chemical Reactions

33. **(a)** $\text{mol } O_2 = 32.8 \text{ g KClO}_3 \times \dfrac{1 \text{ mol KClO}_3}{122.6 \text{ g KClO}_3} \times \dfrac{3 \text{ mol } O_2}{2 \text{ mol KClO}_3} = 0.401 \text{ mol } O_2$

(b) $\text{mass KClO}_3 = 50.0 \text{ g } O_2 \times \dfrac{1 \text{ mol } O_2}{32.00 \text{ g } O_2} \times \dfrac{2 \text{ mol KClO}_3}{3 \text{ mol } O_2} \times \dfrac{122.6 \text{ g KClO}_3}{1 \text{ mol KClO}_3} = 128 \text{ g KClO}_3$

(c) $\text{mass KCl} = 28.3 \text{ g } O_2 \times \dfrac{1 \text{ mol } O_2}{32.00 \text{ g } O_2} \times \dfrac{2 \text{ mol KCl}}{3 \text{ mol } O_2} \times \dfrac{74.55 \text{ g KCl}}{1 \text{ mol KCl}} = 44.0 \text{ g KCl}$

34. **(a)** $\text{amount } H_2 = 42.7\,\text{g Fe} \times \dfrac{1\,\text{mol Fe}}{55.85\,\text{g Fe}} \times \dfrac{4\,\text{mol } H_2}{3\,\text{mol Fe}} = 1.02\,\text{mol } H_2$

(b) $\text{mass } H_2O = 63.5\,\text{g Fe} \times \dfrac{1\,\text{mol Fe}}{55.85\,\text{g Fe}} \times \dfrac{4\,\text{mol } H_2O}{3\,\text{mol Fe}} \times \dfrac{18.02\,\text{g } H_2O}{1\,\text{mol } H_2O} = 27.3\,\text{g } H_2O$

(c) $\text{mass } Fe_3O_4 = 7.36\,\text{mol } H_2 \times \dfrac{1\,\text{mol } Fe_3O_4}{4\,\text{mol } H_2} \times \dfrac{231.54\,\text{g } Fe_3O_4}{1\,\text{mol } Fe_3O_4} = 426\,\text{g } Fe_3O_4$

35. Balance the given equation, and then solve the problem.

$$2\,Ag_2CO_3\,(s) \xrightarrow{\Delta} 4\,Ag(s) + 2\,CO_2\,(g) + O_2\,(g)$$

$\text{mass } Ag_2CO_3 = 75.1\,\text{g Ag} \times \dfrac{1\,\text{mol Ag}}{107.87\,\text{g Ag}} \times \dfrac{2\,\text{mol } Ag_2CO_3}{4\,\text{mol Ag}} \times \dfrac{275.75\,\text{g } Ag_2CO_3}{1\,\text{mol } Ag_2CO_3} = 96.0\,\text{g } Ag_2CO_3$

36. The balanced equation is $Ca_3(PO_4)_2(s) + 4\,HNO_3(aq) \rightarrow Ca(H_2PO_4)_2(s) + 2\,Ca(NO_3)_2(aq)$

$\text{mass } HNO_3 = 125\,\text{kg } Ca(H_2PO_4)_2 \times \dfrac{1\,\text{kmol } Ca(H_2PO_4)_2}{234.05\,\text{kg } Ca(H_2PO_4)_2} \times \dfrac{4\,\text{kmol } HNO_3}{1\,\text{kmol } Ca(H_2PO_4)_2} \times \dfrac{63.01\,\text{kg } HNO_3}{1\,\text{kmol } HNO_3}$

$\text{mass } HNO_3 = 135\,\text{kg } HNO_3$

37. The balanced equation is $Fe_2O_3\,(s) + 3\,C(s) \xrightarrow{\Delta} 2\,Fe(l) + 3\,CO(g)$

$\text{mass } Fe_2O_3 = 523\,\text{kg Fe} \times \dfrac{1\,\text{kmol Fe}}{55.85\,\text{kg Fe}} \times \dfrac{1\,\text{kmol } Fe_2O_3}{2\,\text{kmol Fe}} \times \dfrac{159.7\,\text{kg } Fe_2O_3}{1\,\text{kmol } Fe_2O_3} = 748\,\text{kg } Fe_2O_3$

$\% \; Fe_2O_3 \text{ in ore} = \dfrac{748\,\text{kg } Fe_2O_3}{938\,\text{kg ore}} \times 100\% = 79.7\% \; Fe_2O_3$

38. The following reaction occurs: $2\,Ag_2O(s) \xrightarrow{heat} 4\,Ag(s) + O_2\,(g)$

$\text{mass } Ag_2O = 0.187\,\text{g } O_2 \times \dfrac{1\,\text{mol } O_2}{32.0\,\text{g } O_2} \times \dfrac{2\,\text{mol } Ag_2O}{1\,\text{mol } O_2} \times \dfrac{231.7\,\text{g } Ag_2O}{1\,\text{mol } Ag_2O} = 2.71\,\text{g } Ag_2O$

$\% \; Ag_2O = \dfrac{2.71\,\text{g } Ag_2O}{3.13\,\text{g sample}} \times 100\% = 86.6\% \; Ag_2O$

39. $2\,Al(s) + 6\,HCl(aq) \rightarrow 2\,AlCl_3\,(aq) + 3\,H_2\,(g)$. First determine the mass of Al in the foil.

$\text{mass Al} = (10.25\,\text{cm} \times 5.50\,\text{cm} \times 0.601\,\text{mm}) \times \dfrac{1\,\text{cm}}{10\,\text{mm}} \times \dfrac{2.70\,\text{g}}{1\,\text{cm}^3} = 9.15\,\text{g Al}$

$\text{mass } H_2 = 9.15\,\text{g Al} \times \dfrac{1\,\text{mol Al}}{26.98\,\text{g Al}} \times \dfrac{3\,\text{mol } H_2}{2\,\text{mol Al}} \times \dfrac{2.016\,\text{g } H_2}{1\,\text{mol } H_2} = 1.03\,\text{g } H_2$

40. $2\,Al(s) + 6\,HCl(aq) \rightarrow 2\,AlCl_3\,(aq) + 3\,H_2\,(g)$

$$\text{mass } H_2 = 225\,mL\,soln \times \frac{1.088\,g}{1\,mL} \times \frac{18.0\,g\,HCl}{100.0\,g\,soln} \times \frac{1\,mol\,HCl}{36.46\,g\,HCl} \times \frac{3\,mol\,H_2}{6\,mol\,HCl} \times \frac{2.016\,g\,H_2}{1\,mol\,H_2}$$

$$= 1.22\,g\,H_2$$

41. In each balanced reaction, one mole of $O_2(g)$ is produced from two moles of solid reactant. Thus, the reaction that produces the most $O_2(g)$ per gram of reactant is the one involving the reactant with the smallest molar mass. NH_4NO_3 is 80.04 g/mol; Ag_2O is 231.74 g/mol; HgO is 216.59 g/mol; and $Pb(NO_3)_2$ is 331.2 g/mol. Thus, NH_4NO_3 (reaction 1) produces the most oxygen per gram of reactant.

42. First write the balanced chemical equation for each reaction.

$2\,Na(s) + 2\,HCl(aq) \rightarrow 2\,NaCl(aq) + H_2\,(g)$ $Mg(s) + 2\,HCl(aq) \rightarrow MgCl_2\,(aq) + H_2\,(g)$

$2\,Al(s) + 6\,HCl(aq) \rightarrow 2\,AlCl_3\,(aq) + 3\,H_2\,(g)$ $Zn(s) + 2\,HCl(aq) \rightarrow ZnCl_2\,(aq) + H_2\,(g)$

Three of the reactions—those of Na, Mg, and Zn—produce 1 mole of $H_2(g)$. The one of these three that produces the most hydrogen per gram of metal is the one for which the metal's atomic mass is the smallest, remembering to compare twice the atomic mass for Na. The atomic masses are: 2×23 u for Na, 24.3 u for Mg, and 65.4 u for Zn. Thus, among these three, Mg produces the most H_2 per gram of metal, specifically 1 mol H_2 per 24.3 g Mg. In the case of Al, 3 moles of H_2 are produced by 2 moles of the metal, or 54 g Al. This reduces as follows: 3 mol H_2 / 54 g Al = 1 mol H_2 / 18 g Al. Thus, Al produces the largest amount of H_2 per gram of metal.

Molarity

43. **(a)** $[C_{12}H_{22}O_{11}] = \dfrac{150.0\,g\,C_{12}H_{22}O_{11}}{250.0\,mL\,soln} \times \dfrac{1000\,mL}{1\,L} \times \dfrac{1\,mol\,C_{12}H_{22}O_{11}}{342.3\,g\,C_{12}H_{22}O_{11}} = 1.753\,M$

(b) $[CO(NH_2)_2] = \dfrac{98.3\,mg\,solid}{5.00\,mL\,soln} \times \dfrac{97.9\,mg\,CO(NH_2)_2}{100\,mg\,solid} \times \dfrac{1\,mmol\,CO(NH_2)_2}{60.06\,mg\,CO(NH_2)_2}$

$= 3.20\,M\,CO(NH_2)_2$

(c) $[CH_3OH] = \dfrac{125.0\,mL\,CH_3OH}{15.0\,L\,soln} \times \dfrac{0.792\,g}{1\,mL} \times \dfrac{1\,mol\,CH_3OH}{32.04\,g\,CH_3OH} = 0.206\,M$

44. **(a)** $[H_2C_4H_5NO_4] = \dfrac{0.405\,g\,H_2C_4H_5NO_4}{100.0\,mL} \times \dfrac{1000\,mL}{1\,L} \times \dfrac{1\,mol\,H_2C_4H_5NO_4}{133.10\,g\,H_2C_4H_5NO_4} = 0.0304\,M$

(b) $[C_3H_6O] = \dfrac{35.0\,mL\,C_3H_6O}{425\,mL\,soln} \times \dfrac{1000\,mL}{1\,L} \times \dfrac{0.790\,g\,C_3H_6O}{1\,mL} \times \dfrac{1\,mol}{58.08\,g\,C_3H_6O} = 1.12\,M$

(c) $[(C_2H_5)_2O] = \dfrac{8.8\,mg\,(C_2H_5)_2O}{3.00\,L\,soln} \times \dfrac{1\,g}{1000\,mg} \times \dfrac{1\,mol\,(C_2H_5)_2O}{74.12\,g\,(C_2H_5)_2O} = 4.0 \times 10^{-5}\,M$

45. **(a)** $\text{mass } C_6H_{12}O_6 = 75.0\,\text{mL soln} \times \dfrac{1\,\text{L}}{1000\,\text{mL}} \times \dfrac{0.350\,\text{mol } C_6H_{12}O_6}{1\,\text{L soln}} \times \dfrac{180.16\,\text{g } C_6H_{12}O_6}{1\,\text{mol } C_6H_{12}O_6} = 4.73\,\text{g}$

(b) $V_{CH_3OH} = 2.25\,\text{L soln} \times \dfrac{0.485\,\text{mol}}{1\,\text{L}} \times \dfrac{32.04\,\text{g } CH_3OH}{1\,\text{mol } CH_3OH} \times \dfrac{1\,\text{mL}}{0.792\,\text{g}} = 44.1\,\text{mL } CH_3OH$

46. **(a)** $V_{C_2H_5OH} = 200.0\,\text{L soln} \times \dfrac{1.65\,\text{mol } C_2H_5OH}{1\,\text{L}} \times \dfrac{46.07\,\text{g } C_2H_5OH}{1\,\text{mol } C_2H_5OH} \times \dfrac{1\,\text{mL}}{0.789\,\text{g}} \times \dfrac{1\,\text{L}}{1000\,\text{mL}} = 19.3\,\text{L}$

(b) $V_{HCl} = 12.0\,\text{L} \times \dfrac{0.234\,\text{mol } HCl}{1\,\text{L}} \times \dfrac{36.46\,\text{g } HCl}{1\,\text{mol } HCl} \times \dfrac{100\,\text{g soln}}{36.0\,\text{g } HCl} \times \dfrac{1\,\text{mL soln}}{1.18\,\text{g}} = 241\,\text{mL}$

47. We determine the molar concentration of the 46% by mass sucrose solution.

$$[C_{12}H_{22}O_{11}] = \dfrac{46\,\text{g } C_{12}H_{22}O_{11} \times \dfrac{1\,\text{mol } C_{12}H_{22}O_{11}}{342.3\,\text{g } C_{12}H_{22}O_{11}}}{100\,\text{g soln} \times \dfrac{1\,\text{mL}}{1.21\,\text{g soln}} \times \dfrac{1\,\text{L}}{1000\,\text{mL}}} = 1.6\,\text{M}$$

The 46% by mass sucrose solution is the more concentrated.

48. We calculate the $[C_2H_5OH]$ in the white wine and compare it with $1.71\,\text{M } C_2H_5OH$, the concentration of the solution described in Example 4-8.

$$[C_2H_5OH] = \dfrac{11\,\text{g } C_2H_5OH}{100.0\,\text{g soln}} \times \dfrac{0.95\,\text{g soln}}{1\,\text{mL}} \times \dfrac{1000\,\text{mL}}{1\,\text{L}} \times \dfrac{1\,\text{mol } C_2H_5OH}{46.1\,\text{g } C_2H_5OH} = 2.3\,\text{M } C_2H_5OH$$

The white wine has a greater ethyl alcohol content.

49. $[KNO_3] = \dfrac{10.00\,\text{mL conc'd soln} \times \dfrac{2.05\,\text{mmol } KNO_3}{1\,\text{mL}}}{250.0\,\text{mL}} = 0.0820\,\text{M}$

50. $[HCl] = \dfrac{500.0\,\text{mL dilute soln} \times \dfrac{0.085\,\text{mmol } HCl}{1\,\text{mL soln}}}{25.0\,\text{mL}} = 1.7\,\text{M}$

51. Let us compute how many mL of dilute $(_d)$ solution we obtain from each mL of concentrated $(_c)$ solution. $V_c \times C_c = V_d \times C_d$ becomes $1.00\,\text{mL} \times 0.250\text{M} = x\,\text{mL} \times 0.0125$ M and $x = 20$ Thus, the ratio of the volume of the volumetric flask to that of the pipet would be 20:1. We could use a 100.0-mL flask and a 5.00-mL pipet, a 1000.0-mL flask and a 50.00-mL pipet, or a 500.0-mL flask and a 25.00-mL pipet. There are many combinations that could be used.

52. First determine the amount of solute in the final solution and then the volume of the initial, more concentrated, solution that must be used.

$$\text{volume conc'd soln} = 250.0\,\text{mL} \times \frac{0.175\,\text{mmol KCl}}{1\,\text{mL dil soln}} \times \frac{1\,\text{mL conc'd soln}}{0.496\,\text{mmol KCl}} = 88.2\ \text{mL}$$

Thus the instructions are: Place 88.2 mL of 0.496 M KCl in a 250-mL volumetric flask. Dilute to the mark with distilled water, stopping to mix thoroughly several times during the addition of water.

Chemical Reactions in Solutions

53. **(a)** $\text{mass Na}_2\text{S} = 27.8\,\text{mL} \times \dfrac{1\,\text{L}}{1000\,\text{mL}} \times \dfrac{0.163\,\text{mol AgNO}_3}{1\,\text{L soln}} \times \dfrac{1\,\text{mol Na}_2\text{S}}{2\,\text{mol AgNO}_3}$

$$\times \frac{78.05\,\text{g Na}_2\text{S}}{1\,\text{mol Na}_2\text{S}} = 0.177\,\text{g Na}_2\text{S}$$

(b) $\text{mass Ag}_2\text{S} = 0.177\,\text{g Na}_2\text{S} \times \dfrac{1\,\text{mol Na}_2\text{S}}{78.05\,\text{g Na}_2\text{S}} \times \dfrac{1\,\text{mol Ag}_2\text{S}}{1\,\text{mol Na}_2\text{S}} \times \dfrac{247.80\,\text{g Ag}_2\text{S}}{1\,\text{mol Ag}_2\text{S}} = 0.562\,\text{g Ag}_2\text{S}$

54. **(a)** $\text{mass Ca(OH)}_2 = 415\,\text{mL} \times \dfrac{1\,\text{L}}{1000\,\text{mL}} \times \dfrac{0.477\,\text{mol HCl}}{1\,\text{L soln}} \times \dfrac{1\,\text{mol Ca(OH)}_2}{2\,\text{mol HCl}}$

$$\times \frac{74.1\,\text{g Ca(OH)}_2}{1\,\text{mol Ca(OH)}_2} = 7.33\,\text{g Ca(OH)}_2$$

(b) $\text{mass Ca(OH)}_2 = 324\,\text{L} \times \dfrac{1.12\,\text{kg}}{1\,\text{L}} \times \dfrac{24.28\,\text{kg HCl}}{100.00\,\text{kg soln}} \times \dfrac{1\,\text{kmol HCl}}{36.46\,\text{kg HCl}}$

$$\times \frac{1\,\text{kmol Ca(OH)}_2}{2\,\text{kmol HCl}} \times \frac{74.10\,\text{kg Ca(OH)}_2}{1\,\text{kmol Ca(OH)}_2} = 89.5\,\text{kg Ca(OH)}_2$$

55. **(a)**   We know that the Al forms the AlCl_3.

$$\text{mol AlCl}_3 = 1.87\,\text{g Al} \times \frac{1\,\text{mol Al}}{26.98\,\text{g Al}} \times \frac{1\,\text{mol AlCl}_3}{1\,\text{mol Al}} = 0.0693\,\text{mol AlCl}_3$$

(b) $[\text{AlCl}_3] = \dfrac{0.0693\,\text{mol AlCl}_3}{23.8\,\text{mL}} \times \dfrac{1000\,\text{mL}}{1\,\text{L}} = 2.91\,\text{M AlCl}_3$

56. The balanced chemical reaction indicates that 4 mol NaNO_2 are formed from 2 mol Na_2CO_3.

$$[\text{NaNO}_2] = \frac{138\,\text{g Na}_2\text{CO}_3}{1.42\,\text{L soln}} \times \frac{1\,\text{mol Na}_2\text{CO}_3}{106.0\,\text{g Na}_2\text{CO}_3} \times \frac{4\,\text{mol NaNO}_2}{2\,\text{mol Na}_2\text{CO}_3} = 1.83\,\text{M NaNO}_2$$

57. The molarity unit can be interpreted as millimoles of solute per milliliter of solution.

$$V_{K_2CrO_4} = 415\,mL \times \frac{0.186\,mmol\,AgNO_3}{1\,mL\,soln} \times \frac{1\,mmol\,K_2CrO_4}{2\,mmol\,AgNO_3} \times \frac{1\,mL\,K_2CrO_4\,(aq)}{0.650\,mmol\,K_2CrO_4}$$

$$V_{K_2CrO_4} = 59.4\,mL\,K_2CrO_4$$

58. The volume of solution determines the amount of product.

$$mass\,Ag_2CrO_4 = 415\,mL \times \frac{1\,L}{1000\,mL} \times \frac{0.186\,mol\,AgNO_3}{1\,L\,soln} \times \frac{1\,mol\,Ag_2CrO_4}{2\,mol\,AgNO_3} \times \frac{331.73\,g\,Ag_2CrO_4}{1\,mol\,Ag_2CrO_4}$$

$$mass\,Ag_2CrO_4 = 12.8\,g\,Ag_2CrO_4$$

59.

$$mass\,Na = 155\,mL\,soln \times \frac{1\,L}{1000\,mL} \times \frac{0.175\,mol\,NaOH}{1\,L\,soln} \times \frac{2\,mol\,Na}{2\,mol\,NaOH} \times \frac{22.99\,g\,Na}{1\,mol\,Na}$$

$$= 0.624\,g\,Na$$

60. We determine the amount of HCl present initially, and the amount desired.

$$amount\,HCl\,present = 250.0\,mL \times \frac{1.023\,mmol\,HCl}{1\,mL\,soln} = 255.8\,mmol\,HCl$$

$$amount\,HCl\,desired = 250.0\,mL \times \frac{1.000\,mmol\,HCl}{1\,mL\,soln} = 250.0\,mmol\,HCl$$

$$mass\,Mg = (255.8 - 250.0)\,mmol\,HCl \times \frac{1\,mmol\,Mg}{2\,mmol\,HCl} \times \frac{24.3\,mg\,Mg}{1\,mmol\,Mg} = 70.\,mg\,Mg$$

61. The mass of oxalic acid enables us to determine the amount of NaOH in the solution.

$$[NaOH] = \frac{0.3126\,g\,H_2C_2O_4}{26.21\,mL\,soln} \times \frac{1000\,mL}{1\,L\,soln} \times \frac{1\,mol\,H_2C_2O_4}{90.04\,g\,H_2C_2O_4} \times \frac{2\,mol\,NaOH}{1\,mol\,H_2C_2O_4} = 0.2649\,M$$

62. The total amount of HCl present is the amount that reacted with the $CaCO_3$ plus the amount that reacted with the $Ba(OH)_2$ (aq).

$$\begin{array}{l} moles\,HCl\,from \\ CaCO_3\,reaction \end{array} = 0.1000\,g\,CaCO_3 \times \frac{1\,mol\,CaCO_3}{100.09\,g\,CaCO_3} \times \frac{2\,mol\,HCl}{1\,mol\,CaCO_3} \times \frac{1000\,mmol}{1\,mol}$$

$$= 1.998\,mmol\,HCl$$

$$\begin{array}{l} moles\,HCl\,from \\ Ba(OH)_2\,reaction \end{array} = 43.82\,mL \times \frac{0.01185\,mmol\,Ba(OH)_2}{1\,mL\,soln} \times \frac{2\,mmol\,HCl}{1\,mmol\,Ba(OH)_2}$$

$$= 1.039\,mmol\,HCl$$

The HCl molarity is this total mmol of HCl divided by the total volume of 25.00 mL.

$$[\text{HCl}] = \frac{(1.998 + 1.039)\,\text{mmol HCl}}{25.00\,\text{mL}} = 0.1215\,\text{M}$$

Determining the Limiting Reactant

63. There are equal numbers of moles of each reactant present, but more O_2 is needed than NH_3. Thus, $O_2(g)$ is the limiting reactant, and all of the $O_2(g)$ is consumed. The mass of product produced from 1.00 mol $O_2(g)$ is then calculated.

$$\text{mass NO}(g) = 1.00\,\text{mol } O_2 \times \frac{4\,\text{mol NO}(g)}{5\,\text{mol } O_2(g)} \times \frac{30.01\,\text{g NO}(g)}{1\,\text{mol NO}(g)} = 24.0\,\text{g NO}(g)$$

64. Determine the mass of H_2 produced from each of the reactants. The smaller mass is that produced by the limiting reactant, which is the mass actually formed.

$$\text{mass } H_2 = 1.84\,\text{g Al} \times \frac{1\,\text{mol Al}}{26.98\,\text{g Al}} \times \frac{3\,\text{mol } H_2}{2\,\text{mol Al}} \times \frac{2.016\,\text{g } H_2}{1\,\text{mol } H_2} = 0.206\,\text{g } H_2$$

$$\text{mass } H_2 = 75.0\,\text{mL} \times \frac{1\,\text{L}}{1000\,\text{mL}} \times \frac{2.95\,\text{mol HCl}}{1\,\text{L}} \times \frac{3\,\text{mol } H_2}{6\,\text{mol HCl}} \times \frac{2.016\,\text{g } H_2}{1\,\text{mol } H_2} = 0.223\,\text{g } H_2$$

Thus, 0.206 g H_2 is produced.

65. Determine the amount of Na_2CS_3 produced from each of the reactants.

$$\text{amount } Na_2CS_3 = 92.5\,\text{mL } CS_2 \times \frac{1.26\,\text{g}}{1\,\text{mL}} \times \frac{1\,\text{mol } CS_2}{76.14\,\text{g } CS_2} \times \frac{2\,\text{mol } Na_2CS_3}{3\,\text{mol } CS_2} = 1.02\,\text{mol } Na_2CS_3$$

$$\text{amount } Na_2CS_3 = 2.78\,\text{mol NaOH} \times \frac{2\,\text{mol } Na_2CS_3}{6\,\text{mol NaOH}} = 0.927\,\text{mol } Na_2CS_3$$

Thus, the mass produced is $0.927\,\text{mol } Na_2CS_3 \times \dfrac{154.2\,\text{g } Na_2CS_3}{1\,\text{mol } Na_2CS_3} = 143\,\text{g } Na_2CS_3$

66. Since the two reactants combine in an equimolar basis, the one present with the fewer number of moles is the limiting reactant and determines the mass of the products.

$$\text{mol } ZnSO_4 = 315\,\text{mL} \times \frac{1\,\text{L}}{1000\,\text{mL}} \times \frac{0.275\,\text{mol } ZnSO_4}{1\,\text{L soln}} = 0.0866\,\text{mol } ZnSO_4$$

$$\text{mol BaS} = 285\,\text{mL} \times \frac{1\,\text{L}}{1000\,\text{mL}} \times \frac{0.315\,\text{mol BaS}}{1\,\text{L soln}} = 0.0898\,\text{mol BaS}$$

$ZnSO_4$ is the limiting reactant and 0.0866 mol of each of the products will be produced.

$$\text{mass products} = \left(0.0866\,\text{mol } BaSO_4 \times \frac{233.4\,\text{g } BaSO_4}{1\,\text{mol } BaSO_4}\right) + \left(0.0866\,\text{mol ZnS} \times \frac{97.46\,\text{g ZnS}}{1\,\text{mol ZnS}}\right)$$
$$= 28.7\,\text{g product mixture (lithopone)}$$

67. $Ca(OH)_2(s) + 2NH_4Cl(s) \rightarrow CaCl_2(aq) + 2H_2O(l) + 2NH_3(g)$

Compute the amount of NH_3 formed from each reactant in this limiting reactant problem.

$$\text{amount } NH_3 = 33.0\,g\,NH_4Cl \times \frac{1\,mol\,NH_4Cl}{53.49\,g\,NH_4Cl} \times \frac{2\,mol\,NH_3}{2\,mol\,NH_4Cl} = 0.617\,mol\,NH_3$$

$$\text{amount } NH_3 = 33.0\,g\,Ca(OH)_2 \times \frac{1\,mol\,Ca(OH)_2}{74.09\,g\,Ca(OH)_2} \times \frac{2\,mol\,NH_3}{1\,mol\,Ca(OH)_2} = 0.891\,mol\,NH_3$$

Thus, 0.617 mol NH_3 is produced. $\text{mass } NH_3 = 0.617\,mol\,NH_3 \times \dfrac{17.03\,g\,NH_3}{1\,mol\,NH_3} = 10.5\,g\,NH_3$

Now we determine the mass of reactant in excess, $Ca(OH)_2$.

$$Ca(OH)_2 \text{ used} = 0.617\,mol\,NH_3 \times \frac{1\,mol\,Ca(OH)_2}{2\,mol\,NH_3} \times \frac{74.09\,g\,Ca(OH)_2}{1\,mol\,Ca(OH)_2} = 22.9\,g\,Ca(OH)_2$$

$$\text{excess } Ca(OH)_2 = 33.0\,g\,Ca(OH)_2 - 22.9\,g\,Ca(OH)_2 = 10.1\,g\text{ excess } Ca(OH)_2$$

68. The balanced chemical equation is: $Ca(OH)_2 + 4HCl(g) \rightarrow CaCl_2 + 2H_2O$

$$\text{amount } Cl_2 = 50.0\,g\,Ca(OCl)_2 \times \frac{1\,mol\,Ca(OCl)_2}{142.98\,g\,Ca(OCl)_2} \times \frac{2\,mol\,Cl_2}{1\,mol\,Ca(OCl)_2} = 0.699\,mol\,Cl_2$$

$$\text{amount } Cl_2 = 275\,mL \times \frac{1\,L}{1000\,mL} \times \frac{6.00\,mol\,HCl}{1\,L\,soln} \times \frac{2\,mol\,Cl_2}{4\,mol\,HCl} = 0.825\,mol\,Cl_2$$

Thus, mass Cl_2 produced $= 0.699\,mol\,Cl_2 \times \dfrac{70.91\,g\,Cl_2}{1\,mol\,Cl_2} = 49.6\,g\,Cl_2$

The excess reactant is the one that would produce the most Cl_2; it is HCl(aq). The quantity of excess HCl(aq) is determined from the amount of excess $Cl_2(g)$ it produces.

$$V_{\text{excess HCl}} = (0.825 - 0.699)\,mol\,Cl_2 \times \frac{4\,mol\,HCl}{2\,mol\,Cl_2} \times \frac{1000\,mL}{6.00\,mol\,HCl}$$

$$V_{\text{excess HCl}} = 42.0\,mL\text{ excess HCl}(aq)$$

$$\text{mass excess HCl} = (0.825 - 0.699)\,mol\,Cl_2 \times \frac{4\,mol\,HCl}{2\,mol\,Cl_2} \times \frac{36.46\,g\,HCl}{1\,mol\,HCl}$$

$$\text{mass excess} = 9.19\,g\text{ excess HCl}(aq)$$

Theoretical, Actual, and Percent Yields

69. **(a)** We first need to solve the limiting reactant problem involved here.

$$\text{mol } C_4H_9Br = 15.0\,\text{g } C_4H_9OH \times \frac{1\,\text{mol } C_4H_9OH}{74.12\,\text{g } C_4H_9OH} \times \frac{1\,\text{mol } C_4H_9Br}{1\,\text{mol } C_4H_9OH} = 0.202\,\text{mol } C_4H_9Br$$

$$\text{mol } C_4H_9Br = 22.4\,\text{g NaBr} \times \frac{1\,\text{mol NaBr}}{102.9\,\text{g NaBr}} \times \frac{1\,\text{mol } C_4H_9Br}{1\,\text{mol NaBr}}$$
$$= 0.218\,\text{mol } C_4H_9Br$$

$$\text{mol } C_4H_9Br = 32.7\,\text{g } H_2SO_4 \times \frac{1\,\text{mol } H_2SO_4}{98.1\,\text{g } H_2SO_4} \times \frac{1\,\text{mol } C_4H_9Br}{1\,\text{mol } H_2SO_4} = 0.333\,\text{mol } C_4H_9Br$$

$$\text{Theoretical yield of } C_4H_9Br = 0.202\,\text{mol } C_4H_9Br \times \frac{137.0\,\text{g } C_4H_9Br}{1\,\text{mol } C_4H_9Br} = 27.7\,\text{g } C_4H_9Br$$

(b) The actual yield is the mass obtained, 17.1 g C_4H_9Br.

(c) Then, % yield $= \dfrac{17.1\,\text{g } C_4H_9Br \text{ produced}}{27.7\,\text{g } C_4H_9Br \text{ expected}} \times 100\% = 61.7\%$ yield

70. **(a)** Again, we solve the limiting reactant problem first.

$$\text{amount}\,(C_6H_5N)_2 = 0.10\,\text{L } C_6H_5NO_2 \times \frac{1000\,\text{mL}}{1\,\text{L}} \times \frac{1.20\,\text{g}}{1\,\text{mL}} \times \frac{1\,\text{mol } C_6H_5NO_2}{123.1\,\text{g } C_6H_5NO_2}$$

$$\times \frac{1\,\text{mol}\,(C_6H_5N)_2}{2\,\text{mol } C_6H_5NO_2} = 0.49\,\text{mol}\,(C_6H_5N)_2$$

$$\text{amount}\,(C_6H_5N)_2 = 0.30\,\text{L } C_6H_{14}O_4 \times \frac{1000\,\text{mL}}{1\,\text{L}} \times \frac{1.12\,\text{g}}{1\,\text{mL}} \times \frac{1\,\text{mol } C_6H_{14}O_4}{150.2\,\text{g } C_6H_{14}O_4}$$

$$\times \frac{1\,\text{mol}\,(C_6H_5N)_2}{4\,\text{mol } C_6H_{14}O_4} = 0.56\,\text{mol}\,(C_6H_5N)_2$$

$$\text{theoretical yield}\,(C_6H_5N)_2 = 0.49\,\text{mol}\,(C_6H_5N)_2 \times \frac{182.2\,\text{g}\,(C_6H_5N)_2}{1\,\text{mol}\,(C_6H_5N)_2} = 89\,\text{g}\,(C_6H_5N)_2$$

(b) actual yield $= 55\,\text{g}\,(C_6H_5N)_2$ produced

(c) percent yield $= \dfrac{55\,\text{g}\,(C_6H_5N)_2 \text{ produced}}{89\,\text{g}\,(C_6H_5N)_2 \text{ expected}} \times 100\% = 62\%$ yield

71. Balanced equation: $3C_2H_4O_2 + PCl_3 \rightarrow 3C_2H_3OCl + H_3PO_3$

$$\text{mass acid} = 75\,g\,C_2H_3OCl \times \frac{100.0\,g\,\text{calculated}}{78.2\,g\,\text{produced}} \times \frac{1\,mol\,C_2H_3OCl}{78.5\,g\,C_2H_3OCl} \times \frac{3\,mol\,C_2H_4O_2}{3\,mol\,C_2H_3OCl}$$

$$\times \frac{60.1\,g\,\text{pure}\,C_2H_4O_2}{1\,mol\,C_2H_4O_2} \times \frac{100\,g\,\text{commercial}}{97\,g\,\text{pure}\,C_2H_4O_2} = 76\,g\,\text{commercial}\,C_2H_4O_2$$

72. $$\text{mass}\,CH_2Cl_2 = 112\,g\,CH_4 \times \frac{1\,mol\,CH_4}{16.04\,g\,CH_4} \times \frac{1\,mol\,CH_3Cl}{1\,mol\,CH_4} \times \frac{0.92\,mol\,CH_3Cl\,\text{produced}}{1.00\,mol\,CH_3Cl\,\text{expected}}$$

$$\times \frac{1\,mol\,CH_2Cl_2}{1\,mol\,CH_3Cl} \times \frac{84.93\,g\,CH_2Cl_2}{1\,mol\,CH_2Cl_2} \times \frac{0.92\,g\,CH_2Cl_2\,\text{produced}}{1.00\,g\,CH_2Cl_2\,\text{calculated}}$$

$$= 5.0 \times 10^2\,g\,CH_2Cl_2$$

73. A less-than-100% yield of desired product in synthesis reactions is always the case. This is because of side reactions that yield products other than those desired and because of the loss of material in the various steps of the synthesis procedure. A main criterion for choosing a synthesis reaction is how economically it can be run. In the analysis of a compound, on the other hand, it is essential that all of the material present be detected. Therefore, a 100% yield is required; none of the material present in the sample can be lost during the analysis. Therefore analysis reactions are carefully chosen to meet this criterion; they need not be economical to run.

74. The theoretical yield is 2.07 g Ag_2CrO_4. If the mass actually obtained is less than this, it is likely that some of the pure material was lost in the reaction, perhaps stuck to the walls of the flask in which the reaction occurred, or left suspended in the solution. But the maximum mass of Ag_2CrO_4 that can be produced is 2.07 g. If the precipitate weighs more than 2.07 g, the extra mass must be impurities (e.g., the precipitate was not thoroughly dried).

Consecutive Reactions, Simultaneous Reactions

75. Determine the amount of HCl needed to react with each component of the mixture.

$$Mg(OH)_2\,(s) + 2HCl(aq) \longrightarrow MgCl_2\,(aq) + 2H_2O(l)$$

$$MgCO_3(s) + 2HCl(aq) \longrightarrow MgCl_2\,(aq) + H_2O(l) + CO_2(g)$$

$$\text{mol HCl} = 425\,g\,\text{mixt.} \times \frac{35.2\,g\,MgCO_3}{100.0\,g\,\text{mixt.}} \times \frac{1\,mol\,MgCO_3}{84.3\,g\,MgCO_3} \times \frac{2\,mol\,HCl}{1\,mol\,MgCO_3} = 3.55\,mol\,HCl$$

$$\text{mol HCl} = 425\,g\,\text{mixt.} \times \frac{64.8\,g\,Mg(OH)_2}{100.0\,g\,\text{mixt.}} \times \frac{1\,mol\,Mg(OH)_2}{58.3\,g\,Mg(OH)_2} \times \frac{2\,mol\,HCl}{1\,mol\,MgCO_3} = 9.45\,mol\,HCl$$

$$\text{mass HCl} = (3.55 + 9.45)\,mol\,HCl \times \frac{36.46\,g\,HCl}{1\,mol\,HCl} = 474\,g\,HCl$$

76. Determine the amount of CO_2 produced from each reactant.

$$C_3H_8(g) + 5O_2(g) \longrightarrow 3CO_2(g) + 4H_2O(l)$$
$$2C_4H_{10}(g) + 13O_2(g) \longrightarrow 8CO_2(g) + 10H_2O(l)$$

$$\text{mol } CO_2 = 406 \text{ g mixt.} \times \frac{72.7 \text{ g } C_3H_8}{100.0 \text{ g mixt.}} \times \frac{1 \text{ mol } C_3H_8}{44.10 \text{ g } C_3H_8} \times \frac{3 \text{ mol } CO_2}{1 \text{ mol } C_3H_8} = 20.1 \text{ mol } CO_2$$

$$\text{mol } CO_2 = 406 \text{ g mixt.} \times \frac{27.3 \text{ g } C_4H_{10}}{100.0 \text{ g mixt}} \times \frac{1 \text{ mol } C_4H_{10}}{58.12 \text{ g } C_4H_{10}} \times \frac{8 \text{ mol } CO_2}{2 \text{ mol } C_4H_{10}} = 7.63 \text{ mol } CO_2$$

$$\text{mass } CO_2 = (20.1 + 7.63) \text{ mol } CO_2 \times \frac{44.01 \text{ g } CO_2}{1 \text{ mol } CO_2} = 1.22 \times 10^3 \text{ g } CO_2$$

77. The molar ratios given by the stoichiometric coefficients in the balanced chemical equations are used in the solution.

$$\text{amount } Cl_2 = 2.25 \times 10^3 \text{ g } CCl_2F_2 \times \frac{1 \text{ mol } CCl_2F_2}{120.91 \text{ g } CCl_2F_2} \times \frac{1 \text{ mol } CCl_4}{1 \text{ mol } CCl_2F_2} \times \frac{4 \text{ mol } Cl_2}{1 \text{ mol } CCl_4}$$

$$= 74.4 \text{ mol } Cl_2$$

78. $\text{mass } C_2H_6 = 0.506 \text{ g } BaCO_3 \times \dfrac{1 \text{ mol } BaCO_3}{197.3 \text{ g } BaCO_3} \times \dfrac{1 \text{ mol } CO_2}{1 \text{ mol } BaCO_3} \times \dfrac{2 \text{ mol } C_2H_6}{4 \text{ mol } CO_2} \times \dfrac{30.07 \text{ g } C_2H_6}{1 \text{ mol } C_2H_6}$

$$= 0.0386 \text{ g } C_2H_6$$

79.

$NaI(aq) + AgNO_3(aq)$	$\rightarrow$	$AgI(s) + NaNO_3(aq)$	(multiply by 4)
$2 AgI(s) + Fe(s)$	$\rightarrow$	$FeI_2(aq) + 2 Ag(s)$	(multiply by 2)
$2 FeI_2(aq) + 3 Cl_2(g)$	$\rightarrow$	$2 FeCl_3(aq) + 2 I_2(s)$	(unchanged)

$$4NaI(aq) + 4AgNO_3(aq) + 2Fe(s) + 3Cl_2(g) \rightarrow 4NaNO_3(aq) + 4Ag(s) + 2FeCl_3(aq) + 2I_2(s)$$

For every 4 moles of $AgNO_3$, 2 moles of $I_2(s)$ are produced. The mass of $AgNO_3$ required

$$= 1.00 \text{ kg } I_2(s) \times \frac{1000 \text{ g } I_2(s)}{1 \text{ kg } I_2(s)} \times \frac{1 \text{ mol } I_2(s)}{253.809 \text{ g } I(s)} \times \frac{4 \text{ mol } AgNO_3(s)}{2 \text{ mol } I_2(s)} \times \frac{169.873 \text{ g } AgNO_3(s)}{1 \text{ mol } AgNO_3(s)}$$

$$= 1338.59 \text{ g } AgNO_3 \text{ per kg of } I_2 \text{ produced or } 1.34 \times 10^3 \text{ g } AgNO_3 \text{ per kg of } I_2$$

80.

$Fe + Br_2$	$\rightarrow FeBr_2$	(multiply by 3)
$3 FeBr_2 + Br_2$	$\rightarrow Fe_3Br_8$	
$Fe_3Br_8 + 4 Na_2CO_3$	$\rightarrow 8 NaBr + 4 CO_2 + Fe_3O_4$	

$$3 Fe + 4 Br_2 + 4 Na_2CO_3 \rightarrow 8 NaBr + 4 CO_2 + Fe_3O_4$$

Hence, 3 moles Fe(s) forms 8 mol NaBr

$$\text{mass of Fe} = 2.50 \times 10^3 \text{ kg NaBr} \times \frac{1000 \text{ g NaBr}}{1 \text{ kg NaBr}} \times \frac{1 \text{ mol NaBr}}{102.894 \text{ g NaBr}} \times \frac{3 \text{ mol Fe}}{8 \text{ mol NaBr}} \times \frac{55.847 \text{ g Fe}}{1 \text{ mol Fe}}$$

$$= 509 \times 10^3 \text{ g Fe} \times \frac{1 \text{ kg Fe}}{1000 \text{ g Fe}} = 509 \text{ kg Fe required to produce } 2.5 \times 10^3 \text{ kg KBr}$$

FEATURE PROBLEMS

109. If the sample that was caught is representative of all fish in the lake, there are five marked fish for every 18 fish. Thus, the total number of fish in the lake is determined.

$$\text{total fish} = 100 \text{ marked fish} \times \frac{18 \text{ fish}}{5 \text{ marked fish}} = 360 \text{ fish} \cong 4 \times 10^2 \text{ fish}$$

110. (a) The graph obtained is one of two straight lines, meeting at a peak of about 2.50 g $Pb(NO_3)_2$, corresponding to about 3.5 g PbI_2. Maximum mass of PbI_2 (calculated)

$$= 2.503 \text{ g KI} \times \frac{1 \text{ mol KI}}{166.0 \text{ g KI}} \times \frac{1 \text{ mol PbI}_2}{2 \text{ mol KI}} \times \frac{461.01 \text{ g PbI}_2}{1 \text{ mol PbI}_2} = 3.476 \text{ g PbI}_2$$

(b) The total quantity of reactant is limited to 5.000 g. If either reactant is in excess, the amount in excess will be "wasted," because it cannot be used to form product. Thus, we obtain the maximum amount of product when neither reactant is in excess (i.e., when there is a stoichiometric amount of each present). The balanced chemical equation for this reaction, $2 KI + Pb(NO_3)_2 \rightarrow 2 KNO_3 + PbI_2$, shows that stoichiometric quantities are two moles of KI (166.00 g/mol) for each mole of $Pb(NO_3)_2$ (331.21 g/mol). If we have 5.000 g total, we can let the mass of KI equal x g, so that the mass of

$$Pb(NO_3)_2 = (5.000 - x) \text{ g. and the amount } KI = x \text{ g KI} \times \frac{1 \text{ mol KI}}{166.00 \text{ g}} = \frac{x}{166.00}$$

$$\text{amount } Pb(NO_3)_2 = (5.000 - x) \text{ g } Pb(NO_3)_2 \times \frac{1 \text{ mol Pb}(NO_3)_2}{331.21 \text{ g}} = \frac{5.000 - x}{331.21}$$

At the point of stoichiometric balance, amount $KI = 2 \times$ amount $Pb(NO_3)_2$

$$\frac{x}{166.00} = 2 \times \frac{5.000 - x}{331.21} \quad \text{OR} \quad 331.21 x = 10.00 \times 166.00 - 332.00 x$$

$$x = \frac{1660.0}{331.21 + 332.00} = 2.503 \text{ g KI} \times \frac{1 \text{ mol KI}}{166.00 \text{ g KI}} = 0.01508 \text{ mol KI}$$

$$5.000 - x = 2.497 \text{ g Pb}(NO_3)_2 \times \frac{1 \text{ mol Pb}(NO_3)_2}{331.21 \text{ g Pb}(NO_3)_2} = 0.007539 \text{ mol Pb}(NO_3)_2$$

As a mass ratio we have: $\dfrac{2.503\,g\,KI}{2.497\,g\,Pb(NO_3)_2} = \dfrac{1.002\,g\,KI}{1\,g\,Pb(NO_3)_2}$

As a molar ratio we have: $\dfrac{0.01508\,mol\,KI}{0.007539\,mol\,Pb(NO_3)_2} = \dfrac{2\,mol\,KI}{1\,mol\,Pb(NO_3)_2}$

(c) The molar ratio just determined in part (b) is the same as the ratio of coefficients for KI and $Pb(NO_3)_2$ in the balanced chemical equation. Finally to determine the proportions precisely, we used the balanced chemical equation.

111. (a) For the balanced equation, the order is immaterial; the relative amount of each is important. $20\,rd + 20\,bl + 30\,gr \rightarrow 1\,necklace$

(b) This is similar to a limiting reactant problem. We determine how many necklaces can be made from each quantity of beads.

$$\text{number of necklaces} = 10.0\,kg\,beads \times \dfrac{1000\,g}{1\,kg} \times \dfrac{1\,rd\,bead}{1.98\,g} \times \dfrac{1\,necklace}{20\,rd\,beads} = 252_5 \text{ necklaces}$$

$$\text{number of necklaces} = 10.0\,kg\,beads \times \dfrac{1000\,g}{1\,kg} \times \dfrac{1\,bl\,bead}{3.05\,g} \times \dfrac{1\,necklace}{20\,bl\,beads} = 163_9 \text{ necklaces}$$

$$\text{number of necklaces} = 10.0\,kg\,beads \times \dfrac{1000\,g}{1\,kg} \times \dfrac{1\,gr\,bead}{1.82\,g} \times \dfrac{1\,necklace}{30\,gr\,beads} = 183_1 \text{ necklaces}$$

We have expressed each result with an additional significant figure, written as a subscript, so that we can see the effect of rounding. With the beads available, we can produce 163 necklaces, since we are unable to produce a fraction of a necklace.

(c) Because the mass of a bead, and the total mass available of each type of bead, both are known to just three significant figures, our results are only known that well. The best we can state is that we can make at least 163 necklaces, because 164 is uncertain by one unit. We should not be surprised if we actually made just 161 necklaces, or if we produced 165 of them. More precise masses would help.

112. The more HCl used, the more impure the sample (compared to $NaHCO_3$, twice as much HCl is needed to neutralize Na_2CO_3).

Sample from trona: 6.93 g sample forms 11.89 g AgCl or 1.72 g AgCl per gram sample.

Sample derived from manufactured sodium bicarbonate: 6.78 g sample forms 11.77 g AgCl or 1.74 g AgCl per gram sample.

Thus the trona sample is purer (i.e., it has the greater mass percent $NaHCO_3$).

CHAPTER 5
INTRODUCTION TO REACTIONS
IN AQUEOUS SOLUTIONS
PRACTICE EXAMPLES

1A In determining total $[Cl^-]$, we recall the definition of molarity: moles of solute per liter of solution.

$$\text{from NaCl,}[Cl^-] = \frac{0.438 \text{ mol NaCl}}{1 \text{ L soln}} \times \frac{1 \text{ mol Cl}^-}{1 \text{ mol NaCl}} = 0.438 \text{ M Cl}^-$$

$$\text{from MgCl}_2,[Cl^-] = \frac{0.0512 \text{ mol MgCl}_2}{1 \text{ L soln}} \times \frac{2 \text{ mol Cl}^-}{1 \text{ mol MgCl}_2} = 0.102 \text{ M Cl}^-$$

$$[Cl^-] \text{ total} = [Cl^-] \text{ from NaCl} + [Cl^-] \text{ from MgCl}_2 = 0.438 \text{ M} + 0.102 \text{ M} = 0.540 \text{ M Cl}^-$$

1B **(a)**
$$\frac{1.5 \text{ mg F}^-}{L} \times \frac{1 \text{ g F}^-}{1000 \text{ mg F}^-} \times \frac{1 \text{ mol F}^-}{18.998 \text{ g F}^-} = 7.9 \times 10^{-5} \text{ M F}^-$$

(b) $1.00 \times 10^6 \text{ L} \times \dfrac{7.9 \times 10^{-5} \text{ mol F}^-}{1L} \times \dfrac{1 \text{ mol CaF}_2}{2 \text{ mol F}^-} \times \dfrac{78.075 \text{ g CaF}_2}{1 \text{ mol CaF}_2} \times \dfrac{1 \text{ kg}}{1000 \text{ g}}$

$$= 3.1 \text{ kg CaF}_2$$

2A In each case we use the solubility rules to determine whether either product is insoluble. The ions in each product compound are determined by simply "switching the partners" of the reactant compounds. The designation "(aq)" on each reactant indicates that it is soluble.

(a) Possible products are potassium chloride, KCl, which is soluble, and aluminum hydroxide, $Al(OH)_3$, which is not. The net ionic equation is:
$$Al^{3+}(aq) + 3 OH^-(aq) \rightarrow Al(OH)_3(s)$$

(b) Possible products are iron(III) sulfate, $Fe_2(SO_4)_3$, and potassium bromide, KBr, both of which are soluble. No reaction occurs.

(c) Possible products are calcium nitrate, $Ca(NO_3)_2$, which is soluble, and lead(II) iodide, PbI_2, which is insoluble. The net ionic equation is: $Pb^{2+}(aq) + 2 I^-(aq) \rightarrow PbI_2(s)$

2B **(a)** Possible products are sodium chloride, NaCl, which is soluble, and aluminum phosphate, $AlPO_4$, which is insoluble. The net ionic equation is:
$$Al^{3+}(aq) + PO_4^{3-}(aq) \rightarrow AlPO_4(s)$$

(b) Possible products are aluminum chloride, $AlCl_3$, which is soluble, and barium sulfate, $BaSO_4$, which is insoluble. The net ionic equation is:

$$Ba^{2+}(aq) + SO_4^{2-}(aq) \rightarrow BaSO_4(s)$$

(c) Possible products are ammonium nitrate, NH_4NO_3, which is soluble, and lead (II) carbonate, $PbCO_3$, which is insoluble. The net ionic equation is:

$$Pb^{2+}(aq) + CO_3^{2-}(aq) \rightarrow PbCO_3(s)$$

3A Propionic acid is a weak acid, not dissociated completely in aqueous solution. Ammonia is a weak base. The acid and base react to form a salt solution of ammonium propionate.

$$NH_3(aq) + HC_3H_5O_2(aq) \rightarrow NH_4^+(aq) + C_3H_5O_2^-(aq)$$

3B Since acetic acid is a weak acid, it is not dissociated completely in aqueous solution (except at infinite dilution); it is misleading to write it in ionic form. The products of this reaction are the gas carbon dioxide, the covalent compound water, and the ionic solute calcium acetate. Only the latter exists as ions in aqueous solution.

$$CaCO_3(s) + 2\, HC_2H_3O_2(aq) \rightarrow CO_2(g) + H_2O(l) + Ca^{2+}(aq) + 2\, C_2H_3O_2^-(aq)$$

4A **(a)** This is a metathesis or double displacement reaction. Elements do not change oxidation states during this reaction. It is not an oxidation–reduction reaction.

(b) The presence of $O_2(g)$ as a product indicates that this is an oxidation–reduction reaction. Oxygen is oxidized from O.S.$= -2$ in $Pb(NO_3)_2$ (s) to O.S.$= 0$ in $O_2(g)$. Nitrogen is reduced from O.S.$= +5$ in $Pb(NO_3)_2$ (s) to O.S.$= +4$ in $NO_2(g)$

4B We determine the oxidation state (O.S.) of each element on each side of the equation. The O.S. of H is $+1$ on each side of the equation, and the O.S. of O is -2. For *vanadium*, the O.S. of V is $+4$ in VO^{2+}, and the O.S. of V is $+5$ in VO_2^+; since the oxidation state of V has increased during the reaction, VO^{2+} has been oxidized. For *manganese*, the O.S. of Mn in MnO_4^- is $+7$, and the O.S. of Mn in Mn^{2+} is $+2$; since the oxidation state of Mn has decreased during the reaction, MnO_4^- is the species reduced.

5A Aluminum is oxidized (from an O.S. of 0 to an O.S. of $+3$), while hydrogen is reduced (from an O.S. of $+1$ to an O.S. of 0).

Oxidation : $\{Al(s) \rightarrow Al^{3+}(aq) + 3\, e^-\}\ \times 2$

Reduction: $\{2\, H^+(aq) + 2\, e^- \rightarrow H_2(g)\} \times 3$

Net equation : $2\, Al(s) + 6\, H^+(aq) \rightarrow 2\, Al^{3+}(aq) + 3\, H_2(g)$

5B   Bromide is oxidized (from -1 to 0) while chlorine is reduced (from 0 to -1).

Oxidation : $2\,Br^-\,(aq) \rightarrow Br_2\,(l) + 2\,e^-$

Reduction: $Cl_2\,(g) + 2\,e^- \rightarrow 2\,Cl^-\,(aq)$

Net equation : $2\,Br^-\,(aq) + Cl_2\,(g) \rightarrow Br_2\,(l) + 2\,Cl^-\,(aq)$

6A   Step 1:   Write the two skeleton half-equations.

$$MnO_4^-\,(aq) \rightarrow Mn^{2+}\,(aq)\quad and \quad Fe^{2+}\,(aq) \rightarrow Fe^{3+}\,(aq)$$

Step 2:   Balance each skeleton half-equation for O (with H_2O) and for H atoms (with H^+).

$$MnO_4^-\,(aq) + 8\,H^+\,(aq) \rightarrow Mn^{2+}\,(aq) + 4\,H_2O(l)\quad and \quad Fe^{2+}\,(aq) \rightarrow Fe^{3+}\,(aq)$$

Step 3:   Balance electric charge by adding electrons.

$$MnO_4^-\,(aq) + 8\,H^+\,(aq) + 5\,e^- \rightarrow Mn^{2+}\,(aq) + 4\,H_2O(l)\quad and \quad Fe^{2+}\,(aq) \rightarrow Fe^{3+}\,(aq) + e^-$$

Step 4:   Combine the two ½-reactions

$$\{Fe^{2+}\,(aq) \rightarrow Fe^{3+}\,(aq) + e^-\} \times 5$$

$$MnO_4^-\,(aq) + 8\,H^+\,(aq) + 5\,e^- \rightarrow Mn^{2+}\,(aq) + 4\,H_2O(l)$$

$$MnO_4^-\,(aq) + 8\,H^+\,(aq) + 5\,Fe^{2+}\,(aq) \rightarrow Mn^{2+}\,(aq) + 4\,H_2O(l) + 5\,Fe^{3+}\,(aq)$$

6B   Step 1:   Uranium is oxidized and chromium is reduced in this reaction. The "skeleton" half-equations are: $UO^{2+}\,(aq) \rightarrow UO_2^{2+}\,(aq)\quad and \quad Cr_2O_7^{2-}\,(aq) \rightarrow Cr^{3+}\,(aq)$

Step 2:   First, balance the chromium skeleton half-equation for chromium atoms:

$$Cr_2O_7^{2-}\,(aq) \rightarrow 2\,Cr^{3+}\,(aq)$$

Next, balance oxygen atoms with water molecules in each half-equation:

$$UO^{2+}\,(aq) + H_2O(l) \rightarrow UO_2^{2+}\,(aq)\quad and \quad Cr_2O_7^{2-}\,(aq) \rightarrow 2Cr^{3+}\,(aq) + 7H_2O(l)$$

Then, balance hydrogen atoms with hydrogen ions in each half-equation:

$$UO^{2+}\,(aq) + H_2O(l) \rightarrow UO_2^{2+}\,(aq) + 2\,H^+\,(aq)$$
$$Cr_2O_7^{2-}\,(aq) + 14H^+\,(aq) \rightarrow 2Cr^{3+}\,(aq) + 7H_2O(l)$$

Step 3:   Balance the charge of each half-equation with electrons.

$$UO^{2+}\,(aq) + H_2O(l) \rightarrow UO_2^{2+}\,(aq) + 2\,H^+\,(aq) + 2\,e^-$$
$$Cr_2O_7^{2-}\,(aq) + 14\,H^+\,(aq) + 6\,e^- \rightarrow 2\,Cr^{3+}\,(aq) + 7\,H_2O(l)$$

Step 4: Multiply the uranium half-equation by 3 and add the chromium half-equation to it.

$$\left\{UO^{2+}(aq)+ H_2O(l) \rightarrow UO_2^{2+}(aq)+2\ H^+(aq)+2\ e^-\right\}\times 3$$

$$\underline{Cr_2O_7^{2-}(aq)+14\ H^+(aq)+6\ e^- \rightarrow 2\ Cr^{3+}(aq)+7\ H_2O(l)}$$

$$3\ UO^{2+}(aq)+ Cr_2O_7^{2-}(aq)+14\ H^+(aq)+3\ H_2O(l)$$

$$\rightarrow 3\ UO_2^{2+}(aq)+2\ Cr^{3+}(aq)+7\ H_2O(l)+6\ H^+(aq)$$

Step 5: SIMPLIFY. Subtract 3 H_2O (l) and 6 H^+ (aq) from each side of the equation.

$$3\ UO^{2+}(aq)+ Cr_2O_7^{2-}(aq)+8\ H^+(aq) \rightarrow 3\ UO_2^{2+}(aq)+2\ Cr^{3+}(aq)+4\ H_2O(l)$$

7A Step 1: Write the two skeleton half-equations.

$$S(s) \rightarrow SO_3^{2-}(aq)\ \ and\ \ OCl^-(aq) \rightarrow Cl^-(aq)$$

Step 2: Balance each skeleton half-equation for O (with H_2O) and for H atoms (with H^+).

$$3\ H_2O(l)+ S(s) \rightarrow SO_3^{2-}(aq)+6\ H^+$$

$$OCl^-(aq)+2H^+ \rightarrow Cl^-(aq)+H_2O(l)$$

Step 3: Balance electric charge by adding electrons.

$$3\ H_2O(l)+ S(s) \rightarrow SO_3^{2-}(aq)+6\ H^+(aq)+4\ e^-$$

$$OCl^-(aq)+2H^+(aq)+2e^- \rightarrow Cl^-(aq)+H_2O(l)$$

Step 4: Change from an acidic medium to a basic one by adding OH^- to eliminate H^+.

$$3H_2O(l)+ S(s)+6\ OH^-(aq) \rightarrow SO_3^{2-}(aq)+6\ H^+(aq)+6\ OH^-(aq)+4\ e^-$$

$$OCl^-(aq)+2\ H^+(aq)+2\ OH^-(aq)+2\ e^- \rightarrow Cl^-(aq)+H_2O(l)+2\ OH^-(aq)$$

Step 5: Simplify by removing the items present on both sides of each half-equation, and combine the half-equations to obtain the net redox equation.

$$\left\{S(s)+6\ OH^-(aq) \rightarrow SO_3^{2-}(aq)+3\ H_2O(l)+4\ e^-\right\}\times 1$$

$$\underline{\left\{OCl^-(aq)+ H_2O(l)+2\ e^- \rightarrow Cl^-(aq)+2\ OH^-(aq)\right\}\times 2}$$

$$S(s)+ 6\ OH^-(aq)+2\ OCl^-(aq)+2H_2O(l) \rightarrow SO_3^{2-}(aq)+3\ H_2O(l)+2\ Cl^-(aq)+4OH^-$$

Simplify by removing the species present on both sides.

NET: $S(s)+2\ OH^-(aq)+2\ OCl^-(aq) \rightarrow SO_3^{2-}(aq)+ H_2O(l)+2\ Cl^-(aq)$

7B Step 1: Write the two skeleton half-equations.

$$MnO_4^-(aq) \rightarrow MnO_2(s)\ \ and\ \ SO_3^{2}(aq) \rightarrow SO_4^{2}(aq)$$

Step 2: Balance each skeleton half-equation for O (with H_2O) and for H atoms (with H^+).

$$MnO_4^- (aq) + 4\ H^+ (aq) \rightarrow MnO_2 (s) + 2\ H_2O(l)$$

$$SO_3^{2-} (aq) + H_2O(l) \rightarrow SO_4^{2-} (aq) + 2H^+ (aq)$$

Step 3: Balance electric charge by adding electrons.

$$MnO_4^- (aq) + 4\ H^+ (aq) + 3\ e^- \rightarrow MnO_2 (s) + 2\ H_2O(l)$$

$$SO_3^{2-} (aq) + H_2O(l) \rightarrow SO_4^{2-} (aq) + 2\ H^+ (aq) + 2\ e^-$$

Step 4: Change from an acidic medium to a basic one by adding OH^- to eliminate H^+.

$$MnO_4^- (aq) + 4\ H^+ (aq) + 4\ OH^- (aq) + 3\ e^- \rightarrow MnO_2 (s) + H_2O(l) + 4\ OH^- (aq)$$

$$SO_3^{2-} (aq) + H_2O(l) + 2\ OH^- (aq) \rightarrow SO_4^{2-} (aq) + 2\ H^+ (aq) + 2\ OH^- (aq) + 2\ e^-$$

Step 5: Simplify by removing species present on both sides of each half-equation, and combine the half-equations to obtain the net redox equation.

$$\{MnO_4^- (aq) + 2\ H_2O(l) + 3\ e^- \rightarrow MnO_2 (s) + 4\ OH^- (aq)\} \times 2$$

$$\underline{\{SO_3^{2-} (aq) + 2\ OH^- (aq) \rightarrow SO_4^{2-} (aq) + H_2O(l) + 2\ e^-\} \times 3}$$

$$2\,MnO_4^- (aq) + 3SO_3^{2-} (aq) + 6\ OH^- (aq) + 4\ H_2O(l) \rightarrow$$

$$2\ MnO_2 (s) + 3SO_4^{2-} (aq) + 3\,H_2O(l) + 8\ OH^- (aq)$$

Simplify by removing species present on both sides.

NET: $2\,MnO_4^- (aq) + 3\,SO_3^{2-} (aq) + H_2O(l) \rightarrow 2\ MnO_2 (s) + 3SO_4^{2-} (aq) + 2\ OH^- (aq)$

8A Since the oxidation state of H is 0 in H_2 (g) and is +1 in both NH_3(g) and H_2O(g), hydrogen is oxidized. A substance that is oxidized is called a reducing agent. In addition, the oxidation state of N in NO_2 (g) is +4, while it is −3 in NH_3; the oxidation state of the element N decreases during this reaction, meaning that NO_2 (g) is reduced. The substance that is reduced is called the oxidizing agent.

8B In $\left[Au(CN)_2\right]^-$ (aq), gold has an oxidation state of +1; Au has been oxidized and, thus, Au(s) (oxidization state = 0), is the reducing agent. In OH^- (aq), oxygen has an oxidation state of -2; O has been reduced and thus, O_2(g) (oxidation state = 0) is the oxidizing agent.

9A We first determine the amount of NaOH that reacts with 0.500 g KHP.

$$\text{amount NaOH} = 0.5000\ \text{g KHP} \times \frac{1\ \text{mol KHP}}{204.22\ \text{g KHP}} \times \frac{1\ \text{mol OH}^-}{1\ \text{mol KHP}} \times \frac{1\ \text{mol NaOH}}{1\ \text{mol OH}^-}$$

$$= 0.002448\ \text{mol NaOH}$$

$$[\text{NaOH}] = \frac{0.002448\ \text{mol NaOH}}{24.03\ \text{mL soln}} \times \frac{1000\ \text{mL}}{1\ \text{L}} = 0.1019\ \text{M}$$

9B The net ionic equation when solid hydroxides react with a strong acid is $OH^- + H^+ \rightarrow H_2O$. There are two sources of OH^-: NaOH and $Ca(OH)_2$. We compute the amount of OH^- from each source and add the results.

moles of OH^- from NaOH:

$$= 0.235 \text{ g sample} \times \frac{92.5 \text{ g NaOH}}{100.0 \text{ g sample}} \times \frac{1 \text{ mol NaOH}}{39.997 \text{ g NaOH}} \times \frac{1 \text{ mol OH}^-}{1 \text{ mol NaOH}} = 0.00543 \text{ mol OH}^-$$

moles of OH^- from $Ca(OH)_2$:

$$= 0.235 \text{ g sample} \times \frac{7.5 \text{ g Ca(OH)}_2}{100.0 \text{ g sample}} \times \frac{1 \text{ mol Ca(OH)}_2}{74.093 \text{ g Ba(OH)}_2} \times \frac{2 \text{ mol OH}^-}{1 \text{ mol Ca(OH)}_2} = 0.00048 \text{ mol OH}^-$$

total amount $OH^- = 0.00543$ mol from NaOH $+ 0.00048$ mol from $Ca(OH)_2 = 0.00591$ mol OH^-

$$[\text{HCl}] = \frac{0.00591 \text{ mol OH}^-}{45.6 \text{ mL HCl} \cdot \text{soln}} \times \frac{1 \text{ mol H}^+}{1 \text{ mol OH}^-} \times \frac{1 \text{ mol HCl}}{1 \text{ mol H}^+} \times \frac{1000 \text{ mL soln}}{1 \text{ L soln}} = 0.130 \text{ M}$$

10A First, determine the mass of iron that has reacted as Fe^{2+} with the titrant. The balanced chemical equation provides the essential conversion factor.

$$\text{mass Fe} = 0.04125 \text{ L titrant} \times \frac{0.02140 \text{ mol MnO}_4^-}{1 \text{ L titrant}} \times \frac{5 \text{ mol Fe}^{2+}}{1 \text{ mol MnO}_4^-} \times \frac{55.847 \text{ g Fe}}{1 \text{ mol Fe}^{2+}} = 0.246 \text{ g Fe}$$

Then determine the % Fe in the ore. $\quad \% \text{ Fe} = \dfrac{0.246 \text{ g Fe}}{0.376 \text{ g ore}} \times 100\% = 65.4\% \text{ Fe}$

10B The balanced equation provides stoichiometric coefficients used in the solution.

$$\text{amount MnO}_4^- = 0.2482 \text{ g Na}_2\text{C}_2\text{O}_4 \times \frac{1 \text{ mol Na}_2\text{C}_2\text{O}_4}{134.00 \text{ g Na}_2\text{C}_2\text{O}_4} \times \frac{1 \text{ mol C}_2\text{O}_4^{2-}}{1 \text{ mol Na}_2\text{C}_2\text{O}_4} \times \frac{2 \text{ mol MnO}_4^-}{5 \text{ mol C}_2\text{O}_4^{2-}}$$

$$= 0.0007409 \text{ mol MnO}_4^-$$

$$[\text{KMnO}_4] = \frac{0.0007409 \text{ mol MnO}_4^-}{23.68 \text{ mL soln}} \times \frac{1000 \text{ mL}}{1 \text{ L}} \times \frac{1 \text{ mol KMnO}_4}{1 \text{ mol MnO}_4^-} = 0.03129 \text{ M KMnO}_4$$

REVIEW QUESTIONS

1. **(a)** The symbol "$\rightleftharpoons$" means that a chemical reaction reaches a point of balance or equilibrium where the rate of the forward reaction equals the rate of the reverse reaction.

   (b) The square brackets, [], surrounding the formula of a species, are the symbol for the molarity of that species in solution.

    (c)   A "spectator" ion is one that is present in a solution in which a reaction takes place but is not included in the net ionic equation for that reaction because it does not get involved in the reaction.

    (d)   A weak acid is a species that produces hydrogen ion in aqueous solution (an acid), but does not dissociate completely into its ions (except at infinite dilutions - a weak electrolyte).

2.   **(a)**   In the half-reaction method of balancing redox equations, each species that is oxidized or reduced is the basis for a balanced half-equation. These half-equations then are combined to produce the balanced net ionic equation for the redox reaction.

    (b)   A disproportionation reaction is one in which the same species is both oxidized and reduced.

    (c)   A titration is the procedure of adding a measured amount of one material to a measured amount of another, in such a way that chemically equivalent amounts of substances are present at the end of the titration. The concentration of substance in one of the two materials is known; the technique permits the determination of the other concentration.

    (d)   Standardization of a solution refers to the determination of its concentration by titration.

3.   **(a)**   A strong electrolyte is a substance that dissociates completely into its ions when it is dissolved in aqueous solution. A strong acid is a strong electrolyte that produces hydrogen ions and anions when it dissociates.

    (b)   An oxidizing agent is a species that causes another species to be oxidized: to lose electrons and thereby have the oxidation state of one of its elements increased. A reducing agent is a species that causes another species to be reduced: to accept electrons and have the oxidation state of one of its elements decreased.

    (c)   A precipitation reaction is one in which an insoluble substance is formed when solutions of two soluble substances are mixed. A neutralization reaction is the reaction of an acid with a base; the normal products are water and a salt that has the same cation as the base and the same anion as the acid.

    (d)   A half-reaction refers to just the oxidation or just the reduction aspect of a redox reaction. An overall or "net" reaction refers to the entire chemical reaction, with only spectator ions omitted (or it can be the net result of several reactions that, together, make up a process).

4.   **(a)**   The best electrical conductor is the solution of the strong electrolyte: 0.10 M NaCl. In each liter of this solution, there are 0.10 mol Na^+ ions and 0.10 mol Cl^- ions.

    (b)   The poorest electrical conductor is the solution of the nonelectrolyte: 0.10 M C_2H_5OH. In this solution, the concentration of ions is almost nonexistent.

5. **(a)** Na_2SO_4 is a *salt* of sodium hydroxide, NaOH, and sulfuric acid, H_2SO_4.

(b) $Ba(OH)_2$ is a *strong base*, one of the common strong bases listed in Table 5-1.

(c) $Ba(NO_3)_2$ is a *salt* of barium hydroxide, $Ba(OH)_2$, and nitric acid, HNO_3.

(d) H_3PO_4 is a *weak acid*, (formula begins with hydrogen), but is not listed in Table 5-1.

(e) HBr is a *strong acid*, listed in Table 5-1.

(f) HNO_2 is a *weak acid* (formula begins with hydrogen), but it is not listed in Table 5-1.

(g) NH_3 is a *weak base*, ammonia.

(h) NH_4I is a *salt* of ammonia, NH_3, and hydroiodic acid, HI.

(i) KOH is a *strong base*, one of the common ones listed in Table 5-1.

6. For all these solutes but one— $Al_2(SO_4)_3$ —there is one sulfate ion per formula unit. Consequently, the concentration of the compound and the sulfate ion concentration in that compound's aqueous solution will be the same. This makes 0.22 M $Mg(SO_4)$ the solution with the highest $[SO_4^{2-}]$ among these four. But there are three sulfate ions per formula unit of $Al_2(SO_4)_3$. Thus, $[SO_4^{2-}]$ in the $Al_2(SO_4)_3$ solution is three times the concentration of the solute, or $[SO_4^{2-}] = 3 \times 0.080$ M $= 0.24$ M; therefore this solution has the highest $[SO_4^{2-}]$.

7. **(a)** $[K^+] = \dfrac{0.238 \text{ mol } KNO_3}{1 \text{ L soln}} \times \dfrac{1 \text{ mol } K^+}{1 \text{ mol } KNO_3} = 0.238 \text{ M } K^+$

(b) $[NO_3^-] = \dfrac{0.167 \text{ mol } Ca(NO_3)_2}{1 \text{ L soln}} \times \dfrac{2 \text{ mol } NO_3^-}{1 \text{ mol } Ca(NO_3)_2} = 0.334 \text{ M } NO_3^-$

(c) $[Al^{3+}] = \dfrac{0.083 \text{ mol } Al_2(SO_4)_3}{1 \text{ L soln}} \times \dfrac{2 \text{ mol } Al^{3+}}{1 \text{ mol } Al_2(SO_4)_3} = 0.17 \text{ M } Al^{3+}$

(d) $[Na^+] = \dfrac{0.209 \text{ mol } Na_3PO_4}{1 \text{ L soln}} \times \dfrac{3 \text{ mol } Na^+}{1 \text{ mol } Na_3PO_4} = 0.627 \text{ M } Na^+$

8. The amount of chloride ion in each solution in millimoles is computed.

Cl^- amount $= 200.0 \text{ mL} \times \dfrac{0.35 \text{ mmol NaCl}}{1 \text{ mL soln}} \times \dfrac{1 \text{ mmol } Cl^-}{1 \text{ mmol NaCl}} = 70. \text{ mmol } Cl^-$

Cl^- amount $= 500.0 \text{ mL} \times \dfrac{0.065 \text{ mmol } MgCl_2}{1 \text{ mL soln}} \times \dfrac{2 \text{ mmol } Cl^-}{1 \text{ mmol } MgCl_2} = 65 \text{ mmol } Cl^-$

Cl^- amount $= 1.00 \text{ L} \times \dfrac{1000 \text{ mL}}{1 \text{ L}} \times \dfrac{0.068 \text{ mmol HCl}}{1 \text{ mL}} \times \dfrac{1 \text{ mmol } Cl^-}{1 \text{ mmol HCl}} = 68 \text{ mmol } Cl^-$

The 200.0 mL of 0.035 M NaCl contains the largest amount of Cl^-.

9. $\left[OH^-\right] = \dfrac{0.132 \ g \ Ba(OH)_2 \cdot 8H_2O}{275 \ mL \ soln} \times \dfrac{1000 \ mL}{1 \ L} \times \dfrac{1 \ mol \ Ba(OH)_2 \cdot 8H_2O}{315.5 \ Ba(OH)_2 \cdot 8H_2O} \times \dfrac{2 \ mol \ OH^-}{1 \ mol \ Ba(OH)_2 \cdot 8 \ H_2O}$

$= 3.04 \times 10^{-3} \ M \ OH^-$

10. $\left[K^+\right] = \dfrac{0.126 \ mol \ KCl}{1 \ L \ soln} \times \dfrac{1 \ mol \ K^+}{1 \ mol \ KCl} = 0.126 \ M \ K^+$

$\left[Mg^{2+}\right] = \dfrac{0.148 \ mol \ MgCl_2}{1 \ L \ soln} \times \dfrac{1 \ mol \ Mg^{2+}}{1 \ mol \ MgCl_2} = 0.148 \ M \ Mg^{2+}$

Now determine the amount of Cl^- in 1.00 L of the solution.

$mol \ Cl^- = \left(\dfrac{0.126 \ mol \ KCl}{1 \ L \ soln} \times \dfrac{1 \ mol \ Cl^-}{1 \ mol \ KCl}\right) + \left(\dfrac{0.148 \ mol \ MgCl_2}{1 \ L \ soln} \times \dfrac{2 \ mol \ Cl^-}{1 \ mol \ MgCl_2}\right)$

$= 0.126 \ mol \ Cl^- + 0.296 \ mol \ Cl^- = 0.422 \ mol \ Cl^-$

$\left[Cl^-\right] = \dfrac{0.422 \ mol \ Cl^-}{1 \ L \ soln} = 0.422 \ M \ Cl^-$

11. Determine the amount of I^- in the solution as it now exists, and the amount of I^- in the solution of the desired concentration. The different in these two amounts is the amount of I^- that must be added. Convert this amount to a mass of MgI_2 in grams.

moles of I^- in final solution $= 250.0 \ mL \times \dfrac{1 \ L}{1000 \ mL} \times \dfrac{0.1000 \ mol \ I^-}{1 \ L \ soln} = 0.02500 \ mol \ I^-$

moles of I^- in KI solution $= 250.0 \ mL \times \dfrac{1 \ L}{1000 \ mL} \times \dfrac{0.0876 \ mol \ KI}{1 \ L \ soln} \times \dfrac{1 \ mol \ I^-}{1 \ mol \ KI} = 0.0219 \ mol \ I^-$

mass MgI_2 required $= (0.02500 - 0.0219) \ mol \ I^- \times \dfrac{1 \ mol \ MgI_2}{2 \ mol \ I^-} \times \dfrac{278.11 \ g \ MgI_2}{1 \ mol \ MgI_2} \times \dfrac{1000 \ mg}{1 \ g}$

$= 4.3 \times 10^2 \ mg \ MgI_2$

12. Nitrates, acetates, and alkali metal compounds are water-soluble. $Zn(NO_3)_2$, $Pb(C_2H_3O_2)_2$, and NaI are soluble. Most halides are soluble in water; $CuCl_2$ is soluble in water. Although most sulfates are soluble in water, $BaSO_4(s)$ is not soluble in water. Only a few hydroxides are soluble in water; $Al(OH)_3(s)$ is not soluble in water.

13. HCl(aq) reacts with active metals and some anions to produce a gas.
Ca is an active metal: $Ca(s) + 2 \ HCl(aq) \rightarrow CaCl_2(aq) + H_2(g)$

HSO_3^- produces a gas with an acid:

$KHSO_3(s) + HCl(aq) \rightarrow KCl(aq) + H_2O(l) + SO_2(g)$

14. In each case, each available cation is paired with the available anions, one at a time, to determine if a compound is produced that is insoluble, based on the solubility rules of Chapter 5. Then a net ionic equation is written to summarize this information.

(a) $Pb^{2+}(aq) + 2\ Br^-(aq) \rightarrow\ PbBr_2(s)$ **(b)** No reaction occurs.

(c) $Fe^{3+}(aq) + 3\ OH^-(aq) \rightarrow Fe(OH)_3(s)$ **(d)** $Ca^{2+}(aq) + CO_3^{2-}(aq) \rightarrow\ CaCO_3(s)$

(e) $Ba^{2+}(aq) + SO_4^{2-}(aq) \rightarrow\ BaSO_4(s)$ **(f)** No reaction; CaS(s) is moderately soluble.

15. The type of reaction is given first, followed by the net ionic equation.

(a) Neutralization: $OH^-(aq) +\ HC_2H_3O_2(aq) \rightarrow\ H_2O(l) +\ C_2H_3O_2^-(aq)$

(b) No reaction occurs. This is the mixing of two acids.

(c) Gas evolution: $FeS(s) + 2\ H^+(aq) \rightarrow\ H_2S(g) +\ Fe^{2+}(aq)$

(d) Gas evolution: $HCO_3^-(aq) +\ H^+(aq) \rightarrow "H_2CO_3(aq)" \rightarrow\ H_2O(l) +\ CO_2(g)$

(e) Redox: $Mg(s) + 2\ H^+(aq) \rightarrow\ Mg^{2+}(aq) +\ H_2(g)$

(f) No reaction occurs, based on the information in Table 5-3.

16. Use (b) NH_3(aq): NH_3 affords the OH^- ions necessary to form $Mg(OH)_2$(s)

Applicable reactions: $(NH_3(aq) + H_2O(l) \rightarrow NH_4^+(aq) + OH^-(aq)) \times 2$

$MgCl_2(aq) \rightarrow Mg^{2+}(aq) + 2\ Cl^-(aq)$

$Mg^{2+}(aq) + 2\ OH^-(aq) \rightarrow Mg(OH)_2(s)$

17. The problem is most easily solved with amounts in millimoles.

$$V_{NaOH} = 10.00\ mL\ HCl(aq) \times \frac{0.128\ mmol\ HCl}{1\ mL\ HCl(aq)} \times \frac{1\ mmol\ H^+}{1\ mmol\ HCl} \times \frac{1\ mmol\ OH^-}{1\ mmol\ H^+}$$

$$\times \frac{1\ mmol\ NaOH}{1\ mmol\ OH^-} \times \frac{1\ mL\ NaOH(aq)}{0.0962\ mmol\ NaOH} = 13.3\ mL\ NaOH(aq)\ soln$$

18. $[NaOH] = \dfrac{10.00\ mL\ acid \times \dfrac{0.1012\ mmol\ H_2SO_4}{1\ mL\ acid} \times \dfrac{2\ mmol\ NaOH}{1\ mmol\ H_2SO_4}}{23.31\ mL\ base} = 0.08683\ M$

19. The net ionic equation for the reaction of KOH, a strong base, with HCl, a strong acid, is:

$$OH^-(aq) +\ H^+(aq) \rightarrow\ H_2O(l)$$

Thus, the reactant that produces the smaller amount of ions is the limiting reactant. More to the point, the difference between the larger number of ions and the smaller number, determines whether the resulting solution is acidic or basic. If the difference is zero, the solution is neutral.

$$\text{amount OH}^- = 23.58 \text{ mL KOH(aq)} \times \frac{0.1278 \text{ mmol KOH}}{1 \text{ mL KOH(aq)}} \times \frac{1 \text{ mmol OH}^-}{1 \text{ mmol KOH}} = 3.014 \text{ mmol OH}^-$$

$$\text{amount H}^+ = 25.13 \text{ mL HCl(aq)} \times \frac{0.1264 \text{ mmol HCl}}{1 \text{ mL HCl(aq)}} \times \frac{1 \text{ mmol H}^+}{1 \text{ mmol HCl}} = 3.176 \text{ mmol H}^+$$

excess ion $= 3.176$ mmol $H^+ - 3.014$ mmol $OH^- = 0.162$ mmol H^+. The solution is acidic.

20. The answer is: (a) the missing coefficients are each four.

$(Fe^{2+}(aq) \rightarrow Fe^{3+}(aq) + e^-\ ) \times 4$

$\underline{4 e^- + 4 H^+(aq) + O_2(g) \rightarrow 2 H_2O(l) \qquad\qquad}$

$4 Fe^{2+}(aq) + 4 H^+(aq) + O_2(g) \rightarrow 2 H_2O(l) + 4 Fe^{3+}(aq)$

21. **(a)** The O.S. of H is $+1$, that of O is -2, that of C is $+4$, and that of Mg is $+2$ on each side of this equation. This is not a redox equation.

(b) The O.S. of Cl is 0 on the left and -1 on the right side of this equation. The O.S. of Br is -1 on the left and 0 on the right side of this equation. This is a redox reaction.

(c) The O.S. of Ag is 0 on the left and $+1$ on the right side of this equation. The O.S. of N is $+5$ on the left and $+4$ on the right side of this equation. This is a redox reaction.

(d) The O.S. of O is -2, that of Ag is $+1$, and that of Cr is $+6$ on both sides of this equation. This is not a redox equation.

22. **(a)** The O.S. of O is -2 on both sides of this reaction. The O.S. of H is 0 on the left and $+1$ on the right side of this equation; H is oxidized and thus NO must be an oxidizing agent. The O.S. of N is $+2$ on the left and -3 on the right side of this equation; N is reduced and thus H_2 must be a reducing agent.

(b) The O.S. of O is -2 and that of H is $+1$ on both sides of this equation. The O.S. of Cu is 0 on the left and $+2$ on the right side of this equation; Cu is oxidized and thus NO_3^- must be an oxidizing agent. The O.S. of N is $+5$ on the left and $+2$ on the right side of this equation; N is reduced and thus Cu must be a reducing agent.

(c) The O.S. of O is -2 and that of H is $+1$ on both sides of this equation. The O.S. of Cl is 0 on the left side of this equation; on the right side, the O.S. of Cl is -1 in Cl^- and it is $+5$ in ClO_3^-. Cl is both oxidized and reduced and Cl_2 serves as both an oxidizing agent and as a reducing agent in this disproportionation reaction.

23. **(a)** Reduction: $\qquad 2SO_3^{2-}(aq) + 6 H^+(aq) + 4 e^- \rightarrow S_2O_3^{2-}(aq) + 3 H_2O(l)$

(b) Reduction: $\qquad 2NO_3^-(aq) + 10 H^+(aq) + 8 e^- \rightarrow N_2O(g) + 5 H_2O(l)$

(c) Oxidation: $\qquad I^-(aq) + 3 H_2O(l) \rightarrow IO_3^-(aq) + 6 H^+(aq) + 6 e^-$

(d) Oxidation: $\qquad Al(s) + 4 OH^-(aq) \rightarrow Al(OH)_4^-(aq) + 3 e^-$

24. **(a)** Oxidation: $\{ Zn(s) \rightarrow Zn^{2+}(aq) + 2\ e^- \qquad\qquad\qquad\qquad \} \times 3$

Reduction: $\{ NO_3^-(aq) + 4\ H^+(aq) + 3\ e^- \rightarrow NO(g) + 2\ H_2O(l) \qquad \} \times 2$

Net: $3\ Zn(s) + 2NO_3^-(aq) + 8\ H^+(aq) \rightarrow 3\ Zn^{2+}(aq) + 2\ NO(g) + 4\ H_2O(l)$

(b) Oxidation: $\{ Zn(s) \rightarrow Zn^{2+}(aq) + 2\ e^- \qquad\qquad\qquad\qquad \} \times 4$

Reduction: $NO_3^-(aq) + 10\ H^+(aq) + 8e^- \rightarrow NH_4^+(aq) + 3\ H_2O(l)$

Net: $4\ Zn(s) + NO_3^-(aq) + 10\ H^+(aq) \rightarrow 4\ Zn^{2+}(aq) + NH_4^+(aq) + 3\ H_2O(l)$

(c) Oxidation: $\{ Fe^{2+}(aq) \rightarrow Fe^{3+}(aq) + e^- \qquad\qquad\qquad\qquad \} \times 6$

Reduction: $Cr_2O_7^{2-}(aq) + 14\ H^+(aq) + 6\ e^- \rightarrow 2\ Cr^{3+}(aq) + 7\ H_2O(l)$

Net: $Cr_2O_7^{2-}(aq) + 14\ H^+(aq) + 6\ Fe^{2+}(aq) \rightarrow 6\ Fe^{3+}(aq) + 2\ Cr^{3+}(aq) + 7\ H_2O(l)$

(d) Oxidation: $\{ H_2O_2(aq) \rightarrow O_2(g) + 2\ H^+(aq) + 2\ e^- \qquad\qquad \} \times 5$

Reduction: $\{ MnO_4^-(aq) + 8\ H^+(aq) + 5\ e^- \rightarrow Mn^{2+}(aq) + 4\ H_2O(l) \qquad \} \times 2$

Net: $2MnO_4^-(aq) + 6\ H^+(aq) + 5\ H_2O_2(aq) \rightarrow 2\ Mn^{2+}(aq) + 8\ H_2O(l) + 5\ O_2(g)$

25. **(a)** Oxidation: $\{ MnO_2(s) + 4\ OH^-(aq) \rightarrow MnO_4^-(aq) + 2\ H_2O(l) + 3\ e^- \} \times 2$

Reduction: $ClO_3^-(aq) + 3\ H_2O(l) + 6\ e^- \rightarrow Cl^-(aq) + 6\ OH^-(aq)$

Net: $2\ MnO_2(s) + ClO_3^-(aq) + 2\ OH^-(aq) \rightarrow 2MnO_4^-(aq) + Cl^-(aq) + H_2O(l)$

(b) Oxidation: $\{ Fe(OH)_3(s) + 5\ OH^-(aq) \rightarrow FeO_4^{2-}(aq) + 4\ H_2O(l) + 3\ e^- \quad \} \times 2$

Reduction: $\{ OCl^-(aq) + H_2O(l) + 2\ e^- \rightarrow Cl^-(aq) + 2OH^-(aq) \qquad\qquad \} \times 3$

Net: $2\ Fe(OH)_3(s) + 3\ OCl^-(aq) + 4\ OH^-(aq) \rightarrow 2FeO_4^{2-}(aq) + 3\ Cl^-(aq) + 5\ H_2O(l)$

(c) Oxidation: $\{ ClO_2(aq) + 2\ OH^-(aq) \rightarrow ClO_3^-(aq) + H_2O(l) + e^- \ \} \times 5$

Reduction: $ClO_2(aq) + 2\ H_2O(l) + 5\ e^- \rightarrow Cl^-(aq) + 4\ OH^-(aq)$

Net: $6\ ClO_2(aq) + 6\ OH^-(aq) \rightarrow 5ClO_3^-(aq) + Cl^-(aq) + 3\ H_2O(l)$

26. Oxidation: $\{ C_2O_4^{2-}(aq) \rightarrow 2\ CO_2(g) + 2\ e^- \qquad\qquad\qquad\qquad \} \times 5$

Reduction: $\{ MnO_4^-(aq) + 8\ H^+(aq) + 5\ e^- \rightarrow Mn^{2+}(aq) + 4\ H_2O(l) \quad \} \times 2$

Net: $2MnO_4^-(aq) + \boxed{16}\ H^+(aq) + 5\ C_2O_4^{2-}(aq) \rightarrow 2\ Mn^{2+}(aq) + 8\ H_2O(l) + 10\ CO_2(g)$

$$[MnO_4^-] = \dfrac{0.2879\ g\ Na_2C_2O_4 \times \dfrac{1\ mol\ Na_2C_2O_4}{134.00\ g\ Na_2C_2O_4} \times \dfrac{1\ mol\ C_2O_4^{2-}}{1\ mol\ Na_2C_2O_4} \times \dfrac{2\ mol\ MnO_4^-}{5\ mol\ Cl_2O_4^{2-}}}{25.12\ mL\ soln \times \dfrac{1\ L\ soln}{1000\ mL\ soln}}$$

$= 0.03421\ M$ Thus, the molarity of $KMnO_4$ is 0.03421 M.

EXERCISES

Strong Electrolytes, Weak Electrolytes, and Nonelectrolytes

27. **(a)** Because its formula begins with hydrogen, HC_6H_5O is an acid. It is not listed in Table 5-1, so it is a weak acid. A weak acid is a *weak electrolyte*.

(b) Li_2SO_4 is an ionic compound, that is, a salt. A salt is a *strong electrolyte*.

(c) MgI_2 also is a salt, a *strong electrolyte*.

(d) $(CH_3CH_2)_2 O$ is a covalent compound whose formula does not begin with H. Thus, it is neither an acid nor a salt. It also is not built around nitrogen, and thus it does not behave as a weak base. This is a *nonelectrolyte*.

(e) $Sr(OH)_2$ is a *strong electrolyte*, one of the strong bases listed in Table 5-1.

28. NH_3 (aq) is a weak base; $HC_2H_3O_2$ (aq) is a weak acid. Reaction produces a solution of ammonium acetate, $NH_4C_2H_3O_2 (aq)$, a salt and a strong electrolyte.

$$NH_3(aq) +\ HC_2H_3O_2(aq) \rightarrow NH_4^{+}(aq) +\ C_2H_3O_2^{-}(aq)$$

29. **(a)** Barium bromide-strong electrolyte   **(b)** Propionic acid-weak electrolyte

(c) Ammonia-weak electrolyte

30. Sodium chloride(strong electrolyte)   hypochlorous acid   (weak electrolyte)

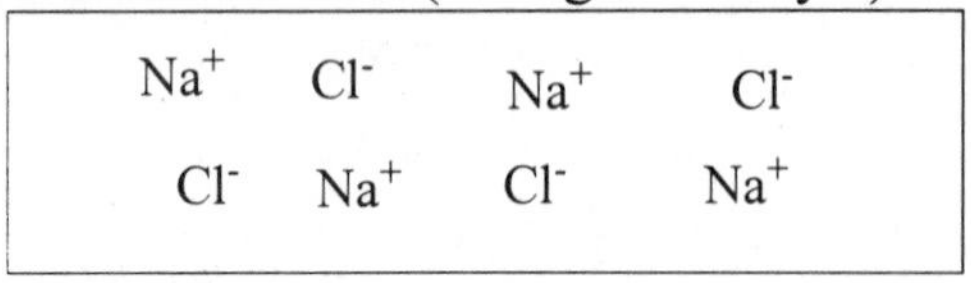

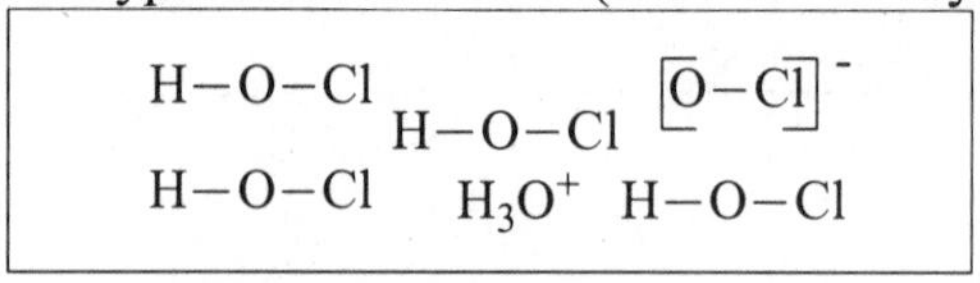

Ammonium chloride   (strong electrolyte) 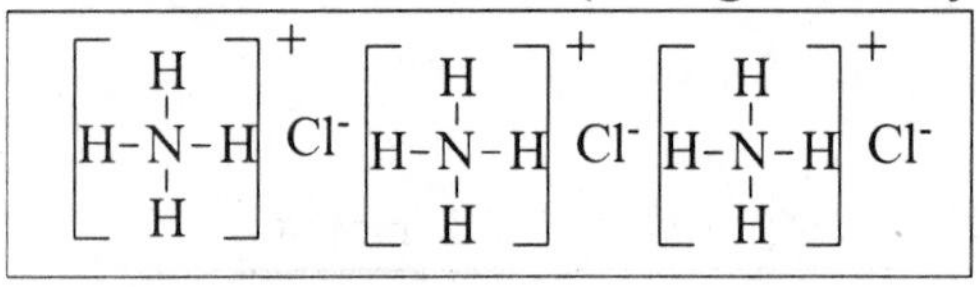   Methanol   (non electrolyte)

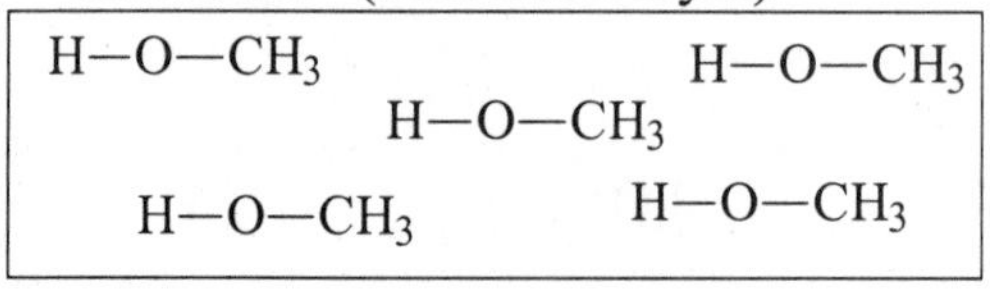

Ion Concentrations

31. **(a)** $\left[Ca^{2+}\right] = \dfrac{35.0\ mg\ Ca^{2+}}{1\ L} \times \dfrac{1\ g}{1000\ mg} \times \dfrac{1\ mol\ Ca^{2+}}{40.078\ g\ Ca^{2+}} = 8.73 \times 10^{-4}\ M\ Ca^{2+}$

(b) $\left[K^{+}\right] = \dfrac{25.6\ mg\ K^{+}}{100\ mL} \times \dfrac{1\ mmol\ K^{+}}{39.098\ mg\ K^{+}} = 6.55 \times 10^{-3}\ M\ K^{+}$ (assumes 3 sig. fig. in 100 mL)

(c) $\left[Zn^{2+}\right] = \dfrac{0.168\ mg\ Zn^{2+}}{1\ mL} \times \dfrac{1\ mmol\ Zn^{2+}}{65.39\ mg\ Zn^{2+}} = 2.57 \times 10^{-3}\ M\ Zn^{2+}$

32. $NaF \ concn = \dfrac{0.9 \ mg \ F^-}{1 \ L} \times \dfrac{1 \ g}{1000 \ mg} \times \dfrac{1 \ mol \ F^-}{19.00 \ g \ F^-} \times \dfrac{1 \ mol \ NaF}{1 \ mol \ F^-} = 5 \times 10^{-5} \ M \ NaF$

<u>**33.**</u> Let us determine the concentration of each solution in $mg \ Na^+ \ / \ mL$. (assume 3 sig. fig. in 100 mL)

(a) $mg \ Na^+/mL = \dfrac{0.208 \ mmol \ Na_2SO_4}{1 \ mL} \times \dfrac{2 \ mmol \ Na^+}{1 \ mmol \ Na_2SO_4} \times \dfrac{23.0 \ mg \ Na^+}{1 \ mmol \ Na^+}$

$mg \ Na^+/mL = 9.57 \ mg \ Na^+/mL$

(b) $mg \ Na^+/mL = \dfrac{1.05 \ g \ NaCl}{100 \ mL} \times \dfrac{1 \ mol \ NaCl}{58.5 \ g \ NaCl} \times \dfrac{1 \ mol \ Na^+}{1 \ mol \ NaCl} \times \dfrac{23.0 \ g \ Na^+}{1 \ mol \ Na^+} \times \dfrac{1000 \ mg}{1 \ g}$

$mg \ Na^+/mL = 4.13 \ mg \ Na^+/mL$

(c) The solution with $14.7 \ mg \ Na^+ \ / \ mL$ has the highest concentration of Na^+.

34. NH_3 is a weak base and would have an exceedingly low $\left[H^+\right]$; the answer is not $1.00 \ M \ NH_3$. $HC_2H_3O_2$ is a very weak acid; $0.011 \ M \ HC_2H_3O_2$ would have a low $\left[H^+\right]$. H_2SO_4 has two ionizable protons per mole while HCl has but one. Thus, H_2SO_4 would have the highest $\left[H^+\right]$ in a 0.010 M aqueous solution.

<u>**35.**</u> Moles of Chloride ion

$= \left(0.225 \ L \times \dfrac{0.625 \ mol \ KCl}{1 \ L \ soln} \times \dfrac{1 \ mol \ Cl^-}{1 \ mol \ KCl} \right) + \left(0.615 \ L \times \dfrac{0.385 \ mol \ MgCl_2}{1 \ L \ soln} \times \dfrac{2 \ mol \ Cl^-}{1 \ mol \ MgCl_2} \right)$

$= 0.141 \ mol \ Cl^- + 0.474 \ mol \ Cl^- = 0.615 \ mol \ Cl^- \quad \left[Cl^-\right] = \dfrac{0.615 \ mol \ Cl^-}{0.225 \ L + 0.615 \ L} = 0.732 \ M$

36. amount of NO_3^- ion $=$

$\left(0.275 \ L \times \dfrac{0.283 \ mol \ KNO_3}{1 \ L \ soln} \times \dfrac{1 \ mol \ NO_3^-}{1 \ mol \ KNO_3} \right) + \left(0.328 \ L \times \dfrac{0.421 \ mol \ Mg(NO_3)_2}{1 \ L \ soln} \times \dfrac{2 \ mol \ NO_3^-}{1 \ mol \ Mg(NO_3)_2} \right)$

$= 0.0778 \ mol \ NO_3^- + 0.276 \ mol \ NO_3^- = 0.354 \ mol \ NO_3^-$

$\left[NO_3^-\right] = \dfrac{0.354 \ mol \ NO_3^-}{0.275 \ L + 0.328 \ L + 0.784 \ L} = 0.255 \ M$

Predicting Precipitation Reactions

37.   Mixture   Result (net ionic equation)

(a) $HI(a) + Zn(NO_3)_2 (aq)$:   No reaction occurs.

(b) $CuSO_4 (aq) + Na_2CO_3 (aq)$: $Cu^{2+}(aq) + CO_3^{2-}(aq) \rightarrow CuCO_3(s)$

(c) $Cu(NO_3)_2 (aq) + Na_3PO_4 (aq)$: $3 Cu^{2+}(aq) + 2 PO_4^{3-}(aq) \rightarrow Cu_3(PO_4)_2(s)$

38.   Mixture   Result (net ionic equation)

(a) $AgNO_3 (aq) + CuCl_2 (aq)$: $Ag^+(aq) + Cl^-(aq) \rightarrow AgCl(s)$

(b) $Na_2S(aq) + FeCl_2 (aq)$: $S^{2-}(aq) + Fe^{2+}(aq) \rightarrow FeS(s)$

(c) $Na_2CO_3 (aq) + AgNO_3 (aq)$: $CO_3^{2-}(aq) + 2 Ag^+(aq) \rightarrow Ag_2CO_3(s)$

39. **(a)** Add $K_2SO_4 (aq)$; $BaSO_4(s)$ will form and $CaSO_4$ will not precipitate.

$BaCl_2 (s) + K_2SO_4 (aq) \rightarrow BaSO_4 (s) + 2 KCl(aq)$

(b) Add $H_2O(l)$; Na_2CO_3 (s) dissolves, $MgCO_3$ (s) will not dissolve (appreciably).

$Na_2CO_3 (s) \xrightarrow{\text{water}} 2 Na^+(aq) + CO_3^{2-}(aq)$

(c) Add KCl(aq); AgCl(s) will form, while $Cu(NO_3)_2$ (s) will dissolve.

$AgNO_3 (s) + KCl(aq) \rightarrow AgCl(s) + KNO_3 (aq)$

40. **(a)** Add H_2O. $Cu(NO_3)_2$ (s) will dissolve, while $PbSO_4(s)$ will not.

$Cu(NO_3)_2 (s) \xrightarrow{\text{water}} Cu^{2+}(aq) + 2NO_3^-(aq)$

(b) Add HCl(aq). $Mg(OH)_2$ (s) will dissolve, but $BaSO_4$ (s) will not.

$Mg(OH)_2 (s) + 2 HCl(aq) \rightarrow MgCl_2 (aq) + 2 H_2O(l)$

(c) Add HCl(aq). Both carbonates dissolve, but $PbCl_2(s)$ will form while $CaCl_2 (aq)$

dissolved. $PbCO_3 (s) + 2 HCl(aq) \rightarrow PbCl_2 (s) + H_2O(l) + CO_2 (g)$

$CaCO_3 (s) + 2 HCl(aq) \rightarrow CaCl_2 (aq) + H_2O(l) + CO_2 (g)$

41.   Mixture   Net ionic equation

(a) $Sr(NO_3)_2 (aq) + K_2SO_4 (aq)$: $Sr^{2+}(aq) + SO_4^{2-}(aq) \rightarrow SrSO_4(s)$

(b) $Mg(NO_3)_2 (aq) + NaOH(aq)$: $Mg^{2+}(aq) + 2 OH^-(aq) \rightarrow Mg(OH)_2(s)$

(c) $BaCl_2 (aq) + K_2SO_4 (aq)$: $Ba^{2+}(aq) + SO_4^{2-}(aq) \rightarrow BaSO_4(s)$

(upon filtering, KCl (aq) is obtained)

42.   Mixture   Net ionic equation

(a) $BaCl_2(aq) + K_2SO_4(aq):$ $Ba^{2+}(aq) + SO_4^{2-}(aq) \rightarrow BaSO_4(s)$

(b) $NaCl(aq) + Ag_2SO_4(s):$ $2 \; AgCl(s) + 2 Na^+(aq) + SO_4^{2-}(aq)$

       $BaCl_2(aq) + Ag_2SO_4(s)$ (a mixed precipitate): $AgCl(s) + BaSO_4(s)$

(c) $Sr(NO_3)_2(aq) + K_2SO_4(aq):$ $Sr^{2+}(aq) + SO_4^{2-}(aq) \rightarrow SrSO_4(s)$

                            (upon filtering, KNO_3 (aq) is obtained)

Acid–Base Reactions

43. (a) $NaHCO_3(s) + H^+(aq) \qquad \rightarrow \; Na^+(aq) + H_2O(l) + CO_2(g)$

(b) $CaCO_3(s) + 2 \; H^+(aq) \qquad \rightarrow \; Ca^{2+}(aq) + H_2O(l) + CO_2(g)$

(c) $Mg(OH)_2(s) + 2 \; H^+(aq) \qquad \rightarrow \; Mg^{2+}(aq) + 2 \; H_2O(l)$

(d) $Mg(OH)_2(s) + 2 \; H^+(aq) \qquad \rightarrow \; Mg^{2+}(aq) + 2 \; H_2O(l)$

       $Al(OH)_3(s) + 3 \; H^+(aq) \qquad \rightarrow \; Al^{3+}(aq) + 3 \; H_2O(l)$

(e) $NaAl(OH)_2 CO_3(s) + 4 \; H^+(aq) \rightarrow \; Al^{3+}(aq) + Na^+(aq) + 3 \; H_2O(l) + CO_2(g)$

44. Dissolve Na_2CrO_4 (s) and $ZnSO_4$ (s) each in H_2O to produce solutions of, respectively, Na^+ and Zn^{2+}. Dissolve each of $BaCO_3$ (s) and $Al(OH)_3$ (s) in $HCl(aq)$ to produce solutions of Ba^{2+} (aq) and Al^{3+} (aq).

45. As a salt: $NaHSO_4(aq) \rightarrow \; Na^+(aq) + HSO_4^-(aq)$

    As an acid: $HSO_4^-(aq) + OH^-(aq) \rightarrow \; H_2O(l) + SO_4^{2-}(aq)$

46. Because all three compounds contain an ammonium cation, all are formed by the reaction of an acid with aqueous ammonia. The identity of the anion determines which acid present.

  $2 \; NH_3(aq) + H_3PO_4(aq) \quad \rightarrow (NH_4)_2 HPO_4(aq)$

  $NH_3(aq) + HNO_3(aq) \qquad \rightarrow NH_4NO_3(aq)$

  $2 \; NH_3(aq) + H_2SO_4(aq) \quad \rightarrow (NH_4)_2 SO_4(aq)$

Oxidation-Reduction (Redox) Equations

47. **(a)** In this reaction, iron is reduced from Fe^{3+} (aq) to Fe^{2+} (aq) *and* manganese is reduced from a +7 O.S. in MnO_4^- (aq) to a +2 O.S. in Mn^{2+} (aq). Thus, there are two reductions and no oxidation, which is an impossibility.

(b) In this reaction, chlorine is oxidized from an O.S. of 0 in Cl_2 (aq) to an O.S. of +1 in ClO^- (aq) *and* oxygen is oxidized from an O.S. of −1 in H_2O_2 (aq) to an O.S. of 0 in O_2 (g). Consequently there are two oxidation reactions and no reduction reactions, also an impossibility.

48. The story is not accurate. The reaction between sodium hydroxide and hydrochloric acid is an acid–base reaction, $NaOH(aq) + HCl(aq) \rightarrow NaCl(aq) + H_2O(l)$. The formation of $Cl_2(g)$ would require an oxidation–reduction reaction.

49. **(a)**
$$2\,H_2O(g) + CH_4(g) \rightarrow CO_2(g) + 8\,H^+(g) + 8\,e^-$$
$$\{2\,e^- + 2\,H^+(g) + NO(g) \rightarrow \tfrac{1}{2}\,N_2(g) + H_2O(g) \quad\}\times 4$$
$$\overline{CH_4(g) + 4\,NO(g) \rightarrow 2\,N_2(g) + CO_2(g) + 2\,H_2O(g)}$$

(b)
$$\{H_2S(g) \rightarrow 1/8\,S_8(s) + 2\,H^+(g) + 2\,e^- \quad\}\times 2$$
$$4\,e^- + 4\,H^+(g) + SO_2(g) \rightarrow 1/8\,S_8(s) + 2\,H_2O(g)$$
$$\overline{2\,H_2S(g) + SO_2(g) \rightarrow 3/8\,S_8(s) + 2\,H_2O(g)} \text{ or}$$
$$16\,H_2S(g) + 8\,SO_2(g) \rightarrow 3\,S_8(s) + 16\,H_2O(g)$$

(c)
$$\{Cl_2O(g) + 2\,NH_4^+(aq) + 2\,H^+(aq) + 4\,e^- \rightarrow 2\,NH_4Cl(s) + H_2O(l) \quad\}\times 3$$
$$\{2\,NH_3(g) \rightarrow N_2(g) + 6\,e^- + 6\,H^+(aq) \quad\}\times 2$$
$$6\,NH_3(g) + 6\,H^+(aq) \rightarrow 6\,NH_4^+(aq)$$
$$\overline{10\,NH_3(g) + 3\,Cl_2O(g) \rightarrow 6\,NH_4Cl(s) + 2\,N_2(g) + 3\,H_2O(l)}$$

50. For the purpose of balancing its redox equation, each of the reactions is treated as if it takes place in acidic aqueous solution.

(a)
$$CH_4(g) + NH_3(g) \rightarrow HCN(g) + 6\,e^- + 6\,H^+$$
$$\{2\,e^- + 2\,H^+(g) + \tfrac{1}{2}\,O_2(g) \rightarrow H_2O(g) \quad\}\times 3$$
$$\overline{CH_4(g) + NH_3(g) + 3/2\,O_2(g) \rightarrow HCN(g) + 3\,H_2O(g)}$$

(b)
$$\{H_2(g) \rightarrow 2\,H^+(aq) + 2\,e^- \quad\}\times 5$$
$$2\,NO(g) + 10\,H^+(aq) + 10\,e^- \rightarrow 2\,NH_3(g) + 2\,H_2O(l)$$
$$\overline{5\,H_2(g) + 2\,NO(g) \rightarrow 2\,NH_3(g) + 2\,H_2O(g)}$$

(c)
$$\{Fe(s) \rightarrow Fe^{3+}(aq) + 3 \ e^- \qquad\qquad\qquad\qquad\}\times 4$$
$$\{4 \ e^- + 2 \ H_2O(l) + O_2(g) \rightarrow 4 \ OH^-(aq) \qquad\qquad\}\times 3$$
$$\overline{4 \ Fe(s) \ + 6 \ H_2O(l) + 3 \ O_2(g) \rightarrow \ 4 \ Fe(OH)_3(s)}$$

51. **(a)** Oxidation: $\{2 \ I^-(aq) \rightarrow \ I_2(s) + 2 \ e^- \qquad\qquad\qquad\qquad\qquad\qquad\}\times 5$

Reduction: $\{MnO_4^-(aq) + 8 \ H^+(aq) + 5 \ e^- \rightarrow \ Mn^{2+}(aq) + 4 \ H_2O(l) \quad\}\times 2$

$\overline{\text{Net: } 10 \ I^-(aq) + 2 \ MnO_4^-(aq) + 16 \ H^+(aq) \rightarrow 5 \ I_2(s) + 2 \ Mn^{2+}(aq) + 8 \ H_2O(l)}$

(b) Oxidation: $\{N_2H_4(l) \rightarrow \ N_2(g) + 4 \ H^+(aq) + 4 \ e^- \qquad\qquad\qquad\qquad\}\times 3$

Reduction: $\{BrO_3^-(aq) + 6 \ H^+(aq) + 6 \ e^- \rightarrow \ Br^-(aq) + 3 \ H_2O(l) \qquad\quad\}\times 2$

$\overline{\text{Net: } \qquad\quad 3 \ N_2H_4(l) + 2 \ BrO_3^-(aq) \rightarrow 3 \ N_2(g) + 2 \ Br^-(aq) + 6 \ H_2O(l)}$

(c) Oxidation: $Fe^{2+}(aq) \rightarrow \ Fe^{3+}(aq) + \ e^-$

Reduction: $VO_4^{3-}(aq) + 6 \ H^+(aq) + \ e^- \rightarrow \ VO^{2+}(aq) + 3 \ H_2O(l)$

$\overline{\text{Net: } \quad Fe^{2+}(aq) + \ VO_4^{3-}(aq) + 6 \ H^+(aq) \rightarrow \ Fe^{3+}(aq) + \ VO^{2+}(aq) + 3 \ H_2O(l)}$

(d) Oxidation: $\{UO^{2+}(aq) + \ H_2O(l) \rightarrow \ UO_2^{2+}(aq) + 2 \ H^+(aq) + 2 \ e^- \quad\}\times 3$

Reduction: $\{NO_3^-(aq) + 4 \ H^+(aq) + 3 \ e^- \rightarrow \ NO(g) + 2 \ H_2O(l) \qquad\quad\}\times 2$

$\overline{\text{Net: } \quad 3 \ UO^{2+}(aq) + 2 \ NO_3^-(aq) + 2 \ H^+(aq) \rightarrow 3 \ UO_2^{2+}(aq) + 2 \ NO(g) + \ H_2O(l)}$

52. **(a)** Oxidation: $\{P_4(s) + 16 \ H_2O(l) \rightarrow 4 \ H_2PO_4^-(aq) + 24 \ H^+(aq) + 20 \ e^-\} \times 3$

Reduction: $\{NO_3^-(aq) + 4 \ H^+(aq) + 3 \ e^- \rightarrow \ NO(g) + 2 \ H_2O(l) \quad\} \times 20$

$\overline{\text{Net: } \quad 3 \ P_4(s) + 20 \ NO_3^-(aq) + 8 \ H_2O + 8 \ H^+(aq) \rightarrow 12 \ H_2PO_4^-(aq) + 20 \ NO(g)}$

(b) Oxidation: $\{S_2O_3^{2-}(aq) + 5 \ H_2O(l) \rightarrow 2 \ SO_4^{2-}(aq) + 10 \ H^+(aq) + 8 \ e^- \quad\} \times 5$

Reduction: $\{MnO_4^-(aq) + 8 \ H^+(aq) + 5 \ e^- \rightarrow \ Mn^{2+}(aq) + 4 \ H_2O(l) \qquad\} \times 8$

$\overline{\text{Net: } \quad 5 \ S_2O_3^{2-}(aq) + 8 \ MnO_4^-(aq) + 14 \ H^+(aq) \rightarrow 10 \ SO_4^{2-}(aq) + 8 \ Mn^{2+}(aq) + 7 \ H_2O(l)}$

(c) Oxidation: $2 \ HS^-(aq) + 3 \ H_2O(l) \rightarrow \ S_2O_3^{2-}(aq) + 8 \ H^+(aq) + 8 \ e^-$

Reduction: $\{2 \ HSO_3^-(aq) + 4 \ H^+(aq) + 4 \ e^- \rightarrow \ S_2O_3^{2-}(aq) + 3 \ H_2O(l) \} \times 2$

$\overline{\text{Net: } \quad 2 \ HS^-(aq) + 4 \ HSO_3^-(aq) \rightarrow 3 \ S_2O_3^{2-}(aq) + 3 \ H_2O(l)}$

(d) Oxidation: $2 \ NH_3OH^+(aq) \rightarrow \ N_2O(g) + \ H_2O(l) + 6 \ H^+(aq) + 4 \ e^-$

Reduction: $\{Fe^{3+}(aq) + \ e^- \rightarrow \ Fe^{2+}(aq) \qquad\qquad\qquad\qquad\qquad\}\times 4$

$\overline{\text{Net: } \quad 4 \ Fe^{3+}(aq) + 2 \ NH_3OH^+(aq) \rightarrow 4 \ Fe^{2+}(aq) + \ N_2O(g) + H_2O(l) + 6 \ H^+(aq)}$

53. **(a)** Oxidation: $\{CN^-(aq) + 2\ OH^-(aq) \rightarrow CNO^-(aq) + H_2O + 2\ e^-$ $\qquad \} \times 3$

Reduction: $\{MnO_4^-(aq) + 2\ H_2O(l) + 3\ e^- \rightarrow MnO_2(s) + 4\ OH^-(aq)$ $\quad \} \times 2$

Net: $3\ CN^-(aq) + 2\ MnO_4^-(aq) + H_2O(l) \rightarrow 3\ CNO^-(aq) + 2\ MnO_2(s) + 2\ OH^-(aq)$

(b) Oxidation: $N_2H_4(l) + 4\ OH^-(aq) \rightarrow N_2(g) + 4\ H_2O(l) + 4\ e^-$

Reduction: $\{[Fe(CN)_6]^{3-}(aq) + e^- \rightarrow [Fe(CN)_6]^{4-}(aq)$ $\qquad \} \times 4$

Net: $4[Fe(CN)_6]^{3-}(aq) + N_2H_4(l) + 4OH^-(aq) \rightarrow 4[Fe(CN)_6]^{4-}(aq) + N_2(g) + 4H_2O(l)$

(c) Oxidation: $\{Fe(OH)_2(s) + OH^-(aq) \rightarrow Fe(OH)_3(s) + e^-$ $\qquad \} \times 4$

Reduction: $O_2(g) + 2\ H_2O(l) + 4\ e^- \rightarrow 4\ OH^-(aq)$

Net: $\qquad 4\ Fe(OH)_2(s) + O_2(g) + 2\ H_2O(l) \rightarrow 4\ Fe(OH)_3(s)$

(d) Oxidation: $\{C_2H_5OH(aq) + 5\ OH^-(aq) \rightarrow C_2H_3O_2^-(aq) + 4\ H_2O(l) + 4\ e^-$ $\quad \} \times 3$

Reduction: $\{MnO_4^-(aq) + 2\ H_2O(l) + 3\ e^- \rightarrow MnO_2(s) + 4\ OH^-(aq)$ $\qquad \} \times 4$

Net: $3\ C_2H_5OH(aq) + 4\ MnO_4^-(aq) \rightarrow 3\ C_2H_3O_2^-(aq) + 4\ MnO_2(s) + OH^-(aq) + 4\ H_2O(l)$

54. **(a)** Oxidation: $Cl_2(g) + 12\ OH^-(aq) \rightarrow 2\ ClO_3^-(aq) + 6\ H_2O(l) + 10\ e^-$

Reduction: $\{Cl_2(g) + 2\ e^- \rightarrow 2\ Cl^-(aq)$ $\qquad \} \times 5$

Net: $\quad 6\ Cl_2(g) + 12\ OH^-(aq) \rightarrow 10\ Cl^-(aq) + 2\ ClO_3^-(aq) + 6\ H_2O(l)$

Or: $\quad 3\ Cl_2(g) + 6\ OH^-(aq) \rightarrow 5\ Cl^-(aq) + ClO_3^-(aq) + 3\ H_2O(l)$

(b) Oxidation: $S_2O_4^{2-}(aq) + 2\ H_2O(l) \rightarrow 2\ HSO_3^-(aq) + 2\ H^+(aq) + 2\ e^-$

Reduction: $S_2O_4^{2-}(aq) + 2\ H^+(aq) + 2\ e^- \rightarrow S_2O_3^{2-}(aq) + H_2O(l)$

Net: $\qquad 2\ S_2O_4^{2-}(aq) + H_2O(l) \rightarrow 2\ HSO_3^-(aq) + S_2O_3^{2-}(aq)$

(c) Oxidation: $\{MnO_4^{2-}(aq) \rightarrow MnO_4^-(aq) + e^-$ $\qquad \} \times 2$

Reduction: $MnO_4^{2-}(aq) + 2\ H_2O(l) + 2\ e^- \rightarrow MnO_2(s) + 4\ OH^-(aq)$

Net: $\quad 3\ MnO_4^{2-}(aq) + 2\ H_2O(l) \rightarrow 2\ MnO_4^-(aq) + MnO_2(s) + 4\ OH^-(aq)$

(d) Oxidation: $\{P_4(s) + 8\ OH^-(aq) \rightarrow 4\ H_2PO_2^-(aq) + 4\ e^-$ $\qquad \} \times 3$

Reduction: $P_4(s) + 12\ H_2O(l) + 12\ e^- \rightarrow 4\ PH_3(g) + 12\ OH^-(aq)$

Net: $\qquad 4\ P_4(s) + 12\ OH^-(aq) + 12\ H_2O(l) \rightarrow 12\ H_2PO_2^-(aq) + 4\ PH_3(g)$

55. **(a)** Oxidation: $S_2O_3^{2-}(aq) + 5\ H_2O(l) \rightarrow 2\ SO_4^{2-}(aq) + 10\ H^+(aq) + 8\ e^-$

Reduction: $\{Cl_2(g) + 2\ e^- \rightarrow 2\ Cl^-(aq)$ $\}\times 4$

Net: $S_2O_3^{2-}(aq) + 5\ H_2O(l) + 4\ Cl_2(g) \rightarrow 2\ SO_4^{2-}(aq) + 8\ Cl^-(aq) + 10\ H^+(aq)$

(b) Oxidation: $\{Sn^{2+}(aq) \rightarrow Sn^{4+}(aq) + 2\ e^-$ $\}\times 3$

Reduction: $Cr_2O_7^{2-}(aq) + 14\ H^+(aq) + 6\ e^- \rightarrow 2\ Cr^{3+}(aq) + 7\ H_2O(l)$

Net: $Cr_2O_7^{2-}(aq) + 14\ H^+(aq) + 3\ Sn^{2+}(aq) \rightarrow 3\ Sn^{4+}(aq) + 2\ Cr^{3+}(aq) + 7\ H_2O(l)$

(c) Oxidation: $S_8(s) + 24\ OH^-(aq) \rightarrow 4\ S_2O_3^{2-}(aq) + 12\ H_2O(l) + 16\ e^-$

Reduction: $S_8(s) + 16\ e^- \rightarrow 8\ S^{2-}(aq)$

Net: $2\ S_8(s) + 24\ OH^-(aq) \rightarrow 8\ S^{2-}(aq) + 4\ S_2O_3^{2-}(aq) + 12\ H_2O(l)$

(d) Oxidation: $As_2S_3(s) + 40\ OH^-(aq) \rightarrow 2\ AsO_4^{3-}(aq) + 3\ SO_4^{2-}(aq) + 20\ H_2O + 28\ e^-$

Reduction: $\{H_2O_2(aq) + 2\ e^- \rightarrow 2\ OH^-(aq)$ $\}\times 14$

Net: $As_2S_3(s) + 12\ OH^-(aq) + 14\ H_2O_2(aq) \rightarrow 2\ AsO_4^{3-}(aq) + 3\ SO_4^{2-}(aq) + 20\ H_2O(l)$

56. **(a)** Oxidation: $\{NO_2^-(aq) + H_2O(l) \rightarrow NO_3^-(aq) + 2\ H^+(aq) + 2\ e^-$ $\}\times 5$

Reduction: $\{MnO_4^-(aq) + 8\ H^+(aq) + 5\ e^- \rightarrow Mn^{2+}(aq) + 4\ H_2O(l)$ $\}\times 2$

Net: $5\ NO_2^-(aq) + 2\ MnO_4^-(aq) + 6\ H^+(aq) \rightarrow 5\ NO_3^-(aq) + 2\ Mn^{2+}(aq) + 3\ H_2O(l)$

(b) Oxidation: $\{Mn^{2+}(aq) + 4\ OH^-(aq) \rightarrow MnO_2(s) + 2\ H_2O(l) + 2\ e^-\} \times 3$

Reduction: $\{MnO_4^-(aq) + 2\ H_2O(l) + 3\ e^- \rightarrow MnO_2(s) + 4\ OH^-(aq)\} \times 2$

Net: $3\ Mn^{2+}(aq) + 2\ MnO_4^-(aq) + 4\ OH^-(aq) \rightarrow 5\ MnO_2(s) + 2\ H_2O(l)$

(c) $2\ Na(s) + 2\ HI(aq) \rightarrow 2NaI(aq) + H_2(g)$

(d) Oxidation: $Zn(s) \rightarrow Zn^{2+}(aq) + 2\ e^-$

Reduction: $\{VO^{2+}(aq) + 2\ H^+(aq) + e^- \rightarrow V^{3+}(aq) + H_2O(l)\} \times 2$

Net: $Zn(s) + 2\,VO^{2+}(aq) + 4\ H^+(aq) \rightarrow Zn^{2+}(aq) + 2\,V^{3+}(aq) + 2\,H_2O(l)$

Oxidizing and Reducing Agents

57. The oxidizing agents experience a decrease in the oxidation state of one of their elements, while the reducing agents experience an increase in the oxidation state of one of their elements.

(a) $SO_3^{2-}(aq)$ is the reducing agent; the O.S. of S $= +4$ in SO_3^{2-} and $= +6$ in SO_4^{2-}.

$MnO_4^{-}(aq)$ is the oxidizing agent; the O.S. of Mn $= +7$ in MnO_4^{-} and $+2$ in Mn^{2+}.

(b) $H_2(g)$ is the reducing agent; the O.S. of H $= 0$ in $H_2(g)$ and $= +1$ in $H_2O(g)$.

$NO_2(g)$ is the oxidizing agent; the O.S. of N $= +4$ in $NO_2(g)$ and -3 in $NH_3(g)$.

(c) $\left[Fe(CN)_6\right]^{4-}(aq)$ is the reducing agent; the O.S. of Fe $= +2$ in $\left[Fe(CN)_6\right]^{4-}$

and $= +3$ in $\left[Fe(CN)_6\right]^{3-}$. $H_2O_2(aq)$ is the oxidizing agent; the O.S. of O $= -1$ in H_2O_2 and $= -2$ in H_2O.

58. (a) $2\ S_2O_3^{2-}(aq) + I_2(s) \rightarrow S_4O_6^{2-}(aq) + 2\ I^{-}(aq)$

(b) $S_2O_3^{2-}(aq) + 4\ Cl_2(g) + 5\ H_2O(l) \rightarrow 2\ HSO_4^{-}(aq) + 8\ Cl^{-}(aq) + 8\ H^{+}(aq)$

(c) $S_2O_3^{2-}(aq) + 4\ OCl^{-}(aq) + 2\ OH^{-}(aq) \rightarrow 2\ SO_4^{2-}(aq) + 4\ Cl^{-}(aq) + H_2O(l)$

Neutralization and Acid–Base Titrations

59. $NaOH(aq) + HCl(aq) \rightarrow NaCl(aq) + H_2O(l)$ is the titration reaction.

$$[NaOH] = \frac{0.02834\,L \times \dfrac{0.1085\,mol\,HCl}{1\,L\,soln} \times \dfrac{1\,mol\,NaOH}{1\,mol\,HCl}}{0.02500\ L\ sample} = 0.1230\ M\ NaOH$$

60. $$[NH_3] = \frac{28.72\ mL\ acid \times \dfrac{1.021\,mmol\,HCl}{1\,mL\,acid} \times \dfrac{1\,mmol\,H^{+}}{1\,mmol\,HCl} \times \dfrac{1\,mmol\,NH_3}{1\,mmol\,H^{+}}}{5.00\ mL\ sample} = 5.86\ M\ NH_3$$

61. The net reaction is $OH^{-}(aq) + HC_3H_5O_2(aq) \rightarrow H_2O(l) + C_3H_5O_2^{-}(aq)$

$$V_{base} = 25.00\ mL\ acid \times \frac{0.3057\ mmol\,HC_3H_5O_2}{1\ mL\ acid} \times \frac{1\ mmol\,KOH}{1\ mmol\,HC_3H_5O_2} \times \frac{1\ mL\ base}{2.155\ mmol\,KOH}$$

$= 3.546\ mL\ KOH\ solution$

62. Titration reaction: $Ba(OH)_2(aq) + 2\ HNO_3(aq) \rightarrow Ba(NO_3)_2(aq) + 2\ H_2O(l)$

$$V_{base} = 50.00\ mL\ acid \times \frac{0.0526\ mol\ HNO_3}{1\ mL\ acid} \times \frac{1\ mmol\ Ba(OH)_2}{2\ mmol\ HNO_3} \times \frac{1\ mL\ base}{0.0844\ mmol\ Ba(OH)_2}$$

$$= 15.6\ mL\ Ba(OH)_2\ solution$$

63. The mass of acetylsalicylic acid is converted to the amount of NaOH, in millimoles, that will react with it.

$$[NaOH] = \frac{0.32\ g\ HC_9H_7O_4}{23\ mL\ NaOH(aq)} \times \frac{1\ mol\ HC_9H_7O_4}{180.2\ g\ HC_9H_7O_4} \times \frac{1\ mol\ NaOH}{1\ mol\ HC_9H_7O_4} \times \frac{1000\ mmol\ NaOH}{1\ mol\ NaOH}$$

$$= 0.077\ M\ NaOH$$

64. **(a)** vol conc. acid $= 20.0\ L \times \dfrac{0.10\ mol\ HCl}{1\ L\ soln} \times \dfrac{36.5\ g\ HCl}{1\ mol\ HCl} \times \dfrac{100\ g\ conc}{38\ g\ HCl} \times \dfrac{1\ mL}{1.19\ g\ conc}$

$$= 1.6 \times 10^2\ mL\ conc.\ acid$$

(b) The titration reaction is $HCl(aq) + NaOH(aq) \rightarrow NaCl(aq) + H_2O(l)$

$$[HCl] = \frac{20.93\ mL\ base \times \dfrac{0.1186\ mmol\ NaOH}{1\ mL\ base} \times \dfrac{1\ mmol\ HCl}{1\ mmol\ NaOH}}{25.00\ mL\ acid} = 0.09929\ M\ HCl$$

(c) First of all, the volume of the dilute solution (20 L) is known at best to a precision of two significant figures. Secondly, HCl is somewhat volatile (we can smell its odor above the solution) and some will likely be lost during the process of preparing the solution.

65. The equation for the reaction is $HNO_3(aq) + KOH(aq) \rightarrow KNO_3(aq) + H_2O(l)$.

This equation shows that equal numbers of moles are needed for a complete reaction. We compute the amount of each reactant.

$$mmol\ HNO_3 = 25.00\ mL\ acid \times \frac{0.132\ mmol\ HNO_3}{1\ mL\ acid} = 3.30\ mmol\ HNO_3$$

$$mmol\ KOH = 10.00\ mL\ acid \times \frac{0.318\ mmol\ KOH}{1\ mL\ base} = 3.18\ mmol\ KOH$$

There is more acid present than base. Thus, the resulting solution is acidic.

66. Compute the amount of acetic acid in the vinegar and the amount of acetic acid needed to react with the sodium carbonate. If there is more than enough acid to react with the solid, the solution will remain acidic.

$$Acetic\ Acid\ in\ vinegar = 125\ mL \times \frac{0.762\ mmol\ HC_2H_3O_2}{1\ mL\ vinegar} = 95.3\ mmol\ HC_2H_3O_2$$

$$Na_2CO_3(s) + 2\ HC_2H_3O_2(aq) \rightarrow 2\ NaC_2H_3O_2(aq) + H_2O(l) + CO_2(g)$$

Acetic acid required for solid:

$$= 7.55 \ g \ Na_2CO_3 \times \frac{1000 \ mmol \ Na_2CO_3}{106.0 \ g \ Na_2CO_3} \times \frac{2 \ mmol \ HC_2H_3O_2}{1 \ mmol \ Na_2CO_3} = 143 \ mmol \ HC_2H_3O_2$$

Clearly there is not enough acetic acid present to react with all of the sodium carbonate. The resulting solution will not be acidic. In fact, the solution will contain only a trace amount of acetic acid ($HC_2H_3O_2$).

67. $V_{base} = 5.00 \ mL \ vinegar \times \dfrac{1.01 \ g \ vinegar}{1 \ mL} \times \dfrac{4.0 \ g \ HC_2H_3O_2}{100.0 \ g \ vinegar} \times \dfrac{1 \ mol \ HC_2H_3O_2}{60.0 \ g \ HC_2H_3O_2}$

$$\times \frac{1 \ mol \ NaOH}{1 \ mol \ HC_2H_3O_2} \times \frac{1 \ L \ base}{0.1000 \ mol \ NaOH} \times \frac{1000 \ mL}{1 \ L} = 34 \ mL \ base$$

68. The titration reaction is $2 \ NaOH(aq) + H_2SO_4(aq) \rightarrow Na_2SO_4(aq) + 2 \ H_2O(l)$

It is most convenient to consider molarity as millimoles per milliliter when solving this problem.

$$[H_2SO_4] = \frac{49.74 \ mL \ base \times \dfrac{0.935 \ mmol \ NaOH}{1 \ mL \ base} \times \dfrac{1 \ mmol \ H_2SO_4}{2 \ mmol \ NaOH}}{5.00 \ mL \ battery \ acid} = 4.65 \ M \ H_2SO_4$$

Thus, the battery acid is *not* sufficiently concentrated.

69. Answer is (d): 120 % of necessary titrant added in titration of NH_3

$$\left.\begin{array}{c} 5 \ NH_3 \\ + \\ 5 \ HCl \end{array}\right\} \begin{array}{c} \text{required for} \\ \text{equivalence} \\ \text{point} \end{array} \quad \longrightarrow \quad \begin{array}{l} 5 \ NH_4^+ + 6 \ Cl^- + H_3O^+ \\ \text{(depicted in question's drawing)} \end{array}$$

$$+$$

$$\left.1 \ HCl \ \right\} \ 20 \ \% \ excess$$

70. **(a)** $H_2O(l) + K^+(aq) + Cl^-(aq)$

(b) $CH_3COOH(aq) + CH_3COO^-(aq) + H_2O(l) + Na^+(aq)$

Stoichiometry of Oxidation–Reduction Reactions

71.

$$[MnO_4^-] = \frac{0.1078 \ g \ As_2O_3 \times \dfrac{1 \ mol \ As_2O_3}{197.84 \ g \ As_2O_3} \times \dfrac{4 \ mol \ MnO_4^-}{5 \ mol \ As_2O_3} \times \dfrac{1 \ mol \ KMnO_4}{1 \ mol \ MnO_4^-}}{22.15 \ mL \times \dfrac{1 \ L}{1000 \ mL}} = 0.01968 \ M \ KMnO_4$$

72. The balanced equation:

$$5\ SO_3^{2-}(aq) + 2\ MnO_4^{-}(aq) + 6\ H^{+}(aq) \rightarrow 5\ SO_4^{2-}(aq) + 2\ Mn^{2+}(aq) + 3\ H_2O(l)$$

$$\left[SO_3^{2-}\right] = \frac{31.46\,mL \times \dfrac{0.02237\,mmol\,KMnO_4}{1\,mL\,soln} \times \dfrac{1\,mmol\,MnO_4^{-}}{1\,mol\,KMnO_4} \times \dfrac{5\,mmol\,SO_3^{2-}}{2\,MnO_4^{-}}}{25.00\ mL\ SO_3^{2-}\ soln} = 0.07038\ M\ SO_3^{2-}$$

73. First, determine the mass of Fe, then the percentage of iron in the ore.

$$mass\ Fe = 28.72\ \ mL \times \frac{1\ L}{1000\ mL} \times \frac{0.05051\ mol\ Cr_2O_7^{2-}}{1\ L\,soln} \times \frac{6\ mol\ Fe^{2+}}{1\ mol\ Cr_2O_7^{2-}} \times \frac{55.85\ g\ Fe}{1\ mol\ Fe^{2+}}$$

$$mass\ Fe = 0.4861\ g\ Fe \qquad\qquad \%\ Fe = \frac{0.4861\,g\,Fe}{0.9132\,g\,ore} \times 100\% = 53.23\%\ Fe$$

74. First balance the equation.

Oxidation: $\{\,Mn^{2+}(aq) + 4\,OH^{-}(aq) \rightarrow MnO_2(s) + 2\,H_2O(l) + 2e^{-} \quad\}\times 3$

Reduction: $\{\,MnO_4^{-}(aq) + 2\,H_2O^{-}(l) + 3e^{-} \rightarrow MnO_2(s) + 4\,OH^{-}(aq)\,\}\times 2$

Net: $\qquad 3\,Mn^{2+}(aq) + 2\,MnO_4^{-}(aq) + 4\,OH^{-}(aq) \rightarrow 5\,MnO_2(s) + 2\,H_2O(l)$

$$\left[Mn^{2+}\right] = \frac{37.21\,mL\ titrant \times \dfrac{0.04162\,mmol\,MnO_4^{-}}{1\,mL\,titrant} \times \dfrac{3\,mmol\,Mn^{2+}}{2\,mmol\,MnO_4^{-}}}{25.00\ mL\ soln} = 0.09292\ M\ Mn^{2+}$$

75. Oxidation: $\{C_2O_4^{2-}(aq) \rightarrow 2\ CO_2(g) + 2\ e^{-} \qquad\qquad\qquad\qquad\ \}\times 5$

Reduction: $\{MnO_4^{-}(aq) + 8\ H^{+}(aq) + 5\ e^{-} \rightarrow \ Mn^{2+}(aq) + 4\ H_2O(l) \quad\}\times 2$

Net: $5\ C_2O_4^{2-}(aq) + 2\ MnO_4^{-}(aq) + 16\ H^{+}(aq) \rightarrow 10\ CO_2(g) + 2\ Mn^{2+}(aq) + 8\ H_2O(l)$

$$mass\ Na_2C_2O_4 = 1.00\ \ L\ satd\ soln \times \frac{1000\ mL}{1\ L} \times \frac{25.8\ mL}{50.0\ mL\ satd\ soln} \times \frac{0.02140\ mol\ KMnO_4}{1000\ mL}$$

$$\times \frac{1\,mol\,MnO_4^{-}}{1\,mol\,KMnO_4} \times \frac{5\,mol\,C_2O_4^{2-}}{2\,mol\,MnO_4^{-}} \times \frac{1\,mol\,Na_2C_2O_4}{1\,mol\,C_2O_4^{2-}} \times \frac{134.0\,g\,Na_2C_2O_4}{1\,mol\,Na_2C_2O_4}$$

$$mass\ Na_2C_2O_4 = 3.70\,g\ Na_2C_2O_4$$

76. Balanced equation:

$$3\ S_2O_4^{2-}(aq) + 2\ CrO_4^{2-}(aq) + 4\ H_2O(l) \rightarrow 6\ SO_3^{2-}(aq) + 2\ Cr(OH)_3(s) + 2\ H^{+}(aq)$$

(a)
$$mass\ Cr(OH)_3 = 100\ L\ soln \times \frac{0.0126\ mol\ CrO_4^{2-}}{1\ L\,soln} \times \frac{2\ mol\ Cr(OH)_3}{2\ mol\ CrO_4^{2-}} \times \frac{103.0\ \ g\ Cr(OH)_3}{1\ mol\ Cr(OH)_3}$$

$$= 130\ g\ Cr(OH)_3 \cong 1.30 \times 10^{2}\,g\ \cong\ 1 \times 10^{2}\,g$$

(b)

$$\text{mass Na}_2\text{S}_2\text{O}_4 = 100.\text{L soln} \times \frac{0.0126 \ \text{mol CrO}_4^{2-}}{1 \ \text{L soln}} \times \frac{3 \ \text{mol S}_2\text{O}_4^{2-}}{2 \ \text{mol CrO}_4^{2-}} \times \frac{1 \ \text{mol Na}_2\text{S}_2\text{O}_4}{1 \ \text{mol S}_2\text{O}_4^{2-}}$$

$$\times \frac{174.1 \ \text{g Na}_2\text{S}_2\text{O}_4}{1 \ \text{mol Na}_2\text{S}_2\text{O}_4} = 329 \ \text{g Na}_2\text{S}_2\text{O}_4 \cong 3.29 \times 10^2 \text{g} \cong 3 \times 10^2 \text{g}$$

FEATURE PROBLEMS

96. From the volume of titrant we can calculate both the amount in moles of NaC_5H_5 and (through its molar mass of 88.08 g/mol) the mass of NaC_5H_5 in a sample. The remaining mass in a sample is that of C_4H_8O (72.11 g/mol) whose amount in moles we calculate. The ratio of the molar amount of C_4H_8O in the sample to the molar amount of NaC_5H_5 is the value of x.

$$\text{moles of NaC}_5\text{H}_5 = 0.01492 \ \text{L} \times \frac{0.1001 \ \text{mol HCl}}{1 \ \text{L soln}} \times \frac{1 \ \text{mol NaOH}}{1 \ \text{mol HCl}} \times \frac{1 \ \text{mol NaC}_5\text{H}_5}{1 \ \text{mol NaOH}}$$

$$= 0.001493 \ \text{mol NaC}_5\text{H}_5$$

$$\text{mass of C}_4\text{H}_8\text{O} = 0.242 \ \text{g sample} - \left(0.001493 \ \text{mol NaC}_5\text{H}_5 \times \frac{88.08 \ \text{g NaC}_5\text{H}_5}{1 \ \text{mol NaC}_5\text{H}_5} \right)$$

$$= 0.110 \ \text{g C}_4\text{H}_8\text{O}$$

$$x = \frac{0.110 \, \text{g C}_4\text{H}_8\text{O} \times \dfrac{1 \, \text{mol C}_4\text{H}_8\text{O}}{72.11 \, \text{g C}_4\text{H}_8\text{O}}}{0.001493 \ \text{mol NaC}_5\text{H}_5} = 1.02$$

For the second sample, parallel calculations give 0.001200 mol NaC_5H_5, 0.093 g C_4H_8, $x = 1.1$. There is rounding error in this second calculation because it is limited to two significant figures. The best answer is from the first run $x \sim 1.02$ or 1. The formula is $NaC_5H_5(THF)_1$.

97. First, we balance the two equations.

Oxidation: $H_2C_2O_4(aq) \rightarrow 2 \ CO_2(g) + 2 \ H^+(aq) + 2 \ e^-$

Reduction: $MnO_2(s) + 4 \ H^+(aq) + 2 \ e^- \rightarrow Mn^{2+}(aq) + 2 \ H_2O(l)$

Net: $H_2C_2O_4(aq) + MnO_2(s) + 2 \ H^+(aq) \rightarrow 2 \ CO_2(g) + Mn^{2+}(aq) + 2 \ H_2O(l)$

Oxidation: $\{ H_2C_2O_4(aq) \rightarrow 2 \ CO_2(g) + 2 \ H^+(aq) + 2 \ e^- \qquad \} \times 5$

Reduction: $\{ MnO_4^-(aq) + 8 \ H^+(aq) + 5 \ e^- \rightarrow Mn^{2+}(aq) + 4 \ H_2O(l) \quad \} \times 2$

Net: $5 \ H_2C_2O_4(aq) + 2 \ MnO_4^-(aq) + 6 \ H^+(aq) \rightarrow 10 \ CO_2(g) + 2 \ Mn^{2+}(aq) + 8 \ H_2O(l)$

Now we determine the mass of the excess oxalic acid.

$$\text{mass } H_2C_2O_4 \cdot 2H_2O = 0.03006\,L \times \frac{0.1000\,\text{mol}\,KMnO_4}{1\,L} \times \frac{1\,\text{mol}\,MnO_4^-}{1\,\text{mol}\,KMnO_4^-} \times \frac{5\,\text{mol}\,H_2C_2O_4}{2\,\text{mol}\,MnO_4^-}$$

$$\times \frac{1\ \text{mol}\,H_2C_2O_4 \cdot 2H_2O}{1\ \text{mol}\,H_2C_2O_4} \times \frac{126.07\ g\,H_2C_2O_4 \cdot 2H_2O}{1\ \text{mol}\,H_2C_2O_4 \cdot 2H_2O}$$

$$= 0.9474\ g\,H_2C_2O_4 \cdot 2H_2O$$

The mass of $H_2C_2O_4 \cdot 2H_2O$ that reacted with MnO_2

$$= 1.651\ g - 0.9474\ g = 0.704\ g\,H_2C_2O_4 \cdot 2H_2O$$

$$\text{mass } MnO_2 = 0.704\ g\,H_2C_2O_4 \cdot 2H_2O \times \frac{1\ \text{mol}\,H_2C_2O_4}{126.07\ g\,H_2C_2O_4 \cdot 2H_2O} \times \frac{1\ \text{mol}\,MnO_2}{1\ \text{mol}\,H_2C_2O_4} \times \frac{86.9\ g\,MnO_2}{1\ \text{mol}\,MnO_2}$$

$$= 0.485\ g\,MnO_2$$

$$\% MnO_2 = \frac{0.485\,g\,MnO_2}{0.533\,g\,\text{sample}} \times 100\% = 91.0\%\ MnO_2$$

98. **(a)**

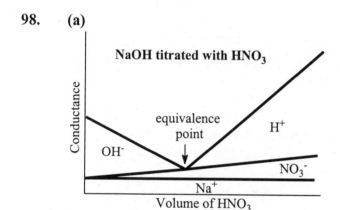

(b)

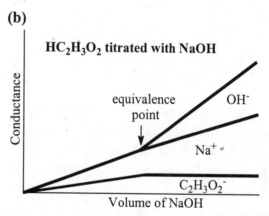

(c)

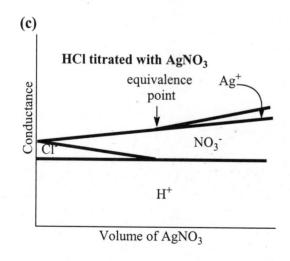

(d)

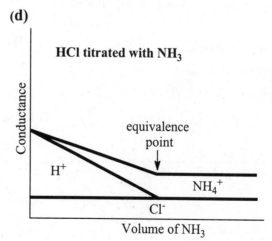

CHAPTER 6
GASES

PRACTICE EXAMPLES

1A   The pressure measured by each liquid must be the same. They are related through $P = g\,h\,d$
Thus, we have the following $g\,h_{DEG}\,d_{DEG} = g\,h_{Hg}\,d_{Hg}$. The g's cancel; we substitute
known values: 9.25 m DEG $\times 1.118$ g/cm^3 DEG $= h_{Hg}\times 13.6$ g/cm^{3}Hg

$$h_{Hg} = 9.25\,\text{m} \times \frac{1.118\,\text{g/cm}^3}{13.6\,\text{g/cm}^3} = 0.760\,\text{m Hg}, \ P = 0.760\,\text{m Hg} = 760.\,\text{mmHg}$$

1B   The solution is found through the expression relating density and height: $h_{TEG}d_{TEG} = h_{Hg}d_{Hg}$
We substitute known values and solve for triethylene glycol's density:

7.39 m TEG $\times d_{TEG} = 757$ mmHg $\times 13.6$ g/cm^3 Hg. Using unit conversions, we get

$$d_{TEG} = \frac{0.757\,\text{m}}{7.39\,\text{m}} \times 13.6\ \text{g/cm}^3 = 1.39\ \text{g/cm}^3$$

2A   We know that $P_{gas} = P_{bar} + \Delta P$ with $P_{bar} = 748.2$ mmHg. We are told that $\Delta P = 7.8$ mmHg.
Thus, $P_{gas} = 748.2$ mmHg $+ 7.8$ mmHg $= 756.0$ mmHg

2B   The difference in pressure between the two levels must be the same just expressed in
different units. Hence, this problem is almost a repetition of Practice Example 6-1.
$h_{Hg} = 748.2$ mmHg $- 739.6$ mmHg $= 8.6$ mmHg. Again we have $g\,h_g\,d_g = g\,h_{Hg}\,d_{Hg}$.
This becomes $h_g \times 1.26$ g/cm^3 glycerol $= 8.6$ mmHg $\times 13.6$ g/cm^3 Hg

$$h_g = 8.6\,\text{mmHg} \times \frac{13.6\,\text{g}/\text{cm}^3\text{Hg}}{1.26\,\text{g}/\text{cm}^3\text{glycerol}} = 93\,\text{mm glycerol}$$

3A $A = \pi r^2$ (here r $= \frac{1}{2}(2.60$ cm $\times \dfrac{1\,\text{m}}{100\,\text{cm}}) = 0.0130$ m)

$A = \pi(0.0130\,\text{m})^2 = 5.31 \times 10^{-4}\,\text{m}^2$

$F = m \times g = (1.000\,\text{kg})(9.81\,\text{m s}^{-2}) = 9.81\,\text{kg m s}^{-2} = 9.81\,\text{N}$

$$P = \frac{F}{A} = \frac{9.81\,\text{N}}{5.31 \times 10^{-4}\,\text{m}^2} = 18475\,\text{N m}^{-2} \ \text{or}\ 1.85 \times 10^4\,\text{Pa}$$

P (torr) $= 1.85 \times 10^4$ Pa $\times = 139$ torr

3B   Final pressure = 100 mb. $100 \text{ mb} \times \dfrac{101,325 \text{ Pa}}{1013.25 \text{ mb}} = 1.000 \times 10^4 \text{ Pa}$

The area of the cylinder is unchanged from that in Example 6-3, $(1.32 \times 10^{-3} \text{ m}^2)$.

$$P = \frac{F}{A} = 1.000 \times 10^4 \text{ Pa} = \frac{F}{1.32 \times 10^{-3} \text{ m}^2}$$

Solving for F, we find $F = 13.2 \text{ (Pa)m}^2 = 13.2 \text{ (N m}^{-2})\text{m}^2 = 13.2 \text{ N}$

$F = m \times g = 13.2 \text{ kg m s}^{-2} = m \times 9.81 \text{ ms}^{-2}$

Total mass = mass of cylinder + mass added weight = $m = \dfrac{F}{g} = \dfrac{13.2 \text{ kg m s}^{-2}}{9.81 \text{ m s}^{-2}} = 1.35 \text{ kg}$

An additional 350 grams must be added to the top of the 1.000 kg (1000 g) red cylinder to increase the pressure to 100 mb. It is not necessary to add a mass with the same cross sectional area. The pressure will only be exerted over the area that is the base of the cylinder on the surface beneath it.

4A   Boyle's Law relates the pressure-volume product. $P_1 V_1 = P_2 V_2$

$$5.25 \text{ atm} \times V_1 = 1.85 \text{ atm} \times 12.5 \text{ L} \quad V_1 = \frac{1.85 \text{ atm} \times 12.5 \text{ L}}{5.25 \text{ atm}} = 4.40 \text{ L}$$

4B   Use Boyle's law, solved for the final pressure. After that, the pressure is converted to mmHg.

$$P_2 = P_1 \times \frac{V_1}{V_2} = 2.25 \text{ atm} \times \frac{1.50 \text{ L}}{8.10 \text{ L}} = 0.417 \text{ atm} \times \frac{760 \text{ mmHg}}{1 \text{ atm}} = 317 \text{ mmHg}$$

5A   Charles's law states that the volume/temperature ratio is constant (temperature in kelvins).

$$\frac{V_1}{T_1} = \frac{V_2}{T_2} = \frac{0.250 \text{ L}}{(25 + 273.15) \text{ K}} = \frac{1.65 \text{ L}}{T_2}; \qquad T_2 = \frac{298 \text{ K} \times 1.65 \text{ L}}{0.250 \text{ L}} = 1.97 \times 10^3 \text{ K} = 1.69 \times 10^3 \text{ °C}$$

5B   For oxygen: The increase in the volume will be directly proportional to the temperature increase (Kelvin scale). Assuming that the pressure remains constant,

$$\frac{V_i}{T_i} = \frac{nR}{P} = \frac{V_f}{T_f} \text{ or } \frac{V_f}{V_i} = \frac{T_f}{T_i} = \frac{(25.0 + 273.15)}{(-13.5 + 273.15)} = 1.15 \text{ (or a 15 \% increase in volume)}$$

For nitrogen: We expect the volume to increase 15% as well.

$V_f = V_i \times 1.15 = 50.5 \text{ mL} \times 1.15 = 58.1 \text{ mL}$

The temperature in Kelvin should also increase by 15%, i.e.

$T_f = T_i \times 1.15 = (33.4 + 273.15) \times 1.15 = 353 \text{ K or } 79 \text{ °C}$

6A The STP molar volume of 22.414 L enables us to determine the amount in moles of propane, from which we find the mass with the use of the molar mass.

$$\text{mass propane} = 30.0 \text{ L} \times \frac{1 \text{ mol}}{22.414 \text{ L}} \times \frac{44.10 \text{ g C}_3\text{H}_8}{1 \text{ mol C}_3\text{H}_8} = 59.0 \text{ g C}_3\text{H}_8$$

6B A gas's STP molar volume is 22.414 L. In addition, we need the molar mass of $CO_2(g)$, 44.01 g/mol.

$$CO_2(g) \text{ volume} = 128 \text{ g} \times \frac{1 \text{ mol CO}_2}{44.01 \text{ g CO}_2} \times \frac{22.414 \text{ L at STP}}{1 \text{ mol CO}_2} = 65.2 \text{ L CO}_2(g) \text{ at STP}$$

7A The ideal gas equation is solved for volume. Conversions are made within the equation.

$$V = \frac{nRT}{P} = \frac{\left(20.2 \text{ g NH}_3 \times \dfrac{1 \text{ mol NH}_3}{17.03 \text{ g NH}_3}\right) \times \dfrac{0.0821 \text{ L} \cdot \text{atm}}{\text{mol} \cdot \text{K}} \times (-25 + 273) \text{K}}{752 \text{ mmHg} \times \dfrac{1 \text{ atm}}{760 \text{ mmHg}}} = 24.4 \text{ L NH}_3$$

7B The amount of $Cl_2(g)$ is 0.193 mol Cl_2 and the pressure is 0.980 atm, as they are in Example 6–7. This information is substituted into the ideal gas equation after it has been solved for temperature.

$$T = \frac{PV}{nR} = \frac{0.980 \text{ atm} \times 7.50 \text{ L}}{0.193 \text{ mol} \times 0.08206 \text{ L atm mol}^{-1} \text{ K}^{-1}} = 464 \text{ K}$$

8A The ideal gas equation is solved for amount and the quantities are substituted.

$$n = \frac{PV}{RT} = \frac{10.5 \text{ atm} \times 5.00 \text{ L}}{\dfrac{0.0821 \text{ L} \cdot \text{atm}}{\text{mol} \cdot \text{K}} \times (30.0 + 273.15) \text{K}} = 2.11 \text{ mol He}$$

8B

$$n = \frac{PV}{RT} = \frac{\left(6.67 \times 10^{-7} \text{ Pa} \times \dfrac{1 \text{ atm}}{101325 \text{ Pa}}\right)\left(3.45 \text{ m}^3 \times \dfrac{1000 \text{ L}}{1 \text{ m}^3}\right)}{(0.08206 \text{ L atm K}^{-1} \text{ mol}^{-1})(25 + 273.15) \text{K}} = 9.28 \times 10^{-10} \text{ moles of N}_2$$

$$\text{molecules of N}_2 = 9.28 \times 10^{-10} \text{ mol N}_2 \times \frac{6.022 \times 10^{23} \text{ molecules of N}_2}{1 \text{ mole N}_2}$$

$$\text{molecules of N}_2 = 5.59 \times 10^{14} \text{ molecules N}_2$$

9A The general gas equation is solved for volume, after the constant amount in moles is cancelled. Temperatures are converted to kelvin.

$$V_2 = \frac{V_1 P_1 T_2}{P_2 T_1} = \frac{1.00 \text{ mL} \times 2.14 \text{ atm} \times (37.8 + 273.2) \text{K}}{1.02 \text{ atm} \times (36.2 + 273.2) \text{K}} = 2.11 \text{ mL}$$

9B The flask has a volume of 1.00 L and initially contains $O_2(g)$ at STP. The mass of $O_2(g)$ that must be released is obtained from the difference in the amount of $O_2(g)$ at the two temperatures, 273 K and 373 K. We also could compute the masses separately and subtract them.

$$\text{mass released} = \left(n_{STP} - n_{100°C}\right) \times M_{O_2} = \left(\frac{PV}{R\,273\,K} - \frac{PV}{R\,373\,K}\right) \times M_{O_2} = \frac{PV}{R}\left(\frac{1}{273\,K} - \frac{1}{373\,K}\right) \times M_{O_2}$$

$$= \frac{1.00\,\text{atm} \times 1.00\,\text{L}}{0.08206\,\text{L atm mol}^{-1}\,\text{K}^{-1}}\left(\frac{1}{273\,K} - \frac{1}{373\,K}\right) \times \frac{32.00\,\text{g}}{1\,\text{mol O}_2} = 0.383\,\text{g O}_2$$

10A The volume of the vessel is 0.09841 L. We substitute other values into the expression for molar mass.

$$M = \frac{mRT}{PV} = \frac{(40.4868\,\text{g} - 40.1305\,\text{g}) \times \dfrac{0.08206\,\text{L} \cdot \text{atm}}{\text{mol} \cdot \text{K}} \times (22.4 \times 273.2)\,\text{K}}{\left(772\,\text{mmHg} \times \dfrac{1\,\text{atm}}{760\,\text{mmHg}}\right) \times 0.09841\,\text{L}} = 86.5\,\text{g / mol}$$

10B The gas's molar mass is its mass (1.27 g) divided by its amount in moles. The amount can be determined from the ideal gas equation.

$$n = \frac{PV}{RT} = \frac{\left(737\,\text{mm Hg} \times \dfrac{1\,\text{atm}}{760\,\text{mm Hg}}\right) \times 1.07\,\text{L}}{\dfrac{0.08206\,\text{L atm}}{\text{mol} \cdot \text{K}} \times (25 + 273)\,\text{K}} = 0.0424\,\text{mol gas}$$

$$M = \frac{1.27\,\text{g}}{0.0424\,\text{mol}} = 30.0\,\text{g/mol} \quad \text{In good aggrement with the molar mass of NO, 30.006 g/mol.}$$

11A The molar mass of He is 4.003 g/mol. This is substituted into the expression for density.

$$d = \frac{MP}{RT} = \frac{4.003\,\text{g mol}^{-1} \times 0.987\,\text{atm}}{0.08206\,\text{L} \cdot \text{atm mol}^{-1}\text{K}^{-1} \times 298\,\text{K}} = 0.162\,\text{g / L}$$

When compared to the density of air under the same conditions (1.16 g/L, based on the "average molar mass of air"=28.8g/mol) the density of He is only about one seventh as much. He is less dense ("lighter") than air.

11B The suggested solution is a simple one; merely solve for the temperature.

$$T = \frac{MP}{Rd} = \frac{\dfrac{32.00\,\text{g O}_2}{1\,\text{mol O}_2}\left(745\,\text{mmHg} \times \dfrac{1\,\text{atm}}{760\,\text{mmHg}}\right)}{\dfrac{0.08206\,\text{L atm}}{\text{mol K}} \times \dfrac{1.00\,\text{g}}{1\,\text{L}}} = 382\,\text{K}$$

However, suppose that you have forgotten the convenient formula for the density of an ideal gas? You can still solve a problem such as this one. The density of 1.00 g/L indicates a mass of 1.00 g of gas in a 1.00-L volume. Of course, the mass of a gas, given its identity (oxygen in this case) enables us to determine the amount in moles of the gas (n in the ideal gas equation). Then, we can solve the ideal gas equation for the desired property, such as temperature, as follows:

$$T = \frac{PV}{nR} = \frac{\left(745\,\text{mmHg} \times \dfrac{1\,\text{atm}}{760\,\text{mmHg}}\right) \times 1.00\,\text{L}}{\left(1.00\,\text{g}\,O_2 \times \dfrac{1\,\text{mol}\,O_2}{32.00\,\text{g}\,O_2}\right) \times \dfrac{0.08206\,\text{L atm}}{\text{mol K}}} = 382\,\text{K}$$

12A The balanced equation is $2\,NaN_3(s) \xrightarrow{\Delta} 2\,Na(l) + 3\,N_2(g)$

$$\text{moles N}_2 = \frac{PV}{RT} = \frac{\left(776\,\text{mmHg} \times \dfrac{1\,\text{atm}}{760\,\text{mmHg}}\right) \times 20.0\,\text{L}}{\dfrac{0.08206\,\text{L atm}}{\text{mol K}} \times (30.0 + 273.2)\,\text{K}} = 0.821\,\text{mol N}_2$$

Now, solve the stoichiometry problem.

$$\text{mass NaN}_3 = 0.821\,\text{mol N}_2 \times \frac{2\,\text{mol NaN}_3}{3\,\text{mol N}_2} \times \frac{65.01\,\text{g NaN}_3}{1\,\text{mol NaN}_3} = 35.6\,\text{g NaN}_3$$

12B Here we are not dealing with gaseous reactants; the law of combining volumes cannot be used. From the ideal gas equation we determine the amount of $N_2(g)$ per liter under the specified conditions. Then we determine the amount of Na(l) produced simultaneously, and finally the mass of that Na(l).

$$\text{mass of Na(l)} = \frac{\left(751\,\text{mmHg} \times \dfrac{1\,\text{atm}}{760\,\text{mmHg}}\right) \times 1.000\,\text{L}}{\dfrac{0.08206\,\text{L atm}}{\text{mol K}} \times (25 + 273)\,\text{K}} \times \frac{2\,\text{mol Na}}{3\,\text{mol N}_2} \times \frac{22.99\,\text{g Na}}{1\,\text{mol Na}} = 0.619\,\text{g Na(l)}$$

13A The law of combining volumes permits us to use stoichiometric coefficients for volume ratios.

$$O_2 \text{ volume} = 1.00\,\text{L NO(g)} \times \frac{5\,\text{L}\,O_2}{4\,\text{L NO}} = 1.25\,\text{L}\,O_2(g)$$

13B The first task is to balance the chemical equation. There must be three moles of hydrogen for every mole of nitrogen in both products (because of the formula of NH_3) and reactants: $N_2(g) + 3\,H_2(g) \rightarrow 2\,NH_3(g)$. The volumes of gaseous reactants and products are related by their stoichiometric coefficients, as long as all gases are measured at the same temperature and pressure.

$$\text{volume } NH_3(g) = 225\,L\,H_2(g) \times \frac{2\,L\,NH_3(g)}{3\,L\,H_2(g)} = 150.\ L\,NH_3$$

14A We can work easily with the ideal gas equation, with a new temperature of $T = (55 + 273)\,K = 328\,K$. The amount of Ne added is readily computed.

$$n_{Ne} = 12.5\,g\,Ne \times \frac{1\,mol\,Ne}{20.18\,g\,Ne} = 0.619\,mol\,Ne$$

$$P = \frac{n_{total}RT}{V} = \frac{(1.75 + 0.619)\,mol \times \dfrac{0.08206\,L\,atm}{mol\,K} \times 328\,K}{5.0\,L} = 13\,atm$$

14B The total volume initially is $2.0\,L + 8.0\,L = 10.0\,L$. These two mixed ideal gases then obey the general gas equation as if they were one gas.

$$P_2 = \frac{P_1V_1T_2}{V_2T_1} = \frac{1.00\,atm \times 10.0\,L \times 298\,K}{2.0\,L \times 273\,K} = 5.5\,atm$$

15A The partial pressures are proportional to the mole fractions.

$$P_{H_2O} = \frac{n_{H_2O}}{n_{tot}} \times P_{tot} = \frac{0.00278\,mol\,H_2O}{0.197\,mol\,CO_2 + 0.00278\,mol\,H_2O} \times 2.50\,atm = 0.0348\,atm\,H_2O(g)$$

$$P_{CO_2} = P_{tot} - P_{H_2O} = 2.50\,atm - 0.0348\,atm = 2.47\,atm\,CO_2(g)$$

15B Expression (6.17) indicates that, in a mixture of gases, the mole percent equals the volume percent, which in turn equals the pressure percent. Thus, we can apply these volume percents—converted to fractions by dividing by 100—directly to the total pressure.

N_2 pressure $= 0.7808 \times 748$ mmHg $= 584$ mmHg,

O_2 pressure $= 0.2095 \times 748$ mmHg $= 157$ mmHg,

CO_2 pressure $= 0.00036 \times 748$ mmHg $= 0.27$ mmHg,

Ar pressure $= 0.0093 \times 748$ mmHg $= 7.0$ mmHg

16A First compute the moles of $H_2(g)$, then use stoichiometry to convert to moles of HCl.

$$\text{amount HCl} = \frac{\left((755-25.2)\,\text{torr} \times \dfrac{1\,\text{atm}}{760\,\text{mmHg}}\right) \times 0.0355\,\text{L}}{\dfrac{0.0821\,\text{L atm}}{\text{mol K}} \times (26+273)\,\text{K}} \times \frac{6\,\text{mol HCl}}{3\,\text{mol H}_2} = 0.00278\,\text{mol HCl}$$

16B The volume occupied by the $O_2(g)$ at its partial pressure is the same as the volume occupied by the mixed gases: water vapor and $O_2(g)$. The partial pressure of $O_2(g)$ is found by difference.

O_2 pressure = 749.2 total pressure $-$ 23.8 mmHg$(H_2O$ pressure$)$ = 725.4 mmHg

The mass of Ag_2O is related to the amount of $O_2(g)$ produced.

$$\text{moles } O_2 = 8.07\,\text{g sample} \times \frac{88.3\,\text{g Ag}_2\text{O}}{100.0\,\text{g sample}} \times \frac{1\,\text{mol Ag}_2\text{O}}{231.74\,\text{g Ag}_2\text{O}} \times \frac{1\,\text{mol O}_2}{2\,\text{mol Ag}_2\text{O}} = 0.0154\,\text{mol O}_2$$

The volume of $O_2(g)$ is found with the ideal gas law.

$$V = \frac{nRT}{P} = \frac{0.0154\,\text{mol O}_2 \times 0.08206 \dfrac{\text{L} \cdot \text{atm}}{\text{mol} \cdot \text{K}} \times (273+25)\,\text{K}}{725.4\,\text{mmHg} \times \dfrac{1\,\text{atm}}{760\,\text{mmHg}}} = 0.395\,\text{L O}_2$$

17A The gas with the smaller molar mass, NH_3 at 17.0 g/mol, has the greater root-mean-square speed

$$u_{\text{rms}} = \sqrt{\frac{3RT}{M}} = \sqrt{\frac{3 \times 8.314\,\text{kg m}^2\,\text{s}^{-2}\,\text{mol}^{-1}\text{K}^{-1} \times 298\,\text{K}}{0.0170\,\text{kg mol}^{-1}}} = 661\,\text{m/s}$$

17B

$$\text{bullet speed } \frac{2180\,\text{mi}}{1\,\text{h}} \times \frac{1\,\text{h}}{3600\,\text{s}} \times \frac{5280\,\text{ft}}{1\,\text{mi}} \times \frac{12\,\text{in.}}{1\,\text{ft}} \times \frac{2.54\,\text{cm}}{1\,\text{in.}} \times \frac{1\,\text{m}}{100\,\text{cm}} = 974.5\,\text{m/s}$$

Solve the rms-speed equation (6.20) for temperature by first squaring both sides.

$$(u_{\text{rms}})^2 = \frac{3RT}{M} \qquad T = \frac{(u_{\text{rms}})^2 M}{3R} = \frac{\left(\dfrac{974.5\,\text{m}}{1\,\text{s}}\right)^2 \times \dfrac{2.016 \times 10^{-3}\,\text{kg}}{1\,\text{mol H}_2}}{3 \times \dfrac{8.3145\,\text{kg m}^2}{\text{s}^2\,\text{mol K}}} = 76.75\,\text{K}$$

We expected the temperature to be lower than 298 K. Note that the speed of the bullet is about half the speed of a H_2 molecule at 298 K. To halve the speed of a molecule, its temperature must be divided by four.

18A The only difference is the molar mass of the gas. 2.2×10^{-4} mol N_2 effuses through the orifice in 105 s.

$$\frac{? \text{ mol O}_2}{2.2 \times 10^{-4} \text{ mol N}_2} = \sqrt{\frac{M_{N_2}}{M_{O_2}}} = \sqrt{\frac{28.014 \text{ g/mol}}{31.999 \text{ g/mol}}} = 0.9357$$

moles $O_2 = 0.9357 \times 2.2 \times 10^{-4} = 2.1 \times 10^{-4}$ mol O_2

18B The two rates of effusion are related as the square root of the ratio of the molar masses of the two gases. The lighter gas, H_2, effuses faster, and thus requires a shorter time for the same amount of gas to effuse.

$$H_2 \text{ time} = N_2 \text{ time} \times \sqrt{\frac{M_{H_2}}{M_{N_2}}} = 105 \text{ s} \times \sqrt{\frac{2.016 \text{ g H}_2/\text{mol H}_2}{28.014 \text{ g N}_2/\text{mol N}_2}} = 28.2 \text{ s}$$

19A The times of effusion are related as the square root of the molar mass. It requires 87.3 s for Kr to effuse.

$$\frac{\text{unknown time}}{\text{Kr time}} = \sqrt{\frac{M_{unk}}{M_{Kr}}} \qquad \text{substitute in values} \qquad \frac{131.3 \text{ s}}{87.3 \text{ s}} = \sqrt{\frac{M_{unk}}{83.80 \text{ g/mol}}} = 1.50$$

$$M_{unk} = (1.504)^2 \times 83.80 \text{ g/mol} = 1.90 \times 10^2 \text{ g/mol}$$

19B This problem is solved in virtually the same manner as Practice Example 18B. The lighter gas is ethane, with a molar mass of 30.07 g/mol.

$$C_2H_6 \text{ time} = \text{Kr time} \times \sqrt{\frac{M(C_2H_6)}{M(Kr)}} = 87.3 \text{ s} \times \sqrt{\frac{30.07 \text{ g C}_2\text{H}_6/\text{mol C}_2\text{H}_6}{83.80 \text{ g Kr/mol Kr}}} = 52.3 \text{ s}$$

20A Because one mole of gas is being considered, the value of n^2a is numerically the same as the value of a, and the value of nb is numerically the same as the value of b.

$$P = \frac{nRT}{V - nb} - \frac{n^2a}{V^2} = \frac{1.00 \text{ mol} \times \dfrac{0.08206 \text{ L atm}}{\text{mol K}} \times 273 \text{ K}}{(2.00 - 0.0427) \text{ L}} - \frac{3.59 \text{ L}^2 \text{ atm}}{(2.00 \text{ L})^2} = 11.4_3 \text{ atm} - 0.898 \text{ atm}$$

$$= 10.5 \text{ atm CO}_2(g) \quad \text{compared with 9.9 atm for Cl}_2(g)$$

$$P_{ideal} = \frac{nRT}{V} = \frac{1.00 \text{ mol} \times \dfrac{0.08206 \text{ L atm}}{\text{mol K}} \times 273 \text{ K}}{2.00 \text{ L}} = 11.2 \text{ atm}$$

$Cl_2(g)$ shows a greater deviation from ideal gas behavior than does $CO_2(g)$.

20B Because one mole of gas is being considered, the value of n^2a is numerically the same as the value of a, and the value of nb is numerically the same as the value of b.

$$P = \frac{nRT}{V-nb} - \frac{n^2a}{V^2} = \frac{1.00 \text{ mol} \times \dfrac{0.08206 \text{ L atm}}{\text{mol K}} \times 273 \text{ K}}{(2.00 - 0.0399) \text{ L}} - \frac{1.49 \text{ L}^2 \text{ atm}}{(2.00 \text{ L})^2} = 11.4_3 \text{ atm} - 0.373 \text{ atm}$$

$$= 11.1 \text{ atm CO}(g)$$

compared to 9.9 atm for $Cl_2(g)$, 11.2 atm for an ideal gas, and 10.5 atm for $CO_2(g)$. Thus, $Cl_2(g)$ displays the greatest deviation from ideality.

REVIEW QUESTIONS

1. **(a)** "atm" is the abbreviation for "atmosphere," a unit of pressure equal to 760 mmHg, 101,325 Pa, or 14.7 $lb/in.^2$.

   (b) "STP" is the abbreviation for "standard temperature and pressure:" $0°C$ and 1 atm pressure.

   (c) R is the symbol for the ideal gas constant. It has a value of 0.08206 L atm $mol^{-1} K^{-1}$.

   (d) Partial pressure is the pressure that one of the gases in a mixture of gases would exert if it were present in the container by itself under the same conditions.

   (e) "u_{rms}" is the abbreviation for the root mean square speed of a number of moving objects, molecules in our considerations. It is the square root of the average of the squares of the speeds. It also is the median speed: half the molecules are traveling faster than this speed, and half are traveling slower.

2. **(a)** The absolute zero of temperature, $-273.15°C$, is the lowest temperature possible. All molecular motion is thought to cease at this temperature.

   (b) A gas is collected over water by bubbling the gas into an upside-down container that is filled with water. The gas rises to the top of the container, displacing the water, and is trapped in the container.

   (c) Effusion of a gas refers to the very slow leakage of the gas out of a small hole in a container and into a vacuum.

   (d) The law of combining volumes states that gases, when present at the same temperature and pressure, react in volumes that are related as small whole numbers. These small whole numbers turn out to be the stoichiometric coefficients.

3. (a) A barometer is a device used to measure absolute pressures; it measures the difference in pressure between some gas and a vacuum. A manometer measures the difference in pressure between two gases.

(b) Celsius and Kelvin temperatures both have the same size degree, large enough that 100 degrees span the temperature range from the boiling point to the freezing point of water. The zero point of Celsius temperature is the freezing point of water; that of Kelvin is $-273.15°C$, absolute zero.

(c) The ideal gas equation relates the properties of pressure, volume, temperature, and amount for one gas. The general gas equation relates the initial and final values of these four properties, or the properties for two gases.

(d) An ideal gas is one that obeys the ideal gas law. On a molecular level, the molecules of such a gas are dimensionless points that exert no forces of attraction or repulsion. The molecules of a real gas occupy some space (but not much compared to the volume of a gas container) and exert weak attractions on each other. At room temperature and pressure the differences in the properties of real and ideal gases are almost insignificant.

4. (a) $P = 736 \text{ mmHg} \times \dfrac{1 \text{ atm}}{760 \text{ mmHg}} = 0.968 \text{ atm}$

(b) $P = 58.2 \text{ cm Hg} \times \dfrac{1 \text{ atm}}{76 \text{ cmHg}} = 0.766 \text{ atm}$

(c) $P = 892 \text{ torr} \times \dfrac{1 \text{ atm}}{760 \text{ torr}} = 1.17 \text{ atm}$

(d) $P = 225 \text{ kPa} \times \dfrac{1000 \text{ Pa}}{1 \text{ kPa}} \times \dfrac{1 \text{ atm}}{101,325 \text{ Pa}} = 2.22 \text{ atm}$

5. (a) $h = 0.984 \text{ atm} \times \dfrac{760 \text{ mmHg}}{1 \text{ atm}} = 748 \text{ mmHg}$

(b) $h = 928 \text{ torr} = 928 \text{ mmHg}$

(c) $h = 142 \text{ ft } H_2O \times \dfrac{12 \text{ in}}{1 \text{ ft}} \times \dfrac{2.54 \text{ cm}}{1 \text{ in}} \times \dfrac{10 \text{ mmH}_2O}{1 \text{ cm H}_2O} \times \dfrac{1 \text{ mmHg}}{13.6 \text{ mmH}_2O} \times \dfrac{1 \text{ mHg}}{1000 \text{ mmHg}} = 3.18 \text{ mHg}$

6. The atmospheric pressure is less than the pressure of the gas.
The difference in pressures is $\Delta P = 16.5 \text{ mmHg} - 7.9 \text{ mmHg} = 8.6 \text{ mmHg}$
$P_{gas} = P_{atm} + \Delta P = 744 \text{ mmHg} + 8.6 \text{ mmHg} = 753 \text{ mmHg}$

7. (a) $V = 26.7 \text{ L} \times \dfrac{762 \text{ mmHg}}{385 \text{ mmHg}} = 52.8 \text{ L}$

(b) $V = 26.7 \text{ L} \times \dfrac{762 \text{ mmHg}}{3.68 \text{ atm} \times \dfrac{760 \text{ mmHg}}{1 \text{ atm}}} = 7.27 \text{ L}$

8. Apply Charles's law: $V = kT$.　　$T_i = 26 + 273 = 299 \text{ K}$

　(a) $T = 273 + 98 = 371 \text{ K}$　　　$V = 886 \text{ mL} \times \dfrac{371 \text{ K}}{299 \text{ K}} = 1.10 \times 10^3 \text{ mL}$

　(b) $T = 273 - 20 = 253 \text{ K}$　　　$V = 886 \text{ mL} \times \dfrac{253 \text{ K}}{299 \text{ K}} = 7.50 \times 10^2 \text{ mL}$

9. Apply Charles's law. $T_f = (22 + 273) \text{ K} \times \dfrac{165 \text{ mL}}{57.3 \text{ mL}} = 849 \text{ K} = 576°\text{C}$

10. $V = n \times V_m = \left(49.6 \text{ g C}_2\text{H}_2 \times \dfrac{1 \text{ mol C}_2\text{H}_2}{26.04 \text{ g}} \right) \times \dfrac{22.414 \text{ L at STP}}{1 \text{ mol}} = 42.7 \text{ L C}_2\text{H}_2$

11. $V = n \times V_m = \left(250.0 \text{ g Cl}_2 \times \dfrac{1 \text{ mol Cl}_2}{70.906 \text{ g}} \right) \times \dfrac{22.414 \text{ L at STP}}{1 \text{ mol}} = 79.03 \text{ L Cl}_2$

12. The gas with the greatest density at STP is the one with the highest molar mass: $\text{Cl}_2 = 70.9 \text{ g/mol}$, $\text{SO}_3 = 80.1 \text{ g/mol}$, $\text{N}_2\text{O} = 44.0 \text{ g/mol}$; and $\text{PF}_3 = 88.0$ g/mol. Thus, PF_3 would have the highest STP density of the four gases listed.

13. Assume that the $\text{CO}_2(g)$ behaves ideally and use the ideal gas law: $PV = nRT$

$$V = \frac{nRT}{P} = \frac{\left(89.2 \text{ g} \times \dfrac{1 \text{ mol CO}_2}{44.01 \text{ g}} \right) 0.08206 \dfrac{\text{L atm}}{\text{mol K}} (37 + 273.2) \text{ K}}{737 \text{ mmHg} \times \dfrac{1 \text{ atm}}{760 \text{ mmHg}}} \times \frac{1000 \text{ mL}}{1 \text{ L}} = 5.32 \times 10^4 \text{ mL}$$

14.

$$P = \frac{nRT}{V} = \frac{\left(285 \text{ g} \times \dfrac{1 \text{ mol SO}_2}{64.07 \text{ g}} \right) 0.08206 \dfrac{\text{L atm}}{\text{mol K}} (27 + 273.2)\text{K}}{40.0 \text{ L}} = 2.74 \text{ atm}$$

15. Use the ideal gas law to determine the amount in moles of the given quantity of gas.

$$n = \frac{PV}{RT} = \frac{\left(743 \text{ mmHg} \times \dfrac{1 \text{ atm}}{760 \text{ mmHg}} \right)\left(115 \text{ mL} \times \dfrac{1 \text{ L}}{1000 \text{ mL}} \right)}{0.08206 \dfrac{\text{L atm}}{\text{mol K}} (273.2 + 66.3) \text{ K}} = 0.00404 \text{ mol gas}$$

$$M = \frac{0.418 \text{ g}}{0.00404 \text{ mol}} = 103 \text{ g} / \text{mol}$$

16. Density, $d(\text{g} / \text{L}) = $ molar mass, $M(\text{g/mol}) \div$ molar volume, V/n (L/mol)

$$V/n = \frac{RT}{P} = \frac{0.08206 \dfrac{\text{L atm}}{\text{mol K}} \times (32.7 + 273.2) \text{ K}}{758 \text{ mmHg} \times \dfrac{1 \text{ atm}}{760 \text{ mmHg}}} = 25.2 \frac{\text{L}}{\text{mol}} \qquad d = \frac{44.01 \dfrac{\text{g}}{\text{mol}}}{25.2 \dfrac{\text{L}}{\text{mol}}} = 1.75 \frac{\text{g}}{\text{L}}$$

17. Each mole of gas occupies 22.4 L at STP.

$$\text{H}_2 \text{ STP volume} = 1.000 \text{ g Al} \times \frac{1 \text{ mol Al}}{26.98 \text{ g Al}} \times \frac{3 \text{ mol H}_2(\text{g})}{2 \text{ mol Al(s)}} \times \frac{22.4 \text{ L H}_2(\text{g}) \text{ at STP}}{1 \text{ mol H}_2}$$

$$= 1.25 \text{ L H}_2(\text{g})$$

18. Determine first the amount of $CO_2(\text{g})$ that can be removed. Then use the ideal gas law.

$$\text{mol CO}_2 = 1.00 \text{ kg LiOH} \times \frac{1000 \text{ g}}{1 \text{ kg}} \times \frac{1 \text{ mol LiOH}}{23.95 \text{ g LiOH}} \times \frac{1 \text{ mol CO}_2}{2 \text{ mol LiOH}} = 20.9 \text{ mol CO}_2$$

$$V = \frac{nRT}{P} = \frac{20.9 \text{ mol} \times 0.08206 \dfrac{\text{L atm}}{\text{mol K}} \times (25.9 + 273.2) \text{ K}}{751 \text{ mmHg} \times \dfrac{1 \text{ atm}}{760 \text{ mmHg}}} = 519 \text{ L CO}_2(\text{g})$$

19. Determine the total amount of gas; then use the ideal gas law, assuming that the gases behave ideally.

$$\text{moles gas} = \left(15.2 \text{ g Ne} \times \frac{1 \text{ mol Ne}}{20.18 \text{ g Ne}} \right) + \left(34.8 \text{ g Ar} \times \frac{1 \text{ mol Ar}}{39.95 \text{ g Ar}} \right)$$

$$= 0.753 \text{ mol Ne} + 0.871 \text{ mol Ar} = 1.624 \text{ mol gas}$$

$$V = \frac{nRT}{P} = \frac{1.624 \text{ mol} \times 0.08206 \dfrac{\text{L atm}}{\text{mol K}} \times (26.7 + 273.2) \text{ K}}{7.15 \text{ atm}} = 5.59 \text{ L gas}$$

20. 2.24 L $H_2(\text{g})$ at STP is 0.100 mol $H_2(\text{g})$. After 0.10 mol He is added, the container holds 0.20 mol gas.

$$V = \frac{nRT}{P} = \frac{0.20 \text{ mol} \times 0.0821 \dfrac{\text{L atm}}{\text{mol K}} (273 + 100) \text{K}}{1.00 \text{ atm}} = 6.1 \text{ L gas}$$

21. **(a)** The total pressure is the sum of all of the partial pressures of $O_2(g)$ and the vapor pressure of water.

$$P_{total} = P_{O_2} + P_{H_2O} = 756 \text{ mmHg} = P_{O_2} + 19 \text{ mmHg}$$
$$P_{O_2} = (756 - 19)\text{mmHg} = 737 \text{ mmHg}$$

(b) The volume percent is equal to the pressure percent.

$$\%O_2(g) \text{ by volume} = \frac{V_{O_2}}{V_{total}} \times 100\% = \frac{P_{O_2}}{P_{total}} \times 100\% = \frac{737 \text{ mmHg of } O_2}{756 \text{ mm Hg total}} \times 100\% = 97.5\%$$

(c) Determine the mass of $O_2(g)$ collected by multiplying the amount of O_2 collected, in moles, by the molar mass of $O_2(g)$.

$$\text{mass } O_2 = \frac{PV}{RT}M = \frac{\left(737. \text{ mmHg} \times \dfrac{1 \text{ atm}}{760 \text{ mmHg}}\right)\left(89.3 \text{ mL} \times \dfrac{1 \text{ L}}{1000 \text{ mL}}\right)}{0.08206 \dfrac{\text{L atm}}{\text{mol K}}(21.3 + 273.2)\text{K}} \times \frac{32.00 \text{ g } O_2}{1 \text{ mol } O_2}$$

$$\text{mass } O_2 = 0.115 \text{ g } O_2$$

22. (1) is the true statement; average molecular kinetic energy depends only on absolute temperature, which is the same for these two gases. (2) is incorrect, since average molecular speed depends also on molar mass, which is different for these two gases. (3) also is incorrect, since volumes at the same temperature and pressure depend on the amount in moles, which is different for these two gases. And (4) is incorrect; the effusion rates at the same temperature and pressure depends inversely on the square root of the molar masses, which differ for these two gases.

23. Effusion time is proportional to the square root of molar mass.

$$\frac{\text{effusion time for NO}}{\text{effusion time for Cl}_2} = \sqrt{\frac{30.01 \text{ g} / \text{mol NO}}{70.91 \text{ g} / \text{mol Cl}_2}} = 0.6505 = \frac{\text{NO effusion time}}{28.6 \text{ s}}$$

NO effusion time $= 0.6505 \times 28.6 \text{ s} = 18.6 \text{ s}$.

24. The best choice for ideal behavior is (3), 200°C and 0.50 atm. Gases are closest to ideal at high temperatures where the molecules move rapidly, and low pressures where molecules are relatively far apart.

EXERCISES

Pressure and Its Measurement

25. We use: $h_{bnz}d_{bnz} = h_{Hg}d_{Hg}$

$$h_{bnz} = 0.970 \text{ atm} \times \frac{0.760 \text{ m Hg}}{1 \text{ atm}} \times \frac{13.6 \text{ g/cm}^3 \text{ Hg}}{0.879 \text{ g/cm}^3 \text{ benzene}} = 11.4 \text{ m benzene}$$

26. We use: $h_{gly}d_{gly} = h_{CCl_4}d_{CCl_4}$

$$h_{gly} = 3.02 \text{ m CCl}_4 \times \frac{1.59 \text{ g/cm}^3 \text{ CCl}_4}{1.26 \text{ g/cm}^3 \text{ glycerol}} = 3.81 \text{ m glycerol}$$

27. The mercury level difference equals the difference in pressure between the container and the atmosphere. This mercury level difference is
$\Delta P = 276 \text{ mmHg} - 49 \text{ mmHg} = 227 \text{ mmHg}$. Since the mercury level in the arm open to the atmosphere is higher than in the arm connected to the container, the pressure in the container is higher than atmospheric pressure.
$P = \Delta P + P_{atm} = 227 \text{ mmHg} + 749 \text{ mmHg} = 976 \text{ mmHg}$

28. The gas inside the container is at a lower pressure than the barometric pressure.

$$\Delta P = P_{bar} - P_{gas} = 4.5 \text{ cmH}_2\text{O} \times \frac{10 \text{ mm H}_2\text{O}}{1 \text{ cm H}_2\text{O}} \times \frac{1 \text{ mmHg}}{13.6 \text{ mm H}_2\text{O}} = 3.3 \text{ mmHg}$$

$P_{gas} = 756.2 \text{ mmHg} - 3.3 \text{ mmHg} = 752.9 \text{ mmHg}$

29. $F = m \times g$ and $1 \text{ atm} = 101325 \text{ Pa} = 101325 \text{ kg m}^{-1} \text{ s}^{-2} = P = \dfrac{F}{A} = \dfrac{m \times 9.81 \text{ m s}^{-2}}{1 \text{ m}^2}$

$$\text{mass (per m}^2) = \frac{101325 \text{ kg m}^{-1}\text{s}^{-2} \times 1 \text{ m}^2}{9.81 \text{ m s}^{-2}} = 10329 \text{ kg}$$

(Note:$1 \text{ m}^2 = (100 \text{ cm})^2 = 10,000 \text{ cm}^2$)

$$P \text{ (kg cm}^{-2}) = \frac{m}{A} = \frac{10329 \text{ kg}}{10,000 \text{ cm}^2} = 1.03 \text{ kg cm}^{-2}$$

30. $\dfrac{1.03 \text{ kg}}{1 \text{ cm}^2} \times \dfrac{(2.54 \text{ cm})^2}{(1 \text{ in})^2} \times \dfrac{2.2046 \text{ lb}}{1 \text{ kg}} = 14.6 \dfrac{\text{lb}}{\text{in}^2}$

The Simple Gas Laws

31. $P_i = P_f \times \dfrac{V_f}{V_i} = \left(721 \text{ mmHg} \times \dfrac{35.8 \text{ L} + 1875 \text{ L}}{35.8 \text{L}}\right) \times \dfrac{1 \text{ atm}}{760 \text{ mm H}_2\text{O}} = 50.6 \text{ atm}$

32. We let P represent barometric pressure, and solve the Boyle's law expression below for P.

$P \times 42.0 \text{ mL} = (P + 85 \text{ mmHg}) \times 37.7 \text{ mL}$ $\qquad$ $42.0P = 37.7P + 3.2 \times 10^3$

$P = \dfrac{3.2 \times 10^3}{42.0 - 37.7} = 7.4 \times 10^2 \text{ mmHg}$

33. Assuming pressure and moles of gas is kept constant, volume is directly proportional to the temperature in kelvins (Charles' Law).

$1 \text{ °C} \rightarrow 2 \text{ °C}$ $\quad$ Volume increase $= \dfrac{V_f}{V_i} = \dfrac{T_f}{T_i} = \dfrac{(2 + 273.15)}{(1 + 273.15)} = 1.0036\underline{5}$ (increase of 0.37 %)

$10 \text{ °C} \rightarrow 20 \text{ °C}$ $\quad$ Volume increase $= \dfrac{(20 + 273.15)}{(10 + 273.15)} = 1.035$ (increase of 3.5 %)

Volume doubles when the temperature in Kelvin doubles (i.e. $273 \text{ K} \rightarrow 546 \text{ K}$)

34. Convert 71 °F to Celsius then to a temperature on the Kelvin scale.

$(71 - 32) \times \dfrac{5}{9} = 22 \text{ °C} + 273.15 = 295 \text{ K}$ $\qquad$ $-196 \text{ °C} = (-196 + 273.15) = 77 \text{ K}$

Volume decrease is proportional to the ratio of the temperatures in Kelvin

Volume contraction $= \dfrac{77 \text{ K}}{295 \text{ K}} = 0.26$ (volume will be about ¼ of its original volume).

35. (a) $\text{mass} = 27.6 \text{ mL} \times \dfrac{1 \text{ L}}{1000 \text{ mL}} \times \dfrac{1 \text{ mol}}{22.414 \text{ L STP}} = 0.00123 \text{ mol PH}_3 \times \dfrac{34.0 \text{ g PH}_3}{1 \text{ mol PH}_3} \times \dfrac{1000 \text{ mg}}{1 \text{ g}}$

$= 41.8 \text{ mg PH}_3$

(b) $\text{number of molecules of PH}_3 = 0.00123 \text{ mol PH}_3 \times \dfrac{6.022 \times 10^{23} \text{ molecules}}{1 \text{ mol PH}_3}$

$\text{number of molecules of PH}_3 = 7.41 \times 10^{20} \text{ molecules}$

36. (a) $\text{mass} = 5.0 \times 10^{17} \text{ atoms} \times \dfrac{1 \text{ mol Rn}}{6.022 \times 10^{23} \text{ atoms}} = 8.30 \times 10^{-7} \text{ mol} \times \dfrac{222 \text{ g Rn}}{1 \text{ mol}} \times \dfrac{10^6 \text{ } \mu\text{g}}{1 \text{ g}}$

$\text{mass} = 1.8 \times 10^2 \text{ } \mu\text{g Rn(g)}$

(b)　　$volume = 8.30 \times 10^{-7} \ mol \times \dfrac{22.414 \ L}{1 \ mol} \times \dfrac{10^{6} \ \mu L}{1 \ L} = 19 \ \mu L \ Rn(g)$

37.　At the higher elevation of the mountains, the atmospheric pressure is lower than at the beach. However, the bag is leak proof; no gas escapes. Thus, the gas inside the bag expands in the lower pressure until the bag is filled to nearly bursting. (It would have been difficult to predict this result. The temperature in the mountains is usually lower than at the beach. The lower temperature would *decrease* the pressure of the gas.)

38.　Based on densities, $1 \ mHg = 13.6 \ mH_2O$. For 30 m of water,

$$30 \ mH_2O \times \frac{1 \ mHg}{13.6 \ mH_2O} = 2.2 \ mHg \quad P_{water} = 2.2 \ mHg \times \frac{1000 \ mm}{1 \ m} \times \frac{1 \ atm}{760 \ mmHg} = 2.9 \ atm$$

To this we add the pressure of the atmosphere above the water:

$P_{total} = 2.9 \ atm + 1.0 \ atm = 3.9 \ atm$.

When the diver rises to the surface, she rises to a pressure of 1.0 atm. Since the pressure is about one fourth of the pressure below the surface, the gas in her lungs attempts to expand to four times the volume of her lungs. It is possible that her lungs would burst.

General Gas Equation

39.　Because the number of moles of gas does not change, $\dfrac{P_i \times V_i}{T_i} = nR = \dfrac{P_f \times V_f}{T_f}$ is obtained

from the ideal gas equation. This expression can be rearranged as follows.

$$V_f = \frac{V_i \times P_i \times T_f}{P_f \times T_i} = \frac{4.25 \ L \times 748 \ mmHg \times (273.2 + 26.8) K}{742 \ mmHg \times (273.2 + 25.6) K} = 4.30 \ L$$

40.　We first compute the pressure at $25°C$ as a result of the additional gas. Of course, the gas pressure increases proportionally to the increase in the mass of gas, because the mass is proportional to the number of moles of gas. $P = \dfrac{12.5 \ g}{10.0 \ g} \times 762 \ mmHg = 953 \ mmHg$.

Now we compute the pressure resulting from increasing the temperature.

$$P = \frac{(62 + 273) \ K}{(25 + 273) \ K} \times 953 \ mmHg = 1.07 \times 10^{3} \ mmHg$$

41.　Volume and Pressure are constant hence: $n_i T_i = \dfrac{PV}{R} = n_f T_f$

$$\frac{n_f}{n_i} = \frac{T_i}{T_f} = \frac{(21 + 273.15)}{(210 + 273.15)} = 0.609 \quad (60.9 \ \% \text{ of the gas remains})$$

Hence, 39.1% of the gas must be released: Mass of gas released $= 12.5 \ g \times \dfrac{39.1}{100} = 4.89 \ g$

42.   First determine the mass of O_2 in the cylinder under the final conditions.

$$\text{mass } O_2 = n \times M = \frac{PV}{RT}M = \frac{1.15 \text{ atm} \times 34.0 \text{ L} \times 32.0 \text{ g / mol}}{0.08206 \dfrac{\text{L atm}}{\text{mol K}}(22+273)\text{K}} = 51.7 \text{ g } O_2$$

mass of O_2 to be released $= 305 \text{ g} - 51.7 \text{ g} = 253 \text{ g } O_2$

Ideal Gas Equation

43. $$P = \frac{nRT}{V} = \frac{\left(35.8 \text{ g } O_2 \times \dfrac{1 \text{ mol } O_2}{32.00 \text{ g } O_2}\right) \times \dfrac{0.08206 \text{ L atm}}{\text{mol K}} \times (46+273.2) \text{ K}}{12.8 \text{ L}} = 2.29 \text{ atm}$$

44. $$\text{mass} = n \times M = \frac{PV}{RT}M = \frac{11.2 \text{ atm} \times 18.5 \text{ L} \times 83.80 \text{ g / mol}}{0.08206 \dfrac{\text{L atm}}{\text{mol K}}(28.2+273.2)\text{K}} = 702 \text{ g Kr}$$

45. $$T = \frac{PV}{nR} = \frac{3.50 \text{ atm} \times 72.8 \text{ L}}{1.85 \text{ mol} \times 0.08206 \dfrac{\text{L atm}}{\text{mol K}}} = 1.68 \times 10^3 \text{K}$$

$$t(^\circ\text{C}) = 1.68 \times 10^3 - 273 = 1.41 \times 10^3 \, ^\circ\text{C}$$

46.   Volume of tank = height × circular area of base

$$\text{Volume} = h \times \pi r^2 = 1.75 \text{ m} \times \pi\left(12.5 \times \frac{1 \text{ m}}{100 \text{ cm}}\right)^2 = 0.0859 \text{ m}^3$$

$1 \text{ m}^3 = 1000 \text{ L}$ Therefore the volume of the tank is 85.9 L

$$P = \frac{nRT}{V} = \frac{(1242 \text{ g CO} \times \dfrac{1 \text{ mol CO}}{28.01 \text{ g CO}})(0.08206 \text{ L atm K}^{-1} \text{ mol}^{-1})(-25 + 273.15)}{85.9 \text{ L}} = 10.5 \text{ atm}$$

$$P(\text{in Pa}) = 10.5 \text{ atm} \times \frac{101325 \text{ Pa}}{1 \text{ atm}} = 1.07 \times 10^6 \text{ Pa}$$

Determining Molar Mass

47.   We first determine the empirical formula of propylene.

$$\text{moles C} = 85.63 \text{ g C} \times \frac{1 \text{ mol C}}{12.01 \text{ g C}} = 7.130 \text{ mol C} \qquad \div 7.130 \rightarrow 1.000 \text{ mol C}$$

$$\text{moles H} = 14.37 \text{ g H} \times \frac{1 \text{ mol H}}{1.008 \text{ g H}} = 14.26 \text{ mol H} \qquad \div 7.130 \rightarrow 2.000 \text{ mol H}$$

The empirical formula is CH_2 and the empirical molar mass is 14.0 g/mol. The molar mass of propylene is 42.08 g/mol, three times the empirical molar mass. The molecular formula is C_3H_6, three times the empirical formula.

48. We first determine the molar mass of the gas.

$$T = 24.3 + 273.2 = 297.5 \text{ K} \qquad\qquad P = 742 \text{ mmHg} \times \frac{1 \text{ atm}}{760 \text{ mmHg}} = 0.976 \text{ atm}$$

$$M = \frac{mRT}{PV} = \frac{2.650 \text{ g} \times 0.08206 \text{ L atm mol}^{-1}\text{K}^{-1} \times 297.5 \text{ K}}{0.976 \text{ atm} \times \left(428 \text{ mL} \times \dfrac{1 \text{ L}}{1000 \text{ mL}}\right)} = 155 \text{ g / mol}$$

Then we determine the empirical formula of the gas, basing our calculations on a 100.0-g sample.

$$\text{mol C} = 15.5 \text{ g C} \times \frac{1 \text{ mol C}}{12.01 \text{ g C}} = 1.29 \text{ mol C} \div 0.649 \rightarrow 1.99 \text{ mol C}$$

$$\text{mol Cl} = 23.0 \text{ g Cl} \times \frac{1 \text{ mol Cl}}{35.45 \text{ g Cl}} = 0.649 \text{ mol Cl} \div 0.649 \rightarrow 1.00 \text{ mol Cl}$$

$$\text{mol F} = 61.5 \text{ g F} \times \frac{1 \text{ mol F}}{19.00 \text{ g F}} = 3.24 \text{ mol F} \div 0.649 \rightarrow 4.99 \text{ mol F}$$

Thus, the empirical formula is C_2ClF_5, which has an empirical molar mass of 154.5 g/mol. This is the same as the experimentally determined molar mass. Hence, the molecular formula is C_2ClF_5.

49. (a) $$M = \frac{mRT}{PV} = \frac{0.231 \text{ g} \times 0.08206 \dfrac{\text{L atm}}{\text{mol K}} \times (23 + 273) \text{ K}}{\left(749 \text{ mmHg} \times \dfrac{1 \text{ atm}}{760 \text{ mmHg}}\right) \times \left(102 \text{ mL} \times \dfrac{1 \text{ L}}{1000 \text{ mL}}\right)} = 55.8 \text{ g/mol}$$

(b) The formula contains 4 atoms of carbon. (5 atoms of carbon gives a molar mass of at least 60-too high-and 3 C atoms gives a molar mass of 36-too low to be made up by adding H's.) To produce a molar mass of 56 with 4 carbons requires the inclusion of 8 atoms of H in the formula of the compound: C_4H_8.

50. (a) First, we obtain the mass of acetylene, and then acetylene's molar mass.

mass of acetylene $= 56.2445 \text{ g} - 56.1035 \text{ g} = 0.1410 \text{ g}$ acetylene

$$M = \frac{mRT}{PV} = \frac{0.1410 \text{ g} \times 0.082057 \dfrac{\text{L atm}}{\text{mol K}} \times (20.02 + 273.15) \text{ K}}{\left(749.3 \text{ mmHg} \times \dfrac{1 \text{ atm}}{760 \text{ mmHg}}\right) \times \left(132.10 \text{ mL} \times \dfrac{1 \text{ L}}{1000 \text{ mL}}\right)} = 26.04 \text{ g/mol}$$

(b)   The formula contains 2 atoms of carbon (3 atoms of carbon gives a molar mass of at least 36-too high and 1C atom gives a molar mass of 12-too low to be made up by adding H's.) To produce a molar mass of 26 with 2 carbons requires the inclusion of 2 atoms of H in the formula for acetylene: C_2H_2.

Gas Densities

51. $d = \dfrac{MP}{RT} \qquad \rightarrow \qquad P = \dfrac{dRT}{M} = \dfrac{1.80 \text{ g/L} \times 0.08206 \dfrac{\text{L atm}}{\text{mol K}} \times (32 + 273) \text{ K}}{28.0 \text{ g/mol}} \times \dfrac{760 \text{ mmHg}}{1 \text{ atm}}$

$P = 1.22 \times 10^3 \text{ mmHg}$

52. $M = \dfrac{dRT}{M} = \dfrac{2.56 \text{ g/L} \times 0.08206 \dfrac{\text{L atm}}{\text{mol K}} \times (22.8 + 273.2) \text{ K}}{756 \text{ mmHg} \times \dfrac{1 \text{ atm}}{760 \text{ mmHg}}} = 62.5 \text{ g/mol}$

53. (a) $d = \dfrac{MP}{RT} = \dfrac{28.96 \text{ g/mol} \times 1.00 \text{ atm}}{0.0821 \dfrac{\text{L atm}}{\text{mol K}} \times (273 + 25)\text{K}} = 1.18 \text{ g/L air}$

(b) $d = \dfrac{MP}{RT} = \dfrac{44.0 \text{ g/mol } CO_2 \times 1.00 \text{ atm}}{0.08206 \dfrac{\text{L atm}}{\text{mol K}} \times (273 + 25)\text{K}} = 1.80 \text{ g/L } CO_2$

Since this density is greater than that of air, the balloon will not rise in air when filled with CO_2 at $25°C$.

54. $d = \dfrac{MP}{RT}$ becomes $T = \dfrac{MP}{RT} = \dfrac{44.0 \text{ g/mol} \times 1.00 \text{ atm}}{0.08206 \dfrac{\text{L atm}}{\text{mol K}} \times 1.18 \text{ g/L}} = 454\text{K} = 181°\text{C}$

55. $d = \dfrac{MP}{RT}$ becomes $M = \dfrac{dRT}{P} = \dfrac{2.64 \text{ g/L} \times 0.0821 \dfrac{\text{L atm}}{\text{mol K}} \times (310 + 273)\text{K}}{775 \text{ mmHg} \times \dfrac{1 \text{ atm}}{760 \text{ mmHg}}} = 124 \text{ g/mol}$

Since the atomic mass of phosphorus is 31.0, the formula of phosphorus molecules in the vapor is P_4. $(4 \text{ atoms/molecule} \times 31.0 = 124)$

56. We first determine the molar mass of the gas, then its empirical formula. These two pieces of information are combined to obtain the molecular formula of the gas.

$$M = \frac{dRT}{P} = \frac{2.33 \text{ g/L} \times 0.08206 \dfrac{\text{L atm}}{\text{mol K}} \times 296 \text{ K}}{746 \text{ mmHg} \times \dfrac{1 \text{ atm}}{760 \text{ mmHg}}} = 57.7 \text{ g/mol}$$

$$\text{mol C} = 82.7 \text{ g C} \times \frac{1 \text{ mol C}}{12.0 \text{ g C}} = 6.89 \text{ mol C} \div 6.89 \rightarrow 1.00 \text{ mol C}$$

$$\text{mol H} = 17.3 \text{ g H} \times \frac{1 \text{ mol H}}{1.01 \text{ g H}} = 17.1 \text{ mol H} \div 6.89 \rightarrow 2.48 \text{ mol H}$$

Multiply both of these mole numbers by 2 to obtain the empirical formula, C_2H_5, which has an empirical molar mass of 29.0 g/mol. Since the molar mass (calculated as 57.7 g/mol above) is twice this empirical molar mass, twice the empirical formula is the molecular formula: C_4H_{10}.

Gases in Chemical Reactions

57. Balanced equation: $C_3H_8(g) + 5\,O_2(g) \rightarrow 3\,CO_2(g) + 4\,H_2O(l)$

Use the law of combining volumes. O_2 volume $= 75.6 \text{ L } C_3H_8 \times \dfrac{5 \text{ L } O_2}{1 \text{ L } C_3H_8} = 378 \text{ L } O_2$

58. We first use the law of combining volumes, and then the general gas law.

$$3\,CO(g) + 7\,H_2(g) \rightarrow C_3H_8(g) + 3\,H_2O(l) \qquad 28.5 \text{ L } CO(g) \times \frac{7 \text{ L } H_2(g)}{3 \text{ L } CO(g)} = 66.5 \text{ L } H_2(g)$$

$$\text{volume } H_2(g) = 66.5 \text{ L } H_2(g) \times \frac{760 \text{ mmHg}}{751 \text{ mmHg}} \times \frac{(26+273)\text{K}}{273 \text{ K}} = 73.7 \text{ L } H_2(g)$$

59. Determine the moles of $SO_2(g)$ produced and then use the ideal gas equation.

$$2.7 \times 10^6 \text{ lb coal} \times \frac{3.28 \text{ lb S}}{100.00 \text{ lb coal}} \times \frac{454 \text{ g S}}{1 \text{ lb S}} \times \frac{1 \text{ mol S}}{32.1 \text{ g S}} \times \frac{1 \text{ mol } SO_2}{1 \text{ mol S}} = 1.3 \times 10^6 \text{ mol } SO_2$$

$$V = \frac{nRT}{P} = \frac{1.3 \times 10^6 \text{ mol } SO_2 \times 0.0821 \dfrac{\text{L atm}}{\text{mol K}} \times 296 \text{ K}}{738 \text{ mmHg} \times \dfrac{1 \text{ atm}}{760 \text{ mmHg}}} = 3.3 \times 10^7 \text{ L } SO_2$$

60. Determine the moles of O_2, and then the mass of $KClO_3$ that produced this amount of O_2.

$$\text{mol } O_2 = \frac{\left(738 \text{ mmHg} \times \dfrac{1 \text{ atm}}{760 \text{ mmHg}}\right) \times \left(119 \text{ mL} \times \dfrac{1 \text{ L}}{1000 \text{ mL}}\right)}{0.08206 \dfrac{\text{L atm}}{\text{mol K}} \times (22.4 + 273.2)\text{K}} = 0.00476 \text{ mol } O_2$$

$$\text{mass } KClO_3 = 0.00476 \text{ mol } O_2 \times \frac{2 \text{ mol } KClO_3}{3 \text{ mol } O_2} \times \frac{122.6 \text{ g } KClO_3}{1 \text{ mol } KClO_3} = 0.389 \text{ g } KClO_3$$

$$\% \text{ } KClO_3 = \frac{0.389 \text{ g } KClO_3}{3.57 \text{ g sample}} \times 100\% = 10.9\% \text{ } KClO_3$$

61. Determine the moles and volume of O_2 liberated. $2 \text{ H}_2\text{O}_2(aq) \rightarrow 2 \text{ H}_2\text{O}(l) + O_2(g)$

$$\text{mol } O_2 = 10.0 \text{ mL soln} \times \frac{1.01 \text{ g}}{1 \text{ mL}} \times \frac{0.0300 \text{ g } H_2O_2}{1 \text{ g soln}} \times \frac{1 \text{ mol } H_2O_2}{34.0 \text{ g } H_2O_2} \times \frac{1 \text{ mol } O_2}{2 \text{ mol } H_2O_2}$$

$$= 0.00446 \text{ mol } O_2$$

$$V = \frac{0.00446 \text{ mol } O_2 \times 0.08206 \dfrac{\text{L atm}}{\text{mol K}} \times (22 + 273) \text{ K}}{752 \text{ mmHg} \times \dfrac{1 \text{ atm}}{760 \text{ mmHg}}} \times \frac{1000 \text{ mL}}{1 \text{ L}} = 109 \text{ mL } O_2$$

62. **(a)**   Volume $NH_3 = 313 \text{ L } H_2 \times \dfrac{2 \text{ L } NH_3}{3 \text{ L } H_2} = 209 \text{ L } NH_3$

(b)   Moles of NH_3 (@ 315 °C and 5.25 atm)

$$= \frac{5.25 \text{ atm} \times 313 \text{ L}}{0.08206 \dfrac{\text{L atm}}{\text{mol K}} \times (315 + 273)\text{K}} = 3.41 \times 10^3 \text{ mol } H_2 \times \frac{2 \text{ mol } NH_3}{3 \text{ mol } H_2}$$

$$= 2.27 \times 10^1 \text{ mol } NH_3$$

$$V(@25°C, 727\text{mmHg}) = \frac{2.27 \times 10^1 \text{ mol } NH_3 \times 0.08206 \dfrac{\text{L atm}}{\text{mol K}} \times 298 \text{ K}}{727 \text{ mmHg} \times \dfrac{1 \text{ atm}}{760 \text{ mmHg}}}$$

$$V(@25°C, 727\text{mmHg}) = 5.80 \times 10^2 \text{ L } NH_3$$

Mixtures of Gases

63. The two pressures are related as are the number of moles of $N_2(g)$ to the total number of moles.

$$\text{moles } N_2 = \frac{PV}{RT} = \frac{28.2 \text{ atm} \times 53.7 \text{ L}}{0.08206 \dfrac{\text{L atm}}{\text{mol K}} \times (26 + 273)\text{K}} = 61.7 \text{ mol } N_2$$

$$\text{total moles of gas} = 61.7 \text{ mol } N_2 \times \frac{75.0 \text{ atm}}{28.2 \text{ atm}} = 164 \text{ mol gas}$$

$$\text{mass Ne} = \left(164 \text{ mol total} - 61.7 \text{ mol } N_2\right) \times \frac{20.18 \text{ g Ne}}{1 \text{ mol Ne}} = 2.06 \times 10^3 \text{ g Ne}$$

64. Solve a Boyle's law problem for each gas and add the resulting partial pressures.

$$P_{H_2} = 762 \text{ mmHg} \times \frac{2.35 \text{ L}}{5.52 \text{ L}} = 324 \text{ mmHg} \qquad P_{He} = 728 \text{ mmHg} \times \frac{3.17 \text{ L}}{5.52 \text{ L}} = 418 \text{ mmHg}$$

$$P_{total} = P_{H_2} + P_{He} = 324 \text{ mmHg} + 418 \text{ mmHg} = 742 \text{ mmHg}$$

65. Initial Pressure of the cylinder

$$P = \frac{nRT}{V} = \frac{(1.60 \text{ g } O_2 \times \dfrac{1 \text{ mol } O_2}{31.998 \text{ g } O_2})(0.08206 \text{ L atm K}^{-1} \text{ mol}^{-1})(273.15 \text{ K})}{2.24 \text{ L}} = 0.500 \text{ atm}$$

We need to quadruple the pressure from 0.500 atm to 2.00 atm.

The mass of O_2 needs to quadruple. $1.60 \text{ g} \rightarrow 6.40 \text{ g}$ or add $4.80 \text{ g } O_2$

(this answer eliminates answer (a) and (b) as being correct)

Could also increase the pressure by adding the same number of another gas (e.g. He)
Mass of He = $n_{He} \times MM_{He}$

(note: moles of O_2 needed $= 4.80 \text{ g} \times \dfrac{1 \text{ mol } O_2}{31.998 \text{ g } O_2} = 0.150 \text{ moles} = 0.150 \text{ moles of He}$)

Mass of He $= 0.150 \text{ moles} \times \dfrac{4.0026 \text{ g He}}{1 \text{ mol He}} = 0.600 \text{ g He}$ ((d) is correct, add 0.600 g of He)

66. (a) First determine the moles of each gas, then the total moles, and then the total pressure.

$$\text{moles } H_2 = 4.0 \text{ g } H_2 \times \frac{1 \text{ mol } H_2}{2.02 \text{ g } H_2} \qquad\qquad \text{moles He} = 10.0 \text{ g He} \times \frac{1 \text{ mol He}}{4.00 \text{ g He}}$$

$$= 2.0 \text{ mol } H_2 \qquad\qquad\qquad\qquad\qquad = 2.50 \text{ mol He}$$

$$P = \frac{nRT}{P} = \frac{(2.0 + 2.50) \text{ mol} \times 0.0821 \dfrac{L \text{ atm}}{\text{mol K}} \times 273 \text{ K}}{4.3 \text{ L}} = 23 \text{ atm}$$

(b) $\quad P_{H_2} = 23 \text{ atm} \times \dfrac{2.0 \text{ mol } H_2}{4.5 \text{ mol total}} = 10 \text{ atm} \qquad\qquad P_{He} = 23 \text{ atm} - 10 \text{ atm} = 13 \text{ atm}$

67. (a) $\quad P_{ben} = \dfrac{nRT}{V} = \dfrac{\left(0.728 \text{ g} \times \dfrac{1 \text{ mol } C_6H_6}{78.11 \text{ g } C_6H_6}\right) \times 0.08206 \dfrac{L \text{ atm}}{\text{mol K}} \times (35 + 273)}{2.00 \text{ L}} \times \dfrac{760 \text{ mmHg}}{1 \text{ atm}}$

$$= 89.5 \text{ mmHg}$$

$$P_{total} = 89.5 \text{ mmHg } C_6H_6(g) + 752 \text{ mmHg Ar}(g) = 842 \text{ mmHg}$$

(b) $\quad P_{benzene} = 89.5 \text{ mmHg} \qquad\qquad P_{Ar} = 752 \text{ mmHg}$

68. (a) The $\%CO_2$ in ordinary air is 0.036%, while from the data of this problem, the $\%CO_2$ in expired air is 3.8%.

$$\frac{P\{CO_2 \text{ expired air}\}}{P\{CO_2 \text{ ordinary air}\}} = \frac{3.8\% \text{ } CO_2}{0.036\% \text{ } CO_2} = 1.1 \times 10^2 \text{ } CO_2 \text{ (expired air to ordinary air)}$$

(b/c) Density should be related to average molar mass. We expect the average molar mass of air to be between the molar masses of its two principal constituents, N_2 (28.0 g/mol) and O_2 (32.0 g/mol). The average molar mass of normal air is approximately 28.9 g/mol. Expired air would be made more dense by the presence of more CO_2 (44.0 g/mol) and less dense by the presence of more H_2O (18.0 g/mol). The change might be minimal. In fact, it is, as the following calculation shows.

$$M_{exp.air} = \left(0.742 \text{ mol } N_2 \times \frac{28.013 \text{ g } N_2}{1 \text{ mol } N_2}\right) + \left(0.152 \text{ mol } O_2 \times \frac{31.999 \text{ g } O_2}{1 \text{ mol } O_2}\right)$$

$$+ \left(0.038 \text{ mol } CO_2 \times \frac{44.01 \text{ g } CO_2}{1 \text{ mol } CO_2}\right) + \left(0.059 \text{ mol } H_2O \times \frac{18.02 \text{ g } H_2O}{1 \text{ mol } H_2O}\right)$$

$$+ \left(0.009 \text{ mol Ar} \times \frac{39.9 \text{ g Ar}}{1 \text{ mol Ar}}\right) = 28.7 \text{ g/mol of expired air}$$

Since the average molar mass of expired air is less than the average molar mass of ordinary air, expired air is less dense than ordinary air. Calculating the densities:

$$d(\text{expired air}) = \frac{(28.7 \text{ g/mol})(1.00 \text{ atm})}{(0.0821 \text{ L} \cdot \text{atm/K} \cdot \text{mol})(310 \text{ K})} = 1.13 \text{ g/L}$$

$$d(\text{ordinary air}) = 1.14 \text{ g/L}$$

69. $1.00 \text{ g H}_2 \approx 0.50 \text{ mol H}_2$ $1.00 \text{ g He} \approx 0.25 \text{ mol He}$

Adding 1.00 g of He to a vessel, which only contains 1.00 g of H_2 results in the number of moles of gas being increased by 50%. Situation (b) best represents the resulting mixture as the volume has increased by 50%

70.

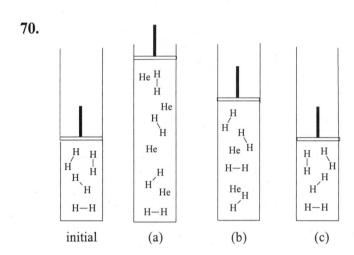

initial   (a)   (b)   (c)

Collecting Gases over Liquids

71.   The pressure of the liberated $H_2(g)$ is 744 mmHg − 23.8 mmHg = 720. mmHg

$$V = \frac{nRT}{P} = \frac{\left(1.65 \text{ g Al} \times \dfrac{1 \text{ mol Al}}{26.98 \text{ g}} \times \dfrac{3 \text{ mol H}_2}{2 \text{ mol Al}}\right) 0.08206 \dfrac{\text{L atm}}{\text{mol K}} (273 + 25)\text{K}}{720. \text{ mmHg} \times \dfrac{1 \text{ atm}}{760 \text{ mmHg}}} = 2.37 \text{ L H}_2(g)$$

This is the total volume of both gases, each with a different partial pressure.

72.   If the gas were measured at the same volume before and after passing it through H_2O, its pressure would increase by 23.8 mmHg from 748 mmHg to 772 mmHg. Since it is measured at the same pressure, we can determine its new volume with Boyle's law. What we are doing is assuming that the gas is collected at the same volume but at 772 mmHg. We then reduced the pressure to 748 mmHg by expanding the volume of the gas.

$$\text{volume} = 367 \text{ mL} \times \frac{772 \text{ mmHg}}{748 \text{ mmHg}} = 379 \text{ mL}$$

73. We first determine the pressure of the gas collected. This would be its "dry gas" pressure and, when added to 22.4 mmHg, gives the barometric pressure.

$$P = \frac{nRT}{V} = \frac{\left(1.46 \text{ g} \times \dfrac{1 \text{ mol O}_2}{32.0 \text{ g O}_2}\right) 0.08206 \dfrac{\text{L atm}}{\text{mol K}} \times 297 \text{ K}}{1.16 \text{ L}} \times \frac{760 \text{ mmHg}}{1 \text{ atm}} = 729 \text{ mmHg}$$

barometric pressure = 729 mm Hg + 22.4 mmHg = 751 mmHg

74. We first determine the "dry gas" pressure of helium. This pressure, subtracted from the barometric pressure of 738.6 mmHg, gives the vapor pressure of hexane at $25°C$.

$$P = \frac{nRT}{V} = \frac{\left(1.072 \text{ g} \times \dfrac{1 \text{ mol He}}{4.003 \text{ g He}}\right) 0.08206 \dfrac{\text{L atm}}{\text{mol K}} \times 298.2 \text{ K}}{8.446 \text{ L}} \times \frac{760 \text{ mmHg}}{1 \text{ atm}} = 589.7 \text{ mmHg}$$

vapor pressure = 738.6 − 589.7 = 148.9 mmHg

Kinetic Molecular Theory

75. $u_{rms} = \sqrt{\dfrac{3RT}{M}} = \sqrt{\dfrac{3 \times 8.3145 \dfrac{\text{J}}{\text{mol K}} \times 303 \text{ K}}{\dfrac{70.91 \times 10^{-3} \text{ kg Cl}_2}{1 \text{ mol Cl}_2}}} = 326 \text{ m/s}$

76. $u_{rms} = \sqrt{\dfrac{3RT}{M}} = 1.84 \times 10^3 \text{ m/s}$ $\qquad$ Solve this equation for temperature with u_{rms} doubled.

$$T = \frac{M u_{rms}^2}{3R} = \frac{2.016 \times 10^{-3} \text{ kg/mol } (2 \times 1.84 \times 10^3 \text{ m/s})^2}{3 \times 8.3145 \dfrac{\text{J}}{\text{mol K}}} = 1.09 \times 10^3 \text{ K}$$

77. $M = \dfrac{3RT}{(u_{rms})^2} = \dfrac{3 \times 8.3145 \dfrac{\text{J}}{\text{mol K}} \times 298 \text{ K}}{\left(2180 \dfrac{\text{mi}}{\text{hr}} \times \dfrac{1 \text{ hr}}{3600 \text{ sec}} \times \dfrac{5280 \text{ ft}}{1 \text{ mi}} \times \dfrac{12 \text{ in.}}{1 \text{ ft}} \times \dfrac{1 \text{ m}}{39.37 \text{ in.}}\right)^2} = 0.00783 \text{ kg/mol}$

$= 7.83$ g/mol. The molecular mass of the gas is $7.83\,u$.

78. A noble gas with molecules having u_{rms} at $25°C$ greater than that of a rifle bullet will have a molar mass less than 7.8 g/mol. Helium is the only possibility. A noble gas with a slower u_{rms} will have a molar mass greater than 7.8 g/mol; any one of the other noble gases will have a slower u_{rms}.

79. We equate the two expressions for root mean square speed, cancel the common factors, and solve for the temperature of Ne. Note that the units of molar masses do not have to be in kg/mol in this calculation; they simply must be expressed in the same units.

$$\sqrt{\frac{3RT}{M}} = \sqrt{\frac{3R \times 300 \text{ K}}{4.003}} = \sqrt{\frac{3R \times T_{Ne}}{20.18}} \quad \text{square both sides:} \quad \frac{300 \text{ K}}{4.003} = \frac{T_{Ne}}{20.18}$$

Solve for T_{Ne}: $T_{Ne} = 300 \text{ K} \times \dfrac{20.18}{4.003} = 1.51 \times 10^3 \text{ K}$

80. u_m, the modal speed, is the speed that occurs most often, 55 mi/h

$$\text{Average speed} = \frac{38 + 44 + 45 + 48 + 50 + 55 + 55 + 57 + 58 + 60}{10} = 51.0 \text{ mi/h} = \bar{u}$$

$$\text{Root mean square speed} = \sqrt{\frac{38^2 + 44^2 + 45^2 + 48^2 + 50^2 + 55^2 + 55^2 + 57^2 + 58^2 + 60^2}{10}}$$

$$\text{Root mean square speed} = \sqrt{\frac{26472}{10}} = 51.5 \frac{\text{mi}}{\text{h}} = u_{rms}$$

Diffusion and Effusion of Gases

81. $\dfrac{\text{rate (NO}_2)}{\text{rate (N}_2\text{O)}} = \sqrt{\dfrac{M(\text{N}_2\text{O})}{M(\text{NO}_2)}} = \sqrt{\dfrac{44.02}{46.01}} = 0.9781 = \dfrac{x \text{ mol NO}_2/t}{0.00484 \text{ mol N}_2\text{O}/t}$

mol $NO_2 = 0.00484$ mol $\times 0.9781 = 0.00473$ mol NO_2

82. $\dfrac{\text{rate (N}_2)}{\text{rate (unknown)}} = \dfrac{\text{mol (N}_2)/38 \text{ s}}{\text{mol (unknown)}/64 \text{ s}} = \dfrac{64 \text{ s}}{38 \text{ s}} = 1.68 = \sqrt{\dfrac{M(\text{unknown})}{M(\text{N}_2)}}$

$M(\text{unknown}) = (1.68)^2 M(\text{N}_2) = (1.68)^2 (28.01 \text{ g/mol}) = 79 \text{ g/mol}$

83.   **(a)** $\dfrac{\text{rate (N}_2)}{\text{rate (O}_2)} = \sqrt{\dfrac{M(\text{O}_2)}{M(\text{N}_2)}} = \sqrt{\dfrac{32.00}{28.01}} = 1.07$

   (b) $\dfrac{\text{rate (H}_2\text{O)}}{\text{rate (D}_2\text{O)}} = \sqrt{\dfrac{M(\text{D}_2\text{O})}{M(\text{H}_2\text{O})}} = \sqrt{\dfrac{20.0}{18.02}} = 1.05$

   (c) $\dfrac{\text{rate (}^{14}\text{CO}_2)}{\text{rate (}^{12}\text{CO}_2)} = \sqrt{\dfrac{M(^{12}\text{CO}_2)}{M(^{14}\text{CO}_2)}} = \sqrt{\dfrac{44.0}{46.0}} = 0.978$

   (d) $\dfrac{\text{rate (}^{235}\text{UF}_6)}{\text{rate (}^{238}\text{UF}_6)} = \sqrt{\dfrac{M(^{238}\text{UF}_6)}{M(^{235}\text{UF}_6)}} = \sqrt{\dfrac{352}{349}} = 1.004$

84. $\dfrac{\text{Rate of effusion O}_2}{\text{Rate of effusion SO}_2} = \dfrac{\sqrt{M(SO_2)}}{\sqrt{M(O_2)}} = \dfrac{\sqrt{64}}{\sqrt{32}} = \sqrt{2} = 1.4$

O_2 will effuse at 1.4 times the rate of effusion of SO_2 The situation depicted in (c) best represents the distribution of the molecules. If 5 molecules of SO_2 effuse, we would expect that 1.4 times as many O_2 molecules to effuse over the same period of time

($1.4 \times 5 = 7$ molecules of O_2). This is depicted in situation (c).

Nonideal Gases

85. For $Cl_2(g), n^2 a = 6.49\,L^2$ atm and $nb = 0.0562\,L.$ $P_{vdw} = \dfrac{nRT}{V - nb} - \dfrac{n^2 a}{V^2}$

At 0°C, $P_{vdw} = 9.90$ atm and $P_{ideal} = 11.2$ atm, off by 1.3 atm or $+13\%$

(a)   At $100°C\,P_{ideal} = \dfrac{nRT}{V} = \dfrac{1.00\,mol \times \dfrac{0.08206\,L\,atm}{mol\,K} \times 373\,K}{2.00\,L} = 15.3\,atm$

$P_{vdw} = \dfrac{1.00\,mol \times \dfrac{0.08206\,L\,atm}{mol\,K} \times T}{(2.00 - 0.0562)\,L} - \dfrac{6.49\,L^2\,atm}{(2.00\,L)^2} = 0.0422\,T\,atm - 1.62\,atm$

$= 0.0422 \times 373\,K - 1.62 = 14.1\,atm$ P_{ideal} is off by 1.2 atm or $+8.5\%$

(b)   At $200°C\,P_{ideal} = \dfrac{nRT}{V} = \dfrac{1.00\,mol \times \dfrac{0.08206\,L\,atm}{mol\,K} \times 473\,K}{2.00\,L} = 19.4\,atm$

$P_{vdw} = 0.0422_2 \times 473\,K - 1.62_3 = 18.3_5\,atm$ P_{ideal} is off by 1.0 atm or $+5.7\%$

(c)   At $400°C\,P_{ideal} = \dfrac{nRT}{V} = \dfrac{1.00\,mol \times \dfrac{0.08206\,L\,atm}{mol\,K} \times 673\,K}{2.00\,L} = 27.6\,atm$

$P_{vdw} = 0.0422 \times 673\,K - 1.62 = 26.8\,atm$ P_{ideal} is off by 0.8 atm or $+3.0\%$

86. **(a)** $V = 100.0$ L,

$$P_{vdw} = \frac{nRT}{V-nb} - \frac{n^2 a}{V^2} = \frac{1.50 \text{ mol} \times \dfrac{0.08206 \text{ L atm}}{\text{mol K}} \times 298 \text{K}}{(100.0 - 1.50 \times 0.0564)\text{L}} - \frac{1.50^2 \times 6.71 \text{ L}^2 \text{ atm}}{(100.0 \text{ L})^2}$$

$$= 0.366 \text{ atm SO}_2\,(g)$$

$$P_{ideal} = \frac{1.50 \text{ mol} \times \dfrac{0.08206 \text{ L atm}}{\text{mol K}} \times 298 \text{K}}{100.0 \text{L}} = 0.367 \text{ atm}$$

The two pressures are almost equal.

(b) $V = 20.0$ L, $P_{vdw} = 1.80$ atm; $P_{ideal} = 1.83$ atm.

Here the two pressures agree within a few percent.

(c) $V = 5.0$ L, $P_{vdw} = 6.86$ atm; $P_{ideal} = 7.34$ atm

(d) $V = 1.0$ L, $P_{vdw} = 25.0$ atm; $P_{ideal} = 36.7$ atm

(e) $V = 0.50$ L, $P_{vdw} = 27.9$ atm; $P_{ideal} = 73.4$ atm

FEATURE PROBLEMS

110. Boyle's Law relates P and V, i.e. $P \times V$ = constant. If V is proportional to the value of A, and $P_{gas} = P_{bar} + P_{Hg}$ (i.e. the pressure of the gas equals the sum of the barometric pressure and the pressure exerted by the mercury column), then a comparison of individual A × P products should show a consistent result or a constant.

A (cm)	P_{bar} (atm)	P_{Hg} (mmHg)	P_{gas} (mmHg)	A × P_{gas} (cm × mmHg)
27.9	739.8	71	810.8	22621
30.5	739.8	0	739.8	22564
25.4	739.8	157	896.8	22779
22.9	739.8	257	996.8	22827
20.3	739.8	383	1123	22793
17.8	739.8	538	1278	22745
15.2	739.8	754	1494	22706
12.7	739.8	1056	1796	22807
10.2	739.8	1476	2216	22601
7.6	739.8	2246	2986	22692

Since consistent A × P_{gas} results are observed, that these data conform reasonably well (within experimental uncertainty) to Boyle's Law.

111.

P_{gas} (mmHg)	$1/A \propto 1/V$ (cm^{-1})
810.8	0.0358
739.8	0.0328
896.8	0.0394
996.8	0.0437
1123	0.0493
1278	0.0562
1494	0.0658
1796	0.0787
2216	0.0980
2986	0.1316

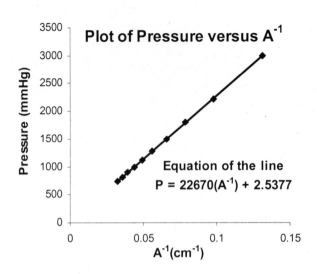

Plot: slope = constant from Feature Problem 110

Factors that would affect the slope of this straight line are related to deviations real gases exhibit from ideality. At higher pressures, real gases tend to interact more, exerting forces of attraction and repulsion that Boyle's Law does not take into account.

112. Nitryl Fluoride

$$65.01 \text{ u} \left(\frac{49.4}{100}\right) = 32.1 \text{ u of X}$$

Nitrosyl Fluoride

$$49.01 \text{ u} \left(\frac{32.7}{100}\right) = 16.0 \text{ u of X}$$

Thionyl Fluoride

$$86.07 \text{ u} \left(\frac{18.6}{100}\right) = 16.0 \text{ u of X}$$

Sulfuryl Fluoride

$$102.07 \text{ u} \left(\frac{31.4}{100}\right) = 32.0 \text{ u of X}$$

The atomic mass of X is 16 u which corresponds to the element oxygen. The number of atoms of X (oxygen) in each compound is given below:

Nitryl Fluoride   = 2 atoms of O

Nitrosyl Fluoride   = 1 atom of O

Thionyl Fluoride   = 1 atom of O

Sulfuryl Fluoride   = 2 atoms of O

113. (a) First convert pressures from mmHg to atm:

density (g/L)	pressure (atm)	density/pressure (g/L·atm)
1.428962	1.0000	$1.428962 \cong 1.4290$
1.071485	0.75000	$1.428647 \cong 1.4286$
0.714154	0.50000	$1.428308 \cong 1.4283$
0.356985	0.25000	$1.42794 \cong 1.4279$

average = 1.4285 g/L·atm

(b) $M_{O_2} = \dfrac{d}{P} RT$

$M_{O_2} = 1.4285 \text{ g/L} \cdot \text{atm} \times 0.082057 \text{ L} \cdot \text{atm/mol} \cdot \text{K} \times 273.15 \text{ K}$

$M_{O_2} = 32.0182 \text{ g/mol}$

Thus, the atomic mass of $O_2 = M_{O_2} / 2 = 16.0009$.

This compares favorably with the value of 15.9994 given in the front of the text.

CHAPTER 7
THERMOCHEMISTRY
PRACTICE EXAMPLES

1A The heat absorbed is the product of the mass of water, its specific heat $\left(4.18\ \mathrm{J\ g^{-1}\ ^{\circ}C^{-1}}\right)$, and the temperature change that occurs.

$$\text{heat energy} = 237\ \mathrm{g} \times \frac{4.18\ \mathrm{J}}{\mathrm{g\ ^{\circ}C}} \times \left(37.0\,^{\circ}\mathrm{C} - 4.0\,^{\circ}\mathrm{C}\right) \times \frac{1\ \mathrm{kJ}}{1000\ \mathrm{J}} = 32.7\ \mathrm{kJ\ of\ heat\ energy}$$

1B The heat absorbed is the product of the amount of mercury, its molar heat capacity, and the temperature change that occurs.

$$\text{heat energy} = \left(2.50\ \mathrm{kg} \times \frac{1000\ \mathrm{g}}{1\ \mathrm{kg}} \times \frac{1\ \mathrm{mol\ Hg}}{200.59\ \mathrm{g\ Hg}}\right) \times \frac{28.0\ \mathrm{J}}{\mathrm{mol\,^{\circ}C}} \times \left[-6.0 - (-20.0)\right]\,^{\circ}\mathrm{C} \times \frac{1\ \mathrm{kJ}}{1000\ \mathrm{J}}$$

$$= 4.89\ \mathrm{kJ\ of\ heat\ energy}$$

2A First calculate the quantity of heat lost by the lead. This heat energy must be absorbed by the surroundings (water). We assume 100% efficiency in the energy transfer.

$$q_{\text{lead}} = 1.00\ \mathrm{kg} \times \frac{1000\ \mathrm{g}}{1\ \mathrm{kg}} \times \frac{0.13\ \mathrm{J}}{\mathrm{g\,^{\circ}C}} \times \left(35.2\,^{\circ}\mathrm{C} - 100.0\,^{\circ}\mathrm{C}\right) = -8.4 \times 10^{3}\ \mathrm{J} = -q_{\text{water}}$$

$$8.4 \times 10^{3}\ \mathrm{J} = m_{\text{water}} \times \frac{4.18\ \mathrm{J}}{\mathrm{g\,^{\circ}C}} \times \left(35.2\,^{\circ}\mathrm{C} - 28.5\,^{\circ}\mathrm{C}\right) = 28 m_{\text{water}} \qquad m_{\text{water}} = \frac{8.4 \times 10^{3}}{28} = 3.0 \times 10^{2}\ \mathrm{g}$$

2B We use the same equation, equating the heat lost by the copper to the heat absorbed by the water, except now we solve for final temperature.

$$q_{\text{Cu}} = 100.0\ \mathrm{g} \times \frac{0.385\ \mathrm{J}}{\mathrm{g\,^{\circ}C}} \times \left(x\,^{\circ}\mathrm{C} - 100.0\,^{\circ}\mathrm{C}\right) = -50.0\ \mathrm{g} \times \frac{4.18\ \mathrm{J}}{\mathrm{g\,^{\circ}C}} \times \left(x\,^{\circ}\mathrm{C} - 26.5\,^{\circ}\mathrm{C}\right) = -q_{\text{water}}$$

$$38.5x - 3850 = -209x + 5539\ \mathrm{J} \qquad 38.5x + 209x = 5539 + 3850 \ \rightarrow\ 247.5x = 9389$$

$$x = \frac{9389}{247.5} = 37.9\,^{\circ}\mathrm{C}$$

3A The molar mass of $C_8H_8O_3$ is 152.15 g/mol. The calorimeter has a heat capacity of 4.90 kJ / $^{\circ}$ C

$$q_{\text{calor}} = \frac{4.90\ \mathrm{kJ\,^{\circ}C^{-1}} \times \left(30.09\,^{\circ}\mathrm{C} - 24.89\,^{\circ}\mathrm{C}\right)}{1.013\ \mathrm{g}} \times \frac{152.15\ \mathrm{g}}{1\ \mathrm{mol}} = 3.83 \times 10^{3}\ \mathrm{kJ\ /\ mol}$$

$$\Delta H_{\text{comb}} = -q_{\text{calor}} = -3.83 \times 10^{3}\ \mathrm{kJ\ /\ mol}$$

3B The heat that is liberated by the benzoic acid's combustion serves to raise the temperature of the assembly. We designate the calorimeter's heat capacity by C.

$$q_{rxn} = 1.176 \text{ g} \times \frac{-26.42 \text{ kJ}}{1 \text{ g}} = -31.07 \text{ kJ} = -q_{calorim}$$

$$q_{calorim} = C\Delta t = 31.07 \text{ kJ} = C \times 4.96°C \qquad C = \frac{31.07 \text{ kJ}}{4.96°C} = 6.26 \text{ kJ} /° C$$

4A The heat that is liberated by the reaction raises the temperature of the reaction mixture. We assume that this reaction mixture has the same density and specific heat as pure water.

$$q_{calorim} = \left(200.0 \text{ mL} \times \frac{1.00 \text{ g}}{1 \text{ mL}} \right) \times \frac{4.18 \text{ J}}{\text{g}°C} \times (30.2 - 22.4)°C = 6.52 \times 10^3 \text{ J} = -q_{rxn}$$

$$\text{moles AgCl} = 100.0 \text{ mL} \times \frac{1 \text{ L}}{1000 \text{ mL}} \times \frac{1.00 \text{ M AgNO}_3}{1 \text{ L}} \times \frac{1 \text{ mol AgCl}}{1 \text{ mol AgNO}_3} = 0.100 \text{ mol AgCl}$$

$$q_{rxn} = \frac{-6.52 \times 10^3 \text{ J}}{0.100 \text{ mol}} \times \frac{1 \text{ kJ}}{1000 \text{ J}} = -65.2 \text{ kJ/mol}$$

Because q_{rxn} is a negative quantity, the precipitation reaction is exothermic.

4B The assumptions include no heat loss to the surroundings or to the calorimeter, a solution density of 1.00 g/mL, a specific heat of $4.18 \text{ J g}^{-1}°C^{-1}$, and that the initial and final solution volumes are the same. The equation for the reaction that occurs is $NaOH(aq) + HCl(aq) \rightarrow NaCl(aq) + H_2O(l)$. Since the two reactants combine in a one to one mole ratio, the limiting reactant is the one present in smaller amount.

$$\text{amount HCl} = 100.0 \text{ mL} \times \frac{1.020 \text{ mmol HCl}}{1 \text{ mL soln}} = 102.0 \text{ mmol HCl}$$

$$\text{amount NaOH} = 50.0 \text{ mL} \times \frac{1.988 \text{ mmol NaOH}}{1 \text{ mL soln}} = 99.4 \text{ mmol NaOH}$$

NaOH is the limiting reactant.

$$q_{neutr} = 99.4 \text{ mmol NaOH} \times \frac{1 \text{ mmol H}_2O}{1 \text{ mmol NaOH}} \times \frac{1 \text{ mol H}_2O}{1000 \text{ mmol H}_2O} \times \frac{-56 \text{ kJ}}{1 \text{ mol H}_2O} = -5.5_7 \text{ kJ}$$

$$q_{calorim} = -q_{neutr} = 5.5_7 \text{ kJ} = (100.0 + 50.0) \text{ mL} \times \frac{1.00 \text{ g}}{1 \text{ mL}} \times \frac{4.18 \text{ J}}{\text{g}°C} \times \frac{1 \text{ kJ}}{1000 \text{ J}} \times (t - 24.52°C)$$

$$= 0.627t - 15.3_7 \qquad t = \frac{5.5_7 + 15.3_7}{0.627} = 33.4°C$$

5A $w = -P\Delta V = -0.750 \text{ atm}(+1.50 \text{ L}) = -1.13 \text{ L atm} \times \frac{101.33 \text{ J}}{1 \text{ L atm}} = -114 \text{ J}$

114 J of work is done by system

5B   Determine the initial number of moles:

$$n = 50.0 \text{ g N}_2 \times \frac{1 \text{ mol N}_2}{28.014 \text{ g N}_2} = 1.78\underline{5} \text{ moles of N}_2$$

$$V = \frac{nRT}{P} = \frac{(1.785 \text{ mol N}_2)(0.08206 \text{ Latm K}^{-1}\text{mol}^{-1})(293.15 \text{ K})}{2.50 \text{ atm}} = 17.2 \text{ L}$$

$$\Delta V = 17.2 - 75.0 \text{ L} = -57.8 \text{ L}$$

$$w = -P\Delta V = -2.50 \text{ atm}(-57.8 \text{ L}) \times \frac{101.33 \text{ J}}{1 \text{ L atm}} \times \frac{1 \text{ kJ}}{1000 \text{ J}} = +14.6 \text{ kJ} \text{ work done on system.}$$

6A   The work is $w = +355$ J. The heat flow is $q = -185$ J. These two are related to the energy change of the system by the first law equation: $\Delta U = q + w$, which becomes

$$\Delta U = +355 \text{ J} - 185 \text{ J} = +1.70 \times 10^2 \text{ J}$$

6B   The internal energy change is $\Delta U = -125$ J. The heat flow is $q = +54$ J. These two are related to the work done on the system by the first law equation: $\Delta U = q + w$, which becomes $-125 \text{ J} = +54 \text{ J} + w$. The solution to this equation is $w = -125 \text{ J} - 54 \text{ J} = -179 \text{ J}$, which means that 179 J of work is done by the system.

7A   Heat that is given off has a negative sign. In addition, we use the molar mass of sucrose, 342.30 g/mol.

$$\text{sucrose mass} = -1.00 \times 10^3 \text{ kJ} \times \frac{1 \text{ mol C}_{12}\text{H}_{22}\text{O}_{11}}{-5.65 \times 10^3 \text{ kJ}} \times \frac{342.30 \text{ g C}_{12}\text{H}_{22}\text{O}_{11}}{1 \text{ mol C}_{12}\text{H}_{22}\text{O}_{11}} = 60.6 \text{ g C}_{12}\text{H}_{22}\text{O}_{11}$$

7B   Although the equation does not say so explicitly, 56 kJ of heat is given off per mole of water formed. The equation then is the source of a conversion factor.

$$\text{heat flow} = 25.0 \text{ mL} \times \frac{1 \text{ L}}{1000 \text{ mL}} \times \frac{0.1045 \text{ mol HCl}}{1 \text{ L soln}} \times \frac{1 \text{ mol H}_2\text{O}}{1 \text{ mol HCl}} \times \frac{56 \text{ kJ evolved}}{1 \text{ mol H}_2\text{O}}$$

heat flow = 0.15 kJ heat evolved

8A $V_{ice} = (2.00 \text{ cm})^3 = 8.00 \text{ cm}^3$

$m_{ice} = m_{water} = 8.00 \text{ cm}^3 \times 0.917 \text{ g cm}^{-3} = 7.34 \text{ g ice} = 7.34 \text{ g H}_2\text{O}$

$$\text{moles of ice} = 7.34 \text{ g ice} \times \frac{1 \text{ mol H}_2\text{O}}{18.015 \text{ g H}_2\text{O}} = 0.407 \text{ moles of ice}$$

$q_{overall} = q_{ice}(-10 \text{ to } 0 \text{ °C}) + q_{fus} + q_{water}(0 \text{ to } 23.2 \text{ °C})$

$q_{overall} = m_{ice}(\text{sp. ht.})_{ice}\Delta T + n_{ice}\Delta H_{fus} + m_{water}(\text{sp. ht.})_{water}\Delta T$

$$q_{overall} = 7.34 \text{ g}(10.0 \text{ °C})(2.01 \frac{\text{J}}{\text{g °C}}) + 0.407 \text{ mol ice}(6.01 \frac{\text{kJ}}{\text{mol}}) + 7.34 \text{ g}(23.2 \text{ °C})(4.184 \frac{\text{J}}{\text{g °C}})$$

$q_{overall} = 0.148 \text{ kJ} + 2.45 \text{ kJ} + 0.712 \text{ kJ}$

$q_{overall} = 3.31 \text{ kJ (absorbs)}$

8B 5.00×10^3 kJ $= q_{ice}(-15$ to 0 °C$) + q_{fus} + q_{water}$ $(0$ to 25 °C$) + q_{vap}$

5.00×10^3 kJ $= m_{ice}$(sp. ht.)$_{ice}\Delta T + n_{ice}\Delta H_{fus} + m_{water}$(sp. ht.)$_{water}\Delta T + n_{water}\Delta H_{vap}$

5.00×10^6 J $= m(15.0$ °C$)(2.01 \dfrac{J}{g\ °C}) + (\dfrac{m}{18.015\ \text{g H}_2\text{O/mol H}_2\text{O}} \times 6.01 \times 10^3 \dfrac{J}{mol})$

$\qquad\qquad + m(25.0$ °C$)(4.184 \dfrac{J}{g\ °C}) + \dfrac{m}{18.015\ \text{g H}_2\text{O/mol}}(44.0 \times 10^3 \dfrac{J}{mol})$

5.00×10^6 J $= m(30.1\underline{5}$ J/g$) + m(333.\underline{6}$ J/g$) + m(104.\underline{5}$ J/g$) + m(2.4\underline{4} \times 10^3$ J/g$)$

5.00×10^6 J $= m(2.91 \times 10^3$ J/g$)$

$m = \dfrac{5.00 \times 10^6\ \text{J}}{2.91 \times 10^3\ \text{J/g}} = 1718$ g or 1.72 kg H_2O

9A   We combine the three combustion reactions to produce the hydrogenation reaction.

$C_3H_6(g) + \frac{9}{2}O_2(g) \rightarrow 3CO_2(g) + 3H_2O(l)$ $\qquad \Delta H_{comb} = \Delta H_1 = -2058$ kJ

$H_2(g) + \frac{1}{2}O_2(g) \rightarrow H_2O(l)$ $\qquad\qquad\qquad \Delta H_{comb} = \Delta H_2 = -285.8$ kJ

$3CO_2(g) + 4H_2O(l) \rightarrow C_3H_8(g) + 5O_2(g)$ $\qquad -\Delta H_{comb} = \Delta H_3 = +2219.9$ kJ

$\overline{C_3H_6(g) + H_2(g) \rightarrow C_3H_8(g)}$ $\qquad\qquad\quad \Delta H_{rxn} = \Delta H_1 + \Delta H_2 + \Delta H_3 = -124$ kJ

9B   The combustion reaction has propanol and $O_{2(g)}$ as reactants; the products are $CO_2(g)$ and $H_2O(l)$. Reverse the reaction given and combine it with the combustion reaction of $C_3H_6(g)$.

$C_3H_7OH(l) \rightarrow C_3H_6(g) + H_2O(l)$ $\qquad\qquad \Delta H_1 = +52.3$ kJ

$\underline{C_3H_6(g) + \frac{9}{2}O_2(g) \rightarrow 3CO_2(g) + 3H_2O(l)} \qquad \Delta H_2 = -2058$ kJ

$C_3H_7OH(l) + \frac{9}{2}O_2(g) \rightarrow 3CO_2(g) + 4H_2O(l) \quad \Delta H_{rxn} = \Delta H_1 + \Delta H_2 = -2006$ kJ

10A The enthalpy of formation is the enthalpy change for the reaction in which one mole of the product, $C_6H_{13}O_2N(s)$, is produced from appropriate amounts of the most stable forms of the elements. $6\ C(\text{graphite}) + \frac{13}{2}H_2(g) + O_2(g) + \frac{1}{2}N_2(g) \rightarrow C_6H_{13}O_2N(s)$

10B The enthalpy of formation is the enthalpy change for the reaction in which one mole of the product, $NH_3(g)$, is produced from appropriate amounts of the most stable forms of the elements, in this case from 0.5 mol $N_2(g)$ and 1.5 mol $H_2(g)$, that is, for the reaction:

$\frac{1}{2}N_2(g) + \frac{3}{2}H_2(g) \rightarrow NH_3(g)$

The specified reaction is twice the reverse of the formation reaction, and its enthalpy change is twice negative of the enthalpy of formation of $NH_3(g)$:

$-2 \times (-46.11$ kJ$) = +92.22$ kJ

11A $\Delta H^{\circ}_{rxn} = 2 \times \Delta H^{\circ}_{f}\left[CO_2(g)\right] + 3 \times \Delta H^{\circ}_{f}\left[H_2O(l)\right] - \Delta H^{\circ}_{f}\left[CH_3CH_2OH(l)\right] - 3 \times \Delta H^{\circ}_{f}\left[O_2(g)\right]$

$$= \left[2 \times (-393.5 \text{ kJ})\right] + \left[3 \times (-285.8 \text{ kJ})\right] - \left[-277.7 \text{ kJ}\right] - \left[3 \times 0.00 \text{ kJ}\right] = -1367 \text{ kJ}$$

11B We write the combustion reaction for each compound, and use that reaction to determine the compound's heat of combustion.

$$C_3H_8(g) + 5O_2(g) \rightarrow 3CO_2(g) + 4H_2O(l)$$

$\Delta H^{\circ}_{combustion} = 3 \times \Delta H^{\circ}_{f}\left[CO_2(g)\right] + 4 \times \Delta H^{\circ}_{f}\left[H_2O(l)\right] - \Delta H^{\circ}_{f}\left[C_3H_8(g)\right] - 5 \times \Delta H^{\circ}_{f}\left[O_2(g)\right]$

$$= \left[3 \times (-393.5 \text{ kJ})\right] + \left[4 \times (-285.8 \text{ kJ})\right] - \left[-103.8\right] - \left[5 \times 0.00 \text{ kJ}\right]$$

$$= -1181 \text{ kJ} - 1143 \text{ kJ} + 103.8 - 0.00 = -2220. \text{ kJ/mol } C_3H_8$$

$$C_4H_{10}(g) + \tfrac{13}{2}O_2(g) \rightarrow 4CO_2(g) + 5H_2O(l)$$

$\Delta H^{\circ}_{combustion} = 4 \times \Delta H^{\circ}_{f}\left[CO_2(g)\right] + 5 \times \Delta H^{\circ}_{f}\left[H_2O(l)\right] - \Delta H^{\circ}_{f}\left[C_4H_{10}(g)\right] - 6.5 \times \Delta H^{\circ}_{f}\left[O_2(g)\right]$

$$= \left[4 \times (-393.5 \text{ kJ})\right] + \left[5 \times (-285.8 \text{ kJ})\right] - \left[-125.6\right] - \left[6.5 \times 0.00 \text{ kJ}\right]$$

$$= -1574 \text{ kJ} - 1429 \text{ kJ} + 125.6 - 0.00 = -2877 \text{ kJ/mol } C_4H_{10}$$

In 1.00 mole of the mixture there are 0.62 mol $C_3H_8(g)$ and 0.38 mol $C_4H_{10}(g)$.

$$\text{heat of combustion} = \left(0.62 \text{ mol } C_3H_8 \times \frac{-2220. \text{ kJ}}{1 \text{ mol } C_3H_8}\right) + \left(0.38 \text{ mol } C_4H_{10} \times \frac{-2877 \text{ kJ}}{1 \text{ mol } C_4H_{10}}\right)$$

$$= -1.4 \times 10^3 \text{ kJ} - 1.1 \times 10^3 \text{ kJ} = -2.5 \times 10^3 \text{ kJ/mole of mixture}$$

12A $6 \, CO_2(g) + 6 \, H_2O(l) \rightarrow C_6H_{12}O_6(s) + 6 \, O_2(g)$

$\Delta H^{\circ}_{rxn} = 2803 \text{ kJ} = \Sigma \Delta H^{\circ}_{f \text{ products}} - \Sigma \Delta H^{\circ}_{f \text{ reactants}}$

$2803 \text{ kJ} = [1 \text{ mol}(\Delta H^{\circ}_{f}[C_6H_{12}O_6(s)]) + 6 \text{ mol}(0 \frac{kJ}{mol})] - [6 \text{ mol}(-393.5 \frac{kJ}{mol}) + 6 \text{ mol}(-285.8 \frac{kJ}{mol})]$

$2803 \text{ kJ} = \Delta H^{\circ}_{f}[C_6H_{12}O_6(s)] - [-4075.8 \text{ kJ}]$. Thus, $\Delta H^{\circ}_{f}[C_6H_{12}O_6(s)] = -1273 \text{ kJ}$

12B $\Delta H^{\circ}_{comb}[CH_3OCH_3(l)] = -31.70 \frac{kJ}{g}$ Molar Mass$_{CH_3OCH_3} = 46.069$ g mol^{-1}

$\Delta H^{\circ}_{comb}[CH_3OCH_3(l)] = -31.70 \frac{kJ}{g} \times 46.069 \frac{g}{mol} = -1460 \frac{kJ}{mol} kJ = \Delta H^{\circ}_{rxn}$

$\Delta H^{\circ}_{rxn} = \Sigma \Delta H^{\circ}_{f \text{ products}} - \Sigma \Delta H^{\circ}_{f \text{ reactants}}$ Reaction: $CH_3OCH_{3(l)} + 3 \, O_2(g) \rightarrow 2 \, CO_2(g) + 3 \, H_2O(l)$

$-1460 \text{ kJ} = [2 \text{ mol}(-393.5 \frac{kJ}{mol}) + 3 \text{ mol}(-285.8 \frac{kJ}{mol})] - [1 \text{ mol}(\Delta H^{\circ}_{f}[CH_3OCH_3(l)]) + 3 \text{ mol}(0 \frac{kJ}{mol})]$

$-1460 \text{ kJ} = -1644.4 \text{ kJ} - \Delta H^{\circ}_{f}[CH_3OCH_3(l)]$ Hence, $\Delta H^{\circ}_{f}[CH_3OCH_3(l)] = -184 \text{ kJ}$

13A The net ionic equation is: $Ag^+(aq) + I^-(aq) \rightarrow AgI(s)$ and we have the following:

$$\Delta H^{\circ}_{rxn} = \Delta H^{\circ}_f[AgI(s)] - \Delta H^{\circ}_f[Ag^+(aq)] - \Delta H^{\circ}_f[I^-(aq)]$$
$$= -61.84 \text{ kJ / mol} - (+105.6 \text{ kJ / mol}) - (-55.19 \text{ kJ / mol}) = -112.3 \text{ kJ / mol}$$

13B $2 Ag^+(aq) + CO_3{}^{2-}(aq) \rightarrow Ag_2CO_3(s)$

$\Delta H^{\circ}_{rxn} = -39.9 \text{ kJ} = \Sigma \Delta H^o_f \text{ products} - \Sigma \Delta H^o_f \text{ reactants} =$

$-39.9 \text{ kJ} = \Delta H^o_f[Ag_2CO_3(s)] - [2 \text{ mol}(105.6 \dfrac{\text{kJ}}{\text{mol}}) + 1 \text{ mol}(-677.1 \dfrac{\text{kJ}}{\text{mol}})]$

$-39.9 \text{ kJ} = \Delta H^o_f[Ag_2CO_3(s)] + 465.9 \text{ kJ}$ Hence, $\Delta H^o_f[Ag_2CO_3(s)] = -505.8 \text{ kJ/mol}$

REVIEW QUESTIONS

1. **(a)** ΔH, the enthalpy change, is the quantity of heat absorbed or released when a process occurs at constant pressure.

   (b) $P\Delta V$, the change in volume multiplied by a constant pressure, is the expression for the pressure-volume work in a process that occurs at constant pressure.

   (c) ΔH°_f, the standard enthalpy of formation, is the heat absorbed or released at constant pressure when 1 mole of product is formed from the most stable form of the elements, with reactants and products in their standard states.

   (d) Standard state is defined as a pressure of exactly 1 bar for a pure substance, or an aqueous solute at a 1 M concentration, each at a temperature of interest.

   (e) A fossil fuel is a material that can be burned for heat and that was produced by material that lived eons ago.

2. **(a)** The law of conservation of energy states that energy is neither created nor destroyed during a process.

   (b) Bomb calorimetry is the technique of running a chemical reaction in a constant-volume container and measuring the heat absorbed or released by the process.

   (c) A function of state is a measurable property that depends only on the initial and final conditions of a process and not on its path.

   (d) An enthalpy diagram represents the enthalpy values of reactants and products by their vertical positions, and the progress of the reaction horizontally.

   (e) Hess's law states that if several reactions can be combined to form a net reaction, the enthalpy changes for those reactions combine in the same way to produce the enthalpy change of the net reaction.

3.   **(a)**   The system is that part of the universe that we are considering, in which we are interested. The surroundings are the rest of the universe, particularly the rest that influences the system.

(b)   Heat is energy in transport that is associated with either a difference in temperature (i.e. sensible heat) or a phase change (such as solid to liquid) of a material (i.e. latent heat). Work is organized energy that has the ability to exert a force through a distance.

(c)   The specific heat of a substance is the quantity of heat needed to raise the temperature of one *gram* of that substance by $1.00°C$. The (molar) heat capacity of a substance is the quantity of heat needed to raise the temperature (of one *mole*) of that substance by $1.00°C$.

(d)   An endothermic reaction is one that absorbs heat from the surroundings. An exothermic reaction is one that evolves heat to the surroundings.

(e)   A constant-volume process takes place in a sealed container with rigid walls where the volume does not change (i.e. a reaction in a bomb calorimeter). A constant-pressure process takes place under a constant external pressure (i.e. an open container or a chamber with a flexible wall or movable piston)

4.   **(a)** $q = 9.25 \text{ L} \times \dfrac{1000 \text{ cm}^3}{1 \text{ L}} \times \dfrac{1.00 \text{ g}}{1 \text{ cm}^3} \times \dfrac{1.00 \text{ cal}}{1 \text{ g }°C} \times \left(29.4°C - 22.0°C\right) \times \dfrac{1 \text{ kcal}}{1000 \text{ cal}} = +68.5 \text{ kcal}$

(b) $q = 5.85 \text{ kg} \times \dfrac{1000 \text{ g}}{1 \text{ kg}} \times \dfrac{0.903 \text{ J}}{\text{g }°C} \times \left(-33.5°C\right) \times \dfrac{1 \text{ kJ}}{1000 \text{ J}} = -177 \text{ kJ}$

5. $\text{heat} = \text{mass} \times \text{sp ht} \times \Delta T \qquad \text{becomes} \qquad \Delta T = \dfrac{\text{heat}}{\text{mass} \times \text{sp. ht.}}$

(a) $\Delta T = \dfrac{+875 \text{ J}}{12.6 \text{ g} \times 4.18 \text{ J g}^{-1}°C^{-1}} = +16.6°C$

$T_f = T_i + \Delta T = 22.9°C + 16.6°C = 39.5°C$

(b) $\Delta T = \dfrac{-1.05 \text{ kcal} \times \dfrac{1000 \text{ cal}}{1 \text{ kcal}}}{\left(1.59 \text{ kg} \times \dfrac{1000 \text{ g}}{1 \text{ kg}}\right) 0.032 \dfrac{\text{cal}}{\text{g }°C}} = -21°C$

$T_f = T_i + \Delta T = 78.2°C - 21°C = 57°C$

6.   **(a)** $\text{sp. ht.} = \dfrac{\text{heat}}{\text{mass} \times \Delta T} = \dfrac{186 \text{ J}}{15.0 \text{ g} \times \left(29.6 - 22.3\right)} = 1.7 \text{ J g}^{-1}°C^{-1}$

(b) $\Delta T = \dfrac{\text{heat}}{\text{mass} \times \text{sp.ht.}} = \dfrac{-2.75 \text{ kcal} \times \dfrac{1000 \text{ cal}}{1 \text{ kcal}}}{\left(2.25 \text{ kg} \times \dfrac{1000 \text{ g}}{1 \text{ kg}}\right) 1.00 \dfrac{\text{cal}}{\text{g} \, ^\circ\text{C}}} = -1.22\,^\circ\text{C}$

$T_f = T_i + \Delta T = 23.1\,^\circ\text{C} - 1.22\,^\circ\text{C} = 21.9\,^\circ\text{C}$

7. The heat capacities of the two substances are added and then multiplied by the temperature change.

$\Delta H = \left(118 \text{ g Cu} \times \dfrac{0.385 \text{ J}}{\text{g} \, ^\circ\text{C}} + 197 \text{ g H}_2\text{O} \times \dfrac{4.18 \text{ J}}{\text{g} \, ^\circ\text{C}}\right)\left(79.2\,^\circ\text{C} - 22.7\,^\circ\text{C}\right) \times \dfrac{1 \text{ kJ}}{1000 \text{ J}} = +49.1 \text{ kJ}$

8. Heat is transferred from the iron to the water.

$q_{\text{water}} = 981 \text{ g} \times \dfrac{4.18 \text{ J}}{\text{g} \, ^\circ\text{C}} \times \left(34.4 - 22.1\right)^\circ\text{C} = 5.04 \times 10^4 \text{ J} = -q_{\text{iron}}$

$q_{\text{iron}} = -5.04 \times 10^4 \text{ J} = 1.22 \text{ kg} \times \dfrac{1000 \text{ g}}{1 \text{ kg}} \times \text{sp. ht.} \times \left(34.4 - 126.5\right)^\circ\text{C} = -1.12 \times 10^5 \times (\text{sp. ht.})$

$\text{sp. ht.} = \dfrac{-5.04 \times 10^4}{-1.12 \times 10^5} = \dfrac{0.450 \text{ J}}{\text{g} \, ^\circ\text{C}}$

9. If the two volumes were the same, a straight average of the temperatures would be the final temperature. But that is not true, so option (3) $\left[50\,^\circ\text{C}\right]$ is incorrect. In fact, there is more of the cooler water (100.0 mL) than of the warmer water (75.0 mL). The cooler side of the straight average should be somewhat favored, but not grossly so; option (4) $\left[28\,^\circ\text{C}\right]$ is incorrect. In order to arrive at $40\,^\circ\text{C}$, the cold water should increase by $20\,^\circ\text{C}$ and the warm water should decrease by $40\,^\circ\text{C}$; their masses, and thus their volume should be in the inverse relationship as their temperature changes: 2:1. That is not the case, so option (1) $\left[40\,^\circ\text{C}\right]$ is incorrect. The correct final temperature is likely to be $46\,^\circ\text{C}$, option (2).

10. **(a)** $\Delta U = q + w = 67 \text{ J heat} -67 \text{ J work} = 0 \text{ J}$

(b) $\Delta U = q + w = 356 \text{ J heat} -592 \text{ J work} = -236 \text{ J}$

(c) $\Delta U = q + w = -38 \text{ J heat} +171 \text{ J work} = +133 \text{ J}$

(d) $\Delta U = q + w = 0 \text{ J heat} - 416 \text{ J work} = -416 \text{ J}$

11. **(a)** $q = \dfrac{-29.4 \text{ kJ}}{0.584 \text{ g C}_3\text{H}_8} \times \dfrac{44.10 \text{ g C}_3\text{H}_8}{1 \text{ mol C}_3\text{H}_8} = -2.22 \times 10^3 \text{ kJ / mol C}_3\text{H}_8$

(b) $q = \dfrac{-1.26 \text{ kcal}}{0.136 \text{ g } C_{10}H_{16}O} \times \dfrac{4.184 \text{ kJ}}{1 \text{ kcal}} \times \dfrac{152.24 \text{ g } C_{10}H_{16}O}{1 \text{ mol } C_{10}H_{16}O} = -5.90 \times 10^3 \text{ kJ / mol } C_{10}H_{16}O$

(c) $q = \dfrac{-58.3 \text{ kJ}}{2.35 \text{ mL } (CH_3)_2 CO} \times \dfrac{1 \text{ mL}}{0.791 \text{ g}} \times \dfrac{58.08 \text{ g} (CH_3)_2 CO}{1 \text{ mol } (CH_3)_2 CO} = -1.82 \times 10^3 \text{ kJ/mol} (CH_3)_2 CO$

12. $\text{heat capacity} = \dfrac{\text{heat absorbed}}{\Delta T} = \dfrac{5228 \text{ cal}}{4.39°C} \times \dfrac{4.184 \text{ J}}{1 \text{ cal}} \times \dfrac{1 \text{ kJ}}{1000 \text{ J}} = 4.98 \text{ kJ/°C}$

13. Heat absorbed by calorimeter $= q_{comb} \times \text{moles} = \text{heat capacity} \times \Delta T$ or $\Delta T = \dfrac{q_{comb} \times \text{moles}}{\text{heat capacity}}$

(a) $\Delta T = \dfrac{\left(1014.2 \dfrac{\text{kcal}}{\text{mol}} \times 4.184 \dfrac{\text{kJ}}{\text{kcal}}\right)\left(0.3268 \text{ g} \times \dfrac{1 \text{ mol } C_8H_{10}O_2N_4}{194.19 \text{ g } C_8H_{10}O_2N_4}\right)}{5.136 \text{ kJ/°C}} = 1.390 °C$

$T_f = T_i + \Delta T = 22.43°C + 1.390°C = 23.82°C$

(b) $\Delta T = \dfrac{2444 \dfrac{\text{kJ}}{\text{mol}}\left(1.35 \text{ mL} \times \dfrac{0.805 \text{ g}}{1 \text{ mL}} \times \dfrac{1 \text{ mol } C_4H_8O}{72.11 \text{ g } C_4H_8O}\right)}{5.136 \text{ kJ/°C}} = 7.17°C$

$T_f = 22.43°C + 7.17°C = 29.60°C$

14. (a) $\dfrac{\text{heat}}{\text{mass}} = \dfrac{\text{heat cap.} \times \Delta t}{\text{mass}} = \dfrac{4.728 \text{ kJ /°C} \times (27.19 - 23.29)°C}{1.183 \text{ g}} = 15.6 \text{ kJ / g xylose}$

$\Delta H = \text{heat given off / g} \times M(\text{g / mol}) = \dfrac{-15.6 \text{ kJ}}{1 \text{ g } C_5H_{10}O_5} \times \dfrac{150.13 \text{ g } C_5H_{10}O_5}{1 \text{ mol}}$

$\Delta H = -2.34 \times 10^3 \text{ kJ / mol } C_5H_{10}O_5$

(b) $C_5H_{10}O_5 (g) + 5O_2 (g) \rightarrow 5CO_2 (g) + 5H_2O(l); \quad \Delta H = -2.34 \times 10^3 \text{ kJ/mol } C_5H_{10}O_5$

15. This is first a limiting reactant problem. There is $0.1000 \text{ L} \times 0.300 \text{ M} = 0.0300 \text{ mol HCl}$ and $1.82 / 65.39 = 0.0278 \text{ mol Zn}$. Stoichiometry demands 2 mol HCl for every 1 mol Zn. HCl is the limiting reactant. The reaction is exothermic. We neglect the slight excess of Zn(s), and assume that the volume of solution remains 100.0 mL and its specific heat, $4.18 \text{ J g}^{-1}°\text{C}^{-1}$. The enthalpy change, in kJ/mol Zn, is

$\Delta H = -\dfrac{100.0 \text{ mL} \times \dfrac{1.00 \text{ g}}{1 \text{ mL}} \times \dfrac{4.18 \text{ J}}{\text{g}°C} \times (30.5 - 20.3)°C}{0.0300 \text{ mol HCl} \times \dfrac{1 \text{ mol Zn}}{2 \text{ mol HCl}}} \times \dfrac{1 \text{ kJ}}{1000 \text{ J}} = -284 \text{ kJ/mol Zn}$

16. **(a)** Because the temperature of the mixture decreases, the reaction (the system) must have absorbed heat from the reaction mixture (the surroundings). Consequently, the reaction must be endothermic.

(b) We assume that the specific heat of the solution is $4.18 \text{ J g}^{-1}\,^\circ\text{C}^{-1}$. The enthalpy change in kJ/mol KCl is obtained by the heat absorbed per gram KCl.

$$\Delta H = -\frac{(0.75+35.0)\,\text{g}\,\dfrac{4.18\,\text{J}}{\text{g}\,^\circ\text{C}}(23.6-24.8)^\circ\text{C}}{0.75\,\text{g KCl}} \times \frac{1\,\text{kJ}}{1000\,\text{J}} \times \frac{74.55\,\text{g KCl}}{1\,\text{mol KCl}} = +18 \text{ kJ / mol}$$

17. As indicated by the negative sign for the enthalpy change, this is an exothermic reaction; the temperature of the system should increase.

$$q_{\text{rxn}} = 0.136 \text{ mol KC}_2\text{H}_3\text{O}_2 \times \frac{-15.3\,\text{kJ}}{1 \text{ mol KC}_2\text{H}_3\text{O}_2} \times \frac{1000\,\text{J}}{1\,\text{kJ}} = -2.08\times10^3 \text{ J} = -q_{\text{calorim}}$$

Now, we assume that the density of water is 1.00 g/mL, the specific heat of the solution in the calorimeter is $4.18 \text{ J g}^{-1}\,^\circ\text{C}^{-1}$, and no heat is lost by the calorimeter.

$$q_{\text{calorim}} = 2.08\times10^3 \text{ J} = \left(\left(525 \text{ mL} \times \frac{1.00\,\text{g}}{1\,\text{mL}}\right) + \left(0.136 \text{ mol KC}_2\text{H}_3\text{O}_2 \times \frac{98.14\,\text{g}}{1 \text{ mol KC}_2\text{H}_3\text{O}_2}\right)\right)$$

$$\times \frac{4.18\,\text{J}}{\text{g}\,^\circ\text{C}} \times \Delta T = 2.25\times10^3 \, \Delta T$$

$$\Delta T = \frac{2.08\times10^3}{2.25\times10^3} = +0.924\,^\circ\text{C} \qquad T_{\text{final}} = T_{\text{initial}} + \Delta T = 25.1\,^\circ\text{C} + 0.924\,^\circ\text{C} = 26.0\,^\circ\text{C}$$

18. **(a)** $N_2(g)+\frac{1}{2}O_2(g)\rightarrow N_2O(g)$ $\Delta H^\circ = +82.05 \text{ kJ / mol}$

(b) $S(\text{rhombic})+O_2(g)+Cl_2(g)\rightarrow SO_2Cl_2(l)$ $\Delta H^\circ = -394.1 \text{ kJ / mol}$

(c) $CH_3CH_2COOH(l)+\frac{7}{2}O_2(g)\rightarrow 3CO_2(g)+3H_2O(l)$ $\Delta H^\circ = -1527 \text{ kJ / mol}$

19. **(a)** $\text{heat evolved} = 1.325 \text{ g C}_4\text{H}_{10} \times \dfrac{1 \text{ mol C}_4\text{H}_{10}}{58.123 \text{ g C}_4\text{H}_{10}} \times \dfrac{2877\,\text{kJ}}{1 \text{ mol C}_4\text{H}_{10}} = 65.59 \text{ kJ}$,

or heat $= -65.59$ kJ

(b) $\text{heat evolved} = 28.4 \text{ L}_{\text{STP}} \text{ C}_4\text{H}_{10} \times \dfrac{1 \text{ mol C}_4\text{H}_{10}}{22.414 \text{ L}_{\text{STP}} \text{ C}_4\text{H}_{10}} \times \dfrac{2877\,\text{kJ}}{1 \text{ mol C}_4\text{H}_{10}} = 3.65\times10^3 \text{ kJ}$,

or heat $= -3.65 \times 10^3$ kJ

(c) Use the ideal gas equation to determine the amount of propane in moles and multiply this amount by 2877 kJ heat produced per mole.

$$\text{heat evolved} = \frac{\left(738\,\text{mmHg} \times \dfrac{1\,\text{atm}}{760\,\text{mmHg}}\right) \times 12.6\,\text{L}}{\dfrac{0.08206\,\text{L atm}}{\text{mol K}} \times (273.2 + 23.6)\,\text{K}} \times \frac{2877\,\text{kJ}}{1\,\text{mol C}_4\text{H}_{10}} = 1.45 \times 10^3\,\text{kJ},$$

or heat $= -1.45 \times 10^3$ kJ

20. The formation reaction for $NH_3(g)$ is $\frac{1}{2}N_2(g) + \frac{3}{2}H_2(g) \rightarrow NH_3(g)$. The given reaction is two-thirds the reverse of the formation reaction. The sign of the enthalpy is changed and it is multiplied by two-thirds. Thus, the enthalpy of the given reaction is $-(-46.11\,\text{kJ}) \times \frac{2}{3} = +30.74$ kJ.

21.

$-(1)$	$CO(g) \rightarrow C(\text{graphite}) + \frac{1}{2}O_2(g)$	$\Delta H^\circ = +110.54$ kJ
$+(2)$	$C(\text{graphite}) + O_2(g) \rightarrow CO_2(g)$	$\Delta H^\circ = -393.51$ kJ
	$CO(g) + \frac{1}{2}O_2(g) \rightarrow CO_2(g)$	$\Delta H^\circ = -282.97$ kJ

22.

$-(3)$	$3\,CO_2(g) + 4\,H_2O(l) \rightarrow C_3H_8(g) + 5\,O_2(g)$	$\Delta H^\circ = +2219.1$ kJ
$+(2)$	$C_3H_4(g) + 4\,O_2(g) \rightarrow 3\,CO_2(g) + 2\,H_2O(l)$	$\Delta H^\circ = -1937$ kJ
$2(1)$	$2\,H_2(g) + O_2(g) \rightarrow 2\,H_2O(l)$	$\Delta H^\circ = -571.6$ kJ
	$C_3H_4(g) + 2H_2(g) \rightarrow C_3H_8(g)$	$\Delta H^\circ = -290.$ kJ

23. The second reaction is the only one in which $NO(g)$ appears; it must be run twice to produce $2NO(g)$.

$2\,NH_3(g) + \frac{5}{2}O_2(g) \rightarrow 2\,NO(g) + 3\,H_2O(l)$ $2 \times \Delta H_2^\circ$

The first reaction is the only one that can eliminate $2NH_3(g)$; it must be run twice to eliminate $2NH_3(g)$.

$N_2(g) + 3\,H_2(g) \rightarrow 2\,NH_3(g)$ $2 \times \Delta H_1^\circ$

We triple and reverse the third reaction to eliminate $3H_2(g)$.

$3\,H_2O(l) \rightarrow 3\,H_2(g) + \frac{3}{2}O_2(g)$ $-3 \times \Delta H_3^\circ$

$\text{Result}: N_2(g) + O_2(g) \rightarrow 2\,NO(g)$ $\Delta H_{\text{rxn}} = 2 \times \Delta H_1 + 2 \times \Delta H_2 - 3 \times \Delta H_3$

24. **(a)** $\Delta H° = \Delta H_f°[C_2H_6(g)] + \Delta H_f°[CH_4(g)] - \Delta H_f°[C_3H_8(g)] - \Delta H_f°[H_2(g)]$

$\Delta H° = -84.68 - 74.81 - (-103.8) - 0.00 = -55.7$ kJ

(b) $\Delta H° = 2\Delta H_f°[SO_2(g)] + 2\Delta H_f°[H_2O(l)] - 2\Delta H_f°[H_2S(g)] - 3\Delta H_f°[O_2(g)]$

$\Delta H° = 2(-296.8) + 2(-285.8) - 2(-20.63) - 3(0.00) = -1124$ kJ

25. $\Delta H° = \Delta H_f°[H_2O(l)] + \Delta H_f°[NH_3(g)] - \Delta H_f°[NH_4^+(aq)] - \Delta H_f°[OH^-(aq)]$

$\Delta H° = -285.8 + (-46.11) - (-132.5) - (-230.0) = +30.6$ kJ

26. $ZnO(s) + SO_2(g) \rightarrow ZnS(s) + \frac{3}{2}O_2(g);$ $\Delta H° = -(-878.2 \text{ kJ})/2 = +439.1$ kJ

$439.1 \text{ kJ} = \Delta H_f°[ZnS(s)] + \frac{3}{2}\Delta H_f°[O_2(g)] - \Delta H_f°[ZnO(s)] - \Delta H_f°[SO_2(g)]$

$439.1 \text{ kJ} = \Delta H_f°[ZnS(s)] + \frac{3}{2}(0.00) - (-348.3) - (-296.8)$

$\Delta H_f°[ZnS(s)] = 439.1 - 348.3 - 296.8 = -206.0$ kJ / mol

EXERCISES

Heat Capacity (Specific Heat)

27. heat gained by the water = heat lost by the metal; heat = mass $\times$ sp.ht. $\times \Delta T$

(a) $50.0 \text{ g} \times 4.18 \dfrac{J}{g°C}(38.9 - 22.0)°C = 3.53 \times 10^3 \text{ J} = -150.0\text{g} \times \text{sp. ht.} \times (38.9 - 100.0)°C$

$\text{sp.ht.} = \dfrac{3.53 \times 10^3 \text{ J}}{150.0 \text{ g} \times 61.1°C} = 0.385 \text{ J g}^{-1} \text{ C}^{°-1} \text{ for Zn}$

(b) $50.0 \text{ g} \times 4.18 \dfrac{J}{g°C}(28.8 - 22.0)°C = 1.4 \times 10^3 \text{J} = -150.0\text{g} \times \text{sp. ht.} \times (28.8 - 100.0)°C$

$\text{sp.ht.} = \dfrac{1.4 \times 10^3 \text{ J}}{150.0 \text{ g} \times 71.2°C} = 0.13 \text{ J g}^{-1} \text{ C}^{°-1} \text{ for Pt}$

(c) $50.0 \text{ g} \times 4.18 \dfrac{J}{g°C}(52.7 - 22.0)°C = 6.42 \times 10^3 \text{J} = -150.0\text{g} \times \text{sp. ht.} \times (52.7 - 100.0)°C$

$\text{sp.ht.} = \dfrac{6.42 \times 10^3 \text{ J}}{150.0 \text{ g} \times 47.3°C} = 0.905 \text{ J g}^{-1} \text{ C}^{°-1} \text{ for Al}$

28. $50.0 \text{ g} \times 4.18 \dfrac{J}{g\,^\circ C}(27.6-23.2)^\circ C = 9.2 \times 10^2 \text{ J} = -75.0 \text{ g} \times \text{sp. ht.} \times (27.6-80.0)^\circ C$

$\text{sp. ht.} = \dfrac{9.2 \times 10^2 \text{ J}}{75.0 \text{ g} \times 52.4\,^\circ C} = 0.23 \text{ J g}^{-1}\ C^{\circ-1} \text{ for Ag}$

29. $q_{water} = 375 \text{ g} \times 4.18 \dfrac{J}{g\,^\circ C}(87-26)^\circ C = 9.5\underline{6} \times 10^4 \text{ J} = -q_{iron}$

$q_{iron} = -9.5\underline{6} \times 10^4 \text{ J} = 465 \text{ g} \times 0.449 \dfrac{J}{g\,^\circ C}(87-T_l) = 1.8_2 \times 10^4 \text{ J} - 2.0_9 \times 10^2\ T_l$

$T_l = \dfrac{-9.5\underline{6} \times 10^4 - 1.8 \times 10^4}{-2.0_9 \times 10^2} = 5.4\underline{4} \times 10^2\,^\circ C \text{ or } 5.4 \times 10^2\ ^\circ C$

The number of significant figures in the final answer is limited by the two significant figures in the temperatures given.

30. heat lost by steel = heat gained by water

$-m \times 0.50 \dfrac{J}{g\,^\circ C}(51.5-183)^\circ C = 66\,m = 125 \text{ mL} \times \dfrac{1.00 \text{ g}}{1 \text{ mL}} \times 4.18 \dfrac{J}{g\,^\circ C}(51.5-23.2)^\circ C$

$66\,m = 1.48 \times 10^4 \text{ J} \qquad m = \dfrac{1.48 \times 10^4}{66} = 2.2 \times 10^2 \text{ g stainless steel.}$

The precision of this method is limited by the assumption that no heat leaks out of the system. When we deal with temperatures far above (or far below) room temperature, this assumption becomes less and less valid. Further, the precision of the method is limited to two significant figures by the specific heat of the steel. If the two specific heats were known more precisely, then the temperature difference would determine the final precision of the method. It is unlikely that we could readily measure temperatures more precisely than $\pm 0.01\,^\circ C$, without expensive equipment. The mass of steel in this case would be measurable to four significant figures, to ± 0.1 g. This is hardly comparable to modern analytical balances which typically measure such masses to ± 0.1 mg.

31. heat lost by Mg = heat gained by water

$-\left(1.00 \text{ kg Mg} \times \dfrac{1000 \text{ g}}{1 \text{ kg}}\right)1.024 \dfrac{J}{g\ C}(T_f - 40.0\ C) = \left(1.00 \text{ L} \times \dfrac{1000 \text{ cm}^3}{1 \text{ L}} \times \dfrac{1.00 \text{ g}}{1 \text{ cm}^3}\right)4.18 \dfrac{J}{g\,^\circ C}(T_f - 20.0\ C)$

$-1.024 \times 10^3\ T_f + 4.10 \times 10^4 = 4.18 \times 10^3\ T_f - 8.36 \times 10^4$

$4.10 \times 10^4 + 8.36 \times 10^4 = (4.18 \times 10^3 + 1.024 \times 10^3)\ T_f \;\rightarrow\; 12.46 \times 10^4 = 5.20 \times 10^3\ T_f$

$T_f = \dfrac{12.46 \times 10^4}{5.20 \times 10^3} = 24.0\,^\circ C$

32. heat gained by the water = heat lost by the brass

$$150.0 \text{ g} \times 4.18 \frac{J}{g\,°C} \times \left(T_f - 22.4\,°C\right) = -\left(15.2 \text{ cm}^3 \times \frac{8.40 \text{ g}}{1 \text{ cm}^3}\right) 0.385 \frac{J}{g\,°C}\left(T_f - 163\,°C\right)$$

$$6.27 \times 10^2\, T_f - 1.40 \times 10^4 = -49.2\, T_f + 8.01 \times 10^3; \quad T_f = \frac{1.40 \times 10^4 + 8.01 \times 10^3}{6.27 \times 10^2 + 49.2} = 32.6\,°C$$

33. heat lost by copper = heat gained by glycerol

$$-74.8 \text{ g} \times \frac{0.385 \text{ J}}{g\,C} \times \left(31.1\,C - 143.2\,C\right) = 165 \text{ mL} \times \frac{1.26 \text{ g}}{1 \text{ mL}} \times \text{sp.ht.} \times \left(31.1\,C - 24.8\,C\right)$$

$$3.23 \times 10^3 = 1.3 \times 10^3 \times (\text{sp.ht.}) \quad \text{sp.ht.} = \frac{3.23 \times 10^3}{1.3 \times 10^3} = 2.5 \text{ J g}^{-1}\,°C^{-1}$$

$$\text{molar heat capacity} = 2.5 \text{ J g}^{-1}\,°C^{-1} \times \frac{92.1 \text{ g}}{1 \text{ mol C}_3\text{H}_8\text{O}_3} = 2.3 \times 10^2 \text{ J mol}^{-1}\,°C^{-1}$$

34. The additional water simply acts as a heat transfer medium. The essential relationship is heat lost by iron = heat gained by water (of unknown mass)

$$-\left(1.23 \text{ kg} \times \frac{1000 \text{ g}}{1 \text{ kg}}\right) 0.449 \frac{J}{g\,C}\left(25.6 - 68.5\right)C = x \text{ g H}_2\text{O} \times 4.18 \frac{J}{g\,C}\left(25.6 - 18.5\right)C$$

$$2.37 \times 10^4 \text{ J} = 29.7\, x \quad x = \frac{2.37 \times 10^4}{29.7} = 798 \text{ g H}_2\text{O} \times \frac{1 \text{ mL H}_2\text{O}}{1.00 \text{ g H}_2\text{O}} = 798 \text{ mL H}_2\text{O}$$

Heats of reaction

35. $\text{heat} = 283 \text{ kg} \times \frac{1000 \text{ g}}{1 \text{ kg}} \times \frac{1 \text{ mol Ca(OH)}_2}{74.09 \text{ g Ca(OH)}_2} \times \frac{65.2 \text{ kJ}}{1 \text{ mol Ca(OH)}_2} = 2.49 \times 10^5 \text{ kJ of heat}$

evolved.

36. $\text{heat energy} = 1.00 \text{ gal} \times \frac{3.785 \text{ L}}{1 \text{ gal}} \times \frac{1000 \text{ mL}}{1 \text{ L}} \times \frac{0.703 \text{ g}}{1 \text{ mL}} \times \frac{1 \text{ mol C}_8\text{H}_{18}}{114.2 \text{ g C}_8\text{H}_{18}} \times \frac{5.48 \times 10^3 \text{ kJ}}{1 \text{ mol C}_8\text{H}_{18}}$

$\text{heat energy} = 1.28 \times 10^5 \text{ kJ}$

37. (a) $\text{mass} = 2.80 \times 10^7 \text{ kJ} \times \frac{1 \text{ mol CH}_4}{890.3 \text{ kJ}} \times \frac{16.04 \text{ g CH}_4}{1 \text{ mol CH}_4} \times \frac{1 \text{ kg}}{1000 \text{ g}} = 504 \text{ kg CH}_4.$

(b) First determine the moles of CH_4 present, with the ideal gas law.

$$\text{mol } CH_4 = \frac{\left(768 \text{ mmHg} \times \dfrac{1 \text{ atm}}{760 \text{ mmHg}}\right) 1.65 \times 10^4 \text{ L}}{0.08206 \dfrac{\text{L atm}}{\text{mol K}} \times (18.6 + 273.2) \text{ K}} = 696 \text{ mol } CH_4$$

$$\text{heat energy} = 696 \text{ mol } CH_4 \times \frac{-890.3 \text{ kJ}}{1 \text{ mol } CH_4} = -6.20 \times 10^5 \text{ kJ of heat energy}$$

(c) $$V_{H_2O} = \frac{6.21 \times 10^5 \text{ kJ} \times \dfrac{1000 \text{ J}}{1 \text{ kJ}}}{4.18 \dfrac{\text{J}}{\text{g} \,^\circ\text{C}} (60.0 - 8.8)\,^\circ\text{C}} \times \frac{1 \text{ mL } H_2O}{1 \text{ g}} = 2.90 \times 10^6 \text{ mL} = 2.90 \times 10^3 \text{ L } H_2O$$

38. The combustion of 1.00 L (STP) of synthesis gas produces 11.13 kJ of heat. The volume of synthesis gas needed to heat 40.0 gal of water is given by first determining the quantity of heat needed to raise the temperature of the water.

$$\text{heat water} = \left(40.0 \text{ gal} \times \frac{3.785 \text{ L}}{1 \text{ gal}} \times \frac{1000 \text{ mL}}{1 \text{ L}} \times \frac{1.00 \text{ g}}{1 \text{ mL}}\right) 4.18 \frac{\text{J}}{\text{g} \,^\circ\text{C}} (65.0 - 15.2)\,^\circ\text{C}$$

$$= 3.15 \times 10^7 \text{ J} \times \frac{1 \text{ kJ}}{1000 \text{ J}} = 3.15 \times 10^4 \text{ kJ}$$

$$\text{gas volume} = 3.15 \times 10^4 \text{ kJ} \times \frac{1 \text{ L (STP)}}{11.13 \text{ kJ of heat}} = 2.83 \times 10^3 \text{ L at STP}$$

39. Since the molar mass of H_2 (2.0 g/mol) is $\frac{1}{16}$ of the molar mass of O_2 (32.0 g/mol) and only twice as many moles of H_2 are needed as O_2, we see that $O_2(g)$ is the limiting reagent in this reaction.

$$\frac{180}{2} \text{ g } O_2 \times \frac{1 \text{ mol } O_2}{32.0 \text{ g } O_2} \times \frac{241.8 \text{ kJ heat}}{0.500 \text{ mol } O_2} = 1.4 \times 10^3 \text{ kJ heat}$$

40. The amounts of the two reactants provided are the same as their stoichiometric coefficients in the balanced equation. Thus 852 kJ of heat is given off by the reaction. We can use this quantity of heat, along with the specific heat of the mixture to determine the temperature change that will occur if all of the heat is retained in the reaction mixture.

$$\text{heat} = 852 \text{ kJ} \times \frac{1000 \text{ J}}{1 \text{ kJ}} = 8.52 \times 10^5 \text{ J} = \text{mass} \times \text{sp.ht.} \times \Delta T$$

$$= \left(\left(1 \text{ mol } Al_2O_3 \times \frac{102 \text{ g } Al_2O_3}{1 \text{ mol } Al_2O_3}\right) + \left(2 \text{ mol Fe} \times \frac{55.8 \text{ g Fe}}{1 \text{ mol Fe}}\right)\right) 0.8 \frac{\text{J}}{\text{g} \,^\circ\text{C}} \times \Delta T$$

$$\Delta T = \frac{8.52 \times 10^5 \text{ J}}{214 \text{ g} \times 0.8 \text{ J g}^{-1}\,^{\circ}\text{C}^{-1}} = 5 \times 10^3\,^{\circ}\text{C}$$

The temperature needs to increase from 25°C to 1530°C or $\Delta T = 1505^{\circ}\text{C} = 1.5 \times 10^3\,^{\circ}\text{C}$. Since the actual ΔT is more than three times as large as this value, the iron indeed will melt, even if a large fraction of the heat evolved is lost to the surroundings and is not retained in the products.

41. **(a)** We first compute the heat produced by this reaction, then determine the value of ΔH in kJ/mol KOH.

$$q_{calorimeter} = (0.205 + 55.9) \text{ g} \times 4.18 \frac{\text{J}}{\text{g}^{\circ}\text{C}} (24.4 - 23.5) = 2 \times 10^2 \text{ J heat} = -q_{rxn}$$

$$\Delta H = -\frac{2 \times 10^2 \text{ J} \times \dfrac{1 \text{ kJ}}{1000 \text{ J}}}{0.205 \text{ g} \times \dfrac{1 \text{ mol KOH}}{56.1 \text{ g KOH}}} = -5 \times 10^1 \text{ kJ / mol}$$

(b) The ΔT here is known to just one significant figure (0.9 °C). Doubling the amount of KOH should give a temperature change known to two significant figures (1.6 °C) and using twenty times the mass of KOH should give a temperature change known to three significant figures (16.0 °C). This would require 4.10 g KOH rather than the 0.205 g KOH actually used, and would increase the precision from one part in five to one part in 500, or ~0.2 %. Note that as the mass of KOH is increased and the mass of H_2O stays constant, the assumption of a constant specific heat becomes less valid.

42. First determine the heat absorbed by the solute during the chemical reaction, q_{rxn}. This is the negative of the heat lost by the solution, q_{soln}. Since the solution (water plus solute) actually gives up heat, the temperature of the solution drops.

$$\text{heat of reaction} = 150.0 \text{ mL} \times \frac{1 \text{ L}}{1000 \text{ mL}} \times \frac{2.50 \text{ mol KI}}{1 \text{ L soln}} \times \frac{20.3 \text{ kJ}}{1 \text{ mol KI}} = 7.61 \text{ kJ} = q_{rxn}$$

$$-q_{rxn} = q_{soln} = \left(150.0 \text{ mL} \times \frac{1.30 \text{ g}}{1 \text{ mL}}\right) \times \frac{2.7 \text{ J}}{\text{g C}} \times \Delta T \quad \Delta T = \frac{-7.61 \times 10^3 \text{J}}{150.0 \text{ mL} \times \dfrac{1.30 \text{ g}}{1 \text{mL}} \times \dfrac{2.7 \text{ J}}{\text{g}^{\circ}\text{C}}} = -14^{\circ}\text{C}$$

final $T = $ initial $T + \Delta T = 23.5^{\circ}\text{C} - 14^{\circ}\text{C} = 10^{\circ}\text{C}$

43. Let x be the mass, (in grams), of NH_4Cl added to the water. heat $= \text{mass} \times \text{sp.ht.} \times \Delta T$

$$x \times \frac{1 \text{ mol NH}_4\text{Cl}}{53.49 \text{ g NH}_4\text{Cl}} \times \frac{14.7 \text{ kJ}}{1 \text{ mol NH}_4\text{Cl}} \times \frac{1000 \text{ J}}{1 \text{ kJ}} = -\left(\left(1400 \text{ mL} \times \frac{1.00 \text{ g}}{1 \text{ mL}}\right) + x\right) 4.18 \frac{\text{J}}{\text{g C}}(10. - 25) \text{ C}$$

$$275 \, x = 8.8 \times 10^4 + 63 \, x; \qquad x = \frac{8.8 \times 10^4}{275 - 63} = 4.2 \times 10^2 \text{ g NH}_4\text{Cl}$$

Our final value is approximate because of the assumed density (1.00 g/mL). The solution's density probably is a bit larger than 1.00 g/mL. Many aqueous solutions are somewhat more dense than water.

44. $\text{heat} = 500 \text{ mL} \times \dfrac{1 \text{ L}}{1000 \text{ mL}} \times \dfrac{7.0 \text{ mol NaOH}}{1 \text{ L soln}} \times \dfrac{-44.5 \text{ kJ}}{1 \text{ mol NaOH}} = -1.6 \times 10^2 \text{ kJ}$

$= \text{heat of reaction} = - \text{ heat absorbed by solution} \quad \text{OR} \quad q_{rxn} = -q_{soln}$

$\Delta T = \dfrac{1.6 \times 10^5 \text{ J}}{500.\text{mL} \times \dfrac{1.08 \text{ g}}{1\text{mL}} \times \dfrac{4.00 \text{ J}}{\text{g} \,^\circ\text{C}}} = 74\,^\circ\text{C} \qquad \text{final } T = 21\,^\circ\text{C} + 74\,^\circ\text{C} = 95 \,^\circ\text{C}$

45. We assume that the solution volumes are additive; that is, that 200.0 mL of solution is formed. Then we compute the heat needed to warm the solution and the cup, and then ΔH for the reaction.

$\text{heat} = \left(200.0 \text{ mL} \times \dfrac{1.02 \text{ g}}{1 \text{ mL}}\right) 4.02 \dfrac{\text{J}}{\text{g}\,^\circ\text{C}} (27.8 - 21.1)\,^\circ\text{C} + 10 \dfrac{\text{J}}{\,^\circ\text{C}} (27.8 - 21.1) = 5.6 \times 10^3 \text{ J}$

$\Delta H_{\text{neutr.}} = \dfrac{-5.6 \times 10^3 \text{ J}}{0.100 \text{ mol}} \times \dfrac{1 \text{ kJ}}{1000 \text{ J}} = -56 \text{ kJ/mol} \; \left(-55.6 \text{ kJ/mol to three significant figures}\right)$

46. Neutralization reaction: $\text{NaOH}(aq) + \text{HCl}(aq) \rightarrow \text{NaCl}(aq) + \text{H}_2\text{O}(l)$

Since NaOH and HCl react in a one-to-one molar ratio, and since there is twice the volume of NaOH solution as HCl solution, but the [HCl] is not twice the [NaOH], the HCl solution is the limiting reagent.

$\text{heat} = 25.00 \text{ mL} \times \dfrac{1 \text{ L}}{1000 \text{ mL}} \times \dfrac{1.86 \text{ mol HCl}}{1 \text{ L}} \times \dfrac{1 \text{ mol H}_2\text{O}}{1 \text{ mol HCl}} \times \dfrac{-55.84 \text{ kJ}}{1 \text{ mol H}_2\text{O}} = -2.60 \text{ kJ}$

$= \text{heat of reaction} = - \text{ heat absorbed by solution or } q_{rxn} = -q_{soln}$

$\Delta T = \dfrac{2.60 \times 10^3 \text{ J}}{75.00 \text{ mL} \times \dfrac{1.02 \text{ g}}{1\text{mL}} \times \dfrac{3.98 \text{ J}}{\text{g} \,^\circ\text{C}}} = 8.54\,^\circ\text{C} \qquad \Delta T = T_{final} - T_i \qquad T_{final} = \Delta T + T_i$

$T_{final} = 8.54\,^\circ\text{C} + 24.72 \,^\circ\text{C} = 33.26 \,^\circ\text{C}$

Enthalpy Changes and States of Matter

47. $q_{\text{H}_2\text{O}(l)} = q_{\text{H}_2\text{O}(s)} \quad m(\text{sp. ht.})_{\text{H}_2\text{O}(l)} \Delta T_{\text{H}_2\text{O}(l)} = \text{mol}_{\text{H}_2\text{O}(s)} \Delta H_{\text{fus H}_2\text{O}(s)}$

$\left(3.50 \text{ mol H}_2\text{O} \times \dfrac{18.015 \text{ g H}_2\text{O}}{1 \text{ mol H}_2\text{O}}\right)\left(4.184 \dfrac{\text{J}}{\text{g} \,^\circ\text{C}}\right)(50.0 \,^\circ\text{C}) = \left(\dfrac{m}{\dfrac{18.015 \text{ g H}_2\text{O}}{1 \text{ mol H}_2\text{O}}} \times 6.01 \times 10^3 \dfrac{\text{J}}{\text{mol}}\right)$

$13.2 \times 10^3 \text{ J} = m(333.\underline{6} \text{ J g}^{-1}) \quad$ Hence, $\text{m} = 39.6 \text{ g}$

48. $-q_{\text{lost by steam}} = q_{\text{gained by water}}$

$$-[(5.00 \text{ g } H_2O \times \frac{1 \text{ mol } H_2O}{18.015 \text{ g } H_2O})(-40.6 \times 10^3 \frac{J}{mol}) + (5.00 \text{ g})(4.184 \frac{J}{g \text{ }°C})(T_f - 100.0 \text{ °C})]$$

$$= (100.0 \text{ g})(4.184 \frac{J}{g \text{ }°C})(T_f - 25.0 \text{ °C})$$

$$11268.4 \text{ J} - 20.92 \frac{J}{°C}(T_f) + 2092 \text{ J} = 418.4 \frac{J}{°C}(T_f) - 10,460 \text{ J}$$

$$11,268.4 \text{ J} + 10,460 \text{ J} + 2092 \text{ J} = 418.4 \frac{J}{°C}(T_f) + 20.92 \frac{J}{°C}(T_f) \quad \text{or} \quad 23.8 \times 10^3 \text{ J} = 439 \text{ J}(T_f)$$

$$T_f = 54.2 \text{ °C}$$

49. Assume $H_2O(l)$ density $= 1.00 \text{ g mL}^{-1}$ (at 28.5 °C) $-q_{\text{lost by ball}} = q_{\text{gained by water}} + q_{\text{vap water}}$

$$-[(125 \text{ g})(0.50 \frac{J}{g \text{ }°C})(100 - 525 \text{ °C})] = [(75.0 \text{ g})(4.184 \frac{J}{g \text{ }°C})(100.0 - 28.5 \text{ °C})] + n_{H_2O}\Delta H°_{vap}$$

$$26562.5 \text{ J} = 22436.7 \text{ J} + n_{H_2O}\Delta H°_{vap} \quad (\text{Note: } n_{H_2O} = \frac{mass_{H_2O}}{molar \text{ } mass_{H_2O}})$$

$$4125.8 \text{ J} = (m_{H_2O})(\frac{1 \text{ mol } H_2O}{18.015 \text{ g } H_2O})(40.6 \times 10^3 \frac{J}{mol})$$

$$m_{H_2O} = 1.83 \text{ g } H_2O \cong 2 \text{ g } H_2O \text{ (1 sig. fig.)}$$

50. $-q_{\text{lost by ball}} = q_{\text{melt ice}}$

$$-[(125 \text{ g})(0.50 \frac{J}{g \text{ }°C})(0 - 525 \text{ °C})] = n_{H_2O}\Delta H°_{fus} = (m_{H_2O})(\frac{1 \text{ mol } H_2O}{18.015 \text{ g } H_2O})(6.01 \times 10^3 \frac{J}{mol})$$

$$32812.5 \text{ J} = m_{H_2O}(333.6 \text{ J g}^{-1}); \quad\quad m_{H_2O} = 98.4 \text{ g } H_2O \cong = 98 \text{ g } H_2O.$$

Bomb Calorimetry

51. To determine the heat capacity of the calorimeter, recognize that the heat evolved by the reaction is the negative of the heat of combustion.

$$\text{heat capacity} = \frac{\text{heat evolved}}{\Delta T} = \frac{1.620 \text{ g } C_{10}H_8 \times \frac{1 \text{ mol } C_{10}H_8}{128.2 \text{ g } C_{10}H_8} \times \frac{5156.1 \text{ kJ}}{1 \text{ mol } C_{10}H_8}}{8.44 \text{ °C}} = 7.72 \text{ kJ/°C}$$

52. Note that the heat evolved is the negative of the heat absorbed.

$$\text{heat capacity} = \frac{\text{heat evolved}}{\Delta T} = \frac{1.201\,g \times \dfrac{1\,mol\,C_7H_6O_3}{138.12\,g\,C_7H_6O_3} \times \dfrac{3023\,kJ}{1\,mol\,C_7H_6O_3}}{(29.82-23.68)\,°C} = 4.28\,kJ/°C$$

53. The temperature should increase as the result of an exothermic combustion reaction.

$$\Delta T = 1.227 \ g\,C_{12}H_{22}O_{11} \times \frac{1\ mol\ C_{12}H_{22}O_{11}}{342.3\ g\ C_{12}H_{22}O_{11}} \times \frac{5.65\times10^3\ kJ}{1\ mol\ C_{12}H_{22}O_{11}} \times \frac{1\,°C}{3.87\ kJ} = 5.23\,°C$$

54. $$q_{comb} = \frac{-11.23\,°C \times 4.68\,kJ/°C}{1.397\,g\,C_{10}H_{14}O \times \dfrac{1\,mol\,C_{10}H_{14}O}{150.2\,g\,C_{10}H_{14}O}} = -5.65\times10^3\ kJ\,/\,mol\,C_{10}H_{14}O$$

Pressure Volume Work

55. **(a)** $P\Delta V = 3.5\ L \times (748\ \text{mmHg})\left(\dfrac{1\,atm}{760\,mmHg}\right) = 3.4\underline{4}\ L\ atm$ or $3.4\ L\ atm$

(b) $1\ L\ kPa = 1\ J$, hence,

$$3.4\underline{4}\ L\ atm \times \left(\frac{101.325\ kPa}{1\ atm}\right) \times \left(\frac{1J}{1\,L\,kPa}\right) = 3.4\underline{9}\times10^2\ J \text{ or } 3.5\times10^2\ J$$

(c) $3.4\underline{9}\times10^2\ J \times \left(\dfrac{1\ cal}{4.184\ J}\right) = 83.\underline{4}\ cal$ or $83\ cal$

56. $w = -P\Delta V = -1.23\ atm \times (3.37\ L - 5.62\ L) \times \left(\dfrac{101.325\ kPa}{1\ atm}\right) \times \left(\dfrac{1J}{1\,L\,kPa}\right) = 280.\ J$

That is, 280. J of work is done on the gas.

57. There is no way to use the expansion process to raise or lower a weight in the surroundings, so there can be no work.

58. Yes, the gas from the aerosol does work. The gas pushes aside the atmosphere.

59. **(a)** No pressure-volume work done (no gases are formed or consumed).

(b) $2\ NO_2(g) \rightarrow N_2O_4(g)$ $\Delta n_{gas} = -1$ mole. Work is done on the system by the surroundings (compression).

(c) $CaCO_3(s) \rightarrow CaO(s) + CO_2(g)$. Formation of a gas, $\Delta n_{gas} = +1$ mole, results in an expansion. The system does work on the surroundings.

60. **(a)** $2 NO(g) + O_2(g) \rightarrow 2 NO_2(g)$ $\Delta n_{gas} = -1$ mole. Work is done on the system by the surroundings (compression).

(b) $MgCl_2(aq) + 2 NaOH(aq) \rightarrow Mg(OH)_2(s) + 2 NaCl(aq)$ $\Delta n_{gas} = 0$, no pressure-volume work is done.

(c) $CuSO_4(s) + 5 H_2O(g) \rightarrow CuSO_4 \bullet 5 H_2O(s)$ $\Delta n_{gas} = -5$ moles. Work is done on the system by the surroundings (compression).

61. **(a)** $\Delta U = q + w = +58 \text{ J} + (-58 \text{ J}) = 0$

(b) $\Delta U = q + w = +125 \text{ J} + (-687 \text{ J}) = -562 \text{ J}$

(c) $280 \text{ cal} \times (4.184 \frac{\text{J}}{\text{cal}}) = 1171.52 \text{ J} = 1.17 \text{ kJ}$ $\Delta U = q + w = -1.17 \text{ kJ} + 1.25 \text{ kJ} = 0.08 \text{ kJ}$

62. **(a)** $\Delta U = q + w = +235 \text{ J} + 128 \text{ J} = 363 \text{ J}$

(b) $\Delta U = q + w = -145 \text{ J} + 98 \text{ J} = -47 \text{ J}$

(c) $\Delta U = q + w = 0 \text{ kJ} + -1.07 \text{ kJ} = -1.07 \text{ kJ}$

63. **(a)** Yes, the gas does work (w = negative value).

(b) Yes, the gas exchanges energy with the surroundings, it absorbs energy.

(c) The temperature of the gas stays the same if the process is isothermal.

(d) ΔU for the gas must equal zero by definition (temperature is not changing).

64. **(a)** Yes, the gas does work (w = negative value).

(b) The internal energy of the gas decreases (energy expended to do work).

(c) The temperature of the gas should decrease, as it cannot attain thermal equilibrium with its surroundings.

65. Impossible, ideal gas expanding isothermally means that $\Delta U = 0 = q + w$, or $w = -q$, not $w = -2q$.

66. Compress the gas adiabatically – the gas will get hotter. Raise the temperature of the surroundings to an even hotter temperature and heat will be transferred to the gas.

67. According the First Law of Thermodynamics, the answer is (c). Both (a) q_v and (b) q_p are heats of chemical reaction carried out under conditions of constant volume and constant pressure respectively. Both ΔU and ΔH incorporate terms related to work as well as heat.

68. **(a)** $C_4H_{10}O(l) + 6 O_2(g) \rightarrow 4CO_2(g) + 5 H_2O(l)$ $\Delta n_{gas} = -2$ mol, $\Delta H < \Delta U$

(b) $C_6H_{12}O_6(s) + 6 O_2(g) \rightarrow 6 CO_2(g) + 6 H_2O(l)$ $\Delta n_{gas} = 0$ mol, $\Delta H = \Delta U$

(c) $NH_4NO_3(s) \rightarrow 2 H_2O(l) + N_2O(g)$ $\Delta n_{gas} = +1$ mol, $\Delta H > \Delta U$

69. $C_3H_8O(l) + 9/2\ O_2(g) \rightarrow 3\ CO_2(g) + 4\ H_2O(l) \quad \Delta n_{gas} = -1.5\ mol$

    (a) $\Delta U = -33.41\ \dfrac{kJ}{g} \times \dfrac{60.096\ g\ C_3H_8O}{1\ mol\ C_3H_8O} = -2008\ \dfrac{kJ}{mol}$

    (b) $\Delta H = \Delta U - w, = \Delta U - (-P\Delta V) = \Delta U - (-\Delta n_{gas}RT) = \Delta U + \Delta n_{gas}RT$

      $\Delta H = -2008\ \dfrac{kJ}{mol} + (-1.5\ mol)(\dfrac{8.3145 \times 10^{-3}\ kJ}{K\ mol})(298.15\ K) = -2012\ \dfrac{kJ}{mol}$

70. $C_{10}H_{14}O(l) + 13\ O_2(g) \rightarrow 10\ CO_2(g) + 7\ H_2O(l) \quad \Delta n_{gas} = -3\ mol$

    $q_{bomb} = q_v = \Delta U = -5.65 \times 10^3\ kJ\ = \Delta U$

    $\Delta H = \Delta U - w;$ where $w = (-\Delta n_{gas}RT) = -(-3\ mol)(\dfrac{8.3145 \times 10^{-3}\ kJ}{K\ mol})(298.15\ K) = +7.4\ kJ$

    $\Delta H = -5.65 \times 10^3\ kJ\ -7.4\ kJ = -5.66 \times 10^3\ kJ$

Hess's Law

71. $2HCl(g) + C_2H_4(g) + \tfrac{1}{2}O_2(g) \rightarrow C_2H_4Cl_2(l) + H_2O(l)\quad \Delta H^{\circ} = -318.7\ kJ$

    $\underline{Cl_2(g) + H_2O(l) \rightarrow 2HCl(g) + \tfrac{1}{2}O_2(g) \hspace{3.5cm} \Delta H^{\circ} = 0.5(+202.4) = +101.2\ kJ}$

    $C_2H_4(g) + Cl_2(g) \rightarrow C_2H_4Cl_2(l) \hspace{4cm} \Delta H^{\circ} = -217.5\ kJ$

72. $N_2H_4(l) + O_2(g) \rightarrow N_2(g) + 2H_2O(l) \hspace{2.5cm} \Delta H^{\circ} = -622.2\ kJ$

    $2H_2O_2(l) \rightarrow 2H_2(g) + 2O_2(g) \hspace{3cm} \Delta H^{\circ} = -2(-187.8\ kJ) = +375.6\ kJ$

    $\underline{2H_2(g) + O_2(g) \rightarrow 2H_2O(l) \hspace{3.5cm} \Delta H^{\circ} = 2(-285.8\ kJ) = -571.6\ kJ}$

    $N_2H_4(l) + 2H_2O_2(l) \rightarrow N_2(g) + 4H_2O(l) \hspace{1.8cm} \Delta H^{\circ} = -818.2\ kJ$

73. $CO(g) + \tfrac{1}{2}O_2(g) \rightarrow CO_2(g) \hspace{4cm} \Delta H^{\circ} = -283.0\ kJ$

    $3C(graphite) + 6H_2(g) \rightarrow 3CH_4(g) \hspace{2.5cm} \Delta H^{\circ} = 3(-74.81) = -224.43\ kJ$

    $2H_2(g) + O_2(g) \rightarrow 2H_2O(l) \hspace{3.2cm} \Delta H^{\circ} = 2(-285.8) = -571.6\ kJ$

    $\underline{3CO(g) \rightarrow \tfrac{3}{2}O_2(g) + 3C(graphite) \hspace{2.7cm} \Delta H^{\circ} = 3(+110.5) = +331.5\ kJ}$

    $4CO(g) + 8H_2(g) \rightarrow CO_2(g) + 3CH_4(g) + 2H_2O(l)\quad \Delta H^{\circ} = -747.5\ kJ$

74. $CS_2(l) + 3O_2(g) \rightarrow CO_2(g) + 2SO_2(g)$ $\qquad \Delta H^\circ = -1077$ kJ

$2S(s) + Cl_2(g) \rightarrow S_2Cl_2(l)$ $\qquad \Delta H^\circ = -58.2$ kJ

$C(s) + 2Cl_2(g) \rightarrow CCl_4(l)$ $\qquad \Delta H^\circ = -135.4$ kJ

$2SO_2(g) \rightarrow 2S(s) + 2O_2(g)$ $\qquad \Delta H^\circ = -2(-296.8 \text{ kJ}) = +593.6$ kJ

$CO_2(g) \rightarrow C(s) + O_2(g)$ $\qquad \Delta H^\circ = -(-393.5 \text{ kJ}) = +393.5$ kJ

$CS_2(l) + 3Cl_2(g) \rightarrow CCl_4(l) + S_2Cl_2(l)$ $\qquad \Delta H^\circ = -284$ kJ

75. $CH_4(g) + CO_2(g) \rightarrow 2CO(g) + 2H_2(g)$ $\qquad \Delta H^\circ = +247$ kJ

$2CH_4(g) + 2H_2O(g) \rightarrow 2CO(g) + 6H_2(g)$ $\qquad \Delta H^\circ = 2(+206 \text{ kJ}) = +412$ kJ

$CH_4(g) + 2O_2(g) \rightarrow CO_2(g) + 2H_2O(g)$ $\qquad \Delta H^\circ = -802$ kJ

$4CH_4(g) + 2O_2(g) \rightarrow 4CO(g) + 8H_2(g)$ $\qquad \Delta H^\circ = -143$ kJ

$\div 4$ produces $CH_4(g) + \frac{1}{2}O_2(g) \rightarrow CO(g) + 2H_2(g)$ $\quad \Delta H^\circ = -35.8$ kJ

76. The thermochemical combustion reactions follow.

$C_4H_6(g) + \frac{11}{2}O_2(g) \rightarrow 4CO_2(g) + 3H_2O(l)$ $\qquad \Delta H^\circ = -2540.2$ kJ

$C_4H_{10}(g) + \frac{13}{2}O_2(g) \rightarrow 4CO_2(g) + 5H_2O(l)$ $\qquad \Delta H^\circ = -2877.6$ kJ

$H_2(g) + \frac{1}{2}O_2(g) \rightarrow H_2O(l)$ $\qquad \Delta H^\circ = -285.85$ kJ

Then these equations are combined in the following manner.

$C_4H_6(g) + \frac{11}{2}O_2(g) \rightarrow 4CO_2(g) + 3H_2O(l)$ $\qquad \Delta H^\circ = -2540.2$ kJ

$4CO_2(g) + 5H_2O(l) \rightarrow C_4H_{10}(g) + \frac{13}{2}O_2(g)$ $\qquad \Delta H^\circ = -(-2877.6 \text{ kJ}) = +2877.6$ kJ

$2H_2(g) + O_2(g) \rightarrow 2H_2O(l)$ $\qquad \Delta H^\circ = 2(-285.8 \text{ kJ}) = -571.6$ kJ

$C_4H_6(g) + 2H_2(g) \rightarrow C_4H_{10}(g)$ $\qquad \Delta H^\circ = -234.2$ kJ

Standard Enthalpies (Heats) of Formation

77. The compounds that have negative standard enthalpies of formation are more stable than the elements from which they are made. Those that have positive standard enthalpies of formation are less stable than the elements. It is unlikely that many compounds would have zero standard enthalpy of formation. Such compounds would be exactly as stable as the elements from which they are made.

78. The correct answer is (b): "The standard enthalpy of formation of $CO_2(g)$ is the standard enthalpy of combustion of C(graphite)." The formation reaction for $CO_2(g)$ is the reaction in which one mol of $CO_2(g)$ is formed from 1 mole of C(graphite) and 1 mole of $O_2(g)$ (i.e.: $C(graphite) + O_2(g) \rightarrow CO_2(g)$). This is the combustion reaction of graphite. (a) This reaction does not have an enthalpy change of zero. (c) *and* (d) Carbon monoxide is not involved at all in the formation reaction.

79. $\Delta H^\circ = 4\Delta H_f^\circ[HCl(g)] + \Delta H_f^\circ[O_2(g)] - 2\Delta H_f^\circ[Cl_2(g)] - 2\Delta H_f^\circ[H_2O(l)]$

$= 4(-92.31) + (0.00) - 2(0.00) - 2(-285.8) = +202.4$ kJ

80. $\Delta H^\circ = 2\Delta H_f^\circ[Fe(s)] + 3\Delta H_f^\circ[CO_2(g)] - \Delta H_f^\circ[Fe_2O_3(s)] - 3\Delta H_f^\circ[CO(g)]$

$= 2(0.00) + 3(-393.5) - (-824.2) - 3(-110.5) = -24.8$ kJ

81. Balanced equation: $\quad C_2H_5OH(l) + 3O_2(g) \rightarrow 2CO_2(g) + 3H_2O(l)$

$\Delta H^\circ = 2\Delta H_f^\circ[CO_2(g)] + 3\Delta H_f^\circ[H_2O(l)] - \Delta H_f^\circ[C_2H_5OH(l)] - 3\Delta H_f^\circ[O_2(g)]$

$= 2(-393.5) + 3(-285.8) - (-277.7) - 3(0.00) = -1366.7$ kJ

82. First determine the value of ΔH_f° for $C_5H_{12}(l)$.

Balanced combustion equation: $\quad C_5H_{12}(l) + 8O_2(g) \rightarrow 5CO_2(g) + 6H_2O(l)$

$\Delta H^\circ = -3509$ kJ $= 5\Delta H_f^\circ[CO_2(g)] + 6\Delta H_f^\circ[H_2O(l)] - \Delta H_f^\circ[C_5H_{12}(l)] - 8\Delta H_f^\circ[O_2(g)]$

$= 5(-393.5) + 6(-285.8) - \Delta H_f^\circ[C_5H_{12}(l)] - 8(0.00) = -3682$ kJ $- \Delta H_f^\circ[C_5H_{12}(l)]$

$\Delta H_f^\circ[C_5H_{12}(l)] = -3682 + 3509 = -173$ kJ

Then use this value to determine the value of ΔH° for the reaction in question.

$\Delta H^\circ = \Delta H_f^\circ[C_5H_{12}(l)] + 5\Delta H_f^\circ[H_2O(l)] - 5\Delta H_f^\circ[CO(g)] - 11\Delta H_f^\circ[H_2(g)]$

$= -173 + 5(-285.8) - 5(-110.5) - 11(0.00) = -1050$ kJ

83. $\Delta H^\circ = -397.3$ kJ $= \Delta H_f^\circ[CCl_4(g)] + 4\Delta H_f^\circ[HCl(g)] - \Delta H_f^\circ[CH_4(g)] - 4\Delta H_f^\circ[Cl_2(g)]$

$= \Delta H_f^\circ[CCl_4(g)] + 4(-92.31) - (-74.81) - 4(0.00) = \Delta H_f^\circ[CCl_4(g)] - 294.4$

$\Delta H_f^\circ[CCl_4(g)] = -397.3 + 294.4 = -102.9$ kJ

84. $\Delta H^\circ = -8326$ kJ $= 12\Delta H_f^\circ[CO_2(g)] + 14\Delta H_f^\circ[H_2O(g)] - 2\Delta H_f^\circ[C_6H_{14}(l)] - 19\Delta H_f^\circ[O_2(g)]$

$\Delta H^\circ = 12(-393.5) + 14(-285.8) - 2\Delta H_f^\circ[C_6H_{14}(l)] - 19(0.00) = -8723 - 2\Delta H_f^\circ[C_6H_{14}(l)]$

$\Delta H_f^\circ[C_6H_{14}(l)] = \dfrac{+8326 - 8723}{2} = -199$ kJ / mol

85. $\Delta H^\circ = \Delta H_f^\circ\left[Al(OH)_3(s)\right] - \Delta H_f^\circ\left[Al^{3+}(aq)\right] - 3\,\Delta H_f^\circ\left[OH^-(aq)\right]$

$\qquad = (-1276) - (-531) - 3\,(-230.0) = -55\ \text{kJ}$

86. $\Delta H^\circ = \Delta H_f^\circ\left[Mg^{2+}(aq)\right] + 2\Delta H_f^\circ\left[NH_3(g)\right] + 2\,\Delta H_f^\circ\left[H_2O(l)\right] - \Delta H_f^\circ\left[Mg(OH)_2(s)\right]$

$\qquad - 2\,\Delta H_f^\circ\left[NH_4^+(aq)\right]$

$\qquad = (-466.9) + 2\,(-46.11) + 2\,(-285.8) - (-924.5) - 2\,(-132.5) = +58.8\ \text{kJ}$

87. Balanced equation: $\qquad CaCO_3(s) \rightarrow CaO(s) + CO_2(g)$

$\Delta H^\circ = \Delta H_f^\circ\left[CaO(s)\right] + \Delta H_f^\circ\left[CO_2(g)\right] - \Delta H_f^\circ\left[CaCO_3(s)\right]$

$\qquad = -635.1 - 393.5 - (-1207) = +178\ \text{kJ}$

$\text{heat} = 1.35\times10^3\ \text{kg CaCO}_3 \times \dfrac{1000\ \text{g}}{1\ \text{kg}} \times \dfrac{1\ \text{mol CaCO}_3}{100.09\ \text{g CaCO}_3} \times \dfrac{178\ \text{kJ}}{1\ \text{mol CaCO}_3} = 2.40\times10^6\ \text{kJ}$

88. Determine the heat of combustion: $\qquad C_4H_{10}(g) + \tfrac{13}{2}O_2(g) \rightarrow 4CO_2(g) + 5H_2O(l)$

$\Delta H^\circ = 4\,\Delta H_f^\circ\left[CO_2(g)\right] + 5\,\Delta H_f^\circ\left[H_2O(l)\right] - \Delta H_f^\circ\left[C_4H_{10}(g)\right] - \tfrac{13}{2}\Delta H_f^\circ\left[O_2(g)\right]$

$\qquad = 4\,(-393.5) + 5\,(-285.8) - (-125.6) - 6.5\,(0.00) = -2877.4\ \text{kJ/mol}$

Now compute the volume of gas needed, with the ideal gas equation, rearranged to: $V = nRT/P$.

$$\text{volume} = \dfrac{\left(5.00\times10^4\,\text{kJ}\times\dfrac{1\,\text{mol}\,C_4H_{10}}{2877.3\,\text{kJ}}\right)0.08206\dfrac{\text{L atm}}{\text{mol K}}(24.6+273.2)\text{K}}{756\,\text{mmHg}\times\dfrac{1\,\text{atm}}{760\,\text{mmHg}}} = 427\,\text{L CH}_4$$

FEATURE PROBLEMS

110. $1\ ^\circ\text{F} = 5/9\ ^\circ\text{C} = 0.555\ ^\circ\text{C}$ $\qquad 1\ \text{lb} = 453.6\ \text{g}$

$E_p = mgh$

$E_p = (772\ \text{lb})(9.80665\ \text{m s}^{-2})(1\ \text{ft}) = 7.57\times10^3\ \dfrac{\text{lb m ft}}{\text{s}^2}$

$E_p = 7.57\times10^3\ \dfrac{\text{lb m ft}}{\text{s}^2} \times \dfrac{0.3048\ \text{m}}{\text{ft}} \times \dfrac{0.4536\ \text{kg}}{\text{lb}}$

$E_p = 1047\ \dfrac{\text{kg m}^2}{\text{s}^2} = 1047\ \text{J} = 1.05\ \text{kJ}$

The statement is validated.

$q = m \times (\text{sp. ht.})\Delta T$

$q = 453.6\ \text{g}(4.184\dfrac{\text{J}}{\text{g}\ ^\circ\text{C}})(0.556\ ^\circ\text{C})$

$q = 1054\ \text{J}$

$q = 1.05\ \text{kJ}$

111. (a) We plot specific heat vs. the inverse of atomic mass.

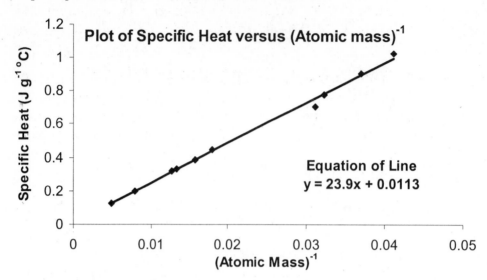

(b) The equation of the line is: Specific Heat = $23.9 \div$(atomic mass) + 0.0113

$$0.23 \text{ J g}^{-1} \,^{\circ}\text{C} = 23.9 \div\text{(atomic mass)} + 0.0113$$

$$\text{atomic mass} = \frac{23.9}{0.23 - 0.0113} = 109 \text{ u or } 110 \text{ u (2 sig fig)};$$

Cadmium's tabulated atomic mass is 112.4 u.

(c) $\text{sp.ht.} = \dfrac{450 \text{ J}}{75.0 \text{ g} \times 15.\,^{\circ}\text{C}} = 0.40 \text{ J g}^{-1}\,^{\circ}\text{C}^{-1} = 0.0113 + 23.9/\text{atomic mass}$

$$\text{atomic mass} = \frac{23.9}{0.40 - 0.0113} = 61.5 \text{ u or } 62 \text{ u} \quad \text{Metal most likely Cu} \,(63.5 \text{ u})$$

112. The plot's maximum is the equivalence point. (assume $\Delta T = 0$ at 0 mL of added NaOH, (i.e. only 60 mL of citric acid) and that $\Delta T = 0$ at 60 mL of NaOH (i.e. no citric acid added)

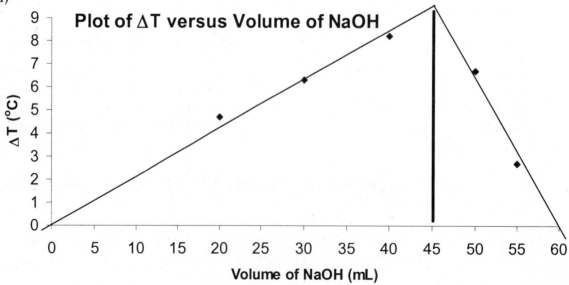

(a) The equivalence point occurs with 45.0 mL of 1.00 M NaOH(aq) [45.0 mmol NaOH] and 15.0 mL of 1.00 M citric acid [15.0 mmol citric acid]. Again, we assume that $\Delta T =$ zero if no NaOH added ($V_{NaOH} = 0$ mL) and $\Delta T = 0$ if no citric acid is added ($V_{NaOH} = 60$).

(b) Heat is a product of the reaction, just as chemical species (products) are. Products are maximized at the exact stoichiometric proportions. Since each reaction mixture has the same volume, and thus about the same mass to heat, the temperature also is a maximum at this point.

(c) $H_3C_6H_5O_7(s) + 3OH^-(aq) \rightarrow 3H_2O(l) + C_6H_5O_7^{3-}(aq)$

113. The reactions, and their temperature changes, are as follows.

(1st) $NH_3(conc.\ aq) + HCl(aq) \rightarrow NH_4Cl(aq)$ $\Delta T = (35.8 - 23.8)°C = 12.0°C$

(2nd,a) $NH_3(conc.\ aq) \rightarrow NH_3(g)$ $\Delta T = (13.2 - 19.3)°C = -6.1°C$

(2nd,b) $NH_3(g) + HCl(aq) \rightarrow NH_4Cl(aq)$ $\Delta T = (42.9 - 23.8)°C = 19.1°C$

The sum of reactions $(2nd, a) + (2nd, b)$ produces the same change as the 1st reaction. We now compute the heat absorbed by the surroundings for each. Hess's law is demonstrated if $\Delta H_1 = \Delta H_{2a} + \Delta H_{2b}$, where in each case $\Delta H = -q$.

$$q_1 = \left[(100.0\ mL + 8.00\ mL) \times 1.00\ g/mL\right]4.18\ J\ g^{-1}°C^{-1} \times 12.0°C = 5.42 \times 10^3\ J = -\Delta H_1$$

$$q_{2a} = \left[(100.0\ mL \times 1.00\ g/mL)\ 4.18\ J\ g^{-1}°C^{-1} \times (-6.1°C)\right] = -2.55 \times 10^3\ J = -\Delta H_{2a}$$

$$q_{2b} = \left[(100.0\ mL \times 1.00\ g/mL)\ 4.18\ J\ g^{-1}°C^{-1} \times (19.1°C)\right] = 7.98 \times 10^3\ J = -\Delta H_{2b}$$

$$\Delta H_{2a} + \Delta H_{2b} = +2.55 \times 10^3\ J - 7.98 \times 10^3\ J = -5.43 \times 10^3\ J \cong -5.42 \times 10^3\ J = \Delta H_1$$

Thus, the argument is complete.

114. According to the kinetic-molecular theory of gases, the internal energy of an ideal gas, U, is proportional to the average translational kinetic energy for the gas particles, $\bar{e}_k$, which in turn is proportional to 3/2 RT. Thus the internal energy for a fixed amount of an ideal gas depends only on its temperature, i.e. $U = 3/2\ nRT$, where U is the internal energy (J), n is the number of moles of gas particles, R is the gas constant (J K^{-1} mol^{-1}), and T is the temperature (K).

If the temperature of the gas sample is changed, the resulting change in internal energy is given by $\Delta U = 3/2\ nR\Delta T$.

(a) At constant volume, $q_v = nC_v\Delta T$. Assuming that no work is done $\Delta U = q_v$ so,
$\Delta U = q_v = 3/2nR\Delta T = nC_v\Delta T$. (divide both sides by $n\Delta T$) $C_v = 3/2\ R = 12.5$ J/K·mol.

(b) The heat flow at constant pressure q_p is the ΔH for the process (i.e., $q_p = \Delta H$) and we know that $\Delta H = \Delta U - w$ and $w = -P\Delta V = -nR\Delta T$

Hence, $q_p = \Delta U - w = \Delta U - (-nR\Delta T) = \Delta U + nR\Delta T$ and $q_p = nC_p\Delta T$ and $\Delta U = 3/2\ nR\Delta T$

Consequently $q_p = nC_p\Delta T = nR\Delta T + 3/2\ nR\Delta T$ (divide both sides by $n\Delta T$)

Now, $C_p = R + 3/2\ R = 5/2\ R = 20.8$ J/K·mol

115. (a) Determine the volume between 2.40 atm and 1.30 atm using $PV = nRT$ that is, plug in the pressure into:

$$V = \frac{0.100\,\text{mol} \times 0.08206\ \dfrac{\text{L atm}}{\text{K mol}} \times 298\ \text{K}}{P} = \frac{2.44\underline{5}\ \text{L atm}}{P}$$

For $P = 2.40$ atm: $V = 1.02$ L
For $P = 2.30$ atm: $V = 1.06$ L $P\Delta V = 2.30$ atm $\times -0.04$ L $= -0.09\underline{2}$ L atm
For $P = 2.20$ atm: $V = 1.11$ L $P\Delta V = 2.20$ atm $\times -0.05$ L $= -0.1\underline{1}$ L atm
For $P = 2.10$ atm: $V = 1.16$ L $P\Delta V = 2.10$ atm $\times -0.05$ L $= -0.1\underline{1}$ L atm
For $P = 2.00$ atm: $V = 1.22$ L $P\Delta V = 2.00$ atm $\times -0.06$ L $= -0.1\underline{2}$ L atm
For $P = 1.90$ atm: $V = 1.29$ L $P\Delta V = 1.90$ atm $\times -0.06$ L $= -0.1\underline{2}$ L atm
For $P = 1.80$ atm: $V = 1.36$ L $P\Delta V = 1.80$ atm $\times -0.07$ L $= -0.1\underline{3}$ L atm
For $P = 1.70$ atm: $V = 1.44$ L $P\Delta V = 1.70$ atm $\times -0.08$ L $= -0.1\underline{4}$ L atm
For $P = 1.60$ atm: $V = 1.53$ L $P\Delta V = 1.60$ atm $\times -0.09$ L $= -0.1\underline{4}$ L atm
For $P = 1.50$ atm: $V = 1.63$ L $P\Delta V = 1.50$ atm $\times -0.10$ L $= -0.1\underline{5}$ L atm
For $P = 1.40$ atm: $V = 1.75$ L $P\Delta V = 1.40$ atm $\times -0.12$ L $= -0.1\underline{7}$ L atm
For $P = 1.30$ atm: $V = 1.88$ L $P\Delta V = 1.30$ atm $\times -0.13$ L $= -0.1\underline{7}$ L atm

Total Work $=$ $-\Sigma\,P\Delta V$ $= -1.4\underline{5}$ L atm

Expressed in joules, the work is $-1.4\underline{5}$ L atm $\times 101.325$ J/L atm $= -14\underline{7}$ J

(b)

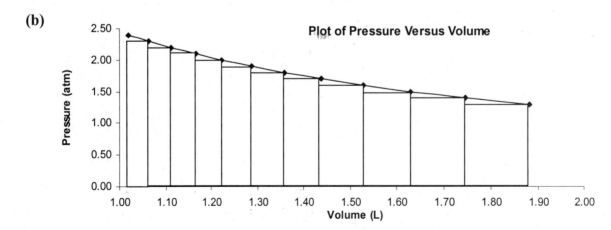

(c)   The total work done in the two-step expansion is minus the total of the area of the two rectangles under the graph is –1.29 L atm or –131 J. In the 11-step expansion in (b), the total area of the rectangles is 1.45 L atm or –14$\underline{7}$ J. If the expansion were divided into a larger number of stages, the total area of the rectangles would be still greater. The maximum amount of work is for an expansion with an infinite number of stages and is equal to the area under the pressure-volume curve between $V = 1.02$ L and 1.88 L. This area is also obtained as the integral obtained from the expression:

$dw = -PdV = -nRT(dV/V)$. The value obtained is:

$w = -nRT \times \ln V_f/Vi = 0.100$ mol $\times$ 8.3145 J mol^{-1} K^{-1} $\times$ 298 K $\times$ ln (1.88 L/1.02 L)

$w = -152$ J

(d)   The maximum work of compression is for a one-stage compression using an external pressure of 2.40 atm and producing a compression in volume of 1.02 L – 1.88 L = –0.86 L:

$w = -P\Delta V = (2.40$ atm $\times 0.86$ L$) \times 101.33$ J/L atm $= 209$ J

The minimum work would be that done in an infinite number of steps and would be the same as the work determined in (c) but with a positive sign, that is, +152 J.

(e)   Because the internal energy of an ideal gas is a function only of temperature, and the temperature remains constant, $\Delta U = 0$. Because $\Delta U = q + w = 0$, q = –w, that is, –209 J corresponding to the maximum work of compression and –152 J corresponding to the minimum work of compression.

(f)   When the expansion described in part in (c),

$q = -w = nRT \ln V_f/Vi$ and $q/T = nR \ln V_f/V_i$

Because the terms on the right side are all constants or functions of state, so too is the term on the left, q/T. In Chapter 20, we learn that q/T is equal to ΔS, the change in a state function called *entropy*.

CHAPTER 8
THE ATMOSPHERIC GASES AND HYDROGEN
PRACTICE EXAMPLES

1A By volume, the atmosphere is 0.934% Ar. $V_{air} = 5.00 \text{ L Ar} \times \dfrac{100.00 \text{ L air}}{0.934 \text{ L Ar}} = 535 \text{ L air}$

1B Calculate the volume of $CO_2(g)$, and of air. The atmosphere is 0.037% $CO_2(g)$ by volume.

$$V_{air} = \frac{nRT}{P} = \frac{\dfrac{5.00 \text{ g } CO_2}{\left(\dfrac{44.01 \text{ g } CO_2}{1 \text{ mol } CO_2}\right)} \times \left(0.08206 \dfrac{\text{L atm}}{\text{mol K}} \times 298 \text{ K}\right)}{1 \text{ atm}} \times \frac{100.00 \text{ L air}}{0.037 \text{ L } CO_2} = 7.5 \times 10^3 \text{ L air}$$

2A First substitute chemical formulas for names:

$PbO_2(s) + HNO_3(aq) \rightarrow Pb(NO_3)_2(aq) + O_2(g) + H_2O(l)$ Balance by inspection.

The oxidation and reduction reactants are in the same substance, $PbO_2(s)$.

$PbO_2(s) + 2 H^+(aq) \rightarrow Pb^{2+}(aq) + H_2O(l) + \dfrac{1}{2} O_2(g)$

$2 PbO_2(s) + 4 HNO_3(aq) \rightarrow 2 Pb(NO_3)_2(aq) + O_2(g) + 2 H_2O(l)$

2B First substitute chemical formulas for names.

$Zn(s) + HNO_3(aq) \rightarrow Zn(NO_3)_2(aq) + NH_4NO_3(aq) + H_2O(l)$

Then write the two skeleton half-equations:

$Zn(s) \rightarrow Zn^{2+}(aq)$ and $NO_3^-(aq) \rightarrow NH_4^+(aq)$

Balance the nitrogen half-equation for oxygen:

$NO_3^-(aq) \rightarrow NH_4^+(aq) + 3 H_2O(l)$ and for hydrogen:

$10 H^+(aq) + NO_3^-(aq) \rightarrow NH_4^+(aq) + 3 H_2O(l)$

Balance the two half-equations for charge with electrons and then combine them to produce the net ionic equation. Finally, add in the nitrate spectator ions.

Oxidation: $(Zn(s) \rightarrow Zn^{2+}(aq) + 2 e^-) \times 4$

Reduction: $10 H^+(aq) + NO_3^-(aq) + 8 e^- \rightarrow NH_4^+(aq) + 3 H_2O(l)$

Net: $4 Zn(s) + 10 H^+(aq) + NO_3^-(aq) \rightarrow 4 Zn^{2+}(aq) + NH_4^+(aq) + 3 H_2O(l)$

Spectators: $9NO_3^-(aq) \rightarrow 9NO_3^-(aq)$

Final: $4 Zn(s) + 10 HNO_3(aq) \rightarrow 4 Zn(NO_3)_2(aq) + NH_4NO_3(aq) + 3 H_2O(l)$

3A As we learned in Chapter 5, half-equations for reactions in basic solution are initially balanced as if they occur in acidic solution. Both of the products in this case are composed of the elements of water. It is easiest to begin with skeleton half-equations that contain only products.

Skeletons: $\rightarrow H_2(g)$ at cathode $\rightarrow O_2(g)$ at anode

Balance oxygens: $\rightarrow H_2(g)$ $2 H_2O \rightarrow O_2(g)$

Balance hydrogens: $2 H^+(aq) \rightarrow H_2(g)$ $2 H_2O \rightarrow O_2(g) + 4 H^+(aq)$

Balance charge: $2 H^+(aq) + 2 e^- \rightarrow H_2(g)$ $2 H_2O \rightarrow O_2(g) + 4 H^+(aq) + 4 e^-$

Add OH^- 's: $2 OH^-(aq) + 2 H^+(aq) + 2 e^- \rightarrow H_2(g) + 2 OH^-(aq)$

$2 H_2O + 4 OH^-(aq) \rightarrow O_2(g) + 4 H^+(aq) + 4 OH^-(aq) + 4 e^-$

Simplify: $2 H_2O + 2 e^- \xrightarrow[\text{reduction}]{\text{cathode}} H_2(g) + 2 OH^-(aq)$

$4 OH^-(aq) \xrightarrow[\text{oxidation}]{\text{anode}} O_2(g) + 2 H_2O(l) + 4 e^-$

3B First substitute chemical formulas for names: $Li_2O_2(s) + CO_2(g) \rightarrow Li_2CO_3(s) + O_2(g)$ Simply use "2" as the stoichiometric coefficient of all species except $O_2(g)$.
$2 Li_2O_2(s) + 2 CO_2(g) \rightarrow 2 Li_2CO_3(s) + O_2(g)$. Oxygen has an oxidation state of −1 in $Li_2O_2(s)$, and −2 in $Li_2CO_3(s)$ and 0 in $O_2(g)$. Thus, $Li_2O_2(s)$ is both an oxidizing agent and a reducing agent.

REVIEW QUESTIONS

1. **(a)** Relative humidity is the ratio of the partial pressure of water vapor to the vapor pressure of water at a given temperature.

    (b) A noble gas is one of the elements in periodic table family 8A: He, Ne, Ar, Kr, Xe, and Rn. These elements are called "noble" because they form few compounds with the other elements. Like "nobility", they do not associate with the other elements.

    (c) A chlorofluorocarbon is a compound, derived from a hydrocarbon that contains the elements carbon, chlorine, and fluorine. These compounds are good-to-excellent refrigerants, but their release into the atmosphere is implicated in the destruction of the ozone layer in the upper atmosphere.

    (d) A nonstoichiometric compound is one in which the ratio of moles of the constituent elements is not a ratio of small whole numbers.

2.   **(a)**   Fractional Distillation is a method used to separate a mixture of volatile components of different boiling points. The mixture is distilled starting at the lowest boiling point where the distillate is collected as one fraction. When the temperature of the vapor rises, the next highest boiling component of the mixture distills. This component is then collected as a separate fraction.

   (b)   Electrolysis is the process by which a non-spontaneous redox reaction is made spontaneous through the passage of electric current from an external power source.

   (c)   Hydrogenation reaction is a reaction where molecular hydrogen ($H_2(g)$) is added to double or triple bonds. Reaction is generally done in the presences of a catalyst and may involve high pressures and temperatures.

   (d)   Dew and frost formation result from fluctuations in the relative humidity with temperature. When water vapor condenses on the earth's surface - this is known as dew; however, if the temperature drops below 0 °C, frost formation results (frozen dew).

3.   **(a)**   The troposphere is the first 12 km of the Earth's atmosphere, the region in which we live and within which weather occurs. The stratosphere is the next layer, 12 to 50 km above the Earth's surface.

   (b)   An allotrope is a form of an element that differs from another form in physical and chemical properties. An isotope is an alternative version of an atom that differs in the neutron number.

   (c)   A "fuel-lean" mixture in an engine is one in which there is more air than is needed to completely burn the fuel. In a "fuel-rich" mixture, there is an abundance of fuel and insufficient air.

   (d)   An ionic hydride is composed of the hydride anion and an active metal cation; it is a stoichiometric compound. Metallic hydrides are composed of hydrogen and less active metals . It actually consists of hydrogen atoms in the voids present in the metallic lattice. It is a nonstoichiometric compound.

4.   **(a)** O_3, ozone   **(b)** N_2O, dinitrogen monoxide (nitrous oxide)

   (c) KO_2, potassium superoxide   **(d)** CaH_2, calcium hydride

   (e) Mg_3N_2, magnesium nitride   **(f)** K_2CO_3, potassium carbonate

   (g) $NH_4H_2PO_4$, ammonium dihydrogen phosphate

5.   **(a)** $C(s) + H_2O(g) \rightarrow CO(g) + H_2(g)$ and $CO(g) + H_2O(g) \rightarrow CO_2(g) + H_2O(g)$

   (b)   coke: almost pure C   **(c)** urea: $CO(NH_2)_2$ **(d)** limestone: principally $CaCO_3$

   (e)   a mixture of $CO(g)$ and $H_2(g)$, produced by steam reforming of a hydrocarbon.

6.   **(a)** NO_2 has N in O.S. $= +4$. **(b)** KO_2 has O in O.S. $= -1/2$.

   (c) N_2O has N in O.S. $= +1$. **(d)** CaH_2 has H in O.S. $= -1$. **(e)** CO has C in O.S. $= +2$.

7. **(a)** Small quantities of $O_2(g)$ can be prepared either by the electrolysis of water, or by the gentle heating of $KClO_3(s)$ in the presence of a $MnO_2(s)$ catalyst.

$$2\,H_2O(l) \xrightarrow{\text{electroysis}} 2H_2(g) + O_2(g) \qquad 2\,KClO_3(s) \xrightarrow{\text{heat, MnO}_2} 2\,KCl(s) + 3\,O_2(g)$$

(b) Small quantities of $N_2O(g)$ can be produced by the thermal decomposition of ammonium nitrate.

$$NH_4NO_3(s) \xrightarrow{\text{200-260°C}} N_2O(g) + 2\,H_2O(g)$$

(c) Small quantities of $H_2(g)$ can be produced by the reaction of a moderately active metal with a strong acid: $Zn(s) + 2\,HCl(aq) \rightarrow ZnCl_2(aq) + H_2(g)$

(d) Small quantities of $CO_2(g)$ can be produced by the reaction of a metal carbonate with a strong acid. $\qquad CaCO_3(s) + 2\,HCl(aq) \rightarrow CaCl_2(aq) + H_2O(l) + CO_2(g)$

8. For a reagent to be an oxidizing agent, it must have atoms that can be easily reduced (to a lower oxidation state). H_2 and NH_3 are not oxidizing reagents as they are in their lowest common oxidation states for the two elements (0 and –3, for H and N, respectively). Both NO_2 and NO have the potential to be oxidizing agents (N in the +4 and +2 oxidation states respectively). NO_2 would be the best oxidizing agent as nitrogen in this compound is in a higher oxidation state and, as we have seen in this chapter, it can be reduced to NO (e.g. see equations 8.8 and 8.9).

9. **(a)** $LiH(s) + H_2O(l) \rightarrow Li^+(aq) + OH^-(aq) + H_2(g)$

(b) $C(s) + H_2O(g) \xrightarrow{\Delta} CO(g) + H_2(g)$

(c) $3\,NO_2(g) + H_2O(l) \rightarrow 2\,HNO_3(aq) + NO(g)$

10. **(a)** $Mg(s) + 2\,HCl(aq) \rightarrow MgCl_2(aq) + H_2(g)$

(b) $NH_3(g) + HNO_3(aq) \rightarrow NH_4NO_3(aq)$

(c) $MgCO_3(s) + 2\,HCl(aq) \rightarrow MgCl_2(aq) + H_2O(l) + CO_2(g)$

(d) $NaHCO_3(s) + HC_2H_3O_2(aq) \rightarrow NaC_2H_3O_2(aq) + H_2O(l) + CO_2(g)$

11. Aqueous sulfuric acid is $H_2SO_4(aq)$; aqueous ammonia is $NH_3(aq)$. The equation for their complete neutralization is: $H_2SO_4(aq) + 2\,NH_3(aq) \rightarrow (NH_4)_2SO_4(aq)$

12. We can identify oxidizing and reducing agents by changes in oxidation state. The oxidation state of one element in an oxidizing agent is lowered, while the oxidation state of one element in a reducing agent is raised.

(8.4) $4NH_3(g) + 5O_2(g) \xrightarrow{\text{850°C, Pt}} 4\ NO(g) + 6H_2O(g)$

The oxidation state of N is -3 on the left- and +2 on the right-hand side of this equation; the O.S. of N increases. $NH_3(g)$ is the reducing agent. The oxidation state of O is 0 on the left- and -2 on the right-hand side of this equation; the O.S. of O decreases. $O_2(g)$ is the oxidizing agent.

(8.8) $3\ NO_2(g) + H_2O(l) \rightarrow 2\ HNO_3(aq) + NO(g)$

The oxidation state of N is +4 on the left- and +5 in $HNO_3(aq)$ on the right-hand side of this equation; the O.S. of N increases. $NO_2(g)$ is the reducing agent. The oxidation state of N is +4 on the left- and +2 in NO(g) on the right-hand side of this equation; the O.S. of O decreases. $NO_2(g)$ is also the oxidizing agent. This is a disproportionation reaction.

(8.16) $C_8H_{18}(l) + \dfrac{25}{2}\ O_2(g) \rightarrow 8\ CO_2(g) + 9\ H_2O(l)$

The oxidation state of C is 2.25 on the left- and +4 on the right-hand side of this equation; the O.S. of C increases. $C_8H_{18}(l)$ is the reducing agent. The oxidation state of O is 0 on the left- and -2 on the right-hand side of this equation; the O.S. of oxygen decreases. $O_2(g)$ is the oxidizing agent.

(8.21) $CaH_2(s) + 2\ H_2O(l) \rightarrow Ca(OH)_2(s) + 2\ H_2(g)$

The oxidation state for H in $CaH_2(s)$ is −1 and in $H_2(g)$ it is 0. Consequently, $CaH_2(s)$ is the reducing agent (H is oxidized in the process). The oxidation state for H in $H_2O(l)$ is +1 and in $H_2(g)$ it is 0. Consequently, $H_2O(l)$ is the oxidizing agent (H is reduced in the process).

13. The third step in the Ostwald process is a disproportionation reaction ($N^{4+} \rightarrow N^{2+}$ and N^{5+}):

$3\ NO_2(g) + H_2O(l) \rightarrow 2\ HNO_3(aq) + NO(g)$

14. The oxide of iron with an oxidation state of +3 is

$Fe_2O_3(s)$: $Fe_2O_3(s) + 3\ H_2(g) \rightarrow 2\ Fe(s) + 3\ H_2O(g)$

15. We assume that copper is oxidized to copper(II) ion. The skeleton half-equations are as follows.

$Cu(s) \rightarrow Cu^{2+}(aq)$ $NO_3^-(aq) \rightarrow NO(g)$

The nitrogen-containing half-reaction is balanced in two steps, first balance for O then for H.

$NO_3^-(aq) \rightarrow NO(g) + 2\ H_2O(l)$ $NO_3^-(aq) + 4\ H^+(aq) \rightarrow NO(g) + 2\ H_2O(l)$

The two half-equations then are balanced for charge using electrons and combined to produce the net ionic equation. Finally, nitrate spectator ions are added to give the balanced equation.

Oxidation: $\{Cu(s) \rightarrow Cu^{2+}(aq) + 2\ e^-$ $\qquad \} \times 3$

Reduction: $\{NO_3^-(aq) + 4\ H^+(aq) + 3\ e^- \rightarrow NO(g) + 2\ H_2O(l) \qquad \} \times 2$

Net: $3\ Cu(s) + 2NO_3^-(aq) + 8\ H^+(aq) \rightarrow 3\ Cu^{2+}(aq) + 2\ NO(g) + 4\ H_2O\ (l)$

16. The gaseous thermal decomposition product is given as the last product in each of these chemical equations.

$2KClO_3(s) \xrightarrow{\text{heat, MnO}_2} 2KCl(s) + 3O_2(g)$ $\qquad$ Oxygen gas is produced.

$CaCO_3(s) \xrightarrow{\Delta} CaO(s) + CO_2(g)$ $\qquad$ Carbon dioxide gas is formed.

$NH_4NO_3(s) \xrightarrow{200-260°C} 2H_2O(g) + N_2O(g)$ $\qquad$ Dinitrogen monoxide gas is formed.

17. **(a)** The basic causes of photochemical smog are high temperature combustion processes, notably those associated with automobile engines (aircraft and industrial processes also contribute to this problem).

(b) Ozone layer depletion results from two major sources: (1) Nitrogen monoxides produced by combustion reactions (supersonic jets operating in the stratosphere) or from natural sources (NO_2 producing bacteria) and (2) chlorofluorocarbons (CFC's) which have been inadvertntly released into the atmosphere. CFCs are particularly worrisome as they have a very long lifetime in the atmosphere.

(c) The major cause of global warming is increased levels of carbon dioxide (CO_2) resulting from the burning of carbon-containing fuels and deforestation of tropical regions.

18. Catalytic converters are basically smog control devices on newer automobiles. Catalytic converters have an oxidation catalyst that oxidizes CO and hydrocarbons to CO_2 and H_2O. It may also have a reduction catalyst that reduces NO to N_2. The catalysts involved with these processes are generally platinum or palladium metal operating at relatively high temperature.

19. He is found in natural gas deposits principally because alpha particles are produced during natural radioactive decay processes. These alpha particles are ^{4}He nuclei; they obtain two electrons from the surrounding material to become helium atoms. This gaseous helium then accumulates with the natural gas trapped beneath the earth. Although other noble gases are produced by radioactive decay—notably ^{40}Ar —they are not produced in the large quantities that helium is.

20. Replacing petroleum products with hydrogen as a fuel for transportation has the following advantages and disadvantages.

Advantages: Eventual decline of supplies of fossil fuels requires finding alternative energy sources. Hydrogen would provide an essentially pollution free exhaust and more efficient fuel. Range of supersonic aircraft could be increased and hypersonic planes become a real possibility.

Disadvantages: Sources of hydrogen gas are relatively expensive along with high storage cost for the gas. Very low temperatures are required for hydrogen storage (bp = -253 °C). Hydrogen must be maintained out of the contact with oxygen or air (explosive mixtures).

EXERCISES

The Atmosphere

21. We assume that the gases in the atmosphere obey the ideal gas law, $PV = nRT$. Then the amount in moles of a particular gas will be given by $n = \dfrac{PV}{RT}$. Since all of the gases in the mixture are at the same temperature and subjected to the same total pressure, their amounts in moles are proportional to their volumes, with the same proportionality constant (P/RT) for each gas in the mixture. Therefore, volume percents and mole percents will be the same.

22. The composition of air on a mass percent basis is not the same as on a volume percent basis. The easiest way to see this is to realize that volume percents and mole percents are equal (Exercise 21), and that different gases have different molar masses. The heavier gases — CO_2, Ar, and O_2 —will have a higher percent by mass than by volume. The calculation for the four most abundant gases in air, based on 100.00 moles of air, follows.

$$\text{mass } N_2 = 78.084 \text{ mol } N_2 \times \frac{28.0135 \text{ g } N_2}{1 \text{ mol } N_2} = 2187.4 \text{ g } N_2$$

$$\text{mass } O_2 = 20.946 \text{ mol } O_2 \times \frac{31.9988 \text{ g } O_2}{1 \text{ mol } O_2} = 670.25 \text{ g } O_2$$

$$\text{mass Ar} = 0.934 \text{ mol Ar} \times \frac{39.95 \text{ g Ar}}{1 \text{ mol Ar}} = 37.3 \text{ g Ar}$$

$$\text{mass } CO_2 = 0.037 \text{ mol } CO_2 \times \frac{44.01 \text{ g } CO_2}{1 \text{ mol } CO_2} = 1.6 \text{ g } CO_2$$

$$\text{mass } \% \, O_2 = \frac{670.25 \, g \, O_2}{2896.6 \, g \, \text{total}} \times 100\% = 23.139 \, \% \, O_2$$

$$\text{mass } \% \, N_2 = \frac{2187.4 \, g \, N_2}{2896.6 \, g \, \text{total}} \times 100\% = 75.516 \% \, N_2$$

$$\text{mass} \% \, Ar = \frac{37.3 \, g \, Ar}{2896.6 \, g \, \text{total}} \times 100\% = 1.29 \, \% \, Ar$$

$$\text{mass} \% \, CO_2 = \frac{1.6 \, g \, CO_2}{2896.6 \, g \, \text{total}} \times 100\% = 0.055 \, \% \, CO_2$$

Nitrogen

23. **(a)** The Haber-Bosch process is the principal artificial method of fixing atmospheric nitrogen. $N_2(g) + 3 \, H_2(g) \rightleftharpoons 2 \, NH_3(g)$

(b) The 1st step of the Ostwald process: $4NH_3(g) + 5 \, O_2(g) \xrightarrow{850°C, \, Pt} 4NO(g) + 6H_2O(g)$

(c) The 2nd and 3rd steps of the process: $2 \, NO(g) + O_2(g) \rightarrow 2 \, NO_2(g)$
$3 \, NO_2(g) + H_2O(l) \rightarrow 2 \, HNO_3(aq) + NO(g)$

24. **(a)** $2NH_4NO_3(s) \xrightarrow{400°C} 2N_2(g) + O_2(g) + 4H_2O(g)$

The equation was balanced by inspection, by realizing first that each mole of $NH_4NO_3(s)$ would produce 1 mol $N_2(g)$ and 2 mol $H_2O(l)$, with 1 mol O [or 1/2 mol $O_2(g)$] remaining.

(b) $NaNO_3(s) \rightarrow NaNO_2(s) + 1/2 \, O_2(g)$ or $2 \, NaNO_3(s) \rightarrow 2 \, NaNO_2(s) + O_2(g)$

(c) $Pb(NO_3)_2(s) \rightarrow PbO(s) + 2 \, NO_2(g) + 1/2 \, O_2(g)$
$2 \, Pb(NO_3)_2(s) \rightarrow 2 \, PbO(s) + 4 \, NO_2(g) + O_2(g)$

25. Decomposition of $HNO_3(l)$: $HNO_3(l) \rightarrow N_2O_4(g) + H_2O(l) + O_2(g)$
Although this equation can be balanced by the half-reaction method, it also can be balanced by inspection. First notice that all of the N is present in $N_2O_4(g)$ in the products. This implies that there are two $HNO_3(aq)$: $2 \, HNO_3(l) \rightarrow N_2O_4(g) + H_2O(l) + O_2(g)$
With this addition, N and H are balanced, and we need to only consider balancing oxygen.

There are 6 O's on the left and 7 O's on the right, an imbalance that can be corrected either by making 1/2 the coefficient of $O_2(g)$ or by doubling all of the other coefficients and leaving the $O_2(g)$ coefficient alone: $4\,HNO_3(l) \rightarrow 2\,N_2O_4(g) + 2\,H_2O(l) + O_2(g)$

26. Begin with the chemical formulas of the species involved:

$$Na_2CO_3(aq) + O_2(g) + NO(g) \rightarrow NaNO_2(aq)$$

Oxygen is reduced and nitrogen (in NO) is oxidized. We use the ion-electron method.

two couples:	$NO(g) \rightarrow NO_2^-(aq)$ and	$O_2 \rightarrow$ products
balance oxygens:	$H_2O + NO \rightarrow NO_2^-$	$O_2 \rightarrow 2\,H_2O$
balance hydrogens:	$H_2O + NO \rightarrow NO_2^- + 2\,H^+$	$4\,H^+ + O_2 \rightarrow 2\,H_2O$
balance charge:	$H_2O + NO \rightarrow NO_2^- + 2\,H^+ + e^-$	$4\,e^- + 4\,H^+ + O_2 \rightarrow 2\,H_2O$

combine: $\qquad 4\,H_2O + 4\,NO + 4\,H^+ + O_2 \rightarrow 2\,H_2O + 4\,NO_2^- + 8\,H^+$

simplify: $\qquad 2\,H_2O + 4\,NO + O_2 \rightarrow 4\,NO_2^- + 4\,H^+$

add spectator ions: $\qquad 4\,Na^+ + 2\,CO_3^{2-} \rightarrow 4\,Na^+ + 2\,CO_3^{2-}$

$\qquad\qquad\qquad\qquad 2\,Na_2CO_3 + 2\,H_2O + 4\,NO + O_2 \rightarrow 4\,NaNO_2 + 2\,H_2O + 2\,CO_2$

simplify: $\qquad 2\,Na_2CO_3(aq) + 4\,NO(g) + O_2(g) \rightarrow 4\,NaNO_2(aq) + 2\,CO_2(g)$

<u>27.</u> We recall that 1 mol of gas occupies 22.414 L at STP.

$$\text{mass } N_2 = 9.39 \times 10^{11}\ ft^3 \times \left(\frac{12\ in.}{1\ ft} \times \frac{2.54\ cm}{1\ in.}\right)^3 \times \frac{1\ L}{1000\ cm^3} \times \frac{1\ mol\ N_2}{22.414\ L} \times \frac{28.01\ g\ N_2}{1\ mol}$$

$$\times \frac{1\ kg}{1000\ g} = 3.32 \times 10^{10}\ kg\ N_2$$

28. $\quad \text{mass \% } HNO_3 = \dfrac{15\ mol\ HNO_3 \times \dfrac{63.01\ g\ HNO_3}{1\ mol\ HNO_3}}{1\ L\ soln \times \dfrac{1000\ mL}{1\ L} \times \dfrac{1.41\ g}{1\ mL\ soln}} \times 100\% = 67\%\ HNO_3$

<u>29.</u> $\quad 75 \times 10^9\ gal \times \dfrac{15\ miles}{1\ gal} \times \dfrac{5\ g}{1\ mile} \times \dfrac{1\ kg}{1000\ g} = 6 \times 10^9\ kg$ of nitrogen oxides released.

30. $2\,NH_{3(g)} + 3\,NO(g) \rightarrow 5/2\,N_2(g) + 3\,H_2O(g)$ (or, multiplying by 2):

$4\,NH_{3(g)} + 6\,NO(g) \rightarrow 5\,N_2(g) + 6\,H_2O(g)$

$\Delta H^{\circ}_{rxn} = \Sigma \Delta H^{\circ}_{f\,products} - \Sigma \Delta H^{\circ}_{f\,reactants}$

$\Delta H^{\circ}_{rxn} = [5\ mol(0\ kJ\ mol^{-1}) + 6\ mol(-241.8\ kJ\ mol^{-1})]$

Oxygen

31. **(a)** $2HgO(s) \xrightarrow{\Delta} 2Hg(l) + O_2(g)$ **(b)** $2KClO_4(s) \xrightarrow{\Delta} 2KClO_3(s) + O_2(g)$

32. **(a)** $O_3(g) + 2\,I^-(aq) + 2\,H^+(aq) \rightarrow I_2(aq) + H_2O(l) + O_2(g)$

(b) $S(s) + H_2O(l) + 3\,O_3(g) \rightarrow H_2SO_4(aq) + 3\,O_2(g)$

(c) $2\left[Fe(CN)_6\right]^{4-}(aq) + O_3(g) + H_2O(l) \rightarrow 2\left[Fe(CN)_6\right]^{3-}(aq) + O_2(g) + 2\,OH^-(aq)$

33. We first write the formulas of the four substances: N_2O_4, Al_2O_3, P_4O_6, CO_2. The one constant in all these substances is oxygen. If we compare amounts of substance with the same amount (in moles) of oxygen, the one with the smallest mass of the other element will have the highest percent oxygen.

3 mol N_2O_4 contains 12 mol O and 6 mol N: $6 \times 14.0 = 84.0$ g N

4 mol Al_2O_3 contains 12 mol O and 8 mol Al: $8 \times 27.0 = 216$ g Al

2 mol P_4O_6 contains 12 mol O and 8 mol P: $8 \times 31.0 = 248$ g P

6 mol CO_2 contains 12 mol O and 6 mol C: $6 \times 12.0 = 72.0$ g C

Thus, of the oxides listed, CO_2 contains the largest percent oxygen by mass.

34. $2NH_4NO_3(s) \xrightarrow{400°C} 2N_2(g) + O_2(g) + 4H_2O(g)$

$2H_2O_2(l) \rightarrow 2H_2O(l) + O_2(g)$

$2KClO_3(s) \xrightarrow{\Delta} 2KCl(s) + 3O_2(g)$

(a) All three reactions have two moles of reactant, and the decomposition of potassium chlorate produces three moles of $O_2(g)$, compared to one mole of $O_2(g)$ for each of the others. The potassium chlorate decomposition produces the most oxygen per mole of reactant.

(b) If each reaction produced the same amount of $O_2(g)$, then the one with the lightest mass of reactant would produce the most oxygen per gram of reactant. When the first two reactions are compared, it is clear that 2 mol H_2O_2 have less mass than 2 mol NH_4NO_3. (Notice that each mol of NH_4NO_3 contains the amounts of elements—2 mol H and 2 mol O—present in each mol H_2O_2 plus some more.) So now we need to compare hydrogen peroxide with potassium chlorate. Notice that 6 H_2O_2 produces the same amount of $O_2(g)$ as 2 $KClO_3$. 6 mol H_2O_2 has a mass of $34.01 \times 6 = 204.1$ g, while 2 mol $KClO_3$ has a mass of $122.2 \times 2 = 244.2$ g. Thus, H_2O_2 produces the most $O_2(g)$ per gram of reactant.

35. Recall that fraction by volume and fraction by pressure are numerically equal. Additionally, one atmosphere pressure is equivalent to 760 mmHg. We combine these two facts.

$$P\{O_3\} = 760 \text{ mmHg} \times \frac{0.04 \text{ mmHg of O}_3}{10^6 \text{ mmHg of atmosphere}} = 3 \times 10^{-5} \text{ mmHg}$$

36. $P = \dfrac{nRT}{V} = \dfrac{\left(5 \times 10^{12} \text{ molecules} \times \dfrac{1 \text{ mol O}_3}{6.022 \times 10^{23} \text{ molec.}}\right) \times \dfrac{0.08206 \text{ L atm}}{\text{mol K}} \times 220 \text{ K}}{1 \text{ cm}^3 \times \dfrac{1 \text{ L}}{100 \text{ cm}^3}}$

$$P = 1._5 \times 10^{-8} \text{ atm} \times \frac{760 \text{ mmHg}}{1 \text{ atm}} = 1._1 \times 10^{-5} \text{ mmHg}$$

37. The electrolysis reaction is $2H_2O(l) \xrightarrow{\text{electroysis}} 2H_2(g) + O_2(g)$. In this reaction, 2 moles of $H_2(g)$ are produced for each mole of $O_2(g)$, by the law of combining volumes, we would expect the volume of hydrogen to be twice the volume of oxygen produced. (Actually the volumes are not exactly in the ratio of 2:1 because of the different solubilities of oxygen and hydrogen in water.)

38. The electrolysis reaction is $2H_2O(l) \xrightarrow{\text{electroysis}} 2H_2(g) + O_2(g)$. In this reaction, 2 moles of water are decomposed to produce each mole of $O_2(g)$. We use the data in the problem to determine the amount of $O_2(g)$ produced, which we then convert to the mass of H_2O needed.

$$\text{mass H}_2\text{O}(l) = \frac{\left(736.7 \text{ mmHg} \times \dfrac{1 \text{ atm}}{760 \text{ mmHg}}\right)\left(22.83 \text{ mL} \times \dfrac{1 \text{ L}}{1000 \text{ mL}}\right)}{\dfrac{0.08206 \text{ L atm}}{\text{mol K}} \times (25.0 + 273.2)\text{K}} \times \frac{2 \text{ mol H}_2\text{O}}{1 \text{ mol O}_2}$$

$$\times \frac{18.02 \text{ g H}_2\text{O}}{1 \text{ mol H}_2\text{O}} = 0.03259 \text{ g H}_2\text{O decomposed}$$

The Noble Gases

39. First we use the ideal gas law to determine the amount in moles of argon.

$$n = \frac{PV}{RT} = \frac{145 \text{ atm} \times 55 \text{ L}}{0.08206 \text{ L atm mol}^{-1} \text{ K}^{-1} \times 299 \text{ K}} = 3.2\underline{5} \times 10^2 \text{ mol Ar}$$

$$\text{L air} = 3.2\underline{5} \times 10^2 \text{ mol Ar} \times \frac{22.414 \text{ L Ar STP}}{1 \text{ mol Ar}} \times \frac{100.000 \text{ L air}}{0.934 \text{ L Ar}} = 7.8 \times 10^5 \text{ L air}$$

40.   First we compute the volume of 5.00 g He at STP, and then the volume of natural gas.

$$\text{natural gas volume} = 5.00 \text{ g He} \times \frac{1 \text{ mol He}}{4.003 \text{ g He}} \times \frac{22.414 \text{ L He}}{1 \text{ mol He}} \times \frac{100 \text{ L air}}{8 \text{ L He}}$$

$$\text{natural gas volume} = 3._5 \times 10^2 \text{ L air at STP} \cong 300 \text{ L}$$

41.   1 mol of the mixture at STP occupies a volume of 22.414 L. It contains 0.79 mol He and 0.21 mol O_2.

$$\text{STP density} = \frac{\text{mass}}{\text{volume}} = \frac{0.79 \text{ mol He} \times \dfrac{4.003 \text{ g He}}{1 \text{ mol He}} + 0.21 \text{ mol O}_2 \times \dfrac{32.0 \text{ g O}_2}{1 \text{ mol O}_2}}{22.414 \text{ L}} = 0.44 \text{ g/L}$$

$25°C$ is a temperature higher than STP. This condition increases the 1.00-L volume that contains 0.44 g of the mixture at STP. We calculate the expanded volume with the combined gas law.

$$V_{\text{final}} = 1.00 \text{ L} \times \frac{(25 + 273.2) \text{ K}}{273.2 \text{ K}} = 1.09 \text{ L} \qquad \text{final density} = \frac{0.44 \text{ g}}{1.09 \text{ L}} = 0.40 \text{ g/L}$$

42.   We determine the apparent molar masses of each mixture by multiplying the mole fraction (numerically equal to the volume fraction) of each gas by its molar mass, and then summing these products for all gases in the mixture.

$$M_{\text{air}} = \left(0.78084 \times \frac{28.01 \text{ g N}_2}{1 \text{ mol N}_2}\right) + \left(0.20946 \times \frac{32.00 \text{ g O}_2}{1 \text{ mol O}_2}\right) + \left(0.00934 \times \frac{39.95 \text{ g Ar}}{1 \text{ mol Ar}}\right)$$

$$= 21.87 \text{ g N}_2 + 6.703 \text{ g O}_2 + 0.373 \text{ g Ar} = 28.95 \text{ g/mol air}$$

$$M_{\text{mix}} = \left(0.79 \times \frac{4.003 \text{ g He}}{1 \text{ mol He}}\right) + \left(0.21 \times \frac{32.00 \text{ g O}_2}{1 \text{ mol O}_2}\right) = 3.2 \text{ g He} + 6.7 \text{ g O}_2 = \frac{9.9 \text{ g mixture}}{\text{mol}}$$

In order to prepare two gases with the same density, the volume of the gas of smaller molar mass must be smaller by a factor equal to the ratio of the molar masses. According to Boyle's law, this means that the pressure on the less dense gas must be larger by a factor equal to a ratio of molar masses.

$$P_{\text{mix}} = \frac{28.95}{9.9} \times 1.00 \text{ atm} = 2.9 \text{ atm}$$

Carbon

43.   **(a)** $2 \text{ C}_6\text{H}_{14}(l) + 19 \text{ O}_2(g) \rightarrow 12 \text{ CO}_2(g) + 14 \text{ H}_2\text{O}(l)$

   (b) $\text{PbO}(s) + \text{CO}(g) \xrightarrow{\text{heat}} \text{Pb}(s) + \text{CO}_2(g)$

   (c) $2 \text{ KOH}(aq) + \text{CO}_2(g) \rightarrow \text{K}_2\text{CO}_3(aq) + \text{H}_2\text{O}(l)$

   (d) $\text{MgCO}_3(s) + 2 \text{ HCl}(aq) \rightarrow \text{MgCl}_2(aq) + \text{H}_2\text{O}(l) + \text{CO}_2(g)$

44. **(a)** $HC_2H_3O_2(aq) + NaHCO_3(s) \rightarrow NaC_2H_3O_2(aq) + H_2O(l) + CO_2(g)$

(b) $ZnO(s) + CO(g) \rightarrow Zn(s) + CO_2(g)$ (The heat from this reaction may melt the Zn.)

(c) $2\,CO(g) + 3\,H_2(g) \rightarrow CH_2OHCH_2OH(l)$

45. The reaction referred to is exemplified by

$CaCO_3(s) + 2\,HCl(aq) \rightarrow CaCl_2(aq) + CO_2(g) + H_2O(l)$,

the action of hydrochloric acid on calcium carbonate. Notice that no oxidation states change in this reaction; it is a double displacement reaction followed by the decomposition of "$H_2CO_3(aq)$". Since the carbon dioxide produced in this reaction is not formed from elemental carbon or by the oxidation of a carbon-containing compound, it cannot be partially oxidized. Thus, no $CO(g)$ will form.

46. Both carbon dioxide and chlorofluorocarbons are greenhouse gases. They absorb infrared radiation, preventing this heat from leaving the Earth. But only the chlorine released by chlorofluorocarbons plays a role in the destruction of ozone in the upper atmosphere.

47. The combustion reaction: $C_8H_{18}(l) + 25/2\,O_2(g) \rightarrow 8\,CO_2(g) + 9\,H_2O(l)$

First compute the enthalpy change for combustion.

$\Delta H° = 8\Delta H°_f[CO_2(g)] + 9\Delta H°_f[H_2O(l)] - [\,\Delta H°_f[C_8H_{18}(l)] + \tfrac{25}{2}\Delta H°_f[O_2(g)]\,]$

$\Delta H° = 8(-393.5) + 9(-285.8) - (-250.0) - 25/2(0.00) = -5470.\ \text{kJ/mol } C_8H_{18}$

Then determine the heat produced from each gallon of gasoline.

$$\text{heat produced} = 1.00\ \text{gal} \times \frac{3.785\ \text{L}}{1\ \text{gal}} \times \frac{1000\ \text{mL}}{1\ \text{L}} \times \frac{0.703\ \text{g}}{1\ \text{mL}} \times \frac{1\ \text{mol } C_8H_{18}}{114.23\ \text{g } C_8H_{18}} \times \frac{5470\ \text{kJ}}{1\ \text{mol } C_8H_{18}}$$

$= 1.27 \times 10^5\ \text{kJ of heat}\ \left(\text{Note that } \Delta H = -1.27 \times 10^5\ \text{kJ}\right)$

48. Let us determine the difference between the two reactions.

combustion: $C_8H_{18}(l) + 25/2\,O_2(g) \rightarrow 8\,CO_2(g) + 9\,H_2O(l)$

$\underline{-(\text{incomplete}): 7\,CO_2(g) + CO(g) + 9\,H_2O(l) \rightarrow C_8H_{18}(l) + 12\,O_2(g)}$

difference: $CO(g) + 1/2\,O_2(g) \rightarrow CO_2(g)$

The heat that accompanies the "difference" reaction is the difference in the heat of the two reactions. To determine the percent of the maximum possible heat this is, first compute the enthalpy change for the combustion reaction.

$\Delta H° = 8\Delta H°_f[CO_2(g)] + 9\Delta H°_f[H_2O(l)] - [\,\Delta H°_f[C_8H_{18}(l)] + \tfrac{25}{2}\Delta H°_f[O_2(g)]\,]$

$= 8(-393.5) + 9(-285.8) - (-250.0) - 25/2(0.00) = -5470.\ \text{kJ / mol } C_8H_{18}$

Then compute the enthalpy change for the "difference" reaction.

$$\Delta H° = \Delta H_f°[CO_2(g)] - \Delta H_f°[CO(g)] - \tfrac{1}{2}\Delta H_f°[O_2(g)] = (-393.5) - (-110.5) - \tfrac{1}{2}(0.00) = -283 \text{ kJ}$$

And finally compute the percent decrease that the difference reaction represents.

$$\% \text{ decrease} = \frac{-283 \text{ kJ}}{-5470 \text{ kJ}} \times 100\% = 5.17\% \text{ decrease}$$

Hydrogen

49. The four reactions of interest are: (Note: $\Delta H°_{combustion} = \Sigma\Delta H°_{f\ products} - \Sigma\Delta H°_{f\ reactants}$)

$CH_4(g) + 2\ O_{2(g)} \rightarrow CO_2(g) + 2\ H_2O(l)$ $\Delta H°_{combustion} = -890.3$ kJ
(Molar mass $CH_4 = 16.0428$ g mol^{-1})

$C_2H_6(g) + 7/2\ O_{2(g)} \rightarrow 2\ CO_2(g) + 3\ H_2O(l)$ $\Delta H°_{combustion} = -1559.7$ kJ
(Molar mass $C_2H_6 = 30.070$ g mol^{-1})

$C_3H_8(g) + 5\ O_{2(g)} \rightarrow 3\ CO_2(g) + 4\ H_2O(l)$ $\Delta H°_{combustion} = -2219.9$ kJ
(Molar mass $C_3H_8 = 44.097$ g mol^{-1})

$C_4H_{10}(g) + 13/2\ O_{2(g)} \rightarrow 4\ CO_2(g) + 5\ H_2O(l)$ $\Delta H°_{combustion} = -2877.4$ kJ
(Molar mass $C_4H_{10} = 58.123$ g mol^{-1})

50. Using the answers obtained in question 49, the per gram energy release is:
$CH_4(g)$ -55.5 kJ; $C_2H_6(g)$ -51.9 kJ; $C_3H_8(g)$ -50.3 kJ; $C_4H_{10}(g)$ -49.5 kJ

(a) $C_4H_{10}(g)$ evolves the most energy on a per mole basis (-2877.4 kJ)

(b) $CH_4(g)$ evolves the most energy on a per gram basis (-55.5 kJ)

(c) $CH_4(g)$ is the most desirable alkane from the standpoint of emission, producing the least $CO_2(g)$ per mole and/or per gram of fuel burned (as well per kJ of energy produced).

51. **(a)** $2\ Al(s) + 6\ HCl(aq) \rightarrow 2\ AlCl_3(aq) + 3\ H_2(g)$

    (b) $C_3H_8(g) + 3\ H_2O(g) \rightarrow 3\ CO(g) + 7\ H_2(g)$

    (c) $MnO_2(s) + 2\ H_2(g) \xrightarrow{\Delta} Mn(s) + 2\ H_2O(g)$

52. **(a)** $2\ H_2O(l) \xrightarrow{electricity} O_2(g) + 2\ H_2(g)$

    (b) $2\ HI(aq) + Zn(s) \rightarrow ZnI_2(aq) + H_2(g)$ Any moderately active metal is suitable.

    (c) $Mg(s) + H_2SO_4(aq) \rightarrow MgSO_4(aq) + H_2(g)$ Any strong acid is suitable.

    (d) $CO(g) + H_2O(g) \rightarrow CO_2(g) + H_2(g)$

53. $CaH_2(s) + 2 H_2O(l) \rightarrow Ca(OH)_2(aq) + 2 H_2(g)$

$\underline{Ca(s) + 2 H_2O(l) \rightarrow Ca(OH)_2(aq) + H_2(g)}$

$2 Na(s) + 2 H_2O(l) \rightarrow 2 NaOH(aq) + H_2(g)$

(a) The reaction that produces the largest volume of $H_2(g)$ per liter of water also produces the largest amount of $H_2(g)$ per mole of water used. All three reactions use two moles of water and the reaction with $CaH_2(s)$ produces the most $H_2(g)$.

(b) We can compare three reactions that produce the same amount of hydrogen; the one that requires the smallest mass of solid produces the greatest amount of H_2 per gram of solid. The amount of hydrogen we choose is 2 moles, which means that we compare 1 mol CaH_2 (42.09 g) with 2 mol Ca (80.16 g) and with 4 mol Na (91.96 g). Clearly CaH_2 produces the greatest amount of H_2 per gram of solid.

54. The balanced equation is $C_{17}H_{33}COOH(l) + H_2(g) \rightarrow C_{17}H_{35}COOH(s)$ One mole of oleic acid requires one mole of $H_2(g)$.

$$V = \frac{nRT}{P} = \frac{1.00 \text{ mol} \times 0.08206 \dfrac{\text{L atm}}{\text{mol K}} \times 298 \text{ K}}{752 \text{ mm Hg} \times \dfrac{1 \text{ atm}}{760 \text{ mmHg}}} \times \frac{0.95 \text{ L produced}}{1 \text{ L calculated}} = 23 \text{ L } H_2(g)$$

55. Greatest mass percent hydrogen:
The atmosphere is mostly $N_2(g)$ and $O_2(g)$ with only a trace of hydrogen containing gas molecules. Seawater is $H_2O(l)$, natural gas is $CH_4(g)$ and ammonia is $NH_3(g)$. Each of these compounds have one non-hydrogen atom, each of which have approximately the same mass ($\sim 14 \pm 2$ g mol^{-1}). Since CH_4 has the highest hydrogen atom to non-hydrogen atom ratio, this molecule has the greatest mass percent hydrogen.

56. The reaction is $CaH_2(s) + 2 H_2O(l) \rightarrow Ca(OH)_2(aq) + 2 H_2(g)$

First calculate the amount of $H_2(g)$ needed and then the mass of $CaH_2(s)$ required.

$$\text{mol } H_2 = \frac{PV}{RT} = \frac{722 \text{ mm Hg} \times \dfrac{1 \text{ atm}}{760 \text{ mmHg}} \times 235 \text{ L}}{0.08206 \text{ L atm mol}^{-1}\text{K}^{-1} \times (273.2 + 19.7)\text{K}} = 9.29 \text{ mol } H_2$$

$$\text{mass } CaH_2 = 9.29 \text{ mol } H_2 \times \frac{1 \text{ mol } CaH_2}{2 \text{ mol } H_2} \times \frac{42.09 \text{ g } CaH_2}{1 \text{ mol } CaH_2} = 196 \text{ g } CaH_2$$

FEATURE PROBLEMS

71. **(a)** An engine with a fuel-lean mixture will produce large proportions of nitrogen oxides since there is a large amount of air (which contains nitrogen and oxygen) compared to the fuel. On the other hand, an engine with a fuel-rich mixture will produce unburned fuel (RH) and carbon monoxide, because of the lack of oxygen.

(b) First we balance the combustion reaction.

$$2\ C_8H_{18}(l) + 25\ O_2(g) \rightarrow 16\ CO_2(g) + 18\ H_2O(l)$$

$$\frac{air}{fuel} = \frac{25\ mol\ O_2}{2\ mol\ C_8H_{18}} \times \frac{100\ mol\ air}{20.946\ mol\ O_2} \times \frac{28.95\ g}{1\ mol\ air} \times \frac{1\ mol\ C_8H_{18}}{114.2\ g\ C_8H_{18}} = 15:1$$

We used the average molar mass of air from Exercise 42.

72. **(a)** The $N_2(g)$ extracted from liquid air has some $Ar(g)$ mixed in. Only $O_2(g)$ was removed from liquid air in the oxygen-related experiments.

(b) Because of the presence of $Ar(g)$ [39.95 g/mol], the $N_2(g)$ [28.01 g/mol] from liquid air will have a greater density than $N_2(g)$ from nitrogen compounds.

(c) Magnesium will react with nitrogen $\left[3\ Mg(s) + N_2(g) \rightarrow Mg_3N_2(s)\right]$ but not with Ar. Thus, magnesium reacts with all the nitrogen in the mixture, but leaves the relatively inert $Ar(g)$ unreacted.

(d) The "nitrogen" remaining after oxygen is extracted from each mole of air (Rayleigh's mixture) contains $(0.78084 + 0.00934 =)\ 0.79018$ mol and has the mass calculated below.

mass of gaseous mixture $= (0.78084 \times 28.013\ g/mol\ N_2) + (0.00934 \times 39.948\ g/mol\ Ar)$

mass of gaseous mixture $= 21.874\ g\ N_2 + 0.373\ g\ Ar = 22.247\ g$ mixture.

Then, the molar mass of the mixture can be computed: 22.247 g mixture / 0.79018 mol = 28.154 g/mol. Since the STP molar volume of an ideal gas is 22.414 L, we can compute the two densities.

$$d(N_2) = \frac{28.013\ g/mol}{22.414\ L/mol} = 1.2498\ g/mol$$

$$d(\text{mixture}) = \frac{28.154\ g/mol}{22.414\ L/mol} = 1.2561\ g/mol$$

These densities differ by 0.50%.

73. The goal is to demonstrate that the three reactions result in the decomposition of water as the net reaction: Net: $2 H_2O \rightarrow 2 H_2 + O_2$ First balance each equation.

(1) $3 FeCl_2 + 4 H_2O \rightarrow Fe_3O_4 + HCl + H_2$ Balance by inspection. Notice that there are 3 Fe and 4 O on the right-hand side. Then balance Cl.

$3 FeCl_2 + 4 H_2O \rightarrow Fe_3O_4 + 6 HCl + H_2$

(2) $Fe_3O_4 + HCl + Cl_2 \rightarrow FeCl_3 + H_2O + O_2$ Try the half-equation method.

$Cl_2 + 2 e^- \rightarrow 2 Cl^-$ But realize that $Fe_3O_4 = Fe_2O_3 \cdot FeO$; only iron(II) needs to be oxidized. $2 FeO \rightarrow 2 Fe^{3+} + O_2 + 6 e^-$ Now combine the two half-equations.

$2 FeO + 3 Cl_2 \rightarrow 2 Fe^{3+} + O_2 + 6 Cl^-$

Add in $2 Fe_2O_3 + 12 H^+ \rightarrow 4 Fe^{3+} + 6 H_2O$

$2 Fe_3O_4 + 3 Cl_2 + 12 H^+ \rightarrow 6 Fe^{3+} + O_2 + 6 Cl^- + 6 H_2O$

And 12 Cl^- spectators: $2 Fe_3O_4 + 3 Cl_2 + 12 HCl \rightarrow 6 FeCl_3 + O_2 + 6 H_2O$

(3) $FeCl_3 \rightarrow FeCl_2 + Cl_2$ by inspection $2 FeCl_3 \rightarrow 2 FeCl_2 + Cl_2$

One strategy is to consider each of the three equations and the net equation. Only equation (1) produces hydrogen; run it twice. Only equation (2) produces oxygen; run it once. Equation (3) can balance out the Cl_2 required by equation (2); run it three times.

$2 \times (1)$ $6 FeCl_2 + 8 H_2O \rightarrow 2 Fe_3O_4 + 12 HCl + 2 H_2$

$1 \times (2)$ $2 Fe_3O_4 + 3 Cl_2 + 12 HCl \rightarrow 6 FeCl_3 + O_2 + 6 H_2O$

$3 \times (3)$ $6 FeCl_3 \rightarrow 6 FeCl_2 + 3 Cl_2$

───

Net: $2 H_2O \rightarrow 2 H_2 + O_2$

CHAPTER 9
ELECTRONS IN ATOMS

PRACTICE EXAMPLES

1A   Use $c = \lambda \nu$, solve for frequency. $\nu = \dfrac{2.9979 \times 10^8 \ \text{m/s}}{690 \ \text{nm}} \times \dfrac{10^9 \ \text{nm}}{1 \ \text{m}} = 4.34 \times 10^{14} \ \text{Hz}$

1B   Wavelength and frequency are related through the equation $c = \lambda \nu$, which can be solved for either one.

$$\lambda = \frac{c}{\nu} = \frac{2.9979 \times 10^8 \ \text{m/s}}{91.5 \times 10^6 \ \text{s}^{-1}} = 3.28 \ \text{m} \quad \text{Note that Hz} = \text{s}^{-1}$$

2A   The relationship $\nu = c / \lambda$ can be substituted into the equation $E = h\nu$ to obtain $E = hc / \lambda$. This energy, in J/photon, can then be converted to kJ/mol.

$$E = \frac{hc}{\lambda} = \frac{6.626 \times 10^{-34} \text{J s photon}^{-1} \times 2.998 \times 10^8 \ \text{m s}^{-1}}{230 \, \text{nm} \times \dfrac{1 \, \text{m}}{10^9 \, \text{nm}}} \times \frac{6.022 \times 10^{23} \ \text{photons}}{1 \, \text{mol}} \times \frac{1 \, \text{kJ}}{1000 \, \text{J}}$$

$$= 520 \ \text{kJ/mol}$$

With a similar calculation one finds that 290 nm corresponds to 410 kJ/mol. Thus, the energy range is from 400 to 520 kJ/mol, respectively.

2B   The equation $E = h\nu$ is solved for frequency and the two frequencies are calculated.

$$\nu = \frac{E}{h} = \frac{3.056 \times 10^{-19} \ \text{J/photon}}{6.626 \times 10^{-34} \ \text{J} \cdot \text{s/photon}} \qquad \nu = \frac{E}{h} = \frac{4.414 \times 10^{-19} \ \text{J/photon}}{6.626 \times 10^{-34} \ \text{J} \cdot \text{s/photon}}$$

$$= 4.612 \times 10^{14} \ \text{Hz} \qquad\qquad = 6.662 \times 10^{14} \ \text{Hz}$$

To determine color, we calculate the wavelength of each frequency and compare it with *text* Figure 9-3.

$$\lambda = \frac{c}{\nu} = \frac{2.9979 \times 10^8 \ \text{m/s}}{4.612 \times 10^{14} \ \text{Hz}} \times \frac{10^9 \ \text{nm}}{1 \ \text{m}} \qquad \lambda = \frac{c}{\nu} = \frac{2.9979 \times 10^8 \ \text{m/s}}{6.662 \times 10^{14} \ \text{Hz}} \times \frac{10^9 \ \text{nm}}{1 \ \text{m}}$$

$$= 650 \ \text{nm} \qquad \text{orange} \qquad\qquad = 450 \ \text{nm} \qquad \text{indigo}$$

The colors of the spectrum that are not absorbed are what we see when we look at a plant: blue, green, and yellow. The plant appears green.

3A We solve the Rydberg equation for n to see if we obtain an integer.

$$n = \sqrt{n^2} = \sqrt{\frac{-R_H}{E_n}} = \sqrt{\frac{-2.179 \times 10^{-18}\ \text{J}}{-2.69 \times 10^{-20}\ \text{J}}} = \sqrt{81.00} = 9.00 \qquad \text{Thus, } E_9 = -2.69 \times 10^{-20}\ \text{J}$$

3B It is not likely that an atomic radius would be precisely equal to an arbitrary unit of length, but let us see how close the radii are. 1 nm = 1000 pm, so we solve the following for n.

$$1000\ \text{pm} = n^2 53\ \text{pm} \qquad\qquad n = \sqrt{\frac{1000}{53}} = 4.3$$

We see that no radius is exactly 1 nm. The closest:

$$r_4 = 4^2 a_0 = 16 \times 0.053\ \text{nm} = 0.85\ \text{nm} \qquad\qquad r_5 = 5^2 a_0 = 25 \times 0.053\ \text{nm} = 1.3\ \text{nm}$$

4A We first determine the energy difference, and then the wavelength of light corresponding to that energy.

$$\Delta E = R_H \left(\frac{1}{2^2} - \frac{1}{4^2} \right) = 2.179 \times 10^{-18}\ \text{J} \left(\frac{1}{2^2} - \frac{1}{4^2} \right) = 4.086 \times 10^{-19}\ \text{J}$$

$$\lambda = \frac{hc}{\Delta E} = \frac{6.626 \times 10^{-34}\ \text{J s} \times 2.998 \times 10^8\ \text{m s}^{-1}}{4.086 \times 10^{-19}\ \text{J}} = 4.862 \times 10^{-7}\ \text{m} = 486.2\ \text{nm}$$

4B The longest wavelength light results from the transition that spans the smallest difference in energy. Since all Lyman series emissions end with $n_f = 1$, the smallest energy transition has $n_i = 2$. From this, we obtain the value of ΔE.

$$\Delta E = R_H \left(\frac{1}{n_i^{\,2}} - \frac{1}{n_f^{\,2}} \right) = 2.179 \times 10^{-18}\ \text{J} \left(\frac{1}{2^2} - \frac{1}{1^2} \right) = -1.634 \times 10^{-18}\ \text{J}$$

From this energy emitted, we can obtain the wavelength of the emitted light: $\Delta E = hc / \lambda$

$$\lambda = \frac{hc}{\Delta E} = \frac{6.626 \times 10^{-34}\ \text{J s} \times 2.998 \times 10^8\ \text{m s}^{-1}}{1.634 \times 10^{-18}\ \text{J}} = 1.216 \times 10^{-7}\ \text{m} = 121.6\ \text{nm} = 1216\ \text{angstroms}$$

5A

$$E_f = \frac{-Z^2 \times R_H}{n_f^{\,2}} = \frac{-4^2 \times 2.179 \times 10^{-18}\ \text{J}}{3^2}$$

$$E_f = -3.874 \times 10^{-18}\ \text{J}$$

$$E_i = \frac{-Z^2 \times R_H}{n_i^{\,2}} = \frac{-4^2 \times 2.179 \times 10^{-18}\ \text{J}}{5^2}$$

$$E_i = -1.395 \times 10^{-18}\ \text{J}$$

$$\Delta E = E_f - E_i$$
$$\Delta E = (-3.874 \times 10^{-18}\ \text{J}) - (-1.395 \times 10^{-18}\ \text{J})$$
$$\Delta E = -2.479 \times 10^{-18}\ \text{J}$$

To determine the wavelength, use $E = h\nu = \dfrac{hc}{\lambda}$; Rearrange for λ:

$$\lambda = \frac{hc}{E} = \frac{(6.626\times10^{-34}\,\text{J s})\left(2.998\times10^{8}\,\dfrac{m}{s}\right)}{2.479\,x10^{-18}\,J} = 8.013\times10^{-8}\,\text{m} = 80.13\,\text{nm}$$

5B   Since $E = \dfrac{-Z^2 \times R_H}{n^2}$, the transitions are related to Z^2, hence, if the frequency is 16 times

greater, then the value of the ratio $\dfrac{Z^2(\text{?-atom})}{Z^2(\text{H-atom})} = \dfrac{Z_?^2}{1^2} = 16$

We can easily see $Z_?^2 = 16$ or $Z = 4$ corresponding to a Be nucleus.
The hydrogen-like ion must be Be^{3+} .

6A   Superman's de Broglie wavelength is given by the relationship $\lambda = h/m\nu$

$$\lambda = \frac{h}{m\nu} = \frac{6.626\times10^{-34}\text{J}\cdot\text{s}}{91\,\text{kg}\times\dfrac{1}{5}\times2.998\times10^{8}\text{m/s}} = 1.21\times10^{-43}\,\text{m}$$

6B   The de Broglie wavelength is given by $\lambda = h/m\nu$ which can be solved for ν.

$$\nu = \frac{h}{m\lambda} = \frac{6.626\times10^{-34}\,\text{J s}}{1.673\times10^{-27}\,\text{kg}\times0.0100\times10^{-9}\,\text{m}} = 3.96\times10^{4}\,\text{m/s}$$

We used the facts that $1\,\text{J} = \text{kg m}^2\text{s}^{-2}$, $1\,\text{nm} = 10^{-9}$ m and $1\,\text{g} = 10^{-3}$ kg

7A $p = (91\,\text{kg})(5.996\times10^{7}\,\text{m s}^{-1}) = 5.4\underline{6}\times10^{9}\,\text{kg m s}^{-1}$

$\Delta p = (0.015)(\,5.4\underline{6}\times10^{9}\,\text{kg m s}^{-1}) = 8.2\times10^{7}\,\text{kg m s}^{-1}$

$$\Delta x = \frac{h}{4\pi\,\Delta p} = \frac{6.626\times10^{-34}\,\text{J s}}{(4\pi)(8.2\times10^{7}\,\dfrac{\text{kg m}}{\text{s}})} = 6.4\times10^{-43}\,\text{m}$$

7B $24\,\text{nm} = 2.4\times10^{-8}\,\text{m} = \Delta x = \dfrac{h}{4\pi\,\Delta p} = \dfrac{6.626\times10^{-34}\,\text{J s}}{(4\pi)(\Delta p)}$

Solve for Δp: $\Delta p = 2.2\times10^{-27}\,\text{kg m s}^{-1}$

$(\Delta\nu)(m) = \Delta p = 2.2\times10^{-27}\,\text{kg m s}^{-1} = (\Delta\nu)(1.67\times10^{-27}\,\text{kg})$ Hence, $\Delta\nu = 1.3\,\text{ m s}^{-1}$

8A   The value of ℓ can range from 0 to $n-1$; in this case, $\ell = 0$ is acceptable. The value of m_ℓ is integral and ranges from $-\ell$ to $+\ell$; in this case, $m_\ell = 0$ is acceptable.
Yes, an orbital can have $n = 3, \ell = 0$, and $m_\ell = 0$.

8B   The first restriction is that ℓ must be an integer smaller than n. This restricts ℓ to the values: 2, 1, and 0. The second restriction is that the absolute value of m_ℓ must be an integer equal to or smaller than ℓ. This further restricts ℓ, and the only allowed values are: $\ell = 2$ and $\ell = 1$.

9A   The magnetic quantum number, m_ℓ, is not reflected in the orbital designation. Because $\ell = 1$, this is a p orbital. Because $n = 3$, the designation is $3p$.

9B   The H-atom orbitals $3s$, $3p$ and $3d$ are degenerate. Therefore, the 9 quantum number combinations are:

	n	ℓ	m_ℓ
$3s$	3	0	0
$3p$	3	1	-1,0,+1
$3d$	3	2	-2,-1, 0,+1,+2

10A We can simply sum the exponents to obtain the number of electrons in the neutral atom and thus the atomic number of the element. $Z = 2 + 2 + 6 + 2 + 6 + 2 + 2 = 22$, which is the atomic number for Ti.

10B Iodine has an atomic number of 53. The first 36 electrons have the same electron configuration as Kr: $1s^2 2s^2 2p^6 3s^2 3p^6 3d^{10} 4s^2 4p^6$. The next two electrons go into the $5s$ subshell $\left(5s^2\right)$, then 10 electrons fill the $4d$ subshell $\left(4d^{10}\right)$, accounting for a total of 48 electrons. The last five electrons partially fill the $5p$ subshell $\left(5p^5\right)$.

The electron configuration of I is therefore $1s^2 2s^2 2p^6 3s^2 3p^6 3d^{10} 4s^2 4p^6 4d^{10} 5s^2 5p^5$. Each iodine atom has ten $3d$ electrons and one unpaired $5p$ electron.

11A Iron has 26 electrons, of which 18 are accounted for by the [Ar] configuration. Beyond [Ar] there are two $4s$ electrons and six $3d$ electrons, as shown in the following orbital diagram.

$$\text{Fe:}\quad [\text{Ar}]\quad \overset{3d}{\boxed{\uparrow\downarrow\,|\,\uparrow\,|\,\uparrow\,|\,\uparrow\,|\,\uparrow}}\quad \overset{4s}{\boxed{\uparrow\downarrow}}$$

11B Bismuth has 83 electrons, of which 54 are accounted for by the [Xe] configuration. Beyond [Xe] there are two $6s$ electrons, fourteen $4f$ electrons, ten $5d$ electrons, and three $6p$ electrons, as shown in the following orbital diagram.

$$\text{Bi:}\quad [\text{Xe}]\quad \overset{4f}{\boxed{\uparrow\downarrow|\uparrow\downarrow|\uparrow\downarrow|\uparrow\downarrow|\uparrow\downarrow|\uparrow\downarrow|\uparrow\downarrow}}\quad \overset{5d}{\boxed{\uparrow\downarrow|\uparrow\downarrow|\uparrow\downarrow|\uparrow\downarrow|\uparrow\downarrow}}\quad \overset{6s}{\boxed{\uparrow\downarrow}}\quad \overset{6p}{\boxed{\uparrow\,|\,\uparrow\,|\,\uparrow}}$$

12A **(a)** Tin is in the 5th period, hence, five electronic shells are filled or partially filled.

(b) The $3p$ subshell was filled with Ar; there are six $3p$ electrons in an atom of Sn.

(c) The electron configuration of Sn is [Kr] $4d^{10}5s^2 5p^2$. There are no $5d$ electrons.

(d) Both of the $5p$ electrons are unpaired; there are two unpaired electrons in a Sn atom.

12B **(a)** The $3d$ subshell was filled at Zn; each Y atom has ten $3d$ electrons.

(b) Ge is in the $4p$ row; each germanium atom has two $4p$ electrons.

(c) We would expect each Au atom to have ten $5d$ electrons and one $6s$ electron. Each Au atom should have one unpaired electron.

REVIEW QUESTIONS

1. **(a)** λ is the symbol for wavelength, which is the distance between like features (two peaks, for instance) of successive waves.

(b) ν is the symbol for frequency, which is the number of times that a particular feature of successive waves (such as a peak) passes a fixed point per unit time, e.g. per second.

(c) h is the symbol for Planck's constant, which relates the energy and frequency of radiation: $E = h\nu$.

(d) Ψ is the symbol for the wavefunction, which is the function that contains all the information known about an electron within a given atomic or molecular system.

(e) The principal quantum number, n, is related to both the distance of an electron from the nucleus and to the electron's energy.

2. **(a)** The atomic (line) spectrum is the result of atoms being energized and then radiating energy as they lose that energy and return to lower energy states. Since only certain energy levels are permitted for each atom, the energies radiated are definite, giving rise to light of only certain wavelengths, or lines in the spectrum.

(b) The photoelectric effect occurs when light (*photo-*) shining on a surface causes electrons (*-electric*) to be emitted from that surface.

(c) A matter wave refers to the wave properties associated with matter.

(d) The Heisenberg uncertainty principle states that it is impossible to simultaneously determine the position and linear momentum of a particle to any desired degree of precision.

(e) Electron spin is that property of electrons that makes them appear as if they were spinning, or rotating, on an axis. Electrons can spin in only one of two states: clockwise (spin up) or counterclockwise (spin down).

(f) The Pauli exclusion principle states that two electrons in an atom do not share the same four quantum numbers.

3.   **(a)**   The frequency of radiation is the number of oscillations that occur per unit time (as the radiation is observed passing a fixed point), the wavelength is the distance between two wave peaks on successive waves.

(b)   Ultraviolet light is radiation that has wavelengths just a bit shorter than those of visible light. Infrared light has wavelengths just a bit longer than visible light.

(c)   A continuous spectrum is visible light of all wavelengths that has been spread out by a prism or a grating. A discontinuous spectrum is light of only certain wavelengths, such as light emitted from an atomic species that has been excited.

(d)   A traveling wave is one whose wave peaks move past a given point as time passes, such as waves in the ocean. A standing wave is one whose nodes remain fixed, such as sound waves in an organ pipe or the waves in a violin string.

(e)   A quantum number is one of the four (principal, orbital, magnetic, spin) definite values that are used to specify the properties of an electron in an atom. An orbital is a region of space in which there is a good chance of finding an electron.

(f)   *spdf* notation specifies the electron configuration of an atom by giving the principal and orbital quantum numbers of electrons, grouped into subshells. An orbital diagram places each electron of an atom, symbolized as an arrow, into a space that represents an orbital.

(g)   An *s*-block element is one in Group 1A or Group 2A. A *p*-block element is one in Group 3A, 4A, 5A, 6A, 7A, or 8A.

(h)   A main-group element is an element in one of the "A" groups, i.e., the *s* block or the *p* block. A transition element is one in a "B" group, that is a *d*-block element. (Sometimes *f*-block elements, also called inner-transition elements, are included as transition elements as well.)

4.   **(a)** $\text{length (nm)} = 1625\,\text{Å} \times \dfrac{1\,\text{m}}{10^{10}\,\text{Å}} \times \dfrac{10^{9}\,\text{nm}}{1\,\text{m}} = 162.5\,\text{nm}$

(b) $\text{length } (\mu\text{m}) = 3880\,\text{Å} \times \dfrac{1\,\text{m}}{10^{10}\,\text{Å}} \times \dfrac{10^{6}\,\mu\text{m}}{1\,\text{m}} = 0.388\,\mu\text{m}$

(c) $\text{length (nm)} = 7.27 \times 10^{-3}\,\text{m} \times \dfrac{10^{9}\,\text{nm}}{1\,\text{m}} = 7.27 \times 10^{6}\,\text{nm}$

(d) $\text{length (m)} = 546\,\text{nm} \times \dfrac{1\,\text{m}}{10^{9}\,\text{nm}} = 5.46 \times 10^{-7}\,\text{m}$

(e) $\text{length (nm)} = 1.12\,\text{cm} \times \dfrac{1\,\text{m}}{10^{2}\,\text{cm}} \times \dfrac{10^{9}\,\text{nm}}{1\,\text{m}} = 1.12 \times 10^{7}\,\text{nm}$

(f) $\text{length (cm)} = 2.6 \times 10^{4}\,\text{Å} \times \dfrac{1\,\text{m}}{10^{10}\,\text{Å}} \times \dfrac{10^{2}\,\text{cm}}{1\,\text{m}} = 2.6 \times 10^{-4}\,\text{cm}$

5.   **(a)** $\lambda = \dfrac{c}{v} = \dfrac{3.00 \times 10^8 \text{ m/s}}{6.8 \times 10^{12} \text{ s}^{-1}} = 4.4 \times 10^{-5}$ m, infrared

(b) $\lambda = \dfrac{c}{v} = \dfrac{3.00 \times 10^8 \text{ m/s}}{9.8 \times 10^{15} \text{ s}^{-1}} = 3.1 \times 10^{-8}$ m, ultraviolet

(c) $\lambda = \dfrac{c}{v} = \dfrac{2.998 \times 10^8 \text{ m/s}}{2.54 \times 10^7 \text{ s}^{-1}} = 11.8$ m, radio

(d) $\lambda = \dfrac{c}{v} = \dfrac{2.998 \times 10^8 \text{ m/s}}{1.07 \times 10^8 \text{ s}^{-1}} = 2.80$ m, radio

6.   The wavelength is the distance between successive peaks. Thus, $4 \times 1.17 \text{ nm} = \lambda = 4.68 \text{ nm}$.

7.   **(a)** $v = \dfrac{c}{\lambda} = \dfrac{3.00 \times 10^8 \text{m/s}}{4.68 \text{ nm} \times \dfrac{1\text{m}}{10^9 \text{ nm}}} = 6.41 \times 10^{16} \text{Hz}$

(b) $E = hv = 6.626 \times 10^{-34} \text{ J s} \times 6.41 \times 10^{16} \text{ Hz} = 4.25 \times 10^{-17}$ J

8.   **(a)**   The velocity of electromagnetic radiation in a vacuum, the speed of light, is a universal constant.

(b)   The wavelength of electromagnetic radiation is inversely proportional to its frequency: $\lambda = c / v$.

(c)   The energy per mole of electromagnetic radiation is directly proportional to its frequency: $E = N_A h v$.

9.   **(a)** $v = 3.2881 \times 10^{15} \text{s}^{-1} \left(\dfrac{1}{2^2} - \dfrac{1}{5^2} \right) = 6.9050 \times 10^{14} \text{s}^{-1}$

(b) $v = 3.2881 \times 10^{15} \text{ s}^{-1} \left(\dfrac{1}{2^2} - \dfrac{1}{7^2} \right) = 7.5492 \times 10^{14} \text{ s}^{-1}$

$\lambda = \dfrac{2.9979 \times 10^8 \text{ m/s}}{7.5492 \times 10^{14} \text{ s}^{-1}} = 3.9711 \times 10^{-7} \text{ m} \times \dfrac{10^9 \text{ nm}}{1 \text{ m}} = 397.11 \text{ nm}$

(c) $v = \dfrac{3.00 \times 10^8 \text{ m}}{380 \text{ nm} \times \dfrac{1\text{m}}{10^9 \text{ nm}}} = 7.89 \times 10^{14} \text{s}^{-1} = 3.2881 \times 10^{15} \text{s}^{-1} \left(\dfrac{1}{2^2} - \dfrac{1}{n^2} \right)$

$0.250 - \dfrac{1}{n^2} = \dfrac{7.89 \times 10^{14} \text{ s}^{-1}}{3.2881 \times 10^{15} \text{ s}^{-1}} = 0.240 \qquad \dfrac{1}{n^2} = 0.250 - 0.240 = 0.010 \quad n = 10$

10. The frequencies of hydrogen emission lines in the infrared region of the spectrum other than the visible region would be predicted by replacing the constant "2" in the Balmer equation by the variable m, where m is an integer smaller than n: $m = 3, 4, ...$

The resulting equation is $v = 3.2881 \times 10^{15} \, s^{-1} \left(\dfrac{1}{m^2} - \dfrac{1}{n^2} \right)$

11. (a) $E = h v = 6.626 \times 10^{-34} \, J\,s \times 8.62 \times 10^{15} \, s^{-1} = 5.71 \times 10^{-18} \, J\,/\,photon$

(b) $E_m = 6.626 \times 10^{-34} \, J\,s \times 1.53 \times 10^{14} \, s^{-1} \times \dfrac{6.022 \times 10^{23} \text{ photons}}{\text{mol}} \times \dfrac{1 \text{ kJ}}{1000 \text{ J}} = 61.0 \text{ kJ/mol}$

12. (a) $v = \dfrac{E}{h} = \dfrac{4.18 \times 10^{-21} \, J}{6.626 \times 10^{-34} \, J\,s} = 6.31 \times 10^{12} \, s^{-1} = 6.31 \times 10^{12} \text{ Hz}$

(b) $E = h v = \dfrac{hc}{\lambda}; \quad \lambda = \dfrac{hc}{E} = \dfrac{6.626 \times 10^{-34} \, J\,s \times 3.00 \times 10^{8} \, m/s}{215 \, kJ/mol \times \dfrac{1000 J}{1 kJ} \times \dfrac{1 mol}{6.022 \times 10^{23} \text{ photons}}} = 5.57 \times 10^{-7} \, m = 557 \text{ nm}$

13. $\Delta E = -2.179 \times 10^{-18} \, J \left(\dfrac{1}{n_f^{\,2}} - \dfrac{1}{n_i^{\,2}} \right) = -2.179 \times 10^{-18} \, J \left(\dfrac{1}{3^2} - \dfrac{1}{5^2} \right) = -1.550 \times 10^{-19} \, J$

$E_{\text{photon emitted}} = 1.550 \times 10^{-19} \, J = h v \qquad v = \dfrac{E}{h} = \dfrac{1.550 \times 10^{-19} \, J}{6.626 \times 10^{-34} \, J\,s} = 2.339 \times 10^{14} \, s^{-1}$

14. The electron (a), has the longest wavelength. De Broglie's equation $(\lambda = h/mv)$ indicates that, if the velocity is constant, the wavelength is inversely proportional to the mass. Thus, the particle with the least mass will have the longest wavelength. Of the four particles given, the *electron* (a) is the least massive.

15. (a) $m_\ell = 0, \pm 1$ Because $\ell = 1$ and $|m_\ell|$ must be $\le \ell$.

(b) $\ell = 1, 2, 3$ ℓ must be less than n (which equals 4) and also must be $\ge |m_\ell|$, and $m_\ell = -1$.

(c) $n = 2, 3, ...$ n must be greater than ℓ (which equals 1 in this case).

16. (a) $4s$ has $n = 4$ $\ell = 0$ **(b)** $3p$ has $n = 3$ $\ell = 1$

(c) $5f$ has $n = 5$ $\ell = 3$ **(d)** $3d$ has $n = 3$ $\ell = 2$

17. **(a)** $n = 3$ $\ell = 2$ $m_\ell = -1$ is permitted.

     (b) $n = 2$ $\ell = 3$ $m_\ell = -1$ is not allowed, because it has $\ell > n$.

     (c) $n = 3$ $\ell = 0$ $m_l = +1$ is not allowed, because it has $|m_\ell| > \ell$.

     (d) $n = 6$ $\ell = 2$ $m_l = -1$ is permitted.

     (e) $n = 4$ $\ell = 4$ $m_\ell = +4$ is not permitted, because it has $\ell = n$.

     (f) $n = 4$ $\ell = 3$ $m_l = -1$ is permitted.

18. The number of orbitals in a subshell is determined by the value of the orbital quantum number, ℓ. The possible m_ℓ values range from $-\ell$ to $+\ell$ in whole number steps. Each m_ℓ value corresponds to a different orbital. Therefore, the number of orbitals in a subshell $= 2\ell + 1$.

     (a) Since an s orbital has $\ell = 0$, there is but one orbital in the $2s$ subshell.

     (b) There is no $3f$ subshell. For $3f$, $n = 3$ and $\ell = 3$, but ℓ must always be less than n.

     (c) A p orbital has $\ell = 1$. There are three orbitals in the $4p$ subshell.

     (d) A d orbital has $\ell = 2$. There are five orbitals in the $5d$ subshell.

     (e) An f orbital has $\ell = 3$. There are seven orbitals in the $5f$ subshell.

     (f) A p orbital has $\ell = 1$. There are three orbitals in the $6p$ subshell.

19. The order of subshell filling is $1s\,2s\,2p\,3s\,3p\,4s\,3d\,4p\,5s\,4d\,5p\,6s\,4f\,5d\,6p\,7s\,5f\,6d$. An s subshell can have a maximum of 2 electrons, a p subshell can have 6 electrons maximum, a d subshell can have 10, and an f subshell can have 14.

     (a) Br has 35 electrons. $1s^2 2s^2 2p^6 3s^2 3p^6 3d^{10} 4s^2 4p^5$

     (b) S has 16 electrons. $1s^2 2s^2 2p^6 3s^2 3p^4$

     (c) Sb has 51 electrons. $1s^2 2s^2 2p^6 3s^2 3p^6 3d^{10} 4s^2 4p^6 4d^{10} 5s^2 5p^3$

     (d) Si has 14 electrons. $1s^2 2s^2 2p^6 3s^2 3p^2$

20. We write the orbital diagram of each element following Hund's rule and count the number of unpaired electrons.

     (a) Mg has 12 electrons, 2 more than the noble gas Ne. Zero unpaired electrons.

     (b) Tl has 81 electrons, 27 more than the noble gas Xe. Tl has one unpaired electron.

     [Xe] $4f$ [↑↓][↑↓][↑↓][↑↓][↑↓][↑↓][↑↓] $5d$ [↑↓][↑↓][↑↓][↑↓][↑↓] $6s$ [↑↓] $6p$ [↑][ ][ ]

(c) Te has <u>52 electrons</u>, 16 more than the noble gas Kr. Te has two unpaired electrons.
[Kr] $_{4d}$ ⇅ ⇅ ⇅ ⇅ ⇅ $_{5s}$ ⇅ $_{4p}$ ⇅ ↑ ↑

(d) Al has <u>13 electrons</u>, 3 more than the noble gas Ne. One unpaired electron.
[Ne] $_{3s}$ ⇅ $_{3p}$ ↑ □ □

21. **(a)** Main group metals include those in Groups 1A and 2A, along with Al, Ga, In, Tl, Sn, Pb, Bi and Po.

(b) Main group nonmetals are: H, F, Cl, Br, I, O, S, Se, N, P, C, and B.

(c) Noble gases are He, Ne, Ar, Kr, Xe, and Rn.

(d) The *d*-block elements are those in Groups 1B, 2B, 3B, 4B, 5B, 6B, 7B, and 8B.

(e) The inner transition elements include those with atomic numbers from $Z = 58$ (Ce) through $Z = 71$ (Lu) and those from $Z = 90$ (Th) through $Z = 103$ (Lr).

22. Base the prediction of each electron configuration upon the configuration of the preceding noble gas.

		prediction		from the text Appendix
(a)	In:	[Kr] $4d^{10}5s^2 5p^1$	in group 13	[Kr] $4d^{10}5s^2 5p^1$
(b)	Cd:	[Kr] $4d^{10}5s^2$	in group 12	[Kr] $4d^{10}5s^2$
(c)	Sb:	[Kr] $4d^{10}5s^2 5p^3$	in group 15	[Kr] $4d^{10}5s^2 5p^3$
(d)	Au:	[Xe] $4f^{14}5d^9 6s^2$	in group 11	[Xe] $4f^{14}5d^{10}6s^1$

The predictions exactly match the given electron configurations for the first three atoms; for Au, the configuration [Xe]$4f^{14}5d^{10}6s^1$ is the one actually adopted because it is a more stable arrangement.

23. **(a)** K is in group 1(1A) and in the fourth period. Elements in group 1 have an s^1 outer electron configuration; those in the fourth period are filling the 4th principal quantum level. Thus K has one $4s$ electron.

(b) I is in group 17(7A)($s^2 p^5$ outer electron configuration) and in the 5th period (5th principal quantum level). I has five $5p$ electrons.

(c) Zn is in group 12[$(n-1)d^{10}n s^2$ outer electron configuration] and in the 4th period $(n = 4)$. Zn has ten $3d$ electrons.

(d) S is in group 16 $\left[n s^2 n p^4 \right]$ and in the 3rd period. The $2p^6$ electron configuration was complete with the preceding noble gas (Ne). S has six $2p$ electrons.

(e) Pb follows the lanthanide series in which fourteen $4f$ electrons were added. Pb has fourteen $4f$ electrons.

(e)   Pb follows the lanthanide series in which fourteen $4f$ electrons were added. Pb has fourteen $4f$ electrons.

(f)   Ni is in the d-block of elements, two from the end. It therefore has eight d electrons. It also is in the fourth period, so these eight d electrons are $3d$ electrons. Ni has eight $3d$ electrons.

24.

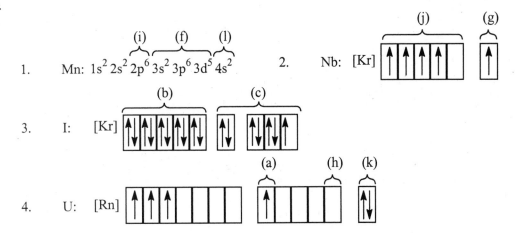

1.   Mn: $1s^2\,2s^2\,2p^6\,3s^2\,3p^6\,3d^5\,4s^2$

2.   Nb: [Kr]

3.   I: [Kr]

4.   U: [Rn]

EXERCISES

Electromagnetic Radiation

25.   **(a)**   TRUE   Since frequency and wavelength are inversely related to each other, radiation of shorter wavelength has higher frequency.

   (b)   FALSE   Light of wavelengths between 390 nm and 790 nm is visible to the eye.

   (c)   FALSE   All electromagnetic radiation has the same speed in vacuum.

   (d)   TRUE   The wavelength of an X-ray is approximately 0.1 nm.

26.   **(a)** $v = \dfrac{c}{\lambda} = \dfrac{2.9979 \times 10^8 \text{ m / s}}{418.7 \text{ nm}} \times \dfrac{10^9 \text{ nm}}{1 \text{ m}} = 7.160 \times 10^{14}$ Hz

   (b)   Light of wavelength 418.7 nm is in the visible region of the spectrum.

   (c)   The light is visible, with a violet color.

27.   Convert all lengths into meters. (a) 5.9×10^{-6} m, (b) 1.13×10^{-3} m, (c) 8.60×10^{-8} m, (d) 6.92×10^{-6} m. The light having the highest frequency is (b) since it has the shortest wavelength.

29. The speed of light is used to convert the distance into an elapsed time.

$$\text{time} = 93 \times 10^6 \text{ mi} \times \frac{5280 \text{ ft}}{1 \text{ mi}} \times \frac{12 \text{ in.}}{1 \text{ ft}} \times \frac{2.54 \text{ cm}}{1 \text{ in}} \times \frac{1 \text{ s}}{3.00 \times 10^{10} \text{ cm}} \times \frac{1 \text{ min}}{60 \text{ s}} = 8.3 \text{ min}$$

30. The speed of light is used to convert the time into a distance spanned by light.

$$1 \text{ light year} = 1 \text{ y} \times \frac{365.25 \text{ d}}{1 \text{ y}} \times \frac{24 \text{ h}}{1 \text{ d}} \times \frac{3600 \text{ s}}{1 \text{ h}} \times \frac{2.9979 \times 10^8 \text{ m}}{1 \text{ s}} \times \frac{1 \text{ km}}{1000 \text{ m}} = 9.4607 \times 10^{12} \text{ km}$$

Atomic Spectra

31. The longest wavelength component has the lowest frequency (and the smallest energy).

$$\nu = 3.2881 \times 10^{15} \text{ s}^{-1} \left(\frac{1}{2^2} - \frac{1}{3^2} \right) = 4.5668 \times 10^{14} \text{ s}^{-1} \qquad \lambda = \frac{c}{\nu} = \frac{2.9979 \times 10^8 \text{ m/s}}{4.5668 \times 10^{14} \text{ s}^{-1}} = 6.5646 \times 10^{-7} \text{m}$$
$$= 656.46 \text{ nm}$$

$$\nu = 3.2881 \times 10^{15} \text{ s}^{-1} \left(\frac{1}{2^2} - \frac{1}{4^2} \right) = 6.1652 \times 10^{14} \text{ s}^{-1} \qquad \lambda = \frac{c}{\nu} = \frac{2.9979 \times 10^8 \text{ m/s}}{6.1652 \times 10^{14} \text{ s}^{-1}} = 4.8626 \times 10^{-7} \text{m}$$
$$= 486.26 \text{ nm}$$

$$\nu = 3.2881 \times 10^{15} \text{ s}^{-1} \left(\frac{1}{2^2} - \frac{1}{5^2} \right) = 6.9050 \times 10^{14} \text{ s}^{-1} \qquad \lambda = \frac{c}{\nu} = \frac{2.9979 \times 10^8 \text{ m/s}}{6.9050 \times 10^{14} \text{ s}^{-1}} = 4.3416 \times 10^{-7} \text{m}$$
$$= 434.16 \text{ nm}$$

$$\nu = 3.2881 \times 10^{15} \text{ s}^{-1} \left(\frac{1}{2^2} - \frac{1}{6^2} \right) = 7.3069 \times 10^{14} \text{ s}^{-1} \qquad \lambda = \frac{c}{\nu} = \frac{2.9979 \times 10^8 \text{ m/s}}{7.3069 \times 10^{14} \text{ s}^{-1}} = 4.1028 \times 10^{-7} \text{m}$$
$$= 410.28 \text{ nm}$$

32. $\lambda = 1880 \text{ nm} \times \dfrac{1 \text{ m}}{10^9 \text{ nm}} = 1.88 \times 10^{-6} \text{ m}$. From Exercise 31, we see that wavelengths in the

Balmer series range downward from $6.5646 \times 10^{-7} \text{ m}$. Since this is less than $1.88 \times 10^{-6} \text{ m}$, we conclude that light with a wavelength of 1880 nm cannot be in the Balmer series.

33. First we determine the frequency of the radiation, and then match it with the Balmer equation.

$$\nu = \frac{c}{\lambda} = \frac{2.9979 \times 10^8 \text{ m s}^{-1} \times \dfrac{10^9 \text{ nm}}{1 \text{m}}}{389 \text{ nm}} = 7.71 \times 10^{14} \text{ s}^{-1} = 3.2881 \times 10^{15} \text{ s}^{-1} \left(\frac{1}{2^2} - \frac{1}{n^2} \right)$$

$$\left(\frac{1}{2^2} - \frac{1}{n^2} \right) = \frac{7.71 \times 10^{14} \text{ s}^{-1}}{3.2881 \times 10^{15} \text{ s}^{-1}} = 0.234 = 0.2500 - \frac{1}{n^2} \qquad \frac{1}{n^2} = 0.016 \qquad n = 7.9 \approx 8$$

34. **(a)** The maximum wavelength occurs when $n = 2$ and the minimum wavelength occurs at the series convergence limit, when n is exceedingly large (and $1/n^2 \approx 0$).

$$v = 3.2881 \times 10^{15} \text{ s}^{-1} \left(\frac{1}{1^2} - \frac{1}{2^2} \right) = 2.4661 \times 10^{15} \text{ s}^{-1} \quad \lambda = \frac{c}{v} = \frac{2.9979 \times 10^8 \text{ m/s}}{2.4661 \times 10^{15} \text{ s}^{-1}} = 1.2156 \times 10^{-7} \text{ m}$$

$$= 121.56 \text{ nm}$$

$$v = 3.2881 \times 10^{15} \text{ s}^{-1} \left(\frac{1}{1^2} - 0 \right) = 3.2881 \times 10^{15} \text{ s}^{-1} \quad \lambda = \frac{c}{v} = \frac{2.9979 \times 10^8 \text{ m/s}}{3.2881 \times 10^{15} \text{ s}^{-1}} = 9.1174 \times 10^{-8} \text{ m}$$

$$= 91.174 \text{ nm}$$

(b) First determine the frequency of this spectral line, and then the value of n to which it corresponds.

$$v = \frac{c}{\lambda} = \frac{2.9979 \times 10^8 \text{ m/s}}{95.0 \text{ nm} \times \dfrac{1 \text{m}}{10^9 \text{nm}}} = 3.16 \times 10^{15} \text{s}^{-1} = 3.2881 \times 10^{15} \text{s}^{-1} \left(\frac{1}{1^2} - \frac{1}{n^2} \right)$$

$$\left(\frac{1}{1^2} - \frac{1}{n^2} \right) = \frac{3.16 \times 10^{15} \text{ s}^{-1}}{3.2881 \times 10^{15} \text{ s}^{-1}} = 0.961 \qquad \frac{1}{n^2} = 1.000 - 0.961 = 0.039 \qquad n = 5$$

(c) Let us pursue the same attack as in part (b).

$$v = \frac{c}{\lambda} = \frac{2.9979 \times 10^8 \text{ m/s}}{108.5 \text{ nm} \times \dfrac{1 \text{m}}{10^9 \text{nm}}} = 2.763 \times 10^{15} \text{s}^{-1} = 3.2881 \times 10^{15} \text{s}^{-1} \left(\frac{1}{1^2} - \frac{1}{n^2} \right)$$

$$\left(\frac{1}{1^2} - \frac{1}{n^2} \right) = \frac{2.763 \times 10^{15} \text{ s}^{-1}}{3.2881 \times 10^{15} \text{ s}^{-1}} = 0.8403 \qquad \frac{1}{n^2} = 1.000 - 0.8403 = 0.1597$$

This gives as a result $n = 2.502$. Since n is not an integer, there is no line in the Lyman spectrum with a wavelength of 108.5 nm.

Quantum Theory

35. **(a)** Combine $E = hv$ and $c = v\lambda$ to obtain $E = hc/\lambda$

$$E = \frac{6.626 \times 10^{-34} \text{J s} \times 2.998 \times 10^8 \text{m/s}}{474 \text{ nm} \times \dfrac{1 \text{m}}{10^9 \text{ nm}}} = 4.19 \times 10^{-19} \text{J / photon}$$

(b) $E_m = 4.19 \times 10^{-19} \dfrac{\text{J}}{\text{photon}} \times 6.022 \times 10^{23} \dfrac{\text{photons}}{\text{mol}} = 2.52 \times 10^5 \text{ J/mol}$

36. First determine the energy of an individual photon, and then its wavelength in nm.

$$E = \frac{1799 \dfrac{kJ}{mol} \times \dfrac{1000J}{1kJ}}{6.022 \times 10^{23} \dfrac{photons}{mol}} = 2.987 \times 10^{-18} \frac{J}{Photon} = \frac{hc}{\lambda} \quad or \quad \frac{hc}{E} = \lambda$$

$$\lambda = \frac{6.626 \times 10^{-34} \ J s \times 2.998 \times 10^{8} \ m/s}{2.987 \times 10^{-18} \ J} \times \frac{10^{9} \ nm}{1 \ m} = 66.5 \ nm \quad \text{This is ultraviolet radiation.}$$

37. The easiest way to answer this question is to convert all (b) through (d) into nanometers. The radiation with the smallest wavelength will have the greatest energy per photon while the radiation with the largest wavelength has the smallest amount of energy per photon.

(a) 6.62×10^{2} nm

(b) 2.1×10^{-5} cm $\times \dfrac{1 \times 10^{7} \ nm}{1 \ cm} = 2.1 \times 10^{2}$ nm

(c) $3.58 \ \mu m \times \dfrac{1 \times 10^{3} \ nm}{1 \ \mu m} = 3.58 \times 10^{3}$ nm

(d) 4.1×10^{-6} m $\times \dfrac{1 \times 10^{9} \ nm}{1 \ m} = 4.1 \times 10^{3}$ nm

So, 2.1×10^{-5} nm radiation, by virtue of possessing the smallest wavelength in the set, has the greatest energy per photon. Conversely, since 4.1×10^{3} nm has the largest wavelength, it possesses the least amount of energy per photon.

38. This time let's express each type of radiation in terms of its frequency (ν).

(a) $\nu = 2.0 \times 10^{15}$ s^{-1}

(b) The maximum frequency for infrared radiation is $\sim 3 \times 10^{14}$ s^{-1}

(c) Here $\lambda = 7000$ Å (where 1 Å $= 1 \times 10^{-10}$ m) so $\lambda = 7000$ Å $\times \dfrac{1 \times 10^{-10} \ m}{1 \overset{\circ}{A}} = 7.000 \times 10^{-7}$ m

Calculate the frequency using the equation: $\nu = c/\lambda = \dfrac{2.998 \times 10^{8} \ m \ s^{-1}}{7.000 \times 10^{-7} \ m}$

$\nu = 4.283 \times 10^{14}$ s^{-1}

(d) X-rays have frequencies that range from 10^{17} s^{-1} to 10^{21} s^{-1}. Recall that for all forms of electromagnetic radiation, the energy per mole of photons increases with increasing frequency. Consequently, the correct order for the energy per mole of photons for the various types of radiation described in this question is:
infrared radiation (b) $< \lambda = 7000$ Å radiation (c) $< \nu = 2.0 \times 10^{15}$ s^{-1} (a) $<$ x-rays (d)
—————————— increasing energy per mole of photons ——————→

39. Notice that energy and wavelength are inversely related: $E = \dfrac{hc}{\lambda}$. Therefore radiation that is 100 times as energetic as radiation with a wavelength of 988 nm will have a wavelength one hundredth as long: 9.88 nm. The frequency of this radiation is:

$$v = \frac{c}{\lambda} = \frac{2.998 \times 10^8 \text{ m/s}}{9.88 \text{ nm} \times \dfrac{1m}{10^9 \text{nm}}} = 3.03 \times 10^{16} \text{s}^{-1} \text{ From Figure 9-3, this is ultraviolet radiation.}$$

40.
$$E_1 = hv = \frac{hc}{\lambda} = \frac{6.62607 \times 10^{-34} \text{J} \cdot \text{s} \times 2.99792 \times 10^8 \text{ m s}^{-1}}{589.00 \text{ nm} \times \dfrac{1m}{10^9 \text{nm}}} = 3.3726 \times 10^{-19} \text{ J/Photon}$$

$$E_2 = \frac{6.62607 \times 10^{-34} \text{J} \cdot \text{s} \times 2.99792 \times 10^8 \text{m s}^{-1}}{589.59 \text{ nm} \times \dfrac{1m}{10^9 \text{ nm}}} = 3.3692 \times 10^{-19} \text{ J/Photon}$$

$$\Delta E = E_1 - E_2 = 3.3726 \times 10^{-19} \text{ J} - 3.3692 \times 10^{-19} \text{ J} = 0.0034 \times 10^{-19} \text{ J/photon}$$
$$\Delta E = 3.4 \times 10^{-22} \text{ J/photon}$$

The Photoelectric Effect

41. **(a)** $E = hv = 6.63 \times 10^{-34} \text{ J s} \times 9.96 \times 10^{14} \text{ s}^{-1} = 6.60 \times 10^{-19} \text{ J/photon}$

(b) Indium will display a photoelectric effect when exposed to ultraviolet light since ultraviolet light has a maximum frequency of $1 \times 10^{16} \text{ s}^{-1}$, which is above the threshold frequency of indium. It will not display a photoelectric effect when exposed to infrared light since the maximum frequency of infrared light is $\sim 3 \times 10^{14} \text{ s}^{-1}$, which is below the threshold frequency of indium.

42. In his explanation of the photoelectric effect, Einstein stated that each photon strikes and is absorbed by only one electron (cannot kill two birds with one stone). He also, and very importantly, stated that no more than one photon could contribute its energy to a given electron (cannot kill one bird with two stones). It is this second principle which explains why photoelectrons are not produced in larger number when the intensity of (sub-threshold frequency) light is increased.

The Bohr Atom

43. **(a)** $\text{radius} = n^2 a_0 = 6^2 \times 0.53 \text{Å} \times \dfrac{1m}{10^{10} \text{ Å}} \times \dfrac{10^9 \text{ nm}}{1m} = 1.9 \text{ nm}$

(b) $E_n = -\dfrac{R_H}{n^2} = -\dfrac{2.179 \times 10^{-18} \text{ J}}{6^2} = -6.053 \times 10^{-20} \text{ J}$

44. **(a)** $r_1 = 1^2 \times 0.53 \text{Å} = 0.53 \text{ Å}$ $r_3 = 3^2 \times 0.53 \text{Å} = 4.8 \text{ Å}$

increase in distance $= r_3 - r_1 = 4.8 \text{Å} - 0.53 \text{Å} = 4.3 \text{ Å}$

(b) $E_1 = \dfrac{-2.179 \times 10^{-18} \text{J}}{1^2} = -2.179 \times 10^{-18} \text{J}$ $E_3 = \dfrac{-2.179 \times 10^{-18} \text{J}}{3^2} = -2.421 \times 10^{-19} \text{J}$

increase in energy $= -2.421 \times 10^{-19} \text{ J} - \left(-2.179 \times 10^{-18} \text{ J}\right) = 1.937 \times 10^{-18} \text{ J}$

45. **(a)** $v = \dfrac{2.179 \times 10^{-18} \text{ J}}{6.626 \times 10^{-34} \text{ J s}}\left(\dfrac{1}{4^2} - \dfrac{1}{7^2}\right) = 1.384 \times 10^{14} \text{ s}^{-1}$

(b) $\lambda = \dfrac{c}{v} = \dfrac{2.998 \times 10^8 \text{ m/s}}{1.384 \times 10^{14} \text{ s}^{-1}} = 2.166 \times 10^{-6} \text{ m} \times \dfrac{10^9 \text{ nm}}{1 \text{ m}} = 2166 \text{ nm}$

(c) This is infrared radiation.

46. The greatest quantity of energy is absorbed in the situation where the difference between the inverses of the squares of the two quantum numbers is the largest, and the system begins with a lower quantum number than it finishes with. The second condition eliminates answer **(d)** from consideration. Now we can consider the difference of the inverses of the squares of the two quantum numbers in each case.

(a) $\left(\dfrac{1}{1^2} - \dfrac{1}{2^2}\right) = 0.75$ **(b)** $\left(\dfrac{1}{2^2} - \dfrac{1}{4^2}\right) = 0.1875$ **(c)** $\left(\dfrac{1}{3^2} - \dfrac{1}{9^2}\right) = 0.0988$

Thus, the largest amount of energy is absorbed in the transition from $n = 1$ to $n = 2$, answer **(a)**, among the four choices given.

47. **(a)** According to the Bohr model, the radii of allowed orbits in a hydrogen atom are given by; where n = 1, 2, 3 . . . and $a_0 = 5.3 \times 10^{-11}$ m (0.53 Å or 53 pm) so,
$r_4 = (4)^2(5.3 \times 10^{-11} \text{ m}) = 8.5 \times 10^{-10} \text{ m}$

(b) Here we want to see if there is an allowed orbit at r = 4.00 Å. To answer this question we will employ the equation $r_n = n^2 a_0$:   4.00 Å $= n^2(0.53$ Å$)$ or $n = 2.75$ Å
Since n is <u>not</u> a whole number, we can conclude that the hydrogen atom does not orbit at a radius of 4.00 Å (i.e., such an orbit is forbidden by selection rules).

(c) The energy level for the n = 8 orbit is calculated using the equation
$E_n = \dfrac{-2.179 \times 10^{-18} \text{J}}{n^2}$ $E_8 = \dfrac{-2.179 \times 10^{-18} \text{J}}{8^2} = -3.405 \times 10^{-20}$ J (relative to $E_\infty = 0$ J)

(d) Here we need to determine if 2.500×10^{-17} J corresponds to an allowed orbit in the hydrogen atom. Once again we will employ the equation $E_n = \dfrac{-2.179 \times 10^{-18} \text{J}}{n^2}$

$2.500 \times 10^{-17} \text{ J} = \dfrac{-2.179 \times 10^{-18} \text{J}}{n^2}$ or $n^2 = \dfrac{-2.179 \times 10^{-18} \text{J}}{-2.500 \times 10^{-17} \text{J}}$ hence, $n = 0.2952$

Because n is not a whole number, 2.500×10^{-17} J is not an allowed energy state for an electron in a hydrogen atom.

48. Only transitions (a) and (d) result in the emission of a photon ((b) and (c) involve absorption, not the emission of light). In the Bohr model, the energy difference between two successive energy levels decreases as the value of n increases. Thus, the $n = 3 \rightarrow n = 2$ electron transition involves a greater loss of energy than the $n = 4 \rightarrow n = 3$ transition. This means that the photon emitted by $n = 4 \rightarrow n = 3$ transition will be lower in energy and hence, longer in wavelength than the photon produced by the $n = 3 \rightarrow n = 2$ transition. Consequently, among the four choices given, the electron transition (a) is the one that will produce light of the longest wavelength.

49. If infrared light is produced, the quantum number of the final state must have a lower value (be of lower energy) than the quantum number of the initial state. First we compute the frequency of the transition being considered (from $v = c / \lambda$), and then solve for the final quantum number.

$$v = \frac{c}{\lambda} = \frac{3.00 \times 10^8 \text{m/s}}{2170 \text{ nm} \times \dfrac{1\text{m}}{10^9 \text{ nm}}} = 1.38 \times 10^{14} \text{s}^{-1} = \frac{2.179 \times 10^{-18} \text{ J}}{6.626 \times 10^{-34} \text{ J s}} \left(\frac{1}{n^2} - \frac{1}{7^2} \right) = 3.289 \times 10^{15} \text{ s}^{-1} \left(\frac{1}{n^2} - \frac{1}{7^2} \right)$$

$$\left(\frac{1}{n^2} - \frac{1}{7^2} \right) = \frac{1.38 \times 10^{14} \text{ s}^{-1}}{3.289 \times 10^{15} \text{ s}^{-1}} = 0.0419_6 \qquad \frac{1}{n^2} = 0.0419_6 + \frac{1}{7^2} = 0.0623_7 \qquad n = 4$$

50. If infrared light is produced, the quantum number of the final state must have a lower value (be of lower energy) than the quantum number of the initial state. First we compute the frequency of the transition being considered (from $v = c / \lambda$), and then solve for the initial quantum number, n.

$$v = \frac{c}{\lambda} = \frac{3.00 \times 10^8 \text{m/s}}{3740 \text{nm} \times \dfrac{1\text{m}}{10^9 \text{nm}}} = 8.02 \times 10^{13} \text{s}^{-1} = \frac{2.179 \times 10^{-18} \text{ J}}{6.626 \times 10^{-34} \text{ J s}} \left(\frac{1}{5^2} - \frac{1}{n^2} \right) = 3.289 \times 10^{15} \text{ s}^{-1} \left(\frac{1}{5^2} - \frac{1}{n^2} \right)$$

$$\left(\frac{1}{5^2} - \frac{1}{n^2} \right) = \frac{8.02 \times 10^{13} \text{ s}^{-1}}{3.289 \times 10^{15} \text{ s}^{-1}} = 0.0243_8 \qquad \frac{1}{n^2} = -0.0243_8 + \frac{1}{5^2} = 0.0156_2 \qquad n = 8$$

Wave-Particle Duality

51. The de Broglie equation is $\lambda = h / mv$. This means that, for a given wavelength to be produced, a lighter particle would have to be moving faster. Thus, electrons would have to move faster than protons to display matter waves of the same wavelength.

52. First, we rearrange the de Broglie equation, solving it for velocity: $v = h / m\lambda$. Then we obtain the velocity of the electron. From Table 2-1, the mass of an electron is 9.109×10^{-28} g.

$$v = \frac{h}{m\lambda} = \frac{6.626 \times 10^{-34} \text{J s}}{\left(9.109 \times 10^{-28} \text{g} \times \dfrac{1\text{kg}}{1000\text{g}} \right) \left(1\mu m \times \dfrac{1\text{m}}{10^6 \mu m} \right)} = 7 \times 10^2 \text{m/s}$$

53.
$$\lambda = \frac{h}{mv} = \frac{6.626 \times 10^{-34}\,\text{J s}}{\left(145\ \text{g} \times \dfrac{1\,\text{kg}}{1000\,\text{g}}\right)\left(168\,\text{km/h} \times \dfrac{1\text{h}}{3600\,\text{s}} \times \dfrac{1000\,\text{m}}{1\,\text{km}}\right)} = 9.79 \times 10^{-35}\,\text{m}$$

The diameter of a nucleus approximates 10^{-15} m, far larger than this wavelength.

54.
$$\lambda = \frac{h}{mv} = \frac{6.626 \times 10^{-34}\,\text{J s}}{1000\ \text{kg} \times 25\ \text{m/s}} = 2.7 \times 10^{-38}\,\text{m}$$

This wavelength is so much smaller than the smallest particle of matter known (nuclear diameter $\approx 10^{-15}$ m) that its experimental measurement is virtually impossible.

The Heisenberg Uncertainty Principle

55. The Bohr model is a determinant model of an atom. It implies that the position of the electron is exactly known at any time in the future, once that position is known at the present. The distance of the electron from the nucleus also is exactly known, as is its energy. And finally, the velocity of the electron in its orbit is exactly known. All of these exactly known quantities—position, distance from nucleus, energy, and velocity—can't, according to the Heisenberg uncertainty principle, be known with great precision simultaneously.

56. Einstein believed very strongly in the law of cause and effect, what is known as a deterministic view of the universe. He felt that the need to use probability and chance ("playing dice") in describing atomic structure resulted because a suitable theory had not yet been developed to permit accurate predictions. He believed that such a theory could be developed, and had good reason for his belief: The developments in the theory of atomic structure came very rapidly during the first thirty years of this century, and those, such as Einstein, who had lived through this period had seen the revision of a number of theories and explanations that were thought to be the final answer. Another viewpoint in this area is embodied in another famous quotation: "Nature is subtle, but not malicious." The meaning of this statement is that the causes of various effects may be obscure but they exist nonetheless. Bohr was stating that we should accept theories as they are revealed by experimentation and logic, rather than attempting to make these theories fit our preconceived notions of what we believe they should be. In other words, Bohr was telling Einstein to keep an open mind.

57.
$$\Delta v = \left(\frac{1}{100}\right)(0.1)\left(2.998 \times 10^{8}\,\frac{\text{m}}{\text{s}}\right) = 2.998 \times 10^{5}\ \text{m/s} \quad m = 1.673 \times 10^{-27}\ \text{kg}$$

$$\Delta p = m\Delta v = (1.673 \times 10^{-27}\ \text{kg})(2.998 \times 10^{5}\ \text{m/s}) = 5.0 \times 10^{-22}\ \text{kg m s}^{-1}$$

$$\Delta x = \frac{h}{4\pi \Delta p} = \frac{6.626 \times 10^{-34}\,\text{J s}}{(4\pi)(5.0 \times 10^{-22}\ \dfrac{\text{kg m}}{\text{s}})} = {\sim}1 \times 10^{-13}\ \text{m} \quad ({\sim}100 \text{ times the diameter of a nucleus})$$

58. Assume a mass of 1000 kg for the mass of an automobile and that its position is known to 1 cm (0.01 m)

$$\Delta v = \frac{h}{4\pi\, m\Delta x}\;\frac{6.626\times 10^{-34}\,\text{J s}}{(4\pi)(1000\,\text{kg})(0.01\text{m})} = 5\times 10^{-36}\,\text{m s}^{-1} \;\text{(undetectable uncertainty in velocity)}$$

59. electron mass $= 9.109\times 10^{-31}$ kg, $\lambda = 0.53$ Å (1 Å $= 1\times 10^{-10}$ m) hence $\lambda = 0.53\times 10^{-10}$ m

$$\lambda = \frac{h}{mv}\;\text{or}\;v = \frac{h}{m\lambda} = \frac{6.626\times 10^{-34}\,\text{J s}}{(9.109\times 10^{-31}\,\text{kg})(0.53\times 10^{-10}\,\text{m})} = 1.4\times 10^{7}\,\text{m s}^{-1}$$

60. $\Delta E = h v = \dfrac{hc}{\lambda}$ or $\lambda = \dfrac{hc}{E} = \dfrac{(6.626\times 10^{-34}\,\text{J s})(2.998\times 10^{8}\,\text{m s}^{-1})}{(-5.45\times 10^{-19}\,\text{J}) - (-2.179\times 10^{-18}\,\text{J})} = 1.216\times 10^{-7}\,\text{m}$

$$v = \frac{h}{m\lambda} = \frac{6.626\times 10^{-34}\,\text{J s}}{(9.109\times 10^{-31}\,\text{kg})(1.216\times 10^{-7}\,\text{m})} = 5.983\times 10^{3}\,\text{m s}^{-1}$$

Wave Mechanics

61. A sketch of this situation is presented at right. We see that 2.50 waves span the space of the 42 cm. Thus, the length of each wave is obtained by equating: $2.50\lambda = 42$ cm, giving $\lambda \approx 17$ cm.

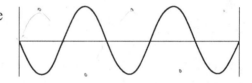

62. If there are four nodes, then there are three half-wavelengths within the string: one between the first and second node, the second half-wavelength between nodes 2 and 3, and the third between nodes 3 and 4. Therefore, $3\times\lambda/2 = $ length $= 3\times 17$ cm/2 $= 26$ cm long string.

63. The second overtone has three half-wavelengths within the string. We can find the wavelength for the second overtone by using the equation 1: string length $= \dfrac{n\lambda}{2}$, which gives the result: 24 in $= \dfrac{3\lambda}{2}$, $\lambda_{\text{second overtone}} = 16$ inches

64. The third harmonic for the wave in the box has five nodes and four antinodes. The total number of nodes is $n + 1$ so that $n = 4$ in this case. Also, we have been told that the box is 100 pm in length, so $L = 100$ pm here. We find the wavelength for the third harmonic by using $\lambda = \dfrac{2L}{n}$. Thus, $\lambda = \dfrac{2(100\text{pm})}{4} = 50$ pm

65. The differences between Bohr orbits and wave mechanical orbitals are given below.

(a) The first difference is that of shape. Bohr orbits, as originally proposed, are circular (later Sommerfeld proposed elliptical orbits). Orbitals, on the other hand, are spherical; or shaped like two tear drops or two squashed spheres; or shaped like four tear drops meeting at their points.

(b) Bohr orbits are planar pathways, while orbitals are three-dimensional regions of space in which there is a high probability of finding electrons.

(c) The electron in a Bohr orbit has a definite trajectory. Its position and velocity are known at all times. The electron in an orbital, however, does not have a well-known position or velocity. In fact, there is a small but definite probability that the electron may be found outside the boundaries generally drawn for the orbital.

Orbits and orbitals are similar in that the radii of Bohr orbits correspond to the distance from the nucleus in an orbital at which the electron is found with high probability.

66. We must be careful to distinguish between probability density—the chance of finding the electron within a definite volume of space—and the probability of finding the electron at a certain distance from the nucleus. The probability density—that is the probability of finding the electron within a small volume element (such as 1 pm^3)—at the nucleus may well be high, in fact higher than the probability density at a distance 0.53 Å from the nucleus. But the probability of finding the electron at a fixed distance from the nucleus is this probability density multiplied by all of the many small volume elements that are located at this distance. (Recall the dart board analogy of Figure 9–19.)

Quantum Numbers and Electron Orbitals

67. Answer (a) is incorrect because the values of m_s may be either $+\dfrac{1}{2}$ or $-\dfrac{1}{2}$. Answers (b) and (d) are incorrect because the value of ℓ may be any integer $\geq |m_\ell|$, and less than n. Thus, answer (c) is correct.

68. **(a)** $n = 3, \ell = 2, m_\ell = 2, m_s = +\dfrac{1}{2}$ (ℓ must be smaller than n and $\geq |m_\ell|$.)

This is a $3d$ orbital.

(b) $n \geq 3, \ell = 2, m_\ell = -1, m_s = -\dfrac{1}{2}$ (n must be larger than ℓ.)

This is a $3d, 4d, 5d, \ldots$ orbital.

(c) $n = 4, \ell = 2, m_\ell = 0, m_s = +\dfrac{1}{2}$ (m_s can be either $+\dfrac{1}{2}$ or $-\dfrac{1}{2}$.)

This is $4d$ orbital.

(d) $n \geq 1, \ell = 0, m_\ell = 0, m_s = +\dfrac{1}{2}$ (n can be any positive integer, m_s could also

equal $-\dfrac{1}{2}$.) This is a $1s$ orbital.

69. **(a)** $n = 5$ $\ell = 1$ $m_\ell = 0$ designates a $5p$ orbital. ($\ell = 1$ for all p orbitals.)

(b) $n = 4$ $\ell = 2$ $m_\ell = -2$ designates a $4d$ orbital. ($\ell = 2$ for all d orbitals.)

(c) $n = 2$ $\ell = 0$ $m_\ell = 0$ designates a $2s$ orbital. ($\ell = 0$ for all s orbitals.)

70. **(a)** TRUE The fourth principal shell has $n = 4$.

(b) TRUE A d orbital has $\ell = 2$. Since $n = 4$, ℓ can be $= 3$, 2, 1, 0. Since $m_\ell = -2$, ℓ can be $= 2, 3$.

(c) FALSE A p orbital has $\ell = 1$. But we demonstrated in part (b) that the only allowed values for ℓ are 2 and 3.

(d) FALSE Either $+\dfrac{1}{2}$ or $-\dfrac{1}{2}$ is permitted as a value of m_s.

71. **(a)** Just one electron can have $n = 3, \ell = 2, m_\ell = 0$, and $m_s = +\dfrac{1}{2}$. Four quantum numbers completely designate an electron.

(b) These three quantum numbers designate an orbital, which can hold two electrons. Two electrons can have the three quantum numbers $n = 3, \ell = 2, m_\ell = 0$.

(c) These two quantum numbers designate the $3d$ subshell, which contains five orbitals, with the possibility of two electrons in each. Ten electrons can have the two quantum numbers $n = 3, \ell = 2$.

(d) $n = 3$ designates the shell that contains the $3d$ subshell that has five orbitals, the $3p$ subshell with three orbitals, and the $3s$ subshell with one orbital. Each of these nine orbitals in the shell can hold two electrons, for a total of 18 electrons.

(e) The first two quantum numbers designate the $3d$ subshell, which has five orbitals. Each orbital can accommodate one electron with spin up. Five electrons can have the quantum numbers $n = 3, \ell = 2, m_s = \dfrac{1}{2}$.

72. **(a)** Each subshell has a different value for ℓ, the orbital quantum number. When $n = 4$, the possible values of ℓ are 0, 1, 2, and 3. There are four subshells in the $n = 4$ level.

(b) In the $n = 3$ level, the possible value of the orbital quantum number are $\ell = 0$, 1, and 2. These correspond to the $3s$ subshell, the $3p$ subshell, and the $3d$ subshell.

(c) In any $\ell = 3$ subshell, the possible values of the magnetic quantum number are $m_\ell = -3, -2, -1$, 0, 1, 2, 3. This means that there are seven orbitals in an $f(\ell = 3)$ subshell.

(d)   The values $n = 4$, $\ell = 3$, and $m_\ell = -2$ completely designate an orbital. Just one orbital has these three quantum numbers.

(e)   Within the $n = 4$ level, there is a $4s$ subshell with one orbital, a $4p$ subshell with three orbitals, a $4d$ subshell with five orbitals, and a $4f$ subshell with seven orbitals. The total number of orbitals in the $n = 4$ level is $1 + 3 + 5 + 7 = 16$ orbitals.

The Shapes of Orbitals and Radial Probabilities

73.   The wavefunction for the 2s orbital of a hydrogen atom is:

$$\psi_{2s} = \frac{1}{4}\left(\frac{1}{2\pi a_o^3}\right)^{1/2}\left(2 - \frac{r}{a_o}\right) e^{-\frac{r}{2a_o}}$$

Where $r = 2a_o$, the $\left(2 - \frac{r}{a_o}\right)$ term becomes zero, thereby making $\psi_{2s} = 0$. At this point the wave function has a radial node (i.e. the electron density is zero). The finite value of r is 2 a_o at the node, which is equal to 2×53 pm or 106 pm. Thus at 106 pm, there is a nodal surface with zero electron density.

74.   The radial part for the 2s wavefunction in the Li^{2+} dication is:

$$R_{2s} = \left(\frac{Z}{2\pi a_o}\right)^{3/2}\left(2 - \frac{Zr}{a_o}\right) e^{-\frac{Zr}{2a_o}}$$

(Z is the atomic number for the element, and, in Li^{2+} $Z = 3$). At $Zr = 2a_o$, the pre-exponential term for the 2s orbital of Li^{2+} is zero. The Li^{2+} 2s orbital has a nodal sphere at $\frac{2a_o}{3}$ or 35 pm.

75.   The angular part of the $2p_y$ wavefunction is $Y(\theta\phi)_{py} = \sqrt{\frac{3}{4\pi}} \sin\theta \sin\phi$. The two lobes of the $2p_y$ orbital lie in the xy plane and perpendicular to this plane is the xz plane. For all points in the xz plane $\phi = 0$, and since the sine of 0° is zero, this means that the entire x z plane is a node. Thus, the probability of finding a $2p_y$ electron in the xz plane is zero.

76.   The angular component of the wavefunction for the $3d_{xz}$ orbital is

$$Y(\theta\phi)_{d_{xz}} = \sqrt{\frac{15}{4\pi}} \sin\theta\cos\theta \cos\phi.$$

The four lobes of the d_{xz} orbital lie in the xz plane. The xy plane is perpendicular to the xz plane, and thus the angle for θ is 90°. The cosine of 90° is zero, so at every point in the xy plane the angular function has a value of zero. In other words, the entire xy plane is a node and, as a result, the probability of finding a $3d_{xz}$ electron in the xy plane is zero.

77./79. The $2p_x$ orbital $Y(\theta\phi) = \sqrt{\dfrac{3}{4\pi}}\sin\theta\cos\phi$; Note: in the xy plane $\theta = 90°$ and $\sin\theta = 1$

Plotting in the xy plane requires that we vary only ϕ

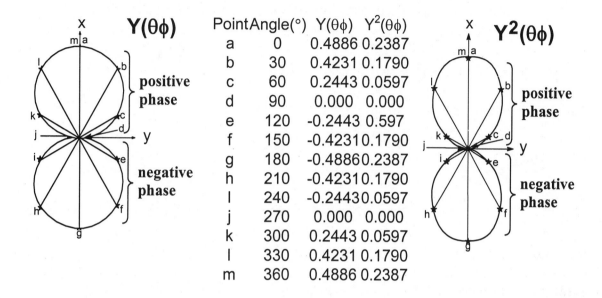

Point	Angle(°)	$Y(\theta\phi)$	$Y^2(\theta\phi)$
a	0	0.4886	0.2387
b	30	0.4231	0.1790
c	60	0.2443	0.0597
d	90	0.000	0.000
e	120	-0.2443	0.597
f	150	-0.4231	0.1790
g	180	-0.4886	0.2387
h	210	-0.4231	0.1790
I	240	-0.2443	0.0597
j	270	0.000	0.000
k	300	0.2443	0.0597
I	330	0.4231	0.1790
m	360	0.4886	0.2387

78./80. The $2p_y$ orbital $Y(\theta\phi) = \sqrt{\dfrac{3}{4\pi}}\sin\theta\sin\phi$, however, in the xy plane $\theta = 90°$ and $\sin\theta = 1$

Plotting in the xy plane requires that we vary only ϕ

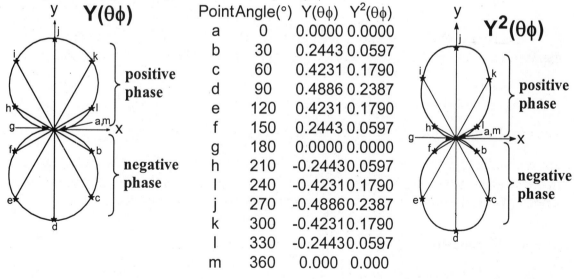

Point	Angle(°)	$Y(\theta\phi)$	$Y^2(\theta\phi)$
a	0	0.0000	0.0000
b	30	0.2443	0.0597
c	60	0.4231	0.1790
d	90	0.4886	0.2387
e	120	0.4231	0.1790
f	150	0.2443	0.0597
g	180	0.0000	0.0000
h	210	-0.2443	0.0597
I	240	-0.4231	0.1790
j	270	-0.4886	0.2387
k	300	-0.4231	0.1790
I	330	-0.2443	0.0597
m	360	0.000	0.000

81. A plot of radial probability distribution versus r/a_o for a H_{1s} orbital shows a maximum at 1.0 (that is, $r = a_o$). The plot is shown below:

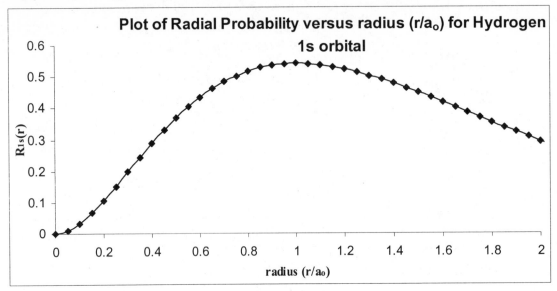

82. A plot of radial probability distribution versus r/a_o for a Li^{2+}_{1s} orbital shows a maximum at 0.33 (that is, $r = a_o/3$). The plot is shown below:

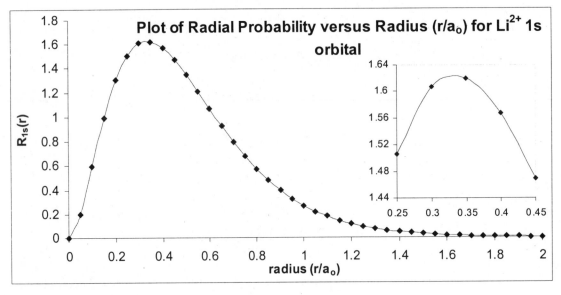

Electron Configurations

83. Configuration **(b)** is correct for phosphorus. The reason why each other configuration is incorrect follows.

(a) The two electrons in the $3s$ subshell must have opposed spins, or different values of m_s.

(c) The three $3p$ orbitals must each contain one electron, before a pair of electrons is placed in any one of these orbitals.

(d) The three unpaired electrons in the $3p$ subshell must all have the same spin, either all spin up or all spin down.

84. The electron configuration of Mo is [Kr] $_{4d}$[↑|↑|↑|↑|↑] $_{5s}$[↑]

(a) [Ar] $_{3d}$[↑↓|↑↓|↑↓|↑↓|↑↓] $_{3f}$[↑↓|↑↓|↑↓|↑↓|↑↓|↑↓|↑↓]   This configuration has the correct number of electrons. But it incorrectly assumes that there is a $3f$ subshell (n = 3, $\ell = 3$).

(b) [Kr] $_{4d}$[↑|↑|↑|↑|↑] $_{5s}$[↑]   This is the correct electron configuration.

(c) [Kr] $_{4d}$[↑|↑|↑|↑|↑] $_{5s}$[↑↓]   This configuration has one electron too many.

(d) [Ar] $_{3d}$[↑↓|↑↓|↑↓|↑↓|↑↓] $_{4s}$[↑↓] $_{4p}$[↑↓|↑|↑]   This configuration incorrectly assumes that electrons enter the 4d subshell before they enter the 5s subshell.

85. We write the correct electron configuration first in each case.

(a) P: $[\text{Ne}]3s^2 3p^3$ There are 3 unpaired electrons in each P atom.

(b) Br: $[\text{Ar}]3d^{10}4s^2 4p^5$ There are ten $3d$ electrons in an atom of Br.

(c) Ge: $[\text{Ar}]3d^{10}4s^2 4p^2$ There are two $4p$ electrons in an atom of Ge.

(d) Ba: $[\text{Xe}]6s^2$ There are two $6s$ electrons in an atom of Ba.

(e) Au: $[\text{Xe}]4f^{14}5d^9 6s^2$ There are fourteen $4f$ electrons in an atom of Au.

86. **(a)** The $4p$ subshell of Br contains 5 electrons   [Ar] $_{3d}$[↑↓|↑↓|↑↓|↑↓|↑↓] $_{4s}$[↑↓] $_{4p}$[↑↓|↑↓|↑]

(b) The $3d$ subshell of Co^{2+} contains 7 electrons   [Ar] $_{3d}$[↑↓|↑↓|↑|↑|↑] $_{4s}$[ ]

(c) The $5d$ subshell of Pb contains 10 electrons:

[Xe] $_{4f}$[↑↓|↑↓|↑↓|↑↓|↑↓|↑↓|↑↓] $_{5d}$[↑↓|↑↓|↑↓|↑↓|↑↓] $_{6s}$[↑↓] $_{6p}$[↑↓|↑|↑]

87. **(a)** N is the third element in the p-block of the second period. It has three $2p$ electrons.

(b) Rb is the first element in the s-block of the *fifth* period. . It has one $4s$ electrons.

(c) As is in the p-block of the fourth period. The $3d$ subshell is filled with ten electrons, but no $4d$ electrons have been added.

(d) Au is in the d-block of the sixth period; the $4f$ subshell is filled. Au has fourteen $4f$ electrons.

(e) Pb is the second element in the p-block of the sixth period; it has two $6p$ electrons. Since these two electrons are placed in separate $6p$ orbitals, they are unpaired. There are two unpaired electrons.

(f) Group 4A of the periodic table is the group with the elements C, Si, Ge, Sn, and Pb. This group currently has five named elements.

(g) The sixth period begins with the element Cs $(Z=55)$ and ends with the element Rn $(Z=86)$. This period is 32 elements long.

88. (a) Sb is in group 15, with an outer shell electron configuration ns^2np^3. Sb has five outer-shell electrons.

(b) Pt has an atomic number of $Z=78$. The fourth principal shell fills as follows: $4s$ from $Z=19$ (K) to $Z=20$ (Ca); $4p$ from $Z=31$ (Ga) to $Z=36$ (Kr); $4d$ from $Z=39$ (Y) to $Z=48$ (Cd); and $4f$ from $Z=58$ (Ce) to $Z=71$ (Lu). Since the atomic number of Pt is greater than $Z=71$, the entire fourth principal shell is filled, with 32 electrons.

(c) The five elements with six outer-shell electrons are those in group 16 (6A): They are O, S, Se, Te, Po.

(d) The outer-shell electron configuration is ns^2np^4, giving the following as a partial orbital diagram: $_{s^2}$ ⊞ $_{p^4}$ ⊞⊞ There are two unpaired electrons in an atom of Te.

(e) The sixth period begins with Cs and ends with Rn. There are 10 outer transition elements in this period (La, and Hf through Hg) and there are 14 inner transition elements in the period (Ce through Lu). Thus, there are $(10+14=)24$ transition elements in the sixth period.

89. Since the periodic table is based on electron structure, two elements in the same group (Pb and element 114) should have similar electron configurations.

(a) Pb: [Xe] $4f^{14}5d^{10}6s^26p^2$ (b)   114: [Rn] $5f^{14}6d^{10}7s^27p^2$

90. (a) The fifth period noble gas in group 18 is the element Xe.

(b) A sixth period element whose atoms have three unpaired electrons is an element in group 15 (5A), which has an outer electron configuration of ns^2np^3, and thus has three unpaired p electrons. This is the element Bi.

(c) One d-block element that has one $4s$ electron is Cu: [Ar] $3d^{10}4s^1$. Another is Cr: [Ar] $3d^54s^1$. Others are Mo, W, Ag, and Au.

(d) There are several p-block elements that are metals: Al, Ga, In, Tl, Sn, Pb, and Bi.

FEATURE PROBLEMS

112. By carefully scanning the diagram, we note that there are no spectral lines in the area of 304 nm and 309 nm, nor at 318 nm, and 327 nm. Likewise, there are none between 435 and 440 nm. We conclude that V is absent.

There are spectral lines that correspond to each of the Cr spectral lines—between 355 nm and 362 nm, and between 425 nm and 430 nm. Cr is present.

There are no spectral lines close to 403 nm; Mn is absent.

There are spectral lines at about 344 nm, 358 nm, 372 nm, 373 nm, and 386 nm. Fe is present.

There are spectral lines at 341 nm, 344 nm to 352 nm, and 362 nm. Ni is present.

There are spectral lines between 310 and 315 nm and at about 415 nm. Another element is present.

Thus, Cr, Fe, and Ni are present. V and Mn are absent. There is an additional element present.

<u>113.</u> The equation of a straight line is $y = mx + b$, where m is the slope of the line and b is its

y-intercept. The Balmer equation is $v = 3.2881 \times 10^{15} \text{ Hz} \left(\dfrac{1}{2^2} - \dfrac{1}{n^2} \right) = \dfrac{c}{\lambda}$. In this equation,

one plots v on the vertical axis, and $1/n^2$ on the horizontal axis. The slope is $b = -3.2881 \times 10^{15}$ Hz and the intercept is $3.2881 \times 10^{15} \text{ Hz} \div 2^2 = 8.2203 \times 10^{14} \text{ Hz}$. The plot of the data of Figure 9.9 follows.

λ	656.3 nm	486.1 nm	434.0 nm	410.1 nm
v	4.568×10^{14} Hz	6.167×10^{14} Hz	6.908×10^{14} Hz	7.310×10^{14} Hz
n	3	4	5	6

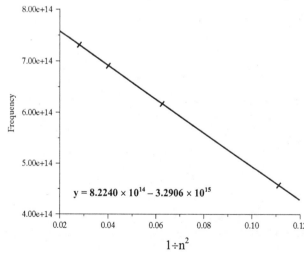

$$y = 8.2240 \times 10^{14} - 3.2906 \times 10^{15}$$

We see the slope (-3.2906×10^{15}) and the y-intercept (8.2240×10^{14}) are almost exactly what we had predicted from the Balmer equation.

114.   This graph differs from the one involving ψ^2 in Figure 9-18(a) because this graph factors in the volume of the thin shell. Figure 9-18(a) simply is a graph of the probability of finding an electron at a distance r from the nucleus. But the graph that accompanies this problem multiplies that radial probability by the volume of the shell that is a distance r from the nucleus. The volume of the shell is the thickness of the shell (a very small value dr) times the area of the shell $(4\pi r^2)$. Close to the nucleus, the area is very small because r is very small. Therefore, the relative probability also is very small near the nucleus; there just isn't enough volume to contain the electrons.

What we plot is the product of the number of darts times the circumference of the outer boundary of the scoring ring. (probability = number × circumference)

darts	200	300	400	250	200
score	"50"	"40"	"30"	"20"	"10"
radius	1.0	2.0	3.0	4.0	5.0
circumference	3.14	6.28	9.42	12.6	15.7
probability	628	1884	3768	3150	3140

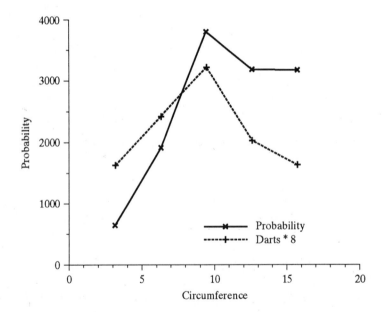

The graph of "probability" is close to the graph that accompanies this problem, except that the dart board is two-dimensional, while the atom is three-dimensional. This added dimension means that the volume close to the nucleus is relatively much smaller than is the area close to the center. The other difference, of course, is that it is harder for darts to get close to the center, while electrons are attracted to the nucleus.

115. **(a)**   First we calculate the range of energies for the incident photons used in the absorption experiment. Remember: $E_{photon} = h\nu$ & $\nu = c/\lambda$. At one end of the range, $\lambda = 100$ nm.

Therefore, $\nu = 2.998 \times 10^8$ m s^{-1} ÷ (1.00×10^{-7} m) $= 2.998 \times 10^{15}$ s^{-1}

So $E_{photon} = 6.626 \times 10^{-34}$ J s(2.998×10^{15} s^{-1}) $= 1.98 \times 10^{-18}$ J

At the other end of the range, $\lambda = 1000$ nm.

Therefore, $\nu = 2.998 \times 10^8$ m s^{-1} ÷ 1.00×10^{-6} m $= 2.998 \times 10^{14}$ s^{-1}

So $E_{photon} = 6.626 \times 10^{-34}$ J s(2.998×10^{14} s^{-1}) $= 1.98 \times 10^{-19}$ J.

Next, we will calculate what excitations are possible using photons with energies between 1.98×10^{-18} J and 1.98×10^{-19} J and the electron initially residing in the $n = 1$ level. These "orbit transitions" can be found with the equation

$$\Delta E = E_f - E_i = -2.179 \times 10^{-18}\left(\frac{1}{(1)^2} - \frac{1}{(n_f)^2}\right) \text{ For the lowest energy photon}$$

$$1.98 \times 10^{-19} \text{ J} = -2.179 \times 10^{-18}\left(\frac{1}{(1)^2} - \frac{1}{(n_f)^2}\right) \quad \text{or} \quad 0.0904 = 1 - \frac{1}{(n_f)^2}$$

From this $-0.9096 = -\dfrac{1}{(n_f)^2}$ and $n_f = 1.05$

Thus, the lowest energy photon is not capable of promoting the electron above the $n = 1$ level. For the highest energy level:

$$1.98 \times 10^{-18} \text{ J} = -2.179 \times 10^{-18}\left(\frac{1}{(1)^2} - \frac{1}{(n_f)^2}\right) \quad \text{or} \quad 0.9114 = 1 - \frac{1}{(n_f)^2}$$

From this $-0.0886 = -\dfrac{1}{(n_f)^2}$ and $n_f = 3.35$ Thus, the highest energy photon can

promote a groundstate electron to both the $n = 2$ and $n = 3$ levels. This means that we would see <u>two</u> lines in the absorption spectrum, one corresponding to the $n = 1 \rightarrow n = 2$ transition and the other to the $n = 1 \rightarrow n = 3$ transition.

Energy for the $n = 1 \rightarrow n = 2$ transition $= -2.179 \times 10^{-18}\left(\dfrac{1}{(1)^2} - \dfrac{1}{(2)^2}\right) = 1.634 \times 10^{-18}$ J

$$\nu = \frac{1.634 \times 10^{-18} \text{ J}}{6.626 \times 10^{-34} \text{ J s}} = 2.466 \times 10^{15} \text{ s}^{-1} \qquad \lambda = \frac{2.998 \times 10^8 \text{ m s}^{-1}}{2.466 \times 10^{15} \text{ s}^{-1}} = 1.215 \times 10^{-7} \text{ m}$$
$\lambda = 121.5$ nm

Thus, we should see a line at 121.5 nm in the absorption spectrum.

Energy for the $n = 1 \rightarrow n = 3$ transition $= -2.179 \times 10^{-18} \left(\dfrac{1}{(1)^2} - \dfrac{1}{(3)^2} \right) = 1.937 \times 10^{-18}$ J

$$v = \frac{1.937 \times 10^{-18} \text{ J}}{6.626 \times 10^{-34} \text{ J s}} = 2.923 \times 10^{15} \text{ s}^{-1} \qquad \lambda = \frac{2.998 \times 10^{8} \text{ m s}^{-1}}{2.923 \times 10^{15} \text{ s}^{-1}} = 1.025 \times 10^{-7} \text{ m}$$

$\lambda = 102.5$ nm

Consequently, the second line should appear at 102.6 nm in the absorption spectrum.

(b)   An excitation energy of 1230 kJ mol^{-1} to 1240 kJ mol^{-1} works out to 2×10^{-18} J per photon. This amount of energy is sufficient to raise the electron to the $n = 4$ level. Consequently, six lines will be observed in the emission spectrum. The calculation for each emission line is summarized below:

$$E_{4 \rightarrow 1} = \frac{2.179 \times 10^{-18} \text{ J}}{6.626 \times 10^{-34} \text{ J s}} \left(\frac{1}{1^2} - \frac{1}{4^2} \right) = 3.083 \times 10^{15} \text{ s}^{-1} \qquad \lambda = \frac{2.998 \times 10^{8} \frac{\text{m}}{\text{s}}}{3.083 \times 10^{15} \text{ s}^{-1}} = 9.724 \times 10^{-8} \text{ m}$$

$\lambda = 97.2$ nm

$$E_{4 \rightarrow 2} = \frac{2.179 \times 10^{-18} \text{ J}}{6.626 \times 10^{-34} \text{ J s}} \left(\frac{1}{2^2} - \frac{1}{4^2} \right) = 6.167 \times 10^{14} \text{ s}^{-1} \qquad \lambda = \frac{2.998 \times 10^{8} \frac{\text{m}}{\text{s}}}{6.167 \times 10^{14} \text{ s}^{-1}} = 4.861 \times 10^{-7} \text{ m}$$

$\lambda = 486.1$ nm

$$E_{4 \rightarrow 3} = \frac{2.179 \times 10^{-18} \text{ J}}{6.626 \times 10^{-34} \text{ J s}} \left(\frac{1}{3^2} - \frac{1}{4^2} \right) = 1.599 \times 10^{14} \text{ s}^{-1} \qquad \lambda = \frac{2.998 \times 10^{8} \frac{\text{m}}{\text{s}}}{1.599 \times 10^{14} \text{ s}^{-1}} = 1.875 \times 10^{-6} \text{ m}$$

$\lambda = 1875$ nm

$$E_{3 \rightarrow 1} = \frac{2.179 \times 10^{-18} \text{ J}}{6.626 \times 10^{-34} \text{ J s}} \left(\frac{1}{1^2} - \frac{1}{3^2} \right) = 2.924 \times 10^{15} \text{ s}^{-1} \qquad \lambda = \frac{2.998 \times 10^{8} \frac{\text{m}}{\text{s}}}{2.924 \times 10^{15} \text{ s}^{-1}} = 1.025 \times 10^{-7} \text{ m}$$

$\lambda = 102.5$ nm

$$E_{3 \rightarrow 2} = \frac{2.179 \times 10^{-18} \text{ J}}{6.626 \times 10^{-34} \text{ J s}} \left(\frac{1}{2^2} - \frac{1}{3^2} \right) = 4.568 \times 10^{14} \text{ s}^{-1} \qquad \lambda = \frac{2.998 \times 10^{8} \frac{\text{m}}{\text{s}}}{4.568 \times 10^{14} \text{ s}^{-1}} = 6.563 \times 10^{-7} \text{ m}$$

$\lambda = 656.3$ nm

$$E_{2 \rightarrow 1} = \frac{2.179 \times 10^{-18} \text{ J}}{6.626 \times 10^{-34} \text{ J s}} \left(\frac{1}{1^2} - \frac{1}{2^2} \right) = 2.467 \times 10^{15} \text{ s}^{-1} \qquad \lambda = \frac{2.998 \times 10^{8} \frac{\text{m}}{\text{s}}}{2.467 \times 10^{15} \text{ s}^{-1}} = 1.215 \times 10^{-7} \text{ m}$$

$\lambda = 121.5$ nm

(c)   The number of lines observed in the two spectra is not the same. The absorption spectrum has two lines while the emission spectrum has six lines. Notice that the 102.5 nm and 1021.5 nm lines are present in both spectra. This is not surprising since the energy difference between each level is the same whether it is probed by emission or absorption spectroscopy.

116. (a)   The wavelength associated with each Helium-4 atom must be close to 100 pm (1.00 $\times$ 10^{-10} m) in order for diffraction to take place. To find the necessary velocity for the He atoms, we need to employ the de Broglie equation: $\lambda = \dfrac{h}{mv}$.

Rearrange to give $v = \dfrac{h}{m\lambda}$ where:

$\lambda = 1.00 \times 10^{-10}$ m, $h = 6.626 \times 10^{-34}$ kg m^{2}s^{-1} and m$_{neutron} = 6.696 \times 10^{-27}$ kg.

So, $v = \dfrac{6.626 \times 10^{-34} \text{ kg m}^2 \text{ s}^{-1}}{6.696 \times 10^{-27} \text{ kg}(1.00 \times 10^{-10} \text{ m})} = 9.90 \times 10^2 \text{ m s}^{-1}$

(b)   The de Broglie wavelength for the beam of protons would be too small for any diffraction to occur. Instead, most of the protons would will simply pass through the film of gold and have little or no interaction with the constituent gold atoms. Keep in mind, however, that some of the protons will end up being deflected or bounced back by passing too close to a nucleus or by colliding with a gold nucleus head on, respectively.

CHAPTER 10
THE PERIODIC TABLE AND SOME ATOMIC PROPERTIES
PRACTICE EXAMPLES

1A   Atomic size decreases left to right through a period, and bottom to top in a family. We expect the smallest elements to be in the upper right corner of the periodic table. S is the element closest to the upper right corner and should have the smallest atom.

S = 104 pm   As = 121 pm   I = 133 pm

1B   From the periodic table inside the front cover, we see that Na is in the same period as Al (period 3), but in a different group from K, Ca, and Br (period 4), which might predict that Na and Al are about the same size. However, there is a substantial decrease in size as one moves from left to right in a period due to an increase in effective nuclear charge, enough such that Ca should be about the same size as Na. Table 10-1 shows:

227 pm for K < 197 pm for Ca ≈ 184 pm for Na < 143 pm for Al < 114 pm for Br

2A Ti^{2+} and V^{3+} are isoelectronic; the one with higher positive charge should be smaller: $V^{3+} < Ti^{2+}$. Sr^{2+} and Br^- are isoelectronic; again, the one with higher positive charge should be smaller: $Sr^{2+} < Br^-$. In addition Ca^{2+} and Sr^{2+} both are ions of Group 2A; the one of lower atomic number should be smaller: $Ca^{2+} < Sr^{2+} < Br^-$. Finally, we know that the size of atoms decreases left to right across a period; we expect size of like-charged ions to follow the same trend: $Ti^{2+} < Ca^{2+}$. Summarized below:

$$V^{3+}(64 \text{ pm}) < Ti^{2+}(86 \text{ pm}) < Ca^{2+}(100 \text{ pm}) < Sr^{2+}(113 \text{ pm}) < Br^-(196 \text{ pm})$$

2B Br^- clearly is larger than As since Br^- is an anion in the same period as As. In turn, As is larger than N since both are in the same group, with As lower down in the group.
As also should be larger than P, which is larger than Mg^{2+}, an ion smaller than N. All that remains is to note that Cs is a truly large atom, one of the largest in the periodic table. The As atom should be in the middle. Data from Table 10-1 are:

65 pm for Mg^{2+} < 70 pm for N < 125 pm for As < 196 pm for Br^- < 265 pm for Cs

3A   Ionization increases from bottom to top of a group and from left to right through a period. The first ionization energy of K is less than that of Mg and the first ionization energy of S is less than that of Cl. We would suppose also that the first ionization energy of Mg is smaller than that of S, because Mg is a metal.

K(418.8 kJ / mol) < Mg(737.7 kJ / mol) < S(1000 kJ / mol) < Cl(1251.1 kJ / mol)

3B We would expect an alkali metal (Rb) or an alkaline earth metal (Sr) to have a low first ionization energy and nonmetals (e.g. Br) to have relatively high first ionization energies. Metalloids (such as Sb and As) should have intermediate ionization energies. Since the first ionization energy for As is larger than that for Sb, the first ionization energy of Sb should be in the middle. Data that were used to produce Figure 10-9 include the following (in kJ/mol): 403 for Rb, 549 for Sr, 834 for Sb, 947 for As, and 1140 for Br.

4A Cl and Al must be paramagnetic, since they each have an odd number of electrons. The electron configurations of K^+ ([Ar]) and O^{2-} ([Ne]) are those of the nearest noble gas; because all electrons are paired, they are diamagnetic. In Zn: [Ar] $3d^{10}4s^2$ all electrons are paired; the atom is diamagnetic.

4B The electron configuration of Cr is [Ar] $3d^5 4s^1$; it has six unpaired electrons. The electron configuration of Cr^{2+} is [Ar] $3d^4$; it has four unpaired electrons. The electron configuration of Cr^{3+} is [Ar] $3d^3$; it has three unpaired electrons. Thus, of the two ions, Cr^{2+} has the greater number of unpaired electrons.

5A We expect the melting point of bromine to be close to the average of those of chlorine and iodine: estimated melting point of $Br_2 = \dfrac{172 \text{ K} + 387 \text{ K}}{2} = 280 \text{ K}$. The actual melting point is 266 K.

5B If the boiling point of I_2 (458 K) is the average of the boiling points of Br_2 (349 K) and At_2, then $458 \text{ K} = (349 \text{ K} + ?)/2$ $? = 2 \times 458 \text{ K} - 349 \text{ K} = 567 \text{ K}$
The estimated boiling point of molecular astatine is about 570 K.

REVIEW QUESTIONS

1. **(a)** Two isoelectronic species (an atom and an ion, or two ions) have the same number of electrons.

    (b) The valence electrons in an atom are those that have the highest principal quantum number. They also are the ones furthest from the nucleus and thus on the outside of the atom. Because they are the electrons that other atoms "see", they are most important in determining the atom's chemical behavior.

    (c) A metal is an element that is relatively easily oxidized. Most elements are metals. In the periodic table, all of the transition elements are metals, as well as the lanthanides and actinides. In addition, metals are the representative elements in group 1(1A) and group 2(2A), as well as Al, Ga, In, Tl, Sn, Pb, Bi and Po.

    (d) A nonmetal is an element that is relatively easily reduced. In the periodic table, nonmetals are at the right and include H, F, Cl, Br, I, O, S, Se, N, P, C and B.

    (e) A metalloid is an element that is relatively easy to oxidize or to reduce. That is, their behavior is between that of metals and of nonmetals. Metalloids include Si, Ge, As, Sb, Te and At.

2. **(a)** The periodic law is, like all natural laws, a summary of experimental observations. It states that a given property of elements varies periodically when the elements are arranged in order of increasing atomic number.

   (b) Ionization energy is the quantity of energy that, when added to a gaseous atom (or ion), will remove an electron.

   (c) Electron affinity is the energy change that occurs when an electron is added to a gaseous atom (or ion). Since the electron affinity process usually is exothermic for atoms, most atomic electron affinities are negative.

   (d) Paramagnetism, the result of a species having one or more unpaired electrons, is the small attraction of that species into a magnetic field. It is not nearly as strong as the more familiar property of ferromagnetism.

3. **(a)** Lanthanide elements are f-block elements in which the $4f$-subshell is filling: Ce through Lu. Actinide elements are in which the $5f$-subshell is filling: Th through Lr.

   (b) A covalent radius is the radius of an atom that is bonded covalently to another. For example, one half the internuclear distance in Cl_2 is the covalent radius of chlorine. A metallic radius is half the shortest internuclear distance in a crystal of solid metal.

   (c) The atomic number of an atom equals the number of protons in the nucleus and thus the positive charge of the nucleus. The effective nuclear charge is the positive charge experienced by the outermost electrons. Intervening electrons of inner shells cancel much of the nuclear charge out.

   (d) Ionization energy is the energy that is needed to remove an electron from a gaseous atom (or ion). Electron affinity is the energy change that occurs when an electron is added to a gaseous atom (or ion).

   (e) A paramagnetic species has one or more unpaired electrons and is drawn towards a magnetic field. A diamagnetic species has all electrons paired and is slightly repelled by a magnetic field.

4. The pairs of elements that are "out of order" based on their atomic masses are presented here, together with their atomic numbers. The periodic table lists elements in order of increasing atomic number, not increasing atomic mass. For one of these pairs there is a further explanation. Most of the Ar in the atmosphere is thought to result from the radioactive decay of ^{40}K, a nuclide of that once was more plentiful than it is now.

Element	Ar	K	Te	I	Co	Ni
Z	18	19	52	53	27	28
At. mass (u)	39.948	39.098	127.60	126.9045	58.9332	58.693

Element	Th	Pa	Pu	Am	Sg	Bh
Z	90	91	94	95	106	107
At. mass (u)	232.0381	231.0359	244	243	263	262

5. **(a)** The number of protons in the nucleus of $^{119}_{50}$Sn equals the atomic number: 50 protons.

   (b) The number of neutrons in the nucleus is the difference between the mass number and the atomic number of the nuclide: $119 - 50 = 69$ neutrons.

   (c) Sn is in Group 14 of the fifth period of the periodic table. The $4d$ subshell completed filling with element 47 (Ag). Thus, there are ten $4d$ electrons in Sn.

   (d) The $3s$ subshell is filled. Sn has two $3s$ electrons.

   (e) The outer (valence) electron configuration of Sn is $5s^2 5p^2$. Sn has two $5p$ electrons.

   (f) There are four electrons in the shell of highest principal quantum number $(n = 5)$ in an atom of Sn.

6. In the literal sense, isoelectronic means having the same number of electrons. (In another sense, not used in the text, it means having the same electron configuration.) We determine the total number of electrons and the electron configuration for each species and make our decisions based on this information.

Fe^{2+}	24 electrons	$[Ar]\,3d^6$	Sc^{3+}	18 electrons	$[Ar]$
Ca^{2+}	18 electrons	$[Ar]$	F^-	10 electrons	$[He]\,2s^2 2p^6$
Co^{2+}	25 electrons	$[Ar]\,3d^7$	Co^{3+}	24 electrons	$[Ar]\,3d^6$
Sr^{2+}	36 electrons	$[Ar]\,3d^{10} 4s^2 4p^6$	Cu^+	28 electrons	$[Ar]\,3d^{10}$
Zn^{2+}	28 electrons	$[Ar]\,3d^{10}$	Al^{3+}	10 electrons	$[He]\,2s^2 2p^6$

Species with the same number of electrons and the same electron configuration are the following. Fe^{2+} and Co^{3+} Sc^{3+} and Ca^{2+} F^- and Al^{3+} Zn^{2+} and Cu^+

7. Isoelectronic species must have the same number of electrons, and each element has a different atomic number, atoms of different elements cannot be isoelectronic. Two different cations may be isoelectronic, as may two different anions, or an anion and a cation. An example would be two anions (or two cations, or an anion and a cation) that have the electron configuration of a nearby noble gas, such as: O^{2-} and F^-, Na^+ and Mg^{2+}, or F^- and Na^+.

8. **(a)** The most nonmetallic element is the one farthest to the right and the top (with the exception of the noble gases). This is the element fluorine.

   (b) The transition elements are those with incomplete d subshells, or those which can form ions with incomplete d subshells. Scandium ($[Ar]\,3d^1 4s^2$) is the transition element with the smallest atomic number.

   (c) The metalloids are defined as the elements in Groups 14 – 17 (4A - 7A) that are adjacent to the stair-step diagonal line in the periodic table, except for Al and Po; that is, Si, Ge, As, Sb, Te and At. Of these 6 elements, only Si $(Z = 14)$ has an atomic number exactly midway between those of two noble gases, Ne $(Z = 10)$ and Ar $(Z = 18)$.

9.   In general, atomic size in the periodic table increases from top to bottom for a group and increases from right to left through a period, as indicated in Figures 10-4 and 10-8. The larger element is indicated first, followed by the reason for making the choice.

   (a)   Te: Te is to the left of Br and also in the period below that of Br in the 4th period.

   (b)   K: K is to the left of Ca within the same period, Period 4.

   (c)   Cs :   Cs is both below and to the left of Ca in the periodic table.

   (d)   N: N is to the left of O within the same period, Period 2.

   (e)   P: P is both below and to the left of O in the periodic table.

   (f)   Au: Au is both below and to the left of Al in the periodic table.

10.   An Al atom is larger than a F atom since Al is both below and to the left of F in the periodic table. As is larger than Al, since As is below Al in the periodic table. (Even though As is to the right of Al, we would not conclude that As is smaller than Al, since increases in size down a group are more pronounced than decreases in size across a period (from left to right). A Cs^+ ion is isoelectronic with an I^- ion, and in an isoelectronic series, anions are larger than cations; thus I^- is larger than Cs^+. I^- also is larger than As, since I is below As in the periodic table (and increases in size down a group are more pronounced than those across a period). Finally, N is larger than F, since N is to the left of F in the periodic table. Therefore, we conclude that F is the smallest species listed and I^- is the largest. In fact, with the exception of Cs^+, we can rank the species in order of decreasing size. $I^- > As > Al > N > F$ and also $I^- > Cs^+$

11.   Ionization energy in the periodic table decreases from top to bottom for a group, and increases from left to right for a period, as summarized in Figure 10-9. Cs has the lowest ionization energy; it is furthest to the left and nearest to the bottom of the periodic table. Next comes Sr, followed by As, then S, and finally F, the most nonmetallic element in the group (and in the periodic table). Thus, the elements listed in order of increasing ionization energy are: $Cs < Sr < As < S < F$

12.   **(a)**   The first element in a group has the smallest atoms. In Group 14 this is C.

   (b)   The atom in a period with the largest atoms is furthest to the left. In the fifth period this is Rb.

   (c)   The bottom element in a group has atoms with the lowest first ionization energy. In Group 17 this is At (or I if we do not wish to consider radioactive elements).

13.   First we convert the mass of Na given to an amount in moles of Na. Then we compute the energy needed to ionize this much Na.

$$\text{Energy} = 1.00 \text{ mg Na} \times \frac{1 \text{ g}}{1000 \text{ mg}} \times \frac{1 \text{ mol Na}}{22.99 \text{ g Na}} \times \frac{495.8 \text{ kJ}}{1 \text{ mol Na}} = 0.0216 \text{ kJ} \times \frac{1000 \text{ J}}{1 \text{ kJ}} = 21.6 \text{ J}$$

14. **(a)** The most metallic element is the one closest to the lower left-hand corner. This is the element Ba.

(b) The most nonmetallic element is the one closest to the upper right-hand corner, the element S.

(c) There are two distinct metals in this group, Ba and Ca; they have the lowest ionization energies. There is one distinct nonmetal, S; it has the highest ionization energy. The remaining two elements, As and Bi, are a metalloid and a metal, respectively. The metalloid has a higher ionization energy than does the metal. Bi has the intermediate value of ionization energy of the five elements listed.

15. Metallic character decreases from left to right and from bottom to top in the periodic table. Thus, in order of decreasing metallic character the elements listed are:
Rb > Ca > Sc > Fe > Te > Br > O > F The difficulty in establishing this series is in placing the elements Te and Br. First, the metal Fe is more metallic than the nonmetal Te. Further, Te clearly is more metallic than the halogen Br. Finally, we only need to recognize that Cl and O have approximately the same nonmetallic character, and Br clearly is more metallic than is Cl.

16. Paramagnetism indicates unpaired electrons, which in turn are often associated with partially filled subshells. First we write the electron configurations of the elements, and then those of the ions. From those electron configurations, we determine whether the species is paramagnetic or diamagnetic.

K	$[Ar]\,4s^1$	K^+	$[Ar]$	diamagnetic, all subshells filled.
Cr	$[Ar]\,3d^5 4s^1$	Cr^{3+}	$[Ar]\,3d^3$	paramagnetic.
Zn	$[Ar]\,3d^{10} 4s^2$	Zn^{2+}	$[Ar]\,3d^{10}$	diamagnetic, all subshells filled.
Cd	$[Kr]\,4d^{10} 5s^2$			diamagnetic, all subshells filled.
Co	$[Ar]\,3d^7 4s^2$	Co^{3+}	$[Ar]\,3d^6$	paramagnetic.
Sn	$[Kr]\,4d^{10} 5s^2 5p^2$	Sn^{2+}	$[Kr]\,4d^{10} 5s^2$	diamagnetic, all subshells filled.
Br	$[Ar]\,3d^{10} 4s^2 4p^5$			paramagnetic

17. **(a)** 6. Tl's electron configuration $[Xe]\,4f^{14} 5d^{10} 6s^2 6p^1$ has one p electron in its outermost shell.

(b) 8. $Z = 70$ identifies the element as Yb, an f-block of the periodic table, an inner transition element.

(c) 5. Ni has the electron configuration $[Ar]\,4s^2 3d^8$. It also is a d-block element.

(d) 1. An s^2 outer electron configuration, with the underlying configuration of a noble gas, is characteristic of elements of group 2(2A), the alkaline earth elements.

(e) 2. The element in the fifth period and Group 15 is Sb, a metalloid.

(f) 4 and 6. The element in the fourth period and Group 16 is Se, a nonmetal. (Note, B is a nonmetal with electron configuration $1s^2 2s^2 2p^1$ (having one p electron in the shell of highest n) 6. is also a possible answer)

18. We expect periodic properties to be functions of atomic number.

Element, atomic number	He, 2	Ne, 10	Ar, 18	Kr, 36	Xe, 54	Rn, 86
Boiling point, K	4.2 K	27.1 K	87.3 K	119.7 K	165 K	
Δ b.p. $/ \Delta Z$		2.86	7.5	1.8	2.5	

With one notable exception, we see that the boiling point increases about 2.4 K per unit of atomic number. The atomic number increases by 32 units from Xe to Rn. We might expect the boiling point to increase by $(2.4 \times 32 =)77$ K to $(165 + 77 =)242$ K for Rn.

A simpler manner is to expect that the boiling point of xenon (165 K) is the average of the boiling points of radon and krypton (120 K).

$$165 \text{ K} = (120 \text{ K} + ?)/2 \quad ? = 2 \times 165 \text{ K} - 120 \text{ K} = 210 \text{ K} = \text{ boiling point of radon}$$

Often the simplest way is best. The tabulated value is 211 K.

EXERCISES

The Periodic Law

19. Element 114 will be a metal in the same group as Pb, element 82 (18 cm^3/mol); Sn, element 50 (18 cm^3/mol); and Ge, element 32 (14 cm^3/mol). We note that the atomic volume of Pb and Sn are essentially equal, probably due to the lanthanide contraction. If there is also an actinide contraction, element 114 will have an atomic volume of $18 \text{ cm}^3 / \text{mol}$. If there is no actinide contraction, we would predict a molar volume of ~ 22 cm^3 / mol. This need to estimate atomic volume is what makes the value for density inaccurate.

$$\text{density} \left(\frac{g}{cm^3} \right) = \frac{298 \ \frac{g}{mol}}{18 \ \frac{cm^3}{mol}} = 16 \ \frac{g}{cm^3} \qquad \text{density} \left(\frac{g}{cm^3} \right) = \frac{298 \ \frac{g}{mol}}{22 \ \frac{cm^3}{mol}} = 14 \ \frac{g}{cm^3}$$

20. Lanthanum has an atomic number of $Z = 57$, and thus its atomic volume is somewhat less than $25 \text{ cm}^3 / \text{mol}$. Let us assume $23 \text{ cm}^3 / \text{mol}$.

$$\text{atomic mass} = \text{density} \times \text{atomic volume} = 6.145 \text{ g}/\text{cm}^3 \times 23 \text{ cm}^3 / \text{mol} = 141 \text{ g}/\text{mol}$$

This compares very well with the listed value of 139 for La.

21. The following data are plotted at right. Density clearly is a periodic property for these two periods of main group elements. It rises, falls a bit, rises again, and falls back to the axis, in both cases.

Element	Atomic Number Z	density $\frac{g}{cm^3}$
Na	11	0.968
Mg	12	1.738
Al	13	2.699
Si	14	2.336
P	15	1.823
S	16	2.069
Cl	17	0.0032
Ar	18	0.0018
K	19	0.856
Ca	20	1.550
Ga	31	5.904
Ge	32	5.323
As	33	5.778
Se	34	4.285
Br	35	3.100
Kr	36	0.0037

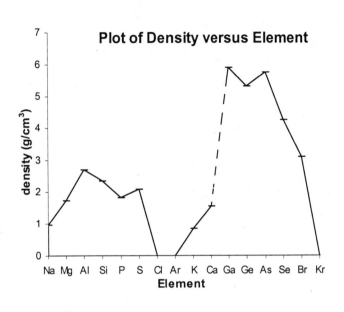

22. The following data are plotted at right below. Melting point clearly is a periodic property for these two periods. It rises to a maximum and then falls off in each case.

Element	Atomic Number Z	Melting Point °C
Li	3	179
Be	4	1278
B	5	2300
C	6	3350
N	7	−210
O	8	−218
F	9	−220
Ne	10	−249
Na	11	98
Mg	12	651
Al	13	660
Si	14	1410
P	15	590
S	16	119
Cl	17	−101
Ar	18	−189

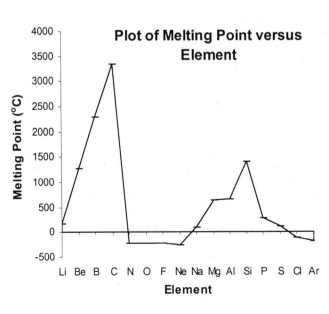

The Periodic Table

23. Mendeleev arranged elements in the periodic table in order of increasing atomic weight. Of course, atomic masses with non-integral values are permissible. Hence, there always is room for an added element between two elements that already are present in the table. On the other hand, Moseley arranged elements in order of increasing atomic number. Only integral (whole number) values of atomic number are permitted. Thus, when elements with all possible integral values in a certain range have been discovered, no new elements are possible in that range.

24. For there to be the same number of elements in each period of the periodic table, each shell of an electron configuration would have to contain the same number of electrons. This is not the case; the shells have 2 (K shell), 8 (L shell), 18 (M shell), and 32 (N shell) electrons each. This is because each of the periods of the periodic table begins with one s electron beyond a noble gas electron configuration, and the noble gas electron configuration corresponds to either s^2 (He) or $s^2 p^6$, with various other full subshells.

25. **(a)** The noble gas following radon $(Z = 86)$ will have an atomic number of $(86 + 32 =)118$.

(b) The alkali metal following francium $(Z = 87)$ will have an atomic number of $(87 + 32 =)119$.

(c) The mass number of radon $(A = 222)$ is $(222 \div 86) = 2.58$ times its atomic number. The mass number of Lr $(A = 262)$ is $(262 \div 103) = 2.54$ times its atomic number. Thus, we would expect the mass numbers, and hence approximate atomic masses of elements 118 and 119 to be about 2.5 times their atomic numbers, that is, $A_{118} \approx 298$ u and $A_{119} \approx 295$ u.

26. **(a)** The $6d$ subshell will be complete with an element in the same group as Hg. This is three elements beyond Une $(Z = 109)$ and thus this element has an atomic number of $Z = 112$.

(b) The element in this period that will most closely resemble bismuth is six elements beyond Une $(Z = 109)$ and thus has $Z = 115$.

(c) The element in this period that would be a noble gas would fall below Rn, nine elements beyond Une $(Z = 109)$ and thus has $Z = 118$.

Atomic Radii and Ionic Radii

27. The reason why the sizes of atoms do not simply increase with atomic number is because electrons often are added successively to the same subshell. These electrons do not fully screen each other from the nuclear charge (they do not effectively get between each other and the nucleus). Consequently, as each electron is added to a subshell and the nuclear charge increases by one unit, all of the electrons in this subshell are drawn more closely into the nucleus.

28. The reason why atomic sizes are uncertain is because the electron cloud that surrounds an atom has no fixed limit. It can be pictured as gradually fading away, rather like the edge of a town. In both cases we pick an arbitrary boundary.

29. **(a)** The smallest atom in Group 13 is the first: B

(b) Po is in the sixth period, and is larger than the others, which are rewritten in the following list from left to right in the fifth period, that is, from largest to smallest: Sr, In, Sb, Te. Thus, Te is the smallest of the elements given.

30. The hydrogen ion contains no electrons, only a nucleus. It is exceedingly tiny, much smaller than any other atom or electron-containing ion. Both H and H^- have a nuclear charge of 1+, but H^- has two electrons to H's one, and thus is larger. Both He and H^- contain two electrons, but He has a nuclear charge of 2+, while H^- has one of only 1+. The smaller nuclear charge of H^- is less effective at attracting electrons than the nuclear charge of He. The only comparison left is between H and He; we expect He to be smaller since atomic size decreases left to right across a period. Thus, ordered by increasing size:

$H^+ < He < H < H^-$.

31. Ions can be isoelectronic without having noble-gas electron configurations. An example is Cu^+ and Zn^{2+}, which both have the electron configuration $[Ar]3d^{10}$.

32. In an isoelectronic series, all of the species have the same number of electrons. The size is determined by the nuclear charge. Those species with the largest (positive) nuclear charge are the smallest. Those with smaller nuclear charges are larger in size. Thus, the more positively charged an ion is in an isoelectronic series, the smaller it will be.

$Y^{3+} < Sr^{2+} < Rb^+ < Kr < Br^- < Se^{2-}$

33. Li^+ is the smallest; it not only is in the second period, but also is a cation. I^- is the largest, an anion in the 5th period. Next largest is Se in the previous (the 4th) period. We expect Br to be smaller than Se because it is both to the right of Se and in the same period.

$Li^+ < Br < Se < I^-$

34. Size decreases from left to right in the periodic table; on this basis I should be smaller than Al. But size increases from top to bottom in the periodic table; on this basis I should be larger than Al. There really is no good way of resolving these conflicting predictions.

Ionization Energies; Electron Affinities

35. The second ionization energy for an atom (I_2) cannot be smaller than the first ionization energy (I_1) for the same atom. The reason is that, when the first electron is removed it is being taken away from a species with a charge of 1+. On the other hand, when the second electron is removed, it is being taken from a species with a charge of 2+. Since the force between two charged particles is proportional to q_+q_-/r^2 (r is the distance between the particles), the higher the charge, the more difficult it will be to remove an electron.

36. In the case of a first electron affinity, a negative electron is being added to a neutral atom. This process may be either exothermic or endothermic. In the case of an ionization potential, however, a negatively charged electron is being removed from a positively charged cation, a process that must always require energy, because unlike charges attract each other.

37. There are four third-shell electrons in each atom of silicon. In Table 10.4, the values for the ionization energies of these four electrons are listed in kJ/mol. Thus, the energy needed to ionize all four electrons from a mole of silicon atoms is given by:
Energy $= (786.5 + 1577 + 3232 + 4356)\,kJ = 9952\,kJ$

38. Data ($I_1 = 375.7$ kJ/mol) are obtained from Table 10.3.

$$\text{no. Cs}^+ \text{ ions} = 1\,J \times \frac{1\,kJ}{1000\,J} \times \frac{1\,\text{mol Cs}^+}{375.7\,kJ} \times \frac{6.022 \times 10^{23}\,\text{Cs}^+\,\text{ions}}{1\,\text{mol Cs}^+\,\text{ions}} = 1.603 \times 10^{18}\,\text{Cs}^+\,\text{ions}$$

39. The electron affinity of bromine is −324.6 kJ/mol (Figure 10-10). We use Hess's law to determine the heat of reaction for $Br_2(g)$ becoming $Br^-(g)$.

$Br_2(g) \rightarrow 2\,Br(g)$ $\Delta H = +193\,kJ$

$2\,Br(g) + 2\,e^- \rightarrow 2\,Br^-(g)$ $2 \times E.A. = 2(-324.6)\,kJ$

$Br_2(g) + 2\,e^- \rightarrow 2\,Br^-(g)$ $\Delta H = -456\,kJ$ Overall process is *exothermic*.

40. The electron affinity of fluorine is −328.0 kJ/mol (Figure 10-10) and the first and second ionization energies of Mg (Table 10.4) are 737.7 kJ/mol and 1451 kJ/mol.

$Mg(g) \rightarrow Mg^+(g) + e^-$ $I_1 = 737.7\,kJ/mol$

$Mg^+(g) \rightarrow Mg^{2+}(g) + e^-$ $I_2 = 1451\,kJ/mol$

$2\,F(g) + 2\,e^- \rightarrow 2\,F^-(g)$ $2E.A. = 2(-328.0)\,kJ/mol$

$Mg(g) + 2\,F(g) \rightarrow Mg^{2+}(g) + 2\,F^-(g)$ $\Delta H = +1533\,kJ/mol$ Endothermic.

41.   The reasoning here is similar to that for the answer to Exercise 35. In both cases, an electron is being removed from an atom with a neon electron configuration. But in the case of Na^+, the electron is being removed from a species with a 2+ charge ,while in the case of Ne, the electron is being removed from a species with a 1+ charge. The more highly charged the resulting species is, the more difficult it is to produce it by removing an electron.

42.   The electron affinity of Li is −59.6 kJ/mol, the smallest ionization energy listed in Table 10.3 (and, except for Fr, displayed in Figure 10-9) is that of Cs, 375.7 kJ/mol. Thus, insufficient energy is produced by the electron affinity of Li to account for the ionization of Cs. We would predict that Li^-Cs^+ will not be stable and hence Li^-Li^+ and Li^-Na^+, owing to the larger ionization energies of Li and Na should be even more unstable. . (One other consideration not delt with here is the energy released when the positive and negative ion combine to form an ion pair. This is an exothermic process, but we have no way of assessing the value of this energy.)

Magnetic Properties

43.   Three of the ions have noble gas electron configurations and thus have no unpaired electrons: F^- is $1s^2 2s^2 2p^6$ Ca^{2+} and S^{2-} are $[Ne]3s^2 3p^6$

Only Fe^{2+} has unpaired electrons. Its electron configuration is $[Ar]3d^6$.

44.   First we write the electron configuration of the element, then that of the ion. In each case, the number of unpaired electrons written beside the configuration agrees with the data given in the statement of the problem.

(a) Ni $[Ar]\,3d^8 4s^2$ $\longrightarrow$ Ni^{2+} $[Ar]\,3d^8$ Two unpaired e^-

(b) Cu $[Ar]\,3d^{10}4s^1$ $\longrightarrow$ Cu^{2+} $[Ar]\,3d^9$ One unpaired e^-

(c) Cr $[Ar]\,3d^5 4s^1$ $\longrightarrow$ Cr^{3+} $[Ar]\,3d^3$ Three unpaired e^-

45.   All atoms with an odd number of electrons must be paramagnetic. There is no way to pair all of the electrons up if there is an odd number of electrons. Many atoms with an even number of electrons are diamagnetic, but some are paramagnetic. The one of lowest atomic number is carbon $(Z=6)$, which has two unpaired p-electrons producing the paramagnetic behavior: $[He]\,2s^2 2p^2$.

46.   The electron configuration of each iron ion can be determined by starting with the electron configuration of the iron atom. $[Fe]=[Ar]3d^6 4s^2$ Then the two ions result from the loss of both $4s$ electrons, followed by the loss of a $3d$ electron in the case of Fe^{3+}.

$$\left[Fe^{2+}\right]=[Ar]3d^6 \qquad \text{four unpaired electrons}$$

$$\left[Fe^{3+}\right]=[Ar]3d^5 \qquad \text{five unpaired electrons}$$

Predictions Based on the Periodic Table

47. **(a)** Elements that one would expect to exhibit the photoelectric effect with visible light should be ones that have a small value of their first ionization energy. Based on Figure 10-9, the alkali metals have the lowest first ionization potentials: Cs, Rb, and K are three suitable metals. Metals that would not exhibit the photoelectric effect with visible light are those that have high values of their first ionization energy. Again from Figure 10-9, Zn, Cd, and Hg seem to be three metals that would not exhibit the photoelectric effect with visible light.

(b) From Figure 10-1, we notice that the atomic (molar) volume increases for the solid forms of the noble gases as we travel down the group (the data points just before the alkali metal peaks). But it seems to increase less rapidly than the molar mass. This means that the density should increase with atomic mass, and Rn should be the densest solid in the group. We expect densities of liquids to follow the same trend as densities of solids.

(c) To estimate the first ionization energy of fermium, we note in Figure 10-9 that the ionization energies of the lanthanides (following the Cs valley) are approximately the same. We expect similar behavior of the actinides, and estimate a first ionization energy of about +600 kJ/mol.

(d) We can estimate densities of solids from the information in Figure 10-1. Radium has $Z = 88$ and an approximate atomic volume of 40 cm^3/mol. Then we use the molar mass of radium to determine its density:

$$\text{density} = \frac{1 \text{ mol}}{40 \text{ cm}^3} \times \frac{226 \text{ g Ra}}{1 \text{ mol}} = 5.7 \text{ g}/\text{cm}^3$$

48. Germanium lies between silicon and tin in Group 14. We expect the heat of atomization of germanium to approximate the average of the heats of atomization of Si and Sn.

$$\text{average} = \frac{452 \text{ kJ/mol Si} + 302 \text{ kJ/mol Sn}}{2} = 377 \text{ kJ/mol Ge}$$

49. **(a)** From Figure 10-1, the atomic (molar) volume of Al is 10 cm^3/mol and that for In is 15 cm^3/mol. Thus, we predict 12.5 cm^3/mol as the molar volume for Ga. Then we compute the expected density for Ga.

$$\text{density} = \frac{1 \text{ mol Ga}}{12.5 \text{ cm}^3} \times \frac{68 \text{ g Ga}}{1 \text{ mol Ga}} = 5.4 \text{ g/cm}^3$$

(b) Since Ga is in group 13 (3A) (Gruppe III on Mendeleev's table), the formula of its oxide should be Ga_2O_3. We use Mendeleev's molar masses to determine the molar mass for Ga_2O_3. Molar mass $= 2 \times 68$ g Ga $+ 3 \times 16$ g O $= 184$ g Ga_2O_3

$$\% \text{ Ga} = \frac{2 \times 68 \text{ g Ga}}{184 \text{ g Ga}_2\text{O}_3} \times 100\% = 74\%(\text{Ga});$$

Using our more recent periodic table, we obtain 74.4% Ga

50. **(a)** The boiling point increases by 52 °C from CH_4 to SiH_4, and by 22 °C from SiH_4 to GeH_4. One expects another increase in boiling point from GeH_4 to SnH_4, probably by about 15 °C. Thus, the predicted boiling point of SnH_4 is $-75°$ C. (The actual boiling point of SnH_4 is $-52°$ C).

(b) The boiling point decreases by 39 C° from H_2Te to H_2Se, and by 20 C° from H_2Se to H_2S. It probably will decrease by about 10 C° to reach H_2O. Therefore, one predicts a value for the boiling point of H_2O of approximately $-71°$ C. Of course, the actual boiling point of water is $100°$ C. The prediction is seriously in error (~ 170 °C) because of hydrogen bonding between water molecules, a topic that is discussed in Chapter 13.

51. **(a)** Size increases down a group and from right to left in a period. Ba is closest to the lower left corner of the periodic table and thus has the largest size.

(b) Ionization energy decreases down a group and from right to left in a period. Although Pb is closest to the bottom of its group, Sr is farthest left in its period (and only one period above Pb). Sr should have the lowest first ionization energy.

(c) Electron affinity becomes more negative from left to right in a period and from bottom to top in a group. Cl is closest to the upper right in the periodic table and has the most negative (smallest) electron affinity.

(d) The number of unpaired electrons can be determined from the orbital diagram for each species.

F	[He] $2s^22p^5$	1 unpaired e^-	N [He] $2s^22p^3$	3 unpaired e^-
S^{2-}	[Ne] $3s^23p^6$	0 unpaired e^-	Mg^{2+} [He] $2s^22p^6$	0 unpaired e^-
Sc^{3+}	[Ne] $3s^23p^6$	0 unpaired e^-	Ti^{3+} [Ar] $3d^1$	1 unpaired e^-

Thus, N has the largest number of unpaired electrons.

52. **(a)** $Z = 32$ 1. This is the element Ge, with an outer electron configuration of $3s^23p^2$. Thus, Ge has two unpaired p electrons.

(b) $Z = 8$ 1. This is the element O. Each atom has an outer electron configuration of $2s^22p^4$. Thus, O has two unpaired p electrons.

(c) $Z = 53$ 3. This is the element I, with an electron affinity more negative than that of the adjacent atoms: Xe and Te.

(d) $Z = 38$ 4. This is the element Sr, which has two $5s$ electrons. It is easier to remove one $5s$ electron than to remove the outermost $4s$ electron of Ca, but harder than removing the outermost $6s$ electron of Cs.

(e) $Z = 48$ 2. This is the element Cd. Its outer electron configuration is s^2 and thus it is diamagnetic.

(f) $Z = 20$ 2. This is the element Ca. Since its electron configuration is [Ar]$4s^2$, all electrons are paired and it is diamagnetic.

FEATURE PROBLEMS

72. **(a)** The work function is the energy required for an electron to escape the solid surface of an element.

   (b) Work functions tend to decrease down a group and increase across a period in the periodic table. The work function increases across the periodic table from left to right following the steady increase in effective nuclear charge. As one proceeds down a given group, the principle quantum number increases and so does the distance of the outer electrons from the nucleus. The further the electrons are from the nucleus, the more easily they can be removed from the solid surface of the element, and hence, the smaller will be the value of the work function for the element.

   (c) Through the process of interpolation, one would predict that the work function for potassium should fall somewhere close to 3.9. The published value for the work function of potassium is 3.69 (CRC handbook). Had we been provided more information on the nature of the bonding in each of these metals, and had we been told what type of crystalline lattice each metal adopts, we would have been able to come up with a more accurate estimate of the work function.

   (d) The periodic trends in work function closely follow those in ionization energy. This should come as no surprise since both ionization and the work function involve the loss of electrons from neutral atoms.

73. The Moseley equation, $\nu = A(Z - B)^2$, where ν is the frequency of the emitted X-ray radiation, Z is the atomic number and A and B are constants, relates the frequency of emitted X-rays to the nuclear charge for the atoms that make up the target of the cathode ray tube. X-rays are emitted by the element after one of its K-level electrons has been knocked out of the atom by collision with a fast moving electron. In this question, we have been asked to determine the values for the constants A and B. The simplest way to find these values is to plot $\sqrt{\nu}$ vs. Z. This plot provides $\sqrt{A}$ as the slope and $-\sqrt{A}\,(B)$ as the y –intercept. Starting with $\nu = A(Z - B)^2$, we first take the square root of both sides. This affords $\sqrt{\nu} = \sqrt{A}\,(Z - B)$. Multiplying out this expression gives $\sqrt{\nu} = \sqrt{A}\,(Z)$ $- \sqrt{A}\,(B)$. This expression follows the equation of a straight line $y = mx + b$, where $y = \sqrt{\nu}$, $m = \sqrt{A}$, $x = Z$ and $b = -\sqrt{A}\,(B)$. So a plot of $\sqrt{\nu}$ vs. Z will provide us with A and B, after a small amount of mathematical manipulation. Before we can construct the plot we need to convert the provided X-ray wavelengths into their corresponding frequencies. For instance, Mg has an X-ray wavelength = 987 pm. The corresponding frequency for this radiation = c/λ, hence,

$$\nu = \frac{2.998 \times 10^8 \text{ m s}^{-1}}{9.87 \times 10^{-10} \text{ m}} = 3.04 \times 10^{17} \text{ s}^{-1}$$

Performing similar conversions on the rest of the data allows for the construction of the following table and plot (below).

(Z)	$\sqrt{v}$
12	5.51×10^8
16	7.48×10^8
20	9.49×10^8
24	1.14×10^9
30	1.45×10^9
37	1.80×10^9

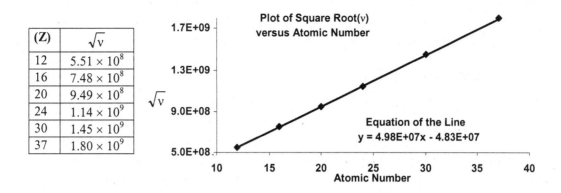

The slope of the line is $4.98 \times 10^7 = \sqrt{A}$ and the y-intercept is $-4.83 \times 10^7 = -\sqrt{A}$ (B).

Thus, $A = 2.30 \times 10^{15}$ Hz and $B = \dfrac{-4.83 \times 10^7}{-4.98 \times 10^7} = 0.970$

According to Bohr's theory, the frequencies that correspond to the lines in the emission spectrum are given by the equation: $(3.2881 \times 10^{15} \text{ s}^{-1}) \left(\dfrac{1}{(n_i)^2} - \dfrac{1}{(n_f)^2} \right)$,

where $(3.2881 \times 10^{15} \text{ s}^{-1})$ represents the frequency for the lowest energy photon that is capable of completely removing (ionizing) an electron from a hydrogen atom in its groundstate. The value of A (calculated in this question) is close to the Rydberg frequency $(3.2881 \times 10^{15} \text{ s}^{-1})$, so it is probably the equivalent term in the Moseley equation. The constant B, which is close to unity, could represent the number of electrons left in the K shell after one K-shell electron has been ejected by a cathode ray. Thus, one can think of b as representing the screening afforded by the remaining electron in the K-shell. Of course screening of the nucleus is only be possible for those elements with Z > 1.

74. **(a)** The table provided in this question shows the energy changes associated with the promotion of the outermost valence electron of sodium into the first four excited states above the highest occupied ground state atomic orbital. In addition, we have been told that the energy needed to completely remove one mole of $3s$ electrons from one mole of sodium atoms in the ground state is 496 kJ. The ionization energy for each excited state can be found by subtracting the "energy quanta" entry for the excited state from 496 kJ mol^{-1}.

e.g., for [Ne]$3p^1$, the 1st ionization energy $= 496 \dfrac{\text{kJ}}{\text{mol}} - 203 \dfrac{\text{kJ}}{\text{mol}} = 293 \dfrac{\text{kJ}}{\text{mol}}$

Thus, the rest of the ionization energies are:

[Ne]$4s^1$, $= 496$ kJ mol^{-1} $- 308$ kJ mol^{-1} $= 188$ kJ mol^{-1}

[Ne]$3d^1$, $= 496$ kJ mol^{-1} $- 349$ kJ mol^{-1} $= 147$ kJ mol^{-1}

[Ne]$4p^1$, $= 496$ kJ mol^{-1} $- 362$ kJ mol^{-1} $= 134$ kJ mol^{-1}

(b) Z_{eff} (the effective nuclear charge) for each state can be found by using the equation:

Ionization Energy in kJ mol^{-1} (I.E.) = $\dfrac{A(Z_{eff})^2}{n^2}$

Where n = starting principle quantum level for the electron that is promoted out of the atom and A = 1.3121×10^3 kJ mol^{-1} (Rydberg constant).

For [Ne]$3p^1$ (n = 3) 2.93×10^2 kJmol^{-1} = $\dfrac{1.3121 \times 10^3 \dfrac{kJ}{mol} (Z_{eff})^2}{3^2}$ $Z_{eff} = 1.42$

For [Ne]$4s^1$ (n = 4) 1.88×10^2 kJmol^{-1} = $\dfrac{1.3121 \times 10^3 \dfrac{kJ}{mol} (Z_{eff})^2}{4^2}$ $Z_{eff} = 1.51$

For [Ne]$3d^1$ (n = 3) 1.47×10^2 kJmol^{-1} = $\dfrac{1.3121 \times 10^3 \dfrac{kJ}{mol} (Z_{eff})^2}{3^2}$ $Z_{eff} = 1.00$

For [Ne]$4p^1$ (n = 4) 1.34×10^2 kJmol^{-1} = $\dfrac{1.3121 \times 10^3 \dfrac{kJ}{mol} (Z_{eff})^2}{4^2}$ $Z_{eff} = 1.28$

(c) $\overline{r}_{nl}$, which is the average distance of the electron from the nucleus for a particular orbital, can be calculated with the equation:

$$\overline{r}_{nl} = \frac{n^2 a_o}{Z_{eff}}\left(1 + \frac{1}{2}\left(1 - \frac{\ell(\ell+1)}{n^2}\right)\right)$$

Where $a_o = 52.9$ pm,

n = principal quantum number

ℓ = angular quantum number for the orbital

For [Ne]$3p^1$ (n = 3, ℓ = 1) $\overline{r}_{3p} = \dfrac{3^2(52.9 \ pm)}{1.42}\left(1+\dfrac{1}{2}\left(1-\dfrac{1(1+1)}{3^2}\right)\right) = 466$ pm

For [Ne]$4s^1$ (n = 4, ℓ = 0) $\overline{r}_{4s} = \dfrac{4^2(52.9 \ pm)}{1.51}\left(1+\dfrac{1}{2}\left(1-\dfrac{0(0+1)}{4^2}\right)\right) = 823$ pm

For [Ne]$3d^1$ (n = 3, ℓ = 2) $\overline{r}_{3d} = \dfrac{3^2(52.9 \ pm)}{1.00}\left(1+\dfrac{1}{2}\left(1-\dfrac{2(2+1)}{3^2}\right)\right) = 555$ pm

For [Ne]$4p^1$ (n = 4, ℓ = 1) $\overline{r}_{4p} = \dfrac{4^2(52.9 \ pm)}{1.28}\left(1+\dfrac{1}{2}\left(1-\dfrac{1(1+1)}{4^2}\right)\right) = 950$ pm

(d)   The results from the Z_{eff} calculations show that the greatest effective nuclear charge is experienced by the 4s orbital ($Z_{eff} = 1.51$). Next are the two p-orbitals, 3p and 4p, which come in at 1.42 and 1.28 respectively. Coming in last is the 3d orbital, which has a $Z_{eff} = 1.00$. These results are precisely in keeping with what we would expect. First of all, only the s-orbital penetrates all the way to the nucleus. Both the p- and d-orbitals have nodes at the nucleus. Also p-orbitals penetrate more deeply than do d-orbitals. Recall that the more deeply an orbital penetrates (i.e. the closer the orbital is to the nucleus), the greater is the effective nuclear charge felt by the electrons in that orbital. It follows then that the 4s orbital will experience the greatest effective nuclear charge and that the Z_{eff} values for the 3p and 4p orbitals should be larger than the Z_{eff} for the 3d orbital.

The results from the $\bar{r}_{nl}$ calculations for the four excited state orbitals show that the largest orbital in the set is the 4p orbital. This is exactly as expected because the 4p orbital is highest in energy and hence, on average furthest from the nucleus. The 4s orbital has an average position closer to the nucleus because it experiences a larger effective nuclear charge. The 3p orbital, being lowest in energy and hence on average closest to the nucleus, is the smallest orbital in the set. The 3p orbital has an average position closer to the nucleus than the 3d orbital (which is in the same principle quantum level), because it penetrates more deeply into the atom.

75. (a)   First of all, we need to find the ionization energy (I.E.) for the process: $F^-(g) \rightarrow F(g) + e^-$. To accomplish this, we need to calculate Z_{eff} for the species in the left hand column and plot the number of protons in the nucleus against Z_{eff}. By extrapolation, we can estimate the first ionization energy for $F^-(g)$. The electron affinity for F is equal to the first ionization energy of F^- multiplied by minus one (i.e., by reversing the ionization reaction, one can obtain the electron affinity).

For $Ne(g) \rightarrow Ne^+(g) + e^-$ (I.E. = 2080 kJ mol^{-1}; n = 2; 10 protons)

$$\text{I.E. (kJ mol}^{-1}) = \frac{1312.1 \ \frac{kJ}{mol}(Zeff)^2}{4} = 2080 \text{ kJ mol}^{-1} \quad Z_{eff} = 2.518$$

For $Na^+(g) \rightarrow Na^{2+}(g) + e^-$ (I.E. = 4565 kJ mol^{-1}; n = 2; 11 protons)

$$\text{I.E. (kJ mol}^{-1}) = \frac{1312.1 \ \frac{kJ}{mol}(Zeff)^2}{4} = 4565 \text{ kJ mol}^{-1} \quad Z_{eff} = 3.730$$

For $Mg^{2+}(g) \rightarrow Mg^{3+}(g) + e^-$ (I.E. = 7732 kJ mol^{-1}; n = 2; 12 protons)

$$\text{I.E. (kJ mol}^{-1}) = \frac{1312.1 \ \frac{kJ}{mol}(Zeff)^2}{4} = 7732 \text{ kJ mol}^{-1} \quad Z_{eff} = 4.855$$

For $Al^{3+}(g) \rightarrow Al^{4+}(g) + e^-$ (I.E. = 11,577 kJ mol^{-1}; n = 2; 13 protons)

$$\text{I.E. (kJ mol}^{-1}) = \frac{1312.1 \ \frac{kJ}{mol}(Zeff)^2}{4} = 11,577 \text{ kJ mol}^{-1} \qquad Z_{eff} = 5.941$$

A plot of the points (10, 2.518). (11, 3.730), (12, 4.855), (13,5.941) gives a straight line that follows the equation: $Z_{eff} = 1.1394(Z) -8.8421$

For F⁻, $Z = 9$; so $Z_{eff} = 1.1394(9) -8.8421 = 1.413$ and $n = 2$. Hence,

$$I.E. (kJ\ mol^{-1}) = \frac{1312.1\ \frac{kJ}{mol}(1.413)^2}{4} = 654.5\ kJ\ mol^{-1}\ for\ F^-$$

The Electron affinity for F must equal the reverse of the first ionization energy or –654.5 kJ. The actual experimental value found for the first electron affinity of F is –328 kJ/mol.

(b) Here, we will use the same method of solution as we did for part (a). To find the electron affinity for the process: $O(g) + e^- \rightarrow O^-(g)$, we first need to calculate the I.E. for $O^-(g)$. This is available from a plot of the number of protons in the nucleus vs. Z_{eff} for the four species in the central column.

For $F(g) \rightarrow F^+(g) + e^-$ (I.E. = 1681 kJ mol⁻¹; n = 2; 9 protons)

$$I.E. (kJ\ mol^{-1}) = \frac{1312.1\ \frac{kJ}{mol}(Zeff)^2}{4} = 1681\ kJ\ mol^{-1}\quad Z_{eff} = 2.264$$

For $Ne^+(g) \rightarrow Ne^{2+}(g) + e^-$ (I.E. = 3963 kJ mol⁻¹; n = 2; 10 protons)

$$I.E. (kJ\ mol^{-1}) = \frac{1312.1\ \frac{kJ}{mol}(Zeff)^2}{4} = 3963\ kJ\ mol^{-1}\quad Z_{eff} = 3.476$$

For $Na^{2+}(g) \rightarrow Na^{3+}(g) + e^-$ (I.E. = 6912 kJ mol⁻¹; n = 2; 11 protons)

$$I.E. (kJ\ mol^{-1}) = \frac{1312.1\ \frac{kJ}{mol}(Zeff)^2}{4} = 6912\ kJ\ mol^{-1}\quad Z_{eff} = 4.590$$

For $Mg^{3+}(g) \rightarrow Mg^{4+}(g) + e^-$ (I.E. = 10,548 kJ mol⁻¹; n = 2; 12 protons)

$$I.E. (kJ\ mol^{-1}) = \frac{1312.1\ \frac{kJ}{mol}(Zeff)^2}{4} = 10,548\ kJ\ mol^{-1}\qquad Z_{eff} = 5.671$$

A plot of the points (9, 2.264), (10, 3.476). (11, 4.590), (12, 5.671), gives a straight line that follows the equation: $Z_{eff} = 1.134(Z) -7.902$
For O⁻, $Z = 8$; so $Z_{eff} = 1.134 (8) -7.902 = 1.170$ and $n = 2$. Hence,

$$I.E. (kJ\ mol^{-1}) = \frac{1312.1\ \frac{kJ}{mol}(1.179)^2}{4} = 449\ kJ\ mol^{-1}\ for\ O^-$$

The Electron affinity for O must equal the reverse of the first ionization energy, or – 449 kJ. Again we will use the same method of solution as was used for part (a). To find the electron affinity for the process: $N(g) + e^- \rightarrow N^-(g)$, we first need to calculate the I.E. for $N^-(g)$. This is accessible from a plot of the number of protons in the nucleus vs. Z_{eff} for the four species in the last column.

For $O(g) \rightarrow O^+(g) + e^-$ (I.E. = 1314 kJ mol^{-1}; n = 2; 8 protons)

$$\text{I.E. (kJ mol}^{-1}) = \frac{1312.1 \frac{kJ}{mol}(Zeff)^2}{4} = 1314 \text{ kJ mol}^{-1} \quad Z_{eff} = 2.001$$

For $F^+(g) \rightarrow F^{2+}(g) + e$ (I.E. = 3375 kJ mol^{-1}; n = 2; 9 protons)

$$\text{I.E. (kJ mol}^{-1}) = \frac{1312.1 \frac{kJ}{mol}(Zeff)^2}{4} = 3375 \text{ kJ mol}^{-1} \quad Z_{eff} = 3.208$$

For $Ne^{2+}(g) \rightarrow Ne^{3+}(g) + e^-$ (I.E. = 6276 kJ mol^{-1}; n = 2; 10 protons)

$$\text{I.E. (kJ mol}^{-1}) = \frac{1312.1 \frac{kJ}{mol}(Zeff)^2}{4} = 6276 \text{ kJ mol}^{-1} \quad Z_{eff} = 4.374$$

For $Na^{3+}(g) \rightarrow Na^{4+}(g) + e^-$ (I.E. = 9,540 kJ mol^{-1}; n = 2; 11 protons)

$$\text{I.E. (kJ mol}^{-1}) = \frac{1312.1 \frac{kJ}{mol}(Zeff)^2}{4} = 9,540 \text{ kJ mol}^{-1} \quad Z_{eff} = 5.393$$

A plot of the points (8, 2.001), (9, 3.204), (10, 4.374). (11, 5.393), gives a straight line that follows the equation: $Z_{eff} = 1.1346(Z) – 7.0357$

For N^-, Z =7; so $Z_{eff} = 1.1346 (7) – 7.0357 = 0.9065$ and $n = 2$. Hence,

$$\text{I.E. (kJ mol}^{-1}) = \frac{1312.1 \frac{kJ}{mol}(0.9065)^2}{4} = 269.6 \text{ kJ mol}^{-1} \text{ for N}^-$$

The Electron affinity for N must equal the reverse of the first ionization energy, or – 269.6 kJ.

(c)   For N, O and F, the additional electron ends up in a 2p orbital. In all three instances the nuclear charge is well shielded by the filled 2s orbital located below the 2p set of orbitals. As we proceed from N to F, electrons are placed, one by one, in the 2p subshell and these electrons do afford some additional shielding, but this extra screening is more than offset by the accompanying increase in nuclear charge. Thus, the increase in electron affinity observed upon moving from N via oxygen to fluorine is the result of the steady increase in Z_{eff} that occurs upon moving further to the right in the periodic table.

76.　**(a)**　An oxygen atom in the ground state has the valence shell configuration $2s^2 2p^4$. Thus there are a total of six electrons in the valence shell. The amount of shielding experienced by one electron in the valence shell is the sum of the shielding provided by the other five electrons in the valence shell and the shielding afforded by the two electrons in the filled 1s orbital below the valence shell. Shielding from electrons in the same shell contribute $5 \times 0.35 = 1.75$ and the shielding from the electrons in the $n = 1$ shell contributes $2 \times 0.85 = 1.70$. The total shielding is 3.45 $(1.75 + 1.70)$. For O, $Z = 8$ hence, $Z_{eff} = 8 - 3.45 = 4.55$.

(b)　A ground state Cu atom has the valence shell configuration $3d^9 4s^2$. According to Slater's rules, the nine 3d electrons do not shield the $4s^2$ electrons from the nucleus. Thus the total amount of shielding experienced by a 4s electron in Cu is:

Shielding from the other $4s$ electron $= 1 \times 0.35$　　　　　 $= \;\; 0.35$
$+$ shielding from the electrons in the $3d$ subshell $= 9 \times 0.85$　 $= \;\; 7.65$
$+$shielding from the electrons in the $3s/3p$ orbitals $= 8 \times 1.00$　 $= \;\; 8.00$
$+$shielding from the electrons in the $2s/2p$ orbitals $= 8 \times 1.00$　 $= \;\; 8.00$
$+$shielding from the electrons in the $1s$ orbital $= 2 \times 1.00$　　 $= \;\; \underline{2.00}$
Total shielding for the $4s$ electrons $= (0.35 + 7.65 + 8 + 8 + 2) = 26.00$

Copper has $Z = 29$, so $Z_{eff} = 29 - 26.00 = 3.00$

(c)　3d electron in a ground state Cu atom will be shielded by the eight other 3d electrons and by the electrons in the lower principle quantum levels. Thus the total amount of shielding for a 3d electron is equal to

shielding from the eight electrons in the $3d$ subshell $= 8 \times 0.35$　 $= \;\; 2.80$

$+$shielding from the electrons in the $3s/3p$ orbitals $= 8 \times 1.00$　 $= \;\; 8.00$

$+$shielding from the electrons in the $2s/2p$ orbitals $= 8 \times 1.00$　 $= \;\; 8.00$

$+$shielding from the electrons in the $1s$ orbital $= 2 \times 1$　　　 $= \;\; \underline{2.00}$

Total shielding for the $3d$ electrons$= (2.80 + 8.00 + 8.00 + 2.00)$　 $= 20.80$

Copper has $Z = 29$, so $Z_{eff} = 29 - 20.80 = 8.2$

(d)　To find Z_{eff} for the valence electron in each Group I element, we first calculate the screening constant for the electron.

For H:　$S = 0$, so $Z_{eff} = Z$ and $Z = 1$; thus $Z_{eff} = 1$

For Li:　$S = 2(0.85) = 1.70$ and $Z = 3$; thus $Z_{eff} = 3 - 1.70 = 1.30$

For Na:　$S = 8(0.85) + 2(1) = 8.80$ and $Z = 11$; thus $Z_{eff} = 11 - 8.80 = 2.20$

For K:　$S = 8(0.85) + 8(1) + 2(1) = 16.80$, $Z = 19$; thus $Z_{eff} = 19 - 16.80 = 2.20$

For Rb:　$S = 8(0.85) + 10(1) + 8(1) + 8(1) + 2(1) = 34.80$, $Z = 37$; thus $Z_{eff} = 37 - 34.80 = 2.20$

For Cs:　$S = 8(0.85) + 10(1) + 8(1) + 10(1) + 8(1) + 8(1) + 2(1) = 52.80$; $Z = 55$, Thus $Z_{eff} = 2.20$

Based upon Slater's rules, we have found that the effective nuclear charge increases sharply between periods one and three and then stays at 2.20 for the rest of the alkali metal group. You may recall that the ionization energy for an element can be calculated by using the equation:

$$\text{I.E. (kJ mol}^{-1}) = \frac{1312.1 \ \frac{\text{kJ}}{\text{mol}} (Z\text{eff})^2}{n^2} \qquad \text{Where } n = \text{principal quantum number.}$$

By plugging the results from our Z_{eff} calculations into this equation, we would find that the ionization energy decreases markedly as we descend the alkali metal group, in spite of the fact that the Z_{eff} remains constant after Li. The reason that the ionization energy drops is, of course, because the value for n becomes larger as we move down the periodic table and, according to the ionization energy equation given above, larger n values translate into smaller ionization energies (this is because n^2 appears in the denominator). Put another way, even though Z_{eff} remains constant throughout most of Group I, the valence s-electrons become progressively easier to remove as we move down the group because they are further and further from the nucleus. Of course, the further away an electron is from the nucleus, the weaker is its attraction to the nucleus and the easier it is to remove.

(e)   As was the case in part (d), to evaluate Z_{eff} for a valence electron in each atom, we must first calculate the screening experienced by the electron with Slater's rules

For Li: $S = 2(0.85) = 1.70$ and $Z = 3$, thus $Z_{eff} = 3 - 1.70 = 1.30$

For Be: $S = 1(0.35) + 2(0.85) = 2.05$ and $Z = 4$; thus $Z_{eff} = 4 - 2.05 = 1.95$

For B: $S = 2(0.35) + 2(0.85) = 2.40$ and $Z = 5$; thus $Z_{eff} = 5 - 2.40 = 2.60$

For C: $S = 3(0.35) + 2(0.85) = 2.75$ and $Z = 6$; thus $Z_{eff} = 6 - 2.75 = 3.25$

For N: $S = 4(0.35) + 2(0.85) = 3.10$ and $Z = 7$; thus $Z_{eff} = 7 - 3.10 = 3.90$

For O: $S = 5(0.35) + 2(0.85) = 3.45$ and $Z = 8$; thus $Z_{eff} = 8 - 3.45 = 4.55$

For F: $S = 6(0.35) + 2(0.85) = 3.80$ and $Z = 9$; thus $Z_{eff} = 9 - 3.80 = 5.20$

For Ne: $S = 7(0.35) + 2(0.85) = 4.15$ and $Z = 10$; thus $Z_{eff} = 10 - 4.15 = 5.85$

The results from these calculations show that the Z_{eff} increases from left to right across the periodic table. Apart from small irregularities, the first ionization energies for the elements within a period also increase with increasing atomic number. Based upon our calculated Z_{eff} values, this is exactly the kind of trend for ionization energies that we would have anticipated. The fact is, a larger effective nuclear charge means that the outer electron(s) is/are held more tightly and this leads to a higher first ionization energy for the atom.

(f)   First we need to calculate the Z_{eff} values for an electron in the 3s, 3p and 3d orbitals of both a hydrogen atom and a sodium atom, by using Slater's rules. Since there is only one electron in a H atom, there is no possibility of shielding, and thus the effective nuclear charge for an electron in a 3s, 3p or 3d orbital of a H atom is one. The picture for Na is more complicated because it contains intervening electrons that shield the outermost electrons from the attractive power of the nucleus. The Z_{eff} calculations(based on Slater's Rules) for an electron in the i) the 3s orbital, ii) the 3p orbital, and iii) the 3d orbital of a Na atom are shown below:

Na (3s electron; n = 3) Z_{eff} = 11.0 –[(8e⁻ in the n = 2 shell × 0.85/e⁻) + (2e⁻ in the n = 1 shell × 1.00/e⁻)] = [11.0 – 8.8] = 2.2

Na (3p electron; n = 3; e- was originally in the 3s orbital)) Z_{eff} = 11.0 – [(8e⁻ in the n = 2 shell × 0.85/e⁻) + (2e⁻ in the n = 1 shell × 1.00/e⁻)] = [11.0 – 8.8] = 2.2

Na (3d electron; n = 3; e- was originally in the 3s orbital)) Z_{eff} = 11.0 – [(8e⁻ in the n = 2 shell × 1.00/e⁻) + (2e⁻ in the n = 1 shell × 1.00/e⁻)] = [11.0 – 10.0] = 1.0

Next, we insert these Z_{eff} values into their appropriate radial functions, which are gathered in Table 9.1, and use the results from these calculations to construct radial probability plots for an electron in the 3s, 3p and 3d orbitals of H and Na. The six plots that result are collected in the two figures presented below:

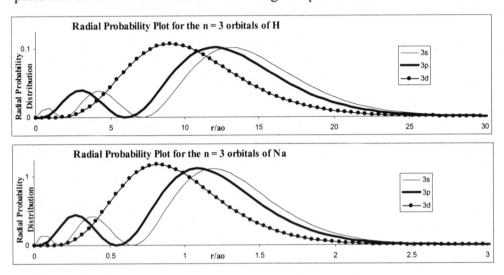

Notice that the 3s and 3p orbitals of sodium are much closer to the nucleus, on average, than the 3s and 3p orbitals of hydrogen. Because they are pulled more strongly towards the nucleus, the 3s and 3p orbitals of sodium end up being much smaller than the corresponding orbitals on hydrogen. Put another way, since the 3s and 3p electrons in sodium experience a larger effective nuclear charge, they are more tightly bound to the nucleus and, hence, are lower in energy than s and p electrons in the third principal shell of a hydrogen atom. If we want to express this in terms of shielding, we can say that the radial probability distributions for the 3s and 3p orbitals of sodium are smaller than those of hydrogen because the 3s and 3p orbitals of sodium are more poorly shielded. The graphs also show that the radial probability plot for the 3d orbital of a hydrogen atom is identical to that for a 3d orbital of a Na atom. This is what one would expect since a 3d electron in H and a 3d electron in Na both experience an effective nuclear charge of one.

CHAPTER 11
CHEMICAL BONDING I: BASIC CONCEPTS
PRACTICE EXAMPLES

1A   Mg is in group 2(2A), and thus has 2 valence electrons and 2 dots in its Lewis symbol. Ge is in group 14(4A), and thus has 4 valence electrons and 4 dots in its Lewis symbol. K is in group 1(1A), and thus has 1 valence electron and 1 dot in its Lewis symbol. Ne is in group 18(8A), and thus has 8 valence electrons and 8 dots in its Lewis symbol.

$$\cdot Mg \cdot \qquad \cdot Ge \cdot \qquad K \cdot \qquad :\overset{\cdot\cdot}{Ne}:$$

1B   Sn is in Family 4A, and thus has 4 electrons and 4 dots in its Lewis symbol. Br is in Family 7A with 7 valence electrons. Adding an electron produces an ion with 8 valence electrons. Tl is in Family 3A with 3 valence electrons. Removing an electron produces a cation with 2 valence electrons.
S is in Family 6A with 6 valence electrons. Adding 2 electrons produces an anion with 8 valence electrons.

$$\cdot \overset{\cdot}{Sn} \cdot \qquad [:\overset{\cdot\cdot}{Br}:]^- \qquad [\cdot Tl\cdot]^+ \qquad [:\overset{\cdot\cdot}{S}:]^{2-}$$

2A   The Lewis structure for the cation, the anion, and the compound follows the explanation.

(a)   Na loses one electron to form Na^+, while S gains two to form S^{2-}.

$$Na\cdot -1\,e^- \rightarrow [\,Na\,]^+ \qquad \cdot\overset{\cdot}{S}\cdot +2e^- \rightarrow [:\overset{\cdot\cdot}{S}:]^{2-} \qquad \text{Lewis Structure: } [Na]^+[:\overset{\cdot\cdot}{S}:]^{2-}[Na]^+$$

(b)   Mg loses two electrons to form Mg^{2+}, while N gains three to form N^{3-}.

$$\cdot Mg\cdot -2\,e^- \rightarrow [\,Mg\,]^{2+} \qquad \cdot\overset{\cdot}{N}\cdot +3e^- \rightarrow [:\overset{\cdot\cdot}{N}:]^{3-}$$

$$\text{Lewis Structure: } [Mg]^{2+}[:\overset{\cdot\cdot}{N}:]^{3-}[Mg]^{2+}[:\overset{\cdot\cdot}{N}:]^{3-}[Mg]^{2+}$$

2B   Below each explanation are the Lewis structures for the cation, for the anion, and for the compound.

(a)   In order to acquire a noble-gas electron configuration, Ca loses two electrons, and I gains one, forming the ions Ca^{2+} and I^-. The formula of the compound is CaI_2.

$$\cdot Ca\cdot -2\,e^- \rightarrow [\,Ca\,]^{2+} \qquad :\overset{\cdot\cdot}{I}\cdot +e^- \rightarrow [:\overset{\cdot\cdot}{I}:]^- \quad \text{Lewis Structure: } \quad [:\overset{\cdot\cdot}{I}:]^-[Ca]^{2+}[:\overset{\cdot\cdot}{I}:]^-$$

(b)   Ba loses two electrons and S gains two to acquire a noble-gas electron configuration, forming the ions Ba^{2+} and S^{2-}. The formula of the compound is BaS.

$$\cdot Ba\cdot -2\,e^- \rightarrow [\,Ba\,]^{2+} \qquad \cdot\overset{\cdot}{S}\cdot +2e^- \rightarrow [:\overset{\cdot\cdot}{S}:]^{2-} \qquad \text{Lewis Structure: } \quad [Ba]^{2+}[:\overset{\cdot\cdot}{S}:]^{2-}$$

(c)   Each Li loses one electron and each O gains two to attain a noble-gas electron configuration, producing the ions Li^+ and O^{2-}. The formula of the compound is Li_2O.

$$\cdot Li - 1\ e^- \rightarrow [\ Li\ ]^+ \quad \cdot \ddot{O} \cdot + 2e^- \rightarrow [:\ddot{O}:]^{2-} \text{ Lewis Structure:} [Li]^+[:\ddot{O}:]^{2-}[Li]^+$$

3A   In the Br_2 molecule, the two Br atoms are joined by a single covalent bond. This bonding arrangement gives each Br atom a closed valence shell configuration that is equivalent to that for a Kr atom.

$$:\ddot{Br} - \ddot{Br}:$$

In CH_4, the carbon atom is covalently bonded to four hydrogen atoms. This arrangement gives the carbon atom a valence shell octet and each H atom a valence shell duet.

$$\begin{array}{c} H \\ | \\ H-C-H \\ | \\ H \end{array}$$

In HOCl, the hydrogen and chlorine atoms are attached to the central oxygen atom through single covalent bonds. This bonding arrangement provides each atom in the molecule with a closed valence shell.

$$H-\ddot{O}-\ddot{Cl}:$$

3B   The Lewis structure for NI_3 is similar to that of NH_3. The central nitrogen atom is attached to each iodine atom by a single covalent bond. All of the atoms in this structure get a closed valence shell.

$$:\ddot{I}-\overset{\ddot{}}{N}-\ddot{I}: \\ | \\ :\ddot{I}:$$

The Lewis diagram for N_2H_4 has each nitrogen with one lone pair of electrons, two covalent bonds to hydrogen atoms and one covalent bond to the other nitrogen atom. With this arrangement, the nitrogen atoms complete their octets while the hydrogen atoms complete their duets.

$$\begin{array}{c} H-\ddot{N}-\ddot{N}-H \\ |\quad | \\ H\quad H \end{array}$$

In the Lewis structure for C_2H_6, each carbon atom shares four pairs of electrons with three hydrogen atoms and the other carbon atom. With this arrangement, the carbon atoms complete their octets while the hydrogen atoms complete their duets.

$$\begin{array}{c} H\quad H \\ |\quad | \\ H-C-C-H \\ |\quad | \\ H\quad H \end{array}$$

4A   The bond with the most ionic character is the one in which the two bonded atoms are the most different in their electronegativities. We find electronegativities in Table 11-2 and calculate ΔEN for each bond.

Electronegativities: H = 2.1   Br = 2.8   N = 3.0   O = 3.5   P = 2.1   Cl = 3.0
Bonds :   H—Br   N—H   N—O   P—Cl
ΔEN values :   0.7   1.4   0.5   0.9
Therefore, the N—H bond is the most polar of the four bonds cited.

4B   The most polar bond is the one with the greatest electronegativity difference.
Electronegativities: C = 2.5   S = 2.5   P = 2.1   O = 3.5   F = 4.0
Bonds: C—S   C—P   P—O   O—F
ΔEN values: 0.0   0.4   1.4   0.5
Therefore, the P—O bond is the most polar of the four bonds cited.

5A **(a)** C has 4 valence electrons and each S has 6 valence electrons: $4 + (2 \times 6) = 16$ valence electrons or 8 pairs of valence electrons. We place C between two S, and use two electron pairs to hold the molecule together, one between C and each S. We complete the octet on each S with three electron pairs for each S. This uses six more electron pairs, for a total of eight electron pairs used. $|\overline{S} - C - \overline{S}|$ But C does not have an octet. We correct this situation by moving one lone pair of each S into a bonding position between C and S. $\overline{S} = C = \overline{S}$

(b) C has 4 valence electrons, N has 5 valence electrons and hydrogen has 1 valence electron: Total number of valence electrons = 4 + 5 + 1 = 10 valence electrons or 5 pairs of valence electrons. We place C between H and N, and use two electron pairs to hold the molecule together, one between C and N, as well as one between C and H. We complete the octet on N using three lone pairs. This uses all five electron pairs. H—C—C—$\overline{N}|$. But C does not have an octet. We correct this situation by moving two lone pair from N into bonding position between C and N. $H - C \equiv N|$

(c) C has 4 valence electrons, each Cl has 7 valence electrons and oxygen has 6 valence electrons: Total number of valence electrons = 4 + 2(7) + 6 = 24 valence electrons or 12 pairs of valence electrons. We choose C as the central atom, and use three electron pairs to hold the molecule together, one between C and O, as well as one between C and each Cl. We complete the octet on Cl and O using three lone pairs. This uses all twelve electron pairs. But C does not have an octet. We correct this situation by moving one lone pair from O into bonding position between C and O.

$$|\overline{O} - C(-\overline{Cl}|)_2 \quad \rightarrow \quad \overline{O} = C(-\overline{Cl}|)_2$$

5B **(a)**

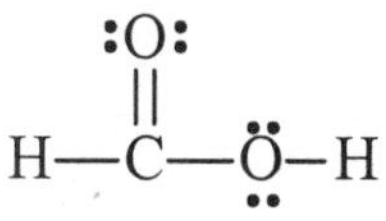

$$\begin{array}{c} :\!O\!: \\ \| \\ H\!-\!C\!-\!\ddot{O}\!-\!H \end{array}$$

(b)

$$\begin{array}{c} H \quad :\!O\!: \\ | \quad\quad \| \\ H\!-\!C\!-\!C\!-\!H \\ | \\ H \end{array}$$

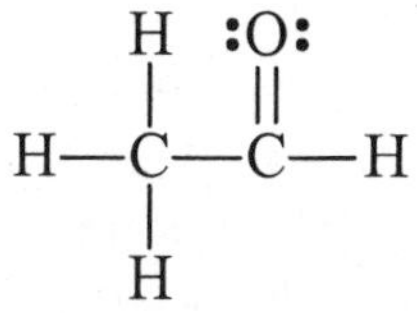

6A **(a)** A plausible Lewis structure for the nitrosonium cation, NO^+ is drawn below:

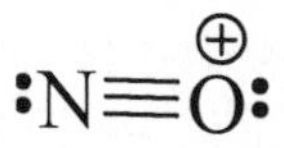

$$:N\!\!\equiv\!\!\overset{\oplus}{O}:$$

The nitrogen atom is triply bonded to the oxygen atom and both atoms in the structure possess a lone pair of electrons. This gives each atom an octet and a positive formal charge appears on the oxygen atom.

(b) A plausible Lewis structure for $N_2H_5^+$ is given below:

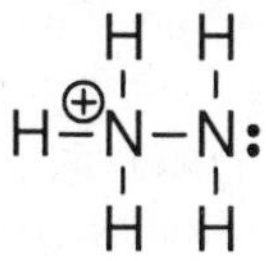

$$\begin{array}{c} H \quad H \\ | \quad | \\ H\!-\!\overset{\oplus}{N}\!-\!N\!: \\ | \quad | \\ H \quad H \end{array}$$

The two nitrogen atoms and the oxygen atom have each achieved an octet by keeping a pair of electrons for themselves while sharing three pairs of electrons with three other atoms.   Each hydrogen atom has completed its duet by sharing a pair of electrons with either a nitrogen or an oxygen atom. A formal 1+ charge of has been assigned to the oxygen atom because it is bonded to three atoms (one more than its usual number) in this structure.

(c)   In order to achieve a noble gas configuration, oxygen gains two electrons, forming the stable dianion. The Lewis structure for O^{2-} is shown below.

$$:\overset{\cdot\cdot}{\underset{\cdot\cdot}{O}}:^{\textcircled{\tiny 2-}}$$

6B   **(a)**   The most likely Lewis structure for BF_4^- is drawn below:

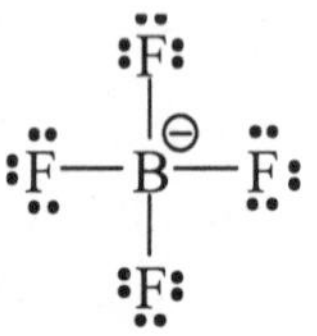

$$\begin{array}{c}
:\overset{\cdot\cdot}{\underset{\cdot\cdot}{F}}: \\
| \\
:\overset{\cdot\cdot}{\underset{\cdot\cdot}{F}}\!\!-\!\!\overset{\ominus}{B}\!\!-\!\!\overset{\cdot\cdot}{\underset{\cdot\cdot}{F}}: \\
| \\
:\overset{\cdot\cdot}{\underset{\cdot\cdot}{F}}:
\end{array}$$

Four bonding pairs of electrons surround the central boron atom in this structure. This arrangement gives the boron atom a complete octet and a formal charge of 1-. By virtue of being surrounded by three lone pairs and one bonding electron pair, each fluorine achieves a full octet.

(b)   A plausible Lewis structural form for NH_3OH^+, the hydroxylammonium ion, has been provided below:

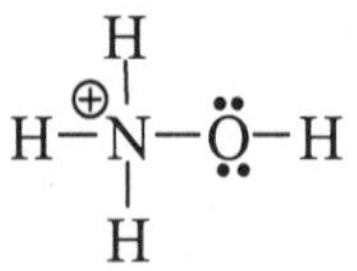

$$\begin{array}{c}
H \\
| \\
H\!\!-\!\!\overset{\oplus}{N}\!\!-\!\!\overset{\cdot\cdot}{O}\!\!-\!\!H \\
| \\
H
\end{array}$$

By sharing bonding electron pairs with three hydrogen atoms and the oxygen atom, the nitrogen atom acquires a full octet and a formal charge of 1+.   The oxygen atom shares one bonding electron pair with the nitrogen and a second bonding pair with a hydrogen atom.

(c)   Three plausible resonance structures can be drawn for the isocyanate ion, NCO^-. The nitrogen contributes five electrons, the carbon four, oxygen six and one more electron is added to account for the negative charge, giving a total of 16 electrons or eight pairs of electrons. In the first resonance contributor, structure 1 below, the carbon atom is joined to the nitrogen and oxygen atoms by two double bonds thereby creating an octet for carbon. To complete the octet of nitrogen and oxygen, each atom is given a lone pair of electrons. Since nitrogen is sharing just two bonding pairs of electrons in this structure, it must be assigned a formal charge of 1-. In structure 2, the carbon atom is again surrounded by four bonding pairs of electrons, but this time, the carbon atom forms a triple bond with oxygen and just a single bond with nitrogen. The octet for the nitrogen atom is closed with three lone pairs of electrons, while that for oxygen is closed with one lone pair of electrons. This bonding arrangement necessitates giving nitrogen a formal charge of 2- and the oxygen atom a formal charge of 1+. In structure 3, which is the dominant contributor because it has a negative formal charge on oxygen (the most electronegative element in the anion), the carbon achieves a full octet by forming a triple bond with the nitrogen atom and a single bond with the oxygen atom. The octet for oxygen is closed with three lone pairs of electrons, while that for nitrogen is closed with one lone pair of electrons.

$$:\ddot{N}=C=\ddot{O}: \qquad \overset{\ominus\ominus}{:\ddot{N}}-C\equiv\overset{\oplus}{O}: \qquad :N\equiv C-\overset{..}{\underset{..}{O}}:^{\ominus}$$
$$\underset{\ominus}{}$$
Structure 1   Structure 2   Structure 3

7A   The total number of valence electrons in NOCl is 18 (5 from nitrogen, 6 from oxygen and 7 from chlorine). Four electrons are used to covalently link the central oxygen atom to the terminal chlorine and nitrogen atoms in the implausible skeletal structure: N—O—Cl.

Next, we need to distribute the remaining electrons to achieve a noble gas electron configuration for each atom. Since four electrons were used to form the two covalent single bonds, fourteen electrons remain to be distributed. By convention, the valence shells for the terminal atoms are filled first. If we follow this convention, we can close the valence shells for both the nitrogen and the chlorine atoms with twelve electrons.

$$:\ddot{N}-O-\ddot{\underset{..}{Cl}}:$$

Oxygen is moved closer to a complete octet by placing the remaining pair of electrons on oxygen as a lone pair.

$$:\ddot{N}-\ddot{\underset{..}{O}}-\ddot{\underset{..}{Cl}}:$$

The valence shell for the oxygen atom can then be closed by forming a double bond between the nitrogen atom and the oxygen atom.

$$:\ddot{N}=\underset{\oplus}{\overset{..}{O}}-\ddot{\underset{..}{Cl}}:$$
$$\underset{\ominus}{}$$

This structure obeys the requirement that all of the atoms end up with a filled valence shell, but is much poorer than the one derived in Example 11-7 because it has a positive formal charge on oxygen, which is the most electronegative atom in the molecule. In other words, this structure can be rejected on the grounds that it does not conform to the third rule for determining plausibility of a Lewis structure based on formal charges which states that "negative formal charges should appear on the most electronegative atom, while any positive formal charge should appear on the least electronegative atom".

7B   There are a total of sixteen valence electrons in the cyanamide molecule (five from each nitrogen atom, four from carbon and one electron from each hydrogen atom). The formula has been written as NH_2CN to remind us that carbon, the most electropositive p-block element in the compound should be selected as the central atom in the skeletal structure.

$$\begin{array}{ccc} H-N-C-N \\ | \\ H \end{array}$$

To construct this skeletal structure we use 8 electrons. Eight electrons remain to be added to the structure. Note: each hydrogen atom at this stage has achieved a duet by forming a covalent bond with the nitrogen atom in the NH_2 group. The octet for the NH_2 nitrogen is completed by giving it a lone pair of electrons.

$$H-\ddot{N}-C-N$$
$$\mid$$
$$H$$

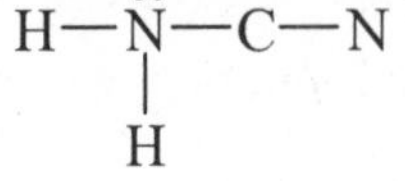

The remaining six electrons can then be given to the terminal nitrogen atom, affording structure 1, shown below. Alternatively, four electrons can be assigned to the terminal nitrogen atom and the last two electrons can be given to the central carbon atom, to produce structure 2 below:

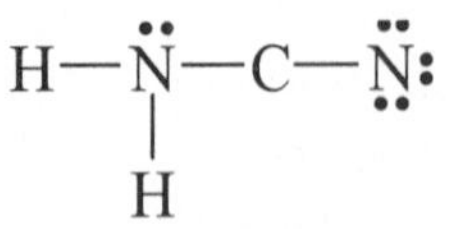

$$H-\ddot{N}-C-\overset{..}{\underset{..}{N}}\text{:}\qquad\qquad H-\ddot{N}-\ddot{C}-\ddot{N}\text{:}$$
$$\mid\qquad\qquad\qquad\qquad\mid$$
$$H\qquad\qquad\qquad\qquad H$$

(Structure 1) 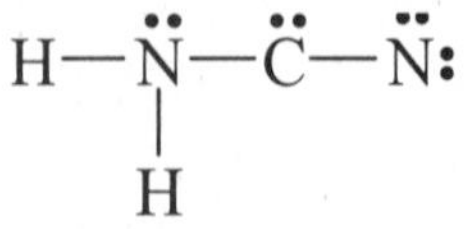   (Structure 2)

The octet for the carbon atom in structure 1 can be completed by converting two lone pairs of electrons on the terminal nitrogen atom into two more covalent bonds to the central carbon atom.

$$H-\ddot{N}-C\equiv N\text{:}$$
$$\mid$$
$$H$$

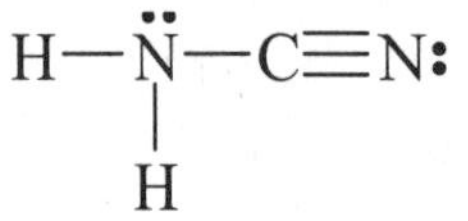

Structure 3

Each atom in structure 3 has a closed-shell electron configuration and a formal charge of zero. We can complete the octet for the carbon and nitrogen atoms in structure 1 by converting a lone pair of electrons on each nitrogen atom into a covalent bond to the central carbon atom.

$$H-\overset{\oplus}{N}=C=\overset{..}{\underset{..}{N}}{}^{\ominus}$$
$$\mid$$
$$H$$

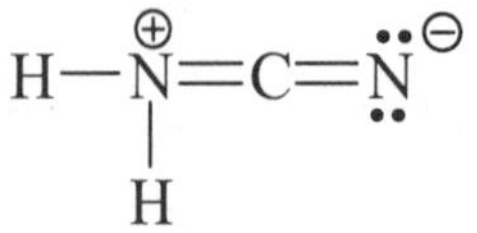

Structure 4

The resulting structure has a formal charge of 1- on the terminal nitrogen atom and a 1+ formal charge on the NH_2 nitrogen atom.

Although both structures 3 and 4 both satisfy the octet and duet rules, structure 3 is the better of the two structures because it has no formal charges. A third structure which obeys the octet rule (depicted below), can be rejected on the grounds that it has formal charges of the same type (two 1+ formal charges) on adjacent atoms, as well as negative formal charges on carbon, which is not the most electronegative element in the molecule.

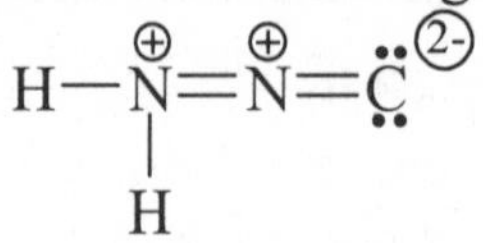

$$H-\overset{\oplus}{N}=\overset{\oplus}{N}=\overset{..}{\underset{..}{C}}{}^{2-}$$
$$\mid$$
$$H$$

8A   The skeletal structure for SO_2 has two terminal oxygen atoms bonded to a central sulfur atom. Sulfur has been selected as the central atom by virtue of it being the most electropositive atom in the molecule. It turns out that two different Lewis structures of identical energy can be drawn on the skeletal structure described above. First we determine that SO_2 has 18 valence electrons (6 from each atom). Four of the valence electrons must be used to covalently bond the three atoms together. The remaining 14 electrons are used to close the valence shell of each atom. Twelve electrons are used to give the terminal oxygen atoms a closed shell. The remaining two electrons (14 -12 = 2) are placed on the sulfur atom, affording the structure depicted below:

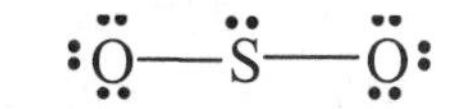

$$:\ddot{\underset{..}{O}}\text{---}\ddot{S}\text{---}\ddot{\underset{..}{O}}:$$

At this stage, the valence shells for the two oxygen atoms are closed, but the sulfur atom is two electrons short of a complete octet. If we complete the octet for sulfur by converting a lone pair of electrons on the right hand side oxygen atom into a sulfur-to-oxygen π-bond, we end up generating the resonance contributor (A) shown below:

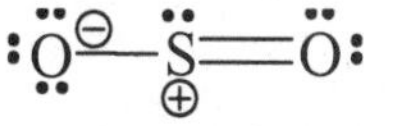

$$:\ddot{\underset{..}{O}}^{\ominus}\text{---}\underset{\oplus}{\ddot{S}}\text{===}\ddot{O}: \quad \text{(A)}$$

Notice that the structure has a positive formal charge on the sulfur atom (most electropositive element) and a negative formal charge on the left-hand oxygen atom. Remember that oxygen is more electronegative than sulfur, so these charges are plausible. The second completely equivalent contributor, (B), is produced by converting a lone pair on the left-most oxygen atom in the structure into a π-bond, resulting in conversion of a S-O single bond into a sulfur-oxygen double bond:

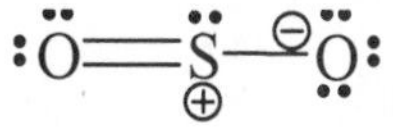

$$:\ddot{O}\text{===}\underset{\oplus}{\ddot{S}}\text{---}^{\ominus}\ddot{\underset{..}{O}}: \quad \text{(B)}$$

Neither structure is consistent with the observation that the two S-O bond lengths in SO_2 are equal, and in fact, the true Lewis structure for SO_2 is neither (A) or (B) but rather an equal blend of the two individual contributors called the resonance hybrid. (see below)

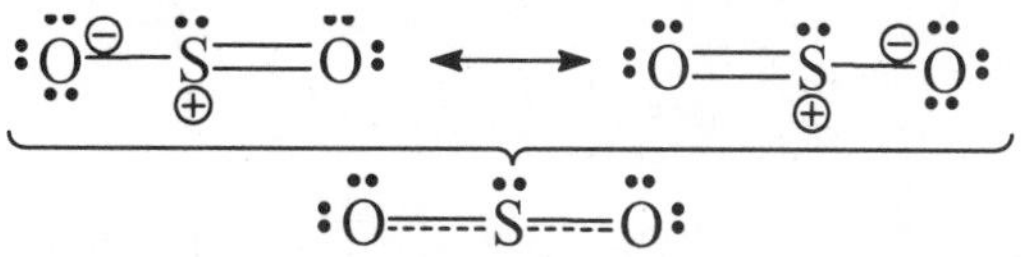

$$:\ddot{\underset{..}{O}}^{\ominus}\text{---}\underset{\oplus}{\ddot{S}}\text{===}\ddot{O}: \quad\longleftrightarrow\quad :\ddot{O}\text{===}\underset{\oplus}{\ddot{S}}\text{---}^{\ominus}\ddot{\underset{..}{O}}:$$

$$\underbrace{\hspace{5cm}}$$

$$:\ddot{\underset{..}{O}}\text{-----}\ddot{S}\text{-----}\ddot{\underset{..}{O}}:$$

8B   The skeletal structure for the NO_3^- ion has three terminal oxygen atoms bonded to a central nitrogen atom. Nitrogen has been chosen as the central atom by virtue of it being the most electropositive atom in the ion. It turns out that three contributing resonance structures of identical energy can be derived from the skeletal structure described here. We begin the process of generating these three structures by counting the total number of valence electrons in the NO_3^- anion. The nitrogen atom contributes five electrons, each oxygen contributes six electrons and an additional electron must be added to account for the 1-charge on the ion. In total, we must account for 24 electrons. Six electrons are used to draw single covalent bonds between the nitrogen atom and each oxygen atom. The remaining 18 electrons are used to complete the octet for the three terminal oxygen atoms:

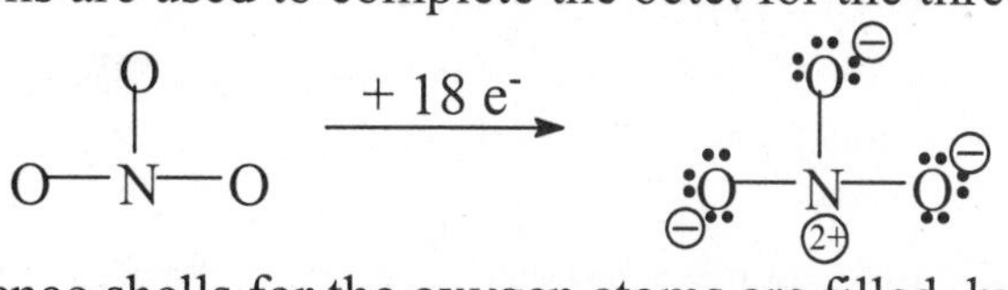

$$
\begin{array}{ccc}
\overset{\displaystyle O}{\underset{\displaystyle |}{}} & \xrightarrow{\;+\;18\;e^-\;} & \overset{\displaystyle :\ddot{\underset{..}{O}}:^{\ominus}}{\underset{\displaystyle |}{}}\\
O\text{---}N\text{---}O & & :\ddot{\underset{..}{O}}\text{---}\underset{\textcircled{\scriptsize 2+}}{N}\text{---}\ddot{\underset{..}{O}}:^{\ominus}\\
& & {}^{\ominus}
\end{array}
$$

At this stage the valence shells for the oxygen atoms are filled, but the nitrogen atom is two electrons short of a complete octet. If we complete the octet for nitrogen by converting a lone pair on O_1 into a nitrogen-to-oxygen π-bond, we end up generating resonance contributor (A):

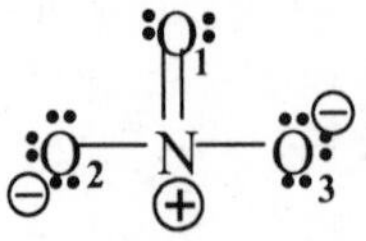

Notice the structure has a 1+ formal charge on the nitrogen atom and a 1- on two of the oxygen atoms (O_2 and O_3). These formal charges are quite reasonable energetically. The second and third equivalent structures are generated similarly, by moving a lone pair from O_2 to form a nitrogen to oxygen (O_2) double bond, we end up generating resonance contributor (B), shown below. Likewise, by converting a lone pair from oxygen (O_3) into a π-bond with the nitrogen atom, we end up generating resonance contributor (C), also shown below.

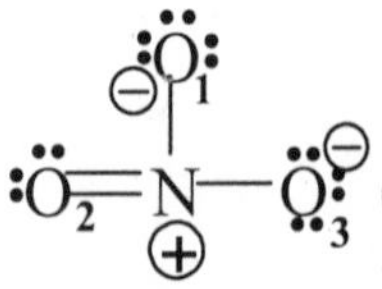

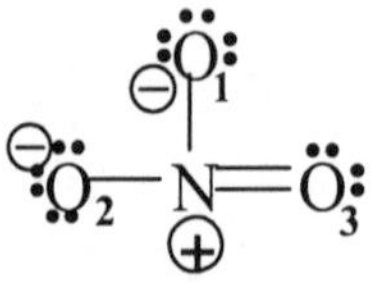

None of these individual structures ((A), (B) or (C)) correctly represents the actual bonding in the nitrate anion. The actual structure, called the resonance hybrid, is an average of all three structures (i.e. 1/3(A) + 1/3(B) + 1/3(C)):

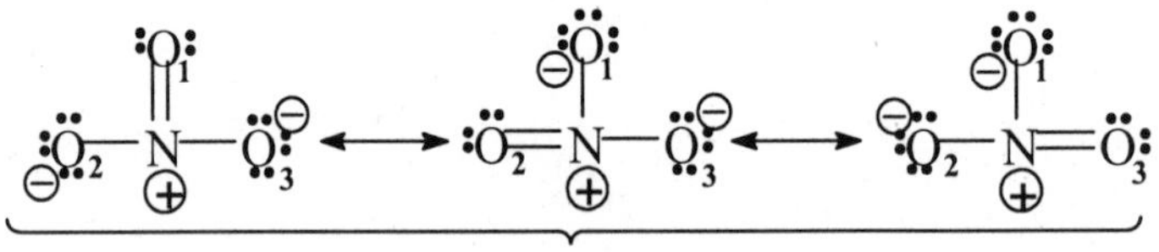

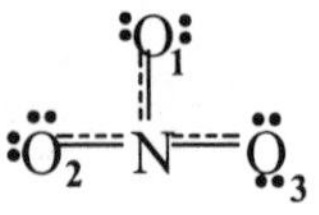

These three resonance forms give bond lengths approximately those of nitrogen-to-nitrogen double bonds.

9A The Lewis structure of NCl_3 has three Cl atoms bonded to N and one lone pair attached to N. These four electron groups around N produce a tetrahedral electron-group geometry. The fact that one of the electron groups is a lone pair means that the molecular geometry is trigonal pyramidal.

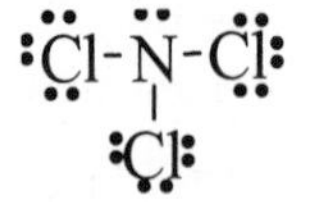

9B The Lewis structure of $POCl_3$ has three single P-Cl bonds and one P-O bond. These four electron groups around P produce a tetrahedral electron-group geometry. No lone pairs are attached to P and thus the molecular geometry is tetrahedral.

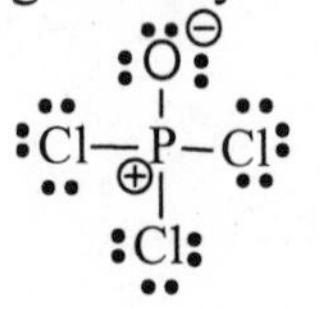

10A The Lewis structure of COS has one S doubly-bonded to C and an O doubly-bonded to C. There are no lone pairs attached to C. The electron-group and molecular geometries are the same: linear. $|\overline{S} = C = \overline{O}|$

10B N is the central atom. $|N \equiv N — \overline{O}\,|$ This gives an octet on each atom, a formal charge of 1+ on the central N, and a 1– on the O atom. There are two bonding pairs of electrons and no lone pairs on the central N atom. The N_2O molecule is linear.

11A In the Lewis structure of methanol, each H atom contributes 1 valence electron, the C atom contributes 4, and the O atom contributes 6, for a total of $(4 \times 1) + 4 + 6 = 14$ valence electrons, or 7 electron pairs. 4 electron pairs are used to connect the H atoms to the C and the O, 1 electron pair is used to connect C to O, and the remaining 2 electron pairs are lone pairs on O, completing its octet.

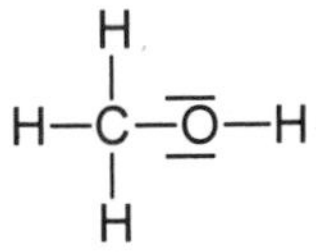

$$\begin{array}{c} H \\ | \\ H—C—\overline{O}—H \\ | \\ H \end{array}$$

The resulting molecule has two central atoms. Around the C there are four bonding pairs, resulting in a tetrahedral electron-group geometry and molecular geometry. The H—C—H bond angles are 109.5° as are the H—C—O bond angles. Around the O there are two bonding pairs of electrons and two lone pairs, resulting in a tetrahedral electron-group geometry and a bent molecular shape around the O atom, with a C—O—H bond angle of slightly less than 109.5°.

11B The Lewis structure is drawn below. With four electron groups surrounding each, the electron-group geometries of N, the central C, and the right-hand O are all tetrahedral. The H—N—H bond angle and the H—N—C bond angles are almost the tetrahedral angle of 109.5°, made a bit smaller by the lone pair.

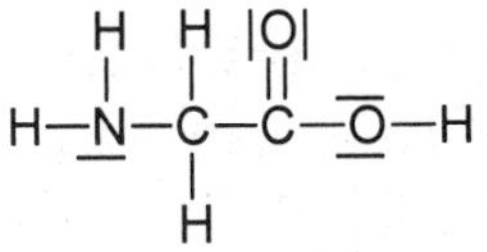

The H—C—N angles, the H—C—H angle and the H—C—C angles all are very close to 109.5°. The C—O—H bond angle is made somewhat smaller than 109.5° by the presence of two lone pairs on O. Three electron groups surround the right-hand C, making its electron-group and molecular geometries trigonal planar. The O—C—O bond angle and the O—C—C bond angles all are very close to 120°.

$$\begin{array}{ccccc} & H & H & |O| & \\ & | & | & || & \\ H—N—C—C—\overline{O}—H \\ & | & | & & \\ & & H & & \end{array}$$

12A Lewis structures of the three molecules are drawn below. Around the S in the SF_6 molecule are six bonding pairs of electrons, and no lone pairs. The molecule is octahedral; each of the S-F bond moments is cancelled by one on the other side of the molecule.

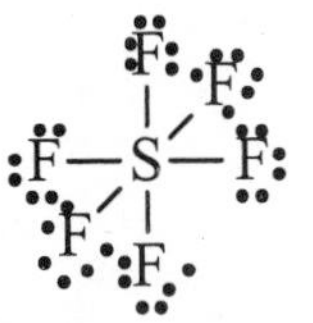

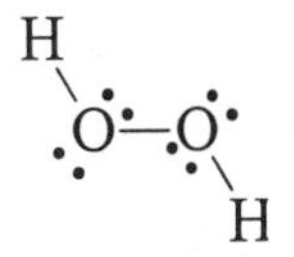

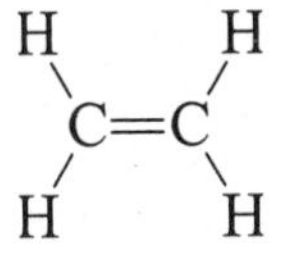

SF_6 is nonpolar. In H_2O_2, the molecular geometry around each O atom is bent; the bond moments do not cancel. H_2O_2 is polar. Around each C in C_2H_4 are three bonding pairs of electron; the molecule is planar around each C and planar overall. The polarity of each —CH_2 group is cancelled by the polarity of the other H_2C— group. C_2H_2 is nonpolar.

12B Lewis structures of the four molecules are drawn below, we can consider the three C—H bonds and the one C—C bond to be non polar. The three C—Cl bonds are tetrahedrally oriented.

$$\underline{|Cl|}-\underset{\underset{\underline{|Cl|}}{|}}{\overset{\overset{\underline{|Cl|}}{|}}{C}}-\underset{\underset{H}{|}}{\overset{\overset{H}{|}}{C}}-H \qquad \underline{|Cl|}-\underset{\underset{\underline{|Cl|}\;\;\underline{|Cl|}}{}}{\overset{\overset{\underline{|Cl|}\;\;\underline{|Cl|}}{}}{P}} \qquad H-\underset{\underset{\underline{|Cl|}}{|}}{\overset{\overset{\underline{|Cl|}}{|}}{C}}-H \qquad \underset{\underset{H}{|}}{\overset{\overset{H}{|}}{|N}}-H$$

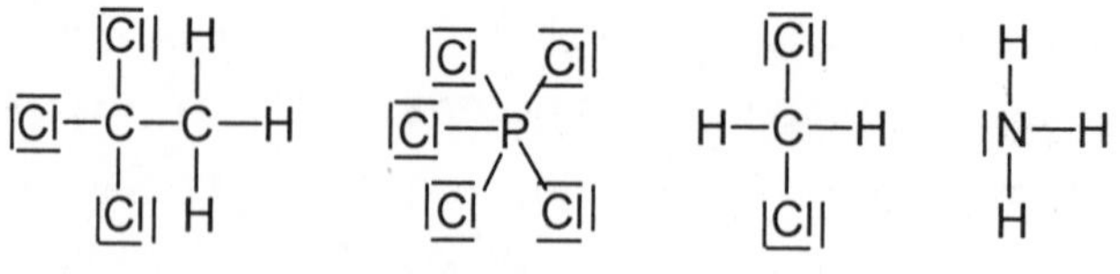

If there were a fourth C—Cl bond on the left-hand C, the bond dipoles would cancel out, producing a nonpolar molecule. Since it is not there, the molecule is polar. A similar argument is made for NH_3, where three tetrahedrally-oriented N—H polar bonds are not balanced by a fourth, and CH_2Cl_2, where two tetrahedrally oriented C—Cl bonds are not balanced by two others. This leaves nonpolar PCl_5, a symmetrical molecule in which bond dipoles cancel.

13A The Lewis structure of CH_3Br has all single bonds. From Table 11.2 the length of a C—H bond is 110 pm. The length of a C—Br bond is not given in the table. A reasonable value is the average of the C—C and Br—Br bond lengths.

$$C-Br = \frac{C-C+Br-Br}{2} = \frac{154 \text{ pm} + 228 \text{ pm}}{2} = 191 \text{ pm} \qquad H-\underset{\underset{H}{|}}{\overset{\overset{H}{|}}{C}}-\underline{Br|}$$

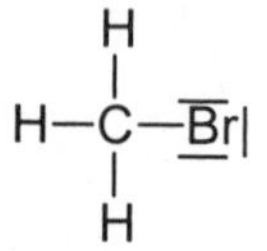

13B In Table 11-2 the C—O bond length is 143 pm, while the C=O bond length is 120 pm, much closer to the experimental bond length in CO_2. In determining the Lewis structure, we begin with 4 valence electrons from C and 6 valence electrons from each of the two O atoms, for a total of $4 + (2 \times 6) = 16$ valence electrons, or 8 pairs of electrons. Two of these eight pairs are used to attach the C atom to each O atom. The remaining 6 pairs complete the octet on each O atom, producing this Lewis structure: $|\overline{O}$—C—$\overline{O}|$ This Lewis structure has two flaws: (1) There is no octet on C and (2) the carbon-to-oxygen bonds are single bonds, in contradiction to experimental evidence. If a lone pair of electrons from each O is shared with C, the following Lewis structure is produced, that overcomes both objections: $|\underline{O} = C = \underline{O}|$

14A   We first draw Lewis structures of all molecules involved.

$$2\ \text{H—H} + \overline{\text{O}} = \overline{\text{O}} \ \rightarrow\ 2\ \text{H—}\overline{\text{O}}\text{—H}$$

Break 1 O=O + 2H—H = 498 kJ/mol + (2 × 436 kJ/mol) = 1370 kJ/mol absorbed
Form 4H—O = (4 × 464 kJ/mol) = 1856 kJ/mol given off
Enthalpy change = 1370 kJ / mol – 1856 kJ / mol = –486 kJ / mol

14B   The chemical equation, with Lewis structures is:

$$1/2\ \text{:N}{\equiv}\text{N:} \quad + \quad 3/2\ \text{H—H} \quad \longrightarrow \quad \text{H—}\overset{\cdot\cdot}{\text{N}}\text{—H}$$
$$\text{H}$$

Energy required to break bonds $= \frac{1}{2}\text{N} \equiv \text{N} + \frac{3}{2}\text{H—H}$

$= (0.5 \times 946\ \text{kJ / mol}) + (1.5 \times 436\ \text{kJ / mol}) = 1.13 \times 10^3\ \text{kJ / mol}$

Energy realized by forming bonds $= 3\ \text{N—H} = 3 \times 389\ \text{kJ / mol} = 1.17 \times 10^3\ \text{kJ / mol}$
$\Delta H = 1.13 \times 10^3\ \text{kJ / mol} - 1.17 \times 10^3\ \text{kJ / mol} = -4 \times 10^1\ \text{kJ / mol of NH}_3.$

Thus, $\Delta H_f = -4 \times 10^1\ \text{kJ / mol NH}_3$ (Appendix D value is $\Delta H_f = -46.11\ \text{kJ / mol NH}_3$)

15A   The reaction with Lewis structures is $\cdot\overline{\text{O}}\text{—N}=\overline{\text{O}} + \cdot\overline{\text{O}}\cdot \rightarrow \text{N}=\overline{\text{O}} + \overline{\text{O}}=\overline{\text{O}}$

The overall result seems to be converting a single N—O bond to a double O=O bond. Since we expect a double bond to be stronger than a single bond, we predict that the products will be more stable than the reactants and this reaction will be exothermic.

15B   Double the chemical equation, as Lewis structures:

$$2\ \text{H—}\overline{\text{O}}\text{—H} + 2|\overline{\text{C}}\text{l—}\overline{\text{C}}\text{l}| \rightarrow |\overline{\text{O}}=\overline{\text{O}}| + 4\text{H—}\overline{\text{C}}\text{l}|$$

Energy required to break bonds = 2 Cl—Cl + 4H—O
$$= (2 \times 243\ \text{kJ / mol}) + (4 \times 464\ \text{kJ / mol}) = 2342\ \text{kJ / mol}$$

Energy realized by forming bonds = 1 O=O + 4 × H—Cl
$$= 498\ \text{kJ / mol} + (4 \times 431\ \text{kJ / mol}) = 2222\ \text{kJ / mol}$$

$\Delta H = \frac{1}{2}(2342\ \text{kJ / mol} - 2222\ \text{kJ / mol}) = +60\ \text{kJ / mol}$ The reaction is endothermic.

REVIEW QUESTIONS

Important Note: In this and subsequent chapters, a lone pair of electrons in a Lewis structure is often shown as a line rather than a pair of dots. Thus, the Lewis structure of Be is Be| rather than Be:

1. **(a)** Valence electrons are those in the highest principal quantum numbers, those in the outermost shell, furthest from the nucleus.

    (b) Electronegativity is a measure of the attraction that an atom in a compound has for the bonding electrons that its shares with other atoms.

    (c) Bond dissociation energy is the energy needed to break a mole of bonds of a given type.

    (d) A double covalent bond results when two pairs of electrons are shared by two atoms.

    (e) A coordinate covalent bond is formed when both electrons of a shared pair come originally from one of the bonded atoms.

2. **(a)** Formal charge is a measure of how many valence electrons surround an atom in a covalently bonded structure, compared with the number of valence electrons of the isolated atom.

    (b) Resonance occurs when the bonding in a molecule cannot be completely represented by one Lewis structure, but requires the "blending" of two or more contributing structures.

    (c) An expanded valence shell results when an atom uses empty, low-lying orbitals to accommodate more than eight electrons in its outermost electron shell.

    (d) Bond energy is the energy required to break a mole of bonds of a given type, averaged over all the compounds in which that type of bond appears.

3. **(a)** An ionic bond is the result of the attraction between a cation and an anion. A covalent bond results when two atoms share one or more pairs of electrons.

    (b) A bonding pair of electrons is a pair that is shared between two atoms. A lone pair resides on one atom only and is not shared between atoms.

    (c) Electron-group geometry describes the orientation of electron groups—bonding and lone pairs—around a central atom. Molecular geometry describes the orientation of bonding groups around the central atom.

    (d) A bond dipole moment is the result of the unequal sharing of bonding electrons between two atoms. The resultant dipole moment of a molecule is the vector sum of all of the individual bond dipole moments.

    (e) A polar molecule is one that has a net dipole moment, a slight overall separation of positive and negative charge. In a nonpolar molecule, there is no such net dipole moment.

4. **(a)** $\left[\text{H:} \right]^-$ **(b)** $:\overset{.}{\text{Kr}}:$ **(c)** $\left[\cdot \text{Sn} \cdot \right]^{2+}$ **(d)** $\left[\text{K} \right]^+$ **(e)** $[:\overset{.}{\text{Br}}:]^-$

    (f) $\cdot \text{G e} \cdot$ **(g)** $:\overset{.}{\text{N}}\cdot$ **(h)** $\cdot \text{Ca} \cdot$ **(i)** $[:\overset{.}{\text{Se}}:]^{2-}$ **(j)** $\left[\text{Sc} \right]^{3+}$

5.   **(a)** $[:\ddot{C}l:]^-\ [Ca]^{2+}\ [:\ddot{C}l:]^-$ **(b)** $[\ Ba\ ]^{2+}\ [:\ddot{S}:]^{2-}$ **(c)** $[\ Li\ ]^+\ [:\ddot{O}:]^{2-}\ [\ Li\ ]^+$

(d) $[\ Na\ ]^+\ [:\ddot{F}:]^-$ **(e)** $[\ Mg\ ]^{2+}[:\ddot{N}:]^{3-}[Mg]^{2+}[:\ddot{N}:]^{3-}[\ Mg\ ]^{2+}$

6.   **(a)** $|\bar{I}-\bar{C}l|$ **(b)** $|\bar{Br}-\bar{Br}|$ **(c)** $|\bar{F}-\bar{O}-\bar{F}|$ **(d)** 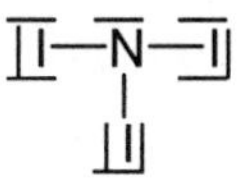 $|\bar{I}-\bar{N}=\bar{I}|$
$$|\underline{I}|$$

(e) $H-\bar{Se}-H$

7.   **(a)** $:\ddot{S}=C=\ddot{S}:$ **(b)**

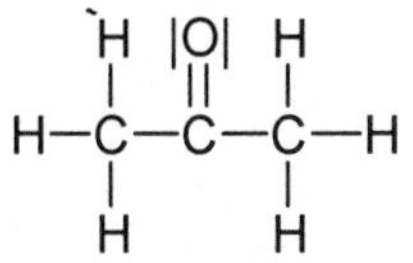

$$
\begin{array}{ccc}
\overset{\cdot}{H} & |\overset{O}{\overset{||}{}}| & H \\
| & || & | \\
H-C- & C- & C-H \\
| & & | \\
H & & H
\end{array}
$$

(c)

$$
\begin{array}{c}
|\overset{O}{\overset{||}{}}| \\
\bar{\underline{C}l}-C-\bar{\underline{C}l}|
\end{array}
$$

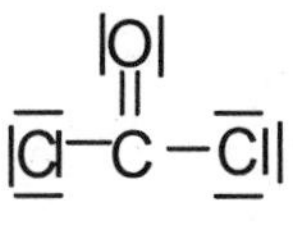

(d) $|\bar{F}-\bar{N}=\bar{O}|$

8.   **(a)** $H-H-\bar{N}-\bar{O}-H$ has two bonds to (four electrons around) the second hydrogen, and only six electrons around the nitrogen. A better Lewis structure is

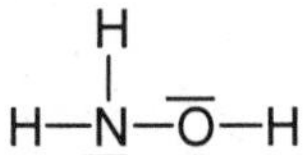

$$
\begin{array}{c}
H \\
| \\
H-\underline{N}-\bar{\underline{O}}-H
\end{array}
$$

(b) $|\bar{O}-\bar{C}l-\bar{O}|$ has 20 valence electrons, whereas the molecule ClO_2 has 19 valence electrons. This is a proper Lewis structure for the chlorite ion, although the brackets and the minus charge are missing. A plausible Lewis structure for the molecule ClO_2 is $|\bar{O}-\underline{C}l-\bar{O}|$

(c) $[\cdot\overset{\cdot}{C}=\ddot{N}:\,]^-$ has only six electrons around the C atom and two too few overall. $[|C\equiv N|]^-$ is a more plausible Lewis structure for the cyanide ion.

(d) $Ca-\bar{O}|$ is improperly written as a covalent Lewis structure, although CaO is an ionic compound. In addition, there are only two electrons around the Ca atom. $[Ca]^{2+}[|\bar{O}|]^{2-}$ is a more plausible Lewis structure for CaO.

9.   **(a)** $[|\bar{O}-H]^-\,[Ca]^{2+}[|\bar{O}-H]^-$ **(b)** 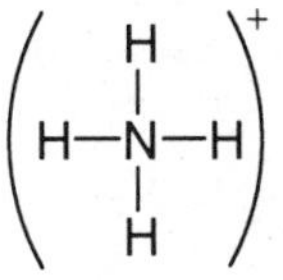 $\left(\begin{array}{c} H \\ | \\ H-N-H \\ | \\ H \end{array}\right)^+ [:\ddot{Br}:]^-$

(c) $[|\bar{O}-\bar{C}l|]^-\,[Ca]^{2+}[|\bar{O}-\bar{C}l|]^-$

10.

(a)

computations for:	H—	—C≡	≡C
no. valence e⁻	1	4	4
- no. lone-pair e⁻	–0	–0	–2
– ½ no. bond-pair e⁻	–1	–4	–3
formal charge	0	0	–1

(b)

computations for:	=O	—O(×2)	C
no. valence e⁻	6	6	4
- no. lone-pair e⁻	–4	–6	–0
– ½ no. bond-pair e⁻	–2	–1	–4
formal charge	–0	–1	0

(c)

computations for:	—H(×7)	side C(×2)	central C
no. valence e⁻	1	4	4
- no. lone-pair e⁻	–0	–0	–0
– ½ no. bond-pair e⁻	–1	–4	–3
formal charge	0	0	+1

(d) The formal charge on each I is 0,
computed as follows:
no. valence e⁻ = 7
–no. lone-pair e⁻ = –6
– ½ no. bond-pair e⁻ = –1
formal charge = 0

(e)

computations for:	=O	—O	=S—
no. valence e⁻	6	6	6
- no. lone-pair e⁻	–4	–6	–2
– ½ no. bond-pair e⁻	–2	–1	–3
formal charge	0	–1	+1

(f)

computations for:	=O	—O	=N—
no. valence e⁻	6	6	5
- no. lone-pair e⁻	–4	–6	–1
– ½ no. bond-pair e⁻	–2	–1	–3
formal charge	0	–1	+1

11. Since electrons pair up (if at all possible) in plausible Lewis structures, we can get a very good indication if a species is paramagnetic if it has an odd number of (valence) electrons.

(a) OH^- $6+1+1=8$ valence electrons   diamagnetic

(b) OH $6+1=7$ valence electrons   paramagnetic

(c) NO_3 $5+(3\times 6)=23$ valence electrons   paramagnetic

(d) SO_3 $6+(3\times 6)=24$ valence electrons   diamagnetic

(e) SO_3^{2-} $6+(3\times 6)+2=26$ valence electrons diamagnetic

(f) HO_2 $1+(2\times 6)=13$ valence electrons   paramagnetic

12. **(a)** 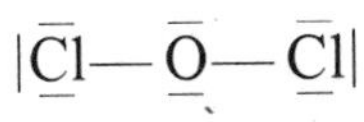 $|\overline{Cl}-\overline{O}-\overline{Cl}|$ **(b)** 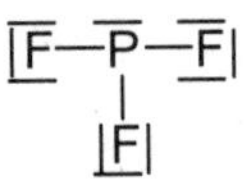 $\overline{F}-\overline{P}-\overline{F}|$ with $|\underline{F}|$ below   **(c)** 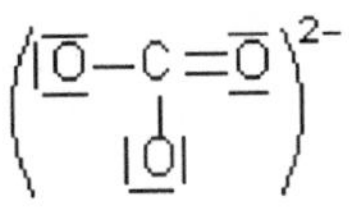 $\left(|\overline{O}-C=\overline{O}\right)^{2-}$ with $|\overline{O}|$ below   **(d)**  $\overline{F}-Br-\overline{F}$ with $\overline{F}$, $\overline{F}$ above and $|\underline{F}|$ below

13. We first give all the Lewis structures. From each, we deduce the electron-group geometry and the molecular shape.

(a)  $\underset{H}{\overset{H}{|S|}}$ **(b)** 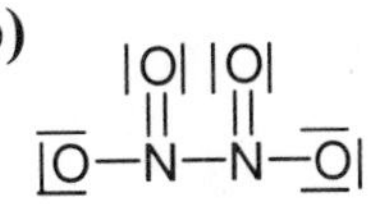 $\overline{O}-N-N-\overline{O}$ with $|\overline{O}|$ $|\overline{O}|$ above   **(c)** $\underset{N}{\overset{H}{|C|}}$ 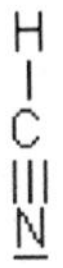   **(d)** 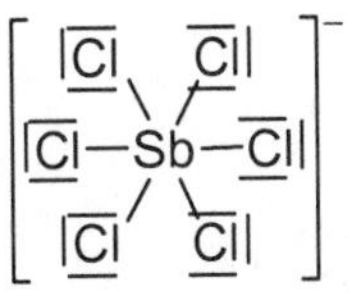 $\left[\begin{array}{c}\overline{Cl}\quad\overline{Cl}|\\ \overline{Cl}-Sb-\overline{Cl}|\\ \overline{Cl}\quad\overline{Cl}|\end{array}\right]^{-}$ **(e)** 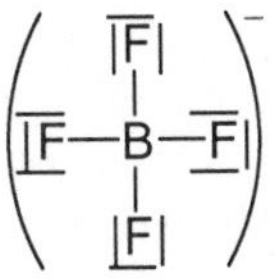 $\left(\begin{array}{c}\overline{F}|\\ \overline{F}-B-\overline{F}|\\ |\underline{F}|\end{array}\right)^{-}$

(a) H_2S tetrahedral electron-group geometry, bent (angular) molecular geometry

(b) N_2O_4 trigonal planar electron-group geometry around each N, planar molecule

(c) HCN   linear electron-group geometry, linear molecular geometry

(d) $SbCl_6^-$ octahedral electron-group geometry, octahedral geometry

(e) BF_4^- tetrahedral electron-group geometry, tetrahedral molecular geometry

14. **(a)** In SO_2, there are a total of $6+(2\times 6)=18$ valence electrons, or 9 electron pairs. A plausible Lewis structure has two resonance forms, of which one is: $\overline{O}=\underline{S}-\overline{O}|$

This molecule is of the type AX_2E; it has a trigonal planar electron-group geometry and a bent molecular shape.

(b) In SO_3^{2-} the total number of valence electrons is $2+(3\times 6)+6=26$ valence electrons, or 13 electron pairs. A plausible Lewis structure is

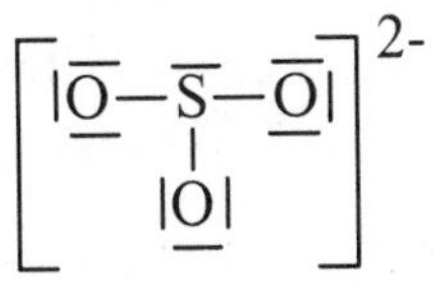

$$\left[\,|\overline{O}-\overline{S}-\overline{O}|\atop |\underline{O}|\right]^{2-}$$

This molecule is of the type AX_3E; it has a tetrahedral electron-group geometry and a trigonal pyramidal molecular shape.

(c)   In SO_4^{2-}, there are a total of $2+6+(4\times 6)=32$ valence electrons, or 16 electron pairs. There are several possible resonance forms, but a simple Lewis structure is provided (right). This ion is of the type AX_4; it has a tetrahedral electron-group geometry and a tetrahedral shape.

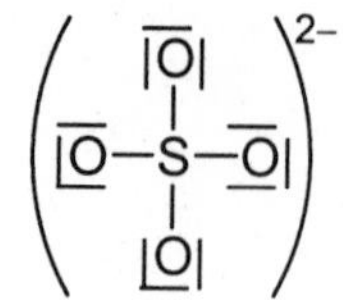

15.   In each case, a plausible Lewis structure is given first, followed by the AX_nE_m notation for each species followed by the electron-group geometry and, finally, the molecular geometry.

(a) $|C \equiv O|$

(b)

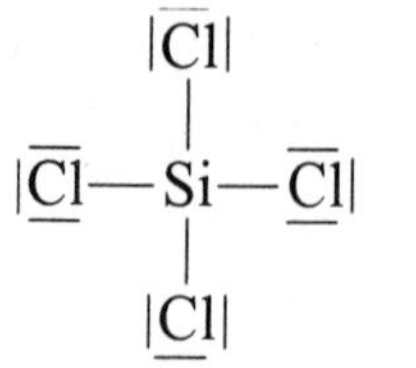

(c)

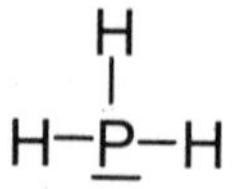

(d)

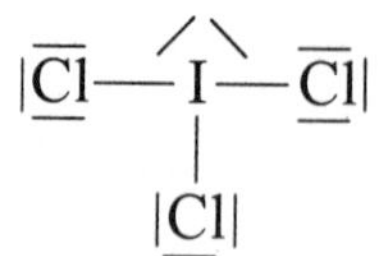

(e)

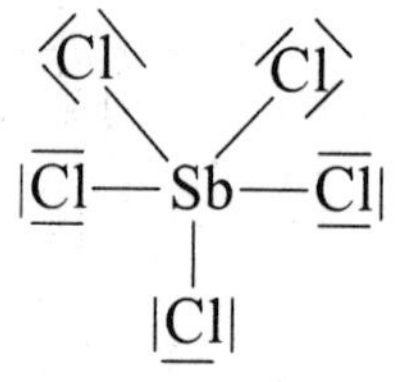

(f) $|\overline{O}-\overline{S}=\overline{O}$

(g)

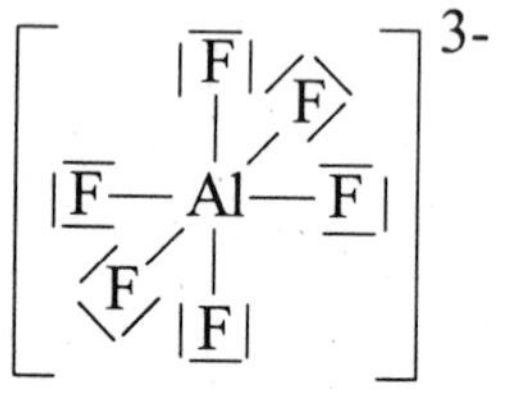

(a)   CO   linear electron-group geometry, linear molecular geometry

(b) $SiCl_4$ tetrahedral electron-group geometry, tetrahedral molecular geometry

(c) PH_3 tetrahedral electron-group geometry, trigonal pyramidal molecular geometry

(d) ICl_3 trigonal bipyramidal electron-group geometry, T-shape molecular geometry

(e) $SbCl_5$ electron-group geometry and molecular geometry, trigonal bipyramidal

(f) SO_2 trigonal planar electron-group geometry, bent molecular geometry

(g) AlF_6^{3-} octahedral electron-group geometry, octahedral geometry

16.

bond	H — C	C=O	C — C	C — Cl	
(a) bond length	110	120	154	178	pm
(b) bond energy	414	736	347	339	kJ/mol

17.   (c) is the longest bond. Single bonds are generally longer than multiple bonds. Of the two molecules with single bonds, Br_2 is expected to have longer bonds than BrCl, since Br is larger than Cl.   (a) $\overline{O}=\overline{O}$ (b) $|N\equiv N|$ (c) $|\overline{Br}-\overline{Br}|$ (d) $|\overline{Br}-\overline{Cl}|$

18.   The reaction $O_2 \rightarrow 2O$ is an endothermic reaction since it requires the breaking of the bond between two oxygen atoms without the formation of any bonds. Since bond breakage is endothermic, the entire process must be endothermic.

19. **(a)** The net result of this reaction involves breaking one mole of C—H bonds (which requires 414 kJ) and forming one mole of H—I bonds (which produces 297 kJ). Thus, this reaction is endothermic; a net infusion of energy is necessary.

(b) The net result of this reaction, involves breaking one mol of H—H bonds (which requires 436 kJ) and one mol of I—I bonds (which requires 151 kJ), and forming two moles of H—I bonds (which produces $2 \times 297 = 594$ kJ). Thus, this reaction is just barely exothermic.

20. Recall that electronegativity increases in the periodic table from lower left to upper right, and specifically that metals have lower electronegativities than do nonmetals, with metalloids having intermediate electronegativities. Based on these principles, the electronegativity of S is greater than that of As, which in turn is greater than that of Bi. The electronegativity of Ba is less than that of Mg. And finally, we would predict the electronegativity of Mg (a definite metal) to be less than that of Bi (somewhat metalloid in character). Thus, ranked in order of increasing electronegativity, these elements are: Ba < Mg < Bi < As < S, with Bi having the intermediate electronegativity of this group of five elements. The actual electronegativities are in parentheses:

Ba (0.9) < Mg (1.2) < Bi (1.9) < As (2.0) < S(2.5)

21. Na—Cl and K—F both are bonds between a metal and a nonmetal; they have the largest ionic character, with the ionic character of K—F being greater than that of Na—Cl, both because K is more metallic (closer to the lower left of the periodic table) than Na and F is more nonmetallic (closer to the upper right) than Cl. The remaining three bonds are covalent bonds to H. Since H and C have about the same electronegativity (a fact you need to memorize), the H—C bond is the most covalent (or the least ionic). Br is somewhat more electronegative than is C, while F is considerably more electronegative than C, making the F—H bond the most polar of the three covalent bonds. Thus, ranked in order of increasing ionic character, these five bonds are:

C—H < Br—H < F—H < Na—Cl < K—F

The actual electronegativity differences follow:

$$C(2.5)\text{-}H(2.1) < Br(2.8)\text{-}H(2.1) < F(4.0)\text{-}H(2.1) < Na(0.9)\text{-}Cl(3.0) < K(0.8)\text{-}F(4.0)$$

| ΔEN | = 0.4 | = 0.7 | = 1.9 | = 2.1 | = 3.2 |

22. **(a)** F_2 cannot possess a dipole moment, since both of the atoms in the diatomic molecule are the same. This means that there is no electronegativity difference between atoms, and hence no polar bonds.

(b) $|\overline{O}\!-\!\overset{\bullet}{N}=\overline{O}$ Each nitrogen-to-oxygen bond in this molecule is polarized toward oxygen, the more electronegative element. The molecule is of the AX_2E category and hence is bent. Therefore the two bond dipoles do not cancel, and the molecule is polar.

(c) Although each B-F bond is polarized toward F in this trigonal planar $|\overline{F}-B-\overline{F}|$
AX$_3$ molecule these bond dipoles cancel. The molecule is nonpolar. $|\underline{F}|$

(d) H-$\overline{\underline{Br}}|$ The H—Br bond is polar toward Br, and this molecule is polar as well.

(e) The H—C bonds are not polar, but the C—Cl bonds are, toward Cl.   H
The molecular shape is tetrahedral (AX_4) and thus these two C—Cl $|\overline{\underline{Cl}}-\underset{|}{\overset{|}{C}}-\overline{\underline{Cl}}|$
dipoles do not cancel each other; the molecule is polar.   H

(f) Although each Si—F bond is polarized toward F, in this tetrahedral $|\overline{F}|$
AX$_4$ molecule these bond dipoles oppose and cancel each other; $|\overline{F}-Si-\overline{F}|$
the molecule is nonpolar. $|\underline{F}|$

(g) $\overline{O}=C=\overline{S}$ In this linear molecule, the two bonds from carbon both are polarized away

from carbon. But the C=O bond is more polar than the C=S bond, and hence the
molecule is polar.

EXERCISES

Lewis theory

23. Hydrogen never has an octet of electrons in any of its compounds, but rather a pair (or
duet, if you prefer). An example is the Lewis structure of H_2O (below). In many
compounds in which the central atom is from the second period or higher, there are more
than eight electrons around the central atom; an example of a compound with such an
"expanded octet" is ICl$_3$ (below). Finally, in some compounds, there are less than eight
electrons around the central atom; one such "electron deficient" compound is BF$_3$.

$$H—\overline{O}—H \qquad |\overline{F}-B-\overline{F}| \qquad |\overline{Cl}-\overset{\frown}{I}-\overline{Cl}| \\ \qquad\qquad\qquad |\underline{F}| \qquad\qquad |\underline{Cl}|$$

24.   NH$_3$ $5+(3\times1)=8$ v.e.$=4$ pairs   BF$_3$ $3+(3\times7)=24$ v.e.$=12$ pairs

SF$_6$ $6+(6\times7)=48$ v.e.$=24$ pairs   SO$_3$ $6+(3\times6)=24$ v.e.$=12$ pairs

NH$_4^+$ $5+(4\times1)-1=8$ v.e.$=4$ pairs   SO$_4^{2-}$ $6+(4\times6)+2=32$ v.e.$=16$ pairs

NO$_2$ $5+(2\times6)=17$ v.e.$=8.5$ pairs

NO$_2$ cannot obey the octet rule; there is no way to pair all electrons when the number of
electrons is odd.

All of these Lewis structures obey the octet rule except for BF$_3$ which is electron deficient
and SF$_6$, which has an expanded octet.

25. **(a)** $Cs^+[:\overset{..}{Br}:]^-$ CsBr, cesium bromide   **(b)** $H—\overline{Sb}—H$ H_3Sb, hydrogen antimonide
$$|$$
$$H$$

(c) $\overline{C}l—B—\overline{C}l|$ BF_3, boron trichloride   **(d)** $Cs^+[:\overset{..}{C}l:]^-$ CsCl, cesium chloride
$$|$$
$$|\underline{C}l|$$

(e) $Li^+[:\overset{..}{O}:]^{2-} Li^+$ Li_2O, lithium oxide   **(f)** $|\overline{I}—\overline{C}l|$ ICl, iodine monochloride

26. **(a)** In order to construct an H_3 molecule, one H would have to bridge the other two. This would place 4 electrons around the central H atom which is more than the stable pair found around H in most Lewis structures.

(b) In HHe there would either three electrons between the two atoms, or three electrons around the He atom. Either of these is not a particularly stable situation.

(c) He_2 would have either a double bond between two He atoms and thus four electrons around each He atom, or three electrons around each He atom. Neither situation achieves the electron configuration of a noble gas.

(d) H_3O has an expanded octet (9 electrons) on oxygen; expanded octets are not found on elements of the second period. Other structures place a multiple bond between O and H. Both situations are unstable.

$$\begin{array}{c} H \\ | \\ H—\underset{..}{O}—H \end{array}$$

27. **(c)** is the correct answer.

(a) $[|\overline{O}—C=\overline{N}]^-$ does not have an octet of electrons around C.

(b) $[C=C:]^-$ does not have an octet around either C, it has only 6 valence electrons, and should have 10, and the sum of the formal charges on the two carbons doesn't equal the charge on the ion.

(d) The total number of valence electrons in $\overset{..}{N}=\overset{..}{\underset{..}{O}}$ is incorrect; this odd-electron species should have 11 valence electrons, not 12.

28. **(a)** $Mg—\overline{O}|$ is written as a covalent structure even though the compound is composed

of a metal (Mg) and a nonmetal (O). One expects an ionic Lewis structure.

$[Mg]^{2+}[|\overline{O}|]^{2-}$

(b) $[|\overline{C}l]^+[|\overline{O}|]^{2-}[|\overline{C}l|]^+$ is written as an ionic structure, even though we expect a

covalent structure between nonmetallic atoms. A more plausible structure is

$|\overline{C}l—\overline{O}—\overline{C}l|$

(c) $[|\bar{O}\!-\!\dot{N}\!=\!\bar{O}|]^{\oplus}$ has too many valence electrons—8.5 electron pairs or 17 valence electrons—it should have $(2\times6)+5-1=16$ valence electrons: 8 electron pairs. A plausible Lewis structure is $[|\bar{O}=N\!=\!\bar{O}|]^{+}$, which has 1+ formal charge on N and 0 formal charge on each oxygen.

(d) In the structure $[|\bar{S}\!-\!C\!=\!\bar{N}]^{-}$ there is not an octet of electrons around either S nor C. In addition, there are only 7 pairs of valence electrons in this structure, 14 valence electrons. There should be $6+4+5+1=16$ valence electrons, or 8 electron pairs. At least two structures are possible. $[\bar{S}=C\!=\!\bar{N}]^{-}$ has a formal charge of 1- on N and is preferred over $[|\bar{S}\!-\!C\equiv N\,|]^{-}$, with its formal charge of 1- on S, which is less electronegative than N.

Ionic bonding

29. **(a)** $\text{Li}\cdot$ forms Li^{+} cations, and $:\!\dot{S}\!:$ forms $[:\!\ddot{S}\!:]^{2-}$ anions. Lithium sulfide: $\text{Li}^{+}[:\!\ddot{S}\!:]^{2-}\,\text{Li}^{+}$

(b) $\text{Na}\cdot$ forms Na^{+} cations, and $:\!\dot{F}\!:$ forms $[:\!\ddot{F}\!:]^{-}$ anions. Sodium fluoride: $\text{Na}^{+}[:\!\ddot{F}\!:]^{-}$

(c) $\text{Ca}\cdot$ forms Ca^{2+} cations, and $:\!\dot{I}\!:$ forms $[:\!\ddot{I}\!:]^{-}$ anions. Calcium iodide: $[:\!\ddot{I}\!:]^{-}\,\text{Ca}^{2+}[:\!\ddot{I}\!:]^{-}$

(d) $\cdot\text{Sc}\cdot$ forms Sc^{3+} cations, and $:\!\dot{C}\!l\!:$ forms $[:\!\ddot{C}l\!:]^{-}$ anions. Scandium chloride is

$$[:\!\ddot{C}l\!:]^{-}\quad\begin{array}{c}[\text{Sc}]^{3+}\\[2pt][:\!\ddot{C}l\!:]^{-}\end{array}\quad[:\!\ddot{C}l\!:]^{-}$$

The formulas of the compounds are Li_{2}S, NaF, CaI_{2}, and ScCl_{3}.

30. The Lewis symbols are $[\text{H}:]^{-}$ for the hydride ion, $[|\bar{N}|]^{3-}$ for the nitride ion.

(a) $[\text{Li}]^{+}\,[\text{H}:]^{-}$

lithium hydride

(b) $[\text{H}:]^{-}\,[\text{Ca}]^{2+}\,[\text{H}:]^{-}$

calcium hydride

(c) $[\text{Mg}]^{2+}\,[|\bar{N}|]^{3-}\,[\text{Mg}]^{2+}[|\,\bar{N}|]^{3-}\,[\text{Mg}]^{2+}$

magnesium nitride

Formal Charge

31. There are three features common to formal charge and oxidation state. First, both indicate how electrons are distributed in the bonding of the compound. Second, negative formal charge (in the most plausible Lewis structure) and negative oxidation state are generally assigned to the more electronegative atoms. And third, both numbers are determined by a set of rules, rather than being determined experimentally. Bear in mind, however, that there are also significant differences. For instance, there are cases where atoms of the same type with the same oxidation state have different formal charges, such as oxygen in ozone, O_3. Another is that formal charges are used to decide between alternative Lewis structures, while oxidation state is used in balancing equations and naming compounds. Also, the oxidation state in a compound is invariant, while the formal charge can change. The most significant difference, though, is that whereas the oxidation state of an element in its compounds is usually not zero, its formal charge usually is.

32. The most common instance in which formal charge is not kept to a minimum occurs is in the case of ionic compounds. For example, in $Mg—\overline{O}|$ the formal charge on Mg is 1+ and on O it is 1-, while in the ionic version $[Mg]^{2+}[|\overline{O}|]^{2-}$, formal charges are 2+ and 2-, respectively. Additionally, in some resonance hybrids, formal charge is not minimized. In order to have bond lengths agree with experimental results, it may not be acceptable to create multiple bonds. Yet a third instance is when double bonds are created to lower formal charge, particularly when this results in the octet rule being violated. Thus, all the $Cl—O$ bonds in ClO_4^- are single bonds, although including some double bonds would minimize formal charges.

33. Formal charge = number of valence electrons $-2\times$ number of lone pairs $-$ number of bonding electron pairs. The formal charge of the central atom is calculated below the Lewis structure of each species.

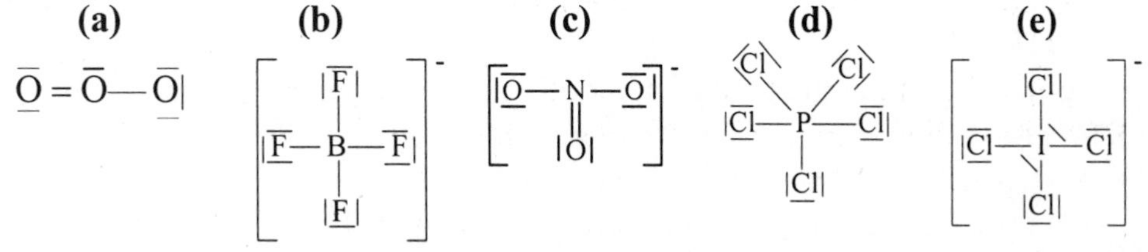

	(a)	**(b)**	**(c)**	**(d)**	**(e)**
valence e⁻	6	3	5	5	7
-2(lone pairs)	-2	0	0	0	-4
-bonding pair	<u>-3</u>	<u>-4</u>	<u>-4</u>	<u>-5</u>	<u>-4</u>
formal charge	+1	-1	+1	0	-1

34. We calculate formal charge = # valence electrons − # lone-pair e⁻ − ½ # bond-pair e⁻

(a)   H−N̄−Ō−H   f.c. of $N = 5 - 2 - 3 = 0$ most plausible
              | f.c. of $O = 6 - 4 - 2 = 0$
              H

           H−Ō−N̄−H   f.c. of $N = 5 - 4 - 2 = -1$
              | f.c. of $O = 6 - 2 - 3 = +1$
              H

(b)   S̄ = C= S̄   f.c. of $C = 4 - 0 - 4 = 0$ most plausible
                         f.c. of $S = 6 - 4 - 2 = 0$

           C̄ = S= S̄   f.c. of $C = 4 - 4 - 2 = -2$
                         f.c. of central $S = 6 - 0 - 4 = +2$
                         f.c. of terminal $S = 6 - 4 - 2 = 0$

(c)   f.c. of $N = 5 - 4 - 2 = -1$
           N̄ = F̄−Ō| f.c. of $O = 6 - 6 - 1 = -1$
                         f.c. of $F = 7 - 3 - 2 = 2+$

           Ō = N̄ − F̄| f.c. of $O = 6 - 4 - 2 = 0$
                         f.c. of $N = 5 - 2 - 3 = 0$ most plausible
                         f.c. of $F = 7 - 6 - 1 = 0$

(d) |S̄—Ō—C̄l| f.c. of $S = 6 - 6 - 1 = -1$ O should have
               | f.c. of $O = 6 - 2 - 3 = +1$ the more negative
             |C̄l| f.c. of $Cl = 7 - 6 - 1 = 0$ formal charge

           |Ō—S̄—C̄l| f.c. of $S = 6 - 2 - 3 = +1$
               | f.c. of $O = 6 - 6 - 1 = -1$ most plausible
             |C̄l| f.c. of $Cl = 7 - 6 - 1 = 0$

           |Ō—C̄l—C̄l—S̄| f.c. of $S = 6 - 6 - 1 = -1$ 26 electrons present;
                         f.c. of $O = 6 - 6 - 1 = -1$ Cl should not have a
                         f.c. of $Cl = 7 - 4 - 2 = +1$ more positive f.c. than S

35. We begin by counting the total number of valence electrons that must appear in the Lewis structure of the ion CO_2H^+: one from hydrogen, four from carbon, and six from each of the two oxygen (12 in all from the oxygen atoms). One electron is lost to establish the 1+ charge on the ion. In all, sixteen electrons. If the usual rules for constructing valid Lewis structures are applied to HCO_2^+, we come up with the following structures:

          Ö̈=C=Ö—H   :O≡C—Ö—H
            (A) ⊕   ⊕   (B)

In structure (A), the internal oxygen atom caries the positive charge, while in structure (B), the positive charge is located on the terminal oxygen atom. Thus, in this case, we cannot use the concept of formal charge to pick one structure over the other because the positive formal charge in both structures is located on the same type of atom, namely, an oxygen atom. In other words, based on formal charge rules alone, we must conclude that structures (A) and (B) are equally plausible.

36. The intention of this question is to make the student aware of the fact that on occasion, one can obtain a better Lewis structure *"from the standpoint of formal charge minimization"* by using chain-like structures rather than the expected compact, symmetrical structures. The two linear Lewis structures for the ClO_4^- (I) and (II) and a compact structure (III) are shown below:

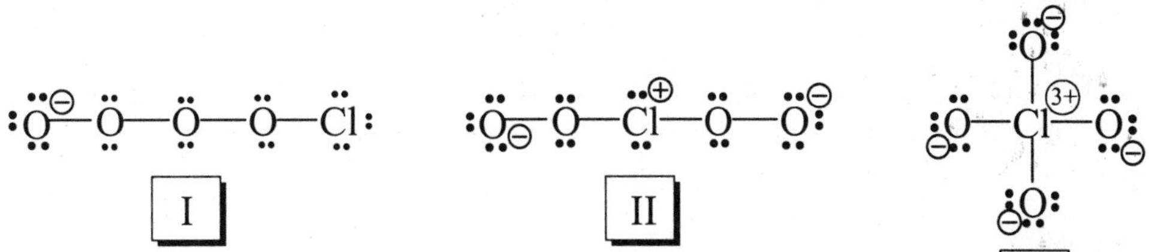

All of these structures have the required 32 valence electrons. Structure I has only one formal charge, 1- on the terminal oxygen atom. Structure II has a total of three formal charges, 1- on the terminal oxygen atoms and 1+ on the central chlorine atom. The compact structure, structure III, has formal charges on all of the atoms, 1- on all oxygen atoms and a formal charge of 3+ on the central chlorine atom.

From the standpoint of minimizing formal charge, structure I would be deemed the most appropriate. Nevertheless, structure III is the one that is actually adopted by the ClO_4^- ion, despite the fact that better minimization is achieved with the linear structure. By using an expanded valence shell with 14 electrons for the central atom, as in structure IV below, one can achieve the same minimum set of formal charges as in structure I:

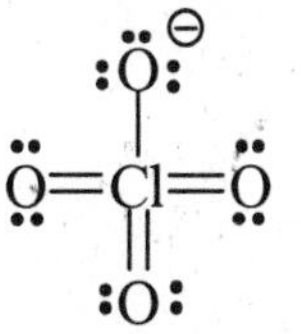

The chlorine atom presumably uses available 3*d* orbitals to accommodate the six additional electrons in its valence shell. Structure IV would appear to be the best structure yet because it has the minimum number of formal charges and is close to the true shape for that seen in the ClO_4^- ion. In light of recent quantum mechanical calculations, however, many chemists now believe that *d*-orbital involvement in expanded octets should only be invoked when there is no way to avoid them, as in PCl_5 or SF_6. This means that whenever possible, octet structures should be used, even though at times they afford unsettlingly large formal charges. Thus, structure III is now considered by most chemists as being superior to structure IV.

Lewis structures

37. **(a)** In H_2NOH, N and O atoms are the central atoms and the terminal atoms are the H atoms. The number of valence electrons in the molecule totals: $(3 \times 1) + 5 + 6 = 14$ valence electrons, or seven electron pairs. A reasonable Lewis structure is:

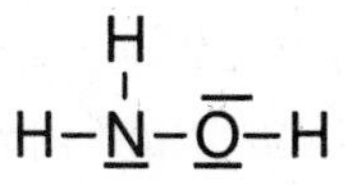

(b)   In $HOClO_2$, two O atoms and one H atom are terminal atoms, and the Cl and O atoms are the central atoms. The total number of valence electrons in the structure is $1+7+(3\times6)=26$ valence electrons, or 13 electron pairs. A reasonable Lewis structure is:

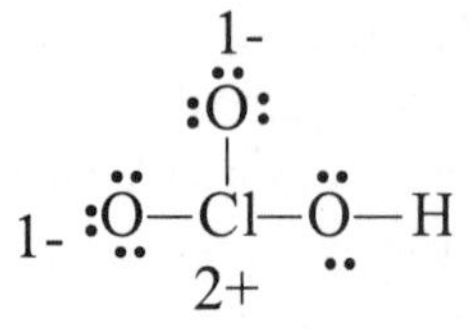

(c)   In HONO, the H atom and one O atom are the terminal atoms; the other O atom and the N atom are the central atoms. The total number of valence electrons in the structure is $1+5+(2\times6)=18$ valence electrons, or nine electron pairs. A plausible Lewis structure is $H—\overline{O}—\overline{N}=\overline{O}$

(d)   In O_2SCl_2, S is the central atom, with Cl and O as terminal atoms. The total number of valence electrons is $(2\times6)+6+(2\times7)=32$ valence electrons, 16 electron pairs. A reasonable Lewis structure is:

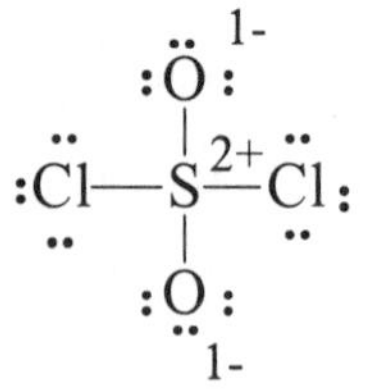

38.   The total number of valence electrons is $(2\times7)+(2\times6)=26$ valence electrons, or 13 pairs of valence electrons. It is unlikely to have F as a central atom; that would require an expanded octet on F. One structure is $|\overline{F}—\overline{S}—\overline{S}—\overline{F}|$; another is $|\overline{S}—\overline{S}\overline{\oplus}\overline{F}|$
$\ominus$ $|\overline{F}|$

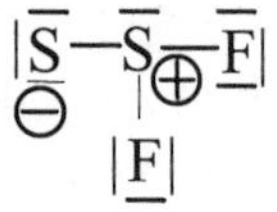

39. **(a)**   The total number of valence electrons in $SO_3{}^{2-}$ is $6+(3\times6)+2=26$, or 13 electron pairs. A plausible Lewis structure is:

(b)   The total number of valence electrons in $NO_2{}^-$ is $5+(2\times6)+1=18$, or 9 electron pairs. There are two resonance forms for the nitrite ion:

$[|\overline{O}—\overline{N}=\overline{O}]^- \leftrightarrow [\overline{O}=\overline{N}—\overline{O}|]^-$

(c)   The total number of valence electrons in $CO_3{}^{2-}$ is $4+(3\times6)+2=24$, or 12 electron pairs. There are three resonance forms for the carbonate ion:

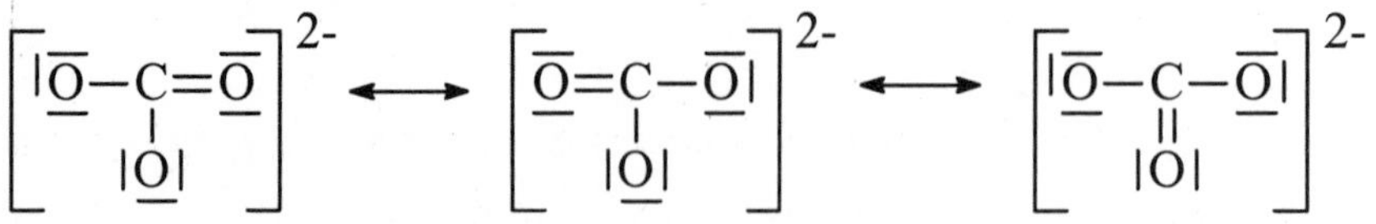

$$\left[\begin{matrix}|\overline{O}—C=\overline{O}\\ |\underline{O}|\end{matrix}\right]^{2-} \longleftrightarrow \left[\begin{matrix}\overline{O}=C—\overline{O}|\\ |\underline{O}|\end{matrix}\right]^{2-} \longleftrightarrow \left[\begin{matrix}|\overline{O}—C—\overline{O}|\\ ||\underline{O}|\end{matrix}\right]^{2-}$$

(d)   The total number of valence electrons in $HO_2{}^-$ is $1+(2\times6)+1=14$, or 7 electron pairs. A plausible Lewis structure is $[H—\overline{O}—\overline{O}]^-$

40.   Each of the cations has an empty valence shell as the result of ionization. The main task is to determine the Lewis structure of each anion.

(a)   The total number of valence electrons in OH^- is $6+1+1=8$, or 4 electron pairs. A plausible Lewis structure for barium hydroxide is $[|\overline{O}-H]^-[Ba]^{2+}[|\overline{O}-H]^-$

(b)   The total number of valence electrons in NO_2^- is $5+(2\times6)+1=18$, or 9 electron pairs. A plausible Lewis structure sodium nitrite is $[Na]^+[|\overline{O}-\overline{N}=\overline{O}]^- \leftrightarrow [\overline{O}=\overline{N}-\overline{O}|]^-$.

(c)   The total number of valence electrons in IO_3^- is $7+(3\times6)+1=26$, or 13 electron pairs. A plausible Lewis structure for magnesium iodate is

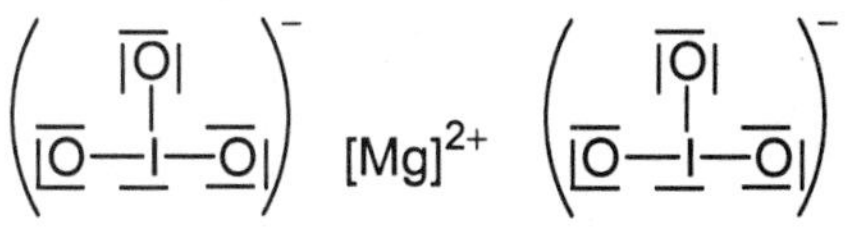

$$\left(\begin{array}{c}\overline{|O|}\\ |\\ |\overline{O}-\underset{}{I}-\overline{O}| \end{array}\right)^- [Mg]^{2+} \left(\begin{array}{c}\overline{|O|}\\ |\\ |\overline{O}-\underset{}{I}-\overline{O}| \end{array}\right)^-$$

(d)   The total number of valence electrons in SO_4^{2-} is $6+(4\times6)+2=32$, or 16 electron pairs. A plausible structure for aluminum sulfate is $[SO_4]^{2-}[Al]^{3+}[SO_4]^{2-}[Al]^{3+}[SO_4]^{2-}$. Because of the ability of S to expand its octet, SO_4^{2-} has several resonance forms, a few of which are:

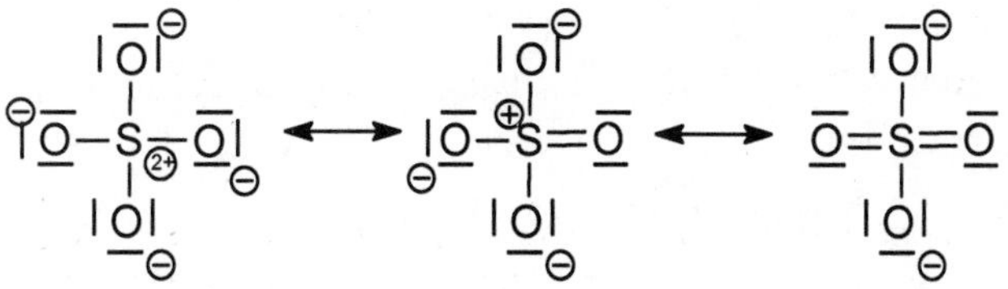

The first structure, without the expanded octet, is preferred.

41.   In $CH_3CHCHCHO$ there are $(4\times4)+(6\times1)+6=28$ valence electrons, or 14 electron pairs. We expect that the carbon atoms bond to each other. A plausible Lewis structure is :

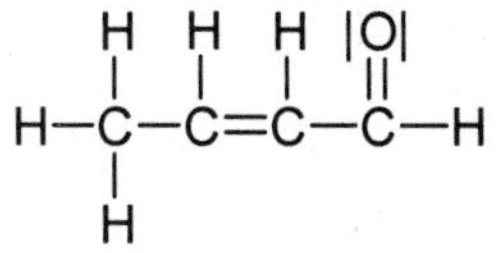

42.   In C_3O_2 there are $(3\times4)+(2\times6)=24$ valence electrons or 12 valence electron pairs. A plausible Lewis structure follows: $\overline{O}=C=C=C=\overline{O}$

43. (a)
```
     :O:
      ‖
  H — C — H
```
(b)
```
      H  :Cl: H
      |   |   |
  H — C — C — C — O — H
      |   |   |   ··
      H   H   H
```

44. (a)
```
      H   H  :O:
      |   |   ‖
  H — C — C — C — H
      |   |
      H   H
```
(b)
```
      H        H   H
      |        |   |
  H — C — O — C — C — H
      |  ··    |   |
      H        H   H
```

45. (a)
```
        H  :O:
        |   ‖
 :Cl — C — C — O — H
  ··    |      ··
        H
```
(b)
```
         H   H  :O:
         |   |   ‖
  H — O — C — C — C — O — H
     ··   |   |       ··
         H   H
```

46. (a)
```
      H   H  :O:  H
      |   |   ··  |
  H — C — C — C — C — H
      |   |       |
      H   H       H
```
(b)
```
         H   H   H
         |   |   |
 :Cl — C — C — N — H
  ··     |   |  ··
         H   H
```

Polar Covalent Bonds

47. The percent ionic character of a bond is based on the difference in electronegativity of its constituent atoms and Figure 11.7.

(a) $S(2.5) — H(2.1)$ (b) $O(3.5) — Cl(3.0)$ (c) $Al(1.5) — O(3.5)$ (d) $As(2.0) - O(3.5)$

Δ EN 0.4	0.5	2.0	1.5
%ionic = 4%	= 5 %	=60 %	=33 %

48.

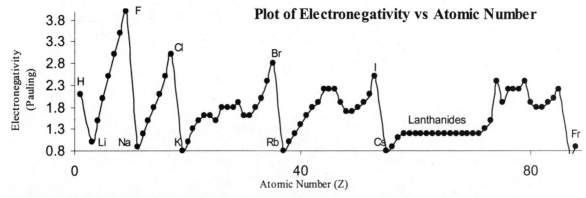

The property of electronegativity does indeed conform to the periodic law. Each of the "low points" corresponds to an alkali metal, and the end of each trend corresponds to a halogen. This is not unexpected.

Resonance

49. In NO_2^-, the total number of valence electrons is $1+5+(2\times6)=18$ valence electrons, or 9 electron pairs. N is the central atom. There are two resonance forms.

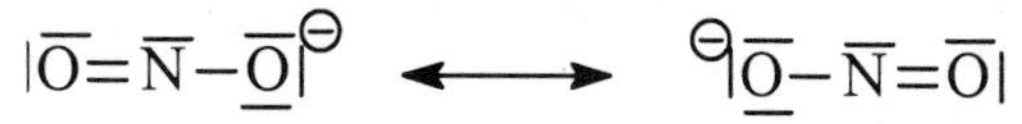

$$\overline{|O}=\overline{N}-\overline{\underline{O}|}^{\ominus} \longleftrightarrow {}^{\ominus}\overline{|\underline{O}}-\overline{N}=\overline{O|}$$

50. The only one of the four species that requires resonance forms for its adequate representation is CO_3^{2-}. In none of the others are there dissimilar electron distributions around like atoms. All four Lewis structures are drawn below.

(a) In CO_2, there are $4+(2\times6)=16$ valence electrons, or 8 electron pairs.

(b) In OCl^-, there are $6+7+1=14$ valence electrons, or 7 electron pairs,

(c) In CO_3^{2-}, there are $4+(3\times6)+2=24$ valence electrons, or 12 electron pairs.

(d) In OH^-, there are $6+1+1=8$ valence electrons, or 4 electron pairs.

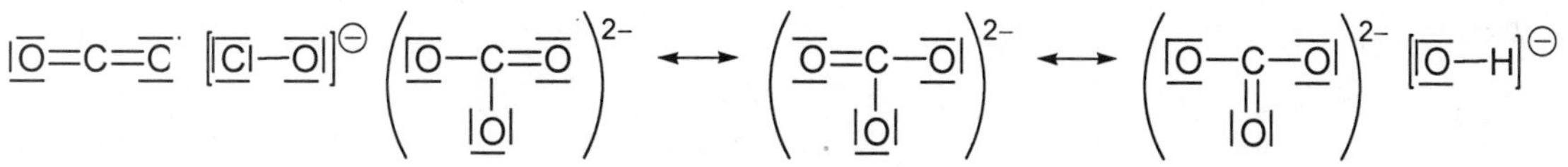

$$\overline{|O}=c=\overline{\underline{C}} \quad \overline{[\underline{Cl}}-\overline{\underline{O}|]}^{\ominus} \quad \left(\overline{|\underline{O}}-c=\overline{\underline{O}}\atop {|\,|\atop |\underline{O}|}\right)^{2-} \longleftrightarrow \left(\overline{|\underline{O}}=c-\overline{\underline{O}|}\atop {|\atop |\underline{O}|}\right)^{2-} \longleftrightarrow \left(\overline{|\underline{O}}-c-\overline{\underline{O}|}\atop {|\,|\atop |\underline{O}|}\right)^{2-} \quad \overline{[\underline{O}}-H]^{\ominus}$$

51. Bond length data from Table 11.2 follow:

$N\equiv N$ 109.8 pm	$N=N$ 123 pm	$N-N$ 145 pm
$N=O$ 120 pm	$N-O$ 136 pm	

The experimental N—N bond length of 113 pm approximates that of the $N\equiv N$ triple bond, which appears in structure (1). The experimental N—O bond length of 119 pm approximates that of the N=O double bond, which appears in structure (2). Structure (4) is highly unlikely because it contains no nitrogen-to-nitrogen bonds, and a N—N bond was found experimentally. Structure (3) also is unlikely, because it contains a very long (145 pm) N—N single bond, which does not agree at all well with the experimental N-to-N bond length. The molecule seems best represented as a resonance hybrid of (1) and (2).

$$\overline{{}^{\ominus}N}=\overset{\oplus}{\overline{N}}=\overline{O} \longleftrightarrow |N\equiv\overset{\oplus}{\overline{N}}-\overline{\underline{O}|}^{\ominus}$$

52. We begin by drawing all three resonance forms of HNO_3 and then analyzing their distribution of formal charge to determine which is the most plausible.

$$\begin{array}{ccc} \textbf{(a)} & \textbf{(b)} & \textbf{(c)} \\ H-\overline{\underline{O}}-\overline{N}-\overline{\underline{O}}| & H-\overline{\underline{O}}-\overline{N}=\overline{O} & H-\overline{\underline{O}}=\overline{N}-\overline{\underline{O}}| \\ {}^{|\,|}_{|\underline{O}|} & {}^{|}_{|\underline{O}|} & {}^{|}_{|\underline{O}|} \end{array}$$

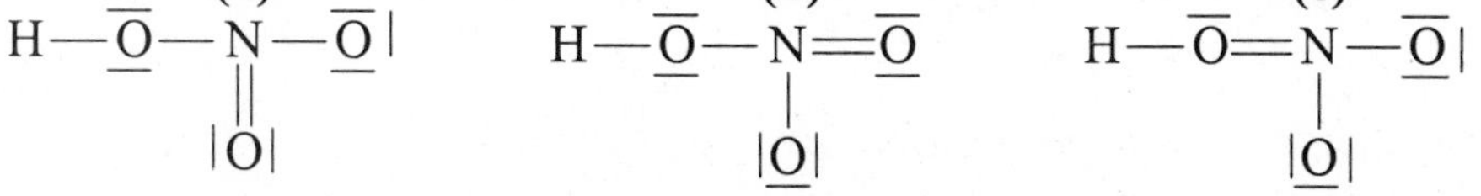

In all three structures, the formal charges of N and H are the same:

f.c. of $H = 1 - 1 - 0 = 0$ f.c. of $N = 5 - 4 - 0 = +1$

For an oxygen that forms two bonds (either 2 single or one double),
f.c. $= 6 - 2 - (2 \times 2) = 0$

For an oxygen that forms only one bond, f.c. $= 6 - 1 - (3 \times 2) = -1$

For the oxygen that forms three bonds (a single and a double), f.c. $= 6 - 3 - (1 \times 2) = +1$

Thus, structures (a) and (b) are equivalent in their distributions of formal charges, zero on all atoms except 1+ on N and 1- on one O. These are resonance forms. Structure (c) is quite different, with formal charges of 1- on two O's, and 1+ on the other, and a formal charge of 1+ on N. Structure (c) is thus the least plausible, especially because of the adjacent like charges.

Odd-electron species

53. **(a)** CH_3 has a total of $(3 \times 1) + 4 = 7$ valence electrons, or 3 electron pairs and a lone electron. C is the central atom. A plausible Lewis structure is

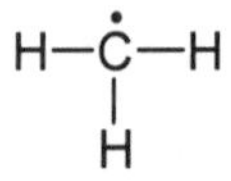

(b) ClO_2 has a total of $(2 \times 6) + 7 = 19$ valence electrons, or 9 electron pairs and a lone electron. Cl is the central atom. A plausible Lewis structure is: $:\ddot{O}\!-\!\ddot{C}l\!-\!\ddot{O}\cdot$

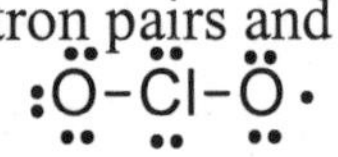

(c) NO_3 has a total of $(3 \times 6) + 5 = 23$ valence electrons, or 11 electron pairs, plus a lone electron. N is the central atom. A plausible Lewis structure is shown to the right. Resonance forms can also be drawn.

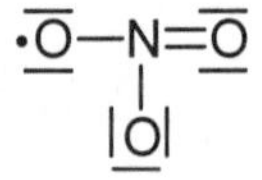

54.   In NO_2, there are $5 + (2 \times 6) = 17$ valence electrons, 8 electron pairs and a lone electron. N is the central atom. A plausible Lewis structure is (A) which has a formal charge of 1+ on N and 1- on the single-bonded O. Another Lewis structure, with zero formal charge on each atom, is (B). Because of the unpaired electron we expect NO_2 to be paramagnetic. We would expect a bond—a pair of electrons—to form between two NO_2 molecules as a result of the pairing of the electrons that are unpaired in the NO_2 molecules. If the second structure for NO_2 is used, the one with zero formal charge on each atom, a plausible structure for N_2O_4 is (C). If the first Lewis structure for NO_2 is used, a plausible Lewis structure contains a $N\!-\!N$ bond (D). Resonance structures can be drawn for this second version of N_2O_4. A $N\!-\!N$ bond is observed experimentally in N_2O_4. In either structure for N_2O_4, all electrons are paired; the molecule is expected to be (and is) diamagnetic. Lewis structures for (A) – (D) are shown on the following page.

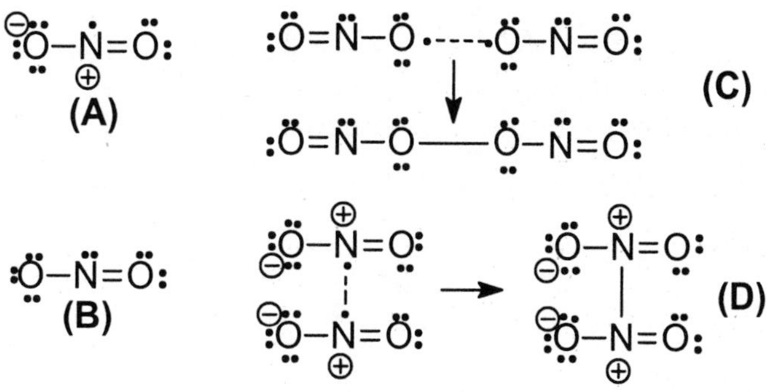

(A)   (B)   (C)   (D)

Expanded octets

55. In PO_4^{3-}; $5 + (4 \times 6) + 3 = 32$ valence electrons or 16 electron pairs.
Expanded octet is not needed.

In PI_3; $5 + (3 \times 7) = 26$ valence electrons or 13 electron pairs.
An expanded octet is not needed.

In ICl_3; are $7 + (3 \times 7) = 28$ valence electrons or 14 electron pairs.
An expanded octet is necessary.

In $OSCl_2$; $6 + 6 + (2 \times 7) = 26$ valence electrons or 13 electron pairs.
Expanded octet is not needed.

In SF_4; $6 + (4 \times 7) = 34$ valence electrons or 17 electron pairs.
An expanded octet is necessary.

In ClO_4^-; $7 + (4 \times 6) + 1 = 32$ valence electrons or 16 electron pairs.
Expanded octet is not needed.

56. Let us draw the Lewis structure of H_2CSF_4. The molecule has $(2 \times 1) + 4 + 6 + (4 \times 7) = 40$ valence electrons, or 20 electron pairs. With only single bonds and all octets complete, there is a 1- formal charge on C and 1+ on S, as in the left structure below . The right structure below avoids an undesirable separation of charge by creating a carbon-to-sulfur double bond.

Molecular shapes

57. The AX_nE_m designations that are cited below are to be found in Table 11.1 of the text, along with a sketch and a picture of a model of each type of structure.

(a) $|N \equiv N|$ is linear, two points define a line.

(b) $H - C \equiv N|$ is linear. The molecule belongs to the AX_2E category, and these species are linear.

(c) NH_4^+ is tetrahedral. The ion is of the AX_4 type, which has a tetrahedral electron-group geometry and a tetrahedral shape.

$$\left(\begin{array}{c} H \\ | \\ H-N-H \\ | \\ H \end{array} \right)^+$$

(d) NO_3^- is trigonal planar. The ion is of the AX_3 type, which has a trigonal planar electron-group geometry and a trigonal planar shape. The other resonance forms are of the same type.

$$\left(\overline{\underline{O}}-N-\overline{\underline{O}} \right)^-$$
$$\underset{|\underline{O}|}{\overset{\|}{}}$$

(e) NSF is bent. The molecule is of the AX_2E type, which has a trigonal planar electron-group geometry and a bent shape. $\underline{N}=\overline{S}-\overline{\underline{F}}|$

58. The AX_nE_m designations that are cited below are to be found in Table 11.1 of the text, along with a sketch and a picture of a model of each type of structure.

(a) PCl_3 is a trigonal pyramid. The molecule is of the AX_3E type, and has a tetrahedral electron-group geometry and a trigonal pyramid shape.

$$|\overline{\underline{Cl}}-P-\overline{\underline{Cl}}|$$
$$\underset{|\underline{Cl}|}{|}$$

(b) SO_4^{2-} has a tetrahedral shape. The ion is of the type AX_4, and has a tetrahedral electron-group geometry and a tetrahedral shape. The other resonance forms of the sulfate ion have the same shape.

$$\left(\begin{array}{c} |\overline{\underline{O}}| \\ | \\ |\overline{\underline{O}}-S-\overline{\underline{O}}| \\ | \\ |\underline{O}| \end{array} \right)^{2-}$$

(c) $SOCl_2$ has a trigonal pyramidal shape. This molecule is of the AX_3E type and has a tetrahedral electron-group geometry and a trigonal pyramidal shape.

$$\overline{\underline{O}}-\overline{S}-\overline{\underline{Cl}}|$$
$$\underset{|\underline{Cl}|}{|}$$

(d) SO_3 has a trigonal planar shape. The molecule is of the AX_3 type, with a trigonal planar electron-group geometry and molecular shape. The other resonance forms have the same shape.

$$\overline{\underline{O}}=S-\overline{\underline{O}}|$$
$$\underset{|\underline{O}|}{|}$$

(e) BrF_4^+ has a see-saw shape. The molecule is of the AX_4E type, with a trigonal bipyramid electron-group geometry and a see-saw molecular shape.

$$\left(\begin{array}{c} |\overline{F}| \\ \backslash \\ |\overline{F}-Br-\overline{F}| \\ | \\ |F| \end{array} \right)^+$$

59. A trigonal planar shape requires that three ligands and no lone pairs be bonded to the central atom. Thus PF_6^- cannot have a trigonal planar shape, since six atoms are attached to the central atom. In addition, PO_4^{3-} cannot have a trigonal planar shape, since four O atoms are attached to the central P atom. We now draw the Lewis structure of each of the remaining ions, as a first step in predicting their shapes. The SO_3^{2-} ion is of the AX_3E type; it has a tetrahedral electron-group geometry and a trigonal pyramidal shape. The CO_3^{2-} ion is of the AX_3 type, and has a trigonal planar electron-group geometry and a trigonal planar shape.

$$\left(\overline{O}-\overset{|}{\underset{|\underline{O}|}{S}}-\overline{O}|\right)^{2-} \quad \left(\overline{O}=\overset{|}{\underset{|\underline{O}|}{C}}-\overline{O}|\right)^{2-}$$

60. We predict shapes by first of all drawing the Lewis structures for the species. Thus, SO_3^{2-} and NI_3 both have the same shape.

$$\overline{I}-\overset{|}{\underset{|\underline{I}|}{N}}-\overline{I}| \qquad H-C\equiv N| \qquad \left(\overline{O}=\overset{|}{\underset{|\underline{O}|}{S}}-\overline{O}|\right)^{2-} \qquad \left(\overline{O}=\overset{|}{\underset{|\underline{O}|}{N}}-\overline{O}|\right)^{-}$$

| AX_nE_m shape | AX_3E_1 trigonal pyramid | AX_2E_0 linear | AX_3E_0 trigonal pyramid | AX_3E_0 trigonal planar |

61. (a) In CO_2 there are a total of $4+(2\times6)=16$ valence electrons, or 8 electron pairs. The following Lewis structure is plausible. $\overline{O}=C=\overline{O}$ This is a molecule of type AX_2; CO_2 has a linear electron-shape geometry and a linear shape.

(b) In Cl_2CO there are a total of $(2\times7)+4+6=24$ valence electrons, or 12 electron pairs. The molecule can be represented by a Lewis structure with C as the central atom. This molecule is of the AX_3 type; it has a trigonal planar electron-group geometry and molecular shape.

$$|\overline{Cl}-\overset{\overset{|\underline{O}|}{||}}{C}-\overline{Cl}|$$

(c) In $ClNO_2$ there are a total of $7+5+(2\times6)=24$ valence electrons, or 12 electron pairs. N is the central atom. A plausible Lewis structure is shown:
This molecule is of the AX_3 type; it has a trigonal planar electron-group geometry and a trigonal planar shape.

$$|\overline{Cl}-\overset{\overset{|\underline{O}|}{||}}{N}-\overline{O}|$$

62. Lewis structures enable us to determine molecular shapes.

(a) N_2O_4 has $(2\times5)+(4\times6)=34$ valence electrons, or 17 electron pairs.

(b) C_2N_2 has $(2\times4)+(2\times5)=18$ valence electrons, or 9 electron pairs.

(c) C_2H_6 has $(2 \times 4) + (6 \times 1) = 14$ valence electrons, or 7 electron pairs.

(d) CH_3OCH_3 has 6 more valence electrons than $C_2H_6 = 20$ valence e⁻ or 10 electron pairs.

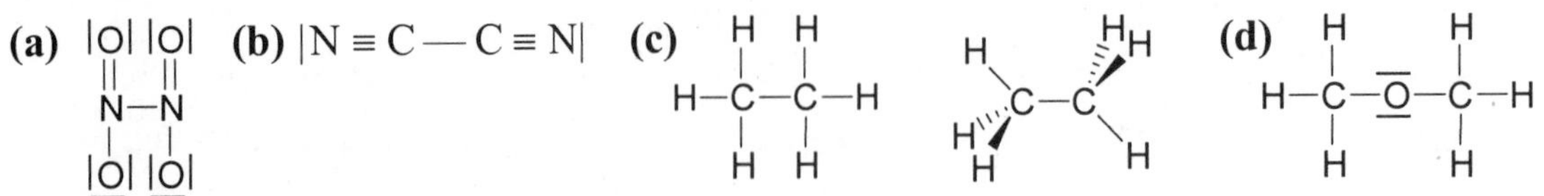

(a)
$$
\overset{|\overline{O}|\;\;|\overline{O}|}{\underset{|\underline{O}|\;\;|\underline{O}|}{N\!-\!N}}
$$
(b) $|N \equiv C \!-\! C \equiv N|$ **(c)**
$$
\overset{H\;\;H}{\underset{H\;\;H}{H\!-\!C\!-\!C\!-\!H}}
$$
(d)
$$
\overset{H\;\;\;\;\;\;H}{\underset{H\;\;\;\;\;\;H}{H\!-\!C\!-\!\overline{O}\!-\!C\!-\!H}}
$$

(a) There are three atoms bound to each N in N_2O_4, making the molecule triangular planar around each N. The entire molecule does not have to be planar, however, since there is free rotation around the N — N bond.

(b) There are two atoms attached to each C, requiring a linear geometry. The entire molecule is linear.

(c) There are four atoms attached to each C. The molecular geometry around each C is tetrahedral, as shown in the sketch above.

(d) Around the central O, the electron-group geometry is tetrahedral. With two of the electron groups being lone pairs, the molecular geometry around the central atom is bent.

63. First we draw the Lewis structure of each species, then use these Lewis structures to predict the molecular shape.

(a) In ClO_4^- there are $7 + (4 \times 6) + 1 = 32$ valence electrons or 16 electron pairs. A plausible Lewis structure follows. Since there are four atoms and no lone pairs bonded to the central atom, the molecular shape and the electron-group geometry are the same: tetrahedral.

(b) In $S_2O_3^{2-}$ there are $(2 \times 6) + (3 \times 6) + 2 = 32$ valence electrons or 16 electron pairs. A plausible Lewis structure follows. Since there are four atoms and no lone pairs bonded to the central atom, both the electron-group geometry and molecular shape are tetrahedral.

(c) In PF_6^- there are $5 + (6 \times 7) + 1 = 48$ valence electrons or 24 electron pairs. A plausible Lewis structure follows. Since there are six atoms and no lone pairs bonded to the central atom, the electron-group geometry and molecular shape are octahedral.

(d) In I_3^- there are $(3 \times 7) + 1 = 22$ valence electrons or 11 electron pairs. There are three lone pairs and two atoms bound to the central atom. The electron-group geometry is trigonal bipyramid; the molecular shape is linear.

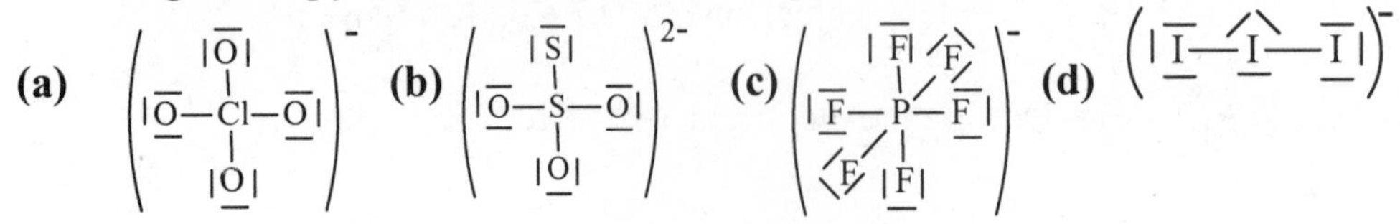

(a)
$$
\left(\overset{|\overline{O}|}{\underset{|\underline{O}|}{|\overline{O}\!-\!Cl\!-\!\overline{O}|}} \right)^{-}
$$
(b)
$$
\left(\overset{|\overline{S}|}{\underset{|\underline{O}|}{|\overline{O}\!-\!S\!-\!\overline{O}|}} \right)^{2-}
$$
(c)
$$
\left(\overset{|\overline{F}|\;\;\diagup F}{\underset{\diagup F\;\;|\overline{F}|}{|\overline{F}\!-\!P\!-\!F|}} \right)^{-}
$$
(d)
$$
\left(|\overline{I}\!-\!I\!-\!\overline{I}| \right)^{-}
$$

64. **(a)** In OSF_2, there are a total of $6+6+(2\times7)=26$ valence electrons, or 13 electron pairs. S is the central atom. A plausible Lewis structure is shown. This molecule is of the type AX_3E_1; it has a tetrahedral electron-group geometry and a trigonal pyramidal shape.

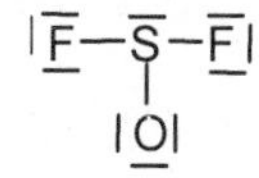

(b) In O_2SF_2, there are a total of $(2\times6)+6+(2\times7)=32$ valence electrons, or 16 electron pairs. S is the central atom. One plausible Lewis structure is shown. The molecule is of the AX_4 type; it has a tetrahedral electron-group geometry and a tetrahedral shape. The structure with all single bonds is preferred because it does not have an expanded octet.

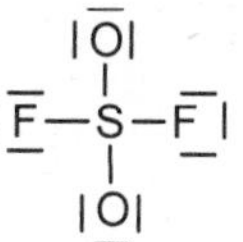

(c) In SF_5^- there are $6+(5\times7)+1=42$ valence electrons, or 21 electron pairs. A plausible Lewis structure is shown. The ion is of the AX_5E type. It has an octahedral electron-group geometry and a square pyramidal molecular shape.

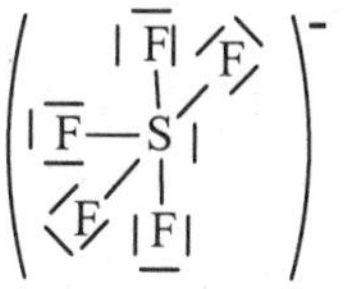

(d) In ClO_4^-, there are a total of $1+7+(4\times6)=32$ valence electrons, or 16 electron pairs. Cl is the central atom. A plausible Lewis structure is shown. This ion is of the AX_4 type; it has a tetrahedral electron-group geometry and a tetrahedral shape.

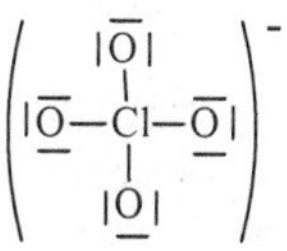

(e) In ClO_3^- the total number of valence electrons is $1+(3\times6)+7=26$ valence electrons, or 13 electron pairs. A plausible Lewis structure is shown. The molecule is of the type AX_3E; it has a tetrahedral electron-group geometry and a trigonal pyramidal molecular shape.

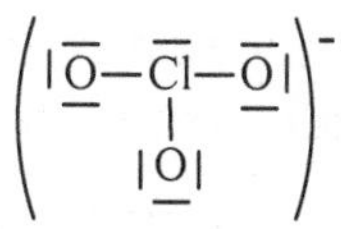

65. In BF_4^-, there are a total of $1+3+(4\times7)=32$ valence electrons, or 16 electron pairs. A plausible Lewis structure has B as the central atom. This ion is of the type AX_4; it has a tetrahedral electron-group geometry and a tetrahedral shape.

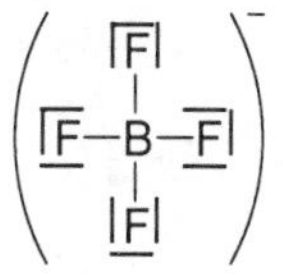

66. The molecular geometry is indicated by the VSEPR notation (i.e. AX_3E_2). Formal charge is reduced by moving lone pairs of electrons from the terminal atoms and forming multiple bonds to the central atom. The VSEPR notation is unchanged when formal charge is increased or decreased. For example, consider SO_2:

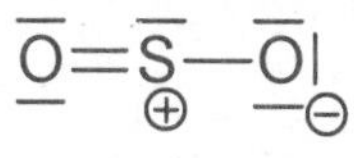

VSEPR notation
AX_2E
shape: bent

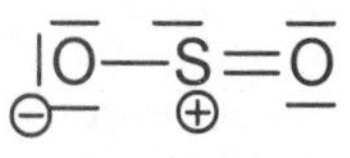

VSEPR notation
AX_2E
shape: bent

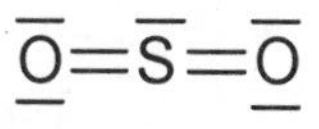

VSEPR notation
AX_2E
shape: bent

Shapes of Molecules with More Than One Central Atom

67.

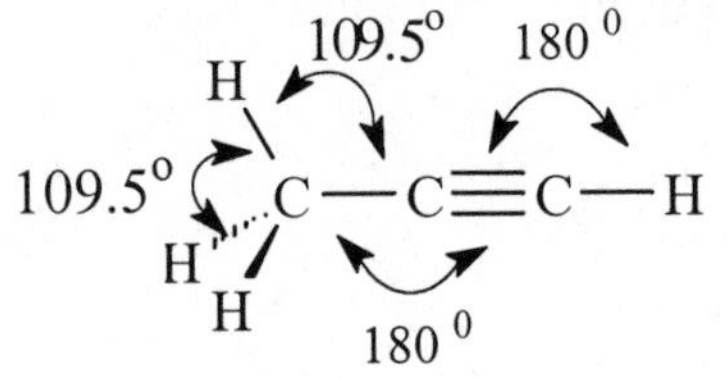

A maximum of 5 atoms can be in the same plane

68.

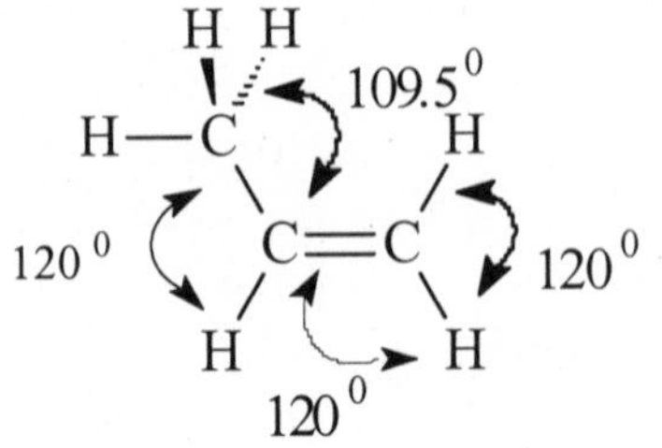

A maximum of 7 atoms can be in the same plane

69.

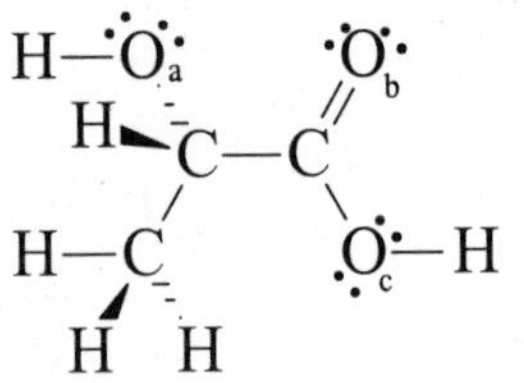

All angles ~ $109.5°$ with the exception of

$$\left.\begin{array}{l} O_c - C - C \\ O_b = C - C \\ O_b = C - O_c \end{array}\right\} \sim 120°$$

70.

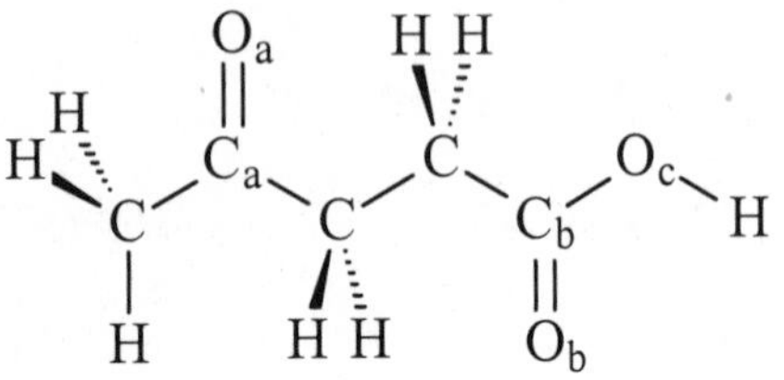

All angles ~ $109.5°$ with the exception of

$$\left.\begin{array}{l} O_c - C_b - C \\ O_b = C_b - O_c \\ O_b = C_b - C \\ O_a = C_a - C \\ C - C_a - C \end{array}\right\} \sim 120°$$

Polar Molecules

71.   For each molecule, we first draw the Lewis structure, which we use to predict the shape.

(a) SO_2 has a total of $6 + (2 \times 6) = 18$ valence electrons, or 9 electron pairs. The molecule has two resonance forms. $\overline{O} = \underset{\oplus}{\overline{S}} - \underset{\ominus}{\overline{O}} \longleftrightarrow \underset{\ominus}{\overline{O}} - \underset{\oplus}{\overline{S}} = \overline{O}$ Each of these resonance forms is of the type AX_2E; it has a triangular planar electron-group geometry and a bent shape. Since each S—O bond is polar toward O, and since the bond dipoles do not point in opposite directions, the molecule has a resultant dipole moment, pointing from S through a point midway between the two O atoms; SO_2 is polar.

(b) NH_3 has a total of $5 + (3 \times 1) = 8$ valence electrons, or 4 electron pairs. N is the central atom. A plausible Lewis structure is shown. The molecule is of the AX_3E type; it has a tetrahedral electron-group geometry and a trigonal pyramidal shape. Each N-H bond is polar toward N. Since the bonds do not symmetrically oppose each other, there is a resultant molecular dipole moment, pointing from the triangular base (formed by the three H atoms) through N. The molecule is polar.

$$H-\underline{N}-H$$
$$\overset{\mid}{H}$$

(c) H_2S has a total of $6 + (2 \times 1) = 8$ valence electrons, or 4 electron pairs. S is the central atom and a plausible Lewis structure is $H\text{-}\underline{\overline{S}}\text{-}H$. This molecule is of the AX_2E_2 type; it has a tetrahedral electron-group geometry and a bent shape. Each H—S bond is polar toward S. Since the bonds do not symmetrically oppose each other, the molecule has a net dipole moment, pointing through S from a point midway between the two H atoms. H_2S is polar.

(d) C_2H_4 consists of atoms that all have about the same electronegativities. Of course, the C—C bond is not polar, but neither are the C—H bonds. The molecule is planar. Thus, the entire molecule is nonpolar.

(e) SF_6 has a total of $6 + (6 \times 7) = 48$ valence electrons, or 24 electron pairs. S is the central atom. All atoms have zero formal charge in the Lewis structure. This molecule is of the AX_6 type; it has an octahedral electron-group geometry and an octahedral shape. Even though each S—F bond is polar toward F, the bonds symmetrically oppose each other resulting in a molecule that is nonpolar.

(f) CH_2Cl_2 has a total of $4 + (2 \times 1) + (2 \times 7) = 20$ valence electrons, or 10 electron pairs. A plausible Lewis structure is shown. The molecule is tetrahedral and polar, since the two polar bonds (C—Cl) do not cancel the effect of each other.

72. (a) HCN is a linear molecule, which can be derived from its Lewis structure. $H—C \equiv N|$. The $C \equiv N$ bond is strongly polar toward N, while the H—C bond is generally considered to be nonpolar. Thus, the molecule has a dipole moment, pointed toward N from C.

(b) SO_3 is a trigonal planar molecule, which can be derived from its Lewis structure. Each sulfur-oxygen bond is polar from S to O, but the three bonds are equally polar and are pointed in symmetrical opposition so that they cancel. The SO_3 molecule has a dipole moment of zero.

(c) CS_2 is a linear molecule, which can be derived from its Lewis structure. $\underline{\overline{S}}=C=\underline{\overline{S}}$. Each carbon-sulfur bond is polar from C to S, but the two bonds are equally polar and are pointed in opposition to each other so that they cancel. The CS_2 molecule has a dipole moment of zero.

(d) OCS also is a linear molecule. Its Lewis structure is $\overline{O}=C=\overline{S}$. But the carbon-oxygen bond is more polar than the carbon-sulfur bond. Although both bond dipoles point from the central atom to the bonded atom, these two bond dipoles are unequal in strength. Thus, the molecule is polar in the direction from C to O.

(e) $SOCl_2$ is a trigonal pyramidal molecule. Its Lewis structure is shown. The lone pair is at one corner of the tetrahedron. Each bond in the molecule is polar, with the dipole moments pointing away from the central atom. The sulfur-chlorine bond is less polar than the sulfur-oxygen bond, and this makes the molecule polar. The dipole moment of the molecule points from the sulfur atom to the base of the trigonal pyramid, not toward the center of the base but slightly toward the O apex of that base.

$$\overline{Cl}-\overline{S}-\overline{Cl}$$
$$|$$
$$\overline{|O|}$$

(f) SiF_4 is a tetrahedral molecule, with the following Lewis structure. Each Si — F bond is polar, with its negative end away from the central atom toward F in each case. These four Si – F bond dipoles oppose each other and thus cancel. SiF_4 has no dipole moment.

$$\overline{|F|}$$
$$|$$
$$\overline{|F}-Si-\overline{F|}$$
$$|$$
$$\underline{|F|}$$

(g) POF_3 is a tetrahedral molecule, with the following Lewis structure All four bonds are polar, with their dipole moments pointing away from the central atom. The P — F bond polarity is greater than that of the P — O bond. Thus, POF_3 is a polar molecule with its dipole moment pointing away from the P towards the center of the triangle formed by the three F atoms.

$$\overline{|F|}$$
$$|$$
$$\overline{|F}-P-\overline{O|}$$
$$|$$
$$\underline{|F|}$$

73. In H_2O_2, there are a total of $(2 \times 1)+(2 \times 6)=14$ valence electrons, 7 electron pairs. The two O atoms are central atoms. A plausible Lewis structure has zero formal charge on each atom. $H-\overline{O}-\overline{O}-H$. In the hydrogen peroxide molecule, the O — O bond is non-polar, while the H — O bonds are polar, toward O. Since the molecule has a resultant dipole moment, it cannot be linear, for, if it were linear the two polar bonds would oppose each other and their polarities would cancel.

74. **(a)** FNO has a total of $7+5+6=18$ valence electrons, or 9 electron pairs. N is the central atom.   A plausible Lewis structure is $\overline{|F}-N=\overline{O}$. The formal charge on each atom in this structure is zero.

(b) FNO is of the AX_2E type; it has a trigonal planar electron-group geometry and a bent shape.

(c) The N — F bond is polar toward F and the N—O bond is polar toward O. In FNO, these two bond dipoles point in the same general direction, producing a polar molecule. In FNO_2, however, the additional N—O bond dipole partially opposes the polarity of the other two bond dipoles, resulting in a smaller net dipole moment.

Bond lengths

75. A heteronuclear bond length (one between two different atoms) is equal to the average of two homonuclear bond lengths (one between two like atoms) of the same order (both single, both double, or both triple).

   (a) I—Cl bond length

   $= [(\text{I—I bond length}) + (\text{Cl—Cl bond length})] \div 2$
   $= [266 \text{ pm} + 199 \text{ pm}] \div 2 = 233 \text{ pm}$

   (b) O—Cl bond length

   $= [(\text{O—O bond length}) + (\text{Cl—Cl bond length})] \div 2$
   $= [145 \text{ pm} + 199 \text{ pm}] \div 2 = 172 \text{ pm}$

   (c) C—F bond length

   $= [(\text{C—C bond length}) + (\text{F—F bond length})] \div 2$
   $= [154 \text{ pm} + 143 \text{ pm}] \div 2 = 149 \text{ pm}$

   (d) C—Br bond length

   $= [(\text{C—C bond length}) + (\text{Br—Br bond length})] \div 2$
   $= [154 \text{ pm} + 228 \text{ pm}] \div 2 = 191 \text{ pm}$

76. First we need to draw the Lewis structure of each of the compounds cited, so that we can determine the order, and thus the relative length, of each O-to-O bond.

   (a) In H_2O_2, there are $(2 \times 1) + (2 \times 6) = 14$ valence electrons or 7 electron pairs. A plausible Lewis structure is $H - \overline{\underline{O}} - \overline{\underline{O}} - H$

   (b) In O_2, the total number of valence electrons is $(2 \times 6 =)12$ valence electrons, or 6 electron pairs. A plausible Lewis structure is $\overline{O} = \overline{O}$

   (c) In O_3, the total number of valence electrons is $(3 \times 6) = 18$ valence electrons, or 9 electron pairs. A plausible Lewis structure is $\overline{O} = \overline{O} - \overline{\underline{O}}| \longleftrightarrow |\overline{\underline{O}} - \overline{O} = \overline{O}$ (Resonance)

   O_2 should have the shortest O-to-O bond, a double bond. The single O—O bond in H_2O_2 should be longest.

77. The N—F bond is a single bond. Its bond length should be the average of the N—N single bond (145 pm) and the F—F single bond (143 pm). Thus, the average N—F bond length $= (145 + 143) \div 2 = 144$ pm

$$\overset{\displaystyle |\overset{\displaystyle O}{\underset{\displaystyle |}{|}}}{|\overline{F} - N - \overline{\underline{O}}|}$$

78. In H_2NOH, there are $(3 \times 1) + 5 + 6 = 14$ valence electrons total, or 7 electron pairs. N and O are the two central atoms. A plausible Lewis structure has zero formal charge on each atom. The N—H bond lengths are 100 pm, the O—H bond is 97 pm, and the N—O bond is 136 pm. All values are taken from Table 11.2. All bond angles approximate the tetrahedral bond angle of 109.5°, but are expected to be somewhat smaller, perhaps by 2° to 4° each.

$$\overset{\displaystyle H}{\underset{\displaystyle |}{H - N - \overline{\underline{O}} - H}}$$

Bond Energies

79.

$$H-\underset{\underset{H}{|}}{\overset{\overset{H}{|}}{C}}-\underset{\underset{H}{|}}{\overset{\overset{H}{|}}{C}}-H \;+\; |\overline{Cl}-\overline{Cl}| \;\longrightarrow\; H-\underset{\underset{H}{|}}{\overset{\overset{H}{|}}{C}}-\underset{\underset{H}{|}}{\overset{\overset{H}{|}}{C}}-\overline{Cl}| \;+\; H-\overline{Cl}|$$

Analysis of the Lewis structures of products and reactants indicates that a $C-H$ bond and a $Cl-Cl$ bond are broken, and a $C-Cl$ and a $H-Cl$ bond are formed.

Energy required to break bonds $= C\!-\!H + Cl\!-\!Cl = 414\,\dfrac{kJ}{mol} + 243\,\dfrac{kJ}{mol} = 657\,\dfrac{kJ}{mol}$

Energy realized by forming bonds $= C\!-\!Cl + H\!-\!Cl = 339\,\dfrac{kJ}{mol} + 431\,\dfrac{kJ}{mol} = 770\,\dfrac{kJ}{mol}$

$\Delta H = 657\ kJ/mol - 770\ kJ/mol = -113\ kJ/mol$

80. The reaction in terms of Lewis structures is $\overline{O}{=}\overline{O}-\overline{O}| + \overline{O}{=}\overline{N}\bullet \longrightarrow \overline{O}{=}N-\overline{O}\bullet + \overline{O}{=}\overline{O}$
The net result is the breakage of an $O-O$ bond (142 kJ/mol) and the formation of an $N-O$ bond (230 kJ/mol). $\Delta H = 142\ kJ/mol - 230\ kJ/mol = -88\ kJ/mol$

81. In each case we write the formation reaction, but specify reactants and products with their Lewis structures. All species are assumed to be gases.

(a) $\dfrac{1}{2}\left(\overline{O}{=}\overline{O}\right) + \dfrac{1}{2}\left(H-H\right) \longrightarrow \bullet\overline{O}-H$

Bonds broken: $\dfrac{1}{2}\,(O=O) + \dfrac{1}{2}\,(H\!-\!H) = 0.5(498\ kJ + 436\ kJ) = 467\ kJ$

Bonds formed: $O\!-\!H = 464\ kJ \qquad \Delta H° = 467\ kJ - 464\ kJ = 3\ kJ/mol$

If the $O-H$ bond dissociation energy of 428.0 kJ/mol from Figure 11-16 is used, $\Delta H_f^{\,o} = 39\ kJ/mol$.

(b) $|N{\equiv}N| + 2\,H-H \longrightarrow H-\underset{\underset{H}{|}}{\overline{N}}-\underset{\underset{H}{|}}{\overline{N}}-H$

Bonds broken $= N\equiv N + 2H\!-\!H = 946\ kJ + 2\times436\ kJ = 1818\ kJ$

Bonds formed $= N\!-\!N + 4N\!-\!H = 163\ kJ + 4\times389\ kJ = 1719\ kJ$

$\Delta H_f^{\,o} = 1818\ kJ - 1719\ kJ = 99\ kJ$

82. The reaction of Example 11-14 is $CH_4(g) + Cl_2(g) \rightarrow CH_3Cl(g) + HCl(g)$

$\Delta H_{rxn} = -113\ kJ$. In Appendix D are the following values:
$\Delta H_f^{\circ}\big[\,CH_4(g)\big] = -74.81\ kJ,\ \Delta H_f^{\circ}\big[\,HCl(g)\big] = -92.31\ kJ,\ \Delta H_f^{\circ}\big[\,Cl_2(g)\big] = 0$

Thus, we have

$\Delta H_{rxn} = \Delta H_f^{\circ}\big[CH_3Cl(g)\big] + \Delta H_f^{\circ}\big[HCl(g)\big] - \Delta H_f^{\circ}\big[CH_4(g)\big] - \Delta H_f^{\circ}\big[Cl_2(g)\big]$

$-113\ kJ = \Delta H_f^{\circ}\big[CH_3Cl(g)\big] - 92.31\ kJ - (-74.81\ kJ) - (0.00)$

$\Delta H_f^{\circ}\big[\,CH_3\,Cl(g)\big] = -113\ kJ + 92.31\ kJ - 74.81\ kJ = -96\ kJ$

83.

$$H-C\equiv C-H_{(g)} \quad + \quad H-H_{(g)} \longrightarrow \quad \underset{(g)}{H-\overset{\overset{H}{|}}{C}=\overset{\overset{H}{|}}{C}-H}$$

Bonds broken	Energy change	Bonds formed	Energy Change
1 mol C≡C	1 × 837 kJ	1 mol C=C	1 × −611 kJ
1 mol H—H	1 × 436 kJ	2 mol C-H	2 × −414 kJ

Energy required to break bonds	+1273 kJ	Energy obtained upon bond formation	-1439 kJ

Overall energy change = 1273 kJ − 1439 kJ = − 166 kJ = $\Delta H°_{rxn}$

84. First determine the value of ΔH for reaction (2).

(1) $C(s) \longrightarrow C(g)$ $\quad\quad\quad\quad\quad \Delta H° = 717$ kJ

(2) $C(g) + 2\,H_2(g) \longrightarrow CH_4(g)$ $\quad \Delta H° = ?$

Net: $\quad C(s) + 2H_2(g) \longrightarrow CH_4(g)$ $\quad \Delta H_f° = -75$ kJ

717 kJ + ? = −75 $\quad$ or $\quad$? = −75 − 717 kJ = −792 kJ

To determine the energy of a C — H bond, we need to analyze reaction (2) in some detail.

In this reaction, 2 H — H bonds are broken and 4 C — H bonds are formed resulting in the production of 792 kJ/mol.

−792 kJ/mol $\quad$ = energy of broken bonds − energy of formed bonds
$\quad\quad\quad\quad\quad$ = $(2 \times 436$ kJ / mol$) - (4 \times$ C — H$)$

4 × C—H $\quad$ = 792 kJ/mol + (2 × 436 kJ/mol) = 1664 kJ/mol

C—H $\quad\quad$ = 1662 ÷ 4 = 416 kJ/mol

This value compares favorably with the value of 414 kJ/mol given in Table 11-3.

FEATURE PROBLEMS

106. (a) The average of the H — H and Cl — Cl bond energies is $(436 + 243) \div 2 = 340$ kJ/mol. The ionic resonance energy is the difference between this calculated value and the measured value of the H — Cl bond energy: $IRE = 431 - 340 = 91 \text{ kJ / mol}$

(b) $\Delta EN = \sqrt{IRE / 96} = \sqrt{91 / 96} = 0.97$

(c) An electronegativity difference of 0.97 gives about a 23% ionic character, read from Figure 11.7. The result of Example 11-4 is that the H — Cl bond is 20% ionic. These values are in good agreement with each other.

107. (a) The two bond moments can be added geometrically, by placing the head of one at the tail of the other, as long as we do not change the direction or the length of the moved dipole. The resultant molecular dipole moment is represented by the arrow drawn from the tail of one bond dipole to the head of the other. This is shown in the figure at right. The 52.0° angle in the figure is one-half of the 104° bond angle in water. The length is given as 1.84 D. We can construct a right angle triangle by bisecting the 76.0° angle. The 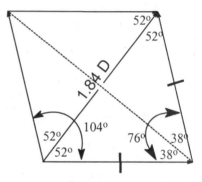 right angle triangle has a hypotenuse = O—H bond dipole and the two other angles are 52 ° and 38 °. The side opposite the bisected 76.0 ° angles is ½ (1.84 D) = 0.92 D. We can calculate the bond dipole using: $\sin 38.0° = \dfrac{0.92 \text{ D}}{\text{O-H bond dipole}} = 0.61566$,

hence O—H bond dipole = 1.49 D.

(b) For H_2S, we do not know the bond angle. We shall represent this bond angle as 2α. Using a similar procedure as above, a diagram can be constructed and the angle 2α calculated as follows:

$$\cos \alpha = \dfrac{\frac{1}{2}(0.93\,\text{D})}{0.67\,\text{D}} = 0.694 \quad \alpha = 46.05\ °$$

or $2\alpha = 92.1$ °

The H—S—H angle is approximately 92 °.

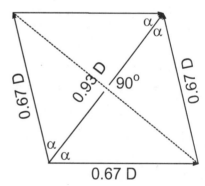

(c)

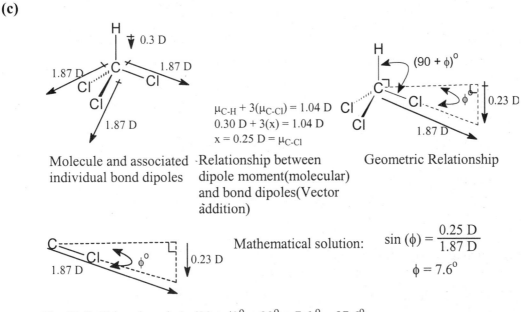

$$\mu_{C\text{-}H} + 3(\mu_{C\text{-}Cl}) = 1.04 \text{ D}$$
$$0.30 \text{ D} + 3(x) = 1.04 \text{ D}$$
$$x = 0.25 \text{ D} = \mu_{C\text{-}Cl}$$

Molecule and associated   ·Relationship between   Geometric Relationship
individual bond dipoles   dipole moment(molecular)
                           and bond dipoles(Vector
                           àddition)

Mathematical solution: $\sin(\phi) = \dfrac{0.25 \text{ D}}{1.87 \text{ D}}$

$$\phi = 7.6^{\circ}$$

The H-C-Cl bond angle is $(90 + \phi)^{\circ} = 90^{\circ} + 7.6^{\circ} = 97.6^{\circ}$

108. Step 1 in the alternate approach is similar to the first step in the method used for drawing Lewis structures. The only significant difference is that "electron pairs" rather than the total number of valence electrons are counted in this alternate approach. The second step in the alternate strategy is also similar to the second step for writing Lewis structures. By counting the number of bonding electron pairs in the alternate method, one is effectively working out the number of bonds present in the skeletal structure of the Lewis diagram. In step 3, the number of electron pairs surrounding the central atom is calculated.

This is basically the same procedure as completing the octets for the terminal atoms and assigning the remaining electrons to the central atom in the Lewis structure. Finally, in step 4 of the alternate method, the number of lone pair electrons on the central atom is calculated. This number together with the result from step 3, allows one to establish the VSEPR class. Consequently, both the alternate strategy and the Lewis diagram provide the number of bonding electron pairs and lone pairs on the central atom in species whose shape is being predicted. Since the shape of the molecule or ion in the VSEPR approach is determined solely by the number and types of electron pairs on the central atom (i.e., the VSEPR class) both methods end up giving the same result.

Included in the "alternative strategy" is the assumption that the central atom does not form double bonds with any of the terminal atoms. This means that in many instances, the central atom does not possess a complete octet. The presence or absence of an octet is, however, of no consequence to the VSEPR method because, according to the tenets of this theory, the shape adopted by the molecule is determined solely by the number and types of electron pairs on the central atom. Examples follow on the next two pages.

(a) PCl_5

   1. Total e⁻ pairs = $\dfrac{(1 \times 5e^- \text{ from P atom}) + (5 \times 7e^- \text{ from the 5 Cl atoms})}{2}$ = 20 pairs of e⁻

   2. Number of bonding e⁻ pairs = 6 atoms (5×Cl + 1×P) − 1 = 5 bonding e⁻ pairs.

   3. Number of e⁻ pairs around the central atom = (20(total) e⁻ pairs) − 3×(5 terminal Cl)
                    = 5 e⁻ pairs around P atom.

   4. Number of lone pair e⁻ = 5 e⁻ pairs around P − 5 bonding pairs of e⁻ = 0

Thus, according to this alternate approach, PCl_5 belongs to the VSEPR class AX_5. Molecules of this type adopt a trigonal bipyramidal structure.

(b) NH_3

   1. Total e⁻ pairs = $\dfrac{(1 \times 5e^- \text{ from N atom}) + (3 \times 1e^- \text{ from the 3 H atoms})}{2}$ = 4 pairs of e⁻

   2. Number of bonding e⁻ pairs = 4 atoms (1×N + 3×H) − 1 = 3 bonding e⁻ pairs.

   3. Number of e⁻ pairs around the central atom = 4(total) e⁻ pairs − 0
                    = 4 e⁻ pairs around N atom.

   4. Number of lone pair e⁻ = 4 e⁻ pairs around N − 3 bonding pairs of e⁻ = 1 lone pair e⁻

Thus, according to this alternate approach, NH_3 belongs to the VSEPR class AX_3E. Molecules of this type adopt a trigonal pyramidal structure.

(c) ClF_3

   1. Total e⁻ pairs = $\dfrac{(1 \times 7e^- \text{ from Cl atom}) + (3 \times 7e^- \text{ from the 3 F atoms})}{2}$ = 14 e⁻ pairs

   2. Number of bonding e⁻ pairs = 4 atoms (1×Cl + 3×F) − 1 = 3 bonding e⁻ pairs.

   3. Number of e⁻ pairs around the central atom = 14(total) pairs − 3×(3 terminal F atoms)
                    = 5 e⁻ pairs around Cl atom.

   4. Number of lone pair e⁻ = 5 e⁻ pairs around Cl − 3 bonding pairs of e⁻ = 2 lone pair e⁻

Thus, according to this alternate approach, ClF_3 belongs to the VSEPR class AX_3E_2. Molecules of this type adopt a T-shaped structure.

(d) SO_2

   1. Total e⁻ pairs = $\dfrac{(1 \times 6e^- \text{ from S atom}) + (2 \times 6e^- \text{ from the 2 O atoms})}{2}$ = 9 e⁻ pairs

   2. Number of bonding e⁻ pairs = 3 atoms (1×S + 2×O) − 1 = 2 bonding e⁻ pairs.

   3. Number of e⁻ pairs around the central atom = 9(total) pairs − 3×(2 terminal O atoms)
                    = 3 e⁻ pairs around S atom.

   4. Number of lone pair e⁻ = 3 e⁻ pairs around S − 2 bonding pairs of e⁻ = 12 lone pair e⁻

Thus, according to this alternate approach, SO_2 belongs to the VSEPR class AX_2E. Molecules of this type adopt a bent structure.

(e) ClF_4^-

1. Total e^- pairs $= \dfrac{(1 \times 7e^- \text{ from Cl}) + (4 \times 7e^- \text{ from the 4 F}) + (1e^- \text{ for charge of -1}))}{2}$

   $= 18$ pairs of e^-

2. Number of bonding e^- pairs $= 5$ atoms $(1 \times Cl + 4 \times F) - 1 = 4$ bonding e^- pairs.

3. Number of e^- pairs around the central atom $= 18$(total) pairs $- 3 \times$(4 terminal F atoms)

   $= 6$ e^- pairs around Cl atom.

4. Number of lone pair $e^- = 6$ e^- pairs around Cl $- 4$ bonding pairs of $e^- = 2$ lone pair e^-

Thus, according to this alternate approach, ClF_4^- belongs to the VSEPR class AX_4E_2. Molecules of this type adopt a square planar structure.

(f) PCl_4^+

1. Total e^- pairs $= \dfrac{(1 \times 5e^- \text{ from P}) + (4 \times 7e^- \text{ from the 4 Cl}) - (1e^- \text{ for +1 charge})}{2}$

   $= 16$ pairs of e^-

2. Number of bonding e^- pairs $= 5$ atoms $(4 \times Cl + 1 \times P) - 1 = 4$ bonding e^- pairs.

3. Number of e^- pairs around the central atom $= 16$(total) pairs $- 3$(4 terminal Cl atoms)

   $= 4$ e^- pairs around P atom.

4. Number of lone pair $e^- = 4$ e^- pairs around P $- 4$ bonding pairs of $e^- = 0$

Thus, according to this alternate approach, ClF_4^+ belongs to the VSEPR class AX_4. Molecules of this type adopt a tetrahedral structure.

CHAPTER 12
CHEMICAL BONDING II:ADDITIONAL ASPECTS
PRACTICE EXAMPLES

1A The valence-shell orbital diagrams of N and I are as follows:

N [He]$_{2s}$⊞ $_{2p}$⊞⊞⊞   I [Kr] 4d^{10}$_{5s}$⊞ $_{5p}$⊞⊞⊞

There are three half-filled $2p$ orbitals on N, and one half-filled $5p$ orbital on I. Each half-filled $2p$ orbital from N will overlap with one half-filled $5p$ orbital of a I. Thus, there will be three N—I bonds. The I atoms will be oriented in the same direction as the three $2p$ orbitals of N: toward the $x-$, $y-$, and z-directions of a Cartesian coordinate system. Thus, the I—N—I angles will be approximately 90° (probably larger because the I atoms will repel each other). The three I atoms will lie in the same plane at the points of a triangle, with the N atom centered above them. The molecule is trigonal pyramidal. (The same molecular shape is predicted if N is assumed to be sp^3 hybridized, but with 109.5° rather than 90° bond angles.)

1B The valence-shell orbital diagrams of N and H are as follows. N: [He] $_{2s}$⊞ $_{2p}$⊞⊞⊞
H: $_{1s}$⊡   There are three half-filled orbitals on N and one half-filled orbital on each H. There will be three N—H bonds, with bond angles of approximately 90°. The molecule is trigonal pyramid. (We obtain the same molecular shape if N is assumed to be sp^3 hybridized, but bond angles are closer to 109.5°, the tetrahedral bond angle.) VSEPR theory begins with the Lewis structure and notes that there are three bond pairs and one lone pair attached to N. This produces a tetrahedral electron pair geometry and a trigonal pyramid molecular shape with bond angles a bit less than the tetrahedral angle of 109.5° because of the lone pair. Since VSEPR theory makes a prediction closer to the experimental bond angle of 107° it seems more appropriate in this case.

2A Following the strategy outlined in the textbook, we begin by drawing a plausible Lewis structure for the cation in question. In this case, the Lewis structure must contain 20 valence electrons. The skeletal structure for the cation has a chlorine atom, the least electronegative element present, in the central position. Next we join the terminal chlorine and fluorine atoms to the central chlorine atom via single covalent bonds and then complete the octets for all three atoms.

$$|\overline{\text{F}}\!\!-\!\!\overset{\oplus}{\underline{\text{Cl}}}\!\!-\!\!\overline{\text{Cl}}|$$

With this bonding arrangement, the central chlorine atom ends up with a 1+ formal charge.

Once the Lewis diagram is complete, we can then use the VSEPR method to establish the geometry for the electron pairs on the central atom. The Lewis structure has two bonding electron pairs and two lone pairs of electrons around the central chlorine atom. These four pairs of electrons assume a tetrahedral geometry to minimize electron-electron repulsions.

The VSEPR notation for the Cl_2F^+ ion is AX_2E_3. According to Table 11.1, molecules of this type exhibit an <u>angular</u> molecular geometry. Our next task is to select a hybridization scheme that is consistent with the predicted shape. It turns out that the only way we can end up with a tetrahedral array of electron groups is if the central chlorine atom is sp^3 hybridized. In this scheme, two of the sp^3 hybrid orbitals are filled, while the remaining two are half occupied.

sp^3 hybridized central chlorine atom (Cl^+)

The Cl—F and Cl—Cl bonds in the cation are then formed by the overlap of the half-filled sp^3 hybrid orbitals of the central chlorine atom with the half-filled p-orbitals of the terminal Cl and F atoms. Thus, by using sp^3 hybridization, we end up with the same <u>bent</u> molecular geometry for the ion as that predicted by VSEPR theory (when the lone pairs on the central atom are ignored)

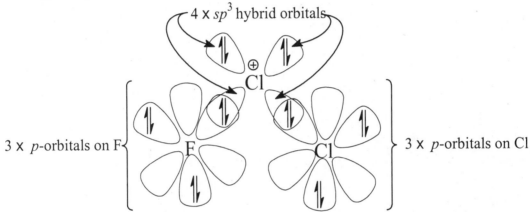

2B   As was the case in 2A, we begin by drawing a plausible Lewis structure for the cation in question. This time the Lewis structure must contain 34 valence electrons. The skeletal structure has bromine, the least electronegative element present, as the central atom. Next, we join the four terminal fluorine atoms to the central bromine atom via single covalent bonds and complete the octets for all of the fluorine atoms. Placing the last two electrons on the central bromine atom completes the diagram.

In order to accommodate ten electrons, the bromine atom is forced to expand its valence shell. Notice that the Br ends up with a 1+ formal charge in this structure. With the completed Lewis structure in hand, we can then use VSEPR theory to establish the geometry for the electron pairs around the central atom. The Lewis structure has four bonding pairs and one lone pair of electrons around the central bromine atom. These five pairs of electrons assume a trigonal bipyramidal geometry to minimize electron-electron repulsions. The VSEPR notation for the BrF_4^+ cation is AX_4E. According to Table 11.1, molecules of this type exhibit a see-saw molecular geometry.

Next we must select a hybridization scheme for the Br atom that is compatible with the predicted shape. It turns out that only sp^3d hybridization will provide the necessary trigonal bipyramidal distribution of electron pairs around the bromine atom. In this scheme, one of the sp^3d hybrid orbitals is filled while the remaining four are half-occupied.

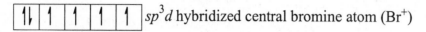 sp^3d hybridized central bromine atom (Br$^+$)

The four Br—F bonds in the cation are then formed by the overlap of the half-filled sp^3d hybrid orbitals of the bromine atom with the half-filled p-orbitals of the terminal fluorine atoms. Thus, by using sp^3d hybridization, we end up with the same see-saw molecular geometry for the cation as that predicted by VSEPR theory (when the lone pair on Br is ignored).

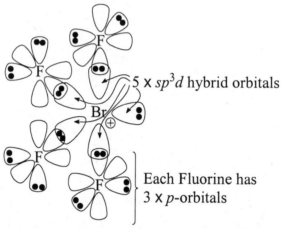

5 × sp^3d hybrid orbitals

Each Fluorine has
3 × p-orbitals

3A We begin by writing the Lewis structure. The H atoms are terminal atoms. There are three central atoms and $(3 \times 1) + 4 + 6 + 4 + (3 \times 1) = 20$ valence electrons, or 10 pairs. A plausible Lewis structure is drawn at right. Each central atom is surrounded by four electron pairs, requiring sp^3 hybridization. The valence-shell orbital diagrams for the atoms follow.

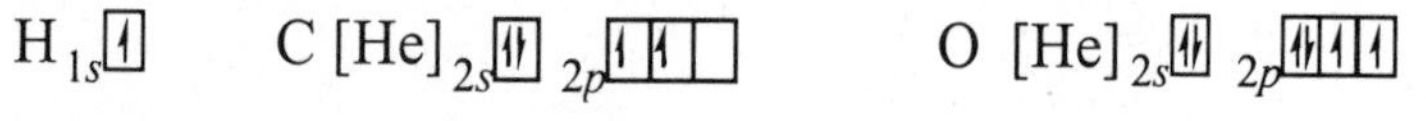

H $_{1s}$☐   C [He]$_{2s}$☐ $_{2p}$☐☐☐   O [He]$_{2s}$☐ $_{2p}$☐☐☐

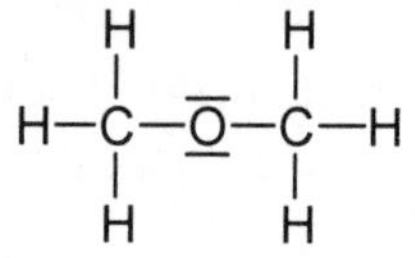

The valence-shell orbital diagrams for the hybridized central atoms then are:

C $_{sp^3}$☐☐☐☐   O $_{sp^3}$☐☐☐☐

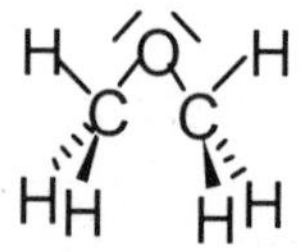

All bonds in the molecule are σ bonds. The H—C—H bond angles are 109.5°, as are the H—C—O bond angles. The C—O—C bond angle is possibly a bit smaller than 109.5° because of the repulsion of the two lone pairs of electrons on O. A wedge-and-dash sketch of the molecule is at right.

3B   The H atoms and one O are terminal atoms in the Lewis structure, which has $3 \times 1 + 4 + 4 + 2 \times 6 + 1 = 24$ valence electrons, or 12 pairs. The left-most C and the right-most O are surrounded by four electron pairs, and thus require sp^3 hybridization.

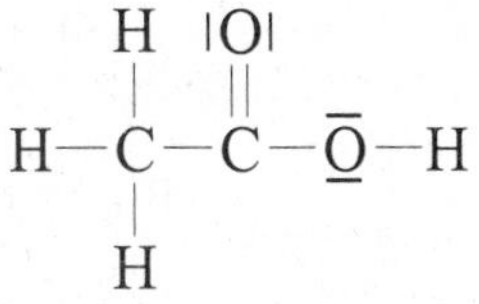

The central carbon is surrounded by three electron groups and is sp^2 hybridized. The orbital diagrams for the un-hybridized atoms are:

H: $1s$ [↓]   C: [He]$_{2s}$[↑↓] $_{2p}$[↑][↑][ ]   O: [He]$_{2s}$[↑↓] $_{2p}$[↑↓][↑][↑]

Hybridized orbital diagrams:

C: [He]$_{sp^3}$[↑][↑][↑][↑]   C: [He]$_{sp^2}$[↑][↑][↑] $_{2p}$[↑]   O:[He]$_{sp^3}$[↑↓][↑↓][↑][↑]   terminal O: [He]$_{sp^2}$[↑↓][↑↓][↑] $_{2p}$[↑]

There is one π bond in the molecule: between the $2p$ on the central C and the $2p$ on the terminal O. The remaining bonds are σ bonds. The H—C—H and H—C—C bond angles are 109.5°. The H—O—C angle is somewhat less, perhaps 105°. The C—C—O bond angles and O—C—O bond angles are all 120°.

4A   There are four bond pairs around the left-hand C, requiring sp^3 hybridization. Three of the bonds that form are C—H sigma bonds resulting from the overlap of a sp^3 hybrid orbital on C with a $1s$ orbital on H. The other C has two attached electron groups, utilizing sp hybridization.

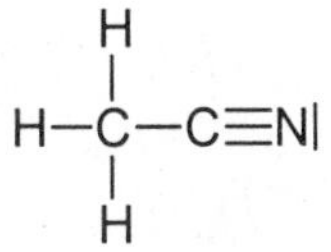

C: [He]$_{sp}$[↑↓][↑↓] $_{2p}$[↑↓][↑↓]   The N atom is sp hybridized. N: [He] $_{2s}$[↑↓] $_{2p}$[↑↓][↑][↑]

The two C atoms join with a sigma bond: overlap of sp^2 on the left-hand C with sp on the right-hand C. The three bonds between C and N consist of a sigma bond (sp on C with sp on N), and two pi bonds ($2p$ on C with $2p$ on N).

4B   The bond lengths that are given indicate that N is the central atom. The molecule has $(2 \times 5) + 6 = 16$ valence electrons, or 8 pairs. Average bond lengths are: N—N = 145 pm, N=N = 123 pm, N≡N = 110 pm, N—O = 136 pm, N=O = 120 pm. Plausible resonance structures are (with subscripts for identification):

structure(1) $|N_a \equiv N_b^{\oplus} - \overline{O}|^{\ominus}$ ↔ $|\underline{N}_a^{\ominus} = N_b^{\oplus} = \overline{O}$ structure(2)

In both structures, the central N is attached to two other atoms, and possesses no lone pairs. The geometry of the molecule thus is linear and the hybridization on this central N is sp. N$_b$ [He]$_{sp}$[↑↓][↑↓] $_{2p}$[↑][↑][↑] We will assume that the terminal atoms are not hybridized in either structure. Their valence-shell orbital diagrams are:
terminal N: N$_a$ [He] $_{2s}$[↑↓] $_{2p}$[↑][↑][↑]   O [He] $_{2s}$[↑↓] $_{2p}$[↑↓][↑][↑]

In structure (1) the N ≡ N bond results from the overlap of three pairs of half-filled orbitals: (1) sp_x on N$_b$ with $2p_x$ on N$_a$ forming a σ bond, (2) $2p_y$ on N$_b$ with $2p_y$ on N$_a$ forming a π bond, and (3) $2p_z$ on N$_b$ with $2p_z$ on N$_a$ also forming a π bond. The N—O bond is a coordinate covalent bond, and requires that the electron configuration of O be written as O $[He]_{2s}[\uparrow\downarrow]$ $_{2p}[\uparrow\downarrow|\uparrow|\uparrow]$ The N—O bond then forms by the overlap of the full sp_x orbital on N_b with the empty $2p$ orbital on O.

In structure (2) the N $=$ O bond results from the overlap of two pairs of half-filled orbitals: (1) sp_y on N_b with $2p_y$ on O forming a σ bond and (2) $2p_z$ on N_b with $2p_z$ on O forming a π bond. The N $=$ N σ bond is a coordinate covalent bond, and requires that the configuration of N_a be written N_a $[He]_{2s}[\uparrow\downarrow]$ $_{2p}[\uparrow\downarrow|\uparrow|\ ]$. The N $=$ N bond is formed by two overlaps: (1) the overlap of the full sp_y orbital on N_b with the empty $2p_y$ orbital on N_a to form a σ bond, and (2) the overlap of the half-filled $2p_x$ orbital on N_b with the half-filled $2p_x$ orbital on N_a to form a π bond. Based on formal charge arguments, structure (1) is preferred, because the negative formal charge is on the more electronegative atom, O.

5A   Removing an electron from Li_2 removes a bonding electron because the valence molecular orbital diagram for Li_2 is the same as that for H_2, only it is just moved up a principal quantum level: $\sigma_{1s}b[\uparrow\downarrow]$ $\sigma_{1s}*[\uparrow\downarrow]$ $\sigma_{2s}b[\uparrow\downarrow]$ $\sigma_{2s}*[\ ]$. The molecular orbital diagram for Li_2^+ is:
$\sigma_{1s}b[\uparrow\downarrow]$ $\sigma_{1s}*[\uparrow\downarrow]$ $\sigma_{2s}b[\uparrow|\ ]$ $\sigma_{2s}*[\ ]$ The bond order is Li_2^+ in 1/2, while that in Li_2 is one. Thus, the bond in Li_2^+ should be one half as strong as the bond in Li_2: 106 kJ/mol $\div 2 = 53$ kJ/mol Li_2^+

5B   The H_2^- ion contains 1 electron from each H plus 1 electron (for the charge) for a total of three electrons. Its molecular orbital diagram is $\sigma_{1s}[\uparrow\downarrow]$ $\sigma_{1s}*[\uparrow]$. There are 2 bonding and 1 anti-bonding electrons. The bond order in H_2^- is:
$$\text{Bond order} = \frac{(2 \text{ bonding } e^- - 1 \text{ antibonding } e^-)}{2} = \frac{1}{2}$$
Thus, we would expect the ion H_2^- to be stable, with a bond strength about half that of a hydrogen molecule.

6A   For each case, the empty molecular-orbital diagram has the following appearance. (KK indicates that the molecular orbitals formed from $1s$ atomic orbitals are full: KK $=\sigma_{1s}[\uparrow\downarrow]$ $\sigma_{1s}*[\uparrow\downarrow]$
KK $\sigma_{2s}[\ ]$ $\sigma_{2s}*[\ ]$ $\sigma_{2p}[\ ]$ $\pi_{2p}[\ |\ ]$ $\pi_{2p}*[\ |\ ]\sigma_{2p}*[\ ]$. We need to simply count up the total number of electrons in each species, and place them in the valence-shell molecular-orbital diagram.

(a) N_2^+ has $(2 \times 5) - 1 = 9$ valence electrons. Its molecular orbital diagram is
N_2^+ KK$\sigma_{2s}[\uparrow\downarrow]$ $\sigma_{2s}*[\uparrow\downarrow]$ $\pi_{2p}[\uparrow\downarrow|\uparrow\downarrow]$ $\sigma_{2p}[\uparrow|\ ]$ $\pi_{2p}*[\ |\ ]$ $\sigma_{2p}*[\ ]$
bond order = (7 bonding electrons -2 antibonding electrons) $\div 2 = 2.5$

(b) Ne_2^+ has $(2 \times 8) - 1 = 15$ valence electrons. Its molecular orbital diagram is
Ne_2^+ KK $\sigma_{2s}[\uparrow\downarrow]$ $\sigma_{2s}*[\uparrow\downarrow]$ $\sigma_{2p}[\uparrow\downarrow]$ $\pi_{2p}[\uparrow\downarrow|\uparrow\downarrow]$ $\pi_{2p}*[\uparrow\downarrow|\uparrow\downarrow]$ $\sigma_{2p}*[\uparrow|\ ]$
bond order = (8 bonding electrons -7 antibonding electrons) $\div 2 = 0.5$

(c) C_2^{2-} has $(2 \times 4) + 2 = 10$ valence electrons. Its molecular orbital diagram is
C_2^{2-} KK $\sigma_{2s}[\uparrow\downarrow]$ $\sigma_{2s}*[\uparrow\downarrow]$ $\pi_{2p}[\uparrow\downarrow|\uparrow\downarrow]$ $\sigma_{2p}[\uparrow\downarrow]$ $\pi_{2p}*[\ |\ ]$ $\sigma_{2p}*[\ ]$
bond order = (8 bonding electrons -2 antibonding electrons) $\div 2 = 3.0$

6B For each case, the empty molecular-orbital diagram has the following appearance. (KK indicates that the molecular orbitals formed from $1s$ atomic orbitals are full: KK $=\sigma_{1s}[\uparrow\downarrow]\;\sigma_{1s}{}^*[\uparrow\downarrow]$ KK $\;\sigma_{2s}[\;]\;\sigma_{2s}{}^*[\;]\;\;\sigma_{2p}[\;]\;\pi_{2p}[\;][\;]\;\pi_{2p}{}^*[\;][\;]\sigma_{2p}[\;]$. All of these species are based on O_2, which has $2 \times 6 = 12$ valence electrons. We simply put the appropriate number of valence electrons in each diagram and determine the bond order.

$O_2{}^+$ 11 v.e. KK $\sigma_{2s}[\uparrow\downarrow]\;\sigma_{2s}{}^*[\uparrow\downarrow]\;\sigma_{2p}[\uparrow\downarrow]\;\pi_{2p}[\uparrow\downarrow][\uparrow\downarrow]\;\pi_{2p}{}^*[\uparrow][\;]\;\sigma_{2p}{}^*[\;]$

bond order $=$ (8 bonding e$^-$ -3 antibonding e$^-$) $\div 2 = 2.5$; bond length $= 112$ pm

O_2 12 v.e. KK $\sigma_{2s}[\uparrow\downarrow]\;\sigma_{2s}{}^*[\uparrow\downarrow]\;\sigma_{2p}[\uparrow\downarrow]\;\pi_{2p}[\uparrow\downarrow][\uparrow\downarrow]\;\pi_{2p}{}^*[\uparrow][\uparrow]\;\sigma_{2p}{}^*[\;]$

bond order $=$ (8 bonding e$^-$ -4 antibonding e$^-$) $\div 2 = 2.0$; bond length $= 121$ pm

$O_2{}^-$ 13 v.e. KK $\sigma_{2s}[\uparrow\downarrow]\;\sigma_{2s}{}^*[\uparrow\downarrow]\;\sigma_{2p}[\uparrow\downarrow]\;\pi_{2p}[\uparrow\downarrow][\uparrow\downarrow]\;\pi_{2p}{}^*[\uparrow\downarrow][\uparrow]\;\sigma_{2p}{}^*[\;]$

bond order $=$ (8 bonding e$^-$ -5 antibonding e$^-$) $\div 2 = 1.5$; bond length $= 128$ pm

$O_2{}^{2-}$ 14 v.e. KK $\sigma_{2s}[\uparrow\downarrow]\;\sigma_{2s}{}^*[\uparrow\downarrow]\;\sigma_{2p}[\uparrow\downarrow]\;\pi_{2p}[\uparrow\downarrow][\uparrow\downarrow]\;\pi_{2p}{}^*[\uparrow\downarrow][\uparrow\downarrow]\;\sigma_{2p}{}^*[\;]$

bond order $=$ (8 bonding electrons $-$ 6 antibonding electrons) $\div 2 = 1.0$; bond length $= 149$ pm
We see that the bond length does indeed increase as the bond order decreases. Longer bonds are weaker bonds.

7A There are 8 valence electrons that must be placed in the molecular orbital diagram for CN^+ (5 electrons from nitrogen, four electrons from carbon and one electron is removed to produce the positive charge). Since both C and N precede oxygen and they are not far apart in atomic number, we must use the modified molecular-orbital energy-level diagram to get the correct configuration. By following the Aufbau orbital filling method, one obtains the ground state diagram asked for in the question.

$$CN^+ \quad \overset{\sigma_{2s}}{[\uparrow\downarrow]}\;\overset{\sigma^*_{2s}}{[\uparrow\downarrow]}\quad \overset{\pi_{2p}}{[\uparrow\downarrow][\uparrow\downarrow]}\quad \overset{\sigma_{2p}}{[\;]}\;\overset{\pi^*_{2p}}{[\;][\;]}\;\overset{\sigma^*_{2p}}{[\;]}$$

The bond order for the C—N bond in CN^+ is $\dfrac{6\text{ bonding e}^- - 2\text{ antibonding e}^-}{2} = 2$

Thus the C and N atoms in CN^+ are joined by a double bond.

7B There are 8 valence electrons that must be placed in the molecular orbital diagram for BN (3 electrons from boron and five electrons from nitrogen). Since both B and N precede oxygen and they are not far apart in atomic numbers, we must use the modified molecular-orbital energy-level diagram to get the correct configuration. By following the Aufbau orbital filling method, one obtains the ground-state diagram asked for in the question.

$$BN \quad \overset{\sigma_{2s}}{[\uparrow\downarrow]}\;\overset{\sigma^*_{2s}}{[\uparrow\downarrow]}\quad \overset{\pi_{2p}}{[\uparrow\downarrow][\uparrow\downarrow]}\quad \overset{\sigma_{2p}}{[\;]}\;\overset{\pi^*_{2p}}{[\;][\;]}\;\overset{\sigma^*_{2p}}{[\;]}$$

The bond order for the B—N bond in BN is $\dfrac{6\text{ bonding e}^- - 2\text{ antibonding e}^-}{2} = 2$

Thus, the B and N atoms in BN are joined by a double bond.

8A   In this exercise we will combine valence-bond and molecular orbital methods to describe the bonding in the SO_3 molecule. By invoking a π-bonding scheme, we can replace the three resonance structures for SO_3 (shown below) with just <u>one</u> structure that exhibits both σ-bonding and delocalized π-bonding.

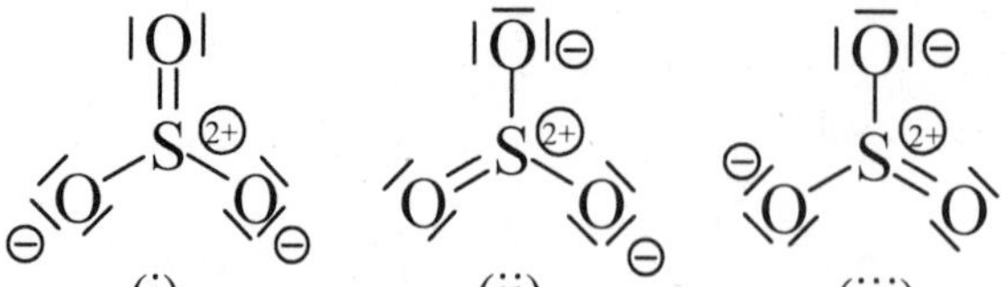

We will use structure (i) to develop a combined localized/delocalized bonding description for the molecule. (Note: Anyone of the three resonance contributors can be used as the starting structure). We begin by assuming that every atom in the molecule is sp^2 hybridized to produce the σ-framework for the molecule. The half-filled sp^2 hybrid orbitals of the oxygen atoms will each be overlapped with a half-filled sp^2 hybrid orbital on sulfur. By contrast, to generate a set of π molecular orbitals, we will combine one unhybridized $2p$ orbital from each of three oxygen atoms with an unhybridized $3p$ orbital on the sulfur atom. This will generate four π molecular orbitals: a bonding molecular orbital, two nonbonding molecular orbitals and an antibonding orbital. Remember that the number of valence electrons assigned to each atom must reflect the formal charge for that atom. Accordingly, sulfur, with a 2+ formal charge, can have only four valence electrons, the two oxygens with a 1- formal charge must each end up with 7 electrons and the oxygen with a zero formal charge must have its customary six electrons

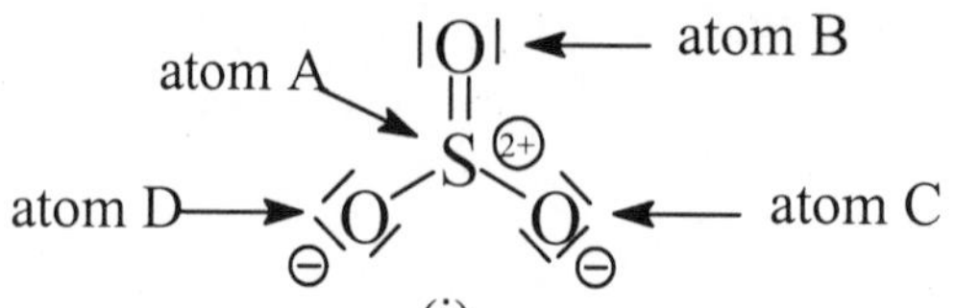

Let's begin assigning valence electrons by half-filling three sp^2 hybrid orbitals on the sulfur atom (atom A) and one sp^2 hybrid orbital on each of the oxygen atoms (atoms B, C and D)

Next, we half-fill the lone unhybridized $3p$ orbital on sulfur and the lone $2p$ orbital on the oxygen atom with a formal charge of zero (atom B). Following this, the $2p$ orbital of the other two oxygen atoms (atoms C and D) , are filled and then lone pairs are placed in the sp^2 hybrid orbitals that are still empty. At this stage, then, all 24 valence electrons have been put into atomic and hybrid orbitals on the four atoms. Now we overlap the six half-filled sp^2 hybrid orbitals to generate the σ-bond framework and combine the three $2p$ orbitals (2 filled, one half-filled) and the $3p$ orbital (half-filled) to form the four π-molecular orbitals, as shown below:

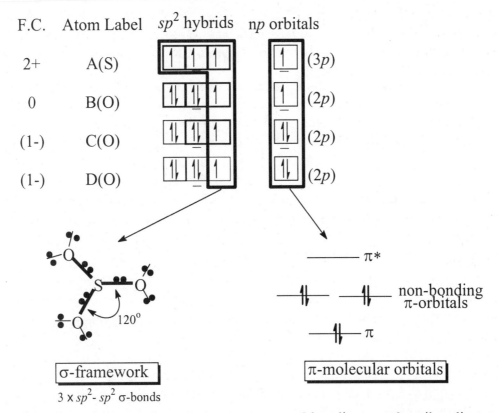

Overall bond order for this set of π-molecular orbitals $\dfrac{2 \text{ bonding e}^- - 0 \text{ antibonding e}^-}{2} = 1$

The π-bond is spread out evenly over the three S—O linkages. This leads to an average bond order of 1.33 for the three S—O bonds in SO_3. By following this *"combined approach"*, we end up with a structure that has the σ-bond framework sandwiched between the delocalized π-molecular orbital framework:

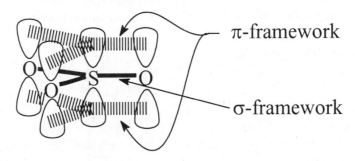

This is a much more accurate description of the bonding in SO_3 than that provided by any one of the three Lewis diagrams given above.

8B We will use the same basic approach to answer this question as was used to solve practice example 12-8A. This time, the bonding in NO_2^- will be described by combining valence-bond and molecular orbital theory. With this approach we will be able to generate a structure that more accurately describes the bonding in NO_2^- than either of the two equivalent Lewis diagrams that can be drawn for the nitrate ion (below).

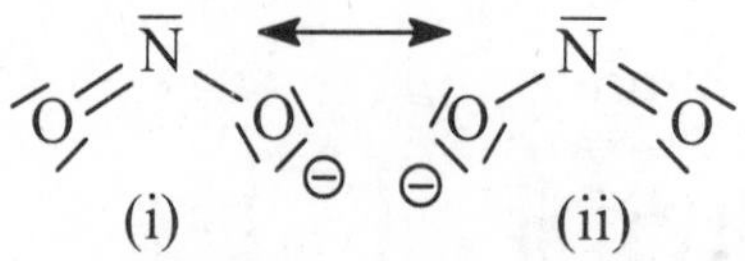

$$
\overline{O}=\overline{N} \longleftrightarrow \overline{O}-\overline{N}=\overline{O}
$$

(i)   (ii)

Structure (i) will be used to develop the combined localized/delocalized bonding description for the anion (either structure could have been used as the starting structure). We begin by assuming that every atom in the molecule is sp^2 hybridized. To produce the σ-framework for the molecule, the half-filled sp^2 orbitals of the oxygen atoms will be overlapped with a half-filled sp^2 orbital of nitrogen. By contrast, to generate a set of π-molecular orbitals, we will combine one unhybridized $2p$ orbital from each of the two oxygen atoms with an unhybridized $2p$ orbital on nitrogen. This will generate three π molecular orbitals: a bonding molecular orbital, a non-bonding molecular orbital and an antibonding molecular orbital.

atom A ⟶ $\overline{N}$

atom B ⟶ O O ⟵ atom C

(i)

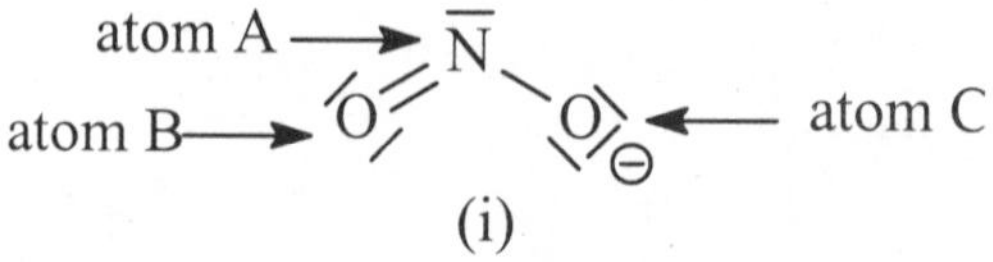

Remember that the number of valence electrons assigned to each atom must reflect the formal charge for that atom. Accordingly, the oxygen with a 1- formal charge (atom C) must end up with 7 electrons, whereas the nitrogen atom (atom A) and the other oxygen atom (atom B) must have their customary 5 and 6 valence electrons, respectively, because they both have a formal charge of zero. Let's begin assigning valence electrons by half-filling two sp^2 hybrid orbitals on the nitrogen atom (atom A) and one sp^2 hybrid orbital on each oxygen atoms (atoms B and C). Next, we half-fill the lone $2p$ orbital on nitrogen (atom A) and the lone $2p$ orbital on the oxygen atom with a formal charge of zero (atom B) Following this, the $2p$ orbital for the remaining oxygen (atom C) is filled and then lone pairs are placed in the sp^2 hybrid orbitals that are still empty. At this stage, then, all 18 valence electrons have been put into atomic and hybrid orbitals on the three atoms. Now, we overlap the four half-filled sp^2 hybrid orbitals to generate the σ-bond framework and combine the three $2p$ orbitals (two half-filled, one filled) to form three π-molecular orbitals as shown below

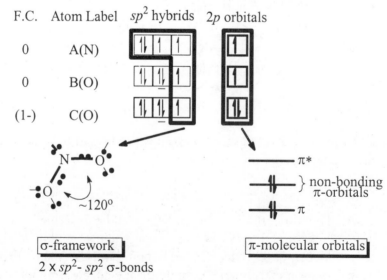

F.C.	Atom Label	sp^2 hybrids	$2p$ orbitals
0	A(N)		
0	B(O)		
(1-)	C(O)		

σ-framework

$2 \times sp^2 - sp^2$ σ-bonds

π-molecular orbitals

Overall bond order for this set of π-molecular orbitals $\dfrac{2\,\text{bonding e}^- - 0\,\text{antibonding e}^-}{2} = 1$

The π bond is spread out evenly over the two N—O linkages. This leads to an average bond order of 1.5 for each of the two N—O bonds in NO_2^-. By following this combined approach, we end up with a structure that has the σ-bond framework sandwiched between the delocalized π-molecular orbitals:

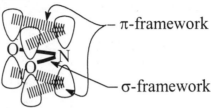

— π-framework

— σ-framework

This is a much more accurate depiction of the bonding in NO_2^- than that provided by any one of the two Lewis diagrams given above.

REVIEW QUESTIONS

Note: In VSEPR theory the term *"bond pair"* is used for a single bond, a double bond, or a triple bond, even though a single bond consists of one pair of electrons, a double bond two pairs of electrons, and a triple bond three pairs of electrons. To avoid any confusion between the number of electron pairs actually involved in the bonding to a central atom, and the number of atoms bonded to that central atom, we shall occasionally use the term *"ligand"* to indicate an atom or a group of atoms attached to the central atom.

1. (a) An sp^2 hybrid orbital is one of a set of three orbitals that are identical in shape and energy and are formed by the combination of an s orbital and two p orbitals. The three orbitals are trigonal planar in orientation.

   (b) σ^*_{2p} is the symbol for the antibonding orbital that arises from the end on overlap of two atomic 2*p* orbitals that are out of phase.

   (c) The bond order is an indication of bond strength. In valence bond theory, it indicates the number of electron pairs shared by two atoms. In molecular orbital theory it is one half of the difference between the number of bonding and the number of antibonding electrons.

   (d) A π bond is a bond that has regions of overlap above and below the internuclear axis.

2. (a) Hybridization of atomic orbitals is that process in which atomic orbitals are blended together to produce new hybrid.

   (b) The σ-bond framework consists of the atoms in the molecule and the single bonds that connect them, before the second (and third) bonds of multiple bonds are in place.

   (c) The Kekulé structures of benzene are the two resonance forms of benzene in which the alternating single and double carbon-to-carbon bonds exchange positions.

   (d) The band theory of metallic bonding describes the bonding between metal atoms in terms of molecular orbitals that extend throughout the crystal.

3.   **(a)**   A σ bond is located along the internuclear axis and is the first bond formed between two atoms. A π bond is located above and below the internuclear axis and only forms after a σ bond has formed.

     (b)   Localized electrons are those confined either to the region around one atom or to the region between two atoms. Delocalized electrons are free to move among three or more atoms.

     (c)   The constructive overlap of two atomic orbitals on two separate atoms produces a bonding molecular orbital. Electrons in bonding molecular orbitals tend to be localized in the region between the two nuclei. The mutual attraction between the positively charged nuclei and the negatively charged electrons hold the nuclei in close proximity. Because the electrons are attracted to <u>two</u> nuclei rather than just a single nucleus, electrons in bonding molecular orbitals are more stable (i.e., have lower energy) than electrons in the parent atomic orbitals. The destructive overlap of two atomic orbitals on two atoms produces an antibonding molecular orbital. In an antibonding molecular orbital, the charge density between the two nuclei decreases to zero. The vast majority of the electron density in an antibonding orbital is concentrated outside the internuclear region at opposite ends of the overlapping atoms. With very little electron density between them, the positively charged nuclei repel each other and consequently, a bond is not formed. Because an electron in an antibonding orbital is, on average, further away from its associated nucleus than an electron in one of the atoms parent atomic orbitals, it is less stable (higher in energy).

     (d)   A metal is a material that readily conducts electricity (due to a small or non-existent band gap). A semiconductor conducts electricity only once it is heated or has a moderate voltage applied to it (due to a moderate band gap).

<u>4.</u>   The Lewis structure of H_2Se, $H—\overline{Se}—H$, indicates that there are two atoms and two lone pairs attached to the central atom. The bond angle of $91°$ indicates that the $1s$ orbitals on H overlap the $4p$ orbitals on Se. This would give an undistorted bond angle of $90°$. But repulsions between the H's open up the bond angle a slight bit.

<u>5.</u>   Determining hybridization is made easier if we begin with Lewis structures. Only one resonance form is drawn for $CO_3{}^{2-}$, SO_2 and $NO_2{}^-$.

$$\left[\overline{|O}{-}C{=}\overline{O}\atop{|\underline{O}|}\right]^{2-}\qquad \overline{|O}{-}\overline{S}{=}\overline{O}\qquad |\overline{Cl}{-}\underset{|\overline{Cl}|}{\overset{|\overline{Cl}|}{C}}{-}\overline{Cl}|\qquad |C{\equiv}O|\qquad [\overline{|O}{-}\overline{N}{=}\overline{O}]^-$$

The C atom is attached to three ligands and no lone pairs and thus is sp^2 hybridized in $CO_3{}^{2-}$. The S atom is attached to two ligands and one lone pair and thus is sp^2 hybridized in SO_2. The C atom is attached to four ligands and no lone pairs and thus is sp^3 hybridized in CCl_4. sp hybridization occurs in CO. The N atom is attached to two ligands and one lone pair in $NO_2{}^-$ and thus is sp^2 hybridized. The central atom is sp^2 hybridized in SO_2, $CO_3{}^{2-}$, and $NO_2{}^-$.

6. Statement (b) is correct. The only consistently single bonds in C—H—O compounds are bonds to H atoms (a H atom can only form one bond), and single bonds must be σ bonds. Thus, both C—H and O—H bonds must be σ bonds, making (a) false and (b) true. Only C=C bonds (or C=O) bonds in these compounds, that is double bonds, consist of a σ bond and a π bond; (c) is false. A σ bond must form before a π bond forms to complete a double bond; (d) is false.

7. **(a)** HI: H $_{1s}$ ☐   I [Kr]$4d^{10}$ $_{5s}$ ☐ $_{5p}$ ☐☐☐

The $1s$ orbital of H overlaps the half-filled $5p$ orbital of I to produce a linear molecule.

(b) BrCl: Br [Ar]$3d^{10}$ $_{4s}$ ☐ $_{4p}$ ☐☐☐   Cl [Ne]$_{3s}$ ☐ $_{3p}$ ☐☐☐

The half-filled $3p$ orbital of Cl overlaps the half-filled $4p$ orbital of Br to produce a linear molecule.

(c) H_2Se: H $_{1s}$ ☐   Se [Ar]$3d^{10}$ $_{4s}$ ☐ $_{4p}$ ☐☐☐

Each half-filled $4p$ orbital of Se overlaps with a half-filled $1s$ orbital of H to produce a bent molecule, with a bond angle of approximately $90°$.

(d) OCl_2: O $_{1s}$ ☐ $_{2s}$ ☐ $_{2p}$ ☐☐☐   Cl [Ne] $_{3s}$ ☐ $_{3p}$ ☐☐☐

Each half-filled $2p$ orbital of O overlaps with a half-filled $3p$ orbital of Cl to produce a bent molecule, with a bond angle of approximately $90°$.

8. The BF_3 molecule is planar with bond angles of $120°$. The $2p$ orbitals are perpendicular to each other, with angles between them of $90°$. Thus, three atoms bound to three $2p$ orbitals would not form a planar structure and would have the wrong bond angles. If two F atoms bonded to two $2p$ orbitals on B and the third F bonded to the $2s$ orbital, the resulting molecule could indeed be planar, but the bond angles would not equal $120°$, and two of the bonds would be different from the other one. By hybridizing the $2s$ and two $2p$ orbitals of B, we create three equivalent orbitals that produce the observed bond angles.

9. **(a)** The central C atom employs four sp^3 hybrid orbitals; these have a tetrahedral geometry. Each of the two H atoms employs a $1s$ orbital, and each of the two Cl atoms employs a half-filled $3p$ orbital. A diagram of this follows:

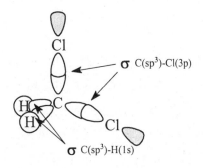

(b)   The ground state C atom has the electron configuration [He] $_{2s}$⊞ $_{2p}$⊞ Based on this ground state electron configuration, we would expect a bent molecule. To produce the two half-filled orbitals required to form two linear bonds a hybridized electron configuration (which corresponds to an excited state for the individual atoms) is required: C* $_{1s}$⊞ $_{2s}$⊞ $_{2p}$⊞ ⟶ C*$_{hyb}$ $_{1s}$⊞ $_{sp}$⊞ $_{2p}$⊞

The overlap of the two half-filled sp orbitals of C, one with a half-filled $2p$ orbital on O and the other with a half-filled $2p$ orbital on N⁻, produces the σ bond framework. O $_{1s}$⊞ $_{2s}$⊞ $_{2p}$⊞   N⁻ $_{1s}$⊞ $_{2s}$⊞ $_{2p}$⊞

The π bonds are produced by the sidewise overlap of the two half-filled $2p$ orbitals on C, one with a half-filled $2p$ orbital on O, the other with a half-filled $2p$ orbital on N⁻. A diagram of this follows:

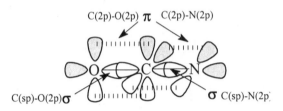

(c)   The ground-state electron configuration of B suggests an ability to form one B—F bond, rather than three. $_{1s}$⊞ $_{2s}$⊞ $_{2p}$⊞ Again, excitation, or promotion of an electron to a higher energy orbital, followed by hybridization is required to form three equivalent half-filled B orbitals.

B* $_{1s}$⊞ $_{2s}$⊞ $_{2p}$⊞ ⟶ B*$_{hyb}$ $_{1s}$⊞ $_{sp^2}$⊞ $_{2p}$☐

Recall that the ground-state electron configuration of F is [He] $_{2s}$⊞ $_{2p}$⊞ . Bonds are produced through overlap of the half-filled sp^2 orbitals on B with the half filled F $2p$ orbitals. The BF_3 molecule is trigonal planar, with F—B—F bond angles of 120° A diagram of this follows:

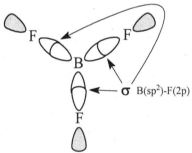

10.   **(a)**   In PF_6^- there are a total of $5+(6\times7)+1=48$ valence electrons, or 24 pairs. A plausible Lewis structure follows. In order to form the six P—F bonds, the hybridization on P must be sp^3d^2.

(b)   In COS there are a total of $4+6+6=16$ valence electrons, or 8 pairs. A plausible Lewis structure follows. In order to bond two atoms to the central C, the hybridization on that C atom must be sp.

(c)   In $SiCl_4$, there are a total of $4+(4\times7)=32$ valence electrons, or 16 pairs. A plausible Lewis structure follows. In order to form four Si—Cl bonds, the hybridization on Si must be sp^3.

(d)   In NO_3^-, there are a total of $5+(3\times6)+1=24$ valence electrons, or 12 pairs. A plausible Lewis structure follows. In order to bond three O atoms to the central N atom, with no lone pairs on that N atom, its hybridization must be sp^2.

(e)   In AsF_5, there are a total of $5+(5\times7)=40$ valence electrons, or 20 pairs. A plausible Lewis structure follows. This is a molecule of the type AX_5. To form five As—F bonds, the hybridization on As must be sp^3d.

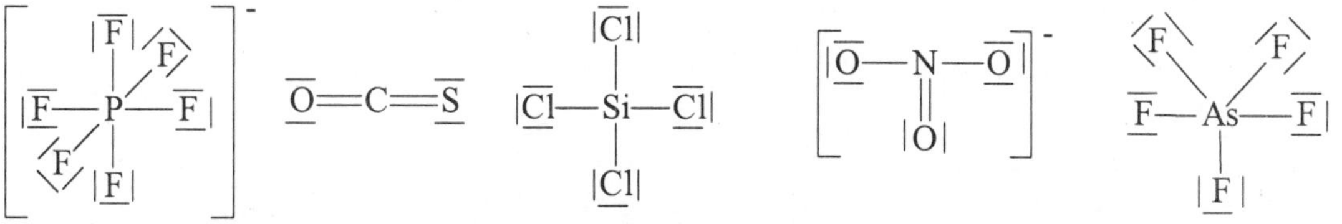

11.   We base each hybridization scheme on the Lewis structure for the molecule.

(a)   H—C≡N| There are two atoms bonded to C and no lone pairs on C. Thus, the hybridization for C must be sp.

(b)   There are four atoms bonded to C and no lone pairs in this molecule. Thus, the hybridization for C must be sp^3. (O atom is also sp^3 hybridized).

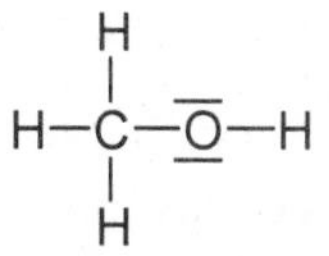

(c)   There are four atoms bonded to each terminal C and no lone pairs on these C atoms. Thus, the hybridization for each terminal C must be sp^3. There are three atoms and no lone pairs bonded to the central C. The hybridization for the central C must be sp^2.

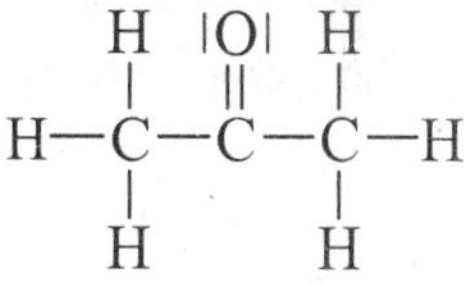

(d)   There are three atoms bonded to C and no lone pairs on C. This requires sp^2 hybridization.

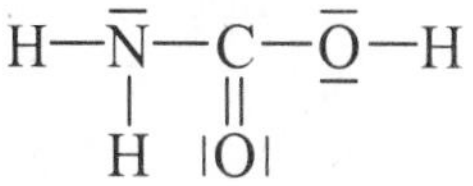

12.   (a)   This is a planar molecule. The hybridization on C is sp^2 (one bond to each of the three attached atoms).

(b) | N≡C—C≡N | is a linear molecule. The hybridization for each C is sp (one bond to each of the two ligands).

(c)   is neither linear nor planar. The shape around the left-hand C is tetrahedral and that C has sp^3 hybridization. The shape around the right-hand carbon is linear and that C has sp hybridization. (N atom is sp hybridized).

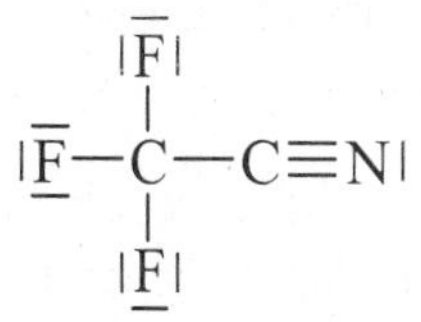

(d) $[|\overline{S}-C≡N|]^-$ is a linear molecule. The hybridization for C is sp.

13. **(a)** In HCN, there are a total of $1+4+5=10$ valence electrons, or 5 pairs. A plausible Lewis structure follows. H—C≡N|. The H—C bond is a σ bond, and the C≡N bond is composed of 1 σ and 2 π bonds.

(b) In C_2N_2, there are a total of $(2 \times 4)+(2 \times 5)=18$ valence electrons, or 9 pairs. A plausible Lewis structure follows: |N≡C—C≡N| The C—C bond is a σ bond, and each C≡N bond is composed of 1 σ and 2 π bonds.

(c) In $CH_3CHCHCCl_3$, there are a total of 42 valence electrons, or 21 pairs. A plausible Lewis structure is shown to the right. All bonds are σ bonds except one of the bonds that comprise the C=C bond. The C=C bond is composed of one σ and one π bond.

$$
\begin{array}{cccc}
\text{H} & & & |\overline{\text{Cl}}| \\
| & & & | \\
\text{H}-\text{C}-\text{C}=\text{C}-\text{C}-\overline{\text{Cl}}| \\
| & | & | & | \\
\text{H} & \text{H} & \text{H} & |\overline{\text{Cl}}|
\end{array}
$$

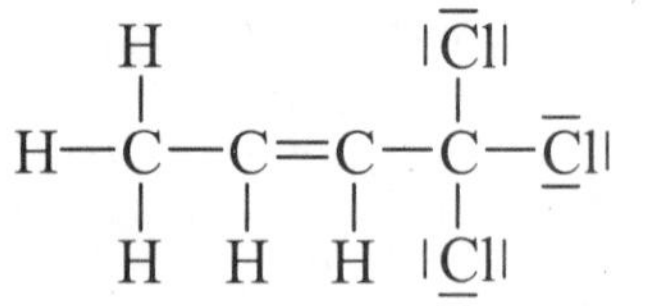

(d) In HONO, there is a total of $1+5+(2 \times 6)=18$ valence electrons, or 9 pairs. A plausible Lewis structure is H—$\overline{\text{O}}$—$\overline{\text{N}}$=$\overline{\text{O}}$ All single bonds in this structure are σ bonds. The double bond is composed of one σ and one π bond.

14. **(a)** In CO_2, there are $4+(2 \times 6)=16$ valence electrons, or 8 pairs. A plausible Lewis structure is |O=C=O|

(b) The molecule is linear and the hybridization on the C atom is sp. Each C=O bond is composed of a σ bond (end-to-end overlap of a half-filled sp orbital on C with a half-filled $2p$ orbital on O) and a π bond (overlap of a half-filled $2p$ orbital on O with a half-filled $2p$ orbital on C).

15. In CH_3NCO, there is a total of $(3 \times 1)+(2 \times 4)+5+6=22$ valence electrons, or 11 pairs. A plausible Lewis structure is

$$
\begin{array}{c}
\text{H} \\
| \\
\text{H}-\text{C}-\overline{\text{N}}=\text{C}=\overline{\text{O}} \\
| \\
\text{H}
\end{array}
$$

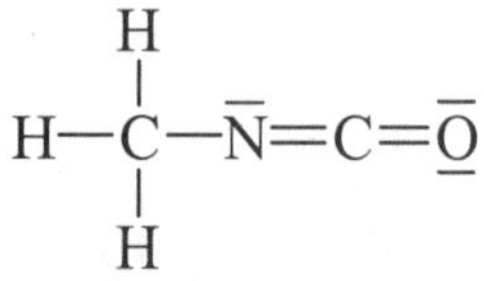

(a) Of course, all single bonds are σ bonds, and a double bond is composed of a σ bond and a π bond. Thus, each bond in the molecule consists of at least a σ bond. There are six (6) σ bonds in this molecule.

(b) Since each double bond consists of a σ bond and a π bond, there are two (2) π bonds in this molecule.

16. **(a)** A σ_{1s} orbital must be lower in energy than a σ_{1s}^* orbital. The bonding orbital always is lower than the antibonding orbital if they are derived from the same atomic orbitals.

(b) A σ_{2s} orbital should be lower in energy than a σ_{2p} orbital, since the $2s$ atomic orbital is lower in energy than the $2p$ atomic orbital, the orbitals from which the molecular orbitals, respectively, are derived.

(c)   A σ_{1s}^* orbital should be lower than a σ_{2s} orbital, since the $1s$ atomic orbital is considerably lower in energy than is the $2s$ orbital.

(d)   A σ_{2p} orbital should be lower in energy than a σ_{2p}^* orbital. Both orbitals are from atomic orbitals in the same subshell but we expect a bonding orbital to be more stable than an antibonding orbital.

17.   Based on Lewis theory, one would expect C_2 to have the greater bond energy due to the formation of quadruple bond, $C\equiv C$ vs. $Li—Li$. The molecular orbital diagrams and bond orders are as follows.

C_2 KK σ_{2s}[↑↓] σ_{2s}^*[↑↓] π_{2p}[↑↓|↑↓] σ_{2p}[ ]   bond order $= (6\text{ bonding } e^- - 2\text{ antibonding } e^-) \div 2 = 2$

Li_2 KK σ_{2s}[↑↓] σ_{2s}^*[ ] π_{2p}[|] σ_{2p}[ ]   bond order $= (2\text{ bonding } e^- - 0\text{ antibonding } e^-) \div 2 = 1$

Thus, molecular orbital theory also predicts that C_2 has a greater bond energy than does Li_2.

18.   **(a)**   In each case, the number of valence electrons in the species is determined first; this is followed by the molecular orbital diagram for each species.

	KK	σ_{2s}	σ_{2s}^*	π_{2p}	σ_{2p}	π_{2p}^*	σ_{2p}^*
C_2^+ no. valence $e^- = (2\times4)-1=7$	KK	[↑↓]	[↑↓]	[↑↓|↑]	[ ]	[	]	[ ]

	KK	σ_{2s}	σ_{2s}^*	σ_{2p}	π_{2p}	π_{2p}^*	σ_{2p}^*
O_2^- no. valence $e^- = (2\times6)+1=13$	KK	[↑↓]	[↑↓]	[↑↓]	[↑↓|↑↓]	[↑↓|↑]	[ ]
F_2^+ no. valence $e^- = (2\times7)-1=13$	KK	[↑↓]	[↑↓]	[↑↓]	[↑↓|↑↓]	[↑↓|↑]	[ ]
NO^+ no. valence $e^- = 5+6-1=10$	KK	[↑↓]	[↑↓]	[↑↓]	[↑↓|↑↓]	[	]	[ ]

(b)   Bond order = (no. bonding electrons – no. antibonding electrons) ÷ 2

C_2^+ bond order $= (5-2) \div 2 = 1.5$ This species is stable.

O_2^- bond order $= (8-5) \div 2 = 1.5$ This species is stable.

F_2^+ bond order $= (8-5) \div 2 = 1.5$ This species is stable.

NO^+ bond order $= (8-2) \div 2 = 3$ This species is stable.

(c) C_2^+ has an odd number of electrons and is paramagnetic, with one unpaired electron. O_2^- has an odd number of electrons and is paramagnetic, with one unpaired electron. F_2^+ has an odd number of electrons and is paramagnetic, with one unpaired electron. NO^+ has an even number of electrons and is diamagnetic.

19.   **(a)**   stainless steel - electrical conductor   **(b)**   solid NaCl - insulator
         (c)   sulfur – insulator   **(d)**   germanium - semiconductor
         (e)   seawater - electrical conductor   **(f)**   solid iodine - insulator

20.   The largest band gap between valence and conduction band is in an insulator. There is no energy gap between the valence and conduction bands in a metal. Metalloids are semiconductors; they have a small band gap between the valence and conduction bands.

EXERCISES

Valence-Bond Theory

21. There are several ways in which valence bond theory is superior to Lewis structures in describing covalent bonds. *First*, valence bond theory clearly distinguishes between sigma and pi bonds. In Lewis theory, a double bond appears to be just two bonds and it is not clear why a double bond is not simply twice as strong as a single bond. In valence bond theory, it is clear that a sigma bond must be stronger than a pi bond, for the orbitals overlap more effectively in a sigma bond (end-to-end) than they do in a pi bond (side-to-side). *Second*, molecular geometries are more directly obtained in valence bond theory than in Lewis theory. Although valence bond theory requires the introduction of hybridization to explain these geometries, Lewis theory does not predict geometries at all; it simply provides the basis from which VSEPR theory predicts geometries. *Third*, Lewis theory does not explain hindered rotation about double bonds. With valence bond theory, any rotation about a double results in poorer overlap of the *p* orbitals that produce the pi bond.

22. The overlap of pure atomic orbitals gives bond angles of $90°$. These bond angles are suitable only for 3- and 4-atom compounds in which the central atom is an atom of the third (or higher) period of the periodic table. For central atoms of the second period of the periodic table, 3- and 4-atom compounds have bond angles closer to $180°$, $120°$, and $109.5°$ than to $90°$. These other bond angles can only be explained well through hybridization. *Second*, hybridization clearly distinguishes between the (hybrid) orbitals that form σ bonds and the *p* orbitals that form π bonds. It places those *p* orbitals in their proper orientation so that they can overlap side-to-side to form π bonds. *Third*, the bond angles of $90°$ and $120°$ that result when the octet of the central atom is expanded cannot be produced with pure atomic orbitals. Hybrid orbitals are necessary. *Finally*, the overlap of pure atomic orbitals does not usually result in all σ bonds being equivalent. Overlaps with hybrid orbitals produce equivalent bonds.

23. (a) Lewis theory does not describe the shape of the water molecule. It does indicate that there is a single bond between each H atom and the O atom, and that there are two lone pairs attached to the O atom, but it says nothing about molecular shape.

(b) In valence-bond theory using simple atomic orbitals, each H—O bond results from the overlap of a $1s$ orbital on H with a $2p$ orbital on O. The angle between $2p$ orbitals is 90° this method initially predicts a $90°$ bond angle. The observed $104°$ bond angle is explained as arising from repulsion between the two slightly positive H atoms.

(c) In VSEPR theory the H_2O molecule is categorized as being of the AX_2E_2 type, with two atoms and two lone pairs attached to the central oxygen atom. The lone pairs repel each other more than do the bond pairs, explaining the smaller than $109.5°$ tetrahedral bond angle.

(d) In valence-bond theory using hybrid orbitals, each H—O bond results from the overlap of a $1s$ orbital on H with a sp^3 orbital on O. The angle between sp^3 orbitals is $109.5°$. The observed bond angle of $104°$ is rationalized based on the greater repulsion of lone pair electrons when compared to bonding pair electrons.

24. (a) Lewis theory really does not describe the shape of the molecule. It does indicate that there is a single bond between each Cl atom and the C atom, but it says nothing about molecular shape.

(b) In valence bond theory using simple atomic orbitals, each C—Cl bond results from the overlap of a $3p$ orbital on Cl with a $2p$ orbital on C. Since the angle between 2p orbitals is $90°$, this method initially predicts a $90°$ bond angle. The observed $109.5°$ bond angle is explained as resulting from the repulsion between the two slightly negative Cl atoms. In addition, the molecule is predicted to have the formula CCl_2, since there are just two half-filled orbitals in the ground state of C.

(c) In VSEPR theory the CCl_4 molecule is categorized as being of the AX_4 type, with four atoms tetrahedrally attached to the central carbon atom.

(d) In valence bond theory, each C—Cl bond results from the overlap of a $3p$ orbital on Cl with a sp^3 orbital on C. The angle between the sp^3 orbitals is $109.5°$.

25. For each species, we first draw the Lewis structure, as an aid to explaining the bonding.

(a) In CO_2, there are a total of $4+(2\times6)=16$ valence electrons, or 8 pairs. C is the central atom. The Lewis structure is $\overline{O}=C=\overline{O}$ The molecule is linear and C is sp hybridized.

(b) In $HONO_2$, there are a total of $1+5+(3\times6)=24$ valence electrons, or 12 pairs. N is the central atom, and a plausible Lewis structure is shown on the right The molecule is trigonal planar around N which is sp^2 hybridized. The O in the H—O—N portion of the molecule is sp^3 hybridized.

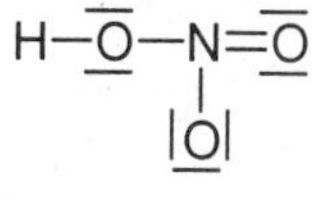

$$H-\overline{O}-N=\overline{\overline{O}}$$
$$\overset{|}{\underset{|}{|\underline{O}|}}$$

(c) In ClO_3^-, there are a total of $7+(3\times6)+1=26$ valence electrons, or 13 pairs. Cl is the central atom, and a plausible Lewis structure is shown on the right The electron-group geometry around Cl is tetrahedral, indicating that Cl is sp^3 hybridized.

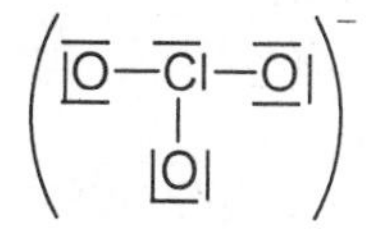

$$\left(\overline{|O}-\overset{|}{\underset{\underset{|\underline{O}|}{|}}{Cl}}-\overline{O|}\right)^-$$

(d) In BF_4^-, there are a total of $3+(4\times7)+1=32$ valence electrons, or 16 pairs. B is the central atom, and a plausible Lewis structure is shown on the right. The electron-group geometry is tetrahedral, indicating that B is sp^3 hybridized.

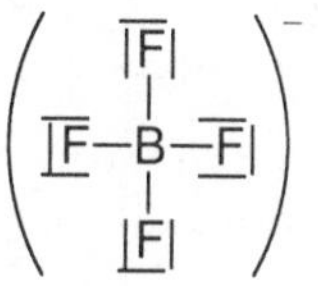

$$\left(\overset{\overline{|F|}}{\underset{|\underline{F}|}{\overline{|F}-\overset{|}{B}-\overline{F|}}}\right)^-$$

26. In NSF there are $5+6+7=18$ valence electrons, or 9 pairs. In the following Lewis structure, $|\overline{N}{=}\overline{S}{-}\underline{F}|$ there are two atoms bonded to the central S, which has a lone pair. Thus, the electron group geometry around the central S is trigonal planar (with sp^2 hybridization) and the molecular shape is bent.

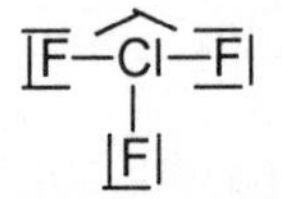

27. The Lewis structure of ClF_3 is shown on the right. There are three atoms and two lone pairs attached to the central atom, its hybridization is sp^3d, which is achieved as follows. $Cl_{unhyb}[Ne]_{3s}$ $\boxed{\uparrow\downarrow}$ $_{3p}\boxed{\uparrow\downarrow|\uparrow\downarrow|\uparrow}$ $_{3d}\boxed{||||}$ $\longrightarrow$ Cl_{hyb} $[Ne]$ $_{dsp^3}\boxed{\uparrow\downarrow|\uparrow\downarrow|\uparrow|\uparrow|\uparrow}$ $_{3d}\boxed{|||}$ The orbital diagram of F is $[He]_{2s}\boxed{\uparrow\downarrow}$ $_{2p}\boxed{\uparrow\downarrow|\uparrow\downarrow|\uparrow}$ Each of the three sigma bonds are formed by the overlap of a $2p$ orbital on F with one of the half-filled dsp^3 orbitals on Cl. Since the dsp^3 hybridization has a trigonal bipyramidal shape, and the two lone pairs occupy the axial positions in the molecule, ClF_3 is T-shaped.

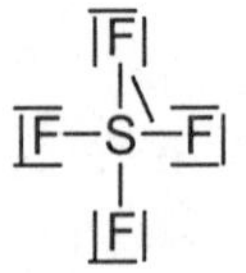

28. In SF_4 there are $6+(4\times7)=34$ valence electrons, or 17 pairs. A plausible Lewis structure has zero formal charge on each atom (structure drawn on right). This molecule has a see-saw shape. There are four atoms and one lone pair attached to S, requiring sp^3d hybridization. Each sigma S—F bond results from the overlap of two half-filled orbitals: sp^3d on S and $2p$ on F. The orbital diagram of F is $[He]_{2s}\boxed{\uparrow\downarrow}$ $_{2p}\boxed{\uparrow\downarrow|\uparrow\downarrow|\uparrow}$. The hybridization diagram for S follows: $S_{unhyb}[Ne]_{3s}\boxed{\uparrow\downarrow}$ $_{3p}\boxed{\uparrow\downarrow|\uparrow|\uparrow}$ $_{3d}\boxed{||||}$ $\longrightarrow$ S_{hyb} $[Ne]$ $_{dsp^3}\boxed{\uparrow\downarrow|\uparrow|\uparrow|\uparrow|\uparrow}$ $_{3d}\boxed{|||}$

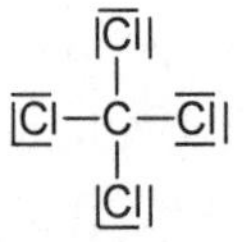

29. (a) In CCl_4, there is a total of $4+(4\times7)=32$ valence electrons, or 8 pairs. C is the central atom. A plausible Lewis structure is shown. The geometry at C is tetrahedral; C is sp^3 hybridized. Cl—C—Cl bond angles are $109.5°$. Each C—Cl bond is represented by $\sigma:$ $C(sp^3)^1$ — $Cl(3p)^1$

(b) In ONCl, there is a total of $6+5+7=18$ valence electrons, or 9 pairs. N is the central atom. A plausible Lewis structure is $\overline{O}=\overline{N}-\overline{C}l|$ The e⁻ group geometry around N is triangular planar, and N is sp^2 hybridized. The O—N—C bond angle is about $120°$. The bonds are: $\sigma:$ $O(2p_y)^1$—$N(sp^2)^1$ $\sigma:$ $N(sp^2)^1$—$Cl(3p_z)^1$ $\pi:$ $O(2p_z)^1$—$N(2p_z)^1$

(c) In HONO, there are a total of $1+(2\times6)+5=18$ valence electrons, or 9 pairs. A plausible Lewis structure is $H-\overline{O}_a-\overline{N}=\overline{O}_b$. The geometry of O_a is tetrahedral, O_a is sp^3 hybridized, the $H-O_a-N$ bond angle is (at least close to) $109.5°$. The e⁻ group geometry at N is trigonal, N is sp^2 hybridized, the O_a-N-O_b bond angle is $120°$ The four bonds are represented as follows. $\sigma:$ $H(1s)^1$ — $O_a(sp^3)^1$ $\sigma:$ $O_a(sp^3)^1$—$N(sp^2)^1$ $\sigma:$ $N(sp^2)^1$—$O_b(2p_y)^1$ $\pi:$ $N(2p_z)^1$—$O_b(2p_z)^1$.

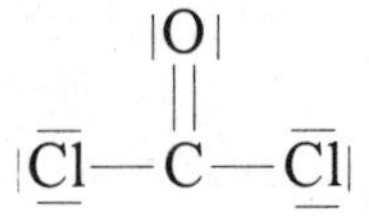

(d) In $COCl_2$, there are a total of $4+6+(2\times7)=24$ valence electrons, or 12 pairs. A plausible Lewis structure is shown to the right. The e⁻ group geometry around C is trigonal planar; all bond angles around C are $120°$, and the hybridization of C is sp^2. The four bonds in the molecule: $2\times\sigma:$ $Cl(3p_x)^1$ —$C(sp^2)^1$ $\sigma:$ $O(2p_y)^1$ —$C(sp^2)^1$ $\pi:$ $O(2p_z)^1$ —$C(2p_z)^1$.

30. **(a)** In NO_2^-, there are a total of $5+(2\times6)+1=18$ valence electrons, or 9 pairs. A plausible Lewis structure follows $\bar{O}_a\!=\!N\!-\!\bar{O}_b|$ The e⁻ group geometry around N is trigonal planar, all bond angles around that atom are $120°$, and the hybridization of N is sp^2. The three bonds in this molecule are:

$$\sigma: O_a\left(2p_y\right)^1-N\left(sp^2\right)^1 \quad \sigma: O_b\left(2p_x\right)^1-N\left(sp^2\right)^1 \text{ and } \pi: O_a\left(2p_z\right)^1-N\left(2p_z\right)^1$$

(b) In I_3^-, there are a total of $(3\times7)+1=22$ valence electrons, or 11 pairs. A plausible Lewis structure follows. $\left[|\bar{I}\!-\!\hat{I}\!-\!\bar{I}|\right]^-$ Since there are three lone pairs and two ligands attached to the central I, the electron-group geometry around that atom is trigonal bipyramidal; its hybridization is sp^3d. Since the two ligand I atoms are located in the axial positions of the trigonal bipyramid, the I—I—I bond angle is $180°$. Each I—I sigma bond is the result of the overlap of a $5p$ orbital on a terminal I with an sp^3d hybrid orbital on the central I.

(c) In $C_2O_4^{2-}$, there are a total of $(2\times4)+(4\times6)+2=34$ valence electrons, or 17 pairs. A plausible Lewis structure is shown on the right. As we see, the ion is symmetrical. There are three atoms or groups of atoms attached to each C, the bond angles around each C are $120°$, and the hybridization on each C is sp^2.

$$\left[\bar{O}_b\!-\!\underset{\|O_a|}{\overset{\|}{C}}\!-\!\underset{\|O_a|}{\overset{\|}{C}}\!-\!\bar{O}_b|\right]^{2-}$$

The bonding to one of these carbons (the right-hand one) is represented as follows.

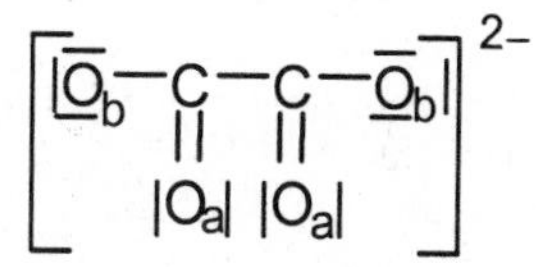

$$\sigma: C\left(sp^2\right)^1-C\left(sp^2\right)^1 \quad \sigma: C\left(sp^2\right)^1-O_a\left(2p_z\right)^1 \quad \sigma: C\left(sp^2\right)^1-O_b\left(2p_x\right)^1$$

$$\pi: C\left(2p_y\right)^1-O_a\left(2p_y\right)^1$$

(d) In HCO_3^-, there are $1+4+(3\times6)+1=24$ valence electrons, or 12 pairs. A plausible Lewis structure is shown on the right. The e⁻ group geometry around C is trigonal planar, with a bond angle of $120°$; the hybridization is sp^2. The geometry around O_a is bent with a bond angle less than $109.5°$; the hybridization is sp^3. The five bonds are represented as follows:

$$\left(H\!-\!\bar{O}_a\!-\!\underset{|O_d}{\overset{\|}{C}}\!-\!\bar{O}_b|\right)^-$$

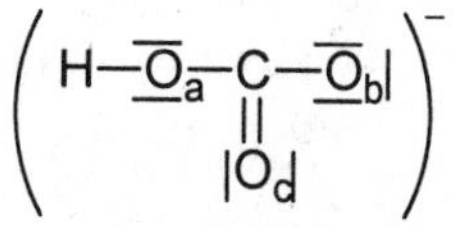

$$\sigma: O_a\left(sp^3\right)^1-H\left(1s\right)^1, \quad \sigma: C\left(sp^2\right)^1-O_a\left(sp^3\right)^1, \quad \sigma: C\left(sp^2\right)^1-O_b\left(2p\right)^1,$$

$$\sigma: C\left(sp^2\right)^1-O_c\left(2p_x\right)^1, \quad \pi: C\left(2p_z\right)^1-O_c\left(2p_z\right)^1$$

31. Citric acid has the molecular structure (shown on right) Using Figure 12-16 as a guide, the flowing hybridization and bonding scheme is obtained for citric acid:

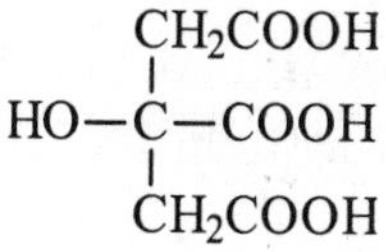

$$CH_2COOH$$
$$HO-\overset{|}{\underset{|}{C}}-COOH$$
$$CH_2COOH$$

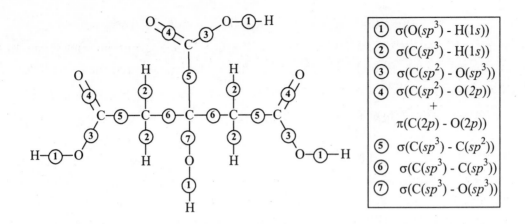

1. $\sigma(O(sp^3) - H(1s))$
2. $\sigma(C(sp^3) - H(1s))$
3. $\sigma(C(sp^2) - O(sp^3))$
4. $\sigma(C(sp^2) - O(2p))$
   $+$
   $\pi(C(2p) - O(2p))$
5. $\sigma(C(sp^3) - C(sp^2))$
6. $\sigma(C(sp^3) - C(sp^3))$
7. $\sigma(C(sp^3) - O(sp^3))$

32. All interior atoms are sp^3 hybridized except C_a and C_d, which are sp^2 hybridized. The overlaps in the sixteen bonds of the molecule are represented as follows.

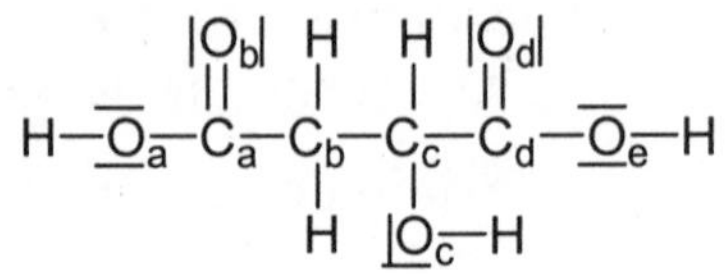

$$\sigma: O_e(sp^3)^1 - H(1s)^1 \quad \sigma: C_d(sp^2)^1 - O_e(sp^3)^1 \quad \sigma: C_a(sp^2)^1 - O_b(2p_x)^1 \quad \pi: C_d - (2p_z)^1 - O_d(2p_z)^1.$$

$$\sigma: O_a(sp^3)^1 - H(1s)^1 \quad \sigma: C_a(sp^2)^1 - O_a(sp^3)^1 \quad \sigma: C_d(sp^2)^1 - O_d(2p_x)^1 \quad \pi: C_a - (2p_z)^1 - O_b(2p_z)^1$$

$$\sigma: C_b(sp^3)^1 - H(1s)^1 \quad \sigma: C_b(sp^3)^1 - H(1s)^1 \quad \sigma: C_a(sp^2)^1 - C_b(sp^3)^1 \quad \sigma: C_b(sp^3)^1 - C_c(sp^3)^1$$

$$\sigma: C_c(sp^3)^1 - H(1s)^1 \quad \sigma: C_c(sp^3)^1 - O_c(sp^3)^1 \quad \sigma: C_c(sp^3)^1 - C_d(sp^2)^1 \quad \sigma: O_c(sp^3)^1 - H(1s)^1$$

33. **(a)**

Br atom is sp^3d hybridized
F atoms are unhybridized

$\sigma(Br(sp^3d) - F(2p))$

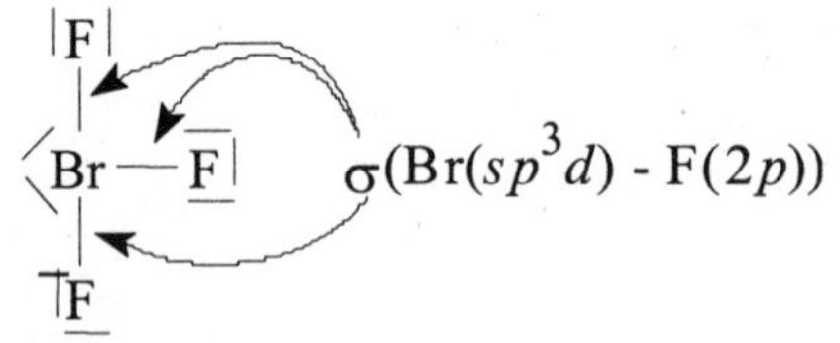

(b)

Central S is sp^2 hybridized and the terrminal O and S atoms are unhybridized

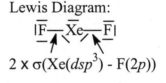

$\sigma(S(sp^2)-S(3p_x))$
$\sigma(S(2p_z)-S(3p_z))$
$\sigma(S(sp^2)-S(2p_z))$

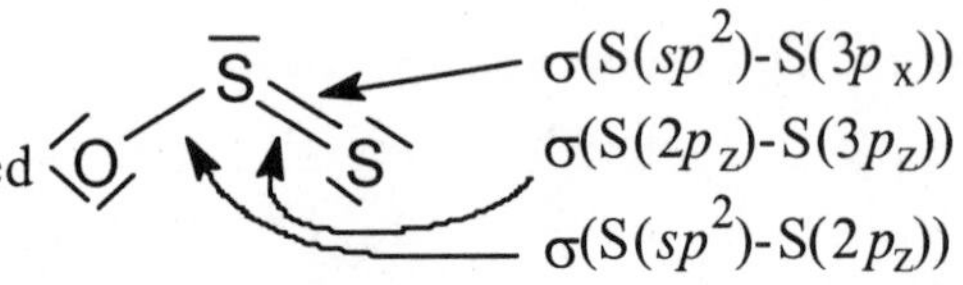

34. Model A is XeF_2
VSEPR class: AX_2E_3
Therefore XeF_2 is a linear molecule
The central Xe atom in XeF_2 is sp^3d or dsp^3

Lewis Diagram:

$|\overline{F} - \overline{X}e - \overline{F}|$

$2 \times \sigma(Xe(dsp^3) - F(2p))$

Model B is IF_6^+
VSEPR class: AX_6
Therefore IF_6^+ is an octahedral molecule
The central I atom in IF_6^+ is sp^3d^2 or d^2sp^3

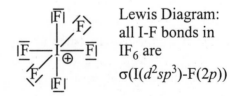

Lewis Diagram:
all I-F bonds in
IF_6 are
$\sigma(I(d^2sp^3)-F(2p))$

35.   The bond lengths are consistent with the left-hand C—C bond being a triple bond (120 pm), the remaining C—C bond being a single bond (154 pm) rather than a double bond (134 pm), and the C—O bond being a double bond (123 pm). Of course, the two C—H bonds are single bonds (110 pm). All of this is depicted in the Lewis structure on the right. The overlap that produces each bond follows:

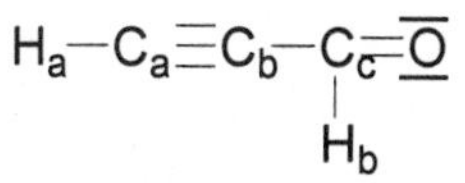

$$\sigma: C_a(sp)^1 - H_a(1s)^1 \qquad \sigma: C_a(sp)^1 - C_b(sp)^1 \qquad \sigma: C_b(sp)^1 - C_c(sp^2)^1$$

$$\sigma: C_c(sp^2)^1 - H_b(1s)^1 \qquad \sigma: C_c(sp^2)^1 - O(2p_y)^1 \qquad \pi: C_a(2p_y)^1 - C_b(2p_y)^1$$

$$\pi: C_a(2p_z)^1 - C_b(2p_z)^1 \qquad \pi: C_c(2p_z)^1 - O(2p_z)^1$$

Lewis structure (upper right):
$$H_a - C_a \equiv C_b - C_c = \overline{\underline{O}}$$
with H_b attached to C_c.

36.   All of the valence electron pairs in the molecule are indicated in the sketch in the problem statement; there are no lone pairs. The two end carbon atoms have three atoms attached to them; they are sp^2 hybridized. This means there is one half-filled $2p$ orbital available on each of these end carbon atoms for the formation of a π bond. The central carbon atom has two atoms attached to it; it is sp hybridized. Thus there are two (perpendicular) half-filled $2p$ orbitals (such as $2p_y$ and $2p_z$) on this central carbon atom available for the formation of π bonds. One π bond is formed from the central carbon atom to each of the terminal carbon atoms. Because the two $2p$ orbitals on the central carbon atom are perpendicular to each other, these two π bonds are also. Thus, the two H—C—H planes are at $90°$ from each other, because of the π bonding in the molecule.

Molecular-Orbital Theory

37.   The valence-bond method describes a covalent bond as the result of the overlap of atomic orbitals. The more complete the overlap, the stronger the bond. Molecular orbital theory describes a bond as a region of space in a molecule where there is a good chance of finding the bonding electrons. The molecular orbital bond does not have to be created from atomic orbitals (although it often is) and the orientations of atomic orbitals do not have to be manipulated to obtain the correct geometric shape. There is little concept of the relative energies of bonding in valence bond theory. In molecular orbital theory, bonds are ordered energetically. These energy orderings, in fact, provide a means of checking the predictions of the theory through the spectroscopic analysis of the molecules.

38. C_2 has $(2 \times 4) = 8$ valence electrons. A plausible Lewis structure, that obeys the octet rule and has zero formal charge for each C atom, incorporates a quadruple bond C::::C or C≡C. Molecular orbital theory, on the other hand, distributes those same 8 electrons as shown below, resulting in a bond order of $(6-2) \div 2 = 2$, that is, a double bond.

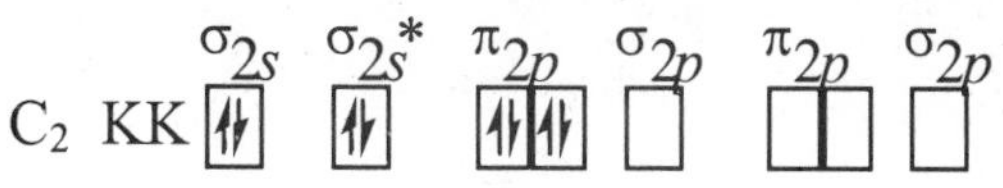

$$\begin{array}{ccccccc} & \sigma_{2s} & \sigma_{2s}{}^* & \pi_{2p} & \sigma_{2p} & \pi_{2p} & \sigma_{2p} \\ C_2 \ KK & \boxed{\uparrow\downarrow} & \boxed{\uparrow\downarrow} & \boxed{\uparrow\downarrow}\,\boxed{\uparrow\downarrow} & \boxed{} & \boxed{}\,\boxed{} & \boxed{} \end{array}$$

39. The two molecular orbital diagrams follow:

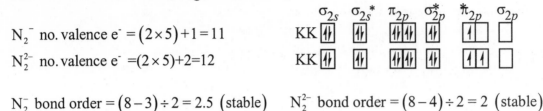

N_2^- no. valence $e^- = (2 \times 5) + 1 = 11$

N_2^{2-} no. valence $e^- = (2 \times 5) + 2 = 12$

N_2^- bond order $= (8 - 3) \div 2 = 2.5$ (stable) N_2^{2-} bond order $= (8 - 4) \div 2 = 2$ (stable)

40. B_2 has $(2 \times 5 =)10$ electrons. We compare the number of unpaired electrons predicted if the order is π_{2p}^b before σ_{2p}^b with the number predicted if the σ_{2p}^b orbital is before the π_{2p}^b orbital. (^{b}indicates a bonding orbital)

$$B_2 \quad \pi_{2p}^b \text{ before } \sigma_{2p}^b$$

$$B_2 \quad \sigma_{2p}^b \text{ before } \pi_{2p}^b$$

The order of orbitals with π_{2p}^b below σ_{2p}^b in energy results in B_2 being paramagnetic, with two unpaired electrons, as observed experimentally. However, if σ_{2p}^b comes before the π_{2p}^b orbital in energy, a diamagnetic B_2 molecule is predicted, in contradiction to experimental evidence.

41. In order to have a bond order higher than three, there would have to be a region in a molecular orbital diagram where four bonding orbitals occur together in order of increasing energy, with no intervening antibonding orbitals. No such region exists in any of the molecular orbital diagrams in Figure 12-23. Alternatively, three bonding orbitals would have to occur together energetically, following an electron configuration which, when full, results in a bond order greater than zero. This arrangement does not occur either.

42. The statement is not true, for there are many instances when the bond is strengthened when a diatomic molecule loses an electron. Specifically, whenever the electron being lost is an antibonding electron (as is the case for $F_2 \rightarrow F_2^+$ and $O_2 \rightarrow O_2^+$), the bond will actually be stronger in the resulting cation than it was in the starting molecule.

43. (^{b}indicates a bonding orbital)

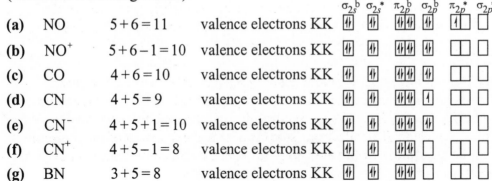

(a)	NO	$5 + 6 = 11$	valence electrons KK
(b)	NO^+	$5 + 6 - 1 = 10$	valence electrons KK
(c)	CO	$4 + 6 = 10$	valence electrons KK
(d)	CN	$4 + 5 = 9$	valence electrons KK
(e)	CN^-	$4 + 5 + 1 = 10$	valence electrons KK
(f)	CN^+	$4 + 5 - 1 = 8$	valence electrons KK
(g)	BN	$3 + 5 = 8$	valence electrons KK

44. In Exercise 43, the species that are isoelectronic are: with 8 electrons: CN^+, BN; with 10 electrons: NO^+, CO, CN^-

<u>45.</u> We first produce the molecular orbital diagram for each species.
  (a) (^{b}indicates a bonding orbital)

$$\sigma_{2s}^b \quad \sigma_{2s}^* \quad \pi_{2p}^b \quad \sigma_{2p}^b \quad \pi_{2p}^* \quad \sigma_{2p}^*$$

  NO^+ $5+6-1=10$ valence electrons KK ⥮ ⥮ ⥮⥮ ⥮ ☐☐ ☐

  N_2^+ $5+5-1=9$ valence electrons KK ⥮ ⥮ ⥮⥮ ↑ ☐☐ ☐

  Bond order = (no. bonding electrons – no. antibonding electrons) ÷ 2
  For NO^+ bond order $= (8-2) \div 2 = 3$; for N_2^+ bond order $= (7-2) \div 2 = 2.5$

  (b) NO^+ is diamagnetic; it has no unpaired electrons. N_2^+ is paramagnetic; it has one unpaired electron.

  (c) The molecule with the lower bond order will also have the greater bond length. Thus, N_2^+ should have the greater bond length.

46. We first produce the molecular orbital diagram for each species.
  (a) (^{b}indicates a bonding orbital)

$$\sigma_{2s}^b \quad \sigma_{2s}^* \quad \pi_{2p}^b \quad \sigma_{2p}^b \quad \pi_{2p}^* \quad \sigma_{2p}^*$$

  CO^+ $4+6-1=9$ valence electrons KK ⥮ ⥮ ⥮⥮ ↑ ☐☐ ☐

  CN^- $4+5+1=10$ valence electrons KK ⥮ ⥮ ⥮⥮ ⥮ ☐☐ ☐

  Bond order = (no. bonding electrons – no. antibonding electrons) ÷ 2
  For CO^+ bond order $= (7-2) \div 2 = 2.5$; for CN^- bond order $= (8-2) \div 2 = 3.0$

  (b) CN^- is diamagnetic; it has no unpaired electrons. CO^+ is paramagnetic; it has one unpaired electron.

  (c) The molecule with the lower bond order will also have the greater bond length. Thus, CO^+ should have the greater bond length.

Delocalized Molecular Orbitals

<u>47.</u> With either Lewis structures or the valence bond method, two structures must be drawn (and "averaged") to explain the π bonding in C_6H_6. The σ bonding is well explained by assuming sp^2 hybridization on each C atom. But the π bonding requires that all six C—C π bonds must be equivalent. This can be achieved by creating six π molecular orbitals—three bonding and three antibonding—into which the 6π electrons are placed. This creates a single structure for the C_6H_6 molecule.

48. Resonance is replaced in molecular orbital theory with delocalized molecular orbitals. These orbitals extend over the entire molecule, rather than being localized between two atoms. Because Lewis theory represents all bonds as being localized, the only way to depict a situation in which a pair of electrons spreads over a larger region of the molecule is to draw several Lewis structures and envision the molecule as a combination, or average of these (localized) Lewis structures.

49. We expect to find delocalized orbitals in those species for which bonding cannot be represented thoroughly by one Lewis structure, that is, for compounds that require several resonance forms.

(a) In C_2H_4, there is a total of $(2 \times 4) + (4 \times 1) = 12$ valence electrons, or 6 pairs. C atoms are the central atoms. The bonding description of C_2H_4 does not require the use of delocalized orbitals.

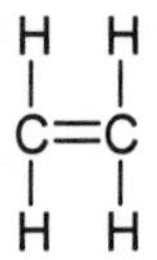

(b) In SO_2, there is a total of $6 + (2 \times 6) = 18$ valence electrons, or 9 pairs. N is the central atom. A plausible Lewis structure has two resonance forms. The bonding description of SO_2 will require the use of delocalized molecular orbitals.

$$\overline{O}=\overline{S}-\overline{O}| \longleftrightarrow |\overline{O}-\overline{S}=\overline{O}$$

(c) In H_2CO, there is a total of $(2 \times 1) + 4 + 6 = 12$ valence electrons, or 6 pairs. C is the central atom. A plausible Lewis structure is shown on the right. Because one Lewis structure adequately represents the bonding in the molecule, the bonding description of H_2CO does not require the use of delocalized molecular orbitals.

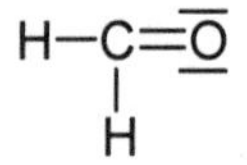

50. We expect to find delocalized orbitals in those species for which bonding cannot be represented thoroughly by one Lewis structure, that is, for species that require several resonance forms.

(a) In HCO_2^-, there are $1 + 4 + (2 \times 6) + 1 = 18$ valence electrons, or 9 pairs. A plausible Lewis structure has two resonance forms. The bonding description of HCO_2^- requires the use of delocalized molecular orbitals.

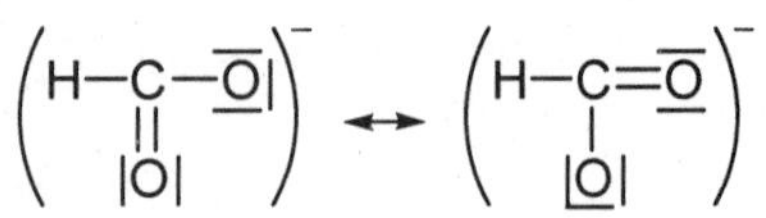

(b) In CO_3^{2-}, there are $4 + (3 \times 6) + 2 = 24$ valence electrons, or 12 pairs. A plausible Lewis structure has three resonance forms. The bonding description of CO_3^{2-} requires the use of delocalized molecular orbitals.

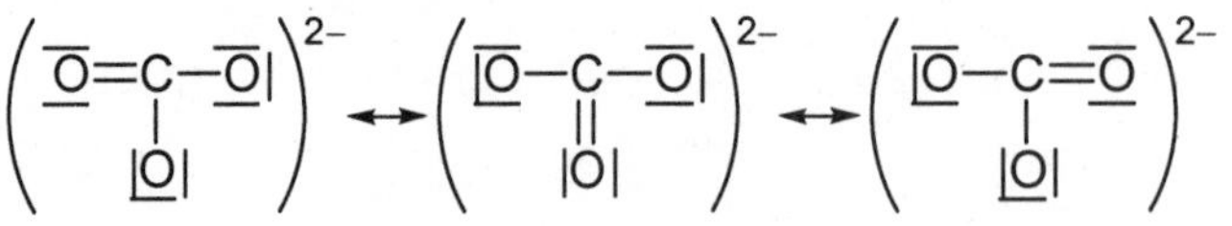

(c) In CH_3^+, there are $4 + (3 \times 1) - 1 = 6$ valence electrons, or 3 pairs. The ion is adequately represented by one Lewis structure; no resonance forms are needed.

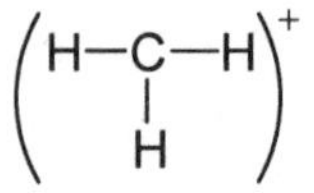

Metallic Bonding

51. **(a)** Atomic number, by itself, is not particularly important in determining whether a substance has metallic properties. However, atomic number determines where an element appears in the periodic table, and the location to the left and toward the bottom of the periodic table is a region where one finds atoms of high metallic character. Therefore atomic number—by locating an element in the periodic table—has some minimal predictive value in determining metallic character.

(b)   The answer for this part is much the same as the answer to part (a), since atomic mass generally parallels atomic number for the elements.

(c)   The number of valence electrons (electrons in the shell of highest principal quantum number) is important in predicting metallic character. Elements with few valence electrons—generally four or less—lose those electrons fairly readily to form cations, chemical behavior characteristic of metals. These elements also have a larger number of valence orbitals, which aids in metallic bond formation.

(d)   The more vacant atomic orbitals there are in the valence shell, the more readily metallic bonding occurs.

(e)   Because metals occur in every period of the periodic table—except the first—there is no particular relationship between the number of electron shells and the metallic behavior of an element. (Remember that one shell begins to be occupied at the start of each period.)

52.   We first determine the ground-state electron configuration of Na, Fe, and Zn

$$[\,Na\,] = [\,Ne\,]3s^1 \quad [\,Fe\,] = [\,Ar\,]3d^6 4s^2 \quad [\,Zn\,] = [\,Ar\,]3d^{10} 4s^2.$$

Based on the number of electrons in the valence shell, we would expect Na to be the softest of these three metals. The one $4s$ electron per Na atom simply will not contribute as strongly to bonding as will two $4s$ electrons in the other two metals. If just the number of valence electrons is considered, then one would conclude that Zn should exhibit stronger metallic bonding than iron (Zn has 12 valence electrons, while Fe has 8). The opposite is true, however (i.e. Fe has stronger metallic bonding), because the ten 3d-electrons in zinc are deep in the atom and thus contribute very little to the metallic bonding. Thus, Fe should be harder than Zn. Melting point should also reflect the strengths of bonding between atoms, with higher melting points occurring in metals that are more strongly bonded. Thus in order, melting point: $\text{Fe}\,(1530^\circ\text{C}) > \text{Zn}\,(420^\circ\text{C}) > \text{Na}\,(98^\circ\text{C})$ and

hardness: $\text{Fe}\,(4.5) > \text{Zn}\,(2.5) > \text{Na}\,(0.4)$. Parenthesized numbers are actual values, with hardness on a scale (MOHS) of 0 (talc) to 10 (diamond).

53.   We first determine the number of Na atoms in the sample.

$$\text{no. Na atoms} = 26.8 \text{ mg Na} \times \frac{1 \text{ g Na}}{1000 \text{ mg Na}} \times \frac{1 \text{ mol Na}}{22.99 \text{ g Na}} \times \frac{6.022 \times 10^{23} \text{ Na atoms}}{1 \text{ mol Na}}$$

$$= 7.02 \times 10^{20} \text{ Na atoms}$$

Because there is one $3s$ orbital per Na atom, and since the number of energy levels (molecular orbitals) created is equal to the number of atomic orbitals initially present, there are 7.02×10^{20} energy levels present in the conduction band of this sample. Also, there is one $3s$ electron contributed by each Na atom, for a total of 7.02×10^{20} electrons. Because each energy level can hold two electrons, the conduction band is half full.

54.   Although the band that results from the combination of 3*s* orbitals in magnesium metal is full, there is another empty band that overlaps this 3*s* band. This empty band results from the combination of 3*p* orbitals. Although it does not overlap the 3*s* band in the Mg_2 molecule or in atom clusters, once the metal crystal has grown to visible size, the overlap is complete.

55.   A semiconducting element is one that displays poor conductivity when pure, but attains much higher levels of conductivity when doped with small quantities of selected elements or when heated. The best semiconducting materials are made from the Group 14 elements Si and Ge. *P*-type semiconductors result when Group 14 elements are doped with small quantities of an element that has fewer than four valence electrons. For instance, when Si is doped with B, each boron atom ends us forming one silicon bond that has just one electron in it. The transfer of valence electrons from adjacent atoms into these electron-deficient bonds creates a domino effect that results in the movement of an electron-deficient hole through the semiconductor in a direction opposite to the movement of the electrons. Thus, in order to produce a *p*-type semiconductor, the added dopant atom must each have at least one less valence electron than the individual atoms that make up the bulk of the material.

   (a)   Sulfur has six valence electrons, which is three too many, so doping Si with sulfur will not produce a *p*-type semiconductor.

   (b)   Arsenic has five valence electrons, which is two electrons too many, so doping Si with arsenic will not produce a *p*-type semiconductor.

   (c)   Lead has four valence electrons which is one too many, so doping with lead will not produce a *p*-type semiconductor.

   (d)   Boron, with one less valence electron than silicon, has the requisite number of electrons needed to form a *p*-type semiconductor.

   (e)   Gallium arsenide is an *n*-type semiconductor with an excess of electrons, so doping Si with GaAs will not produce a *p*-type semiconductor.

   (f)   Like boron, gallium has three valence electrons. Thus, doping Si with gallium will produce a *p*-type semiconductor.

56.   *N*-type semiconductors are formed when the parent element is doped with atoms of an element that has more valence electrons. For instance, when silicon is doped with phosphorus, one extra electron is left over after each phosphorus forms four covalent bonds to four neighboring silicon atoms. The extra electrons can be forced to move through the solid and conduct electric current in the process by applying a small voltage or thermal excitation. Thus, to create an *n*-type semiconductor, the added dopant atoms must each have at least one more valence electron than the individual atoms that make up the bulk of the material.

   (a)   Since sulfur has two more valence electrons than germanium, doping germanium with sulfur will produce an *n*-type semiconductor.

   (b)   Aluminum has just three valence electrons which is two short of what is required, so doping germanium with aluminum will not produce an *n*-type semiconductor.

(c)   Tin has four electrons, which is one short of the number required, so doping with tin will not produce an *n*-type semiconductor.

(d)   Cadmium sulfide is composed of Cd^{2+} and S^{2-} ions, neither ion has electrons that it can donate to germanium and so doping germanium with cadmium sulfide will not produce an *n*-type semiconductor.

(e)   Arsenic had one more valence electron than germanium, doping germanium with arsenic will produce an *n*-type semiconductor.

(f)   Since GaAs is itself an *n*-type semiconductor, the addition of small amounts of gallium arsenide to pure germanium should produce an n-type semiconductor.

57.   In ultra pure crystalline silicon, there are no extra electrons in the lattice that can conduct an electric current. If however, the silicon becomes contaminated with arsenic atoms, then there will be one additional electron added to the silicon crystal lattice for each arsenic atom that is introduced. Upon heating, some of those *"extra"* electrons will be promoted into the conduction band of the solid. The electrons that end up in the conduction band are able to move freely through the structure. In other words, the arsenic atoms increase the conductivity of the solid by providing additional electrons that can carry a current when they are promoted into the conduction band by thermal excitation. Thus, by virtue of having extra electrons in the lattice, silicon contaminated with arsenic will exhibit greater electrical conductance than pure silicon at elevated temperatures.

58.   When a semiconductor is doped with donor atoms (e.g., Si doped with P), the extra electron is promoted to the conduction band of the solid where it is able to move freely (good electrical conductor). When a semiconductor is doped with an acceptor atom (e.g., Si doped with Al), electron-deficient bonding results. The transfer of valence electrons from adjacent atoms into these electron-deficient bonds creates a domino effect that results in the movement of an electron-deficient hole through the semiconductor in a direction opposite to the movement of the electrons. If there are equal numbers of donor and acceptor atoms present, electrons will be redistributed from donor to acceptor atoms. There will be no excesses or deficiencies , and conductivity of the solid will not be enhanced.

59. $\Delta E_{Si} = 110.$ kJ mol^{-1}. $\Delta E_{Si\ atom} = \dfrac{110.\ kJ\ mol^{-1}}{6.022 \times 10^{23}} = 1.83 \times 10^{-19}$ J $E = h\nu = \dfrac{hc}{\lambda}$ or $\lambda = \dfrac{hc}{E}$

$\lambda = \dfrac{(6.626 \times 10^{-34}\ J\,s)(2.998 \times 10^{8}\ \frac{m}{s})}{1.83 \times 10^{-19}\ J} = 1.09 \times 10^{-6}$ m or 1090 nm. This is IR-radiation.

60.   Electrons are being promoted into the conduction band, which is comprised of several energy levels. The conduction band for a solid has a large number of energy levels with extremely small energy separations between each energy level (this collection of energy levels is termed an energy band). A solar cell operates over a broad range of wavelengths because electrons can be promoted to numerous energy levels in the conduction band, rather than to just one level.

FEATURE PROBLEMS

84. **(a)** $C_6H_6(l) + 3 H_2(g) \rightarrow C_6H_{12}(l)$ $\Delta H° = \Sigma \Delta H°_{f, products} - \Sigma \Delta H°_{f, reactants}$
$\Delta H° = -156.4 \text{ kJ} - [3 \text{ mol} \times 0 \text{ kJ mol}^{-1} + 49.0 \text{ kJ}] = -205.4 \text{ kJ} = \Delta H°(a)$

(b) $C_6H_{10}(l) + H_2(g) \rightarrow C_6H_{12}(l)$ $\Delta H° = \Sigma \Delta H°_{f \, products} - \Sigma \Delta H°_{f \, reactants}$
$\Delta H° = -156.4 \text{ kJ} - [1 \text{ mol} \times 0 \text{ kJ mol}^{-1} + (-38.5 \text{ kJ})] = -117.9 \text{ kJ} = \Delta H°(b)$

(c) Enthalpy of hydrogenation for 1,3,5-cyclohexatriene = $3 \times \Delta H°(b)$
Enthalpy of hydrogenation = $3 (-117.9 \text{ kJ}) = -353.7 \text{ kJ} = \Delta H°(c)$

$\Delta H°_{f, cyclohexene} = -38.5$ kJ/mole (given in part b of this question).

Resonance energy is the difference between $\Delta H°(a)$ and $\Delta H°(c)$.
Resonance energy = $-353.7 \text{ kJ} - (-205.4 \text{ kJ}) = -148.3 \text{ kJ}$

(d) Using bond energies:
$\Delta H°_{atomization} = 6(C—H) + 3(C—C) + 3(C=C)$
$\Delta H°_{atomization} = 6(414 \text{ kJ}) + 3(347 \text{ kJ}) + 3(611 \text{ kJ})$
$\Delta H°_{atomization} = 5358 \text{ kJ}$ (per mole of C_6H_6)

$C_6H_6(g) \rightarrow 6 C(g) + 6 H(g)$
$\Delta H° = [6(716.7 \text{ kJ}) + 6(218 \text{ kJ})] - 82.6 \text{ kJ} = 5525.6 \text{ kJ}$

Resonance energy = $5358 - 5525.6 \text{ kJ} = -168 \text{ kJ}$

85. Consider the semiconductor device below which is hooked up to a battery (direct current). The *n*-type semiconductor (a) is connected to the negative terminal of a battery, the *p*-type to the positive terminal. This has the effect of pushing conduction electrons from right to left and positive holes from left to right.

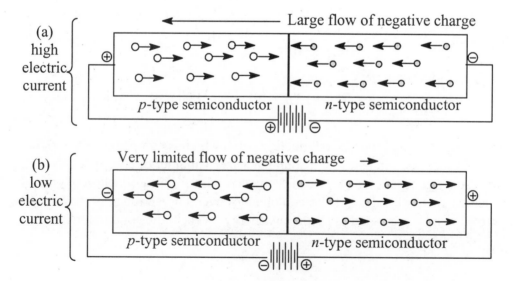

A large current flows across the *p-n* junction ad through the electric circuit.
Note: the flow of positive holes in one direction is, in effect, a flow of electrons in the opposite direction. So, the *p-n* junction should still have the same orientation in (b) as in (a). Now the conduction electrons are pulled to the right and the positive holes to the left. Because there are very few conduction electrons in the *p*-type semiconductor and very few positive holes in the *n*-type semiconductor, there are very few carriers of electric charge across the *p-n* junction. Very little electric current flows.

When the *p-n* junction rectifier is connected to 60-cycle alternating current, each terminal of the of the electric generator switches back and forth between being a positive and a negative terminal 120 times per second. If the electrical contact to the *p-n* junction rectifier is made through an electric generator rather than a battery, half of the time the situation is that depicted in (a) and half the time it is that depicted in (b). Thus, half the time there is a large current flow, always in the same direction, and half the time there is essentially no current. The alternating current is converted to direct current: it is rectified, and the *p-n* junction device shown above is called a rectifier.

86. (a) Five valid resonance forms can be drawn for furan:

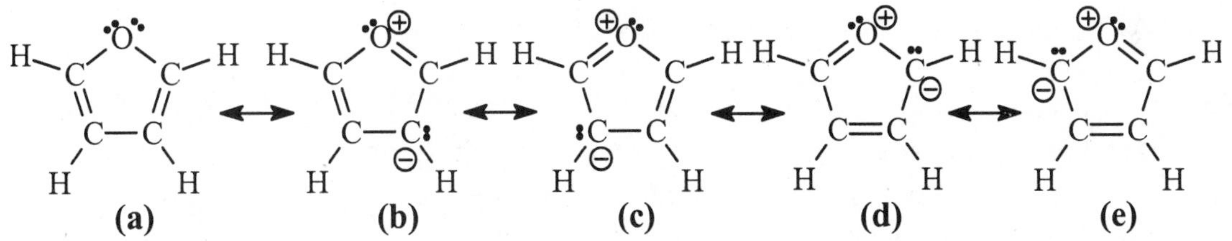

These individual resonance forms do not exist. The actual structure is a hybrid possessing characteristics of all five individual contributors.

(b) Furan is an aromatic five-membered herterocycle. Its four carbon atoms and lone oxygen atom are sp^2 hybridized in the classical bonding description. One of the lone pairs on oxygen occupies an unhybridized p-orbital, and this lone pair overlaps with the half-filled $2p$ orbitals on the four carbon atoms to form an aromatic sextet:

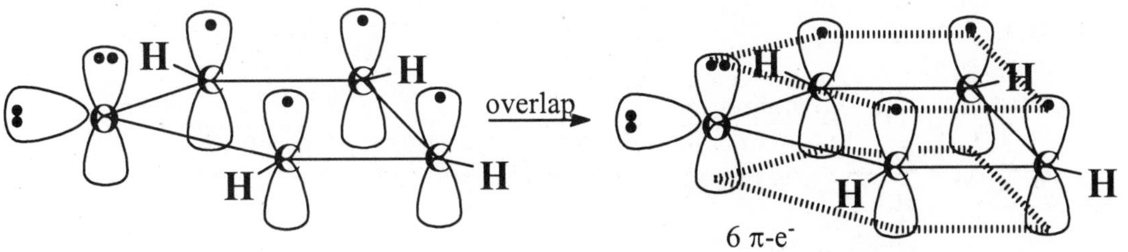

$6 \pi\text{-e}^-$
$(4n + 2) \pi\text{-e}^-$ where n = 1
Therefore, aromatic ring

The groundstate molecular orbital diagram for the π-system in furan is depicted below:

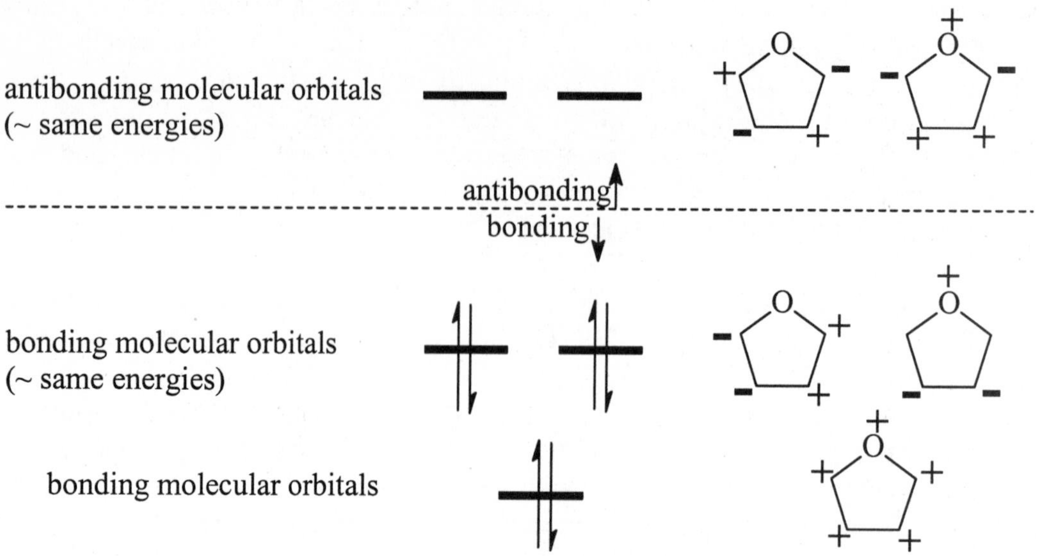

antibonding molecular orbitals
(~ same energies)

antibonding ↑
bonding ↓

bonding molecular orbitals
(~ same energies)

bonding molecular orbitals

The diagram shows that the six π-electrons fill the three bonding molecular orbitals for the π-system. Keep in mind that the ensemble of π-molecular orbitals is superimposed on the sp^2 framework for the molecule.

(c) All five resonance structures for furan (structures a-e) have six electrons involved in π-bonding. In structure (a), a lone pair in a $2p$ orbital on the oxygen atom along with the four π-electrons in unhybridized $2p$ orbitals which form the two C=C bonds are combined to give a π-system. In structures (b) - (e) inclusive, a π-system is produced by the combination of a lone pair in a 2p orbital on a carbon atom with the four π-electrons from the C=O and C=C bonds.

87. (a) In order to see the shape of the sp hybrid the simplest approach is to combine just the angular parts of the s and p orbitals. Including the radial part makes the plot more challenging (see below). The angular parts for the 2s and $2p_z$ orbitals are

$$Y(s) = \left(\frac{1}{4\pi}\right)^{1/2} \quad \text{and} \quad Y\left(p_z\right) = \left(\frac{3}{4\pi}\right)^{1/2} \cos\theta$$

Combining the two angular parts:

$$\psi_1(sp) = \frac{1}{\sqrt{2}}\left[Y(2s) + Y(2p)\right]$$

$$\psi_1(sp) = \frac{1}{\sqrt{2}}\left[\left(\frac{1}{4\pi}\right)^{1/2} + \left(\frac{3}{4\pi}\right)^{1/2} \cos\theta\right]$$

$$\psi_1(sp) = \frac{1}{\sqrt{2}}\left(\frac{1}{4\pi}\right)^{1/2}\left[1 + \sqrt{3}\cos\theta\right]$$

We now evaluate this function (ignoring the constants in front of the square brackets) for various values of θ

Theta(deg)	Theta(rad)	sp_1	sp_2
0	0	2.732	-0.732
30	0.5236	2.500	-0.500
45	0.7854	2.225	-0.225
60	1.0472	1.866	0.134
90	1.5708	1.000	1.000
120	2.0944	0.134	1.866
135	2.3562	-0.225	2.225
150	2.6180	-0.500	2.500
170	2.9670	-0.706	2.706
180	3.1416	-0.732	2.732
190	3.3161	-0.706	2.706
210	3.6652	-0.500	2.500
225	3.9270	-0.225	2.225
240	4.1888	0.134	1.866
270	4.7124	1.000	1.000
300	5.2360	1.866	0.134
315	5.4978	2.225	-0.225
330	5.7596	2.500	-0.500
360	6.2832	2.732	-0.732

A plot of these values in the form of Figure 9.24 gives the general shape of the *sp* hybrid.

This is a plot in the xz-plane, for $\Psi_1(sp)$.

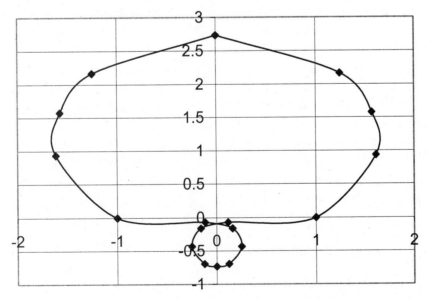

The other *sp* hybrid is

$$\psi_2(sp) = \frac{1}{\sqrt{2}}\left[Y(2s) - Y(2p)\right]$$

$$\psi_2(sp) = \frac{1}{\sqrt{2}}\left[\left(\frac{1}{4\pi}\right)^{1/2} - \left(\frac{3}{4\pi}\right)^{1/2}\cos\theta\right]$$

$$\psi_2(sp) = \frac{1}{\sqrt{2}}\left(\frac{1}{4\pi}\right)^{1/2}\left[1 - \sqrt{3}\cos\theta\right]$$

Graphically, $\Psi_2(sp)$ is generated as follows:

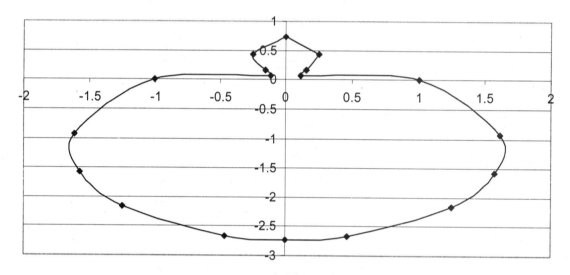

We now see that the second *sp* hybrid ($\Psi_2(sp)$) points in a direction opposite to the first ($\Psi_1(sp)$)

(b)   Figure 12-10 pictorially shows the hybridization of a *2s* orbital and a *2p* (there is no mention of whether it is a $2p_x$, $2p_y$ or $2p_z$ orbital). When one of the degenerate *2p*-orbitals and the 2s orbital are hybridized, we expect that there should be no difference in shape or energy of the resulting *sp*-hybrid orbitals. The only difference between the $2p_x$, $2p_y$ and $2p_z$ orbitals is the direction in which these atomic orbitals are oriented. Similarly, the only difference in the resulting hybrid orbitals should be their directional properties. Mathematically, we have shown (part (a)), that an sp_z hybrid orbital is proportional to $1 - \sqrt{3}\cos\theta$. Similar calculations for an sp_x hybrid orbital result in an analogous relationship, namely one proportional to to $1 - \sqrt{3}\sin\theta\cos\phi$. (Note: in the xz plane, $\phi = 90°$, therefore $\cos\phi = 1$). In the xz plane, this expression simplifies to $1 - \sqrt{3}\sin\theta$. (Note: $\sin\theta = \cos(90° - \theta)$). These two relationships result in similar answers (i.e., same shape), the only difference being a shift by an expect 90°. Similar arguments can be made for an sp_y hybrid orbital expression.

(c)   The sp^2 hybrids. To show the spatial distribution of the sp^2 hybrids we will again use the angular functions only. Thus

$$Y(p_x) = \left(\frac{3}{4\pi}\right)^{1/2} \sin\theta\cos\phi$$

$$Y(s) = \left(\frac{1}{4\pi}\right)^{1/2}$$

Combining these functions

$$\psi_1(sp^2) = \frac{1}{\sqrt{3}}\psi(2s) + \frac{\sqrt{2}}{\sqrt{3}}\psi(2p_x)$$

$$\psi_1(sp^2) = \frac{1}{\sqrt{3}}\left(\frac{1}{4\pi}\right)^{1/2} + \frac{\sqrt{2}}{\sqrt{3}}\left(\frac{3}{4\pi}\right)^{1/2}\sin\theta\cos\phi$$

$$\psi_1(sp^2) = \left(\frac{1}{4\pi}\right)^{1/2}\left[\frac{1}{\sqrt{3}} + \sqrt{2}\sin\theta\cos\phi\right]$$

In the xy plane $\theta = 90°$ so that the function becomes $(\sin(90°) = 1)$

$$\psi_1(sp^2) = \left(\frac{1}{4\pi}\right)^{1/2}\left[\frac{1}{\sqrt{3}} + \sqrt{2}\cos\phi\right]$$

We now evaluate the functional form of one of the other sp^2 hybrids

$$Y(p_y) = \left(\frac{3}{4\pi}\right)^{1/2}\sin\theta\sin\phi$$

$$\psi_2(sp^2) = \frac{1}{\sqrt{3}}\psi(2s) - \frac{1}{\sqrt{6}}\psi(2p_x) + \frac{1}{\sqrt{2}}\psi(2p_y)$$

$$\psi_2(sp^2) = \frac{1}{\sqrt{3}}\left(\frac{1}{4\pi}\right)^{1/2} - \frac{1}{\sqrt{6}}\left(\frac{3}{4\pi}\right)^{1/2}\sin\theta\cos\phi + \frac{1}{\sqrt{2}}\left(\frac{3}{4\pi}\right)^{1/2}\sin\theta\sin\phi$$

$$\psi_2(sp^2) = \left(\frac{1}{4\pi}\right)^{1/2}\left[\frac{1}{\sqrt{3}} - \frac{1}{\sqrt{6}}\sqrt{3}\sin\theta\cos\phi + \frac{1}{\sqrt{2}}\sqrt{3}\sin\theta\sin\phi\right]$$

$$\psi_2(sp^2) = \left(\frac{1}{4\pi}\right)^{1/2}\left[\frac{1}{\sqrt{3}} - \frac{1}{\sqrt{2}}\sin\theta\cos\phi + \frac{\sqrt{3}}{\sqrt{2}}\sin\theta\sin\phi\right]$$

Again in the xy plane we have

$$\psi_2\left(sp^2\right) = \left(\frac{1}{4\pi}\right)^{1/2}\left[\frac{1}{\sqrt{3}} - \frac{1}{\sqrt{2}}\cos\phi + \frac{\sqrt{3}}{\sqrt{2}}\sin\phi\right]$$

The third sp^2 hybrid is

$$\psi_3\left(sp^2\right) = \left(\frac{1}{4\pi}\right)^{1/2}\cdot\left[\frac{1}{\sqrt{3}} - \frac{1}{\sqrt{2}}\sin\theta\cos\phi - \frac{\sqrt{3}}{\sqrt{2}}\sin\theta\sin\phi\right]$$

and in the xy plane

$$\psi_3\left(sp^2\right) = \left(\frac{1}{4\pi}\right)^{1/2}\left[\frac{1}{\sqrt{3}} - \frac{1}{\sqrt{2}}\cos\phi - \frac{\sqrt{3}}{\sqrt{2}}\sin\phi\right]$$

We can evaluate these functions and obtain sp^2 angular values:

Phi(deg)	Phi(rad)	$sp^2(1)$	$sp^2(2)$	$sp^2(3)$
0	0	1.992	-0.130	-0.647
30	0.5236	1.802	0.577	-0.837
45	0.7854	1.577	0.943	-0.789
60	1.0472	1.284	1.284	-0.647
90	1.5708	0.577	1.802	-0.130
120	2.0944	-0.123	1.992	0.577
135	2.3562	-0.423	1.943	0.943
150	2.6180	-0.647	1.802	1.284
170	2.9671	-0.815	1.486	1.661
180	3.1416	-0.837	1.284	1.802
190	3.3161	-0.815	1.061	1.906
210	3.6652	-0.647	0.577	1.992
225	3.9270	-0.423	0.211	1.943
240	4.1888	-0.130	-0.130	1.802
270	4.7124	0.577	-0.647	1.284
300	5.2360	1.284	-0.837	0.577
315	5.4978	1.577	-0.789	0.211
330	5.7596	1.802	-0.647	-0.130
360	6.2832	1.992	-0.130	-0.647

The graphs are as follows:

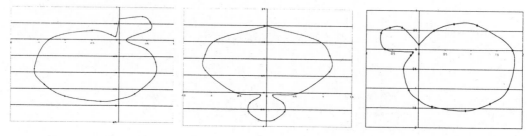

Part (a) INCLUDING THE RADIAL PART

If we combine the angular and radial parts of the $2s$ and $2p$ orbitals the situation becomes more complicated as we have two variables to deal with. The angular and radial parts of the $2s$ and $2p$ orbitals are:

$$Y(s) = \left(\frac{1}{4\pi}\right)^{1/2} \qquad R(2s) = \frac{1}{2\sqrt{2}}\left(\frac{Z}{a_0}\right)^{3/2}(2-\sigma)e^{-\sigma/2}$$

$$Y(p_z) = \left(\frac{3}{4\pi}\right)^{1/2}\cos\theta \qquad R(2p) = \frac{1}{2\sqrt{6}}\left(\frac{Z}{a_0}\right)^{3/2}\sigma e^{-\sigma/2}$$

Combining these

$$\psi(2s) = \left(\frac{1}{4\pi}\right)^{1/2} \times \frac{1}{2\sqrt{2}}\left(\frac{Z}{a_0}\right)^{3/2}(2-\sigma)e^{-\sigma/2}$$

$$\psi(2p_z) = \left(\frac{3}{4\pi}\right)^{1/2}\cos\theta \times \frac{1}{2\sqrt{6}}\left(\frac{Z}{a_0}\right)^{3/2}\sigma e^{-\sigma/2}$$

We now combine these to form the sp-hybrids

$$\psi_1(sp) = \frac{1}{\sqrt{2}}\left[\psi(2s) + \psi(2p)\right]$$

$$\psi_1(sp) = \frac{1}{\sqrt{2}}\left[\left(\frac{1}{4\pi}\right)^{1/2} \times \frac{1}{2\sqrt{2}}\left(\frac{Z}{a_0}\right)^{3/2}(2-\sigma)e^{-\sigma/2} + \left(\frac{3}{4\pi}\right)^{1/2}\cos\theta \times \frac{1}{2\sqrt{6}}\left(\frac{Z}{a_0}\right)^{3/2}\sigma e^{-\sigma/2}\right]$$

$$\psi_1(sp) = \frac{1}{\sqrt{2}}\left(\frac{1}{4\pi}\right)^{1/2}\left[\frac{1}{2\sqrt{2}}\left(\frac{Z}{a_0}\right)^{3/2}(2-\sigma)e^{-\sigma/2} + \sqrt{3} \times \cos\theta \times \frac{1}{2\sqrt{6}}\left(\frac{Z}{a_0}\right)^{3/2}\sigma e^{-\sigma/2}\right]$$

$$\psi_1(sp) = \frac{1}{\sqrt{2}}\left(\frac{1}{4\pi}\right)^{1/2}\left(\frac{Z}{a_0}\right)^{3/2}\frac{1}{2\sqrt{2}}\left[(2-\sigma)e^{-\sigma/2} + \sqrt{3} \times \cos\theta \times \frac{1}{\sqrt{3}}\sigma e^{-\sigma/2}\right]$$

$$\psi_1(sp) = \frac{1}{\sqrt{2}}\left(\frac{1}{4\pi}\right)^{1/2}\left(\frac{Z}{a_0}\right)^{3/2}\frac{e^{-\sigma/2}}{2\sqrt{2}}\left[(2-\sigma) + \cos\theta \times \sigma\right]$$

$$\psi_1(sp) = \frac{1}{\sqrt{2}}\left(\frac{1}{4\pi}\right)^{1/2}\left(\frac{Z}{a_0}\right)^{3/2}\frac{e^{-\sigma/2}}{2\sqrt{2}}\left[2 + \sigma(\cos\theta - 1)\right]$$

Similarly

$$\psi_2(sp) = \frac{1}{\sqrt{2}}\left[\psi(2s) - \psi(2p)\right]$$

$$\psi_2(sp) = \frac{1}{\sqrt{2}}\left(\frac{1}{4\pi}\right)^{1/2}\left(\frac{Z}{a_0}\right)^{3/2}\frac{e^{-\sigma/2}}{2\sqrt{2}}\left[2 - \sigma(\cos\theta + 1)\right]$$

We are dealing with orbitals with $n=2$, so that

$$\sigma = \frac{2Zr}{na_0} = \frac{Zr}{a_0}$$

In order to proceed further we have to choose values of r and θ, but how do we plot this? The way to do this is to evaluate the function for various values of (r/a_0) and θ and plot points in the xz plane calculating the values of x and z from the polar coordinates

$$z = r\cos\theta$$

$$x = r\sin\theta\cos\phi$$

choosing $\phi=0$ in the xz plane so that $x = r\sin\theta$

Thus, we evaluate the probability as the square of the functions

$$\psi_1(sp) = \frac{1}{\sqrt{2}}\left(\frac{1}{4\pi}\right)^{1/2}\left(\frac{Z}{a_0}\right)^{3/2}\frac{e^{-Zr/2a_0}}{2\sqrt{2}}\left[2 + \frac{Zr}{a_0}(\cos\theta - 1)\right]$$

$$\psi_2(sp) = \frac{1}{\sqrt{2}}\left(\frac{1}{4\pi}\right)^{1/2}\left(\frac{Z}{a_0}\right)^{3/2}\frac{e^{-Zr/2a_0}}{2\sqrt{2}}\left[2 - \frac{Zr}{a_0}(\cos\theta + 1)\right]$$

and plot the value of the function at the x,z coordinate and join points of equal value of the probability in the manner of a contour plot.. The shape of the sp hybrid will then be revealed. Choose a value of Z (or Z_{eff}) that is convenient.

This is suitable for a class team project using *Microsoft Excel*. A primitive spreadsheet is available upon request.

CHAPTER 13
LIQUIDS, SOLIDS, AND INTERMOLECULAR FORCES
PRACTICE EXAMPLES

1A d = 0.701 g/L at 25 °C for C_6H_{14} (Molar Mass = 86.177 g mol^{-1})
Consider a 1.00 L sample. This contains 0.701 g C_6H_{14}

$$\text{Moles } C_6H_{14} \text{ in 1.00 L sample} = 0.701 \text{ g } C_6H_{14} \times \frac{1 \text{ mol } C_6H_{14}}{86.177 \text{ g } C_6H_{14}} = 8.13 \times 10^{-3} \text{ mol } C_6H_{14}$$

Find pressure using the ideal gas law: $P = \dfrac{nRT}{V} = \dfrac{(8.31 \times 10^{-3} \text{ mol})(\frac{0.08206 \text{ L atm}}{\text{K mol}})(298\text{K})}{1.00 \text{ L}}$

$$P = 0.199 \text{ atm}$$

$$P(\text{torr}) = 0.199 \text{ atm} \times \frac{760 \text{ torr}}{1 \text{ atm}} = 151 \text{ torr}$$

1B From Figure 13-9, the vapor pressure is ≈ 420 mmHg or 420 mmHg $\times \dfrac{1 \text{ atm}}{760 \text{ torr}} = 0.55_3$ atm

Molar mass = 74.123 g mol^{-1}. $P = \dfrac{nRT}{V} = \dfrac{\left(\dfrac{\text{mass}}{\text{molar mass}}\right)RT}{V} = \dfrac{(\text{density})RT}{\text{molar mass}}$

or $d = \dfrac{(\text{molar mass})P}{RT} = \dfrac{\left(74.123 \dfrac{\text{g}}{\text{mol}}\right)(0.55_3 \text{ atm})}{\left(0.08206 \dfrac{\text{L atm}}{\text{K mol}}\right)(293\text{K})} = 1.70 \text{ g L}^{-1} \approx 1.7 \text{ g/L}$

2A We first calculate pressure created by the water at $80.0° \text{ C}$, assuming all 0.132 g H_2O vaporizes.

$$P_2 = \frac{nRT}{V} = \frac{\left(0.132 \text{ g } H_2O \times \dfrac{1 \text{ mol } H_2O}{18.02 \text{ g } H_2O}\right) \times 0.08206 \dfrac{\text{L atm}}{\text{mol K}} \times 353.2 \text{ K}}{0.525 \text{ L}} \times \frac{760 \text{ mmHg}}{1 \text{ atm}}$$

$$P_2 = 307 \text{ mmHg}$$

At $80.0°\text{C}$, the vapor pressure of water is 355.1 mmHg. Thus, all the water can exist as vapor at $80.0°\text{C}$.

2B   The result of Example 13-2 is that 0.132 g H_2O would exert a pressure of 281 mmHg if it all existed as a vapor. Since that 281 mmHg is greater than the vapor pressure of water at this temperature, some of the water must exist as liquid. The calculation of the example is based on the equation $P = nRT/V$, which means that the pressure of water is proportional to its mass. Thus, the mass of water needed to produce a pressure of 92.5 mmHg under this situation is

$$\text{mass of water vapor} = 92.5 \text{ mmHg} \times \frac{0.132 \text{ g } H_2O}{281 \text{ mmHg}} = 0.0435 \text{ g } H_2O$$

mass of liquid water = 0.132 g H_2O total $-$ 0.0435 g H_2O vapor = 0.089 g liquid water

3A   From Table 13-1 we know that $\Delta H_{vap} = 38.0 \text{ kJ}/\text{mol}$ for methyl alcohol. We now can use the Clausius-Clapeyron equation to determine the vapor pressure at $25.0°\text{C} = 298.2 \text{ K}$.

$$\ln \frac{P}{100 \text{ mmHg}} = \frac{38.0 \times 10^3 \text{ J mol}^{-1}}{8.3145 \text{ J mol}^{-1} \text{ K}^{-1}} \left(\frac{1}{(273.2 + 21.2) \text{ K}} - \frac{1}{298.2 \text{K}} \right) = +0.198$$

$$\frac{P}{100 \text{ mmHg}} = e^{+0.198} = 1.22 \qquad P = 1.22 \times 100 \text{ mmHg} = 121 \text{ mmHg}$$

3B   The vapor pressure at the normal boiling point $\left(99.2°\text{C} = 372.4 \text{ K} \right)$ is 760 mmHg precisely. We can use the Clausius-Clapeyron equation to determine the vapor pressure at $25°\text{C} = 298 \text{ K}$.

$$\ln \frac{P}{760 \text{ mmHg}} = \frac{35.76 \times 10^3 \text{ J mol}^{-1}}{8.3145 \text{ J mol}^{-1} \text{ K}^{-1}} \left(\frac{1}{372.4 \text{ K}} - \frac{1}{298.2 \text{ K}} \right) = -2.874$$

$$\frac{P}{760 \text{ mmHg}} = e^{-2.874} = 0.0565 \qquad P = 0.0565 \times 760 \text{ mmHg} = 42.9 \text{ mmHg}$$

4A   Moving from point R to P we begin with $H_2O(g)$ at high temperature (>100°C). When the temperature reaches the point on the vaporization curve, OC, water condenses at constant temperature (100°C). Once all of the water is in the liquid state, the temperature drops. When the temperature reaches the point on the fusion curve, OD, ice begins to form at constant temperature (0°C). Once all of the water has been converted to $H_2O(s)$, the temperature of the sample decreases slightly until point P is reached.

Since solids are not very compressible, very little change occurs until the pressure reaches the point on the fusion curve OD. Here, melting begins. A significant decrease in the volume occurs ($\approx$10%) as ice is converted to liquid water. After melting, additional pressure produces very little change in volume because liquids are not very compressible.

4B

1.00 mol H_2O. At Point R, T = 374.1 °C or 647.3 K

$$V_{\text{point R}} = \frac{nRT}{P} = \frac{(1.00\,\text{mol})\left(0.08206\,\dfrac{L\,atm}{K\,mol}\right)(647.3\,K)}{1.00\,atm} = 53.1\,L$$

1.00 mol H_2O on P-R line, if 1/2 of water is vaporized, T = 100 °C(273.015 K)

$$V_{\text{1/2 vap(PR)}} = \frac{nRT}{P} = \frac{(0.500\,\text{mol})\left(0.08206\,\dfrac{L\,atm}{K\,mol}\right)(373.15\,K)}{1.00\,atm} = 15.3\,L$$

A much smaller volume results when just 1/2 of the sample is vaporized (moles of gas smaller as well, temperature is smaller). 53.1 L vs 15.3 L (about 28.8 % of the volume as that seen at point R).

51. 3 L

At

Point

R

15.3 L
at 100C
1/2 vap

5A The substance with the highest boiling point will have the strongest intermolecular forces. The weakest of van der Waals forces are London forces, which depend on molar mass (and surface area): C_3H_8 is 44 g/mol, CO_2 is 44 g/mol, and CH_3CN is 41 g/mol. Thus, the London forces are approximately equal for these three compounds. Next to consider are dipole-dipole forces. C_3H_8 is nonpolar; its bonds are not polar bonds. CO_2 is nonpolar; its two bond moments cancel each other. CH_3CN is polar and thus has the strongest intermolecular forces and should have the highest boiling point. Boiling points are $-78.44°C$ for CO_2, $-42.1°C$ for C_3H_8, and $81.6°C$ for CH_3CN.

5B Dispersion forces, which depend on molar mass and structure, are one of the determinants of boiling point. The molar masses are: C_8H_{18} (114.2 g/mol), $CH_3CH_2CH_2CH_3$ (58.1 g/mol), $(CH_3)_3CH$ (58.1 g/mol), C_6H_5CHO (106.1 g/mol), and SO_3 (80.1 g/mol). We would expect $(CH_3)_3CH$ to have the lowest boiling point because it has the lowest molar mass and the most compact (ball-like) shape, whereas $CH_3CH_2CH_2CH_3$ which has the same mass, but is longer (and so has more chances for intermolecular interactions), should have the second highest boiling point. Then should follow polar SO_3. C_6H_5CHO should have a boiling point higher than the more massive C_8H_{18} because benzaldehyde is polar while octane is not. Actual boiling points are given in parentheses in the following ranking. $(CH_3)_3CH(-11.6°C) < CH_3CH_2CH_2CH_3(-0.5°C) < SO_3(44.8°C) < C_8H_{18}(125.7°C) < C_6H_5CHO(178°C)$

6A We first look to molar masses: Ne (20.2 g/mol), He (4.0 g/mol), Cl_2 (70.9 g/mol), $(CH_3)_2CO$ (58.1 g/mol), O_2 (32.0 g/mol), and O_3 (48.0 g/mol). Both $(CH_3)_2CO$ and O_3 are polar, O_3 weakly so (because of its uneven distribution of electrons). We expect $(CH_3)_2CO$ to be the highest boiling, followed by Cl_2, O_3, O_2, Ne, and He. In the following ranking, actual boiling points are given in parentheses. He (-268.9 °C), Ne (-245.9 °C), O_2 (-183.0 °C), O_3 (-111.9 °C), Cl_2 (-34.6 °C), and $(CH_3)_2CO$ (56.2°C)

6B The magnitude of the enthalpy of vaporization is strongly related to the strength of intermolecular forces: the stronger those forces are, the more endothermic the vaporization process. The first three substances all are nonpolar and, therefore, their only intermolecular forces are London forces, whose strength primarily depends on molar mass. The substances are arranged in order of increasing molar mass: $H_2 = 2.0$ g / mol, $CH_4 = 16.0$ g / mol, $C_6H_6 = 78.1$ g / mol, and also in order of increasing heat of vaporization. The last substance has a molar mass of 61.0 g/mol, which would produce intermolecular forces smaller than those of C_6H_6 if CH_3NO_2 were nonpolar. But the molecule is definitely polar. Dipole-dipole forces are strong enough that the intermolecular forces are stronger and the enthalpy of vaporization is larger for CH_3NO_2 than for C_6H_6.

7A Strong interionic forces lead to high melting points. Strong interionic forces are created by ions with high charge and of small size. Thus, a compound with a lower melting point than KI would have ions of larger size, such as RbI or CsI. A compound with a melting point higher than CaO would have either smaller ions, such as MgO, or more highly charged ions, such as Ga_2O_3 or Ca_3N_2, or both, such as AlN or Mg_3N_2.

7B Mg^{2+} has a higher charge and a smaller size than does Na^+. In addition, Cl^- has a smaller size than I^-. Thus, interionic forces should be stronger in $MgCl_2$ than in NaI. We expect $MgCl_2$ to have lower solubility and, in fact, 12.3 mol (1840 g) of NaI dissolves in a liter of water, compared to just 5.7 mol (543 g) of $MgCl_2$, confirming our prediction.

8A The length (l) of a bcc unit cell and the radius (r) of the atom involved are related as $4r = l\sqrt{3}$. For potassium, $r = 227$ pm. Then $l = 4 \times 227$ pm $/ \sqrt{3} = 524$ pm

8B Consider just the face of Figure 13.42c. Note that it is composed of one atom at each of four corners and one in the center. The four corner atoms touch the atom in the center, but not each other. Thus, the atoms are in contact across the diagonal of the face. If each atomic radius is designated r, then the length of the diagonal is $4r (= r$ for one corner atom $+2r$ for the center atom $+r$ for the other corner atom). The diagonal also is related to the length of a side, l, by the Pythagorean theorem: $d^2 = l^2 + l^2 = 2l^2$ or $d = \sqrt{2}l$. We have two quantities equal to the diagonal, and thus to each other.

$$\sqrt{2}l = \text{diagonal} = 4r = 4 \times 143.1 \text{ pm} = 572.4 \text{ pm}$$

$$l = \frac{572.4}{\sqrt{2}} = 404.7 \text{ pm}$$

The cubic unit cell volume, V, is equal to the cube of one side.

$$V = l^3 = (404.7 \text{ pm})^3 = 6.628 \times 10^7 \text{ pm}^3$$

9A In a bcc unit cell, there are eight corner atoms, of which $\frac{1}{8}$ of each is apportioned to the unit cell. There is also one atom in the center. The total number of atoms per unit cell is:

$$=1 \text{ center} + 8 \text{ corners} \times \frac{1}{8} = 2 \text{ atoms. The density, in } g/cm^3, \text{ for this cubic cell:}$$

$$\text{density} = \frac{2 \text{ atoms}}{(524 \text{ pm})^3} \times \left(\frac{10^{12} \text{ pm}}{10^2 \text{ cm}}\right)^3 \times \frac{1 \text{ mol}}{6.022 \times 10^{23} \text{ atoms}} \times \frac{39.10 \text{ g K}}{1 \text{ mol K}} = 0.903 \text{ g}/cm^3$$

The tabulated density of potassium at $20°C$ is $0.86 \text{ g}/cm^3$.

9B In a fcc unit cell the number of atoms is computed as 1/8 atom for each of the eight corner atoms (since each is shared among eight unit cells) plus 1/2 atom for each of the six face atoms (since each is shared between two unit cells). This gives the total number of atoms per unit cell as: atoms/unit cell = (1/8 corner atom × 8 corner atoms/unit cell) + (1/2 face atom × 6 face atoms/unit cell) = 4 atoms / unit cell

Now we can determine the mass per Al atom, and a value for the Avogadro constant.

$$\frac{\text{mass}}{\text{Al atom}} = \frac{2.6984 \text{ g Al}}{1 \text{ cm}^3} \times \left(\frac{100 \text{ cm}}{1 \text{ m}} \times \frac{1 \text{ m}}{10^{12} \text{ pm}}\right)^3 \times \frac{6.628 \times 10^7 \text{ pm}^3}{1 \text{ unit cell}} \times \frac{1 \text{ unit cell}}{4 \text{ Al atoms}}$$

$$= 4.471 \times 10^{-23} \text{ g} / \text{Al atom}$$

$$N_A = \frac{26.9815 \text{ g Al}}{1 \text{ mol Al}} \times \frac{1 \text{ Al atom}}{4.471 \times 10^{-23} \text{g Al}} = 6.035 \times 10^{23} \frac{\text{atoms}}{\text{mol}}$$

10A Across the diagonal of a CsCl unit cell are Cs^+ and Cl^- ions, so that the body diagonal equals $2r(Cs^+) + 2r(Cl^-)$. This body diagonal equals $\sqrt{3}l$, where l is the length of the unit cell.

$$l = \frac{2r(Cs^+) + 2r(Cl^-)}{\sqrt{3}} = \frac{2(167 + 181) \text{ pm}}{\sqrt{3}} = 402 \text{ pm}$$

10B Since NaCl is fcc, the Na^+ ions are in the same locations as were the Al atoms in Practice Example 13-9B, and there are 4 Na^+ ions per unit cell. For stoichiometric reasons, there must also be 4 Cl^- ions per unit cell. These are accounted for as follows: There is one Cl^- along each edge, and each of these edge Cl^- ions are shared among four unit cells and there is one Cl^- precisely in the body center of the unit cell, not shared with any other unit cells. Thus, the number of Cl^- ions is given by: Cl^- ions/unit cell = $\left(1/4 \text{ } Cl^- \text{ on edge} \times 12 \text{ edges per unit cell}\right) + 1 \text{ } Cl^- \text{ in body center} = 4 \text{ } Cl^-/\text{unit cell}$.

The volume of this cubic unit cell is the cube of its length. The density is:

$$\text{NaCl density} = \frac{4 \text{ formula units}}{1 \text{ unit cell}} \times \frac{1 \text{ unit cell}}{(560 \text{ pm})^3} \times \left(\frac{10^{12} \text{ pm}}{1 \text{ m}} \times \frac{1 \text{ m}}{100 \text{ cm}}\right)^3 \times \frac{1 \text{ mol NaCl}}{6.022 \times 10^{23} \text{ f.u.}}$$

$$\times \frac{58.44 \text{ g NaCl}}{1 \text{ mol NaCl}} = 2.21 \text{ g/cm}^3$$

11A

Sublimation of Cs(g):	$Cs(s) \rightarrow Cs(g)$	$\Delta H_{sub} = +78.2 \text{ kJ / mol}$
Ionization of Cs(g):	$Cs(g) \rightarrow Cs^+(g) + e^-$	$\Delta I_1 = +375.7 \text{ kJ / mol}$
		(Table 10.3)

$\frac{1}{2}$ Dissociation of $Cl_2(g)$: $\quad \frac{1}{2} Cl_2(g) \rightarrow Cl(g) \qquad DE = \frac{1}{2} \times 243 = 121.5 \text{ kJ / mol}$

(Table 11.3)

Cl(g) electron affinity:	$Cl(g) + e^- \rightarrow Cl^-(g)$	$EA_1 = -349.0 \text{ kJ / mol}$
		(Figure 10–10)
Lattice energy:	$Cs^+(g) + Cl^-(g) \rightarrow CsCl(s)$	L.E.

Enthalpy of formation: $\quad Cs(s) + \frac{1}{2} Cl_2(s) \rightarrow CsCl(s) \qquad \Delta H_f^{\,o} = -442.8 \text{ kJ / mol}$

$-442.8 \text{ kJ/mol} = +78.2 \text{ kJ/mol} + 375.7 \text{ kJ/mol} + 121.5 \text{ kJ/mol} - 349.0 \text{ kJ/mol} + \text{ L.E.}$
$\qquad\qquad = +226.4 \text{ kJ/mol} + \text{ L.E.}$

L.E. $= -442.8 \text{ kJ} - 226.4 \text{ kJ} = -669.2 \text{ kJ/mol}$

11B

Sublimation:	$Ca(s) \rightarrow Ca(g)$	$\Delta H_{sub} = +178.2 \text{ kJ / mol}$
First ionization energy:	$Ca(g) \rightarrow Ca^+(g) + e^-$	$I_1 = +590 \text{ kJ / mol}$
Second ionization energy:	$Ca^+(g) \rightarrow Ca^{2+}(g) + e^-$	$I_2 = +1145 \text{ kJ / mol}$
Dissociation energy:	$Cl_2(g) \rightarrow 2Cl(g)$	$D.E. = (2 \times 122) \text{ kJ / mol}$
Electron Affinity:	$2 Cl(g) + 2 e^- \rightarrow 2Cl^-(g)$	$2 \times E.A. = 2(-349) \text{ kJ/mol}$
Lattice energy:	$Ca^{2+}(g) + 2 Cl^-(g) \rightarrow CaCl_2(s)$	L.E. $= -2223 \text{ kJ/mol}$

Enthalpy of formation: $\quad Ca(s) + Cl_2(s) \rightarrow CaCl_2(s) \qquad \Delta H_f^{\,o} = ?$

$\Delta H_f^{\,o} = \Delta H_{sub} + I_1 + I_2 + D.E. + (2 \times E.A.) + \text{ L.E.}$
$\qquad = 178.2 \text{ kJ/mol} + 590 \text{ kJ/mol} + 1145 \text{ kJ/mol} + 244 \text{ kJ/mol} - 698 \text{ kJ/mol} - 2223 \text{ kJ/mol}$
$\qquad = -764 \text{ kJ/mol}$

REVIEW QUESTIONS

1. **(a)** ΔH_{vap} symbolizes the molar enthalpy of vaporization, the quantity of heat needed to convert one mole of a substance from liquid to vapor.

    (b) T_c symbolizes the critical temperature, that temperature above which a gas cannot be condensed to a liquid, no matter how high the applied pressure.

    (c) An instantaneous dipole is a momentary imbalance of the positive and negative charges in an atom or a molecule, caused when the electron cloud moves off center in a random manner.

    (d) Coordination number refers to the number of atoms of one type surrounding another atom at the same distance, often those touching the central atom.

    (e) A unit cell is the smallest portion of a crystal that will replicate the crystal through simple translations.

2. **(a)** Capillary Action: A liquid with strong adhesive forces with the glass in tubes of small diameter rises noticeably higher inside the capillary tube than the liquid level outside of the tube. Inversely proportional to the diameter of the tube and proportional to the surface tension (simplest test for γ).

   (b) Polymorphism: Existence of a solid substance in more than one form (i.e., ice I, ice II, ice III, …)

   (c) Sublimation is the physical change in which a substance is converted directly from a solid to a vapor.

   (d) Supercooling refers to the cooling of a liquid below its freezing point without the formation of any solid. The supercooled liquid is in a metastable state.

   (e) The freezing point of a liquid is the point on the cooling curve of a liquid where the temperature remains constant while heat is withdrawn from the sample. During that period, solid forms as liquid freezes.

3. **(a)** Adhesive forces are those between different types of substances, usually forces between two condensed phases. Cohesive forces are those within a substance, again usually referring to liquids or solids.

   (b) Vaporization is the physical process of transforming a liquid into a vapor. Condensation is the reverse process, namely, converting a vapor to liquid.

   (c) A triple point of a substance is the point where three phases coexist, often solid, liquid, and vapor. The critical point is that temperature and pressure above which liquid and vapor cannot be distinguished; just a fluid exists, called a supercritical fluid.

   (d) Face-centered and body-centered cubic unit cells both have atoms at the vertices of a cube. In a face-centered unit cell there also is an atom on the center of each of the six faces of the cube. In contrast, in a body-centered cubic unit cell there is an atom exactly in the center of the three-dimensional cubic cell.

   (e) A tetrahedral hole is one surrounded by four atoms at the vertices of a tetrahedron. An octahedral hole is surrounded by six atoms: above and below, in front and behind, and to the right and the left.

4. *Instantaneous dipole-induced dipole* forces occur after a chance distortion of the electron cloud of an atom or molecule creates an instantaneous imbalance of charge in that particle: an instantaneous dipole. This imbalance creates a similar imbalance in an adjacent particle: an induced dipole. The two dipoles reinforce each other and can induce dipoles in other nearby particles. The dipoles thus created attract each other and hold the particles together. *Dipole-dipole forces* exist between molecules that have permanent resultant dipole moments, which attract each other and hold these molecules together. *Hydrogen bonds* result from the attraction between a highly electronegative atom, usually N, O, or F, and a hydrogen atom bonded to another highly electronegative atom. *Relative strengths* of these forces are hydrogen bonds > dipole-dipole forces > instantaneous dipole-induced dipole forces. Each force in the preceding ordering is approximately ten times stronger than the one that follows it.

5. **(a/ b)** It is possible for a substance not to have a normal melting point or a normal boiling point. Since normal is defined as the boiling point at 760 mmHg, if the solid-liquid-vapor triple point occurs at a pressure higher than 1 atm, for instance, then melting and vaporization must occur at pressures greater than 1 atm. In such a case, however, there will be a normal sublimation point.

(c) Every substance should have a critical point. There should always be a temperature above which molecules are moving so rapidly that forces of attraction between them are ineffective in producing a condensed phase.

6. One would expect the enthalpy of sublimation (d) to be the largest of the four quantities cited. Molar heat capacities are quite small, on the order of fractions of a kilojoule per mole-degree. (Remember that specific heats have values of joules per gram-degree.) All of the heats of transition (or latent heats) are positive numbers and on the order of kilojoules per mole. Since the heat of sublimation is the sum of the heat of fusion and the heat of vaporization, ΔH_{subl} must be the largest of the three.

7. **(a)** Intermolecular forces in a liquid *do* affect its vapor pressure. The stronger these forces are, the more difficult it is for molecules to vaporize and the lower the vapor pressure is at a given temperature.

(b) The volume of liquid present *does not* affect its vapor pressure. Vaporization occurs at the surface of the liquid and thus volume of liquid is of no consequence.

(c) The volume of vapor present also *does not* affect the vapor pressure of a liquid, for similar reasons to those given in part (b).

(d) The size of the container also *does not* affect a liquid's vapor pressure, again following the reasoning in part (b). Of course, both liquid and vapor must be present at the point of equilibrium.

(e) The temperature of the liquid *does* affect that liquid's vapor pressure. Higher temperature means that more energy is present per molecule, each molecule has a larger fraction of the energy needed to overcome cohesive forces, and thus it is more likely that it will be able to overcome those cohesive forces and vaporize.

8. **(a)** $\text{mass } CHCl_3 \text{ vaporized} = 6.62 \text{ kJ} \times \dfrac{1000 \text{ J}}{1 \text{ kJ}} \times \dfrac{1 \text{ g } CHCl_3}{247 \text{ J}} = 26.8 \text{ g } CHCl_3$

(b) $\Delta H_{vap} = \dfrac{247 \text{ J}}{1 \text{ g } CHCl_3} \times \dfrac{1 \text{ kJ}}{1000 \text{ J}} \times \dfrac{119.38 \text{ g } CHCl_3}{1 \text{ mol } CHCl_3} = 29.5 \text{ kJ/mol chloroform}$

(c) $\text{heat evolved} = 19.6 \text{ g } CHCl_3 \times \dfrac{247 \text{ J}}{1 \text{ g } CHCl_3} \times \dfrac{1 \text{ kJ}}{1000 \text{ J}} = 4.84 \text{ kJ} \text{ or } \Delta H = -4.84 \text{ kJ.}$

9. **(a)** We read up the $100°C$ line until we arrive at C_6H_7N curve (e). This occurs at about 45 mmHg.

(b) We read across the 760 mmHg line until we arrive at the C_7H_8 curve (d). This occurs at about $110°C$.

10. **(a)** The normal boiling point occurs where the vapor pressure is 760 mmHg, and thus $\ln P = 6.63$. For aniline, this occurs at about the uppermost data point (open circle) on the aniline line. This corresponds to $1/T = 2.18 \times 10^{-3}$ K^{-1}. Thus,

$$T_{nbp} = \frac{1}{2.18 \times 10^{-3} \text{ K}^{-1}} = 459 \text{ K}.$$

(b) $25°$ C $= 298$ K $= T$ and thus $1/T = 3.36 \times 10^{-3}$ K^{-1}. This occurs at about $\ln P = 6.25$. Thus, $P = e^{6.25} = 518$ mmHg.

11. Use the ideal gas equation, $n = $ moles Br$_2$ $= 0.486$ g Br$_2 \times \dfrac{1 \text{ mol Br}_2}{159.8 \text{ g Br}_2} = 3.04 \times 10^{-3}$ mol Br$_2$.

$$P = \frac{nRT}{V} = \frac{3.04 \times 10^{-3} \text{ mol Br}_2 \times 0.03206 \text{ L atm mol}^{-1} \text{ K}^{-1} \times 298.2 \text{ K}}{0.2500 \text{ L}} \times \frac{760 \text{ mmHg}}{1 \text{ atm}}$$

$= 226$ mmHg

12. We use the Clausius-Clapeyron equation (13.2).

$$T_1 = (56.0 + 273.2) \text{ K} = 329.2 \text{ K} \qquad T_2 = (103.7 + 273.2) \text{ K} = 376.9 \text{ K}$$

$$\ln\frac{10.0 \text{ mmHg}}{100.0 \text{ mmHg}} = \frac{\Delta H_{vap}}{8.3145 \text{ J mol}^{-1} \text{ K}^{-1}}\left(\frac{1}{376.9 \text{ K}} - \frac{1}{329.2 \text{ K}}\right) = -2.30 = -4.624 \times 10^{-5} \Delta H_{vap}$$

$$\Delta H_{vap} = 4.97 \times 10^4 \text{ J/mol} = 49.7 \text{ kJ/mol}$$

13. We use the Clausius-Clapeyron equation (13.2). $T_1 = (5.0 + 273.2) \text{ K} = 278.2 \text{ K}$

$$\ln\frac{760.0 \text{ mmHg}}{40.0 \text{ mmHg}} = 2.944 = \frac{38.0 \times 10^3 \text{ J/mol}}{8.3145 \text{ J mol}^{-1} \text{ K}^{-1}}\left(\frac{1}{278.2 \text{ K}} - \frac{1}{T_{nbp}}\right)$$

$$\left(\frac{1}{278.2 \text{ K}} - \frac{1}{T_{nbp}}\right) = 2.944 \times \frac{8.3145 \text{ K}^{-1}}{38.0 \times 10^3} = 6.44 \times 10^{-4} \text{ K}^{-1}$$

$$\frac{1}{T_{nbp}} = \frac{1}{278.2 \text{ K}} - 6.44 \times 10^{-4} \text{ K}^{-1} = 2.95 \times 10^{-3} \text{ K}^{-1} \qquad T_{nbp} = 339 \text{ K}$$

WRONG!

14. heat needed $= (5.08 \text{ cm})^3 \times \dfrac{0.92 \text{ g}}{1 \text{ cm}^3} \times \dfrac{1 \text{ mol}}{18.0 \text{ g}} \times \dfrac{6.01 \text{ kJ}}{1 \text{ mol}} = 40 \text{ .kJ}$

15. **(a)** heat evolved $= 3.78$ kg Cu $\times \dfrac{1000 \text{ g}}{1 \text{ kg}} \times \dfrac{1 \text{ mol Cu}}{63.55 \text{ g Cu}} \times \dfrac{13.05 \text{ kJ}}{1 \text{ mol Cu}} = 776$ kJ evolved or

$\Delta H = -776$ kJ

(b) heat absorbed $= (75 \text{ cm} \times 15 \text{ cm} \times 12 \text{ cm}) \times \dfrac{8.92 \text{ g}}{1 \text{ cm}^3} \times \dfrac{1 \text{ mol Cu}}{63.55 \text{ g Cu}} \times \dfrac{13.05 \text{ kJ}}{1 \text{ mol Cu}} = 2.5 \times 10^4$ kJ

16. Let us use the ideal gas law to determine the final pressure in the container, assuming that all of the dry ice vaporizes. We then locate this pressure, at a temperature of $25°$ C, on the phase diagram of Figure 13-19.

$$P = \frac{nRT}{V} = \frac{\left(80.0\,g\,CO_2 \times \dfrac{1\,mol\,CO_2}{44.0\,g\,CO_2}\right) \times 0.08206 \dfrac{L\,atm}{mol\,K} \times 298\,K}{0.500\,L} = 88.9\,atm$$

Although this point ($25°$ C and 88.9 atm) is most likely in the region labeled "liquid" in Figure 13–19, we computed its pressure assuming the CO_2 is a gas. Some of this gas should condense to a liquid. Thus, both liquid and gas are present in the container.

17. (After each formula is that substance's boiling point)

(a) $C_{10}H_{22}\left(174.1°\ C\right)$ has a higher boiling point than $C_7H_{16}\left(98.4°\ C\right)$. All else being equal, the substance with the higher molar mass has the higher boiling point.

(b) $CH_3OCH_3\left(M = 46.0\ g/mol\right)\left(-25°C\right)$ has a higher boiling point than $C_3H_8\left(M = 44.0\ g/mol\right)\left(-42.1°C\right)$. For two substances with nearly the same molar masses, the polar substance has a higher boiling point than the nonpolar substance.

(c) $CH_3CH_2OH\left(M = 46.0\ g/mol\right)$ has a higher boiling point than $CH_3CH_2SH\left(M = 62.1\ g/mol\right)$. Even though $CH_3CH_2SH\left(35°C\right)$ is the heavier molecule, hydrogen bonds form between $CH_3CH_2OH\left(78.5°C\right)$ molecules, requiring more energy to vaporize them.

18. All of these substances are homonuclear molecules, composed of the same type of atoms. Thus, boiling point should increase with increasing molar mass: $N_2\left(M = 28.0\ g/mol\right)$, $O_3\left(M = 48.0\ g/mol\right)$, $F_2\left(M = 38.0\ g/mol\right)$, $Ar\left(M = 39.9\ g/mol\right)$, and $Cl_2\left(M = 70.9\ g/mol\right)$. Thus, O_3 is seen to be out of order. The correct order is:

$$N_2\left(-195.8°\ C\right) < \ F_2\left(-188.1°C\right) < \ Ar\left(-185.7°C\right) < O_3\left(-111.9°C\right) \ < \ Cl_2\left(-34.6°C\right)$$

19. Both Ne and C_3H_8 are nonpolar; their solids are held together by relatively weak London forces, which are stronger for substances with higher molecular masses (or larger surface areas). Ne has a lower melting point than C_3H_8. Next in melting are CH_3CH_2OH and $CH_2OHCHOHCH_2OH$, which are held together with hydrogen bonds in addition to London forces. There are more hydrogen bonds possible for a molecule of $CH_2OHCHOHCH_2OH$ than for one of CH_3CH_2OH , and $CH_2OHCHOHCH_2OH$ molecules are heavier; $CH_2OHCHOHCH_2OH$ has the higher melting point. KI, K_2SO_4, and MgO are ionic solids and have the highest melting points. For ionic solids the melting point is high for small, highly charged ions. Thus, MgO has the highest melting point of these three, KI the lowest. Arranged in order of increasing melting point, the seven substances are as follows:

$$Ne < C_3H_8 < CH_3CH_2OH < CH_2OHCHOHCH_2OH < KI < K_2SO_4 < MgO$$
$$\left(-248.6°\ C < -187.7°\ C < -97.8°\ C < 18.2°C < 681°C < 1067°C < 1067°C < 2825°C\right)$$

20. The polarity of hydrazine (34.0 g/mol) might explain its high boiling point. But HCl (36.5 g/mol) also is polar, and it is a gas at room temperature. The distinction between the two substances is that hydrazine can form strong hydrogen bonds, while HCl cannot. This raises the boiling point of hydrazine significantly above that of $HCl\left(-114.18°\ C\right)$.

21. **(a)** Si(s) is a network covalent solid. Its structure is similar to that of diamond; both elements are in the same family of the periodic table.

  (b) $SiCl_4(s)$ is a molecular solid. $SiCl_4$ exists as discrete nonpolar molecules. The molecules are attracted to each other by London forces.

  (c) $CaCl_2(s)$ is an ionic solid. Calcium chloride consists of Ca^{2+} cations and Cl^- anions.

  (d) Ag(s) is a metallic solid. Silver is one of the metals in the periodic table.

  (e) $SO_2(s)$ will form a molecular solid. SO_2 exists as discrete polar molecules. The molecules are attracted to each other by London forces and dipole-dipole attractions.

22. Answer (c) is correct; atoms at the corners, along the edges, and on the faces of a unit cell are shared with adjacent unit cells. **(a)** is incorrect; often there are several formula units within a unit cell, as in the case for NaCl. **(b)** is incorrect; the unit cell need not be cubic; that of hexagonal close packing is not. **(d)** is incorrect; a unit cell will not contain the same number of cations as anions if their numbers are not equal in the formula of the compound.

23. In the unit cell of Figure 13-49, there is one Cs^+ ion completely contained within the unit cell. There are eight "corner" Cl^- ions, each shared by eight other unit cells. Thus, the number of Cl^- ions $= (8 \times 1/8) = 1$ Cl^- ions. Thus, the "formula of cesium chloride" is CsCl.

24. **(a)** The atoms touch along the face diagonal, d. The length of that diagonal thus is $4r$ (r from one corner atom, $+2r$ from the center atom, $+r$ from the other corner atom). The Pythagorean theorem relates the length of the unit cell, l, to the face diagonal: $d^2 = l^2 + l^2$ (the cell is square; both sides are equal) or $d = \sqrt{2}l$. Two quantities equal to the face diagonal are equal to each other: $d = \sqrt{2}l = 4r = 4 \times 128$ pm

  $$l = \frac{4 \times 128\ \text{pm}}{1.414} = 362\ \text{pm}$$

  (b) The volume of the unit cell is that of a cube: $V = l^3 = (362\ \text{pm})^3 = 4.74 \times 10^7\ \text{pm}^3$.

  (c) In a face-centered unit cell, there are eight "corner" Cu atoms, each shared by eight unit cells; there are six "face" Cu atoms, each shared by two unit cells.

$$\text{No. Cu atoms} = \left(8 \times \frac{1}{8}\right) + \left(6 \times \frac{1}{2}\right) = 4 \text{ Cu atoms.}$$

(d) Volume of 4 Cu atoms $= 4\left(\frac{4}{3}\pi r^3\right) = 4\left(\frac{4}{3}\pi(128\,pm)^3\right) = 3.51 \times 10^7\,pm^3$

Percentage of unit cell occupied by Cu $= \dfrac{3.51 \times 10^7\,pm^3}{4.74 \times 10^7\,pm^3} \times 100\% = 74.1\%$

(e) The mass of 4 Cu atoms is determined from the atomic mass.

$$\frac{mass}{unit\ cell} = \frac{4\ Cu\ atoms}{unit\ cell} \times \frac{1\ mol\ Cu\ atoms}{6.022 \times 10^{23}\ Cu\ atoms} \times \frac{63.55\ g\ Cu}{1\ mol\ Cu\ atoms} = \frac{4.221 \times 10^{-22}\ g\ Cu}{1\ unit\ cell}$$

(f) $density = \dfrac{mass}{volume} = \dfrac{4.221 \times 10^{-21}\ g}{4.74 \times 10^7\ pm^3} \times \left(\dfrac{10^{12}\ pm}{1\ m} \times \dfrac{1\ m}{100\ cm}\right)^3 = 8.91\ g/cm^3$

EXERCISES

Surface Tension; Viscosity

25. Since both the silicone oil and the cloth or leather are composed of relatively nonpolar molecules, they attract each other. The oil thus adheres well to the material. Water, on the other hand is polar and adheres very poorly to the silicone oil (actually, the water is repelled by the oil), much more poorly, in fact, than it adheres to the cloth or leather. This is because the oil is more nonpolar than is the cloth or the leather. Thus, water is repelled from the silicone-treated cloth or leather.

26. Both surface tension and viscosity deal with the work needed to overcome the attractions between molecules. Increasing the temperature of a liquid sample causes the molecules to move around faster. Some of the work has been done by adding thermal energy (heat) and less work needs to be done by the experimenter; both surface tension and viscosity decrease. The vapor pressure is a measure of the concentration of molecules that have come free of the surface. As thermal energy is added to the liquid sample, more and more molecules have enough energy to break free of the surface, and the vapor pressure increases.

27. (a) Molasses, like honey, is a very viscous liquid (high resistance to flow)
(b) Temperatures in January (Northern Hemisphere) are generally at their lowest point during the year
(c) Viscosity generally increases as the temperature decreaseHence, molasses at low temperature is a very slow flowing liquid - there is indeed a scientific basis for "slower than molasses" in January.

28. The product can lower the surface tension of water. Then the water can more easily wet a solid substance, because a greater surface area of water can be created with the same energy. (Surface tension equals the work needed to create a given quantity of surface area.)

This greater water surface area means a greater area of contact with a solid object, such as a piece of fabric. More of the fabric being in contact with the water means that the water indeed is wetter.

Vaporization

29. The process of evaporation is an endothermic one; it requires energy. If evaporation occurs from an uninsulated container, this energy is obtained from the surroundings, through the walls of the container. However, if the evaporation occurs from an insulated container, the only source of the needed energy is the liquid that is evaporating. Therefore, the temperature of the liquid decreases.

30. Vapor cannot form throughout the liquid at temperatures below the boiling point because for vapor to form, it must overcome the atmospheric pressure (≈ 1 atm) or slightly more due to the pressure of the liquid. Formation of a bubble of vapor in the liquid, requires that it must push the liquid out of the way. This is not true at the surface. The vapor molecules simply move into the gas phase at the surface, which is mostly empty space.

31. We use the quantity of heat to determine the number of moles of benzene that vaporize.

$$V = \frac{nRT}{P} = \frac{\left(1.54\,kJ \times \dfrac{1\,mol}{33.9\,kJ}\right) \times 0.08206\,\dfrac{L\,atm}{mol\,K} \times 298\,K}{95.1\,mmHg \times \dfrac{1\,atm}{760\,mmHg}} = 8.88\,L\,C_6H_6(l)$$

32. $n_{acetonitrile} = \dfrac{PV}{RT} = \dfrac{1.00\,atm \times 1.17\,L}{0.08206\,L\,atm\,mol^{-1}\,K^{-1} \times (273.2 + 81.6)\,K} = 0.0402$ mol acetonitrile

$\Delta H_{vap} = \dfrac{1.00\,kJ}{0.0402\,mol} = 24.9$ kJ / mol acetonitrile

33. 25.00 mL of N_2H_4 (25 °C) density(25°C) = 1.0036 g mL^{-1} (Molar mass = 32.0452 g mol^{-1})
Mass of N_2H_4 = (volume)(density) = (25.00 mL)(1.0036 g mL^{-1}) = 25.09 g N_2H_4

$n_{N2H4} = 25.09\,g\,N_2H_4 \times \dfrac{1\,mol\,N_2H_4}{32.0452\,g\,N_2H_4} = 0.7830$ mol

Energy required to increase temperature from 25.0 °C to 113.5 °C ($\Delta t = 88.5$ °C)

$q_{heating} = (n)(C)(\Delta t) = (0.78295\,mol\,N_2H_4)\left(\dfrac{98.84\,J}{1\,mol\,N_2H_4\,°C}\right)(88.5\,°C) = 6848.7$ J or 6.85 kJ

$q_{vap} = (n_{N_2H_4})(\Delta H_{vap}) = (0.78295\,mol\,N_2H_4)\left(\dfrac{43.0\,kJ}{1\,mol\,N_2H_4}\right) = 33.7$ kJ

$q_{overall} = q_{heating} + q_{vap} = 6.85\,kJ + 33.7\,kJ = 40.5$ kJ

34. ΔH_{vap} for $CH_3OH(l) = 38.0$ kJ mol^{-1} at 298 K (assume constant)

$\Delta t = 30.0\ °C - 20.0\ °C = 10.0\ °C$

$$n_{CH_3OH} = 215\ g\ CH_3OH \times \left(\frac{1\,mol\,CH_3OH}{32.0422\,g\,CH_3OH}\right) = 6.71\ mol\ CH_3OH$$

Raise temperature of liquid from 20.0 °C to 30.0 °C

$$q = (n)(C)(\Delta t) = (6.71\ mol\ CH_3OH)\left(\frac{81.1\ J}{1\ mol\ CH_3OH\ °C}\right)(10.0\ °C) = 5441.8\ J\ or\ 5.44\ kJ$$

Vaporize liquid at 30°C (Use ΔH_{vap} at 25 °C and assume constant)

$$q_{vap} = (n_{CH_3OH})(\Delta H_{vap}) = (6.71\ mol\ CH_3OH)\left(\frac{38.0\,kJ}{1\ mol\,CH_3OH}\right) = 255\ kJ$$

$q_{overall} = q_{heating} + q_{vap} = 5.44\ kJ + 255\ kJ = 260.\ kJ$

35. heat needed $= 3.78\ L\ H_2O \times \dfrac{1000\ cm^3}{1\ L} \times \dfrac{0.958\ g\,H_2O}{1\ cm^3} \times \dfrac{1\ mol\ H_2O}{18.02\ g\,H_2O} \times \dfrac{40.7\ kJ}{1\ mol\ H_2O} = 8.18 \times 10^3\ kJ$

amount CH_4 needed $= 8.18 \times 10^3\ kJ \times \dfrac{1\ mol\ CH_4}{890\ kJ} = 9.19\ mol\ CH_4$

$V = \dfrac{nRT}{P} = \dfrac{9.19\,mol \times 0.08206\,L\ atm\ mol^{-1}K^{-1} \times 296.6\,K}{768\,mmHg \times \dfrac{1\,atm}{760\,mmHg}} = 221\ L$ methane

36. If not all of the water vaporizes, the final temperature of the system will be $100.00°$ C. Let us proceed on that assumption and modify our final state if it is not true. First we determine the heat available from the iron in cooling down, then the heat needed to warm the water to boiling, and finally the mass of water that vaporizes.

heat from Fe $=$ mass $\times$ sp.ht. $\times \Delta t = 50.0\ g \times \dfrac{0.45\ J}{g°\ C} \times (100.00°\ C - 152°\ C) = -1.17 \times 10^3\ J$

heat to warm water $= 20.0\ g \times \dfrac{4.21\ J}{g°C} \times (100.00°C - 89°C) = 9.3 \times 10^2\ J$

mass of water vaporized:

$= (11.7 \times 10^2\ J\ available - 9.3 \times 10^2\ J\ used) \times \dfrac{1\ mol\ H_2O\ vaporized}{40.7 \times 10^3\ J} \times \dfrac{18.02\ g\ H_2O}{1\ mol\ H_2O}$

$= 0.11$ g of water vaporize

Clearly, all of the water does not vaporize and our initial assumption was true.

Vapor Pressure and Boiling Point

37. **(a)** When the water vaporizes in the outer container, heat is required; vaporization is an endothermic process. When this vapor (steam) condenses on the outside walls of the inner container, that same heat is liberated; condensation is an exothermic process.

(b) Liquid water, condensed on the outside wall, is in equilibrium with the water vapor that fills the space between the two containers. This equilibrium exists at the boiling point of water. We assume that the pressure is 1.000 atm, and thus, the temperature of the equilibrium must be 373.15 K or 100.00° C. This is the maximum temperature that can be realized without pressurizing the apparatus.

38. As the can cools, the water vapor in the can condenses to liquid and the pressure inside the can lowers to the vapor pressure of water at room temperature ($\approx$ 25 mmHg). The pressure on the outside of the can is still near 760 mmHg. It is this huge difference in pressure that is responsible for the can being crushed.

39. With the Clausius-Clapeyron equation, we use the vapor pressure of water at $100.0°C = 373.2$ K and $120.0°C = 393.2$ K to determine ΔH_{vap} of water near its boiling point. We then use the equation again, to determine the temperature where water's vapor pressure is 2.00 atm.

$$\ln \frac{1489.1 \text{ mmHg}}{760.0 \text{ mmHg}} = \frac{\Delta H_{vap}}{8.3145 \text{ J mol}^{-1} \text{ K}^{-1}} \left(\frac{1}{373.2 \text{ K}} - \frac{1}{393.2 \text{ K}} \right) = 0.6726 = 1.639 \times 10^{-5} \Delta H_{vap}$$

$$\Delta H_{vap} = 4.104 \times 10^4 \text{ J / mol} = 41.04 \text{ kJ / mol}$$

$$\ln \frac{1520.0 \text{ mmHg}}{760.0 \text{ mmHg}} = 0.6931 = \frac{41.04 \times 10^3 \text{ J mol}}{8.3145 \text{ J mol}^{-1} \text{ K}^{-1}} \left(\frac{1}{373.2 \text{ K}} - \frac{1}{T} \right)$$

$$\left(\frac{1}{373.2 \text{ K}} - \frac{1}{T_{bp}} \right) = 0.6931 \times \frac{8.1345 \text{ K}^{-1}}{41.03 \times 10^3} = 1.404 \times 10^{-4} \text{ K}^{-1}$$

$$\frac{1}{T_{bp}} = \frac{1}{373.2 \text{ K}} - 1.404 \times 10^{-4} \text{ K}^{-1} = 2.539 \times 10^{-3} \text{ K}^{-1} \quad T_{bp} = 393.9 \text{ K} = 120.7° \text{ C}$$

40. **(a)** We need the temperature at which the vapor pressure of water is 640 mmHg. This is a temperature between 95.0°C (633.9 mmHg) and 96.0°C (657.6 mmHg), about one-third between, in fact. We estimate a boiling point of 95.3°C.

(b) If the observed boiling point is 94°C, the atmospheric pressure equals the vapor pressure of water at 94°C, that is, 611 mmHg.

41. The 25.0 L of He becomes saturated with aniline vapor, at a pressure equal to the vapor pressure of aniline.

$$n_{aniline} = (6.220 \text{ g} - 6.108 \text{ g}) \times \frac{1 \text{ mol aniline}}{93.13 \text{ g aniline}} = 0.00120 \text{ mol aniline}$$

$$P = \frac{nRT}{V} = \frac{0.00120 \text{ mol} \times 0.08206 \text{ L atm mol}^{-1} \text{ K}^{-1} \times 303.2 \text{ K}}{25.0 \text{ L}} = 0.00119 \text{ atm} = 0.907 \text{ mmHg}$$

42. This is essentially a Boyle's Law problem. The initial pressure is the pressure of the $N_2(g)$ and of the CCl_4 vapor: $P_i = 742 \text{ mmHg} + 261 \text{ mmHg} = 1003 \text{ mmHg}$, according to Dalton's law of partial pressures. The final pressure is 742 mmHg. The initial volume is 7.53 L.

$$V_f = V_i \times \frac{P_i}{P_f} = 7.53 \text{ L} \times \frac{1003 \text{ mmHg}}{742 \text{ mmHg}} = 10.18 \text{ L} \cong 10.2 \text{ L}$$

43. 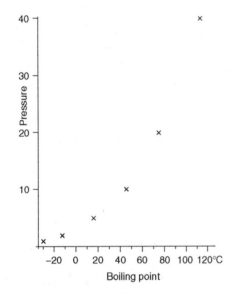 The graph of pressure vs. boiling point for Freon-12 is shown.

At a temperature of $25°$ C the vapor pressure is approximately 6.5 atm for Freon-12. Thus the compressor must be capable of producing a pressure greater than 6.5 atm.

44. At $27°$ C, the vapor pressure of water is 26.7 mmHg. We use this value to determine the mass of water that could exist within the container as vapor.

$$\text{mass of water vapor} = \frac{\left(26.7 \text{ mmHg} \times \dfrac{1 \text{ atm}}{760 \text{ mmHg}}\right) \times \left(1515 \text{ mL} \times \dfrac{1 \text{ L}}{1000 \text{ mL}}\right)}{\dfrac{0.08206 \text{ L atm}}{\text{mol K}} \times (27 + 273) \text{ K}}$$

$$= 0.00216 \text{ mol H}_2\text{O}(g)$$

$$\text{mass of water vapor} = 0.00216 \text{ mol H}_2\text{O}(g) \times \frac{18.02 \text{ g H}_2\text{O}}{1 \text{ mol H}_2\text{O}} = 0.0389 \text{ g H}_2\text{O}(g)$$

The Clausius-Clapeyron Equation

45. $T = 56.2°$ C is $T = 329.4$ K

$$\ln\frac{760 \text{ mmHg}}{375 \text{ mmHg}} = \frac{25.5 \times 10^3 \text{ J/mol}}{8.3145 \text{ J mol}^{-1} \text{ K}^{-1}}\left(\frac{1}{T} - \frac{1}{329.4 \text{ K}}\right) = 0.706$$

$$\left(\frac{1}{T} - \frac{1}{329.4 \text{ K}}\right) = \frac{0.706 \times 8.3145}{25.5 \times 10^3} \text{ K}^{-1} = 2.30 \times 10^{-4} \text{ K}^{-1} = 1/T - 3.03_6 \times 10^{-3} \text{ K}^{-1}$$

$$1/T = (3.03_6 + 0.230) \times 10^{-3} \text{ K}^{-1} = 3.266 \times 10^{-3} \text{ K}^{-1} \qquad T = 306 \text{ K} = 33° \text{ C}$$

46. $P_1 = 40.0$ torr $T_1 = -7.1$ °C (266 K) and $\Delta H_{vap} = 29.2$ kJ mol^{-1}
$P_2 = 760.0$ torr $T_2 = ?$

$$\ln\left(\frac{760.0}{40.0}\right) = \frac{29,200 \text{ J}}{8.31451 \dfrac{\text{J}}{\text{K mol}}}\left(\frac{1}{266 \text{ K}} - \frac{1}{T_2}\right) \qquad T_2 = 342.\underline{3} \text{ K or } \sim 69 \text{ °C (literature 61 °C)}$$

47. normal boiling point = 179 °C and critical point = 422 °C and 45.9 atm

$$\ln\left(\frac{P_2}{P_1}\right) = \frac{\Delta H_{vap}}{R}\left(\frac{1}{T_1} - \frac{1}{T_2}\right) \qquad \ln\left(\frac{45.9}{1}\right) = \frac{\Delta H_{vap}}{8.3145 \text{ J K}^{-1}\text{mol}^{-1}}\left(\frac{1}{452.2 \text{ K}} - \frac{1}{695.2 \text{ K}}\right)$$

$\Delta H_{vap} = 41.2$ kJ mol^{-1}

$$\ln\left(\frac{1}{P}\right) = \frac{41,200 \text{ J mol}^{-1}}{8.3145 \text{ J K}^{-1}\text{mol}^{-1}}\left(\frac{1}{373.2} - \frac{1}{452.2 \text{ K}}\right) \qquad P = 0.0981 \text{ atm or 74.6 torr}$$

48. **(a)** The plot of ln P vs 1/T for benzene is located to the right of that for toluene. This tells us that for a given temperature, the vapor pressure for toluene is lower than that for benzene, and thus benzene is a more volatile liquid than toluene

(b) According to Figure 13-13, at 65 °C, the vapor pressure for benzene is ~ 450 mm Hg or 0.60 atm. We have been asked to estimate the temperature at which toluene has a vapor pressure of 0.60 atm. First, we find ln 450 = 6.11 on the benzene curve of Figure 13.13. Then, we move horizontally to the left along ln P = 6.11 until we reach the toluene curve. At this point on the toluene curve, we observe the value of 1/T = 0.00273. This corresponds to T = 366 K.

Critical Point

49. A substance that can exist as a liquid at room temperature (about $20°$ C) is one whose critical temperature is above $20°$ C, 293 K. Of the substances listed in Table 13.3, this includes $CO_2 (T_c = 304.2$ K), $HCl(T_c = 324.6$ K), $NH_3 (T_c = 405.7$ K), $SO_2 (T_c = 431.0$ K), and $H_2O(T_c = 647.3$ K). In fact, CO_2 exists as a liquid in CO_2 fire extinguishers.

50.   The critical temperature of SO_2 is 431.0 K, which is above the temperature of $0°C$, 273 K. The critical pressure of SO_2 is 77.7 atm, which is below the pressure of 100 atm. Thus, SO_2 can be maintained as a liquid at $0°$ C and 100 atm. The critical temperature of methane, CH_4, is 191.1 K, which is below the temperature of $0°$ C, 273 K. Thus, CH_4 cannot exist as a liquid at $0°$ C, no matter what pressure is applied.

States of Matter and Phase Diagrams

51. **(a)**   As heat is added initially, the temperature of the ice rises from $-20°$ C to $0°$ C. At (or just slightly below) $0°$ C, ice begins to melt to liquid water. The temperature remains at $0°$ C until all of the ice has melted. Adding heat then warms the liquid until a temperature of about $93.5°$ C is reached, where the liquid begins to vaporize to steam. The temperature remains fixed until all the water is converted to steam. Adding heat then warms the steam to $200°$ C. (Data for this part are taken from Figure 13-21 and Table 13.2.)

(b)   As the pressure is raised, initially gaseous iodine is compressed. At about 91 mmHg, liquid iodine appears; the system remains at a fixed pressure with further compression until all vapor is converted to liquid. Increasing the pressure further simply compresses the liquid until a high pressure is reached, perhaps 50 atm, where solid iodine appears. Again the pressure remains fixed with further compression until all iodine is converted to solid. After this has occurred further compression raises the pressure of the system until 100 atm is reached. (Data for this part are from Figure 13-18 and the surrounding text.)

(c)   Cooling of gaseous CO_2 simply lowers the temperature until a temperature of perhaps $20°$ C is reached. At this point, liquid CO_2 appears. The temperature remains constant as more heat is removed until all the gas is converted to liquid. Further cooling then lowers the temperature of the liquid until a temperature of slightly higher than $-56.7°$ C is reached, where solid CO_2 appears. At this point, further cooling simply converts liquid to solid at constant temperature, until all liquid has been converted to solid. From this point, further cooling lowers the temperature of the solid. (Data for this part are taken from Figure 13-19 and Table 13.3.)

52. **(a)**   The upper-right region of the phase diagram is the liquid region, while the lower-right region is the region of gas. One way to figure this out is to imagine moving from left to right (toward higher temperatures) at constant pressure. (Assume 45 atm for this case.) From experience, we known that the progression of states is solid (low temperature) $\rightarrow$ liquid (intermediate temperature) $\rightarrow$ gas (high temperature).

(b)   Melting means that we convert the solid to a liquid. As the phase diagram shows, the lowest pressure at which liquid exists is at the triple point pressure, 43 atm. 1.00 atm is far below 43 atm; liquid cannot exist at this temperature.

(c)   As we move from point A to point B by lowering the pressure, initially nothing happens. At a certain pressure, the solid liquefies. The pressure continues to drop, with the entire sample being liquid while it does, until another, lower pressure is reached. At this lower pressure the entire sample vaporizes. The pressure then continues to drop, with the gas becoming less dense as it does so, until point B is reached.

53.   0.240 g H_2O corresponds to 0.0133 mol H_2O. If the water does not vaporize completely, the pressure of the vapor in the flask equals the vapor pressure of water at the indicated temperature. However, if the water vaporizes completely, the pressure of the vapor is determined by the ideal gas law.

(a)   30.0° C, vapor pressure of H_2O = 31.8 mmHg = 0.0418 atm

$$n = \frac{PV}{RT} = \frac{0.0418 \text{ atm} \times 3.20 \text{ L}}{0.08206 \text{ L atm mol}^{-1} \text{ K}^{-1} \times 303.2 \text{ K}}$$

n = 0.00538 mol H_2O vapor, this is less than 0.0133 mol H_2O;

This represents a non-eqilibrium condition, as not all the H_2O vaporizes.

The pressure in the flask is 0.0418 atm. (from tables)

(b)   50.0° C, vapor pressure of H_2O = 92.5 mmHg = 0.122 atm

$$n = \frac{PV}{RT} = \frac{0.122 \text{ atm} \times 3.20 \text{ L}}{0.08206 \text{ L atm mol}^{-1} \text{ K}^{-1} \times 323.2 \text{ K}}$$
= 0.0147 mol H_2O vapor > 0.0133 mol H_2O; all the H_2O vaporizes. Thus,

$$P = \frac{nRT}{V} = \frac{0.0133 \text{ mol} \times 0.08206 \text{ L atm mol}^{-1} \text{ K}^{-1} \times 323.2 \text{ K}}{3.20 \text{ L}} = 0.110 \text{ atm} = 83.8 \text{ mmHg}$$

(c)   70.0° C All the H_2O must vaporize, as this temperature is higher than that of part (b). Thus,

$$P = \frac{nRT}{V} = \frac{0.0133 \text{ mol} \times 0.08206 \text{ L atm mol}^{-1} \text{ K}^{-1} \times 343.2 \text{ K}}{3.20 \text{ L}} = 0.117 \text{ atm} = 89.0 \text{ mmHg}$$

54.   **(a)**   If the pressure exerted by 2.50 g of H_2O (vapor) is less than the vapor pressure of water at 120. °C (393 K), then the water must exist entirely as a vapor.

$$P = \frac{nRT}{V} = \frac{\left(2.50 \text{ g} \times \dfrac{1 \text{ mol } H_2O}{18.02 \text{ g } H_2O}\right) \times 0.08206 \text{ L atm mol}^{-1}\text{K}^{-1} \times 393 \text{ K}}{5.00 \text{ L}} = 0.895 \text{ atm}$$

Since this is less than the 1.00 atm vapor pressure at 100. °C, it must be less than the vapor pressure of water at 120 °C; the water exists entirely as a vapor.

(b) 0.895 atm $= 680$ mmHg. From Table 13.2, we see that this corresponds to a temperature of $97.0°$ C (at which temperature the vapor pressure of water is 682.1 mmHg). Thus, at temperatures slightly less than $97.0°$ C the water will begin to condense to liquid. But we have forgotten that, in this constant-volume container, the pressure (of water) will decrease as the temperature decreases. Calculating the precise decrease in both involves linking the Clausius-Clapeyron equation with the expression $P = kT$, but we can estimate the final temperature. First, we determine the pressure if we lower the temperature from 393 K to $97°$ C (370 K).

$$P_f = \frac{370 \text{ K}}{393 \text{ K}} \times 680 \text{ mmHg} = 640 \text{ mmHg}$$ From Table 13.2, this pressure occurs at a

temperature of about $95°$ C (368 K). We now determine the final pressure at this temperature. $P_f = \dfrac{368 \text{ K}}{393 \text{ K}} \times 680 \text{ mmHg} = 637 \text{ mmHg}$ This is just slightly above the

vapor pressure of water (633.9 mmHg) at $95°$ C, which we conclude must be the temperature at which liquid water appears.

55. **(a)** According to Figure 13-19, $CO_2(s)$ exists at temperatures below $-78.5°C$ when the pressure is 1 atm or less. We do not expect to find temperatures this low and partial pressures of $CO_2(g)$ of 1 atm on the surface of the earth.

(b) According to Table 13.3, the critical temperature of CH_4, the maximum temperature at which $CH_4(l)$ can exist, is $191.1 \text{ K} = -82.1°$ C. We do not expect to find temperatures this low on the surface of the earth.

(c) Since, according to Table 13.3, the critical temperature of SO_2 is $431.0 \text{ K} = 157.8 °C$, $SO_2(g)$ can be found on the surface of the earth.

(d) According to Figure 13–18, $I_2(l)$ can exist at pressures less than 1.00 atm between the temperatures of $114°$ C and $184°$ C. There are very few places on the Earth that reach temperatures this far above the boiling point of water at pressures below 1 atm, places such as the mouths of volcanoes high above sea level Essentially, $I_2(l)$ is not found on the surface of the earth.

(e) According to Table 13.3, the critical temperature–the maximum temperature at which $O_2(l)$ exists, is $154.8 \text{ K} = -118.4 °C$. Temperatures this low do not exist on the surface of the earth.

56. The final pressure specified (100. atm) is below the ice III-ice I-liquid water triple point, according to the text adjacent to Figure 13–21; no ice III is formed. The starting point is that of $H_2O(g)$. When the pressure is increased to about 4.5 mmHg, the vapor condenses to ice I. As the pressure is raised, the melting point of ice I decreases–by $1°$ C per 125 atm pressure increase. Thus, somewhere between 10 and 15 atm, the ice I melts to liquid water, in which state it remains until the final pressure is reached.

57. **(a)** Heat lost by water = q_{water} = (m)(C)(Δt)

$$q_{water} = (100.0 \text{ g})\left(4.184\frac{\text{J}}{\text{g} \, ^\circ\text{C}}\right)(0.00 \, ^\circ\text{C} - 20.00 \, ^\circ\text{C})\left(\frac{1\text{kJ}}{1000\text{J}}\right) = -8.37 \text{ kJ}$$

Using $\Delta H_{cond} = -\Delta H_{vap}$ and heat lost by system = heat loss of condensation + cooling

$$q_{steam} = (175 \text{ g H}_2\text{O})\left(4.184\frac{\text{J}}{\text{g} \, ^\circ\text{C}}\right)(0.0 \, ^\circ\text{C} - 100.0 \, ^\circ\text{C})\left(\frac{1\text{kJ}}{1000\text{J}}\right)$$

$$+(175 \text{ g H}_2\text{O})\left(\frac{1\text{mol H}_2\text{O}}{18.015\text{g H}_2\text{O}}\right)\left(\frac{-40.7\text{kJ}}{1\text{mol H}_2\text{O}}\right)$$

q_{steam} = -395.4 kJ + -73.2 kJ = -468.6 kJ or ~ -469 kJ

Total energy to melt the ice = q_{water} + q_{steam} = -8.37 kJ + $-$469 kJ = $-$477 kJ

$$\text{Moles of ice melted} = (477 \text{ kJ})\left(\frac{1\text{mol ice}}{6.01\text{kJ}}\right) = 79.4 \text{ mol ice melted}$$

$$\text{Mass of ice melted} = (79.4 \text{ mol H}_2\text{O})\left(\frac{18.015\text{g H}_2\text{O}}{1\text{mol H}_2\text{O}}\right)\left(\frac{1\text{kg H}_2\text{O}}{1000\text{g H}_2\text{O}}\right) = 1.43 \text{ kg}$$

Mass of unmelted ice = 1.65 kg $-$ 1.43 kg = 0.22 kg

(b) Mass of unmelted ice = 0.22 kg (above)
Heat required to melt ice = $n \, \Delta H_{fusion}$

$$\text{Heat required} = (0.22 \text{ kg ice})\left(\frac{1000\text{g H}_2\text{O}}{1\text{kg H}_2\text{O}}\right)\left(\frac{1\text{mol H}_2\text{O}}{18.015\text{g H}_2\text{O}}\right)\left(\frac{6.01\text{kJ}}{1\text{mol H}_2\text{O}}\right) = 73.\underline{4} \text{ kJ}$$

Need to determine heat produced when 1 mole of steam (18.015 g) condenses and cools from 100.°C to 0.0 °C. Heat evolved,

$$= (1 \text{ mol H}_2\text{O})\left(\frac{-40.7\text{kJ}}{1\text{mol H}_2\text{O}}\right) + (18.015 \text{ g})\left(4.184\frac{\text{J}}{\text{g} \, ^\circ\text{C}}\right)(0.0 \, ^\circ\text{C} - 100. \, ^\circ\text{C})\left(\frac{1\text{kJ}}{1000\text{J}}\right)$$

= -40.7 kJ + -7.54 kJ = -48.2 kJ per mole of H_2O(g) or per 18.015 g H_2O(g)

$$\text{Mass of steam required} = (73.\underline{4} \text{ kJ})\left(\frac{1\text{mol H}_2\text{O(g)}}{48.2\text{kJ}}\right)\left(\frac{18.015\text{g H}_2\text{O}}{1\text{mol H}_2\text{O}}\right) = 27 \text{ g steam}$$

58. The heat gained by the ice equals the negative of the heat lost by the water. Let us use this fact in a step-by-step approach. We first compute the heat needed to raise the temperature of the ice to 0.0 °C and then the heat given off when the temperature of the water is lowered to 0.0 °C.

$$\text{to heat the ice} = 54 \text{ cm}^3 \times \frac{0.917 \text{ g}}{1 \text{ cm}^3} \times 2.01 \text{ J g}^{-1} \, ^\circ\text{C}^{-1} \times (0 \, ^\circ\text{C} + 25.0 \, ^\circ\text{C}) = 2.5 \times 10^3 \text{ J}$$

to cool the water $= -400.0 \text{ cm}^3 \times \dfrac{0.998 \text{ g}}{1 \text{ cm}^3} \times 4.18 \text{ J g}^{-1} \text{ °C}^{-1} \times (0 \text{ °C} - 32.0 \text{ °C}) = 53.4 \times 10^3 \text{ J}$

Thus, at 0 °C, we have 50. g ice, 399 g water, and $(53.4 - 2.5) \times 10^3 \text{ J} = 50.9 \text{ kJ}$ of heat available. Since 50.0 g of ice is a bit less than 3 mol of ice and 50.9 kJ is enough heat to melt at least 8 mole of ice, all of the ice will melt. The heat needed to melt the ice is

$50. \text{ g ice} \times \dfrac{1 \text{ mol H}_2\text{O}}{18.0 \text{ g H}_2\text{O}} \times \dfrac{6.01 \text{ kJ}}{1 \text{ mol ice}} = 17 \text{ kJ}$

We now have $50.9 \text{ kJ} - 17 \text{ kJ} = 34 \text{ kJ}$ of heat, and $399 \text{ g} + 50. \text{ g} = 449 \text{ g}$ of water at 0 °C. We compute the temperature change that is produced by adding the heat to the water.

$\Delta T = \dfrac{34 \times 10^3 \text{ J}}{449 \text{ g} \times 4.18 \text{ J g}^{-1} \text{ °C}^{-1}} = 18 \text{ °C}$ The final temperature is 18 °C

59. The liquid in the can is supercooled. When the can is opened, gas bubbles released from the carbonated beverage serve as nuclei for the formation of ice crystals. The condition of supercooling is destroyed and the liquid reverts to the solid phase instantly.

An alternative explanation follows. The process of the gas coming out of solution is endothermic (heat is required). (We know this to be true because the reaction solution of gas in water $\rightarrow$ gas + liquid water proceeds to the right as the temperature is raised, a characteristic direction of an endothermic reaction.) The required heat is taken from the cooled liquid, causing it to freeze.

60. Both the melting point of ice and the boiling point of water are temperatures that vary as the pressure changes, the boiling point more substantially than the melting point. The triple point, however, does not vary with pressure. Solid, liquid, and vapor coexist only at one fixed temperature and pressure.

Intermolecular Forces

61. **(a)** HCl is not a very heavy diatomic molecule; London forces are expected to be relatively weak. Hydrogen bonding is weak in the case of H — Cl bonds; Cl is not one of the three atoms (F, O, N) that form strong hydrogen bonds. Finally, because Cl is an electronegative atom, and H is only moderately electronegative, dipole-dipole interactions should be relatively strong.

(b) In Br_2 neither hydrogen bonds nor dipole-dipole attractions are important; there are no H atoms in the molecule, and homonuclear molecules are nonpolar. London forces are more important than in HCl since Br_2 is heavier.

(c) In ICl there are no hydrogen bonds since there are no H atoms in the molecule. The London forces are as strong as in Br_2 since the two molecules have the same number of electrons. However, dipole-dipole interactions are important in ICl; the molecule is polar toward Cl.

(d) In HF London forces are not very important; the molecule has only 10 electrons. Hydrogen bonding is quite important and definitely overshadows even the strong dipole-dipole interactions.

(e) In CH_4, H bonds are not important; the H atoms are not bonded to F, O, or N. In addition the molecule is not polar, so there are no dipole-dipole interactions. Finally, London forces are quite weak since the molecule contains only 10 electrons, and this is why CH_4 is a gas with a low critical temperature.

62. Substituting Cl for H both makes the molecule heavier and thus increases London forces, and makes it polar which increases dipole-dipole interactions. Both of these effects make it more difficult to disrupt the forces of attraction between molecules, increasing the boiling point. A substitution of Br for Cl increases London forces, but makes the molecule less polar. Since London forces are relatively more important than dipole-dipole interactions, the boiling point increases yet again. Finally, substituting OH for Br decreases London forces but both increases the dipole-dipole interactions and creates opportunities for hydrogen bonding. Since hydrogen bonds are much stronger than London forces, the boiling point increases even further.

63.   (c)   <   (b)   <   (d)   <   (a)
          (ethane thiol)   (ethanol)   (butanol)   (acetic acid)

Viscosity will depend on the intermolecular forces, e.g. varying H-bonding forces.

64.   (d)   <   (b)   <   (a)   <   (c)
          (butane)   (carbon disulfide)   (ethanol)   (1,2-dihydroxyethane)

The boiling point is dependent on the intermolecular forces; hence, hydrogen bonding produces the strongest interactions (highest boiling point) and non polar molecules like butane and CS_2 have the lowest boiling point. We must also consider the effect of Van der Waals forces.

65. We expect CH_3OH to be a liquid from among the four substances listed. Of these four molecules, C_3H_8 has the most electrons and should have the strongest London forces. However, only CH_3OH satisfies the conditions for hydrogen bonding (H bonded to and attracted to N, O, or F) and thus its intermolecular attractions should be much stronger than those of the other substances.

66. **(a)** Intramolecular hydrogen bonding cannot occur in C_4H_{10} since the conditions for hydrogen bonding (H bonded to and also attracted to N, O, or F) are not satisfied in this molecule. There is no N, O, or F atom in this molecule.

(b) Intramolecular hydrogen bonding is important in $HOOCCH_2CH_2CH_2CH_2COOH$. The H of one end –COOH group can be attracted to one of the O atoms of the other end –COOH group to cause ring closure.

(c) Intramolecular hydrogen bonding is not important in H_3CCOOH. Although there is another O atom to which the H of —OH can hydrogen bond, the resulting configurations will create a four membered ring $\left(\begin{array}{c} O\text{-}\text{-}H \\ \| \quad | \\ -C-O \end{array}\right)$ with bond angles of $90°$, quite different, and therefore strained, compared to the normal bond angles of $109.5°$ and $120°$.

(d) In orthophthalic acid, intramolecular hydrogen bonds can be important. The H of one —COOH group can be attracted to one of the O atoms of the other —COOH group. The resulting ring is seven atoms around and thus should not cause bond angle strain.

Network Covalent Solids

<u>67.</u> One would expect diamond to have a greater density than graphite. Although the bond distance in graphite, "one-and-a-half" bonds, would be expected to be shorter than the single bonds in diamond, the large spacing between the layers of C atoms in graphite makes its crystals much less dense than those of diamond.

68. Diamond works well in glass cutters because of its extreme hardness, a hardness that is due to the crystal being held together entirely by covalent bonds. Graphite will not function effectively in a glass cutter, since it is quite soft, soft enough to flake off in microscopic pieces when used in pencils. In fact, graphite is so soft that pure graphite is rarely used in common wooden pencils. Often clay or some other substance is mixed with the graphite to produce a mechanically strong pencil "lead".

<u>69.</u> **(a)** We expect Si and C atoms to alternate in the structure, as shown at right. The C atoms are on the corners ($8 \times 1/8 = 1$ C atom) and on the faces ($6 \times 1/2 = 3$ C atoms), a total of four C atoms/unit cell. The Si atoms are each totally within the cell, a total of four Si atoms/unit cell.

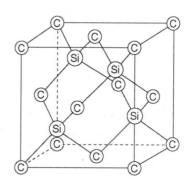

(b) To have a graphite structure, we expect sp^2 hybridization for each atom. The hybridization schemes for B and N atoms are shown to the right. The half-filled sp^2 hybrid orbitals overlap to form the σ bonding structure, and a hexagonal array of atoms. The $2p_z$ orbitals then overlap to form the π bonding orbitals. There are as many π electrons in a sample of BN as there are in a sample of graphite, assuming both samples have the same number of atoms.

B [He] $\underset{sp^2}{\boxed{\uparrow}\boxed{\uparrow}\boxed{\uparrow}}$ $\underset{2p}{\boxed{}}$

N [He] $\underset{sp^2}{\boxed{\uparrow}\boxed{\uparrow}\boxed{\uparrow}}$ $\underset{2p}{\boxed{\uparrow\downarrow}}$

70. Buckminsterfullerene is composed of C_{60} spheres. These molecules of carbon are rather like the N_2, P_4 and S_8 molecules and produce a nonpolar molecular solid. The chain of alternating single and triple bonds could be a network covalent solid in the sense that graphite is. The long carbon chain should be linear and these rods of carbon $(-C \equiv C - C \equiv C - C \equiv C - C \equiv C - C \equiv C - C \equiv C -)$ should fit together rather like spaghetti in a box, held together by the π electrons of the triple bonds from adjacent rods attracting each other.

Ionic Bonding and Properties.

71. We expect forces in ionic compounds to increase as sizes of ions become smaller and as ionic charges become greater. As the forces between ions become stronger, a higher temperature is required to melt the crystal. In the series of compounds NaF, NaCl, NaBr, and NaI, the anions are progressively larger, and thus the ionic forces become weaker. We expect the melting points to decrease in this series from NaF to NaI. This is precisely what we see.

72. Coulomb's law (Appendix B) states that the force between two particles of charges Q_1 and Q_2 that are separated by a distance r is given by $F = Q_1 Q_2 / \& r^2$ where $\&$, the dielectric constant, equals 1 for a vacuum. Since we are comparing forces, we can use $+1, -1, +2$, and -2 for the charges on ions. We take r to equal the sum of the cation and anion radii, which are taken from Figure 13-36.

For NaCl, $r_+ = 99$ pm, $r_- = 181$ pm　　$F = (+1)(-1) / (99 + 181)^2 = -1.3 \times 10^{-5}$

For MgO, $r_+ = 72$ pm, $r_- = 140$ pm　　$F = (+2)(-2) / (72 + 140)^2 = -8.9 \times 10^{-5}$

Thus, it is clear that interionic forces are about seven times stronger in MgO than in NaCl.

Crystal Structures

73. In each layer of a closest packing arrangement of spheres, there are six spheres surrounding and touching any given sphere. A second similar layer then is placed on top of this first layer so that its spheres fit into the indentations in the layer below. The two different closest packing arrangements arise from two different ways of placing the third layer on top of these two, with its spheres fitting into the indentations of the layer below. In one case, one can look down into these indentations and see a sphere of the bottom (first) layer. If these indentations are used, the closest packing arrangement *abab* results (hexagonal closest packing). In the other case, no first layer sphere is visible through the indentation; the closest packing arrangement *abcabc* results (cubic closest packing).

74. Physical properties are determined by the type of bonding between atoms in a crystal or the types of interactions between ions or molecules in the crystal. Different types of interactions can produce the same geometrical relationships of unit cell components. In both Ar and CO_2, London forces hold the particles in the crystal. In NaCl, the forces are interionic attractions, while metallic bonds hold Cu atoms in their crystals. It is the type of force, not the geometric arrangement of the components, that largely determines the physical properties of the crystalline material.

75. **(a)** We naturally tend to look at crystal structures in right-left, up-down terms. If we do that here, we might assign a unit cell as a square, with its corners at the centers of the light-colored squares. But the crystal does not "know" right and left or top and bottom. If we look at this crystal from the lower right corner, we see a unit cell that has its corners at the centers of dark-colored diamonds. These two types of unit cells are outlined at the top of the diagram below.

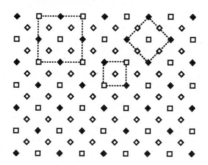

(b) The unit cell has one light-colored square fully inside it. It has four light-colored "circles" (which the computer doesn't draw as very round) on the edges, each shared with one other unit cell. So a total of $4 \times 1/2 = 2$ circles per unit cell. And the unit cell has four dark-colored diamonds, one at each corner, and each shared with four other unit cells, for a total of $4 \times 1/4 = 1$ diamond per unit cell.

(c) One example of an erroneous unit cell is the small square outlined near the center of the figure drawn in part (a). But notice that simply repeatedly translating this unit cell toward the right, so that its left edge sits where its right edge is now, will not generate the lattice.

76. This question reduces to asking what percentage of the area of a square is covered by a circle inscribed within it. The diameter of the circle equals the side of the square. Since diameter $= 2 \times$ radius, we have

$$\frac{\text{area of circle}}{\text{area of square}} = \frac{\pi r^2}{(2r)^2} = \frac{\pi}{4} = \frac{3.14159}{4} = 0.7854 \quad or \quad 78.54\%$$

Area uncovered $= 100.00\% - 78.54\% = 21.46\%$

77. In Figure 13–45 we see that the body diagonal of a cube has a length of $\sqrt{3}l$, where l is the length of one edge of the cube. The length of this body diagonal also equals $4r$, where r is the radius of the atom in the structure. Hence $4r = \sqrt{3}l$ or $l = 4r \div \sqrt{3}$ Recall that the volume of a cube is l^3, and $\sqrt{3} = 1.732$

$$\text{density} = \frac{\text{mass}}{\text{volume}} = \frac{\dfrac{2\ \text{W atoms}}{1\ \text{unit cell}} \times \dfrac{1\ \text{mol W}}{6.022 \times 10^{23}\ \text{W atoms}} \times \dfrac{183.85\ \text{g W}}{1\ \text{mol W}}}{\left(\dfrac{4 \times 139\ \text{pm}}{1.732} \times \dfrac{1\ \text{m}}{10^{12}\ \text{pm}} \times \dfrac{100\ \text{cm}}{1\ \text{m}}\right)^3} = 18.5\ \text{g}/\text{cm}^3$$

This compares well with a tabulated density of $19.25\ \text{g}/\text{cm}^3$.

78. One atom lies entirely within the hcp unit cell and there are eight corner atoms, each of which are shared among eight unit cells. Thus, there are a total of two atoms per unit cell. The volume of the unit cell is given by the product of its height and the area of its base. The base is a parallelogram, which can be subdivided along its shorter diagonal into two triangles. The height of either triangle, equivalent to the perpendicular distance between the parallel sides of the parallelogram, is given by $320. \text{ pm} \times \sin(60°) = 277 \text{ pm}$.

$$\text{Area of base} = 2 \times \text{area of triangle} = 2 \times (1/2 \times \text{ base} \times \text{ height})$$
$$= 277 \text{ pm} \times 320. \text{ pm} = 8.86 \times 10^4 \text{ pm}^2$$

$$\text{Volume of unit cell} = 8.86 \times 10^4 \text{ pm}^2 \times 520. \text{ pm} = 4.61 \times 10^7 \text{ pm}^3$$

$$\text{Density} = \frac{\text{mass}}{\text{volume}} = \frac{2 \text{ Mg atoms} \times \dfrac{1 \text{ mol Mg}}{6.022 \times 10^{23} \text{ Mg atoms}} \times \dfrac{24.305 \text{ g Mg}}{1 \text{ mol Mg}}}{4.61 \times 10^7 \text{ pm}^3 \times \left(\dfrac{1 \text{ m}}{10^{12} \text{ pm}} \times \dfrac{100 \text{ cm}}{1 \text{ m}} \right)^3} = 1.75 \text{ g} / \text{ cm}^3$$

This is in reasonable agreement with the experimental value of $1.738 \text{ g} / \text{cm}^3$.

79. **(a)** 335 pm = 2 radii or 1 diameter. Hence Po diameter = 335 pm

(b) 1 Po unit cell = $(335 \text{ pm})^3 = 3.76 \times 10^7 \text{ pm}^3$ $(3.76 \times 10^{-23} \text{ cm}^3)$ per unit cell.

$$\text{density} = \frac{m}{V} = \frac{3.47 \times 10^{-22} \text{ g}}{3.76 \times 10^{-23} \text{ cm}^3} = 9.23 \text{ g cm}^{-3}$$

(c) $n = 1$, d = 335 pm and $\lambda = 1.785 \times 10^{-10}$ or 178.5 pm
solve for $\sin \theta$ then determine θ

$$\sin \theta = \frac{n\lambda}{2d} = \frac{(1)(1.785 \times 10^{-10})}{2(335 \times 10^{-12})} = 0.266\underline{4} \text{ or } \theta = 15.4\underline{5}°$$

80. Ge unit cell has a length of 565 pm. volume = (length)^3 = $(565 \text{ pm})^3 = 1.80 \times 10^8 \text{ pm}^3$

$$\text{Convert to cm}^3\text{: } 1.80 \times 10^8 \text{ pm}^3 \left(\frac{1 \times 10^{-12} \text{ m}}{1 \text{ pm}} \right)^3 \left(\frac{100 \text{ cm}}{1 \text{ m}} \right)^3 = 1.80\underline{4} \times 10^{-22} \text{ cm}^3$$

Determine the number of Ge atoms per unit cell
Method: Find number of atoms of Ge per cm^3 then using the unit cell volume, (calculation above), determine number of Ge atoms in the unit cell

$$\frac{\text{atoms Ge}}{\text{cm}^3} = \left(\frac{5.36 \text{ g Ge}}{\text{cm}^3} \right) \left(\frac{1 \text{ mol Ge}}{72.61 \text{ g Ge}} \right) \left(\frac{6.022 \times 10^{23}}{1 \text{ mol Ge}} \right) = 4.44\underline{5} \times 10^{22} \frac{\text{atoms Ge}}{\text{cm}^3}$$

$$\frac{\text{atoms Ge}}{\text{unit cell}} = (1.80\underline{4} \times 10^{-22} \text{ cm}^3) \left(4.44\underline{5} \times 10^{22} \frac{\text{atoms Ge}}{\text{cm}^3} \right) = 8.02 \frac{\text{atoms Ge}}{\text{unit cell}}$$

There are 8 atoms of Ge per unit cell. Ge must adopt a face centered cubic structure with the 4 tetrahedral holes filled (diamond structure)

Ionic Crystal Structures

81. CaF_2. There are eight Ca^{2+} ions on the corners, each shared among eight unit cells, for a total of $(8 \times 1/8)$ one corner ion per unit cell. There are six Ca^{2+} ions on the faces, each shared between two unit cells, for a total of $(6 \times 1/2)$ three face ions per unit cell. This gives a total of four Ca^{2+} ions per unit cell. There are eight F^- ions, each wholly contained within the unit cell. The ratio of Ca^{2+} ions to F^- ions is 4 Ca^{2+} ions per 8 F^- ions: Ca_4F_8 or CaF_2.

TiO_2. There are eight Ti^{4+} ions on the corners, each shared among eight unit cells, for a total of $(8 \times 1/8)$ one Ti^{4+} corner ion per unit cell. There is one Ti^{4+} ion in the center, wholly contained within the unit cell. Thus, there are a total of two Ti^{4+} ions per unit cell. There are four O^{2-} ions on the faces of the unit cell, each shared between two unit cells, for a total of $(4 \times 1/2)$ two face atoms per unit cells. There are two O^{2-} ions totally contained within the unit cell. This gives a total of four O^{2-} ions per unit cell. The ratio of Ti^{4+} ions to O^{2-} ions is 2 Ti^{4+} ions per 4 O^{2-} ion: Ti_2O_4 or TiO_2.

82. The unit cell of CsCl is pictured in Figure 13–49. The length of the body diagonal equals $2r_+ + 2r_-$. From Figure 10-8, for Cl^-, $r_- = 181$ pm. Thus, the length of the body diagonal equals $2(169 \text{ pm} + 181 \text{ pm}) = 700.$ pm. From Figure 13–45, the length of this body diagonal is $\sqrt{3}l = 700.$ pm. The volume of the unit cell is $V = l^3$. ($\sqrt{3} = 1.732$).

$$V = \left(\frac{700. \text{ pm}}{1.732} \times \frac{1 \text{ m}}{10^{12} \text{ pm}} \times \frac{100 \text{ cm}}{1 \text{ m}} \right)^3 = 6.60 \times 10^{-23} \text{ cm}^3$$

Per unit cell, there is one Cs^+ and one Cl^- (8 corner ions $\times 1/8$ corner ion per unit cell).

$$\text{density} = \frac{\text{mass}}{\text{volume}} = \frac{\left(1 \text{ f.u. CsCl} \times \dfrac{1 \text{ mol CsCl}}{6.022 \times 10^{23} \text{ CsCl f.u.'s}} \times \dfrac{168.4 \text{ g CsCl}}{1 \text{ mol CsCl}} \right)}{6.60 \times 10^{-23} \text{ cm}^3} = 4.24 \text{ g}/\text{cm}^3$$

83. (a) In a sodium chloride type of lattice, there are six cations around each anion and six anions around each cation. These oppositely charged ions are arranged as follows: one above, one below, one in front, one in back, one to the right, and one to the left. Thus the coordination number of Mg^{2+} is 6 and that of O^{2-} is 6 also.

(b) In the unit cell, there is an oxide ion at each of the eight corners; each of these is shared between eight unit cells. There also is an oxide ion at the center of each of the six faces; each of these oxide ions is shared between two unit cells. Thus, the total number of oxide ions is computed as follows.

$$\text{total \# of oxide ions} = 8 \text{ corners} \times \frac{1 \text{ oxide ion}}{8 \text{ unit cells}} + 6 \text{ faces} \times \frac{1 \text{ oxide ion}}{2 \text{ unit cells}} = 4 \text{ } O^{2-} \text{ ions}$$

There is a magnesium ion on each of the twelve edges; each of these is shared between four unit cells. There also is a magnesium ion in the center which is not shared with another unit cell.

$$\text{total \# of Mg}^{2+}\text{ ions} = 12 \text{ adjoining cells} \times \frac{1 \text{ magnesium ion}}{4 \text{ unit cells}} + 1 \text{ central Mg}^{2+}\text{ ion}$$

$$= 4 \text{ Mg}^{2+}\text{ ions}$$

Thus, there are four formula units per unit cell of MgO.

(c)   Along the edge of the unit cell, Mg^{2+} and O^{2-} ions are in contact. The length of the edge is equal to the radius of one O^{2-}, plus the diameter of Mg^{2+}, plus the radius of another O^{2-}

edge length$= 2 \times O^{2-}$ radius$+2 \times Mg^{2+}$ radius$= 2 \times 140$ pm$+2 \times 72$ pm$= 424$ pm

The unit cell is a cube; its volume is the cube of its length.

$$\text{volume} = (424 \text{ pm})^3 = 7.62 \times 10^7 \text{ pm}\left(\frac{1 \text{ m}}{10^{12} \text{ pm}} \times \frac{100 \text{ cm}}{1 \text{ m}}\right)^3 = 7.62 \times 10^{-23} \text{ cm}^3$$

(d) $\text{density} = \dfrac{\text{mass}}{\text{volume}} = \dfrac{4 \text{ MgO f.}u}{7.62 \times 10^{-23} \text{ cm}^3} \times \dfrac{1 \text{ mol MgO}}{6.022 \times 10^{23} \text{ f.}u.} \times \dfrac{40.30 \text{ g MgO}}{1 \text{ mol MgO}} = 3.51 \text{ g/cm}^3$

84.   According to Figure 13-48, there are four formula units of KCl in a unit cell and the length of the edge of that unit cell is twice the internuclear distance between K^+ and Cl^- ions. (The analysis is given in more detail in the answer to Exercise 83.) The unit cell is a cube; its volume is the cube of its length.   length $= 2 \times 314.54$ pm $= 629.08$ pm

$$\text{volume} = (629.08 \text{ pm})^3 = 2.4895 \times 10^8 \text{ pm}^3 \left(\frac{1 \text{ m}}{10^{12} \text{ pm}} \times \frac{100 \text{ cm}}{1 \text{ m}}\right)^3 = 2.4895 \times 10^{-22} \text{ cm}^3$$

unit cell mass $=$ volume $\times$ density $= 2.4895 \times 10^{-22} \text{ cm}^3 \times 1.9893 \text{ g}/\text{cm}^3 = 4.9524 \times 10^{-22}$ g
$= $ mass of 4 KCl formula units

$$N_A = \frac{4 \text{ KCl formula unit}}{4.9524 \times 10^{-22} \text{ g KCl}} \times \frac{74.5513 \text{ g KCl}}{1 \text{ mol KCl}} = 6.0214 \times 10^{23} \text{ formula unit/mol}$$

<u>85</u>.   CaO $\rightarrow$ radius ratio $= \dfrac{r_{Ca^{2+}}}{r_{O^{2-}}} = \dfrac{100 \text{ pm}}{140 \text{ pm}} = 0.714$

Cations occupies octahedral holes of a face centered cubic array of anions

CuCl $\rightarrow$ radius ratio $= \dfrac{r_{Cu^+}}{r_{Cl^-}} = \dfrac{96 \text{ pm}}{181 \text{ pm}} = 0.530$

Cations occupy octahedral holes of a face centered cubic array of anions

$$LiO_2 \rightarrow \text{radius ratio} = \frac{r_{Li^+}}{r_{O_2^-}} = \frac{59\,pm}{128\,pm} = 0.461$$

Cations occupy octahedral holes of a face centered cubic array of anions

86. $$BaO \rightarrow \text{radius ratio} = \frac{r_{Ba^{2+}}}{r_{O^{2-}}} = \frac{135\,pm}{140\,pm} = 0.964$$

Cubic hole of simple cubic array of anions occupied by the cations

$$CuI \rightarrow \text{radius ratio} = \frac{r_{Cu^+}}{r_{I^-}} = \frac{96\,pm}{220\,pm} = 0.436$$

Cations occupy octahedral holes of a face centered cubic array of anions

$$LiS_2 \rightarrow \text{radius ratio} = \frac{r_{Li^+}}{r_{S_2^-}} = \frac{59\,pm}{198\,pm} = 0.298$$

Cations occupy tetrahedral holes of a face centered cubic array of anions

Lattice Energy

87.   Whether or not enthalpies of sublimation of the alkali metals are approximately the same, lattice energies of a series such as LiCl(s), NaCl(s), KCl(s), RbCl(s), and CsCl(s) will vary approximately with the size of the cation. A smaller cation will produce a more exothermic lattice energy. Thus, the lattice energy for LiCl(s) should be the most exothermic and CsCl(s) the least in this series, with NaCl(s) falling in the middle of the series.

88.   The cycle of reactions is shown. Recall that Hess's law states that the enthalpy change is the same, whether a chemical change is produced by one reaction or several.

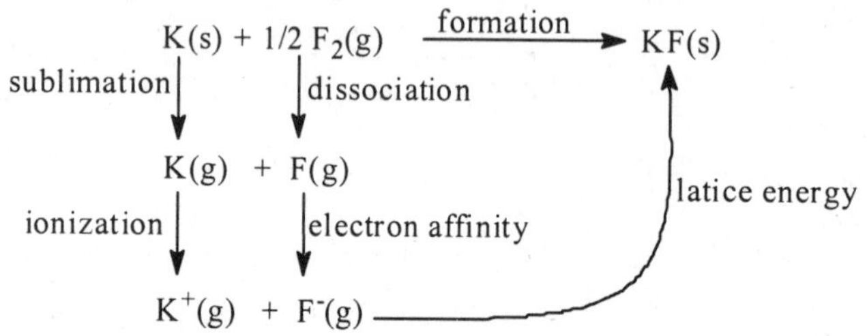

Formation reaction:	$K(s) + \frac{1}{2} F_2(g) \rightarrow KF(s)$	$\Delta H_f^\circ = -567.3$ kJ / mol
Sublimation:	$K(s) \rightarrow K(g)$	$\Delta H_{sub} = 89.24$ kJ / mol
Ionization:	$K(s) \rightarrow K^+(g) + e^-$	$I_1 = 418.9$ kJ/mol
Dissociation:	$\frac{1}{2} F_2(g) \rightarrow F(g)$	$D.E. = (159 / 2)$ kJ / mol F
Electron Affinity:	$F(g) + e^- \rightarrow F^-(g)$	$E.A. = -328$ kJ / mol

$\Delta H_f^\circ = \Delta H_{sub} + I_1 + D.E. + E.A. + \text{lattice energy (L.E.)}$

-567.3 kJ / mol $= 89.24$ kJ / mol $+ 418.9$ kJ / mol $+ (159 / 2)$ kJ / mol $- 328$ kJ / mol $+$ L.E.

$L.E. = -827$ kJ / mol

89.

Second ionization energy:	$Mg^+(g) \rightarrow Mg^{2+}(g) + e^-$	$I_2 = 1451$ kJ/mol
Lattice energy:	$Mg^{2+}(g) + 2\ Cl^-(g) \rightarrow MgCl_2(s)$	L.E. $= -2526$ kJ / mol
Sublimation:	$Mg(s) \rightarrow Mg(g)$	$\Delta H_{sub} = 146$ kJ/mol
First ionization energy	$Mg(g) \rightarrow Mg^+(g) + e^-$	$I_1 = 738$ kJ/mol
Dissociation energy:	$Cl_2(g) \rightarrow 2\ Cl(g)$	D.E. $= (2 \times 122)$ kJ/mol
Electron Affinity:	$2\ Cl(g) + 2\ e^- \rightarrow 2\ Cl^-(g)$	$2 \times$ E.A. $= 2(-349)$ kJ/mol

$$\Delta H_f^o = \Delta H_{sub} + I_1 + I_2 + \text{D.E.} + (2 \times \text{E.A.}) + \text{L.E.}$$

$$= 146\ \frac{kJ}{mol} + 738\ \frac{kJ}{mol} + 1451\ \frac{kJ}{mol}1 + 244\ \frac{kJ}{mol} - 698\ \frac{kJ}{mol} - 2526\ \frac{kJ}{mol} = -645\ \frac{kJ}{mol}$$

In Example 13-11, the value of ΔH_f^o for MgCl is calculated as -19 kJ/mol. Therefore, $MgCl_2$ is much more stable than MgCl, since considerably more energy is released when it forms. We expect $MgCl_2(s)$ to be more stable than MgCl(s) because a 2+ cation more strongly attracts anions than does a 1+ cation.

90.

Formation reaction:	$Na(s) + \dfrac{1}{2} H_2(g) \rightarrow NaH(s)$	$\Delta H_f^o = -57$ kJ/mol
Heat of sublimation:	$Na(s) \rightarrow Na(g)$	$\Delta H_{sub} = +107$ kJ/mol
Ionization energy:	$Na(g) \rightarrow Na^+(g) + e^-$	$I_1 = +496$ kJ/mol
Dissociation energy:	$\dfrac{1}{2} H_2(g) \rightarrow H(g)$	D.E. $= +218$ kJ/mol H
Lattice energy:	$Na^+(g) + H^-(g) \rightarrow NaH(s)$	L.E. $= -812$ kJ/mol

$$\Delta H_f^o = \Delta H_{sub} + I_1 + \text{D.E.} + \text{E.A.} + \text{lattice energy (L.E.)}$$

-57 kJ/mol $= 107$ kJ/mol $+ 496$ kJ/mol $+$ E.A. $+ 218$ kJ/mol $- 812$ kJ/mol

E.A. $= -66$ kJ/mol

FEATURE PROBLEMS

120. We obtain the surface tension by substituting the experimental values into the equation for surface tension.

$$h = \frac{2\gamma}{dgr} \quad \gamma = \frac{hdgr}{2} = \frac{1.1 \text{ cm} \times 0.789 \text{ g cm}^{-3} \times 981 \text{ cm s}^{-2} \times 0.050 \text{ cm}}{2} = 21 \text{ g/s}^2 = 0.021 \text{ J/m}^2$$

121. (a)

$$\frac{dP}{dT} = \frac{\Delta H_{vap}}{T(V_g - V_l)} = \frac{\Delta H_{vap}}{T(V_g)} \quad Note : V_l \approx 0 \quad \text{Rearrange expression, Use } V_g = \frac{nRT}{P}$$

$$\frac{dP}{dT} = \frac{\Delta H_{vap}}{T(\dfrac{nRT}{P})} = \frac{\Delta H_{vap}}{\dfrac{nRT^2}{P}} \quad or \quad \frac{dP}{P} = \frac{dT \Delta H_{vap}}{nRT^2} \quad \text{Integrate expression from } P_1 \to P_2 \text{ and } T_1 \to T_2$$

$$\ln\left(\frac{P_2}{P_1}\right) = \frac{\Delta H_{vap}}{nR}\left(\frac{1}{T_1} - \frac{1}{T_2}\right) \quad \text{Plug in } \Delta H_{vap} = 15{,}971 + 14.55 \text{ T} - 0.10 \text{ T}^2$$

$$\ln\left(\frac{P}{P_1}\right) = \frac{(15{,}971 + 14.55T - 0.160T^2)}{nR}\left(\frac{1}{T_1} - \frac{1}{T}\right)$$

Where P = 1 atm and T = normal boiling point

(b) $n = 1$ P = 10.16 torr (0.01337 atm) at 120 K, find boiling point at 1 atm.

$$\ln\left(\frac{1}{0.01337}\right) = \frac{(15{,}971 + 14.55 \text{ T} - 0.160 \text{ T}^2)}{8.3145 \text{ JK}^{-1} \text{ mol}^{-1}}\left(\frac{1}{120 \text{ K}} - \frac{1}{T}\right) \quad \text{solve for T: } T \approx 175 \text{ K}$$

122. (a) 1 NaCl unit missing from the NaCl unit cell → overall stoichiometry is the same. The unit cell usually has 4 Na$^+$ and 4 Cl$^-$ in the unit cell. Now the unit cell will have 3 Na$^+$ and 3 Cl$^-$. Accordingly, the density will decrease by a factor of 25% (1/4) if 1 Na$^+$ and 1 Cl$^-$ are consistently absent throughout the structure. Thus, the density will be 0.75(d$_{NaCl, normal}$).

(b) No change in stoichiometry or density, as this is just a simple displacement of an ion within the unit cell.

(c) Unit cell should contain 4 Ti^{2+} and 4 O^{2-} ions (same as in the NaCl unit cell). 4 TiO ions have a mass of :

$$4 \text{ formula units} \times \left(\frac{1 \text{ mol TiO}}{6.022 \times 10^{23} \text{ formula units}}\right)\left(\frac{63.88 \text{ g TiO}}{1 \text{ mol TiO}}\right) = 4.243 \times 10^{-22} \text{ g TiO}$$

$$V = (418 \text{ pm})^3 \times \left(\frac{1 \times 10^{-12} \text{ m}}{1 \text{ pm}}\right)^3 \left(\frac{100 \text{ cm}}{1 \text{ m}}\right)^3 = 7.30 \times 10^{-23} \text{ cm}^3$$

$$\text{Calculated density} = \frac{mass}{V} = \frac{4.243 \times 10^{-22} \text{ g}}{7.30 \times 10^{-23} \text{ cm}^3} = 5.81 \text{ g cm}^{-3} (\text{actual density} = 4.92 \text{ g cm}^{-3})$$

This indicates the presence of vacancies, possibly an example of the Schottky defect.

CHAPTER 14
SOLUTIONS AND THEIR PHYSICAL PROPERTIES
PRACTICE EXAMPLES

1A   To determine mass percent, we need both the mass of ethanol and the mass of solution. From volume percent, we know that 100.0 mL of solution contains 20.0 mL pure ethanol. The density of pure ethanol is 0.789 g/mL. We now can determine the mass of solute (ethanol) and solution. We perform the calculation in one set-up.

$$\text{mass percent ethanol} = \frac{20.0 \text{ mL ethanol} \times \dfrac{0.789 \text{ g}}{1 \text{ mL ethanol}}}{100.0 \text{ mL soln} \times \dfrac{0.977 \text{ g}}{1 \text{ mL soln}}} \times 100\% = 16.2\% \text{ ethanol by mass}$$

1B   In each case, we use the definition of the concentration unit, making sure that numerator and denominator are converted to the correct units.

(a)   We first determine the mass in grams of the solute and of the solution.

$$\text{mass } CH_3OH = 11.3 \text{ mL } CH_3OH \times \frac{0.793 \text{ g}}{1 \text{ mL}} = 8.96 \text{ g } CH_3OH$$

$$\text{mass soln} = 75.0 \text{ mL soln} \times \frac{0.980 \text{ g}}{1 \text{ mL}} = 73.5 \text{ g soln}$$

$$\text{amount of } H_2O = \left(73.5 \text{ g soln} - 8.96 \text{ g } CH_3OH\right) \times \frac{1 \text{ mol } H_2O}{18.02 \text{ g } H_2O}$$

$$\text{amount of } H_2O = 3.58 \text{ mol } H_2O \left(= 64.5 \text{ g } H_2O\right)$$

$$\text{amount of } CH_3OH = 8.96 \text{ g } CH_3OH \times \frac{1 \text{ mol } CH_3OH}{32.04 \text{ g } CH_3OH} = 0.280 \text{ mol } CH_3OH$$

$$\frac{H_2O \text{ mole}}{\text{fraction}} = \frac{\text{amount of } H_2O \text{ in moles}}{\text{amount of soln in moles}} = \frac{3.58 \text{ mol } H_2O}{3.58 \text{ mol } H_2O + 0.280 \text{ mol } CH_3OH} = 0.927$$

(b) $$[CH_3OH] = \frac{\text{amount of } CH_3OH \text{ in moles}}{\text{volume of soln in L}} = \frac{0.280 \text{ mol } CH_3OH}{75.0 \text{ mL soln} \times \dfrac{1 \text{ L}}{1000 \text{ mL}}} = 3.73 \text{ M}$$

(c) $$\text{molality of } CH_3OH = \frac{\text{amount of } CH_3OH \text{ in moles}}{\text{mass of } H_2O \text{ in kg}} = \frac{0.280 \text{ mol } CH_3OH}{64.5 \text{ g } H_2O \times \dfrac{1 \text{ kg}}{1000 \text{ g}}} = 4.34 \text{ } m$$

2A   First we need the amount of each component in solution. We use a 100.00-g sample of solution, in which there are 16.00 g glycerol and 84.00 g water.

$$\text{amount of glycerol} = 16.00 \text{ g } C_3H_5(OH)_3 \times \frac{1 \text{ mol } C_3H_5(OH)_3}{92.10 \text{ g } C_3H_5(OH)_3} = 0.1737 \text{ mol } C_3H_5(OH)_3$$

$$\text{amount of water} = 84.00 \text{ g } H_2O \times \frac{1 \text{ mol } H_2O}{18.02 \text{ g } H_2O} = 4.661 \text{ mol } H_2O$$

$$\text{mole fraction of } C_3H_5(OH)_3 = \frac{\text{amount } C_3H_5(OH)_3}{\text{amount } [C_3H_5(OH)_3 + H_2O]}$$

$$= \frac{0.1737 \text{ mol } C_3H_5(OH)_3}{0.1737 \text{ mol } C_3H_5(OH)_3 + 4.661 \text{ mol } H_2O} = 0.03593$$

2B   First we need the amount of sucrose in solution. We use a 100.00-g sample of solution, in which there are 10.00 g sucrose and 90.00 g water.

$$\text{amount } C_{12}H_{22}O_{11} = 10.00 \text{ g } C_{12}H_{22}O_{11} \times \frac{1 \text{ mol } C_{12}H_{22}O_{11}}{342.30 \text{ g } C_{12}H_{22}O_{11}} = 0.02921 \text{ mol } C_{12}H_{22}O_{11}$$

(a)   Molarity is amount of solute in moles per liter of solution. Convert the 100.00 g of solution to L with density as a conversion factor.

$$C_{12}H_{22}O_{11} \text{ molarity} = \frac{0.02921 \text{ mol } C_{12}H_{22}O_{11}}{1000. \text{ g soln}} \times \frac{1.040 \text{ g soln}}{1 \text{ mL}} \times \frac{1000 \text{ mL}}{1 \text{ L}} = 0.3038 \text{ M}$$

(b)   Molality is amount of solute in moles per kilogram of solvent. Convert 90.00 g of solute to kg.

$$C_{12}H_{22}O_{11} \text{ molality} = \frac{0.02921 \text{ mol } C_{12}H_{22}O_{11}}{90.00 \text{ g } H_2O} \times \frac{1000 \text{ g}}{1 \text{ kg}} = 0.3246 \, m$$

(c)   Mole fraction is amount of solute per amount of solution. First compute the amount in 90.00 g H_2O.

$$\text{amount } H_2O = 90.00 \text{ g} \times \frac{1 \text{ mol } H_2O}{18.02 \text{ g } H_2O} = 4.994 \text{ mol } H_2O$$

$$\text{mole fraction } C_{12}H_{22}O_{11} = \frac{0.02921 \text{ mol } C_{12}H_{22}O_{11}}{0.02921 \text{ mol } C_{12}H_{22}O_{11} + 4.994 \text{ mol } H_2O} = 0.005815$$

3A   Water is a polar, hydrogen bonding compound. It should mix well with other polar, hydrogen bonding compounds. (a) Toluene is nonpolar and should not be very soluble in water. (c) Benzaldehyde can form hydrogen bonds to water through its O atom. However, most of the molecule is non-polar and, as a result, it has limited solubility in water. (b) Oxalic acid is polar and can form hydrogen bonds. Oxalic acid should be the most readily soluble of these three compounds in water. Actual solubilities (w/w%) are: toluene (0.067%) < benzaldehyde (0.28%) < oxalic acid (14%).

3B   Both I_2 and CCl_4 are nonpolar molecules. It does not take much energy to break the attractions among I_2 molecules, or among CCl_4 molecules. Also, there is not a strong I_2-CCl_4 attraction created when a solution forms. The result is that I_2 should dissolve well in CCl_4 by simple mixing. H_2O is extensively hydrogen bonded with strong intermolecular forces requiring significant energy to break, but there is not a strong I_2-H_2O attraction created when a solution forms. We expect I_2 to dissolve poorly in water. Actual solubilities are 2.603 g $I_2/100$ g CCl_4 and 0.033 g $I_2/100$ g H_2O.

4A   The two suggestions are quoted first, followed by the means of achieving each one.
(1)   Dissolve the 95 g NH_4Cl in just enough water to produce a saturated solution (55 g $NH_4Cl/100$ g H_2O) at $60°C$.

$$\text{mass of water needed} = 95 \text{ g } NH_4Cl \times \frac{100 \text{ g } H_2O}{55 \text{ g } NH_4Cl} = 173 \text{ g } H_2O$$

The mass of NH_4Cl in the saturated solution at $20°C$ now is smaller.

$$\text{mass } NH_4Cl \text{ dissolved} = 173 \text{ g } H_2O \times \frac{37 \text{ g } NH_4Cl}{100 \text{ g } H_2O} = 64 \text{ g } NH_4Cl \text{ dissolved}$$

$$\text{crystallized mass } NH_4Cl = 95 \text{ g } NH_4Cl \text{ total} - 64 \text{ g } NH_4Cl \text{ dissolved at } 20°C$$
$$= 31 \text{ g } NH_4Cl \text{ crystallized}$$

(2)   Lower the final temperature to $0°C$, rather than $20°C$. From Figure 14-8, at $0°C$ the solubility of NH_4Cl is 28.5 g $NH_4Cl/100$ g H_2O. From this (and knowing that there is 173 g H_2O present in the solution) we calculate the mass of NH_4Cl dissolved at this lower temperature.

$$\text{mass dissolved } NH_4Cl = 173 \text{ g } H_2O \times \frac{28.5 \text{ g } NH_4Cl}{100 \text{ g } H_2O} = 49.3 \text{ g } NH_4Cl \text{ dissolved}$$

The mass of NH_4Cl recrystallized is 95 g $-$ 49.3 g $=$ 46 g

$$\text{yield} = (46/95) \times 100\% = 48\%$$

4B   Percent yield for the recrystallization can be defined as:

$$\% \text{ yield} = \frac{\text{mass crystallized}}{\text{mass dissolved}(40°C)} \times 100\%$$

$$\% \text{ yield} = \frac{\text{mass dissolved}(40°C) - \text{mass dissolved}(20°C)}{\text{mass dissolved}(40°C)} \times 100\%$$

Figure 14-8 solubilities per 100 g H_2O are followed by percent yield calculations.
Solubility of $KClO_4$: 4.84 g at 40°C and 3.0 g at 20° C

$$\text{Percent yield of } KClO_4 = \frac{4.8 \text{ g}(40°C) - 3.0 \text{ g } (20°C)}{4.8 \text{ g } (40°C)} \times 100\% = 38\% \text{ } KClO_4$$

Solubility of KNO_3: 60.7 g at 40°C and 32.3 g at 20°C

$$\text{Percent Yield of } KNO_3 = \frac{60.7 \text{ g}(40°C)\text{-}32.3 \text{ g }(20°C)}{60.7 \text{ g }(40°C)} \times 100\% = 47\% \; KNO_3$$

Solubility of K_2SO_4: 15.1 g at 40°C and 11.9 g at 20°C

$$\text{Percent yield of } K_2SO_4 = \frac{15.1 \text{ g}(40°C)\text{-}11.9 \text{ g }(20°C)}{15.1 \text{ g }(40°C)} \times 100\% = 21\% \; K_2SO_4$$

Ranked in order from highest to lowest percent yield:
$KNO_3 (47\%) > KClO_4 (38\%) > K_2SO_4 (21\%)$

5A   From Example 14-5, we know that the Henry's law constant for O_2 dissolved in water is
$k = 2.18 \times 10^{-3}$ M atm^{-1}. Consequently,

$$P_{gas} = \frac{C}{k} = \frac{8.23 \times 10^{-4} \text{ M}}{2.18 \times 10^{-3} \text{ M atm}^{-1}} = 0.378 \text{ atm } O_2 \text{ pressure}$$

5B   The pressure required must allow 0.0100 mol CO will dissolve in each liter of solution,
which we take as 1000 mL H_2O. We know that at 1 atm pressure, 35.4 mL CO dissolves
in each liter of solution (0.0354 mL CO/mL solution). Thus, [CO] at 1 atm CO pressure is

$$[CO] = \frac{35.4 \text{ mL CO at STP} \times \dfrac{1 \text{ mol CO}}{22,414 \text{ mL CO at STP}}}{1 \text{ L soln}} = 0.00158 \text{ M}$$

Now, Henry's law:

$$k = \frac{c}{P_{gas}} = \frac{0.00158 \text{ M}}{1 \text{ atm}} = \frac{0.0100 \text{ M}}{P} \qquad P = \frac{0.0100 \text{ M} \times 1 \text{ atm}}{0.00158 \text{ M}} = 6.33 \text{ atm CO}$$

6A   Raoult's law enables us to determine the vapor pressure of each component.
$P_{hex} = \chi_{hex} P°_{hex} = 0.750 \times 149.1 \text{ mmHg} = 112 \text{ mmHg}$
$P_{pen} = \chi_{pen} P°_{pen} = 0.250 \times 508.5 \text{ mmHg} = 127 \text{ mmHg}$.
We use Dalton's law to determine the total vapor pressure:
$P_{total} = P_{hex} + P_{pen} = 112 \text{ mmHg} + 127 \text{ mmHg} = 239 \text{ mmHg}$

6B   Masses of solution components need to be converted to amounts in moles through the use
of molar masses. Let us choose as our mass precisely 1.0000 mole
of $C_6H_6 = 78.11$ g C_6H_6 and an equal mass of toluene.

$$\text{amount of toluene} = 78.11 \text{ g } C_7H_8 \times \frac{1 \text{ mol } C_7H_8}{92.14 \text{ g } C_7H_8} = 0.8477 \text{ mol } C_7H_8$$

$$\text{mole fraction toluene} = \chi_{tol} = \frac{0.8477 \text{ mol } C_7H_8}{0.8477 \text{ mol } C_7H_8 + 1.0000 \text{ mol } C_6H_6} = 0.4588$$

toluene vapor pressure $= \chi_{tol} P°_{tol} = 0.4588 \times 28.4 \text{ mmHg} = 13.0 \text{ mmHg}$
benzene vapor pressure $= \chi_{benz} P°_{benz} = (1.0000 - 0.4588) \times 95.1 \text{ mmHg} = 51.5 \text{ mmHg}$.
total vapor pressure $= 13.0 \text{ mmHg} + 51.5 \text{ mmHg} = 64.5 \text{ mmHg}$

7A   The mole fraction composition of each component is that component's partial pressure divided by the total pressure. Again, we note that the vapor is richer in the more volatile component.

$$y_{hexane} = \frac{P_{hexane}}{P_{total}} = \frac{112 \text{ mmHg hexane}}{239 \text{ mmHg total}} = 0.469 \quad y_{pentane} = \frac{P_{pentane}}{P_{total}} = \frac{127 \text{ mmHg pentane}}{239 \text{ mmHg total}} = 0.531$$

or simply $1.000 - 0.469 = 0.531$

7B   The mole fraction composition of each component is that component's partial pressure divided by the total pressure. Again we note that the vapor is richer in the more volatile component.

$$y_t = \frac{P_{toluene}}{P_{total}} = \frac{13.0 \text{ mmHg toluene}}{64.5 \text{ mmHg total}} = 0.202 \quad y_b = \frac{P_{benzene}}{P_{total}} = \frac{51.5 \text{ mmHg benzene}}{64.5 \text{ mmHg total}} = 0.798$$

or simply $1.000 - 0.202 = 0.798$

8A   We use the osmotic pressure equation, converting the mass of solute to amount in moles, the temperature to Kelvin, and the solution volume to liters.

$$\pi = \frac{nRT}{V} = \frac{\left(1.50 \text{ g C}_{12}\text{H}_{22}\text{O}_{11} \times \dfrac{1 \text{ mol C}_{12}\text{H}_{22}\text{O}_{11}}{342.3 \text{ g C}_{12}\text{H}_{22}\text{O}_{11}}\right) \times 0.08206 \dfrac{\text{L atm}}{\text{mol K}} \times 298 \text{ K}}{125 \text{ mL} \times \dfrac{1 \text{ L}}{1000 \text{ mL}}} = 0.857 \text{ atm}$$

8B   We use the osmotic pressure equation to determine the molarity of the solution.

$$\frac{n}{V} = \frac{\pi}{RT} = \frac{0.015 \text{ atm}}{0.08206 \text{ L atm mol}^{-1} \text{ K}^{-1} \times 298 \text{ K}} = 6.1 \times 10^{-4} \text{ M}$$

Now, we can calculate the mass of urea.

$$\text{urea mass} = 0.225 \text{ L} \times \frac{6.1 \times 10^{-4} \text{ mol urea}}{1 \text{ L soln}} \times \frac{60.06 \text{ g CO(NH}_2)_2}{1 \text{ mol CO(NH}_2)_2} = 8.24 \times 10^{-3} \text{ g}$$

9A   We could substitute directly into the equation for molar mass derived in Example 14-9, but let us rather think our way through each step of the process. First, we find the concentration of the solution, by rearranging $\pi = \dfrac{n}{V} RT$. We need to convert the osmotic pressure to atmospheres.

$$\frac{n}{V} = \frac{\pi}{RT} = \frac{8.73 \text{ mmHg} \times \dfrac{1 \text{ atm}}{760 \text{ mmHg}}}{0.08206 \dfrac{\text{L atm}}{\text{mol K}} \times 298 \text{ K}} = 4.70 \times 10^{-4} \text{ M}$$

Next we determine the amount in moles of dissolved solute.

$$\text{amount of solute} = 100.0 \text{ mL} \times \frac{1 \text{ L}}{1000 \text{ mL}} \times \frac{4.70 \times 10^{-4} \text{ mol solute}}{1 \text{ L solution}} = 4.70 \times 10^{-5} \text{ mol solute}$$

We use the mass of solute, 4.04 g, to determine the molar mass. $\rightarrow$ $M = \dfrac{4.04 \text{ g}}{4.70 \times 10^{-5} \text{ mol}} = 8.60 \times 10^4 \text{ g/mol}$

9B We use the osmotic pressure equation along with the molarity of the solution.

$$\pi = \frac{n}{V} RT = \frac{2.12 \text{ g} \times \dfrac{1 \text{ mol}}{6.86 \times 10^4 \text{ g}}}{75.00 \text{ mL} \times \dfrac{1 \text{ L}}{1000 \text{ mL}}} \times \frac{0.08206 \text{ L atm}}{\text{mol K}} \times (310.2)\text{K} = 0.0105 \text{ atm} = 7.97 \text{ mmHg}$$

10A (a) The freezing point depression constant for water is $K_f = 1.86°C \, m^{-1}$.

$$\text{molality} = \frac{\Delta T_f}{-K_f} = \frac{-0.227 °C}{-1.86 °C \, m^{-1}} = 0.122 \ m$$

(b) Use the definition of molality to determine the number of moles of riboflavin in 0.833 g of dissolved riboflavin.

$$\text{amount of riboflavin} = 18.1 \text{ g solvent } H_2O \times \frac{1 \text{ kg solvent}}{1000 \text{ g}} \times \frac{0.122 \text{ mol solute}}{1 \text{ kg solvent}}$$

amount of riboflavin = 2.21×10^{-3} mol riboflavin

$$M = \frac{0.833 \text{ g riboflavin}}{2.21 \times 10^{-3} \text{ mol}} = 377 \text{ g/mol}$$

(c) Use the method of Chapter 3 to find riboflavin's empirical formula, starting with a 100.00-g sample.

$$54.25 \text{ g C} \times \frac{1 \text{ mol C}}{12.01 \text{ g C}} = 4.517 \text{ mol C} \div 1.063 \quad \rightarrow \ = 4.249 \text{ mol C}$$

$$5.36 \text{ g H} \times \frac{1 \text{ mol H}}{1.008 \text{ g H}} = 5.32 \text{ mol H} \ \div 1.063 \quad \rightarrow \ = 5.00 \text{ mol H}$$

$$25.51 \text{ g O} \times \frac{1 \text{ mol O}}{16.00 \text{ g O}} = 1.594 \text{ mol O} \div 1.063 \quad \rightarrow \ = 1.500 \text{ mol O}$$

$$14.89 \text{ g N} \times \frac{1 \text{ mol N}}{14.01 \text{ g N}} = 1.063 \text{ mol N} \div 1.063 \rightarrow \ = 1.000 \text{ mol N}$$

If we multiply each of these amounts by 4 (because 4.249 is almost equal to $4\frac{1}{4}$), the empirical formula is $C_{17}H_{20}O_6N_4$ with a molar mass of 376 g/mol. The molecular formula is $C_{17}H_{20}O_6N_4$.

10B The boiling point of pure water at 760.0 mmHg is $100.000°C$. For higher pressures, the boiling point occurs at a higher temperature; for lower pressures, a lower boiling point is observed. The boiling point elevation of this urea solution is calculated as follows.

$\Delta T_b = K_b \times m = 0.512 °C \, m^{-1} \times 0.205 m = 0.105 °C$

We would expect this urea solution to boil at $(100.00 + 0.105 =)100.105\,°C$ under 760.0 mmHg atmospheric pressure. Since it boils at a lower temperature, the atmospheric pressure must be lower than 760.0 mmHg.

<u>11A</u>   We assume a van't Hoff factor of $i = 3.00$ and convert the temperature to Kelvin, 298 K.

$$\pi = iMRT = \frac{3.00 \text{ mol ions}}{1 \text{ mol MgCl}_2} \times \frac{0.0530 \text{ mol MgCl}_2}{1 \text{ L soln}} \times 0.08206 \frac{\text{L atm}}{\text{mol K}} \times 298 \text{ K} = 3.89 \text{ atm}$$

<u>11B</u>   We first determine the molality of the solution, and assume a van't Hoff factor of $i = 2.00$.

$$m = \frac{\Delta T_f}{-K_f \times i} = \frac{-0.100\,°C}{-1.86\,°C\,m^{-1} \times 2.00} = 0.0269\,m \approx 0.0269 \text{ M}$$

volume of $HCl(aq) = 250.0 \text{ mL final soln} \times \frac{0.0269 \text{ mmol HCl}}{1 \text{ mL soln}} \times \frac{1 \text{ mL conc soln}}{12.0 \text{ mmol HCl}}$

volume of $HCl(aq) = 0.560 \text{ mL conc soln}$

REVIEW QUESTIONS

1.   **(a)** χ_B is the symbol for the mole fraction of component B: the amount in moles of B divided by the total amount in moles in the solution.

   (b) $P_A^°$ represents the vapor pressure of pure substance A.

   (c) K_f is the freezing point depression constant, the amount by which the freezing point of a solvent is lowered by having one mole of dissolved particles present in a kilogram of solvent.

   (d) i is the symbol for the van't Hoff factor, the measured value of a colligative property divided by the value calculated assuming that each mole of solute produces one mole of particles in solution.

   (e)   The activity of a solute is its effective concentration, the concentration it would have to have if solute particles did not interact with each other in solution.

2.   **(a)**   Henry's law states that the solubility of a gas in a solution is directly proportional to the pressure of that gas above the solution.

   (b)   Freezing point depression is the change in the freezing point of a liquid due to the presence of dissolved solute.

   (c)   Recrystallization is a purification process. An impure solute is dissolved in a solvent. As solvent evaporates away, the less soluble pure solute crystallizes first. When this process is repeated with the crystals obtained they become purer still.

   (d)   A hydrated ion is an ion surrounded by and attracted to water molecules.

   (e)   Deliquescence is the absorbance by a solid of water vapor from the air to such an extent that a saturated solution is formed.

3. **(a)**   Molarity is the amount of dissolved solute in moles per liter of solution, while molality is the amount of dissolved solute in moles per kilogram of solvent.

(b)   In an ideal solution the attractions among particles of all types are approximately the same in strength. In a nonideal solution, there is a disparity in the strengths of these attractive forces.

(c)   More solute can be dissolved in an unsaturated solution. The attempt to dissolve more solute in a supersaturated solution results in the precipitation of some solute.

(d)   Fractional crystallization is the process of first producing a concentrated and warm solution and then chilling it to a point of lower solute solubility to precipitate out some of that solute. Repetition of this process produces very pure solute. Fractional distillation is the repeated vaporization and condensation of a liquid mixture. This also produces quite pure components.

(e)   Osmosis is the motion of solute through a membrane permeable only to it. This creates an imbalance of pressures, if the solvent and solution volumes are limited, known as osmotic pressure. Reverse osmosis describes obtaining pure solvent by pushing a solution through a semipermeable membrane. This, of course, leaves the solute behind.

4. $$\% \text{ NaBr} = \frac{116 \text{ g NaBr}}{116 \text{ g NaBr} + 100 \text{ g H}_2\text{O}} \times 100\% = 53.7\% = 53.7 \text{ g NaBr}/100 \text{ g solution}$$

5.   **(a)** $$\% \text{ by volume} = \frac{12.8 \text{ mL CH}_3\text{CH}_2\text{CH}_2\text{OH}}{75.0 \text{ mL soln}} \times 100\% = 17.1\% \text{ CH}_3\text{CH}_2\text{CH}_2\text{OH}$$

(b) $$\text{percent by mass} = \frac{12.8 \text{ mL CH}_3\text{CH}_2\text{CH}_2\text{OH} \times \dfrac{0.803 \text{ g}}{1 \text{ mL}}}{75.0 \text{ mL} \times \dfrac{0.988 \text{ g}}{1 \text{ mL}}} \times 100\%$$

percent by mass $= 13.9\% \text{ CH}_3\text{CH}_2\text{CH}_2\text{OH}$ by mass

(c) $$\text{percent}(\text{mass}/\text{vol}) = \frac{12.8 \text{ mL CH}_3\text{CH}_2\text{CH}_2\text{OH} \times \dfrac{0.803 \text{ g}}{1 \text{ mL}}}{75.0 \text{ mL}} \times 100\%$$

$$\text{percent}(\text{mass}/\text{vol}) = 13.7 \frac{\text{mass}}{\text{vol}} \% \text{ CH}_3\text{CH}_2\text{CH}_2\text{OH}$$

6. $$\text{soln. volume} = 725 \text{ kg NaCl} \times \frac{1000 \text{ g NaCl}}{1 \text{ kg NaCl}} \times \frac{100.00 \text{ g soln}}{3.87 \text{ g NaCl}} \times \frac{75.0 \text{ mL soln}}{76.9 \text{ g soln}} \times \frac{1 \text{ L soln}}{1000 \text{ mL soln}}$$

$$= 1.83 \times 10^4 \text{ L sol'n}$$

7. $$\text{mass AgNO}_3 = 0.1250 \text{ L soln} \times \frac{0.0321 \text{ mol AgNO}_3}{1 \text{ L soln}} \times \frac{169.9 \text{ g AgNO}_3}{1 \text{ mol AgNO}_3} \times \frac{100.00 \text{ g mixt.}}{99.81 \text{ g AgNO}_3}$$

$$\text{mass AgNO}_3 = 0.683 \text{ g mixture}$$

8. $\text{molality} = \dfrac{2.65 \text{ g } C_6H_4Cl_2 \ \times \dfrac{1 \text{ mol } C_6H_4Cl_2}{147.0 \text{ g } C_6H_4Cl_2}}{50.0 \text{ mL} \times \dfrac{0.879 \text{ g}}{1 \text{ mL}} \times \dfrac{1 \text{ kg}}{1000 \text{ g}}} = 0.410 \ m$

9. $\text{molarity} = \dfrac{6.00 \text{ g } CH_3OH \ \times \dfrac{1 \text{ mol } CH_3OH}{32.04 \text{ g } CH_3OH}}{100.00 \text{ g soln} \times \dfrac{1 \text{ mL}}{0.988 \text{ g}} \times \dfrac{1 \text{ L}}{1000 \text{ mL}}} = 1.85 \text{ M} = [CH_3OH]$

10. total amount $= 1.28 \text{ mol } C_7H_{16} + 2.92 \text{ mol } C_8H_{18} + 2.64 \text{ mol } C_9H_{20} = 6.84 \text{ moles}$

(a) $\chi_{C_7H_{16}} = \dfrac{1.28 \text{ mol } C_7H_{16}}{6.84 \text{ moles total}} = 0.187$ **(b)** $\times 100\% = 18.7 \text{ mol\% } C_7H_{16}$

$\chi_{C_8H_{18}} = \dfrac{2.92 \text{ mol } C_8H_{18}}{6.84 \text{ moles total}} = 0.427$ $\times 100\% = 42.7 \text{ mol\% } C_8H_{18}$

$\chi_{C_9H_{20}} = \dfrac{2.64 \text{ mol } C_9H_{20}}{6.84 \text{ moles total}} = 0.386$ $\times 100\% = 38.6 \text{ mol\% } C_9H_{20}$

or $1.00 - 0.187 - 0.427 = 0.386$ or $100 - 18.7 - 42.7 = 38.6 \%$

11. **(a)** $[C_3H_8O_3] = \dfrac{62.0 \text{ g } C_3H_8O_3 \ \times \dfrac{1 \text{ mol } C_3H_8O_3}{92.09 \text{ g } C_3H_8O_3}}{100.0 \text{ g soln} \times \dfrac{1 \text{ mL soln}}{1.159 \text{ g soln}} \times \dfrac{1 \text{ L soln}}{1000 \text{ mL soln}}} = 7.80 \text{ M}$

(b) $[H_2O] = \dfrac{38.0 \text{ g } H_2O \ \times \dfrac{1 \text{ mol } H_2O}{18.02 \text{ g } H_2O}}{100.0 \text{ g soln} \times \dfrac{1 \text{ mL soln}}{1.159 \text{ g soln}} \times \dfrac{1 \text{ L soln}}{1000 \text{ mL soln}}} = 24.4 \text{ M}$

(c) $H_2O \text{ molality} = \dfrac{38.0 \text{ g } H_2O \ \times \dfrac{1 \text{ mol } H_2O}{18.02 \text{ g } H_2O}}{62.0 \text{ g } C_3H_8O_3 \ \times \dfrac{1 \text{ kg}}{1000 \text{ g}}} = 34.0 \ m$

(d) $C_3H_8O_3 \text{ mole fraction} = \dfrac{62.0 \text{ g } C_3H_8O_3 \ \times \dfrac{1 \text{ mol } C_3H_8O_3}{92.09 \text{ g } C_3H_8O_3}}{\dfrac{62.0 \text{ g } C_3H_8O_3}{92.09 \text{ g mol}^{-1} C_3H_8O_3} + \dfrac{38.0 \text{ g } H_2O}{18.02 \text{ g mol}^{-1} H_2O}}$

$= \dfrac{0.674 \text{ mol } C_3H_8O_3}{0.674 \text{ mol } C_3H_8O_3 + 2.11 \text{ mol } H_2O} = 0.242$

(e) H_2O mole percent $= \dfrac{2.11 \text{ mol } H_2O}{0.674 \text{ mol } C_3H_8O_3 + 2.11 \text{ mol } H_2O} \times 100\% = 75.8 \text{ mol \% } H_2O$

or $(1.000 - 0.242) \times 100\% = 75.8 \text{ \% } H_2O$

12.   Solubilities in Figure 14-8 are given as grams of solute per 100 g of solvent.

mass $NH_4Cl = 1.12 \text{ mol } NH_4Cl \times \dfrac{53.49 \text{ g } NH_4Cl}{1 \text{ mol } NH_4Cl} = 59.9 \text{ g } NH_4Cl$

Mass of $NH_4Cl = \dfrac{59.9 \text{ g } NH_4Cl}{150.0 \text{ g } H_2O} \times 100 \text{ g } H_2O = 39.9 \text{ g } NH_4Cl$

From Figure 14-8, the solubility of NH_4Cl in water at $30°C$ is 42 g/100 g water. The solution described in the problem is unsaturated.

13.   Answer (d) is correct. An ideal solution is most likely to form when the solute and the solvent have similar polarities, and thus the forces between solvent and solute molecules are about the same strength as those among solute or among solvent molecules. In (a), NaCl is an ionic solid, while H_2O is a polar liquid. In (b), C_2H_5OH is a polar liquid, while $C_6H_6(l)$ is nonpolar. In (c), C_7H_{16} is nonpolar , while H_2O is highly polar. In (d), both C_7H_{16} and C_8H_{18} are nonpolar liquids.

14. $NH_2OH(s)$ should be the most water soluble. Both $C_6H_6(l)$ and $C_{10}H_8(s)$ are composed of nonpolar molecules, which are barely (if at all) soluble in water. Both $NH_2OH(s)$ and $CaCO_3(s)$ should be able to be attracted by water molecules. But $CaCO_3(s)$ contains ions of high charge, difficult to dissolve because of the high lattice energy. (Recall the solubility rules of Chapter 5: most carbonates are insoluble in water.)

15.   Butyl alcohol should be moderately soluble in both water and benzene. A solute that is moderately soluble in both solvents will have some properties in common with each solvent. Both naphthalene and hexane are nonpolar molecules, like benzene, but have no properties in common with water molecules; they are soluble in benzene but not in water Sodium chloride consists of charged ions, similar to the charges in the polar bonds of water. Thus, as expected NaCl is very soluble in water. Butyl alcohol, on the other hand, possesses both a nonpolar part $(C_4H_9 -)$ like benzene, and a polar bond $(-O-H)$ like water. In fact, water and butyl alcohol can mutually hydrogen bond.

16.   We first determine the number of moles of O_2 that have dissolved.

amount $O_2 = \dfrac{PV}{RT} = \dfrac{1.00 \text{ atm} \times 0.02831 \text{ L}}{0.08206 \text{ L atm mol}^{-1} \text{ K}^{-1} \times 298 \text{ K}} = 1.16 \times 10^{-3} \text{ mol } O_2$

$[O_2] = \dfrac{1.16 \times 10^{-3} \text{ mol } O_2}{1.00 \text{ L soln}} = 1.16 \times 10^{-3} \text{ M}$

The oxygen concentration now is computed at the higher pressure.

$[O_2] = \dfrac{1.16 \times 10^{-3} \text{ M}}{1 \text{ atm } O_2} \times 3.86 \text{ atm } O_2 = 4.48 \times 10^{-3} \text{ M}$

17. First determine the number of moles of each component, then its mole fraction in the solution, then the partial pressure due to that component above the solution, and finally the total pressure.

$$\text{amount benzene} = n_b = 35.8 \text{ g } C_6H_6 \times \frac{1 \text{ mol } C_6H_6}{78.11 \text{ g } C_6H_6} = 0.458 \text{ mol } C_6H_6$$

$$\text{amount toluene} = n_t = 56.7 \text{ g } C_7H_8 \times \frac{1 \text{ mol } C_7H_8}{92.14 \text{ g } C_7H_8} = 0.615 \text{ mol } C_7H_8$$

$$\chi_b = \frac{0.458 \text{ mol } C_6H_6}{(0.458 + 0.615) \text{ total moles}} = 0.427 \qquad \chi_t = \frac{0.615 \text{ mol } C_7H_8}{(0.458 + 0.615) \text{ total moles}} = 0.573$$

$$P_b = 0.427 \times 95.1 \text{ mmHg} = 40.6 \text{ mmHg} \qquad P_t = 0.573 \times 28.4 \text{ mmHg} = 16.3 \text{ mmHg}$$

$$\text{total pressure} = 40.6 \text{ mmHg} + 16.3 \text{ mmHg} = 56.9 \text{ mmHg}$$

18.
$$\text{vapor fraction of benzene} = f_b = \frac{40.6 \text{ mmHg}}{56.9 \text{ mmHg}} = 0.714$$

$$\text{vapor fraction of toluene} = f_t = \frac{16.3 \text{ mmHg}}{56.9 \text{ mmHg}} = 0.286$$

These are both pressure fractions, as calculated, and also are the mole fractions in the vapor phase.

19. The freezing point depression is proportional to the product of the solute's molality and the van't Hoff factor. For nonelectrolytes, such as C_2H_5OH (0.050 m), $i = 1$, and thus $i \cdot m = 0.050 \ m$. For 1:1 electrolytes, such as $MgSO_4$ and NaCl, $i = 2$, and thus $i \cdot m = 0.020 \ m$ for (0.010 m) $MgSO_4$ and $im = 0.022 \ m$ for (0.011 m) NaCl. For 1:2 electrolytes, such as MgI_2 (0.010 m), $i = 3$, and thus $i \cdot m = 0.030 \ m$. The $i \times m$ product is largest for C_2H_5OH; thus it has the largest freezing point depression and the freezing point.

20. The solution with the largest freezing-point depression will have the lowest freezing point. Since C_2H_5OH is a nonelectrolyte, it will produce 1 mol of particles per mole of solute, and give the smallest freezing point depression, hence the highest freezing point. $HC_2H_3O_2$ is a weak electrolyte and produces slightly more than 1 mole of particles per mole of solute; its freezing point is the second lowest. NaCl, $MgBr_2$, and $Al_2(SO_4)_3$ all are strong electrolytes, producing 2, 3, and 5 moles of particles per mole of solute. They are listed in order of decreasing value of the freezing points of their solutions. Thus, the solutes, listed in order of decreasing freezing points of their 0.01 m aqueous solutions, are: $C_2H_5OH > HC_2H_3O_2 > NaCl > MgBr_2 > Al_2(SO_4)_3$

21. **(a)** $$\left[Na^+ \right] = \frac{0.92 \text{ g NaCl}}{100 \text{ mL soln}} \times \frac{1 \text{ mol NaCl}}{58.44 \text{ g NaCl}} \times \frac{1 \text{ mol } Na^+}{1 \text{ mol NaCl}} \times \frac{1000 \text{ mL}}{1 \text{ L soln}} = 0.16 \text{ M } Na^+$$

(b) $$\left[\text{ions} \right] = \left[Na^+ \right] + \left[Cl^- \right] = 2 \times \left[Na^+ \right] = 2 \times 0.16 \text{ M} = 0.32 \text{ M}$$

(c) $\pi = \dfrac{n}{V}RT = \dfrac{0.32\ \text{mol ions}}{1\ \text{L}} \times \dfrac{0.08206\ \text{L atm}}{\text{mol K}} \times 310\ \text{K} = 8.1\ \text{atm}$

(d) To determine freezing point depression, we first need molality, which means we need mass of solvent.

$$\text{molality} = \dfrac{0.92\ \text{g NaCl} \times \dfrac{1\ \text{mol NaCl}}{58.44\ \text{g NaCl}} \times \dfrac{2\ \text{mol ions}}{1\ \text{mol NaCl}}}{\left(\left(100\ \text{mL soln} \times \dfrac{1.005\ \text{g}}{1\ \text{mL soln}}\right) - 0.92\ \text{g NaCl}\right) \times \dfrac{1\ \text{kg}}{1000\ \text{g}}} = 0.32\ m$$

$\Delta T_f = -K_f \times m = -1.86°\text{C}/m \times 0.32 m = -0.60°\text{C} \qquad T_f = -0.60°\text{C}$

22. We compute the concentration of the solution. Then, assuming that the solution volume is the same as that of the solvent (0.2500 L), we determine the amount of solute dissolved, and finally the molar mass.

$$\dfrac{n}{V} = \dfrac{\pi}{RT} = \dfrac{1.67\ \text{mmHg} \times \dfrac{1\ \text{atm}}{760\ \text{mmHg}}}{0.0821\ \text{L atm mol}^{-1}\ \text{K}^{-1} \times 298\ \text{K}} = 8.98 \times 10^{-5}\ \text{M}$$

$\dfrac{\text{solute}}{\text{amount}} = 0.2500\ \text{L} \times \dfrac{8.98 \times 10^{-5}\ \text{mol}}{1\ \text{L}} = 2.25 \times 10^{-5}\ \text{mol}$

$M = \dfrac{0.72\ \text{g}}{2.25 \times 10^{-5}\ \text{mol}} = 3.2 \times 10^{4}\ \text{g/mol}$

23. First compute the molality of the benzene solution, then the number of moles of solute dissolved, and finally the molar mass of the unknown compound.

$m = \dfrac{\Delta T_f}{-K_f} = \dfrac{4.92°\text{C} - 5.53°\text{C}}{-5.12°\text{C}/m} = 0.12\ m$

$\text{amount solute} = 0.07522\ \text{kg benzene} \times \dfrac{0.12\ \text{mol solute}}{1\ \text{kg benzene}} = 9.0 \times 10^{-3}\ \text{mol solute}$

$M = \dfrac{1.10\ \text{g unknown compound}}{9.0 \times 10^{-3}\ \text{mol}} = 1.2 \times 10^{2}\ \text{g/mol}$

24. We compute the value of the van't Hoff factor, then use this value to compute the boiling point elevation.

$i = \dfrac{\Delta T_f}{-K_f m} = \dfrac{-0.072°\text{C}}{-1.86°\text{C}/m \times 0.010\ m} = 3.9$

$\Delta T_b = i K_b m = 3.9 \times 0.512°\text{C}\ m^{-1} \times 0.010\ ; m = 0.020°\text{C}$

The solution begins to boil at 100.02 °C.

EXERCISES

Homogeneous and Heterogeneous Mixtures

25. (b) salicyl alcohol probably is moderately soluble in both benzene and water. The reason for this assertion is that salicyl alcohol contains a benzene ring, which would make it soluble in benzene, and also can use its −OH groups to hydrogen bond to water molecules. On the other hand, (c) diphenyl contains only nonpolar benzene rings; it should be soluble in benzene but not in water. (a) *para*-dichlorobenzene contains a benzene ring, making it soluble in benzene, and two polar C–Cl bonds, which oppose each other, producing a nonpolar—and thus water-insoluble—molecule. (d) hydroxyacetic acid is a very polar molecule with many opportunities for hydrogen bonding. Its polar nature would make it insoluble in benzene, while the prospective hydrogen bonds will enhance aqueous solubility.

26. Vitamin C is a water-soluble vitamin. Its molecules contain a number of polar −OH groups, capable of hydrogen bonding with water, making these molecules soluble in water. Vitamin E is fat-soluble. Its molecules are composed of largely non-polar hydrocarbon chains, with few polar groups and thus should be soluble in nonpolar solvents.

27. (c) formic acid and (f) propylene glycol are soluble in water. They both can form hydrogen bonds with water, and they both have small nonpolar portions. (b) benzoic acid and (d) butyl alcohol are only slightly soluble in water. Although they both can form hydrogen bonds with water, both molecules contain reasonably large nonpolar portions, which will not interact strongly with water. (a) iodoform and (e) chlorobenzene are insoluble in water. Although both molecules have polar groups, their influence is too small to enable the molecules to disrupt the hydrogen bonds in water and form a homogeneous liquid mixture.

28. Benzoic acid will react with NaOH to form a solution of sodium ions and benzoate ions. (In this equation, ϕ represents the benzene ring with one hydrogen omitted, C_6H_5-.)

$$Na^+(aq) + OH^-(aq) + \phi-COOH \rightarrow Na^+(aq) + \phi-COO^-(aq) + H_2O(l)$$

The result of this reaction is that the concentration of molecular benzoic acid decreases, and thus more of the acid can be dissolved before the solution is saturated.

29. We expect small, highly charged ions to form crystals with large lattice energies, a factor that decreases their solubility. Based on this information, we would expect MgF_2 to be insoluble and KF to be soluble. It is probable that CaF_2 is insoluble due to a high lattice energy, but that NaF, with a smaller lattice energy, is soluble. KF is probably the most water soluble. Actual solubilities at 25 °C are: 0.00020 M CaF_2 < 0.0021 M MgF_2 < 0.95 M NaF < 16 M KF.

30. The nitrate ion is reasonably large and has a small negative charge. It thus should produce crystals with small lattice energies. In contrast, S^{2-} is highly charged, small and easily polarized. It should form crystals with relatively high lattice energies. In fact, partial covalent bonding occurs in many metal sulfides, as predicted from the polarizability of the sulfide ion to be most polarized by small, highly charged cations; these metal sulfides should be the least soluble. The most soluble sulfides thus should be those of large cations of low charge: K_2S, Rb_2S, Cs_2S, for example.

Percent Concentration

31. For water, the mass in grams and the volume in mL are about equal; the density of water is close to 1.0 g/mL. For ethanol, on the other hand, the density is about 0.8 g/mL. As long as the final solution volume after mixing is close to sum of the volumes for the two pure liquids, the percent by volume of ethanol will have to be larger than its percent by mass. This would not necessarily be true of other ethanol solutions. It would only be true in those cases where the density of the other component is greater than the density of ethanol.

32. Volume percent is not usually independent of temperature. As temperature increases, the volume of a liquid also usually increases. Unless the volume of the solution and the volume of the solute increase in the same proportion, the volume percent will be altered by these temperature increases. Mass, on the other hand, is unaffected by temperature, and thus mass percent is independent of temperature.

33.
$$\text{mass } HC_2H_3O_2 = 355 \text{ mL vinegar} \times \frac{1.01 \text{ g vinegar}}{1 \text{ mL}} \times \frac{6.02 \text{ g } HC_2H_3O_2}{100.00 \text{ g vinegar}}$$

$$\text{mass } HC_2H_3O_2 = 21.6 \text{ g } HC_2H_3O_2$$

34. We start with the molarity and convert both solute amount (numerator) and solution volume (denominator) to mass in order to obtain the mass percent.

$$\text{mass \% } H_2SO_4 = \frac{6.00 \text{ mol } H_2SO_4 \times \dfrac{98.08 \text{ g } H_2SO_4}{1 \text{ mol } H_2SO_4}}{1 \text{ L soln} \times \dfrac{1000 \text{ mL}}{1 \text{ L}} \times \dfrac{1.338 \text{ g}}{1 \text{ mL}}} \times 100\% = 44.0\% \ H_2SO_4$$

35. $46.1 \text{ ppm} = \dfrac{46.1 \text{ mg } SO_4^{2-}}{1 \text{ L solution}}$ (Assumes density of water ~1.00 g mL^{-1})

$$[SO_4^{2-}] = \frac{46.1 \text{ mg } SO_4^{2-}}{1 \text{ L solution}} \times \frac{1 \text{ g } SO_4^{2-}}{1000 \text{ mg } SO_4^{2-}} \times \frac{1 \text{ mol } SO_4^{2-}}{96.06 \text{ g } SO_4^{2-}} = 4.80 \times 10^{-4} \text{ M}$$

36. $9.4 \text{ ppb } CHCl_3 = \dfrac{9.4 \text{ µg } CHCl_3}{1 \text{ L solution}}$ (Assumes density of water ~1.00 g mL^{-1})

$$\text{mass } CHCl_3 = \frac{9.4 \text{ µg } CHCl_3}{1 \text{ L solution}} \times \frac{1 \text{ g } CHCl_3}{1 \times 10^6 \text{ µg } CHCl_3} \times \frac{1 \text{ L solution}}{1000 \text{ g solution}} \times 250 \text{ g solution}$$

$$\text{mass } CHCl_3 = 2.4 \times 10^{-6} \text{ g}$$

Molarity

37. The solution of Example 14-1 is 1.71 M C_2H_5OH, or 1.71 mmol C_2H_5OH in each mL of solution.

$$\text{volume conc. soln} = 825 \text{ mL} \times \frac{0.235 \text{ mmol } C_2H_5OH}{1 \text{ mL soln}} \times \frac{1 \text{ mL conc. soln}}{1.71 \text{ mmol } C_2H_5OH}$$

$$\text{volume conc. soln} = 113 \text{ mL conc. soln}$$

38. HNO_3 molarity $= \dfrac{30.00 \text{ g } HNO_3 \times \dfrac{1 \text{ mol } HNO_3}{63.01 \text{ g } HNO_3}}{100.0 \text{ g soln} \times \dfrac{1 \text{ mL}}{1.18 \text{ g}} \times \dfrac{1 \text{ L}}{1000 \text{ mL}}} = 5.62$ M at 20 °C

Molality

39. The mass of solvent in kg multiplied by the molality gives the amount in moles of the solute.

$$\text{mass } I_2 = \left(725.0 \text{ mL } CS_2 \times \frac{1.261 \text{ g}}{1 \text{ mL}} \times \frac{1 \text{ kg}}{1000 \text{ g}} \right) \times \frac{0.236 \text{ mol } I_2}{1 \text{ kg } CS_2} \times \frac{253.8 \text{ g } I_2}{1 \text{ mol } I_2} = 54.8 \text{ g } I_2$$

40. In the original solution the mass of CH_3OH associated with each 1.00 kg $= 1000$ g of water is:

$$\text{mass } CH_3OH = 1.00 \text{ kg } H_2O \times \frac{1.38 \text{ mol } CH_3OH}{1 \text{ kg } H_2O} \times \frac{32.04 \text{ g } CH_3OH}{1 \text{ mol } CH_3OH} = 44.2 \text{ g } CH_3OH$$

Then we compute the mass of methanol in 1.0000 kg $= 1000.0$ g of the original solution.

$$\text{mass } CH_3OH = 1000.0 \text{ g soln} \times \frac{44.2 \text{ g } CH_3OH}{1044.2 \text{ g soln}} = 42.3 \text{ g } CH_3OH$$

Thus, 1.0000 kg of this original solution contains $(1000.0 - 42.3 =)957.7$ g H_2O
Now, a 1.00 *m* CH_3OH solution contains 32.04 g CH_3OH (one mole) for every 1000.0 g H_2O. We can compute the mass of H_2O associated with 42.3 g CH_3OH in such a solution.

$$\text{mass } H_2O = 42.3 \text{ g } CH_3OH \times \frac{1000.0 \text{ g } H_2O}{32.04 \text{ g } CH_3OH} = 1320 \text{ g } H_2O$$

(We have temporarily retained an extra significant figure in the calculation.) Since we already have 957.7 g H_2O in the original solution, the mass of water that must be added is

mass of H_2O to be added $= 1320 \text{ g} - 957.7 \text{ g} = 362 \text{ g } H_2O = 3.6 \times 10^2$ g H_2O

41. H_3PO_4 molarity $= \dfrac{34.0 \text{ g H}_3PO_4 \times \dfrac{1 \text{ mol H}_3PO_4}{98.00 \text{ g H}_3PO_4}}{100.0 \text{ g soln} \times \dfrac{1 \text{ mL}}{1.209 \text{ g}} \times \dfrac{1 \text{ L}}{1000 \text{ mL}}} = 4.19 \text{ M}$

molality $= \dfrac{34.0 \text{ g H}_3PO_4 \times \dfrac{1 \text{ mol H}_3PO_4}{98.00 \text{ g H}_3PO_4}}{66.0 \text{ g solvent} \times \dfrac{1 \text{ kg}}{1000 \text{ g}}} = 5.26 \ m$

42. C_2H_5OH molarity $= \dfrac{10.00 \text{ g C}_2H_5OH \times \dfrac{1 \text{ mol C}_2H_5OH}{46.069 \text{ g C}_2H_5OH}}{90.00 \text{ g H}_2O \times \dfrac{1 \text{ kg}}{1000 \text{ g}}} = 2.412 \ m$

The molality of this solution does not vary with temperature. We would, however, expect the solution's molarity to vary with temperature because the solution's density, and thus the volume of solution that contains one mole, varies with temperature. But, unlike the volume, the mass of the solution does not vary with temperature, and so the solution's molality doesn't vary with temperature either.

Mole Fraction, Mole Percent

43. **(a)** amount $C_2H_5OH = 21.7 \text{ g C}_2H_5OH \times \dfrac{1 \text{ mol C}_2H_5OH}{46.07 \text{ g C}_2H_5OH} = 0.471 \text{ mol C}_2H_5OH$

amount $H_2O = 78.3 \text{ g H}_2O \times \dfrac{1 \text{ mol H}_2O}{18.02 \text{ g H}_2O} = 4.35 \text{ mol H}_2O$

$\chi_{ethanol} = \dfrac{0.471 \text{ mol C}_2H_5OH}{(0.471 + 4.34) \text{ total moles}} = 0.0979$

(b) amount $H_2O = 1000. \text{ g} \times \dfrac{1 \text{ mol H}_2O}{18.02 \text{ g H}_2O} = 55.49 \text{ mol H}_2O$

$\chi_{urea} = \dfrac{0.684 \text{ mol urea}}{(55.49 + 0.684) \text{ total moles}} = 0.0122$

44. **(a)** The amount of solvent is found after the solute's mass is subtracted from that of the solution.

$\text{solvent amount} = \left(\left(1 \text{ L soln} \times \dfrac{1000 \text{ mL}}{1 \text{ L}} \times \dfrac{1.006 \text{ g}}{1 \text{ mL}} \right) - \left(0.112 \text{ mol C}_6H_{12}O_6 \times \dfrac{180.2 \text{ g C}_6H_{12}O_6}{1 \text{ mol C}_6H_{12}O_6} \right) \right)$

$= \left[1006 \text{ g solution} - 20.2 \text{ g C}_6H_{12}O_6 \right] \times \dfrac{1 \text{ mol H}_2O}{18.02 \text{ g H}_2O} = 54.7 \text{ mol H}_2O$

$\chi_{solute} = \dfrac{0.112 \text{ mol C}_6H_{12}O_6}{0.112 \text{ mol C}_6H_{12}O_6 + 54.7 \text{ mol H}_2O} = 0.00204$

(b) First determine the mass and the amount of ethanol. The amount of solvent is calculated after the ethanol's mass is subtracted from the solution's mass. Use a 100.00-mL sample of solution for computation. This 100 mL sample would contain 3.20 mL of C_2H_5OH. We calculate the mass of ethanol first, followed by the number of moles:

$$V_{C_2H_5OH} = 3.20 \text{ mL } C_2H_5OH \times \frac{0.789 \text{ g}}{1 \text{ mL } C_2H_5OH} = 2.52 \text{ g } C_2H_5OH$$

$$\text{moles } C_2H_5OH = 2.52 \text{ g } C_2H_5OH \times \frac{1 \text{ mol } C_2H_5OH}{46.07 \text{ g } C_2H_5OH} = 0.0547 \text{ mol } C_2H_5OH$$

$$\text{mass of } H_2O = \left(\left(100.0 \text{ mL soln} \times \frac{0.993 \text{ g}}{1 \text{ mL}}\right) - 2.52 \text{ g } C_2H_5OH\right) = 96.8 \text{ g } H_2O$$

$$\text{amount of } H_2O = 96.8 \text{ g } H_2O \times \frac{1 \text{ mol } H_2O}{18.02 \text{ g } H_2O} = 5.37 \text{ mol } H_2O$$

$$\chi_{\text{solute}} = \frac{0.0547 \text{ mol } C_2H_5OH}{0.0547 \text{ mol } C_2H_5OH + 5.37 \text{ mol } H_2O} = 0.0101$$

45. Solve the following relationship for n_{ethanol}, the number of moles of ethanol, C_2H_5OH. The number of moles of water calculated in the Example is 5.01 moles, that of ethanol is 0.171 moles.

$$\chi_{\text{ethanol}} = \frac{n_{\text{ethanol}}}{n_{\text{ethanol}} + 5.01 \text{ mol } H_2O} = 0.0525 \qquad n_{\text{ethanol}} = 0.0525 \, n_{\text{ethanol}} + 0.263$$

$$n_{\text{ethanol}} = \frac{0.263}{1.0000 - 0.0525} = 0.278 \text{ mol ethanol}$$

$$\text{moles of added ethanol} = (0.278 - 0.171) \text{ mol } C_2H_5OH = 0.107 \text{ mol } C_2H_5OH$$

$$\text{mass added ethanol} = 0.107 \text{ mol } C_2H_5OH \times \frac{46.07 \text{ g } C_2H_5OH}{1 \text{ mol } C_2H_5OH} = 4.93 \text{ g } C_2H_5OH$$

46. The amount of water present in 1 kg is

$$n_{\text{water}} = 1000 \text{ g } H_2O \times \frac{1 \text{ mol } H_2O}{18.02 \text{ g } H_2O} = 55.49 \text{ mol } H_2O \text{ Now solve the following expression}$$

for n_{gly}, the amount of glycerol. $4.85\% = 0.0485$ mole fraction.

$$\chi_{\text{gly}} = 0.0485 = \frac{n_{\text{gly}}}{n_{\text{gly}} + 55.49} \qquad n_{\text{gly}} = 0.0485 \, n_{\text{gly}} + 2.69$$

$$n_{\text{gly}} = \frac{2.69}{(1.0000 - 0.0485)} = 2.83 \text{ mol glycerol}$$

$$\text{volume glycerol} = 2.83 \text{ mol } C_3H_8O_3 \times \frac{92.09 \text{ g } C_3H_8O_3}{1 \text{ mol } C_3H_8O_3} \times \frac{1 \text{ mL}}{1.26 \text{ g}} = 207 \text{ mL glycerol}$$

Solubility Equilibrium

47. At 40 °C the solubility of NH_4Cl is 46.3 g per 100 g of H_2O. To determine molality, we calculate amount in moles of the solute and the solvent mass in kg.

$$\text{molarity} = \frac{46.3 \text{ g} \times \dfrac{1 \text{ mol } NH_4Cl}{53.49 \text{ g } NH_4Cl}}{100 \text{ g } H_2O \times \dfrac{1 \text{ kg}}{1000 \text{ g}}} = 8.66 \; m$$

48. The data in Figure 14-8 are given in mass of solute per 100 g H_2O. 0.200 m $KClO_4$ contains 0.200 mol $KClO_4$ associated with each 1 kg($= 1000$) g of H_2O. The mass of $KClO_4$ dissolved in 100 g H_2O is:

$$\text{Mass } KClO_4 = 100 \text{ g } H_2O \times \frac{0.200 \text{ mol } KClO_4}{1000 \text{ g } H_2O} \times \frac{138.5 \text{ g } KClO_4}{1 \text{ mol } KClO_4} = 2.77 \text{ g } KClO_4.$$

This concentration is realized at a temperature of about 32 °C. At this temperature a saturated solution of $KClO_4$ is 0.200 m.

49. **(a)** The concentration for $KClO_4$ in this mixture is calculated first.

$$\frac{\text{mass solute}}{100 \text{ g } H_2O} = 100 \text{ g } H_2O \times \frac{20.0 \text{ g } KClO_4}{500.0 \text{ g water}} = 4.00 \text{ g } KClO_4$$

At 40 °C a saturated $KClO_4$ solution has a concentration of about 4.6 g $KClO_4$ dissolved in 100 g water. Thus, the solution is unsaturated.

(b) The mass of $KClO_4$ that must be added is the difference between the mass now present in the mixture and the mass that is dissolved in 500 g H_2O to produce a saturated solution.

$$\text{mass to be added} = \left(500.0 \text{ g } H_2O \times \frac{4.6 \text{ g } KClO_4}{100 \text{ g } H_2O} \right) - 20.0 \text{ g } KClO_4 = 3.0 \text{ g } KClO_4$$

50. From Figure 14-8, the solubility of KNO_3 in water is 38 g KNO_3 per 100 g water at 25.0 °C, while at 0.0 °C its solubility is 15 g KNO_3 per 100 g water. At 25.0 °C, every 138 g solution contains 38 g KNO_3 and 100.0 g water; and at 0.0° C every 115 g solution contains 15 g KNO_3 and 100.0 g water. We calculate the mass of water and of KNO_3 present at 25.0 °C. mass water $= 335$ g soln $\times \dfrac{100.0 \text{ g water}}{138 \text{ g soln}} = 243$ g water

mass $KNO_3 = 335$ g soln $- 243$ g water $= 92$ g KNO_3

At 0.0 °C the mass of water has been reduced 55 g by evaporation to $(243 \text{ g} - 55 \text{ g} =)188$ g. The mass of KNO_3 soluble in this mass of water at 0.0°C is

mass KNO_3 dissolved $= 188$ g $H_2O \times \dfrac{15 \text{ g } KNO_3}{100.0 \text{ g } H_2O} = 28$ g KNO_3

Thus, of the 92 g KNO_3 originally present in solution, the mass that recrystallizes is

KNO_3 recrystallized (g) $= 92$ g KNO_3 (orig. dissol.) $- 28$ g KNO_3 (still dissol.) $= 64$ g KNO_3

Solubility of Gases

51. mass of $CH_4 = 1.00 \times 10^3$ kg $H_2O \times \dfrac{0.02 \text{ g } CH_4}{1 \text{ kg } H_2O \cdot \text{ atm}} \times 20 \text{ atm} = 4 \times 10^2$ g CH_4 (natural gas)

52. We first compute the amount of O_2 dissolved in 515 mL $= 0.515$ L at each temperature.

amount $O_2 = 0.515$ L $\times \dfrac{2.18 \times 10^{-3} \text{ mol } O_2}{1 \text{ L}} = 1.12 \times 10^{-3}$ mol O_2 at 0 °C

amount $O_2 = 0.515$ L $\times \dfrac{1.26 \times 10^{-3} \text{ mol } O_2}{1 \text{ L}} = 6.49 \times 10^{-4}$ mol O_2 at 25 °C

Determine the difference between these amounts, and the corresponding volume.

O_2 volume $= \dfrac{(11.2 - 6.49) \times 10^{-4} \text{ mol } O_2 \times \dfrac{0.08206 \text{ L atm}}{\text{mol 1 K}} \times (25 + 273) \text{ K}}{1 \text{ atm}}$

$= 1.2 \times 10^{-2}$ L $= 12$ mL expelled

53. We use the STP molar volume $(22.414$ L $= 22,414$ mL$)$ to determine the molarity of Ar under 1 atom pressure and then the Henry's law constant.

$k_{Ar} = \dfrac{C}{P_{Ar}} = \dfrac{\dfrac{33.7 \text{ mL Ar}}{1 \text{ L soln}} \times \dfrac{1 \text{ mol Ar}}{22,414 \text{ mL at STP}}}{1 \text{ atm pressure}} = \dfrac{0.00150 \text{ M}}{\text{atm}}$

In the atmosphere, the partial pressure of argon is $P_{Ar} = 0.00934$ atm. (Recall that pressure fractions equal volume fractions for ideal gases.) We now compute the concentration of argon in aqueous solution.

$C = k_{Ar} P_{Ar} = \dfrac{0.00150 \text{ M}}{\text{atm}} \times 0.00934 \text{ atm} = 1.40 \times 10^{-5}$ M

54. We use the STP molar volume $(22.414$ L $= 22,414$ mL$)$ to determine the molarity of CO_2 under 1 atm pressure and then the Henry's law constant.

$k_{CO_2} = \dfrac{C}{P_{CO_2}} = \dfrac{\dfrac{87.8 \text{ mL } CO_2}{1 \text{ L soln}} \times \dfrac{1 \text{ mol } CO_2}{22,414 \text{ mL at STP}}}{1 \text{ atm pressure}} = \dfrac{0.00392 \text{ M}}{\text{atm}}$

In the atmosphere, the partial pressure of CO_2 is $P_{CO_2} = 0.000360$ atm. (Recall that pressure fractions equal volume fractions for ideal gases.) We now compute the concentration of carbon dioxide in aqueous solution.

$C = k_{CO_2} P_{CO_2} = \dfrac{0.00392 \text{ M } CO_2}{\text{atm}} \times 0.000360 \text{ atm} = 1.41 \times 10^{-6}$ M CO_2

55. Equation 14.2 is $c = kP_{gas}$. Concentration, c, is the number of moles of gas dissolved in a fixed volume of solution. But gas solubilities are quite low and thus the volume of solution is almost exactly the same as volume of solvent. Of course the amount of gas in moles is directly proportional to its mass. Therefore the two statements are equivalent.

56. Although this statement initially may seem quite different, we need to remember that the volume of a gas decreases as its partial pressure increases, as described by Boyle's law. Thus increasing the partial pressure means that more moles of gas are forced into the same volume. Therefore this statement is equivalent to equation 14.2.

$$kP_{gas} = c_{gas} = \frac{n_{gas}}{V_{solvent}} = \frac{P_{gas}V_{gas}}{RT_{gas}V_{solvent}} \quad or \quad V_{gas} = kRT_{gas}V_{solvent}$$

Thus, V_{gas} is constant at fixed temperature for a given volume of solvent. This statement is not valid if the ideal gas law is not obeyed or if the gas reacts with the solvent.

Raoult's Law and Liquid-Vapor Equilibrium

57. We determine the mole fraction of water in this solution.

$$n_{glucose} = 165 \text{ g C}_6\text{H}_{12}\text{O}_6 \times \frac{1 \text{ mol C}_6\text{H}_{12}\text{O}_6}{180.2 \text{ g C}_6\text{H}_{12}\text{O}_6} = 0.916 \text{ mol C}_6\text{H}_{12}\text{O}_6$$

$$n_{water} = 685 \text{ g H}_2\text{O} \times \frac{1 \text{ mol H}_2\text{O}}{18.02 \text{ g H}_2\text{O}} = 38.0 \text{ mol H}_2\text{O} \qquad \chi_{water} = \frac{38.0 \text{ mol H}_2\text{O}}{(38.0 + 0.916) \text{ total moles}} = 0.976$$

$$P_{sol'n} = \chi_{water}P_{water}^* = 0.976 \times 23.8 \text{ mmHg} = 23.2 \text{ mmHg}$$

58. We determine the mole fraction of methanol in this solution.

$$n_{urea} = 17 \text{ g CO(NH}_2)_2 \times \frac{1 \text{ mol CO(NH}_2)_2}{60.06 \text{ g CO(NH}_2)_2} = 0.283 \text{ mol CO(NH}_2)_2$$

$$n_{methanol} = 100 \text{ mL CH}_3\text{OH} \times \frac{0.792 \text{ g}}{1 \text{ mL}} \times \frac{1 \text{ mol CH}_3\text{OH}}{32.04 \text{ g CH}_3\text{OH}} = 2.47 \text{ mol CH}_3\text{OH}$$

$$\chi_{methanol} = \frac{2.47 \text{ mol CH}_3\text{OH}}{(2.47 + 0.283) \text{ total moles}} = 0.897$$

$$P_{sol'n} = \chi_{methanol}P_{methanol}^* = 0.897 \times 95.7 \text{ mmHg} = 85.8 \text{ mmHg}$$

59. We consider a sample of 100.0 g of the solution and determine the number of moles of each component in this sample. From this information and the given vapor pressures, we determine the vapor pressure of each component.

$$\text{amount styrene} = n_s = 38 \text{ g styrene} \times \frac{1 \text{ mol C}_8\text{H}_8}{104 \text{ g C}_8\text{H}_8} = 0.37 \text{ mol C}_8\text{H}_8$$

$$\text{amount ethylbenzene} = n_e = 62 \text{ g ethylbenzene} \times \frac{1 \text{ mol C}_8\text{H}_{10}}{106 \text{ g C}_8\text{H}_{10}} = 0.58 \text{ mol C}_8\text{H}_{10}$$

$$\chi_s = \frac{n_s}{n_s + n_e} = \frac{0.37 \text{ mol styrene}}{(0.37 + 0.58) \text{ total moles}} = 0.39; \quad P_s = 0.39 \times 134 \text{ mmHg} = 52 \text{ mmHg for } C_8H_8$$

$$\chi_e = \frac{n_e}{n_s + n_e} = \frac{0.58 \text{ mol ethylbenzene}}{(0.37 + 0.58) \text{ total moles}} = 0.61; \quad P_e = 0.61 \times 182 \text{ mmHg} = 111 \text{ mmHg for } C_8H_{10}$$

Then the mole fraction in the vapor can be determined.

$$y_e = \frac{P_e}{P_e + P_s} = \frac{111 \text{ mmHg}}{(111 + 52) \text{ mmHg}} = 0.68 \qquad\qquad y_s = 1.00 - 0.68 = 0.32$$

60. The vapor pressures of the pure liquids are: for benzene $P_b^\circ = 95.1$ mmHg and for toluene $P_t^\circ = 28.4$ mmHg. We know that $P_t = \chi_t P_t^\circ$ and $P_b = \chi_b P_b^\circ$. We also know that $\chi_t + \chi_b = 1.000$. We solve for χ_b in the following expression.

$$0.620 = \frac{P_b}{P_b + P_t} = \frac{\chi_b P_b^\circ}{\chi_b P_b^\circ + \chi_t P_t^\circ} = \frac{\chi_b P_b^\circ}{\chi_b P_b^\circ + (1 - \chi_b) P_t^\circ} = \frac{\chi_b 95.1}{\chi_b 95.1 + (1 - \chi_b) 28.4}$$

$$95.1\chi_b = 0.620 \left[\chi_b 95.1 + (1 - \chi_b) 28.4 \right] = 59.0\chi_b + 17.6 - 17.6\chi_b$$

$$17.6 = (95.1 - 59.0 + 17.6)\chi_b = 53.7\chi_b \qquad \chi_b = \frac{17.6}{53.7} = 0.328$$

61. The total vapor pressure above the solution at its normal boiling point is 760 mm Hg. The vapor pressure due to toluene is given by the following equation.

$$P_{toluene} = \chi_{toluene} \cdot P^\circ_{toluene} = 0.700 \times 533 \text{ mm Hg} = 373 \text{ mm Hg}$$

Next, the vapor pressure due to benzene is determined, followed by the vapor pressure of pure benzene.

$$P_{benzene} = P_{total} - P_{toluene} = 760 \text{ mm Hg} - 373 \text{ mm Hg} = 387 \text{ mm Hg} = \chi_{benzene} \cdot P^\circ_{benzene}$$
$$387 \text{ mm Hg} = 0.300 \times P^\circ_{benzene} \text{ and hence, } P^\circ_{benzene} = 1.29 \times 10^3 \text{ mm Hg}$$

62. (a) The vapor pressure of water is greater above the solution that has the higher mole fraction of water, according to Raoult's Law. This also is the solution that has the smaller mole fraction of NaCl, namely, the unsaturated solution.

  (b) As water evaporates from the unsaturated solutionm its concentration changes and so does its vapor pressure, according to Raoult's Law. This does not happen with the saturated solution; its concentration and thus its vapor pressure remains constant as the solvent evaporates.

  (c) The solution with the higher boiling point is the more concentrated solution (the saturated solution) because the boiling point increases as the amount of dissolved solute increases.

Osmotic Pressure

63. Both the flowers and the cucumber contain solutions (plant sap), but both of these solutions are less concentrated than the salt solution. Thus, the solution in the plant

material moves across the semipermeable membrane in an attempt to dilute the salt solution, leaving behind wilted flowers and shriveled pickles (less water in their tissues).

64.   The main purpose of drinking water is to provide water for living tissues. In fresh water, water should move across the semipermeable membranes of a fish's organs as water moves past the gills for breathing. Thus, it should not be necessary for freshwater fish to drink. In contrast, the concentration of solutes in salt water should be about the same as their concentration in a fish's blood plasma. In this case, osmotic pressure will not be effective in moving water into a fish's tissues. The purpose of drinking for a salt water fish is to isolate the ingested water within the fish. Then mechanisms within the fish's body can move water across membranes into the fish's tissues. Or possibly, the fish has a means of desalinating the ingested water, in which case osmotic pressure would carry water into the fish's tissues.

<u>65.</u>   We first determine the molarity of the solution. Let's work the problem with three significant figures.

$$\frac{n}{V} = \frac{\pi}{RT} = \frac{1.00 \text{ atm}}{0.08206 \text{ L atm mol}^{-1} \text{ K}^{-1} \times 273 \text{ K}} = 0.0446 \text{ M}$$

$$\text{volume} = 1 \text{ mol} \times \frac{1 \text{ L}}{0.0446 \text{ mol solute}} = 22.4 \text{ L solution} \approx 22.4 \text{ L solvent}$$

We have assumed that the solution is so dilute that its volume closely approximates the volume of the solvent constituting it. Note that this volume corresponds to the STP molar volume of an ideal gas. The osmotic pressure equation also resembles the ideal gas equation.

66.   First calculate the concentration of solute in the solution, then the mass of solute in 100.0 mL of that solution.

$$[\text{solute}] = \frac{\pi}{RT} = \frac{7.25 \text{ mmHg} \times \left(\frac{1 \text{ atm}}{760 \text{ mmHg}}\right)}{0.08206 \text{ L atm mol}^{-1} \text{ K}^{-1} \times 298 \text{ K}} = 3.90 \times 10^{-4} \text{ M}$$

$$\text{mass of hemoglobin} = 1.000 \text{ L} \times \frac{3.90 \times 10^{-4} \text{ mol hemoglobin}}{1 \text{ L soln}} \times \frac{6.86 \times 10^4 \text{ g}}{1 \text{ mol hemoglobin}}$$

$$= 2.68 \text{ g hemoglobin}$$

<u>67.</u>   First determine the concentration of the solution from the osmotic pressure, then the amount of solute dissolved, and finally the molar mass of that solute.

$$\pi = 5.1 \text{ mm soln} \times \frac{0.88 \text{ mmHg}}{13.6 \text{ mm soln}} \times \frac{1 \text{ atm}}{760 \text{ mmHg}} = 4.3 \times 10^{-4} \text{ atm}$$

$$\frac{n}{V} = \frac{\pi}{RT} = \frac{4.3 \times 10^{-4} \text{ atm}}{0.0821 \text{ L atm mol}^{-1} \text{ K}^{-1} \times 298 \text{ K}} = 1.8 \times 10^{-5} \text{ M}$$

$$\text{amount solute} = 100.0 \text{ mL} \times \frac{1 \text{ L}}{1000 \text{ mL}} \times 1.8 \times 10^{-5} \text{ M} = 1.8 \times 10^{-6} \text{ mol solute}$$

$$M = \frac{0.50 \text{ g}}{1.8 \times 10^{-6} \text{ mol}} = 2.8 \times 10^{5} \text{ g / mol}$$

68. We need the molar concentration of ions in the solution to determine its osmotic pressure. We assume that the solution has a density of 1.00 g/mL.

$$[\text{ions}] = \frac{0.92 \text{ g NaCl} \times \dfrac{1 \text{ mol NaCl}}{58.4 \text{ g NaCl}} \times \dfrac{2 \text{ mol ions}}{1 \text{ mol NaCl}}}{100.0 \text{ mL soln} \times \dfrac{1 \text{ L soln}}{1000 \text{ mL}}} = 0.32 \text{ M}$$

$$\pi = \frac{n}{V} RT = 0.32 \frac{\text{mol}}{\text{L}} \times 0.0821 \text{ L atm mol}^{-1} \text{ K}^{-1} \times (37.0 + 273.2) \text{ K} = 8.1 \text{ atm}$$

69. The reverse osmosis process requires a pressure equal to or slightly greater than the osmotic pressure of the solution. We assume that this solution has a density of 1.00 g/mL. First, we determine the molar concentration of ions in the solution.

$$[\text{ions}] = \frac{2.5 \text{ g NaCl} \times \dfrac{1 \text{ mol NaCl}}{58.4 \text{ g NaCl}} \times \dfrac{2 \text{ mol ions}}{1 \text{ mol NaCl}}}{100.0 \text{ mL soln} \times \dfrac{1 \text{ mL soln}}{1.00 \text{ g}} \times \dfrac{1 \text{ L soln}}{1000 \text{ mL}}} = 0.86 \text{ M}$$

$$\pi = \frac{n}{V} RT = 0.86 \frac{\text{mol}}{\text{L}} \times 0.0821 \text{ L atm mol}^{-1} \text{ K}^{-1} \times (25 + 273.2) \text{ K} = 21 \text{ atm}$$

70. We need to determine the molarity of each solution. The solution with the higher molarity has the higher osmotic pressure, and water will flow from the other solution into that more concentrated solution, in an attempt to create two solutions of equal concentrations.

$$[C_3H_8O_3] = \frac{14.0 \text{ g } C_3H_8O_3 \times \dfrac{1 \text{ mol } C_3H_8O_3}{92.09 \text{ g } C_3H_8O_3}}{55.2 \text{ mL soln} \times \dfrac{1 \text{ L}}{1000 \text{ mL}}} = 2.75 \text{ M}$$

$$[C_{12}H_{22}O_{11}] = \frac{17.2 \text{ g } C_{12}H_{22}O_{11} \times \dfrac{1 \text{ mol } C_{12}H_{22}O_{11}}{342.3 \text{ g } C_{12}H_{22}O_{11}}}{62.5 \text{ mL soln} \times \dfrac{1 \text{ L}}{1000 \text{ mL}}} = 0.804 \text{ M}$$

Thus, water will move from the $C_{12}H_{22}O_{11}$ solution into the $C_3H_8O_3$ solution, that is, from right to left.

Freezing Point Depression and Boiling Point Elevation

71. **(a)** First determine the molality of the solution, then the value of the freezing-point depression constant.

$$m = \frac{1.00 \text{ g C}_6\text{H}_6 \times \dfrac{1 \text{ mol C}_6\text{H}_6}{78.11 \text{ g C}_6\text{H}_6}}{80.00 \text{ g solvent} \times \dfrac{1 \text{ kg}}{1000 \text{ g}}} = 0.160 \ m \qquad K_f = \frac{\Delta T_f}{-m} = \frac{3.3°\text{C} - 6.5°\text{C}}{-0.160 \ m} = 20.°\text{C}/m$$

(b)   For benzene, $K_f = 5.12 \ °\text{C} m^{-1}$. Cyclohexane is the better solvent for freezing point depression determinations of molar mass, since a less concentrated solution will still give a substantial freezing-point depression. For the same concentration, cyclohexane solutions will show a freezing point depression approximately four times that of benzene. (Another factor is that benzene is labeled as a carcinogen and should be avoided.)

72.   The solution's boiling point is $0.40°\text{C}$ above the boiling point of pure water. First compute the molality of the solution, then the mass of sucrose that must be added to 1 kg (1000 g) H_2O, and finally the mass% sucrose in the solution.

$$m = \frac{\Delta T_b}{K_b} = \frac{0.40 \ °\text{C}}{0.512 \ °\text{C}/m} = 0.78 \ m \rightarrow \frac{0.78 \text{ mol C}_{12}\text{H}_{22}\text{O}_{11}}{1 \text{kg H}_2\text{O}} \times \frac{342.3 \text{ g C}_{12}\text{H}_{22}\text{O}_{11}}{1 \text{ mol C}_{12}\text{H}_{22}\text{O}_{11}} = \frac{26_7 \text{ g C}_6\text{H}_{12}\text{O}_6}{1 \text{ kg H}_2\text{O}}$$

$$\text{mass \% C}_{12}\text{H}_{22}\text{O}_{11} = \frac{26_7 \text{ g C}_{12}\text{H}_{22}\text{O}_{11}}{(1000. + 26_7) \text{ g total}} \times 100\% = 21\% \ \text{C}_{12}\text{H}_{22}\text{O}_{11}$$

73.   We determine the molality of the solution, then the number of moles of solute present, and then the molar mass of the solute. Then we determine the compound's empirical formula, and combine this with the molar mass to determine the molecular formula.

$$m = \frac{\Delta T_f}{-K_f} = \frac{1.37 \ °\text{C} - 5.53 \ °\text{C}}{-5.12°\text{C}/m} = 0.813 \ m$$

$$\text{amount} = \left(50.0 \text{ mL C}_6\text{H}_6 \times \frac{0.879 \text{ g}}{1 \text{ mL}} \times \frac{1 \text{ kg}}{1000 \text{ g}} \right) \times \frac{0.813 \text{ mol solute}}{1 \text{ kg C}_6\text{H}_6} = 3.57 \times 10^{-2} \text{ mol}$$

$$M = \frac{6.45 \text{ g}}{3.57 \times 10^{-2} \text{ mol}} = 181 \text{ g}/\text{mol}$$

$$42.9 \text{ g C} \times \frac{1 \text{ mol C}}{12.01 \text{ g C}} = 3.57 \text{ mol C} \qquad\qquad \div 1.19 \rightarrow 3.00 \text{ mol C}$$

$$2.4 \text{ g H} \times \frac{1 \text{ mol H}}{1.01 \text{ g H}} = 2.4 \text{ mol H} \qquad\qquad \div 1.19 \rightarrow 2.0 \text{ mol H}$$

$$16.7 \text{ g N} \times \frac{1 \text{ mol N}}{14.01 \text{ g N}} = 1.19 \text{ mol N} \qquad\qquad \div 1.19 \rightarrow 1.00 \text{ mol N}$$

$$38.1 \text{ g O} \times \frac{1 \text{ mol O}}{16.00 \text{ g O}} = 2.38 \text{ mol O} \qquad\qquad \div 1.19 \rightarrow 2.00 \text{ mol O}$$

The empirical formula is $\text{C}_3\text{H}_2\text{NO}_2$, with a formula mass of 84.0 g/mol. This is one-half the experimentally determined molar mass. Thus, the molecular formula is $\text{C}_6\text{H}_4\text{N}_2\text{O}_4$.

74. We determine the molality of the nitrobenzene $(K_f = 8.1\,°C\,/\,m)$ solution first, then the solute's molar mass.

$$m = \frac{\Delta T_f}{-K_f} = \frac{-1.4\,°C - 5.7\,°C}{-8.1\,°C/m} = 0.88\,m$$

$$\text{amount solute} = 30.0\text{ mL nitrobenzene} \times \frac{1.204\text{ g}}{1\text{ mL}} \times \frac{1\text{ kg}}{1000\text{ g}} \times \frac{0.88\text{ mol solute}}{1\text{ kg nitrobenzene}}$$

$$= 0.032\text{ mol solute}$$

$$M = \frac{3.88\text{ g nicotinamide}}{0.032\text{ mol nicotinamide}} = 1.2 \times 10^2\text{ g}\,/\,\text{mol}$$

Then we determine the compound's empirical formula, and combine this with the molar mass to determine the molecular formula.

$$59.0\text{ g C} \times \frac{1\text{ mol C}}{12.01\text{ g C}} = 4.91\text{ mol C} \qquad \div 0.819 \rightarrow 6.00\text{ mol C}$$

$$5.0\text{ g H} \times \frac{1\text{ mol H}}{1.01\text{ g H}} = 5.0\text{ mol H} \qquad \div 0.819 \rightarrow 6.0\text{ mol H}$$

$$22.9\text{ g N} \times \frac{1\text{ mol N}}{14.01\text{ g N}} = 1.63\text{ mol N} \qquad \div 0.819 \rightarrow 2.00\text{ mol N}$$

$$13.1\text{ g O} \times \frac{1\text{ mol O}}{16.00\text{ g O}} = 0.819\text{ mol O} \qquad \div 0.819 \rightarrow 1.00\text{ mol O}$$

A reasonable empirical formula is $C_6H_6N_2O$, which has an empirical mass of 122. Since this is the same as the experimentally determined molar mass, the molecular formula of nicotinamide is $C_6H_6N_2O$.

75. We determine the molality of the benzene solution first, then the molar mass of the solute.

$$m = \frac{\Delta T_f}{-K_f} = \frac{-1.183\,°C}{-5.12\,°C/m} = 0.231\,m$$

$$\text{amount solute} = 0.04456\text{ kg benzene} \times \frac{0.231\text{ mol solute}}{1\text{ kg benzene}} = 0.0103\text{ mol solute}$$

$$M = \frac{0.867\text{ g thiophene}}{0.0103\text{ mol thiophene}} = 84.2\text{ g}\,/\,\text{mol}$$

Next, we determine the empirical formula from the masses of the combustion products.

$$\text{amount C} = 4.913\text{ g CO}_2 \times \frac{1\text{ mol CO}_2}{44.010\text{ g CO}_2} \times \frac{1\text{ mol C}}{1\text{ mol CO}_2} = 0.1116\text{ mol C} \div 0.02791 \rightarrow 4.000\text{ mol C}$$

$$\text{amount H} = 1.005\text{ g H}_2O \times \frac{1\text{ mol H}_2O}{18.015\text{ g H}_2O} \times \frac{2\text{ mol H}}{1\text{ mol H}_2O} = 0.1116\text{ mol H} \div 0.02791 \rightarrow 4.000\text{ mol H}$$

$$\text{amount S} = 1.788\text{ g SO}_2 \times \frac{1\text{ mol SO}_2}{64.065\text{ g SO}_2} \times \frac{1\text{ mol S}}{1\text{ mol SO}_2} = 0.02791\text{ mol S} \div 0.02791 \rightarrow 1.000\text{ mol S}$$

A reasonable empirical formula is C_4H_4S, which has an empirical mass of 84.1 g/mol. Since this is the same as the experimentally determined molar mass, the molecular formula of thiophene if C_4H_4S.

76. First we determine the empirical formula of coniferin from the combustion data.

$$0.698 \text{ g } H_2O \times \frac{1 \text{ mol } H_2O}{18.015 \text{ g } H_2O} \times \frac{2 \text{ mol } H}{1 \text{ mol } H_2O} = 0.0775 \text{ mol } H \times \frac{1.008 \text{ g } H}{1 \text{ mol } H} = 0.0781 \text{ g } H$$

$$2.479 \text{ g } CO_2 \times \frac{1 \text{ mol } CO_2}{44.010 \text{ g } CO_2} \times \frac{1 \text{ mol } C}{1 \text{ mol } CO_2} = 0.05633 \text{ mol } C \times \frac{12.011 \text{ g } C}{1 \text{ mol } C} = 0.6766 \text{ g } C$$

$$(1.205 \text{ g} - 0.0781 \text{ g } H - 0.6766 \text{ g } C) \times \frac{1 \text{ mol } O}{16.00 \text{ g } O} = 0.0281 \text{ mol } O$$

0.0775 mol H ÷ 0.0281 = 2.76 mol H $\qquad$ ×4 → 11.04 mol H
0.05633 mol C ÷ 0.0281 = 2.00 mol C $\qquad$ ×4 → 8.00 mol C
0.0281 mol O ÷ 0.0281 = 1.00 mol O $\qquad$ ×4 → 4.00 mol O

Empirical formula = $C_8H_{11}O_4$ with a formula weight of 171 g/mol.

Now we determine the molality of the boiling solution, the number of moles of solute present in that solution, and the molar mass of the solute.

$$m = \frac{\Delta T_b}{K_b} = \frac{0.068°C}{0.512°C/m} = 0.13 \text{ } m$$

$$\text{amount solute} = 48.68 \text{ g } H_2O \times \frac{1 \text{ kg solvent}}{1000 \text{ g}} \times \frac{0.13 \text{ mol solute}}{1 \text{ kg solvent}} = 6.3 \times 10^{-3} \text{ mol solute}$$

$$M = \frac{2.216 \text{ g solute}}{6.3 \times 10^{-3} \text{ mol solute}} = 3.5 \times 10^2 \text{ g}/\text{mol}$$

This molar mass is twice the formula weight of the empirical formula. Thus, the molecular formula is twice the empirical formula. Molecula formula = $C_{16}H_{22}O_8$.

77. The boiling point must go up by 2 degrees, so $\Delta T_b = 2$ °C. We know that $K_b = 0.512$ °C/m for water. We assume that the mass of a litre of water is 1.000 kg and that the van't Hoff factor for NaCl is $i = 2.00$. We first determine the molality of the saltwater solution and then the mass of solute needed.

$$m = \frac{\Delta T_b}{i \text{ } K_b} = \frac{2°C}{2.00 \times 0.512 \text{ } °C/m} = 2 \text{ } m$$

$$\text{solute mass} = 1.00 \text{ L } H_2O \times \frac{1 \text{ kg } H_2O}{1 \text{ L } H_2O} \times \frac{2 \text{ mol NaCl}}{1 \text{ kg } H_2O} \times \frac{58.4 \text{ g NaCl}}{1 \text{ mol NaCl}} = 120 \text{ g NaCl}$$

This is at least ten times the amount of salt one would typically add to a liter of water for cooking purposes.

78. (a) The impure compound begins to melt at a temperature lower than that of the pure compound because the compound acts as a solvent for its impurity. The impurity thus causes the depression of the freezing point of the compound. Actually, the compound freezes (or melts) over a range of temperatures because once some of the pure compound freezes out, the remaining solution is a more concentrated solution of the impurity and thus has an even lower freezing point. It is this range of freezing points (or melting points) rather than a single freezing point- a so called "sharp" melting point- that is the true characteristic of an impure solid. (We have been referring to melting points in parentheses above because melting is, of course, simply the reverse of freezing).

(b) the melting point of the impure compound can be higher than the melting point of the pure compound if two conditions are met. First, the impurity must have a higher melting point than does the pure compound. Second, the impurity must be soluble in the solid solvent. Then we have a solution of the solvent in the impurity, and the impurity's melting point can determine the melting point of the solution.

Strong Electrolytes, Weak Electrolytes, and Nonelectrolytes

79. The freezing point depression is given by $\Delta T_f = -iK_f m$. Since $K_f = 1.86°C/m$ for water, $\Delta T_f = -i0.186°C$ for this group of $0.10\,m$ solutions.

(a) $T_f = -0.19°C$ Urea is a nonelectrolyte, and $i = 1$.

(b) $T_f = -0.37°C$ NH_4NO_3 is a strong electrolyte, composed of two ions per formula unit; $i = 2$.

(c) $T_f = -0.37°C$ HCl is a strong electrolyte, composed of two ions per formula unit; $i = 2$.

(d) $T_f = -0.56°C$ $CaCl_2$ is a strong electrolyte, composed of three ions per formula unit; $i = 3$.

(e) $T_f = -0.37°C$ $MgSO_4$ is a strong electrolyte, composed of two ions per formula unit; $i = 2$.

(f) $T_f = -0.19°C$ Ethanol is a nonelectrolyte; $i = 1$.

(g) $T_f < -0.19°C$ $HC_2H_3O_2$ is a weak electrolyte; i is somewhat larger than 1.

80. (a) First calculate the freezing point for a $0.050\,m$ solution of a nonelectrolyte.

$$\Delta T_{f,calc} = -K_f m = -1.86°C/m \times 0.050\,m = -0.093°C \qquad i = \frac{\Delta T_{f,obs}}{\Delta T_{f,calc}} = \frac{-0.0986°C}{-0.093°C} = 1.1$$

(b) $\left[H^+\right] = \left[NO_2^-\right] = 6.91 \times 10^{-3}\,M = 0.00691\,M$

$\left[HNO_2\right] = 0.100\,M - 0.00691\,M = 0.093\,M$

$[\text{particles}] = 0.00691\,M + 0.00691\,M + 0.093\,M = 0.107\,M \qquad i = \frac{0.107\,M}{0.100\,M} = 1.07$

81. The combination of $NH_3(aq)$ with $HC_2H_3O_2(aq)$, results in the formation of $NH_4C_2H_3O_2(aq)$, a solution of an ionic substance and a strong electrolyte.

$$NH_3(aq) + HC_2H_3O_2(aq) \rightarrow NH_4C_2H_3O_2(aq) \rightarrow NH_4^+(aq) + C_2H_3O_2^-(aq)$$

This solution of strong electrolyte conducts a current very well.

82. For two solutions to be isotonic, they must contain the same concentration of particles, For the solution to also have the same % mass/volume would mean that each gram of each substance produces the same number of particles. This would be true if they had the same molar mass and the same van't Hoff factor. It would also be true if the quotient of molar mass and van't Hoff factor were the same. While this condition is not impossible to meet, it is pretty unlikely.

FEATURE PROBLEMS

105. **(a)** The temperature at which the total vapor pressure is equal to 760 mm Hg corresponds to the boiling point for the cinnamaldehyde/H_2O mixture. Consequently, to find the temperature at which steam distillation begins under 1 atm pressure, we must construct a plot of P_{total} against the temperature. The data needed for this graph and a separate graph required for the answer to part (c) are given in the table below.

Temperature (°C)	P_{H_2O} (mmHg)	$P_{C_9H_8O}$ (mmHg)	P_{total} (mmHg)	Mole fraction of C_9H_8O
76.1	308	1	309	3.24×10^{-3}
105.8	942	5	947	5.28×10^{-3}
120.0	1489	10	1499	6.67×10^{-3}

The graph of total vapor pressure vs temperature is drawn below:

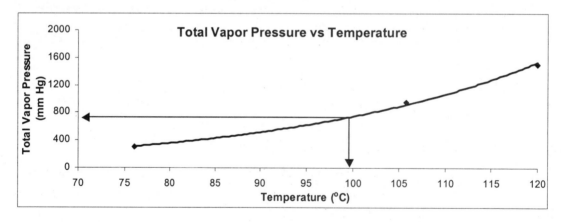

According to this plot, when the barometric pressure is 1 atm. (i.e. 760 mmHg), distillation occurs at 99.6 °C. Thus, as one would expect for a steam distillation, the boiling point for the mixture is below the boiling point for the lower boiling point component, viz. water. Moreover, the graph shows that the boiling point for this mixture is close to that for pure water and, once again, this is the anticipated result since the mixture contains very little cinnamaldehyde.

(b) The condensate that is collected in a steam distillation is the liquefied form of the gaseous mixture generated in the distillation pot. The composition of the gaseous mixture and that of the derived condensate is determined not by the composition of the liquid mixture but rather by the vapor pressure for each component at the boiling point. The boiling point of the cinnamaldehyde/H_2O mixture will stay fixed at 99.6 °C and the composition will remain at 99.5$\underline{4}$ mole % H_2O and 0.46$\underline{4}$ % cinnamaldehyde as long as there is at least a modicum of each component present.

(c) To answer this part of the question we must construct a plot of the mole fraction of cinnamaldehyde against temperature. The graph, which is drawn below, shows that the mole fraction for cinnamaldehyde at the boiling point (99.6 °C) is 0.0046$\underline{4}$.

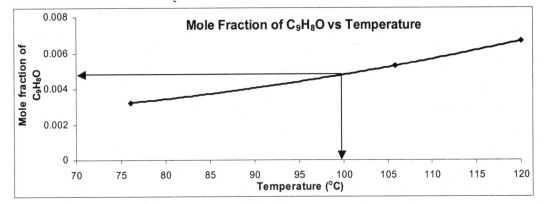

Because the concentration of each component in the vapor is directly proportional to its vapor pressure, the number of moles of component A divided by the moles of component B is equal to the partial pressure for component A divided by the partial pressure of component B.

$$\frac{\#\ \text{moles component A}}{\#\ \text{moles component B}} = \frac{P_{\text{component A}}}{P_{\text{component B}}}$$

Expressed another way, $\dfrac{\text{Mass A}}{\text{Mass B}} = \dfrac{P_{\text{component A}}}{P_{\text{component B}}} \times \dfrac{\text{Molar Mass component A}}{\text{Molar Mass component B}}$

If we make cinnamaldehyde component A and water component B, and then plug in the appropriate numbers into the second equation involving masses, we obtain

$P_{\text{cinn}} = 0.00464 \times (760 \text{ mmHg}) = {\sim}3.53 \text{ mmHg};$
$P_{H_2O} = (760\text{-}3.53) \text{ mmHg} = {\sim}\ 756 \text{ mmHg}$

$$\frac{\text{Mass C}_9\text{H}_8\text{O}}{\text{Mass H}_2\text{O}} = \frac{3.53 \text{ mmHg}}{756 \text{ mm Hg}} \times \frac{1\ \text{mol C}_9\text{H}_8\text{O} \cdot \frac{132\ \text{g C}_9\text{H}_8\text{O}}{1\ \text{mol C}_9\text{H}_8\text{O}}}{18.0\ \text{g H}_2\text{O} \cdot \frac{1\ \text{mol H}_2\text{O}}{1\ \text{mol H}_2\text{O}}} = 3.42 \times 10^{-2}$$

Clearly then, water condenses in the greater quantity by mass. In fact, every gram of cinnamaldehyde that condenses is accompanied by ~29 g of water in the collection flask!

106. (a) A solution with $\chi_{HCl} = 0.50$ begins to boil at about 18° C. At that temperature the composition of the vapor is about $\chi_{HCl} = 0.63$, reading directly across the tie line at 18°C. The vapor has $\chi_{HCl} > 0.50$.

(b) The composition of HCl(aq) changes as the solution boils in an open container because the vapor has a different composition than does the liquid. Thus, the component with the lower boiling point is depleted as the solution boils.

(c) The azeotrope occurs at the maximum of the curve: at $\chi_{HCl} = 0.12$ and a boiling temperature of $110\,^{\circ}C$.

(d) We first determine the amount of HCl in the sample.

$$\text{amount HCl} = 30.32 \text{ mL NaOH} \times \frac{1\text{ L}}{1000\text{ mL}} \times \frac{1.006 \text{ mol NaOH}}{1\text{ L}} \times \frac{1 \text{ mol HCl}}{1 \text{ mol NaOH}}$$

amount HCl = 0.03050 mol HCl

The mass of water is the difference between the mass of solution and that of HCl.

$$\text{mass } H_2O = \left(5.00 \text{ mL soln} \times \frac{1.099 \text{ g}}{1 \text{ mL}} \right) - \left(0.03050 \text{ mol HCl} \times \frac{36.46 \text{ g HCl}}{1 \text{ mol HCl}} \right) = 4.38 \text{ g}$$

Now we determine the amount of H_2O and then the mole fraction of HCl.

$$\text{amount } H_2O = 4.38 \text{ g } H_2O \times \frac{1 \text{ mol } H_2O}{18.02 \text{ g } H_2O} = 0.243 \text{ mol } H_2O$$

$$\chi_{HCl} = \frac{0.03050 \text{ mol HCl}}{0.03050 \text{ mol HCl} + 0.243 \text{ mol } H_2O} = 0.111$$

107. (a) At 20 °C, the solubility of NaCl is 35.9 g NaCl / 100 g H_2O. We determine the mole fraction of H_2O in this solution

$$\text{amount } H_2O = 100 \text{ g } H_2O \times \frac{1 \text{ mol } H_2O}{18.02 \text{ g } H_2O} = 5.549 \text{ mol } H_2O$$

$$\text{amount NaCl} = 35.9 \text{ g NaCl} \times \frac{1 \text{ mol NaCl}}{58.44 \text{ g NaCl}} = 0.614 \text{ mol NaCl}$$

$$\chi_{water} = \frac{5.549 \text{ mol } H_2O}{0.614 \text{ mol NaCl} + 5.549 \text{ mol } H_2O} = 0.9004$$

The approximate relative humidity then will be 90% (90.04 %), because the water vapor pressure above the NaCl saturated solution will be 90.04% of the vapor pressure of pure water at 20° C.

(b) $CaCl_2 \cdot 6 H_2O$ deliquesces if the relative humidity is over 32%. Thus, $CaCl_2 \cdot 6 H_2O$ will deliquesce.

(c)   If the substance in the bottom of the dessicator has a high water solubility its saturated solution will have a low χ_{water}, which in turn will produce a low relative humidity. Relative humidity lower than 32% is needed to keep $CaCl_2 \cdot 6\,H_2O$ dry.

108. (a)   We first compute the molality of a 0.92% mass/volume solution, assuming the solution's density is about 1.00 g/mL, meaning that 100.0 mL solution has a mass of 100.0 g.

$$molality = \frac{0.92 \text{ g NaCl} \times \dfrac{1 \text{ mol NaCl}}{58.44 \text{ g NaCl}}}{(100.0 \text{ g soln} - 0.92 \text{ g NaCl}) \times \dfrac{1 \text{ kg solvent}}{1000 \text{ g}}} = 0.16\,m$$

Then we compute the freezing point depression of this solution.

$$\Delta T_f = -iK_f m = \frac{-2.0 \text{ mol ions}}{\text{mol NaCl}} \times \frac{1.86°C}{m} \times 0.16\,m = -0.61\ °C$$

The van't Hoff factor of NaCl most likely is not equal to 2.0, but a bit less and thus the two definitions are in fair agreement.

(b)   We calculate the amount of each solute, assume 1.00 L of solution has a mass of 1000 g, and subtract the mass of all solutes to determine the mass of solvent.

$$amount \text{ NaCl ions} = 3.5 \text{ g NaCl} \times \frac{1 \text{ mol NaCl}}{58.44 \text{ g NaCl}} \times \frac{2 \text{ mol ions}}{1 \text{ mol NaCl}} = 0.12 \text{ mol ions}$$

$$amount \text{ KCl ions} = 1.5 \text{ g} \times \frac{1 \text{ mol KCl}}{74.55 \text{ g KCl}} \times \frac{2 \text{ mol ions}}{1 \text{ mol KCl}} = 0.040 \text{ mol ions}$$

$$amount \text{ Na}_3\text{C}_6\text{H}_5\text{O}_7 \text{ ions} = 2.9 \text{ g} \times \frac{1 \text{ mol Na}_3\text{C}_6\text{H}_5\,\text{O}_7}{258.07 \text{ g Na}_3\text{C}_6\text{H}_5\text{O}_7} \times \frac{4 \text{ mol ions}}{1 \text{ mol Na}_3\text{C}_6\text{H}_5\text{O}_7}$$

$$amount \text{ Na}_3\text{C}_6\text{H}_5\text{O}_7 \text{ ions} = 0.045 \text{ mol ions}$$

$$amount \text{ C}_6\text{H}_{12}\text{O}_6 = 20.0 \text{ g C}_6\text{H}_{12}\text{O}_6 \times \frac{1 \text{ mol C}_6\text{H}_{12}\text{O}_6}{180.2 \text{ g C}_6\text{H}_{12}\text{O}_6} = 0.111 \text{ mol C}_6\text{H}_{12}\text{O}_6$$

$$solvent \text{ mass} = 1000.0 \text{ g} - (3.5 \text{ g} + 1.5 \text{ g} + 2.9 \text{ g} + 20.0 \text{ g}) = 972.1 \text{ g H}_2\text{O}$$

$$solvent \text{ mass} = 0.9721 \text{ kg H}_2\text{O}$$

$$solution \text{ molality} = \frac{(0.120 + 0.040 + 0.045 + 0.111) \text{ mol}}{0.9721 \text{ kg H}_2\text{O}} = 0.325\,m$$

$$\Delta T_f = -K_f m = -1.86\ °C/m \times 0.325\,m = -0.60°C$$

This again is close to the defined freezing point of $-0.52\ °C$, with the error most likely arising from the van't Hoff factors not being integral.

CHAPTER 15
CHEMICAL KINETICS

PRACTICE EXAMPLES

1A The rate of reaction for a reactant is expressed as the negative of the change in molarity divided by the time interval. Thus rate of reaction of A

$$= \frac{-\Delta[A]}{\Delta t} = \frac{-(0.3187\ M - 0.3629\ M)}{8.25\ min} \times \frac{1\ min}{60\ sec} = 8.93 \times 10^{-5}\ M\ sec^{-1}$$

1B We use the rate of reaction of A to determine the rate of formation of B, noting from the balanced equation that 3 moles of B form (+3 moles B) when 2 moles of A react (–2 moles A). (Recall that "M" means "moles per liter.")

$$rate\ of\ B\ formation = \frac{0.5522\ M\ A - 0.5684\ M\ A}{2.50\ min \times \frac{60\ s}{1\ min}} \times \frac{+3\ moles\ B}{-2\ moles\ A} = 1.62 \times 10^{-4}\ M\ s^{-1}$$

2A **(a)** The 2400-s tangent line intersects the 1200-s vertical line at 0.75 M and reaches 0 M at 3500 s. The slope of that tangent line is thus

$$slope = \frac{0\ M - 0.75\ M}{3500\ s - 1200\ s} = -3.3 \times 10^{-4}\ M\ s^{-1} = -\ instantaneous\ rate\ of\ reaction$$

The instantaneous rate of reaction $= 3.3 \times 10^{-4}\ M\ s^{-1}$

(b) At 2400 s, $[H_2O_2] = 0.39$ M. At 2450 s, $[H_2O_2] = 0.39\ M +\ rate \times \Delta t$

$$At\ 2450\ s, [H_2O_2] = 0.39\ M + \left[-3.3 \times 10^{-4}\ mol\ H_2O_2\ L^{-1}s^{-1} \times 50s \right]$$

$$= 0.39\ M - 0.017\ M = 0.37\ M$$

2B With only the data of Table 15.2 we can use only the reaction rate during the first 400 s, $-\Delta[H_2O_2]/\Delta t = 15.0 \times 10^{-4}\ M\ s^{-1}$, and the initial concentration, $[H_2O_2]_0 = 2.32$ M. We calculate the change in $[H_2O_2]$ and add it to $[H_2O_2]_0$ to determine $[H_2O_2]_{100}$.

$$\Delta[H_2O_2] =\ rate\ of\ reaction\ of\ H_2O_2 \times \Delta t = -15.0 \times 10^{-4}\ M\ s^{-1} \times 100\ s = -0.15\ M$$

$$[H_2O_2]_{100} = [H_2O_2]_0 + \Delta[H_2O_2] = 2.32\ M + (-0.15\ M) = 2.17\ M$$

This value differs from the value of 2.15 M determined in *text* Example 15-2b because the *text* used the initial rate of reaction $(17.1 \times 10^{-4}\ M\ s^{-1})$, which is a bit faster than the average rate over the first 400 seconds.

3A   We write the equation for each rate, divide them into each other, and solve for n.

$$R_1 = k \times [N_2O_5]_1^n = 5.45 \times 10^{-5} \text{ M s}^{-1} = k(3.15 \text{ M})^n$$

$$R_2 = k \times [N_2O_5]_2^n = 1.35 \times 10^{-5} \text{ M s}^{-1} = k(0.78 \text{ M})^n$$

$$\frac{R_1}{R_2} = \frac{5.45 \times 10^{-5} \text{ M s}^{-1}}{1.35 \times 10^{-5} \text{ M s}^{-1}} = 4.04 = \frac{k \times [N_2O_5]_1^n}{k \times [N_2O_5]_2^n} = \frac{k(3.15 \text{ M})^n}{k(0.78 \text{ M})^n} = \left(\frac{3.15}{0.78}\right)^n = (4.0_4)^n$$

We kept an extra significant figure $(_4)$ to emphasize that the value of $n = 1$; the reaction is first-order in N_2O_5.

3B   For the reaction, we know that $\text{rate} = k[\text{HgCl}_2]^1[C_2O_4^{2-}]^2$. Compare Expt. 4 to Expt. 1:

$$\frac{\text{rate}_4}{\text{rate}_1} = \frac{k[\text{HgCl}_2]^1[C_2O_4^{2-}]^2}{k[\text{HgCl}_2]^1[C_2O_4^{2-}]^2} = \frac{0.025 \text{ M} \times (0.045 \text{ M})^2}{0.105 \text{ M} \times (0.150 \text{ M})^2} = 0.0214 = \frac{\text{rate}_4}{1.8 \times 10^{-5} \text{ M min}^{-1}}$$

The desired rate is $\text{rate}_4 = 0.0214 \times 1.8 \times 10^{-5} \text{ M min}^{-1} = 3.9 \times 10^{-7} \text{ M min}^{-1}$

4A   We substitute into the rate law and solve for k.

$$\text{rate} = k[A]^2[B] = 4.78 \times 10^{-2} \text{ M s}^{-1} = k(1.12 \text{ M})^2(0.87 \text{ M})$$

$$k = \frac{4.78 \times 10^{-2} \text{ M s}^{-1}}{(1.12 \text{ M})^2 \, 0.87 \text{ M}} = 4.4 \times 10^{-2} \text{ M}^{-2} \text{ s}^{-1}$$

4B   We already know that $\text{rate} = k[\text{HgCl}_2]^1[C_2O_4^{2-}]^2$ and $k = 7.6 \times 10^{-3} \text{ M}^{-2}\text{min}^{-1}$.
Thus, substitution yields:

$$\text{Rate} = 7.6 \times 10^{-3} \text{ M}^{-2} \text{ min}^{-1}(0.050 \text{ M})^1(0.025 \text{ M})^2 = 2.4 \times 10^{-7} \text{ M min}^{-1}$$

5A   We substitute directly into the integrated rate equation.

$$\ln[A]_t = -kt + \ln[A]_0 = -3.02 \times 10^{-3} \text{ s}^{-1} \times 325 \text{ s} + \ln(2.80) = -0.982 + 1.030 = 0.048$$

$$[A]_t = e^{0.048} = 1.0 \text{ M}$$

5B   We substitute the suggested values into text equation (15.12).

$$\ln\frac{[H_2O_2]_t}{[H_2O_2]_0} = -kt = -k \times 600 \text{ s} = \ln\frac{1.49 \text{ M}}{2.32 \text{ M}} = -0.443 \qquad k = \frac{-0.443}{-600 \text{ s}} = 7.38 \times 10^{-4} \text{ s}^{-1}$$

Now we choose $[H_2O_2]_0 = 1.49$ M, $[H_2O_2]_t = 0.62$, $t = 1800 \text{ s} - 600 \text{ s} = 1200 \text{ s}$

$$\ln\frac{[H_2O_2]_t}{[H_2O_2]_0} = -kt = -k \times 1200 \text{ s} = \ln\frac{0.62 \text{ M}}{1.49 \text{ M}} = -0.88 \qquad k = \frac{-0.88}{-1200 \text{ s}} = 7.3 \times 10^{-4} \text{ s}^{-1}$$

These two values agree within the limits of the experimental measurement.

6A We can use the integrated rate equation to find the ratio of the final and initial concentrations. This ratio equals the fraction of the initial concentration that remains.

$$\ln\frac{[A]_t}{[A]_0} = -kt = -2.95\times10^{-3} \text{ s}^{-1}\times150 \text{ s} = -0.443$$

$$\frac{[A]_t}{[A]_0} = e^{-0.443} = 0.642; \quad 64.2\% \text{ of } [A]_0 \text{ remains.}$$

6B After two-thirds of the sample has decomposed, one-third of the sample remains. Thus $[H_2O_2]_t = [H_2O_2]_0 \div 3$, and we have

$$\ln\frac{[H_2O_2]_t}{[H_2O_2]_0} = -kt = \ln\frac{[H_2O_2]_0 \div 3}{[H_2O_2]_0} = \ln(1/3) = -1.099 = -7.30\times10^{-4} \text{ s}^{-1}t$$

$$t = \frac{-1.099}{-7.30\times10^{-4} \text{ s}^{-1}} = 1.51\times10^3 \text{ s}\times\frac{1 \text{ min}}{60 \text{ s}} = 25.2 \text{ min}$$

7A At the end of one half-life the pressure of DTBP will have halved, to 400 mmHg. At the end of another half-life, at 160 min, the pressure of DTBP will have halved again, to 200 mmHg. Thus, the pressure of DTBP at 125 min will be intermediate between the pressure at 80.0 min (400 mmHg) and that at 160 min (200 mmHg). To obtain an exact answer, first we determine the value of the rate constant from the half-life.

$$k = \frac{0.693}{t_{1/2}} = \frac{0.693}{80.0 \text{ min}} = 0.00866 \text{ min}^{-1}$$

$$\ln\frac{(P_{DTBP})_t}{(P_{DTBP})_0} = -kt = -0.00866 \text{ min}^{-1}\times125 \text{ min} = -1.08$$

$$\frac{(P_{DTBP})_t}{(P_{DTBP})_0} = e^{-1.08} = 0.340$$

$$(P_{DTBP})_t = 0.340\times(P_{DTBP})_0 = 0.340\times800 \text{ mmHg} = 272 \text{ mmHg}$$

7B **(a)** We use partial pressures in place of concentrations in the integrated first-order rate equation. Notice first that more than 30 half-lives have elapsed, and thus the ethylene oxide pressure has declined to at most $(0.5)^{30} = 9\times10^{-10}$ of its initial value.

$$\ln\frac{P_{30}}{P_0} = -kt = -2.05\times10^{-4} \text{ s}^{-1}\times30.0 \text{ h}\times\frac{3600 \text{ s}}{1 \text{ h}} = -22.1 \quad \frac{P_{30}}{P_0} = e^{-22.1} = 2.4\times10^{-10}$$

$$P_{30} = 2.4\times10^{-10} P_0 = 2.4\times10^{-10}\times782 \text{ mmHg} = 1.9\times10^{-7} \text{ mmHg}$$

(b) $P_{\text{ethylene oxide}}$ initially 782 mmHg $\rightarrow 1.9\times10^{-7}$ mmHg (~ 0). Essentially all of the ethylene oxide is converted to CH_4 and CO. Since pressure is proportional to moles, the final pressure will be twice the initial pressure (1 mole $\rightarrow$ 2 moles; 782 mmHg $\rightarrow$ 1564 mmHg). The final pressure will be 1.56×10^3 mmHg.

8A We first begin by looking for a constant rate, indicative of a zero-order reaction. If the rate is constant, the concentration will decrease by the same quantity during the same time period. If we choose a 25-s time period, we note that the concentration decreases $(0.88\text{ M} - 0.74\text{ M} =)0.14\text{ M}$ during the first 25 s, $(0.74\text{ M} - 0.62\text{ M} =)0.12\text{ M}$ during the second 25 s, $(0.62\text{ M} - 0.52\text{ M} =)0.10\text{ M}$ during the third 25 s, and $(0.52\text{ M} - 0.44\text{ M} =)0.08\text{ M}$ during the fourth 25 s period. This is hardly a constant rate and we conclude that the reaction is not zero-order.

We next look for a constant half-life, indicative of a first-order reaction. The initial concentration of 0.88 M decreases to one half of that value, 0.44 M, during the first 100 s, indicating a 100-s half-life. The concentration halves again to 0.22 M in the second 100 s, another 100-s half-life. Finally, we note that the concentration halves also from 0.62 M at 50 s to 0.31 M at 150 s, yet another 100-s half-life. The rate is established as first-order.

The rate constant is $k = \dfrac{0.693}{t_{1/2}} = \dfrac{0.693}{100\text{ s}} = 6.93 \times 10^{-3}\text{ s}^{-1}$

We can easily see that the reaction is first-order in the plots below, $k = 6.85 \times 10^{-3}\text{ s}^{-1}$

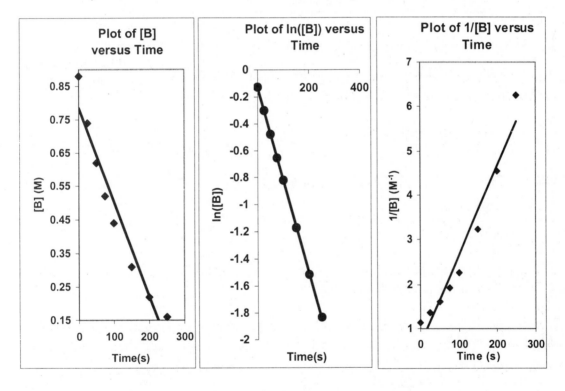

8B We plot the data in three ways to determine the order. (1) A plot of [A] vs. time is linear if the reaction is zero-order. (2) A plot of ln [A] vs. time is linear if the reaction is first-order. (3) A plot of 1/[A] vs. time is linear if the reaction is second-order. It is obvious from the plots below that the reaction is zero-order. The negative of the slope of the line equals $k = -(0.083\text{ M} - 0.250\text{ M}) \div 18.00\text{ min} = 9.28 \times 10^{-3}\text{ M/min}$ ($k = 9.30 \times 10^{-3}$ M/min using a graphical approach (next page)).

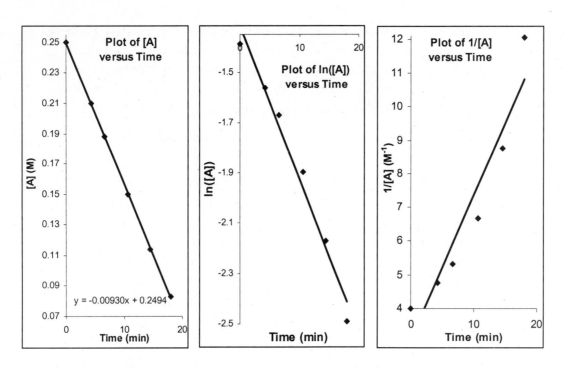

9A First we compute the value of the rate constant at $75.0\,^{\circ}C$ with the Arrhenius equation. We know that the activation energy is $E_a = 1.06 \times 10^5$ J/mol, and that $k = 3.46 \times 10^{-5}$ s^{-1} at 298 K. The temperature of $75.0\,^{\circ}C = 348.2$ K.

$$\ln\frac{k_2}{k_1} = \ln\frac{k_2}{3.46 \times 10^{-5}\ \text{s}^{-1}} = \frac{E_a}{R}\left(\frac{1}{T_1} - \frac{1}{T_2}\right) = \frac{1.06 \times 10^5\ \text{J/mol}}{8.3145\ \text{J mol}^{-1}\ \text{K}^{-1}}\left(\frac{1}{298.2\ \text{K}} - \frac{1}{348.2\ \text{K}}\right) = 6.14$$

$$k_2 = 3.46 \times 10^{-5}\ \text{s}^{-1} \times e^{+6.14} = 3.46 \times 10^{-5}\ \text{s}^{-1} \times 4.6 \times 10^2 = 0.016\ \text{s}^{-1}$$

$$t_{1/2} = \frac{0.693}{k} = \frac{0.693}{0.016\ \text{s}^{-1}} = 43\ \text{s at } 75\ ^{\circ}C$$

9B We use the integrated rate equation to determine the rate constant, realizing that one-third remains when two-thirds have decomposed.

$$\ln\frac{\left[N_2O_5\right]_t}{\left[N_2O_5\right]_0} = \ln\frac{\left[N_2O_5\right]_0 \div 3}{\left[N_2O_5\right]_0} = \ln\frac{1}{3} = -kt = -k(1.50\ \text{h}) = -1.099$$

$$k = \frac{1.099}{1.50\ \text{h}} \times \frac{1\ \text{h}}{3600\ \text{s}} = 2.04 \times 10^{-4}\ \text{s}^{-1}$$

Now use the Arrhenius equation to determine the temperature at which the rate constant is 2.04×10^{-4} s^{-1}.

$$\ln\frac{k_2}{k_1} = \ln\frac{2.04 \times 10^{-4}\ \text{s}^{-1}}{3.46 \times 10^{-5}\ \text{s}^{-1}} = 1.77 = \frac{E_a}{R}\left(\frac{1}{T_1} - \frac{1}{T_2}\right) = \frac{1.06 \times 10^5\ \text{J/mol}}{8.3145\ \text{J mol}^{-1}\ \text{K}^{-1}}\left(\frac{1}{298\ \text{K}} - \frac{1}{T_2}\right)$$

$$\frac{1}{T_2} = \frac{1}{298\ \text{K}} - \frac{1.77 \times 8.3145\ \text{K}^{-1}}{1.06 \times 10^5} = 3.22 \times 10^{-3}\ \text{K}^{-1} \quad T_2 = 311\ \text{K}$$

10A The two steps of the mechanism must add, in a Hess's law fashion, to produce the overall reaction.

overall reaction : $CO + NO_2 \longrightarrow CO_2 + NO$ or $CO + NO_2 \longrightarrow CO_2 + NO$

$-$ second step:$-(NO_3 + CO \longrightarrow NO_2 + CO_2)$ or $+(NO_2 + CO_2 \longrightarrow NO_3 + CO)$

first step: $2\,NO_2 \longrightarrow NO + NO_3$

If the first step is the slow step, then it will be the rate-determining step, and the rate of that step will be the rate of the reaction: rate of reaction $= k_1 [NO_2]^2$

10B **(1)** The steps of the mechanism must add, in a Hess's law fashion, to produce the overall reaction. This is done at right. The two intermediates, $NO_2F_2(g)$ and $F(g)$, are each produced in one step and consumed in the next one.

Fast: $NO_2(g) + F_2(g) \rightleftharpoons NO_2F_2(g)$

Slow: $NO_2F_2(g) \rightarrow NO_2F(g) + F(g)$

Fast: $F(g) + NO_2(g) \rightarrow NO_2F(g)$

Net: $2\,NO_2(g) + F_2(g) \rightarrow 2\,NO_2F(g)$

(2) The proposed mechanism must agree with the rate law. We expect the rate-determining step to determine the reaction rate: Rate $= k_3 [NO_2F_2]$. To eliminate $[NO_2F_2]$, we recognize that the first elementary reaction is very fast and will have the same rate forward as reverse: $R_f = k_1 [NO_2][F_2] = k_2 [NO_2F_2] = R_r$. We solve for the concentration of intermediate:$[NO_2F_2] = k_1 [NO_2][F_2]/k_2$. We now substitute this expression for $[NO_2F_2]$ into the rate equation: Rate $= (k_1 k_3 / k_2)[NO_2][F_2]$. The predicted rate law agrees with the experimental rate law.

REVIEW QUESTIONS

1.   **(a)** $[A]_0$ symbolizes the concentration of species A, usually a reactant, at time $= 0$.

   (b) k is the symbol for the specific rate constant of a reaction, the speed of the reaction if all rate-determining species were present at unit molarity.

   (c) $t_{1/2}$ is the symbol for the half-life of a reaction, the time during which the concentration of the reactant drops to one-half of its initial value.

   (d)   A zero-order reaction is one whose rate is independent of the concentration of reactant.

   (e)   A catalyst is a substance that is added to increase the rate of a chemical reaction. The catalyst takes part in the chemical reaction, without being consumed by the reaction (it is regenerated). Catalysts work by providing an alternate reaction pathways with a lower activation energy than the uncatalyzed reaction.

2.   **(a)**   The method of initial rates is used to determine the overall reaction order and the rate constant for a reaction by measuring the effect of different reactant concentrations on the initial rate of reaction.

(b)   An activated complex is a high-energy species that is formed as a result of the collision of two reacting particles.

(c)   The reaction mechanism is the series of molecular processes that accounts for the overall reaction and its observed kinetics.

(d)   Heterogeneous catalysis refers to catalytic activity that occurs at the interface between two phases, usually a liquid or a gas in contact with a solid catalytic surface.

(e)   The rate-determining step is the slowest step of a reaction mechanism (i.e. the step with the highest activation energy).

3.   **(a)**   The rate of a first-order reaction depends on the first power of the concentration of a reactant; a second-order reaction's rate depends on the second power of the concentration of a reactant, or on the first power of each of the two reactants.

(b)   A rate equation or rate law is the relationship between the instantaneous rate of a reaction and the powers, (orders), of concentrations of the reactants. An integrated rate equation provides the mathematical relationship between the concentration of reactant and elapsed time.

(c)   The activation energy is the energy that must be supplied to energize reactants to the level of the activated complex; it is almost always endothermic. The enthalpy of reaction is the energy released or absorbed as the overall reaction occurs, i.e., it may be exothermic or endothermic, respectively.

(d)   An elementary process is a description of how molecules interact with each other. The net reaction is the result of the several elementary processes that constitute it.

(e)   An enzyme is a biological catalyst. A substrate is the reactant whose reaction the catalyst promotes.

4.   **(a)** $\text{Rate} = -\dfrac{\Delta[A]}{\Delta t} = -\dfrac{(0.1832\ \text{M} - 0.2643\ \text{M})}{35\ \text{min}} = 2.3 \times 10^{-3}\ \text{M min}^{-1}$

(b) $\text{Rate} = 2.3 \times 10^{-3}\ \dfrac{\text{mol}}{\text{L min}} \times \dfrac{1\ \text{min}}{60\ \text{s}} = 3.8 \times 10^{-5}\ \text{M sec}^{-1}$

5. $2A + B \rightarrow C + 3D \qquad -\dfrac{\Delta[A]}{\Delta t} = 6.2 \times 10^{-4}\ \text{M s}^{-1}$

(a) $\text{Rate} = -\dfrac{1}{2}\dfrac{\Delta[A]}{\Delta t} = 1/2(6.2 \times 10^{-4}\ \text{M s}^{-1}) = 3.1 \times 10^{-4}\ \text{M}^{-1}\text{s}^{-1}$

(b) $\text{Rate of disappearance of B} = -\dfrac{1}{2}\dfrac{\Delta[A]}{\Delta t} = 1/2(6.2 \times 10^{-4}\ \text{M s}^{-1}) = 3.1 \times 10^{-4}\ \text{M}^{-1}\text{s}^{-1}$

(c) $\text{Rate of appearance of D} = -\dfrac{3}{2}\dfrac{\Delta[A]}{\Delta t} = 3(6.2 \times 10^{-4}\ \text{M s}^{-1}) = 9.3 \times 10^{-4}\ \text{M}^{-1}\text{s}^{-1}$

6. In each case, we draw the tangent line to the plotted curve.

    (a) The slope of the line is $\dfrac{\Delta[H_2O_2]}{\Delta t} = \dfrac{1.7\text{ M} - 0.6\text{ M}}{400\text{ s} - 1600\text{ s}} = -9.2 \times 10^{-4}\text{ M s}^{-1}$

       Reaction rate $= -\dfrac{\Delta[H_2O_2]}{\Delta t} = 9.2 \times 10^{-4}\text{ M s}^{-1}$

    (b) Read the value where the horizontal line $[H_2O_2] = 0.50$ M, intersects the curve ≈ 2150 s or ≈ 36 minutes

7. Statement **(b)** is correct. After each half-life—that is, after each 75 s—the amount of reactant remaining is half of the amount that was present at the beginning of that half-life. Statement **(a)** is incorrect; the quantity of A remaining after 150 s is half of what was present after 75 s. Statement **(c)** is incorrect because different quantities of A are consumed in each 75 s of the reaction: 1/2 of the original amount in the first 75 s, 1/4 of the original amount in the second 75 s, 1/8 of the original amount in the third 75 s, and so on. Statement **(d)** is incorrect; it implies a constant rate during the first half-life. The rate of a first-order reaction actually decreases as time passes and reactant is consumed.

8. Substitute the given values into the rate equation to obtain the rate of reaction.

$$\text{Rate} = k[A]^2[B]^0 = \left(0.0103\text{ M}^{-1}\text{min}^{-1}\right)(0.116\text{ M})^2(3.83\text{ M})^0 = 1.39 \times 10^{-4}\text{ M / min}$$

Recall that $(\text{any quantity})^0 = 1$.

9. The rate constant is determined as $k = 0.693 / t_{1/2} = 0.693 / 19.8\text{ min} = 0.0350\text{ min}^{-1}$

Then the rate can be determined. $\text{Rate} = k[A] = 0.0350\text{ min}^{-1} \times 0.632\text{ M} = 0.0221\text{ M min}^{-1}$

10. **(a)** A first-order reaction has a constant half-life. Thus, half of the initial concentration remains after 30.0 minutes, and at the end of another half-life—60.0 minutes total—half of the concentration present at 30.0 minutes will have reacted: the concentration has decreased to one-quarter of its initial value. Or, we could say that the reaction is 75% complete after two half-lives—60.0 minutes.

    (b) A zero-order reaction proceeds at a constant rate. Thus, if the reaction is 50% complete in 30.0 minutes, in twice the time—60.0 minutes—the reaction will be 100% complete. (And in one-fifth the time—6.0 minutes—the reaction will be 10% complete. Alternatively, we can say that the rate of reaction is 10%/6.0 min.) Time to be 75% complete $= 75\% \times \dfrac{60.0\text{ min}}{100\%} = 45\text{ min}$

11. **(a)** Although we might be tempted to apply equation (15.12) to the given data, and calculate k first, there is an easier method. During one half-life, [A] decreases from 2.00 M to 1.00 M. Then, during the second half-life, [A] decreases from 1.00 M to 0.500 M. And finally, during the third half-life, [A] decreases 0.500 M to 0.250 M. Consequently, three half-lives must have elapsed while [A] has decreased from 2.00 M to 0.250 M. $3t_{1/2} = 126$ min $t_{1/2} = 126$ min $\div 3 = 42.0$ min

(b) $k = \dfrac{0.693}{t_{1/2}} = \dfrac{0.693}{42.0 \text{ min}} = 0.0165 \text{ min}^{-1}$

12. **(a)** The mass of A has decreased to one fourth of its original value, from 1.60 g to 0.40 g. Since $\dfrac{1}{4} = \dfrac{1}{2} \times \dfrac{1}{2}$, we see that two half-lives have elapsed.

Thus, $2 \times t_{1/2} = 38$ min, or $t_{1/2} = 19$ min.

(b) $k = 0.693/t_{1/2} = \dfrac{0.693}{19 \text{ min}} = 0.036 \text{ min}^{-1}$ $\ln \dfrac{[A]_t}{[A]_0} = -kt = -0.036 \text{ min}^{-1} \times 60 \text{ min} = -2.2$

$\dfrac{[A]_t}{[A]_0} = e^{-2.2} = 0.11$ or $[A]_t = [A]_0 \, e^{-kt} = 1.60 \text{ g A} \times 0.11 = 0.2 \text{ g A}$

13. **(a)** $\ln \dfrac{[A]_t}{[A]_0} = -kt = \ln \dfrac{0.632 \text{ M}}{0.816 \text{ M}} = -0.256$ $k = -\dfrac{-0.256}{16.0 \text{ min}} = 0.0160 \text{ min}^{-1}$

(b) $t_{1/2} = \dfrac{0.693}{k} = \dfrac{0.693}{0.0160 \text{ min}^{-1}} = 43.3$ min

(c) Solve the integrated rate equation for the elapsed time.

$\ln \dfrac{[A]_t}{[A]_0} = -kt = \ln \dfrac{0.235 \text{ M}}{0.816 \text{ M}} = -1.245 = -0.0160 \text{ min}^{-1} \times t$

$t = \dfrac{-1.245}{-0.0160 \text{ min}^{-1}} = 77.8 \text{ min}$

(d) $\ln \dfrac{[A]}{[A]_0} = -kt$ becomes $\dfrac{[A]}{[A]_0} = e^{-kt}$ which in turn becomes

$[A] = [A]_0 \, e^{-kt} = 0.816 \text{ M} \exp\left(-0.0160 \text{ min}^{-1} \times 2.5 \text{ h} \times \dfrac{60 \text{ min}}{1 \text{ h}} \right) = 0.816 \times 0.0907 = 0.074 \text{ M}$

14. **(a)** From Expt. 1 to Expt. 3, when [B] remains constant and [A] doubles, the rate

increases by a factor of $\dfrac{6.75 \times 10^{-4}\ \text{M s}^{-1}}{3.35 \times 10^{-4}\ \text{M s}^{-1}} = 2.01 \approx 2$

Thus, the reaction is first-order with respect to A.

From Expt 1 to Expt. 2, when [A] remains constant and [B] doubles, the rate

increases by a factor of $\dfrac{1.35 \times 10^{-3}\ \text{M s}^{-1}}{3.35 \times 10^{-4}\ \text{M s}^{-1}} = 4.03 \approx 4$

Thus, the reaction is second-order with respect to B.

(b) Overall reaction order = order with respect to A + order with respect to B = $1 + 2 = 3$
The reaction is third-order overall.

(c) $\text{Rate} = 3.35 \times 10^{-4}\ \text{M s}^{-1} = k(0.185\ \text{M})(0.133\ \text{M})^2$

$k = \dfrac{3.35 \times 10^{-4}\ \text{M s}^{-1}}{(0.185\ \text{M})(0.133\ \text{M})^2} = 0.102\ \text{M}^{-2}\ \text{s}^{-1}$

15. In the first 500 s, [A] decreases from 2.00 M to 1.00 M; this is the first half-life. From 500 s to 1500 s, an elapsed period of 1000 s, [A] decreases by half again, from 1.00 M to 0.50 M; this is the second half-life. Since the half-life is not constant, the reaction is not first-order.

During the first 500 s, $\text{Rate} = -\dfrac{1.00\ \text{M} - 2.00\ \text{M}}{500\ \text{s}} = 0.00200\ \text{M/s}$. Then during the first

1500 s, the rate is computed as $\text{Rate} = -\dfrac{0.50\ \text{M} - 2.00\ \text{M}}{1500\ \text{s}} = 0.00100\ \text{M/s}$.

Since the rate is not constant, this reaction is not zero-order.

We conclude that the reaction is second-order, by the process of elimination. We confirm this conclusion by computing values of the second-order rate constant with the equation

$1/[\text{A}]_t - 1/[\text{A}]_0 = kt$

$\dfrac{1}{1.00\ \text{M}} - \dfrac{1}{2.00\ \text{M}} = 0.500\ \text{M}^{-1} = k \times 500\ \text{s} \quad k = \dfrac{0.500\ \text{M}^{-1}}{500\ \text{s}} = 0.0010\ \text{M}^{-1}\ \text{s}^{-1}$

$\dfrac{1}{0.50\ \text{M}} - \dfrac{1}{2.00\ \text{M}} = 1.50\ \text{M}^{-1} = k \times 1500\ \text{s} \quad k = \dfrac{1.50\ \text{M}^{-1}}{1500\ \text{s}} = 0.0010\ \text{M}^{-1}\ \text{s}^{-1}$

$\dfrac{1}{0.25\ \text{M}} - \dfrac{1}{2.00\ \text{M}} = 3.50\ \text{M}^{-1} = k \times 3500\ \text{s} \quad k = \dfrac{3.50\ \text{M}^{-1}}{3500\ \text{s}} = 0.0010\ \text{M}^{-1}\ \text{s}^{-1}$

The constant value of the second-order rate constant indicates that this reaction indeed is second-order.

16. Statement (d) is correct; the activation energy of an endothermic reaction must at least equal the value of ΔH for that reaction. However, if $E_a = \Delta H_{rxn}$, the products will readily "slip back down the hill" to reactants; that is, no net reaction will occur. (There will be no activation energy for the reverse reaction, nothing to prevent that reaction from occurring regardless of the energy of its reactants.) Thus, E_a should be at least a slight bit greater than ΔH_{rxn}.

17. **(a)** $\ln\dfrac{k_1}{k_2} = \dfrac{E_a}{R}\left(\dfrac{1}{T_2} - \dfrac{1}{T_1}\right) = \ln\dfrac{5.4\times10^{-4}\ \text{L mol}^{-1}\ \text{s}^{-1}}{2.8\times10^{-2}\ \text{L mol}^{-1}\ \text{s}^{-1}} = \dfrac{E_a}{R}\left(\dfrac{1}{683\ \text{K}} - \dfrac{1}{599\ \text{K}}\right)$

$-3.95R = -E_a \times 2.05\times10^{-4}$

$E_a = \dfrac{3.95\ R}{2.05\times10^{-4}} = 1.93\times10^4\ \text{K}^{-1} \times 8.3145\ \text{J mol}^{-1}\ \text{K}^{-1} = 1.60\times10^5\ \text{J/mol} = 160\ \text{kJ/mol}$

(b) $\ln\dfrac{k_1}{k_2} = \dfrac{E_a}{R}\left(\dfrac{1}{T_2} - \dfrac{1}{T_1}\right) = \ln\dfrac{5.0\times10^{-3}\ \text{L mol}^{-1}\ \text{s}^{-1}}{2.8\times10^{-2}\ \text{L mol}^{-1}\ \text{s}^{-1}} = \dfrac{1.60\times10^5\ \text{J/mol}}{8.3145\ \text{J mol}^{-1}\ \text{K}^{-1}}\left(\dfrac{1}{683\ \text{K}} - \dfrac{1}{T}\right)$

$-1.72 = 1.92\times10^4\left(\dfrac{1}{683\ \text{K}} - \dfrac{1}{T}\right)$ $\quad\left(\dfrac{1}{683\ \text{K}} - \dfrac{1}{T}\right) = \dfrac{-1.72}{1.92\times10^4} = -8.96\times10^{-5}$

$\dfrac{1}{T} = 8.96\times10^{-5} + 1.46\times10^{-3} = 1.55\times10^{-3}$ $\quad T = 645\ \text{K}$

18. **(a)** Set II is data from a zero-order reaction. We know this because the rate of set II is constant. 0.25 M/25 s = 0.010 M s^{-1}. A zero-order reaction has a constant rate.

(b) A first-order reaction has a constant half-life. In set I, the first half-life is slightly less than 75 sec, since the concentration decreases by slightly more than half (from 1.00 M to 0.47 M) in 75 s. Again, from 75 s to 150 s the concentration decreases from 0.47 M to 0.22 M, again by slightly more than half, in a time of 75 s. Finally, two half-lives should see the concentration decrease to one-fourth of its initial value. This is what we see: from 100 s to 250 sec, 150 s of elapsed time, the concentration decreases from 0.37 M to 0.08 M; that is, to slightly less than one-fourth of its initial value. Notice that we cannot make the same statement of constancy of half-life regarding set III. The first half-life is 100 s, but it takes more than 150 s (from 100 s to 250 s) for [A] to again decrease by half.

(c) For a second-order reaction, $1/[A]_t - 1/[A]_0 = kt$. For the initial 100 s in set III, we have

$\dfrac{1}{0.50\ \text{M}} - \dfrac{1}{1.00\ \text{M}} = 1.0\ \text{L mol}^{-1} = k\,100\ \text{s},$ $\quad k = 0.010\ \text{L mol}^{-1}\ \text{s}^{-1}$

For the initial 200 s, we have

$$\frac{1}{0.33\ \text{M}} - \frac{1}{1.00\ \text{M}} = 2.0\ \text{L mol}^{-1} = k\ 200\ \text{s}, \qquad k = 0.010\ \text{L mol}^{-1}\ \text{s}^{-1}$$

Since we obtain the same value of the rate constant using the equation for second-order kinetics, set III must be second-order.

19. For a zero-order reaction (set II), the slope equals the rate constant:

$$k = -\Delta[A]/\Delta t = 1.00\ \text{M}/100\ \text{s} = 0.0100\ \text{M/s}$$

20. Set I is the data for a first-order reaction; we can analyze those items of data to determine the half-life. In the first 75 s, the concentration decreases by a bit more than half. This implies a half-life slightly less than 75 s, perhaps 70 s. This is consistent with the other time periods noted in the answer to Review Question 18 (b) and also to the fact that in the 150 s from 50 s to 200 s, the concentration decreases from 0.61 M to 0.14 M, a bit more than a factor-of-four decrease. The factor-of-four decrease, to one-fourth of the initial value, is what we would expect of two successive half-lives. We can determine the half-life more accurately, by obtaining a value of k from the relation $\ln\left([A]_t/[A]_0\right) = -kt$ followed by $t_{1/2} = 0.693/k$

21. We can determine an approximate initial rate by using data from the first 25 s.

$$\text{Rate} = -\frac{\Delta[A]}{\Delta t} = -\frac{0.80\ \text{M} - 1.00\ \text{M}}{25\ \text{s} - 0\ \text{s}} = 0.0080\ \text{M s}^{-1}$$

22. The approximate rate at 75 s can be taken as the rate over the time period from 50 s to 100 s.

(a) $\displaystyle \text{Rate}_{II} = -\frac{\Delta[A]}{\Delta t} = -\frac{0.00\ \text{M} - 0.50\ \text{M}}{100\ \text{s} - 50\ \text{s}} = 0.010\ \text{M s}^{-1}$

(b) $\displaystyle \text{Rate}_{I} = -\frac{\Delta[A]}{\Delta t} = -\frac{0.37\ \text{M} - 0.61\ \text{M}}{100\ \text{s} - 50\ \text{s}} = 0.0048\ \text{M s}^{-1}$

(c) $\displaystyle \text{Rate}_{III} = -\frac{\Delta[A]}{\Delta t} = -\frac{0.50\ \text{M} - 0.67\ \text{M}}{100\ \text{s} - 50\ \text{s}} = 0.0034\ \text{M s}^{-1}$

Alternatively we can use [A] at 75 s (the values given in the table) in the relationship $\text{Rate} = k[A]^m$, where $m = 0$, 1, or 2.

(a) $\text{Rate}_{II} = 0.010\ \text{M s}^{-1} \times (0.25\ \text{mol/L})^0 = 0.010\ \text{M s}^{-1}$

(b) Since $t_{1/2} = 70\ \text{s}$, $k = 0.693/70\ \text{s} = 0.0099\ \text{s}^{-1}$

$\text{Rate}_{I} = 0.0099\ \text{s}^{-1} \times (0.47\ \text{mol/L})^1 = 0.0047\ \text{M s}^{-1}$

(c) $\text{Rate}_{III} = 0.010\ \text{L mol}^{-1}\ \text{s}^{-1} \times (0.57\ \text{mol/L})^2 = 0.0032\ \text{M s}^{-1}$

23. We can combine the approximate rates from Review Question 22, with the fact that 10 s have elapsed, and the concentration at 100 s.

  (a) $[A]_{II} = 0.00$ M   There is no reactant left after 100 s.

  (b) $[A]_{I} = [A]_{100} - (10 \text{ s} \times \text{ rate}) = 0.37$ M $- (10 \text{ s} \times 0.0047 \text{ M s}^{-1}) = 0.32$ M

  (c) $[A]_{III} = [A]_{100} - (10 \text{ s} \times \text{ rate}) = 0.50$ M $- (10 \text{ s} \times 0.0032 \text{ M s}^{-1}) = 0.47$ M

24. (a) We can demonstrate consistency with the stoichiometry by adding the two reactions. Sum of the reactions: $A + B + I + B \longrightarrow I + C + D$ We then cancel the I that appears on both sides of the resultant equation: $A + 2B \longrightarrow C + D$ We determine the rate of the reaction from the mechanism by assuming that the rate of the elementary slow step equals the reaction rate: Reaction rate $= k_{slow}[A][B]$. This is equivalent to the observed rate law.

  (b) Again, we demonstrate consistency with the stoichiometry by adding the two reactions to obtain: $2B + A + B_2 \longrightarrow B_2 + C + D$ We then cancel the "B_2" that appears on both sides of the resultant equation, and finally have: $2B + A \longrightarrow C + D$ The reaction rate is the rate of the elementary slow step: Reaction rate $= k_{slow}[A][B_2]$ But B_2 is a reaction intermediate whose concentration is difficult to determine. We can determine that concentration by assuming that the fast initial step goes equally rapidly in the forward and reverse directions. $k_f[B]^2 = k_r[B_2]$ $[B_2] = (k_f/k_r)[B]^2$ The resulting expression for $[B_2]$ is substituted into the expression for reaction rate, and we see that the experimentally determined rate law is not recovered.
  Reaction rate $= k_{slow}[A][B_2] = k_{slow}[A](k_f/k_r)[B]^2 = k'[A][B]^2$

EXERCISES

Rates of Reactions

25. Rate $= -\dfrac{\Delta[A]}{\Delta t} = -\dfrac{(0.474 \text{ M} - 0.485 \text{ M})}{82.4 \text{ s} - 71.5 \text{ s}} = 1.0 \times 10^{-3} \text{ M s}^{-1}$

26. (a) Rate $= -\dfrac{\Delta[A]}{\Delta t} = -\dfrac{0.1498 \text{ M} - 0.1565 \text{ M}}{1.00 \text{ min} - 0.00 \text{ min}} = 0.0067 \text{ M min}^{-1}$

  Rate $= -\dfrac{\Delta[A]}{\Delta t} = -\dfrac{0.1433 \text{ M} - 0.1498 \text{ M}}{2.00 \text{ min} - 1.00 \text{ min}} = 0.0065 \text{ M min}^{-1}$

(b) The rates are not equal because, in all except zero-order reactions, the rate depends on the concentration of reactant. And, of course, as the reaction proceeds reactant is consumed and its concentration decreases, changing the rate of the reaction.

27. **(a)** $[A] = [A]_i + \Delta[A] = 0.588\ M - 0.013\ M = 0.575\ M$

(b) $\Delta[A] = 0.565\ M - 0.588\ M = -0.023\ M$

$$\Delta t = \Delta[A]\frac{\Delta t}{\Delta[A]} = \frac{-0.023\ M}{-2.2 \times 10^{-2}\ M/min} = 1.0\ min$$

$$\text{time} = t + \Delta t = (4.40 + 1.0)\ min = 5.4\ min$$

28. Initial concentrations are $[HgCl_2] = 0.105\ M$ and $[C_2O_4^{2-}] = 0.300\ M$.

The initial rate of the reaction is $7.1 \times 10^{-5}\ M\ min^{-1}$. Of course, the reaction is

$$2\ HgCl_2(aq) + C_2O_4^{2-}(aq) \rightarrow 2\ Cl^-(aq) + 2\ CO_2(g) + Hg_2Cl_2(aq).$$

The rate of reaction equals the rate of disappearance of $C_2O_4^{2-}$. Then, after 1 hour, assuming that the rate is the same as the initial rate,

(a) $[HgCl_2] = 0.105\ M - \left(7.1 \times 10^{-5}\ \dfrac{mol\ C_2O_4^{2-}}{L \cdot s} \times \dfrac{2\ mol\ HgCl_2}{1\ mol\ C_2O_4^{2-}} \times 1\ h \times \dfrac{60\ min}{1\ h}\right)$

$[HgCl_2] = 0.096\ M$

(b) $[C_2O_4^{2-}] = 0.300\ M - \left(7.1 \times 10^{-5}\ \dfrac{mol}{L \cdot min} \times 1\ h \times \dfrac{60\ min}{1\ h}\right) = 0.296\ M$

29. **(a)** $\text{Rate} = \dfrac{-\Delta[A]}{\Delta t} = \dfrac{\Delta[C]}{2\Delta t} = 1.76 \times 10^{-5}\ M\ s^{-1}$

$$\frac{\Delta[C]}{\Delta t} = 2 \times 1.76 \times 10^{-5}\ M\ s^{-1} = 3.52 \times 10^{-5}\ M/s$$

(b) $\dfrac{\Delta[A]}{\Delta t} = -\dfrac{\Delta[C]}{2\Delta t} = -1.76 \times 10^{-5}\ M\ s^{-1}$ Assume this rate is constant.

$$[A] = 0.3580\ M + \left(-1.76 \times 10^{-5}\ M\ s^{-1} \times 1.00\ min \times \frac{60\ s}{1\ min}\right) = 0.357\ M$$

(c) $\dfrac{\Delta[A]}{\Delta t} = -1.76 \times 10^{-5}\ M\ s^{-1}$

$$\Delta t = \frac{\Delta[A]}{-1.76 \times 10^{-5}\ M/s} = \frac{0.3500\ M - 0.3580\ M}{-1.76 \times 10^{-5}\ M/s} = 4.5 \times 10^2\ s$$

30. **(a)** $\dfrac{\Delta n[O_2]}{\Delta t} = 1.00 \text{ L soln} \times \dfrac{5.7 \times 10^{-4} \text{ mol } H_2O_2}{1 \text{ L soln} \cdot \text{s}} \times \dfrac{1 \text{ mol } O_2}{2 \text{ mol } H_2O_2} = 2.9 \times 10^{-4} \text{ mol } O_2/\text{s}$

(b) $\dfrac{\Delta n[O_2]}{\Delta t} = 2.9 \times 10^{-4} \dfrac{\text{mol } O_2}{\text{s}} \times \dfrac{60 \text{ s}}{1 \text{min}} = 1.7 \times 10^{-2} \text{ mol } O_2 / \text{min}$

(c) $\dfrac{\Delta V[O_2]}{\Delta t} = 1.7 \times 10^{-2} \dfrac{\text{mol } O_2}{\text{min}} \times \dfrac{22{,}414 \text{ mL } O_2 \text{ at STP}}{1 \text{ mol } O_2} = \dfrac{3.8 \times 10^2 \text{ mL } O_2 \text{ at STP}}{\text{min}}$

31. Notice that, for every 1000 mmHg drop in the pressure of A(g), there will be a corresponding 2000 mmHg rise in the pressure of B(g) plus a 1000 mmHg rise in the pressure of C(g).

(a) We set up the calculation with three lines of information below the balanced equation: (1) the initial conditions, (2) the changes that occur, which are related to each other by reaction stoichiometry, and (3) the final conditions, which simply are initial conditions + changes.

	A(g)	→	2B(g)	+	C(g)
Initial	1000. mmHg		0. mmHg		0. mmHg
Changes	−1000. mmHg		+2000. mmHg		+1000. mmHg
Final	0. mmHg		2000. MmHg		1000. mmHg

Total final pressure = 0. mmHg + 2000. mmHg + 1000. mmHg = 3000. mmHg

(b)

	A(g)	→	2B(g)	+	C(g)
Initial	1000. mmHg		0. mmHg		0. mmHg
Changes	−200. mmHg		+400. mmHg		+200. mmHg
Final	800 mmHg		400. mmHg		200. mmHg

Total pressure = 800. mmHg + 400. mmHg + 200. mmHg = 1400. mmHg

32. **(a)** We use the ideal gas law to determine N_2O_5 pressure

$$P\{N_2O_5\} = \dfrac{nRT}{V} = \dfrac{\left(1.00 \text{ g} \times \dfrac{1 \text{ mol } N_2O_5}{108.0 \text{ g}}\right) \times 0.08206 \dfrac{\text{L atm}}{\text{mol K}} \times (273 + 65) \text{ K}}{15 \text{ L}} \times \dfrac{760 \text{ mmHg}}{1 \text{ atm}}$$

$$= 13 \text{ mmHg}$$

(b) After 2.38 min, one half-life has passed, so the initial pressure of N_2O_5 has decreased by half to 6.5 mmHg.

(c)   From the balanced chemical equation, the reaction of 2 mol $N_2O_5(g)$ produces 4 mol $NO_2(g)$ and 1 mol $O_2(g)$. That is, the consumption of 2 mol of reactant gas produces 5 mol of product gas. When measured at the same temperature and confined to the same volume, pressures will behave as amounts: the reaction of 2 mmHg of reactant produces 5 mmHg of product.

$$P_{total} = 13 \text{ mmHg } N_2O_{5 \text{ (initially)}} - 6.5 \text{ mmHg } N_2O_{5 \text{ (reactant)}} + \left(6.5 \text{ mmHg}_{(reactant)} \times \frac{5 \text{ mmHg}_{(product)}}{2 \text{ mmHg}_{(reactant)}} \right)$$

$$= (13 - 6.5 + 16) \text{ mmHg} = 23 \text{ mmHg}$$

Method of Initial Rates

33.   From Expt. 1 and Expt. 2 we see that [B] remains fixed while [A] triples. As a result, the initial rate increases from 4.2×10^{-3} M/min to 1.3×10^{-2} M/min, that is, the initial reaction rate triples. Therefore, the reaction is first-order in [A]. Between Expt. 2 and Expt. 3, we see that [A] doubles, which would double the rate, and [B] doubles. As a consequence, the initial rate goes from 1.3×10^{-2} M/min to 5.2×10^{-2} M/min, that is, the rate quadruples. Since an additional doubling of the rate is due to the change in [B], the reaction is first-order in [B]. Now we determine the value of the rate constant.

$$\text{Rate} = k[A]^1[B]^1 \qquad k = \frac{\text{Rate}}{[A][B]} = \frac{5.2 \times 10^{-2} \text{ M / min}}{3.00 \text{ M} \times 3.00 \text{ M}} = 5.8 \times 10^{-3} \text{ L mol}^{-1}\text{min}^{-1}$$

The rate law is $\text{Rate} = \left(5.8 \times 10^{-3} \text{ L mol}^{-1}\text{min}^{-1} \right)[A]^1[B]^1$

34.   From Experiment 1 and 2 we see [NO] remains constant while $[Cl_2]$ doubles. At the same time the initial rate of reaction is found to double. Thus, the reaction is first-order with respect to $[Cl_2]$. Between experiment 1 and 3 we see the $[Cl_2]$ remains constant, while [NO] is doubled, resulting in a quadrupling of the initial rate of reaction. Thus, the reaction must be second-order in [NO]. Overall the reaction is third-order: $\text{Rate} = k[NO]^2[Cl_2]$. The rate constant may be calculated from any one of the experiments. Using data from exp. 1,

$$k = \frac{\text{Rate}}{[NO]^2[Cl_2]} = \frac{2.27 \times 10^{-5} \text{ M s}^{-1}}{(0.0125 \text{ M})^2(0.0255 \text{ M})} = 5.70 \text{ M}^{-2}\text{s}^{-1}$$

35.   (a)   From Expt. 1 to Expt. 2, [B] remains constant at 1.40 M and [C] remains constant at 1.00 M, but [A] is halved $(\times 0.50)$. At the same time the rate is halved $(\times 0.50)$. Thus, the reaction is first-order with respect to A, since $0.50^x = 0.50$ when $x = 1$.
From Expt. 2 to Expt. 3, [A] remains constant at 0.70 M and [C] remains constant at 1.00 M, but [B] is halved $(\times 0.50)$, from 1.40 M to 0.70 M. At the same time, the rate is quartered $(\times 0.25)$. Thus, the reaction is second-order with respect to B, since $0.50^y = 0.25$ when $y = 2$.

From Expt. 1 to Expt. 4, [A] remains constant at 1.40 M and [B] remains constant at 1.40 M, but [C] is halved $(\times 0.50)$, from 1.00 M to 0.50 M. At the same time, the rate is increased by a factor of 2.0.

$$\left[rate_4 = 16 \; rate_3 = 16 \times \frac{1}{4} rate_2 = 4 \; rate_2 = 4 \times \frac{1}{2} rate_1 = 2 \times rate_1. \right]$$

Thus, the order of the reaction with respect to C is -1, since $0.5^z = 2.0$ when $z = -1$.

(b) $rate_5 = k \left(0.70 \text{ M}\right)^1 \left(0.70 \text{ M}\right)^2 \left(0.50 \text{ M}\right)^{-1} = k \left(\frac{1.40 \text{ M}}{2}\right)^1 \left(\frac{1.40 \text{ M}}{2}\right)^2 \left(\frac{1.00 \text{ M}}{2}\right)^{-1}$

$= k \frac{1}{2}^1 \left(1.40 \text{ M}\right)^1 \frac{1}{2}^2 \left(1.40 \text{ M}\right)^2 \frac{1}{2}^{-1} \left(1.00 \text{ M}\right)^{-1} = rate_1 \left(\frac{1}{2}\right)^{1+2-1} = rate_1 \left(\frac{1}{2}\right)^2 = \frac{1}{4} rate_1$

This is based on $rate_1 = k \left(1.40 \text{ M}\right)^1 \left(1.40 \text{ M}\right)^2 \left(1.00 \text{ M}\right)^{-1}$

36. $CH_3CHO(g) \rightarrow CH_4(g) + CO(g)$ Rate $= k[CH_3CHO]^x$. Rate increases by 2.8 times when initial concentration of CH_3CHO is doubled. Hence: $2.8 = (2.0)^x$. Solve by taking the log of both sides. $\log 2.8 = x \log 2.0$ $x = 1.5$. The reaction order is 1.5 or 3/2.

Rate $= k[CH_3CHO]^{1.5}$

First-Order Reactions

37. **(a)** TRUE The rate of the reaction does decrease as more and more of B and C are formed, but not because more and more of B and C are formed. Rather the rate decreases because the concentration of A must decrease to form more and more of B and C.

(b) FALSE The time required for one half of substance A to react—the half-life—is independent of the quantity of "A" present.

38. **(a)** FALSE A plot of ln [A] or log [A] vs. time yields a straight line. One of [A] vs. time yields a curved line.

(b) TRUE The rate of formation of "C" is related to the rate of disappearance of "A" by the stoichiometry of the reaction.

39. **(a)** Since the half-life is 180 s, after 900 s five half-lives have elapsed, and the original quantity of A has been cut in half five times.

final quantity of A $= (0.5)^5 \times$ initial quantity of A $= 0.03125 \times$ initial quantity of A

About 3.13% of the original quantity of A remains unreacted after 900 s.

(b) For a first-order reaction $k = \dfrac{0.693}{t_{1/2}} = \dfrac{0.693}{180 \text{ s}} = 0.00385 \text{ s}^{-1}$

Rate $= k[A] = 0.00385 \text{ s}^{-1} \times 0.50 \text{ M} = 0.00193 \text{ M} / \text{s}$

40. **(a)** We note that the final concentration is one-eighth of the initial concentration. Thus, three half-lives have elapsed, the first to reduce [A] from 0.800 M to 0.400 M, the second to reduce [A] from 0.400 M to 0.200 M, and the third half-life to reduce [A] from 0.200 M to 0.100 M.

$$54 \text{ min} = 3 \, t_{1/2} \qquad t_{1/2} = \frac{54 \text{ min}}{3} = 18 \text{ min}$$

To reduce the concentration even further requires two additional half-lives: the first of these to lower [A] from 0.100 M to 0.050 M and the second of them to reduce [A] from 0.050 M to 0.025 M. These two half-lives equal $2 \times 18 \text{ min} = 36 \text{ min}$. Thus, at $54 \text{ min} + 36 \text{ min} = 90 \text{ min}$ from the start of the reaction, $[A] = 0.025$ M.

(b) We determine the rate constant for this first-order reaction:

$$k = \frac{0.693}{t_{1/2}} = \frac{0.693}{18 \text{ min}} = 0.039 \text{ min}^{-1}$$

Then we determine the rate:

$$\text{Rate} = k[A]^1 = 0.039 \text{ min}^{-1} \times 0.025 \text{ M} = 9.8 \times 10^{-4} \text{ M/min}$$

41. We determine the value of the first-order rate constant and from that we determine the half-life. If the reactant is 99% decomposed in 137 min, then only 1% (0.010) of the initial concentration remains.

$$\ln\frac{[A]_t}{[A]_0} = -kt = \ln\frac{0.010}{1.000} = -4.61 = -k \times 137 \text{min} \qquad k = \frac{4.61}{137 \text{ min}} = 0.0336 \text{ min}^{-1}$$

$$t_{1/2} = \frac{0.0693}{k} = \frac{0.693}{0.0336 \text{ min}^{-1}} = 20.6 \text{ min}$$

42. If 99% of the radioactivity of ^{32}P is lost, 1% (0.010) of that radioactivity remains. First we compute the value of the rate constant from the half-life. $k = \dfrac{0.693}{t_{1/2}} = \dfrac{0.693}{14.3 \text{ d}} = 0.0485 \text{ d}^{-1}$

Then we use the integrated rate equation to determine the elapsed time.

$$\ln\frac{[A]_t}{[A]_0} = -kt \quad t = -\frac{1}{k}\ln\frac{[A]_t}{[A]_0} = -\frac{1}{0.0485 \text{ d}^{-1}}\ln\frac{0.010}{1.000} = 95 \text{ days}$$

43. **(a)** $\ln\left(\dfrac{\dfrac{35}{100}[A]_0}{[A]_0}\right) = \ln(0.35) = -kt = (-4.81 \times 10^{-3} \text{ min}^{-1})t \qquad t = 218 \text{ min.}$

Note: We did not need to know the initial concentration of acetoacetic acid to answer the question.

(b) Assume the reaction takes placed in a 1.00 L container.

$$10.0 \text{ g acetoacetic acid} \times \frac{1 \text{ mol acetoacetic acid}}{102.090 \text{ g acetoacetic acid}} = 0.09795 \text{ mol acetoacetic acid.}$$

After 575 min. (~ 4 half lives, hence, we expect ~ 6.25% remains), use integrated form of the rate law:

$$\ln\left(\frac{[A]_t}{[A]_o}\right) = -kt = (-4.81 \times 10^{-3} \text{ min}^{-1})(575 \text{ min}) = -2.766$$

$$\frac{[A]_t}{[A]_o} = e^{-2.766} = 0.06293 \ (\sim 6.3\% \text{ remains}) \qquad \frac{[A]_t}{0.09795 \text{ M}} = 0.063$$

$$[A]_t = 6.2 \times 10^{-3} \text{ moles.}$$

$[A]_{reacted} = [A]_o - [A]_t = (0.098 - 6.2 \times 10^{-3})$ moles $= 0.092$ moles acetoacetic acid. Stoichiometry is such that for every mole of acetoacetic acid, one mole of CO_2 forms. Hence, we need to determine the volume of 0.0918 moles CO_2 at 24.5 °C (297.65 K) and 748 torr (0.984 atm).

$$V = \frac{nRT}{P} = \frac{0.0918 \text{ mol}\left(0.08206 \dfrac{\text{L atm}}{\text{K mol}}\right) 297.65 \text{ K}}{0.984 \text{ atm}} = 2.3 \text{ L } CO_2$$

44. (a)
$$\ln\frac{[A]_t}{[A]_o} = -kt = \ln\frac{2.5 \text{ g}}{80.0 \text{ g}} = -3.47 = -6.2\times10^{-4} \text{ s}^{-1}t, \qquad t = \frac{3.47}{6.2\times10^{-4} \text{ s}^{-1}} = 5.6\times10^3 \text{ s}$$

We substituted masses for concentrations, because the same substance (with the same molar mass) is present initially and at time t, and because it is present in the same volume.

(b)
$$\text{amount } O_2 = 77.5 \text{ g } N_2O_5 \times \frac{1 \text{ mol } N_2O_5}{108.0 \text{ g } N_2O_5} \times \frac{1 \text{ mol } O_2}{2 \text{ mol } N_2O_5} = 0.359 \text{ mol } O_2$$

$$V = \frac{nRT}{P} = \frac{0.359 \text{ mol } O_2 \times 0.08206 \dfrac{\text{L atm}}{\text{mol K}} \times (45+273) \text{ K}}{745 \text{ mmHg} \times \dfrac{1 \text{ atm}}{760 \text{ mmHg}}} = 9.56 \text{ L } O_2$$

45. **(a)** If the reaction is first-order, we will obtain the same value of the rate constant from several sets of data.

$$\ln \frac{[A]_t}{[A]_0} = -kt = \ln \frac{0.497 \text{ M}}{0.600 \text{ M}} = -k \times 100 \text{ s} = -0.188, \quad k = \frac{0.188}{100 \text{ s}} = 1.88 \times 10^{-3} \text{ s}^{-1}$$

$$\ln \frac{[A]_t}{[A]_0} = -kt = \ln \frac{0.344 \text{ M}}{0.600 \text{ M}} = -k \times 300 \text{ s} = -0.556, \quad k = \frac{0.556}{300 \text{ s}} = 1.85 \times 10^{-3} \text{ s}^{-1}$$

$$\ln \frac{[A]_t}{[A]_0} = -kt = \ln \frac{0.285 \text{ M}}{0.600 \text{ M}} = -k \times 400 \text{ s} = -0.744, \quad k = \frac{0.744}{400 \text{ s}} = 1.86 \times 10^{-3} \text{ s}^{-1}$$

$$\ln \frac{[A]_t}{[A]_0} = -kt = \ln \frac{0.198 \text{ M}}{0.600 \text{ M}} = -k \times 600 \text{ s} = -1.109, \quad k = \frac{1.109}{600 \text{ s}} = 1.85 \times 10^{-3} \text{ s}^{-1}$$

The virtual constancy of the rate constant throughout the time of the reaction confirms that the reaction is first-order.

(b) We assume that the rate constant equals the average of the values obtained in part (a).

$$k = \frac{1.88 + 1.85 + 1.86 + 1.85}{4} \times 10^{-3} \text{ s}^{-1} = 1.86 \times 10^{-3} \text{ s}^{-1}$$

(c) We use the integrated first-order rate equation:

$$[A]_{750} = [A]_0 \exp(-kt) = 0.600 \text{ M} \exp\left(-1.86 \times 10^{-3} \text{ s}^{-1} \times 750 \text{ s}\right)$$

$$[A]_{750} = 0.600 \text{ M } e^{-1.40} = 0.148 \text{ M}$$

46. **(a)** If the reaction is first-order, we will obtain the same value of the rate constant from several sets of data.

$$\ln \frac{P_t}{P_0} = -kt = \ln \frac{264 \text{ mmHg}}{312 \text{ mmHg}} = -k \times 390 \text{ s} = -0.167, \quad k = \frac{0.167}{390 \text{ s}} = 4.28 \times 10^{-4} \text{ s}^{-1}$$

$$\ln \frac{P_t}{P_0} = -kt = \ln \frac{224 \text{ mmHg}}{312 \text{ mmHg}} = -k \times 777 \text{ s} = -0.331, \quad k = \frac{0.331}{777 \text{ s}} = 4.26 \times 10^{-4} \text{ s}^{-1}$$

$$\ln \frac{P_t}{P_0} = -kt = \ln \frac{187 \text{ mmHg}}{312 \text{ mmHg}} = -k \times 1195 \text{ s} = -0.512, \quad k = \frac{0.512}{1195 \text{ s}} = 4.28 \times 10^{-4} \text{ s}^{-1}$$

$$\ln \frac{P_t}{P_0} = -kt = \ln \frac{78.5 \text{ mmHg}}{312 \text{ mmHg}} = -k \times 3155 \text{ s} = -1.38, \quad k = \frac{1.38}{3155 \text{ s}} = 4.37 \times 10^{-4} \text{ s}^{-1}$$

The virtual constancy of the rate constant confirms that the reaction is first-order.

(b) We assume the rate constant to be the average of the values in part (a): 4.3×10^{-4} s^{-1}.

(c) At 390 s, the pressure of dimethyl ether has dropped to 264 mmHg. Thus, an amount of dimethyl ether equivalent to a pressure of $(312 \text{ mmHg} - 264 \text{ mmHg} =)$ 48 mmHg has decomposed. For each 1 mmHg pressure of dimethyl ether that decomposes, 3 mmHg of pressure of the products is produced. Thus, the increase in the pressure of the products is $3 \times 48 = 144$ mmHg. The total pressure at this point is 264 mmHg $+144$ mmHg $= 408$ mmHg. Done in a more systematic fashion:

$(CH_3)_2O(g)$	$\rightarrow$	$CH_4(g)$	$+$	$H_2(g)$	$+$	$CO(g)$
Initial	312 mmHg	0 mmHg		0 mmHg		0 mmHg
Changes	−48 mmHg	+48 mmHg		+48 mmHg		+48 mmHg
Final	264 mmHg	48 mmHg		48 mmHg		48 mmHg

$$P_{total} = P_{DME} + P_{methane} + P_{hydrogen} + P_{CO}$$
$$= 264 \text{ mmHg} + 48 \text{ mmHg} + 48 \text{ mmHg} + 48 \text{ mmHg} = 408 \text{ mmHg}$$

(d) In the same manner as we solved part (c) we have the following.

$(CH_3)_2O(g)$	$\rightarrow$	$CH_4(g)$	$+$	$H_2(g)$	$+$	$CO(g)$
Initial	312 mmHg	0 mmHg		0 mmHg		0 mmHg
Changes	−312 mmHg	+312 mmHg		+312 mmHg		+312 mmHg
Final	0 mmHg	312 mmHg		312 mmHg		312 mmHg

$$P_{total} = P_{DME} + P_{methane} + P_{hydrogen} + P_{CO}$$
$$= 0 \text{ mmHg} + 312 \text{ mmHg} + 312 \text{ mmHg} + 312 \text{ mmHg} = 936 \text{ mmHg}$$

(e) We first determine P_{DME} at 1000 s. $\ln \dfrac{P_{1000}}{P_0} = -kt = -4.3 \times 10^{-4}$ s$^{-1} \times 1000$ s $= -0.43$

$$\frac{P_{1000}}{P_0} = e^{-0.43} = 0.65 \qquad P_{1000} = 312 \text{ mmHg} \times 0.65 = 203 \text{ mmHg}$$

Then we use the technique of parts (c) and (d)

$(CH_3)_2O(g)$	$\rightarrow$	$CH_4(g)$	$+$	$H_2(g)$	$+$	$CO(g)$
Initial	312 mmHg	0 mmHg		0 mmHg		0 mmHg
Changes	−109 mmHg	+109 mmHg		+109 mmHg		+109 mmHg
Final	203 mmHg	109 mmHg		109 mmHg		109 mmHg

$$P_{total} = P_{DME} + P_{methane} + P_{hydrogen} + P_{CO}$$
$$= 203 \text{ mmHg} + 109 \text{ mmHg} + 109 \text{ mmHg} + 109 \text{ mmHg} = 530. \text{ mmHg}$$

Reactions of Various Orders

47. For reaction: $HI(g) \rightarrow 1/2\ H_2(g) + 1/2\ I_2(g)$ (700 K)

Time (s)	[HI] (M)	ln[HI]	1/[HI](M^{-1})
0	1.00	0	1.00
100	0.90	−0.105	1.11
200	0.81	−0.211	1.23<u>5</u>
300	0.74	−0.301	1.35
400	0.68	−0.386	1.47

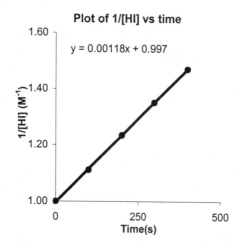

From data above, a plot of 1/[HI] vs. t yields a straight line. The reaction is second-order in HI at 700 K. Rate = $k[HI]^2$. The slope of the line = $k = 0.00118\ M^{-1}s^{-1}$

48. (a) We can graph 1/[ArSOOH] vs. time and obtain a straight line. We may also graph [ArSOOH] *vs.* time and ln([ArSOOH]) *vs.* time to demonstrate that they do not yield a straight line. (only the plot of 1/[ArSO$_2$H] versus time is shown.

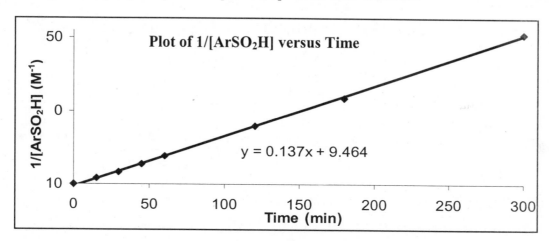

The straightness of the line indicates that the reaction is second-order.

(b) We solve the rearranged integrated second-order rate law for the rate constant, using the longest time interval.
$$\frac{1}{[A]_t} - \frac{1}{[A]_0} = kt \qquad \frac{1}{t}\left(\frac{1}{[A]_t} - \frac{1}{[A]_0}\right) = k$$

$$k = \frac{1}{300\ \text{min}}\left(\frac{1}{0.0196\ M} - \frac{1}{0.100\ M}\right) = 0.137\ \text{L mol}^{-1}\text{min}^{-1}$$

(c) We use the same equation as in part (b), but solved for t, rather than k.

$$t = \frac{1}{k}\left(\frac{1}{[A]_t} - \frac{1}{[A]_0}\right) = \frac{1}{0.137\ \text{L mol}^{-1}\text{min}^{-1}}\left(\frac{1}{0.0500\ M} - \frac{1}{0.100\ M}\right) = 73.0\ \text{min}$$

(d) We use the same equation as in part (b), but solved for t, rather than k.

$$t = \frac{1}{k}\left(\frac{1}{[A]_t} - \frac{1}{[A]_0}\right) = \frac{1}{0.137 \text{ L mol}^{-1}\text{min}^{-1}}\left(\frac{1}{0.0250 \text{ M}} - \frac{1}{0.100 \text{ M}}\right) = 219 \text{ min}$$

(e) We use the same equation as in part (b), but solved for t, rather than k.

$$t = \frac{1}{k}\left(\frac{1}{[A]_t} - \frac{1}{[A]_0}\right) = \frac{1}{0.137 \text{ L mol}^{-1}\text{min}^{-1}}\left(\frac{1}{0.0350 \text{ M}} - \frac{1}{0.100 \text{ M}}\right) = 136 \text{ min}$$

49. (a) In the first 22 s, [A] decreases from 0.715 M to 0.605 M, that is, $\Delta[A] = -0.110\,M$.

The rate in these 22 s is then determined. $\text{Rate} = \dfrac{-\Delta[A]}{\Delta t} = \dfrac{0.110 \text{ M}}{22 \text{ s}} = 5.0 \times 10^{-3}\,M/s$

In the first 74 s, $\Delta[A] = 0.345\,M - 0.715\,M$, and the rate is determined.

$$\text{Rate} = \frac{-\Delta[A]}{\Delta t} = \frac{0.370 \text{ M}}{74 \text{ s}} = 5.0 \times 10^{-3}\,M/s$$

Finally, in the first 132 s, $\Delta[A] = 0.055\,M - 0.715\,M$, and the rate is determined as

follows. $\text{Rate} = \dfrac{-\Delta[A]}{\Delta t} = \dfrac{0.660 \text{ M}}{132 \text{ s}} = 5.0 \times 10^{-3}\,M/s$. Since the rate is constant for this

reaction, it must be zero-order.

(b) The half-life of this reaction is the time needed for one half of the initial [A] to react.

Thus, $\Delta[A] = 0.715\,M \div 2 = 0.358\,M$ and $t_{1/2} = \dfrac{0.358 \text{ M}}{5.0 \times 10^{-3} \text{ M/s}} = 72$ s.

50. (a) We can either graph $1/[C_4H_6]$ vs. time and obtain a straight line, or we can determine the second-order rate constant from several data points. Then, if k indeed is a constant, the reaction is demonstrated to be second-order. We shall use the second technique in this case. First we do a bit of algebra.

$$\frac{1}{[A]_t} - \frac{1}{[A]_0} = kt \qquad \frac{1}{t}\left(\frac{1}{[A]_t} - \frac{1}{[A]_0}\right) = k$$

$$k = \frac{1}{12.18 \text{ min}}\left(\frac{1}{0.0144 \text{ M}} - \frac{1}{0.0169 \text{ M}}\right) = 0.843 \text{ L mol}^{-1}\text{min}^{-1}$$

$$k = \frac{1}{24.55 \text{ min}}\left(\frac{1}{0.0124 \text{ M}} - \frac{1}{0.0169 \text{ M}}\right) = 0.875 \text{ L mol}^{-1}\text{min}^{-1}$$

$$k = \frac{1}{42.50 \min}\left(\frac{1}{0.0103 \text{ M}} - \frac{1}{0.0169 \text{ M}}\right) = 0.892 \text{ L mol}^{-1}\text{min}^{-1}$$

$$k = \frac{1}{68.05 \min}\left(\frac{1}{0.00845 \text{ M}} - \frac{1}{0.0169 \text{ M}}\right) = 0.870 \text{ L mol}^{-1}\text{min}^{-1}$$

The nearly constant value for the rate constant indicates that the reaction is second-order.

(b) The rate constant is the average of the values obtained in part (a).

$$k = \frac{0.843 + 0.875 + 0.892 + 0.870}{4} \text{ L mol}^{-1}\text{min}^{-1} = 0.87 \text{ L mol}^{-1}\text{min}^{-1}$$

(c) We use the same equation as in part (a), but solved for t, rather than k.

$$t = \frac{1}{k}\left(\frac{1}{[\text{A}]_t} - \frac{1}{[\text{A}]_0}\right) = \frac{1}{0.870 \text{ L mol}^{-1}\text{min}^{-1}}\left(\frac{1}{0.00423 \text{ M}} - \frac{1}{0.0169 \text{ M}}\right) = 2.0 \times 10^2 \text{ min}$$

(d) We use the same equation as in part (a), but solved for t, rather than k.

$$t = \frac{1}{k}\left(\frac{1}{[\text{A}]_t} - \frac{1}{[\text{A}]_0}\right) = \frac{1}{0.870 \text{ L mol}^{-1}\text{min}^{-1}}\left(\frac{1}{0.0050 \text{ M}} - \frac{1}{0.0169 \text{ M}}\right) = 1.6 \times 10^2 \text{ min}$$

51. **(a)** $\text{Initial rate} = -\dfrac{\Delta[\text{A}]}{\Delta t} = -\dfrac{1.490 \text{ M} - 1.512 \text{ M}}{1.0 \min - 0.0 \min} = +0.022 \text{ M/min}$

$\text{Initial rate} = -\dfrac{\Delta[\text{A}]}{\Delta t} = -\dfrac{2.935 \text{ M} - 3.024 \text{ M}}{1.0 \min - 0.0 \min} = +0.089 \text{ M/min}$

(b) When the initial concentration is doubled $(\times 2.0)$, from 1.512 M to 3.024 M, the initial rate quadruples $(\times 4.0)$. Thus, the reaction is second-order in A, since $2.0^x = 4.0$ when $x = 2$.

52. **(a)** Let us assess the possibilities. If the reaction is zero-order, its rate will be constant. During the first 8 min, the rate is $-(0.60 \text{ M} - 0.80 \text{ M})/8 \min = 0.03 \text{ M/min}$. Then, during the first 24 min, the rate is $-(0.35 \text{ M} - 0.80 \text{ M})/24 \min = 0.019 \text{ M/min}$.
Thus, the reaction is not zero-order.
If the reaction is first-order, it will have a constant half-life, that is consistent with its rate constant. The half-life can be assessed from the fact that 40 min elapse while the concentration drops from 0.80 M to 0.20 M, that is, to one-fourth of its initial value. Thus, 40 min equals two half-lives and $t_{1/2} = 20 \min$.

This gives $k = 0.693 / t_{1/2} = 0.693 / 20 \text{ min} = 0.035 \text{ min}^{-1}$. Also

$$kt = -\ln \frac{[A]_t}{[A]_0} = -\ln \frac{0.35 \text{ M}}{0.80 \text{ M}} = 0.827 = k \times 24 \text{ min} \qquad k = \frac{0.827}{24 \text{ min}} = 0.034 \text{ min}^{-1}$$

The constancy of the value of k indicates that the reaction is first-order.

(b) The value of the rate constant is $k = 0.034 \text{ min}^{-1}$.

(c) Reaction rate $= \frac{1}{2}$ (rate of formation of B) $= k[A]^1$ First we need [A] at $t = 30$. min

$$\ln \frac{[A]}{[A]_0} = -kt = -0.034 \text{ min}^{-1} \times 30. \text{ min} = -1.0_2 \qquad \frac{[A]}{[A]_0} = e^{-1.02} = 0.36$$

$$[A] = 0.36 \times 0.80 \text{ M} = 0.29 \text{ M}$$

rate of formation of B $= 2 \times 0.034 \text{ min}^{-1} \times 0.29 \text{ M} = 2.0 \times 10^{-2} \text{ M min}^{-1}$

53. The half-life of the reaction depends on the concentration of "A" and, thus, this reaction cannot be first-order. For a second-order reaction, the half-life varies inversely with the rate: $t_{1/2} = 1 / (k[A]_0)$ or $k = 1 / (t_{1/2}[A]_0)$. Let us attempt to verify the second-order nature of this reaction by seeing if the rate constant is fixed.

$$k = \frac{1}{1.00 \text{ M} \times 50 \text{ min}} = 0.020 \text{ L mol}^{-1}\text{min}^{-1}$$

$$k = \frac{1}{2.00 \text{ M} \times 25 \text{ min}} = 0.020 \text{ L mol}^{-1}\text{min}^{-1}$$

$$k = \frac{1}{0.50 \text{ M} \times 100 \text{ min}} = 0.020 \text{ L mol}^{-1} \text{ min}^{-1}$$

The constancy of the rate constant demonstrates that this reaction indeed is second-order. The rate equation is Rate $= k[A]^2$ and $k = 0.020 \text{ L mol}^{-1}\text{min}^{-1}$.

54. (a) The half-life depends on the initial $[NH_3]$ and, thus, the reaction cannot be first-order. Let us attempt to verify second-order kinetics.

$$k = \frac{1}{[NH_3]_0 t_{1/2}} \text{ for a second-order reaction} \qquad k = \frac{1}{0.0031 \text{ M} \times 7.6 \text{ min}} = 42 \text{ M}^{-1}\text{min}^{-1}$$

$$k = \frac{1}{0.0015 \text{ M} \times 3.7 \text{ min}} = 180 \text{ M}^{-1}\text{min}^{-1} \qquad k = \frac{1}{0.00068 \text{ M} \times 1.7\text{min}} = 865 \text{ M}^{-1}\text{min}^{-1}$$

The reaction is not second-order. But, if the reaction is zero-order, its rate will be constant.

$$\text{Rate} = \frac{[A]_o/2}{t_{1/2}} = \frac{0.0031 \text{ M} \div 2}{7.6 \text{ min}} = 2.0 \times 10^{-4} \text{ M/min}$$

$$\text{Rate} = \frac{0.0015 \text{ M} \div 2}{3.7 \text{ min}} = 2.0 \times 10^{-4} \text{ M/min}$$

$$\text{Rate} = \frac{0.00068 \text{ M} \div 2}{1.7 \text{ min}} = 2.0 \times 10^{-4} \text{ M/min} \qquad \text{Zero-order reaction}$$

(b) The constancy of the rate indicates that the decomposition of ammonia under these conditions is zero-order, and the rate constant is $k = 2.0 \times 10^{-4}$ M/min.

55. Zeroth-order: $t_{1/2} = \dfrac{[A]_o}{2k}$ \qquad Second-order: $t_{1/2} = \dfrac{1}{k[A]_o}$

A zero-order reaction has a half life that varies proportionally to $[A]_o$, therefore, increasing $[A]_o$ increases the half-life for the reaction. A second-order reaction's half-life varies inversely proportional to $[A]_o$, that is, as $[A]_o$ increases, the half-life decreases. The reason for the difference is that a zero-order reaction has a constant rate of reaction (independent of $[A]_o$). The larger the value of $[A]_o$, the longer it will take to react. In a second-order reaction, the rate of reaction increases as the square of the $[A]_o$, hence, for high $[A]_o$, the rate of reaction is large and for very low $[A]_o$, the rate of reaction is very slow. If we consider a bimolecular elementary reaction, we can easily see that a reaction will not take place unless two molecules of reactants collide. This is more likely when the $[A]_o$ is large than when it is small.

56. **(a)** $\dfrac{[A]_o}{2k} = \dfrac{0.693}{k}$ \qquad Hence, $\dfrac{[A]_o}{2} = 0.693$ or $[A]_o = 1.39$ M

(b) $\dfrac{[A]_o}{2k} = \dfrac{1}{k[A]_o}$ \qquad Hence, $\dfrac{[A]_o^2}{2} = 1$ or $[A]_o^2 = 2.00$ M \qquad $[A]_o = 1.41\underline{4}$ M

(c) $\dfrac{0.693}{k} = \dfrac{1}{k[A]_o}$ \qquad Hence, $0.693 = \dfrac{1}{[A]_o}$ or $[A]_o = 1.44$ M

Collision Theory; Activation Energy

57. **(a)** The rate of a reaction depends on at least two factors other than the frequency of collisions. The first of these is whether each collision possesses sufficient energy to get over the energy barrier to products. This depends on the activation energy of the reaction; the higher it is, the smaller will be the fraction of successful collisions. The second factor is whether the molecules in a given collision are properly oriented for a successful reaction. The more complex the molecules are, the smaller will be the the fraction of collisions that are correctly oriented.

(b) Although the collision frequency increases relatively slowly with temperature, the fraction of those collisions that have sufficient energy to overcome the activation energy increases much more rapidly. Therefore, the rate of reaction will increase dramatically with temperature.

(c) The addition of a catalyst has the net effect of decreasing the activation energy of the overall reaction, by enabling an alternate mechanism. The lower activation energy of the alternate mechanism, (compared to the uncatalyzed mechanism), means that a larger fraction of molecules have sufficient energy to react. Thus the rate increases, even though the temperature does not.

58. (a) The activation energy for the reaction of hydrogen with oxygen is quite high, too high to be supplied by the energy ordinarily available in a mixture of the two gases at ambient temperatures. However, the spark supplies a suitably concentrated form of energy to initiate the reaction of at least a few molecules. Since the reaction is highly exothermic, the reaction of these first few molecules supplies sufficient energy for yet other molecules to react and the reaction proceeds to completion or to the elimination of the limiting reactant.

(b) A larger spark simply means that a larger number of molecules react initially. But the eventual course of the reaction remains the same, with the initial reaction producing enough energy to initiate still more molecules, and so on.

59. (a) The products are 21 kJ/mol closer in energy to the constant energy activated complex than are the reactants. Thus, the activation energy for the reverse reaction is
$84 \text{ kJ} / \text{mol} - 21 \text{ kJ} / \text{mol} = 63 \text{ kJ} / \text{mol}$.

(b) The reaction profile for this reaction is sketched at below.

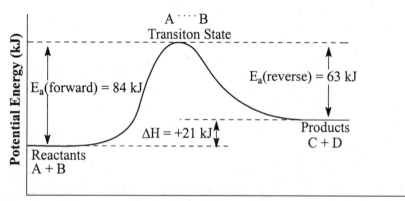

Progress of Reaction

60.	In an endothermic reaction (right), E_a must be larger than ΔH for the reaction. For an exothermic reaction (left), the magnitude of E_a may be either larger or smaller than that of ΔH. That is, a small E_a may be associated with a large decrease in enthalpy, or a large E_a with a small decrease in enthalpy.

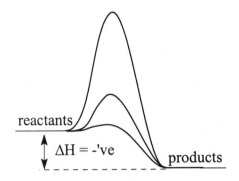 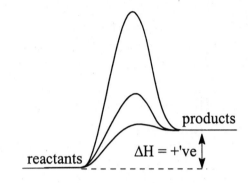

<u>61.</u>	**(a)**	There are two intermediates (B and C).

(b)	There are three transition states (peaks/maxima) in the energy diagram.

(c)	Fastest step has the smallest E_a, hence, step 3 is the fastest step in the reaction with step 2 a close second.

(d)	Reactant A (step 1) is the reactant in the rate-limiting step

(e)	endothermic, need energy to go from A→B

(f)	exothermic, energy released when going from A→D.

62.	**(a)**	There are two intermediates (B and C).

(b)	There are three transition states (peaks/maxima) in the energy diagram.

(c)	Fastest step has the smallest E_a, hence, step 2 is the fastest step in the reaction.

(d)	Reactant A (step 1) is the reactant in the rate-limiting step.

(e)	endothermic, need energy to go from A→B.

(f)	endothermic, need energy to go from A→D.

Effect of Temperature on Rates of Reaction

63. **(a)** First we need to compute values of $\ln k$ and $1/T$. Then we plot the graph of $\ln k$ versus 1/T.

$T, °C$	0 °C	10 °C	20 °C	30 °C
T, K	273 K	283 K	293 K	303 K
$1/T, K^{-1}$	0.00366	0.00353	0.00341	0.00330
k, s^{-1}	5.6×10^{-6}	3.2×10^{-5}	1.6×10^{-4}	7.6×10^{-4}
$\ln k$	-12.09	-10.35	-8.74	-7.18

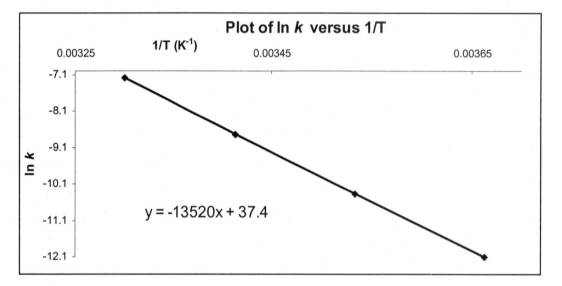

Plot of ln **k** versus 1/T

$y = -13520x + 37.4$

(b) The slope $= -E_a / R$. $E_a = -R \times$ slope $= -8.3145 \dfrac{J}{mol\,K} \times -1.35_2 \times 10^4\ K = 112\ kJ/mol$

(c) We apply the Arrhenius equation, with $k = 5.6 \times 10^{-6}\ s^{-1}$ at $0\,°C$ (273 K), $k = ?$ at $40°C$ (313 K), and $E_a = 113 \times 10^3$ J/mol.

$$\ln\frac{k}{5.6 \times 10^{-6}\ s^{-1}} = \frac{E_a}{R}\left(\frac{1}{T_1} - \frac{1}{T_2}\right) = \frac{113 \times 10^3\ J/mol}{8.3145\ J\,mol^{-1}\,K^{-1}}\left(\frac{1}{273\ K} - \frac{1}{313\ K}\right) = 6.36$$

$$e^{6.36} = 578 = \frac{k}{5.6 \times 10^{-6}\ s^{-1}} \qquad\qquad k = 578 \times 5.6 \times 10^{-6}\ s^{-1} = 3.2 \times 10^{-3}\ s^{-1}$$

$$t_{1/2} = \frac{0.693}{k} = \frac{0.693}{3.2 \times 10^{-3}\ s^{-1}} = 2.2 \times 10^2\ s$$

64. **(a)** We plot $\ln k$ vs. $1/T$. The slope of the straight line equals $-E_a/R$. First we tabulate the data to plot. (plot on next page).

T, °C	15.83	32.02	59.75	90.61
T, K	288.98	305.17	332.90	363.76
$1/T$, K^{-1}	0.0034604	0.0032769	0.0030039	0.0027491
k, M^{-1}s^{-1}	5.03×10^{-5}	3.68×10^{-4}	6.71×10^{-3}	0.119
$\ln k$	-9.898	-7.907	-5.004	-2.129

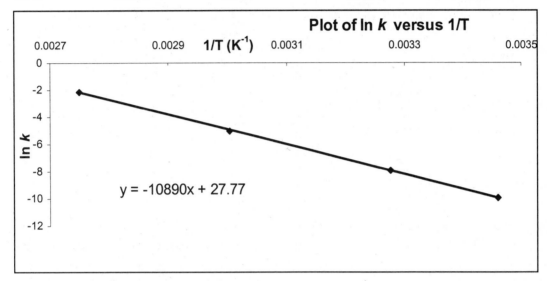

The slope of this graph $= -1.09\times10^4\ \text{K} = E_a/R$

$$E_a = -\left(-1.089\times10^4\ \text{K}\right)\times8.3145\ \text{J mol}^{-1}\ \text{K}^{-1} = 9.054\times10^4\ \text{J/mol} = 90.5\ \text{kJ/mol}$$

(b) We calculate the activation energy with the Arrhenius equation using the two extreme data points.

$$\ln\frac{k_2}{k_1} = \ln\frac{0.119}{5.03\times10^{-5}} = +7.77 = \frac{E_a}{R}\left(\frac{1}{T_1} - \frac{1}{T_2}\right) = \frac{E_a}{R}\left(\frac{1}{288.98\ \text{K}} - \frac{1}{363.76\ \text{K}}\right)$$

$$= 7.1138\times10^{-4}\ \text{K}^{-1}\frac{E_a}{R} \qquad E_a = \frac{7.769\times8.3145\ \text{J mol}^{-1}\ \text{K}^{-1}}{7.1138\times10^{-4}\ \text{K}^{-1}} = 9.08\times10^4\ \text{J/mol}$$

$E_a = 91\,\text{kJ/mol}$ The values are in quite good agreement, within experimental limits.

(d) We apply the Arrhenius equation, with $E_a = 9.080\times10^4$ J/mol,

$k = 5.03\times10^{-5}$ M^{-1} s^{-1} at 15.83 °C (288.98 K), and $k = ?$ at 100.0 °C (373.2 K).

$$\ln\frac{k}{5.03\times10^{-5}\ \text{M}^{-1}\ \text{s}^{-1}} = \frac{E_a}{R}\left(\frac{1}{T_1} - \frac{1}{T_2}\right) = \frac{90.80\times10^3\ \text{J/mol}}{8.3145\ \text{J mol}^{-1}\ \text{K}^{-1}}\left(\frac{1}{288.98\ \text{K}} - \frac{1}{373.2\ \text{K}}\right)$$

$$\ln\frac{k}{5.03\times10^{-5}\ \text{M}^{-1}\ \text{s}^{-1}} = 8.528$$

$$e^{8.528} = 5.05 \times 10^3 = \frac{k}{5.03 \times 10^{-5} \text{ M}^{-1} \text{ s}^{-1}}$$

$$k = 5.05 \times 10^3 \times 5.03 \times 10^{-5} \text{ M}^{-1} \text{ s}^{-1} = 0.254 \text{ M}^{-1} \text{ s}^{-1}$$

65. The half-life of a first-order reaction is inversely proportional to its rate constant: $k = 0.693 / t_{1/2}$. Thus we can apply a modified version of the Arrhenius equation.

(a) $\ln \dfrac{k_2}{k_1} = \ln \dfrac{(t_{1/2})_1}{(t_{1/2})_2} = \dfrac{E_a}{R}\left(\dfrac{1}{T_1} - \dfrac{1}{T_2}\right) = \ln \dfrac{46.2 \text{ min}}{2.6 \text{ min}} = \dfrac{E_a}{R}\left(\dfrac{1}{298 \text{ K}} - \dfrac{1}{(102 + 273) \text{ K}}\right)$

$2.88 = \dfrac{E_a}{R} 6.89 \times 10^{-4} \quad E_a = \dfrac{2.88 \times 8.3145 \text{ J mol}^{-1} \text{ K}^{-1}}{6.89 \times 10^{-4} \text{ K}^{-1}} \times \dfrac{1 \text{ kJ}}{1000 \text{ J}} = 34.8 \text{ kJ/mol}$

(b) $\ln \dfrac{10.0 \text{ min}}{46.2 \text{ min}} = \dfrac{34.8 \times 10^3 \text{ J / mol}}{8.3145 \text{ J mol}^{-1} \text{ K}^{-1}}\left(\dfrac{1}{T} - \dfrac{1}{298}\right) = -1.53 = 4.19 \times 10^3 \left(\dfrac{1}{T} - \dfrac{1}{298}\right)$

$\left(\dfrac{1}{T} - \dfrac{1}{298}\right) = \dfrac{-1.53}{4.19 \times 10^3} = -3.65 \times 10^{-4} \quad \dfrac{1}{T} = 2.99 \times 10^{-3} \quad T = 334 \text{ K} = 61 \text{ °C}$

66. The half-life of a first-order reaction is inversely proportional to its rate constant: $k = 0.693 / t_{1/2}$. Thus, we can apply a modified version of the Arrhenius equation.

(a) $\ln \dfrac{k_2}{k_1} = \ln \dfrac{(t_{1/2})_1}{(t_{1/2})_2} = \dfrac{E_a}{R}\left(\dfrac{1}{T_1} - \dfrac{1}{T_2}\right) = \ln \dfrac{22.5 \text{ h}}{1.5 \text{ h}} = \dfrac{E_a}{R}\left(\dfrac{1}{293 \text{ K}} - \dfrac{1}{(40 + 273) \text{ K}}\right)$

$2.71 = \dfrac{E_a}{R} 2.18 \times 10^{-4}, \quad E_a = \dfrac{2.71 \times 8.3145 \text{ J mol}^{-1} \text{ K}^{-1}}{2.18 \times 10^{-4} \text{ K}^{-1}} \times \dfrac{1 \text{ kJ}}{1000 \text{ J}} = 103 \text{ kJ/mol}$

(b) The relationship is $k = A \exp(-E_a / RT)$

$k = 2.05 \times 10^{13} \text{ s}^{-1} \exp\left(\dfrac{-103 \times 10^3 \text{ J mol}^{-1}}{8.3145 \text{ J mol}^{-1} \text{ K}^{-1} \times (273 + 30) \text{ K}}\right) = 2.05 \times 10^{13} \text{ s}^{-1} \text{ e}^{-40.9}$

$= 3.5 \times 10^{-5} \text{ s}^{-1}$

67. (a) It is the change in the value of the rate constant that causes the reaction to go faster. Let k_1 be the rate constant at room temperature, $20°$ C or 293 K. Then $k_2 = 2k_1$ is the rate constant ten degrees higher, at $30°$ C or 303 K.

$\ln \dfrac{k_2}{k_1} = \ln \dfrac{2k_1}{k_1} = 0.693 = \dfrac{E_a}{R}\left(\dfrac{1}{T_1} - \dfrac{1}{T_2}\right) = \dfrac{E_a}{R}\left(\dfrac{1}{293} - \dfrac{1}{303 \text{ K}}\right) = 1.13 \times 10^{-4} \text{ K}^{-1} \dfrac{E_a}{R}$

$E_a = \dfrac{0.693 \times 8.3145 \text{ J mol}^{-1} \text{ K}^{-1}}{1.13 \times 10^{-4} \text{ K}^{-1}} = 5.1 \times 10^4 \text{ J / mol} = 51 \text{ kJ / mol}$

(b) Since the activation energy for the depicted reaction (i.e., $N_2O + NO \rightarrow N_2 + NO_2$) is 209 kJ/mol, we would not expect this reaction to follow the rule of thumb.

68. Under a pressure of 2.00 atm the boiling point of water is approximately 121 °C or 394 K. Under a pressure of 1 atm, the boiling point of water is 100° C or 373 K. We assume an activation energy of 5.1×10^4 J / mol and compute the ratio of the two rates.

$$\ln\frac{\text{Rate}_2}{\text{Rate}_1} = \frac{E_a}{R}\left(\frac{1}{T_1} - \frac{1}{T_2}\right) = \frac{5.1 \times 10^4 \text{ J / mol}}{8.3145 \text{ J mol}^{-1} \text{ K}^{-1}}\left(\frac{1}{373} - \frac{1}{394 \text{ K}}\right) = 0.88$$

$\text{Rate}_2 = e^{0.88} \text{ Rate}_1 = 2.4 \text{ Rate}_1$.

Cooking will occur 2.4 times faster in the pressure cooker.

Catalysis

69. **(a)** Although a catalyst is *recovered unchanged from the reaction mixture*, it does "take part in the reaction." Some catalysts actually slow down the rate of a reaction. Usually, however, these negative catalysts are called inhibitors.

(b) The function of a catalyst is to *change the mechanism of a reaction*. The new mechanism is one that has a different (lower) activation energy than the original reaction.

70. If the reaction is first-order, its half-life is 100 min, for in this time period [S] decreases from 1.00 M to 0.50 M, that is, by one half. This gives a rate constant of

$$k = 0.693 / t_{1/2} = 0.693 / 100 \text{ min} = 0.00693 \text{ min}^{-1} .$$

The rate constant also can be determined from any two of the other sets of data.

$$kt = \ln\frac{[A]_0}{[A]_t} = \ln\frac{1.00 \text{ M}}{0.70 \text{ M}} = 0.357 = k \times 60 \text{ min} \qquad k = \frac{0.357}{60 \text{ min}} = 0.00595 \text{ min}^{-1}$$

This is not a particularly constant rate constant. Let's try zero-order, where the rate should be constant.

$$\text{Rate} = -\frac{0.90 \text{ M} - 1.00 \text{ M}}{20 \text{ min}} = 0.0050 \text{ M/min} \qquad \text{Rate} = -\frac{0.50 \text{ M} - 1.00 \text{ M}}{100 \text{ min}} = 0.0050 \text{ M/min}$$

$$\text{Rate} = -\frac{0.20 \text{ M} - 0.90 \text{ M}}{160 \text{ min} - 20 \text{ min}} = 0.0050 \text{ M/min} \qquad \text{Rate} = -\frac{0.50 \text{ M} - 0.90 \text{ M}}{100 \text{ min} - 20 \text{ min}} = 0.0050 \text{ M/min}$$

Thus, this reaction is zero-order with respect to [S].

71. Both platinum and an enzyme can be considered heterogeneous catalysts in the sense that the catalytic activity occurs on their surfaces or near them. Even so, some would classify an enzyme as a homogeneous catalyst. The most important difference, however, is one of specificity. Platinum is a rather nonspecific catalyst, catalyzing many different reactions. An enzyme, however, is quite specific, usually catalyzing only one reaction rather than all reactions of a given class.

72. In both the enzyme and the metal surface cases, the reaction occurs in a specialized location: either with the enzyme or on the surface of the catalyst. At high concentrations of reactant, the limiting factor in determining the rate is not the concentration of reactant present but how rapidly active sites become available for reaction to occur. Thus, the rate of the reaction depends on either the quantity of enzyme present or the surface area of the catalyst, rather than on how much reactant is present; the reaction is zero-order. At low concentrations or gas pressures the reaction rate depends on how rapidly molecules can reach the available active sites. Thus, the rate depends on concentration or pressure of reactant and is first-order.

73. For the straight-line graph of Rate versus [Enzyme], an excess of substrate must be present.

74. For human enzymes, we would expect the maximum in the curve to appear around 37°C, i.e., normal body temperature. (possibly slightly elevated temperatures to aid in the control of diseases (37 - 41 °C). At lower temperatures, the reaction rate of enzyme-activated reactions decreases with decreasing temperature, following the Arrhenius equation. However, at higher temperatures, these temperature sensitive biochemical processes become inhibited, probably by temperature-induced structural modifications to the enzyme or the substrate, which prevent formation of the enzyme-substrate complex.

Reaction Mechanisms

75. The molecularity of an elementary process is the number of reactant molecules in that process. This molecularity is equal to the order of the overall reaction only if the elementary process in question is the slowest and, thus, the rate-determining step of the overall reaction. In addition, the elementary process in question should be the only elementary step that influences the rate of the reaction.

76. If the type of molecule that is expressed in the rate law as being first-order collides with other molecules that are present in much larger concentrations, the reaction will seem to depend only on the concentration of those types of molecules present in smaller concentration, since the larger concentration will be essentially unchanged during the course of the reaction. Such a situation is quite common, and has been given the name pseudo first-order. Another situation is that molecules which do not participate directly in the reaction— including product molecules—strike the reactant molecules and impart to them sufficient energy to react.

77. The three elementary steps must sum to the overall reaction. That is, the overall reaction is the sum of step 1 + step 2 + step 3. Hence, step 2 = overall reaction −step 1 −step 3. All species in the equations below are gases.

overall: $2\,NO + 2H_2 \rightarrow N_2 + 2\,H_2O$ $2\,NO + 2\,H_2 \rightarrow N_2 + 2\,H_2O$

−first: $-\left(2\,NO \rightleftharpoons N_2O_2\right)$ $N_2O_2 \rightleftharpoons 2NO$

−third $-\left(N_2O + H_2 \rightarrow N_2 + H_2O\right)$ or $N_2 + H_2O \rightarrow N_2O + H_2$

The result is the second step, which is slow: $H_2 + N_2O_2 \rightarrow H_2O + N_2O$

The rate of this rate-determining step is: Rate $= k_2\left[H_2\right]\left[N_2O_2\right]$

Since N_2O_2 does not appear in the overall reaction, we need to replace its concentration with the concentrations of species that do appear in the overall reaction. To do this, recall that the first step is rapid, with the forward reaction occurring at the same rate as the reverse reaction. $k_1\left[NO\right]^2 =$ forward rate $=$ reverse rate $= k_{-1}\left[N_2O_2\right]$

This expression is solved for $\left[N_2O_2\right]$, which then is substituted into the rate equation for the overall reaction.

$$\left[N_2O_2\right] = \frac{k_1\left[NO\right]^2}{k_{-1}} \qquad\qquad \text{Rate} = \frac{k_2 k_1}{k_{-1}}\left[H_2\right]\left[NO\right]^2$$

The reaction is first-order in $\left[H_2\right]$ and second-order in [NO]. This result conforms to the experimentally determined reaction order.

78.   Proposed mechanism: $I_2(g) \underset{k_{-1}}{\overset{k_1}{\rightleftharpoons}} 2\,I(g)$ Observed rate law:

$$\underline{2\,I(g) + H_2(g) \xrightarrow{k_2} 2\,HI(g)} \qquad \text{Rate} = k[I_2][H_2]$$

$$I_2(g) + H_2(g) \rightarrow 2\,HI(g)$$

The first step is a fast equilibrium reaction and step 2 is slow. Thus, the predicted rate law is Rate $= k_2[I]^2[H_2]$

In the first step, set the rate in the forward direction for the equilibrium equal to the rate in the reverse direction. Then solve for $[I]^2$.

Rate$_{forward}$ = Rate$_{reverse}$ Use: Rate$_{forward} = k_1[I_2]$ and Rate$_{reverse} = k_{-1}[I]^2$

From this we see: $k_1[I_2] = k_{-1}[I]^2$. Rearranging (solving for $[I]^2$)

$$[I]^2 = \frac{k_1[I_2]}{k_{-1}} \quad \text{Substitute into Rate} = k_2[I]^2[H_2] = k_2\frac{k_1[I_2]}{k_{-1}}\,[H_2] = k_{obs}[I_2][H_2]$$

since the predicted rate law is the same as the experimental rate law, this mechanism is plausible and it is supported by the observed rate law.

79. Proposed mechanism: $\quad Cl_2(g) \underset{k_{-1}}{\overset{k_1}{\rightleftharpoons}} 2\,Cl(g) \qquad$ Observed rate law:

$$2\,Cl(g) + 2\,NO(g) \xrightarrow{k_2} 2\,NOCl(g) \qquad \text{Rate} = k[Cl_2][NO_2]^2$$

$$Cl_2(g) + 2NO(g) \rightarrow 2\,NOCl(g)$$

The first step is a fast equilibrium reaction and step 2 is slow. Thus, the predicted rate law is Rate $= k_2[Cl]^2[NO]^2$

In the first step, set the rate in the forward direction for the equilibrium equal to the rate in the reverse direction. Then solve for $[Cl]^2$.

Rate$_{\text{forward}}$ = Rate$_{\text{reverse}}$ $\qquad$ Use: Rate$_{\text{forward}} = k_1[Cl_2]$ and Rate$_{\text{reverse}} = k_{-1}[Cl]^2$
From this we see: $k_1[Cl_2] = k_{-1}[Cl]^2$. Rearranging (solving for $[Cl]^2$)

$$[Cl]^2 = \frac{k_1[Cl_2]}{k_{-1}} \quad \text{Substitute into Rate} = k_2[Cl]^2[NO]^2 = k_2\frac{k_1[Cl_2]}{k_{-1}}[NO]^2 = k_{obs}[Cl_2][NO_2]^2$$

Since the predicted rate law is the same as the experimental rate law, this mechanism is plausible..

80. A possible mechanism is: $\quad$ Step 1: $\quad O_3 \underset{k_2}{\overset{k_1}{\rightleftharpoons}} O_2 + O\,(\text{fast})$

$$\text{Step 2:} \quad O + O_3 \xrightarrow{k_3} 2\,O_2\,(\text{slow})$$

The overall rate is that of the slow step: Rate $= k_3[O][O_3]$. But O is a reaction intermediate, whose concentration is difficult to determine. An expression for [O] can be found by assuming that the forward and reverse "fast" steps proceed equally rapidly.

Rate$_1$ = Rate$_2$ $\qquad k_1[O_3] = k_2[O_2][O] \quad [O] = \dfrac{k_1[O_3]}{k_2[O_2]}$

Then substitute this expression into the rate law for the reaction.

$$\text{Rate} = k_3\frac{k_1[O_3]}{k_2[O_2]}[O_3] = \frac{k_3 k_1}{k_2}\frac{[O_3]^2}{[O_2]}$$

This rate equation has the same form as the experimentally determined rate law and thus the proposed mechanism is plausible.

FEATURE PROBLEMS

104. (a) To determine the order of the reaction, we need $[C_6H_5N_2Cl]$ at each time. To determine this value, note that 58.3 mL $N_2(g)$ evolved corresponds to total depletion of $C_6H_5N_2Cl$, to $[C_6H_5N_2Cl] = 0.000$ M.

Thus, at any point in time,

$$[C_6H_5N_2Cl] = 0.071 \text{ M} - \left(\text{volume } N_2(g) \times \frac{0.071 \text{ M } C_6H_5N_2Cl}{58.3 \text{ mL } N_2(g)} \right)$$

Consider 21 min:

$$[C_6H_5N_2Cl] = 0.071 \text{ M} - \left(44.3 \text{ mL } N_2 \times \frac{0.071 \text{ M } C_6H_5N_2Cl}{58.3 \text{ mL } N_2(g)} \right) = 0.017 \text{ M}$$

The numbers in the following table are determined with this method.

time, min	0	3	6	9	12	15	18	21	24	27	30	∞
V_{N_2}, mL	0	10.8	19.3	26.3	32.4	37.3	41.3	44.3	46.5	48.4	50.4	58.3
$[C_6H_5N_2Cl]$, mM	71	58	47	39	32	26	21	17	14	12	10	0

[The concentration is given in thousandths of a mole per liter (mM).]

(b)

time(min)	0	3	6	9	12	15	18	21	24	27	30	∞
ΔT(min)		3	3	3	3	3	3	3	3	3	3	
$[C_6H_5N_2Cl]$(mM)	71	58	47	39	32	26	21	17	14	12	10	∞
$\Delta[C_6H_5N_2Cl]$(mM)		-13	-11	-8	-7	-6	-5	-4	-3	-2	-2	
Reaction Rate (mM min^{-1})		4.3	3.7	2.7	2.3	2.0	1.7	1.3	1.0	0.7	0.7	

(c) The two graphs are drawn on the same axes.

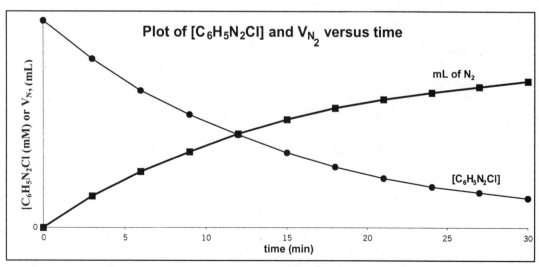

(d) The rate of the reaction at $t = 21$ min is the slope of the tangent line to the $[C_6H_5N_2Cl]$ curve. The tangent line intercepts the vertical axis at about $[C_6H_5N_2Cl] = 39$ mM and the horizontal axis at about 37 min

$$\text{Rate} = \frac{39 \times 10^{-3} \text{ M}}{37 \text{ min}} = 1.0_5 \times 10^{-3} \text{ M min}^{-1} = 1.1 \times 10^{-3} \text{ M min}^{-1}$$

The agreement with the reported value is very good.

(e) The initial rate is the slope of the tangent line to the $[C_6H_5N_2Cl]$ curve at $t = 0$. The intercept with the vertical axis is 71 mM, of course. That with the horizontal axis is about 13 min.

$$\text{Rate} = \frac{71 \times 10^{-3} \text{ M}}{13 \text{ min}} = 5.5 \times 10^{-3} \text{ M min}^{-1}$$

(f) The first-order rate law is $\text{Rate} = k[C_6H_5N_2Cl]$, which we solve for k:

$$k = \frac{\text{Rate}}{[C_6H_5N_2Cl]}$$

$$k_0 = \frac{5.5 \times 10^{-3} \text{ M min}^{-1}}{71 \times 10^{-3} \text{ M}} = 0.077 \text{ min}^{-1}$$

$$k_{21} = \frac{1.1 \times 10^{-3} \text{ M min}^{-1}}{17 \times 10^{-3} \text{ M}} = 0.065 \text{ min}^{-1}$$

An average value would be a good estimate: $k_{avg} = 0.071 \text{ min}^{-1}$

(g) The estimated rate constant gives one value of the half-life:

$$t_{1/2} = \frac{0.693}{k} = \frac{0.693}{0.071 \text{ min}^{-1}} = 9.8 \text{ min}$$

The first half-life occurs when $[C_6H_5N_2Cl]$ drops from 0.071 M to 0.0355 M. This seems to occur at about 10.5 min.

(h) The reaction should be three-fourths complete in two half-lives, about 20 minutes.

(i) The graph plots $\ln\left[C_6H_5N_2Cl\right]$ (in millimoles/L) vs. time in minutes.

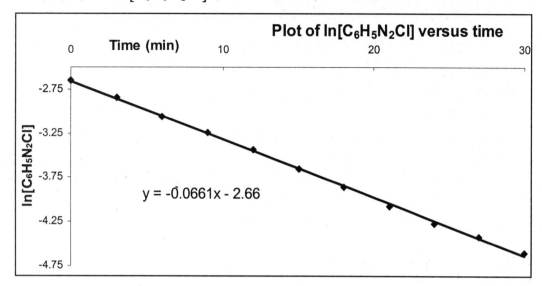

The linearity of the graph demonstrates that the reaction is first-order.

(j) $k = -\text{slope} = -\left(-6.61 \times 10^{-2}\right) \text{min}^{-1} = 0.0661 \text{ min}^{-1}$

$$t_{1/2} = \frac{0.693}{0.0661 \text{ min}^{-1}} = 10.5 \text{ min},$$

in good agreement with our previously determined values.

105. (a) In Experiments 1 & 2, [KI] is the same (0.20 M), while $\left[(NH_4)_2S_2O_8\right]$ is halved, from 0.20 M to 0.10 M. As a consequence, the time to produce a color change doubles, that is, the rate is halved. This indicates that reaction (a) is first-order in $S_2O_8^{2-}$. Experiments 2 and 3 produce a similar conclusion. In Experiments 4 and 5, $\left[(NH_4)_2S_2O_8\right]$ is the same (0.20 M) while [KI] is halved, from 0.10 to 0.050 M. As a consequence, the time to produce a color change nearly doubles, that is, the rate is halved. This indicates that reaction (a) is also first-order in I^-. Reaction (a) is $(1 + 1)$ second-order overall.

(b) The blue color appears when all the $S_2O_3^{2-}$ has been consumed, for only then does reaction (b) cease. The same amount of $S_2O_3^{2-}$ is placed in each reaction mixture.

$$\text{amount } S_2O_3^{2-} = 10.0 \text{ mL} \times \frac{1 \text{ L}}{1000 \text{ mL}} \times \frac{0.010 \text{ mol Na}_2S_2O_3}{1 \text{ L}} \times \frac{1 \text{ mol } S_2O_3^{2-}}{1 \text{ mol Na}_2S_2O_3}$$

$$= 1.0 \times 10^{-4} \text{ mol}$$

Through stoichiometry, we determine the amount of each reactant that reacts before this amount of $S_2O_3^{2-}$ will be consumed.

$$\text{amount } S_2O_8^{2-} = 1.0 \times 10^{-4} \text{ mol } S_2O_3^{2-} \times \frac{1 \text{ mol } I_3^-}{2 \text{ mol } S_2O_3^{2-}} \times \frac{1 \text{ mol } S_2O_8^{2-}}{1 \text{ mol } I_3^-}$$

$$= 5.0 \times 10^{-5} \text{ mol } S_2O_8^{2-}$$

$$\text{amount } I^- = 5.0 \times 10^{-5} \text{ mol } S_2O_8^{2-} \times \frac{2 \text{ mol } I^-}{1 \text{ mol } S_2O_8^{2-}} = 1.0 \times 10^{-4} \text{ mol } I^-$$

Note that we do not use "3 mol I^-" from equation (a) since one mole has not been oxidized; it simply complexes with the product I_2. The total volume of each solution is $(25.0 \text{ mL} + 25.0 \text{ mL} + 10.0 \text{ mL} + 5.0 \text{ mL} =)65.0 \text{ mL}$, or 0.0650 L.

The amount of $S_2O_8^{2-}$ that reacts in each case is 5.0×10^{-5} mol and thus

$$\Delta[S_2O_8^{2-}] = \frac{-5.0 \times 10^{-5} \text{ mol}}{0.0650 \text{ L}} = -7.7 \times 10^{-4} \text{ M}$$

$$\text{Thus, } \text{Rate}_1 = \frac{-\Delta[S_2O_8^{2-}]}{\Delta t} = \frac{+7.7 \times 10^{-4} \text{ M}}{21 \text{ s}} = 3.7 \times 10^{-5} \text{ M s}^{-1}$$

(c)   For Experiment 2, $\text{Rate}_2 = \dfrac{-\Delta[S_2O_8^{2-}]}{\Delta t} = \dfrac{+7.7 \times 10^{-4} \text{ M}}{42 \text{ s}} = 1.8 \times 10^{-5} \text{ M s}^{-1}$

To determine the value of k, we need initial concentrations, as altered by dilution.

$$[S_2O_8^{2-}]_1 = 0.20 \text{ M} \times \frac{25.0 \text{ mL}}{65.0 \text{ mL total}} = 0.077 \text{ M} \qquad [I^-]_1 = 0.20 \text{ M} \times \frac{25.0 \text{ mL}}{65.0 \text{ mL}} = 0.077 \text{ M}$$

$$\text{Rate}_1 = 3.7 \times 10^{-5} \text{ M s}^{-1} = k[S_2O_8^{2-}]^1[I^-]^1 = k(0.077 \text{ M})^1(0.077 \text{ M})^1$$

$$k = \frac{3.7 \times 10^{-5} \text{ M s}^{-1}}{0.077 \text{ M} \times 0.077 \text{ M}} = 6.2 \times 10^{-3} \text{ M}^{-1} \text{ s}^{-1}$$

$$[S_2O_8^{2-}]_2 = 0.10 \text{ M} \times \frac{25.0 \text{ mL}}{65.0 \text{ mL total}} = 0.038 \text{ M} \qquad [I^-]_2 = 0.20 \text{ M} \times \frac{25.0 \text{ mL}}{65.0 \text{ mL}} = 0.077 \text{ M}$$

$$\text{Rate}_2 = 1.8 \times 10^{-5} \text{ M s}^{-1} = k[S_2O_8^{2-}]^1[I^-]^1 = k(0.038 \text{ M})^1(0.077 \text{ M})^1$$

$$k = \frac{1.8 \times 10^{-5} \text{ M s}^{-1}}{0.038 \text{ M} \times 0.077 \text{ M}} = 6.2 \times 10^{-3} \text{ M}^{-1} \text{ s}^{-1}$$

(d) First we determine concentrations for Experiment 4.

$$\left[S_2O_8^{2-} \right]_4 = 0.20 \text{ M} \times \frac{25.0 \text{ mL}}{65.0 \text{ mL total}} = 0.077 \text{ M} \qquad \left[I^- \right]_4 = 0.10 \times \frac{25.0 \text{ mL}}{65.0 \text{ mL}} = 0.038 \text{ M}$$

We have two expressions for Rate; let us equate them and solve for the rate constant.

$$\text{Rate}_4 = \frac{-\Delta\left[S_2O_8^{2-} \right]}{\Delta t} = \frac{+7.7 \times 10^{-4} \text{ M}}{\Delta t} = k\left[S_2O_8^{2-} \right]_4^1 \left[I^- \right]_4^1 = k\left(0.077 \text{ M} \right)\left(0.038 \text{ M} \right)$$

$$k = \frac{7.7 \times 10^{-4} \text{ M}}{\Delta t \times 0.077 \text{ M} \times 0.038 \text{ M}} = \frac{0.26 \text{ M}^{-1}}{\Delta t}$$

$$k_3 = \frac{0.26 \text{ M}^{-1}}{189 \text{ s}} = 0.0014 \text{ M}^{-1} \text{ s}^{-1}$$

$$k_{13} = \frac{0.26 \text{ M}^{-1}}{88 \text{ s}} = 0.0030 \text{ M}^{-1} \text{ s}^{-1}$$

$$k_{24} = \frac{0.26 \text{ M}^{-1}}{42 \text{ s}} = 0.0062 \text{ M}^{-1} \text{ s}^{-1}$$

$$k_{33} = \frac{0.26 \text{ M}^{-1}}{21 \text{ s}} = 0.012 \text{ M}^{-1} \text{ s}^{-1}$$

(e) We plot $\ln k$ vs. $1/T$ The slope of the line $= -E_a / R$.

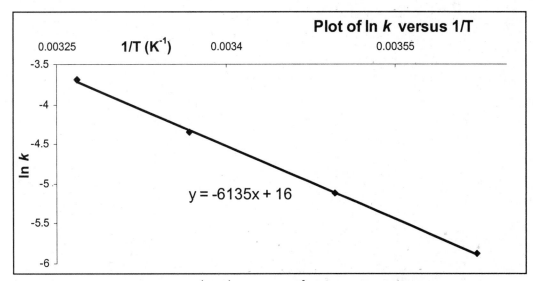

$E_a = +6135 \text{ K} \times 8.3145 \text{ J mol}^{-1} \text{ K}^{-1} = 51.0 \times 10^3 \text{ J/mol} = 51.0 \text{ kJ/mol}$

The scatter of the data permits only a two significant figure result: 51 kJ/mol

(f) For the mechanism to agree with the reaction stoichiometry, the steps of the mechanism must sum to the overall reaction, in the manner of Hess's law.

(slow) $I^- + S_2O_8^{2-} \rightarrow IS_2O_8^{3-}$

(fast) $IS_2O_8^{3-} \rightarrow 2\,SO_4^{2-} + I^+$

(fast) $I^+ + I^- \rightarrow I_2$

(fast) $I_2 + I^- \rightarrow I_3^-$

(net) $3\,I^- + S_2O_8^{2-} \rightarrow 2\,SO_4^{2-} + I_3^-$

Each of the intermediates cancels: $IS_2O_8^{3-}$ is produced in the first step and consumed in the second, I^+ is produced in the second step and consumed in the third, I_2 is produced in the third step and consumed in the 4th. The mechanism is consistent with the stoichiometry. The rate of the slow step of the mechanism is

$$\text{Rate}_1 = k_1 \left[S_2O_8^{2-} \right]^1 \left[I^- \right]^1$$

This is exactly the same as the experimental rate law. It is reasonable that the first step be slow since it involves two negatively charged species coming together. We know that like charges repel, and thus this should not be an easy or rapid process.

CHAPTER 16
PRINCIPLES OF CHEMICAL EQUILIBRIUM
PRACTICE EXAMPLES

1A For the cited reaction $CO(g) + 2 H_2(g) \rightleftharpoons CH_3OH(g)$ $\qquad K = \dfrac{[CH_3OH]}{[CO][H_2]^2} = 14.5$

Given that $[CO] = [CH_3OH]$. Thus, one may be substituted for the other, and the resulting equation solved for $[H_2]$. $\quad K = \dfrac{[CO]}{[CO][H_2]^2} = 14.5 = \dfrac{1}{[H_2]^2} \qquad [H_2] = \sqrt{\dfrac{1}{14.5}} = 0.263 \text{ M}$

1B Substitute the values we have into the equilibrium constant expression, and solve for $[H_2]$.

$$K_c = \frac{[NH_3]^2}{[N_2][H_2]^3} = 1.8 \times 10^4 = \frac{(2.00 \text{ M})^2}{0.015 \text{ M}[H_2]^3} \qquad [H_2] = \sqrt[3]{[H_2]^3} = \sqrt[3]{\frac{(2.00)^2}{1.8 \times 10^4 \times 0.015}} = 0.25 \text{ M}$$

2A The Example gives $K_c = 3.6 \times 10^8$ for the reaction $N_2(g) + 3 H_2(g) \rightleftharpoons 2 NH_3(g)$. The reaction we are considering is one-third of this reaction. If we divide the reaction by 3, we should take the cube root of the equilibrium constant to obtain the value of the equilibrium constant for the "divided" reaction. $\quad K_{c3} = \sqrt[3]{K_c} = \sqrt[3]{3.6 \times 10^8} = 7.1 \times 10^2$

2B First we reverse the given reaction to put $NO_2(g)$ on the reactant side. The new equilibrium constant is the inverse of the given one.

$$NO_2(g) \rightleftharpoons NO(g) + \tfrac{1}{2} O_2(g) \qquad K_c' = 1/(7.5 \times 10^2) = 0.0013$$

Then we double the reaction to obtain 2 moles of $NO_2(g)$ as reactant. The equilibrium constant is then raised to the second power.

$$2 NO_2(g) \rightleftharpoons 2 NO(g) + O_2(g) \qquad K_c = (0.0013)^2 = 1.7 \times 10^{-6}$$

3A We use the expression $K_p = K_c (RT)^{\Delta n_g}$. In this case, $\Delta n_{gas} = 3 + 1 - 2 = 2$ and thus we have

$$K_p = K_c (RT)^2 = 2.8 \times 10^{-9} \times (0.08206 \times 298)^2 = 1.7 \times 10^{-6}$$

3B We begin by writing the K_p expression. We then substitute $(n/V)RT = [\ ]RT$ for each pressure. We collect terms to obtain an expression relating K_c and K_p, into which we substitute to find the value of K_c.

$$K_p = \frac{\{P(H_2)\}^2 \{P(S_2)\}}{\{P(H_2S)\}^2} = \frac{([H_2]RT)^2 ([S_2]RT)}{([H_2S]RT)^2} = \frac{[H_2]^2[S_2]}{[H_2S]^2} RT = K_c RT$$

Same result can be obtained by using $K_p = K_c (RT)^{\Delta n_{gas}}$, since $\Delta n_{gas} = 2 + 1 - 2 = +1$.

$$K_c = \frac{K_p}{RT} = \frac{1.2 \times 10^{-2}}{0.0806 \times (1065 + 273)} = 1.1 \times 10^{-4}$$

But the reaction has been reversed and halved. Thus

$$K_{final} = \sqrt{\frac{1}{K_c}} = \sqrt{\frac{1}{1.1 \times 10^{-4}}} = \sqrt{9090} = 95.3$$

4A We remember that neither solids, such as $Ca_5(PO_4)_3OH(s)$, nor liquids, such as $H_2O(l)$, appear in the equilibrium constant expression. Concentrations of products appear in the numerator, those of reactants in the denominator. $K_c = \dfrac{\left[Ca^{2+} \right]^5 \left[HPO_4^{2-} \right]^3}{\left[H^+ \right]^4}$

4B First we write the balanced chemical equation for the reaction. Then we write the equilibrium constant expressions, remembering that gases and solutes in aqueous solution appear in the K_c expression, but pure liquids and pure solids do not.

$$3\,Fe(s) + 4\,H_2O(g) \rightleftharpoons Fe_3O_4(s) + 4\,H_2(g)$$

$$K_p = \frac{\{P(H_2)\}^4}{\{P(H_2O)\}^4} \qquad\qquad K_c = \frac{[H_2]^4}{[H_2O]^4} \qquad\qquad \text{Because } \Delta n_{gas} = 4 - 4 = 0,\ K_p = K_c$$

5A We compare the value of the reaction quotient, Q_c, with that of K_c.

$$Q_c = \frac{[PCl_3][Cl_2]}{[PCl_5]} = \frac{0.50\,M \times 0.20\,M}{4.50\,M} = 0.022 < 0.0454 = K_c$$

Because $Q_c < K_c$, the net reaction will proceed to the right, forming products.

5B We compute the value of Q_c. Each concentration equals the mass (m) of the substance divided by its molar mass (which quotient is the amount of the substance in moles) and further divided by the volume of the container.

$$Q_c = \frac{[CO_2][H_2]}{[CO][H_2O]} = \frac{\dfrac{m \times \dfrac{1\,mol\,CO_2}{44.0\,g\,CO_2}}{V} \times \dfrac{m \times \dfrac{1\,mol\,H_2}{2.0\,g\,H_2}}{V}}{\dfrac{m \times \dfrac{1\,mol\,CO}{28.0\,g\,CO}}{V} \times \dfrac{m \times \dfrac{1\,mol\,H_2O}{18.0\,g\,H_2O}}{V}} = \frac{\dfrac{1}{44.0 \times 2.0}}{\dfrac{1}{28.0 \times 18.0}} = \frac{28.0 \times 18.0}{44.0 \times 2.0} = 5.7 > 1.00 = K_c$$

(In evaluating the expression above, we cancelled the equal values of V, and we also cancelled the equal values of m.) Because the value of Q_c is larger than the value of K_c, reaction will proceed to the left to reach a state of equilibrium. Thus, at equilibrium there will be larger amounts of reactants, and smaller amounts of products than there are initially.

6A $O_2(g)$ is a reactant. The equilibrium system will shift right forming product in an attempt to consume some of the added reactant. Looked at in another way, $[O_2]$ is increased above its equilibrium value by the addition of oxygen. This makes Q_c smaller than K_c. (The $[O_2]$ is in the denominator of the expression.) And the system shifts right to compensate.

6B **(a)** The position of an equilibrium mixture is affected only by changing the concentration of substances that appear in the equilibrium constant expression, $K_c = [CO_2]$. Since CaO(s) is a pure solid, its concentration does not appear in the equilibrium constant expression and thus its addition will have no direct effect on the position of equilibrium.

(b) The addition of $CO_2(g)$ will increase $[CO_2]$ above its equilibrium value. The reaction will shift left to alleviate this increase; some $CaCO_3(s)$ will form.

(c) Since $CaCO_3$(s) is a pure solid like CaO(s), its concentration does not appear in the equilibrium constant expression and thus the addition of any solid $CaCO_3$ to an equilibrium mixture will not have an effect upon the position of equilibrium

7A We know that a decrease in volume or an increase in pressure of an equilibrium mixture of gases causes a net reaction in the direction producing the smaller number of moles of gas. In the reaction in question, that direction is to the left: one mole of $N_2O_4(g)$ is formed when two moles of $NO_2(g)$ combine. Thus, decreasing the cylinder volume would have the initial effect of doubling both $[N_2O_4]$ and $[NO_2]$. In order to reestablish equilibrium, some NO_2 will then be converted into N_2O_4. Note, however, that the NO_2 concentration will still ultimately end up being higher than it was prior to pressurization.

7B In the balanced chemical equation for the chemical reaction $\Delta n_{gas} = (1+1)-(1+1)=0$. As a consequence, a change in overall volume or total gas pressure will have no effect on the position of equilibrium. In the equilibrium constant expression, the two partial pressures in the numerator will be affected to exactly the same degree as the two partial pressures in the denominator, and Q_p will continue to equal K_p.

8A The cited reaction is endothermic. Raising the temperature on an equilibrium mixture favors the endothermic reaction. Thus, $N_2O_4(g)$ should decompose more completely at higher temperatures and the amount of $NO_2(g)$ formed from a given amount of $N_2O_4(g)$ will be greater at high temperatures than at low ones.

8B The NH_3(g) formation reaction is $\frac{1}{2}N_2(g) + \frac{3}{2}H_2(g) \rightarrow NH_3(g)$ $\Delta H° = -46.11$ kJ/mol
This reaction is an exothermic reaction. Lowering temperature causes a shift in the direction of this exothermic reaction. The reaction will shift right at a lower temperature. $[NH_3(g)]$ will be greater at $-100°C$.

9A   We write the expression for K_c and then substitute expressions for molar concentrations.

$$K_c = \frac{[H_2]^2[S_2]}{[H_2S]^2} = \frac{\left(\dfrac{0.22 \text{ mol } H_2}{3.00 \text{ L}}\right)^2 \dfrac{0.11 \text{ mol } S_2}{3.00 \text{ L}}}{\left(\dfrac{2.78 \text{ mol } H_2S}{3.00 \text{ L}}\right)^2} = 2.3 \times 10^{-4}$$

9B   We write the equilibrium constant expression and solve it for $[N_2O_4]$.

$$K_c = 4.61 \times 10^{-3} = \frac{[NO_2]^2}{[N_2O_4]} \qquad [N_2O_4] = \frac{[NO_2]^2}{4.61 \times 10^{-3}} = \frac{(0.0236 \text{ M})^2}{4.61 \times 10^{-3}} = 0.121 \text{ M}$$

Then we determine the mass of N_2O_4 present in 2.26 L.

$$N_2O_4 \text{ mass} = 2.26 \text{ L} \times \frac{0.121 \text{ mol } N_2O_4}{1 \text{ L}} \times \frac{92.01 \text{ g } N_2O_4}{1 \text{ mol } N_2O_4} = 25.2 \text{ g } N_2O_4$$

10A   We use the initial-change-equilibrium setup to establish the amount of each substance at equilibrium. We label each entry in the table in the order of its determination (1st, 2nd, 3rd, 4th, 5th), to better illustrate the technique. We know the initial amounts of all substances (1st). (There are zero moles of each product at the start).

Because "initial"+ "change"= "equilibrium", the equilibrium amount (2nd) of $Br_2(g)$ enables us to determine "change" (3rd) for $Br_2(g)$. We then use stoichiometry to write other entries (4th) on the "change" line. And finally, we determine the remaining equilibrium amounts (5th).

reaction:	$2 \text{ NOBr}(g)$	$\rightleftharpoons$	$2 \text{ NO}(g)$	$+$	$Br_2(g)$
initial:	1.86 mol (1st)		0.00 mol (1st)		0.00 mol (1st)
change:	−0.164 mol (4th)		+0.164 mol (4th)		+0.082 mol (3rd)
equil.:	1.70 mol (5th)		0.164 mol (5th)		0.082 mol (2nd)

$$K_c = \frac{[NO]^2[Br_2]}{[NOBr]^2} = \frac{\left(\dfrac{0.164 \text{ mol NO}}{5.00 \text{ L}}\right)^2 \left(\dfrac{0.082 \text{ mol } Br_2}{5.00 \text{ L}}\right)}{\left(\dfrac{1.70 \text{ mol NOBr}}{5.00 \text{ L}}\right)^2} = 1.5 \times 10^{-4}$$

In this case,

$$\Delta n_{gas} = 2 + 1 - 2 = +1. \qquad K_p = K_c (RT)^{+1} = 1.5 \times 10^{-4} \times 0.08206 \times 298 = 3.7 \times 10^{-3}$$

10B We use the amounts stated in the problem to determine the equilibrium amount of each substance.

reaction:	$2 SO_3(g)$	$\rightleftharpoons$	$2 SO_2(g)$	$+$	$O_2(g)$
initial:	0 mol		0.100 mol		0.100 mol
changes:	+0.0916 mol		−0.0916 mol		−0.0916/2 mol
equil.:	0.0916 mol		0.0084 mol		0.0542 mol
concentrations:	$\dfrac{0.916 \text{ mol}}{1.52 \text{ L}}$		$\dfrac{0.0084 \text{ mol}}{1.52 \text{ L}}$		$\dfrac{0.0542 \text{ mol}}{1.52 \text{ L}}$
concentrations:	0.0603 M		0.0055 M		0.0357 M

We use these values to compute K_c for the reaction and then the relationship

$K_p = K_c (RT)^{\Delta n_{gas}}$ (with $\Delta n_{gas} = 1 + 1 - 1 = +1$) to determine the value of K_p.

$$K_c = \frac{[SO_2]^2 [O_2]}{[SO_3]^2} = \frac{(0.0055)^2 (0.0357)}{(0.0603)^2} = 3.0 \times 10^{-4} \quad K_p = 3.0 \times 10^{-4} \times 0.08206 \times 900 \approx 0.022$$

11A The equilibrium constant expression is $K_p = P\{H_2O\} P\{CO_2\} = 0.231$ at $100\,^{\circ}C$. From the balanced chemical equation, we see that one mole of $H_2O(g)$ is formed for each mole of $CO_2(g)$. Consequently, $P\{H_2O\} = P\{CO_2\}$ and $K_p = (P\{CO_2\})^2$. We solve this expression for $P\{CO_2\}$: $P\{CO_2\} = \sqrt[2]{(P\{CO_2\})^2} = \sqrt[2]{K_p} = \sqrt[2]{0.231} = 0.481$ atm

11B The equation for the reaction is $NH_4HS(s) \rightleftharpoons NH_3(g) + H_2S(g)$ $\quad K_p = 0.108$ at $25\,^{\circ}C$
The two partial pressures do not have to be equal at equilibrium. The only instance in which they must be equal is when all of the two gases come from the decomposition of $NH_4HS(s)$. In this case, some of the $NH_3(g)$ comes from another source. We can obtain the pressure of $H_2S(g)$ by substitution into the equilibrium constant expression, since we are given the equilibrium pressure of $NH_3(g)$.

$$K_p = P\{H_2S\} P\{NH_3\} = 0.108 = P\{H_2S\} \times 0.500 \text{ atm } NH_3 \quad P\{H_2S\} = \frac{0.108}{0.500} = 0.216 \text{ atm}$$

So, $P_{total} = P_{H_2S} + P_{NH_3} = 0.216$ atm $+ 0.500$ atm $= 0.716$ atm

12A We set up this problem in the same manner we have previously employed, designating the equilibrium amount of HI as $2x$. (Note that we use the same multipliers for x as the stoichiometric coefficients.)

Equation:	$H_2(g)$	$+$	$I_2(g)$	$\rightleftharpoons$	$2 HI(g)$
Initial:	0.150 mol		0.200 mol		0 mol
Changes:	$-x$ mol		$-x$ mol		$+2x$ mol
Equil:	$(0.150-x)$ mol		$(0.200-x)$ mol		$2x$ mol

$$K_c = \frac{\left(\dfrac{2x}{15.0}\right)^2}{\dfrac{0.150-x}{15.0} \times \dfrac{0.200-x}{15.0}} = \frac{(2x)^2}{(0.150-x)(0.200-x)} = 50.2$$

We substitute these expressions into the equilibrium constant expression and solve for x

$$4x^2 = (0.150-x)(0.200-x)50.2 = 50.2(0.0300 - 0.350x + x^2) = 1.51 - 17.6x + 50.2x^2$$

$$0 = 46.2x^2 - 17.6x + 1.51 \qquad \text{Now we use the quadratic formula to obtain a value of } x.$$

$$x = \frac{-b \pm \sqrt{b^2 - 4ac}}{2a} = \frac{17.6 \pm \sqrt{(17.6)^2 - 4 \times 46.2 \times 1.51}}{2 \times 46.2} = \frac{17.6 \pm 5.54}{92.4} = 0.250 \text{ or } 0.131$$

The first root cannot be used because we cannot have $(0.150 - 0.250 = -0.100)$ a negative amount of H_2. Thus, we have $2 \times 0.131 = 0.262$ mol HI at equilibrium. We check by substituting the amounts into the K_c expression. (Notice that the volumes cancel.) The slight disagreement in the two values (52 compared to 50.2) is the result of rounding error.

$$K_c = \frac{(0.262)^2}{(0.150 - 0.131)(0.200 - 0.131)} = \frac{0.0686}{0.019 \times 0.069} = 52$$

12B **(a)** The equation for the reaction is $N_2O_4(g) \rightleftharpoons 2\,NO_2(g)$ $\qquad K_c = 4.61 \times 10^{-3}$ at

$25°C$. In the example, this reaction is conducted in a 0.372 L flask. The effect of moving the mixture to the larger, 10.0 L container is that the reaction will shift to produce a greater number of moles of gas; $NO_2(g)$ will be produced and $N_2O_4(g)$ will dissociate. Consequently, the amount of N_2O_4 will decrease.

(b) The equilibrium constant expression substituting 10.0 L for 0.372 L, follows.

$$K_c = \frac{[NO_2]^2}{[N_2O_4]} = \frac{\left(\dfrac{2x}{10.0}\right)^2}{\dfrac{0.0240-x}{10.0}} = \frac{4x^2}{10.0(0.0240-x)} = 4.61 \times 10^{-3}$$

This can be solved with the quadratic equation, and the result is $x = 0.0118$ moles. We can attempt the method of successive approximations. *First*, assume that $x \ll 0.0240$. We obtain:

$$x = \frac{\sqrt{4.61 \times 10^{-3} \times 10.0\,(0.0240 - 0)}}{4} = \sqrt{4.61 \times 10^{-3} \times 2.50\,(0.0240 - 0)} = 0.0166$$

Clearly x is not much smaller than 0.0240. So, *second,* assume $x \approx 0.0166$. We obtain:

$$x = \sqrt{4.61 \times 10^{-3} \times 2.50(0.0240 - 0.0166)} = 0.00925$$

This assumption is not valid either. So, *third,* assume $x \approx 0.00925$. We obtain:

$$x = \sqrt{4.61 \times 10^{-3} \times 2.50(0.0240 - 0.0925)} = 0.0130$$

Notice that after each cycle the value we obtain for x gets closer to the value obtained from the quadratic equation. The values from the next several cycles follow.

cycle	4th	5th	6th	7th	8th	9th	10th	11th
x value	0.0112	0.0121	0.0117	0.0119	0.0118_1	0.0118_6	0.0118_3	0.0118_4

The amount of N_2O_4 at equilibrium is 0.0118 mol, less than the 0.0210 mol N_2O_4 at equilibrium in the 0.372 L flask, as predicted.

13A Again we organize our solution on the balanced chemical equation.

equation: $Ag^+ (aq) + Fe^{2+} (aq) \rightleftharpoons Fe^{3+} (aq) + Ag(s) \qquad K_c = 2.98$

initial: 0 M   0 M   1.20 M

changes: $+x$ M $+x$ M $-x$ M

equil: x M x M $(1.20 - x)$ M

$$K_c = \frac{[Fe^{3+}]}{[Ag^+][Fe^{2+}]} = 2.98 = \frac{1.20 - x}{x^2} \qquad 2.98 \ x^2 = 1.20 - x \qquad 0 = 2.98 x^2 + x - 1.20$$

We use the quadratic formula to obtain a solution.

$$x = \frac{-b \pm \sqrt{b^2 - 4ac}}{2a} = \frac{-1.00 \pm \sqrt{(1.00)^2 + 4 \times 2.98 \times 1.20}}{2 \times 2.98} = \frac{-1.00 \pm 3.91}{5.96} = 0.488 \text{ or} -0.824$$

A negative root makes no physical sense. We obtain the equilibrium concentrations from x.

$$\left[Ag^+ \right] = \left[Fe^{2+} \right] = 0.488 \text{ M} \qquad\qquad \left[Fe^{3+} \right] = 1.20 - 0.488 = 0.71 \text{ M}$$

13B We first calculate the value of Q_c to determine the direction of the reaction.

$$Q_c = \frac{[V^{2+}][Cr^{3+}]}{[V^{3+}][Cr^{2+}]} = \frac{0.150 \times 0.150}{0.0100 \times 0.0100} = 225 < 7.2 \times 10^2 = K_c$$

Because the reaction quotient has a smaller value than the equilibrium constant, a net reaction will occur to the right. We now set up this solution as we have others, based on the balanced chemical equation.

            $V^{3+} (aq) \quad + \quad Cr^{2+} (aq) \rightleftharpoons V^{2+} (aq) \quad + \quad Cr^{3+} (aq)$

initial   0.0100 M   0.0100 M   0.150 M   0.150 M

changes $-x$ M $-x$ M $+x$ M $+x$ M

equil $(0.0100 - x)$M $(0.0100 - x)$M $(0.150 + x)$M $(0.150 + x)$M

$$K_c = \frac{[V^{2+}][Cr^{3+}]}{[V^{3+}][Cr^{2+}]} = \frac{(0.150 + x) \times (0.150 + x)}{(0.0100 - x) \times (0.0100 - x)} = 7.2 \times 10^2 = \left(\frac{0.150 + x}{0.0100 - x} \right)^2$$

If we take the square root of both sides of this expression, we have

$$\sqrt{7.2 \times 10^2} = \frac{0.150 + x}{0.0100 - x} = 27$$

$0.150 + x = 0.27 - 27x$ which becomes $28x = 0.12$ and yields $x = 0.0043$ M. Then the equilibrium concentrations are: $\left[V^{3+} \right] = \left[Cr^{2+} \right] = 0.0100$ M $- 0.0043$ M $= 0.0057$ M

$\left[V^{2+} \right] = \left[Cr^{3+} \right] = 0.150$ M $+ 0.0043$ M $= 0.154$ M

REVIEW QUESTIONS

1. **(a)** K_p is the symbol for the equilibrium constant in terms of partial pressures: the quotient of the equilibrium partial pressures of the products divided by those of the reactants, with each reactant or product partial pressure (in atmospheres) raised to a power equal to its stoichiometric coefficient in the balanced equation.

   (b) Q_c is the symbol for the reaction quotient in terms of molarities: the quotient of the molarities of the products divided by those of the reactants, each concentration raised to a power equal to its stoichiometric coefficient. These concentrations need not be those at equilibrium.

   (c) Δn_{gas} is the difference between the sum of the stoichiometric coefficients of the gaseous products and the similar sum for gaseous reactants, using the coefficients from the balanced chemical equation.

2. **(a)** A dynamic equilibrium is a balanced situation that is created by opposing processes proceeding at such rates that the net result is no apparent change, although individual molecules may indeed be undergoing change, and probably are.

   (b) The direction of net reaction describes whether a chemical reaction produces more product (to the right) or forms more reactant (to the left), generally as the result of the alteration of reaction conditions.

   (c) Le Châtelier's principle states that if a stress is applied to a system that is at equilibrium, the system will react in such a way as to relieve the stress and attain a new equilibrium position.

   (d) The effect of a catalyst on the position of equilibrium is nonexistent. A catalyst does, however, speed up the rate at which equilibrium is established.

3. **(a)** A reaction that goes to completion is one that reacts until the amount of one of the reactants is gone. A reversible reaction is one that comes to a point of balance before all reactants are depleted.

(b) The equilibrium constant K_c contains the equilibrium concentrations of products and reactants, while the K_p expression contains their partial pressures. The values of these two numbers are related by the expression $K_p = K_c (RT)^{\Delta n_g}$.

(c) The reaction quotient, Q_c, and the equilibrium constant expression, K_c, have the same functional form, with the concentrations of products over those of reactants, each raised to a power equal to the stoichiometric coefficient of that species. However, the concentrations that are used differ. In the K_c expression, these concentrations are equilibrium concentrations, while in the Q_c expression they are not necessarily the concentrations at equilibrium—often they are the initial concentrations. Similar statements can be made concerning Q_p and K_p, except that partial pressures replace concentrations.

(d) A homogeneous equilibrium is one that takes place in a solution, either entirely gas or entirely liquid. A heterogeneous equilibrium is one that is established when two phases of matter are present, generally a solid with either a liquid or a gaseous solution.

4. **(a)** $K_c = \dfrac{[NO_2]^2}{[NO]^2[O_2]}$ **(b)** $K_c = \dfrac{[Zn^{2+}]}{[Ag^+]^2}$ **(c)** $K_c = \dfrac{[OH^-]^2}{[CO_3^{2-}]}$

5. **(a)** $K_p = \dfrac{P\{CH_4\}P\{H_2S\}^2}{P\{CS_2\}P\{H_2\}^4}$ **(b)** $K_p = P\{O_2\}^{1/2}$ **(c)** $K_p = P\{CO_2\}P\{H_2O\}$

6. In each case we write the equation for the formation reaction and then the equilibrium constant expression for that reaction.

(a) $N_2(g) + \tfrac{1}{2} O_2(g) \rightleftharpoons N_2O(g)$ $\qquad K_c = \dfrac{[N_2O]}{[N_2][O_2]^{1/2}}$

(b) $\tfrac{1}{2}H_2(g) + \tfrac{1}{2} Br_2(g) \rightleftharpoons HBr(g)$ $\qquad K_c = \dfrac{[HBr]}{[H_2]^{1/2}[Br_2]^{1/2}}$

(c) $\tfrac{1}{2}N_2(g) + \tfrac{3}{2} H_2(g) \rightleftharpoons NH_3(g)$ $\qquad K_c = \dfrac{[NH_3]}{[N_2]^{1/2}[H_2]^{3/2}}$

(d) $\tfrac{1}{2}Cl_2(g) + \tfrac{3}{2}F_2(g) \rightleftharpoons ClF_3(g)$ $\qquad K_c = \dfrac{[ClF_3]}{[Cl_2]^{1/2}[F_2]^{3/2}}$

(e) $\tfrac{1}{2}N_2(g) + \tfrac{1}{2} O_2(g) + F_2(g) \rightleftharpoons NOF_2(g)$ $\qquad K_c = \dfrac{[NOF_2]}{[N_2]^{1/2}[O_2]^{1/2}[F_2]}$

7. (a) This is the reverse of the first reaction; K_c is the inverse value.

$$K_c = \frac{[CO][H_2O]}{[CO_2][H_2]} = \frac{1}{23.2} = 0.0431$$

(b) This is twice the second reaction; K_c is its square.

$$K_c = \frac{[SO_3]^2}{[SO_2]^2[O_2]} = (56)^2 = 3.1 \times 10^3$$

(c) This is half of the reverse of the third reaction; K_c is the the inverse of the square

root. $K_c = \dfrac{[H_2S]}{[H_2][S_2]^{1/2}} = \dfrac{1}{\sqrt{2.3 \times 10^{-4}}} = 66$

8. We combine one-half of the reversed first reaction with the second reaction to obtain the desired reaction.

$\frac{1}{2} N_2(g) + \frac{1}{2} O_2(g) \rightleftharpoons NO(g)$ $K_c = \dfrac{1}{\sqrt{2.1 \times 10^{30}}}$

$NO(g) + \frac{1}{2} Br_2(g) \rightleftharpoons NOBr(g)$ $K_c = 1.4$

net : $\frac{1}{2} N_2(g) + \frac{1}{2} O_2(g) + \frac{1}{2} Br_2(g) \rightleftharpoons NOBr(g)$ $K_c = \dfrac{1.4}{\sqrt{2.1 \times 10^{30}}} = 9.7 \times 10^{-16}$

9. Answer (d) is correct. I_2 (1 mol) is the limiting reagent in this reaction. If the reaction were to go to completion, 2 mol HI would be produced. Thus, the only way that 2 mol HI could be produced (answer b) would be if K_c were infinite. 1 mol HI could be produced (answer a), but without knowing the value of K_c for this reaction, we cannot be sure. The only certain answer is that something less than 2 mol HI are produced, answer (d).

10. Simply substitute into the K_c expression. $K_c = \dfrac{[C]^2}{[A]^2[B]} = \dfrac{(0.43 \text{ M})^2}{(0.55 \text{ M})^2 \times 0.33 \text{ M}} = 1.9$

11. The formation of 2.0×10^{-11} mol S(g) does not significantly change the amount of $S_2(g)$

from its original 0.0040 mol. Thus, $K_c = \dfrac{[S]^2}{[S_2]} = \dfrac{\left(\dfrac{2.0 \times 10^{-11} \text{ mol}}{0.500 \text{ L}}\right)^2}{\dfrac{0.0040 \text{ mol}}{0.500 \text{ L}}} = 2.0 \times 10^{-19}$

12. (a) If the amounts of SO_2 and SO_3 are equal, then $[SO_2] = [SO_3]$.

$$K_c = 35.5 = \frac{[SO_3]^2}{[SO_2]^2[O_2]} = \frac{1}{[O_2]} \quad ; [O_2] = \frac{1}{35.5} = 0.0282 \text{ M}$$

mol $O_2 = 0.0282$ M $\times 2.05$ L $= 0.0578$ mol O_2

(b) If there are twice as many moles of SO_3 in the flask as SO_2, then $[SO_3] = 2 \times [SO_2]$

$$K_c = 35.5 = \frac{[SO_3]^2}{[SO]^2[O_2]} = \frac{2^2[SO_3]^2}{[SO_2]^2[O_2]} = \frac{4}{[O_2]} \quad [O_2] = \frac{4}{35.5} = 0.113 \text{ M}$$

mol $O_2 = 0.113 \text{ M} \times 2.05 \text{ L} = 0.232 \text{ mol } O_2$

13. In each case we use the relationship $K_p = K_c(RT)^{\Delta n \text{ gas}}$

(a) For $CO_2(g) + H_2(g) \rightleftharpoons CO(g) + H_2O(g)$ $K_p = K_c = 0.0431$

(b) For $2\,SO_2(g) + O_2(g) \rightleftharpoons 2\,SO_3(g)$ $\quad K_p = K_c(RT)^{-1} = \dfrac{3.1 \times 10^3}{0.0821 \times 900} = 42$

(c) For $H_2(g) + \frac{1}{2}\,S_2(g) \rightleftharpoons H_2S(g)$ $\quad K_p = K_c(RT)^{-1/2} = \dfrac{66}{\sqrt{0.0821 \times 1405}} = 6.1$

14. We first find the value of K_p for the reaction

$2\,NO_2(g) \rightleftharpoons 2\,NO(g) + O_2(g)$, $K_c = 1.8 \times 10^{-6}$ at $184\,°C = 457\,K$.

For this reaction $\Delta n_{gas} = 2 + 1 - 2 = +1$.

$K_p = K_c(RT)^{\Delta n_g} = 1.8 \times 10^{-6}(0.08206 \times 457)^{+1} = 6.8 \times 10^{-5}$

To obtain the final reaction $NO(g) + \frac{1}{2}O_2(g) \rightleftharpoons NO_2(g)$ from the initial reaction, that initial reaction must be reversed and then divided by two. Thus, in order to determine the value of the equilibrium constant for the final reaction, the value of K_p for the initial reaction must be inverted, and the square root taken of the result.

$$K_{p,\text{ final}} = \sqrt{\frac{1}{6.8 \times 10^{-5}}} = 1.2 \times 10^2$$

15. (a) $K_c = \dfrac{[CO][H_2O]}{[CO_2][H_2]} = \dfrac{\dfrac{n\{CO\}}{V} \times \dfrac{n\{H_2O\}}{V}}{\dfrac{n\{CO_2\}}{V} \times \dfrac{n\{H_2\}}{V}}$

(b) Note that $K_p = K_c$ for this reaction, since $\Delta n_{gas} = 0$.

$$K_c = K_p = \frac{0.224 \text{ mol CO} \times 0.224 \text{ mol H}_2O}{0.276 \text{ mol CO}_2 \times 0.276 \text{ mol H}_2} = 0.659$$

16. (a) $K_c = \dfrac{[PCl_5]}{[PCl_3][Cl_2]} = \dfrac{\dfrac{0.105 \text{ g PCl}_5}{2.50 \text{ L}} \times \dfrac{1 \text{ mol PCl}_5}{208.2 \text{ g}}}{\left(\dfrac{0.220 \text{ g PCl}_3}{2.50 \text{ L}} \times \dfrac{1 \text{ mol PCl}_3}{137.3 \text{g}}\right) \times \left(\dfrac{2.12 \text{ g Cl}_2}{2.50 \text{ L}} \times \dfrac{1 \text{ mol Cl}_2}{70.9 \text{ g}}\right)} = 26.3$

(b) $K_p = K_c(RT)^{\Delta n} = 26.3\,(0.08206 \times 523)^{-1} = 0.613$

17. **(a)** If 5% of the 1.00 mol I_2 is dissociated into atoms, then 0.05 mol I_2 has dissociated.

Equation:	$I_2(g)$	$\rightleftharpoons$	$2\,I(g)$
Initial:	1.00 mol		0.00 mol
Changes	-0.05 mol		+0.10 mol
Equil:	0.95 mol		0.10 mol

$$K_c = \frac{[I^2]}{[I_2]} = \frac{\left(\dfrac{0.10\,\text{mol I}}{1.00\,\text{L}}\right)^2}{\dfrac{0.95\,\text{mol I}_2}{1.00\,\text{L}}} = 0.011$$

(b) $K_p = K_c(RT)^{+1} = 0.011\,(0.0821 \times 1473) = 1.3$

18. **(a)** We determine the concentration of each species in the gaseous mixture, use these concentrations to determine the value of the reaction quotient, and compare this value of Q_c with the value of K_c.

$$[SO_2] = \frac{0.455\ \text{mol SO}_2}{1.90\ \text{L}} = 0.239\ \text{M} \qquad [O_2] = \frac{0.183\ \text{mol O}_2}{1.90\ \text{L}} = 0.0963\ \text{M}$$

$$[SO_3] = \frac{0.568\ \text{mol SO}_3}{1.90\ \text{L}} = 0.299\ \text{M} \qquad Q_c = \frac{[SO_3]^2}{[SO_2]^2[O_2]} = \frac{(0.299)^2}{(0.239)^2\,0.0963} = 16.3$$

Since $Q_c = 16.3 \neq 2.8 \times 10^2 = K_c$, this mixture is not at equilibrium.

(b) Since the value of Q_c is smaller than that of K_c, the reaction will proceed to the right, forming products to reach equilibrium.

19. Increasing the volume of an equilibrium mixture causes that mixture to shift toward the side (reactants or products) where the sum of the stoichiometric coefficients of the gaseous species is the larger. That is: shifts to the right if $\Delta n_{gas} > 0$, shifts to the left if $\Delta n_{gas} < 0$, and does not shift if $\Delta n_{gas} = 0$

(a) $C(s) + H_2O(g) \rightleftharpoons CO(g) + H_2(g), \Delta n_{gas} > 0$, shift right, toward products

(b) $Ca(OH)_2(s) + CO_2(g) \rightleftharpoons CaCO_3(s) + H_2O(g), \quad \Delta n_{gas} = 0$, no shift, no change in equilibrium position.

(c) $4\,NH_3(g) + 5\,O_2(g) \rightleftharpoons 4\,NO(g) + 6\,H_2O(g), \quad \Delta n_{gas} > 0$, shift right, form products

20. The equilibrium position for a reaction that is exothermic shifts to the left (favors reactants) when temperature is raised. For one that is endothermic, it shifts right (favors products) when temperature is raised.

(a) $NO(g) \rightleftharpoons \frac{1}{2} N_2(g) + \frac{1}{2} O_2(g)$ $\Delta H^\circ = -90.2$ kJ shifts left, % dissociation $\downarrow$

(b) $SO_3(g) \rightleftharpoons SO_2(g) + \frac{1}{2} O_2(g)$ $\Delta H^\circ = +98.9$ kJ shifts right, % dissociation $\uparrow$

(c) $N_2H_4(g) \rightleftharpoons N_2(g) + 2\,H_2(g)$ $\Delta H^\circ = -95.4$ kJ shifts left, % dissociation $\downarrow$

(d) $COCl_2(g) \rightleftharpoons CO(g) + Cl_2(g)$ $\Delta H^\circ = +108.3$ kJ shifts right, % dissociation $\uparrow$

21. $4 HCl(g) + O_2(g) \rightleftharpoons 2 H_2O(g) + 2 Cl_2(g)$ $\Delta H° = -114$ kJ

(a) Adding $O_2(g)$ to the equilibrium mixture at constant volume will cause the position of equilibrium to shift to the right, increasing the equilibrium amount of $Cl_2(g)$.

(b) Removing HCl(g) from the equilibrium mixture at constant volume will cause the position of equilibrium to shift to the left, decreasing the amount of $Cl_2(g)$.

(c) Because $\Delta n_{gas} < 0$, transferring the equilibrium mixture to a container of twice the volume will cause the position of equilibrium to shift to the left, decreasing the amount of $Cl_2(g)$.

(d) Adding a catalyst to a mixture at equilibrium has no effect on the equilibrium position.

(e) Raising the temperature of this exothermic reaction will cause the equilibrium position to shift to the left, decreasing the amount of $Cl_2(g)$.

22. The information for the calculation is organized around the chemical equation. Let $x =$ mol H_2 (or I_2) that reacts. Then use stoichiometry to determine the amount of HI formed, in terms of x, and finally solve for x.

Equation: $H_2(g)$ + $I_2(g) \rightleftharpoons$ $2 HI(g)$

Initial: 0.150 mol 0.150 mol 0.000 mol

Changes: $-x$ mol $-x$ mol $+2x$ mol

Equil: $0.150-x$ $0.150-x$ $2x$

$$K_c = \frac{[HI]^2}{[H_2][I_2]} = \frac{\left(\dfrac{2x}{3.25\ L}\right)^2}{\dfrac{0.150-x}{3.25\,L} \times \dfrac{0.150-x}{3.25\ L}}$$

Then take the square root of both sides: $\sqrt{K_c} = \sqrt{50.2} = \dfrac{2x}{0.150-x} = 7.09$

$2x = 1.06 - 7.09x$ $x = \dfrac{1.06}{9.09} = 0.117$ mol, amount HI $= 2x = 2 \times 0.117$ mol $= 0.234$ mol HI

amount $H_2 =$ amount $I_2 = (0.150 - x)$ mol $= (0.150 - 0.117)$ mol $= 0.033$ mol H_2 or I_2

23. This calculation is set up in a manner similar to the calculation of Review Question 22. In this case, we let x equal the amount of I_2 that is formed.

Equation: $H_2(g)$ + $I_2(g)$ $\rightleftharpoons$ $2 HI(g)$

Initial: 0.000 mol 0.000 mol 0.150 mol

Changes: $+x$ mol $+x$ mol $-2x$ mol

Equil. x x $0.150-2x$

$$K_c = 50.2 = \frac{[HI]^2}{[H_2][I_2]} = \frac{\left(\dfrac{0.150-2x}{3.25\,L}\right)^2}{\dfrac{x}{3.25\,L} \times \dfrac{x}{3.25\,L}} = \frac{(0.150-2x)^2}{x^2} \qquad \sqrt{50.2} = \frac{0.150-2x}{x} = 7.09$$

$0.150 - 2x = 7.09x$ $x = \dfrac{0.150}{2 + 7.09} = 0.0165$ amount $I_2 = x = 0.0165$ mol I_2

24. We first determine the initial pressure of NH_3.

$$P\{NH_3(g)\} = \frac{nRT}{V} = \frac{0.100 \text{ mol } NH_3 \times 0.08206 \text{ L atm mol}^{-1} \text{ K}^{-1} \times 298 \text{ K}}{2.58 \text{ L}} = 0.948 \text{ atm}$$

Equation:	$NH_4HS(s)$	$\rightleftharpoons$	$NH_3(g)$	+	$H_2S(g)$
Initial:			0.948 atm		0 atm
Changes:			$+x$ atm		$+x$ atm
Equil:			$(0.948+x)$ atm		x atm

$$K_p = P\{NH_3\} P\{H_2S\} = 0.108 = (0.948+x)x = 0.948x + x^2 \quad 0 = x^2 + 0.948x - 0.108$$

$$x = \frac{-b \pm \sqrt{b^2 - 4ac}}{2a} = \frac{-0.948 \pm \sqrt{0.899 + 0.432}}{2} = 0.103 \text{ atm}, -1.05 \text{ atm}$$

The negative root makes no physical sense. The total gas pressure is obtained as follows.

$$P_{tot} = P\{NH_3\} + P\{H_2 S\} = (0.948+x) + x = 0.948 + 2x = 0.948 + 2 \times 0.103 = 1.154 \text{ atm}$$

EXERCISES

Writing Equilibrium Constant Expressions

25. **(a)** $2 COF_2(g) \rightleftharpoons CO_2(g) + CF_4(g)$ $\qquad K_c = \dfrac{[CO_2][CF_4]}{[COF_2]^2}$

(b) $Cu(s) + 2 Ag^+(aq) \rightleftharpoons Cu^{2+}(aq) + 2 Ag(s)$ $\qquad K_c = \dfrac{[Cu^{2+}]}{[Ag^+]^2}$

(c) $S_2O_8^{2-}(aq) + 2 Fe^{2+}(aq) \rightleftharpoons 2 SO_4^{2-}(aq) + 2 Fe^{3+}(aq)$ $\qquad K_c = \dfrac{[SO_4^{2-}]^2[Fe^{3+}]^2}{[S_2O_8^{2-}][Fe^{2+}]^2}$

26. **(a)** $4 NH_3(g) + 3 O_2(g) \rightleftharpoons 2 N_2(g) + 6 H_2O(g)$ $\qquad K_c = \dfrac{[N_2]^2[H_2O]^6}{[NH_3]^4[O_2]^3}$

(b) $7 H_2(g) + 2 NO_2(g) \rightleftharpoons 2 NH_3(g) + 4 H_2O(g)$ $\qquad K_c = \dfrac{[NH_3]^2[H_2O]^4}{[H_2]^7[NO_2]^2}$

(c) $N_2(g) + Na_2CO_3(s) + 4 C(s) \rightleftharpoons 2 NaCN(s) + 3 CO(g)$ $\qquad K_c = \dfrac{[CO]^3}{[N_2]}$

27. Since $K_p = K_c(RT)^{\Delta n_g}$, it is also true that $K_c = K_p(RT)^{-\Delta n_g}$.

(a) $K_c = \dfrac{[SO_2(g)][Cl_2(g)]}{[SO_2Cl_2(g)]} = K_p(RT)^{-(+1)} = 2.9 \times 10^{-2}(0.0821 \times 303)^{-1} = 0.0012$

(b) $K_c = \dfrac{[NO_2]^2}{[NO]^2[O_2]} = K_p(RT)^{-(-1)} = (1.48 \times 10^4)(0.0821 \times 457) = 5.55 \times 10^5$

(c) $K_c = \dfrac{[H_2S]^3}{[H_2]^3} = K_p(RT)^0 = K_p = 0.429$

28. $K_p = K_c(RT)^{\Delta n_g}$, with $R = 0.0821$ L $\cdot$ atm mol^{-1} K^{-1}

(a) $K_p = \dfrac{P\{NO_2\}^2}{P\{N_2O_4\}} = K_c(RT)^{+1} = 4.61 \times 10^{-3}(0.0821 \times 298)^1 = 0.113$

(b) $K_p = \dfrac{P\{C_2H_2\}P\{H_2\}^3}{P\{CH_4\}^2} = K_c(RT)^{(+2)} = (0.154)(0.0821 \times 2000)^2 = 4.15 \times 10^3$

(c) $K_p = \dfrac{P\{H_2\}^4 P\{CS_2\}}{P\{H_2S\}^2 P\{CH_4\}} = K_c(RT)^{(+2)} = (5.27 \times 10^{-8})(0.0821 \times 973)^2 = 3.36 \times 10^{-4}$

29. The equilibrium reaction is $H_2O(l) \rightleftharpoons H_2O(g)$ with $\Delta n_{gas} = +1$. $K_p = K_c(RT)^{\Delta n_g}$ gives

$K_c = K_p(RT)^{-\Delta n_g}$.

$K_p = P\{H_2O\} = 23.8 \text{ mmHg} \times \dfrac{1 \text{ atm}}{760 \text{ mmHg}} = 0.0313$

$K_c = K_p(RT)^{-+1} = \dfrac{K_p}{RT} = \dfrac{0.0313}{0.0821 \times 298} = 1.28 \times 10^{-3}$

30. The equilibrium rxn is $C_6H_6(l) \rightleftharpoons C_6H_6(g)$ with $\Delta n_{gas} = +1$. Using $K_p = K_c(RT)^{\Delta n_g}$,

$K_p = K_c(RT) = 5.12 \times 10^{-3}(0.08206 \times 298) = 0.125 = P\{C_6H_6\}$

$P\{C_6H_6\} = 0.125 \text{ atm} \times \dfrac{760 \text{ mmHg}}{1 \text{ atm}} = 95.0 \text{ mmHg}$

31. We combine the several given reactions to obtain the net reaction.

$2\,N_2O(g) \rightleftharpoons 2\,N_2(g) + O_2(g) \qquad K_c = \dfrac{1}{(2.7 \times 10^{-18})^2}$

$4\,NO_2(g) \rightleftharpoons 2\,N_2O_4(g) \qquad K_c = \dfrac{1}{(4.6 \times 10^{-3})^2}$

$2\,N_2(g) + 4\,O_2(g) \rightleftharpoons 4\,NO_2(g) \qquad K_c = (4.1 \times 10^{-9})^4$

net : $2\,N_2O(g) + 3\,O_2(g) \rightleftharpoons 2\,N_2O_4(g) \quad K_{c(Net)} = \dfrac{(4.1 \times 10^{-9})^4}{(2.7 \times 10^{-18})^2 (4.6 \times 10^{-3})^2} = 1.8 \times 10^6$

32. We combine the K_c values to obtain the value of K_c for the overall reaction, and then convert this to a value for K_p.

$$2\,CO_2\,(g)+2H_2\,(g)\rightleftharpoons 2\,CO(g)+2\,H_2O(g) \qquad K_c = (1.4)^2$$

$$2\,C(graphite)+O_2\,(g)\rightleftharpoons 2\,CO(g) \qquad K_c = (1\times10^8)^2$$

$$4\,CO(g)\rightleftharpoons 2\,C(graphite)+2\,CO_2\,(g) \qquad K_c = \dfrac{1}{(0.64)^2}$$

net: $\quad 2H_2\,(g)+O_2\,(g)\rightleftharpoons 2H_2O(g) \qquad K_{c(Net)} = \dfrac{(1.4)^2(1\times10^8)^2}{(0.64)^2}=5\times10^6$

$$K_p = K_c\,(RT)^{\Delta n} = \dfrac{K_c}{RT} = \dfrac{5\times10^{16}}{0.08206\times1200} = 5\times10^{14}$$

Experimental Determination of Equilibrium Constants

33. First, we determine the concentration of PCl_5 and of Cl_2 present initially and at equilibrium, respectively. Then we use the balanced equation to help us determine the concentration of each species present at equilibrium.

$$[PCl_5] = \dfrac{1.00\times10^{-3}\text{ mol } PCl_5}{0.250\text{ L}} = 0.00400\text{ M} \qquad [Cl_2] = \dfrac{9.65\times10^{-4}\text{ mol } Cl_2}{0.250\text{ L}} = 0.00386\text{ M}$$

Equation: $PCl_5\,(g)\rightleftharpoons PCl_3\,(g) + Cl_2\,(g)$

$$K_c = \dfrac{[PC_3][Cl_2]}{[PCl_5]} = \dfrac{(0.00386\,M)(0.00386\,M)}{0.00014\,M} = 0.106$$

34. First we determine the partial pressure of each gas.

$$P_{initial}\{H_2\,(g)\} = \dfrac{nRT}{V} = \dfrac{1.00\,g\,H_2\times\dfrac{1\,mol\,H_2}{2.016\,g\,H_2}\times\dfrac{0.08206\,L\,atm}{mol\,K}\times1670\,K}{0.500\,L} = 136\,atm$$

$$P_{initial}\{H_2S(g)\} = \dfrac{nRT}{V} = \dfrac{1.06\,g\,H_2S\times\dfrac{1\,mol\,H_2S}{34.08\,g\,H_2S}\times\dfrac{0.08206\,L\,atm}{mol\,K}\times1670\,K}{0.500\,L} = 8.52\,atm$$

$$P_{equil}\{S_2\,(g)\} = \dfrac{nRT}{V} = \dfrac{8.00\times10^{-6}\,mol\,S_2\times\dfrac{0.08206\,L\,atm}{mol\,K}\times1670\,K}{0.500\,L} = 2.19\times10^{-3}\,atm$$

Equation:	$2\,H_2\,(g)$	$+\quad S_2\,(g)$	$\rightleftharpoons$	$2\,H_2S(g)$
Initial :	136 atm	0 atm		8.52 atm
Changes :	+0.00438 atm	0.00219 atm		−0.00438 atm
Equil :	136 atm	0.00219 atm		8.52 atm

$$K_p = \frac{P\{H_2S(g)\}^2}{P\{H_2(g)\}^2\,P\{S_2(g)\}} = \frac{(8.52)^2}{(136)^2\,0.00219} = 1.79$$

Equilibrium Relationships

35. $K_c = 281 = \dfrac{[SO_3]^2}{[SO_2]^2[O_2]} = \dfrac{[SO_3]^2}{[SO_2]^2} \times \dfrac{0.185\ L}{0.00247\ mol}$ $\quad \dfrac{[SO_2]}{[SO_3]} = \sqrt{\dfrac{0.185}{0.00247 \times 281}} = 0.516$

36. $K_c = 0.011 = \dfrac{[I]^2}{[I_2]} = \dfrac{\left(\dfrac{0.37\ mol\ L}{V}\right)^2}{\dfrac{1.00\ mol\ I_2}{V}} = \dfrac{1}{V} \times 0.14$ $\quad V = \dfrac{0.14}{0.011} = 13\ L$

37. **(a)** A possible equation for the oxidation of $NH_3(g)$ to $NO_2(g)$ follows.

$$NH_3(g) + \tfrac{7}{4}O_2(g) \rightleftharpoons NO_2(g) + \tfrac{3}{2}H_2O(g)$$

(b) We obtain K_p for the reaction in part (a) by appropriately combining the values of K_p given in the problem.

$$NH_3(g) + \tfrac{5}{4}O_2(g) \rightleftharpoons NO(g) + \tfrac{3}{2}H_2O(g) \qquad\qquad K_p = 2.11 \times 10^{19}$$

$$NO(g) + \tfrac{1}{2}O_2(g) \rightleftharpoons NO_2(g) \qquad\qquad K_p = \dfrac{1}{0.524}$$

$$\text{net: } NH_3(g) + \tfrac{7}{4}O_2(g) \rightleftharpoons NO_2(g) + \tfrac{3}{2}H_2O(g) \quad K_p = \dfrac{2.11 \times 10^{19}}{0.524} = 4.03 \times 10^{19}$$

38. **(a)** We first determine $[H_2]$ and $[CH_4]$, and then $[C_2H_2]$. $[CH_4] = [H_2] = \dfrac{0.10\ mol}{1.00\ L} = 0.10\ M$

$$K_c = \dfrac{[C_2H_2][H_2]^3}{[CH_4]^2} \quad [C_2H_2] = \dfrac{K_c[CH_4]^2}{[H_2]^3} = \dfrac{0.154 \times 0.10^2}{0.10^3} = 1.5\ M$$

In a 1.00 L container, each concentration numerically equals the amount of substance.

$$\chi\{C_2H_2\} = \dfrac{1.5\ mol\ C_2H_2}{1.5\ mol\ C_2H_2 + 0.10\ mol\ CH_4 + 0.10\ mol\ H_2} = 0.88$$

(b) The conversion of $CH_4(g)$ to $C_2H_2(g)$ is favored at low pressures, since the conversion reaction has a larger sum of the stoichiometric coefficients of gaseous products (4) than of reactants (2).

(c) Initially, all concentrations are halved when the mixture is transferred to a flask that is twice as large. In re-establishing equilibrium the system reacts to the right, forming more moles of gas. We base our solution on the balanced chemical equation, in the manner we have used before.

Equation:	$2\,CH_4(g)$	$\rightleftharpoons$	$C_2H_2(g)$	$+$	$3\,H_2$
Initial:	$\dfrac{0.10\,mol}{2.00\,L}$		$\dfrac{1.5\,mol}{2.00\,L}$		$\dfrac{0.10\,mol}{2.00\,L}$
	$=0.050\ M$		$=0.75\ M$		$=0.050\ M$
Changes:	$-2x\ M$		$+x\ M$		$+3x\ M$
Equil:	$(0.050-2x)\ M$		$(0.0750+x)\ M$		$(0.050+3x)\ M$

$$K_c = \frac{[C_2H_2][H_2]^3}{[CH_4]^2} = \frac{(0.050+3x)^3(0.750+x)}{(0.050-2x)^2} = 0.154$$

We solve this fourth-order equation by successive approximations. First we guess that $x=0.010\ M$.

$$x=0.010 \quad Q_c = \frac{(0.050+3(0.010))^3(0.750+0.010)}{(0.050-2(0.010))^2} = \frac{(0.080)^3(0.760)}{(0.030)^2} = 0.433 > 0.154$$

$$x=0.020 \quad Q_c = \frac{(0.050+3(0.020))^3(0.750+0.020)}{(0.050-2(0.020))^2} = \frac{(0.110)^3(0.770)}{(0.010)^2} = 10.2 > 0.154$$

$$x=0.005 \quad Q_c = \frac{(0.050+3(0.005))^3(0.750+0.005)}{(0.050-2(0.005))^2} = \frac{(0.065)^3(0.755)}{(0.040)^2} = 0.129 < 0.154$$

$$x=0.006 \quad Q_c = \frac{(0.050+3(0.006))^3(0.750+0.006)}{(0.050-2(0.006))^2} = \frac{(0.068)^3(0.756)}{(0.038)^2} = 0.165 > 0.154$$

This is the maximum number of significant figures our system permits. We have $x=0.006\ M$. $[CH_4]=0.038\ M$; $[C_2H_2]=0.756\ M$; $[H_2]=0.068\ M$
Because we have a 2.00 L container, molar amounts are double the numerical value of molarities.

$$2.00\ L \times \frac{0.756\ mol\ C_2H_2}{1\ L} = 1.51\ mol\ C_2H_2 \qquad 2.00\ L \times \frac{0.038\ mol\ CH_4}{1\ L} = 0.076\ mol\ CH_4$$

$$2.00\ L \times \frac{0.068\ mol\ H_2}{1\ L} = 0.14\ mol\ H_2$$

Thus, the increased volume allows for the production of some additional C_2H_2.

Direction and Extent of Chemical Change

39. We compute the value of Q_c for the given amounts of product and reactants.

$$Q_c = \frac{[SO_3]^2}{[SO_2]^2[O_2]} = \frac{\left(\dfrac{1.8\,\text{mol }SO_3}{7.2\,\text{L}}\right)^2}{\left(\dfrac{3.6\,\text{mol }SO_2}{7.2\,\text{L}}\right)^2 \dfrac{2.2\,\text{mol }O_2}{7.2\,\text{L}}} = 0.82 < K_c = 100$$

The mixture described cannot be maintained indefinitely. In fact, because $Q_c < K_c$, the reaction will proceed to the right, that is, toward products, until equilibrium is established.

40. We compute the value of Q_c for the given amounts of product and reactants.

$$Q_c = \frac{[NO_2]^2}{[N_2O_2]} = \frac{\left(\dfrac{0.0205\,\text{mol }NO_2}{5.25\,\text{L}}\right)^2}{\dfrac{0.750\,\text{mol }N_2O_4}{5.25\,\text{L}}} = 1.07 \times 10^{-4} < K_c = 4.61 \times 10^{-3}$$

The mixture described cannot be maintained indefinitely. In fact, because $Q_c < K_c$, the reaction will proceed to the right, that is, toward products, until equilibrium is established.

41. We use the balanced chemical equation as a basis to organize the information we have about the reaction.

Equation:	$SbCl_5(g)$	$\rightleftharpoons$	$SbCl_3(g)$	+	$Cl_2(g)$
Initial:	$\dfrac{0.00\,\text{mol}}{2.50\,\text{L}}$		$\dfrac{0.280\,\text{mol}}{2.50\,\text{L}}$		$\dfrac{0.160\,\text{mol}}{2.50\,\text{L}}$
Initial:	0.000 M		0.112 M		0.0640 M
Changes:	$+x$ M		$-x$ M		$-x$ M
Equil:	x M		$(0.112-x)$ M		$(0.0640-x)$ M

$$K_c = 0.025 = \frac{[SbCl_3][Cl_2]}{[SbCl_5]} = \frac{(0.112-x)(0.0640-x)}{x} = \frac{0.00717 - 0.176x + x^2}{x}$$

$$0.025x = 0.00717 - 0.176x + x^2 \qquad x^2 - 0.201x + 0.00717 = 0$$

$$x = \frac{-b \pm \sqrt{b^2 - 4ac}}{2a} = \frac{0.201 \pm \sqrt{0.0404 - 0.0287}}{2} = 0.0464 \text{ or } 0.155$$

The second of the two values for x gives a negative value of $[Cl_2](= -0.091\,\text{M})$, and thus is physically meaningless. Thus, concentrations and amounts follow.

$[SbCl_5] = x = 0.0464$ M   amount $SbCl_5 = 2.50$ L $\times$ 0.0464 M = 0.116 mol $SbCl_5$

$[SbCl_3] = 0.112 - x = 0.066$ M   amount $SbCl_3 = 2.50$ L $\times$ 0.066 M = 0.17 mol $SbCl_3$

$[Cl_2] = 0.0640 - x = 0.0176$ M   amount $Cl_2 = 2.50$ L $\times$ 0.0176 M = 0.0440 mol Cl_2

Or, in the first line labeled "Initial," we could have set $y =$ number of moles of $SbCl_5$ produced. Then our final answer would be $y = 0.116$ mol $SbCl_5$. That is, we would have obtained the answer more directly.

42. Use the chemical equation as a basis to organize the information we have about the reaction, and then determine the final number of moles of $Cl_2(g)$ present.

Equation: $CO(g)$ + $Cl_2(g)$ $\rightleftharpoons$ $COCl_2(g)$

Initial: 0.3500 mol   0.0000 mol   0.05500 mol

Changes: $+x$ mol $+x$ mol $-x$ mol

Equil.: $(0.3500+x)$ mol x mol $(0.05500-x)$ mol

$$K_c = 1.2\times10^3 = \frac{[COCl_2]}{[CO][Cl_2]} = \frac{\dfrac{(0.0550-x)\,\text{mol}}{3.050\,\text{L}}}{\dfrac{(0.3500+x)\text{mol}}{3.050\,\text{L}}\times\dfrac{x\,\text{mol}}{3.050\,\text{L}}}$$

$$\frac{1.2\times10^3}{3.050} = \frac{0.05500-x}{(0.3500+x)x}$$

We assume that $x \ll 0.0550$ This produces the following expression.

$$\frac{1.2\times10^3}{3.050} = \frac{0.05500}{0.3500\,x} \qquad x = \frac{3.050\times0.05500}{0.3500\times1.2\times10^3} = 4.0\times10^{-4} \text{ mol } Cl_2$$

We use the first value we obtained 4.0×10^{-4} $(= 0.00040)$ to arrive at a second value.

$$x = \frac{3.050\times(0.0550-0.00040)}{(0.3500+0.00040)\times1.2\times10^3} = 4.0\times10^{-4} \text{ mol } Cl_2$$

Because the value did not change on the second iteration, we have arrived at a solution.

43. We first compute the initial concentration of each species present. Then we determine the equilibrium concentrations of all species. Finally, we compute the mass of CO_2 present at equilibrium.

$$[CO]_{int} = \frac{1.00\text{ g}}{1.41\text{ L}}\times\frac{1\text{ mol CO}}{28.01\text{ g CO}} = 0.0253 \text{ M} \qquad [H_2O]_{int} = \frac{1.00\text{ g}}{1.41\text{ L}}\times\frac{1\text{ mol }H_2O}{18.02\text{ g }H_2O} = 0.0394 \text{ M}$$

$$[H_2]_{int} = \frac{1.00\text{ g}}{1.41\text{ L}}\times\frac{1\text{ mol }H_2}{2.016\text{ g }H_2} = 0.352 \text{ M}$$

Equation: $CO(g)$ + $H_2O(g)$ $\rightleftharpoons$ $CO_2(g)$ + $H_2(g)$

Initial: 0.0253 M   0.0394 M   0.0000 M   0.352 M

Changes: $-x$ M $-x$ M $+x$ M $+x$ M

Equil: $(0.0253-x)$ M $(0.0394-x)$ M x M $(0.352+x)$ M

$$K_c = \frac{[CO_2][H_2]}{[CO][H_2O]} = 23.2 = \frac{x(0.352+x)}{(0.0253-x)(0.0394-x)} = \frac{0.352x+x^2}{0.000997-0.0647x+x^2}$$

$$0.0231-1.50x+23.2x^2 = 0.352x+x^2 \qquad 22.2x^2 -1.852x+0.0231 = 0$$

$$x = \frac{-b\pm\sqrt{b^2-4ac}}{2a} = \frac{1.852\pm\sqrt{3.430-2.051}}{44.4} = 0.0682 \text{ M}, \ 0.0153 \text{ M}$$

The first value of x gives a negative $[CO](=-0.0429 \text{ M})$ and a negative

$[H_2O](=-0.0288 \text{ M})$. Thus, $x=0.0153 \text{ M} =[CO_2]$. Now we find the mass of CO_2.

$$1.41 \text{ L}\times\frac{0.0153 \text{ mol } CO_2}{1 \text{ L mixture}}\times\frac{44.01 \text{ g } CO_2}{1 \text{ mol } CO_2} = 0.949 \text{ g } CO_2$$

44.   We base each of our solutions on the balanced chemical equation.

(a)   Equation : $PCl_5(g) \rightleftharpoons PCl_3(g) + Cl_2(g)$

Initial : $\dfrac{0.550 \text{ mol}}{2.50 \text{ L}}$ $\dfrac{0.550 \text{ mol}}{2.50 \text{ L}}$ $\dfrac{0 \text{ mol}}{2.50 \text{ L}}$

Changes : $\dfrac{-x \text{ mol}}{2.50 \text{ L}}$ $\dfrac{+x \text{ mol}}{2.50 \text{ L}}$ $\dfrac{+x \text{ mol}}{2.50 \text{ L}}$

Equil : $\dfrac{(0.550-x) \text{ mol}}{2.50 \text{ L}}$ $\dfrac{(0.550+x) \text{ mol}}{2.50 \text{ L}}$ $\dfrac{x \text{ mol}}{2.50 \text{ L}}$

$$K_c = \frac{[PCl_3][Cl_2]}{[PCl_5]} = 3.8\times10^{-2} = \frac{\dfrac{(0.550+x)\text{ mol}}{2.50\text{ L}}\times\dfrac{x\text{ mol}}{2.50\text{ L}}}{\dfrac{(0.550-x)\text{ mol}}{2.50\text{ L}}}$$

$$\frac{x(0.550+x)}{2.50(0.550-x)} = 3.8\times10^{-2}$$

$$x^2+0.550x = 0.052-0.095x \qquad x^2+0.645x-0.052 = 0$$

$$x = \frac{-b\pm\sqrt{b^2-4ac}}{2a} = \frac{-0.645\pm\sqrt{0.416+0.208}}{2} = 0.0725 \text{ mol}, \ -0.717 \text{ mol}$$

The second answer gives a negative amt. of Cl_2, and that makes no physical sense.

amount $PCl_5 = (0.550-0.0725)=0.478$ mol

amount $PCl_3 = (0.550+0.0725)=0.623$ mol

amount $Cl_2 = x = 0.0725$ mol

(b) Equation : $PCl_5(g) \rightleftharpoons PCl_3(g) + Cl_2(g)$

Initial : $\dfrac{0.610 \text{ mol}}{2.50 \text{ L}}$ 0 M   0 M

Changes : $\dfrac{-x \text{ mol}}{2.50 \text{ L}}$ $\dfrac{+x \text{ mol}}{2.50 \text{ L}}$ $\dfrac{+x \text{ mol}}{2.50 \text{ L}}$

Equil : $\dfrac{0.610 - x \text{ mol}}{2.50 \text{ L}}$ $\dfrac{(x \text{ mol})}{2.50 \text{ L}}$ $\dfrac{(x \text{ mol})}{2.50 \text{ L}}$

$$K_c = \frac{[PCl_3][Cl_2]}{[PCl_5]} = 3.8 \times 10^{-2} = \frac{\dfrac{(x \text{ mol})}{2.50 \text{ L}} \times \dfrac{(x \text{ mol})}{2.50 \text{ L}}}{\dfrac{0.610 - x \text{ mol}}{2.50 \text{ L}}}$$

$$2.50 \times 3.8 \times 10^{-2} = \frac{x^2}{0.610 - x} = 0.095$$

$$0.058 - 0.095x = x^2 \quad x^2 + 0.095x - 0.058 = 0$$

$$x = \frac{-b \pm \sqrt{b^2 - 4ac}}{2a} = \frac{-0.095 \pm \sqrt{0.0090 + 0.23}}{2} = 0.20 \text{ mol}, \; -0.29 \text{ mol}$$

amount $PCl_3 = 0.20$ mol $=$ amount Cl_2; amount $PCl_5 = 0.610 - 0.20 = 0.41$ mol

45. **(a)** We use the balanced chemical equation as a basis to organize the information we have about the reactants and products.

Equation: $2\,COF_2(g) \rightleftharpoons CO_2(g) + CF_4(g)$

Initial: $\dfrac{0.145 \text{ mol}}{5.00 \text{ L}}$ $\dfrac{0.262 \text{ mol}}{5.00 \text{ L}}$ $\dfrac{0.074 \text{ mol}}{5.00 \text{ L}}$

Initial: 0.0290 M   0.0524 M   0.0148 M

And we now compute a value of Q_c to compare with the given value of K_c.

$$Q_c = \frac{[CO_2][CF_4]}{[COF_2]^2} = \frac{(0.0524)(0.0148)}{(0.0290)^2} = 0.922 < 2.00 = K_c$$

Because Q_c is not equal to K_c, the mixture is not at equilibrium.

(b) Because Q_c is smaller than K_c, the reaction will shift right, that is, products will be formed, in reaching a state of equilibrium.

(c) We continue the organization of information about reactants and products.

Equation:	$2 COF_2(g)$	$\rightleftharpoons$	$CO_2(g)$	$+$	$CF_4(g)$
Initial:	0.0290 M		0.0524 M		0.0148 M
Changes:	$-2x$ M		$+x$ M		$+x$ M
Equil:	$(0.0290-2x)$ M		$(0.0524+x)$ M		$(0.0148+x)$ M

$$K_c = \frac{[CO_2][CF_4]}{[COF_2]^2} = \frac{(0.0524+x)(0.0148+x)}{(0.0290-2x)^2} = 2.00 = \frac{0.000776+0.0672x+x^2}{0.000841-0.1160x+4x^2}$$

$$0.00168-0.232x+8x^2 = 0.000776+0.0672x+x^2 \quad 7x^2-0.299x+0.000904 = 0$$

$$x = \frac{-b\pm\sqrt{b^2-4ac}}{2a} = \frac{0.299\pm\sqrt{0.0894-0.0253}}{14} = 0.0033 \text{ M}, 0.0394 \text{ M}$$

The second of these values for x (0.0394) gives a negative $[COF_2](=-0.0498 \text{ M})$, clearly a nonsensical result. We now compute the concentration of each species at equilibrium, and check to ensure that the equilibrium constant is satisfied.

$$[COF_2] = 0.0290-2x = 0.0290-2(0.0033) = 0.0224 \text{ M}$$

$$[CO_2] = 0.0524+x = 0.0524+0.0033 = 0.0557 \text{ M}$$

$$[CF_4] = 0.0148+x = 0.0148+0.0033 = 0.0181 \text{ M}$$

$$K_c = \frac{[CO_2][CF_4]}{[COF_2]^2} = \frac{0.0557 \text{ M} \times 0.0181 \text{ M}}{(0.0224 \text{ M})^2} = 2.01$$

The agreement of this value of K_c with the cited value (2.00) indicates that this solution is correct. Now we determine the number of moles of each species at equilibrium.

$$\text{mol } COF_2 = 5.00 \text{ L} \times 0.0224 \text{ M} = 0.112 \text{ mol } COF_2$$

$$\text{mol } CO_2 = 5.00 \text{ L} \times 0.0557 \text{ M} = 0.279 \text{ mol } CO_2$$

$$\text{mol } CF_4 = 5.00 \text{ L} \times 0.0181 \text{ M} = 0.0905 \text{ mol } CF_4$$

But suppose we had incorrectly concluded, in part (b), that reactants would be formed in reaching equilibrium. What result would we obtain? The set up follows.

Equation :	$2 COF_2(g)$	$\rightleftharpoons$	$CO_2(g)$	$+$	$CF_4(g)$
Initial :	0.0290 M		0.0524 M		0.0148 M
Changes :	$+2y$ M		$-y$ M		$-y$ M
Equil :	$(0.0290+2y)$ M		$(0.0524-y)$ M		$(0.0148-y)$ M

$$K_c = \frac{[CO_2][CF_4]}{[COF_2]^2} = \frac{(0.0524-y)(0.0148-y)}{(0.0290+2y)^2} = 2.00 = \frac{0.000776-0.0672y+y^2}{0.000841+0.1160y+4y^2}$$

$$0.00168 + 0.232y + 8y^2 = 0.000776 - 0.0672y + y^2 \quad 7y^2 + 0.299y + 0.000904 = 0$$

$$y = \frac{-b \pm \sqrt{b^2 - 4ac}}{2a} = \frac{-0.299 \pm \sqrt{0.0894 - 0.0253}}{14} = -0.0033 \text{ M}, -0.0394 \text{ M}$$

The second of these values for $x(-0.0394)$ gives a negative $[COF_2](= -0.0498 \text{ M})$, clearly a nonsensical result. We now compute the concentration of each species at equilibrium, and check to ensure that the equilibrium constant is satisfied.

$$[COF_2] = 0.0290 + 2y = 0.0290 + 2(-0.0033) = 0.0224 \text{ M}$$

$$[CO_2] = 0.0524 - y = 0.0524 + 0.0033 = 0.0557 \text{ M}$$

$$[CF_4] = 0.0148 - y = 0.0148 + 0.0033 = 0.0181 \text{ M}$$

These are the same equilibrium concentrations that we obtained by making the correct decision regarding the direction that the reaction would take. Thus, you can be assured that, if you perform the algebra correctly, it will guide you even if you make the incorrect decision about the direction of the reaction.

46. **(a)** We calculate the initial amount of each substance.

$$n\ \{C_2H_5OH\ \} = 17.2 \text{ g } C_2H_5OH \times \frac{1 \text{ mol } C_2H_5OH}{46.07 \text{ g } C_2H_5OH} = 0.373 \text{ mol } C_2H_5OH$$

$$n\{CH_3CO_2H\ \} = 23.8 \text{ g } CH_3CO_2H \times \frac{1 \text{ mol } CH_3CO_2H}{60.05 \text{ g } CH_3CO_2H} = 0.396 \text{ mol } CH_3CO_2H$$

$$n\{CH_3CO_2C_2H_5\} = 48.6 \text{ g } CH_3CO_2C_2H_5 \times \frac{1 \text{ mol } CH_3CO_2C_2H_5}{88.11 \text{ g } CH_3CO_2C_2H_5}$$

$$n\{CH_3CO_2C_2H_5\} = 0.552 \text{ mol } CH_3CO_2C_2H_5$$

$$n\{\ H_2O\ \} = 71.2 \text{ g } H_2O \times \frac{1 \text{ mol } H_2O}{18.02 \text{ g } H_2O} = 3.95 \text{ mol } H_2O$$

Since we would divide each amount by the total volume, and since there are the same numbers of product and reactant stoichiometric coefficients, we can use amounts rather than concentrations in the Q_c expression.

$$Q_c = \frac{n\{CH_3CO_2C_2H_5\ \}n\{\ H_2O\ \}}{n\{C_2H_5OH\ \}n\{\ CH_3CO_2H\ \}} = \frac{0.552 \text{ mol} \times 3.95 \text{ mol}}{0.373 \text{ mol} \times 0.396 \text{ mol}} = 14.8 > K_c = 4.0$$

Since $Q_c > K_c$ the reaction will shift to the left, forming reactants, as it attains equilibrium.

(b)   Equation: C_2H_5OH + CH_3CO_2H $\rightleftharpoons$ $CH_3CO_2C_2H_5$ + H_2O

Initial   0.373 mol   0.396 mol   0.552 mol   3.95 mol

Changes   +x mol   +x mol   −x mol   −x mol

Equil   (0.373+x) mol   (0.396+x) mol   (0.552−x) mol   (3.95−x) mol

$$K_c = \frac{(0.552-x)(3.95-x)}{(0.373+x)(0.396+x)} = \frac{2.18-4.50x+x^2}{0.148+0.769x+x^2} = 4.0$$

$$x^2-4.50x+2.18 = 4x^2+3.08x+0.59 \qquad 3x^2+7.58x-1.59=0$$

$$x = \frac{-b\pm\sqrt{b^2-4ac}}{2a} = \frac{-7.58\pm\sqrt{57+19}}{6} = 0.19 \text{ moles}, -2.72 \text{ moles}$$

Negative amounts do not make physical sense. We compute the equilibrium amount of each substance.

$$n\{C_2H_5OH\} = 0.373 \text{ mol} + 0.19 \text{ mol} = 0.56 \text{ mol } C_2H_5OH$$

$$\text{mass } C_2H_5OH = 0.56\,\text{mol } C_2H_5OH \times \frac{46.07\text{ g }C_2H_5OH}{1\text{ mol }C_2H_5OH} = 26\,\text{g }C_2H_5OH$$

$$n\{CH_3CO_2H\} = 0.396 \text{ mol} + 0.19 \text{ mol} = 0.59 \text{ mol } CH_3CO_2H$$

$$\text{mass } CH_3CO_2H = 0.59\,\text{mol } CH_3CO_2H \times \frac{60.05\text{ g }CH_3CO_2H}{1\text{ mol }CH_3CO_2H} = 35\,\text{g }CH_3CO_2H$$

$$n\{CH_3CO_2C_2H_5\} = 0.552 \text{ mol} - 0.19 \text{ mol} = 0.36 \text{ } CH_3CO_2C_2H_5$$

$$\text{mass } CH_3CO_2C_2H_5 = 0.36\,\text{mol } CH_3CO_2C_2H_5 \times \frac{88.10\text{ g }CH_3CO_2C_2H_5}{1\text{ mol }CH_3CO_2C_2H_5} = 32\,\text{g }CH_3CO_2C_2H_5$$

$$n\{H_2O\} = 3.95 \text{ mol} - 0.19 \text{ mol} = 3.76 \text{ mol } H_2O$$

$$\text{mass } H_2O = 3.76\,\text{mol } H_2O \times \frac{18.02\text{ g }H_2O}{1\text{ mol }H_2O} = 68\,\text{g }H_2O$$

$$\text{To check } K_c = \frac{n\{CH_3CO_2C_2H_5\}n\{H_2O\}}{n\{C_2H_5OH\}n\{CH_3CO_2H\}} = \frac{0.36\text{ mol}\times 3.76\text{ mol}}{0.56\text{ mol}\times 0.59\text{ mol}} = 4.1$$

47.   We organize the known data around the balanced chemical equation.

Equation: $H_2(g)$ + $I_2(g)$ $\rightleftharpoons$ $2\,HI(g)$

Initial :   0.125 mol   0.125 mol   0.000 mol

Changes: −x mol   −x mol   +2x mol

Equil :   (0.125−x) mol   (0.125−x) mol   2x mol

$$K_c = \frac{[HI]^2}{[H_2][I_2]} = \frac{\left(\frac{2x\,\text{mol}}{6.14\,\text{L}}\right)^2}{\frac{(0.125-x)\,\text{mol}}{6.14\,\text{L}} \times \frac{(0.125-x)\,\text{mol}}{6.14\,\text{L}}} = \left(\frac{2x}{0.125-x}\right)^2 = 50.2$$

$$\sqrt{50.2} = \frac{2x}{0.125-x} = 7.09 \quad 2x = 0.886 - 7.09\,x \quad 9.09\,x = 0.886 \quad x = \frac{0.886}{9.09} = 0.0975\,\text{mol}$$

amount $HI = 2x = 2 \times 0.0975$ mol $= 0.195$ mol

amount $H_2 =$ amount $I_2 = 0.125 - x = 0.125 - 0.0975 = 0.0275$ mol

Total amount $= 0.195$ mol HI $+ 0.0275$ mol $H_2 + 0.0275$ mol $I_2 = 0.250$ mol

mol% $HI = \dfrac{0.195\text{ mol HI}}{0.250\text{ mol total}} \times 100\% = 78.0\%$ HI

48. The final volume of the mixture is $0.750\text{ L} + 2.25\text{ L} = 3.00\text{ L}$. Then use the balanced chemical equation to organize the data we have concerning the reaction. The reaction should shift to the right, that is form products, in reaching a new equilibrium, since the volume is greater.

Equation:	$N_2O_4\,(g)$	$\rightleftharpoons$	$2\,NO_2\,(g)$
Initial:	$\dfrac{0.971\text{ mol}}{3.00\text{ L}}$		$\dfrac{0.0580\text{ mol}}{3.00\text{ L}}$
Initial:	0.324 M		0.0193 M
Changes:	$-x\,$M		$+2x\,$M
Equil :	$(0.324 - x)$M		$(0.0193 + 2x)$M

$$K_c = \frac{[NO_2]^2}{[N_2O_4]} = \frac{(0.0193 + 2x)^2}{0.324 - x} = \frac{0.000372 + 0.0772x + 4x^2}{0.324 - x} = 4.61 \times 10^{-3}$$

$$0.000372 + 0.0772x + 4x^2 = 0.00149 - 0.00461x \qquad 4x^2 + 0.0818x - 0.00112 = 0$$

$$x = \frac{-b \pm \sqrt{b^2 - 4ac}}{2a} = \frac{-0.0818 \pm \sqrt{0.00669 + 0.0179}}{8} = 0.00938 \text{ M}, \ -0.0298 \text{ M}$$

$[NO_2] = 0.0193 + (2 \times 0.00938) = 0.0381$ M

amount $NO_2 = 0.0381$ M $\times 3.00$ L $= 0.114$ mol NO_2

$[N_2O_4] = 0.324 - 0.00938 = 0.314_6$ M

amount $N_2O_4 = 0.314_6$ M $\times 3.00$ L $= 0.944$ mol N_2O_4

49. $[HCONH_2]_{init} = \dfrac{0.186 \text{ mol}}{2.16 \text{ L}} = 0.0861 \text{ M}$

Equation: $HCONH_2(g) \rightleftharpoons NH_3(g) + CO(g)$

Initial :   0.0861 M   0 M   0 M

Changes: $-x$ M $+x$ M $+x$ M

Equil : $(0.0861 \text{-} x)$ M x M x M

$K_c = \dfrac{[NH_3][CO]}{[HCONH_2]} = \dfrac{x \cdot x}{0.0861 - x} = 4.84$ $x^2 = 0.417 - 4.84x$ $0 = x^2 + 4.84x - 0.417$

$x = \dfrac{-b \pm \sqrt{b^2 - 4ac}}{2a} = \dfrac{-4.84 \pm \sqrt{23.4 + 1.67}}{2} = 0.084 \text{ M}, \; -4.92 \text{ M}$

The negative concentration obviously is physically meaningless. We determine the total concentration of all species, and then the total pressure.

$[\text{total}] = [NH_3] + [CO] + [HCONH_2] = x + x + 0.0861 - x = 0.0861 + 0.084 = 0.170 \text{ M}$

$P_{tot} = 0.170 \text{ mol L}^{-1} \times 0.08206 \text{ L atm mol}^{-1} \text{ K}^{-1} \times 400. \text{ K} = 5.58 \text{ atm}$

50. We determine the value of Q_p for this situation and compare it to K_p. We assume that the added solids are of negligible volume so that the initial partial pressures of $CO_2(g)$ and $H_2O(g)$ do not significantly change.

$P\{H_2O\} = \left(715 \text{ mmHg} \times \dfrac{1 \text{ atm}}{760 \text{ mmHg}}\right) = 0.941 \text{ atm H}_2O$

$Q_p = P\{CO_2\}P\{H_2O\} = 2.10 \text{ atm CO}_2 \times 0.941 \text{ atm H}_2O = 1.98 > 0.23 = K_p$

Because Q_p is larger than K_p, the reaction will proceed left toward reactants in establishing equilibrium; the partial pressures of the two gases will decrease.

51. **(a)** We organize the solution around the balanced chemical equation.

Equation: $2 Cr^{3+}(aq) + Cd(s) \rightleftharpoons 2 Cr^{2+}(aq) + Cd^{2+}(aq)$

Initial :   1.00 M   0 M   0 M

Changes: $-2x$ M $+2x$ $+x$

Equil: $(1.00 - 2x)$ M $2x$ x

$K_c = \dfrac{[Cr^{2+}]^2[Cd^{2+}]}{[Cr^{3+}]^2} = \dfrac{(2x)^2(x)}{(1.00 - 2x)^2} = 0.288$ $x = 0.257 \text{ M}$

Therefore, at equilibrium, $[Cd^{2+}] = 0.257 \text{ M}$, $[Cr^{2+}] = 0.514 \text{ M}$ and $[Cr^{3+}] = 0.486 \text{ M}$

(b) Minimum mass of $Cd(s) = 0.350L \times 0.257$ M $\times 112.41$ g/mol = 10.1 of Cd metal

52. Again we base the set up of the problem around the balanced chemical equation.

Equation: $Pb(s) + 2\,Cr^{3+}(aq) \xrightleftharpoons{K_c=3.2\times10^{-10}} Pb^{2+}(aq) + 2\,Cr^{2+}(aq)$

Initial:	0.100 M		0 M	0 M
Changes:	$-2x$ M		$+x$M	$+2x$ M
Equil :	$(0.100-2x)$M		xM	$2x$ M

$$K_c = \frac{x(2x)^2}{(0.100)^2} = 3.2\times10^{-10} \qquad 4x^3 = (0.100)^2 \times 3.2\times10^{-10} = 3.2\times10^{-12}$$

$$x = \sqrt[3]{\frac{3.2\times10^{-12}}{4}} = 9.3\times10^{-5} \text{ M}$$

Our assumption, that $2x \ll 0.100$, is valid and thus also is our result:

$\left[Pb^{2+}\right] = x = 9.3\times10^{-5}$ M, $[Cr^{2+}] = 1.9 \times 10^{-4}$ M and $[Cr^{3+}] = 0.100$ M

53. We are told in this question that the reaction $SO_2(g) + Cl_2(g) \rightleftharpoons SO_2Cl_2(g)$ has $K_c = 4.0$ at a certain temperature T. This means that at the temperature T, $[SO_2Cl_2] = 4.0 \times [Cl_2] \times [SO_2]$. Careful scrutiny of the three diagrams reveals that sketch (b) is the best representation because it contains numbers of SO_2Cl_2, SO_2 and Cl_2 molecules that are consistent with the K_c for the reaction. In other words, sketch (b) is the best choice because it contains 12 SO_2Cl_2 molecules (per unit volume), 1 Cl_2 molecule (per unit volume) and 3 SO_2 molecules (per unit volume) which, is the requisite number of each type of molecule needed to generate the expected K_c value for the reaction at temperature T.

54. In this question we are told that the reaction $2\,NO(g) + Br_2(g) \rightleftharpoons 2\,NOBr(g)$ has $K_c = 3.0$ at a certain temperature T. This means that at the temperature T, $[NOBr]^2 = 3.0 \times [Br_2][NO]^2$. Sketch (c) is the most accurate representation because it contains 18 NOBr molecules (per unit volume), 6 NO molecules (per unit volume) and 3 Br_2 molecules (per unit volume) which is the requisite number of each type of molecule needed to generate the expected K_c value for the reaction at temperature T.

Partial Pressure Equilibrium Constant, K_p

55. We substitute the given equilibrium pressure into the equilibrium constant expression and solve for the other equilibrium pressure. $K_p = \dfrac{P\{O_2\}^3}{P\{CO_2\}^2} = 28.5 = \dfrac{P\{O_2\}^3}{(0.0721 \text{ atm CO}_2)^2}$

$P\{O_2\} = \sqrt[3]{P\{O_2\}^3} = \sqrt[3]{28.5(0.0712 \text{ atm})^2} = 0.529 \text{ atm O}_2$

$P_{total} = P\{CO_2\} + P\{O_2\} = 0.0721 \text{ atm CO}_2 + 0.529 \text{ atm O}_2 = 0.601 \text{ atm total}$

56. The composition of dry air is given in volume percent. Division of these percents by 100 gives the volume fraction, which equals the mole fraction and also the partial pressure in atmospheres, if the total pressure is 1.00 atm. Thus, we have $P\{O_2\} = 0.20946$ atm and $P\{CO_2\} = 0.00036$ atm. We substitute these two values into the expression for Q_p.

$$Q_p = \frac{P\{O_2\}^3}{P\{CO_2\}^2} = \frac{(0.20946 \text{ atm } O_2)^3}{(0.00036 \text{ atm } CO_2)^2} = 7.1 \times 10^4 > 28.5 = K_p$$

The value of Q_p in the atmosphere is much larger than the value of K_p. This reaction should proceed to the left. It should only occur to the right when the pressure of O_2 drops or that of CO_2 rises (as would be the case in self-contained breathing devices).

57. **(a)** We first determine the initial pressure of each gas.

$$P\{CO\} = P\{Cl_2\} = \frac{nRT}{V} = \frac{1.00 \text{ mol} \times 0.08206 \text{ L atm mol}^{-1} \text{ K}^{-1} \times 668 \text{ K}}{1.75 \text{ L}} = 31.3 \text{ atm}$$

Then we calculate equilibrium partial pressures, organizing our calculation around the balanced chemical equation. We see that the equilibrium constant is quite large, meaning that the equilibrium position favors the product. Thus, we begin by forming as much product as possible; then proceed to equilibrium. (Significant figure rules produce a somewhat erroneous answer if we do not initially form product.)

Equation	$CO(g)$	$+ Cl_2(g)$	$\rightleftharpoons$	$COCl_2(g)$	$K_p = 22.5$
Initial:	31.3 atm	31.3 atm		0 atm	
To right:	0 atm	0 atm		31.3 atm	
Changes:	$+x$ atm	$+x$ atm		$-x$ atm	
Equil:	x atm	x atm		$(31.3 - x)$ atm	

$$K_p = \frac{P\{COCl_2\}}{P\{CO\} \, P\{Cl_2\}} = 22.5 = \frac{31.3 - x}{x^2}$$

$$22.5 \, x^2 = 31.3 - x$$

$$22.5 x^2 + x - 31.3 = 0$$

$$x = \frac{-b \pm \sqrt{b^2 - 4ac}}{2a} = \frac{-1 \pm \sqrt{1 + 2817}}{45.0} = 1.16 \text{ atm}, -1.20 \text{ atm}$$

$$P\{CO\} = P\{Cl_2\} = 1.16 \text{ atm} \qquad P\{COCl_2\} = 31.3 \text{ atm} - 1.16 \text{ atm} = 30.1 \text{ atm}$$

(b) $P_{total} = P\{CO\} + P\{Cl_2\} + P\{COCl_2\} = 1.16 \text{ atm} + 1.16 \text{ atm} + 30.1 \text{ atm} = 32.4 \text{ atm}$

58. The $I_2(s)$ maintains the presence of I_2 in the flask until it has all vaporized. Thus, if enough HI(g) is produced to completely consume the $I_2(s)$, equilibrium will not be achieved.

$$P\{H_2S\} = 747.6 \text{ mmHg} \times \frac{1 \text{ atm}}{760 \text{ mmHg}} = 0.9837 \text{ atm}$$

Equation:
$$H_2S(g) + \quad I_2(s) \rightleftharpoons \quad 2\,HI(g) + S(s)$$

Initial:	0.9837 atm	0 atm
Changes:	$-x$ atm	$+2x$ atm
Equil:	$(0.9837 - x)$ atm	$2x$ atm

$$K_p = \frac{P\{HI\}^2}{P\{H_2S\}} = \frac{(2x)^2}{(0.9837-x)} = 1.34\times10^{-5} = \frac{4x^2}{0.9837}$$

$$x = \sqrt{\frac{1.34\times10^{-5}\times0.9837}{4}} = 1.82\times10^{-3} \text{ atm}$$

Our assumption, that $0.9837 \gg x$, clearly is valid. We now verify that sufficient I_2 (s) is present by computing the mass of I_2 needed to produce the predicted pressure of HI(g). There is 1.85 g I_2 available.

$$\text{mass } I_2 = \frac{1.82\times10^{-3} \text{ atm}\times0.725 \text{ L}}{0.08206 \text{ L atm mol}^{-1}\text{ K}^{-1}\times333\text{ K}} \times \frac{1 \text{ mol } I_2}{2 \text{ mol HI}} \times \frac{253.8 \text{ g } I_2}{1 \text{ mol } I_2} = 0.00613 \text{ g } I_2$$

$$P_{tot} = P\{H_2S\} + P\{HI\} = (0.9837 - x) + 2x = 0.9837 + x = 0.9837 + 0.00182 = 0.9855 \text{ atm}$$

$$P_{tot} = 749.0 \text{ mmHg}$$

Le Châtelier's Principle

59. Continuous removal of the product, of course, has the effect of decreasing the concentration of the products below their equilibrium values. Thus, the equilibrium system is disturbed by removing the products and the system will attempt (in vain, as it turns out) to re-establish the equilibrium by shifting toward the right, that is, by producing more products.

60. We notice that the density of the solid ice is smaller than is that of liquid water. This means that the same mass of liquid water is present in a smaller volume than an equal mass of ice. Thus, if pressure is placed on ice, attempting to force it into a smaller volume, the ice will be transformed into the less-space-occupying water at $0°C$. Thus, at $0°C$ under pressure, $H_2O(s)$ will melt to form $H_2O(l)$. This behavior is *not* expected in most cases because generally a solid is *more* dense than its liquid phase.

61. (a) This reaction is exothermic with $\Delta H° = -150$. kJ. Thus, high temperatures favor the reverse reaction. The amount of H_2 (g) present at high temperatures will be less than that present at low temperatures.

(b) $H_2O(g)$ is one of the reactants involved. Introducing more will cause the equilibrium position to shift to the right, favoring products. The amount of H_2 (g) will increase.

(c) Doubling the volume of the container will favor the side of the reaction with the largest sum of gaseous stoichiometric coefficients. The sum of the stoichiometric coefficients of gaseous species is the same (4) on both sides of this reaction.

Therefore, increasing the volume of the container will have no effect on the amount of $H_2(g)$ present at equilibrium.

(d)   A catalyst merely speeds up the rate at which a reaction reaches the equilibrium position. The addition of a catalyst has no effect on the amount of $H_2(g)$ present at equilibrium.

62.   (a)   This reaction is endothermic, with $\Delta H^\circ = +92.5$ kJ. Thus, a higher temperature will favor the forward reaction and increase the amount of HI(g) present at equilibrium.

(b)   The introduction of more product will favor the reverse reaction and decrease the amount of HI(g) present at equilibrium.

(c)   The sum of the stoichiometric coefficients of gaseous products is larger than that for gaseous reactants. Increasing the volume of the container will favor the forward reaction and increase the amount of HI(g) present at equilibrium.

(d)   A catalyst merely speeds up the rate at which a reaction reaches the equilibrium position. The addition of a catalyst has no effect on the amount of HI(g) present at equilibrium.

(e)   The addition of an inert gas to the constant-volume reaction mixture will not change any partial pressures. It will have no effect on the amount of HI(g) present at equilibrium.

63.   (a)   The formation of NO(g) from the elements is an endothermic reaction ($\Delta H^\circ = +181$ kJ/mol). Since the equilibrium position of endothermic reactions is shifted toward products at higher temperatures, we expect the formation of NO(g) from the elements to be enhanced at higher temperatures.

(b)   Reaction rates always are enhanced by higher temperatures, since a larger fraction of the collisions will have an energy which surmounts the activation energy. This enhancement of rates affects both the forward and the reverse reactions. Thus, the position of equilibrium is reached more rapidly at higher temperatures than at lower temperatures.

64.   If the reaction is endothermic ($\Delta H^\circ > 0$), the forward reaction is favored at high temperatures. If the reaction is exothermic ($\Delta H^\circ < 0$), the forward reaction is favored at low temperatures.

(a) $\Delta H^\circ = \Delta H_f^\circ \left[PCl_5(g) \right] - \Delta H_f^\circ \left[PCl_3(g) \right] - \Delta H_f^\circ \left[Cl_2(g) \right]$

$\Delta H^\circ = -374.9 - (-287.0) - 0.00 = -87.9$ kJ/mol   (at low temperatures)

(b) $\Delta H^\circ = 2\Delta H_f^\circ [H_2O(g)] + 3\Delta H_f^\circ [S(\text{rhombic})] - \Delta H_f^\circ [SO_2(g)] - 2\Delta H_f^\circ [H_2S(g)]$

$\Delta H^\circ = 2(-241.8) + 3(0.00) - (-296.8) - 2(-20.63)$

$\Delta H^\circ = -145.5$ kJ/mol (at low temperatures)

(c) $\Delta H^\circ = 4\Delta H_f^\circ \left[NOCl(g) \right] + 2\Delta H_f^\circ \left[H_2O(g) \right]$

$-2\Delta H_f^\circ \left[N_2(g) \right] - 3\Delta H_f^\circ \left[O_2(g) \right] - 4\Delta H_f^\circ \left[HCl(g) \right]$

$\Delta H^\circ = 4(51.71) + 2(-241.8) - 2(0.00) - 3(0.00) - 4(-92.31)$

$\Delta H^\circ = +92.5$ kJ/mol (at higher temperatures)

65. If the total pressure of a mixture of gases at equilibrium is doubled by compression, the equilibrium will shift to the side with fewer moles of gas to counteract the increase in pressure. Thus, if the pressure of an equilibrium mixture of $N_2(g)$, $H_2(g)$ and $NH_3(g)$ is doubled, the reaction involving these three gases, i.e. $N_2(g) + 3 H_2(g) \rightleftharpoons 2 NH_3(g)$

will proceed in the forward direction to produce a new equilibrium mixture that contains additional ammonia and less molecular nitrogen and molecular hydrogen. In other words, the partial pressure of $N_2(g)$ will have decreased when equilibrium is re-established. It is important to note, however, that the final equilibrium partial pressure for the N_2 will, nevertheless, be higher than its original partial pressure prior to the doubling of the total pressure.

66. (a) Since $\Delta H^\circ = 0$, the position of the equilibrium for this reaction will not be affected by temperature. Because the position of equilibrium is expressed by the value of the equilibrium constant, we expect K_p to be unaffected by, or to remain constant with, temperature.

(b) From part (a), we know that the value of K_p will not change when the temperature is changed. The pressures of the gases, however, will change with temperature. (Recall the ideal gas law: $P = nRT/V$.) In fact, all pressures will increase. The stoichiometric coefficients in the reaction are such that higher pressure favor the formation of reactant in reattaining equilibrium. Thus, the amount of $D(g)$ will be smaller when equilibrium is reestablished at the higher temperature for the cited reaction.

$$A(s) \rightleftharpoons B(s) + 2\ C(g) + \frac{1}{2}\ D(g)$$

FEATURE PROBLEMS

88. We first determine the amount in moles of acetic acid in the equilibrium mixture.

$$\text{amount } CH_3CO_2H = 28.85 \text{ mL} \times \frac{1 \text{ L}}{1000 \text{ mL}} \times \frac{0.1000 \text{ mol } Ba(OH)_2}{1 \text{ L}} \times \frac{2 \text{ mol } CH_3CO_2H}{1 \text{ mol } Ba(OH)_2}$$

$$\times \frac{\text{complete equilibrium mixture}}{0.01 \text{ of equilibrium mixture}} = 0.5770 \text{ mol } CH_3CO_2H$$

Equation: C_2H_5OH + CH_3CO_2H $\rightleftharpoons$ $CH_3CO_2C_2H_5$ + H_2O

Initial: 0.500 mol   1.000 mol   0.000 mol   0.000 mol

Changes: −0.423 mol   −0.423 mol   +0.423 mol   +0.423 mol

Equil: 0.077 mol   0.577 mol   0.423 mol   0.423 mol

$$K_c = \frac{[CH_3CO_2C_2H_5][H_2O]}{[C_2H_5OH][CH_3CO_2H]} = \frac{\dfrac{0.423\,mol}{V} \times \dfrac{0.423\,mol}{V}}{\dfrac{0.077\,mol}{V} \times \dfrac{0.577\,mol}{V}} = \frac{0.423 \times 0.423}{0.077 \times 0.577} = 4.0$$

89. In order to determine whether or not equilibrium has been established in each bulb, we need to calculate the concentrations for all three species at the time of opening. The results from these calculations are tabulated below and a typical calculation is given beneath this table.

Bulb #	Time Bulb opened (hours)	initial amount HI(g) (in mmol)	amount of $I_2(g)$ and $H_2(g)$ at time of opening (in mmol)	amount HI(g) at time of opening (in mmol)	[HI] (mM)	$[I_2]$ & $[H_2]$ (mM)	$\dfrac{[H_2][I_2]}{[HI]^2}$
1	2	2.345	0.1572	2.03	5.08	0.393	0.00599
2	4	2.518	0.2093	2.10	5.25	0.523	0.00992
3	12	2.463	0.2423	1.98	4.95	0.606	0.0150
4	20	3.174	0.3113	2.55	6.38	0.778	0.0149
5	40	2.189	0.2151	1.76	4.40	0.538	0.0150

Consider for instance bulb #4 (opened after 20 hours)

Initial moles of HI(g) = 0.406 g HI(g) $\times \dfrac{1\ \text{mole HI}}{127.9\ \text{g HI}}$ = 0.003174 mol HI(g) or 3.174 mmol

moles of $I_2(g)$ present in bulb when opened.

= 0.04150 L $Na_2S_2O_3 \times \dfrac{0.0150\ \text{mol } Na_2S_2O_3}{1\ \text{L } Na_2S_2O_3} \times \dfrac{1\ \text{mol } I_2}{2\ \text{mol } Na_2S_2O_3}$ = 3.113 × 10^{-4} mol I_2

millimoles of $I_2(g)$ present in bulb when opened = 3.113 × 10^{-4} mol I_2

moles of H_2 present in bulb when opened = moles of $I_2(g)$ present in bulb when opened.

HI reacted = 3.113 × 10^{-4} mol $I_2 \times \dfrac{2\ \text{mole HI}}{1\ \text{mol } I_2}$ = 6.226 × 10^{-4} mol HI (0.6226 mmol HI)

moles of HI(g) in bulb when opened= 3.174 mmol HI − 0.6226 mmol HI = 2.55 mmol HI

Concentrations of HI, I_2 and H_2

[HI] = 2.55 mmol HI ÷ 400. mL = 6.38 mM

$[I_2] = [H_2] = 0.3113$ mmol $\div$ 400 mL = 0.778 mM

Ratio: $\dfrac{[H_2][I_2]}{[HI]^2} = \dfrac{(0.778 \text{ mM})(0.778 \text{ mM})}{(6.38 \text{ mM})^2} = 0.0149$

As the time increases, the ratio $\dfrac{[H_2][I_2]}{[HI]^2}$ initially climbs sharply, but then plateaus at

0.0150 somewhere between 4 and 12 hours. Consequently, it seems quite reasonable to conclude that the reaction $2HI(g) \rightleftharpoons H_2(g) + I_2(g)$ has a $K_c \sim 0.015$ at 623 K

90.   We first need to determine the number of moles of ammonia that were present in the sample of gas that left the reactor. This will be accomplished by using the data from the titrations involving HCl(aq).

Original number of moles of HCl(aq) in the 20.00 mL sample

$= 0.01872$ L of KOH $\times \dfrac{0.0523 \text{ mol KOH}}{1 \text{ L KOH}} \times \dfrac{1 \text{ mol HCl}}{1 \text{ mol KOH}}$

$= 9.79\underline{06} \times 10^{-4}$ moles of HCl$_{(initially)}$

Moles of unreacted HCl(aq)

$= 0.01542$ L of KOH $\times \dfrac{0.0523 \text{ mol KOH}}{1 \text{ L KOH}} \times \dfrac{1 \text{ mol HCl}}{1 \text{ mol KOH}} =$

$8.06\underline{47} \times 10^{-4}$ moles of HCl$_{(unreacted)}$

Moles of HCl that reacted and /or moles of NH_3 present in the sample of reactor gas
$= 9.79\underline{06} \times 10^{-4}$ moles $- 8.06\underline{47} \times 10^{-4}$ moles $= 1.73 \times 10^{-4}$ mole of NH_3 (or HCl)

The remaining gas, which is a mixture of N_2 and H_2 gases, was found to occupy 1.82 L at 273.2 K and 1.00 atm. Thus, the total number of moles of N_2 and H_2 can be found via the

ideal gas law: $n_{H_2 + N_2} = \dfrac{PV}{RT} = \dfrac{(1.00 \text{ atm})(1.82 \text{ L})}{(0.08206 \dfrac{\text{L atm}}{\text{K mol}})(273.2 \text{ K})} = 0.0811\underline{8}$ moles of $(N_2 + H_2)$

According to the stoichiometry for the reaction, 2 parts NH_3 decompose to give 3 parts H_2 and 1 part N_2. Thus the non-reacting mixture must be 75% H_2 and 25% N_2.

So, the number of moles of $N_2 = 0.25 \times 0.0811\underline{8}$ moles $= 0.0203$ moles N_2 and the number of moles of $H_2 = 0.75 \times 0.0811\underline{8}$ moles $= 0.0609$ moles H_2

Before we can calculate K_c, we need to determine the volume that the NH_3, N_2 and H_2 molecules occupied in the reactor. Once again, the ideal gas law ($PV = nRT$) will be employed. $n_{gas} = 0.0811\underline{8}$ moles ($N_2 + H_2$)$+ 1.73 \times 10^{-4}$ moles $NH_3 = 0.0813\underline{5}$ moles

$$V_{gases} = \frac{nRT}{P} = \frac{(0.08135 \text{ mol})(0.08206 \frac{L \text{ atm}}{K \text{ mol}})(1174.2 \text{ K})}{30.0 \text{ atm}} = 0.261\underline{3} \text{ L}$$

So, $K_c = \dfrac{\left[\dfrac{1.73 \times 10^{-4} \text{ moles}}{0.2613 \text{ L}}\right]^2}{\left[\dfrac{0.0609 \text{ moles}}{0.2613 \text{ L}}\right]^3 \left[\dfrac{0.0203 \text{ moles}}{0.2613 \text{ L}}\right]^1} = 4.46 \times 10^{-4}$

To calculate K_p at 901 °C, we need to employ the equation $K_p = K_c(RT)^{\Delta n_{gas}}$; $\Delta n_{gas} = -2$

$K_p = 4.46 \times 10^{-4} [(0.08206 \text{ L atm K}^{-1}\text{mol}^{-1})] \times (1174.2 \text{ K})]^{-2} = 4.80 \times 10^{-8}$ at 901°C for the reaction $N_2(g) + 3 H_2(g) \rightleftharpoons 2 NH_3(g)$

91. For step 1, rate of the forward reaction = rate of the reverse reaction, so,

$k_1[I_2] = k_{-1}[I]^2$ or $\dfrac{k_1}{k_{-1}} = \dfrac{[I]^2}{[I_2]} = K_c$ (step 1)

Like the first step, the rates for the forward and reverse reactions are equal in the second step and thus,

$k_2[I]^2[H_2] = k_{-2}[HI]^2$ or $\dfrac{k_2}{k_{-2}} = \dfrac{[HI]^2}{[I]^2[H_2]} = K_c$ (step 2)

Now we combine the two elementary steps to obtain the overall equation and its associated equilibrium constant.

$I_2(g) \rightleftharpoons 2 I(g)$ $\qquad K_c = \dfrac{k_1}{k_{-1}} = \dfrac{[I]^2}{[I_2]}$ (STEP 1)

and

$H_2(g) + 2 I(g) \rightleftharpoons 2 HI(g)$ $\qquad K_c = \dfrac{k_2}{k_{-2}} = \dfrac{[HI]^2}{[I]^2[H_2]}$ (STEP 2)

$H_2(g) + I_2(g) \rightleftharpoons 2 HI(g)$ $\qquad K_{c(overall)} = K_{c(step\ 1)} \times K_{c(step\ 2)}$

$$K_{c(overall)} = \frac{k_1}{k_{-1}} \times \frac{k_2}{k_{-2}} = \frac{[I]^2}{[I_2]} \times \frac{[HI]^2}{[I]^2[H_2]}$$

$$K_{c(overall)} = \frac{k_1 k_2}{k_{-1} k_{-2}} = \frac{[I]^2[HI]^2}{[I]^2[I_2][H_2]} = \frac{[HI]^2}{[I_2][H_2]}$$

CHAPTER 17
ACIDS AND BASES
PRACTICE EXAMPLES

1A **(a)** In the forward direction, HF is the acid, (proton donor; forms F^-), and H_2O is the base (proton acceptor; forms H_3O^+). In the reverse direction F^- is the base (forms HF), accepting a proton from H_3O^+, which is the acid, (forms H_2O).

   (b) In the forward direction, HSO_4^- is the acid, (proton donor; forms SO_4^{2-}), and NH_3 is the base (proton acceptor; forms NH_4^+). In the reverse direction SO_4^{2-} is the base (forms HSO_4^-), accepting a proton from H_3O^+, which is the acid, (forms H_2O).

   (c) In the forward direction, HCl is the acid, (proton donor; forms Cl^-), and $C_2H_3O_2^-$ is the base (proton acceptor; forms $HC_2H_3O_2$). In the reverse direction Cl^- is the base (forms HCl), accepting a proton from $HC_2H_3O_2$, which is the acid, (forms $C_2H_3O_2^-$).

1B We know that the formulas of most acids begin with H. Thus, we identify HNO_2 and HCO_3^- as acids.

$$HNO_2(aq) + H_2O(l) \rightleftharpoons NO_2^-(aq) + H_3O^+(aq);$$
$$HCO_3^-(aq) + H_2O(l) \rightleftharpoons CO_3^{2-}(aq) + H_3O^+(aq)$$

A negatively charged species will attract a positively charged proton and act as a base. Thus PO_4^{3-} and HCO_3^- can act as bases. We also know that PO_4^{3-} is a base because it cannot act as an acid—it has no protons to donate—and we know that all three species have acid-base properties.

$$PO_4^{3-}(aq) + H_2O(l) \rightleftharpoons HPO_4^{2-}(aq) + OH^-(aq);$$
$$HCO_3^-(aq) + H_2O(l) \rightleftharpoons H_2CO_3(aq) \rightarrow CO_2 \cdot H_2O(aq) + OH^-(aq)$$

Notice that HCO_3^- is the amphiprotic species, acting as both an acid and as a base.

2A $[H_3O^+]$ is readily computed from pH: $[H_3O^+] = 10^{-pH}$ $[H_3O^+] = 10^{-2.85} = 1.4 \times 10^{-3}$ M.

$[OH^-]$ can be found in two ways: (1) from $K_w = [H_3O^+][OH^-]$, giving

$$[OH^-] = \frac{K_w}{[H_3O^+]} = \frac{1.0 \times 10^{-14}}{1.4 \times 10^{-3}} = 7.1 \times 10^{-12} \text{ M, or (2) from pH} + pOH = 14.00, \text{ giving}$$

$pOH = 14.00 - pH = 14.00 - 2.85 = 11.15$, and then $[OH^-] = 10^{-pOH} = 10^{-11.15} = 7.1 \times 10^{-12}$ M.

2B $[H_3O^+]$ is computed from pH in each case: $[H_3O^+]=10^{-pH}$

$$[H_3O^+]_{conc.}=10^{-2.50}=3.2\times10^{-3}\,M \qquad [H_3O^+]_{dil.}=10^{-3.10}=7.9\times10^{-4}\,M$$

All of the H_3O^+ in the dilute solution comes from the concentrated solution.

$$\text{amount } H_3O^+ =1.00\text{ L conc. soln}\times\frac{3.2\times10^{-3}\,\text{mol }H_3O^+}{1\text{ L conc. soln}}=3.2\times10^{-3}\,\text{mol }H_3O^+$$

Next we calculate the volume of the dilute solution.

$$\text{volume of dilute solution}=3.2\times10^{-3}\,\text{mol }H_3O^+\times\frac{1\text{ L dilute soln}}{7.9\times10^{-4}\,\text{mol }H_3O^+}=4.1\text{ L dilute soln}$$

Thus, the volume of water to be added $= 3.1$ L

Infinite dilution does not lead to infinitely small hydrogen ion concentrations. Since dilution is done with water, the pH of an infinitely dilute solution will approach that of pure water, approximately pH = 7.

3A pH is computed directly from the $[H_3O^+]$, $pH=-\log[H_3O^+]=-\log(0.0025)=2.60$.

We know that HI is a strong acid and, thus, is completely dissociated into H_3O^+ and I^-. The consequence is that $[I^-]=[H_3O^+]=0.0025$ M. $[OH^-]$ is most readily computed from pH: $pOH=14.00-pH=14.00-2.60=11.40$; $[OH^-]=10^{-pOH}=10^{-11.40}=4.0\times10^{-12}$ M

3B The number of moles of HCl(g) is calculated from the ideal gas law. Then $[H_3O^+]$ is calculated, based on the fact that HCl(aq) is a strong acid (1 mol H_3O^+ is produced from each mol of HCl).

$$\text{moles HCl(g)}=\frac{\left(747\,\text{mmHg}\times\dfrac{1\,\text{atm}}{760\,\text{mmHg}}\right)\times0.535\,\text{L}}{\dfrac{0.08206\,\text{L atm}}{\text{mol K}}\times(26.5+273.2)\,\text{K}}=0.0214\,\text{mol HCl(g)}$$

$$[H_3O^+]=\frac{0.0214\text{ mol HCl}}{0.625\text{ L soln}}\times\frac{1\text{ mol }H_3O^+}{1\text{ mol HCl}}=0.0342\,M \qquad pH=-\log(0.0342)=1.466$$

4A pH is most readily determined from $pOH=-\log[OH^-]$. Assume $Mg(OH)_2$ is a strong base.

$$[OH^-]=\frac{9.63\text{ mg }Mg(OH)_2}{100.0\text{ mL soln}}\times\frac{1000\text{ mL}}{1\text{ L}}\times\frac{1\,g}{1000\text{ mg}}\times\frac{1\text{ mol }Mg(OH)_2}{58.32\,g\,Mg(OH)_2}\times\frac{2\text{ mol }OH^-}{1\text{ mol }Mg(OH)_2}$$

$[OH^-]=0.00330\,M$; $pOH=-\log(0.00330)=2.481$

$pH=14.000-pOH=14.000-2.481=11.519$

4B Notice that KOH is a strong base, which means that each mole of KOH produces one mole of $OH^-(aq)$. First we calculate $[OH^-]$ and the pOH. We then use $pH + pOH = 14.00$ to determine pH.

$$[OH^-] = \frac{3.00\,g\,KOH}{100.00\,g\,soln} \times \frac{1\,mol\,KOH}{56.11\,g\,KOH} \times \frac{1\,mol\,OH^-}{1\,mol\,KOH} \times \frac{1.0242\,g\,soln}{1\,mL\,soln} \times \frac{1000\,mL}{1\,L} = 0.548\,M$$

$$pOH = -\log(0.548) = 0.261 \quad pH = 14.000 - pOH = 14.000 - 0.261 = 13.739$$

5A First we determine $[H_3O^+] = 10^{-pH} = 10^{-4.18} = 6.6 \times 10^{-5}$ M. Then we organize the solution around the balanced chemical equation.

Equation: $HOCl(aq) + H_2O(l) \rightleftharpoons H_3O^+(aq) + OCl^-(aq)$

Initial:	0.150 M	≈ 0 M	0 M
Changes:	-6.6×10^{-5} M	$+6.6 \times 10^{-5}$ M	$+6.6 \times 10^{-5}$ M
Equil:	0.150 M	6.6×10^{-5} M	6.6×10^{-5} M

$$K_a = \frac{[H_3O^+][OCl^-]}{[HOCl]} = \frac{(6.6 \times 10^{-5})(6.6 \times 10^{-5})}{0.150} = 2.9 \times 10^{-8}$$

5B First, we use pH to determine $[OH^-]$. $pOH = 14.00 - pH = 14.00 - 10.08 = 3.92$.

$[OH^-] = 10^{-pOH} = 10^{-3.92} = 1.2 \times 10^{-4}$ M. We determine the initial concentration of cocaine and then organize the solution around the balanced equation in the manner we have used before.

$$[C_{17}H_{21}O_4N] = \frac{0.17\,g\,C_{17}H_{21}O_4N}{100\,mL\,soln} \times \frac{1000\,mL}{1\,L} \times \frac{1\,mol\,C_{17}H_{21}O_4N}{303.36\,g\,C_{17}H_{21}O_4N} = 0.0056\,M$$

Equation: $C_{17}H_{21}O_4N(aq) + H_2O(l) \rightleftharpoons C_{17}H_{21}O_4NH^+(aq) + OH^-(aq)$

Initial:	0.0056 M	0 M	≈ 0 M
Changes:	-1.2×10^{-4} M	$+1.2 \times 10^{-4}$ M	$+1.2 \times 10^{-4}$ M
Equil:	0.0055 M	1.2×10^{-4} M	1.2×10^{-4} M

$$K_b = \frac{[C_{17}H_{21}O_4NH^+][OH^-]}{[C_{17}H_{21}O_4N]} = \frac{(1.2 \times 10^{-4})(1.2 \times 10^{-4})}{0.0055} = 2.6 \times 10^{-6}$$

6A Again we organize our solution around the balanced chemical equation.

Equation: $HC_2H_2FO_2(aq) + H_2O(l) \rightleftharpoons H_3O^+(aq) + C_2H_2FO_2^-(aq)$

Initial:	0.100 M	≈ 0M	0 M
Changes:	$-x$ M	$+x$ M	$+x$ M
Equil:	$(0.100 - x)$ M	x M	x M

$$K_a = \frac{\left[H_3O^+\right]\left[C_2H_2FO_2^-\right]}{\left[HC_2H_2FO_2^-\right]} \qquad K_a = 2.6\times10^{-3}$$

With simplifying assumption $\left[H_3O^+\right] = 0.016$ M, pH $= 1.80$

With the quadratic formula $\left[H_3O^+\right] = 0.0149$M, pH $= 1.83$

Thus, the calculated pH is considerably lower than 2.89. (Example 17-6).

6B We first determine the concentration of undissociated acid. We then use this value in a set up that is based on the balanced chemical equation.

Equation: $\quad HC_9H_7O_4\,(aq) \;+\; H_2O(l) \;\rightleftharpoons\; H_3O^+(aq) \;+\; C_9H_7O_4^-\,(aq)$
Initial: $\qquad$ 0.0171 M $\qquad\qquad\qquad\qquad\qquad$ 0 M $\qquad\qquad$ 0 M
Changes: $\qquad$ $-x$ M $\qquad\qquad\qquad\qquad\qquad$ $+x$ M $\qquad\qquad$ $+x$ M
Equil: $\qquad$ $(0.0171-x)$ M $\qquad\qquad\qquad\qquad$ x M $\qquad\qquad$ x M

$$K_a = \frac{\left[H_3O^+\right]\left[C_9H_7O_4^-\right]}{\left[HC_9H_7O_4\right]} = 3.3\times10^{-4} = \frac{x\cdot x}{0.0171-x}$$

$x^2 + 3.3\times10^{-4} - 5.6\times10^{-6} = 0$ (Solve quadratic for the positive root)

$$x = \frac{-3.3\times10^{-4} \pm \sqrt{1.1\times10^{-7} + 2.3\times10^{-5}}}{2} = 0.0022\,\text{M}; \quad \text{pH} = -\log(0.0022) = 2.66$$

7A Again we organize our solution around the balanced chemical equation.

Equation: $\quad HC_2H_2FO_2\,(aq) \;+\; H_2O(l) \;\rightleftharpoons\; H_3O^+(aq) +\; C_2H_2FO_2^-\,(aq)$
Initial: $\qquad$ 0.015 M $\qquad\qquad\qquad\qquad\quad$ ≈ 0 M $\qquad\qquad$ 0 M
Changes $\qquad$ $-x$ M $\qquad\qquad\qquad\qquad\qquad$ $+x$ M $\qquad\qquad$ $+x$ M
Equil: $\qquad$ $(0.015-x)$ M $\qquad\qquad\qquad\qquad$ x M $\qquad\qquad$ x M

$$K_a = \frac{\left[H_3O^+\right]\left[C_2H_2FO_2^-\right]}{\left[HC_2H_2FO_2\right]} = 2.6\times10^{-3} = \frac{(x)(x)}{0.015-x} \approx \frac{x^2}{0.015}$$

$x = \sqrt{x^2} = \sqrt{0.015 \times 2.6\times10^{-3}} = 0.0062\,\text{M} = [H_3O^+]$ $\qquad$ Our assumption is invalid:

0.0062 is not quite small compared to 0.015. Thus we use another cycle of successive approximations.

$$K_a = \frac{(x)(x)}{0.015-0.0062} = 2.6\times10^{-3} \quad x = \sqrt{(0.015-0.0062)\times2.6\times10^{-3}} = 0.0048\ \text{M} = [H_3O^+]$$

$$K_a = \frac{(x)(x)}{0.015-0.0048} = 2.6\times10^{-3} \quad x = \sqrt{(0.015-0.0048)\times2.6\times10^{-3}} = 0.0051\ \text{M} = [H_3O^+]$$

$$K_a = \frac{(x)(x)}{0.015 - 0.0051} = 2.6 \times 10^{-3} \quad x = \sqrt{(0.015 - 0.0051) \times 2.6 \times 10^{-3}} = 0.0051 \ M = [H_3O^+]$$

Two successive identical results is a signal that we have the solution.

$pH = -\log[H_3O^+] = -\log(0.0051) = 2.29$. The quadratic equation gives the same result (0.0051 M) as this method of successive approximations.

7B We first determine the concentration of undissociated piperidine. We then use that value in a set up based on the balanced chemical equation.

$$[C_5H_{11}N] = \frac{114 \ mg \ C_5H_{11}N}{315 \ mL \ soln} \times \frac{1 \ mmol \ C_5H_{11}N}{85.15 \ mg \ C_5H_{11}N} = 0.00425 \ M$$

Equation:	$C_5H_{11}N(aq)$	$+$	$H_2O(l)$	$\rightleftharpoons$	$C_5H_{11}NH^+(aq)$	$+$	$OH^-(aq)$
Initial:	0.00425 M				0 M		≈ 0 M
Changes:	$-x$ M				$+x$ M		$+x$ M
Equil:	$(0.00425 - x)$ M				x M		x M

$$K_b = \frac{[C_5H_{11}NH^+][OH^-]}{[C_5H_{11}N]} = 1.6 \times 10^{-3} = \frac{x \cdot x}{0.00425 - x} \approx \frac{x \cdot x}{0.00425}$$

We assumed that $x \ll 0.00425$

$x = \sqrt{0.0016 \times 0.00425} = 0.0026 \ M$ The assumption is not valid. Let's assume

$x \approx 0.0026$

$x = \sqrt{0.0016(0.00425 - 0.0026)} = 0.0016$ Let's try again, with $x \approx 0.0016$

$x = \sqrt{0.0016(0.00425 - 0.0016)} = 0.0021$ Yet another try, with $x \approx 0.0021$

$x = \sqrt{0.0016(0.00425 - 0.0021)} = 0.0019$ The last time, with $x \approx 0.0019$

$x = \sqrt{0.0016(0.00425 - 0.0019)} = 0.0019 \ M = [OH^-]$

$pOH = -\log[H_3O^+] = -\log(0.0019) = 2.72$

$pH = 14.00 - pOH = 14.00 - 2.72 = 11.28$

We could have solved the problem with the quadratic equation rather than by successive approximations. The same answer is obtained. In fact, if we substitute $x = 0.0019$ into the K_b expression, we obtain $(0.0019)^2 / (0.00425 - 0.0019) = 1.5 \times 10^{-3}$ compared to $K_b = 1.6 \times 10^{-3}$. The error is due to rounding, not to an incorrect value. Using $x = 0.0020$ gives a value of 1.8×10^{-3}, while using $x = 0.0018$ gives 1.3×10^{-3}.

8A We organize the solution around the balanced chemical equation; a M is the initial concentration of HF.

Equation: $HF(aq)$ + $H_2O(l)$ $\rightleftharpoons$ $H_3O^+(aq)$ + $F^-(aq)$

Initial: a M ≈ 0 M 0 M
Changes: $-x$ M $+x$ M $+x$ M
Equil: $(a-x)$ M x M x M

$$K_a = \frac{[H_3O^+][F^-]}{[HF]} = \frac{(x)(x)}{a-x} \approx \frac{x^2}{a} = 6.6 \times 10^{-4}$$

$$x = \sqrt{a \times 6.6 \times 10^{-4}}$$

For 0.20 M HF, $x = \sqrt{0.20 \times 6.6 \times 10^{-4}} = 0.011$ M $\%\text{dissoc} = \dfrac{0.011 \text{ M}}{0.20 \text{ M}} \times 100\% = 5.5\%$

For 0.020 M HF, $x = \sqrt{0.020 \times 6.6 \times 10^{-4}} = 0.0036$ M

We need another cycle of approximation: $x = \sqrt{(0.020 - 0.0036) \times 6.6 \times 10^{-4}} = 0.0033$ M

Yet another cycle with $x \approx 0.0033$ M : $x = \sqrt{(0.020 - 0.0033) \times 6.6 \times 10^{-4}} = 0.0033$ M

$\%\text{dissoc} = \dfrac{0.0033 \text{M}}{0.020 \text{M}} \times 100\% = 17\%$

As expected, the weak acid is more dissociated when diluted.

8B Since both H_3O^+ and $C_3H_5O_3^-$ come from the same source in equimolar amounts, their concentrations are equal. $[H_3O^+] = [C_3H_5O_3^-] = 0.067 \times 0.0284 \text{M} = 0.0019 \text{M}$

$$K_a = \frac{[H_3O^+][C_3H_5O_3^-]}{[HC_3H_5O_3]} = \frac{(0.0019)(0.0019)}{0.0284 - 0.0019} = 1.4 \times 10^{-4}$$

9A For an aqueous solution of a diprotic acid, the concentration of the divalent anion equals the second ionization constant: $[^-OOCCH_2COO^-] = K_{a_2} = 2.0 \times 10^{-6} \text{M}$. We organize around the chemical equation.

Equation: $CH_2(COOH)_2 (aq)$ + $H_2O(l)$ $\rightleftharpoons$ $H_3O^+(aq)$ + $HCH_2(COO)_2^- (aq)$

Initial: 1.0 M ≈ 0M   0 M
Changes: $-x$ M $+x$ M $+x$ M
Equil: (1.0 $-x$) M x M xM

$$K_a = \frac{[H_3O^+][HCH_2(COO)_2^-]}{[CH_2(COOH)_2]} = \frac{(x)(x)}{1.0-x} \approx \frac{x^2}{1.0} = 1.4 \times 10^{-3}$$

$$x = \sqrt{1.0 \times 1.4 \times 10^{-3}} = 3.7 \times 10^{-2} \text{ M} = [H_3O^+] = [HOOCCH_2COO^-]$$

The assumption that $x \ll 1.0 \text{M}$ is valid.

9B   We already know that $K_{a_2} = \left[\text{doubly charged anion}\right]$ for a polyprotic acid. Thus

$K_{a_2} = 5.3 \times 10^{-5} = \left[C_2O_4^{2-}\right]$. From the pH we find the $\left[H_3O^+\right] = 10^{-pH} = 10^{-0.67} = 0.21$ M.

We also recognize that $\left[HC_2O_4^-\right] = \left[H_3O^+\right]$, since the second ionization occurs to only a

very small extent. We realize that $HC_2O_4^-$ is produced by the ionization of $H_2C_2O_4$.

Each mole of $HC_2O_4^-$ present results from the ionization of 1 mole of $H_2C_2O_4$.

Now we have sufficient information to determine the value of K_{a_1}.

$$K_{a_1} = \frac{\left[H_3O^+\right]\left[HC_2O_4^-\right]}{\left[H_2C_2O_4\right]} = \frac{0.21 \times 0.21}{1.05 - 0.21} = 5.3 \times 10^{-2}$$

10A H_2SO_4 is a strong acid in its first ionization, and somewhat weak in its second, with

$K_{a_2} = 1.1 \times 10^{-2} = 0.011$. Because of the strong first ionization step, this problem is one of

determining concentrations in a solution that initially is 0.20 M H_3O^+ and 0.20 M HSO_4^-.

We base the setup on the balanced chemical equation.

Equation:	$HSO_4^-\,(aq) + H_2O\,(l)$	$\rightleftharpoons$	$H_3O^+(aq)$	+	$SO_4^{2-}(aq)$
Initial:	0.20 M		0.20 M		0 M
Changes:	$-x$ M		$+x$ M		$+x$ M
Equil:	$(0.20 - x)$ M		$(0.20 + x)$ M		x M

$$K_{a_2} = \frac{\left[H_3O^+\right]\left[SO_4^{2-}\right]}{\left[HSO_4^-\right]} = \frac{(0.20 + x)x}{0.20 - x} = 0.011 \approx \frac{0.20 \times x}{0.20}, \text{ assuming that } x \ll 0.20\text{M}.$$

$x = 0.011$M

Try one cycle of approximation:

$$0.011 \approx \frac{(0.20 + 0.011)x}{(0.20 - 0.011)} = \frac{0.21x}{0.19} \qquad x = \frac{0.19 \times 0.011}{0.21} = 0.010 \text{ M}$$

The next cycle of approximation produces the same answer $0.010\text{M} = \left[SO_4^{2-}\right]$,

$\left[H_3O^+\right] = 0.010 + 0.20 \text{ M} = 0.21 \text{ M}, \qquad \left[HSO_4^-\right] = 0.20 - 0.010 \text{ M} = 0.19 \text{ M}$

10B   We know that H_2SO_4 is a strong acid in its first ionization, and a somewhat weak acid in

its second, with $K_{a_2} = 1.1 \times 10^{-2} = 0.011$. Because of the strong first ionization step, the

problem essentially reduces to determining concentrations in a solution that initially is

0.020 M H_3O^+ and 0.020 M HSO_4^-. We base the setup on the balanced chemical

equation. The result is solved with the quadratic equation.

Equation:	$HSO_4^-\,(aq) + H_2O\,(l)$	$H_3O^+(aq)$	+	$SO_4^{2-}(aq)$
Initial:	0.020 M	0.020 M		0 M

Changes: $-x$ M $+x$ M $+x$ M

Equil: $(0.020-x)$ M $(0.020+x)$ M x M

$$K_{a_2} = \frac{\left[H_3O^+\right]\left[SO_4^{2-}\right]}{\left[HSO_4^-\right]} = \frac{(0.020+x)x}{0.020-x} = 0.011 \qquad\qquad 0.020x + x^2 = 2.2\times10^{-4} - 0.011x$$

$$x^2 + 0.031x - 0.00022 = 0 \qquad x = \frac{-0.031\pm\sqrt{0.00096+0.00088}}{2} = 0.0060 \text{ M} = \left[SO_4^{2-}\right]$$

$$\left[HSO_4^-\right] = 0.020 - 0.0060 = 0.014 \text{ M} \qquad\qquad \left[H_3O^+\right] = 0.020 + 0.0060 = 0.026 \text{ M}$$

(The method of successive approximations converges to $x = 0.006$ M in 8 cycles.)

11A (a) $CH_3NH_3^+NO_3^-$ is the salt of the cation of a weak base that will hydrolyze to form an acidic solution $\left(CH_3NH_3^+ + H_2O \rightleftharpoons CH_3NH_2 + H_3O^+\right)$, and the anion of a strong acid that will not hydrolyze. The aqueous solutions of this compound will be acidic.

(b) NaI is the salt of the cation of a strong base and the anion of a strong acid, neither of which hydrolyzes in water. Solutions of this compound will be pH neutral.

(c) $NaNO_2$ is the salt of the cation of a strong base that will not hydrolyze in water and the cation of a weak acid that will hydrolyze to form an alkaline solution $\left(NO_2^- + H_2O \rightleftharpoons HNO_2 + OH^-\right)$. Aqueous solutions of this compound will be basic (alkaline).

11B Without comparing values of K we can predict that the reaction that produces H_3O^+ occurs to the greater extent. The reason, of course, is that the solution produced is acidic, since its pH is less than 7.00. We write the two reactions of $H_2PO_4^-$ with water, along with the values of their equilibrium constants.

$$H_2PO_4^-\,(aq) + H_2O(l) \rightleftharpoons H_3O^+\,(aq) + HPO_4^-\,(aq) \quad K_{a_2} = 6.3\times10^{-8}$$

$$H_2PO_4^-\,(aq) + H_2O(l) \rightleftharpoons OH^-\,(aq) + H_3PO_4\,(aq) \quad K_b = \frac{K_w}{K_{a_1}} = \frac{1.0\times10^{-14}}{7.1\times10^{-3}} = 1.4\times10^{-12}$$

As predicted, the acid ionization occurs to the greater extent.

12A From the value of pK_b we determine the value of K_b and then K_a for the cation.

cocaine: $K_b = 10^{-pK} = 10^{-8.41} = 3.9\times10^{-9}$ $\qquad K_a = \dfrac{K_w}{K_b} = \dfrac{1.0\times10^{-14}}{3.9\times10^{-9}} = 2.6\times10^{-6}$

codeine: $K_b = 10^{-pK} = 10^{-7.95} = 1.1\times10^{-8}$ $\qquad K_a = \dfrac{K_w}{K_b} = \dfrac{1.0\times10^{-14}}{1.1\times10^{-8}} = 9.1\times10^{-7}$

(This method may be a bit easier:

$pK_a = 14.00 - pK_b = 14.00 - 8.41 = 5.59, K_a = 10^{-5.59} = 2.6 \times 10^{-6}$) The acid with the larger K_a will produce the higher $[H^+]$, and that solution will have the lower pH. Thus, the solution of codeine hydrochloride will have the higher pH.

12B Both of the ions of $NH_4CN(aq)$ react with water in hydrolysis reactions.

$$NH_4^+(aq) + H_2O(l) \rightleftharpoons NH_3(aq) + H_3O^+(aq) \qquad K_a = \frac{K_w}{K_b} = \frac{1.0 \times 10^{-14}}{1.8 \times 10^{-5}} = 5.6 \times 10^{-10}$$

$$CN^-(aq) + H_2O(l) \rightleftharpoons HCN(aq) + OH^-(aq) \qquad K_b = \frac{K_w}{K_a} = \frac{1.0 \times 10^{-14}}{6.2 \times 10^{-10}} = 1.6 \times 10^{-5}$$

Since the value of the equilibrium constant for the hydrolysis reaction of cyanide ion is larger than that for the hydrolysis of ammonium ion, the cyanide ion hydrolysis reaction will proceed to a greater extent. The solution of $NH_4CN(aq)$ will be basic (alkaline).

13A NaF dissociates completely into sodium ions and fluoride ions. Fluoride ion hydrolyzes in aqueous solution to form hydroxide ion. The solution is organized around the balanced equation.

Equation:	$F^-(aq)$	+	$H_2O(l)$	$\rightleftharpoons$	$HF(aq)$	+	$OH^-(aq)$
Initial:	0.10 M				0 M		0 M
Changes:	$-x$ M				$+x$ M		$+x$ M
Equil:	$(0.10-x)$ M				x M		x M

$$K_b = \frac{K_w}{K_a} = \frac{1.0 \times 10^{-14}}{6.6 \times 10^{-4}} = 1.5 \times 10^{-11} = \frac{[HF][OH^-]}{[F^-]} = \frac{(x)(x)}{(0.10-x)} = \frac{x^2}{0.10}$$

$x = \sqrt{0.10 \times 1.5 \times 10^{-11}} = 1.2 \times 10^{-6}$ M=$[OH^-]$; pOH = $-\log(1.2 \times 10^{-6}) = 5.92$

pH = 14.00 − pOH = 14.00 − 5.92 = 8.08

13B The cyanide ion hydrolyzes in solution, as shown in the solution to Practice Example 17-12B. As a consequence of that hydrolysis, $[OH^-] = [HCN]$. $[OH^-]$ can be found from the pH of the solution and then values are substituted into the K_b expression for CN^-, which is solved for $[CN^-]$.

pOH = 14.00 − pH = 14.00 − 10.38 = 3.62

$$[OH^-] = 10^{-pOH} = 10^{-3.62} = 2.4 \times 10^{-4} M = [HCN]$$

$$K_b = \frac{[HCN][OH^-]}{[CN^-]} = 1.6 \times 10^{-5} = \frac{(2.4 \times 10^{-4})^2}{[CN^-]} \qquad [CN^-] = \frac{(2.4 \times 10^{-4})^2}{1.6 \times 10^{-5}} = 3.6 \times 10^{-3} M$$

14A First we draw the Lewis structures of the four acids. Lone pairs are not depicted since we are interested only in the arrangements of atoms.

$$\underset{\text{H—O—N—O}}{\overset{\displaystyle O}{\|}} \qquad \underset{\displaystyle\underset{O}{|}}{\overset{\displaystyle O}{\underset{|}{\text{H—O—Cl—O}}}} \qquad \underset{\underset{H}{|}}{\overset{H\ \ O}{\underset{|}{\text{F—C—C—O—H}}}} \qquad \underset{\underset{H}{|}}{\overset{H\ \ O}{\underset{|}{\text{Br—C—C—O—H}}}}$$

$HClO_4$ should be stronger than HNO_3. Although Cl and N have similar electronegativities, there are more terminal oxygen atoms attached to the chlorine in perchloric acid than to the nitrogen in nitric acid. By virtue of having more terminal oxygens, perchloric acid, when ionized, affords a more stable conjugate base. The more stable the anion, the more easily it is formed and hence the stronger is the conjugate acid from which it is derived. CH_2FCOOH will be a stronger acid than $CH_2BrCOOH$ because F is a more electronegative atom than Br. The F atom withdraws additional electron density from the O—H bond, making that bond more easily broken.

14B First we draw the Lewis structures of the first two acids. Lone pairs are not depicted since we are interested in the arrangements of atoms.

$$\underset{\underset{\text{O—H}}{|}}{\overset{\displaystyle O}{\underset{|}{\text{H—O—P—O—H}}}} \qquad \overset{\displaystyle O}{\underset{|}{\text{H—O—S—O—H}}}$$

H_3PO_4 and H_2SO_3 both have one terminal oxygen atom, but S is more electronegative than P. This indicates that $H_2SO_3 \left(K_{a_1} = 1.3 \times 10^{-2}\right)$ should be a stronger acid that $H_3PO_4 \left(K_{a_1} = 7.1 \times 10^{-3}\right)$, and it is. The only difference between CCl_3CH_2COOH and CCl_2FCH_2COOH is the replacement of Cl by F. Since F is more electronegative than Cl, CCl_3CH_2COOH should be a weaker acid than CCl_2FCH_2COOH.

15A We draw Lewis structures to help us decide.

$$\underset{\underset{|\overline{F}|}{|}}{\overset{|\overline{F}|}{\underset{|}{|\overline{F}\text{—B}}}} + \underset{\underset{H}{|}}{\overset{H}{\underset{|}{|\text{N—H}}}} \longrightarrow \underset{\underset{|\overline{F}|\ \ H}{|\ \ |}}{\overset{|\overline{F}|\ \ H}{\underset{|\ \ |}{|\overline{F}\text{—B—N—H}}}} \qquad \underset{\underset{H}{|}}{\overset{}{\underset{|}{|\overline{O}\text{—H}}}}$$

(a) Clearly, BF_3 is an electron pair acceptor, a Lewis acid, and NH_3 is an electron pair donor, a Lewis base.

(b) H_2O certainly has electron pairs (lone pairs) to donate and thus it can be a Lewis base. It is unlikely that the cation Cr^{3+} has any accessible valence electrons that can be donated to another atom; it is the Lewis acid.

15B The Lewis structures of the six species follow.

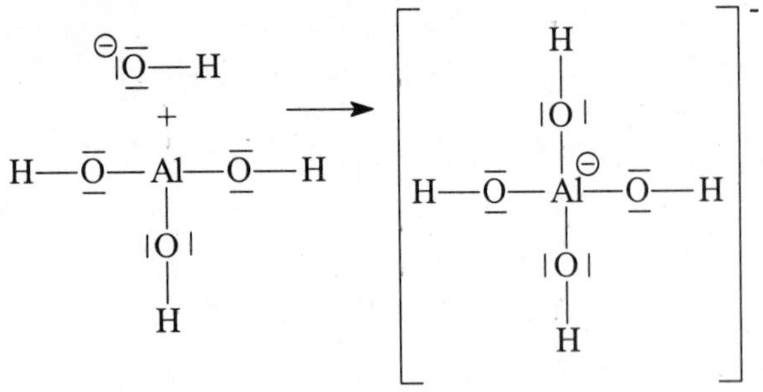

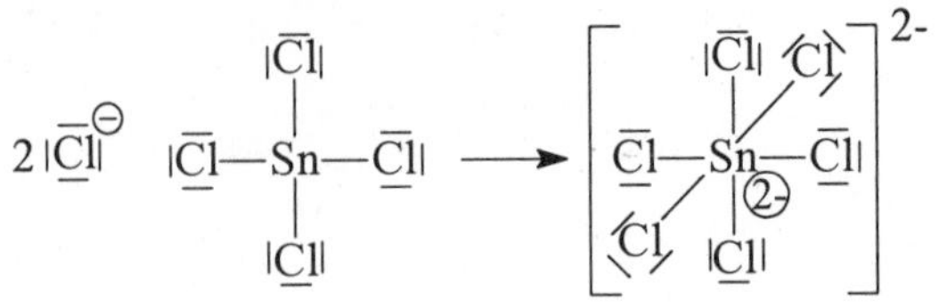

Both hydroxide ion and chloride ion have extra lone pairs of electrons; these two are the electron pair donors, the Lewis bases. $Al(OH)_3$ and $SnCl_4$ have additional spaces in their structures to accept pairs of electrons, which is what occurs when they form complex anions. $Al(OH)_3$ and $SnCl_4$ are the Lewis acids in these reactions.

REVIEW QUESTIONS

1. **(a)** K_w is the symbol for the ion product of water: $K_w = \left[H_3O^+\right]\left[OH^-\right] = 1.0 \times 10^{-14}$ ($25°C$).

    (b) pH symbolizes the negative of the (base 10) logarithm of the hydrogen ion (actually hydronium ion, H_3O^+) concentration: $pH = -\log\left[H_3O^+\right]$. High pH indicates an alkaline solution; low pH indicates an acidic one.

    (c) pK_a symbolizes the negative of the logarithm of the ionization constant of an acid: $pK_a = -\log K_a$.

    (d) Hydrolysis refers to the process where an ion reacts (as an acid or a base) with water to produce either hydrogen ion or hydroxide ion.

    (e) Lewis acids are electron pair acceptors (i.e., the recipient's pairs of e^- from Lewis bases)

2. **(a)** A conjugate base is the species produced when another species (molecule or ion) donates, and thus loses a proton, H^+.

    (b) Percent ionization is the ratio of the concentration of the ion formed by a species divided by the original concentration of that species prior to ionization multiplied by 100%..

    (c) Self-ionization: The process in which pure solvents partially dissociate into cations and anions. (e.g. $2H_2O(l) \rightleftharpoons H_3O^+(aq) + OH^-(aq)$)

    (d) Amphiprotic behavior refers to the ability of some substances, often oxides, to behave as acids in the presence of strong bases, and behave as bases in the presence of strong acids.

3.   **(a)**   A Brønsted-Lowry acid is a proton (H^+) donor; a Brønsted-Lowry base is a proton acceptor.

(b) $pH = -\log[H_3O^+]$; the one is a logarithmic expression of the other, necessary because of the wide range of $[H_3O^+]$ values commonly encountered.

(c) K_a for NH_4^+ and K_b for NH_3 are inversely related to each other: $K_a \times K_b = K_w$.

The leveling effect refers the inability of a moderatelybasic solvent to distinguish between the strengths of two strong acids; they both donate protons equally well to the solvent. The electron-withdrawing effect refers to the attraction of electrons to an electronegative atom or group within a molecule. If the electrons are drawn away from an $O-H$ bond, that bond will become more polar and hence, more easily broken, thereby producing a stronger acid.

4.   **(a)** HNO_2 is an acid, a proton donor. It's conjugate base is NO_2^-.

(b) OCl^- is a base, a proton acceptor. It's conjugate acid is $HOCl$.

(c) NH_2^- is a base, a proton acceptor. It's conjugate acid is NH_3.

(d) NH_4^+ is an acid, a proton donor. It's conjugate base is NH_3.

(e) $CH_3NH_3^+$ is an acid, a proton donor. It's conjugate base is CH_3NH_2.

5.   We write the conjugate base as the first product of the equilibrium in which the acid is the first reactant.

(a) $HIO_3(aq) + H_2O(l) \rightleftharpoons IO_3^-(aq) + H_3O^+(aq)$

(b) $C_6H_5COOH(aq) + H_2O(l) \rightleftharpoons C_6H_5COO^-(aq) + H_3O^+(aq)$

(c) $HPO_4^{2-}(aq) + H_2O(l) \rightleftharpoons PO_4^{3-}(aq) + H_3O^+(aq)$

(d) $C_2H_5NH_3^+(aq) + H_2O(l) \rightleftharpoons C_2H_5NH_2(aq) + H_3O^+(aq)$

6.   All of the solutes are strong acids or strong bases.

(a) $[H_3O^+] = 0.00165 \text{ M HNO}_3 \times \dfrac{1 \text{ mol } H_3O^+}{1 \text{ mol HNO}_3} = 0.00165 \text{ M}$

$[OH^-] = \dfrac{K_w}{[H_3O^+]} = \dfrac{1.0 \times 10^{-14}}{0.00165\text{M}} = 6.1 \times 10^{-12} \text{ M}$

(b) $[OH^-] = 0.0087 \text{ M KOH} \times \dfrac{1 \text{ mol } OH^-}{1 \text{ mol KOH}} = 0.0087 \text{ M}$

$[H_3O^+] = \dfrac{K_w}{[OH^-]} = \dfrac{1.0 \times 10^{-14}}{0.0087\text{M}} = 1.1 \times 10^{-12} \text{ M}$

(c) $\left[OH^-\right] = 0.00213 \text{ M } Sr(OH)_2 \times \dfrac{2 \text{ mol } OH^-}{1 \text{ mol } Sr(OH)_2} = 0.00426 \text{ M}$

$\left[H_3O^+\right] = \dfrac{K_w}{\left[OH^-\right]} = \dfrac{1.0 \times 10^{-14}}{0.00426 \text{ M}} = 2.3 \times 10^{-12} \text{ M}$

(d) $\left[H_3O^+\right] = 5.8 \times 10^{-4} \text{M HI} \times \dfrac{1 \text{mol } H_3O^+}{1 \text{mol HI}} = 5.8 \times 10^{-4} \text{ M}$

$\left[OH^-\right] = \dfrac{K_w}{\left[H_3O^+\right]} = \dfrac{1.0 \times 10^{-14}}{5.8 \times 10^{-4} \text{M}} = 1.7 \times 10^{-11} \text{ M}$

7. Again, all of the solutes are strong acids or strong bases.

(a) $\left[H_3O^+\right] = 0.0045 \text{ M HCl} \times \dfrac{1 \text{ mol } H_3O^+}{1 \text{ mol HCl}} = 0.0045 \text{ M}$

$pH = -\log(0.0045) = 2.35$

(b) $\left[H_3O^+\right] = 6.14 \times 10^{-4} \text{M HNO}_3 \times \dfrac{1 \text{ mol } H_3O^+}{1 \text{ mol HNO}_3} = 6.14 \times 10^{-4} \text{M}$

$pH = -\log(6.14 \times 10^{-4}) = 3.212$

(c) $\left[OH^-\right] = 0.00683 \text{ M NaOH} \times \dfrac{1 \text{ mol } OH^-}{1 \text{ mol NaOH}} = 0.00683 \text{M}$

$pOH = -\log(0.00683) = 2.166$ and $pH = 14.000 - pOH = 14.000 - 2.166 = 11.834$

(d) $\left[OH^-\right] = 4.8 \times 10^{-3} \text{M } Ba(OH)_2 \times \dfrac{2 \text{ mol } OH^-}{1 \text{ mol } Ba(OH)_2} = 9.6 \times 10^{-3} \text{M}$

$pOH = -\log(9.6 \times 10^{-3}) = 2.02 \qquad pH = 14.00 - 2.02 = 11.98$

8. Most probably, $0.11 \text{ M} = \left[H_3O^+\right]$ in 0.10 M H_2SO_4. Although H_2SO_4 is completely ionized in dilute solution, this solution is not sufficiently dilute. (Recall the paragraph before Example 17-10 in the text.) Thus, 0.20 M is incorrect. However, the first ionization of H_2SO_4 is complete; 0.05 M cannot be correct. And the second ionization of H_2SO_4 is at least partially complete; 0.10 M incorrectly assumes that the only source of H_3O^+ is the first ionization.

9. We determine the amounts of H_3O^+ and OH^- and then the amount of the one that is in excess. We express molar concentration in millimoles/milliliter, equivalent to mol/L.

$24.80 \text{ mL} \times \dfrac{0.248 \text{ mmol HNO}_3}{1 \text{ mL soln}} \times \dfrac{1 \text{ mmol } H_3O^+}{1 \text{ mmol HNO}_3} = 6.15 \text{ mmol } H_3O^+$

$$15.40 \text{ mL} \times \frac{0.394 \text{ mmol KOH}}{1 \text{ mL soln}} \times \frac{1 \text{ mmol OH}^-}{1 \text{ mmol KOH}} = 6.07 \text{ mmol OH}^-$$

The net reaction is $H_3O^+(aq) + OH^-(aq) \rightarrow 2H_2O(l)$. There is an excess amount of H_3O^+ of $(6.15 - 6.07 \text{ mmol} =) 0.08 \text{ mmol } H_3O^+$. This is an acidic solution. The total solution volume is $(24.80 + 15.40 \text{ mL} =) 40.20 \text{ mL}$.

$$\left[H_3O^+ \right] = \frac{0.08 \text{ mmol } H_3O^+}{40.20 \text{mL}} = 0.002M \qquad pH = -\log(0.002) = 2.7$$

10. Answer (4) is correct, $pH < 13$ ..If this were a strong base, its $\left[OH^- \right] = 0.10$, giving $pOH = 1.0$, and $pH = 13.0$. But since methylamine is a weak base, its $\left[OH^- \right] < 0.10$, giving $pOH > 1.0$, and $pH < 13.0$.

11. $$\left[HC_3H_5O_2 \right]_{initial} = \left(\frac{0.275 \text{ mol}}{625 \text{ mL}} \right) \left(\frac{1000 \text{ mL}}{1 \text{ L}} \right) = 0.440 \text{ M}$$

$$\left[HC_3H_5O_2 \right]_{equilibrium} = 0.440 \text{ M} - 0.00239 \text{ M} = 0.438 \text{ M}$$

$$K_a = \frac{\left[H_3O^+ \right] \left[C_3H_5O_2^- \right]}{\left[HC_3H_5O_2 \right]} = \frac{(0.00239)^2}{0.438} = 1.30 \times 10^{-5}$$

12. **(a)** The set up is based on the balanced chemical equation.

Equation: $\quad HC_8H_7O_2(aq) \quad + \quad H_2O(l) \quad \rightleftharpoons \quad H_3O^+(aq) + \quad C_8H_7O_2^-(aq)$

Initial:	0.186 M	≈ 0 M	0 M
Changes:	$-x$ M	$+x$ M	$+x$ M
Equil:	$(0.186 - x)$ M	x M	x M

$$K_a = 4.9 \times 10^{-5} = \frac{[H_3O^+][C_8H_7O_2^-]}{[HC_8H_7O_2]} = \frac{x \cdot x}{0.186 - x} \approx \frac{x^2}{0.186}$$

$$x = \sqrt{0.186 \times 4.9 \times 10^{-5}} = 0.0030 \text{ M} = [H_3O^+] = [C_8H_7O_2^-]$$

(b) The set up is based on the balanced chemical equation.

Equation: $\quad HC_8H_7O_2(aq) \quad + \quad H_2O(l) \quad \rightleftharpoons \quad H_3O^+(aq) \quad + \quad C_8H_7O_2(aq)$

Initial:	0.121 M	≈ 0 M	0 M
Changes:	$-x$ M	$+x$ M	$+x$ M
Equil:	$(0.121 - x)$ M	x M	x M

$$K_a = 4.9 \times 10^{-5} = \frac{\left[H_3O^+ \right] \left[C_8H_7O_2^- \right]}{\left[HC_8H_7O_2 \right]} = \frac{x^2}{0.121 - x} \approx \frac{x^2}{0.121} \qquad x = 0.0024 \text{ M} = \left[H_3O^+ \right]$$

We have assumed $x \ll 0.121$, an assumption that clearly is correct.
pH $= -\log(0.0024) = 2.62$

13. We need first to determine the molarity S of the acid needed.

$$\left(\left[H_3O^+\right] = 10^{-2.85} = 1.4 \times 10^{-3} M\right)$$

Equation: $HC_7H_5O_2(aq) + H_2O(l) \rightleftharpoons H_3O^+(aq) + C_7H_5O_2^-(aq)$

Initial:	S	0 M	0 M
Changes:	-0.0014 M	$+0.0014$M	$+0.0014$M
Equil:	$S - 0.0014$M	0.0014 M	0.0014 M

$$K_a = \frac{\left[H_3O^+\right]\left[C_7H_5O_2^-\right]}{\left[HC_7H_5O_2\right]} = \frac{(0.0014)^2}{S - 0.0014} = 6.3 \times 10^{-5} \qquad S - 0.0014 = \frac{(0.0014)^2}{6.3 \times 10^{-5}} = 0.031$$

$$S = 0.031 + 0.0014 = 0.032 \text{ M} = \left[HC_7H_5O_2\right]$$

$$350.0 \text{ mL} \times \frac{1 \text{ L}}{1000 \text{ mL}} \times \frac{0.032 \text{ mol } HC_7H_5O_2}{1 \text{ L soln}} \times \frac{122.1 \text{ g } HC_7H_5O_2}{1 \text{ mol } HC_7H_5O_2} = 1.4 \text{ g } HC_7H_5O_2$$

14. First determine $\left[OH^-\right]$, and thus $\left[(CH_3)_3NH^+\right]$, in the solution. Then use the K_b expression to find $\left[(CH_3)_3N\right]$

pOH $= 14.00 - $ pH $= 14.00 - 11.12 = 2.88 \qquad \left[OH^-\right] = 10^{-pOH} = 10^{-2.88} = 0.0013$ M

$$K_b = 6.3 \times 10^{-5} = \frac{\left[(CH_3)_3NH^+\right]\left[OH^-\right]}{\left[(CH_3)_3N\right]_{equil}} = \frac{(0.0013)^2}{\left[(CH_3)_3N\right]_{equil}}$$

$$\left[(CH_3)_3N\right]_{equil} = \frac{(0.0013)^2}{6.3 \times 10^{-5}} = 0.027 \text{ M } (CH_3)_3N$$

Since some of the $(CH_3)_3N$ has ionized, $\left[(CH_3)_3N\right]_{initial} = 0.027 \text{ M} + 0.0013 \text{ M} = 0.028$ M

15. For H_2CO_3, $K_1 = 4.4 \times 10^{-7}$ and $K_2 = 4.7 \times 10^{-11}$

(a) The first acid ionization proceeds to a far greater extent than does the second and determines the value of $\left[H_3O^+\right]$.

Equation: $H_2CO_3(aq) + H_2O(l) \rightleftharpoons H_3O^+(aq) + HCO_3^-(aq)$

Initial:	0.045 M	≈ 0 M	0 M
Changes:	$-x$ M	$+x$ M	$+x$ M
Equil:	$(0.045 - x)$ M	x M	x M

$$K_1 = \frac{\left[H_3O^+\right]\left[HCO_3^-\right]}{\left[H_2CO_3\right]} = \frac{x^2}{0.045 - x} = 4.4 \times 10^{-7} \approx \frac{x^2}{0.045} \qquad x = 1.4 \times 10^{-4} M = \left[H_3O^+\right]$$

(b) Since the second ionization occurs to only a limited extent,

$$\left[HCO_3^-\right] = 1.4 \times 10^{-4} \ M = 0.00014 \ M$$

(c) We use the second ionization to determine $\left[CO_3^{2-}\right]$.

Equation: $HCO_3^-(aq) \ + \ H_2O(l) \ \rightleftharpoons \ H_3O^+(aq) \ + \ CO_3^{2-}(aq)$

Initial:	0.00014 M		0.00014 M	0 M
Changes:	$-x$ M		$+x$ M	$+x$ M
Equil:	$(0.00014-x)$ M		$(0.00014+x)$ M	x M

$$K_2 = \frac{\left[H_3O^+\right]\left[CO_3^{2-}\right]}{\left[HCO_3^-\right]} = \frac{(0.00014+x)x}{0.00014-x} = 4.7 \times 10^{-11} \approx \frac{(0.00014) \ x}{0.00014}$$

$$x = 4.7 \times 10^{-11} M = \left[CO_3^{2-}\right]$$

Again we assumed that $x \ll 0.00014$ M, which clearly is the case. We also note that our original assumption, that the second ionization is much less significant than the first, also is a valid one. As an alternative to all of this, we could have recognized that the concentration of the divalent anion equals the second ionization constant:

$$\left[CO_3^{2-}\right] = K_2 .$$

16. The species that hydrolyze are the cations of weak bases—NH_4^+ and $C_6H_5NH_3^+$—and the anions of weak acids—NO_2^- and $C_7H_5O_2^-$.

(a) $NH_4^+(aq) + NO_3^-(aq) + H_2O(l) \rightleftharpoons NH_3(aq) + NO_3^-(aq) + H_3O^+(aq)$

(b) $Na^+(aq) + NO_2^-(aq) + H_2O(l) \rightleftharpoons Na^+(aq) + HNO_2(aq) + OH^-(aq)$

(c) $K^+(aq) + C_7H_5O_2^-(aq) + H_2O(l) \rightleftharpoons K^+(aq) + HC_7H_5O_2(aq) + OH^-(aq)$

(d) $K^+(aq) + Cl^-(aq) + Na^+(aq) + I^-(aq) + H_2O(l) \rightarrow$ no reaction

(e) $C_6H_5NH_3^+(aq) + Cl^-(aq) + H_2O(l) \rightleftharpoons C_6H_5NH_2(aq) + Cl^-(aq) + H_3O^+(aq)$

17. Equation: $NH_4^+(aq) \ + \ H_2O(l) \ \rightleftharpoons \ H_3O^+(aq) \ + \ NH_3(aq)$

Initial:	1.68 M		≈ 0 M	0 M
Changes:	$-x$ M		$+x$ M	$+x$ M
Equil:	$(1.68-x)$ M		x M	x M

$$K_a = \frac{\left[H_3O^+\right]\left[NH_3\right]}{\left[NH_4^+\right]} = \frac{x^2}{1.68-x} = 5.6 \times 10^{-10} \approx \frac{x^2}{1.68}$$

$$x = 3.0_7 \times 10^{-5} \ M = [H_3O^+] \qquad pH = -\log(3.07 \times 10^{-5}) = 4.51$$

18. Recall that, for a conjugate weak acid-weak base pair, $K_a \times K_b = K_w$

(a) $K_a = \dfrac{K_w}{K_b} = \dfrac{1.0 \times 10^{-14}}{1.5 \times 10^{-9}} = 6.7 \times 10^{-6}$ for $C_5H_5NH^+$

(b) $K_b = \dfrac{K_w}{K_a} = \dfrac{1.0 \times 10^{-14}}{1.8 \times 10^{-4}} = 5.6 \times 10^{-11}$ for CHO_2^-

(c) $K_b = \dfrac{K_w}{K_a} = \dfrac{1.0 \times 10^{-14}}{1.0 \times 10^{-10}} = 1.0 \times 10^{-4}$ for $C_6H_5O^-$

19. We base the calculation on the balanced chemical equation for the hydrolysis of the anion.

Equation: $\quad C_2H_2ClO_2^-(aq) + H_2O(l) \rightleftharpoons HC_2H_2ClO_2(aq) + OH^-(aq)$

Initial:	2.05 M	0 M	≈ 0 M
Changes:	$-x$ M	$+x$ M	$+x$ M
Equil:	$(2.05-x)$ M	x M	x M

$$K_b = \frac{K_w}{K_a} = \frac{1.0 \times 10^{-14}}{1.4 \times 10^{-3}} = 7.1 \times 10^{-12} = \frac{[HC_2H_2ClO_2][OH^-]}{[C_2H_2ClO_2^-]} = \frac{x \cdot x}{2.05 - x} \approx \frac{x^2}{2.05}$$

$x = \sqrt{2.05 \times 7.1 \times 10^{-12}} = 3.8 \times 10^{-6} = [OH^-]$ $\quad$ $pOH = -\log(3.8 \times 10^{-6}) = 5.42$

$pH = 14.00 - pOH = 14.00 - 5.42 = 8.58$

20. CCl_3COOH is a stronger acid than CH_3COOH because in CCl_3COOH there are electronegative (electron withdrawing) Cl atoms bonded to the carbon atom adjacent to the COOH group. The electron withdrawing Cl atoms further polarize the OH bond, resulting in a stronger acid.

21. (a) HI is the stronger acid because the $H-I$ bond length is longer than the $H-Br$ bond length and the $H-I$ bond is weaker.

(b) HOClO is a stronger acid than HOBr because there is a terminal O in HOClO but not in HOBr and Cl is more electronegative than Br.

(c) $H_3CCH_2CCl_2COOH$ is a stronger acid than $I_3CCH_2CH_2COOH$ both because Cl is more electronegative than is I and also because the Cl atoms are closer to the acidic hydrogen in the COOH group and thus can exert a stronger e^- withdrawing effect on the O–H bond than can the more distant I atoms.

22. A Lewis base is an electron pair donor, while a Lewis acid is an electron pair acceptor. We draw Lewis structures to assist our interpretation.

(a) $\left[|\overline{\underline{O}}-H \right]^-$

(b)
```
      H   H        H   H
      |   |        |   |
   H—C—C —B— C—C—H
      |   |   |    |   |
      H   H  CH₂   H   H
              |
             CH₃
```

(c)
```
      H   H
      |   |
   H—N—C—H
      |   |
          H
```

(a) The lone pairs on oxygen can readily be donated; this is a Lewis base.

(b) The incomplete octet of B provides a site for acceptance of an electron pair; this is a Lewis acid.

(c) The lone pair on nitrogen can be donated; this is a Lewis base.

23. A Lewis base is an electron pair donor, while a Lewis acid is an electron pair acceptor. We draw Lewis structures to assist our interpretation.

(a) According to the following Lewis structures, SO_3 appears to be the electron pair acceptor (the Lewis acid), and H_2O is the electron pair donor (the Lewis base). Note that an additional sulfur-to-oxygen bond is formed.

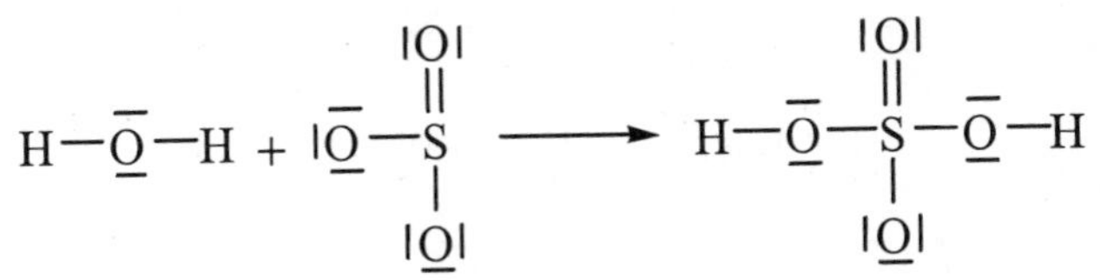

$$H-\overline{O}-H + I\overline{O}-S \longrightarrow H-\overline{O}-S-\overline{O}-H$$

(b) Zn in $Zn(OH)_2$ accepts a pair of electrons; $Zn(OH)_2$ is a Lewis acid. OH^- donates the pair of electrons that form the covalent bond; OH^- is a Lewis base. We have assumed that Zn has sufficient covalent character to form coordinate covalent bonds with hydroxide ions. (Lewis structure drawn below).

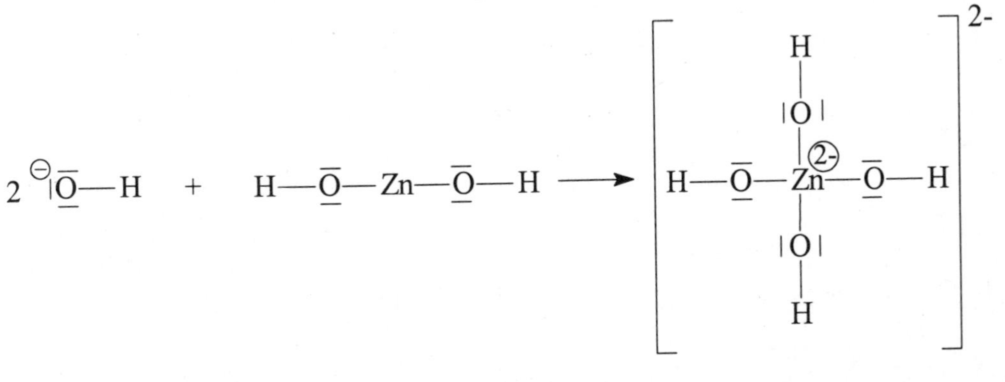

$$2\ ^\ominus I\overline{O}-H\ +\ H-\overline{O}-Zn-\overline{O}-H \longrightarrow \left[H-\overline{O}-Zn-\overline{O}-H \right]^{2-}$$

24. (a) $NH_4Cl(aq)$ contains a cation, , that is a weak acid; it should have a pH slightly below 7.00.

(b) $NH_3(aq)$ is a weak base; it should have a reasonably high pH.

(c) $NaC_2H_3O_2(aq)$ contains an anion, $C_2H_3O_2^-$ (aq), that is a weak base; it should have a pH slightly above 7.00.

(d) Neither $K^+(aq)$ nor $Cl^-(aq)$ is an acid or a base in aqueous solution; KCl(aq) should have a pH of 7.00. In order of decreasing pH:
$NH_3(aq) > NaC_2H_3O_2(aq) > KCl(aq) > NH_4Cl(aq)$

EXERCISES

Brønsted-Lowry Theory of Acids and Bases

25. The acids (proton donors) and bases (proton acceptors) are labeled below their formulas. Remember that a proton, in Brønsted-Lowry acid-base theory, is H^+.

(a) $HOBr(aq) + H_2O(l) \rightleftharpoons H_3O^+(aq) + OBr^-(aq)$
   acid   base   acid   base

(b) $HSO_4^-(aq) + H_2O(l) \rightleftharpoons H_3O^+(aq) + SO_4^{2-}(aq)$
   acid   base   acid   base

(c) $HS^-(aq) + H_2O(l) \rightleftharpoons H_2S(aq) + OH^-(aq)$
   base   acid   acid   base

(d) $C_6H_5NH_3^+(aq) + OH^-(aq) \rightleftharpoons C_6H_5NH_2(aq) + H_2O(l)$
   acid   base   base   acid

26. For each amphiprotic substance, we write its reactions with water: with the substance acting as an acid, and then with it acting as a base. Even for the substances that are not usually considered amphiprotic, both reactions are written, but one of them is labeled as unlikely. In some instances we have written an oxygen as $\varnothing$ to keep track of it through the reaction.

(a) $\varnothing H^- + H_2O \rightleftharpoons \varnothing^{2-} + H_3O^+$ (unlikely) $\quad \varnothing H^- + H_2O \rightleftharpoons H_2\varnothing + OH^-$

(b) $NH_4^+ + H_2O \rightleftharpoons NH_3 + H_3O^+$ $\quad NH_4^+$ cannot add a proton without

   N-aton expanding its octet

(c) $H_2\varnothing + H_2O \rightleftharpoons \varnothing H^- + H_3O^+$ $\quad H_2\varnothing + H_2O \rightleftharpoons H_3\varnothing^+ + OH^-$

(d) $HS^- + H_2O \rightleftharpoons S^{2-} + H_3O^+$ $\quad HS^- + H_2O \rightleftharpoons H_2S + OH^-$

(e) NO_2^- cannot act as an acid, no protons $\quad NO_2^- + H_2O \rightleftharpoons HNO_2 + OH^-$

(f) $HCO_3^- + H_2O \rightleftharpoons CO_3^{2-} + H_3O^+$ $\quad HCO_3^- + H_2O \rightleftharpoons H_2CO_3 + OH^-$

(g) $HBr + H_2O \rightleftharpoons H_3O^+ + Br^-$ $\quad HBr + H_2O \rightleftharpoons H_2Br^+ + OH^-$ (unlikely)

27. Answer (2), NH_3, is correct. $HC_2H_3O_2$ will react most completely with the strongest base. NO_3^- and Cl^- are very weak bases. H_2O is a weak base, but it is amphiprotic, acting as an acid (donating protons), as in the presence of NH_3. Thus, NH_3 must be the strongest base and the most effective deprotonating $HC_2H_3O_2$.

28. Lewis structures are given below each equation.

(a) $2NH_3(l) \rightleftharpoons NH_4^+ + NH_2^-$

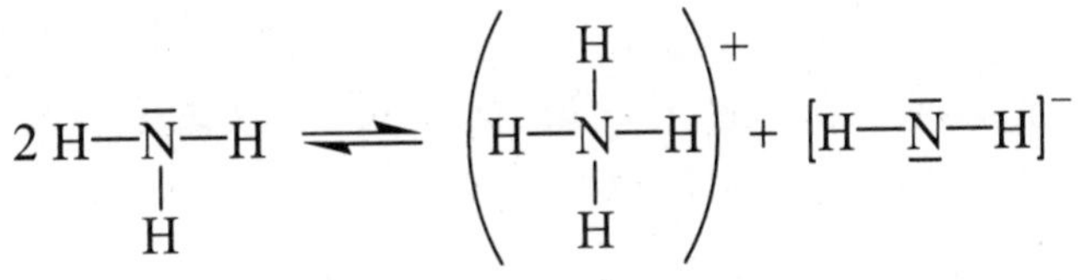

$$2 \; H-\overline{N}-H \;\; \rightleftharpoons \;\; \left(H-\overset{H}{\underset{H}{N}}-H \right)^+ + \left[H-\overline{\overline{N}}-H \right]^-$$

(b) $2HF(l) \rightleftharpoons H_2F^+ + F^-$

(c) $2CH_3OH(l) \rightleftharpoons CH_3OH_2^+ + CH_3O^-$

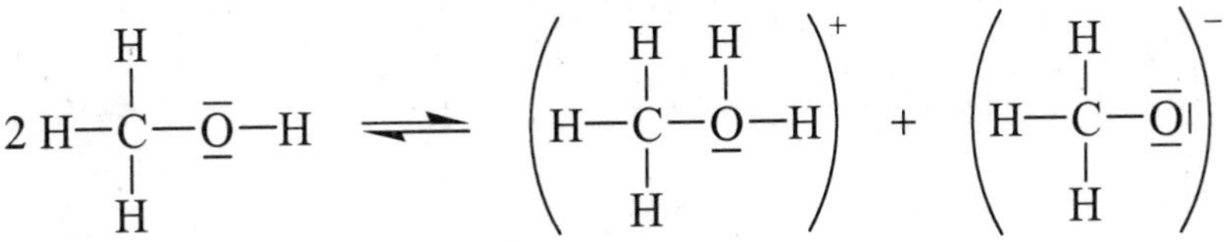

$$2 \; H-\overset{H}{\underset{H}{C}}-\overline{O}-H \;\; \rightleftharpoons \;\; \left(H-\overset{H}{\underset{H}{C}}-\overset{H}{O}-H \right)^+ + \left(H-\overset{H}{\underset{H}{C}}-\overline{O}| \right)^-$$

(d) $2HC_2H_3O_2(l) \rightleftharpoons H_2C_2H_3O_2^+ + C_2H_3O_2^-$

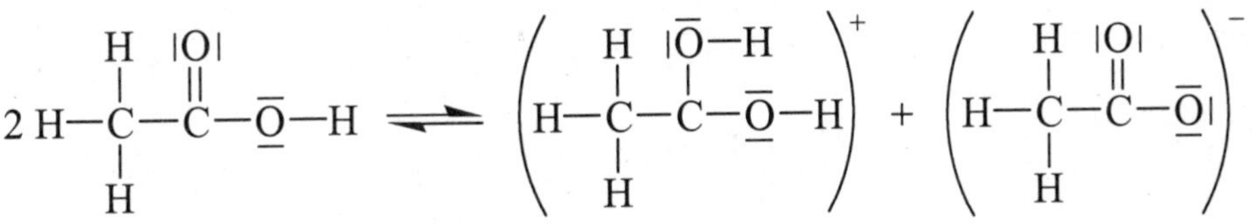

$$2 \; H-\overset{H}{\underset{H}{C}}-\overset{|O|}{C}-\overline{O}-H \;\; \rightleftharpoons \;\; \left(H-\overset{H}{\underset{H}{C}}-\overset{|\overline{O}-H}{C}-\overline{O}-H \right)^+ + \left(H-\overset{H}{\underset{H}{C}}-\overset{|O|}{C}-\overline{O}| \right)^-$$

(e) $2\,H_2SO_4 \rightleftharpoons 2\,H_3SO_4^+ + HSO_4^-$

$$H-\overline{O}-\overset{|\overline{O}|}{\underset{|O|}{S}}-\overline{O}-H + H-\overline{O}-\overset{|\overline{O}|}{\underset{|O|}{S}}-\overline{O}-H \;\; \rightleftharpoons \;\; \left[\overset{\ominus}{|O}-\overset{|\overline{O}|}{\underset{|O|}{S}}-\overline{O}-H \right]^- + \left[H-\overline{O}-\overset{|\overline{O}|}{\underset{\oplus}{S}}-\overline{O}-H \atop |O-H \right]^+$$

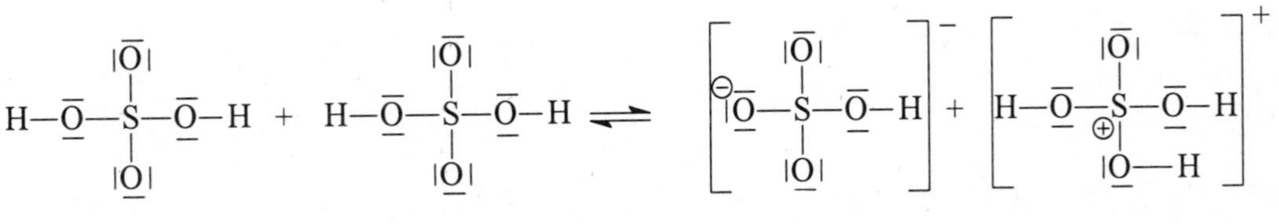

29. The principle here is that the weaker acid and the weaker base will predominate at equilibrium. The reason is that a strong acid will do a good job of donating its protons and, having done so, its conjugate base will be left behind. The preferred direction is:

Strong acid + strong base → weak (conjugate) base + weak (conjugate) acid.

(a) The reaction will favor the forward direction because OH^- (a strong base) $> NH_3$ as bases and $NH_4^+ > H_2O$ as acids.

(b) The reaction will favor the reverse direction because $HNO_3 > HSO_4^-$ (a weak acid in the second ionization) as acids, and $SO_4^{2-} > NO_3^-$ as bases.

(c) The reaction will favor the reverse direction because $HC_2H_3O_2 > CH_3OH$ (not usually thought of as an acid) as acids, and $CH_3O^- > C_2H_3O_2^-$ as bases.

30. The principle here is that the weaker acid and the weaker base predominate at equilibrium. This is because a strong acid will do a good job of donating its protons and, having done so, its conjugate base will remain. The preferred direction is:

Strong acid + strong base → weak (conjugate) base + weak (conjugate) acid.

(a) The reaction will favor the forward direction because $HC_2H_3O_2$ (a moderate acid) $> HCO_3^-$ (a rather weak acid) as acids and $CO_3^{2-} > C_2H_3O_2^-$ as bases.

(b) The reaction will favor the reverse direction because $HClO_4$ (a strong acid) $> HNO_2$ as acids, and $NO_2^- > ClO_4^-$ as bases.

(c) The reaction will favor the forward direction, because $H_2CO_3 > HCO_3^-$ as acids (because $K_1 > K_2$) and $CO_3^{2-} > HCO_3^-$ as bases.

Strong Acids, Strong Bases, and pH

31.

$$\left[OH^-\right] = \frac{3.9 \text{ g Ba}(OH)_2 \cdot 8H_2O}{100 \text{ mL soln}} \times \frac{1000 \text{ mL}}{1 \text{ L}} \times \frac{1 \text{ mol Ba}(OH)_2 \cdot 8H_2O}{315.5 \text{ g Ba}(OH)_2 \cdot 8H_2O} \times \frac{2 \text{ mol OH}^-}{1 \text{ mol Ba}(OH)_2 \cdot 8H_2O}$$

$$= 0.25 \text{ M}$$

$$\left[H_3O^+\right] = \frac{K_w}{\left[OH^-\right]} = \frac{1.0 \times 10^{-14}}{0.25 \text{ M OH}^-} = 4.0 \times 10^{-14} \text{ M} \qquad pH = -\log(4.0 \times 10^{-14}) = 13.40$$

32. The dissolved $Ca(OH)_2$ is completely dissociated into ions.

$$pOH = 14.00 - pH = 14.00 - 12.35 = 1.65$$

$$\left[OH^-\right] = 10^{-pOH} = 10^{-1.65} = 2.2 \times 10^{-2} \text{ M OH}^- = 0.022 \text{ M OH}^-$$

$$\text{solubility} = \frac{0.022 \text{ mol OH}^-}{1 \text{ L soln}} \times \frac{1 \text{ mol Ca}(OH)_2}{2 \text{ mol OH}^-} \times \frac{74.09 \text{ g Ca}(OH)_2}{1 \text{ mol Ca}(OH)_2} \times \frac{1000 \text{ mg Ca}(OH)_2}{1 \text{ g Ca}(OH)_2}$$

$$= \frac{8.1 \times 10^2 \text{ mg Ca}(OH)_2}{1 \text{ L soln}}$$

In 100 mL the solubility is $100 \text{ mL} \times \dfrac{1 \text{ L}}{1000 \text{ mL}} \times \dfrac{8.1 \times 10^2 \text{ mg Ca}(OH)_2}{1 \text{ L soln}} = 81 \text{ mg Ca}(OH)_2$

33. First determine the moles of HCl, and then its concentration.

$$\text{moles HCl} = \frac{PV}{RT} = \frac{\left(751 \text{ mmHg} \times \dfrac{1 \text{ atm}}{760 \text{ mmHg}}\right) \times 0.205 \text{ L}}{0.08206 \text{ L atm mol}^{-1} \text{ K}^{-1} \times 296 \text{ K}} = 8.34 \times 10^{-3} \text{ mol HCl}$$

$$\left[H_3O^+\right] = \frac{8.34 \times 10^{-3} \text{ mol HCl}}{4.25 \text{ L soln}} \times \frac{1 \text{ mol H}_3O^+}{1 \text{ mol HCl}} = 1.96 \times 10^{-3} \text{ M}$$

34. First determine the concentration of NaOH, then of OH^-.

$$\left[OH^-\right] = \frac{0.125 \text{ L} \times \dfrac{0.606 \text{ mol NaOH}}{1 \text{ L}} \times \dfrac{1 \text{ mol OH}^-}{1 \text{ mol NaOH}}}{15.0 \text{ L final solution}} = 0.00505 \text{ M}$$

$$\text{pOH} = -\log(0.00505 \text{ M}) = 2.297 \qquad \text{pH} = 14.00 - 2.297 = 11.703$$

35. First determine the amount of HCl, and then the volume of the concentrated solution.

$$\text{amount HCl} = 12.5 \text{ L} \times \frac{10^{-2.10} \text{ mol H}_3O^+}{1 \text{ L soln}} \times \frac{1 \text{ mol HCl}}{1 \text{ mol H}_3O^+} = 0.099 \text{ mol HCl}$$

$$V_{\text{solution}} = 0.099 \text{ mol HCl} \times \frac{36.46 \text{ g HCl}}{1 \text{ mol HCl}} \times \frac{100.0 \text{ g soln}}{36.0 \text{ g HCl}} \times \frac{1 \text{ mL soln}}{1.18 \text{ g soln}} = 8.5 \text{ mL soln}$$

36. First determine the amount of KOH, and then the volume of the concentrated solution.

$$\text{pOH} = 14.00 - \text{pH} = 14.00 - 11.55 = 2.45 \qquad \left[OH^-\right] = 10^{-\text{pOH}} = 10^{-2.45} = 0.0035 \text{ M}$$

$$\text{amount KOH} = 25.0 \text{ L} \times \frac{0.0035 \text{ M mol OH}^-}{1 \text{ L soln}} \times \frac{1 \text{ mol KOH}}{1 \text{ mol OH}^-} = 0.088 \text{ mol KOH}$$

$$V_{\text{solution}} = 0.088 \text{ mol KOH} \times \frac{56.11 \text{ g KOH}}{1 \text{ mol KOH}} \times \frac{100.0 \text{ g soln}}{15.0 \text{ g KOH}} \times \frac{1 \text{ mL soln}}{1.14 \text{ g soln}} = 29 \text{ mL soln}$$

37. The volume of HCl(aq) needed is determined by first finding the amount of $NH_3(aq)$ present, and then realizing that acid and base react in a 1:1 molar ratio.

$$V_{\text{HCl}} = 1.25 \text{ L base} \times \frac{0.265 \text{ mol NH}_3}{1 \text{ L base}} \times \frac{1 \text{ mol H}_3O^+}{1 \text{ mol NH}_3} \times \frac{1 \text{ mol HCl}}{1 \text{ mol H}_3O^+} \times \frac{1 \text{ L acid}}{6.15 \text{ mol HCl}}$$

$$= 0.0539 \text{ L acid}$$

38. NH_3 and HCl react in a 1:1 molar ratio. Since, by Avogadro's hypothesis, equal volumes of gases measured at the same temperature and pressure contain equal numbers of moles, we need to determine the volume at 762 mmHg and $21.0°C$ that is equivalent to 28.2 L at 742 mmHg and $25.0°C$.

$$V_{NH_3(g)} = 28.2 \text{ L HCl(g)} \times \frac{742 \text{ mmHg}}{762 \text{ mmHg}} \times \frac{(273.2+21.0) \text{ K}}{(273.2+25.0) \text{ K}} \times \frac{1 \text{ L NH}_3(g)}{1 \text{ L HCl(g)}} = 27.1 \text{ L NH}_3(g)$$

Alternatively, we can make sure that we get the ratios correct with the approach of setting up and solving the ideal gas equation for the amount of each gas, and then equating these two expressions and solving for the volume of NH_3.

$$n\{HCl\} = \frac{742 \text{ mmHg} \cdot 28.2 \text{ L}}{R \cdot 298.2 \ K} \qquad n\{NH_3\} = \frac{762 \text{ mmHg} \cdot V\{NH_3\}}{R \cdot 294.2 \ K}$$

$$\frac{742 \text{ mmHg} \cdot 28.2 \text{ L}}{R \cdot 298.2 \ K} = \frac{762 \text{ mmHg} \cdot V\{NH_3\}}{R \cdot 294.2 \ K} \quad \text{This yields } V\{NH_3\} = 27.1 \text{ L NH}_3(g)$$

39. We determine the amounts of H_3O^+ and OH^- and then the amount of the one that is in excess. We express molar concentration in millimoles/milliliter, equivalent to mol/L.

$$50.00 \text{ mL} \times \frac{0.0155 \text{ mmol HI}}{1 \text{ mL soln}} \times \frac{1 \text{ mmol H}_3O^+}{1 \text{ mmol HI}} = 0.775 \text{ mmol H}_3O^+$$

$$75.00 \text{ mL} \times \frac{0.0106 \text{ mmol KOH}}{1 \text{ mL soln}} \times \frac{1 \text{ mmol OH}^-}{1 \text{ mmol KOH}} = 0.795 \text{ mmol OH}^-$$

The net reaction is $H_3O^+ (aq) + OH^- (aq) \rightarrow 2H_2O(l)$.

There is an excess amount of OH^- of $(0.795 - 0.775 =)$ 0.020 mmol OH^-.
This is a basic solution. The total solution volume is $(50.00 + 75.00 =)$ 125.00 mL.

$$[OH^-] = \frac{0.020 \text{ mmol OH}^-}{125.00 \text{ mL}} = 1.6 \times 10^{-4} \text{ M,}$$

$$pOH = -\log(1.6 \times 10^{-4}) = 3.80, \quad pH = 14.00 - 3.80 = 10.20$$

40. In each case, we need to determine the $[H_3O^+]$ or $[OH^-]$ of each solution being mixed, and then the amount of H_3O^+ or OH^-, so that we can determine the amount in excess. We express molar concentration in millimoles/milliliter, equivalent to mol/L.

$[H_3O^+] = 10^{-2.12} = 7.6 \times 10^{-3}$ M   moles $H_3O^+ = 25.00 \text{ mL} \times 7.6 \times 10^{-3}$ M $= 0.19 \text{ mmol H}_3O^+$

$pOH = 14.00 - 12.65 = 1.35$ $[OH^-] = 10^{-1.35} = 4.5 \times 10^{-2}$ M

amount $OH^- = 25.00 \text{ mL} \times 4.5 \times 10^{-2}$ M $= 1.13 \text{ mmol OH}^-$

There is excess OH^- in the amount of 0.94 mmol $(= 1.13 \text{ mmol OH}^- - 0.19 \text{ mmol H}_3O^+)$

$$[OH^-] = \frac{0.94 \text{ mmol OH}^-}{25.00 \text{ mL} + 25.00 \text{ mL}} = 1.9 \times 10^{-2} \text{ M} \qquad pOH = -\log(1.9 \times 10^{-2}) = 1.72$$

$pH = 14.00 - 1.72 = 12.28$

Weak Acids, Weak Bases, and pH

41. We organize the solution around the balanced chemical equation.

Equation: $HNO_2(aq) + H_2O(l) \rightleftharpoons H_3O^+(aq) + NO_2^-(aq)$

Initial: 0.143 M ≈ 0 M 0 M

Changes: $-x$ M $+x$ M $+x$ M

Equil: $(0.143-x)$ M x M x M

$$K_a = \frac{[H_3O^+][NO_2^-]}{[HNO_2]} = \frac{x^2}{0.143-x} = 7.2\times10^{-4} \gg \frac{x^2}{0.143} \text{ (if } x \ll 0.143)$$

$$x = \sqrt{0.143\times7.2\times10^{-4}} = 0.010 \text{ M}$$

We have assumed that $x \ll 0.143$ M, an almost-correct assumption. Another cycle of approximations yields:

$$x = \sqrt{(0.143-0.010)\times7.2\times10^{-4}} = 0.0098 \text{ M} = [H_3O^+] \qquad pH = -\log(0.0098) = 2.01$$

This is the same result as is determined with the quadratic equation.

42. We organize the solution around the balanced chemical equation, and solve first for $[OH^-]$

Equation: $C_2H_5NH_2(aq) + H_2O(l) \rightleftharpoons OH^-(aq) + C_2H_5NH_3^+(aq)$

Initial: 0.085 M ≈ 0 M 0 M

Changes: $-x$ M $+x$ M $+x$ M

Equil: $(0.085-x)$ M x M x M

$$K_b = \frac{[OH^-][C_2H_5NH_3^+]}{[C_2H_5NH_2]} = \frac{x^2}{0.085-x} = 4.3\times10^{-4} \approx \frac{x^2}{0.085} \text{ assuming } x \ll 0.085$$

$$x = \sqrt{0.085\times4.3\times10^{-4}} = 0.0060 \text{ M}$$

We have assumed that $x \ll 0.085$ M, an almost-correct assumption. Another cycle of approximations yields:

$$x = \sqrt{(0.085-0.0060)\times4.3\times10^{-4}} = 0.0058 \text{ M} = [OH^-] \qquad \text{yet another cycle produces:}$$

$$x = \sqrt{(0.085-0.0058)\times4.3\times10^{-4}} = 0.0058 \text{ M} = [OH^-] \qquad pOH = -\log(0.0058) = 2.24$$

$$pH = 14.00 - 2.24 = 11.76 \qquad [H_3O^+] = 10^{-pH} = 10^{-11.76} = 1.7\times10^{-12} \text{ M}$$

This is the same result as determined with the quadratic equation.

43. We base our set up on the balanced chemical equation. $[H_3O^+] = 10^{-1.56} = 2.8 \times 10^{-2}$ M

Equation :	$CH_2FCOOH(aq)$	$+ H_2O(l)$	$\rightleftharpoons$	$CH_2FCOO^-(aq)$	$+ H_3O^+(aq)$
Initial :	0.318 M			0 M	0 M
Changes :	-0.028 M			$+0.028$ M	$+0.028$ M
Equil :	0.290 M			0.028 M	0.028 M

$$K_a = \frac{[H_3O^+][CH_2FCOO^-]}{[CH_2FCOOH]} = \frac{(0.028)(0.028)}{0.290} = 2.7 \times 10^{-3}$$

44. $[HC_6H_{11}O_2] = \dfrac{11 \text{ g}}{1 \text{ L}} \times \dfrac{1 \text{ mol } HC_6H_{11}O_2}{116.2 \text{ g } HC_6H_{11}O_2} = 9.5 \times 10^{-2}$ M $= 0.095$ M

$[H_3O^+] = 10^{-2.94} = 1.1 \times 10^{-3}$ M $= 0.0011$ M

The stoichiometry of the reaction, written below, indicates that $[H_3O^+] = [C_6H_{11}O_2^-]$

Equation :	$HC_6H_{11}O_2(aq) + H_2O(l)$	$\rightleftharpoons$	$C_6H_{11}O_2^-(aq)$	$+$	$H_3O^+(aq)$
Initial :	0.095 M		0 M		≈ 0 M
Changes :	-0.0011 M		$+0.0011$ M		$+0.0011$ M
Equil :	0.094 M		0.0011 M		0.0011 M

$$K_a = \frac{[C_6H_{11}O_2^-][H_3O^+]}{[HC_6H_{11}O_2]} = \frac{(0.0011)^2}{0.094} = 1.3 \times 10^{-5}$$

45. We use the balanced chemical equation, then solve using the quadratic formula.

Equation :	$HClO_2(aq)$	$+$	$H_2O(l)$	$\rightleftharpoons$	$H_3O^+(aq)$	$+$	$ClO_2^-(aq)$
Initial :	0.55 M				≈ 0 M		0 M
Changes :	$-x$ M				$+x$ M		$+x$ M
Equil :	$(0.55-x)$ M				x M		x M

$$K_a = \frac{[H_3O^+][ClO_2^-]}{[HClO_2]} = \frac{x^2}{0.55-x} = 1.1 \times 10^{-2} = 0.011 \qquad x^2 = 0.0061 - 0.011x$$

$x^2 + 0.011x - 0.0061 = 0$

$$x = \frac{-b \pm \sqrt{b^2 - 4ac}}{2a} = \frac{-0.011 \pm \sqrt{0.000121 + 0.0244}}{2} = 0.073 \text{ M} = [H_3O^+]$$

The method of successive approximations converges to the same answer in four cycles.

$pH = -\log[H_3O^+] = -\log(0.073) = 1.14 \qquad pOH = 14.00 - pH = 14.00 - 1.14 = 12.86$

$[OH^-] = 10^{-pOH} = 10^{-12.86} = 1.4 \times 10^{-13}$ M

46. Organize the solution around the balanced chemical equation, and solve first for $[OH^-]$.

Equation: $CH_3NH_2(aq)$ + $H_2O(l)$ $\rightleftharpoons$ $OH^-(aq)$ + $CH_3NH_3^+(aq)$

Initial: 0.386 M ≈ 0 M 0 M

Changes: $-x$ M $+x$ M $+x$ M

Equil: $(0.386-x)$ M x M x M

$$K_b = \frac{[OH^-][CH_3NH_3^+]}{[CH_3NH_2]} = \frac{x^2}{0.386-x} = 4.2\times10^{-4} \approx \frac{x^2}{0.386} \quad \text{assuming } x \ll 0.386$$

$$x = \sqrt{0.386 \times 4.2\times10^{-4}} = 0.013 \text{ M} = [OH^-] \quad pOH = -\log(0.013) = 1.89$$

$$pH = 14.00 - 1.89 = 12.11 \qquad [H_3O^+] = 10^{-pH} = 10^{-12.11} = 7.8\times10^{-13} \text{ M}$$

This is the same result as is determined with the quadratic equation.

47.
$$[C_{10}H_7NH_2] = \frac{1 \text{ g } C_{10}H_7NH_2}{590 \text{ g } H_2O} \times \frac{1.00 \text{ g } H_2O}{1 \text{ mL}} \times \frac{1000 \text{ mL}}{1 \text{ L}} \times \frac{1 \text{ mol } C_{10}H_7NH_2}{143.2 \text{ g } C_{10}H_7NH_2}$$
$$= 0.012 \text{ M } C_{10}H_7NH_2$$

$$K_b = 10^{-pK} = 10^{-3.92} = 1.2\times10^{-4}$$

Equation: $C_{10}H_7NH_2(aq)$ + $H_2O(l)$ $\rightleftharpoons$ $OH^-(aq)$ + $C_{10}H_7NH_3^+(aq)$

Initial: 0.012 M ≈ 0 M 0 M

Changes: $-x$ M $+x$ M $+x$ M

Equil: $(0.012-x)$ M x M x M

$$K_b = \frac{[OH^-][C_{10}H_7NH_3^+]}{[C_{10}H_7NH_2]} = \frac{x^2}{0.012-x} = 1.2\times10^{-4} \approx \frac{x^2}{0.012} \qquad \text{assuming } x \ll 0.012$$

$x = \sqrt{0.012 \times 1.2\times10^{-4}} = 0.0012$ This is an almost-correct assumption. Another approximation cycle gives:

$x = \sqrt{(0.012-0.0012) \times 1.2\times10^{-4}} = 0.0011$ Yet another cycle seems necessary.

$x = \sqrt{(0.012-0.0011) \times 1.2\times10^{-4}} = 0.0011 \text{ M} = [OH^-]$

$$pOH = -\log[OH^-] = -\log(0.0011) = 2.96 \qquad pH = 14.00 - pOH = 11.04$$

$$H_3O^+ = 10^{-pH} = 10^{-11.04} = 9.1\times10^{-12} \text{ M}$$

48. The stoichiometry of the ionization reaction indicates that

$$\left[H_3O^+\right]=\left[OC_6H_4NO_2^-\right], K_a = 10^{-pK} = 10^{-7.23} = 5.9 \times 10^{-8} \text{ and}$$

$$\left[H_3O^+\right]=10^{-4.53} = 3.0 \times 10^{-5} M. \text{ We let } S = \text{ the molar solubility of } o\text{-nitrophenol.}$$

Equation:	$HOC_6H_4NO_2(aq) + H_2O(l)$	$\rightleftharpoons$	$OC_6H_4NO_2^-(aq)$ +	$H_3O^+(aq)$
Initial:	S		$0\ M$	$\approx 0\ M$
Changes:	-3.0×10^{-5} M		$+3.0 \times 10^{-5}$ M	$+3.0 \times 10^{-5}$ M
Equil:	$(S - 3.0 \times 10^{-5})$ M		3.0×10^{-5} M	3.0×10^{-5} M

$$K_a = 5.9 \times 10^{-8} = \frac{\left[OC_6H_4NO_2^-\right]\left[H_3O^+\right]}{\left[HOC_6H_5NO_2\right]} = \frac{\left(3.0 \times 10^{-5}\right)^2}{S - 3.0 \times 10^{-5}}$$

$$S - 3.0 \times 10^{-5} = \frac{\left(3.0 \times 10^{-5}\right)^2}{5.9 \times 10^{-8}} = 1.5 \times 10^{-2} M \quad S = 1.5 \times 10^{-2} M + 3.0 \times 10^{-5} M = 1.5 \times 10^{-2} M$$

Hence,

$$\text{solubility} = \frac{1.5 \times 10^{-2}\ \text{mol HOC}_6\text{H}_4\text{NO}_2}{1\ \text{L soln}} \times \frac{139.1\ \text{g HOC}_6\text{H}_4\text{NO}_2}{1\ \text{mol HOC}_6\text{H}_4\text{NO}_2} = 2.1\ \text{g HOC}_6\text{H}_4\text{NO}_2\ /\ \text{L soln}$$

49. We determine $\left[H_3O^+\right]$ which, because of the stoichiometry of the reaction, equals $\left[C_2H_3O_2^-\right]$.

$$\left[H_3O^+\right]=10^{-pH} = 10^{-4.52} = 3.0 \times 10^{-5} M = \left[C_2H_3O_2^-\right]$$

We solve for S, the concentration of $HC_2H_3O_2$ in the 0.750 L solution before it dissociates.

Equation:	$HC_2H_3O_2(aq)$ +	$H_2O(l)$	$\rightleftharpoons$	$C_2H_3O_2^-(aq)$ +	$H_3O^+(aq)$
initial:	S M			$0\ M$	$\approx 0M$
changes:	$-3.0 \times 10^{-5} M$			$+3.0 \times 10^{-5} M$	$+3.0 \times 10^{-5} M$
equil:	$\left(S - 3.0 \times 10^{-5}\right)$ M			$3.0 \times 10^{-5} M$	$3.0 \times 10^{-5} M$

$$K_a = \frac{[H_3O^+][C_2H_3O_2^-]}{[HC_2H_3O_2]} = 1.8 \times 10^{-5} = \frac{(3.0 \times 10^{-5})^2}{(S - 3.0 \times 10^{-5})}$$

$$(3.0 \times 10^{-5})^2 = 1.8 \times 10^{-5}(S - 3.0 \times 10^{-5}) = 9.0 \times 10^{-10} = 1.8 \times 10^{-5} S - 5.4 \times 10^{-10}$$

$$S = \frac{9.0 \times 10^{-10} + 5.4 \times 10^{-10}}{1.8 \times 10^{-5}} = 8.0 \times 10^{-5} M$$

Now we determine the mass of vinegar needed.

$$\text{mass vinegar} = 0.750\ L \times \frac{8.0 \times 10^{-5}\ \text{mol HC}_2\text{H}_3\text{O}_2}{1\ \text{L soln}} \times \frac{60.05\ \text{g HC}_2\text{H}_3\text{O}_2}{1\ \text{mol HC}_2\text{H}_3\text{O}_2} \times \frac{100.0\ \text{g vinegar}}{5.7\ \text{g HC}_2\text{H}_3\text{O}_2}$$

$$= 0.063\ \text{g vinegar}$$

50. We determine $[OH^-]$ which, because of the stoichiometry of the reaction, equals $[NH_4^+]$.

$$pOH = 14.00 - pH = 14.00 - 11.55 = 2.45$$

$$[OH^-] = 10^{-pOH} = 10^{-2.45} = 3.5 \times 10^{-3} M = [NH_4^+]$$

We solve for S, the concentration of NH_3 in the 0.625 L solution before it dissociates.

Equation: $NH_3(aq) \ + \ H_2O(l) \ \rightleftharpoons \ NH_4^+(aq) \ + \ OH^-(aq)$

initial:	S M	0 M	≈ 0 M
changes:	-0.0035 M	$+0.0035$ M	$+0.0035$ M
equil:	$(S-0.0035)$ M	0.0035 M	0.0035 M

$$K_b = \frac{[NH_4^+][OH^-]}{[NH_3]} = 1.8 \times 10^{-5} = \frac{(0.0035)^2}{(S-0.0035)}$$

$$(0.0035)^2 = 1.8 \times 10^{-5}(S-0.0035) = 1.2 \times 10^{-5} = 1.8 \times 10^{-5}S - 6.3 \times 10^{-8}$$

$$S = \frac{1.2 \times 10^{-5} + 6.3 \times 10^{-8}}{1.8 \times 10^{-5}} = 0.67 \text{ M}$$

Now we determine the vol. of household ammonia needed

$$V_{ammonia} = 0.625 \text{ L soln} \times \frac{0.67 \text{ mol } NH_3}{1 \text{ L soln}} \times \frac{17.03 \text{ g } NH_3}{1 \text{ mol } NH_3} \times \frac{100.0 \text{ g soln}}{6.8 \text{ g } NH_3} \times \frac{1 \text{ mL soln}}{0.97 \text{ g soln}}$$

$$= 1.1 \times 10^2 \text{ mL household ammonia solution}$$

51. **(a)** $n_{proplyamine} = \dfrac{PV}{RT} = \dfrac{(316 \text{ torr} \times \dfrac{1 \text{ atm}}{760 \text{ torr}})(0.275 \text{ L})}{0.08206 \text{ L atm K}^{-1} \text{ mol}^{-1})(298.15 \text{ K})} = 4.67 \times 10^{-3}$ mol propylamine

$$[propylamine] = \frac{n}{V} = \frac{4.67 \times 10^{-3} \text{ moles}}{0.500 \text{ L}} = 9.35 \times 10^{-3} \text{ M}$$

$$K_b = 10^{-pK_b} = 10^{-3.43} = 3.7 \times 10^{-4}$$

$CH_3CH_2CH_2NH_2(aq) + H_2O(l) \ \rightleftharpoons \ CH_3CH_2CH_2NH_3^+(aq) \ + \ OH^-(aq)$

Initial	9.35×10^{-3} M	0 M	≈ 0 M
Change	$-x$ M	$+x$ M	$+x$ M
Equil.	$(9.35 \times 10^{-3} - x)$ M	x M	x M

$$K_b = 3.7 \times 10^{-4} = \frac{x^2}{9.35 \times 10^{-3} - x} \text{ or } 3.5 \times 10^{-6} - 3.7 \times 10^{-4}(x) = x^2$$

$x^2 + 3.7 \times 10^{-4}x - 3.5 \times 10^{-6} = 0$ solve quadratic:

$$x = \frac{-3.7 \times 10^{-4} \pm \sqrt{(3.7 \times 10^{-4})^2 - 4(1)(-3.5 \times 10^{-6})}}{2(1)}$$

Therefore $x = 1.695 \times 10^{-3}$ M $= [OH^-]$ $pOH = 2.77$ and $pH = 11.23$

(b) $[OH^-] = 1.7 \times 10^{-3}\ M = [NaOH]$ $(MM_{NaOH} = 39.997\ g\ mol^{-1})$
$n_{NaOH} = (C)(V) = 1.7 \times 10^{-3}\ M \times 0.500\ L = 8.5 \times 10^{-4}$ moles NaOH
mass of NaOH $= (n)(MM_{NaOH}) = 8.5 \times 10^{-4}$ mol NaOH $\times\ 39.997\ g\ NaOH/mol\ NaOH$
mass of NaOH $= 0.034\ g$ NaOH(34 mg of NaOH)

52. $K_b = 10^{-pK_b} = 10^{-9.5} = 3.\underline{2} \times 10^{-10}$ $MM_{quinoline} = 129.16\ g\ mol^{-1}$

Solubility(25 °C) $= \dfrac{0.6\ g}{100\ mL}$

Molar solubility(25 °C) $= \dfrac{0.6\ g}{100\ mL} \times \dfrac{1\ mol}{129.16\ g} \times \dfrac{1000\ mL}{1\ L} = 0.04\underline{6}\ M$

(note: final answer only 1 sig fig)

$$C_9H_7N(aq) \quad + H_2O(l) \;\rightleftharpoons\; C_9H_7NH^+(aq) \quad + \quad OH^-(aq)$$

Initial	0.046 M	0 M	≈ 0 M
Change	$-x$ M	$+x$ M	$+x$ M
Equil.	$(0.046 - x)$ M	x M	x M

$3.\underline{2} \times 10^{-10} = \dfrac{x^2}{0.04_6 - x} \approx \dfrac{x^2}{0.04_6},$ $x = 3.\underline{8} \times 10^{-6}\ M$ $(x \ll 0.04\underline{6}$, valid assumption)

Therefore, $[OH^-] = 4 \times 10^{-6}\ M$ pOH $= 5.4$ and pH $= 8.6$

53. If the molarity of acetic acid is doubled, we expect a lower initial pH (more $H_3O^+(aq)$ in solution) and a lower percent ionization as the concentration increases. Therefore (b) (the diagram in the center) best represents the conditions $(\sim(2)^{1/2}$ times greater).

54. If NH_3 is diluted to half its original molarity, we expect a lower pH (a lower $[OH^-]$) and a higher percent ionization in the diluted sample. The diagram that is next to the one furthest to the right best represents the hydroxide ion concentration ($\sim (2)^{-1/2}$ times the original $[OH^-]$).

Percent Ionization

55. Let us first compute the $\left[H_3O^+\right]$ in this solution.

Equation: $HC_3H_5O_2(aq) \quad + \quad H_2O(l) \;\rightleftharpoons\; H_3O^+(aq) \quad + \quad C_3H_5O_2^-(aq)$

Initial:	0.45 M	≈ 0 M	≈ 0 M
Changes:	$-x$ M	$+x$ M	$+x$ M
Equil:	$(0.45 - x)$ M	x M	x M

$K_a = \dfrac{\left[H_3O^+\right]\left[C_3H_5O_2^-\right]}{\left[HC_3H_5O_2\right]} = \dfrac{x^2}{0.45 - x} = 10^{-4.89} = 1.3 \times 10^{-5} \approx \dfrac{x^2}{0.45}$

$x = 2.4 \times 10^{-3}\ M$; We have assumed that $x \ll 0.45\ M$, an assumption that clearly is correct.

(a) $\alpha = \dfrac{\left[H_3O^+\right]_{equil}}{\left[HC_3H_5O_2\right]_{initial}} = \dfrac{2.4\times10^{-3}\,M}{0.45\,M} = 0.0053 = $ degree of ionization

(b) % ionization $= \alpha \times 100\% = 0.0053 \times 100\% = 0.53\%$

56. For $C_2H_5NH_2$ (ethylamine), $K_b = 4.3 \times 10^{-4}$

$$C_2H_5NH_2(aq) + H_2O(l) \rightleftharpoons C_2H_5NH_3^+(aq) + OH^-(aq)$$

Initial	0.85 M		0 M	$\approx$ 0 M
Change	$-x$ M		$+x$ M	$+x$ M
Equil.	(0.85 $-x$) M		x M	x M

$$4.3 \times 10^{-4} = \frac{x^2}{(0.85-x)\,M} \approx \frac{x^2}{0.85\,M} \qquad x = 0.019\,M \ \ (x \ll 0.85, \text{ valid assumption})$$

Degree of ionization $\alpha = \dfrac{0.019\,M}{0.85\,M} = 0.022$ Percent ionization $= \alpha \times 100\% = 2.2\,\%$

57. The fact that the NH_3 is 4.2% ionized means that $\left[NH_4^+\right] = 0.042\left[NH_3\right]_{initial} = \left[OH^-\right]$. Expressed another way, this is $\left[NH_3\right]_{initial} = 23.8\left[OH^-\right]$. $(23.8 = 1 \div 0.042.)$ Of course, we also know that $\left[NH_3\right]_{equil} = \left[NH_3\right]_{initial} - \left[OH^-\right]$. We use the base dissociation constant expression for NH_3 to organize our information.

$$K_b = 1.8\times10^{-5} = \frac{\left[NH_4^+\right]\left[OH^-\right]}{\left[NH_3\right]_{equil}} = \frac{\left[OH^-\right]^2}{23.8\times\left[OH^-\right]-\left[OH^-\right]} = \frac{\left[OH^-\right]}{22.8}$$

$\left[OH^-\right] = 22.8\times1.8\times10^{-5} = 4.1\times10^{-4}\,M \qquad \left[NH_3\right]_{initial} = 23.8\times4.1\times10^{-4} = 0.0098\,M$

58.
$$HC_3H_5O_2(aq) + H_2O(l) \xrightarrow{K_a = 1.3 \times 10^{-5}} C_3H_5O_2^-(aq) + H_3O^+(aq)$$

Initial	0.100 M		0 M	$\approx$ 0 M
Change	$-x$ M		$+x$ M	$+x$ M
Equil.	(0.100 $-x$) M		x M	x M

$$1.3 \times 10^{-5} = \frac{x^2}{0.100\,M - x} \approx \frac{x^2}{0.100} \qquad x = 1.1 \times 10^{-3} \ \ (x \ll 0.100, \text{ valid assumption})$$

Percent ionization $= \dfrac{0.0011}{0.100} \times 100\% = 1.1\underline{4}\,\%$

Need to find molarity of acetic acid that is 1.1% ionized

$$HC_2H_3O_2(aq) \ + \ H_2O(l) \ \xrightleftharpoons{K_a = 1.8 \times 10^{-5}} \ C_2H_3O_2^-(aq) \ + \ H_3O^+(aq)$$

Initial	X M	0 M	≈ 0 M
Change	$-\dfrac{1.1\underline{4}}{100} X$ M	$+\dfrac{1.1\underline{4}}{100} X$ M	$+\dfrac{1.1\underline{4}}{100} X$ M
Equil.	$X - \dfrac{1.1\underline{4}}{100} X$	$\dfrac{1.1\underline{4}}{100} X$ M	$\dfrac{1.1\underline{4}}{100} X$ M
	$= 0.98\underline{86}\, X$ M	$= 0.011\underline{4}$ M	$= 0.011\underline{4}$ M

$$1.8 \times 10^{-5} = \frac{\left(\dfrac{1.1\underline{4}}{100} X\right)^2}{0.9886 \ X} = 1.3 \times 10^{-4}(X); \qquad X = 0.13\underline{7} \text{ M}$$

Approximately 0.14 moles of acetic acid must be dissolved in one liter of water in order to have the same percent ionization as 0.100 M propionic acid.

59. We would not expect these ionizations to be correct because the calculated degree of ionization is based on the assumption that the $[HC_2H_3O_2]_{\text{initial}} \approx [HC_2H_3O_2]_{\text{initial}} - [HC_2H_3O_2]_{\text{equil.}}$, which is invalid at the 13 and 42 percent levels of ionization seen here.

60. $HC_2Cl_3O_2$ (trichloroacetic acid) $pK_a = 0.52$ $K_a = 10^{-0.52} = 0.30$
For a 0.035 M solution, the assumption will not work (concentration small and K_a large) Must solve the quadratic equation.

$$HC_2Cl_3O_2(aq) \ + H_2O(l) \ \rightleftharpoons \ C_2Cl_3O_2^-(aq) \ + \ H_3O^+(aq)$$

Initial	0.035 M	0 M	≈ 0 M
Change	$-x$ M	$+x$ M	$+x$ M
Equil.	$(0.035 - x)$ M	x M	x M

$$0.30 = \frac{x^2}{0.035 - x} \quad \text{or} \quad 0.010\underline{5} - 0.30(x) = x^2 \quad \text{or} \quad x^2 + 0.30x - 0.010\underline{5} = 0 \text{ ; solve quadratic:}$$

$$x = \frac{-0.30 \pm \sqrt{(0.30)^2 - 4(1)(0.010\underline{5})}}{2(1)} \qquad \text{Therefore } x = 0.032 \text{ M} = [H_3O^+]$$

$$\text{Degree of ionization } \alpha = \frac{0.032 \text{ M}}{0.035 \text{ M}} = 0.91 \quad \text{Percent ionization} = \alpha \times 100\% = 91. \%$$

Polyprotic Acids

61. Because H_3PO_4 is a weak acid, there is little HPO_4^{2-} (produced in the second ionization) compared to the H_3O^+ (produced in the first ionization). In turn, there is little PO_4^{3-} (produced in the third ionization) compared to the HPO_4^{2-}, and very little compared to the H_3O^+.

62. The main estimate involves assuming that the mass percents can be expressed as 0.057 g of 75% H_3PO_4 per 100. mL of solution and 0.084 g of 75% H_3PO_4 per 100. mL of solution. That is, that the density of the aqueous solution is essentially 1.00 g/mL. Based on this assumption, we can compute the initial concentrations of H_3PO_4.

$$[H_3PO_4] = \frac{0.057 \, g \, impure \, H_3PO_4 \times \dfrac{75 \, g \, H_3PO_4}{100 \, g \, impure \, H_3PO_4} \times \dfrac{1 \, mol \, H_3PO_4}{98.00 \, g \, H_3PO_4}}{100 \, mL \, soln \times \dfrac{1L}{1000 \, mL}} = 0.0044 \, M$$

$$[H_3PO_4] = \frac{0.084 \, g \, impure \, H_3PO_4 \times \dfrac{75 \, g \, H_3PO_4}{100 \, g \, impure \, H_3PO_4} \times \dfrac{1 \, mol \, H_3PO_4}{98.00 \, g \, H_3PO_4}}{100 \, mL \, soln \times \dfrac{1L}{1000 \, mL}} = 0.0064 \, M$$

Equation:	$H_3PO_4(aq)$	+	$H_2O(l)$	$\rightleftharpoons$	$H_2PO_4^-(aq)$	+	$H_3O^+(aq)$
Initial:	0.0044 M				0 M		≈ 0 M
Changes:	$-x$ M				$+x$ M		$+x$ M
Equil:	$(0.0044 - x)$ M				x M		x M

$$K_{a_1} = \frac{[H_2PO_4^-][H_3O^+]}{[H_3PO_4]} = \frac{x^2}{0.0044 - x} = 7.1 \times 10^{-3} \qquad x^2 + 0.0071x - 3.1 \times 10^{-5} = 0$$

$$x = \frac{-b \pm \sqrt{b^2 - 4ac}}{2a} = \frac{-0.0071 \pm \sqrt{5.0 \times 10^{-5} + 1.2 \times 10^{-4}}}{2} = 3.0 \times 10^{-3} \, M = [H_3O^+]$$

The set up for the second concentration is the same as for the first, with the exception of substitution 0.0064 M for 0.0044 M.

$$K_{a_1} = \frac{[H_2PO_4^-][H_3O^+]}{[H_3PO_4]} = \frac{x^2}{0.0064 - x} = 7.1 \times 10^{-3} \qquad x^2 + 0.0071x - 4.5 \times 10^{-5} = 0$$

$$x = \frac{-b \pm \sqrt{b^2 - 4ac}}{2a} = \frac{-0.0071 \pm \sqrt{5.0 \times 10^{-5} + 1.8 \times 10^{-4}}}{2} = 4.0 \times 10^{-3} \, M = [H_3O^+]$$

The two values of pH now are determined, representing the pH range in a cola drink.

$$pH = -\log(3.0 \times 10^{-3}) = 2.52 \qquad pH = -\log(4. \times 10^{-3}) = 2.40$$

63. **(a)** Equation: $H_2S(aq)$ $+$ $H_2O(l)$ $\rightleftharpoons$ $HS^-(aq)$ $+$ $H_3O^+(aq)$

Initial:	0.075 M		0 M	≈ 0 M
Changes:	$-x$ M		$+x$ M	$+x$ M
Equil:	$(0.075-x)$ M		x M	$+x$ M

$$K_{a_1} = \frac{[HS^-][H_3O^+]}{[H_2S]} = 1.0 \times 10^{-7} = \frac{x^2}{0.075-x} \approx \frac{x^2}{0.075} \quad x = 8.7 \times 10^{-5} M = [H_3O^+]$$

$$[HS^-] = 8.7 \times 10^{-5} M \text{ and } [S^{2-}] = K_{a_2} = 1 \times 10^{-19} M$$

(b) The set up for this problem is the same as for part (a), with the substitution of 0.0050 M for 0.075 M as the initial value of $[H_2S]$.

$$K_{a_1} = \frac{[HS^-][H_3O^+]}{[H_2S]} = 1.0 \times 10^{-7} = \frac{x^2}{0.0050-x} \approx \frac{x^2}{0.0050} \quad x = 2.2 \times 10^{-5} M = [H_3O^+]$$

$$[HS^-] = 2.2 \times 10^{-5} M \text{ and } [S^{2-}] = K_{a_2} = 1 \times 10^{-19} M$$

(c) The set up for this part is the same as for part (a), with the substitution of 1.0×10^{-5} M for 0.075 M as the initial value of $[H_2S]$. The solution differs in that we cannot assume $x \ll 1.0 \times 10^{-5}$. Rather than solve the quadratic equation, however, we shall use the method of successive approximations.

$$K_{a_1} = \frac{[HS^-][H_3O^+]}{[H_2S]} = 1.0 \times 10^{-7} = \frac{x^2}{1.0 \times 10^{-5} - x} \approx \frac{x^2}{1.0 \times 10^{-5}} \quad x = 1.0 \times 10^{-6} M = [H_3O^+]$$

$$K_{a_1} = \frac{[HS^-][H_3O^+]}{[H_2S]} = 1.0 \times 10^{-7} = \frac{x^2}{1.0 \times 10^{-5} - 1.0 \times 10^{-6}} = \frac{x^2}{9.0 \times 10^{-6}} \quad x = 9.5 \times 10^{-7} M$$

$$K_{a_1} = \frac{[HS^-][H_3O^+]}{[H_2S]} = 1.0 \times 10^{-7} = \frac{x^2}{1.0 \times 10^{-5} - 9.5 \times 10^{-7}} = \frac{x^2}{9.0 \times 10^{-6}} \quad x = 9.5 \times 10^{-7} M$$

$$x = 9.5 \times 10^{-7} M = [H_3O^+] \quad [HS^-] = 9.5 \times 10^{-7} M \quad [S^{2-}] = K_{a_2} = 1 \times 10^{-19} M$$

64. In all cases, of course, the first ionization of H_2SO_4 is complete, and establishes the initial values of $[H_3O^+]$ and $[HSO_4^-]$. We deal with the second ionization in each case.

(a) Equation: $HSO_4^-(aq)$ $+$ $H_2O(l)$ $\rightleftharpoons$ $SO_4^{2-}(aq)$ $+$ $H_3O^+(aq)$

Initial:	0.75 M		0 M	0.75 M
Changes:	$-x$ M		$+x$ M	$+x$ M
Equil:	$(0.75-x)$ M		x M	$(0.75+x)$ M

$$K_{a_2} = \frac{\left[SO_4^{2-}\right]\left[H_3O^+\right]}{\left[HSO_4^-\right]} = 0.011 = \frac{x(0.75+x)}{0.75-x} \approx \frac{0.75x}{0.75} \qquad x = 0.011\,M = \left[SO_4^{2-}\right]$$

We have assumed that $x \ll 0.75$ M, an assumption that clearly is correct.

$$\left[HSO_4^-\right] = 0.75 - 0.011 = 0.74\,M \qquad \left[H_3O^+\right] = 0.75 + 0.011 = 0.76\,M$$

(b)   The set up for this part is similar to part (a), with the exception of substituting 0.075 M for 0.75 M.

$$K_{a_2} = \frac{\left[SO_4^{2-}\right]\left[H_3O^+\right]}{\left[HSO_4^-\right]} = 0.011 = \frac{x(0.075+x)}{0.075-x} \qquad 0.011(0.075-x) = 0.075x + x^2$$

$$x^2 + 0.086x - 8.3\times10^{-4} = 0 , \ x = \frac{-b \pm \sqrt{b^2-4ac}}{2a} = \frac{-0.086 \pm \sqrt{0.0074+0.0033}}{2} = 0.0087\,M$$

$$x = 0.0087\,M = \left[SO_4^{2-}\right]$$

$$\left[HSO_4^-\right] = 0.075 - 0.0087 = 0.066\,M \qquad \left[H_3O^+\right] = 0.075 + 0.0088\,M = 0.084\,M$$

(c)   Again, the set up is the same as for part (a), with the exception of substituting 0.00075 M for 0.75 M.

$$K_{a_2} = \frac{[SO_4^{2-}][H_3O^+]}{[HSO_4^-]} = 0.011 = \frac{x(0.00075+x)}{0.00075-x} \qquad \begin{array}{l} 0.011(0.00075-x) = 0.00075\,x + x^2 \\ x^2 + 0.0118x - 8.3\times10^{-6} = 0 \end{array}$$

$$x = \frac{-b \pm \sqrt{b^2-4ac}}{2a} = \frac{-0.0118 \pm \sqrt{1.39\times10^{-4}+3.3\times10^{-5}}}{2} = 6.6\times10^{-4}$$

$$x = 6.6\times10^{-4}\,M = \left[SO_4^{2-}\right] \qquad \left[HSO_4^-\right] = 0.00075 - 0.00066 = 9\times10^{-5}\,M$$

$\left[H_3O^+\right] = 0.00075 + 0.00066\,M = 1.41\times10^{-3}\,M$ $\left[H_3O^+\right]$ is almost twice the initial value of $\left[H_2SO_4\right]$. The second ionization of H_2SO_4 is nearly complete in this dilute solution.

65.   First determine $\left[H_3O^+\right]$, with calculation based, on the balanced chemical equation.

Equation: $HOOC(CH_2)_4 COOH(aq) + H_2O(l) \rightleftharpoons HOOC(CH_2)_4 COO^-(aq) + H_3O^+(aq)$

Initial:	0.10 M	0 M	≈ 0 M
Changes:	$-x$ M	$+x$ M	$+x$ M
Equil:	$(0.10-x)$ M	x M	x M

$$K_{a_1} = \frac{[H_3O^+][HOOC(CH_2)_4COO^-]}{[HOOC(CH_2)_4COOH]} = 3.9 \times 10^{-5} = \frac{x \cdot x}{0.10 - x} \approx \frac{x^2}{0.10}$$

$$x = \sqrt{0.10 \times 3.9 \times 10^{-5}} = 2.0 \times 10^{-3} \, M = [H_3O^+]$$

We see that our simplifying assumption, that $x \ll 0.10$ M, is indeed valid. Now we consider the second ionization. We shall see that very little H_3O^+ is produced in this ionization because of the small size of two numbers: K_{a_2} and $[HOOC(CH_2)_4COO^-]$. Again we base our calculation on the balanced chemical equation.

Equation: $HOOC(CH_2)_4COO^-(aq) + H_2O(l) \rightleftharpoons {}^-OOC(CH_2)_4COO^-(aq) + H_3O^+(aq)$

Initial:	2.0×10^{-3} M	0 M	2.0×10^{-3} M
Changes:	$-y$ M	$+y$ M	$+y$ M
Equil:	$(0.0020 - y)$ M	y M	$(0.0020 + y)$ M

$$K_{a_2} = \frac{[H_3O^+][{}^-OOC(CH_2)_4COO^-]}{[HOOC(CH_2)_4COO^-]} = 3.9 \times 10^{-6} = \frac{y(0.0020 + y)}{0.0020 - y} \approx \frac{0.0020y}{0.0020}$$

$$y = 3.9 \times 10^{-6} \, M = [{}^-OOC(CH_2)_4COO^-]$$

Again, we see that our assumption, that $y \ll 0.0020$ M, is valid. In addition, we also see that virtually no H_3O^+ is created in this second ionization. The concentrations of all species have been calculated above, with the exception of $[OH^-]$.

$$[OH^-] = \frac{K_w}{[H_3O^+]} = \frac{1.0 \times 10^{-14}}{2.0 \times 10^{-3}} = 5.0 \times 10^{-12} \, M; \qquad [HOOC(CH_2)_4COOH] = 0.10 \, M$$

$$[H_3O^+] = [HOOC(CH_2)_4COO^-] = 2.0 \times 10^{-3} \, M; \qquad [{}^-OOC(CH_2)_4COO^-] = 3.9 \times 10^{-6} \, M$$

66. (a) Recall that a base is a proton acceptor, in this case, accepting from H_2O.

First ionization: $C_{20}H_{24}O_2N_2 + H_2O \rightleftharpoons C_{20}H_{24}O_2N_2H^+ + OH^- \qquad pK_{b_1} = 6.0$

Second ionization: $C_{20}H_{24}O_2N_2H^+ + H_2O \rightleftharpoons C_{20}H_{24}O_2N_2H_2^{2+} + OH^- \quad pK_{b_2} = 9.8$

(b)

$$[C_{20}H_{24}O_2N_2] = \frac{1.00 \, g \, quinine \times \dfrac{1 \, mol \, quinine}{324.4 \, g \, quinine}}{1900 \ mL \times \dfrac{1 \, L}{1000 \, mL}} = 1.62 \times 10^{-3} \, M$$

$$K_{b_1} = 10^{-6.0} = 1 \times 10^{-6}$$

As always, set up the table and solve for the $[OH^-]$

Equation: $C_{20}H_{24}O_2N_2 (aq)\ +\ H_2O(l) \rightleftharpoons\ C_{20}H_{24}O_2N_2H^+(aq)\ +\ OH^-(aq)$

Initial:	0.00162 M	0 M	≈ 0 M
Changes:	$-x$ M	$+x$ M	$+x$ M
Equil:	$(0.00162-x)$ M	x M	x M

$$K_{b_1} = \frac{\left[C_{20}H_{24}O_2N_2H^+\right]\left[OH^-\right]}{\left[C_{20}H_{24}O_2N_2\right]} = 1\times10^{-6} = \frac{x^2}{0.00162-x} = \frac{x^2}{0.00162} \qquad x = 4\times10^{-5} M$$

Our assumption, that $x \ll 0.00162$, clearly is valid.

$$pOH = -\log\left(4\times10^{-5}\right) = 4.4 \quad pH = 14.00 - 4.4 = 9.6$$

Ions as Acids and Bases (Hydrolysis)

67. **(a)** KCl forms a neutral solution, being composed of the cation of a strong base, and the anion of a strong acid; no hydrolysis occurs.

(b) KF forms an alkaline (basic) solution, being composed of the cation of a strong base and the anion of a weak acid; the fluoride ion hydrolyzes.

$$F^-(aq) + H_2O(l) \rightleftharpoons HF(aq) + OH^-(aq)$$

(c) $NaNO_3$ forms a neutral solution, being composed of the cation of a strong base and the anion of a strong acid. No hydrolysis occurs.

(d) $Ca(OCl)_2$ forms an alkaline (basic) solution, being composed of the cation of a strong base, and the anion of a weak acid; the hypochlorite ion hydrolyzes.

$$OCl^-(aq) + H_2O(l) \rightleftharpoons HOCl(aq) + OH^-(aq)$$

(e) NH_4NO_2 forms an acidic solution. The salt is composed of the cation of a weak base and the anion of a weak acid; the ammonium ion hydrolyzes:

$$NH_4^+(aq) + H_2O(l) \rightleftharpoons NH_3(aq) + H_3O^+(aq) \qquad K_a = 5.6\times10^{-10}$$

So does the nitrite ion:

$$NO_2^-(aq) + H_2O(l) \rightleftharpoons HNO_2(aq) + OH^-(aq) \qquad K_b = 1.4\times10^{-11}$$

Since NH_4^+ is a stronger acid than NO_2^- is a base, the solution will be acidic. The ionization constants were computed from data in Table 17-3 and with the relationship $K_w = K_a \times K_b$.

For NH_4^+, $K_a = \dfrac{1.0\times10^{-14}}{1.8\times10^{-5}} = 5.6\times10^{-10}$ For NO_2^-, $K_b = \dfrac{1.0\times10^{-14}}{7.2\times10^{-4}} = 1.4\times10^{-11}$

Where $1.8\times10^{-5} = K_b$ of NH_3 and $7.2\times10^{-4} = K_a$ of HNO_2

68. Our list in order of increasing pH is in order of decreasing acidity. First we look for the strong acids; there is just HNO_3. Next we look for the weak acids; there is only one, $HC_2H_3O_2$. Next in order of decreasing acidity come salts with cations from weak bases and anions from strong acids; NH_4ClO_4 is in this category. Then come salts in which both ions hydrolyze to the same degree; $NH_4C_2H_3O_2$ is an example, forming a pH-neutral solution. Next come salts that have the cation of a strong base and the anion of a weak acid; $NaNO_2$ is in this category. Then come weak bases, of which $NH_3(aq)$ is an example. And finally, strong bases: NaOH is the only one. Thus, in order of increasing pH of their 0.010 M aqueous solutions, the solutes are:

$$HNO_3 < HC_2H_3O_2 < NH_4ClO_4 < NH_4C_2H_3O_2 < NaNO_2 < NH_3 < NaOH$$

69. NaOCl dissociates completely in aqueous solution into $Na^+(aq)$, which does not hydrolyze, and $OCl^-(aq)$, which does hydrolyze. We determine $[OH^-]$ in a 0.089 M solution of OCl^-, finding the value of the hydrolysis constant from the ionization constant of HOCl, $K_a = 2.9 \times 10^{-8}$.

Equation:	$OCl^-(aq) + H_2O(l)$	$\rightleftharpoons$	$HOCl(aq)$	$+$	$OH^-(aq)$
Initial:	0.089 M		0 M		≈ 0 M
Changes:	$-x$ M		$+x$ M		$+x$ M
Equil:	$(0.089 - x)$ M		x M		x M

$$K_b = \frac{K_w}{K_a} = \frac{1.0 \times 10^{-14}}{2.9 \times 10^{-8}} = 3.4 \times 10^{-7} = \frac{[HOCl][OH^-]}{[OCl^-]} = \frac{x^2}{0.089 - x} \approx \frac{x^2}{0.089}$$

$$x = 1.7 \times 10^{-4} M = [OH^-]; \quad pOH = -\log(1.7 \times 10^{-4}) = 3.77, \quad pH = 14.00 - 3.77 = 10.23$$

70. NH_4Cl dissociates completely in aqueous solution into $NH_4^+(aq)$, which hydrolyzes, and $Cl^-(aq)$, which does not. We determine $[H_3O^+]$ in a 0.123 M solution of NH_4^+, finding the value of the hydrolysis constant from the ionization constant of NH_3, $K_b = 1.8 \times 10^{-5}$.

Equation:	$NH_4^+(aq)$	$+ H_2O(l) \rightleftharpoons$	$NH_3(aq)$	$+$	$H_3O^+(aq)$
Initial:	0.123 M		0 M		≈ 0 M
Changes:	$-x$ M		$+x$ M		$+x$ M
Equil:	$(0.123 - x)$ M		x M		x M

$$K_b = \frac{K_w}{K_a} = \frac{1.0 \times 10^{-14}}{1.8 \times 10^{-5}} = 5.6 \times 10^{-10} = \frac{[NH_3][H_3O^+]}{[NH_4^+]} = \frac{x^2}{0.123 - x} \approx \frac{x^2}{0.123}$$

$$x = 8.3 \times 10^{-6} M = [H_3O^+]; \quad pH = -\log(8.3 \times 10^{-6}) = 5.08$$

71. $KC_6H_7O_2$ dissociates completely in aqueous solution into $K^+(aq)$, which does not hydrolyze, and the ion $C_6H_7O_2^-(aq)$, which does hydrolyze. We determine $[OH^-]$ in 0.37 M $KC_6H_7O_2$ solution. Note: $K_a = 10^{-pK} = 10^{-4.77} = 1.7 \times 10^{-5}$

Equation : $C_6H_7O_2^-(aq) + H_2O(l) \rightleftharpoons HC_6H_7O_2(aq) + OH^-(aq)$

Initial :   0.37 M   0 M ≈ 0 M

Changes : $-x$ M $+x$ M $+x$ M

Equil : $(0.37-x)$ M x M x M

$$K_b = \frac{K_w}{K_a} = \frac{1.0 \times 10^{-14}}{1.7 \times 10^{-5}} = 5.9 \times 10^{-10} = \frac{[HC_6H_7O_2][OH^-]}{[C_6H_7O_2^-]} = \frac{x^2}{0.37-x} \approx \frac{x^2}{0.37}$$

$x = 1.5 \times 10^{-5}$ M $= [OH^-]$, $pOH = -\log(1.5 \times 10^{-5}) = 4.82$, $pH = 14.00 - 4.82 = 9.18$

72. Equation : $C_5H_5NH^+(aq) + H_2O(l) \rightleftharpoons C_5H_5N(aq) + H_3O^+(aq)$

Initial :   0.0482 M   0 M ≈ 0 M

Changes : $-x$ M $+x$ M $+x$ M

Equil : $(0.0482-x)$ M x M x M

$$K_a = \frac{K_w}{K_b} = \frac{1.0 \times 10^{-14}}{1.5 \times 10^{-9}} = 6.7 \times 10^{-6} = \frac{[C_5H_5N][H_3O^+]}{[C_5H_5NH^+]} = \frac{x^2}{0.0482-x} \approx \frac{x^2}{0.0482}$$

$x = 5.7 \times 10^{-4}$ M $= [H_3O^+]$, $pH = -\log(5.7 \times 10^{-4}) = 3.24$

73. (a) $HSO_3^- + H_2O \rightleftharpoons H_3O^+ + SO_3^{2-}$ $K_a = K_{a_2,\text{Sulfurous}} = 6.2 \times 10^{-8}$

$HSO_3^- + H_2O \rightleftharpoons OH^- + H_2SO_3$ $K_b = \dfrac{K_w}{K_{a_1,\text{Sulfurous}}} = \dfrac{1.0 \times 10^{-14}}{1.3 \times 10^{-2}} = 7.7 \times 10^{-13}$

Since $K_a > K_b$, a solution of HSO_3^- is acidic.

(b) $HS^- + H_2O \rightleftharpoons H_3O^+ + S^{2-}$ $K_a = K_{a_2,\text{Hydrosulfuric}} = 1 \times 10^{-19}$

$HS^- + H_2O \rightleftharpoons OH^- + H_2S$ $K_b = \dfrac{K_w}{K_{a_1,\text{Hydrosulfuric}}} = \dfrac{1.0 \times 10^{-14}}{1.0 \times 10^{-7}} = 1.0 \times 10^{-7}$

Since $K_a < K_b$, a solution of HS^- is alkaline, or basic.

(c) $HPO_4^{2-} + H_2O \rightleftharpoons H_3O^+ + PO_4^{3-}$ $K_a = K_{a_3,\text{Phosphoric}} = 4.2 \times 10^{-13}$

$HPO_4^{2-} + H_2O \rightleftharpoons OH^- + H_2PO_4^-$ $K_b = \dfrac{K_w}{K_{a_2,\text{Phosphoric}}} = \dfrac{1.0 \times 10^{-14}}{6.3 \times 10^{-8}} = 1.6 \times 10^{-7}$

Since $K_a < K_b$, a solution of HPO_4^{2-} is alkaline, or basic.

74. pH = 8.65 is a basic solution. To produce a solution with this pH, we need the salt of the cation of a strong base and the anion of a weak acid, which will hydrolyze to produce this basic solution. This must be (c) KNO_2. (a) NH_4Cl is the salt of the cation of a weak base and the anion of a strong acid, and should form an acidic solution. (b) $KHSO_4$ and (d) $NaNO_3$ both are the salts of cations of strong bases with anions of strong acids; they form pH-neutral solutions. $pOH = 14.00 - 8.65 = 5.35$ $[OH^-] = 10^{-5.35} = 4.5 \times 10^{-6} M$

Equation: $NO_2^-(aq) + H_2O(l) \rightleftharpoons HNO_2(aq) + OH^-(aq)$

Initial:	S	0 M	≈ 0 M
Changes:	$-4.5 \times 10^{-6} M$	$+4.5 \times 10^{-6} M$	$+4.5 \times 10^{-6} M$
Equil:	$(S - 4.5 \times 10^{-6} M)$	$4.5 \times 10^{-6} M$	$4.5 \times 10^{-6} M$

$$K_b = \frac{K_w}{K_a} = \frac{1.0 \times 10^{-14}}{7.2 \times 10^{-4}} = 1.4 \times 10^{-11} = \frac{[HNO_2][OH^-]}{[NO_2^-]} = \frac{(4.5 \times 10^{-6})^2}{S - 4.5 \times 10^{-6}}$$

$$S - 4.5 \times 10^{-6} = \frac{(4.5 \times 10^{-6})^2}{1.4 \times 10^{-11}} = 1.4 M \quad S = 1.4 M + 4.5 \times 10^{-6} = 1.4 M = [KNO_2]$$

Molecular Structure and Acid Strength

75. **(a)** $HClO_3$ should be a stronger acid than is $HClO_2$. In each acid there is an $H-O-Cl$ grouping. The remaining oxygen atoms are bonded directly to Cl as terminal O atoms. Thus, there are two terminal O atoms in $HClO_3$ and only one in $HClO_2$. Of oxoacids of the same element, the one with the higher number of terminal oxygen atoms is the stronger. $HClO_3 : K_{a_1} = 5 \times 10^2$ f and $HClO_2 : K_a = 1.1 \times 10^{-2}$.

(b) HNO_2 and H_2CO_3 have the same number (one) of terminal oxygen atoms. They differ in N being more electronegative than C. $HNO_2 (K_a = 7.2 \times 10^{-4})$ is stronger than $H_2CO_3 (K_{a_1} = 4.4 \times 10^{-7})$.

(c) H_3PO_4 and H_2SiO_3 have the same number (one) of terminal oxygen atoms. They differ in P being more electronegative than Si. $H_3PO_4 (K_{a_1} = 7.1 \times 10^{-3})$ is stronger than $H_2SiO_3 (K_{a_1} = 1.7 \times 10^{-10})$.

76. The weakest of the five acids is CH_3CH_2COOH. The reasoning is as follows. HBr is a strong acid, stronger than the carboxylic acids. A carboxylic acid—such as $CH_2ClCOOH$ and CH_2FCH_2COOH—with a strongly electronegative atom attached to the hydrocarbon chain will be stronger than one in which no such group is attached. But the I atom is so weakly electronegative that it barely influences the acid strength. Acid strengths of some of these acids (as values of pK_a) follow. (Larger values of pK_a indicate weaker acids.)

Strength: $CH_3CH_2COOH (4.89) < CI_3COOH < CH_2FCH_2COOH < CH_2ClCOOH < HBr (-8.72)$

77. Largest K_b(most basic) belongs to $CH_3CH_2CH_2NH_2$ (hydrocarbon chains have the lowest electronegativity). Smallest K_b (least basic) is that of o-chlorolaniline (the nitrogen lone pair is delocalized (spread out over the ring), hence, less available to accept a proton (poorer base)).

78. Most basic: CH_3O^- (methoxide ion). Methanol is the weakest acid, thus its anion is the strongest base.

Most acidic: ortho-chlorophenol (only acidic compound)

Lewis Theory of Acids and Bases

79. **(a)**

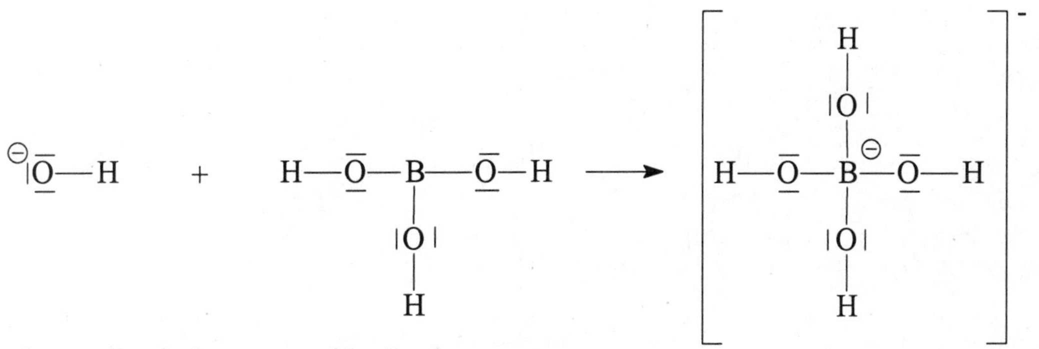

base: e⁻ pair donor   acid: e⁻ pair acceptor

(b)

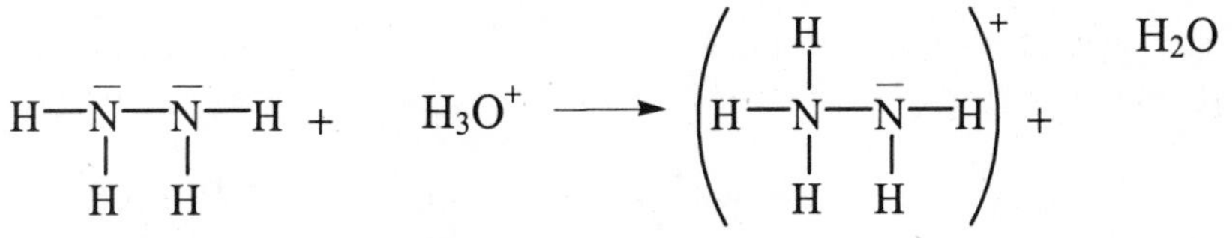

The actual Lewis acid is H^+, which is supplied by H_3O^+

(c)

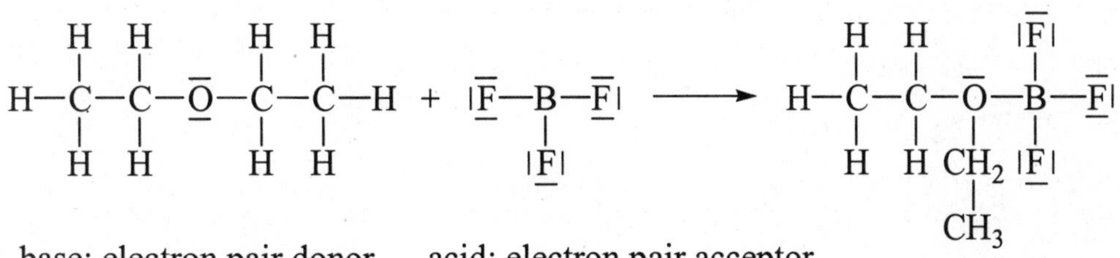

base: electron pair donor   acid: electron pair acceptor

80. The C in a CO_2 molecule can accept of a pair of electrons: it is the Lewis acid. The hydroxide anion is the Lewis base: it has pairs of electrons it can donate.

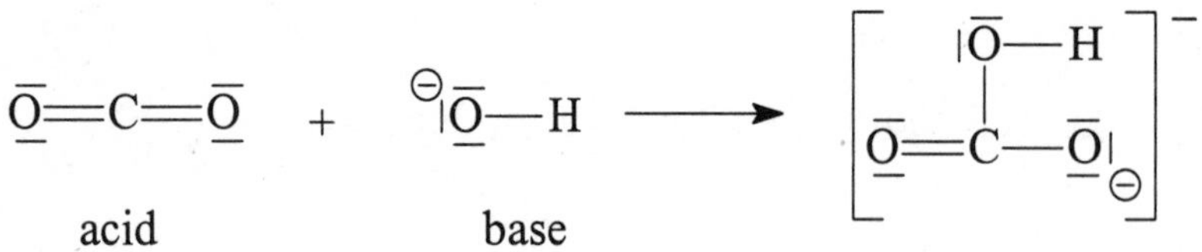

$$\overline{O}=C=\overline{O} \quad + \quad {}^{\ominus}|\overline{O}-H \quad \longrightarrow \quad \left[\begin{array}{c} |\overline{O}-H \\ | \\ \overline{O}=C-\overline{O}| \\ {}_{\ominus} \end{array} \right]^{-}$$

       acid   base

81. $I_2(aq) + I^-(aq) \rightleftharpoons I_3^-(aq)$

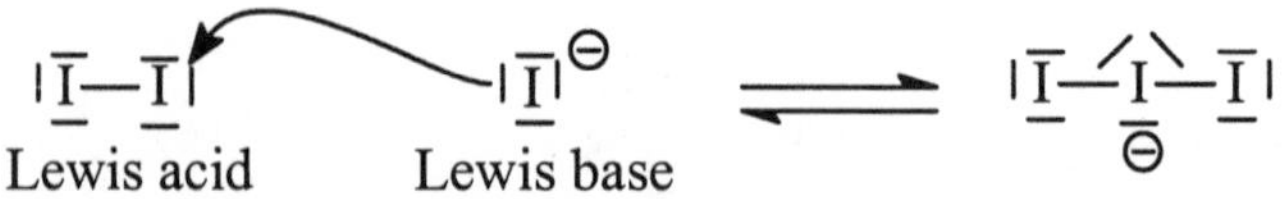

$|\overline{I}-\overline{I}|$ $|\overline{I}|^{\ominus}$ $\rightleftharpoons$ $|\overline{I}-\overset{\frown}{I}-\overline{I}|$
  Lewis acid   Lewis base $\ominus$
(e⁻ pair acceptor)   (e⁻ pair donor)

82. (a)

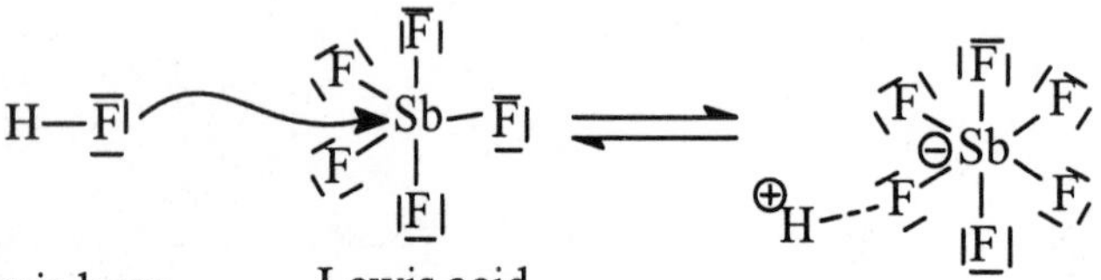

  Lewis base   Lewis acid
(e⁻ pair donor)   (e⁻ pair acceptor)

(b)

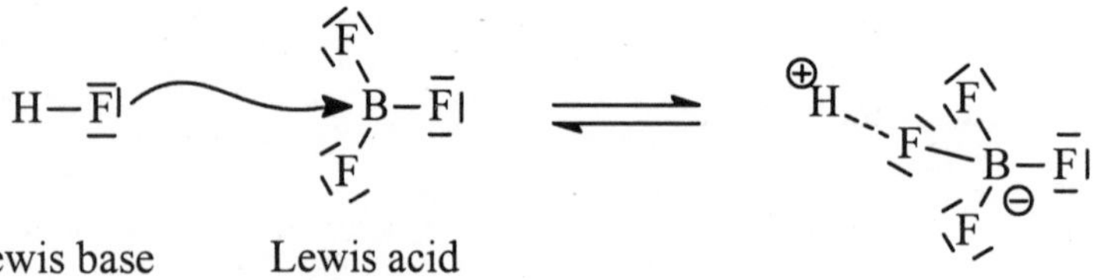

  Lewis base   Lewis acid
(e⁻ pair donor)   (e⁻ pair acceptor)

83. $H-\ddot{O}-H$ $:\ddot{O}-S=\ddot{O}:$ $\longrightarrow$ $H-\overset{\oplus}{O}-S:$ $\longrightarrow$ $H-\ddot{O}-\overset{\oplus}{S}:$

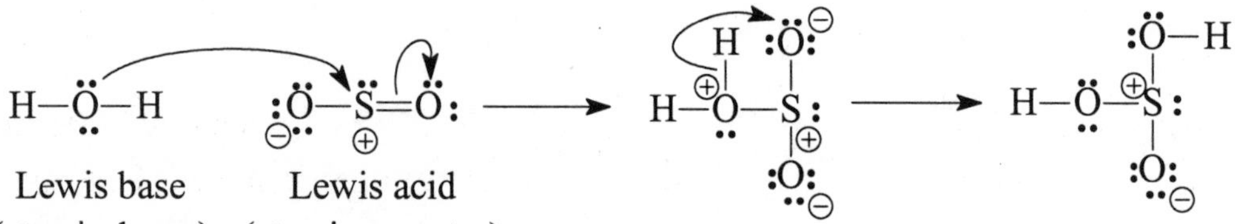

  Lewis base   Lewis acid
(e⁻ pair donor)   (e⁻ pair acceptor)

84. $2\ \ddot{N}H_3(aq) \quad + \quad Ag^+(aq) \longrightarrow \left[H_3N-Ag-NH_3\right]^+$
  Lewis base   Lewis acid
(e⁻ pair donor)(e⁻ pair acceptor)

FEATURE PROBLEMS

105. (a) From the combustion analysis we can determine the empirical formula. Note that the mass of oxygen is determined by difference.

$$\text{amount C} = 1.599 \text{ g CO}_2 \times \frac{1 \text{ mol CO}_2}{44.01 \text{ g CO}_2} \times \frac{1 \text{ mol C}}{1 \text{ mol CO}_2} = 0.03633 \text{ mol C}$$

$$\text{mass of C} = 0.03633 \text{ mol C} \times \frac{12.011 \text{ g C}}{1 \text{ mol C}} = 0.4364 \text{ g C}$$

$$\text{amount H} = 0.327 \text{ g H}_2\text{O} \times \frac{1 \text{ mol H}_2\text{O}}{18.02 \text{ g H}_2\text{O}} \times \frac{2 \text{ mol H}}{1 \text{ mol H}_2\text{O}} = 0.03629 \text{ mol H}$$

$$\text{mass of H} = 0.03629 \text{ mol H} \times \frac{1.008 \text{ g H}}{1 \text{ mol H}} = 0.03658 \text{ g H}$$

$$\text{amount O} = (1.054 \text{ g sample} - 0.4364 \text{ g C} - 0.03568 \text{ g H}) \times \frac{1 \text{ mol O}}{16.00 \text{ g O}} = 0.0364 \text{ mol O}$$

There are equal moles of the three elements. The empirical formula is CHO. The freezing-point depression data are used to determine the molar mass.

$$\Delta T_f = -K_f m \quad m = \frac{\Delta T_f}{-K_f} = \frac{-0.82 \text{ °C}}{-3.90 \text{ °C}/m} = 0.21 \, m$$

$$\text{amount of solute} = 25.10 \text{ g solvent} \times \frac{1 \text{ kg}}{1000 \text{ g}} \times \frac{0.21 \text{ mol solute}}{1 \text{ kg solvent}} = 0.0053 \text{ mol}$$

$$M = \frac{0.615 \text{ g solute}}{0.0053 \text{ mol solute}} = 1.2 \times 10^2 \text{ g/mol}$$

The formula mass of the empirical formula is: $12.0 \text{ g C} + 1.0 \text{ g H} + 16.0 \text{ g O} = 29.0 \text{ g/mol}$ Thus, there are four empirical units in a molecule, the molecular formula is $C_4H_4O_4$, and the molar mass is 116.1 g/mol.

(b) We determine the mass of maleic acid that reacts with one mol OH^-,

$$\frac{\text{mass}}{\text{mol OH}^-} = \frac{0.4250 \text{ g maleic acid}}{34.03 \text{ mL base} \times \frac{1 \text{ L}}{1000 \text{ mL}} \times \frac{0.2152 \text{ mol KOH}}{1 \text{ L base}}} = 58.03 \text{ g/mol OH}^-$$

This means that one mole of maleic acid (116.1 g/mol) reacts with two moles of hydroxide ion. Maleic acid is a diprotic acid: $H_2C_4H_2O_4$.

(c) Maleic acid has two—COOH groups joined together by a bridging C_2H_2 group. A plausible Lewis structure is shown below:

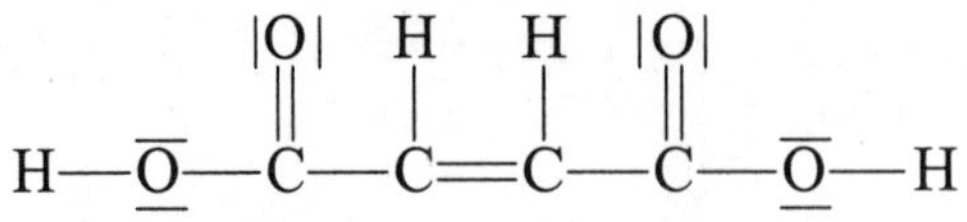

$$H-\overline{\underline{O}}-\overset{\displaystyle |\overline{O}|}{\overset{\|}{C}}-\overset{\displaystyle H}{\overset{|}{C}}=\overset{\displaystyle H}{\overset{|}{C}}-\overset{\displaystyle |\overline{O}|}{\overset{\|}{C}}-\overline{\underline{O}}-H$$

(d) We first determine $\left[H_3O^+\right]=10^{-pH}=10^{-1.80}=0.016 \text{ M}$ and then the initial concentration of acid.

$$\left[(CHCOOH)_2\right]_{initial}=\frac{0.215\,g}{50.00\,mL}\times\frac{1000\,mL}{1L}\times\frac{1\,mol}{116.1\,g}=0.0370 \text{ M}$$

We use the first ionization to determine the value of K_{a_1}

Equation: $(CHCOOH)_2 \text{ (aq)} + H_2O(l) \rightleftharpoons H(CHCOO)_2^- \text{ (aq)}+ \quad H_3O^+\text{(aq)}$

Initial:	0.0370 M	0 M	≈ 0 M
Changes:	−0.016 M	+0.016 M	+0.016 M
Equil:	0.021 M	0.016 M	0.016 M

$$K_a=\frac{\left[H(CHCOO)_2^-\right]\left[H_3O^+\right]}{\left[(CHCOOH)_2\right]}=\frac{(0.016)(0.016)}{0.021}=1.2\times10^{-2}$$

K_{a_2} could be determined if we had some way to measure the total concentration of all ions in solution, or if we could determine $\left[(CHCOO)_2^{2-}\right]=K_{a_2}$

(e) Practically all the $\left[H_3O^+\right]$ arises from the first ionization.

Equation: $(CHCOOH)_2 \text{ (aq)}+ H_2O(l) \rightleftharpoons H(CHCOO)_2^- \text{ (aq)}+ H_3O^+\text{(aq)}$

Initial:	0.0500 M	0 M	≈ 0 M
Changes:	$-x$ M	$+x$ M	$+x$ M
Equil:	$(0.0500-x)$ M	x M	x M

$$K_a=\frac{[H(CHCOO)_2^-][H_3O^+]}{[(CHCOOH)_2]}=\frac{x^2}{0.0500-x}=1.2\times10^{-2}$$

$$x^2=0.00060-0.012\,x \qquad x^2+0.012\,x-0.00060=0$$

$$x=\frac{-b\pm\sqrt{b^2-4ac}}{2a}=\frac{-0.012\pm\sqrt{0.00014+0.0024}}{2}=0.019 \text{ M}=\left[H_3O^+\right]$$

$$pH=-\log(0.019)=1.72$$

106. (a)

$$\frac{x^2}{0.00250-x} \simeq \frac{x^2}{0.00250} \simeq 4.2 \times 10^{-4} \qquad x_1 \simeq 0.00102$$

$$\frac{x^2}{0.00250-0.00102} = \frac{x^2}{0.00148} \simeq 4.2 \times 10^{-4} \qquad x_2 \simeq 0.000788$$

$$\frac{x^2}{0.00250-0.000788} = \frac{x^2}{0.00171} \simeq 4.2 \times 10^{-4} \qquad x_3 \simeq 0.000848$$

$$\frac{x^2}{0.00250-0.000848} = \frac{x^2}{0.00165} \simeq 4.2 \times 10^{-4} \qquad x_4 \simeq 0.000833$$

$$\frac{x^2}{0.00250-0.000833} = \frac{x^2}{0.00167} \simeq 4.2 \times 10^{-4} \qquad x_5 \simeq 0.000837$$

$$\frac{x^2}{0.00250-0.000837} = \frac{x^2}{0.00166} \simeq 4.2 \times 10^{-4} \qquad x_6 \simeq 0.000836$$

$$x_6 \simeq 0.000836 \qquad \text{or} \simeq 8.4 \times 10^{-4},$$

which is the same as the value obtained with the quadratic equation on page 681.

(b) We organize the solution around the balanced chemical equation, as we have done before.

Equation: $HClO_2(aq) + H_2O(l) \rightleftharpoons ClO_2^-(aq) + H_3O^+(aq)$

Initial: 0.500 M $\qquad\qquad$ 0 M $\qquad$ ≈0 M

Changes: $-x$ M $\qquad\qquad$ $+x$ M $\qquad$ $+x$ M

Equil: x M $\qquad\qquad$ x M $\qquad$ x M

$$K_a = \frac{\left[ClO_2^-\right]\left[H_3O^+\right]}{\left[HClO_2\right]} = \frac{x^2}{0.500-x} = 1.1 \times 10^{-2} \approx \frac{x^2}{0.500} \qquad \text{Assuming } x \ll 0.500 \text{ M},$$

$$x = \sqrt{x^2} = \sqrt{0.500 \times 1.1 \times 10^{-2}} = 0.074 \text{ M} \qquad \text{This is not greatly smaller than 0.500 M.}$$

Assume $x = 0.074$ $x = \sqrt{(0.500-0.0774) \times 1.1 \times 10^{-2}} = 0.068 \text{ M}$ Try once more.

Assume $x = 0.068$ $x = \sqrt{(0.500-0.068) \times 1.1 \times 10^{-2}} = 0.069 \text{ M}$ One more time.

Assume $x = 0.069$ $x = \sqrt{(0.500-0.069) \times 1.1 \times 10^{-2}} = 0.069 \text{ M}$ Final result!

$$\left[H_3O^+\right] = 0.069 \text{ M}, \quad pH = -\log\left[H_3O^+\right] = -\log(0.069) = 1.16$$

107. (a) Consider *formic* acid ionization first. Then add acetic acid contribution.

$$HCHO_2(aq) \quad + H_2O(l) \quad \underset{K_a = 1.8 \times 10^{-4}}{\rightleftharpoons} \quad CHO_2^-(aq) \quad + \quad H_3O^+(aq)$$

Initial	0.250 M	0 M	≈ 0 M
Change	$-x$ M	$+x$ M	$+x$ M
Equil.	(0.250 M $-x$) M	x M	x M

$$1.8 \times 10^{-4} = \frac{x^2}{0.250 - x} \approx \frac{x^2}{0.250\ M} \qquad x = 0.0067 \quad (x << 0.250, \text{ valid assumption})$$

Now consider acetic acid dissociation $K_a = 1.8 \times 10^{-5}$

$$HC_2H_3O_2(aq) \quad + H_2O(l) \quad \rightleftharpoons \quad C_2H_3O_2^-(aq) \quad + \quad H_3O^+(aq)$$

Initial	0.315 M	0 M	0.0067 M
Change	$-x$ M	$+x$ M	$+x$ M
Equil.	(0.315 $-x$) M	x M	(0.0067 $+x$) M

$$1.8 \times 10^{-5} = \frac{x(0.0067 + x)}{(0.315 - x)} \quad \text{Solve: quadratic equation (or successive approx.):}$$

$$5.6\underline{7} \times 10^{-6} - 1.8 \times 10^{-5}x = x(0.0067 + x) = 0.0067x + x^2$$

$$\text{or } x^2 + 6.7\underline{2} \times 10^{-3}x - 5.6\underline{7} \times 10^{-6} = 0$$

$$x = \frac{-6.72 \times 10^{-3} \pm \sqrt{(6.72 \times 10^{-3})^2 - 4(1)(-5.67 \times 10^{-6})}}{2(1)}$$

Therefore $x = 7.6 \times 10^{-4}$ M

At equilibrium, the $[H_3O^+] = 0.0067 + 0.00076 = 7.5 \times 10^{-3}$ M $\qquad$ pH = 2.13

(b) $[NH_3] = \dfrac{(12.5\ g)\dfrac{1\ mol}{17.031\ g}}{0.375\ L} = 1.96$ M; $\quad [CH_3NH_2] = \dfrac{(1.55\ g)\dfrac{1\ mol}{31.0575\ g}}{0.375\ L} = 0.133$ M

Consider ammonia ionization first: $K_b = 1.8 \times 10^{-5}$

$$NH_3(aq) \quad + H_2O(l) \quad \rightleftharpoons \quad NH_4^+(aq) \quad + \quad OH^-(aq)$$

Initial	1.96 M	0 M	≈ 0 M
Change	$-x$ M	$+x$ M	$+x$ M
Equil.	(1.96 $-x$) M	x M	x M

$$1.8 \times 10^{-5} = \frac{x^2}{1.96 - x} \approx \frac{x^2}{1.96} \qquad x = 5.9\underline{4} \times 10^{-3} = [OH^-] \quad (x << 1.9\underline{6}, \text{ valid assumption})$$

Now consider methylamine ionization: $K_b = 4.2 \times 10^{-4}$

Equation:
$$CH_3NH_2(aq) \quad + H_2O(l) \quad \rightleftharpoons \quad CH_3NH_3^+(aq) \quad + \quad OH^-(aq)$$

Initial	0.133 M	0 M	0.0059\underline{4} M
Change	$-x$ M	$+x$ M	$+x$ M
Equil.	(0.133 $-x$) M	x M	(0.0059\underline{4} $+ x$) M

$$4.2 \times 10^{-4} = \frac{x(0.00594 + x)}{0.133 - x} \quad \text{or} \quad 5.59 \times 10^{-5} - 4.2 \times 10^{-4} x = x^2 + 0.00594x$$

$$x^2 + 6.36 \times 10^{-3} x - 5.59 \times 10^{-5} = 0 \quad \text{Solve using quadratic formula:}$$

$$x = \frac{-6.36 \times 10^{-3} \pm \sqrt{(6.36 \times 10^{-3})^2 - 4(1)(-5.59 \times 10^{-5})}}{2(1)} \qquad x = 4.94 \times 10^{-3} \text{ M}$$

$[OH^-] = (0.00594 + 4.94 \times 10^{-3}) \text{ M} = 0.011 \text{ M}, \quad pOH = 1.96 \text{ and } pH = 12.04$

(c)   1.0 M NH_4^+(aq) $K_a = 5.6 \times 10^{-10}$ (very weak acid)

1.0 M CN^-(aq) $K_b = 1.6 \times 10^{-5}$ (weak base)

We can neglect the effect of NH_4^+ ions because K_a for NH_4^+ is very small compared to the K_b for CN^- ($K_b \gg K_a$)

Equation:	CN^-(aq)	+ H_2O(l)	$\rightleftharpoons$	HCN(aq)	+	OH^-(aq)
Initial:	1.0 M			0 M		$\approx$ 0 M
Change:	$-x$ M			$+x$ M		$+x$ M
Equil.:	(1.0 $-x$) M			x M		x M

$$K_b = 1.6 \times 10^{-5} = \frac{x^2}{1.0 - x} \approx \frac{x^2}{1.0} \qquad x = 4.0 \times 10^{-3} \quad (x \ll 1.0, \text{ valid assumption})$$

$[OH^-] = 4.0 \times 10^{-3}$, pOH = 2.40 and pH = 11.60

CHAPTER 18
ADDITIONAL ASPECTS OF ACID— BASE EQUILIBRIA
PRACTICE EXAMPLES

1A Organize the solution around the balanced chemical equation, as we have done before.

Equation: $HF(aq) + H_2O(l) \rightleftharpoons H_3O^+(aq) + F^-(aq)$

Initial: 0.500 M ≈ 0 M   0 M

Changes: $-x$ M $+x$ M $+x$ M

Equil: $(0.500 - x)$ M x M x M

$$K_a = \frac{[H_3O^+][F^-]}{[HF]} = \frac{(x)(x)}{0.500-x} = 6.6 \times 10^{-4} \approx \frac{x^2}{0.500} \qquad \text{assuming } x \ll 0.500$$

$x = \sqrt{0.500 \times 6.6 \times 10^{-4}} = 0.018\,\text{M}$ One further cycle of approximations gives:

$x = \sqrt{(0.500 - 0.018) \times 6.6 \times 10^{-4}} = 0.018\,\text{M} = [H_3O^+]$

$[HF] = 0.500\,\text{M} - 0.018\,\text{M} = 0.482\,\text{M}$

Recognize that 0.100 M HCl means $\left[H_3O^+\right]_{\text{initial}} = 0.100\,\text{M}$, since HCl is a strong acid.

Equation: $HF(aq) + H_2O(l) \rightleftharpoons H_3O^+(aq) + F^-(aq)$

Initial: 0.500 M   0.100 M   0 M

Changes: $-x$ M $+x$ M $+x$ M

Equil: $(0.500 - x)$ M $(0.100 + x)$ M x M

$$K_a = \frac{[H_3O^+][F^-]}{[HF]} = \frac{(x)(0.100+x)}{0.500-x} = 6.6 \times 10^{-4} \approx \frac{0.100\,x}{0.500} \quad \text{assuming } x \ll 0.100$$

$x = \dfrac{6.6 \times 10^{-4} \times 0.500}{0.100} = 3.3 \times 10^{-3}\,\text{M} = \left[H_3O^+\right]$ The assumption is valid.

$\left[HF\right] = 0.500\,\text{M} - 0.003\,\text{M} = 0.497\,\text{M}$

1B From Example 17-6 *in the text*, we know that $[H_3O^+] = [C_2H_3O_2^-] = 1.3 \times 10^{-3}$ M in 0.100 M $HC_2H_3O_2$. We base our calculation, as usual, on the balanced chemical equation. The concentration of H_3O^+ from the added HCl is represented by x.

Equation: $HC_2H_3O_2(aq) + H_2O(l) \rightleftharpoons H_3O^+(aq) + C_2H_3O_2^-(aq)$

Initial: 0.100 M ≈ 0 M   0 M

Changes: -0.00010 M $+0.00010$ M $+0.00010$ M

From HCl: $+x$ M

Equil: 0.100 M $(0.00010 + x)$ M $+0.00010$ M

$$K_a = \frac{[H_3O^+][C_2H_3O_2^-]}{[HC_2H_3O_2]} = \frac{(0.00010 + x)0.00010}{0.100} = 1.8 \times 10^{-5}$$

$$0.00010 + x = \frac{1.8 \times 10^{-5} \times 0.100}{0.00010} = 0.018 \, M \qquad x = 0.018 \text{ M} - 0.00010 \text{ M} = 0.018 \text{ M}$$

$$V_{12 \, M \, HCl} = 1.00 \, L \times \frac{0.018 \text{ mol } H_3O^+}{1 \, L} \times \frac{1 \text{ mol HCl}}{1 \text{ mol } H_3O^+} \times \frac{1 \text{ L soln}}{12 \text{ mol HCl}} \times \frac{1000 \text{ mL}}{1 \, L} \times \frac{1 \text{ drop}}{0.050 \text{ mL}} = 30. \text{ drops}$$

Since 30. drops corresponds to 1.5 mL of 12 M solution, we see that the volume of solution does indeed remain 1.00 L after addition of the 12 M HCl.

2A   We again organize the solution around the balanced chemical equation.

Equation: $HCHO_2(aq) + H_2O(l) \rightleftharpoons CHO_2^-(aq) + H_3O^+(aq)$

Initial: 0.100 M   0.150 M ≈ 0 M

Changes: $-x$ M $+x$ M $+x$ M

Equil: $(0.100 - x)$ M $(0.150 + x)$ M x M

$$K_a = \frac{[CHO_2^-][H_3O^+]}{[HCHO_2]} = \frac{(0.150 + x)(x)}{0.100 - x} = 1.8 \times 10^{-4} \approx \frac{0.150x}{0.100} \qquad \text{assuming } x \ll 0.100$$

$$x = \frac{0.100 \times 1.8 \times 10^{-4}}{0.150} = 1.2 \times 10^{-4} M = [H_3O^+], \text{ valid assumption}$$

$$[CHO_2^-] = 0.150 \text{ M} + 0.00012 \text{ M} = 0.150 \text{ M}$$

2B   This time a solid sample of a weak base is being added to a solution of its conjugate acid. We let x represent the concentration of acetate ion from the added sodium acetate. Notice that sodium acetate is a strong electrolyte, completely dissociated in aqueous solution.

$$[H_3O^+] = 10^{-pH} = 10^{-5.00} = 1.0 \times 10^{-5} M = 0.000010 \text{ M}$$

Equation: $HC_2H_3O_2(aq) + H_2O(l) \rightleftharpoons C_2H_3O_2^-(aq) + H_3O^+(aq)$

Initial: 0.100 M   0 M ≈ 0 M

Changes: -0.000010 M $+0.000010$ M $+0.000010$ M

From NaAc: $+x$ M

Equil: 0.100 M $(0.000010 + x)$ M   0.000010 M

$$K_a = \frac{\left[H_3O^+\right]\left[C_2H_3O_2^-\right]}{\left[HC_2H_3O_2\right]} = \frac{0.000010\left(0.000010+x\right)}{0.100} = 1.8\times10^{-5}$$

$$0.000010+x = \frac{1.8\times10^{-5}\times0.100}{0.000010} = 0.18\ M \qquad x = 0.18\ M - 0.000010\ M = 0.18\ M$$

$$\text{mass of } NaC_2H_3O_2 = 1.00\,L\times\frac{0.18\ mol\ C_2H_3O_2^-}{1L}\times\frac{1\,mol\ NaC_2H_3O_2}{1\,mol\ C_2H_3O_2^-}\times\frac{82.03\,g\ NaC_2H_3O_2}{1\,mol\ NaC_2H_3O_2}$$

$$= 15\,g\ NaC_2H_3O_2$$

3A A strong acid dissociates completely, and essentially is a source of H_3O^+. $NaC_2H_3O_2$ also dissociates completely in solution. The hydronium ion and the acetate ion react to form acetic acid: $H_3O^+(aq) + C_2H_3O_2^-(aq) \rightleftharpoons HC_2H_3O_2(aq) + H_2O(l)$
All that is necessary to form a buffer is to have approximately equal amounts of a weak acid and its conjugate base in solution. This will be achieved if we add an amount of HCl equal to approximately half the original amount of acetate ion.

3B HCl dissociates completely and serves as a source of hydronium ion. This reacts with ammonia to form ammonium ion: $NH_3(aq) + H_3O^+(aq) \rightleftharpoons NH_4^+(aq) + H_2O(l)$.
Because a buffer contains approximately equal amounts of a weak base (NH_3) and its conjugate acid (NH_4^+), to prepare a buffer we simply add an amount of HCl equal to approximately half the amount of $NH_3(aq)$ initially present.

4A We first find the formate ion concentration, remembering that $NaCHO_2$ is a strong electrolyte, existing in solution as $Na^+(aq)$ and $CHO_2^-(aq)$.

$$[CHO_2^-] = \frac{23.1\,g\ NaCHO_2}{500.0\ mL\ soln}\times\frac{1000\ mL}{1L}\times\frac{1\ mol\ NaCHO_2}{68.01g\ NaCHO_2}\times\frac{1\ mol\ CHO_2^-}{1\ mol\ NaCHO_2} = 0.679\,M$$

As usual, the solution to the problem is organized around the balanced chemical equation.

Equation: $HCHO_2(aq)\ +\ H_2O(l)\ \rightleftharpoons\ CHO_2^-(aq)\ +\ H_3O^+(aq)$

Initial:	0.432 M	0.679 M	$\approx 0\,M$
Changes:	$-x$ M	$+x$ M	$+x$ M
Equil:	$(0.432-x)$ M	$(0.679+x)$ M	x M

$$K_a = \frac{\left[H_3O^+\right]\left[CHO_2^-\right]}{\left[HCHO_2\right]} = \frac{x\left(0.679+x\right)}{0.432-x} = 1.8\times10^{-4} \approx \frac{0.679x}{0.432} \qquad x = \frac{0.432\times1.8\times10^{-4}}{0.679}$$

This gives $\left[H_3O^+\right] = 0.0892\ M$. The assumption that $x \ll 0.432$ is clearly correct.

$$pH = -\log\left[H_3O^+\right] = -\log\left(1.1\times10^{-4}\right) = 3.96$$

4B The concentrations of the components in the 100.0 mL of buffer solution are found by dilution. Remember that $NaC_2H_3O_2$ is a strong electrolyte, existing in solution as $Na^+(aq)$ and $C_2H_3O_2^-(aq)$.

$$[HC_2H_3O_2] = 0.200 \text{ M} \times \frac{63.0 \text{ mL}}{100.0 \text{ mL}} = 0.126 \text{ M} \quad [C_2H_3O_2^-] = 0.200 \text{ M} \times \frac{37.0 \text{ mL}}{100.0 \text{ mL}} = 0.0740 \text{ M}$$

As usual, the solution to the problem is organized around the balanced chemical equation.

Equation: $HC_2H_3O_2(aq) + H_2O(l) \rightleftharpoons C_2H_3O_2^-(aq) + H_3O^+(aq)$

Initial: 0.126 M   0.0740 M ≈ 0 M

Changes: $-x$ M $+x$ M $+x$ M

Equil: $(0.126 - x)$ M $(0.0740 + x)$ M x M

$$K_a = \frac{[H_3O^+][C_2H_3O_2^-]}{[HC_2H_3O_2]} = \frac{x(0.0740 + x)}{0.126 - x} = 1.8 \times 10^{-5} \approx \frac{0.0740\,x}{0.126}$$

$$x = \frac{1.8 \times 10^{-5} \times 0.126}{0.0740} = 3.1 \times 10^{-5}\text{M} = [H_3O^+]; \quad pH = -\log[H_3O^+] = -\log 3.1 \times 10^{-5} = 4.51$$

Note: assumption valid: $x \ll 0.0740 < 0.126$; x is neglected when added or subtracted

5A We know the initial concentration of NH_3 in the buffer solution and can use the pH to find the equilibrium $[OH^-]$. The rest of the solution is organized around the balanced chemical equation. Our first goal is to determine the initial concentration of NH_4^+.

$$pOH = 14.00 - pH = 14.00 - 9.00 = 5.00 \qquad [OH^-] = 10^{-pOH} = 10^{-5.00} = 1.0 \times 10^{-5} \text{ M}$$

Equation: $NH_3(aq) + H_2O(l) \rightleftharpoons NH_4^+(aq) + OH^-(aq)$

Initial: 0.35 M x M ≈ 0 M

Changes: -1.0×10^{-5} M $+1.0 \times 10^{-5}$ M $+1.0 \times 10^{-5}$ M

Equil: $(0.35 - 1.0 \times 10^{-5})$ M $(x + 1.0 \times 10^{-5})$ M 1.0×10^{-5} M

$$K_b = \frac{[NH_4^+][OH^-]}{[NH_3]} = 1.8 \times 10^{-5} = \frac{(x + 1.0 \times 10^{-5})(1.0 \times 10^{-5})}{0.35 - 1.0 \times 10^{-5}} = \frac{1.0 \times 10^{-5} \times x}{0.35}$$

Assume $x \gg 1.0 \times 10^{-5}$ $x = \dfrac{0.35 \times 1.8 \times 10^{-5}}{1.0 \times 10^{-5}} = 0.63$ M = initial NH_4^+ concentration

$$\text{mass}(NH_4)_2SO_4 = 0.500\,L \times \frac{0.63 \text{ mol } NH_4^+}{1 \text{ L soln}} \times \frac{1 \text{ mol } (NH_4)_2SO_4}{2 \text{ mol } NH_4^+} \times \frac{132.1 \text{ g } (NH_4)_2SO_4}{1 \text{ mol } (NH_4)_2SO_4}$$

Mass of $(NH_4)_2SO_4 = 21$ g

5B   The solution is composed of 33.05 g $NaC_2H_3O_2 \cdot 3\,H_2O$ dissolved in 300.0 mL of 0.250 M HCl. $NaC_2H_3O_2 \cdot 3\,H_2O$, a strong electrolyte, exists in solution as $Na^+(aq)$, $C_2H_3O_2^-(aq)$ ions and H_2O. First calculate the moles of $NaC_2H_3O_2 \cdot 3\,H_2O$, which based on the 1:1 stoichiometry will also be the number of moles of), $C_2H_3O_2^-$. From this, we may calculate the initial), $[C_2H_3O_2^-]$, assuming the solution's volume remains at 300. mL.

moles of $NaC_2H_3O_2 \cdot 3\,H_2O$ (and moles of $C_2H_3O_2^-$)

$$= \frac{33.05 \text{ g } NaC_2H_3O_2 \cdot 3H_2O}{\dfrac{1 \text{ mole } NaC_2H_3O_2 \cdot 3H_2O}{138.08 \text{ g } NaC_2H_3O_2 \cdot 3H_2O}} = 0.243 \text{ moles } NaC_2H_3O_2 \cdot 3H_2O = \text{moles } C_2H_3O_2^-$$

$$[C_2H_3O_2^-] = \frac{0.243 \text{ mol } C_2H_3O_2^-}{0.300 \text{ L soln}} = 0.810\,M \quad \text{(Note: [HCl] assume unchanged at 0.250 M)}$$

We organize this information around the balanced chemical equation, as before. We recognize that as much hydronium ion (a strong acid) as possible will react to produce the weak acid, acetic acid.

Equation:	$HC_2H_3O_2(aq) + H_2O(l)$	$\rightleftharpoons$	$C_2H_3O_2^-(aq)$	$+$	$H_3O^+(aq)$
Initial:	0 M		0.810 M		0.250 M
Form HAc:	+0.250 M		−0.250 M		−0.250 M
	0.250 M		0.560 M		≈ 0 M
Changes:	$-x$ M		$+x$ M		$+x$ M
Equil:	$(0.250 - x)$ M		$(0.560 + x)$ M		$+x$ M

$$K_a = \frac{[H_3O^+][C_2H_3O_2^-]}{[HC_2H_3O_2]} = \frac{x(0.560+x)}{0.250-x} = 1.8\times10^{-5} \approx \frac{0.560\,x}{0.250}$$

$$x = \frac{1.8\times10^{-5}\times0.250}{0.560} = 8.0\times10^{-6} \text{ M} = [H_3O^+]$$

$$pH = -\log[H_3O^+] = -\log 8.0\times10^{-6} = 5.09 \approx 5.1$$

6A   **(a)**   For formic acid, $pK_a = -\log(1.8\times10^{-4}) = 3.74$. The Henderson-Hasselbalch equation provides the pH of the original buffer solution:

$$pH = pK_a + \log\frac{[CHO_2^-]}{[HCHO_2]} = 3.74 + \log\frac{0.350}{0.550} = 3.54$$

(b)   The added acid completely reacts with formate ion and produces formic acid. Each mole/L of added acid consumes one 1 M of formate ion and forms 1 M of formic acid: $CHO_2^-(aq) + H_3O^+(aq) \longrightarrow HCHO_2(aq) + H_2O(l)$.

Thus, $\left[CHO_2^-\right] = 0.350$ M $- 0.0050$ M $= 0.345$ M and

$[HCHO_2] = 0.550$ M $+ 0.0050$ M $= 0.555$ M. Use Henderson-Hasselbalch equation.

$$pH = pK_a + \log\frac{[CHO_2^-]}{[HCHO_2]} = 3.74 + \log\frac{0.345}{0.555} = 3.53$$

(c) Added base reacts completely with formic acid and produces formate ion. Each mole/L of added base consumes 1 M of formic acid and forms 1 M of formate ion:

$HCHO_2 + OH^- \longrightarrow CHO_2^- + H_2O$. Thus, $\left[CHO_2^- \right] = 0.350\,M + 0.0050\ M = 0.355\ M$

and $\left[HCHO_2 \right] = 0.550\ M - 0.0050\ M = 0.545\ M$.

With the Henderson-Hasselbalch equation

$$pH = pK_a + \log\frac{\left[CHO_2^- \right]}{\left[HCHO_2 \right]} = 3.74 + \log\frac{0.355}{0.545} = 3.55$$

6B   The buffer cited has the same concentration of weak acid and its anion as does the buffer of Example 18-6. Our goal is $pH = 5.03$ or $\left[H_3O^+ \right] = 10^{-pH} = 10^{-5.03} = 9.3 \times 10^{-6}\ M$.

Adding strong acid (H_3O^+), of course, produces $HC_2H_3O_2$ at the expense of $C_2H_3O_2^-$; it drives the reaction to the left. Again, we use the data around the balanced chemical equation.

Equation:	$HC_2H_3O_2\,(aq)$	$+ H_2O(l) \rightleftharpoons$	$C_2H_3O_2^-\,(aq)$	$+$	$H_3O^+(aq)$
Initial:	0.250 M		0.560 M		8.0×10^{-6} M
Add acid:					$+y$ M
Form HAc:	$+y$ M		$-y$ M		$-y$ M
	$(0.250 + y)$ M		$(0.560 - y)$ M		≈ 0 M
Changes:	$-x$ M		$+x$ M		$+x$ M
Equil:	$(0.250 + y - x)$ M		$(0.560 - y + x)$ M		9.3×10^{-6} M

$$K_a = \frac{[H_3O^+][C_2H_3O_2^-]}{[HC_2H_3O_2]} = \frac{9.3 \times 10^{-6}(0.560 - y + x)}{0.250 + y - x} = 1.8 \times 10^{-5} \approx \frac{9.3 \times 10^{-6}(0.560 - y)}{0.250 + y}$$

(Assume that x is negligible compared to y)

$$\frac{1.8 \times 10^{-5} \times (0.250 + y)}{9.3 \times 10^{-6}} = 0.484 + 1.94\ y = 0.560 - y \qquad y = \frac{0.560 - 0.484}{1.94 + 1.00} = 0.026\ M$$

Notice that our assumption is valid: $x \ll 0.250 + y\ (= 0.276) < 0.560 - y\ (= 0.534)$.

$$V_{HNO_3} = 300.0\ \text{mL buffer} \times \frac{0.026\ \text{mmol } H_3O^+}{1\ \text{mL buffer}} \times \frac{1\ \text{mL } HNO_3\,(aq)}{6.0\ \text{mmol } H_3O^+} = 1.3\ \text{mL of 6.0 M } HNO_3$$

Instead of the algebraic solution, we use the Henderson-Hasselbalch equation, since the final pH falls within one pH unit of the pK_a of acetic acid. We let z indicate the increase in $[HC_2H_3O_2]$, and also the decrease in $\left[C_2H_3O_2^- \right]$

$$pH = pK_a + \log\frac{\left[C_2H_3O_2^- \right]}{\left[HC_2H_3O_2 \right]} = 4.74 + \log\frac{0.560 - z}{0.250 + z} = 5.03 \qquad \frac{0.560 - z}{0.250 + z} = 10^{5.03 - 4.74} = 1.95$$

$$0.560 - z = 1.95\ (0.250 + z) = 0.488 + 1.95\ z \qquad z = \frac{0.560 - 0.488}{1.95 + 1.00} = 0.024\ M$$

This is—and should be—almost exactly the value of y we obtained by the other method. The differences are due to imprecision: rounding errors.

<u>7A</u> **(a)** The initial pH is the pH of 0.150 M HCl, which we obtain from $\left[H_3O^+\right]$ of that strong acid solution.

$$[H_3O^+] = \frac{0.150 \text{ mol HCl}}{1 \text{ L soln}} \times \frac{1 \text{ mol } H_3O^+}{1 \text{ mol HCl}} = 0.150 \text{ M},$$

$$pH = -\log[H_3O^+] = -\log(0.150) = 0.824$$

(b) To determine $\left[H_3O^+\right]$ and then pH at the 50.0% point, we need the volume of the solution and the amount of H_3O^+ unreacted. First we calculate the amount of hydronium ion present and then the volume of base solution needed for its complete neutralization.

$$\text{amount } H_3O^+ = 25.00 \text{ mL} \times \frac{0.150 \text{ mmol HCl}}{1 \text{ mL soln}} \times \frac{1 \text{ mmol } H_3O^+}{1 \text{ mmol HCl}} = 3.75 \text{ mmol } H_3O^+$$

$$V_{acid} = 3.75 \text{ mmol } H_3O^+ \times \frac{1 \text{ mmol } OH^-}{1 \text{ mmol } H_3O^+} \times \frac{1 \text{ mmol NaOH}}{1 \text{ mmol } OH^-} \times \frac{1 \text{ mL titrant}}{0.250 \text{ mmol NaOH}}$$

$$= 15.0 \text{ mL titrant}$$

At the 50.0% point, half (1.88 mmol H_3O^+) will remain unreacted and only half (7.50 mL titrant) of the titrant solution will be added. From this information, and the original 25.00-mL volume of the solution, we calculate $\left[H_3O^+\right]$ and then pH.

$$\left[H_3O^+\right] = \frac{1.88 \text{ mmol } H_3O^+ \text{ left}}{25.00 \text{ mL original} + 7.50 \text{ mL titrant}} = 0.0578 \text{ M}$$

$$pH = -\log(0.0578) = 1.238$$

(c) Since this is the titration of a strong acid by a strong base, at the neutralization point $pH = 7.00$. This is a solution of NaCl(aq), in which neither ion hydrolyzes.

(d) Beyond the equivalence point, the added titrant contains the only species that determine solution pH. The volume of the solution is $25.00 \text{ mL} + 1.00 \text{ mL} = 26.00$ mL. The amount of hydroxide ion in the excess titrant is calculated and used to determine $\left[OH^-\right]$, from which pH is computed.

$$\text{amount of } OH^- = 1.00 \text{ mL} \times \frac{0.250 \text{ mmol NaOH}}{1 \text{ mL}} = 0.250 \text{ mmol } OH^-$$

$$\left[OH^-\right] = \frac{0.250 \text{ mmol } OH^-}{26.00 \text{ mL}} = 0.00962 \text{ M}$$

$$pOH = -\log(0.00962) = 2.017; \quad pH = 14.00 - 2.017 = 11.983$$

7B **(a)** The initial pH is simply the pH of 0.00812 M $Ba(OH)_2$, which we obtain from $\left[OH^-\right]$ for the solution.

$$\left[OH^-\right] = \frac{0.00812\,mol\,Ba(OH)_2}{1\,L\,soln} \times \frac{2\,mol\,OH^-}{1\,mol\,Ba(OH)_2} = 0.01624\,M$$

$$pOH = -\log\left[OH^-\right] = -\log(0.0162) = 1.790; \quad pH = 14.00 - pOH = 14.00 - 1.790 = 12.21$$

(b) To determine $\left[OH^-\right]$ and then pH at the 50.0% point, we need the volume of the solution and the amount of OH^- unreacted. First we calculate the amount of hydroxide ion present and then the volume of acid solution needed for its complete neutralization.

$$amount\,OH^- = 50.00\,mL \times \frac{0.00812\,mmol\,Ba(OH)_2}{1\,mL\,soln} \times \frac{2\,mmol\,OH^-}{1\,mmol\,Ba(OH)_2} = 0.812\,mmol\,OH^-$$

$$V_{acid} = 0.812\,mmol\,OH^- \times \frac{1\,mmol\,H_3O^+}{1\,mmol\,OH^-} \times \frac{1\,mmol\,HCl}{1\,mmol\,H_3O^+} \times \frac{1\,mL\,titrant}{0.0250\,mmol\,HCl}$$

$$= 32.48\,mL\,titrant$$

At the 50.0 % point, half (0.406 mmol OH^-) will remain unreacted and only half (16.24 mL titrant) of the titrant solution will be added. From this information, and the original 50.00-mL volume of the solution, we calculate $\left[OH^-\right]$ and then pH.

$$\left[OH^-\right] = \frac{0.406\,mmol\,OH^-\,left}{50.00\,mL\,original + 16.24\,mL\,titrant} = 0.00613\,M$$

$$pOH = -\log(0.00613) = 2.213; \quad pH = 14.00 - pOH = 11.79$$

(c) Since this is the titration of a strong base by a strong acid, at the neutralization point, $pH = 7.00$. This is a solution of $BaCl_2(aq)$, in which neither ion hydrolyzes.

8A **(a)** Initial pH is just that of 0.150 M HF ($pK_a = -\log(6.6 \times 10^{-4}) = 3.18$).

[Initial solution contains $20.00\,mL \times \dfrac{0.150\,mmol\,HF}{1\,mL} = 3.00$ mmol HF].

$$Equation: HF(aq) + H_2O(l) \rightleftharpoons H_3O^+(aq) + F^-(aq)$$

Initial: $\quad$ 0.150 M $\qquad\qquad\qquad \approx 0\,M \qquad\quad 0\,M$

Changes: $\quad -x\,M \qquad\qquad\qquad\quad +x\,M \qquad +x\,M$

Equil: $\quad (0.150 - x)\,M \qquad\qquad\quad x\,M \qquad\quad x\,M$

$$K_a = \frac{[H_3O^+][F^-]}{[HF]} = \frac{x \cdot x}{0.150 - x} \approx \frac{x^2}{0.150} = 6.6 \times 10^{-4}$$

$$x = \sqrt{0.150 \times 6.6 \times 10^{-4}} = 9.9 \times 10^{-3} \text{ M}$$

$x > 0.05(0.150)$. The assumption is invalid. After a 2nd cycle of approximation

cycle, $[H_3O^+] = 9.6 \times 10^{-3} \text{M}$; $\text{pH} = -\log(9.6 \times 10^{-3}) = 2.02$

(b) When the titration is 25.0% complete, there are $(0.25 \times 3.00 =)0.75$ mmol F^- for every 3.00 mmol HF that were present initially. $(3.00 - 0.75 =)2.25$ mmol HF remain untitrated. We designate the solution volume (the volume holding these 3.00 mmol total) as V and use the Henderson-Hasselbalch equation.

$$\text{pH} = pK_a + \log \frac{[F^-]}{[HF]} = 3.18 + \log \frac{0.75 \text{ mmol}/V}{2.25 \text{ mmol}/V} = 2.70$$

(c) At the midpoint of the titration of a weak base, $\text{pH} = pK_a = 3.18$

(d) At the endpoint of the titration, the anion of the weak acid hydrolyzes. We calculate the amount of that anion and the volume of the solution in order to calculate its initial concentration.

$$\text{amount } F^- = 20.00 \text{ mL} \times \frac{0.150 \text{ mmol HF}}{1 \text{ mL soln}} \times \frac{1 \text{ mmol } F^-}{1 \text{ mmol HF}} = 3.00 \text{ mmol } F^-$$

$$\text{volume titrant} = 3.00 \text{ mmol HF} \times \frac{1 \text{ mmol OH}^-}{1 \text{ mmol HF}} \times \frac{1 \text{ mL titrant}}{0.250 \text{ mmol OH}^-} = 12.0 \text{ mL titrant}$$

$$[F^-] = \frac{3.00 \text{ mmol } F^-}{25.00 \text{ mL original volume} + 12.0 \text{ mL titrant}} = 0.0811 \text{M}$$

We organize the solution of the hydrolysis problem around its balanced equation.

Equation: $\quad F^-(aq) \quad + \quad H_2O(l) \quad \rightleftharpoons \quad HF(aq) \quad + \quad OH^-(aq)$

Initial: $\quad\quad 0.0811 \text{M} \quad\quad\quad\quad\quad\quad\quad\quad\quad 0 \text{M} \quad\quad \approx 0 \text{M}$

Changes: $\quad\quad -x \text{M} \quad\quad\quad\quad\quad\quad\quad\quad\quad\quad +x \text{M} \quad\quad +x \text{M}$

Equil: $\quad\quad (0.0811 - x)\text{M} \quad\quad\quad\quad\quad\quad x \text{M} \quad\quad\quad x \text{ M}$

$$K_b = \frac{[HF][OH^-]}{[F^-]} = \frac{K_w}{K_a} \frac{1.0 \times 10^{-14}}{6.6 \times 10^{-4}} = 1.5 \times 10^{-11} = \frac{x \cdot x}{0.0811 - x} \approx \frac{x^2}{0.0811}$$

$$x = \sqrt{0.0811 \times 1.5 \times 10^{-11}} = 1.1 \times 10^{-6} \text{ M} = [OH^-] \quad\quad \text{The assumption is valid.}$$

$$\text{pOH} = -\log(1.1 \times 10^{-6}) = 5.96; \quad \text{pH} = 14.00 - \text{pOH} = 14.00 - 5.96 = 8.04$$

8B **(a)** The initial pH is simply that of $0.106\ \text{M}\ NH_3$.

Equation: $NH_3(aq) + H_2O(l) \rightleftharpoons NH_4^+(aq) + OH^-(aq)$

Initial:	0.106 M	0 M	≈ 0 M
Changes:	$-x$ M	$+x$ M	$+x$ M
Equil:	$(0.106-x)$ M	x M	x M

$$K_b = \frac{\left[NH_4^+\right]\left[OH^-\right]}{\left[NH_3\right]} = \frac{x \cdot x}{0.106-x} \approx \frac{x^2}{0.106} = 1.8 \times 10^{-5}$$

$$x = \sqrt{0.106 \times 1.8 \times 10^{-4}} = 1.4 \times 10^{-3}\ \text{M}$$

$\left[OH^-\right] = 1.4 \times 10^{-3}\ \text{M}$ Assumption is valid. $pOH = -\log(0.0014) = 2.85$

$pH = 14.00 - pOH = 14.00 - 2.85 = 11.15$

(b) When the titration is 25.0% complete, there are 25.0 mmol NH_4^+ for every 100.0 mmol of NH_3 that were present initially (i.e. there are 1.33 mmol of NH_4^+ in solution) 3.98 mmol NH_3 remain untitrated. We designate the solution volume (the volume holding these 5.30 mmol total) as V and use a version of the Henderson-Hasselbalch equation.

$$pOH = pK_b + \log\frac{\left[NH_4^+\right]}{\left[NH_3\right]} = 4.74 + \log\frac{\dfrac{1.33\ \text{mmol}}{V}}{\dfrac{3.98\ \text{mmol}}{V}} = 4.26$$

$pH = 14.00 - 4.26 = 9.74$

(c) At the midpoint of the titration of a weak base,
$pOH = pK_b = 4.74$ and $pH = 14.00 - 4.74 = 9.26$

(d) At the endpoint of the titration, the cation of the weak base hydrolyzes. We calculate the amount of that cation and the volume of the solution in order to calculate its initial concentration.

$$\text{amount } NH_4^+ = 50.00\ \text{mL} \times \frac{0.106\ \text{mmol } NH_3}{1\ \text{mL soln}} \times \frac{1\ \text{mmol } NH_4^+}{1\ \text{mmol } NH_3}$$

$$\text{amount } NH_4^+ = 5.30\ \text{mmol } NH_4^+$$

$$V_{\text{titrant}} = 5.30\ \text{mmol } NH_3 \times \frac{1\ \text{mmol } H_3O^+}{1\ \text{mmol } NH_3} \times \frac{1\ \text{mL titrant}}{0.225\ \text{mmol } H_3O^+} = 23.6\ \text{mL titrant}$$

$$\left[NH_4^+\right] = \frac{5.30 \text{ mmol } NH_4^+}{50.00 \text{ mL original volume} + 23.6 \text{ mL titrant}} = 0.0720 \text{ M}$$

We organize the solution of the hydrolysis problem around its balanced chemical equation.

Equation: $NH_4^+ (aq) + H_2O(l) \rightleftharpoons NH_3(aq) + H_3O^+(aq)$

Initial:	0.0720 M	0 M	≈ 0 M
Changes:	$-x$ M	$+x$ M	$+x$ M
Equil:	$(0.0720-x)$ M	x M	x M

$$K_b = \frac{[NH_3][H_3O^+]}{[NH_4^+]} = \frac{K_w}{K_b} = \frac{1.0 \times 10^{-14}}{1.8 \times 10^{-5}} = 5.6 \times 10^{-10} = \frac{x \cdot x}{0.0720 - x} \approx \frac{x^2}{0.0720}$$

$$x = \sqrt{0.0720 \times 5.6 \times 10^{-10}} = 6.3 \times 10^{-6} \text{ M} = [H_3O^+] \quad \text{The assumption is valid.}$$

$$pH = -\log(6.3 \times 10^{-6}) = 5.20$$

9A   The acidity of the solution is principally due to the hydrolysis of the carbonate ion, considered first.

Equation: $CO_3^{2-}(aq) + H_2O(l) \rightleftharpoons HCO_3^-(aq) + OH^-(aq)$

Initial:	1.0 M	0 M	≈ 0 M
Changes:	$-x$ M	$+x$ M	$+x$ M
Equil:	$(1.0-x)$ M	x M	x M

$$K_b = \frac{K_w}{K_a(HCO_3^-)} = \frac{1.0 \times 10^{-14}}{4.7 \times 10^{-11}} = 2.1 \times 10^{-4} = \frac{[HCO_3^-][OH^-]}{[CO_3^{2-}]} = \frac{x \cdot x}{1.0-x} \approx \frac{x^2}{1.0}$$

$$x = \sqrt{1.0 \times 2.1 \times 10^{-4}} = 1.5 \times 10^{-2} \text{ M} = 0.015 \text{ M} = [OH^-] \quad \text{The assumption is valid.}$$

Now we consider the hydrolysis of the bicarbonate ion.

Equation: $HCO_3^-(aq) + H_2O(l) \rightleftharpoons H_2CO_3(aq) + OH^-(aq)$

Initial:	0.015 M	0 M	0.015 M
Changes:	$-y$ M	$+y$ M	$+y$ M
Equil:	$(0.015-y)$ M	y M	$(0.015+y)$ M

$$K_b = \frac{K_w}{K_a(H_2CO_3)} = \frac{1.0 \times 10^{-14}}{4.4 \times 10^{-7}} = 2.3 \times 10^{-8} = \frac{[H_2CO_3][OH^-]}{[HCO_3^-]} = \frac{y(0.015+y)}{0.015-x} \approx \frac{0.015y}{0.015} = y$$

The assumption is valid and $y = [H_2CO_3] = 2.3 \times 10^{-8}$ M. The second hydrolysis makes a negligible contribution to the acidity of the solution. For the entire solution, then

$$pOH = -\log\left[OH^-\right] = -\log(0.015) = 1.82 \qquad\qquad pH = 14.00 - 1.82 = 12.18$$

9B The acidity of the solution is principally due to the hydrolysis of the sulfite ion.

Equation: $SO_3^{2-}(aq) + H_2O(l) \rightleftharpoons HSO_3^-(aq) + OH^-(aq)$

Initial:	0.500 M	0 M	≈ 0 M
Changes:	$-x$ M	$+x$ M	$+x$ M
Equil:	$(0.500 - x)$ M	x M	x M

$$K_b = \frac{K_w}{K_a HSO_3^-} = \frac{1.0\times10^{-14}}{6.2\times10^{-8}} = 1.6\times10^{-7} = \frac{\left[HSO_3^-\right]\left[OH^-\right]}{[SO_3^{2-}]} = \frac{x\cdot x}{0.500 - x} \approx \frac{x^2}{0.500}$$

$$x = \sqrt{0.500\times1.6\times10^{-7}} = 2.8\times10^{-4}\ M = 0.00028\ M = [OH^-] \qquad \text{The assumption is valid.}$$

Next we consider the hydrolysis of the bisulfite ion.

Equation: $HSO_3^-(aq) + H_2O(l) \rightleftharpoons H_2SO_3(aq) + OH^-(aq)$

Initial:	0.00028 M	0 M	0.00028 M
Changes:	$-y$ M	$+y$ M	$+y$ M
Equil:	$(0.00028 - y)$ M	y M	$(0.00028 + y)$ M

$$K_b = \frac{K_w}{K_a H_2SO_3} = \frac{1.0\times10^{-14}}{1.3\times10^{-2}} = 7.7\times10^{-13}$$

$$K_b = 7.7\times10^{-13} = \frac{\left[H_2SO_3\right]\left[OH^-\right]}{\left[HSO_3^-\right]} = \frac{y(0.00028 + y)}{0.00028 - y} \approx \frac{0.00028\,y}{0.00028} = y$$

The assumption is valid and $y = \left[H_2CO_3\right] = 7.7\times10^{-13}$ M. The second hydrolysis makes a negligible contribution to the acidity of the solution. For the entire solution, then

$$pOH = -\log\left[OH^-\right] = -\log(0.00028) = 3.55 \qquad\qquad pH = 14.00 - 3.55 = 10.45$$

REVIEW QUESTIONS

1. (a) The abbreviation "mmol" stands for millimole, one thousandth of a mole. This amount of material often is a convenient one to use in calculations involving solutions where volumes are measured in mL.

    (b) HIn is the generic symbol for the formula of the acid form of an acid/base indicator.

    (c) The equivalence point of a titration is that point when sufficient added titrant is present to react with all of the substance being titrated, with no titrant left over.

    (d) A titration curve is a plot of the pH of solution versus the volume of added titrant (or sometimes versus the percentage of substance that has been titrated).

2. **(a)** The common-ion effect refers to the suppression of an ionization equilibrium caused by the presence (or addition) of one or more of the ions produced by the ionization.

   (b) A buffer solution maintains constant pH by consuming added strong acid or added strong base. The buffer solution contains components that react with each.

   (c) The value of pK_a can be determined from the titration curve of a monoprotic weak acid since the pH at the half equivalence point is equal to the pK_a of the weak acid.

   (d) pH is measured with a series of acid-base indicators that change color at different pH values. By determining the color of each indicator in a sample of the solution being tested, the pH of that solution can be placed within a narrow range.

3. **(a)** Buffer capacity is the amount of strong acid or base that can be added to a specified volume of a buffer before a significant change in pH occurs. Buffer range is the range of pH values within which a buffer solution will resist changes in pH.

   (b) Hydrolysis refers to the reaction of an ion with water to produce either $H_3O^+(aq)$ or $OH^-(aq)$. Neutralization refers to the reaction of an acid with a base to produce a solution with a pH close to 7.0.

   (c) The first and second equivalence points of a weak diprotic acid are the points where, respectively, the first proton has been completely ionized, and the second proton has been completely ionized.

   (d) The equivalence point of a titration is the point where stoichiometric amounts of acid and base have been combined. The end point is when the indicator changes color. Careful selection of the indicator can ensure that these two points nearly coincide.

4. **(a)** Note that HI is a strong acid and the initial $[H_3O^+] = [HI] = 0.0892 M$

Equation : $\quad HC_3H_5O_2 + H_2O \rightleftharpoons C_3H_5O_2^- + H_3O^+$

Initial :	0.275 M	0M	0.0892M
Changes :	$-x$ M	$+x$ M	$+x$ M
Equil :	$(0.275 - x)$ M	x M	$(0.0892 + x)$ M

$$K_a = \frac{[C_3H_5O_2^-][H_3O^+]}{[HC_3H_5O_2]} = 1.3 \times 10^{-5} = \frac{x(0.0892 + x)}{0.275 - x} \approx \frac{0.0892x}{0.275} \qquad x = 4.0 \times 10^{-5} \text{ M}$$

The assumption that $x \ll 0.0892$ M is correct. $[H_3O^+] = 0.0892 M$

   (b) $[OH^-] = \dfrac{K_w}{[H_3O^+]} = \dfrac{1.0 \times 10^{-14}}{0.0892} = 1.1 \times 10^{-13}$ M

   (c) $[C_3H_5O_2^-] = x = 4.0 \times 10^{-5}$ M

   (d) $[I^-] = [HI]_{int} = 0.0892$ M

5. **(a)** The NH_4Cl dissociates completely, and thus, $\left[NH_4^+\right]_{int} = \left[Cl^-\right]_{int} = 0.102$ M

Equation: $NH_3(aq) + H_2O(l) \rightleftharpoons NH_4^+(aq) + OH^-(aq)$

Initial:	0.164 M	0.102 M	≈ 0 M
Changes:	$-x$ M	$+x$ M	$+x$ M
Equil:	$(0.164-x)$M	$(0.102+x)$M	x M

$$K_b = \frac{\left[NH_4^+\right]\left[OH^-\right]}{\left[NH_3\right]} = \frac{(0.102+x)x}{0.164-x} = 1.8\times10^{-5} \approx \frac{0.102x}{0.164}; \quad x = 2.9\times10^{-5}\,M$$

Assumed $x \ll 0.102$ M, a clearly valid assumption. $\left[OH^-\right] = x = 2.9\times10^{-5}$ M

(b) $\left[NH_4^+\right] = 0.102 + x = 0.102$ M   **(c)** $\left[Cl^-\right] = 0.102$ M

(d) $\left[H_3O^+\right] = \dfrac{1.0\times10^{-14}}{2.9\times10^{-5}} = 3.4\times10^{-10}$ M

6. Adding $H_3O^+(aq)$ represents reaction with strong acid; adding $OH^-(aq)$ represents reaction with strong base.

(a) $CHO_2^-(aq) + H_3O^+(aq) \rightarrow HCHO_2(aq) + H_2O(l)$

$HCHO_2(aq) + OH^-(aq) \rightarrow CHO_2^-(aq) + H_2O(l)$

(b) $C_6H_5NH_2(aq) + H_3O^+(aq) \rightarrow C_6H_5NH_3^+(aq) + H_2O(l)$

$C_6H_5NH_3^+(aq) + OH^-(aq) \rightarrow C_6H_5NH_2(aq) + H_2O(l)$

(c) $HPO_4^{2-}(aq) + H_3O^+(aq) \rightarrow H_2PO_4^-(aq) + H_2O(l)$

$H_2PO_4^-(aq) + OH^-(aq) \rightarrow HPO_4^{2-}(aq) + H_2O(l)$

7. **(a)** Equation: $HC_7H_5O_2(aq) + H_2O(l) \rightleftharpoons C_7H_5O_2^-(aq) + H_3O^+(aq)$

Initial:	0.012 M	0.033 M	≈ 0 M
Changes:	$-x$ M	$+x$ M	$+x$ M
Equil:	$(0.012-x)$M	$(0.033+x)$M	x M

$$K_a = \frac{\left[H_3O^+\right]\left[C_7H_5O_2^-\right]}{\left[HC_7H_5O_2\right]} = 6.3\times10^{-5} = \frac{x(0.033+x)}{0.012-x} \approx \frac{0.033x}{0.012} \quad x = 2.3\times10^{-5}\,M$$

To determine the value of x, we assumed $x \ll 0.012$ M, an assumption that clearly is correct. $\left[H_3O^+\right] = 2.3\times10^{-5}$ M   pH $= -\log(2.3\times10^{-5}) = 4.64$

(b) Equation: $NH_3(aq) + H_2O(l) \rightleftharpoons NH_4^+(aq) + OH^-(aq)$

Initial: $\quad$ 0.408 M $\qquad\qquad$ 0.153 M $\qquad$ ≈ 0 M

Changes: $\quad -x$ M $\qquad\qquad\quad +x$ M $\qquad +x$ M

Equil: $\quad (0.408 - x)$ M $\qquad (0.153 + x)$ M $\qquad x$ M

$$K_b = \frac{[NH_4^+][OH^-]}{[NH_3]} = 1.8 \times 10^{-5} = \frac{x(0.153 + x)}{0.408 - x} \approx \frac{0.153x}{0.408} \qquad x = 4.8 \times 10^{-5} M$$

To determine the value of x, we assumed $x \ll 0.153$, a clearly valid assumption.

$$[OH^-] = 4.8 \times 10^{-5} M; \quad pOH = -\log(4.8 \times 10^{-5}) = 4.32; \quad pH = 14.00 - 4.32 = 9.68$$

8. $[H_3O^+] = 10^{-4.06} = 8.7 \times 10^{-5} M$. We let $S = [CHO_2^-]_{int}$

Equation: $HCHO_2(aq) + H_2O(l) \rightleftharpoons CHO_2^-(aq) + H_3O^+(aq)$

Initial: $\quad$ 0.366 M $\qquad\qquad\qquad$ S M $\qquad\qquad$ ≈ 0 M

Changes: -8.7×10^{-5} M $\qquad +8.7 \times 10^{-5}$ M $\qquad +8.7 \times 10^{-5}$ M

Equil: $\quad$ 0.366 M $\qquad (S + 8.7 \times 10^{-5})$ M $\qquad 8.7 \times 10^{-5}$ M

$$K_a = \frac{[H_3O^+][CHO_2^-]}{[HCHO_2]} = 1.8 \times 10^{-4} = \frac{(S + 8.7 \times 10^{-5})8.7 \times 10^{-5}}{0.366} \approx \frac{8.7 \times 10^{-5}S}{0.366}; \quad S = 0.76 M$$

To determine S, we assumed $S \gg 8.7 \times 10^{-5} M$, clearly a valid assumption. Or, we could have used the Henderson-Hasselbalch equation. $pK_a = -\log(1.8 \times 10^{-4}) = 3.74$

$$4.06 = 3.74 + \log\frac{[CHO_2^-]}{[HCHO_2]}; \qquad \frac{[CHO_2^-]}{[HCHO_2]} = 2.1; \qquad [CHO_2^-] = 2.1 \times 0.366 = 0.77 M$$

The difference in the two answers is due simply to rounding.

9. We use the Henderson-Hasselbalch equation. $pK_b = -\log(1.8 \times 10^{-5}) = 4.74$

$$pK_a = 14.00 - pK_b = 14.00 - 4.74 = 9.26 \qquad\qquad pH = 9.12 = 9.26 + \log\frac{[NH_3]}{[NH_4^+]}$$

$$\frac{[NH_3]}{[NH_4^+]} = 10^{-0.14} = 0.72 \qquad\qquad [NH_3] = 0.72 \times [NH_4^+] = 0.72 \times 0.732 M = 0.53 M$$

10. 0.60 mol $NaC_2H_3O_2$ will raise the pH of 1.00 L of 0.50 M HCl to the greatest extent. Both OH^- and $C_2H_3O_2^-$ are bases that will react with the strong acid (H_3O^+) in HCl(aq), resulting in an increase in the pH of the solution. Since there is more acetate ion available than hydroxide ion, 0.60 mol $NaC_2H_3O_2$ will consume more H_3O^+ than will 0.40 mol NaOH, causing a greater increase in the pH of the solution. In fact, 0.60 mol $NaC_2H_3O_2$ will consume all of the H_3O^+ in 1.00 L of 0.50 M HCl, leaving only a solution of acetic acid and sodium acetate, while 0.40 mol NaOH will leave a solution of HCl (with a concentration of 0.10 M) and NaCl. With regard to the other two possibilities, 0.70 mol NaCl will not affect the pH and 0.50 mol $HC_2H_3O_2$ will lower the pH because $HC_2H_3O_2$ is an acid.

11. We use the Henderson-Hasselbalch equation to determine K_a of lactic acid.

$$\left[C_3H_5O_3^-\right] = \frac{1.00 \text{ g } NaC_3H_5O_3}{100.0 \text{ mL soln}} \times \frac{1000 \text{ mL}}{1 \text{ L soln}} \times \frac{1 \text{ mol } NaC_3H_5O_3}{112.1 \text{ g } NaC_3H_5O_3} \times \frac{1 \text{ mol } C_3H_5O_3^-}{1 \text{ mol } NaC_3H_5O_3} = 0.0892 \text{ M}$$

$$pH = 4.11 = pK_a + \log\frac{\left[C_3H_5O_3^-\right]}{\left[HC_3H_5O_3\right]} = pK_a + \log\frac{0.0892 \text{ M}}{0.0500 \text{ M}} = pK_a + 0.251$$

$$pK_a = 4.11 - 0.251 = 3.86; \quad K_a = 10^{-3.86} = 1.4 \times 10^{-4}$$

12. (a) Use the Henderson-Hasselbalch equation to determine $\left[CHO_2^-\right]$ in the buffer solution.

$$pH = pK_a + \log\frac{\left[CHO_2^-\right]}{\left[HCHO_2\right]}; \quad 3.82 = 3.74 + \log\frac{\left[CHO_2^-\right]}{\left[HCHO_2\right]}; \quad \log\frac{\left[CHO_2^-\right]}{\left[HCHO_2\right]} = 3.82 - 3.74 = 0.08$$

$$\frac{\left[CHO_2^-\right]}{\left[HCHO_2\right]} = 1.2; \quad \left[CHO_2^-\right] = 1.2\left[HCHO_2\right] = 1.2 \times 0.465 \text{ M} = 0.56 \text{ M}$$

$$\text{mass } NaCHO_2 = 0.250 \text{ L} \times \frac{0.56 \text{ mol } CHO_2^-}{1 \text{ L soln}} \times \frac{1 \text{ mol } NaCHO_2}{1 \text{ mol } CHO_2^-} \times \frac{68.0 \text{ g } NaCHO_2}{1 \text{ mol } NaCHO_2}$$

$$= 9.5 \text{ g } NaCHO_2 \qquad \left[\text{Note that } pK_a = -\log K_a = -\log\left(1.8 \times 10^{-4}\right) = 3.74\right]$$

(b) $[OH^-]_{int} = \dfrac{0.20 \text{ g NaOH}}{0.250 \text{ L}} \times \dfrac{1 \text{ mol NaOH}}{40.0 \text{ g NaOH}} \times \dfrac{1 \text{ mol } OH^-}{1 \text{ mol NaOH}} = 0.020 \text{ M } OH^-$

Thus, $\left[HCHO_2\right]$ will decrease by 0.020 M and $\left[CHO_2^-\right]$ will increase by 0.020 M because of the reaction with the added hydroxide ion, as a result of:

$$HCHO_2 + OH^- \rightarrow CHO_2^- + H_2O \qquad pH = 3.74 + \log\frac{0.56 + 0.02}{0.465 - 0.02} = 3.86 \quad \text{The}$$

pH has increased 0.04 units due to the base addition.

13. **(a)** The pH color change range is 1.00 pH unit on either side of pK_{HIn}. If the pH color change range is below $pH = 7.00$, the indicator changes color in acidic solution. If it is above $pH = 7.00$, the indicator changes color in alkaline solution. If $pH = 7.00$ falls within the pH color change range, the indicator changes color near the neutral point.

indicator	K_{HIn}	pK_{HIn}	pH color change range	changes color in?
bromophenol blue	1.4×10^{-4}	3.85	2.9 (yellow) to 4.9 (blue)	acidic solution
bromocresol green	2.1×10^{-5}	4.68	3.7 (yellow) to 5.7 (blue)	acidic solution
bromothymol blue	7.9×10^{-8}	7.10	6.1 (yellow) to 8.1 (blue)	neutral solution
2,4-dinitrophenol	1.3×10^{-4}	3.89	2.9 (clrless) to 4.9 (yellow)	acidic solution
chlorophenol red	1.0×10^{-6}	6.00	5.0 (yellow) to 7.0 (red)	acidic solution
thymolphthalein	1.0×10^{-10}	10.00	9.0 (clrless) to 11.0 (blue)	basic solution

(b) If bromcresol green is green, the pH is between 3.7 and 5.7, probably about $pH = 4.7$.

If chlorophenol red is orange, the pH is between 5.0 and 7.0, probably about $pH = 6.0$.

14. We first determine the pH of each solution, and then use the answer in Exercise 13(a) to predict the color of the indicator. (The weakly acidic or basic character of the indicator does not affect the pH of the solution, since the very little indicator is added.)

(a) $\left[H_3O^+\right] = 0.100 \text{ M HCl} \times \dfrac{1 \text{ mol } H_3O^+}{1 \text{ mol HCl}} = 0.100 \text{ M}; \quad pH = -\log(0.100 \text{ M}) = 1.000$

2,4-dinitrophenol assumes its acid color in a solution with $pH = 1.000$. The solution is colorless.

(b) Solutions of NaCl(aq) are pH neutral, with $pH = 7.000$. Chlorophenol red assumes its neutral color in such a solution; the solution is red/orange.

(c) Equation: $NH_3(aq) + H_2O(l) \rightleftharpoons NH_4^+(aq) + OH^-(aq)$

Initial:	1.00 M	0 M	≈ 0 M
Changes:	$-x$ M	$+x$ M	$+x$ M
Equil:	$(1.00 - x)$ M	x M	x M

$$K_b = \frac{\left[NH_4^+\right]\left[OH^-\right]}{\left[NH_3\right]} = 1.8 \times 10^{-5} = \frac{x^2}{1.00 - x} \approx \frac{x^2}{1.00} \quad x = 4.2 \times 10^{-3} \text{ M} = \left[OH^-\right]$$

$pOH = -\log(4.2 \times 10^{-3}) = 2.38 \qquad pH = 14.00 - 2.38 = 11.62$

Thymolphthalein assumes its basic color in a solution with $pH = 11.62$; the solution is blue.

(d) From Figure 17-6, seawater has $pH = 7.00$ to 8.50. Bromcresol green assumes its basic color in this solution; the solution is blue.

15. **(a)** The titration reaction is $KOH(aq) + HI(aq) \rightarrow KI(aq) + H_2O(l)$

vol. KOH soln $= 25.00 \text{ mL} \times \dfrac{0.212 \text{ mmol HI}}{1 \text{ mL soln}} \times \dfrac{1 \text{ mmol KOH}}{1 \text{ mmol HI}} \times \dfrac{1 \text{ mL soln}}{0.146 \text{ mmol KOH}}$

$= 36.3 \text{ mL KOH soln}$

(b) The titration reaction is $2KOH(aq) + H_2SO_4(aq) \rightarrow K_2SO_4(aq) + 2H_2O(l)$

vol. KOH soln $= 20.00 \text{ mL} \times \dfrac{0.0942 \text{ mol } H_2SO_4}{1 \text{L soln}} \times \dfrac{2 \text{ mol KOH}}{1 \text{ mol } H_2SO_4} \times \dfrac{1 \text{ mL soln}}{0.146 \text{ mmol KOH}}$

$= 25.8 \text{ mL KOH soln}$

16. We can just sketch approximate titration curves of pH vs. percent of titration, since we do not have the concentration of the acid or the base, or the volume of solution being titrated. We can, however, precisely determine the pH at the half-equivalence point [mid-way between untitrated and completely titrated for a weak acid (or base)]; this is equal to the pK_a of the weak acid (or $pOH = pK_b$ of the weak base). Further, if we assume all solutions are 1.00 M, we can determine the pH at each equivalence point. For a weak base, the pH at the equivalence point equals $-\log\sqrt{0.50(K_a)}$. For a weak acid, the pH at the equivalence point equals $14.00 + \log\sqrt{0.50(K_b)}$ or $14.00 + \log\sqrt{0.50(K_w / K_a)}$ [Indicator choices are given in square brackets.]

(a) In this titration (assuming 1.00 M NaOH and 1.00 M HNO_3), the initial $pH = 14.00$ and at the equivalence point $pH = 7.00$. The pH drops rapidly after the equivalence point, rapidly reaching $pH = 1.00$. [Bromothymol blue changes from blue at $pH = 8$ to yellow at $pH = 6$].

(b) The initial pH is that of 1.00 M NH_3, $pH = 11.6$. The half-equivalence point is $pOH = pK_b = 4.74$ and thus $pH = 9.26$. That of the equivalence point is $-\log\sqrt{0.50(K_a)} = 4.78$. The pH then drops rapidly with added strong acid to $pH = 1.00$. [Methyl red changes from yellow at $pH = 6.2$ to red at $pH = 4.5$]

(c) The initial pH is that of 1.00 M $HC_2H_3O_2$, $pH = 2.38$. That of the half-equivalence point is $pH = pK_a = 4.74$. The pH at the equivalence point is $pH = 14.00 + \log\sqrt{0.50(K_w / K_a)}$ or $pH = 9.22$. The pH then rises rapidly with the addition of strong base, rapidly reaching $pH = 13.00$. [Phenolphthalein changes from colorless at $pH = 8$ to red at $pH = 10$.]

(d) The initial pH is that of a 1.00 M $H_2PO_4^-$, $pH = 3.60$. The first half-equivalence point is $pH = pK_{a_2} = 7.20$. The pH of the first equivalence point is about $pH = 10.4$. (the OH- ions are formed via hydrolysis of HPO_4^{2-}) That for the second half equivalence point would be $pH = pK_{a_3} = 12.37$, but this will be difficult to reach; the solution is becoming pretty dilute. Finally, the pH for the second equivalence point is about $pH = 12.9$ (here OH- is formed via hydrolysis of PO_4^{3-}) which again will be hard to reach without adding excessively concentrated base. [Alizarin yellow R changes from yellow at $pH = 10$ to violet at $pH = 12$. These is no suitable indicator given in Figure 18-8 for the second equivalence point.] The curves are sketched below.

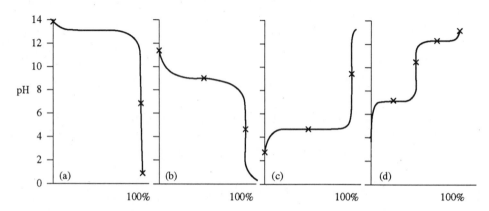

17. First we calculate the amount of HCl. The titration reaction is

$$HCl(aq) + KOH(aq) \rightarrow KCl(aq) + H_2O(l)$$

$$\text{amount HCl} = 25.00 \text{ mL} \times \frac{0.160 \text{ mmol HCl}}{1 \text{ mL soln}} = 4.00 \text{ mmol HCl} = 4.00 \text{ mmol } H_3O^+ \text{present}$$

Then, in each case, we calculate the amount of OH^- that has been added, determine which ion, $OH^-(aq)$ or $H_3O^+(aq)$, is in excess, compute the concentration of that ion, and determine the pH.

(a) amount $OH^- = 10.00 \text{ mL} \times \dfrac{0.242 \text{ mmol } OH^-}{1 \text{ mL soln}} = 2.42 \text{ mmol } OH^-$; H_3O^+ is in excess.

$$[H_3O^+] = \frac{4.00 \text{ mmol } H_3O^+ - \left(2.42 \text{ mmol } OH^- \times \dfrac{1 \text{ mmol } H_3O^+}{1 \text{ mmol } OH^-}\right)}{25.00 \text{ mL originally} + 10.00 \text{ mL titrant}} = 0.0451 \text{ M}$$

$$pH = -\log(0.0451) = 1.346$$

(b) amount $OH^- = 15.00 \text{ mL} \times \dfrac{0.242 \text{ mmol } OH^-}{1 \text{ mL soln}} = 3.63 \text{ mmol } OH^-$; H_3O^+ is in excess.

$$[H_3O^+] = \frac{4.00 \text{ mmol } H_3O^+ - \left(3.63 \text{ mmol } OH^- \times \dfrac{1 \text{ mmol } H_3O^+}{1 \text{ mmol } OH^-}\right)}{25.00 \text{ mL originally} + 15.00 \text{ mL titrant}} = 0.00925 \text{ M}$$

$$pH = -\log(0.00925) = 2.034$$

18.   The titration reaction is $KOH(aq) + HCl(aq) \rightarrow KCl(aq) + H_2O(l)$

mmol of $KOH = 20.00 \text{ mL} \times \dfrac{0.275 \text{ mmol KOH}}{1 \text{ mL soln}} = 5.50 \text{ mmol KOH}$

(a)   The total volume of the solution is $V = 20.00 \text{ mL} + 15.00 \text{ mL} = 35.00 \text{ mL}$

mmol $HCl = 15.00 \text{ mL} \times \dfrac{0.350 \text{ mmol HCl}}{1 \text{ mL soln}} = 5.25 \text{ mmol HCl}$

mmol excess $OH^- = (5.50 \text{ mmol KOH} - 5.25 \text{ mmol HCl}) \times \dfrac{1 \text{ mmol OH}^-}{1 \text{ mmol KOH}} = 0.25 \text{ mmol OH}^-$

$\left[OH^-\right] = \dfrac{0.25 \text{ mmol OH}^-}{35.00 \text{ mL soln}} = 0.0071 M \qquad pOH = -\log(0.0071) = 2.15$

$pH = 14.00 - 2.15 = 11.85$

(b)   The total volume of solution is $V = 20.00 \text{ mL} + 20.00 \text{ mL} = 40.00 \text{ mL}$

mmol $HCl = 20.00 \text{ mL} \times \dfrac{0.350 \text{ mmol HCl}}{1 \text{ mL soln}} = 7.00 \text{ mmol HCl}$

mmol excess $H_3O^+ = (7.00 \text{ mmol HCl} - 5.50 \text{ mmol KOH}) \times \dfrac{1 \text{ mmol H}_3O^+}{1 \text{ mmol HCl}}$

$= 1.50 \text{ mmol H}_3O^+$

$\left[H_3O^+\right] = \dfrac{1.50 \text{ mmol H}_3O^+}{40.00 \text{ mL}} = 0.0375 M \qquad pH = -\log(0.0375) = 1.426$

19.   The titration reaction is $HNO_2(aq) + NaOH(aq) \rightarrow NaNO_2(aq) + H_2O(l)$

amount $HNO_2 = 25.00 \text{ mL} \times \dfrac{0.132 \text{ mmol HNO}_2}{1 \text{ mL soln}} = 3.30 \text{ mmol HNO}_2$

(a)   The volume of the solution is $25.00 \text{ mL} + 10.00 \text{ mL} = 35.00 \text{ mL}$

amount $NaOH = 10.00 \text{ mL} \times \dfrac{0.116 \text{ mmol NaOH}}{1 \text{ mL soln}} = 1.16 \text{ mmol NaOH}$

1.16 mmol $NaNO_2$ are formed in this reaction and there is an excess of (3.30 mmol $HNO_2 - 1.16$ mmol NaOH) $= 2.14$ mmol HNO_2. We can use the Henderson-Hasselbalch equation to determine the pH of the solution.

$pK_a = -\log(7.2 \times 10^{-4}) = 3.14$

$pH = pK_a + \log\dfrac{\left[NO_2^-\right]}{\left[HNO_2\right]} = 3.14 + \log\dfrac{1.16 \text{ mmol NO}_2^- / 35.00 \text{ mL}}{2.14 \text{ mmol HNO}_2 / 35.00 \text{ mL}} = 2.87$

(b) The volume of the solution is $25.00 \text{ mL} + 20.00 \text{ mL} = 45.00 \text{ mL}$

$$\text{amount NaOH} = 20.00 \text{ mL} \times \frac{0.116 \text{ mmol NaOH}}{1 \text{ mL soln}} = 2.32 \text{ mmol NaOH}$$

2.32 mmol $NaNO_2$ are formed in this reaction and there is an excess of
$(3.30 \text{ mmol } HNO_2 - 2.32 \text{ mmol NaOH} =) \ 0.98 \text{ mmol } HNO_2$.

$$\text{pH} = \text{p}K_a + \log\frac{\left[NO_2^-\right]}{\left[HNO_2\right]} = 3.14 + \log\frac{2.32 \text{ mmol } NO_2^- / 45.00 \text{ mL}}{0.98 \text{ mmol } HNO_2 / 45.00 \text{ mL}} = 3.51$$

20. The calculation is very similar to that of Review Question 19. In this case, however, the titration reaction is $NH_3(aq) + HCl(aq) \rightarrow NH_4Cl(aq) + H_2O(l)$

$$\text{amount } NH_3 = 20.00 \text{ mL} \times \frac{0.318 \text{ mmol } NH_3}{1 \text{ mL soln}} = 6.36 \text{ mmol } NH_3$$

(a) The volume of the solution is $20.00 \text{ mL} + 10.00 \text{ mL} = 30.00 \text{ mL}$

$$\text{amount HCl} = 10.00 \text{ mL} \times \frac{0.475 \text{ mmol NaOH}}{1 \text{ mL soln}} = 4.75 \text{ mmol HCl}$$

4.75 mmol NH_4Cl is formed in this reaction and there is an excess of (6.36 mmol $NH_3 - 4.75 \text{ mmol HCl} =) 1.61 \text{ mmol } NH_3$. We can use the Henderson-Hasselbalch equation to determine the pH of the solution.

$$\text{p}K_b = -\log\left(1.8 \times 10^{-5}\right) = 4.74; \quad \text{p}K_a = 14.00 - \text{p}K_b = 14.00 - 4.74 = 9.26$$

$$\text{pH} = \text{p}K_a + \log\frac{\left[NH_3\right]}{\left[NH_4^+\right]} = 9.26 + \log\frac{1.61 \text{ mmol } NH_3 / 30.00 \text{ mL}}{4.75 \text{ mmol } NH_4^+ / 30.00 \text{ mL}} = 8.79$$

(b) The volume of the solution is $20.00 \text{ mL} + 15.00 \text{ mL} = 35.00 \text{ mL}$

$$\text{amount HCl} = 15.00 \text{ mL} \times \frac{0.475 \text{ mmol NaOH}}{1 \text{ mL soln}} = 7.13 \text{ mmol HCl}$$

6.36 mmol NH_4Cl is formed in this reaction and there is an excess of (7.13 mmol $HCl - 6.36 \text{ mmol } NH_3 =) 0.77 \text{ mmol HCl}$; this excess HCl determines the pH of the solution.

$$\left[H_3O^+\right] = \frac{0.77 \text{ mmol HCl}}{35.00 \text{ mL soln}} \times \frac{1 \text{ mmol } H_3O^+}{1 \text{ mmol HCl}} = 0.022 \text{ M} \quad \text{pH} = -\log\left(0.022\right) = 1.66$$

21. **(a)** First we calculate the pH of a 0.01000 M $HC_7H_5O_2$ solution with an ICE table.

Equation : $HC_7H_5O_2$ (aq) + H_2O(l) $\rightleftharpoons$ $C_7H_5O_2^-$ (aq) + H_3O^+ (aq)

Initial :   0.01000 M   0 M ≈ 0 M

Changes : $-x$ M $+x$ M $+x$ M

Equil : $(0.01000 - x)$ M x M x M

$$K_a = \frac{\left[C_7H_5O_2^-\right]\left[H_3O^+\right]}{\left[HC_7H_5O_2\right]} = 6.3 \times 10^{-5} = \frac{x^2}{0.0100 - x} \approx \frac{x^2}{0.0100}$$

$$x = 7.9 \times 10^{-4}\,M = \left[H_3O^+\right] \qquad pH = -\log(7.9 \times 10^{-4}) = 3.10$$

(Note: x is about 8% of the value of 0.0100, thus it is not much smaller than 0.0100) A more accurate calculation requires solution of the quadratic:

$x^2 + 6.3 \times 10^{-5}x - 6.3 \times 10^{-7} = 0$ where $x = 7.6 \times 10^{-4}$ M and pH = 3.12

(b) amount of $HC_7H_5O_2 = 25.0\,mL \times 0.01000\,M = 0.250\,mmol\ HC_7H_5O_2$

We determine the volume of 0.01000 M $Ba(OH)_2$ to reach the equivalence point. The titration reaction is:

$$Ba(OH)_2\,(aq) + 2HC_7H_5O_2\,(aq) \rightarrow Ba(C_7H_5O_2)_2\,(aq) + 2H_2O(l)$$

$$V_{base} = 0.250\,mmol\ HC_7H_5O_2 \times \frac{1\,mmol\ Ba(OH)_2}{2\,mmol\ HC_7H_5O_2} \times \frac{1\,mL\ soln}{0.01000\,mmol\ Ba(OH)_2}$$

$$V_{base} = 12.5\ mL$$

Thus, the addition of 6.25 mL of 0.01000 M $Ba(OH)_2$ brings us to the half-equivalence point, where $pH = pK_a = 4.20$.

(c) At the equivalence point, there is 0.250 mmol $C_7H_5O_2^-$ in $25.00 + 12.50 = 37.50$ mL solution. It is the hydrolysis of this anion that determines the pH of the solution.

$$\left[C_7H_5O_2^-\right] = \frac{0.250\,mmol\ C_7H_5O_2^-}{37.50\,mL\ soln} = 6.67 \times 10^{-3}\,M$$

Equation: $C_7H_5O_2^-$ (aq) + H_2O(l) $\rightleftharpoons$ $HC_7H_5O_2$ (aq) + OH^- (aq)

Initial :   0.00667 M   0 M ≈ 0 M

Changes : $-x$ M $+x$ M $+x$ M

Equil : $(0.00667 - x)$ M x M x M

$$K_b = \frac{K_w}{K_a} = \frac{1.0 \times 10^{-14}}{6.3 \times 10^{-5}} = \frac{x^2}{0.00667 - x} \approx \frac{x^2}{0.00667} \qquad x = 1.0 \times 10^{-6}\,M = \left[OH^-\right]$$

$$pOH = -\log(1.0 \times 10^{-6}) = 6.00 \qquad pH = 14.00 - 6.00 = 8.00$$

(d) The excess added base determines the pH of the solution.

$$\text{excess OH}^- = 2.50\,\text{mL} \times \frac{0.0100\,\text{mmol Ba(OH)}_2}{1\,\text{mL Ba(OH)}_2} \times \frac{2\,\text{mmol OH}^-}{1\,\text{mmol Ba(OH)}_2}$$

$$\text{excess OH}^- = 0.0500\,\text{mmol OH}^-$$

$$\left[\text{OH}^-\right] = \frac{0.0500\,\text{mmol OH}^-}{25.00\,\text{mL} + 15.00\,\text{mL}} = 1.25 \times 10^{-3}\,\text{M}$$

$$\text{pOH} = -\log\left(1.25 \times 10^{-3}\right) = 2.903 \qquad \text{pH} = 14.000 - 2.903 = 11.097$$

22. The 0.10 M solutions of $NaHSO_4$ is the most acidic. The Na^+ cation does not contribute to the acidity of the solution; it does not hydrolyze. Of the anions involved, S^{2-} hydrolyzes to produce an alkaline solution. Each of the other anions $\left(HSO_4^-, HCO_3^-, HPO_4^{2-}\right)$ can ionize further to yield H_3O^+, but HSO_4^- does so to a far greater extent than do the other two; it is quite strong for a weak acid. In fact, HPO_4^{2-} is expected to hydrolyze to form a basic solution.

EXERCISES

The Common Ion Effect

23. **(a)** We first determine the pH of 0.100 M HNO_2.

Equation	$HNO_2\,(aq)$	+	$H_2O(l)$	$\rightleftharpoons$	$NO_2^-\,(aq)$	+	$H_3O^+\,(aq)$
Initial:	0.100 M				0 M		≈ 0 M
Changes:	$-x$ M				$+x$ M		$+x$ M
Equil:	$(0.100 - x)$ M				x M		x M

$$K_a = \frac{\left[NO_2^-\right]\left[H_3O^+\right]}{\left[HNO_2\right]} = 7.2 \times 10^{-4} = \frac{x^2}{0.100 - x} \approx \frac{x^2}{0.100}$$

Via successive approximations, $x = 8.1 \times 10^{-3}\,\text{M} = \left[H_3O^+\right]$.

Then $\text{pH} = -\log\left(8.1 \times 10^{-3}\right) = 2.09$

When 0.100 mol $NaNO_2$ is added to 1.00 L of a 0.100 M HNO_2, a solution with $\left[NO_2^-\right] = 0.100\,\text{M} = \left[HNO_2\right]$ is produced. The answer obtained with the Henderson-Hasselbalch equation, is $\text{pH} = pK_a = -\log\left(7.2 \times 10^{-4}\right) = 3.14$. Thus, the addition has caused a pH change of 1.05 units.

(b) $NaNO_3$ contributes nitrate ion, NO_3^-, to the solution. Since, however, there is no molecular $HNO_3(aq)$ in equilibrium with hydrogen and nitrate ions, there is no equilibrium to be shifted by the addition of nitrate ions. The $[H_3O^+]$ and the pH are thus unaffected by the addition of $NaNO_3$ to a solution of nitric acid. The pH changes are not the same because there is an equilibrium system to be shifted in the first solution, whereas there is no equilibrium, just total ionization, in the second solution.

24. The explanation for the different result is that each of these solutions has present an ion, acetate ion, $C_2H_3O_2^-$, that is produced in the ionization of acetic acid. The presence of this ion suppresses the ionization of acetic acid, thus minimizing the increase in $[H_3O^+]$. All three solutions are buffer solutions and their pH can be found with the aid of the Henderson-Hasselbalch equation.

(a)
$$pH = pK_a + \log\frac{[C_2H_3O_2^-]}{[HC_2H_3O_2]} = 4.74 + \log\frac{0.10}{1.0} = 3.74 \qquad [H_3O^+] = 10^{-3.74} = 1.8 \times 10^{-4}\,M$$

$$\% \text{ ionization} = \frac{[H_3O^+]}{[HC_2H_3O_2]} \times 100\% = \frac{1.8 \times 10^{-4}\,M}{1.0\,M} \times 100\% = 0.018\%$$

(b)
$$pH = pK_a + \log\frac{[C_2H_3O_2^-]}{[HC_2H_3O_2]} = 4.74 + \log\frac{0.10}{0.10} = 4.74$$

$$[H_3O^+] = 10^{-4.74} = 1.8 \times 10^{-5}\,M$$

$$\% \text{ ionization} = \frac{[H_3O^+]}{[HC_2H_3O_2]} \times 100\% = \frac{1.8 \times 10^{-5}\,M}{0.10\,M} \times 100\% = 0.018\%$$

(c)
$$pH = pK_a + \log\frac{[C_2H_3O_2^-]}{[HC_2H_3O_2]} = 4.74 + \log\frac{0.10}{0.010} = 5.74$$

$$[H_3O^+] = 10^{-5.74} = 1.8 \times 10^{-6}\,M$$

$$\% \text{ ionization} = \frac{[H_3O^+]}{[HC_2H_3O_2]} \times 100\% = \frac{1.8 \times 10^{-6}\,M}{0.010\,M} \times 100\% = 0.018\%$$

25. **(a)** The strong acid HCl suppresses the ionization of the weak acid HOCl so much that a negligible concentration of H_3O^+ is contributed to the solution by HOCl. Thus,
$$[H_3O^+] = [HCl] = 0.035\,M$$

(b) This is a buffer solution. We can use the Henderson-Hasselbalch equation to determine its pH. $pK_a = -\log(7.2 \times 10^{-4}) = 3.14$;

$$pH = pK_a + \log\frac{[NO_2^-]}{[HNO_2]} = 3.14 + \log\frac{0.100\,M}{0.0550\,M} = 3.40$$

$$[H_3O^+] = 10^{-3.40} = 4.0 \times 10^{-4}\,M$$

(c) This also is a buffer solution, as we see by an analysis of the reaction between the components.

Equation: $\quad H_3O^+\,(aq,\ \text{from HCl}) + C_2H_3O_2^-\,(aq,\ \text{from NaC}_2\text{H}_3\text{O}_2) \rightarrow HC_2H_3O_2\,(aq) + H_2O(l)$

In soln:	0.0525 M	0.0768 M	0 M
Produce HAc:	−0.0525 M	−0.0525 M	+0.0525 M
Initial:	≈ 0M	0.0243 M	0.0525 M

Now the Henderson-Hasselbalch equation can be used.

$$pK_a = -\log(1.8 \times 10^{-5}) = 4.74$$

$$pH = pK_a + \log\frac{[C_2H_3O_2^-]}{[HC_2H_3O_2]} = 4.74 + \log\frac{0.0243\,M}{0.0525\,M} = 4.41$$

$$[H_3O^+] = 10^{-4.41} = 3.9 \times 10^{-5}\,M$$

26. (a) Neither $Ba^{2+}(aq)$ nor $Cl^-(aq)$ hydrolyzes or ionizes to affect the acidity of the solution. $[OH^-]$ is determined entirely by the $Ba(OH)_2$ solute.

$$[OH^-] = \frac{0.0062\ \text{mol Ba}(OH)_2}{1\,\text{L soln}} \times \frac{2\,\text{mol OH}^-}{1\,\text{mol Ba}(OH)_2} = 0.012\,M$$

(b) We use the Henderson-Hasselbalch equation for this buffer solution.

$$[NH_4^+] = 0.315\,M\,(NH_4)_2\,SO_4 \times \frac{2\ \text{mol NH}_4^+}{1\ \text{mol }(NH_4)_2\,SO_4} = 0.630\,M \quad pK_a = 9.26\ \text{for NH}_4^+.$$

$$pH = pK_a + \log\frac{[NH_3]}{[NH_4^+]} = 9.26 + \log\frac{0.486\,M}{0.630\,M} = 9.15 \quad pOH = 14.00 - 9.15 = 4.85$$

$$[OH^-] = 10^{-4.85} = 1.4 \times 10^{-5}\,M$$

(c) This solution also is a buffer solution, as analysis of the reaction between its components shows.

Equation: NH_4^+ (aq, from NH_4Cl) $+ OH^-$ (aq, from NaOH) $\rightarrow NH_3$ (aq) $+ H_2O(l)$

In soln:	0.264 M	0.196 M	0 M
Form NH_3:	−0.196 M	−0.196 M	+0.196 M
Initial:	0.068 M	≈ 0M	0.196 M

$$pH = pK_a + \log \frac{[NH_3]}{[NH_4^+]} = 9.26 + \log \frac{0.196\,M}{0.068\,M} = 9.72 \qquad pOH = 14.00 - 9.72 = 4.28$$

$$[OH^-] = 10^{-4.28} = 5.2 \times 10^{-5}\,M$$

Buffer Solutions

27. **(a)** 0.100 M NaCl is not a buffer solution. Neither a weak acid nor a weak base present.

(b) 0.100 M NaCl—0.100 M NH_4Cl is not a buffer solution. Although a weak acid, NH_4^+, is present, its conjugate base is not.

(c) 0.100 M CH_3NH_2 and 0.150 M $CH_3NH_3^+Cl^-$ is a buffer solution. Both a weak base, CH_3NH_2, and its conjugate acid, $CH_3NH_3^+$, are present in approximately equal concentrations.

(d) 0.100 M HCl—0.050 M $NaNO_2$ is not a buffer solution. All the NO_2^- is converted to HNO_2 and thus the solution is a mixture of a strong acid and a weak acid.

(e) 0.100 M HCl—0.200 M $NaC_2H_3O_2$ is a buffer solution. All of the HCl reacts with half of the $C_2H_3O_2^-$ to form a solution with 0.100 M $HC_2H_3O_2$, a weak acid, and 0.100 M $C_2H_3O_2^-$, its conjugate base.

(f) 0.100 M $HC_2H_3O_2$ and 0.125 M $NaC_3H_5O_2$ is not a buffer in the strict sense because it does not contain a weak acid and its conjugate base, but rather the conjugate base of another weak acid. These two weak acids (acetic, $K_a = 1.8 \times 10^{-5}$ and propionic, $K_a = 1.35 \times 10^{-5}$) have approximately the same strength, however, this solution would resist changes in its pH on the addition of strong acid or strong base and thus, it could be argued that this system should also be called a buffer.

28. **(a)** Reaction with added acid: $HPO_4^{2-} + H_3O^+ \rightarrow H_2PO_4^- + H_2O$

Reaction with added base: $H_2PO_4^- + OH^- \longrightarrow HPO_4^{2-} + H_2O$

(b) We assume initially that the buffer has equal concentrations of the two ions, $[H_2PO_4^-] = [HPO_4^{2-}]$

$$pH = pK_{a_2} + \log \frac{\left[HPO_4^{2-}\right]}{\left[H_2PO_4^{-}\right]} = 7.20 + 0.00 = 7.20 \text{ (pH where the buffer is most effective)}.$$

(c) $\quad pH = 7.20 + \log \dfrac{\left[HPO_4^{2-}\right]}{\left[H_2PO_4^{-}\right]} = 7.20 + \log \dfrac{0.150 \text{ M}}{0.050 \text{ M}} = 7.20 + 0.48 = 7.68$

29. amount of solute $= 1.15 \text{ mg} \times \dfrac{1 \text{ g}}{1000 \text{ mg}} \times \dfrac{1 \text{ mol } C_6H_5NH_3{}^+Cl^-}{129.6 \text{ g}} \times \dfrac{1 \text{ mol } C_6H_5NH_3{}^+}{1 \text{mol } C_6H_5NH_3{}^+Cl^-}$

$$= 8.87 \times 10^{-6} \text{ mol } C_6H_5NH_3{}^+$$

$$\left[C_6H_5NH_3{}^+\right] = \frac{8.87 \times 10^{-6} \text{ mol } C_6H_5NH_3{}^+}{3.18 \text{ L soln}} = 2.79 \times 10^{-6} \text{ M}$$

Equation: $\quad C_6H_5NH_2 (aq) + H_2O(l) \rightleftharpoons C_6H_5NH_3{}^+ (aq) + OH^- (aq)$

Initial:	0.105 M	2.79×10^{-6} M	≈ 0 M
Changes:	$-x$ M	$+x$ M	$+x$ M
Equil:	$(0.105 - x)$M	$(2.79 \times 10^{-6} + x)$M	x M

$$K_b = \frac{\left[C_6H_5NH_3{}^+\right]\left[OH^-\right]}{\left[C_6H_5NH_2\right]} = 7.4 \times 10^{-10} = \frac{(2.79 \times 10^{-6} + x)x}{0.105 - x}$$

$$7.4 \times 10^{-10} (0.105 - x) = (2.79 \times 10^{-6} + x)x; \quad 7.8 \times 10^{-11} - 7.4 \times 10^{-10} x = 2.79 \times 10^{-6} x + x^2$$

$$x^2 + (2.79 \times 10^{-6} + 7.4 \times 10^{-10})x - 7.8 \times 10^{-11} = 0; \quad x^2 + 2.79 \times 10^{-6} x - 7.8 \times 10^{-11} = 0$$

$$x = \frac{-b \pm \sqrt{b^2 - 4ac}}{2a} = \frac{-2.79 \times 10^{-6} \pm \sqrt{7.78 \times 10^{-12} + 3.1 \times 10^{-10}}}{2} = 7.5 \times 10^{-6} \text{ M} = \left[OH^-\right]$$

$$pOH = -\log(7.5 \times 10^{-6}) = 5.12 \qquad pH = 14.00 - 5.12 = 8.88$$

30. We determine the concentration of the cation of the weak base.

$$\left[C_6H_5NH_3{}^+\right] = \frac{8.50 \text{ g} \times \dfrac{1 \text{ mmol } C_6H_5NH_3{}^+Cl^-}{129.6 \text{ g}} \times \dfrac{1 \text{ mmol } C_6H_5NH_3{}^+}{1 \text{ mmol } C_6H_5NH_3{}^+Cl^-}}{750 \text{ mL} \times \dfrac{1 \text{ L}}{1000 \text{ mL}}} = 0.0874 \text{ M}$$

In order to be an effective buffer, each concentration must exceed the ionization constant $(K_b = 7.4 \times 10^{-10})$ by a factor of at least 100, which clearly is true. Also, the ratio of the

two concentrations must fall between 0.1 and 10: $\dfrac{\left[C_6H_5NH_3^+\right]}{\left[C_6H_5NH_2\right]} = \dfrac{0.0874M}{0.215M} = 0.407$.

This solution will be an effective buffer.

31. **(a)** First use the Henderson-Hasselbalch equation

$\left[pK_b = -\log(1.8 \times 10^{-5}) = 4.74, pK_a = 14.00 - 4.74 = 9.26\right]$ to determine $\left[NH_4^+\right]$ in the buffer solution.

$$pH = 9.45 = pK_a + \log\frac{\left[NH_3\right]}{\left[NH_4^+\right]} = 9.26 + \log\frac{\left[NH_3\right]}{\left[NH_4^+\right]}; \quad \log\frac{\left[NH_3\right]}{\left[NH_4^+\right]} = 9.45 - 9.26 = +0.19$$

$$\frac{\left[NH_4^+\right]}{\left[NH_3\right]} = 10^{-0.19} = 0.65 \qquad \left[NH_4^+\right] = 0.65 \times \left[NH_3\right] = 0.65 \times 0.258M = 0.17M$$

We now assume that the volume of the solution does not change significantly when the solid is added.

$$\text{mass}\left(NH_4\right)_2SO_4 = 425 \text{ mL} \times \frac{1\text{ L soln}}{1000\text{ mL}} \times \frac{0.17 \text{ mol } NH_4^+}{1\text{ L soln}} \times \frac{1 \text{ mol}\left(NH_4\right)_2SO_4}{2 \text{ mol } NH_4^+}$$

$$\times \frac{132.2 \text{ g}\left(NH_4\right)_2SO_4}{1 \text{ mol}\left(NH_4\right)_2SO_4} = 4.8 \text{ g}\left(NH_4\right)_2SO_4$$

(b) Let's use the Henderson-Hasselbalch equation to determine the ratio of concentrations of cation and weak base in the altered solution.

$$pH = 9.30 = pK_a + \log\frac{\left[NH_3\right]}{\left[NH_4^+\right]} = 9.26 + \log\frac{\left[NH_3\right]}{\left[NH_4^+\right]} \qquad \log\frac{\left[NH_3\right]}{\left[NH_4^+\right]} = 9.30 - 9.26 = +0.04$$

$$\frac{\left[NH_4^+\right]}{\left[NH_3\right]} = 10^{-0.04} = 0.91 = \frac{0.17 \text{ M} + x \text{ M}}{0.258} \qquad 0.17 + x = 0.91 \times 0.258 = 0.235$$

The reason we decided to add x to the numerator follows. (Notice we cannot remove a component.) A pH of 9.30 is more acidic than a pH of 9.45 and therefore the conjugate acid's $\left(NH_4^+\right)$ concentration must increase. Additionally, mathematics tells us that for the concentration ratio to increase from 0.65 to 0.91, its numerator must increase. We solve this expression for x. $x = 0.235 - 0.17 = 0.06_5$. We need to add NH_4^+ to increase its concentration by 0.06_5 M in 750 mL of solution.

$$(NH_4)_2 SO_4 \ mass = 0.100\,L \times \frac{0.06_5 \ mol \ NH_4^+}{1\,L} \times \frac{1\,mol\,(NH_4)_2\,SO_4}{2\,mol\,NH_4^+} \times \frac{132.1g\ (NH_4)_2\,SO_4}{1\,mol\ (NH_4)_2\,SO_4}$$

$$= 0.4_3\,g\ (NH_4)_2\,SO_4 \ to \ add \simeq 0.4\,g$$

32. **(a)** $\quad$ amount $HC_7H_5O_2 = 2.00$ g $HC_7H_5O_2 \times \dfrac{1\,mol\ HC_7H_5O_2}{122.1g\ HC_7H_5O_2} = 0.0164\,mol\ HC_7H_5O_2$

amount $C_7H_5O_2^- = 2.00$ g $NaC_7H_5O_2 \times \dfrac{1\,mol\ NaC_7H_5O_2}{144.1g\ NaC_7H_5O_2} \times \dfrac{1\,mol\ C_7H_5O_2^-}{1\,mol\ NaC_7H_5O_2}$

$$= 0.0139\,mol\ C_7H_5O_2^-$$

$$pH = pK_a + \log\frac{\left[C_7H_5O_2^-\right]}{\left[HC_7H_5O_2\right]} = -\log\left(6.3\times10^{-5}\right) + \log\frac{0.0139\,mol\ C_7H_5O_2^-/0.7500\,L}{0.0164\,mol\ HC_7H_5O_2/0.7500\,L}$$

$$= 4.20 - 0.0718 = 4.13$$

(b) $\quad$ To lower the pH of this buffer solution, that is, to make it more acidic, benzoic acid must be added. The quantity is determined as follows. We use moles rather than concentrations because all components are present in the same volume of solution.

$$4.00 = 4.20 + \log\frac{0.0139\,mol\ C_7H_5O_2^-}{x\,mol\ HC_7H_5O_2} \qquad \log\frac{0.0139\,mol\ C_7H_5O_2^-}{x\,mol\ HC_7H_5O_2} = -0.20$$

$$\frac{0.0139\,mol\ C_7H_5O_2^-}{x\,mol\ HC_7H_5O_2} = 10^{-0.20} = 0.63 \qquad x = \frac{0.0139}{0.63} = 0.022\,mol\ HC_7H_5O_2 \ (required)$$

$HC_7H_5O_2$ that must be added = amount required $-$ amount already in solution

$HC_7H_5O_2$ that must be added = $0.022\,mol\ HC_7H_5O_2 - 0.0164\,mol\ HC_7H_5O_2$

$HC_7H_5O_2$ that must be added = $0.006\,mol\ HC_7H_5O_2$

added mass $HC_7H_5O_2 = 0.006\,mol\ HC_7H_5O_2 \times \dfrac{122.1g\ HC_7H_5O_2}{1\,mol\ HC_7H_5O_2} = 0.7g\ HC_7H_5O_2$

33. The added HCl will react with the ammonia, and the pH of the buffer solution will decrease. The original buffer solution has $\left[NH_3\right] = 0.258$ M and $\left[NH_4^+\right] = 0.17$ M . We first calculate the [HCl] in solution, reduced from 12 M because of dilution. [HCl] added $= 12\,M \times \dfrac{0.55\,mL}{100.6\,mL} = 0.066\,M$ We determine pK_a for ammonium ion:

$$pK_b = -\log\left(1.8\times10^{-5}\right) = 4.74 \quad pK_a = 14.00 - 4.74 = 9.26$$

Equation: $NH_3(aq) + H_3O^+(aq) \rightleftharpoons NH_4^+(aq) + H_2O(l)$

Buffer:	0.258 M		0.17 M
Added:		+0.066 M	
Changes:	−0.066 M	−0.066 M	+0.066 M
Final:	0.192 M	0 M	0.24 M

$$pH = pK_a + \log \frac{[NH_3]}{[NH_4^+]} = 9.26 + \log \frac{0.192}{0.24} = 9.16$$

34. The added NH_3 will react with the benzoic acid, and the pH of the buffer solution will increase. Original buffer solution has $[C_7H_5O_2^-] = 0.0139$ mol $C_7H_5O_2^-/0.750$ L = 0.0185 M and $[HC_7H_5O_2] = 0.0164$ mol $HC_7H_5O_2/0.7500$ L = 0.0219 M. We first calculate the $[NH_3]$ in solution, reduced from 15 M because of dilution.

$$[NH_3] \text{ added} = 15 \text{ M} \times \frac{0.35 \text{ mL}}{750.4 \text{ mL}} = 0.0070 \text{ M} \quad \text{For benzoic acid, } pK_a = -\log(6.3 \times 10^{-5}) = 4.20$$

Equation: $NH_3(aq) + HC_7H_5O_2(aq) \rightleftharpoons NH_4^+(aq) + C_7H_5O_2^-(aq)$

Buffer:		0.0219 M		0.0185 M
Added:	0.0070 M			
Changes:	−0.0070M	−0.0070 M	+0.0070	+0.0070 M
Final:	0.000 M	0.0149 M	0.0070	0.0255 M

$$pH = pK_a + \log \frac{[C_7H_5O_2^-]}{[HC_7H_5O_2]} = 4.20 + \log \frac{0.0255}{0.0149} = 4.43$$

35. The pK_a's of the three acids help us choose the one to be used in the buffer. It is the acid with a pK_a within 1.00 pH unit of 3.50. $pK_a = 3.74$ for $HCHO_2$, $pK_a = 4.74$ for $HC_2H_3O_2$, and $pK_1 = 2.15$ for H_3PO_4. Thus, we choose $HCHO_2$ and $NaCHO_2$ to prepare a buffer with pH = 3.50. The Henderson-Hasselbalch equation is used to determine the relative amounts of each component present in the buffer solution.

$$pH = 3.50 = 3.74 + \log \frac{[CHO_2^-]}{[HCHO_2]} \qquad \log \frac{[CHO_2^-]}{[HCHO_2]} = 3.50 - 3.74 = -0.24$$

$$\frac{[CHO_2^-]}{[HCHO_2]} = 10^{-0.24} = 0.58$$

This ratio of concentrations is also the ratio of the number of moles of each component in the buffer solution, since both concentrations are a number of moles in a certain volume, and the volumes are the same (the two solutes are in the same solution). This ratio also is the ratio of the volumes of the two solutions, since both solutions being mixed contain the same concentration of solute. If we assume 100. mL of acid solution, $V_{acid} = 100.$ mL.

Then the volume of salt solution is $V_{salt} = 0.58 \times 100.$ mL = 58 mL 0.100 M $NaCHO_2$

36. We can lower the pH of the 0.250 M $HC_2H_3O_2$ — 0.560 M $C_2H_3O_2^-$ buffer solution by increasing $[HC_2H_3O_2]$ or lowering $[C_2H_3O_2^-]$. Small volumes of NaCl solutions will have no effect, and the addition of NaOH(aq) or $NaC_2H_3O_2$(aq) will raise the pH. The addition of 0.150 M HCl will raise $[HC_2H_3O_2]$ and lower $[C_2H_3O_2^-]$ through the reaction

$$H_3O^+(aq) + C_2H_3O_2^-(aq) \rightleftharpoons HC_2H_3O_2(aq) + H_2O(l)$$

We first use the Henderson-Hasselbalch equation to determine the ratio of the

concentration of acetate ion and acetic acid. $pH = 5.00 = 4.74 + \log\dfrac{[C_2H_3O_2^-]}{[HC_2H_3O_2]}$

$\log\dfrac{[C_2H_3O_2^-]}{[HC_2H_3O_2]} = 5.00 - 4.74 = 0.26;\quad \dfrac{[C_2H_3O_2^-]}{[HC_2H_3O_2]} = 10^{0.26} = 1.8$

Now we compute the amount of each component in the original buffer solution.

amount of $C_2H_3O_2^-$ = 300. mL $\times \dfrac{0.560 \text{ mmol } C_2H_3O_2^-}{1 \text{ mL soln}} = 168 \text{ mmol } C_2H_3O_2^-$

amount of $HC_2H_3O_2$ = 300. mL $\times \dfrac{0.250 \text{ mmol } HC_2H_3O_2}{1 \text{ mL soln}} = 75.0 \text{ mmol } HC_2H_3O_2$

Now let x represent the amount of H_3O^+ added in mmol.

$1.8 = \dfrac{168 - x}{75.0 + x};\quad 168 - x = 1.8(75 + x) = 13_5 + 1.8x \quad 168 - 13_5 = 2.8x$

$x = \dfrac{168 - 13_5}{2.7} = 12 \text{ mmol } H_3O^+$

mL 0.150 M HCl $= 12 \text{ mmol } H_3O^+ \times \dfrac{1 \text{ mmol HCl}}{1 \text{ mmol } H_3O^+} \times \dfrac{1 \text{ mL soln}}{0.150 \text{ mmol HCl}}$

$= 80 \text{ mL } 0.150 \text{ M HCl solution}$

37. **(a)** The pH of the buffer is determined via the Henderson-Hasselbalch equation.

$$pH = pK_a + \log\dfrac{[C_3H_5O_2^-]}{[HC_3H_5O_2]} = 4.89 + \log\dfrac{0.100M}{0.100M} = 4.89$$

The effective pH range is the same for every propionate buffer: from pH = 3.89 to pH = 5.89, one pH unit on either side of pK_a for propionic acid.

(b) To each liter of 0.100 M $HC_3H_5O_2$ — 0.100M $NaC_3H_5O_2$ we can add 0.100 mol OH^- before all of the $HC_3H_5O_2$ is consumed, and we can add 0.100 mol H_3O^+ before all of the $C_3H_5O_2^-$ is consumed. The buffer capacity thus is 100. millimoles (0.100 mol) of acid or base per liter of buffer solution.

38. (a) The solution will be an effective buffer one pH unit on either sides of the pK_a of methylammonium ion, $CH_3NH_3^+$, $K_b = 4.2 \times 10^{-4}$ for methylamine,

$pK_b = -\log(4.2 \times 10^{-4}) = 3.38$. For methylammonium cation,

$pK_a = 14.00 - 3.38 = 10.62$. Thus, this buffer will be effective from a pH of 9.62 to a pH of 11.62.

(b) The capacity of the buffer is reached when all of the weak base or all of the conjugate acid has been neutralized by added strong acid or strong base. Because their concentrations are the same, the number of moles of base are equal to the number of moles of conjugate acid in the same volume of solution.

$$\text{amount of weak base} = 125 \text{ mL} \times \frac{0.0500 \text{ mmol}}{1 \text{ mL}} = 6.25 \text{ mmol } CH_3NH_2 \text{ or } CH_3NH_3^+$$

The buffer capacity is 6.25 millimoles of acid or base per 125 mL buffer solution.

39. (a) pH of this buffer solution is determined with the Henderson-Hasselbalch equation.

$$pH = pK_a + \log\frac{\left[CHO_2^-\right]}{\left[HCHO_2\right]} = -\log\left(1.8 \times 10^{-4}\right) + \log\frac{8.5 \text{ mmol}/75.0 \text{ mL}}{15.5 \text{ mmol}/75.0 \text{ mL}}$$

$$= 3.74 - 0.26 = 3.48$$

[Note: solution is not a good buffer, as $\left[CHO_2^-\right] = 1.1 \times 10^{-1} \sim 600$ times K_a]

(b) Amount of added $OH^- = 0.25 \text{ mmol } Ba(OH)_2 \times \dfrac{2 \text{ mmol } OH^-}{1 \text{ mmol } Ba(OH)_2} = 0.50 \text{ mmol } OH^-$

The OH^- added reacts with the formic acid and produces formate ion.

Equation:	$HCHO_2$ (aq)	+	OH^- (aq)	$\rightleftharpoons$	CHO_2^- (aq)	+	H_2O(l)
Buffer:	15.5 mmol		≈ 0 M		8.5 mmol		
Add base:			+0.50 mmol				
React:	−0.50 mmol		−0.50 mmol		+0.50 mmol		
Final:	15.0 mmol		0 mmol		9.0 mmol		

$$pH = pK_a + \log\frac{\left[CHO_2^-\right]}{\left[HCHO_2\right]} = -\log\left(1.8\times10^{-4}\right) + \log\frac{9.0\,mmol/75.0\,mL}{15.0\,mmol/75.0\,mL}$$

$$= 3.74 - 0.22 = 3.52$$

(c) Amount of added $H_3O^+ = 1.05\,mL$ acid $\times\dfrac{12\,mmol\,HCl}{1\,mL\,acid}\times\dfrac{1\,mmol\,H_3O^+}{1\,mmol\,HCl} = 13\,mmol\,H_3O^+$

The H_3O^+ added reacts with the formate ion and produces formic acid.

Equation:	CHO_2^- (aq)	+	H_3O^+ (aq)	$\rightleftharpoons$	$HCHO_2$ (aq) + $H_2O(l)$
Buffer :	8.5 mmol		≈ 0 mmol		15.5 mmol
Add acid :			+13 mmol		
React :	− 8.5 mmol		− 8.5 mmol		+8.5 mmol
Final :	0 mmol		4.5 mmol		24.0 mmol

The buffer's capacity has been exceeded. The pH of the solution is determined by the strong acid present.

$$\left[H_3O^+\right] = \frac{4.5\,mmol}{75.0\,mL + 1.05\,mL} = 0.059\,M; \quad pH = -\log\left(0.059\right) = 1.23$$

40. For $NH_3, pK_b = -\log\left(1.8\times10^{-5}\right)$ For $NH_4^+, pK_a = 14.00 - pK_b = 14.00 - 4.74 = 9.26$

(a) $\left[NH_3\right] = \dfrac{1.68\,g\,NH_3}{0.500\,L}\times\dfrac{1\,mol\,NH_3}{17.03\,g\,NH_3} = 0.197\,M$

$$\left[NH_4^+\right] = \frac{4.05\,g\,(NH_4)_2\,SO_4}{0.500\,L}\times\frac{1\,mol(NH_4)_2\,SO_4}{132.1\,g\,(NH_4)_2\,SO_4}\times\frac{2\,mol\,NH_4^+}{1\,mol\,(NH_4)_2\,SO_4} = 0.123\,M$$

$$pH = pK_a + \log\frac{\left[NH_3\right]}{\left[NH_4^+\right]} = 9.26 + \log\frac{0.197\,M}{0.123\,M} = 9.46$$

(b) The OH^-(aq) reacts with the NH_4^+(aq) to produce an equivalent amount of NH_3(aq).

$$\left[OH^-\right]_i = \frac{0.88\,g\,NaOH}{0.500\,L}\times\frac{1mol\,NaOH}{40.00\,g\,NaOH}\times\frac{1mol\,OH^-}{1mol\,NaOH} = 0.044\,M$$

Equation:	NH_4^+ (aq) +	OH^- (aq)	$\rightleftharpoons$	NH_3 (aq) +	$H_2O(l)$
Initial :	0.123 M	≈ 0 M		0.197 M	
Add NaOH :		+0.044 M			
React :	−0.044 M	−0.044 M		+0.044 M	
Final :	0.079 M	0.0000 M		0.241 M	

$$pH = pK_a + \log \frac{[NH_3]}{[NH_4^+]} = 9.26 + \log \frac{0.241\,M}{0.079\,M} = 9.74$$

(c) Equation: $\quad NH_3(aq) + H_3O^+(aq) \rightleftharpoons NH_4^+(aq) + H_2O(l)$

Initial: $\qquad 0.197\,M \qquad \approx 0\,M \qquad\qquad 0.123\,M$

Add HCl: $\qquad\qquad\qquad +x\,M$

React: $\qquad -x\,M \qquad -x\,M \qquad\qquad +x\,M$

Final: $\qquad (0.197-x)\,M \quad 1\times10^{-9}\,M \qquad (0.123+x)\,M$

$$pH = 9.00 = pK_a + \log \frac{[NH_3]}{[NH_4^+]} = 9.26 + \log \frac{(0.197-x)\,M}{(0.123+x)\,M}$$

$$\log \frac{(0.197-x)\,M}{(0.123+x)\,M} = 9.00 - 9.26 = -0.26 \qquad \frac{(0.197-x)\,M}{(0.123+x)\,M} = 10^{-0.26} = 0.55$$

$$0.197 - x = 0.55(0.123 + x) = 0.068 + 0.55x \quad 1.55x = 0.197 - 0.068 = 0.129$$

$$x = \frac{0.129}{1.55} = 0.0832\,M$$

$$\text{volume HCl} = 0.500\,L \times \frac{0.0832\,mol\,H_3O^+}{1L\,soln} \times \frac{1\,mol\,HCl}{1\,mol\,H_3O^+} \times \frac{1000\,mL\,HCl}{12\,mol\,HCl} = 3.5\,mL$$

41. **(a)** We use the Henderson-Hasselbalch equation to determine the pH of the solution. The total solution volume is

$$36.00\,mL + 64.00\,mL = 100.00\,mL. \quad pK_a = 14.00 - pK_b = 14.00 + \log(1.8\times10^{-5}) = 9.26$$

$$[NH_3] = \frac{36.00\,mL \times 0.200\,M\,NH_3}{100.00\,mL} = \frac{7.20\,mmol\,NH_3}{100.0\,mL} = 0.0720\,M$$

$$[NH_4^+] = \frac{64.00\,mL \times 0.200\,M\,NH_4^+}{100.00\,mL} = \frac{12.8\,mmol\,NH_4^+}{100.0\,mL} = 0.128\,M$$

$$pH = pK_a + \log \frac{[NH_3]}{[NH_4^+]} = 9.26 + \log \frac{0.0720}{0.128\,M} = 9.01$$

(b) The solution has $[OH^-] = 10^{-4.99} = 1.0\times10^{-5}\,M$

The Henderson-Hasselbalch equation depends on the assumption:

$$[NH_3] \gg 1.8\times10^{-5}\,M \ll [NH_4^+]$$

If the solution is diluted to 1.00 L, $[NH_3] = 7.20 \times 10^{-3}$ M, and

$[NH_4^+] = 1.28 \times 10^{-2}$ M. These concentrations are consistent with the assumption.

However, if the solution is diluted to 1000. L, $[NH_3] = 7.2 \times 10^{-6}$ M, and

$[NH_3] = 1.28 \times 10^{-5}$ M, and these two concentrations are not consistent with the assumption. Thus, in 1000. L of solution, the given quantities of NH_3 and NH_4^+ will not produce a solution with pH = 9.00. With sufficient dilution, the solution will become indistinguishable from pure water; its pH will equal 7.00.

(c)   The 0.20 mL of added 1.00 M HCl does not significantly affect the volume of the solution, but it does add $0.20 \, \text{mL} \times 1.00 \, \text{M HCl} = 0.20 \, \text{mmol} \, H_3O^+$. This added H_3O^+ reacts with NH_3, decreasing its amount from 7.20 mmol NH_3 to 7.00 mmol NH_3, and increasing the amount of NH_4^+ from 12.8 mmol NH_4^+ to 13.0 mmol NH_4^+, through the reaction: $NH_3 + H_3O^+ \rightarrow NH_4^+ + H_2O$

$$pH = 9.26 + \log \frac{7.00 \, \text{mmol} \, NH_3 \, / \, 100.20 \, \text{mL}}{13.0 \, \text{mmol} \, NH_4^+ \, / \, 100.20 \, \text{mL}} = 8.99$$

(d)   We see in the calculation of part (c) that the total volume of the solution does not affect the pOH of the solution, at least as long as the Henderson-Hasselbalch equation is obeyed. We let x represent the number of millimoles of H_3O^+ added, through 1.00 M HCl. This increases the amount of NH_4^+ and decreases the amount of NH_3, through the reaction $NH_3 + H_3O^+ \rightarrow NH_4^+ + H_2O$

$$pH = 8.90 = 9.26 + \log \frac{7.20 - x}{12.8 + x}; \quad \log \frac{7.20 - x}{12.8 + x} = 8.90 - 9.26 = -0.36$$

Inverting, we have:

$$\frac{12.8 + x}{7.20 - x} = 10^{0.36} = 2.29; \quad 12.8 + x = 2.29(7.20 - x) = 16.5 - 2.29x$$

$$x = \frac{16.5 - 12.8}{1.00 + 2.29} = 1.1 \, \text{mmol} \, H_3O^+$$

$$\text{vol 1.00 M HCl} = 1.1 \, \text{mmol} \, H_3O^+ \times \frac{1 \, \text{mmol HCl}}{1 \, \text{mmol} \, H_3O^+} \times \frac{1 \, \text{mL soln}}{1.00 \, \text{mmol HCl}} = 1.1 \, \text{mL 1.00 M HCl}$$

42.   (a) $$[C_2H_3O_2^-] = \frac{12.0 \, \text{g NaC}_2\text{H}_3\text{O}_2}{0.300 \, \text{L soln}} \times \frac{1 \, \text{mol NaC}_2\text{H}_3\text{O}_2}{82.03 \, \text{g NaC}_2\text{H}_3\text{O}_2} \times \frac{1 \, \text{mol C}_2\text{H}_3\text{O}_2^-}{1 \, \text{mol NaC}_2\text{H}_3\text{O}_2}$$

$$= 0.488 \, \text{M C}_2\text{H}_3\text{O}_2^-$$

Equation : $\qquad HC_2H_3O_2 + H_2O \rightleftharpoons C_2H_3O_2^- + H_3O^+$

Initial : $\qquad\qquad\qquad\qquad\qquad\qquad\qquad\qquad\qquad$ 0.200 M

Add $NaC_2H_3O_2$ $\qquad\qquad\qquad\qquad\qquad$ 0.488 M

Consume H_3O^+ $\quad +0.200$ M $\qquad\qquad\quad$ -0.200 M $\quad -0.200$ M

Buffer : $\qquad\quad$ 0.200 M $\qquad\qquad\quad$ 0.288 M $\qquad \approx 0$ M

Then use the Henderson-Hasselbalch equation.

$$pH = pK_a + \log\frac{\left[C_2H_3O_2^-\right]}{\left[HC_2H_3O_2\right]} = 4.74 + \log\frac{0.288\,M}{0.200\,M} = 4.74 + 0.16 = 4.90$$

(b) We can calculate the initial $\left[OH^-\right]$ due to the added $Ba(OH)_2$.

$$\left[OH^-\right] = \frac{1.00\,g\,Ba(OH)_2}{0.300\,L} \times \frac{1\,mol\,Ba(OH)_2}{171.3\,g\,Ba(OH)_2} \times \frac{2\,mol\,OH^-}{1\,mol\,Ba(OH)_2} = 0.0389\,M$$

Then $HC_2H_3O_2$ is consumed.

Equation: $\qquad HC_2H_3O_2 + OH^- \rightleftharpoons C_2H_3O_2^- + H_2O$

Initial: $\qquad\quad$ 0.200 M $\qquad$ 0.0389 M $\qquad$ 0.288 M

Consume OH^-: $\quad$ 0.161 M $\qquad \sim 0$M $\qquad$ 0.327 M

Then use the Henderson-Hasselbalch equation.

$$pH = pK_a + \log\frac{\left[C_2H_3O_2^-\right]}{\left[HC_2H_3O_2\right]} = 4.74 + \log\frac{0.327\,M}{0.161\,M} = 4.74 + 0.31 = 5.05$$

(c) $Ba(OH)_2$ can be added until all of the $HC_2H_3O_2$ is consumed.

$$Ba(OH)_2 + 2HC_2H_3O_2 \longrightarrow Ba(C_2H_3O_2)_2 + 2H_2O$$

$$\text{mass of } Ba(OH)_2 = 0.300\,L \times \frac{0.200\,mol\,HC_2H_3O_2}{1\,L\,soln} \times \frac{1\,mol\,Ba(OH)_2}{2\,mol\,HC_2H_3O_2} = 0.0300\,mol\,Ba(OH)_2$$

$$\text{moles of } Ba(OH)_2 = 0.0300\,mol\,Ba(OH)_2 \times \frac{171.3\,g\,Ba(OH)_2}{1\,mol\,Ba(OH)_2} = 5.14\,g\,Ba(OH)_2$$

(d) This is an excess of 0.36 g $Ba(OH)_2$ and it is this excess that determines the pOH of the solution.

$$\left[OH^-\right] = \frac{0.36\,g\,Ba(OH)_2}{0.300\,L\,soln} \times \frac{1\,mol\,Ba(OH)_2}{171.3\,g\,Ba(OH)_2} \times \frac{2\,mol\,OH^-}{1\,mol\,Ba(OH)_2} = 1.4 \times 10^{-2}\,M\,OH^-$$

$$pOH = -\log(1.4 \times 10^{-2}) = 1.85 \qquad pH = 14.00 - 1.85 = 12.15$$

Acid-Base Indicators

43. **(a)** In an acid-base titration, the pH of the solution changes sharply at a definite pH that is known prior to titration. (This pH change occurs during the addition of a very small volume of titrant.) Determining the pH of a solution, on the other hand, is more difficult because the pH of the solution is not known precisely in advance. Since each indicator only serves to fix the pH over a quite small region, often less than 2.0 pH units, several indicators—carefully chosen to span the entire range of 14 pH units—must be employed to narrow the pH to ±1 pH unit or possibly lower.

(b) An indicator is, after all, a weak acid. Its addition to a solution will affect the acidity of that solution. Thus, one adds only enough indicator to show a color change and not enough to affect solution acidity.

44. **(a)** We use an equation similar to the Henderson-Hasselbalch equation to determine the relative concentrations of HIn, and its anion, In^-, in this solution.

$$pH = pK_{HIn} + \log\frac{[In^-]}{[HIn]}; \quad 4.55 = 4.95 + \log\frac{[In^-]}{[HIn]}; \quad \log\frac{[In^-]}{[HIn]} = 4.55 - 4.95 = -0.40$$

$$\frac{[In^-]}{[HIn]} = 10^{-0.40} = 0.40 = \frac{x}{100-x} \quad x = 40 - 0.40x \quad x = \frac{40}{1.40} = 29\% \text{ In and } 71\% \text{ HIn}$$

(b) When the indicator is in a solution whose pH equals its pK_a (4.95), the ratio $[In^-]/[HIn] = 1.00$. And yet, at the midpoint of its color change range (about pH = 5.3), the ratio $[In^-]/[HIn]$ is greater than 1.00. Even though $[HIn] < [In^-]$ at this midpoint, the contribution of HIn to establishing the color of the solution is about the same as the contribution of In^-. This must mean that HIn (red) is more strongly colored than In^- (yellow).

45. **(a)** 0.10 M KOH is an alkaline solution and phenol red will display its basic color in such a solution; the solution will be red.

(b) 0.10 M $HC_2H_3O_2$ is an acidic solution, although that of a weak acid, and phenol red will display its acidic color in such a solution; the solution will be yellow.

(c) 0.10 M NH_4NO_3 is an acidic solution due to the hydrolysis of the ammonium ion. Phenol red will display its acidic color, that is, yellow, in this solution.

(d) 0.10 M HBr is an acidic solution, the aqueous solution of a strong acid. Phenol red will display its acidic color in this solution; the solution will be yellow.

(e) 0.10 M NaCN is an alkaline solution because of the hydrolysis of the cyanide ion. Phenol red will display its basic color, red, in this solution.

(f)   An equimolar acetic acid–potassium acetate buffer has $pH = pK_a = 4.74$ for acetic acid. In this solution phenol red will display its acid color, yellow.

46. **(a)** $pH = -\log(0.205) = 0.688$ The indicator is red in this solution.

(b)   The total volume of the solution is 600.0 mL. We compute the amount of each solute.

amount $H_3O^+ = 350.0 \text{ mL} \times 0.205 \text{ M} = 71.8 \text{ mmol } H_3O^+$

amount $NO_2^- = 250.0 \text{ mL} \times 0.500 \text{ M} = 125 \text{ mmol } NO_2^-$

$$[H_3O^+] = \frac{71.8 \text{ mmol}}{600.0 \text{ mL}} = 0.120 \text{ M} \qquad [NO_2^-] = \frac{125 \text{ mmol}}{600.0 \text{ mL}} = 0.208 \text{ M}$$

The H_3O^+ and NO_2^- react to produce a buffer solution in which $[HNO_2] = 0.120 \text{ M}$ and $[NO_2^-] = 0.208 - 0.120 = 0.088 \text{ M}$. We use the Henderson-Hasselbalch equation to determine the pH of this solution. $pK_a = -\log(7.2 \times 10^{-4}) = 3.14$

$$pH = pK_a + \log \frac{[NO_2^-]}{[HNO_2]} = 3.14 + \log \frac{0.088 \text{ M}}{0.120 \text{ M}} = 3.01 \text{ Indicator is yellow in this solution.}$$

(c)   The total volume of the solution is 750. mL. We compute the amount and then the concentration of each solute. Amount $OH^- = 150 \text{ mL} \times 0.100 \text{ M} = 15.0 \text{ mmol } OH^-$ This OH^- reacts with HNO_2 in the buffer solution to neutralize some of it and leave 56.8 mmol $(= 71.8 - 15.0)$ unneutralized.

$$[HNO_2] = \frac{56.8 \text{ mmol}}{750. \text{ mL}} = 0.0757 \text{ M} \qquad [NO_2^-] = \frac{(125 + 15) \text{ mmol}}{750. \text{ mL}} = 0.187 \text{ M}$$

We use the Henderson-Hasselbalch equation to determine the pH of this solution.

$$pH = pK_a + \log \frac{[NO_2^-]}{[HNO_2]} = 3.14 + \log \frac{0.187 \text{ M}}{0.0757 \text{ M}} = 3.53$$

The indicator is yellow in this solution.

(d)   We determine the $[OH^-]$ due to the added $Ba(OH)_2$.

$$[OH^-] = \frac{5.00 \text{ g Ba(OH)}_2}{0.750 \text{ L}} \times \frac{1 \text{ mol Ba(OH)}_2}{171.34 \text{ g Ba(OH)}_2} \times \frac{2 \text{ mol OH}^-}{1 \text{ mol Ba(OH)}_2} = 0.0778 \text{ M}$$

This is sufficient $[OH^-]$ to react with the existing $[HNO_2]$ and leave an excess

$[OH^-] = 0.0778 \text{ M} - 0.0757 \text{ M} = 0.0021 \text{ M}. \quad pOH = -\log(0.0021) = 2.68.$

$pH = 14.00 - 2.68 = 11.32$ The indicator is blue in this solution.

47.   Moles of HCl = $C \times V$ = 0.04050 M $\times$ 0.01000 L = 4.050×10^{-4} moles

Moles of Ba(OH)$_2$ at endpoint = $C \times V$ = 0.01120 M $\times$ 0.01790 L = 2.005×10^{-4} moles.

Moles of HCl that react with Ba(OH)$_2$ = 2 $\times$ moles Ba(OH)$_2$

Moles of HCl in excess 4.050×10^{-4} moles $- 4.010 \times 10^{-4}$ moles = $4.0\underline{4} \times 10^{-6}$ moles

Total volume at the equivalence point = (10.00 mL + 17.90 mL) = 27.90 mL

$$[\text{HCl}]_{\text{excess}} = \frac{4.04 \times 10^{-6} \text{ mole HCl}}{0.02790 \text{ L}} = 1.45 \times 10^{-4} \text{ M}; \quad \text{pH} = -\log(1.45 \times 10^{-4}) = 3.84$$

a)   The approximate pK_{HIn} = 3.84 (generally $\pm$ 1 pH unit)

b)   This is a relatively good indicator (with $\approx$ 1 % of the equivalence point volume), however, pK_{HIn} is not very close to the theoretical pH at the equivalence point (pH = 7.000) For very accurate work, a better indicator is needed (i.e. bromothymol blue (pK_{HIn} = 7.1) Note: 2,4-dinitrophenol works relatively well here because the pH near the equivalence point of a strong acid/strong base titration rises very sharply ($\approx$ 6 pH units for an addition of only 2 drops (0.10 mL))

48.   Solution (a): 100.0 mL of 0.100 M HCl, [H$_3$O$^+$] = 0.100 M and pH = 1.000 (yellow)

Solution (b): 150 mL of 0.100 M NaC$_2$H$_3$O$_2$

K_a of HC$_2$H$_3$O$_3$ = 1.8×10^{-5} K_b of C$_2$H$_3$O$_2^-$ = 5.6×10^{-10}

	C$_2$H$_3$O$_2^-$(aq)	+	H$_2$O(l)	$\xrightarrow{K_b = 5.6 \times 10^{-10}}$	HC$_2$H$_3$O$_2$(aq)	+	OH$^-$(aq)
initial	0.100 M				0 M		~0 M
change	$-x$				$+x$		$+x$
equil.	0.100 $-x$				x		x

Assume x is small: $5.6 \times 10^{-11} = x^2$; $x = 7.4\underline{8} \times 10^{-6}$ M (assumption valid by inspection)

[OH$^-$] = x = $7.4\underline{8} \times 10^{-6}$ M, pOH = 5.13 and pH = 8.87 (green-blue)

Mixture of solution (a) and (b). Total volume = 250.0 mL

$n_{\text{HCl}} = C \times V$ = 0.1000 L $\times$ 0.100 M = 0.0100 mol HCl

$n_{\text{C}_2\text{H}_3\text{O}_2^-} = C \times V$ = 0.1500 L $\times$ 0.100 M = 0.0150 mol C$_2$H$_3$O$_2^-$

HCl is the limiting reagent. Assume 100% reaction.

Therefore, 0.0050 mole C$_2$H$_3$O$_2^-$ left unreacted, and 0.0100 moles of HC$_2$H$_3$O$_2$ form.

$$[\text{C}_2\text{H}_3\text{O}_2^-] = \frac{n}{V} = \frac{0.0050 \text{ mol}}{0.250 \text{ L}} = 0.020 \text{ M} \quad [\text{HC}_2\text{H}_3\text{O}_2] = \frac{n}{V} = \frac{0.0100 \text{ mol}}{0.250 \text{ L}} = 0.0400 \text{ M}$$

$$HC_2H_3O_2(aq) \ + \ H_2O(l) \ \underset{}{\overset{K_a = 1.8 \times 10^{-5}}{\rightleftharpoons}} \ C_2H_3O_2^-(aq) \ + \ H_3O^+(aq)$$

initial	0.0400 M	0.020 M	~ 0 M
change	$-x$	$+x$	$+x$
equil.	$0.0400 - x$	$0.020 + x$	x

$$1.8 \times 10^{-5} = \frac{x(0.020 + x)}{0.0400 - x} \approx \frac{x(0.020)}{0.0400} \quad x = 3.6 \times 10^{-5} \text{ (proof 0.18 \% } < 5\%, \text{ assumption valid)}$$

$[H_3O^+] = 3.6 \times 10^{-5}$ pH = 4.44 Color of thymol blue at various pHs:

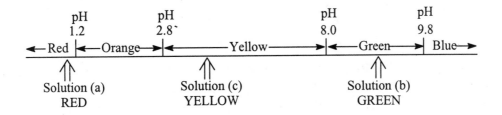

pH 1.2	pH 2.8	pH 8.0	pH 9.8

←—Red—→|←—Orange—→|←———Yellow———→|←—Green—→| Blue—→

⇑	⇑	⇑
Solution (a)	Solution (c)	Solution (b)
RED	YELLOW	GREEN

Neutralization Reactions

49. The reaction (to 2nd equiv. pt.) is: $H_3PO_4(aq) + 2KOH(aq) \longrightarrow K_2HPO_4(aq) + 2H_2O(l)$.

The molarity of the H_3PO_4 solution is determined in the following manner.

$$H_3PO_4 \text{ molarity} = \frac{31.15 \, \text{mL KOH soln} \times \dfrac{0.2420 \, \text{mmol KOH}}{1 \, \text{mL KOH soln}} \times \dfrac{1 \, \text{mmol } H_3PO_4}{2 \, \text{mmol KOH}}}{25.00 \, \text{mL } H_3PO_4 \text{ soln}} = 0.1508 \, \text{M}$$

50. The reaction (1st to 2nd equiv. pt.) is:

$NaH_2PO_4(aq) + NaOH(aq) \longrightarrow Na_2HPO_4(aq) + H_2O(l)$. The molarity of the H_3PO_4 solution is determined in the following manner.

$$H_3PO_4 \text{ molarity} = \frac{18.67 \, \text{mL NaOH soln} \times \dfrac{0.1885 \, \text{mmol NaOH}}{1 \, \text{mL NaOH soln}} \times \dfrac{1 \, \text{mmol } H_3PO_4}{1 \, \text{mmol NaOH}}}{20.00 \, \text{mL } H_3PO_4 \text{ soln}} = 0.1760 \, \text{M}$$

51. Determine amount of H_3O^+ or OH^- in each solution, and the amount of excess reagent.

A. $\text{amount } H_3O^+ = 50.00 \, \text{mL} \times \dfrac{0.0150 \, \text{mmol } H_2SO_4}{1 \, \text{mL soln}} \times \dfrac{2 \, \text{mmol } H_3O^+}{1 \, \text{mmol } H_2SO_4} = 1.50 \, \text{mmol } H_3O^+$

(assuming complete ionization of H_2SO_4 and HSO_4^- in the presence of OH^-)

B. $\text{amount } OH^- = 50.00 \, \text{mL} \times \dfrac{0.0385 \, \text{mmol NaOH}}{1 \, \text{mL soln}} \times \dfrac{1 \, \text{mmol } OH^-}{1 \, \text{mmol NaOH}} = 1.93 \, \text{mmol } OH^-$

Result: Titration reaction: $OH^-(aq) + H_3O^+(aq) \rightleftharpoons 2H_2O(l)$

Initial amounts: 1.93 mmol   1.50 mmol

After reaction: 0.43 mmol   0 mmol

$$[OH^-] = \frac{0.43\,\text{mmol OH}^-}{100.0\,\text{mL soln}} = 4.3 \times 10^{-3}\,M$$

$$pOH = -\log(4.3 \times 10^{-3}) = 2.37 \qquad\qquad pH = 14.00 - 2.37 = 11.63$$

52. Determine the amount of solute in each solution, followed by the amount excess reagent.

   A. $[H_3O^+] = 10^{-2.50} = 0.0032\,M$

   $$\text{mmol HCl} = 100.0\,\text{mL} \times \frac{0.0032\,\text{mmol H}_3O^+}{1\,\text{mL soln}} \times \frac{1\,\text{mmol HCl}}{1\,\text{mmol H}_3O^+} = 0.32\,\text{mmol HCl}$$

   B. $pOH = 14.00 - 11.00 = 3.00 \quad [OH^-] = 10^{-3.00} = 1.0 \times 10^{-3}\,M$

   $$\text{mmol NaOH} = 100.0\,\text{mL} \times \frac{0.0010\,\text{mmol OH}^-}{1\,\text{mL soln}} \times \frac{1\,\text{mmol NaOH}}{1\,\text{mmol OH}^-} = 0.10\,\text{mmol NaOH}$$

   Result:

   Titration reaction: $NaOH(aq) + HCl(aq) \rightleftharpoons NaCl(aq) + H_2O(l)$

   Initial amounts: 0.10 mmol   0.32 mmol

   After reaction: 0.00 mmol   0.22 mmol

   $$[H_3O^+] = \frac{0.22\,\text{mmol HCl}}{200.0\,\text{mL soln}} \times \frac{1\,\text{mmol H}_3O^+}{1\,\text{mmol HCl}} = 1.1 \times 10^{-3}\,M$$

   $$pH = -\log(1.1 \times 10^{-3}) = 2.96$$

Titration Curves

53. In each case, the volume of acid and its molarity are the same. Thus, also the amount of acid is the same in each case. The volume of titrant needed to reach the equivalence point will also be the same in both cases, since the titrant has the same concentration in each case, and it is the same amount of base that reacts with a given amount (in moles) of acid. Realize that, as the titration of a weak acid proceeds, the weak acid will ionize, replenishing the H_3O^+ in solution. This will occur until all of the weak acid has ionized and subsequently reacted with the strong base. At the equivalence point in the titration of a strong acid with a strong base is an aqueous solution of ions that do not hydrolyze. But the equivalence point solution of the titration of a weak acid with a strong base contains the anion of a weak acid, which will hydrolyze to produce a basic (alkaline) solution. (Don't forget, however, that the inert cation of a strong base is also present.)

54. **(a)** This equivalence point is the result of the titration of a weak acid with a strong base. The species present in solution is CO_3^{2-} which, through its hydrolysis, will form an alkaline, or basic solution. The other ionic species in solution, Na^+, will not hydrolyze. Thus, $pH > 7.0$

(b) This is the titration of a strong acid with a weak base. The species present in solution at the equivalent point is NH_4^+ which hydrolyzes to form an acidic solution. Cl^- does not hydrolyze. Thus, $pH < 7.0$

(c) This is the titration of a strong acid with a strong base. Two ions are present in the solution at the equivalence point, K^+ and Cl^-, and neither of these hydrolyze. The solution will have a pH of 7.00.

55. **(a)** Initial $[OH^-] = 0.100\,M\ OH^-$ $pOH = -\log(0.100) = 1.000$ $pH = 13.00$

Since this is the titration of a strong base with a strong acid, KI is the solute present at the equivalence point and thus, the $pH = 7.00$. The titration reaction is:

$$KOH(aq) + HI(aq) \longrightarrow KI(aq) + H_2O(l)$$

$$V_{HI} = 25.0\,mL\ KOH\ soln \times \frac{0.100\,mmol\ KOH\ soln}{1\,mL\ soln} \times \frac{1\,mmol\ HI}{1\,mmol\ KOH} \times \frac{1\,mL\ HI\ soln}{0.200\,mmol\ HI}$$

$$= 12.5\,mL\ HI\ soln$$

Initial amount of KOH present $= 25.0\,mL\ KOH\ soln \times 0.100\,M = 2.50\,mmol\ KOH$
At the 40% titration point: $5.00\,mL\ HI\ soln \times 0.200\,M\ HI = 1.00\,mmol\ HI$
excess KOH $= 2.50\,mmol\ KOH - 1.00\,mmol\ HI = 1.50\,mmol\ KOH$

$$[OH^-] = \frac{1.50\,mmol\ KOH}{30.0\,mL\ total} \times \frac{1\,mmol\ OH^-}{1\,mmol\ KOH} = 0.0500\,M \quad pOH = -\log(0.0500) = 1.30$$
$$pH = 14.00 - 1.30 = 12.70$$

At the 80% titration point: $10.00\,mL\ HI\ soln \times 0.200\,M\ HI = 2.00\,mmol\ HI$
excess KOH $= 2.50\,mmol\ KOH - 2.00\,mmol\ HI = 0.50\ mmol\ KOH$

$$[OH^-] = \frac{0.50\,mmol\ KOH}{35.0\,mL\ total} \times \frac{1\,mmol\ OH^-}{1\,mmol\ KOH} = 0.0143\,M \quad pOH = -\log(0.0143) = 1.84$$
$$pH = 14.00 - 1.84 = 12.16$$

At the 110% titration point: $13.75\,mL\ HI\ soln \times 0.200\,M\ HI = 2.75\,mmol\ HI$
excess HI $= 2.75\ mmol\ HI - 2.50\,mmol\ HI = 0.25\,mmol\ HI$

$$[H_3O^+] = \frac{0.25\,mmol\ HI}{38.8\,mL\ total} \times \frac{1\,mmol\ H_3O^+}{1\,mmol\ HI} = 0.0064\,M; \quad pH = -\log(0.0064) = 2.19$$

Since the pH changes very rapidly at the equivalence point, from about $pH = 10$ to about $pH = 4$, most of the indicators in Figure 18-6 can be used. The main exceptions are alizarin yellow R, bromophenol blue, thymol blue (in its acid range), and methyl violet.

(b) *Initial pH*:

Equation: $\quad NH_3(aq) + H_2O(l) \rightleftharpoons NH_4^+(aq) + OH^-(aq)$

Initial: $\qquad$ 1.00 M $\qquad\qquad\qquad$ 0 M $\qquad$ ≈ 0 M

Changes: $\qquad -x$ M $\qquad\qquad\qquad$ $+x$ M $\qquad$ $+x$ M

Equil: $\qquad (1.00-x)$ M $\qquad\qquad$ x M $\qquad\quad$ x M

$$K_b = \frac{[NH_4^+][OH^-]}{[NH_3]} = 1.8\times10^{-5} = \frac{x^2}{1.00-x} \approx \frac{x^2}{1.00}$$

$x = 4.2\times10^{-3}$ M $= [OH^-]$, $pOH = -\log(4.2\times10^{-3}) = 2.38$,

$pH = 14.00 - 2.38 = 11.62 =$ initial pH

Volume of titrant: $\quad NH_3 + HCl \longrightarrow NH_4Cl + H_2O$

$$V_{HCl} = 10.0mL \times \frac{1.00\,mmol\,NH_3}{1\,mL\,soln} \times \frac{1\,mmol\,HCl}{1\,mmol\,NH_3} \times \frac{1\,mL\,HCl\,soln}{0.250\,mmol\,HCl} = 40.0\,mL\,soln$$

pH at equivalence point: The total solution volume at the equivalence point is $10.0 + 40.0 = 50.0$ mL

Also at the equivalence point, all of the NH_3 has reacted to form NH_4^+. It is this NH_4^+ that hydrolyzes to determine the pH of the solution.

$$[NH_4^+] = \frac{10.0\,mL \times \dfrac{1.00\,mmol\,NH_3}{1\,mL\,soln} \times \dfrac{1\,mmol\,NH_4^+}{1\,mmol\,NH_3}}{50.0\,mL\,total\,solution} = 0.200\,M$$

Equation: $NH_4^+(aq) + H_2O(l) \rightleftharpoons NH_3(aq) + H_3O^+(aq)$

Initial: $\quad$ 0.200 M $\qquad\qquad\quad$ 0 M $\qquad$ ≈ 0 M

Changes: $\quad -x$ M $\qquad\qquad\qquad$ $+x$ M $\qquad$ $+x$ M

Equil: $\quad (0.200-x)$ M $\qquad\qquad$ x M $\qquad\quad$ x M

$$K_a = \frac{K_w}{K_b} = \frac{1.0\times10^{-14}}{1.8\times10^{-5}} = \frac{[NH_3][H_3O^+]}{[NH_3]} = \frac{x^2}{0.200-x} \approx \frac{x^2}{0.200}$$

$x = 1.1\times10^{-5}$ M; $\quad [H_3O^+] = 1.1\times10^{-5}$ M; $\quad pH = -\log(1.1\times10^{-5}) = 4.96$

Of the indicators in Figure 18-6, one that has the pH of the equivalence point within its pH color change range is methyl red (yellow at $pH = 6.2$ and red at $pH = 4.5$);

Bromcresol green would be another choice. At the 50 % titration point,

$$\left[NH_3\right] = \left[NH_4^+\right] \text{ and } pOH = pK_b = 4.74 \qquad pH = 14.00 - 4.74 = 9.26$$

The titration curves for parts **(a)** and **(b)** follow.

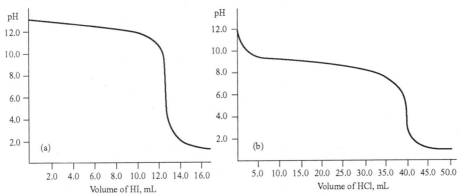

56. **(a)**   This part simply involves calculating the pH of a 0.275 M NH_3 solution.

Equation: $NH_3(aq) + H_2O(l) \rightleftharpoons NH_4^+(aq) + OH^-(aq)$

Initial:	0.275 M	0 M	≈ 0 M
Changes:	$-x$ M	$+x$ M	$+x$ M
Equil:	$(0.275 - x)$ M	x M	x M

$$K_b = \frac{\left[NH_4^+\right]\left[OH^-\right]}{\left[NH_3\right]} = 1.8 \times 10^{-5} = \frac{x^2}{0.275 - x} \approx \frac{x^2}{0.275} \qquad x = 2.2 \times 10^{-3} \, M = \left[OH^-\right]$$

$$pOH = -\log\left(2.2 \times 10^{-3}\right) = 2.66 \qquad pH = 11.34 \, M$$

(b)   This is the volume of titrant needed to reach the equivalence point.
The titration reaction is $NH_3(aq) + HI(aq) \longrightarrow NH_4I(aq)$

$$V_{HI} = 20.00 \text{ mL } NH_3(aq) \times \frac{0.275 \text{ mmol } NH_3}{1 \text{ mL } NH_3 \text{ soln}} \times \frac{1 \text{ mmol } HI}{1 \text{ mmol } NH_3} \times \frac{1 \text{ mL } HI \text{ soln}}{0.325 \text{ mmol } HI}$$

$$V_{HI} = 16.9 \text{ mL } HI \text{ soln}$$

(c)   The pOH at the half-equivalence point of the titration of a weak base with a strong acid
is equal to the pK_b of the weak base. $pOH = pK_b = 4.74$; $pH = 14.00 - 4.74 = 9.26$

(d) NH_4^+ is formed during the titration, and its hydrolysis determines the pH of the
solution. Total volume of solution $= 20.00 \text{ mL} + 16.9 \text{ mL} = 36.9 \text{ mL}$

$$\text{mmol } NH_4^+ = 20.00 \text{ mL } NH_3(aq) \times \frac{0.275 \text{ mmol } NH_3}{1 \text{ mL } NH_3 \text{ soln}} \times \frac{1 \text{ mmol } NH_4^+}{1 \text{ mmol } NH_3} = 5.50 \text{ mmol } NH_4^+$$

$$\left[NH_4^+\right] = \frac{5.50 \text{ mmol } NH_4^+}{36.9 \text{ mL soln}} = 0.149 \, M$$

Equation: $NH_4^+(aq) + H_2O(l) \rightleftharpoons NH_3(aq) + H_3O^+(aq)$

Initial:	0.149 M	0 M	≈ 0 M
Changes:	$-x$ M	$+x$ M	$+x$ M
Equil:	$(0.149 - x)$ M	x M	x M

$$K_a = \frac{K_w}{K_b} = \frac{1.0 \times 10^{-14}}{1.8 \times 10^{-5}} = \frac{[NH_3][H_3O^+]}{[NH_4^+]} = \frac{x^2}{0.149 - x} \approx \frac{x^2}{0.149}$$

$$x = 9.1 \times 10^{-6} \text{ M} = [H_3O^+] \quad pH = -\log(9.1 \times 10^{-6}) = 5.04$$

57. A pH greater than 7.00 in the titration of a strong base with a strong acid means that the base is not completely titrated. A pH less than 7.00 means that excess acid has been added.

(a) We can determine $[OH^-]$ of the solution from the pH. $[OH^-]$ is also the quotient of the amount of hydroxide ion in excess divided by the volume of the solution: 20.00 mL base $+x$ mL added acid.

$$pOH = 14.00 - pH = 14.00 - 12.55 = 1.45 \quad [OH^-] = 10^{-pOH} = 10^{-1.45} = 0.035 \text{ M}$$

$$[OH^-] = \frac{\left(20.00 \text{ mL base} \times \dfrac{0.175 \text{ mmol OH}^-}{1 \text{ mL base}}\right) - \left(x \text{ mL acid} \times \dfrac{0.200 \text{ mmol H}_3\text{O}^+}{1 \text{ mL acid}}\right)}{20.00 \text{ mL} + x \text{ mL}} = 0.035 \text{ M}$$

$$3.50 - 0.200x = 0.70 + 0.035x; \quad 3.50 - 0.70 = 0.035x + 0.200x; \quad 2.80 = 0.235x$$

$$x = \frac{2.80}{0.235} = 11.9 \text{ mL acid added.}$$

(b) The set-up here is the same as for part (a).

$$pOH = 14.00 - pH = 14.00 - 10.80 = 3.20 \quad [OH^-] = 10^{-pOH} = 10^{-3.20} = 0.00063 \text{ M}$$

$$[OH^-] = \frac{\left(20.00 \text{ mL base} \times \dfrac{0.175 \text{ mmol OH}^-}{1 \text{ mL base}}\right) - \left(x \text{ mL acid} \times \dfrac{0.200 \text{ mmol H}_3\text{O}^+}{1 \text{ mL acid}}\right)}{20.00 \text{ mL} + x \text{ mL}}$$

$$[OH^-] = 0.00063 \frac{\text{mmol}}{\text{mL}} = 0.00063 \text{ M}$$

$$3.50 - 0.200x = 0.0126 + 0.00063x; \quad 3.50 - 0.0126 = 0.00063x + 0.200x; \quad 3.49 = 0.201x$$

$$x = \frac{3.49}{0.201} = 17.4 \text{ mL acid added. This is close to the equivalence point at } 17.5 \text{ mL.}$$

(c) Here the acid is in excess, so we reverse the set-up of part (a). We are just slightly beyond the equivalence point. This is close to the "mirror image" of part **(b)**.

$$\left[H_3O^+\right] = 10^{-pH} = 10^{-4.25} = 0.000056 \text{ M}$$

$$\left[H_3O^+\right] = \frac{\left(x \text{ mL acid} \times \dfrac{0.200 \text{ mmol } H_3O^+}{1 \text{ mL acid}}\right) - \left(20.00 \text{ mL base} \times \dfrac{0.175 \text{ mmol } OH^-}{1 \text{ mL base}}\right)}{20.00 \text{ mL} + x \text{ mL}}$$

$$= 5.6 \times 10^{-5} \text{ M}$$

$$0.200\,x - 3.50 = 0.0011 + 5.6 \times 10^{-5}\,x; \quad 3.50 + 0.0011 = -5.6 \times 10^{-5}\,x + 0.200\,x;$$

$$3.50 = 0.200\,x$$

$$x = \frac{3.50_1}{0.200} = 17.5_1 \text{ mL acid added. Just beyond the equivalence point at } 17.5 \text{ mL.}$$

58. In the titration of a weak acid with a strong base, the middle range of the titration, with the pH within one unit of pK_a ($= 4.74$ for acetic acid), is known as the buffer region. The Henderson-Hasselbalch equation can be used to determine the ratio of weak acid and anion concentrations. The amount of weak acid then is used in these calculations to determine the amount of base to be added.

(a) $$pH = pK_a + \log\frac{\left[C_2H_3O_2^-\right]}{\left[HC_2H_3O_2\right]} = 3.85 = 4.74 + \log\frac{\left[C_2H_3O_2^-\right]}{\left[HC_2H_3O_2\right]} \qquad \log\frac{\left[C_2H_3O_2^-\right]}{\left[HC_2H_3O_2\right]} = 3.85 - 4.74$$

$$\frac{\left[C_2H_3O_2^-\right]}{\left[HC_2H_3O_2\right]} = 10^{-0.89} = 0.13; \quad \begin{array}{l}\text{amount of}\\ HC_2H_3O_2\end{array} = 25.00 \text{ mL} \times \frac{0.100 \text{ mmol } HC_2H_3O_2}{1 \text{ mL acid}} = 2.50 \text{ mmol}$$

Since acetate ion and acetic acid are in the same solution, we can use their amounts in millimoles in place of their concentrations. The amount of acetate ion is the amount created by the addition of strong base, one millimole of acetate ion for each millimole of strong based added. The amount of acetic acid is reduced by the same number of millimoles. $HC_2H_3O_2 + OH^- \rightarrow C_2H_3O_2^- + H_2O$

$$0.13 = \frac{x \text{ mL base} \times \dfrac{0.200 \text{ mmol } OH^-}{\text{mL base}} \times \dfrac{1 \text{ mmol } C_2H_3O_2^-}{1 \text{ mmol } OH^-}}{2.50 \text{ mmol } HC_2H_3O_2 - \left(x \text{ mL base} \times \dfrac{0.200 \text{ mmol } OH^-}{\text{mL base}}\right)} = \frac{0.200\,x}{2.50 - 0.200\,x}$$

$$0.200\,x = 0.13(2.50 - 0.200\,x) = 0.33 - 0.026x \qquad 0.33 = 0.200\,x + 0.026\,x = 0.226\,x$$

$$x = \frac{0.33}{0.226} = 1.5 \text{ mL base}$$

(b) This is the same set-up as part (a), except for a different ratio of concentrations.

$$\text{pH} = 5.25 = 4.74 + \log\frac{\left[C_2H_3O_2^-\right]}{\left[HC_2H_3O_2\right]} \qquad \log\frac{\left[C_2H_3O_2^-\right]}{\left[HC_2H_3O_2\right]} = 5.25 - 4.74 = 0.51$$

$$\frac{\left[C_2H_3O_2^-\right]}{\left[HC_2H_3O_2\right]} = 10^{0.51} = 3.2 \qquad 3.2 = \frac{0.200x}{2.50 - 0.200x}$$

$$0.200x = 3.2(2.50 - 0.200x) = 8.0 - 0.64x \qquad 8.0 = 0.200x + 0.64x = 0.84x$$

$$x = \frac{8.0}{0.84} = 9.5 \, \text{mL base.} \quad \text{This is close to the equivalence point, 12.5 mL base.}$$

(c) This is after the equivalence point, where the pH is determined by the excess added base.

$$\text{pOH} = 14.00 - \text{pH} = 14.00 - 11.10 = 2.90 \qquad \left[OH^-\right] = 10^{-\text{pOH}} = 10^{-2.90} = 0.0013 \, \text{M}$$

$$\left[OH^-\right] = 0.0013 \, \text{M} = \frac{x \, \text{mL} \times \dfrac{0.200 \, \text{mmol OH}^-}{1 \, \text{mL base}}}{x \, \text{mL} + (12.50 \, \text{mL} + 25.00 \, \text{mL})} = \frac{0.200x}{37.50 + x}$$

$$0.200x = 0.0013(37.50 + x) = 0.049 + 0.0013x \qquad x = \frac{0.049}{0.200 - 0.0013} = 0.25 \, \text{mL excess}$$

Total base added $= 12.5$ mL to equivalence point $+ 0.25$ mL excess $= 12.8$ mL

59. For each of the titrations, the pH at the half-equivalence point equals the pK_a of the acid.

The initial pH is that of 0.1000 M weak acid: $K_a = \dfrac{x^2}{0.1000} \qquad x = \sqrt{0.1000 \times K_a} = [H_3O^+]$

The pH at the equivalence point is that of 0.05000 M anion of the weak acid, for which the

$\left[OH^-\right]$ is determined as follows. $K_b = \dfrac{K_w}{K_a} = \dfrac{x^2}{0.05000} \qquad x = \sqrt{\dfrac{K_w}{K_a} 0.05000} = [OH^-]$

And finally when 0.100 mL of base has been added beyond the equivalence point, the pH is determined by the excess added base, as follows (for all three titrations).

$$\left[OH^-\right] = \frac{0.100 \, \text{mL} \times \dfrac{0.1000 \, \text{mmol NaOH}}{1 \, \text{mL NaOH soln}} \times \dfrac{1 \, \text{mmol OH}^-}{1 \, \text{mmol NaOH}}}{20.1 \, \text{mL soln total}} = 4.98 \times 10^{-4} \, \text{M}$$

$$\text{pOH} = -\log(4.98 \times 10^{-4}) = 3.303 \qquad \text{pH} = 14.000 - 3.303 = 10.697$$

(a) Initial: $\quad [H_3O^+] = \sqrt{0.1000 \times 7.0 \times 10^{-3}} = 0.026 \, \text{M} \qquad \text{pH} = 1.59$

Equiv: $\quad \left[OH^-\right] = \sqrt{\dfrac{1.0 \times 10^{-14}}{7.0 \times 10^{-3}} \times 0.05000} = 2.7 \times 10^{-7} \qquad \begin{array}{l} \text{pOH} = 6.57 \\ \text{pH} = 14.00 - 6.57 = 7.43 \end{array}$

Indicator: bromothymol blue, yellow at pH $= 6.2$ and blue at pH $= 7.8$

(b) Initial: $[H_3O^+] = \sqrt{0.1000 \times 3.0 \times 10^{-4}} = 0.0055$ M   pH = 2.26

Equiv: $\left[OH^-\right] = \sqrt{\dfrac{1.0 \times 10^{-14}}{3.0 \times 10^{-4}}} \times 0.05000 = 1.3 \times 10^{-6}$ pOH = 5.89

pH = 14.00 − 5.89 = 8.11

Indicator: thymol blue, yellow at pH = 8.0 and blue at pH = 9.6

(c) Initial: $[H_3O^+] = \sqrt{0.1000 \times 2.0 \times 10^{-8}} = 0.000045$ M pH = 4.35

Equiv: $\left[OH^-\right] = \sqrt{\dfrac{1.0 \times 10^{-14}}{2.0 \times 10^{-8}}} \times 0.0500 = 1.6 \times 10^{-4}$ pOH = 3.80

pH = 14.00 − 3.80 = 10.20

Indicator: alizarin yellow R, yellow at pH = 10.0 and violet at pH = 12.0

The three titration curves are drawn with respect to the same axes in the diagram below.

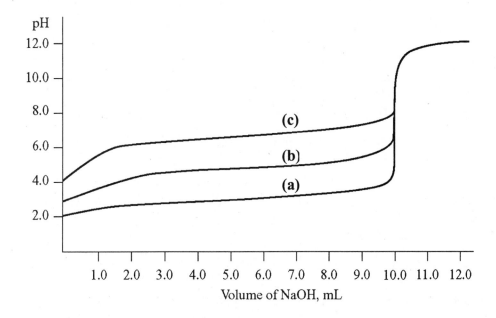

60. For each of the titrations, the pOH at the half-equivalence point equals the pK_b of the base. The initial pOH is that of 0.1000 M weak base, determined as follows.

$$K_b = \frac{x^2}{0.1000} \qquad x = \sqrt{0.1000 \times K_a} = [OH^-]$$

The pH at the equivalence point is that of 0.05000 M cation of the weak base, for which the $[H_3O^+]$ is determined as follows.

$$K_a = \frac{K_w}{K_b} = \frac{x^2}{0.05000} \qquad x = \sqrt{\frac{K_w}{K_b} 0.05000} = [H_3O^+]$$

And finally when 0.100 mL of acid has been added beyond the equivalence point, the pH for all three titrations is determined by the excess added acid, as follows.

$$\left[H_3O^+\right] = \frac{0.100 \text{ mL HCl} \times \dfrac{0.1000 \text{ mmol HCl}}{1 \text{ mL HCl soln}} \times \dfrac{1 \text{ mmol } H_3O^+}{1 \text{ mmol HCl}}}{20.1 \text{ mL soln total}} = 4.98 \times 10^{-4} \text{ M}$$

$$pH = -\log\left(4.98 \times 10^{-4}\right) = 3.303$$

(a) Initial: $[OH^-] = \sqrt{0.1000 \times 1 \times 10^{-3}} = 0.01$ M pOH = 2.0 pH = 12.0

Half-equiv: $pOH = -\log\left(1 \times 10^{-3}\right) = 3.0$ pH = 11.0

Equiv: $[H_3O^+] = \sqrt{\dfrac{1.0 \times 10^{-14}}{1 \times 10^{-3}} \times 0.05000} = 7 \times 10^{-7}$ pH=6.2

Indicator: methyl red, yellow at pH = 6.3 and red at pH = 4.5

(b) Initial: $[OH^-] = \sqrt{0.1000 \times 3 \times 10^{-6}} = 5 \times 10^{-4}$ M pOH = 3.3 pH = 10.7

Half-equiv: $pOH = -\log\left(3 \times 10^{-6}\right) = 5.5$ pH = 8.5

Equiv: $[H_3O^+] = \sqrt{\dfrac{1.0 \times 10^{-14}}{3 \times 10^{-6}} \times 0.05000} = 1 \times 10^{-5}$ pH=5.0

Indicator: methyl red, yellow at pH = 6.3 and red at pH = 4.5

(c) Initial: $[OH^-] = \sqrt{0.1000 \times 7 \times 10^{-8}} = 8 \times 10^{-5}$ M pOH = 4.1 pH = 9.9

Half-equiv: $pOH = -\log\left(7 \times 10^{-8}\right) = 7.2$ pH = 6.8

Equiv: $[H_3O^+] = \sqrt{\dfrac{1.0 \times 10^{-14}}{7 \times 10^{-8}} \times 0.05000} = 8 \times 10^{-5}$ pH=4.1

Indicator: bromcresol green, blue at pH = 5.5 and yellow at pH = 4.0

The titration curves are drawn with respect to the same axes in the diagram below.

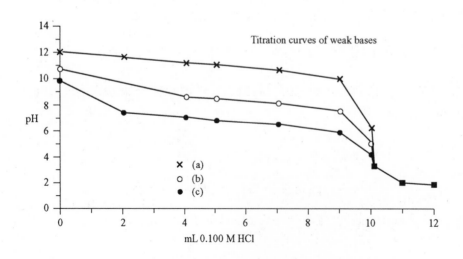

61. 25.00 mL of 0.100 M NaOH titrated with 0.100 M HCl

(i) Initial pOH for 0.100 M NaOH: $[OH^-] = 0.100$ M, pOH = 1.000 or pH = 13.000

(ii) After addition of 24 mL: $[NaOH] = 0.100$ M $\times \dfrac{25.00 \text{ mL}}{49.00 \text{ mL}} = 0.0510$ M

$$[HCl] = 0.100 \text{ M} \times \dfrac{24.00 \text{ mL}}{49.00 \text{ mL}} = 0.0490 \text{ M}$$

NaOH is in excess by 0.0020 M = $[OH^-]$ pOH = 2.70

(iii) At the equivalence point (25.00 mL), the pOH should be 7.000 and pH = 7.000

(iv) After addition of 26 mL: $[NaOH] = 0.100$M $\times \dfrac{25.00}{51.00} = 0.0490$ M

$$[HCl] = 0.100 \text{ M} \times \dfrac{26.00 \text{ mL}}{51.00 \text{ mL}} = 0.0510 \text{ M}$$

HCl is in excess by 0.0020 M = $[H_3O^+]$ pH = 2.70 or pOH = 11.30

(v) After addition of 33.00 mL HCl(xs) $[NaOH] = 0.100$ M $\times \dfrac{25.00 \text{ mL}}{58.00 \text{ mL}} = 0.0431$ M

$[HCl] = 0.100$ M $\times \dfrac{33.00 \text{ mL}}{58.00 \text{ mL}} = 0.0569$ M $[HCl]_{excess} = 0.0138$ M

pH = 1.860 and pOH = 12.140

The graphs look to be mirror images of one another. In reality, one must reflect about a horizontal line centered at pH or pOH = 7 to obtain the other curve.

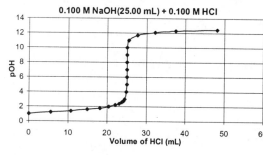

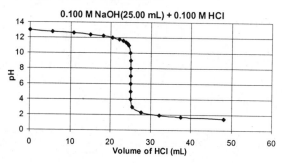

62. 25.00 mL of 0.100 M NH$_3$ with 0.100 M HCl $K_a = \dfrac{K_w}{K_b} = \dfrac{1 \times 10^{-14}}{1.8 \times 10^{-5}} = 5.6 \times 10^{-10}$

(i)   For initial pOH, use I.C.E.(initial, change, equilibrium) table.

	NH$_3$(aq)	+ H$_2$O(l)	$\xrightarrow{K_b = 1.8 \times 10^{-5}}$	NH$_4^+$(aq)	+ OH$^-$(aq)
initial	0.100 M			0 M	~ 0 M
change	$-x$			$+x$	$+x$
equil.	$0.100 - x$			x	x

$1.8 \times 10^{-5} = \dfrac{x^2}{0.100 - x} \approx \dfrac{x^2}{0.100}$ (Assume $x \sim 0$) $x = 1.3 \times 10^{-3}$

($x < 5\%$ of 0.100, assumption valid).

Hence, $x = [OH^-] = 1.3 \times 10^{-3}$ pOH = 2.87, pH = 11.13

(ii) After 2 mL of HCl added: $[HCl] = 0.100 \text{ M} \times \dfrac{2.00 \text{ mL}}{27.00 \text{ mL}} = 0.00741$ M (after dilution)

$[NH_3] = 0.100 \text{ M} \times \dfrac{25.00 \text{ mL}}{27.00 \text{ mL}} = 0.0926$ M (after dilution)

The equilibrium constant for ht neutralization reaction is large and thus the reaction goes to nearly 100% completion. Assume that the limiting reagent is used up (100% reaction in the reverse direction) and re-establish the equilibrium by a shift in the forward direction. Here H_3O^+ (HCl) is the limiting reagent.

	NH_4^+(aq) + H_2O(l)	$K_a = 5.6 \times 10^{-10}$	NH_3(aq) +	H_3O^+(aq)
initial	0 M		0.0926 M	0.00741 M
change	+x	$x = 0.00741$	$-x$	$-x$
100% rxn	0.00741		0.0852	0
change	$-y$	re-establish equilibrium	+y	+y
equil	0.00741−y		0.0852 + y	y

$5.6 \times 10^{-10} = \dfrac{y(0.0852 + y)}{(0.00741 - y)} = \dfrac{y(0.0852)}{(0.00741)}$ (set $y \sim 0$) $y = 4.8 \times 10^{-11}$ (valid)

$y = [H_3O^+] = 4.8 \times 10^{-11}$; pH = 10.32 and pOH = 3.68

(iii) pH at 1/2 equivalence point = $pK_a = -\log 5.6 \times 10^{-10} = 9.26$ and pOH = 4.74

(iv) After addition of 24 mL of HCl:

$[HCl] = 0.100 \text{ M} \times \dfrac{24.00 \text{ mL}}{49.00 \text{ mL}} = 0.0490 \text{ M}; \quad [NH_3] = 0.100 \text{ M} \times \dfrac{25.00 \text{ mL}}{49.00 \text{ mL}} = 0.0510$ M

The equilibrium constant for the neutralization reaction is large and thus the reaction goes to nearly 100% completion. Assume that the limiting reagent is used up (100% reaction in the reverse direction) and re-establish the equilibrium in the reverse direction. Here H_3O^+ (HCl) is the limiting reagent.

	NH_4^+(aq) + H_2O(l)	$K_a = 5.6 \times 10^{-10}$	NH_3(aq) +	H_3O^+(aq)
initial	0 M		0.0541 M	0.0490 M
change	+x	$x = 0.0490$	$-x$	$-x$
100% rxn	0.0490		0.0020	0
change	$-y$	re-establish equilibrium	+y	+y
equil	0.0490−y		0.0020 + y	y

$5.6 \times 10^{-10} = \dfrac{y(0.0020 + y)}{(0.0490 - y)} = \dfrac{y(0.0020)}{(0.0490)}$ (Assume $y \sim 0$) $y = 1.3 \times 10^{-8}$ (valid)

$y = [H_3O^+] = 1.3 \times 10^{-8}$; pH = 7.89 and pOH = 6.11

(v) Equiv. point: 100% reaction of $NH_3 \rightarrow NH_4^+$: $[NH_4^+] = 0.100 \times \dfrac{25.00 \text{ mL}}{50.00 \text{ mL}} = 0.0500 \text{ M}$

$$NH_4^+(aq) \quad + \quad H_2O(l) \quad \xrightarrow{K_a = 5.6 \times 10^{-10}} \quad NH_3(aq) \quad + \quad H_3O^+(aq)$$

initial	0.0500 M		0 M	~ 0 M
change	−x		+x	+x
equil	0.0500−x		x	x

$5.6 \times 10^{-10} = \dfrac{x^2}{(0.0500-x)} = \dfrac{x^2}{0.0500}$ (Assume $x \sim 0$) $x = 5.3 \times 10^{-6}$ (valid)

$x = [H_3O^+] = 5.3 \times 10^{-6}$; pH = 5.28 and pOH = 8.72

(vi) After addition of 26 mL of HCl, HCl is in excess. The pH and pOH should be the same as those in question 61. pH = 2.70 and pOH = 11.30

(vii) After addition of 33 mL of HCl, HCl is in excess. The pH and pOH should be the same as those in question 61. pH = 1.860 and pOH = 12.140

This time the curves are not mirror images of one another, but rather they are related through a reflection of a horizontal line centered at pH or pOH = 7.

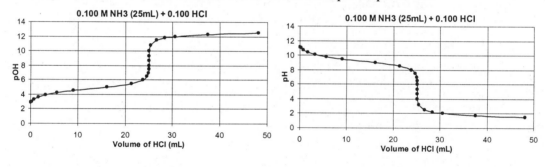

pH of Salts of Polyprotic Acids

63. We expect a solution of Na_2S to be alkaline, or basic. This alkalinity is created by the hydrolysis of the sulfide ion, the anion of a very weak acid ($K_2 = 1 \times 10^{-19}$ for H_2S).

$$S^{2-}(aq) + H_2O(l) \rightleftharpoons HS^-(aq) + OH^-(aq)$$

64. We expect the pH of a solution of sodium dihydrogen citrate, NaH_2Cit , to be acidic because pK_a values of first and second ionization constants of polyprotic acids are reasonably large. The pH of a solution of the salt is the average of pK_1 and pK_2 . For citric acid, in fact, this average is $(3.13 + 4.77) \div 2 = 3.95$. This is an acidic solution.

65. **(a)** $H_3PO_4(aq) + CO_3^{2-}(aq) \rightleftharpoons H_2PO_4^-(aq) + HCO_3^-(aq)$

$H_2PO_4^-(aq) + CO_3^{2-}(aq) \rightleftharpoons HPO_4^{2-}(aq) + HCO_3^-(aq)$

$HPO_4^{2-}(aq) + OH^-(aq) \rightleftharpoons PO_4^{3-}(aq) + H_2O(l)$

(b) The pH values of 1.00 M solutions of the three ions are;

$1.0\ M\ OH^- \rightarrow pH = 14.00$ $1.0\ M\ CO_3^{2-} \rightarrow pH = 12.16$ $1.0\ M\ PO_4^{3-} \rightarrow pH = 13.15$

Thus, we see that CO_3^{2-} is not a strong enough base to remove the third proton from H_3PO_4. As an alternative method of solving this problem, we can compute the equilibrium constant the reactions of carbonate ion with H_3PO_4, $H_2PO_4^-$ and HPO_4^{2-}.

$$H_3PO_4 + CO_3^{2-} \rightleftharpoons H_2PO_4^- + HCO_3^- \quad K = \frac{K_{a1}\{H_3PO_4\}}{K_{a2}\{H_2CO_3\}} = \frac{7.1 \times 10^{-3}}{4.7 \times 10^{-11}} = 1.5 \times 10^8$$

$$H_2PO_4^- + CO_3^{2-} \rightleftharpoons HPO_4^{2-} + HCO_3^- \quad K = \frac{K_{a2}\{H_3PO_4\}}{K_{a2}\{H_2CO_3\}} = \frac{6.3 \times 10^{-8}}{4.7 \times 10^{-11}} = 1.3 \times 10^3$$

$$HPO_4^{2-} + CO_3^{2-} \rightleftharpoons PO_4^{3-} + HCO_3^- \quad K = \frac{K_{a3}\{H_3PO_4\}}{K_{a2}\{H_2CO_3\}} = \frac{4.2 \times 10^{-13}}{4.7 \times 10^{-11}} = 8.9 \times 10^{-3}$$

Since the equilibrium constant for the third reaction is much smaller than 1.00, we conclude that it proceeds to the right to only a negligible extent and thus is not a practical method of producing PO_4^{3-}. The other two reactions have large equilibrium constants, and products are expected to strongly predominate. They have the advantage of involving an inexpensive base and, even if they do not go to completion, they will be drawn to completion by reaction with OH^- in the last step of the process.

66. We expect the bicarbonate ion to hydrolyze and the hydrolysis products to determine the pH of the solution.

Equation: $HCO_3^-(aq) + H_2O(l) \rightleftharpoons$ "H_2CO_3"(aq)   + $OH^-(aq)$

Initial	1.00 M	0 M	≈ 0 M
Changes:	$-x$ M	$+x$ M	$+x$ M
Equil:	$(1.00 - x)$ M	x M	x M

$$K_b = \frac{K_w}{K_1} = \frac{1.00 \times 10^{-14}}{4.4 \times 10^{-7}} = 2.3 \times 10^{-8} = \frac{[H_2CO_3][OH^-]}{[HCO_3^-]} = \frac{(x)(x)}{1.00 - x} \approx \frac{x^2}{1.00}$$

$x = \sqrt{1.00 \times 2.3 \times 10^{-8}} = 1.5 \times 10^{-4}$ M $= [OH^-]$; pOH $= -\log(1.5 \times 10^{-4}) = 3.82$ pH $= 10.18$

For 1.00 M NaOH, $[OH^-] = 1.00$ pOH $= -\log(1.00) = 0.00$ pH $= 14.00$

Both 1.00 M $NaHCO_3$ and 1.00 M NaOH have an equal capacity to neutralize acids since one mole of each neutralizes one mole of strong acid.

$$NaOH(aq) + H_3O^+(aq) \rightarrow Na^+(aq) + 2H_2O(l)$$

$$NaHCO_3(aq) + H_3O^+(aq) \rightarrow Na^+(aq) + CO_2(g) + 2H_2O(l).$$

But on a per gram basis, the one with the smaller molar mass is the more effective. Because the molar mass of NaOH is 40.0 g/mol, while that of $NaHCO_3$ is 84.0 g/mol, NaOH(s) is more than twice as effective than $NaHCO_3(s)$ on a per gram basis. $NaHCO_3$ is preferred in laboratories for safety and expense reasons. NaOH is not a good choice because it can cause severe burns. $NaHCO_3$, baking soda, by comparison, is relatively non-hazardous. It also is much cheaper than NaOH.

67. $H_2C_3H_2O_4$ MM $= 104.06 \dfrac{g}{mol}$ Moles of $H_2C_3H_2O_4 = 19.5 \text{ g} \times \dfrac{1 \text{ mol}}{104.06 \text{ g}} = 0.187$ mol

Concentration of $H_2C_3H_2O_4 = \dfrac{\text{moles}}{V} = \dfrac{0.187 \text{ moles}}{0.250 \text{ L}} = 0.748$ M

The second proton that can dissociate has a negligible effect on pH (K_{a_2} very small)

	$H_2A(aq)$	$+$	$H_2O(l)$	$\rightleftharpoons$	$HA^-(aq)$	$+$	$H_3O^+(aq)$
initial	0.748 M				0 M		~0 M
change	$-x$				$+x$		$+x$
equil	$0.748-x$				x		x

pH $= 1.47$, therefore, $[H_3O^+] = 0.034$ M $= x$, $K_{a_1} = \dfrac{(0.034)^2}{0.748 - 0.034} = 1.6 \times 10^{-3}$

(1.5×10^{-3} in tables, difference owing to ionization of the second proton)

	$HA^-(aq)$	$+$	$H_2O(l)$	$\rightleftharpoons$	$A^{2-}(aq)$	$+$	$H_3O^+(aq)$
initial	0.300 M				0 M		~0 M
change	$-x$				$+x$		$+x$
equil	$0.300-x$				x		x

pH $= 4.26$, therefore, $[H_3O^+] = 5.5 \times 10^{-5}$ M $= x$, $K_{a_2} = \dfrac{(5.5 \times 10^{-5})^2}{0.300 - 5.5 \times 10^{-5}} = 1.0 \times 10^{-8}$

68. Ortho-phthalic acid. $K_{a_1} = 1.1 \times 10^{-3}$, $K_{a_2} = 3.9 \times 10^{-6}$

(a) We have solution of NaHA.

	$HA^-(aq)$	$+$	$H_2O(l)$	$\rightleftharpoons$	$A^{2-}(aq)$	$+$	$H_3O^+(aq)$
initial	0.350 M				0 M		~0 M
change	$-x$				$+x$		$+x$
equil	$0.350-x$				x		x

$$3.9 \times 10^{-6} = \frac{x^2}{0.350 - x} \simeq \frac{x^2}{0.350} \ , x = 1.2 \times 10^{-3} \text{ (assumption valid)}$$

$$x = [H_3O^+] = 1.2 \times 10^{-3}, \ pH = 2.93$$

(b) 36.35 g of potassium ortho-phthalate $(MM = 242.314 \text{ g mol}^{-1})$

moles of potassium ortho-phthalate $= 36.35 \text{ g} \times \dfrac{1 \text{ mol}}{242.314 \text{ g}} = 0.150 \text{ mol in 1 L}$

$$A^{2-}(aq) \ + \ H_2O(l) \ \rightleftharpoons \ HA^-(aq) \ + \ OH^-(aq)$$

initial	0.150 M	0 M	~0 M
change	$-x$	$+x$	$+x$
equil	$0.150-x$	x	x

$$K_{b,A^{2-}} = \frac{K_w}{K_{a_2}} = \frac{1.0 \times 10^{-14}}{3.9 \times 10^{-6}} = 2.6 \times 10^{-9} = \frac{x^2}{0.150 - x} = \frac{x^2}{0.150}$$

$$x = 2.0 \times 10^{-5} \text{ (assumption valid)} = [OH^-]$$

$$pOH = -\log 2.0 \times 10^{-5} = 4.70; \quad pH = 9.30$$

General Acid-Base Equilibria

69. (a) $Ba(OH)_2$ is a strong base. $pOH = 14.00 - 11.88 = 2.12 \quad \left[OH^-\right] = 10^{-2.12} = 0.0076 \text{ M}$

$$\left[Ba(OH)_2\right] = \frac{0.0076 \text{ mol OH}^-}{1 \text{ L}} \times \frac{1 \text{ mol } Ba(OH)_2}{2 \text{ mol OH}^-} = 0.0038 \text{ M}$$

(b) $pH = 4.52 = pK_a + \log \dfrac{\left[C_2H_3O_2^-\right]}{\left[HC_2H_3O_2\right]} = 4.74 + \log \dfrac{0.294 \text{ M}}{\left[HC_2H_3O_2\right]}$

$$\log \frac{0.294 \text{ M}}{\left[HC_2H_3O_2\right]} = 4.52 - 4.74 \qquad \frac{0.294 \text{ M}}{\left[HC_2H_3O_2\right]} = 10^{-0.22} = 0.60$$

$$\left[HC_2H_3O_2\right] = \frac{0.294 \text{ M}}{0.60} = 0.49 \text{ M}$$

70. (a) $pOH = 14.00 - 8.95 = 5.05 \qquad \left[OH^-\right] = 10^{-5.05} = 8.9 \times 10^{-6} \text{ M}$

Equation: $\quad C_6H_5NH_2(aq) + H_2O(l) \ \rightleftharpoons \ C_6H_5NH_3^+(aq) \ + \ OH^-(aq)$

Initial	x M	0 M	≈ 0 M
Changes:	-8.9×10^{-6} M	$+8.9 \times 10^{-6}$ M	$+8.9 \times 10^{-6}$ M
Equil:	$(x - 8.9 \times 10^{-6})$ M	8.9×10^{-6} M	8.9×10^{-6} M

$$K_b = \frac{[C_6H_5NH_3^+][OH^-]}{[C_6H_5NH_2]} = 7.4 \times 10^{-10} = \frac{(8.9 \times 10^{-6})^2}{x - 8.9 \times 10^{-6}}$$

$$x - 8.9 \times 10^{-6} = \frac{(8.9 \times 10^{-6})^2}{7.4 \times 10^{-10}} = 0.11 \text{ M} \qquad x = 0.11 \text{ M} = \text{molarity of aniline}$$

(b) $[H_3O^+] = 10^{-5.12} = 7.6 \times 10^{-6}$ M

Equation:	$NH_4^+(aq) + H_2O(l)$	$\rightleftharpoons$	$NH_3(aq)$	$+$	$H_3O^+(aq)$
Initial	x M		0 M		≈ 0 M
Changes:	-7.6×10^{-6} M		$+7.6 \times 10^{-6}$ M		$+7.6 \times 10^{-6}$ M
Equil:	$(x - 7.6 \times 10^{-6})$ M		7.6×10^{-6} M		7.6×10^{-6} M

$$K_a = \frac{[NH_3][H_3O^+]}{[NH_4^+]} = \frac{K_w}{K_b \text{ for } NH_3} = \frac{1.0 \times 10^{-14}}{1.8 \times 10^{-5}} = 5.6 \times 10^{-10} = \frac{(7.6 \times 10^{-6})^2}{x - 7.6 \times 10^{-6}}$$

$$x - 7.6 \times 10^{-6} = \frac{(7.6 \times 10^{-6})^2}{5.6 \times 10^{-10}} = 0.10 \text{ M} \qquad x = [NH_4^+] = [NH_4Cl] = 0.10 \text{ M}$$

71. **(a)** $\quad$ pOH $= 14.00 - 6.07 = 7.93 \qquad [OH^-] = 10^{-7.93} = 1.2 \times 10^{-8}$ M

$$Q_b = \frac{[NH_4^+][OH^-]}{[NH_3]} = \frac{(0.10 \text{ M})(1.2 \times 10^{-8} \text{ M})}{0.10} = 1.2 \times 10^{-8}$$

Since the value of Q_b does not equal the tabulated value of $K_b = 1.8 \times 10^{-5}$, the solution described cannot exist.(The assumption here is that equilibrium has been established)

(b) These solutes can be added to the same solution, but the final solution will have an appreciable $[HC_2H_3O_2]$ because of the reaction of $H_3O^+(aq)$ with $C_2H_3O_2^-(aq)$

Equation:	$H_3O^+(aq)$	$+$	$C_2H_3O_2^-(aq)$	$\rightleftharpoons$	$HC_2H_3O_2(aq) + H_2O(l)$
Initial	0.058 M		0.10 M		0 M
Changes:	−0.058 M		−0.058 M		+0.058 M
Equil:	0.000 M		0.04 M		0.058 M

Of course, some H_3O^+ will exist in the final solution, but not equivalent to 0.058 M HI.

(c) Both 0.10 M KNO_2 and 0.25 M KNO_3 can exist together. Some hydrolysis of the $NO_2^-(aq)$ ion will occur, forming $HNO_2(aq)$.

(d) $Ba(OH)_2$ is a strong base and will react as much as possible with the weak conjugate acid NH_4^+, to form $NH_3(aq)$. We will end up with a solution of $BaCl_2(aq)$, $NH_3(aq)$, and unreacted $NH_4Cl(aq)$.

(e)   This will be a benzoic acid-benzoate ion buffer solution. Since the two components have the same concentration, the buffer solution will have $pH = pK_a = -\log(6.3 \times 10^{-5}) = 4.20$. This solution can indeed exist.

(f)   The first three components contain no ions that will hydrolyze. But $C_2H_3O_2^-$ is the anion of a weak acid and will hydrolyze to form a slightly basic solution. Since $pH = 6.4$ is an acidic solution, the solution described cannot exist.

72.   **(a)**   When $[H_3O^+]$ and $[HC_2H_3O_2]$ are high and $[C_2H_3O_2^-]$ is very low, a common ion H_3O^+ has been added to a solution of acetic acid, suppressing its ionization.

(b)   When $[C_2H_3O_2^-]$ is high and $[H_3O^+]$ and $[HC_2H_3O_2]$ are very low, we are dealing with a solution of acetate ion, which hydrolyzes to produce a small concentration of $HC_2H_3O_2$.

(c)   When $[HC_2H_3O_2]$ is high and both $[H_3O^+]$ and $[C_2H_3O_2^-]$ are low, the solution is an acetic acid solution, in which the solute is partially ionized.

(d)   When both $[HC_2H_3O_2]$ and $[C_2H_3O_2^-]$ are high while $[H_3O^+]$ is low, the solution is a buffer solution, in which the presence of acetate ion suppresses the ionization of acetic acid.

FEATURE PROBLEMS

98.   **(a)**   The two curves cross the point at which half of the total acetate is present as acetic acid and half is present as acetate ion. This is the half equivalence point in a titration, where $pH = pK_a = 4.74$.

(b)   For carbonic acid, there are three carbonate containing species: "H_2CO_3" which predominates at low pH, HCO_3^-, and CO_3^{2-} which predominates in alkaline solution. The points of intersection should occur at the half-equivalence points in each step-wise titration: at $pH = pK_{a_1} = -\log(4.4 \times 10^{-7}) = 6.36$ and at $pH = pK_{a_2} = -\log(4.7 \times 10^{-11}) = 10.33$. The following graph was computer-calculated (and then drawn) from these equations. f in each instance represents the fraction of the species whose formula is in parentheses.

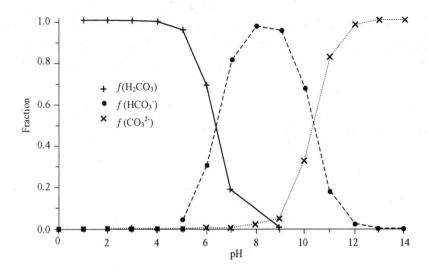

$$\frac{1}{f\left(H_{2}A\right)}=1+\frac{K_{1}}{\left[H^{+}\right]}+\frac{K_{1}K_{2}}{\left[H^{+}\right]^{2}};\quad\frac{1}{f\left(HA^{-}\right)}=\frac{\left[H^{+}\right]}{K_{1}}+1+\frac{K_{2}}{\left[H^{+}\right]};\quad\frac{1}{f\left(A^{2-}\right)}=\frac{\left[H^{+}\right]^{2}}{K_{1}\ K_{2}}+\frac{\left[H^{+}\right]}{K_{2}}+1$$

(c) For phosphoric acid, there are four phosphate containing species: H_3PO_4 under acidic conditions, $H_2PO_4^-$, HPO_4^{2-}, and PO_4^{3-} which predominates in alkaline solution. The points of intersection should occur at $pH = pK_{a_1} = -\log(7.1\times10^{-3}) = 2.15$, $pH = pK_{a_2} = -\log(6.3\times10^{-8}) = 7.20$, and $pH = pK_{a_3} = -\log(4.2\times10^{-13}) = 12.38$, a quite alkaline solution. The graph that follows was computer-calculated and drawn.

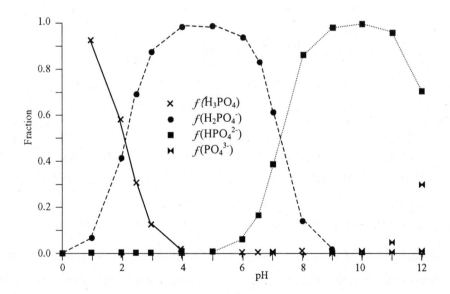

99. (a) This is exactly the same titration curve we would obtain for the titration of 25.00 mL of 0.200 M HCl with 0.200 M NaOH, because the acid species being titrated is H_3O^+. Both acids are strong acids and have ionized completely before titration begins. The initial pH is that of 0.200 M $H_3O^+ = [HCl] + [HNO_3]$; $pH = -\log(0.200) = 0.70$. At the equivalence point, $pH = 7.000$. We treat this problem in the same we would for the titration of a single strong acid with a strong base.

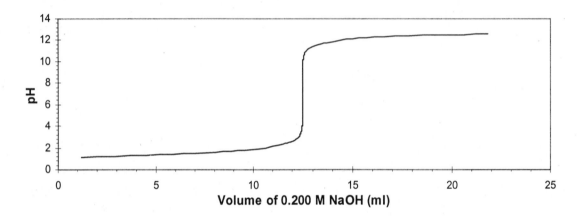

(b) In Figure 18-9, we note that the equivalence point of the titration of a strong acid occurs at $pH = 7.00$, but that the strong acid is essentially completely neutralized at $pH = 4$. In Figure 18-13, we see that the first equivalence point of H_3PO_4 occurs at about $pH = 4.6$. Thus, the first equivalence point represents the complete neutralization of HCl and the neutralization of H_3PO_4 to $H_2PO_4^-$. Then, the second equivalence point represents the neutralization of $H_2PO_4^-$ to HPO_4^{2-}. To reach the first equivalence point requires about 20.0 mL of 0.216 M NaOH, while to reach the second one requires a total of 30.0 mL of 0.216 M NaOH, or an additional 10.0 mL of base beyond the first equivalence point. The equations for the two titration reactions are as follows.

To the first equivalence point: $NaOH + H_3PO_4 \longrightarrow NaH_2PO_4 + H_2O$

$NaOH + HCl \longrightarrow NaCl + H_2O$

To the second equivalence point: $NaOH + NaH_2PO_4 \longrightarrow Na_2HPO_4 + H_2O$

There is a third equivalence point, not shown in the figure, that would require an additional 10.0 mL of base to reach. Its titration reaction is represented by the following equation.

To the third equivalence point: $NaOH + Na_2HPO_4 \longrightarrow Na_3PO_4 + H_2O$

We determine the molar concentration of H_3PO_4 and then of HCl. Notice that only 10.0 mL of the NaOH needed to reach the first equivalence point reacts with the HCl(aq); the rest reacts with H_3PO_4.

$$\frac{(30.0-20.0) \text{ mL NaOH(aq)} \times \dfrac{0.216 \text{ mmol NaOH}}{1 \text{ mL NaOH soln}} \times \dfrac{1 \text{ mmol H}_3\text{PO}_4}{1 \text{ mmol NaOH}}}{10.00 \text{ mL acid soln}} = 0.216 \text{ M H}_3\text{PO}_4$$

$$\frac{(20.0-10.0) \text{ mL NaOH(aq)} \times \dfrac{0.216 \text{ mmol NaOH}}{1 \text{ mL NaOH soln}} \times \dfrac{1 \text{ mmol HCl}}{1 \text{ mmol NaOH}}}{10.00 \text{ mL acid soln}} = 0.216 \text{ M HCl}$$

(c) We start with a phosphoric acid-dihydrogen phosphate buffer solution and titrate until all of the H_3PO_4 is consumed. We begin with

$$10.00 \text{ mL} \times \frac{0.0400 \text{ mmol H}_3\text{PO}_4}{1 \text{ mL}} = 0.400 \text{ mmol H}_3\text{PO}_4 \text{ and the diprotic anion,}$$

$$10.00 \text{ mL} \times \frac{0.0150 \text{ mmol H}_2\text{PO}_4^{-}}{1 \text{ mL}} = 0.150 \text{ mmol H}_2\text{PO}_4^{2-} . \text{ The volume of } 0.0200 \text{ M}$$

NaOH needed is $0.400 \text{ mmol H}_3\text{PO}_4 \times \dfrac{1 \text{ mmol NaOH}}{1 \text{ mmol H}_3\text{PO}_4} \times \dfrac{1 \text{ mL NaOH}}{0.0200 \text{ mmol NaOH}} = 20.00 \text{ mL}$

to reach the first equivalence point. The pH of points during this titration are computed with the Henderson-Hasselbalch equation.

$$\text{Initially: pH} = pK_1 + \log\frac{\left[\text{H}_2\text{PO}_4^{-}\right]}{\left[\text{H}_3\text{PO}_4\right]} = -\log\left(7.1\times10^{-3}\right) + \log\frac{0.0150}{0.0400} = 2.15 - 0.43 = 1.62$$

$$\text{At } 5.00 \text{ mL : pH} = 2.15 + \log\frac{0.150+0.100}{0.400-0.100} = 2.15 - 0.08 = 2.07$$

$$\text{At } 10.0 \text{ mL, pH} = 2.15 + \log\frac{0.150+0.200}{0.400-0.200} = 2.15 + 0.24 = 2.39$$

$$\text{At } 15.0 \text{ mL, pH} = 2.15 + \log\frac{0.150+0.300}{0.400-0.300} = 2.15 + 0.65 = 2.80$$

This is the first equivalence point, a solution of 30.00 mL ($=10.00$ mL originally $+$ 20.00 mL titrant), containing 0.400 mmol $H_2PO_4^{-}$ from the titration and the 0.150 mmol $H_2PO_4^{2-}$ originally present. This is a solution with

$$\left[\text{H}_2\text{PO}_4^{-}\right] = \frac{(0.400+0.150) \text{ mmol H}_2\text{PO}_4^{-}}{30.00 \text{ mL}} = 0.0183 \text{ M , which has}$$

$$\text{pH} = \frac{1}{2}\left(pK_1 + pK_2\right) = 0.50\left(2.15 - \log\left(6.3\times10^{-8}\right)\right) = 0.50\left(2.15 + 7.20\right) = 4.68$$

To reach the second equivalence point means titrating 0.550 mmol $H_2PO_4^{-}$, which requires an additional volume of titrant given by

$$0.550 \text{ mmol H}_2\text{PO}_4^- \times \frac{1 \text{ mmol NaOH}}{1 \text{ mmol H}_2\text{PO}_4^-} \times \frac{1 \text{ mL NaOH}}{0.0200 \text{ mmol NaOH}} = 27.5 \text{ mL}.$$

To determine pH during this titration, we divide the region into five equal portions of 5.5 mL and use the Henderson-Hasselbalch equation.

At $(20.0 + 5.5)$ mL,

$$pH = pK_2 + \log\frac{\left[\text{HPO}_4^{2-}\right]}{\left[\text{H}_2\text{PO}_4^-\right]} = 7.20 + \log\frac{(0.20 \times 0.550) \text{ mmol HPO}_4^{2-} \text{ formed}}{(0.80 \times 0.550) \text{ mmol H}_2\text{PO}_4^- \text{ remaining}}$$

$$pH = 7.20 - 0.60 = 6.60$$

At $(20.0 + 11.0)$ mL $= 31.0$ mL, $pH = 7.20 + \log\dfrac{0.40 \times 0.550}{0.60 \times 0.550} = 7.02$

At 36.5 mL, $pH = 7.38$ At 42.0 mL, $pH = 7.80$

The pH at the second equivalence point is given by

$$pH = \frac{1}{2}\left(pK_2 + pK_3\right) = 0.50\left(7.20 - \log\left(4.2 \times 10^{-13}\right)\right) = 0.50 \ (7.20 + 12.38) = 9.79.$$

Another 27.50 mL of 0.020 M NaOH would be required to reach the third equivalence point. pH values at each of four equally spaced volumes of 5.50 mL additional 0.0200 M NaOH are computed as before, assuming the Henderson-Hasselbalch equation is valid.

At $(47.50 + 5.50)$ mL $= 53.00$ mL, $pH = pK_3 + \log\dfrac{\left[\text{PO}_4^{3-}\right]}{\left[\text{HPO}_4^{2-}\right]} = 12.38 + \log\dfrac{0.20 \times 0.550}{0.80 \times 0.550}$

$$= 12.38 - 0.60 = 11.78$$

At 58.50 mL, $pH = 12.20$ At 64.50 mL, $pH = 12.56$ At 70.00 mL, $pH = 12.98$
But at infinite dilution with 0.0200 M NaOH, the pH = 12.30, so this point can't be reached.

At the last equivalence point, the solution will contain 0.550 mmol PO_4^{3-} in a total of $10.00 + 20.00 + 27.50 + 27.50$ mL $= 85.00$ mL of solution, with

$$\left[\text{PO}_4^{3-}\right] = \frac{0.550 \text{ mmol}}{85.00 \text{ mL}} = 0.00647 \text{ M. But we can never reach this point, because the}$$

pH of the 0.0200 M NaOH titrant is 12.30. Moreover, the titrant is diluted by its addition to the solution. Thus, our titration will cease sometime shortly after the second equivalence point. We never will see the third equivalence point, largely because the titrant is too dilute. Our results are plotted on the next page.

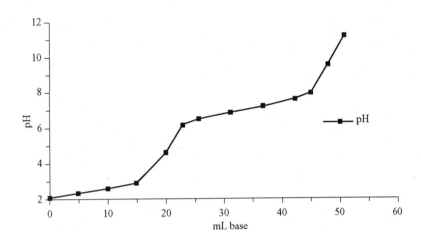

100. $pK_{a_1} = 2.34$; $K_{a_1} = 4.6 \times 10^{-3}$ and $pK_{a_2} = 9.69$; $K_{a_2} = 2.0 \times 10^{-10}$

(a)   Since the K_a values are so different, we can treat alanine (H_2A^+) as a monoprotic acid with $K_{a_1} = 4.6 \times 10^{-3}$. Hence:

	H_2A^+(aq)	+	H_2O(l)	$\rightleftharpoons$	HA(aq)	+	H_3O^+(aq)
Initial	0.500 M		-		0 M		≈ 0 M
Change	−x M		-		+x M		+x M
Equilibrium	(0.500 −x) M		-		x M		x M

$$K_{a_1} = \frac{[HA][H_3O^+]}{[H_2A^+]} = \frac{(x)(x)}{(0.500 - x)} = 4.6 \times 10^{-3} \approx \frac{x^2}{0.500}$$

$x = 0.048 \, M = [H_3O^+]$ ($x = 0.0457$ solving the quadratic equation)

$pH = -\log[H_3O^+] = -\log(0.046) = 1.34$

(b)   At the first half-neutralization point a buffer made up of H_2A^+/HA is formed, where $[H_2A^+] = [HA]$. The Henderson Hasselbalch equation gives $pH = pK_a = 2.34$

(c)   At the 1st equivalence point all of the H_2A^+(aq) is converted to HA(aq). HA(aq) is involved in both K_{a_1} and K_{a_2}, both ionizations must be considered.

   If we assume that the solution is converted to 100% HA, we must consider two reactions. HA may act as a weak acid (HA→A^- + H^+) or HA may act as a base (HA + H^+→H_2A^+). See Diagram on the next page:

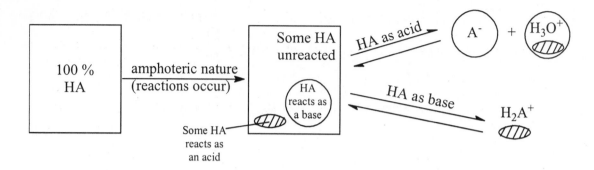

Using the diagram above, we see the following relations must hold true.

$[A^-] = [H_3O^+] + [H_2A^+]$

$$K_{a_2} = \frac{[A^-][H_3O^+]}{[HA]} \quad \text{or} \quad [A^-] = \frac{K_{a_2}[HA]}{[H_3O^+]} \quad \& \quad K_{a_1} = \frac{[HA][H_3O^+]}{[H_2A^+]} \quad \text{or} \quad [H_2A^+] = \frac{[H_3O^+][HA]}{K_{a_1}}$$

Substitute for $[A^-]$ and $[H_2A^+]$ in $[A^-] = [H_3O^+] + [H_2A^+]$

$$\frac{K_{a_2}[HA]}{[H_3O^+]} = [H_3O^+] + \frac{[H_3O^+][HA]}{K_{a_1}} \quad \text{(multiply both sides by } K_{a_1}[H_3O^+]\text{)}$$

$$K_{a_1}K_{a_2}[HA] = K_{a_1}[H_3O^+][H_3O^+] + [H_3O^+][H_3O^+][HA]$$

$$K_{a_1}K_{a_2}[HA] = [H_3O^+]^2(K_{a_1} + [HA])$$

$$[H_3O^+]^2 = \frac{K_{a_1}K_{a_2}[HA]}{(K_{a_1} + [HA])} \quad \text{Usually, } [HA] \gg K_{a_1} \quad \text{(Here, } 0.500 \gg 4.6 \times 10^{-3}\text{)}$$

Make the assumption that $K_{a_1} + [HA] \approx [HA]$

$$[H_3O^+]^2 = \frac{K_{a_1}K_{a_2}[HA]}{[HA]} = K_{a_1}K_{a_2} \qquad \text{Take } -\log \text{ of both sides}$$

$$-\log[H_3O^+]^2 = -2\log[H_3O^+] = 2(\text{pH}) = -\log K_{a_1}K_{a_2} = -\log K_{a_1} - \log K_{a_2} = pK_{a_1} + pK_{a_2}$$

$$2(\text{pH}) = pK_{a_1} + pK_{a_2}$$

$$\text{pH} = \frac{pK_{a_1} + pK_{a_2}}{2} = \frac{2.34 + 9.69}{2} = 6.02$$

(d) Half way between the 1st and 2nd equivalence point , half of HA(aq) is converted to A$^-$(aq). We have a HA/A$^-$ buffer solution where [HA] = [A$^-$]. The Henderson-Hasselbalch equation yields pH = pK_{a_2} = 9.69

(e) At the second equivalence point, all of the $H_2A^+(aq)$ is converted to $A^-(aq)$. We may treat this simply as a weak base in water having $K_b = \dfrac{K_w}{K_{a_2}} = \dfrac{1 \times 10^{-14}}{2.0 \times 10^{-10}} = 5.0 \times 10^{-5}$

Note: There has been a 1:3 dilution, hence the $[A^-] = 0.500 \text{ M} \times \dfrac{1 \text{ V}}{3 \text{ V}} = 0.167 \text{ M}$

	$A^-(aq)$	+	$H_2O(l)$	$\rightleftharpoons$	$HA(aq)$	+	$OH^-(aq)$
Initial	0.167 M		-		0 M		≈ 0 M
Change	$-x$ M		-		$+x$ M		$+x$ M
Equilibrium	$(0.167 - x)$ M		-		x M		x M

$$K_b = \frac{[HA][OH^-]}{[A^-]} = \frac{(x)(x)}{(0.167 - x)} = 5.0 \times 10^{-5} \approx \frac{x^2}{0.167}$$

$x = 0.0029 \text{ M} = [OH^-]; \quad pOH = -\log[OH^-] = 2.54;$

$pH = 14.00 - pOH = 14.00 - 2.54 = 11.46$

(f) All of the points required in (f) can be obtained using the Henderson-Hasselbalch equation (the chart below shows that the buffer ratio for each point is within the acceptable range (0.25 to 4.0))

mL NaOH →	0.0	10.0	20.0	30.0	40.0	50.0	60.0	70.0	80.0	90.0	100.0	110.0
%H_2A^+ →	100	80	60	40	20	0	0	0	0	0	0	0
%HA →	0	20	40	60	80	100	80	60	40	20	0	0
%A^- →	0	0	0	0	0	0	20	40	60	80	100	100
buffer ratio = base/acid →	0	0.25	0.67	1.5	4.0	∞	0.25	0.67	1.5	4.0	∞	∞

May use Henderson-Hasselbalch equation   May use Henderson-Hasselbalch equation

(i) After 10.0 mL

Here we will show how to obtain the answer using both the Henderson-Hasselbalch equation and setting up the I. C. E. (Initial, Change, Equilibrium) table. The results will differ within accepted experimental limitation of the experiment ($\pm$ 0.01 pH units)

$n_{H_2A^+} = (C \times V) = (0.500 \text{ M})(0.0500 \text{ L}) = 0.0250 \text{ moles } H_2A^-$

$n_{OH^-} = (C \times V) = (0.500 \text{ M})(0.0100 \text{ L}) = 0.00500 \text{ moles } OH^-$

$V_{total} = (50.0 + 10.0) \text{ mL} = 60.0 \text{ mL or } 0.0600 \text{ L}$

$[H_2A^+] = \dfrac{n_{H_2A^+}}{V_{total}} = \dfrac{0.0250 \text{ mol}}{0.0600 \text{ L}} = 0.417 \text{ M} \qquad [OH^-] = \dfrac{n_{OH^-}}{V_{total}} = \dfrac{0.00500 \text{ mol}}{0.0600 \text{ L}} = 0.0833 \text{ M}$

$$\text{K}_{eq} \text{ for titration reaction} = \frac{1}{K_{b(HA)}} = \frac{1}{\left(\dfrac{K_w}{K_{a_1}}\right)} = \frac{K_{a_1}}{K_w} = \frac{4.6\times10^{-3}}{1.00\times10^{-14}} = 4.6\times10^{11}$$

	H_2A^+(aq)	+	OH^-(aq)	$\rightleftharpoons$	HA(aq)	+	H_2O(l)
Initial:	0.417 M		0.0833 M		0 M		-
100% rxn:	-0.0833		-0.0833 M		+0.0833 M		-
New initial:	0.334 M		0 M		0.0833 M		-
Change:	+x M		+x M		−x M		-
Equilibrium:	≈0.334 M		x M		≈0.0833 M		-

$$4.6\times10^{11} = \frac{(0.0833)}{(0.334)(x)} \; ; \; x = \frac{(0.0833)}{(0.334)(4.6\times10^{11})} = 5.4\times10^{-13} \text{ (valid assumption)}$$

$x = 5.4\times10^{-13} = [OH^-]; \; pOH = -\log(5.4\times10^{-13}) = 12.27;$
$pH = 14.00 - pOH = 14.00 - 12.27 = 1.73$

Alternate method using the Henderson-Hasselbalch equation:

(i) After 10.0 mL, 20% of H_2A^+ reacts forming the conjugate base HA
Hence the buffer solution is 80% H_2A^+ (acid) and 20% HA (base)

$$pH = pK_{a_1} + \log\frac{base}{acid} = 2.34 + \log\frac{20.0}{80.0} = 2.34 + (-0.602) = 1.74 \text{ (within } \pm 0.01)$$

For the remainder of the calculations we will employ the Henderson-Hasselbalch equation with the understanding that using the method that employs the I.C.E. table gives the same result within the limitation of the data.

(ii) After 20.0 mL, 40% of H_2A^+ reacts forming the conjugate base HA
Hence the buffer solution is 60% H_2A^+ (acid) and 40% HA (base)

$$pH = pK_{a_1} + \log\frac{base}{acid} = 2.34 + \log\frac{40.0}{60.0} = 2.34 + (-0.176) = 2.16$$

(iii) After 30.0 mL, 60% of H_2A^+ reacts forming the conjugate base HA
Hence the buffer solution is 40% H_2A^+ (acid) and 60% HA (base)

$$pH = pK_{a_1} + \log\frac{base}{acid} = 2.34 + \log\frac{60.0}{40.0} = 2.34 + (+0.176) = 2.52$$

(iv) After 40.0 mL, 80% of H_2A^+ reacts forming the conjugate base HA
Hence the buffer solution is 20% H_2A^+ (acid) and 80% HA (base)

$$pH = pK_{a_1} + \log\frac{base}{acid} = 2.34 + \log\frac{80.0}{20.0} = 2.34 + (0.602) = 2.94$$

(v)   After 50 mL all of the $H_2A^+(aq)$ has reacted and we begin with essentially 100% HA(aq), which is a weak acid. Addition of base results in the formation of the conjugate base (buffer system) $A^-(aq)$. We employ a similar solution, however, now we must use $pK_{a_2} = 9.69$

(vi)   After 60.0 mL, 20% of HA reacts forming the conjugate base A^-
        Hence the buffer solution is 80% HA (acid) and 20% A^- (base)

$$pH = pK_{a_2} + \log \frac{base}{acid} = 9.69 + \log \frac{20.0}{80.0} = 9.69 + (-0.602) = 9.09$$

(vii)   After 70.0 mL, 40% of HA reacts forming the conjugate base A^-
        Hence the buffer solution is 60% HA (acid) and 40% A^- (base)

$$pH = p\,pK_{a_2} + \log \frac{base}{acid} = 9.69 + \log \frac{40.0}{60.0} = 9.69 + (-0.176) = 9.51$$

(viii) After 80.0 mL, 60% of HA reacts forming the conjugate base A^-
        Hence the buffer solution is 40% HA (acid) and 60% A^- (base)

$$pH = pK_{a_2} + \log \frac{base}{acid} = 9.69 + \log \frac{60.0}{40.0} = 9.69 + (+0.176) = 9.87$$

(ix)   After 90.0 mL, 80% of HA reacts forming the conjugate base A^-
        Hence the buffer solution is 20% HA (acid) and 80% A^- (base)

$$pH = pK_{a_2} + \log \frac{base}{acid} = 9.69 + \log \frac{80.0}{20.0} = 9.69 + (0.602) = 10.29$$

(x)   After the addition of 110.0 mL, NaOH is in excess. (10.0 mL of 0.500 M NaOH is in excess, or, 0.00500 moles of NaOH remains unreacted). The pH of a solution which has NaOH in excess is determined by the $[OH^-]$ that is in excess. (For a diprotic acid this occurs after the 2nd equivalence point).

$$[OH^-]_{excess} = \frac{n_{OH^-}}{V_{total}} = \frac{0.00500\,mol}{0.1600\,L} = 0.03125\,M \; ; \; pOH = -\log(0.03125) = 1.51$$

$$pH = 14.00 - pOH = 14.00 - 1.51 = 12.49$$

(g) A sketch of the titration curve for the 0.500 M solution of alanine hydrochloride, with some significant points labeled on the plot is shown on the next page.

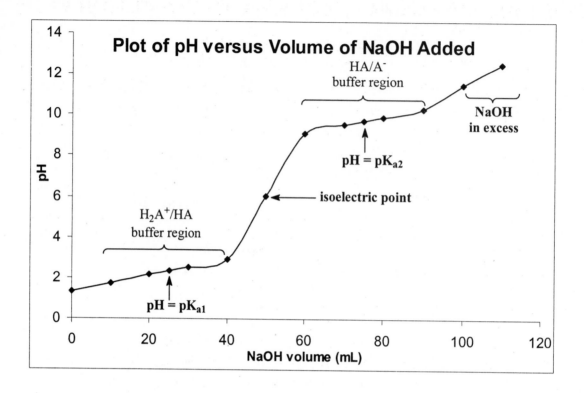

CHAPTER 19
SOLUBILITY AND COMPLEX-ION EQUILIBRIA
PRACTICE EXAMPLES

1A   In each case, we first write the balanced equation for the solubility equilibrium and then the equilibrium constant expression for that equilibrium, the K_{sp} expression:

    (a) $MgCO_3(s) \rightleftharpoons Mg^{2+}(aq) + CO_3^{2-}(aq)$ $K_{sp} = [Mg^{2+}][CO_3^{2-}]$

    (b) $Ag_3PO_4(s) \rightleftharpoons 3Ag^+(aq) + PO_4^{3-}(aq)$ $K_{sp} = [Ag^+]^3[PO_4^{3-}]$

1B   **(a)**   Provided $[OH^-]$ is not too high, the hydrogen phosphate ion is not expected to ionize in aqueous solution to a significant extent because of the quite small values for the second and third ionization constants of phosphoric acid.

          $CaHPO_4(s) \rightleftharpoons Ca^{2+}(aq) + HPO_4^{2-}(aq)$

    (b)   The ionization constant is written in the manner of a K_c expression:

        $K_{sp} = [Ca^{2+}][HPO_4^{2-}] = 1. \times 10^{-7}$

2A   We calculate the solubility of silver cyanate, s, as a molarity. We then use the solubility equation to (1) relate the concentrations of the ions and (2) write the K_{sp} expression.

$$s = \frac{7 \text{ mg AgOCN}}{100 \text{ mL}} \times \frac{1000 \text{ mL}}{1 \text{ L}} \times \frac{1 \text{ g}}{1000 \text{ mg}} \times \frac{1 \text{ mol AgOCN}}{149.9 \text{ g AgOCN}} = 5 \times 10^{-4} \text{ moles/L}$$

Equation : $AgOCN(s) \rightleftharpoons Ag^+(aq) + OCN^-(aq)$

Solubility Product : s s

$$K_{sp} = [Ag^+][OCN^-] = (s) \times (s) = s^2 = (5 \times 10^{-4})^2 = 3 \times 10^{-7}$$

2B   We calculate the solubility of lithium phosphate, s, as a molarity. We then use the solubility equation to (1) relate the concentrations of the ions and (2) write the K_{sp} expression.

$$s = \frac{0.034 \text{ g Li}_3PO_4}{100 \text{ mL soln}} \times \frac{1000 \text{ mL}}{1 \text{ L}} \times \frac{1 \text{ mol Li}_3PO_4}{115.79 \text{ g Li}_3PO_4} = 0.0029 \text{ moles/L}$$

Equation: $Li_3PO_4(s) \rightleftharpoons 3Li^+(aq) + PO_4^{3-}(aq)$

Solubility Product: $(3s)^3$ s

$$K_{sp} = [Li^+]^3[PO_4^{3-}] = (3s)^3 \cdot (s) = 27s^4 = 27(0.0029)^4 = 1.9 \times 10^{-9}$$

3A We use the solubility equilibrium to write the K_{sp} expression, which we then solve to obtain the molar solubility, s, of $Fe(OH)_3$.

$$Fe(OH)_3(s) \rightleftharpoons Fe^{3+}(aq) + 3OH^-(aq) \quad K_{sp} = [Fe^{3+}][OH^-]^3 = (s)(3s)^3 = 27s^4 = 4 \times 10^{-38}$$

Solubility: $s = \sqrt[4]{\dfrac{4 \times 10^{-38}}{27}} = 2 \times 10^{-10} M$

3B First we determine the solubility of $BaSO_4$, and then find the mass dissolved.

$$BaSO_4(aq) \rightleftharpoons Ba^{2+}(aq) + SO_4^{2-}(aq) \qquad\qquad K_{sp} = [Ba^{2+}][SO_4^{2-}] = s^2$$

The last relationship is true because $[Ba^{2+}] = [SO_4^{2-}]$ in a solution produced by dissolving $BaSO_4$ in pure water. Thus, $s = \sqrt{K_{sp}} = \sqrt{1.1 \times 10^{-10}} = 1.0_5 \times 10^{-5} M$

$$\text{mass } BaSO_4 = 225 \text{ mL} \times \frac{1.0_5 \times 10^{-5} \text{mmol } BaSO_4}{1 \text{ mL sat'd soln}} \times \frac{233.39 \text{ mg } BaSO_4}{1 \text{ mmol } BaSO_4} = 0.55 \text{ mg } BaSO_4$$

4A For PbI_2, $K_{sp} = [Pb^{2+}][I^-]^2 = 7.1 \times 10^{-9}$. The solubility equilibrium is the basis of the calculation.

Equation:	$PbI_2(s)$	$\rightleftharpoons$	$Pb^{2+}(aq)$	$+$	$2I^-(aq)$
Initial:			0.10 M		0 M
Changes:			$+s$ M		$+2s$ M
Equil:			$(0.10 + s)$ M		$2s$ M

$$K_{sp} = [Pb^{2+}][I^-]^2 = 7.1 \times 10^{-9} = (0.10 + s)(2s)^2 \approx 0.40 s^2 \qquad s = \sqrt{\frac{7.1 \times 10^{-9}}{0.40}} = 1.3 \times 10^{-4} M$$

(assumption $0.10 \gg s$ valid)

This value of s is the solubility of PbI_2 in $0.10 \text{ M } Pb(NO_3)_2(aq)$.

4B We find pOH from the given pH:

$$pOH = 14.00 - 8.20 = 5.80; \quad [OH^-] = 10^{-pOH} = 10^{-5.80} = 1.6 \times 10^{-6} M.$$

We assume that pOH remains constant, and use the K_{sp} expression for $Fe(OH_3)$.

$$K_{sp} = [Fe^{3+}][OH^-]^3 = 4 \times 10^{-38} = [Fe^{3+}](1.6 \times 10^{-6})^3 \quad [Fe^{3+}] = \frac{4 \times 10^{-38}}{(1.6 \times 10^{-6})^3} = 1 \times 10^{-20} M$$

Therefore, the molar solubility of $Fe(OH)_3$ is 1×10^{-20} M.

The dissolved $Fe(OH)_3$ does not significantly affect $[OH^-]$.

5A First determine $[I^-]$ as altered by dilution. We then compute Q_{sp} and compare it with K_{sp}.

$$[I^-] = \frac{3\,\text{drops} \times \dfrac{0.05\,\text{mL}}{1\,\text{drop}} \times \dfrac{0.20\,\text{mmol KI}}{1\,\text{mL}} \times \dfrac{1\,\text{mmol I}^-}{1\,\text{mmol KI}}}{100.0\,\text{mL soln}} = 3 \times 10^{-4}\,\text{M}$$

$$Q_{sp} = [Ag^+][I^-] = (0.010)(3 \times 10^{-4}) = 3 \times 10^{-6}$$

$$Q_{sp} > 8.5 \times 10^{-17} = K_{sp} \quad \text{Precipitation should occur.}$$

5B We first use the solubility product constant expression for PbI_2 to determine the $[I^-]$ needed in solution to just form a precipitate when $[Pb^{2+}] = 0.010$ M. We assume that the volume of solution added is small and that $[Pb^{2+}]$ remains at 0.010 M throughout.

$$K_{sp} = [Pb^{2+}][I^-]^2 = 7.1 \times 10^{-9} = (0.010)[I^-]^2 \qquad [I^-] = \sqrt{\frac{7.1 \times 10^{-9}}{0.010}} = 8.4 \times 10^{-4}\,\text{M}$$

We determine the volume of 0.20 M KI needed.

$$\text{volume of KI(aq)} = 100.0\,\text{mL} \times \frac{8.4 \times 10^{-4}\,\text{mmol I}^-}{1\,\text{mL}} \times \frac{1\,\text{mmol KI}}{1\,\text{mmol I}^-} \times \frac{1\,\text{mL KI(aq)}}{0.20\,\text{mmol KI}} \times \frac{1\,\text{drop}}{0.050\,\text{mL}}$$

$$= 8.4\,\text{drops} = 9\,\text{drops}$$

Since one additional drop is needed, 10 drops will be required. This is an insignificant volume compared to the original solution, so that $[Pb^{2+}]$ remains constant.

6A We find $[Ca^{2+}]$ that can coexist with $[OH^-] = 0.040$ M. $K_{sp} = 5.5 \times 10^{-6}$ for $Ca(OH)_2$.

$$K_{sp} = 5.5 \times 10^{-6} = [Ca^{2+}][OH^-]^2 = [Ca^{2+}](0.040)^2 ; \quad [Ca^{2+}] = \frac{5.5 \times 10^{-6}}{(0.040)^2} = 3.4 \times 10^{-3}\,\text{M}$$

For precipitation to be considered complete, $[Ca^{2+}]$ should be less than 0.1% of its original value. 3.4×10^{-3} M is 34% of 0.010 M and therefore precipitation of $Ca(OH)_2$ is not complete under these conditions.

6B We begin by finding $[Mg^{2+}]$ that corresponds to $1\,\mu g\ Mg^{2+}/L$.

$$[Mg^{2+}] = \frac{1\,\mu g\ Mg^{2+}}{1\,\text{L soln}} \times \frac{1\,g}{10^6\,\mu g} \times \frac{1\,\text{mol Mg}^{2+}}{24.3\,g\ Mg^{2+}} = 4 \times 10^{-8}\,\text{M}$$

Now we use the K_{sp} expression for $Mg(OH)_2$ to determine $[OH^-]$.

$$K_{sp} = 1.8 \times 10^{-11} = [Mg^{2+}][OH^-]^2 = (4 \times 10^{-8})[OH^-]^2 \qquad [OH^-] = \sqrt{\frac{1.8 \times 10^{-11}}{4 \times 10^{-8}}} = 0.02\,\text{M}$$

7A Let us first determine $\left[Ag^+\right]$ when AgCl(s) just begins to precipitate. At that point, Q_{sp} and K_{sp} are equal.

$$K_{sp} = 1.8 \times 10^{-10} = \left[Ag^+\right]\left[Cl^-\right] = Q_{sp} = \left[Ag^+\right] \times 0.115M \quad \left[Ag^+\right] = \frac{1.8 \times 10^{-10}}{0.115} = 1.6 \times 10^{-9}\,M$$

Now let us determine $\left[Br^-\right]$ that can coexist with this $\left[Ag^+\right]$.

$$K_{sp} = 5.0 \times 10^{-13} = \left[Ag^+\right]\left[Br^-\right] = 1.6 \times 10^{-9}\,M \times \left[Br^-\right]; \quad \left[Br^-\right] = \frac{5.0 \times 10^{-13}}{1.6 \times 10^{-9}} = 3.1 \times 10^{-4}\,M$$

The remaining bromide ion has precipitated as AgBr(s) with the addition of $AgNO_3(aq)$.

$$\text{Percent of } Br^- \text{ remaining} = \frac{[Br^-]_{final}}{[Br^-]_{initial}} \times 100\% = \frac{3.1 \times 10^{-4}\,M}{0.264\,M} \times 100\% = 0.12\%$$

7B Since the ions have the same charge and the same concentrations, we look for two K_{sp} values with the same anion that are as far apart as possible. The K_{sp} values for the carbonates are very close, while those for the sulfates and fluorides are quite different. However, the difference in the K_{sp} values is greatest for the chromates; K_{sp} for $BaCrO_4 \left(1.2 \times 10^{-10}\right)$ is so much smaller than K_{sp} for $SrCrO_4 \left(2.2 \times 10^{-5}\right)$, $BaCrO_4$ will precipitate first and $SrCrO_4$ will begin to precipitate when $\left[CrO_4^{2-}\right]$ has the value:

$$\left[CrO_4^{2-}\right] = \frac{K_{sp}}{\left[Sr^{2+}\right]} = \frac{2.2 \times 10^{-5}}{0.10} = 2.2 \times 10^{-4}\,M\,.$$

At this point $\left[Ba^{2+}\right]$ is found as follows.

$$\left[Ba^{2+}\right] = \frac{K_{sp}}{\left[CrO_4^{2-}\right]} = \frac{1.2 \times 10^{-10}}{2.2 \times 10^{-4}} = 5.5 \times 10^{-7}\,M\,;$$

$[Ba^{2+}]$ has dropped to 0.00055% of its initial value and therefore is considered to be completely precipitated, before $SrCrO_4$ begins to precipitate. The two ions are thus effectively separated as chromates. The best precipitating agent is a group 1 chromate salt.

8A First determine $\left[OH^-\right]$ resulting from the hydrolysis of acetate ion.

Equation:	$C_2H_3O_2^- (aq) + H_2O(l)$	$\rightleftharpoons$	$HC_2H_3O_2(aq)$	$+$	$OH^-(aq)$
Initial:	0.10 M		0 M		$\approx 0M$
Changes:	$-x$ M		$+x$ M		$+x$ M
Equil:	$(0.10 - x)$ M		x M		x M

$$K_b = \frac{1.0 \times 10^{-14}}{1.8 \times 10^{-5}} = 5.6 \times 10^{-10} = \frac{[HC_2H_3O_2][OH^-]}{[C_2H_3O_2^-]} = \frac{x \cdot x}{0.10-x} \approx \frac{x^2}{0.10}$$

$x = [OH^-] = \sqrt{0.10 \times 5.6 \times 10^{-10}} = 7.5 \times 10^{-6} M$ (the assumption $x \ll 0.10$ was valid)

Now compute the value of the ion product in this solution and compare it with the value of K_{sp} for $Mg(OH)_2$.

$$Q_{sp} = [Mg^{2+}][OH^-]^2 = (0.010M)(7.5 \times 10^{-6}M)^2 = 5.6 \times 10^{-13} < 1.8 \times 10^{-11} = K_{sp}[Mg(OH)_2]$$

Because Q_{sp} is smaller than K_{sp}, this solution is unsaturated and precipitation of $Mg(OH)_2(s)$ will not occur.

8B Use the Henderson–Hasselbalch equation to determine the pH of the buffer.

$$pH = pK_a + \log\frac{[C_2H_3O_2^-]}{[HC_2H_3O_2]} = -\log(1.8 \times 10^{-5}) + \log\frac{0.250M}{0.150M} = 4.74 + 0.22 = 4.96$$

$$pOH = 14.00 - pH = 14.00 - 4.96 = 9.04 \qquad [OH^-] = 10^{-pOH} = 10^{-9.04} = 9.1 \times 10^{-10} M$$

Now we determine Q_{sp} to see if precipitation will occur.

$$Q_{sp} = [Fe^{3+}][OH^-]^3 = (0.013M)(9.1 \times 10^{-10})^3 = 9.8 \times 10^{-30}$$

$$Q_{sp} > 4 \times 10^{-38} = K_{sp}; \ Fe(OH)_3 \text{ precipitation should occur.}$$

9A Determine $[OH^-]$, and then the pH necessary to prevent the precipitation of $Mn(OH)_2$.

$$K_{sp} = 1.9 \times 10^{-13} = [Mn^{2+}][OH^-]^2 = (0.0050M)[OH^-]^2$$

$$[OH^-] = \sqrt{\frac{1.9 \times 10^{-13}}{0.0050}} = 6.2 \times 10^{-6} M$$

$$pOH = -\log(6.2 \times 10^{-6}) = 5.21 \qquad pH = 14.00 - 5.21 = 8.79$$

We use this pH in the Henderson–Hasselbalch equation to determine $[NH_4^+]$.

$pK_b = 4.74$ for NH_3.

$$pH = pK_a + \log\frac{[NH_3]}{[NH_4^+]} = 8.79 = (14.00 - 4.74) + \log\frac{0.025M}{[NH_4^+]}$$

$$\log\frac{0.025M}{[NH_4^+]} = 8.79 - (14.00 - 4.74) = -0.47 \qquad \frac{0.025}{[NH_4^+]} = 10^{-0.47} = 0.34$$

$$[NH_4^+] = \frac{0.025}{0.34} = 0.074M$$

9B   First we must calculate the $[H_3O^+]$ in the buffer solution that is being employed to dissolve the magnesium hydroxide:

$$NH_3(aq) + H_2O(l) \rightleftharpoons NH_4^+(aq) + OH^-(aq) \; ; \; K_b = 1.8 \times 10^{-5}$$

$$K_b = \frac{[NH_4^+][OH^-]}{[NH_3]} = \frac{[0.100M][OH^-]}{[0.250M]} = 1.8 \times 10^{-5}$$

$$[OH^-] = 4.5 \times 10^{-5} \, M \; ; [H_3O^+] = \frac{1.00 \times 10^{-14} \, M^2}{4.5 \times 10^{-5} \, M} = 2.2_2 \times 10^{-10} \, M$$

Now we can employ equation 19.4 to calculate the molar solubility of $Mg(OH)_2$ in the buffer solution; Molar Solubility $Mg(OH)_2 = [Mg^{2+}]_{equil}$

$$Mg(OH)_2(s) + 2\,H_3O^+(aq) \xrightarrow{K = 1.8 \times 10^{17}} Mg^{2+}(aq) + 4H_2O(l) \; ;$$

Equilibrium   – 2.2×10^{-10} x –

$$K = \frac{[Mg^{2+}]}{[H_3O^+]^2} = \frac{x}{[2.2 \times 10^{-10}M]^2} = 1.8 \times 10^{17} \qquad x = 8.7 \times 10^{-3} \, M = [Mg^{2+}]_{equil}$$

So, the Molar Solubility for $Mg(OH)_2 = 8.7 \times 10^{-3}M$

10A (a)   In solution are $Cu^{2+}(aq)$, $SO_4^{\,2-}(aq)$, $Na^+(aq)$, and $OH^-(aq)$.

$$Cu^{2+}(aq) + 2\,OH^-(aq) \rightarrow Cu(OH)_2(s)$$

(b)   In solution above $Cu(OH)_2(s)$ is $Cu^{2+}(aq)$:

$$Cu(OH)_2(s) \rightleftharpoons Cu^{2+}(aq) + 2\,OH^-(aq)$$

This $Cu^{2+}(aq)$ reacts with the added $NH_3(aq)$:

$$Cu^{2+}(aq) + 4\,NH_3(aq) \rightleftharpoons \left[Cu(NH_3)_4\right]^{2+}(aq)$$

The overall result is:

$$Cu(OH)_2(s) + 4NH_3(aq) \rightleftharpoons \left[Cu(NH_3)_4\right]^{2+}(aq) + 2\,OH^-(aq)$$

(c) $HNO_3(aq)$ (strong acid), forms $H_3O^+(aq)$, which reacts with $OH^-(aq)$ and $NH_3(aq)$.

$$OH^-(aq) + H_3O^+(aq) \rightarrow 2\,H_2O(l); \qquad NH_3(aq) + H_3O^+(aq) \rightarrow NH_4^+(aq) + H_2O(l)$$

As $NH_3(aq)$ is consumed, this reaction shifts to the left.

$$Cu(OH)_2(s) + 4\,NH_3(aq) \rightleftharpoons \left[Cu(NH_3)_4\right]^{2+}(aq) + 2\,OH^-(aq)$$

But as $OH^-(aq)$ is consumed, this reaction shifts to the right:

$$Cu(OH)_2(s) \rightleftharpoons Cu^{2+}(aq) + 2\,OH^-(aq)$$

The species in solution at the end of all this are

$Cu^{2+}(aq)$, $NO_3^-(aq)$, $NH_4^+(aq)$, excess $H_3O^+(aq)$, $Na^+(aq)$; and $SO_4^{\,2-}(aq)$

(Probably $HSO_4^-(aq)$ as well)

10B **(a)**   In solution are $Zn^{2+}(aq)$, $SO_4^{2-}(aq)$, and $NH_3(aq)$,

$$Zn^{2+}(aq) + 4 NH_3(aq) \rightleftharpoons \left[Zn(NH_3)_4\right]^{2+}(aq)$$

(b) $HNO_3(aq)$, a strong acid, produces $H_3O^+(aq)$, which reacts with $NH_3(aq)$.

$NH_3(aq) + H_3O^+(aq) \rightarrow NH_4^+(aq) + H_2O(l)$ As $NH_3(aq)$ is consumed, the complex ion is destroyed.

$$\left[Zn(NH_3)_4\right]^{2+}(aq) + 4 H_3O^+(aq) \rightleftharpoons \left[Zn(H_2O)_4\right]^{2+}(aq) + 4 NH_4^+(aq)$$

(c)   NaOH(aq) is a strong base that produces $OH^-(aq)$, forming a hydroxide precipitate.

$$\left[Zn(H_2O)_4\right]^{2+}(aq) + 2 OH^-(aq) \rightleftharpoons Zn(OH)_2(s) + 4 H_2O(l)$$

Another possibility is reversing the effect of the reaction of part **(b)**.

$$\left[Zn(H_2O)_4\right]^{2+}(aq) + 4 NH_4^+(aq) + 4 OH^-(aq) \rightleftharpoons \left[Zn(NH_3)_4\right]^{2+}(aq) + 8 H_2O(l)$$

(d)   The precipitate dissolves in excess base.

$$Zn(OH)_2(s) + 2 OH^-(aq) \rightleftharpoons \left[Zn(OH)_4\right]^{2-}(aq)$$

11A   We first determine $\left[Ag^+\right]$ in a solution that is 0.100 M $Ag^+(aq)$ (from $AgNO_3$) and 0.225 M $NH_3(aq)$. Because of the large value of $K_f = 1.6 \times 10^7$, we start by having the reagents form as much complex ion as possible, and approach equilibrium from that point.

Equation:	$Ag^+(aq)$ +	$2 NH_3(aq)$	$\rightleftharpoons$	$\left[Ag(NH_3)_2\right]^+(aq)$
In soln:	0.100 M	0.225 M		0 M
Form complex:	−0.100 M	−0.200 M		+0.100 M
Initial	0 M	0.025 M		0.100 M
Changes	+x M	+2x M		−x M
Equil :	x M	(0.025 + x) M		(0.100 − x) M

$$K_f = 1.6 \times 10^7 = \frac{\left[Ag(NH_3)_2\right]^+}{\left[Ag^+\right]\left[NH_3\right]^2} = \frac{0.100 - x}{x(0.025 + 2x)^2} \approx \frac{0.100}{x(0.025)^2}$$

$$x = \frac{0.100}{(0.025)^2 \, 1.6 \times 10^7} = 1.0 \times 10^{-5} \, M = \left[Ag^+\right] = \text{concentration of free silver ion}$$

The $\left[Cl^-\right]$ is diluted: $\left[Cl^-\right]_{final} = \left[Cl^-\right]_{initial} \times \dfrac{1.00 \, mL_{initial}}{1,500 \, mL_{final}} = 3.50 \, M \div 1500 = 0.00233 \, M$

Finally we compare Q_{sp} with K_{sp} to determine if precipitation of AgCl(s) will occur.

$$Q_{sp} = \left[Ag^+\right]\left[Cl^-\right] = \left(1.0 \times 10^{-5} \, M\right)\left(0.00233 \, M\right) = 2.3 \times 10^{-8} > 1.8 \times 10^{-10} = K_{sp}$$

Because the value of the Q_{sp} is larger than the value of the K_{sp}, precipitation should occur.

11B We organize the solution around the balanced equation of the formation reaction.

Equation: $Pb^{2+}(aq)$ + $EDTA^{4-}(aq)$ $\rightleftharpoons$ $[PbEDTA]^{2-}(aq)$

Initial: 0.100 M   0.250 M   0 M
Form Complex: −0.100 M $(0.250 - 0.100)$ M   0.100 M
Equil : x M $(0.150 + x)$ M $(0.100 - x)$ M

$$K_f = \frac{\left[[PbEDTA]^{2-}\right]}{\left[Pb^{2+}\right]\left[EDTA^{4-}\right]} = 2 \times 10^{18} = \frac{0.100 - x}{x(0.150 + x)} \approx \frac{0.100}{0.150\, x}$$

$$x = \frac{0.100}{0.150 \times 2 \times 10^{18}} = 3 \times 10^{-19}\,M$$

We calculate Q_{sp} to determine if precipitation should occur.

$$Q_{sp} = \left[Pb^{2+}\right]\left[I^-\right]^2 = \left(3 \times 10^{-19}\,M\right)\left(0.10\,M\right)^2 = 3 \times 10^{-21}\ .$$

$Q_{sp} < 7.1 \times 10^{-9} = K_{sp}$ Precipitation will not occur.

12A We first determine the maximum concentration of free Ag^+.

$$K_{sp} = \left[Ag^+\right]\left[Cl^-\right] = 1.8 \times 10^{-10} \qquad \left[Ag^+\right] = \frac{1.8 \times 10^{-10}}{0.0075} = 2.4 \times 10^{-8}\,M.$$

This is so small that we assume that all the Ag^+ in solution is present as complex ion:

$\left[\left[Ag(NH_3)_2\right]^+\right] = 0.13$ M. We use K_f to determine the concentration of free NH_3.

$$K_f = \frac{\left[\left[Ag(NH_3)_2\right]^+\right]}{\left[Ag^+\right]\left[NH_3\right]^2} = 1.6 \times 10^7 = \frac{0.13M}{2.4 \times 10^{-8}\left[NH_3\right]^2}$$

$$\left[NH_3\right] = \sqrt{\frac{0.13}{2.4 \times 10^{-8} \times 1.6 \times 10^7}} = 0.58\,M.$$

If we combine this with the ammonia present in the complex ion, the total ammonia concentration is $0.58\,M + (2 \times 0.13\,M) = 0.84\,M$.

12B We use the solubility product constant expression to determine the maximum $\left[Ag^+\right]$ that is possible in 0.010 M Cl^- without precipitation occurring.

$$K_{sp} = 1.8 \times 10^{-10} = \left[Ag^+\right]\left[Cl^-\right] = \left[Ag^+\right](0.010\,M) \qquad \left[Ag^+\right] = \frac{1.8 \times 10^{-10}}{0.010} = 1.8 \times 10^{-8}\,M$$

This is also the concentration of free silver ion in the K_f expression. Because of the large value of K_f, practically all of the silver ion in solution is present as the complex ion, and

$[Ag(S_2O_3)_2]^{3-}]$. We solve the expression for $[S_2O_3^{2-}]$ and then add the $[S_2O_3^{2-}]$ "tied up" in the complex ion.

$$K_f = 1.7 \times 10^{13} = \frac{\left[\left[Ag(S_2O_3)_2\right]^{3-}\right]}{\left[Ag^+\right]\left[S_2O_3^{2-}\right]^2} = \frac{0.10 \text{ M}}{1.8 \times 10^{-8} \text{ M} \left[S_2O_3^{2-}\right]^2}$$

$$\left[S_2O_3^{2-}\right] = \sqrt{\frac{0.10}{1.8 \times 10^{-8} \times 1.7 \times 10^{13}}} = 5.7 \times 10^{-4} \text{ M} = \text{concentration of free } S_2O_3^{2-}$$

$$\text{total } \left[S_2O_3^{2-}\right] = 5.7 \times 10^{-4} \text{ M} + 0.10 \text{ M} \left[Ag(S_2O_3)_2\right]^{3-} \times \frac{2 \text{ mol } S_2O_3^{2-}}{1 \text{ mol} \left[Ag(S_2O_3)_2\right]^{3-}}$$

$$= 0.20 + 0.00057 \text{ M} = 0.20 \text{ M}$$

13A We combine the two equilibrium expressions, for K_f and for K_{sp}.

$$Fe(OH)_3(s) \rightleftharpoons Fe^{3+}(aq) + 3OH^-(aq) \qquad\qquad K_{sp} = 4 \times 10^{-38}$$

$$Fe^{3+}(aq) + 3C_2O_4^{2-}(aq) \rightleftharpoons \left[Fe(C_2O_4)_3\right]^{3-}(aq) \qquad K_f = 2 \times 10^{20}$$

$$Fe(OH)_3(s) + 3C_2O_4^{2-}(aq) \rightleftharpoons \left[Fe(C_2O_4)_3\right]^{3-}(aq) + 3OH^-(aq) \quad K_{overall} = 8 \times 10^{-18}$$

initial:	0.100 M	0 M	≈ 0 M
changes:	$-3x$ M	$+x$ M	$+3x$ M
equil:	$(0.100 - 3x)$ M	x M	$3x$ M

$$K_{overall} = \frac{\left[\left[Fe(C_2O_4)_3\right]^{3-}\right]\left[OH^-\right]^3}{\left[C_2O_4^{2-}\right]^3} = \frac{(x)(3x)^3}{(0.100 - 3x)^3} = 8 \times 10^{-18} \approx \frac{27x^4}{(0.100)^3} \qquad \text{assuming } 3x \ll 0.100$$

$$x = \sqrt[4]{\frac{(0.100)^3 8 \times 10^{-18}}{27}} = 4 \times 10^{-6} \text{ M The assumption is valid.}$$

The solubility of $Fe(OH)_3$ in 0.100 M $C_2O_4^{2-}$ is 4×10^{-6} M.

13B In Example 19-13 we saw that the expression for the solubility, s, of a silver halide in an aqueous ammonia solution, where $[NH_3]$ is the concentration of aqueous ammonia, is given by:

$$K_{sp} \times K_f = \left(\frac{s}{[NH_3] - 2s}\right)^2 \qquad or \qquad \sqrt{K_{sp} \times K_f} = \frac{s}{[NH_3] - 2s}$$

We are comparing situations of constant $[NH_3] = 0.100$ M and $K_f = 1.6 \times 10^7$.

We see that s will decrease as does K_{sp}. The values are:

$$K_{sp}(AgCl) = 1.8 \times 10^{-10}, \quad K_{sp}(AgBr) = 5.0 \times 10^{-13}, \quad K_{sp}(AgI) = 8.5 \times 10^{-17}.$$

14A For FeS, we know that $K_{spa} = 6 \times 10^2$; for Ag_2S, $K_{spa} = 6 \times 10^{-30}$.

We compute the value of Q_{spa} in each case, with $[H_2S] = 0.10\,M$ and $[H_3O^+] = 0.30\,M$.

For FeS, $Q_{spa} = \dfrac{0.020 \times 0.10}{(0.30)^2} = 0.022 < 6 \times 10^2 = K_{spa}$; Precipitation of FeS should not occur.

For Ag_2S, $Q_{spa} = \dfrac{(0.010)^2 \times 0.10}{(0.30)^2} = 1.1 \times 10^{-4}$

$Q_{spa} > 6 \times 10^{-30} = K_{spa}$; Precipitation of Ag_2S should occur.

14B The $[H_3O^+]$ needed to just form a precipitate can be obtained by direct substitution into the K_{spa} expression. When that expression is satisfied, a precipitate will just form.

$$K_{spa} = \frac{[Fe^{2+}][H_2S]}{[H_3O^+]^2} = 6 \times 10^2 = \frac{(0.015\,M\,Fe^{2+})(0.10\,M\,H_2S)}{[H_3O^+]^2}, \quad [H_3O^+] = \sqrt{\frac{0.015 \times 0.10}{6 \times 10^2}} = 0.002\,M$$

$$pH = -\log[H_3O^+] = -\log(0.002) = 2.7$$

REVIEW QUESTIONS

1. **(a)** K_{sp} is the symbol for the solubility product constant, representing the equilibrium constant for the dissolution of a sparingly soluble (ionic) compound in waer.

   (b) K_f is the symbol for the formation constant of a complex ion: the complex ion is formed (the product) from combination of a simple cation with the appropriate set of ligands.

   (c) Q_{sp}, or ion product, is reaction quotient for the dissolution of an ionic solid in water. The ion product has the same format as the solubility product expression, but it involves initial rather than equilibrium concentrations.

   (d) A complex ion is charged species consisting of a central metal ion attached to one or more groups of neutral molecules or ions called ligands.

2. **(a)** The common-ion effect refers to the phenomenon whereby less of a sparingly soluble ionic compound will dissolve in solution because of the presence of one of its constituent ions in solution.

   (b) Fractional precipitation refers to adding a precipitating agent (such as an anion) to a solution containing two other ions (such as cations) gradually so that only one compound precipitates.

   (c) Ion-pair formation refers to the result of the attraction between solvated ions of unlike charges in solution. These solvated ions cluster together and behave to some extent as one particle.

(d) Qualitative cation analysis is a scheme of alternating selective precipitation and dissolution, with the goal of isolating each type of cation into its own sample, where a definitive test for it can be performed.

3. **(a)** The solubility of a compound refers to the concentration of that compound in solution, either as a molarity or as a mass per unit volume. The solubility product constant is the equilibrium constant in terms of concentrations of ions, for the dissolution equilibrium, raised to their appropriate coefficients.

(b) The common-ion effect describes the lowering of the solubility of a compound in a solution due to the presence of one of the ions of that compound in the solution. The salt effect describes the enhancement of the solubility of a compound in a solution due to the presence of a different type of ion in solution.

(c) An ion pair is a close association of a cation and an anion in solution, whereas the ion product is the value obtained when the initial concentrations for the dissolved ions involved in the solubility equilibrium are inserted into the equilibrium constant expression.

4. **(a)** $Ag_2SO_4(s) \rightleftharpoons 2\,Ag^+(aq) + SO_4^{2-}(aq)$ $\qquad K_{sp} = \left[Ag^+\right]^2 \left[SO_4^{2-}\right]$

(b) $Ra(IO_3)_2(s) \rightleftharpoons Ra^{2+}(aq) + 2\,IO_3^-(aq)$ $\qquad K_{sp} = \left[Ra^{2+}\right]\left[IO_3^-\right]^2$

(c) $Ni_3(PO_4)_2(s) \rightleftharpoons 3\,Ni^{2+}(aq) + 2\,PO_4^{3-}(aq)$ $\quad K_{sp} = \left[Ni^{2+}\right]^3 \left[PO_4^{3-}\right]^2$

(d) $PuO_2CO_3(s) \rightleftharpoons PuO_2^{2+}(aq) + CO_3^{2-}(aq)$ $\qquad K_{sp} = \left[PuO_2^{2+}\right]\left[CO_3^{2-}\right]$

5. **(a)** $K_{sp} = \left[Fe^{3+}\right]\left[OH^-\right]^3$ $\qquad Fe(OH)_3(s) \rightleftharpoons Fe^{3+}(aq) + 3\,OH^-(aq)$

(b) $K_{sp} = \left[BiO^+\right]\left[OH^-\right]$ $\qquad BiOOH(s) \rightleftharpoons BiO^+(aq) + OH^-(aq)$

(c) $K_{sp} = \left[Hg_2^{2+}\right]\left[I^-\right]^2$ $\qquad Hg_2I_2(s) \rightleftharpoons Hg_2^{2+}(aq) + 2\,I^-(aq)$

(d) $K_{sp} = \left[Pb^{2+}\right]^3 \left[AsO_4^{3-}\right]^2$ $\qquad Pb_3(AsO_4)_2(s) \rightleftharpoons 3\,Pb^{2+}(aq) + 2\,AsO_4^{3-}(aq)$

6. **(a)** $CrF_3(s) \rightleftharpoons Cr^{3+}(aq) + 3\,F^-(aq)$ $\qquad K_{sp} = \left[Cr^{3+}\right]\left[F^-\right]^3 = 6.6 \times 10^{-11}$

(b) $Au_2(C_2O_4)_3(s) \rightleftharpoons 2\,Au^{3+}(aq) + 3\,C_2O_4^{2-}(aq)$ $\quad K_{sp} = \left[Au^{3+}\right]^2 \left[C_2O_4^{2-}\right]^3 = 1 \times 10^{-10}$

(c) $Cd_3(PO_4)_2(s) \rightleftharpoons 3\,Cd^{2+}(aq) + 2\,PO_4^{3-}(aq)$ $\quad K_{sp} = \left[Cd^{2+}\right]^3 \left[PO_4^{3-}\right]^2 = 2.1 \times 10^{-33}$

(d) $SrF_2(s) \rightleftharpoons Sr^{2+}(aq) + 2\,F^-(aq)$ $\qquad K_{sp} = \left[Sr^{2+}\right]\left[F^-\right]^2 = 2.5 \times 10^{-9}$

7.   Let $s = $ solubility of each compound in moles of compound per liter of solution.

(a) $K_{sp} = \left[Ba^{2+}\right]\left[CrO_4^{2-}\right] = (s)(s) = s^2 = 1.2 \times 10^{-10}$ $\qquad$ $s = 1.1 \times 10^{-5}\,M$

(b) $K_{sp} = \left[Pb^{2+}\right]\left[Br^-\right]^2 = (s)(2s)^2 = 4s^3 = 4.0 \times 10^{-5}$ $\qquad$ $s = 2.2 \times 10^{-2}\,M$

(c) $K_{sp} = \left[Ce^{3+}\right]\left[F^-\right]^3 = (s)(3s)^3 = 27s^4 = 8 \times 10^{-16}$ $\qquad$ $s = 7 \times 10^{-5}\,M$

(d) $K_{sp} = \left[Mg^{2+}\right]^3\left[AsO_4^{3-}\right]^2 = (3s)^3(2s)^2 = 108s^5 = 2.1 \times 10^{-20}$ $\qquad$ $s = 4.5 \times 10^{-5}\,M$

8.   Again, let $s = $ solubility of each compound in moles of solute per liter of solution.

(a) $K_{sp} = \left[Cs^+\right]\left[MnO_4^-\right] = (s)(s) = s^2 = \left(3.8 \times 10^{-3}\right)^2 = 1.4 \times 10^{-5}$

(b) $K_{sp} = \left[Pb^{2+}\right]\left[ClO_2^-\right]^2 = (s)(2s)^2 = 4s^3 = 4\left(2.8 \times 10^{-3}\right)^3 = 8.8 \times 10^{-8}$

(c) $K_{sp} = \left[In^{3+}\right]\left[IO_3^-\right]^3 = (s)(3s)^3 = 27s^4 = 27\left(1.0 \times 10^{-3}\right)^4 = 2.7 \times 10^{-11}$

9.   Statement (d) is correct. Consider the solubility equation: $PbI_2\,(s) \rightleftharpoons Pb^{2+}\,(aq) + 2I^-\,(aq)$. The stoichiometry of the dissolving reaction indicates that two I^- ions are formed for each Pb^{2+} ion, thus, $\left[Pb^{2+}\right] = 0.5\left[I^-\right]$. The relationship between K_{sp} and $\left[Pb^{2+}\right]$ is $\left[Pb^{2+}\right] = \sqrt[3]{K_{sp}/4}$.

10.   We let $s = $ molar solubility of $Mg(OH)_2$ in moles solute per liter of solution.

(a) $K_{sp} = \left[Mg^{2+}\right]\left[OH^-\right]^2 = (s)(2s)^2 = 4s^3 = 1.8 \times 10^{-11}$ $\qquad$ $s = 1.7 \times 10^{-4}\,M$

(b) Equation : $Mg(OH)_2\,(s) \;\rightleftharpoons\; Mg^{2+}\,(aq)$ $\quad + \quad$ $2OH^-\,(aq)$

Initial: 0.0862 M $\approx 0\,M$

Changes: $+s\,M$ $+2s\,M$

Equil : $(0.0862 + s)\,M$ $2s\,M$

$K_{sp} = (0.0862 + s)(2s)^2 = 1.8 \times 10^{-11} \approx (0.0862)(2s)^2 = 0.34\,s^2$ $\qquad$ $s = 7.3 \times 10^{-6}\,M$

(c) $\left[OH^-\right] = \left[KOH\right] = 0.0355\,M$

Equation : $Mg(OH)_2\,(s) \;\rightleftharpoons\; Mg^{2+}\,(aq)$ $\quad + \quad$ $2OH^-\,(aq)$

Initial: 0 M   0.0355 M

Changes : $+s\,M$ $+2s\,M$

Equil: $s\,M$ $(0.0355 + 2s)\,M$

$K_{sp} = (s)(0.0355 + 2s)^2 = 1.8 \times 10^{-11} \approx (s)(0.0355)^2 = 0.0013\,s$ $\qquad$ $s = 1.4 \times 10^{-8}\,M$

11. The solubility equilibrium is $CaCO_3(s) \rightleftharpoons Ca^{2+}(aq) + CO_3^{2-}(aq)$

(a) The addition of $Na_2CO_3(aq)$ produces $CO_3^{2-}(aq)$ in solution. This common ion suppresses the solubility of $CaCO_3(s)$.

(b) HCl(aq) is a strong acid that reacts with carbonate ion:
$CO_3^{2-}(aq) + 2H_3O^+(aq) \rightarrow CO_2(g) + 3H_2O(l)$. This decreases $\left[CO_3^{2-}\right]$ in the solution, allowing more $CaCO_3(s)$ to dissolve.

(c) $HSO_4^-(aq)$ is a moderately weak acid. It is strong enough to protonate carbonate ion, decreasing $\left[CO_3^{2-}\right]$ and enhancing the solubility of $CaCO_3(s)$, as the value of K_c indicates.

$HSO_4^-(aq) + CO_3^{2-}(aq) \rightleftharpoons SO_4^{2-}(aq) + HCO_3^-(aq)$

$$K_c = \frac{K_a(HSO_4^-)}{K_a(HCO_3^-)} = \frac{0.011}{4.7 \times 10^{-11}} = 2.3 \times 10^8$$

12. In each case, compute Q_{sp} and compare its value with the value of K_{sp}. If $Q_{sp} > K_{sp}$, a precipitate should form.

(a) $Q_{sp} = \left[Mg^{2+}\right]\left[CO_3^{2-}\right] = (0.0037)(0.0068) = 2.5 \times 10^{-5} > 3.5 \times 10^{-8} = K_{sp}$
Precipitation should occur.

(b) $Q_{sp} = \left[Ag^+\right]^2\left[SO_4^{2-}\right] = (0.018)^2(0.0062) = 2.0 \times 10^{-6} < 1.4 \times 10^{-5} = K_{sp}$
Precipitation is not expected to occur.

(c) $pOH = 14.00 - 3.20 = 10.80$ $\left[OH^-\right] = 10^{-10.80} = 1.6 \times 10^{-11} M$

$Q_{sp} = \left[Cr^{3+}\right]\left[OH^-\right]^3 = (0.038)(1.6 \times 10^{-11})^3 = 1.6 \times 10^{-34} < 6.3 \times 10^{-31} = K_{sp}$
Precipitation is not expected to occur.

13. We use the K_{sp} expression to determine $\left[Ca^{2+}\right]$ that can coexist with $\left[SO_4^{2-}\right] = 0.750 M$.

$$K_{sp} = 9.1 \times 10^{-6} = \left[Ca^{2+}\right]\left[SO_4^{2-}\right] = \left[Ca^{2+}\right](0.750 M); \quad \left[Ca^{2+}\right] = \frac{9.1 \times 10^{-6}}{0.750 M} = 1.2 \times 10^{-5} M$$

% unprecipitated $= \dfrac{1.2 \times 10^{-5} M}{0.103 M} \times 100\% = 0.012\%$ unprecipitated which is essentially

complete precipitation. Of course, $\left[SO_4^{2-}\right]$ actually does decrease because some is

consumed in the precipitation reaction. One mole of sulfate ion is removed from solution for each mole of calcium ion that precipitates, which amounts to 0.103 moles from each liter of solution. So, in this case $\left[SO_4^{2-}\right] = 0.750 M - 0.103 M = 0.647 M$. But this increases

the final $\left[Ca^{2+}\right]$ only very slightly, to $1.4 \times 10^{-5} M$, or 0.014% unprecipitated.

14. **(a)** Determine $[I^-]$ when AgI just begins to precipitate, and $[I^-]$ when PbI_2 just begins to precipitate.

$$K_{sp} = [Ag^+][I^-] = 8.5 \times 10^{-17} = (0.10)[I^-] \qquad [I^-] = 8.5 \times 10^{-16}\,M$$

$$K_{sp} = [Pb^{2+}][I^-]^2 = 7.1 \times 10^{-9} = (0.10)[I^-]^2 \qquad [I^-] = \sqrt{\dfrac{7.1 \times 10^{-9}}{0.10}} = 2.7 \times 10^{-4}\,M$$

Since $8.5 \times 10^{-16}\,M$ is less than 2.7×10^{-4} M, AgI will precipitate before PbI_2.

(b) $[I^-] = 2.7 \times 10^{-4}$ M before the second cation, Pb^{2+}, begins to precipitate.

(c) $K_{sp} = [Ag^+][I^-] = 8.5 \times 10^{-17} = [Ag^+](2.7 \times 10^{-4}) \qquad [Ag^+] = 3.1 \times 10^{-13}\,M$

(d) Since $[Ag^+]$ has decreased to much less than 0.1% of its initial value before PbI_2 begins to precipitate, we conclude that Ag^+ and Pb^{2+} can be separated by precipitation with iodide ion.

15. $Mg(OH)_2(s)$ will be the most soluble in a solution of $NH_4Cl(aq)$. NH_4Cl will form an acidic solution, via the reaction $NH_4^+(aq) + H_2O(l) \rightleftharpoons H_3O^+(aq) + NH_3(aq)$ and this H_3O^+ will react with OH^- ion: $OH^-(aq) + H_3O^+(aq) \rightleftharpoons 2H_2O(l)$ causing the solubility equilibrium to shift right. $Mg(OH)_2(s) \rightleftharpoons Mg^{2+}(aq) + 2OH^-(aq)$. Since NaOH has an ion in common with $Mg(OH)_2$, its use will actually decrease the solubility of $Mg(OH)_2$. Addition of Na_2CO_3 will also decrease $Mg(OH)_2$ solubility, through hydrolysis of the carbonate ion: $H_2O + CO_3^{2-} \rightleftharpoons HCO_3^- + OH^-$ followed by the common ion effect of OH^-.

16. **(a)** $Ag^+(aq) + NO_3^-(aq) + Na^+(aq) + Br^-(aq) \longrightarrow AgBr(s) + Na^+(aq) + NO_3^-(aq)$

(b) $Cu^{2+}(aq) + NO_3^-(aq) + H_3O^+(aq) + Cl^-(aq) \longrightarrow$ no reaction

(c) $Fe^{2+}(aq) + H_2S$ (aq, in 0.3 M HCl) $\longrightarrow$ no reaction

(d) $Cu(OH)_2(s) + 4NH_3(aq) \longrightarrow [Cu(NH_3)_4]^{2+}(aq) + 2OH^-(aq)$

(e) $Fe^{3+}(aq) + 3NH_3(aq) + 3H_2O(l) \longrightarrow Fe(OH)_3(s) + 3NH_4^+(aq)$

(f) $Ag_2SO_4(s) + 4NH_3(aq) \longrightarrow 2[Ag(NH_3)_2]^+(aq) + SO_4^{2-}(aq)$

(g) $CaSO_3(s) + 2H_3O^+(aq) \longrightarrow Ca^{2+}(aq) + 3H_2O(l) + SO_2(g)$

17. $Cu(OH)_2$ dissolves readily in acidic solutions [HCl(aq) and HNO_3 (aq)] and ammoniacal solutions $[NH_3(aq)]$. The net ionic equations for these two reactions are, respectively:

$$Cu(OH)_2(s) + 2H_3O^+(aq) \longrightarrow Cu^{2+}(aq) + 4H_2O(l)$$

$$Cu(OH)_2(s) + 4NH_3(aq) \longrightarrow [Cu(NH_3)_4]^{2+}(aq) + 2OH^-(aq)$$

18. The best choice is HCl(aq). AgCl is insoluble in water, while $CuCl_2$ is water soluble. On the other hand, both Cu^{2+} and Ag^+ form insoluble hydroxides and carbonates, so neither NaOH(aq) nor $(NH_4)_2CO_3(aq)$ would be capable of separating these two cations.

19. Both Cu^{2+} and Mg^{2+} form hydroxides, although $Cu(OH)_2 (K_{sp} = 2.2 \times 10^{-20})$ is much less soluble than $Mg(OH)_2 (K_{sp} = 1.8 \times 10^{-11})$. Thus, NaOH(aq) might be one choice to produce a precipitate of $Cu(OH)_2(s)$ while leaving $Mg^{2+}(aq)$ in solution. A better choice, however, seems to be $NH_3(aq)$ in which $[Cu(NH_3)_4]^{2+}$ forms, $(K_f = 1.1 \times 10^{13})$, while simultaneously creating an alkaline solution from which $Mg(OH)_2(s)$ precipitates.

20. First we determine $[OH^-]$ in this buffer solution.

$$pH = pK_a + \log \frac{[C_2H_3O_2^-]}{[HC_2H_3O_2]} = 4.74 + \log \frac{0.35\,M}{0.45\,M} = 4.63 \qquad pOH = 14.00 - 4.63 = 9.37$$

$[OH^-] = 10^{-9.37} = 4.3 \times 10^{-10}\,M$. Now, we compute the value of Q_{sp} for $Al(OH)_3$.

$$Q_{sp} = [Al^{3+}][OH^-]^3 = (0.275)(4.3 \times 10^{-10})^3 = 2.2 \times 10^{-29}$$

$Q_{sp} > 1.3 \times 10^{-33} = K_{sp}$ Precipitation should occur from this solution.

21. We first find the concentration of free metal ion. Then we determine the value of Q_{sp} for the precipitation reaction, and compare that value with the value of K_{sp} to determine whether precipitation will occur.

Equation:	$Ag^+(aq)$	+	$2CN^-(aq)$	$\rightleftharpoons$	$[Ag(CN)_2]^-(aq)$.
Initial:	0 M		1.05 M		0.012 M
Changes:	$+x\,M$		$+2x\,M$		$-x\,M$
Equil:	$x\,M$		$(1.05 + 2x)\,M$		$(0.012 - x)\,M$

$$K_f = \frac{[[Ag(CN)_2]^-]}{[Ag^+][CN^-]^2} = 5.6 \times 10^{18} = \frac{0.012 - x}{x(1.05 + 2x)^2} \approx \frac{0.012}{(1.05)^2 x} \qquad x = 1.9 \times 10^{-21}\,M = [Ag^+]$$

$$Q_{sp} = [Ag^+][I^-] = (1.9 \times 10^{-21})(2.0) = 3.8 \times 10^{-21}.$$

$Q_{sp} < 8.5 \times 10^{-17} = K_{sp}$ Precipitation should not occur.

22. We use the K_f expression to determine the concentration of free silver ion, $\left[Ag^+\right]$.

$$K_f = 1.6\times10^7 = \frac{[[Ag(NH_3)_2]^+]}{[Ag^+][NH_3]^2} = \frac{1.6\,M}{[Ag^+](1.25\,M)^2}; \quad [Ag^+] = \frac{1.6}{1.6\times10^7(1.25)^2} = 6.4\times10^{-8}\ M$$

The K_{sp} expression is used to find the maximum $\left[Cl^-\right]$ that can coexist with this $\left[Ag^+\right]$.

$$K_{sp} = 1.8\times10^{-10} = \left[Ag^+\right]\left[Cl^-\right] = \left(6.4\times10^{-8}\,M\right)\left[Cl^-\right]; \quad \left[Cl^-\right] = \frac{1.8\times10^{-10}}{6.4\times10^{-8}} = 2.8\times10^{-3}\,M.$$

EXERCISES

K_{sp} and Solubility

23. We use the value of K_{sp} for each compound in determining the molar solubility in a saturated solution. In each case, s represents the molar solubility of the compound.

AgCN $\quad K_{sp} = \left[Ag^+\right]\left[CN^-\right] \quad = (s)(s) = s^2 = 1.2\times10^{-16} \quad s = 1.1\times10^{-8}\,M$

AgIO$_3$ $\quad K_{sp} = \left[Ag^+\right]\left[IO_3^-\right] \quad = (s)(s) = s^2 = 3.0\times10^{-8} \quad s = 1.7\times10^{-4}\,M$

AgI $\quad K_{sp} = \left[Ag^+\right]\left[I^-\right] \quad = (s)(s) = s^2 = 8.5\times10^{-17} \quad s = 9.2\times10^{-9}\,M$

AgNO$_2$ $\quad K_{sp} = \left[Ag^+\right]\left[NO_2^-\right] \quad = (s)(s) = s^2 = 6.0\times10^{-4} \quad s = 2.4\times10^{-2}\,M$

Ag$_2$SO$_4$ $\quad K_{sp} = \left[Ag^+\right]^2\left[SO_4^{2-}\right] \quad = (2s)^2(s) = 4s^3 = 1.4\times10^{-5} \quad s = 1.5\times10^{-2}\,M$

In order of increasing molar solubility, smallest to largest:

$AgI < AgCN < AgIO_3 < Ag_2SO_4 < AgNO_2$

24. We use the value of K_{sp} for each compound in determining $\left[Mg^{2+}\right]$ in its saturated solution. In each case, s represents the molar solubility of the compound.

(a) MgCO$_3$ $\qquad K_{sp} = \left[Mg^{2+}\right]\left[CO_3^{2-}\right] = (s)(s) = s^2 = 3.5\times10^{-8}$

$\qquad\qquad\qquad s = 1.9\times10^{-4}\,M \qquad \left[Mg^{2+}\right] = 1.9\times10^{-4}\,M$

(b) MgF$_2$ $\qquad K_{sp} = \left[Mg^{2+}\right]\left[F^-\right]^2 = (s)(2s)^2 = 4s^3 = 3.7\times10^{-8}$

$\qquad\qquad\qquad s = 2.1\times10^{-3}\,M \qquad \left[Mg^{2+}\right] = 2.1\times10^{-3}\,M$

(c) Mg$_3$(PO$_4$)$_2$ $\qquad K_{sp} = \left[Mg^{2+}\right]^3\left[PO_4^{3-}\right]^2 = (3s)^3(2s)^2 = 108s^5 = 2.1\times10^{-25}$

$\qquad\qquad\qquad s = 5\times10^{-6}\,M \qquad \left[Mg^{2+}\right] = 1\times10^{-5}\ M$

A saturated solution of MgF$_2$ has the highest $\left[Mg^{2+}\right]$.

25. We determine $[F^-]$ in saturated CaF_2, and from that value the concentration of F^- in ppm.

For CaF_2 $\quad K_{sp} = [Ca^{2+}][F^-]^2 = (s)(2s)^2 = 4s^3 = 5.3 \times 10^{-9}$ $\quad\quad s = 1.1 \times 10^{-3} M$

The solubility in ppm is the number of grams of CaF_2 in 10^6 g of solution. We assume a solution density of $1.00 \, g/mL$.

$$\text{mass of } F^- = 10^6 \, g \, soln \times \frac{1 \, mL}{1.00 \, g \, soln} \times \frac{1 \, L \, soln}{1000 \, mL} \times \frac{1.1 \times 10^{-3} \, mol \, CaF_2}{1 \, L \, soln}$$

$$\times \frac{2 \, mol \, F^-}{1 \, mol \, CaF_2} \times \frac{19.0 \, g \, F^-}{1 \, mol \, F^-} = 42 \, g \, F^-$$

This is 42 times more concentrated than the optimum concentration of fluoride ion. CaF_2 is, in fact, more soluble than is necessary. Its uncontrolled use might lead to excessive F^- in solution.

26. We determine $[OH^-]$ in a saturated solution. From this $[OH^-]$ we determine pH.

$$K_{sp} = [BiO^+][OH^-] = 4 \times 10^{-10} = s^2 \quad\quad\quad s = 2 \times 10^{-5} M = [OH^-] = [BiO^+]$$

$$pOH = -\log(2 \times 10^{-5}) = 4.7 \quad\quad\quad\quad pH = 9.3$$

27. We first assume that the volume of the solution does not change appreciably when its temperature is lowered. Then we determine the mass of $Mg(C_{16}H_{31}O_2)_2$ dissolved in each solution, recognizing that the molar solubility of $Mg(C_{16}H_{31}O_2)_2$ equals the cube root of one fourth of its solubility product constant, since it is the only solute in the solution.

$$K_{sp} = 4s^3 \quad\quad\quad\quad s = \sqrt[3]{K_{sp}/4}$$

At $50°C$: $s = \sqrt[3]{4.8 \times 10^{-12}/4} = 1.1 \times 10^{-4}$ M; At $25°C$: $s = \sqrt[3]{3.3 \times 10^{-12}/4} = 9.4 \times 10^{-5}$ M

$$\text{amount of } Mg(C_{16}H_{31}O_2)_2 \, (50°C) = 0.965 \, L \times \frac{1.1 \times 10^{-4} \, mol \, Mg(C_{16}H_{31}O_2)_2}{1 \, L \, soln} = 1.1 \times 10^{-4} \, mol$$

$$\text{amount of } Mg(C_{16}H_{31}O_2)_2 \, (25°C) = 0.965 \, L \times \frac{9.4 \times 10^{-5} \, mol \, Mg(C_{16}H_{31}O_2)_2}{1 \, L \, soln} = 0.91 \times 10^{-4} \, mol$$

mass of $Mg(C_{16}H_{31}O_2)_2$ precipitated:

$$= (1.1 - 0.91) \times 10^{-4} \, mol \times \frac{535.15 \, g \, Mg(C_{16}H_{31}O_2)_2}{1 \, mol \, Mg(C_{16}H_{31}O_2)_2} \times \frac{1000 \, mg}{1 \, g} = 11 \, mg$$

28. We first assume that the volume of the solution does not change appreciably when its temperature is lowered. Then we determine the mass of CaC_2O_4 dissolved in each solution, recognizing that the molar solubility of CaC_2O_4 equals the square root of its solubility product constant, since it is the only solute in the solution.

At 95 °C $s = \sqrt{1.2 \times 10^{-8}} = 1.1 \times 10^{-4}\,M$; At 13 °C: $s = \sqrt{2.7 \times 10^{-9}} = 5.2 \times 10^{-5}\,M$

mass of CaC_2O_4 (95 °C) $= 0.725\,L \times \dfrac{1.1 \times 10^{-4}\,mol\ CaC_2O_4}{1\,L\ soln} \times \dfrac{128.1\,g\ CaC_2O_4}{1\,mol\ CaC_2O_4} = 0.010\,g\ CaC_2O_4$

mass of CaC_2O_4 (13 °C) $= 0.725\,L \times \dfrac{5.2 \times 10^{-5}\,mol\ PbSO_4}{1\,L\ soln} \times \dfrac{128.1\,g\ CaC_2O_4}{1\,mol\ CaC_2O_4} = 0.0048\,g\ CaC_2O_4$

mass of CaC_2O_4 precipitated $= (0.010\ g - 0.0048\ g) \times \dfrac{1000\ mg}{1g} = 5\,mg$

29. First we determine $\left[I^-\right]$ in the saturated solution.

$K_{sp} = \left[Pb^{2+}\right]\left[I^-\right]^2 = 7.1 \times 10^{-9} = (s)(2s)^2 = 4s^3$ $\qquad s = 1.2 \times 10^{-3}\,M$

The $AgNO_3$ reacts with the I^- in this saturated solution in the titration.
$Ag^+(aq) + I^-(aq) \rightarrow AgI(s)$ We determine the amount of Ag^+ needed for this titration, and then $\left[AgNO_3\right]$ in the titrant.

moles $Ag^+ = 0.02500\,L \times \dfrac{1.2 \times 10^{-3}\,mol\ PbI_2}{1\,L\ soln} \times \dfrac{2\,mol\ I^-}{1\,mol\ PbI_2} \times \dfrac{1\,mol\ Ag^+}{1\,mol\ I^-} = 6.0 \times 10^{-5}\,mol\ Ag^+$

$AgNO_3$ molarity $= \dfrac{6.0 \times 10^{-5}\,mol\ Ag^+}{0.0133\,L\ soln} \times \dfrac{1\,mol\ AgNO_3}{1\,mol\ Ag^+} = 4.5 \times 10^{-3}\,M$

30. We determine $\left[C_2O_4^{2-}\right] = s$, the solubility of the saturated solution.

$\left[C_2O_4^{2-}\right] = \dfrac{4.8\,mL \times \dfrac{0.00134\,mmol\ KMnO_4}{1\,mL\ soln} \times \dfrac{5\,mmol\ C_2O_4^{2-}}{2\,mmol\ MnO_4^-}}{250.0\,mL} = 6.4 \times 10^{-5}\,M = s = \left[Ca^{2+}\right]$

$K_{sp} = \left[Ca^{2+}\right]\left[C_2O_4^{2-}\right] = (s)(s) = s^2 = \left(6.4 \times 10^{-5}\right)^2 = 4.1 \times 10^{-9}$

31. We use the ideal gas law to determine the moles of H_2S gas used.

$n = \dfrac{PV}{RT} = \dfrac{\left(748\ mmHg \times \dfrac{1\,atm}{760\ mmHg}\right) \times \left(30.4\ mL \times \dfrac{1\,L}{1000\ mL}\right)}{0.08206\ L\ atm\ mol^{-1}\ K^{-1} \times (23 + 273)\,K} = 1.23 \times 10^{-3}\ moles$

If we assume that all the H_2S is consumed in forming Ag_2S, we can compute the $\left[Ag^+\right]$ in the $AgBrO_3$ solution. This assumption is valid if the equilibrium constant for the cited reaction is large, which is the case, as shown below:

$$2Ag^+(aq)+HS^-(aq)+OH^-(aq)\rightleftharpoons Ag_2S(s)+H_2O(l) \quad K_{a_2}/K_{sp}=\frac{1.0\times10^{-19}}{2.6\times10^{-51}}=3.8\times10^{31}$$

$$H_2S(aq)+H_2O(l)\rightleftharpoons HS^-(aq)+H_3O^+(aq) \qquad\qquad K_1=1.0\times10^{-7}$$

$$2H_2O(l)\rightleftharpoons H_3O^+(aq)+OH^-(aq) \qquad\qquad\qquad K_w=1.0\times10^{-14}$$

$$\overline{2Ag^+(aq)+H_2S(aq)+2H_2O(l)\rightleftharpoons Ag_2S(s)+2H_3O^+(aq)}$$

$$K_{overall}=(K_{a_2}/K_{sp})(K_1)(K_w)=(3.8\times10^{31})(1.0\times10^{-7})(1.0\times10^{-14})=3.8\times10^{10}$$

$$\left[Ag^+\right]=\frac{1.23\times10^{-3}\text{ mol }H_2S}{338\text{ mL soln}}\times\frac{1000\text{ mL}}{1\text{ L soln}}\times\frac{2\text{ mol }Ag^+}{1\text{ mol }H_2S}=7.28\times10^{-3}\text{ M}$$

Then, for $AgBrO_3$, $K_{sp}=\left[Ag^+\right]\left[BrO_3^-\right]=\left(7.28\times10^{-3}\right)^2=5.30\times10^{-5}$

32. The titration reaction is $Ca(OH)_2(aq)+2HCl(aq)\longrightarrow CaCl_2(aq)+2H_2O(l)$

$$\left[OH^-\right]=\frac{10.7\text{ mL HCl}\times\dfrac{1\text{ L}}{1000\text{ mL}}\times\dfrac{0.1032\text{ mol HCl}}{1\text{ L}}\times\dfrac{1\text{ mol }OH^-}{1\text{ mol HCl}}}{50.00\text{ mL }Ca(OH)_2\text{ soln}\times\dfrac{1\text{ L}}{1000\text{ mL}}}=0.0221\text{ M}$$

In a saturated solution of $Ca(OH)_2$, $\left[Ca^{2+}\right]=\left[OH^-\right]\div2$

$K_{sp}=\left[Ca^{2+}\right]\left[OH^-\right]^2=(0.0221\div2)(0.0221)^2=5.40\times10^{-6}$

Compare with 5.5×10^{-6} in Appendix D.

The Common-Ion Effect

<u>33.</u> The presence of KI in a solution produces a significant $\left[I^-\right]$ in the solution. Not as much AgI can dissolve in such a solution as in pure water, since the ion product, $\left[Ag^+\right]\left[I^-\right]$, cannot exceed the value of K_{sp} (i.e. the I^- from the KI that dissolves represses the dissociation of AgI(s). In similar fashion, $AgNO_3$ produces a significant $\left[Ag^+\right]$ in solution, again influencing the value of the ion product; not as much AgI can dissolve as in pure water.

34. If the solution contains KNO_3, more AgI will end up dissolving than in pure water, because the activity of each ion will be less than its molarity. On an ionic level, the reason is that ion pairs, such as $Ag^+NO_3^-(aq)$ and $K^+I^-(aq)$, form in the solution, preventing Ag^+ and I^- ions from coming together and precipitating.

35. Equation:

$$Ag_2SO_4(s) \rightleftharpoons 2Ag^+(aq) + SO_4^{2-}(aq)$$

Original:	0 M	0.150 M
Add solid:	$+x$ M	$+x/2$ M
Equil:	x M	$(0.150+x/2)$M

$x = \left[Ag^+\right] = 9.7\times10^{-3}M$; $K_{sp} = \left[Ag^+\right]^2\left[SO_4^{2-}\right] = \left(9.7\times10^{-3}\right)^2(0.150+0.0049) = 1.5\times10^{-5}$

36. Equation:

$$CaSO_4(s) \rightleftharpoons Ca^{2+}(aq) + SO_4^{2-}(aq).$$

Soln:	0 M	0.0025 M
Add $CaSO_4(s)$	$+x$ M	$+x$ M
Equil:	x M	$(0.0025+x)$ M

$K_{sp} = \left[Ca^{2+}\right]\left[SO_4^{2-}\right] = 9.1\times10^{-6} = x(0.0025+x) = 0.0025x + x^2$

$x^2 + 0.0025x - 9.1\times10^{-6} = 0$

$x = \dfrac{-b \pm \sqrt{b^2-4ac}}{2a} = \dfrac{-0.0025 \pm \sqrt{6.3\times10^{-6}+3.6\times10^{-5}}}{2} = 2.0\times10^{-3}M = \left[CaSO_4\right]$

mass $CaSO_4 = 0.1000\,L \times \dfrac{2.0\times10^{-3}\text{ mol }CaSO_4}{1\text{ L soln}} \times \dfrac{136.1\text{ g }CaSO_4}{1\text{ mol }CaSO_4} = 0.027\text{ g }CaSO_4$

37. For PbI_2, $K_{sp} = 7.1\times10^{-9} = \left[Pb^{2+}\right]\left[I^-\right]^2$

In a solution where 1.5×10^{-4} mol PbI_2 is dissolved, $\left[Pb^{2+}\right] = 1.5\times10^{-4}$ M, and

$\left[I^-\right] = 2\left[Pb^{2+}\right] = 3.0\times10^{-4}$ M

Equation:

$$PbI_2(s) \rightleftharpoons Pb^{2+}(aq) + 2I^-(aq)$$

Initial:	1.5×10^{-4} M	3.0×10^{-4} M
Add lead(II):	$+x$ M	
Equil:	$(0.00015+x)$ M	0.00030 M

$K_{sp} = 7.1\times10^{-9} = (0.00015+x)(0.00030)^2$; $(0.00015+x) = 0.079$; $x = 0.079$ M $= \left[Pb^{2+}\right]$

38. For PbI_2, $K_{sp} = 7.1 \times 10^{-9} = \left[Pb^{2+}\right]\left[I^-\right]^2$

In a solution where 1.5×10^{-5} mol PbI_2/L is dissolved, $\left[Pb^{2+}\right] = 1.5 \times 10^{-5}$ M, and

$\left[I^-\right] = 2\left[Pb^{2+}\right] = 3.0 \times 10^{-5}$ M

Equation: $PbI_2(s) \rightleftharpoons$ $Pb^{2+}(aq) + 2I^-(aq)$

Initial: 1.5×10^{-5} M 3.0×10^{-5} M
Add KI: $+x$ M
Equil: 1.5×10^{-5} M $(3.0 \times 10^{-5} + x)$ M

$K_{sp} = 7.1 \times 10^{-9} = \left(1.5 \times 10^{-5}\right)\left(3.0 \times 10^{-5} + x\right)^2$ $\left(3.0 \times 10^{-5} + x\right) = \sqrt{\dfrac{1 \times 10^{-9}}{1.5 \times 10^{-5}}} = 2.2 \times 10^{-2}$

$x = 2.2 \times 10^{-2} - 3.0 \times 10^{-5} = 2.2 \times 10^{-2}$ M $= \left[I^-\right]$

39. For Ag_2CrO_4, $K_{sp} = 1.1 \times 10^{-12} = \left[Ag^+\right]^2\left[CrO_4^{2-}\right]$

In a solution where 5.0×10^{-8} mol Ag_2CrO_4/L is dissolved, $\left[CrO_4^{2-}\right] = 5.0 \times 10^{-8}$ M and

$\left[Ag^+\right] = 1.0 \times 10^{-7}$ M

Equation: $Ag_2CrO_4(s) \rightleftharpoons 2Ag^+(aq) + CrO_4^{2-}(aq)$

Initial: 1.0×10^{-7} M 5.0×10^{-8} M
Add chromate: $+x$ M
Equil: 1.0×10^{-7} M $(5.0 \times 10^{-8} + x)$ M

$K_{sp} = 1.1 \times 10^{-12} = \left(1.0 \times 10^{-7}\right)^2\left(5.0 \times 10^{-8} + x\right);$ $\left(5.0 \times 10^{-8} + x\right) = 1.1 \times 10^2$

$x = 1.1 \times 10^2$ M $= \left[CrO_4^{2-}\right].$

This is an impossibly high concentration to reach; we cannot lower the solubility of
Ag_2CrO_4 to 5.0×10^{-8} mol Ag_2CrO_4 / L with CrO_4^{2-} as the common ion.
Let's consider using Ag^+ as the common ion.

Equation: $Ag_2CrO_4(s) \rightleftharpoons 2Ag^+(aq) + CrO_4^{2-}(aq)$

Initial: 1.0×10^{-7} M 5.0×10^{-8} M
Add silver(I) ion: $+x$ M
Equil: $(1.0 \times 10^{-7} + x)$ M 5.0×10^{-8} M

$K_{sp} = 1.1 \times 10^{-12} = \left(1.0 \times 10^{-7} + x\right)^2\left(5.0 \times 10^{-8}\right)$ $\left(1.0 \times 10^{-7} + x\right) = \sqrt{\dfrac{1.1 \times 10^{-12}}{5.0 \times 10^{-8}}} = 4.7 \times 10^{-3}$

$x = 4.7 \times 10^{-3} - 1.0 \times 10^{-7} = 4.7 \times 10^{-3}$ M $= \left[I^-\right];$ This is an easy-to-reach concentration.

40. Even though $BaCO_3$ is more soluble than $BaSO_4$, it will still precipitate when 0.50 M $Na_2CO_3(aq)$ is added to a saturated solution of $BaSO_4$ because there is a sufficient $[Ba^{2+}]$ in such a solution for the ion product $[Ba^{2+}][CO_3^{2-}]$ to exceed the value of K_{sp} for the compound. An example will demonstrate this phenomenon. Let us assume that the two solutions being mixed are of equal volume and, to make the situation even more unfavorable, that the saturated $BaSO_4$ solution is not in contact with solid $BaSO_4$, meaning that it does not maintain its saturation when it is diluted. First we determine $[Ba^{2+}]$ in saturated $BaSO_4(aq)$.

$$K_{sp} = [Ba^{2+}][SO_4^{2-}] = 1.1 \times 10^{-10} = s^2 \qquad s = \sqrt{1.1 \times 10^{-10}} = 1.0 \times 10^{-5} \text{ M}$$

Mixing solutions of equal volumes means that the concentration of solutes not common to the two solutions are halved by dilution.

$$[Ba^{2+}] = \frac{1}{2} \times \frac{1.0 \times 10^{-5} \text{ mol } BaSO_4}{1 \text{ L}} \times \frac{1 \text{ mol } Ba^{2+}}{1 \text{ mol } BaSO_4} = 5.0 \times 10^{-6} \text{ M}$$

$$[CO_3^{2-}] = \frac{1}{2} \times \frac{0.50 \text{ mol } Na_2CO_3}{1 \text{ L}} \times \frac{1 \text{ mol } CO_3^{2-}}{1 \text{ mol } Na_2CO_3} = 0.25 \text{ M}$$

$$Q_{sp}\{BaCO_3\} = [Ba^{2+}][CO_3^{2-}] = (5.0 \times 10^{-6})(0.25) = 1.3 \times 10^{-6} > 5.0 \times 10^{-9} = K_{sp}\{BaCO_3\}$$

Thus, precipitation of $BaCO_3$ indeed should occur under the conditions described.

41. $$[Ca^{2+}] = \frac{115 \text{ g } Ca^{2+}}{10^6 \text{ g soln}} \times \frac{1 \text{ mol } Ca^{2+}}{40.08 \text{ g } Ca^{2+}} \times \frac{1000 \text{ g soln}}{1 \text{ L soln}} = 2.87 \times 10^{-3} \text{ M}$$

$$[Ca^{2+}][F^-]^2 = K_{sp} = 5.3 \times 10^{-9} = (2.87 \times 10^{-3})[F^-]^2 \qquad [F^-] = 1.4 \times 10^{-3} \text{ M}$$

$$\text{ppm } F^- = \frac{1.4 \times 10^{-3} \text{ mol } F^-}{1 \text{ L soln}} \times \frac{19.00 \text{ g } F^-}{1 \text{ mol } F^-} \times \frac{1 \text{ L soln}}{1000 \text{ g}} \times 10^6 \text{ g soln} = 27 \text{ ppm}$$

42. We first calculate the $[Ag^+]$ and the $[Cl^-]$ in the saturated solution.

$$K_{sp} = [Ag^+][Cl^-] = 1.8 \times 10^{-10} = (s)(s) = s^2 \qquad s = 1.3 \times 10^{-5} \text{ M} = [Ag^+] = [Cl^-]$$

Both of these concentrations are marginally diluted by the addition of 1 mL of NaCl(aq)

$$[Ag^+] = [Cl^-] = 1.3 \times 10^{-5} \text{ M} \times \frac{100.0 \text{ mL}}{100.0 \text{ mL} + 1.0 \text{ mL}} = 1.3 \times 10^{-5} \text{ M}$$

The $[Cl^-]$ in the NaCl(aq) also is diluted. $[Cl^-] = 1.0 \text{ M} \times \dfrac{1.0 \text{ mL}}{100.0 \text{ mL} + 1.0 \text{ mL}} = 9.9 \times 10^{-3} \text{ M}$

Let us use this $[Cl^-]$ to determine the $[Ag^+]$ that can exist in this solution.

$$\left[\text{Ag}^+\right]\left[\text{Cl}^-\right]=1.8\times10^{-10}=\left[\text{Ag}^+\right]\left(9.9\times10^{-3}\text{ M}\right) \qquad \left[\text{Ag}^+\right]=\frac{1.8\times10^{-10}}{9.9\times10^{-3}}=1.8\times10^{-8}\text{ M}$$

We compute the amount of AgCl in this final solution, and in the initial solution.

$$\text{mmol AgCl final}=101.0\text{ mL}\times\frac{1.8\times10^{-8}\text{mol Ag}^+}{1\text{ L soln}}\times\frac{1\text{ mmol AgCl}}{1\text{ mmol Ag}^+}=1.8\times10^{-6}\text{mmol AgCl}$$

$$\text{mmol AgCl initial}=100.0\text{ mL}\times\frac{1.3\times10^{-5}\text{mol Ag}^+}{1\text{ L soln}}\times\frac{1\text{ mmol AgCl}}{1\text{ mmol Ag}^+}=1.3\times10^{-3}\text{mmol AgCl}$$

The difference between these two amounts is the amount of AgCl that precipitates. Next we compute its mass.

$$\text{mass AgCl}=\left(1.3\times10^{-3}-1.8\times10^{-6}\right)\text{ mmol AgCl}\times\frac{143.3\text{ mg AgCl}}{1\text{ mmol AgCl}}=0.19\text{ mg}$$

We conclude that the precipitate will not be visible to the unaided eye, since its mass is less than 1 mg.

Criterion for Precipitation from Solution

43. We first determine $\left[\text{Mg}^{2+}\right]$, and then determine the value of Q_{sp} in order to compare it to the value of K_{sp}. We express molarity in millimoles per milliliter, entirely equivalent to moles per liter.

$$[\text{Mg}^{2+}]=\frac{22.5\text{ mg MgCl}_2}{325\text{ mL soln}}\times\frac{1\text{ mmol MgCl}_2\cdot6\text{H}_2\text{O}}{203.3\text{ mg MgCl}_2\cdot6\text{H}_2\text{O}}\times\frac{1\text{ mmol Mg}^{2+}}{1\text{ mmol MgCl}_2}=3.41\times10^{-4}\text{ M}$$

$$Q_{\text{sp}}=[\text{Mg}^{2+}][\text{F}^-]^2=(3.41\times10^{-4})(0.035)^2=4.1\times10^{-7}>3.7\times10^{-8}=K_{\text{sp}}$$

Thus, precipitation of $\text{MgF}_2(\text{s})$ should occur from this solution.

44. The solutions mutually dilute each other.

$$\left[\text{Cl}^-\right]=0.016\text{ M}\times\frac{155\text{ mL}}{155\text{ mL}+245\text{ mL}}=6.2\times10^{-3}\text{ M}$$

$$\left[\text{Pb}^{2+}\right]=0.175\text{ M}\times\frac{245\text{ mL}}{245\text{ mL}+155\text{ mL}}=0.107\text{ M}$$

Then we compute the value of the ion product and compare it to the solubility product constant value.

$$Q_{\text{sp}}=\left[\text{Pb}^{2+}\right]\left[\text{Cl}^-\right]^2=(0.107)\left(6.2\times10^{-3}\right)^2=4.1\times10^{-6}<1.6\times10^{-5}=K_{\text{sp}}$$

Thus, precipitation of $\text{PbCl}_2(\text{s})$ will not occur from these mixed solutions.

45. We determine the $\left[OH^-\right]$ needed to just initiate precipitation of $Cd(OH)_2$

$$K_{sp} = \left[Cd^{2+}\right]\left[OH^-\right]^2 = 2.5 \times 10^{-14} = (0.0055 M)\left[OH^-\right]^2 \qquad \left[OH^-\right] = \sqrt{\frac{2.5 \times 10^{-14}}{0.0055}} = 2.1 \times 10^{-6} M$$

$$pOH = -\log\left(2.1 \times 10^{-6}\right) = 5.68 \qquad pH = 14.00 - 5.68 = 8.32$$

$Cd(OH)_2$ will precipitate from a solution with $pH > 8.32$.

46. We determine the $\left[OH^-\right]$ needed to just initiate precipitation of $Cr(OH)_3$

$$K_{sp} = [Cr^{3+}][OH^-]^3 = 6.3 \times 10^{-31} = (0.086\ M)\ [OH^-]^3; \quad [OH^-] = \sqrt{\frac{6.3 \times 10^{-31}}{0.086}} = 1.9 \times 10^{-10}\ M$$

$$pOH = -\log\left(1.9 \times 10^{-10}\right) = 9.72 \qquad pH = 14.00 - 9.72 = 4.28$$

$Cr(OH)_3$ will precipitate from a solution with $pH > 4.28$.

47. (a) First we determine $\left[Cl^-\right]$ due to the added NaCl.

$$\left[Cl^-\right] = \frac{0.10\ \text{mg NaCl}}{1.0\ \text{L soln}} \times \frac{1\ \text{g}}{1000\ \text{mg}} \times \frac{1\ \text{mol NaCl}}{58.4\ \text{g NaCl}} \times \frac{1\ \text{mol } Cl^-}{1\ \text{mol NaCl}} = 1.7 \times 10^{-6}\ M$$

Then we determine the value of the ion product and compare it to the solubility product constant value.

$$Q_{sp} = \left[Ag^+\right]\left[Cl^-\right] = (0.10)\left(1.7 \times 10^{-6}\right) = 1.7 \times 10^{-7} > 1.8 \times 10^{-10} = K_{sp} \text{ for AgCl}$$

Precipitation of AgCl(s) should occur.

(b) The KBr(aq) is diluted on mixing, but the $\left[Ag^+\right]$ and $\left[Cl^-\right]$ are barely affected by dilution.

$$\left[Br^-\right] = 0.10\ M \times \frac{0.05\ \text{mL}}{0.05\ \text{mL} + 250\ \text{mL}} = 2 \times 10^{-5}\ M$$

Now we determine $\left[Ag^+\right]$ in a saturated AgCl solution.

$$K_{sp} = \left[Ag^+\right]\left[Cl^-\right] = (s)(s) = s^2 = 1.8 \times 10^{-10} \qquad s = 1.3 \times 10^{-5}\ M$$

Then we determine the value of the ion product for AgBr and compare it to the solubility product constant value.

$$Q_{sp} = \left[Ag^+\right][Br] = \left(1.3 \times 10^{-5}\right)\left(2 \times 10^{-5}\right) = 3 \times 10^{-10} > 5.0 \times 10^{-13} = K_{sp} \text{ for AgBr}$$

Precipitation of AgBr(s) should occur.

(c)   The hydroxide ion is diluted by mixing the two solutions.

$$\left[OH^-\right] = 0.0150\ M \times \frac{0.05\ mL}{0.05\ mL + 3000\ mL} = 2 \times 10^{-7}\ M$$

But the $\left[Mg^{2+}\right]$ does not change significantly.

$$\left[Mg^{2+}\right] = \frac{2.0\ mg\ Mg^{2+}}{1.0\ L\ soln} \times \frac{1\ g}{1000\ mg} \times \frac{1\ mol\ Mg^{2+}}{24.3\ g\ Mg} = 8.2 \times 10^{-5}\ M$$

Then we determine the value of the ion product and compare it to the solubility product constant value.

$$Q_{sp} = \left[Mg^{2+}\right]\left[OH^-\right]^2 = \left(3 \times 10^{-7}\right)\left(8.2 \times 10^{-5}\right)^2 = 2 \times 10^{-15}$$

$$Q_{sp} < 1.8 \times 10^{-11} = K_{sp}\ \text{for}\ Mg(OH)_2\ \ \text{Thus, no precipitate forms.}$$

48.   Determine the moles of H_2 produced during the electrolysis, and then determine $\left[OH^-\right]$.

$$\text{moles}\ H_2 = \frac{PV}{RT} = \frac{752\ mmHg \times \dfrac{1\ atm}{760\ mmHg} \times 0.652\ L}{0.08206\ L\ atm\ mol^{-1}\ K^{-1} \times 295\ K} = 0.0267\ mol\ H_2$$

$$\left[OH^-\right] = \frac{0.0267\ mol\ H_2 \times \dfrac{2\ mol\ OH^-}{1\ mol\ H_2}}{0.315\ L\ sample} = 0.170\ M$$

$$Q_{sp} = \left[Mg^{2+}\right]\left[OH^-\right]^2 = (0.185)(0.170)^2 = 5.35 \times 10^{-3} > 1.8 \times 10^{-11} = K_{sp}$$

Yes, precipitation of $Mg(OH)_2(s)$ should occur during the electrolysis.

49.   We determine $\left[C_2O_4^{2-}\right]$ in this solution. From key idea 3 in Section 17-6,

$$\left[C_2O_4^{2-}\right] = K_{a_2} = 5.4 \times 10^{-5}$$

$$Q_{sp} = \left[Ca^{2+}\right]\left[C_2O_4^{2-}\right] = (0.150)(5.4 \times 10^{-5}) = 8.1 \times 10^{-6} > 4 \times 10^{-9} = K_{sp}$$

Thus, CaC_2O_4 should precipitate from this solution.

50.   The solutions mutually dilute each other. We first determine the solubility of each compound in its saturated solution and then its concentration after dilution.

$$K_{sp} = \left[Ag^+\right]^2\left[SO_4^{2-}\right] = 1.4 \times 10^{-5} = (2s)^2 s = 4s^3 \qquad s = \sqrt[3]{\frac{1.4 \times 10^{-5}}{4}} = 1.5 \times 10^{-2}\ M$$

$$\left[SO_4^{2-}\right] = 0.015\ M \times \frac{100.0\ mL}{100.0\ mL + 250.0\ mL} = 0.0043\ M \qquad \left[Ag^+\right] = 0.0086\ M$$

$$K_{sp} = [Pb^{2+}][CrO_4^{2-}] = 2.8 \times 10^{-13} = (s)(s) = s^2 \qquad s = \sqrt{2.8 \times 10^{-13}} = 5.3 \times 10^{-7} \text{ M}$$

$$\left[Pb^{2+} \right] = \left[CrO_4^{\ 2-} \right] = 5.3 \times 10^{-7} \times \frac{250.0 \text{ mL}}{250.0 \text{ mL} + 100.0 \text{ mL}} = 3.8 \times 10^{-7} \text{ M}$$

From the balanced chemical equation, we see that the two possible precipitates are $PbSO_4$ and Ag_2CrO_4. (Neither $PbCrO_4$ nor Ag_2SO_4 can precipitate because they have been diluted below their saturated concentrations.) $PbCrO_4 + Ag_2SO_4 \rightleftharpoons PbSO_4 + Ag_2CrO_4$

Thus, we compute the value of Q_{sp} for each of these compounds and compare those values with the solubility constant product value.

$$Q_{sp} = \left[Pb^{2+} \right]\left[SO_4^{\ 2-} \right] = (3.8 \times 10^{-7})(0.0043) = 1.6 \times 10^{-9} < 1.6 \times 10^{-8} = K_{sp} \text{ for } PbSO_4$$

Thus, $PbSO_4(s)$ will not precipitate.

$$Q_{sp} = \left[Ag^+ \right]^2 \left[CrO_4^{\ 2-} \right] = (0.0086)^2 (3.8 \times 10^{-7}) = 2.8 \times 10^{-11} > 1.1 \times 10^{-12} = K_{sp} \text{ for } Ag_2CrO_4$$

Thus, $Ag_2CrO_4(s)$ should precipitate.

Completeness of Precipitation

51. First determine that a precipitate forms. The solutions mutually dilute each other.

$$\left[CrO_4^{\ 2-} \right] = 0.350 \text{ M} \times \frac{200.0 \text{ mL}}{200.0 \text{ mL} + 200.0 \text{ mL}} = 0.175 \text{ M}$$

$$\left[Ag^+ \right] = 0.0100 \text{ M} \times \frac{200.0 \text{ mL}}{200.0 \text{ mL} + 200.0 \text{ mL}} = 0.00500 \text{ M}$$

We determine the value of the ion product and compare it to the solubility product constant value.

$$Q_{sp} = \left[Ag^+ \right]^2 \left[CrO_4^{\ 2-} \right] = (0.00500)^2 (0.175) = 4.4 \times 10^{-6} > 1.1 \times 10^{-12} = K_{sp} \text{ for } Ag_2CrO_4$$

Ag_2CrO_4 should precipitate.

Now, we assume that as much solid forms as possible, and then we approach equilibrium by dissolving that solid in a solution that contains the ion in excess.

Equation:	$Ag_2CrO_4(s) \rightleftharpoons$	$2Ag^+(aq)$	$+ \ CrO_4^{\ 2-}(aq)$
Orig. soln:		0.00500 M	0.175 M
Form solid:		−0.00500 M	−0.00250 M
Not at equilibrium		0 M	0.173 M
Changes:		+2x M	+x M
Equil:		2x M	(0.173 + x) M

$$K_{sp} = \left[Ag^+ \right]^2 \left[CrO_4^{\ 2-} \right] = 1.1 \times 10^{-12} = (2x)(0.173 + x) \approx (4x^2)(0.173)$$

$$x = \sqrt{\frac{1.1 \times 10^{-12}}{4 \times 0.173}} = 1.3 \times 10^{-6} \, \text{M} \qquad\qquad [Ag^+] = 2x = 2.6 \times 10^{-6} \, \text{M}$$

$$\% \, Ag^+ \text{ unprecipitated} = \frac{2.6 \times 10^{-6} \text{ M final}}{0.00500 \text{ M initial}} \times 100\% = 0.052\% \text{ unprecipitated}$$

52. $[Ag^+]_{\text{diluted}} = 0.0208 \text{ M} \times \dfrac{175 \text{ mL}}{425 \text{ mL}} = 0.008565 \text{ M}$

$[CrO_4^{2-}]_{\text{diluted}} = 0.0380 \text{ M} \times \dfrac{250 \text{ mL}}{425 \text{ mL}} = 0.02235 \text{ M}$

$Q_{sp} = [Ag^+]^2 [CrO_4^{2-}] = 1.6 \times 10^{-6} > K_{sp}$

The equilibrium constant for this precipitation reaction suggests that the reaction nearly goes to 100% completion. Assume that the limiting reagent is used up (100% reaction in the reverse direction) and re-establish the equilibrium in the reverse direction. Here Ag^+ is the limiting reagent.

	$Ag_2CrO_4(s)$	$\xrightarrow{K_{sp(Ag_2CrO_4)} = 1.1 \times 10^{-12}}$	$2 \, Ag^+(aq)$	$+$	$CrO_4^{2-}(aq)$
initial	0 M		0.008565 M		0.02235 M
change	$+x$	$x = 0.00428$ M	$-2x$		$-x$
100% rxn	0.00428		0		0.0181
change	$-y$	**re-establish equil**	$+2y$		$+y$
equil	$0.00428 - y$	**(assume $y \sim 0$)**	$2y$		$0.0181 + y$

$1.1 \times 10^{-12} = (2y)^2(0.0181 + y) \approx (2y)^2(0.0181) \qquad y = 3.9 \times 10^{-6}$ (valid assumption)

$y = [Ag^+] = 3.9 \times 10^{-6}$ M after precipitation is complete.

$\% \, [Ag^+]$ left in solution $= \dfrac{3.9 \times 10^{-6}}{0.00856} \times 100 \% = 0.046 \%$ (quantitative)

53. We first use the solubility product constant expression to determine $[Pb^{2+}]$ in a solution with 0.100 M Cl^-.

$$K_{sp} = [Pb^{2+}][Cl^-]^2 = 1.6 \times 10^{-5} = [Pb^{2+}](0.100)^2 \qquad [Pb^{2+}] = \frac{1.6 \times 10^{-5}}{(0.100)^2} = 1.6 \times 10^{-3} \text{ M}$$

$$\% \text{ unprecipitated} = \frac{1.6 \times 10^{-3} \text{ M}}{0.065 \text{ M}} \times 100\% = 2.5\%$$

Now, we want to determine what $[Cl^-]$ must be maintained to keep $[Pb^{2+}]_{\text{final}} = 1\%$;
$[Pb^{2+}]_{\text{initial}} = 0.010 \times 0.065 \, \text{M} = 6.5 \times 10^{-4} \text{ M}$

$$K_{sp} = [Pb^{2+}][Cl^-]^2 = 1.6 \times 10^{-5} = (6.5 \times 10^{-4})[Cl^-]^2 \qquad\qquad [Cl^-] = \sqrt{\frac{6 \times 10^{-5}}{6.5 \times 10^{-4}}} = 0.16 \text{ M}$$

54. Let's start by assuming that the concentration of Pb^{2+} in the untreated wine is no higher than 1.5×10^{-4} M (this assumption is not unreasonable). As long as the Pb^{2+} concentration is less than 1.5×10^{-4} M, then the final sulfate ion concentration in the $CaSO_4$ treated wine should be virtually the same as the sulfate ion concentration in a saturated solution of $CaSO_4$ formed by dissolving solid $CaSO_4$ in pure water (i.e., with $[Pb^{2+}]$ less than or equal to 1.5×10^{-4} M, the $[SO_4^{2-}]$ will not drop significantly below that for a saturated solution, $\approx 3.0 \times 10^{-3}$ M). Thus, the addition of $CaSO_4$ to the wine would result in the precipitation of solid $PbSO_4$, which would continue until the concentration of Pb^{2+} was equal to the K_{sp} for $PbSO_4$ divided by the concentration of dissolved sulfate ion, i.e., $[Pb^{2+}]_{max} = 1.6 \times 10^{-8} M^2 / 3.0 \times 10^{-3}$ M $= 5.3 \times 10^{-6}$ M.

Fractional Precipitation

55. The concentrations of silver ion that are cited in Example 19-7 range from 5.0×10^{-11} M to 1.0×10^{-5} M. These are incredibly small concentrations, especially the first. Virtually any $AgNO_3(aq)$ solution that we would prepare by usual means would have at least these concentrations. However, there is the matter of dilution to be considered. If one drop (0.05 mL) of $AgNO_3(aq)$ is added to 500.0 mL of solution, the $[Ag^+]$ will decrease by a factor of 10^4 Thus, we would have to begin with 0.15 M $AgNO_3$ for this dilution to produce 1.5×10^{-5} M. So we cannot be too careless and use extremely dilute $AgNO_3(aq)$.

56. **(a)** 0.10 M NaCl will not work at all, since both $BaCl_2$ and $CaCl_2$ are soluble in water.

   (b) $K_{sp} = 1.1 \times 10^{-10}$ for $BaSO_4$ and $K_{sp} = 9.1 \times 10^{-6}$ for $CaSO_4$. Since these values differ by more than 1000, 0.05 M Na_2SO_4 would effectively separate Ba^{2+} from Ca^{2+}.

   We first compute $[SO_4^{2-}]$ when $BaSO_4$ begins to precipitate.

   $$[Ba^{2+}][SO_4^{2-}] = (0.050)[SO_4^{2-}] = 1.1 \times 10^{-10}; \quad [SO_4^{2-}] = \frac{1.1 \times 10^{-10}}{0.050} = 2.2 \times 10^{-9} \text{ M}$$

   And then we calculate $[SO_4^{2-}]$ when $[Ba^{2+}]$ has decreased to 0.1% of its initial value, that is, to 5.0×10^{-5} M

   $$[Ba^{2+}][SO_4^{2-}] = (5.0 \times 10^{-5})[SO_4^{2-}] = 1.1 \times 10^{-10}; \quad [SO_4^{2-}] = \frac{1.1 \times 10^{-10}}{5.0 \times 10^{-5}} = 2.2 \times 10^{-6} \text{ M}$$

   And finally $[SO_4^{2-}]$ when $CaSO_4$ begins to precipitate.

   $$[Ca^{2+}][SO_4^{2-}] = (0.050)[SO_4^{2-}] = 9.1 \times 10^{-6}; \quad [SO_4^{2-}] = \frac{9.1 \times 10^{-6}}{0.050} = 1.8 \times 10^{-4} \text{ M}$$

(c) Now, $K_{sp} = 5 \times 10^{-3}$ for $Ba(OH)_2$ and $K_{sp} = 5.5 \times 10^{-6}$ for $Ca(OH)_2$. The fact that these two K_{sp} values differ by almost a factor of 1000 does not tell the entire story, because $[OH^-]$ is squared in both K_{sp} expressions. We compute $[OH^-]$ when $Ca(OH)_2$ begins to precipitate.

$$[Ca^{2+}][OH^-]^2 = 5.5 \times 10^{-6} = (0.050)[OH^-]^2 \quad [OH^-] = \sqrt{\frac{5.5 \times 10^{-6}}{0.050}} = 1.0 \times 10^{-2} \text{ M}$$

Precipitation will not proceed, as we only have 0.001 M NaOH, which has $[OH^-] = 1 \times 10^{-3}$ M

(d) $K_{sp} = 5.1 \times 10^{-9}$ for $BaCO_3$ and $K_{sp} = 2.8 \times 10^{-9}$ for $CaCO_3$. Since these two values differ by less than a factor of 2, 0.50 M Na_2CO_3 would not effectively separate Ba^{2+} from Ca^{2+}.

57. Normally we would worry about the mutual dilution of the two solutions, but the values of the solubility product constants are so small that only a very small volume of 0.50 M $Pb(NO_3)_2$ solution needs to be added, as we shall see.

(a) Since the two anions are present at the same concentration and they have the same type of formula (one anion per cation), the one forming the compound with the smallest K_{sp} value will precipitate first. Thus, CrO_4^{2-} is the first anion to precipitate.

(b) At the point where SO_4^{2-} begins to precipitate, we have

$$K_{sp} = [Pb^{2+}][SO_4^{2-}] = 1.6 \times 10^{-8} = [Pb^{2+}](0.010M); [Pb^{2+}] = \frac{1.6 \times 10^{-8}}{0.010} = 1.6 \times 10^{-6} \text{ M}$$

Now we can test our original assumption, that only a very small volume of 0.50 M $Pb(NO_3)_2$ solution has been added. We assume that we have 1.00 L of the original solution, the one with the two anions dissolved in it, and compute the volume of 0.50 M $Pb(NO_3)_2$ that has to be added to achieve $[Pb^{2+}] = 1.6 \times 10^{-6}$ M

$$V_{added} = 1.00 \text{ L} \times \frac{1.6 \times 10^{-6} \text{ mol } Pb^{2+}}{1 \text{ L soln}} \times \frac{1 \text{ mol } Pb(NO_3)_2}{1 \text{ mol } Pb^{2+}} \times \frac{1 \text{ L } Pb^{2+}\text{soln}}{0.50 \text{ mol } Pb(NO_3)_2}$$

$$V_{added} = 3.2 \times 10^{-5} \text{ L } Pb^{2+}\text{soln} = 0.0032 \text{mL } Pb^{2+}\text{soln}$$

This is less than one drop (0.05 mL) of the Pb^{2+} solution, clearly a very small volume.

(c) The two anions are effectively separated if $\left[Pb^{2+}\right]$ has not reached 1.6×10^{-6} M when $\left[CrO_4{}^{2-}\right]$ is reduced to 0.1% of its original value, that is, to

$$0.010 \times 10^{-3} \text{ M} = 1.0 \times 10^{-5} \text{ M} = \left[CrO_4{}^{2-}\right]$$

$$K_{sp} = \left[Pb^{2+}\right]\left[CrO_4{}^{2-}\right] = 2.8 \times 10^{-13} = \left[Pb^{2+}\right]\left(1.0 \times 10^{-5}\right)$$

$$\left[Pb^{2+}\right] = \frac{2.8 \times 10^{-13}}{1.0 \times 10^{-5}} = 2.8 \times 10^{-8} \text{ M}$$

Thus, the two anions can be effectively separated by fractional precipitation.

58. (a) We answer this question by determining the $\left[Ag^+\right]$ needed to initiate precipitation of each compound.

AgCl: $K_{sp} = \left[Ag^+\right]\left[Cl^-\right] = 1.8 \times 10^{-10} = \left[Ag^+\right](0.250);$ $\left[Ag^+\right] = \dfrac{1.8 \times 10^{-10}}{0.250} = 7.2 \times 10^{-10}$ M

AgBr: $K_{sp} = \left[Ag^+\right]\left[Br^-\right] = 5.0 \times 10^{-13} = \left[Ag^+\right](0.0022);$ $\left[Ag^+\right] = \dfrac{5.0 \times 10^{-13}}{0.0022} = 2.3 \times 10^{-10}$ M

Thus, Br^- precipitates first, as AgBr, because it requires a lower $\left[Ag^+\right]$.

(b) $\left[Ag^+\right] = 7.2 \times 10^{-10}$ M when chloride ion, the second anion, begins to precipitate.

(c) Cl^- and Br^- cannot be separated by this fractional precipitation. $\left[Ag^+\right]$ will have to rise to 1000 times its initial value, to 2.3×10^{-7} M, before AgBr is completely precipitated. But as soon as $\left[Ag^+\right]$ reaches 7.2×10^{-10} M, AgCl will begin to precipitate.

Solubility and pH

59. In each case we indicate whether the compound is more soluble in acid than in water. We write the net ionic equation for the reaction in which the solid dissolves in acid. Substances are more soluble in acid if either (1) an acid-base reaction occurs [as in (b-d)] or (2) a gas is produced, since escape of the gas from the reaction mixture causes the reaction to shift to the right.

(a) Same: KCl

(b) Acid: $MgCO_3(s) + 2H^+(aq) \longrightarrow Mg^{2+}(aq) + H_2O(l) + CO_2(g)$

(c) Acid: $FeS(s) + 2H^+(aq) \longrightarrow Fe^{2+}(aq) + H_2S(g)$

(d) Acid: $Ca(OH)_2(s) + 2H^+(aq) \longrightarrow Ca^{2+}(aq) + 2H_2O(l)$

(e) Water: C_6H_5COOH is less soluble in acid, because of the H_3O^+ common ion.

60. In each case we indicate whether the compound is more soluble in base than in water. We write the net ionic equation for the reaction in which the solid dissolves in base. Substances are soluble in base if either (1) acid-base reaction occurs [as in (b)] or (2) a gas is produced, since escape of the gas from the reaction mixture causes the reaction to shift to the right.

(a) Water: $BaSO_4$ is less soluble in base, hydrolysis of SO_4^{2-} will be repressed.

(b) Base: $H_2C_2O_4(s) + 2OH^-(aq) \longrightarrow C_2O_4^{2-}(aq) + 2H_2O(l)$

(c) Water: $Fe(OH)_3$ is less soluble in base because of the OH^- common ion.

(d) Same: $NaNO_3$

(e) Water: MnS is less soluble in base because hydrolysis of S^{2-} will be repressed.

61. We determine $[Mg^{2+}]$ in the solution.

$$[Mg^{2+}] = \frac{0.65 \text{ g Mg(OH)}_2}{1 \text{ L soln}} \times \frac{1 \text{ mol Mg(OH)}_2}{58.3 \text{ g Mg(OH)}_2} \times \frac{1 \text{ mol Mg}^{2+}}{1 \text{ mol Mg(OH)}_2} = 0.011 M$$

Then we determine $[OH^-]$ in the solution, and its pH.

$$K_{sp} = [Mg^{2+}][OH^-]^2 = 1.8 \times 10^{-11} = (0.011)[OH^-]^2; \quad [OH^-] = \sqrt{\frac{1.8 \times 10^{-11}}{0.011}} = 4.0 \times 10^{-5} \text{ M}$$

$$pOH = -\log(4.0 \times 10^{-5}) = 4.40 \qquad\qquad pH = 14.00 - 4.40 = 9.60$$

62. First determine the $[Mg^{2+}]$ and $[NH_3]$ that result from dilution to a total volume of 0.500 L.

$$[Mg^{2+}] = 0.100 M \times \frac{0.150 \text{ L}_{initial}}{0.500 \text{ L}_{final}} = 0.0300 M; \quad [NH_3] = 0.150 \text{ M} \times \frac{0.350 \text{ L}_{initial}}{0.500 \text{ L}_{final}} = 0.105 \text{ M}$$

Then determine the $[OH^-]$ that will allow $[Mg^{2+}] = 0.0300$ M in this solution.

$$K_{sp} = 1.8 \times 10^{-11} = [Mg^{2+}][OH^-]^2 = (0.0300)[OH^-]^2; \quad [OH^-] = \sqrt{\frac{1.8 \times 10^{-11}}{0.0300}} = 2.4 \times 10^{-5} \text{ M}$$

This $[OH^-]$ is maintained by the NH_3 / NH_4^+ buffer, for which we use the Henderson–Hasselbalch equation.

$$pH = 14.00 - pOH = 14.00 + \log(2.4 \times 10^{-5}) = 9.38 = pK_a + \log\frac{[NH_3]}{[NH_4^+]} = 9.26 + \log\frac{[NH_3]}{[NH_4^+]}$$

$$\log\frac{[NH_3]}{[NH_4^+]} = 9.38 - 9.26 = +0.12; \quad \frac{[NH_3]}{[NH_4^+]} = 10^{+0.12} = 1.3; \quad [NH_4^+] = \frac{0.105 \text{ M NH}_3}{1.3} = 0.081 \text{ M}$$

$$\text{mass } (NH_4)_2SO_4 = 0.500 \text{ L} \times \frac{0.081 \text{ mol NH}_4^+}{\text{L soln}} \times \frac{1 \text{ mol } (NH_4)_2SO_4}{2 \text{ mol NH}_4^+} \times \frac{132.1 \text{ g } (NH_4)_2SO_4}{1 \text{ mol } (NH_4)_2SO_4} = 2.7 \text{ g}$$

63. (a) We calculate $[OH^-]$ needed for precipitation.

$$K_{sp} = [Al^{3+}][OH^-]^3 = 1.3 \times 10^{-33} = (0.075 \text{ M})[OH^-]^3$$

$$[OH^-] = \sqrt[3]{\frac{1.3 \times 10^{-33}}{0.075}} = 2.6 \times 10^{-11} \quad pOH = -\log(2.6 \times 10^{-11}) = 10.59$$

$$pH = 14.00 - 10.59 = 3.41$$

(b) We use the Henderson–Hasselbalch equation to determine $[C_2H_3O_2^-]$.

$$pH = 3.41 = pK_a + \log \frac{[C_2H_3O_2^-]}{[HC_2H_3O_2]} = 4.74 + \log \frac{[C_2H_3O_2^-]}{1.00 \text{ M}}$$

$$\log \frac{[C_2H_3O_2^-]}{1.00 \text{ M}} = 3.41 - 4.74 = -1.33; \quad \frac{[C_2H_3O_2^-]}{1.00 \text{ M}} = 10^{-1.33} = 0.047; \quad [C_2H_3O_2^-] = 0.047 \text{ M}$$

This situation does not quite obey the guideline that the ratio of concentrations must fall in the range 0.10 to 10.0, but the resulting error is a small one in this circumstance.

$$\text{mass NaC}_2\text{H}_3\text{O}_2 = 0.2500 \text{ L} \times \frac{0.047 \text{ mol C}_2\text{H}_3\text{O}_2^-}{1 \text{ L soln}} \times \frac{1 \text{ mol NaC}_2\text{H}_3\text{O}_2}{1 \text{ mol C}_2\text{H}_3\text{O}_2^-} \times \frac{82.03 \text{ g NaC}_2\text{H}_3\text{O}_2}{1 \text{ mol NaC}_2\text{H}_3\text{O}_2}$$

$$= 0.96 \text{ g NaC}_2\text{H}_3\text{O}_2$$

64. (a) Since HI is a strong acid, $[I^-] = 1.05 \times 10^{-3} \text{ M} + 1.05 \times 10^{-3} \text{ M} = 2.10 \times 10^{-3} \text{ M}$

We determine the value of the ion product and compare it to the solubility product constant value.

$$Q_{sp} = [Pb^{2+}][I^-]^2 = (1.1 \times 10^{-3})(2.10 \times 10^{-3})^2 = 4.9 \times 10^{-9} < 7.1 \times 10^{-9} = K_{sp} \text{ for PbI}_2$$

A precipitate of PbI_2 will not form under these conditions.

(b) We compute the $[OH^-]$ needed for precipitation.

$$K_{sp} = [Mg^{2+}][OH^-]^2 = 1.8 \times 10^{-11} = (0.0150)[OH^-]^2; \quad [OH^-] = \sqrt{\frac{1.8 \times 10^{-11}}{0.0150}} = 3.5 \times 10^{-5} \text{ M}$$

Then we compute $[OH^-]$ in this solution, resulting from the ionization of NH_3.

$$[NH_3] = 6.00 \text{ M} \times \frac{0.05 \times 10^{-3} \text{ L}}{2.50 \text{ L}} = 1 \times 10^{-4} \text{ M}$$

Even though NH_3 is a weak base, $[OH^-]$ produced from NH_3 will approximate 4×10^{-5} M in this very dilute solution. (Recall that degree of ionization is high in dilute solution.) And since $[OH^-] = 3.5 \times 10^{-5}$ M is needed for precipitation to occur, we conclude that $Mg(OH)_2$ will not precipitate from this solution.

(c)   0.010 M $HC_2H_3O_2$ and 0.010 M $NaC_2H_3O_2$ is a buffer solution with $pH = pK_a$ of acetic acid, since the acid and its anion are present in equal concentrations. From this, we determine the $[OH^-]$.

$$pH = 4.74 \qquad pOH = 14.00 - 4.74 = 9.26 \qquad [OH^-] = 10^{-9.26} = 5.5 \times 10^{-10}$$

$$Q = [Al^{3+}][OH^-]^3 = (0.010)(5.5 \times 10^{-10})^3 = 1.7 \times 10^{-30} > 1.3 \times 10^{-33} = K_{sp} \text{ of } Al(OH)_3$$

Thus, $Al(OH)_3(s)$ should precipitate from this solution.

Complex-Ion Equilibria

65. Lead(II) ion forms a complex ion with chloride ion. It forms no such complex ion with nitrate ion. The formation of this complex ion decreases the concentrations of free $Pb^{2+}(aq)$ and free $Cl^-(aq)$. Thus, ~~solid or of~~ $PbCl_2$ will dissolve in the HCl(aq) up until the value of the solubility product is exceeded. $Pb^{2+}(aq) + 3Cl^-(aq) \rightleftharpoons [PbCl_3]^-(aq)$

66. $Zn^{2+}(aq) + 4NH_3(aq) \rightleftharpoons [Zn(NH_3)_4]^{2+}(aq) \qquad K_f = 4.1 \times 10^8$

$NH_3(aq)$ will be least effective in reducing the concentration of the complex ion. In fact, the addition of $NH_3(aq)$ will increase the concentration of the complex ion by favoring a shift of the equilibrium to the right. $NH_4^+(aq)$ will have a similar effect, but not as direct. $NH_3(aq)$ is formed by the hydrolysis of $NH_4^+(aq)$ and, thus, increasing $[NH_4^+]$ will eventually increase $NH_3(aq)$: $NH_4^+(aq) + H_2O(l) \rightleftharpoons NH_3(aq) + H_3O^+(aq)$. The addition of HCl(aq) will cause the greatest decrease in the concentration of the complex ion. HCl(aq) will react with $NH_3(aq)$ to decrease its concentration (by forming NH_4^+) and this will cause the complex ion equilibrium reaction to shift left.

67. We substitute the given concentrations directly into the K_f expression.

$$K_f = \frac{[[Cu(CN)_4^{3-}]]}{[Cu^+][CN^-]^4} = \frac{0.0500}{(6.1 \times 10^{-32})(0.80)^4} = 2.0 \times 10^{30}$$

68. The solution is organized around the balanced chemical equation. Free $[NH_3]$ is 6.0 M at equilibrium. The size of the equilibrium constant indicates that most copper(II) is present as the complex ion.

Equation:	$Cu^{2+}(aq) + 4NH_3(aq)$		$\rightleftharpoons$	$[Cu(NH_3)_4]^{2+}(aq)$
Initial:	0.10 M			0 M
Change(100 % rxn):	−0.10 M			0.10 M
Changes:	+x M			−x M
Equil:	x M	6.0 M		(0.10−x) M

$$K_f = \frac{\left[\left[Cu(NH_3)_4\right]^{2+}\right]}{\left[Cu^{2+}\right]\left[NH_3\right]^4} = 1.1 \times 10^{13} = \frac{0.10 - x}{x(6.0)^4} \approx \frac{0.10}{1.3 \times 10^3 x}$$

$$x = \frac{0.10}{1.3 \times 10^3 \times 1.1 \times 10^{13}} = 7.0 \times 10^{-18} \text{ M} = \left[Cu^{2+}\right]$$

69. We first find the concentration of free metal ion. Then we determine the value of Q_{sp} for the precipitation reaction, and compare its value with the value of K_{sp} to determine whether precipitation should occur.

Equation: $\quad Ag^+(aq) + \qquad 2\,S_2O_3^{2-}(aq) \quad \rightleftharpoons \quad \left[Ag(S_2O_3)_2\right]^{3-}(aq)$

Initial:	0 M	0.76 M	0.048 M
Changes:	$+x$ M	$+2x$ M	$-x$ M
Equil:	x M	$(0.76 + 2x)$ M	$(0.048 - x)$ M

$$K_f = \frac{\left[\left[Ag(S_2O_3)_2\right]^{3-}\right]}{\left[Ag^+\right]\left[S_2O_3^{2-}\right]^2} = 1.7 \times 10^{13} = \frac{0.048 - x}{x(0.76 + 2x)^2} \approx \frac{0.048}{(0.76)^2 x}; x = 4.9 \times 10^{-15} \text{ M} = \left[Ag^+\right]$$

$$Q_{sp} = \left[Ag^+\right]\left[I^-\right] = \left(4.9 \times 10^{-15}\right)(2.0) = 9.8 \times 10^{-15} > 8.5 \times 10^{-17} = K_{sp}.$$

Because $Q_{sp} > K_{sp}$, precipitation of AgI(s) should occur.

70. We determine $\left[OH^-\right]$ in this solution, and also the free $\left[Cu^{2+}\right]$.

$$pH = pK_a + \log\frac{\left[NH_3\right]}{\left[NH_4^+\right]} = 9.26 + \log\frac{0.10 \text{ M}}{0.10 \text{ M}} = 9.26 \quad pOH = 14.00 - 9.26 = 4.74$$

$$\left[OH^-\right] = 10^{-4.74} = 1.8 \times 10^{-5} \text{ M}$$

$$Cu^{2+}(aq) + 4NH_3(aq) \rightleftharpoons \left[Cu(NH_3)_4\right]^{2+}(aq)$$

$$K_f = \frac{\left[\left[Cu(NH_3)_4\right]^{2+}\right]}{\left[Cu^{2+}\right]\left[NH_3\right]^4} = 1.1 \times 10^{13} = \frac{0.015}{\left[Cu^{2+}\right]0.10^4}; \left[Cu^{2+}\right] = \frac{0.015}{1.1 \times 10^{13} \times 0.10^4} = 1.4 \times 10^{-11} \text{ M}$$

Now we determine the value of Q_{sp} and compare it with the value of K_{sp} for $Cu(OH)_2$.

$$Q_{sp} = \left[Cu^{2+}\right]\left[OH^-\right]^2 = \left(1.4 \times 10^{-11}\right)\left(1.8 \times 10^{-5}\right)^2 = 4.5 \times 10^{-21} < 2.2 \times 10^{-20} = K_{sp} \text{ for } Cu(OH)_2$$

Precipitation of $Cu(OH)_2(s)$ from this solution should not occur.

71. We first compute the free $\left[Ag^+\right]$ in the original solution. The size of the complex ion formation equilibrium constant indicates that the reaction lies to th efar right, so we form as much complex ion as possible stoichiometrically.

Equation:	$Ag^+(aq) +$	$2NH_3(aq)$	$\rightleftharpoons$	$\left[Ag(NH_3)_2\right]^+(aq)$
In soln:	0.10 M	1.00 M		0 M
Form complex:	−0.10 M	−0.20 M		+0.10 M
	0 M	0.80 M		0.10 M
Changes:	+x M	+2x M		−x M
Equil:	x M	$(0.80 + 2x)$ M		$(0.10 - x)$ M

$$K_f = 1.6 \times 10^7 = \frac{\left[Ag(NH_3)_2\right]^+}{\left[Ag^+\right]\left[NH_3\right]^2} = \frac{0.10 - x}{x(0.80 + 2x)^2} \approx \frac{0.10}{x(0.80)^2} \qquad x = \frac{0.10}{1.6 \times 10^7 (0.80)^2} = 9.8 \times 10^{-9} \text{ M}.$$

Thus, $\left[Ag^+\right] = 9.8 \times 10^{-9}$ M. We next determine the $\left[I^-\right]$ that can coexist in this solution without precipitation.

$$K_{sp} = \left[Ag^+\right]\left[I^-\right] = 8.5 \times 10^{-17} = \left(9.8 \times 10^{-9}\right)\left[I^-\right]; \qquad \left[I^-\right] = \frac{8.5 \times 10^{-17}}{9.8 \times 10^{-9}} = 8.7 \times 10^{-9} \text{ M}$$

Finally, we determine the mass of KI needed to produce this $\left[I^-\right]$

$$\text{mass KI} = 1.00 \text{ L soln} \times \frac{8.7 \times 10^{-9} \text{ mol } I^-}{1 \text{ L soln}} \times \frac{1 \text{ mol KI}}{1 \text{ mol } I^-} \times \frac{166.0 \text{ g KI}}{1 \text{ mol KI}} = 1.4 \times 10^{-6} \text{ g KI}$$

72. First we determine $\left[Ag^+\right]$ that can exist with this $\left[Cl^-\right]$. We know that $\left[Cl^-\right]$ will be unchanged because precipitation will not be allowed to occur.

$$K_{sp} = \left[Ag^+\right]\left[Cl^-\right] = 1.8 \times 10^{-10} = \left[Ag^+\right]0.100 \text{ M}; \qquad \left[Ag^+\right] = \frac{1.8 \times 10^{-10}}{0.100} = 1.8 \times 10^{-9} \text{ M}$$

We now consider the complex ion equilibrium. If the complex ion's final concentration is x, then the decrease in $\left[NH_3\right]$ is $2x$, because 2 mol NH_3 react to form each mol of complex ion, as follows. $Ag^+(aq) + 2NH_3(aq) \rightleftharpoons \left[Ag(NH_3)_2\right]^+(aq)$ We can solve the K_f expression for x.

$$K_f = 1.6 \times 10^7 = \frac{\left[Ag(NH_3)_2\right]^+}{\left[Ag^+\right]\left[NH_3\right]^2} = \frac{x}{1.8 \times 10^{-9}(1.00 - 2x)^2}$$

$$x = \left(1.6 \times 10^7\right)\left(1.8 \times 10^{-9}\right)(1.00 - 2x)^2 = 0.029\left(1.00 - 4.00x + 4.00x^2\right) = 0.029 - 0.12x + 0.12x^2$$

$$0 = 0.029 - 1.12x + 0.12x^2 \quad \text{We use the quadratic formula to solve for } x.$$

$$x = \frac{-b \pm \sqrt{b^2 - 4ac}}{2a} = \frac{1.12 \pm \sqrt{(1.12)^2 - 4 \times 0.029 \times 0.12}}{2 \times 0.12} = \frac{1.12 \pm 1.114}{0.24} = 9.3,\ 0.025$$

Thus, we can add 0.025 mol $AgNO_3$ (~ 4.4 g $AgNO_3$) to this solution before we see a precipitate of $AgCl(s)$ form.

Precipitation and Solubilities of Metal Sulfides

73. We know that $K_{spa} = 3 \times 10^7$ for MnS and $K_{spa} = 6 \times 10^2$ for FeS. The metal sulfide will begin to precipitate when $Q_{spa} = K_{spa}$. Let us determine $\left[H_3O^+\right]$ just necessary to form each precipitate. We assume that the solution is saturated with H_2S, $\left[H_2S\right] = 0.10$ M.

$$K_{spa} = \frac{[M^{2+}][H_2S]}{[H_3O^+]^2} \quad [H_3O^+] = \sqrt{\frac{[M^{2+}][H_2S]}{K_{spa}}} = \sqrt{\frac{(0.10\ \text{M})(0.10\ \text{M})}{3 \times 10^7}} = 1.8 \times 10^{-5}\ \text{M for MnS}$$

$$\left[H_3O^+\right] = \sqrt{\frac{(0.10\ \text{M})(0.10\ \text{M})}{6 \times 10^2}} = 4.1 \times 10^{-3}\ \text{M for FeS}$$

Thus, if the $[H_3O^+]$ is maintained just a bit higher than 1.8×10^{-5} M, FeS will precipitate and $Mn^{2+}(aq)$ will remain in solution. To determine if the separation is complete, we see whether $\left[Fe^{2+}\right]$ has decreased to 0.1% or less of its original value when the solution is held at the aforementioned acidity. Let $\left[H_3O^+\right] = 2.0 \times 10^{-5}$ M and calculate $\left[Fe^{2+}\right]$.

$$K_{spa} = \frac{\left[Fe^{2+}\right][H_2S]}{\left[H_3O^+\right]^2} = 6 \times 10^2 = \frac{\left[Fe^{2+}\right](0.10\ \text{M})}{\left(2.0 \times 10^{-5}\ \text{M}\right)^2};\ \left[Fe^{2+}\right] = \frac{\left(6 \times 10^2\right)\left(2.0 \times 10^{-5}\right)^2}{0.10} = 2.4 \times 10^{-6}\ \text{M}$$

$$\%\,Fe^{2+}(aq)\ \text{remaining} = \frac{2.4 \times 10^{-6}\ \text{M}}{0.10\ \text{M}} \times 100\% = 0.0024\% \quad \therefore \text{Separation is complete.}$$

74. Since the cation concentrations are identical, the value of Q_{spa} is the same for each one. It is this value of Q_{spa} that we compare with K_{spa} to determine if precipitation occurs.

$$Q_{spa} = \frac{\left[M^{2+}\right][H_2S]}{\left[H_3O^+\right]^2} = \frac{0.05\,\text{M} \times 0.10\,\text{M}}{\left(0.010\,\text{M}\right)^2} = 5 \times 10^1$$

If $Q_{spa} > K_{spa}$, precipitation of the metal sulfide should occur. But, if $Q_{spa} < K_{spa}$, precipitation will not occur.

For CuS, $K_{spa} = 6 \times 10^{-16} < Q_{spa} = 5 \times 10^1$ Precipitation of $CuS(s)$ should occur.

For HgS, $K_{spa} = 2 \times 10^{-32} < Q_{spa} = 5 \times 10^1$ Precipitation of $HgS(s)$ should occur.

For MnS, $K_{spa} = 3 \times 10^7 > Q_{spa} = 5 \times 10^1$ Precipitation of $MnS(s)$ will not occur.

75. **(a)** Calculate $[H_3O^+]$ in the buffer solution with the Henderson–Hasselbalch equation.

$$pH = pK_a + \log\frac{\left[C_2H_3O_2^-\right]}{\left[HC_2H_3O_2\right]} = 4.74 + \log\frac{0.15\,M}{0.25\,M} = 4.52 \quad \left[H_3O^+\right] = 10^{-4.52} = 3.0\times10^{-5}\,M$$

We use this information to calculate a value of Q_{spa} for MnS in this solution and then comparison of Q_{spa} with K_{spa} will allow us to decide if a precipitate will form.

$$Q_{spa} = \frac{\left[Mn^{2+}\right]\left[H_2S\right]}{\left[H_3O^+\right]^2} = \frac{(0.15)(0.10)}{\left(3.0\times10^{-5}\right)^2} = 1.7\times10^7 < 3\times10^7 = K_{spa} \text{ for MnS}$$

Precipitation of MnS(s) will not occur.

(b) We need to change $[H_3O^+]$ so that

$$Q_{spa} = 3\times10^7 = \frac{(0.15)(0.10)}{\left[H_3O^+\right]^2}; \quad \left[H_3O^+\right] = \sqrt{\frac{(0.15)(0.10)}{3\times10^7}}$$

$$[H_3O^+] = 2.2\times10^{-5}\,M \qquad pH = 4.66$$

This is a more basic solution, which we can achieve by increasing the basic component of the buffer solution, the acetate ion. We find out the new acetate ion concentration with the Henderson–Hasselbalch equation.

$$pH = pK_a + \log\frac{[C_2H_3O_2^-]}{[HC_2H_3O_2]} = 4.66 = 4.74 + \log\frac{[C_2H_3O_2^-]}{0.25\,M}$$

$$\log\frac{[C_2H_3O_2^-]}{0.25\,M} = 4.66 - 4.74 = -0.08$$

$$\frac{\left[C_2H_3O_2^-\right]}{0.25\,M} = 10^{-0.08} = 0.83 \quad \left[C_2H_3O_2^-\right] = 0.83\times0.25\,M = 0.21\,M$$

76. **(a)** CuS is in the hydrogen sulfide group of qualitative analysis. Its precipitation occurs when 0.3 M HCl is saturated with H_2S. It will certainly precipitate from a (non-acidic) saturated solution of H_2S, which has a much higher $\left[S^{2-}\right]$.

$$Cu^{2+}\left(aq\right) + H_2S\left(satd\ aq\right) \longrightarrow CuS(s) + 2H^+\left(aq\right)$$

This reaction proceeds to a significant extent in the forward direction.

(b) MgS is soluble, according to the solubility rules listed in Chapter 5.
$$Mg^{2+}\left(aq\right) + H_2S\left(satd\ aq\right) \xrightarrow{\ 0.3\,M\ HCl\ } \text{no reaction}$$

(c) As in part (a), PbS is in the qualitative analysis hydrogen sulfide group, which precipitates from a 0.3 M HCl solution saturated with H_2S. Therefore, PbS does not dissolve appreciably in 0.3 M HCl. $PbS(s) + HCl\left(0.3M\right) \longrightarrow$ no reaction

(d) Since ZnS(s) does not precipitate in the hydrogen sulfide group, we conclude that it is soluble in acidic solution.

$$ZnS(s) + 2HNO_3(aq) \longrightarrow Zn(NO_3)_2(aq) + H_2S(g)$$

Qualitative Analysis

77. The purpose of adding hot water is to separate Pb^{2+} from AgCl and Hg_2Cl_2. Thus, the most important consequence of the absence of a valid test for the presence or absence of Pb^{2+}. In addition, if we add NH_3 first, $PbCl_2$ may form $Pb(OH)_2$. If $Pb(OH)_2$ does form, it will be present with Hg_2Cl_2 in the solid, although $Pb(OH)_2$ will not darken with added NH_3. Thus, we might falsely conclude that Ag^+ is present, but not falsely conclude that Hg_2^{2+} is present.

78. For $PbCl_2(aq)$, $2\left[Pb^{2+}\right] = \left[Cl^-\right]$ and we let $s = $ molar solubility of $PbCl_2$. Thus $s = \left[Pb^{2+}\right]$.

$$K_{sp} = [Pb^{2+}][Cl^-]^2 = (s)(2s)^2 = 4s^3 = 1.6 \times 10^{-5}; \quad s = \sqrt[3]{1.6 \times 10^{-5} \div 4} = 1.6 \times 10^{-2} \text{ M} = [Pb^{2+}]$$

Both $\left[Pb^{2+}\right]$ and $\left[CrO_4^{2-}\right]$ are diluted by mixing the two solutions.

$$\left[Pb^{2+}\right] = 0.016 \text{ M} \times \frac{1.00 \text{ mL}}{1.05 \text{ mL}} = 0.015 \text{ M} \qquad \left[CrO_4^{2-}\right] = 1.0 \text{ M} \times \frac{0.05 \text{ mL}}{1.05 \text{ mL}} = 0.048 \text{ M}$$

$$Q_{sp} = \left[Pb^{2+}\right]\left[CrO_4^{2-}\right] = (0.015 \text{ M})(0.048 \text{ M}) = 7.2 \times 10^{-4} > 2.8 \times 10^{-13} = K_{sp}$$

Thus, precipitation should occur from the solution described.

79. **(a)** Ag^+ and/or Hg_2^{2+} are probably present. Both of these cations form precipitates from an acidic solution of chloride ion.

(b) We cannot tell whether Mg^{2+} is present or not. Both MgS and $MgCl_2$ are water soluble.

(c) Pb^{2+} possibly is absent; it is the only cation of those given which forms a precipitate in an acidic solution that is treated with H_2S, and no sulfide precipitate was formed.

(d) We cannot tell whether Fe^{2+} is present or not. FeS will not precipitate from an acidic solution that is treated with H_2S; the solution must be alkaline for a FeS precipitate to form.

(a) and **(c)** are the valid conclusions.

80. **(a)** $Pb^{2+}(aq) + 2Cl^-(aq) \longrightarrow PbCl_2(s)$

(b) $Zn(OH)_2(s) + 2OH^-(aq) \longrightarrow \left[Zn(OH)_4\right]^{2-}(aq)$

(c) $Fe(OH)_3(s) + 3H_3O^+(aq) \longrightarrow Fe^{3+}(aq) + 6H_2O(l)$ or $\left[Fe(H_2O)_6\right]^{3+}(aq)$

(d) $Cu^{2+}(aq) + H_2S(aq) \longrightarrow CuS(s) + 2H^+(aq)$

Feature Problems

100. $\left[Ca^{2+}\right] = \left[SO_4^{2-}\right]$ in the saturated solution. Let us first determine the amount of H_3O^+ in the 100.0 mL diluted effluent. $H_3O^+(aq) + NaOH(aq) \longrightarrow 2H_2O(l) + Na^+(aq)$

$$\text{mmol } H_3O^+ = 100.0 \text{ mL} \times \frac{8.25 \text{ mL base}}{10.00 \text{ mL sample}} \times \frac{0.0105 \text{ mmol NaOH}}{1 \text{ mL base}} \times \frac{1 \text{ mmol } H_3O^+}{1 \text{ mmol NaOH}}$$

$$= 0.866 \text{ mmol } H_3O^+(aq)$$

Now we determine $\left[Ca^{2+}\right]$ in the original 25.00 mL sample, remembering that $2H_3O^+$ were produced for each Ca^{2+}.

$$\left[Ca^{2+}\right] = \frac{0.866 \text{ mmol } H_3O^+(aq) \times \dfrac{1 \text{ mmol } Ca^{2+}}{2 \text{ mmol } H_3O^+}}{25.00 \text{ mL}} = 0.0173 \text{ M}$$

$K_{sp} = \left[Ca^{2+}\right]\left[SO_4^{2-}\right] = (0.0173)^2 = 3.0 \times 10^{-4}$; Compared with 9.1×10^{-6} in Appendix D.

101. (a) We assume that there is little of each ion present in solution at equilibrium; that this is a simple stoichiometric calculation. This is true because we are titrating: we stop when just enough silver ion has been added. $Ag^+(aq) + Cl^-(aq) \longrightarrow AgCl(s)$

$$V = 100.0 \text{ mL} \times \frac{29.5 \text{ mg } Cl^-}{1000 \text{ mL}} \times \frac{1 \text{ mmol } Cl^-}{35.45 \text{ mg } Cl^-} \times \frac{1 \text{ mmol } Ag^+}{1 \text{ mmol } Cl^-} \times \frac{1 \text{ mL}}{0.01000 \text{ mmol } AgNO_3}$$

$$= 8.32 \text{ mL}$$

(b) We first calculate the concentration of each ion as the consequence of dilution. Then we determine the $\left[Ag^+\right]$ from the value of K_{sp}

$$\text{initial } \left[Ag^+\right] = 0.01000 \text{ M} \times \frac{8.32 \text{ mL added}}{108.3 \text{ mL final volume}} = 7.68 \times 10^{-4} \text{ M}$$

$$\text{initial} \left[Cl^- \right] = \frac{29.5 \text{ mg } Cl^- \times \dfrac{1 \text{ mmol } Cl^-}{35.45 \text{ mg } Cl^-}}{1000 \text{ mL}} \times \frac{100.0 \text{ mL taken}}{108.3 \text{ mL final volume}} = 7.68 \times 10^{-4} M$$

The slight excess of each ion will precipitate until the solubility constant is satisfied.

$$\left[Ag^+ \right] = \left[Cl^- \right] = \sqrt{K_{sp}} = \sqrt{1.8 \times 10^{-10}} = 1.3 \times 10^{-5} M$$

(c) If we want Ag_2CrO_4 to appear just when AgCl has completed precipitation,

$\left[Ag^+ \right] = 1.3 \times 10^{-5}$ M. We determine $\left[CrO_4^{2-} \right]$ from the K_{sp} expression.

$$K_{sp} = \left[Ag^+ \right]^2 \left[CrO_4^{2-} \right] = 1.1 \times 10^{-12} = \left(1.3 \times 10^{-5} \right)^2 \left[CrO_4^{2-} \right];$$

$$\left[CrO_4^{2-} \right] = \frac{1.1 \times 10^{-12}}{\left(1.3 \times 10^{-5} \right)^2} = 0.0065 \text{ M}$$

(d) If $\left[CrO_4^{2-} \right]$ were greater than just computed, Ag_2CrO_4 would appear before all Cl^- had precipitated, leading to a false early endpoint. We would calculate a falsely low $\left[Cl^- \right]$ of the original solution.

If $\left[CrO_4^{2-} \right]$ were less than computed in part 3, Ag_2CrO_4 would appear somewhat after all Cl^- had precipitated, leading one to conclude there was more Cl^- in solution than actually was the case.

(e) If Ag^+ were in the sample being titrated, it would react immediately with the CrO_4^{2-} in that sample, forming a red-orange precipitate. This precipitate would not be likely to dissolve so that AgCl could form. There would be no visual indication of the endpoint.

102. (a) We need to calculate the $\left[Mg^{2+} \right]$ in a solution that is saturated with $Mg(OH)_2$.

$$K_{sp} = 1.8 \times 10^{-11} = \left[Mg^{2+} \right] \left[OH^- \right]^2 = (s)(2s)^2 = 4s^3$$

$$s = \sqrt[3]{\frac{1.8 \times 10^{-11}}{4}} = 1.7 \times 10^{-4} \text{ M} = [Mg^{2+}]$$

(b) Even though water has been added to the original solution, it still is saturated (it is in equilibrium with the undissolved solid $Mg(OH)_2$). $\left[Mg^{2+} \right] = 1.7 \times 10^{-4} M.$

(c) Although HCl(aq) reacts with OH^-, it will not react with Mg^{2+}. The solution simply

is diluted. $\left[Mg^{2+}\right] = 1.7 \times 10^{-4} \, M \times \dfrac{100.0 \text{ mL initial volume}}{(100.0 + 500.) \text{ mL final volume}} = 2.8 \times 10^{-5} \, M$

(d) In this instance, we have a dual dilution to a 275.0 mL total volume, followed by a common-ion problem.

$$\text{initial } \left[Mg^{2+}\right] = \frac{\left(25.00 \text{ mL} \times \dfrac{1.7 \times 10^{-4} \text{ mmol Mg}^{2+}}{1 \text{ mL}}\right) + \left(250.0 \text{ mL} \times \dfrac{0.065 \text{ mmol Mg}^{2+}}{1 \text{ mL}}\right)}{275.0 \text{ mL total volume}}$$

$$= 0.059 \, M$$

$$\text{initial } \left[OH^-\right] = \frac{25.00 \text{ mL} \times \dfrac{1.7 \times 10^{-4} \text{ mmol Mg}^{2+}}{1 \text{ mL}} \times \dfrac{2 \text{ mmol OH}^-}{1 \text{ mmol Mg}^{2+}}}{275.0 \text{ mL total volume}} = 3.1 \times 10^{-5} \, M$$

Let's see if precipitation occurs.

$$Q_{sp} = \left[Mg^{2+}\right]\left[OH^-\right]^2 = (0.059)(3.1 \times 10^{-5})^2 = 5.7 \times 10^{-11} > 1.8 \times 10^{-11} = K_{sp}$$

Precipitation barely occurs. If $\left[OH^-\right]$ goes down by 1.4×10^{-5} M (which means that

$\left[Mg^{2+}\right]$ drops by $0.7 \times 10^{-5} \, M$) then $\left[OH^-\right] = 1.7 \times 10^{-5} \, M$ and

$\left[Mg^{2+}\right] = (0.059 \text{ M} - 0.7 \times 10^{-5} \text{ M} =)0.059 \text{ M}$, then $Q_{sp} < K_{sp}$ and precipitation will

stop. Thus, $\left[Mg^{2+}\right] = 0.059 \text{ M}$.

(e) Again we have a dual dilution, now to a 200.0 mL final volume, followed by a common-ion problem.

$$\text{initial } \left[Mg^{2+}\right] = \frac{50.00 \text{ mL} \times \dfrac{1.7 \times 10^{-4} \text{ mmol Mg}^{2+}}{1 \text{ mL}}}{200.0 \text{ mL total volume}} = 4.3 \times 10^{-5} \, M$$

$$\text{initial } \left[OH^-\right] = 0.150 \text{ M} \times \frac{150.0 \text{ mL initial volume}}{200.0 \text{ mL total volume}} = 0.113 \text{ M}$$

Now it is evident that precipitation will occur. We determine the $\left[Mg^{2+}\right]$ that can

exist in solution with 0.113 M OH^-. It is clear that $\left[Mg^{2+}\right]$ will drop dramatically to

satisfy the K_{sp} expression but the larger value of $\left[OH^-\right]$ will scarcely be affected.

$$K_{sp} = \left[Mg^{2+}\right]\left[OH^-\right]^2 = 1.8 \times 10^{-11} = \left[Mg^{2+}\right](0.0113 \text{ M})^2$$

$$\left[Mg^{2+}\right] = \frac{1.8 \times 10^{-11}}{(0.113)^2} = 1.4 \times 10^{-9} \, M$$

CHAPTER 20
SPONTANEOUS CHANGE:
ENTROPY AND FREE ENERGY
PRACTICE EXAMPLES

1A In general, $\Delta S > 0$ if $\Delta n_{gas} > 0$. This is because gases are very disordered compared to liquids or solids; (gases possess large entropies). Recall that Δn_{gas} is the difference between the sum of the stoichiometric coefficients of the gaseous products and a similar sum for the reactants.

(a) $\Delta n_{gas} = 2 + 0 - (2 + 1) = -1$. One mole of gas is consumed here. We predict $\Delta S < 0$.

(b) $\Delta n_{gas} = 1 + 0 - 0 = +1$. Since one mole of gas is produced, we predict $\Delta S > 0$.

1B **(a)** The outcome is uncertain in the reaction between $ZnS(s)$ and $Ag_2O(s)$. We have used Δn_{gas} to estimate the sign of entropy change. There is no gas involved in this reaction and thus our prediction is uncertain.

(b) In the chlor-alkali process we are confident that entropy increases because two moles of gas have formed where none where originally present.

2A For a vaporization, $\Delta G^{\circ}_{vap} = 0 = \Delta H^{\circ}_{vap} - T\Delta S^{\circ}_{vap}$. Thus, $\Delta S^{\circ}_{vap} = \Delta H^{\circ}_{vap} / T_{vap}$.

We substitute the given values. $\Delta S^{\circ}_{vap} = \dfrac{\Delta H^{\circ}_{vap}}{T_{vap}} = \dfrac{20.2 \text{ kJ mol}^{-1}}{(-29.79 + 273.15) \text{ K}} = 83.0 \text{ J mol}^{-1} \text{ K}^{-1}$

2B For a phase change, $\Delta G^{\circ}_{tr} = 0 = \Delta H^{\circ}_{tr} - T\Delta S^{\circ}_{tr}$. Thus, $\Delta H^{\circ}_{tr} = T\Delta S^{\circ}_{tr}$. We substitute the given values. $\Delta H^{\circ}_{tr} = T\Delta S^{\circ}_{tr} = (95.5 + 273.2) \text{ K} \times 1.09 \text{ J mol}^{-1} \text{ K}^{-1} = 402 \text{ J / mol}$

3A The entropy change for the reaction is expressed in terms of the standard entropies of the reagents.

$$\Delta S^{\circ} = 2S^{\circ}\left[NH_3(g)\right] - S^{\circ}\left[N_2(g)\right] - 3S^{\circ}\left[H_2(g)\right]$$
$$= 2 \times 192.5 \text{ J mol}^{-1}\text{K}^{-1} - 191.6 \text{ J mol}^{-1} \text{ K}^{-1} - 3 \times 130.7 \text{ J mol}^{-1} \text{ K}^{-1} = -198.7 \text{ J mol}^{-1} \text{ K}^{-1}$$

To form *one* mole of $NH_3(g)$, the standard entropy change is -99.4 J mol^{-1} K^{-1}

3B   The entropy change for the reaction is expressed in terms of the standard entropies of the reagents.

$$\Delta S^\circ = S^\circ \left[NO(g) \right] + S^\circ \left[NO_2(g) \right] - S^\circ \left[N_2O_3(g) \right]$$

$$138.5 \text{ J mol}^{-1} \text{ K}^{-1} = 210.8 \text{ J mol}^{-1} \text{ K}^{-1} + 240.1 \text{ J mol}^{-1} \text{ K}^{-1} - S^\circ \left[N_2O_3(g) \right]$$

$$= 450.9 \text{ J mol}^{-1} \text{ K}^{-1} - S^\circ \left[N_2O_3(g) \right]$$

$$S^\circ \left[N_2O_3(g) \right] = 450.9 \text{ J mol}^{-1} \text{ K}^{-1} - 138.5 \text{ J mol}^{-1} \text{ K}^{-1} = 312.4 \text{ J mol}^{-1} \text{ K}$$

4A   **(a)**   Because $\Delta n_{gas} = 2 - (1+3) = -2$ for the synthesis of ammonia, we would predict $\Delta S < 0$ for the reaction. We already know that $\Delta H < 0$. Thus, the reaction falls into case 2, spontaneous at low temperatures and nonspontaneous at high temperatures.

   (b)   For the formation of ethylene $\Delta n_{gas} = 1 - (2+0) = -1$ and thus $\Delta S < 0$. We are given that $\Delta H > 0$ and, thus, this reaction corresponds to case 4, a reaction that is nonspontaneous at all temperatures.

4B   **(a)**   Because $\Delta n_{gas} = +1$ for the decomposition of calcium carbonate, we would predict $\Delta S > 0$ for the reaction, favoring the reaction at high temperatures. High temperatures also favor this endothermic $(\Delta H^\circ > 0)$ reaction.

   (b)   Roasting ZnS(s) has $\Delta n_{gas} = 2 - 3 = -1$ and, thus, $\Delta S < 0$. We are given that $\Delta H < 0$; thus, this reaction corresponds to case 2, a reaction spontaneous at low temperatures, and nonspontaneous at high ones.

5A   The expression $\Delta G^\circ = \Delta H^\circ - T\Delta S^\circ$ is used with $T = 298.15 \text{ K}$.

$$\Delta G^\circ = \Delta H^\circ - T\Delta S^\circ = -1648 \text{ kJ} - 298.15 \text{ K} \times \left(-549.3 \text{ J K}^{-1} \right) \times \left(1 \text{ kJ} / 1000 \text{ J} \right)$$

$$= -1648 \text{ kJ} + 163.8 \text{ kJ} = -1484 \text{ kJ}$$

5B   We just need to substitute values from Appendix D into the supplied expression.

$$\Delta G^\circ = 2\Delta G_f^\circ \left[NO_2(g) \right] - 2\Delta G_f^\circ \left[NO(g) \right] - \Delta G_f^\circ \left[O_2(g) \right]$$

$$= 2 \times 51.31 \text{ kJ mol}^{-1} - 2 \times 86.55 \text{ kJ mol}^{-1} - 0.00 \text{ kJ mol}^{-1} = -70.48 \text{ kJ mol}^{-1}$$

6A   Pressures of gases and molarities of solutes in aqueous solution appear in thermodynamic equilibrium constant expressions. Pure solids and liquids (including solvents) do not appear.

   (a) $K_{eq} = \dfrac{P\{SiCl_4\}}{P\{Cl_2\}^2} = K_p$ **(b)** $K_{eq} = \dfrac{[HOCl][H^+][Cl^-]}{P\{Cl_2\}}$

$K_{eq} = K_p$ for (a) because all factors in the K_{eq} expression are gas pressures.

6B   We need the balanced chemical equation in order to write the equilibrium constant expression. We start by translating names into formulas.

$$PbS(s) + HNO_3(aq) \rightarrow Pb(NO_3)_2(aq) + S(s) + NO(g)$$

The equation then is balanced with the ion-electron method.

$$\text{oxidation}: \{PbS(s) \rightarrow Pb^{2+}(aq) + S(s) + 2e^- \qquad\qquad\} \times 3$$

$$\text{reduction}: \{NO_3^-(aq) + 4H^+(aq) + 3e^- \rightarrow NO(g) + 2H_2O(l)\ \} \times 2$$

$$\text{net ionic}: 3\,PbS(s) + 2NO_3^-(aq) + 8H^+(aq) \rightarrow 3Pb^{2+}(aq) + 3S(s) + 2NO(g) + 4H_2O(l)$$

In writing the thermodynamic equilibrium constant, recall that neither pure solids ($PbS(s)$ and $S(s)$) nor pure liquids $(H_2O(l))$ appear in the thermodynamic equilibrium constant expression. Note also that we have written $H^+(aq)$ here for brevity even though we understand that $H_3O^+(aq)$ is the acidic species in aqueous solution.

$$K = \frac{\left[Pb^{2+}\right]^3 P\{NO\}^2}{\left[NO_3^-\right]^2\left[H^+\right]^8}$$

7A   Since the reaction is taking place at 298.15 K, we can use standard free energies of formation to calculate the standard free energy change for the reaction:

$$N_2O_4 \ \rightleftharpoons\ 2\,NO_2$$

$$\Delta G^\circ = 2\Delta G_f^\circ\left[NO_2(g)\right] - \Delta G_f^\circ\left[N_2O_4(g)\right] = 2 \times 51.31\ \text{kJ/mol} - 97.89\ \text{kJ/mol} = +4.73\,\text{kJ}$$

$\Delta G^{\circ}_{rxn} = +4.73$ kJ . Thus, the reaction is non-spontaneous as written at 298.15 K.

7B   In order to answer this question we must calculate the reaction quotient and compare it to the K_p value for the reaction:

$$N_2O_4 \ \rightleftharpoons\ 2\,NO_2 \qquad\qquad Q_p = \frac{(0.5)^2}{0.5} = 0.5$$

0.5 bar   0.5 bar

$$\Delta G^{\circ}_{rxn} = +4.73\ \text{kJ} = -RT\ln K_p; \quad 4.73\ \text{kJ/mol} = -(8.3145 \times 10^{-3}\ \text{kJ/K·mol})(298.15\ \text{K})\ln K_p$$

Therefore, $K_p = 0.148$. Since Q_p is greater than K_p, we can conclude that the reverse reaction will proceed spontaneously, i.e. NO_2 will spontaneously convert into N_2O_4.

8A   We first determine the value of ΔG° and then set $\Delta G^\circ = -RT\ln K_{eq}$ to determine K_{eq}.

$$\Delta G = \Delta G_f^\circ\left[Ag^+(aq)\right] + \Delta G_f^\circ[I^-(aq)] - \Delta G_f^\circ\left[AgI(s)\right]$$

$$= [(77.11 - 51.57) - (-66.19)]\ \text{kJ/mol} = +91.73$$

$$\ln K_{eq} = \frac{-\Delta G^\circ}{RT} = -\frac{-91.73\ \text{kJ/mol}}{8.3145\ \text{J mol}^{-1}\,\text{K}^{-1} \times 298.15\ \text{K}} \times \frac{1000\ \text{J}}{1\ \text{kJ}} = -37.00$$

$$K_{eq} = e^{-37.00} = 8.5 \times 10^{-17}$$

This is precisely equal to the value for the K_{sp} listed in Appendix D.

8B   We begin by translating names into formulas. $MnO_2(s) + HCl(aq) \rightarrow Mn^{2+}(aq) + Cl_2(aq)$
Then we produce a balanced net ionic equation with the ion-electron method.

$$\text{oxidation}: 2\,Cl^-(aq) \rightarrow Cl_2(g) + 2e^-$$

$$\text{reduction}: MnO_2(s) + 4H^+(aq) + 2e^- \rightarrow Mn^{2+}(aq) + 2H_2O(l)$$

$$\text{net ionic}: MnO_2(s) + 4H^+(aq) + 2Cl^-(aq) \rightarrow Mn^{2+}(aq) + Cl_2(g) + 2H_2O(l)$$

Next we determine the value of $\Delta G°$ for the reaction and then the value of the equilibrium constant

$$\Delta G° = \Delta G_f°\left[Mn^{2+}(aq)\right] + \Delta G_f°\left[Cl_2(g)\right] + 2\Delta G_f°\left[H_2O(l)\right]$$

$$- \Delta G_f°\left[MnO_2(s)\right] - 4\Delta G_f°\left[H^+(aq)\right] - 2\Delta G_f°\left[Cl^-(aq)\right]$$

$$= -228.1\text{ kJ} + 0.0\text{ kJ} + 2 \times (-237.1\text{ kJ})$$

$$-(-465.1\text{ kJ}) - 4 \times 0.0\text{ kJ} - 2 \times (-131.2\text{ kJ}) = +25.2\text{ kJ}$$

$$\ln K_{eq} = \frac{-\Delta G°}{RT} = \frac{-(+25.2 \times 10^3\text{ J mol}^{-1})}{8.3145\text{ J mol}^{-1}\text{ K}^{-1} \times 298.15\ K} = -10.1\underline{7} \quad K_{eq} = e^{-10.2} = 4 \times 10^{-5}$$

Because the value of K_{eq} is so much smaller than unity, we do not expect an appreciable reaction.

9A   We set equal the two expressions for $\Delta G°$ and solve for the absolute temperature.

$$\Delta G° = \Delta H° - T\Delta S° = -RT\ln K_{eq} \qquad \Delta H° = T\Delta S° - RT\ln K_{eq} = T\left(\Delta S° - R\ln K_{eq}\right)$$

$$T = \frac{\Delta H°}{\Delta S° - R\ln K_{eq}} = \frac{-114.1 \times 10^3\text{ J / mol}}{\left[-146.4 - 8.3145\ \ln(150)\right]\text{J mol}^{-1}\text{ K}^{-1}} = 607\text{ K}$$

9B   We expect the value of the equilibrium constant to increase as temperature decreases since this is an exothermic reaction and it should become more spontaneous (shift right) at lower temperatures. Thus, we expect K_{eq} to be larger than 1000, which is its value at 4.3×10^2 K.

(a)   The value of the equilibrium constant at $25°$ C is obtained directly from the value of $\Delta G°$, since that value is also for $25°C$. Note:
$$\Delta G° = \Delta H° - T\Delta S° = -77.1\text{ kJ/mol} - 298.15\text{ K}(-0.1213\text{ kJ/mol}\cdot\text{K}) = -40.9\text{ kJ/mol}$$

$$\ln K_{eq} = \frac{-\Delta G°}{RT} = \frac{-(-40.9 \times 10^3\text{ J mol}^{-1})}{8.3145\text{ J mol}^{-1}\text{ K}^{-1} \times 298.15\text{ K}} = 16.5 \quad K_{eq} = e^{+16.5} = 1.5 \times 10^7$$

(b)   First, we solve for $\Delta G°$ at $75°C = 348$ K

$$\Delta G° = \Delta H° - T\Delta S° = -77.1\ \frac{\text{kJ}}{\text{mol}} \times \frac{1000\text{ J}}{1\text{ kJ}} - \left(348.15\text{ K} \times \left(-121.3\frac{\text{J}}{\text{mol K}}\right)\right)$$

$$= -34.87 \times 10^3\text{ J/mol}$$

Then we use this value to obtain the value of the equilibrium constant, as in part (a).

$$\ln K_{eq} = \frac{-\Delta G°}{RT} = \frac{-(-34.87 \times 10^3 \text{ J mol}^{-1})}{8.3145 \text{ J mol}^{-1}\text{ K}^{-1} \times 348.15 \text{ K}} = 12.05 \qquad K_{eq} = e^{+12.05} = 1.7 \times 10^5$$

10A   We use the value of $K_p = 9.1 \times 10^2$ at 800 K and $\Delta H° = -1.8 \times 10^5$ J / mol in the van't Hoff equation.

$$\ln \frac{5.8 \times 10^{-2}}{9.1 \times 10^2} = \frac{-1.8 \times 10^5 \text{ J/mol}}{8.3145 \text{ J mol}^{-1}\text{ K}^{-1}} \left(\frac{1}{800 \text{ K}} - \frac{1}{T \text{ K}} \right) = -9.66; \quad \frac{1}{T} = \frac{1}{800} - \frac{9.66 \times 8.3145}{1.8 \times 10^5}$$

$$1/T = 1.25 \times 10^{-3} - 4.5 \times 10^{-4} = 8.0 \times 10^{-4} \qquad T = 1240 \text{ K} \approx 970°\text{C}$$

This temperature is an estimate because it is an extrapolated point beyond the range of the data supplied.

10B   The temperature we are considering is $235°\text{C} = 508$ K. We use the value of $K_p = 9.1 \times 10^2$ at 800 K and $\Delta H° = -1.8 \times 10^5$ J/mol in the van't Hoff equation.

$$\ln \frac{K_p}{9.1 \times 10^2} = \frac{-1.8 \times 10^5 \text{ J/mol}}{8.3145 \text{ J mol}^{-1}\text{ K}^{-1}} \left(\frac{1}{800 \text{ K}} - \frac{1}{508 \text{ K}} \right) = +15_{.6}; \quad \frac{K_p}{9.1 \times 10^2} = e^{+15.6} = 6 \times 10^6$$

$$K_p = 6 \times 10^6 \times 9.1 \times 10^2 = 5 \times 10^9$$

REVIEW QUESTIONS

1.   **(a)** ΔS_{univ} is the symbol for the entropy change of the universe. If this quantity is positive, the associated process is spontaneous.

   (b) $\Delta G_f°$ is the symbol for the standard free energy of formation, the free energy change of the standard state reaction when one mole of a substance is produced from stable forms of its elements.

   (c) K_{eq} is the symbol for the thermodynamic equilibrium constant; in this expression gases are represented by pressures and solutes in aqueous solution by molarities.

2.   **(a)**   *Absolute* molar entropy refers to the entropy of a substance relative to pure perfect crystals at 0 K, which have zero entropy.

   (b)   A reversible process is one occurring in a state of dynamic equilibrium, one that can be made to reverse its direction when an infinitesimal change is made to some property of the system.

   (c)   Trouton's rule states that the entropy change of vaporization approximates 87 J mol^{-1} K^{-1}. It works well for nonpolar liquids.

   (d)   An equilibrium constant is evaluated from tabulated thermodynamic data by combining values of $\Delta G_f°$ to obtain $\Delta G_{rxn}°$ and then using $\Delta G_{rxn}° = -RT \ln K_{eq}$.

3. (a) A spontaneous process is one that occurs without (or sometimes in spite of) external intervention. A nonspontaneous process will only occur when some external agency operates on it.

    (b) The second law of thermodynamics places limitations on converting heat into work and on the spontaneous directions of processes. The third law of thermodynamics establishes a zero point for entropy.

    (c) ΔG refers to the free energy change for any process, while $\Delta G°$ requires that both the initial and final states of the process are standard ones: gases at 1 atm pressure, aqueous solutes at 1 molar concentration, solids and liquids are pure and the temperature is 298 K.

4. (a) Increase in entropy because a gas has been created from a liquid, a condensed phase.

    (b) Decrease in entropy as a condensed phase, a solid, is created from a solid and a gas.

    (c) For this reaction we cannot be certain of the entropy change; even though the number of moles of gas produced is the same as the number that reacted, we cannot conclude that the entropy change is zero because not all gases have the same molar entropy.

    (d) $2H_2S(g) + 3O_2(g) \rightarrow 2H_2O(g) + 2SO_2(g)$ Decrease in entropy since five moles of gas with high entropy become only four moles of gas, with about the same quantity of entropy per mole.

5. (a) At $75°\,C$, 1 mol H_2O (g, 1 atm) has a greater entropy than 1 mol H_2O (liq., 1 atm) since a gas is much more disordered than a liquid.

    (b) $50.0 \text{ g Fe} \times \dfrac{1 \text{ mol Fe}}{55.8 \text{ g Fe}} = 0.896$ mol Fe has a higher entropy than 0.80 mol Fe, both (s)

    at 1 atm and $5°\,C$, because entropy is an extensive property that depends on the amount of substance present.

    (c) 1 mol Br_2 (liq., 1 atm, $8°\,C$) has a higher entropy than 1 mol Br_2 (s, 1atm, $-8°\,C$) because solids are more ordered substances than are liquids, and the temperature is higher.

    (d) 0.312 mol SO_2 (g, 0.110 atm, $32.5°\,C$) has a higher entropy than 0.284 mol O_2 (g, 15.0 atm, $22.3°\,C$) for at least three reasons. First, entropy is an extensive property that depends on the amount of substance present. Second, entropy increases with temperature. Third, entropy is greater at lower pressures. Furthermore, entropy generally is higher per mole for more complicated molecules.

6. We predict the sign of ΔS based on the number of moles of gas in the balanced chemical equation; if $\Delta n_{gas} < 0$, ΔS is smaller than zero. If $\Delta n_{gas} > 0$, ΔS is greater than zero.

   (a) $\Delta S > 0$ and $\Delta H > 0$ This is case 3 in Table 20-1.

   (b) $\Delta S < 0$ and $\Delta H < 0$ This is case 2 in Table 20-1.

   (c) $\Delta S > 0$ and $\Delta H < 0$ This is case 1 in Table 20-1.

   (d) $\Delta S < 0$ and $\Delta H > 0$ This is case 4 in Table 20-1.

7. Answer (b) is correct. Br—Br bonds are broken in this reaction, meaning that it is endothermic, with $\Delta H > 0$. Since the number of moles of gas increases during the reaction, $\Delta S > 0$. And, because $\Delta G = \Delta H - T \Delta S$, this reaction is nonspontaneous at low temperatures where the ΔH term predominates, with $\Delta G > 0$, and spontaneous at high temperatures where the $T \Delta S$ term predominates, with $\Delta G < 0$.

8. Answer (d) is correct. A reaction that proceeds only through electrolysis is a reaction that is nonspontaneous. Such a reaction has $\Delta G > 0$.

9. Answer (d) is correct. Because $\Delta G^\circ = -RT \ln K_{eq}$, if $\Delta G^\circ = 0, \ln K_{eq} = 0$, which means that $K_{eq} = 1$.

10. $\Delta H^\circ = \Delta H_f^\circ \left[NH_4Cl(s) \right] - \Delta H_f^\circ \left[NH_3(g) \right] - \Delta H_f^\circ \left[HCl(g) \right]$

   $= -314.4 \, kJ/mol - (-46.11 \, kJ/mol - 92.31 \, kJ/mol) = -176.0 \, kJ/mol$

   $\Delta G^\circ = \Delta G_f^\circ \left[NH_4Cl(s) \right] - \Delta G_f^\circ \left[NH_3(g) \right] - \Delta G_f^\circ \left[HCl(g) \right]$

   $= -202.9 \, kJ/mol - (-16.48 \, kJ/mol - 95.30 \, kJ/mol) = -91.1 \, kJ/mol$

   $\Delta G^\circ = \Delta H^\circ - T \Delta S^\circ \quad \Delta S^\circ = \dfrac{\Delta H^\circ - \Delta G^\circ}{T} = \dfrac{-176.0 \, kJ/mol + 91.1 \, kJ/mol}{298 \, K} \times \dfrac{1000 \, J}{1 \, kJ} = -285 \, J \, mol^{-1}$

11. (a) $\Delta G^\circ = \Delta G_f^\circ \left[C_2H_6(g) \right] - \Delta G_f^\circ \left[C_2H_2(g) \right] - 2\Delta G_f^\circ \left[H_2(g) \right]$

   $= -32.82 \, kJ/mol - 209.2 \, kJ/mol - 2(0.00 \, kJ/mol) = -242.0 \, kJ/mol$

(b) $\Delta G^\circ = 2\Delta G_f^\circ \left[SO_2(g) \right] + \Delta G_f^\circ \left[O_2(g) \right] - 2\Delta G_f^\circ \left[SO_3(g) \right]$

   $= 2(-300.2 \, kJ / mol) + 0.00 \, kJ / mol - 2(-371.1 \, kJ / mol) = +141.8 \, kJ / mol$

(c) $\Delta G^\circ = 3\Delta G_f^\circ \left[Fe(s) \right] + 4\Delta G_f^\circ \left[H_2O(g) \right] - \Delta G_f^\circ \left[Fe_3O_4(s) \right] - 4\Delta G_f^\circ \left[H_2(g) \right]$

   $= 3(0.00 \, kJ/mol) + 4(-228.6 \, kJ/mol) - (-1015 \, kJ/mol) - 4(0.00 \, kJ/mol)$

   $= 101 \, kJ/mol$

(d) $\Delta G^\circ = 2\Delta G_f^\circ \left[Al^{3+}(aq) \right] + 3\Delta G_f^\circ \left[H_2(g) \right] - 2\Delta G_f^\circ \left[Al(s) \right] - 6\Delta G_f^\circ \left[H^+(aq) \right]$

   $= 2(-485 \, kJ/mol) + 3(0.00 \, kJ/mol) - 2(0.00 kJ/mol) - 6(0.00 \, kJ/mol)$

   $= -970. \, kJ/mol$

12. **(a)** This temperature is the melting point of $I_2(s)$ at 1 atm pressure. From the phase diagram for iodine, we see that this process occurs at $113.6°C$.

(b) $\Delta G° = 0$ for $I_2(s, 1\text{ atm}) \rightleftharpoons I_2(l, 1\text{ atm})$. The point of the crossing of the lines in Figure 20-9 is the point where $\Delta G° = 0$. In addition the melting point of a substance is the temperature where liquid and solid have the same free energy.

13. **(a)** $\Delta H°_{vap} = \dfrac{\Delta H°_{vap}}{T_{vap}} = \dfrac{3.86\text{ kcal/mol}}{-85.05°C + 273.15\text{ K}} \times \dfrac{1000\text{ cal}}{1\text{ kcal}} \times \dfrac{4.184\text{ J}}{1\text{ cal}} = 85.9\text{ J mol}^{-1}\text{K}^{-1}$

(b) $\Delta S°_{fus} = \dfrac{\Delta H°_{fus}}{T_{fus}} = \dfrac{27.05\text{ cal/g}}{97.82°C + 273.15\text{ K}} \times \dfrac{22.99\text{ g Na}}{1\text{ mol Na}} \times \dfrac{4.184\text{ J}}{1\text{ cal}} = 7.014\text{ J mol}^{-1}\text{K}^{-1}$

14. **(a)** $K_{eq} = \dfrac{P\{NO_2(g)\}^2}{P\{NO(g)\}^2\, P\{O_2\}} = K_p$ **(c)** $K_{eq} = \dfrac{\left[H_3O^+\right]\left[C_2H_3O_2^-\right]}{\left[HC_2H_3O_2\right]} = K_c$

(b) $K_{eq} = P\{SO_2(g)\} = K_p$ **(d)** $K_{eq} = P\{H_2O(g)\}\, P\{CO_2(g)\} = K_p$

(e) $K_{eq} = \dfrac{\left[Mn^{2+}(aq)\right]P\{Cl_2(g)\}}{\left[H^+\right]^4\left[Cl^-\right]^2}$, neither K_p nor K_c

15. $\Delta G° = -RT \ln K_p = -\left(8.3145\text{ J mol}^{-1}\text{K}^{-1}\right)(1000.\text{K})\left(\dfrac{1\text{ kJ}}{1000\text{ J}}\right)\ln\left(2.45\times10^{-7}\right) = 127\text{ kJ/mol}$

16. $\Delta G° = 2\Delta G°_f\left[NO(g)\right] - \Delta G°_f\left[N_2O(g)\right] - 0.5\,\Delta G°_f\left[O_2(g)\right]$

$\qquad = 2(86.55\text{ kJ/mol}) - (104.2\text{ kJ/mol}) - 0.5(0.00\text{ kJ/mol}) = 68.9\text{ kJ/mol}$

$\qquad = -RT \ln K_p = -\left(8.3145\times10^{-3}\text{ kJ mol}^{-1}\text{ K}^{-1}\right)(298\text{ K})\ln K_p$

$\ln K_p = -\dfrac{68.9\text{ kJ/mol}}{8.3145\times10^{-3}\text{ kJ mol}^{-1}\text{ K}^{-1}\times 298\text{ K}} = -27.8 \qquad K_p = e^{-27.8} = 8\times10^{-13}$

17. We first balance the chemical equation and then calculate the value of $\Delta G°$ with data from Appendix D, and finally calculate the value of K_{eq} with the use of $\Delta G° = -RT \ln K$.

(a) $4HCl(g) + O_2(g) \rightleftharpoons 2H_2O(g) + 2Cl_2(s)$

$\Delta G° = 2\Delta G°_f\left[H_2O(g)\right] + 2\Delta G°_f\left[Cl_2(g)\right] - 4\Delta G°_f\left[HCl(g)\right] - \Delta G°_f\left[O_2(g)\right]$

$\qquad = 2\times(-228.6\text{ kJ/mol}) + 2\times0\text{ kJ/mol} - 4\times(-95.30\text{ kJ/mol}) - 0\text{ kJ/mol}$

$\Delta G° = -76.0\text{ kJ/mol}$

$$\ln K_{eq} = \frac{-\Delta G^{\circ}}{RT} = \frac{+76.0 \times 10^3 \text{ J/mol}}{8.3145 \text{ J mol}^{-1} \text{ K}^{-1} \times 298 \text{ K}} = +30.7 \qquad K_{eq} = e^{+30.7} = 2 \times 10^{13}$$

(b) $3Fe_2O_3(s) + H_2(g) \rightleftharpoons 2Fe_3O_4(s) + H_2O(g)$

$$\Delta G^{\circ} = 2\Delta G_f^{\circ}\left[Fe_3O_4(s)\right] + \Delta G_f^{\circ}\left[H_2O(g)\right] - 3\Delta G_f^{\circ}\left[Fe_2O_3(s)\right] - \Delta G_f^{\circ}\left[H_2(g)\right]$$

$$= 2 \times (-1015 \text{ kJ/mol}) - 228.6 \text{ kJ/mol} - 3 \times (-742.2 \text{ kJ/mol}) - 0.00 \text{ kJ/mol}$$

$$= -32 \text{ kJ/mol}$$

$$\ln K_{eq} = \frac{-\Delta G^{\circ}}{RT} = \frac{32 \times 10^3 \text{ J/mol}}{8.3145 \text{ J mol}^{-1} \text{ K}^{-1} \times 298 \text{ K}} = 13; \quad K_{eq} = e^{+13} = 4 \times 10^5$$

(c) $2Ag^+(aq) + SO_4^{2-}(aq) \rightleftharpoons Ag_2SO_4(s)$

$$\Delta G^{\circ} = \Delta G_f^{\circ}\left[Ag_2SO_4(s)\right] - 2\Delta G_f^{\circ}\left[Ag^+(aq)\right] - \Delta G_f^{\circ}\left[SO_4^{2-}(aq)\right]$$

$$= -618.4 \text{ kJ/mol} - 2 \times 77.11 \text{ kJ/mol} - (-744.5 \text{ kJ/mol}) = -28.1 \text{ kJ/mol}$$

$$\ln K_{eq} = \frac{-\Delta G^{\circ}}{RT} = \frac{28.1 \times 10^3 \text{ J/mol}}{8.3145 \text{ J mol}^{-1} \text{ K}^{-1} \times 298 \text{ K}} = 11.3; \quad K_{eq} = e^{+11.3} = 8 \times 10^4$$

18. $\Delta S^{\circ} = S^{\circ}\{CO_2(g)\} + S^{\circ}\{H_2(g)\} - S^{\circ}\{CO(g)\} - S^{\circ}\{H_2O(g)\}$

$$= 213.7 \text{ J mol}^{-1} \text{ K}^{-1} + 130.7 \text{ J mol}^{-1} \text{ K}^{-1} - 197.7 \text{ J mol}^{-1} \text{ K}^{-1} - 188.8 \text{ J mol}^{-1}\text{K}^{-1}$$

$$= -42.1 \text{ J mol}^{-1} \text{ K}^{-1}$$

19. We only compute a value of ΔG° for each reaction. If this value is significantly less than zero, we conclude that the reaction is expected to occur to some extent at 298.15 K.

(a) $\Delta G^{\circ} = 2\Delta G_f^{\circ}\left[O_3(g)\right] - 3\Delta G_f^{\circ}\left[O_2(g)\right] = 2(163.2 \text{ kJ/mol}) - 3(0.00 \text{ kJ/mol})$

$$= 326.4 \text{ kJ/mol} \qquad \text{Does not occur to a significant extent.}$$

(b) $\Delta G^{\circ} = 2\Delta G_f^{\circ}\left[NO_2(g)\right] - \Delta G_f^{\circ}\left[N_2O_4(g)\right] = 2(51.31 \text{ kJ/mol}) - 97.89 \text{ kJ/mol}$

$$= 4.73 \text{ kJ/mol} \text{ Occurs only to a very small extent.}$$

(c) $\Delta G^{\circ} = 2\Delta G_f^{\circ}\left[BrCl(g)\right] - \Delta G_f^{\circ}\left[Cl_2(g)\right] - \Delta G_f^{\circ}\left[Br_2(l)\right]$

$$= 2(-0.98 \text{ kJ/mol}) - 0.00 \text{ kJ/mol} - 0.00 \text{ kJ/mol}$$

$$= -2.0 \text{ kJ/mol} \qquad \text{Occurs to some small extent.}$$

We could have predicted the first and last result by simply looking at the free energy of formation of product. The small size of the result in the other case indicates that some calculation is necessary.

20. $\Delta G^\circ = \Delta H^\circ - T\Delta S^\circ$ $T\Delta S^\circ = \Delta H^\circ - \Delta G^\circ$ $T = \dfrac{\Delta H^\circ - \Delta G^\circ}{\Delta S^\circ}$

$$T = \frac{-24.8 \times 10^3 \text{ J} - \left(-45.5 \times 10^3 \text{ J}\right)}{15.2 \text{ J/K}} = 1.36 \times 10^3 \text{ K}$$

21.   **(a)** $\Delta G^\circ = 2\Delta G_f^\circ\left[N_2O_5\,(g)\right] - 2\Delta G_f^\circ\left[N_2O_4\,(g)\right] - \Delta G_f^\circ\left[O_2\,(g)\right]$

$\qquad\qquad = 2(115.1 \text{ kJ/mol}) - 2(97.89 \text{ kJ/mol}) - (0.00 \text{ kJ/mol}) = 34.4 \text{ kJ/mol}$

(b) $\Delta G^\circ = -RT \ln K_p$ $\ln K_p = -\dfrac{\Delta G^\circ}{RT} = -\dfrac{34.4 \times 10^3 \text{ J/mol}}{8.3145 \text{ J mol}^{-1}\text{ K}^{-1} \times 298\text{K}} = -13.9$

$\qquad K_p = e^{-13.9} = 9 \times 10^{-7}$

22   **(a)** $\Delta S^\circ = S^\circ\left[Na_2CO_3\,(s)\right] + S^\circ\left[H_2O(l)\right] + S^\circ\left[CO_2\,(g)\right] - 2S^\circ\left[NaHCO_3\,(s)\right]$

$\qquad\qquad = 135.0\,\dfrac{\text{J}}{\text{K mol}} + 69.91\,\dfrac{\text{J}}{\text{K mol}} + 213.7\,\dfrac{\text{J}}{\text{K mol}} - 2\left(101.7\,\dfrac{\text{J}}{\text{K mol}}\right) = 215.2\,\dfrac{\text{J}}{\text{K mol}}$

(b) $\Delta H^\circ = \Delta H_f^\circ\left[Na_2CO_3\,(s)\right] + \Delta H_f^\circ\left[H_2O(l)\right] + \Delta H_f^\circ\left[CO_2\,(g)\right] - 2\Delta H_f^\circ\left[NaHCO_3\,(s)\right]$

$\qquad\qquad = -1131\dfrac{\text{kJ}}{\text{mol}} - 285.8\dfrac{\text{kJ}}{\text{mol}} - 393.5\dfrac{\text{kJ}}{\text{mol}} - 2\left(-950.8\dfrac{\text{kJ}}{\text{mol}}\right) = +91\dfrac{\text{kJ}}{\text{mol}}$

(c) $\Delta G^\circ = \Delta H^\circ - T\Delta S^\circ = 91 \text{ kJ/mol} - (298 \text{ K})\left(215.2 \times 10^{-3} \text{ kJ mol}^{-1}\text{K}^{-1}\right)$

$\qquad\qquad = 91 \text{ kJ/mol} - 64.13 \text{ kJ/mol} = 27 \text{ kJ/mol}$

(d) $\Delta G^\circ = -RT \ln K_{eq}$ $\ln K_{eq} = -\dfrac{\Delta G^\circ}{RT} = -\dfrac{27 \times 10^3 \text{ J/mol}}{8.3145 \text{ J mol}^{-1}\text{K}^{-1} \times 298\text{K}} = -10._9$

$\qquad K_{eq} = e^{-10._9} = 2 \times 10^{-5}$

23.   **(a)** $\Delta S^\circ = S^\circ\left[CH_3CH_2OH(g)\right] + S^\circ\left[H_2O(g)\right] - S^\circ\left[CO(g)\right] - 2S^\circ\left[H_2\,(g)\right] - S^\circ\left[CH_3OH(g)\right]$

$\qquad\qquad = 282.7\,\dfrac{\text{J}}{\text{K mol}} + 188.8\,\dfrac{\text{J}}{\text{K mol}} - 197.7\,\dfrac{\text{J}}{\text{K mol}} - 2\left(130.7\,\dfrac{\text{J}}{\text{K mol}}\right) - 239.8\,\dfrac{\text{J}}{\text{K mol}}$

$\qquad\qquad = -227.4\,\dfrac{\text{J}}{\text{K mol}}$

$\Delta H^\circ = \Delta H_f^\circ[CH_3CH_2OH(g)] + \Delta H_f^\circ\left[H_2O(g)\right] - \Delta H_f^\circ\left[CO(g)\right] - 2\Delta H_f^\circ[H_2\,(g)] - \Delta H_f^\circ[CH_3OH(g)]$

$\qquad\qquad = -235.1\dfrac{\text{kJ}}{\text{mol}} - 241.8\dfrac{\text{kJ}}{\text{mol}} - \left(-110.5\dfrac{\text{kJ}}{\text{mol}}\right) - 2\left(0.00\dfrac{\text{kJ}}{\text{mol}}\right) - \left(-200.7\dfrac{\text{kJ}}{\text{mol}}\right) = -165.7\dfrac{\text{kJ}}{\text{mol}}$

$\Delta G^\circ = -165.7\dfrac{\text{kJ}}{\text{mol}} - (298\text{K})\left(-227.4 \times 10^{-3}\dfrac{\text{kJ}}{\text{K mol}}\right) = -165.4\dfrac{\text{kJ}}{\text{mol}} + 67.8\dfrac{\text{kJ}}{\text{mol}} = -97.9\dfrac{\text{kJ}}{\text{mol}}$

(b) $\Delta H^\circ < 0$ for this reaction; it is favored at low temperatures. And because $\Delta n_{gas} = +2 - 4 = -2$, which is less than zero, the reaction is favored at high pressures.

(c) We assume that neither ΔS° nor ΔH° varies significantly with temperature. Then we compute a value for ΔG° at 750 K. From this value of ΔG°, we compute a value for K_p.

$$\Delta G^\circ = \Delta H^\circ - T\Delta S^\circ = -165.7\,\text{kJ/mol} - (750.\text{K})\left(-227.4 \times 10^{-3}\,\text{kJ mol}^{-1}\,\text{K}^{-1}\right)$$

$$= -165.7\,\text{kJ/mol} + 170.6\,\text{kJ/mol} = +4._9\,\text{kJ/mol} = -RT \ln K_p$$

$$\ln K_p = -\frac{\Delta G^\circ}{RT} = -\frac{4._9 \times 10^3\,\text{J/mol}}{8.3145\,\text{J mol}^{-1}\,\text{K}^{-1} \times 750.\text{K}} = -0.7_9 \qquad K_p = e^{-0.7_9} = 0.5$$

24. We use the van't Hoff equation with $\Delta H^\circ = -1.8 \times 10^5\,\text{J/mol}$, $T_1 = 800.\,\text{K}$, $T_2 = 100.^\circ\text{C} = 373\,\text{K}$, and $K_1 = 9.1 \times 10^2$.

$$\ln\frac{K_2}{K_1} = \frac{\Delta H^\circ}{R}\left(\frac{1}{T_1} - \frac{1}{T_2}\right) = \frac{-1.8 \times 10^5\,\text{J/mol}}{8.3145\,\text{J mol}^{-1}\,\text{K}^{-1}}\left(\frac{1}{800\,\text{K}} - \frac{1}{373\,\text{K}}\right) = 31$$

$$\frac{K_2}{K_1} = e^{31} = 2.\underline{9} \times 10^{13} = \frac{K_2}{9.1 \times 10^2} \qquad K_2 = \left(2.\underline{9} \times 10^{13}\right)\left(9.1 \times 10^2\right) = 3 \times 10^{16}$$

EXERCISES

Spontaneous Change, Entropy, and Disorder

25. **(a)** The freezing of ethanol involves a *decrease* in the entropy of the system, since solids are more ordered than liquids, in general.

(b) The sublimation of dry ice involves converting a quite ordered solid into a very disordered vapor. Thus, the entropy of the system *increases* substantially.

(c) The burning of rocket fuel involves converting a somewhat ordered liquid fuel into the highly disordered mixture of the gaseous combustion products. The entropy of the system *increases* substantially.

26. Although there is a substantial change in entropy involved in **(a)** changing H_2O (liq., 1 atm) to H_2O (g, 1 atm), it is not as large as **(c)** converting the liquid to a gas at 10 mmHg. The gas is less ordered at lower pressures. In turn, **(b)** if we start with a solid and convert it to a gas at the lower pressure, the entropy change should be even larger, since a solid is more ordered than a liquid. Thus, in order of increasing ΔS, the processes are: **(a)** $<$ **(c)** $<$ **(b)**.

27. The first law of thermodynamics states that energy is neither created nor destroyed (thus, "The energy of the universe is constant"). A consequence of the second law of thermodynamics is that entropy of the universe increases for all spontaneous, that is, naturally occurring, processes (and therefore, "the entropy of the universe increases toward a maximum").

28. When environmental pollutants are produced they are dispersed throughout the environment. These pollutants thus start in a relatively compact form and end up dispersed throughout a large volume. In this large volume, because they are mixed with many other substances, they are highly disordered and thus have a high entropy. Returning them to their original compact form requires reducing this entropy, a highly nonspontaneous process. If we have had enough foresight to retain these pollutants in a reasonably compact form, such as disposing of them in a *secure* landfill, rather than dispersing them in the atmosphere or in rivers and seas, the task of permanently removing them from the environment, and perhaps even converting them to useful forms, would be considerably easier.

29. **(a)** Negative; A liquid (moderate entropy) combines with a solid to form another solid.

(b) Positive; One mole of high entropy gas forms where no gas was present before.

(c) Positive; One mole of high entropy gas forms where no gas was present before.

(d) Uncertain; Same number of moles of gaseous products as of gaseous reactants.

(e) Negative; Two moles of gas (and a solid) combine to form just one mole of gas.

30. The entropy of formation of a compound would be the difference between the absolute entropy of one mole of the compound and the sum of the absolute entropies of the appropriate amounts of the elements constituting the compound, each in its most stable form. It seems as though $CS_2(l)$ would have the highest molar entropy of formation of the compounds listed, since it is the only substance whose formation does not involve the consumption of high entropy gaseous reactants.

(a) $C(graphite) + 2H_2(g) \rightleftharpoons CH_4(g)$

$$\Delta S_f^\circ [CH_4(g)] = S^\circ [CH_4(g)] - S^\circ [C(graphite)] - 2S^\circ [H_2(g)]$$

$$= 186.3 \text{ J mol}^{-1}\text{K}^{-1} - 5.74 \text{ J mol}^{-1}\text{K}^{-1} - 2 \times 130.7 \text{ J mol}^{-1}\text{K}^{-1}$$

$$= -80.8 \text{ J mol}^{-1}\text{K}^{-1}$$

(b) $2C(graphite) + 3H_2(g) + \frac{1}{2}O_2(g) \rightleftharpoons CH_3CH_2OH(l)$

$$\Delta S_f^\circ [CH_3CH_2OH(l)] = S^\circ [CH_3CH_2OH(l)] - 2S^\circ [C(graphite)] - 3S^\circ [H_2(g)] - \frac{1}{2}S^\circ [O_2(g)]$$

$$= 160.7 \text{ J mol}^{-1}\text{ K}^{-1} - 2 \times 5.74 \text{ J mol}^{-1}\text{ K}^{-1} - 3 \times 130.7 \text{ J mol}^{-1}\text{ K}^{-1} - \frac{1}{2} \times 205.1 \text{ J mol}^{-1}\text{K}^{-1}$$

$$= -345.4 \text{ J mol}^{-1}\text{ K}^{-1}$$

(c) $C(graphite) + 2S(rhombic) \rightleftharpoons CS_2(l)$

$$\Delta S_f^\circ \left[CS_2(g)\right] = S^\circ \left[CS_2(l)\right] - S^\circ \left[C(graphite)\right] - 2S^\circ \left[S(rhombic)\right]$$

$$= 151.3 \text{ J mol}^{-1}\text{ K}^{-1} - 5.74 \text{ J mol}^{-1}\text{ K}^{-1} - 2 \times 31.80 \text{ J mol}^{-1}\text{ K}^{-1}$$

$$= 82.0 \text{ J mol}^{-1}\text{ K}^{-1}$$

Phase Transitions

31. **(a)** $\Delta H_{vap}^\circ = \Delta H_f^\circ[H_2O(g)] - \Delta H_f^\circ[H_2O(l)] = -241.8 \text{ kJ/mol} - (-285.8 \text{ kJ/mol})$

$$= +44.0 \text{ kJ/mol}$$

$$\Delta S_{vap}^\circ = S^\circ\left[H_2O(g)\right] - S^\circ\left[H_2O(l)\right] = 188.8 \text{ J mol}^{-1}\text{ K}^{-1} - 69.91 \text{ J mol}^{-1}\text{K}^{-1}$$

$$= 118.9 \text{ J mol}^{-1}\text{K}^{-1}$$

There is an alternate, but incorrect, method of obtaining ΔS_{vap}°.

$$\Delta S_{vap}^\circ = \frac{\Delta H_{vap}^\circ}{T} = \frac{44.0 \times 10^3 \text{ J/mol}}{298.15 \text{ K}} = 148 \text{ J mol}^{-1}\text{ K}^{-1}$$

This method is invalid because the temperature in the denominator of the equation must be the temperature at which the liquid-vapor transition is at equilibrium. Liquid water and water vapor at 1 atm pressure (standard state, indicated by $^\circ$) are in equilibrium only at 100° C $= 373$ K.

(b) The reason why ΔH_{vap}° is different from its value at 100° C has to do with the heat required to bring the reactants and products down to 298 K from 373 K. The specific heat of liquid water is higher than the heat capacity of steam. Thus, more heat is given off (this is negative heat, an exothermic process) by lowering the temperature of the liquid water from 100° C to 25° C than is given off by lowering the temperature of the same amount of steam. Another way to think of this is that hydrogen bonding is more disrupted in water at 100° C than at 25° C (because the molecules are in rapid—thermal—motion), and hence, there is not as much energy needed to convert liquid to vapor.

The reason why ΔS_{vap}° has a larger value at 25° C than at 100° C has to do with disorder. A vapor at 1 atm pressure (the case at both temperatures) has about the same entropy. On the other hand, liquid water is more disordered at higher temperatures since more of the hydrogen bonds are disrupted by thermal motion. (The hydrogen bonds are totally disrupted in the two vapors).

32. **(a)** We use Trouton's rule $\left(\Delta S_{vap}^\circ \approx 87 \text{ J mol}^{-1}\text{ K}^{-1}\right)$ to estimate the normal boiling point.

$$T_{nbp} = \frac{\Delta H_{vap}^\circ}{\Delta S_{vap}^\circ} = \frac{\left[-146.9 \text{ kJ/mol} - (-173.5) \times \dfrac{1000 \text{ J}}{1 \text{ kJ}}\right]}{87 \text{ J mol}^{-1}\text{K}^{-1}} = 30_6 \text{ K} \approx 3.0 \times 10^2 \text{ K}$$

(b) $\Delta G_{vap}^{\circ} = \Delta H_{vap}^{\circ} - T\Delta S_{vap}^{\circ}$

$= [-146.9 - (-173.5)] \text{ kJ/mol} - \dfrac{87 \text{ J mol}^{-1}\text{K}^{-1} \times 298\,\text{K}}{\dfrac{1000\,\text{J}}{1\,\text{kJ}}} = 26.6 \text{ kJ/mol} - 25.9 \text{ kJ/mol}$

$= +0.7 \text{ kJ/mol}$

(c)   A positive value of ΔG_{vap}° indicates that normal boiling (having a vapor pressure of 1.00 atm) should be nonspontaneous (will not occur) at 298 K for pentane. The vapor pressure of pentane at 298 K should be less than 1.00 atm.

33. Trouton's rule is obeyed most closely by liquids that do not have a high degree of order within the liquid. In both HF and CH_3OH, hydrogen bonds create considerable order within the liquid. In $C_6H_5CH_3$, the only attractive forces are non-directional London forces, which cause the molecules to attract each other, but have no preferred orientation as hydrogen bonds do. Thus, of the three choices, liquid $C_6H_5CH_3$ would most closely follow Trouton's rule.

34. $\Delta H_{vap}^{\circ} = \Delta H_f^{\circ}[Br_2(g)] - \Delta H_f^{\circ}[Br_2(l)] \approx 30.91 \text{ kJ/mol} - 0.00 \text{ kJ/mol} = 30.91 \text{ kJ/mol}$

$\Delta S_{vap}^{\circ} = \dfrac{\Delta H_{vap}^{\circ}}{T_{vap}} \approx 87 \text{ J mol}^{-1} \text{ K}^{-1}$ or $T_{vap} = \dfrac{\Delta H_{vap}^{\circ}}{\Delta S_{vap}^{\circ}} \approx \dfrac{30.91 \times 10^3 \text{ J/mol}}{87 \text{ J mol}^{-1}\text{K}^{-1}} = 3.5 \times 10^2 \text{ K}$

The accepted value of the boiling point of bromine is $58.8^{\circ}\text{C} = 332 \text{ K} = 3.32 \times 10^2 \text{ K}$. Thus, our estimate is in reasonable agreement with the measured value.

35. The liquid water-gaseous water equilibrium H_2O (l, 0.50 atm) $\rightleftharpoons H_2O$ (g, 0.50 atm) can only be established at <u>one temperature</u>, namely the boiling point for water under 0.50 atm external pressure. We can estimate the boiling point for water under 0.50 atm external pressure by using the Clausius-Clapeyron equation:

$$\ln \frac{P_2}{P_1} = \frac{\Delta H_{vap}^{o}}{R} \left(\frac{1}{T_1} - \frac{1}{T_2} \right)$$

We know that at 373 K, the pressure of water vapor is 1.00 atm. Let's make $P_1 = 1.00$ atm, $P_2 = 0.50$ atm and $T_1 = 373$ K. Thus, the boiling point under 0.50 atm pressure is T_2. To find T_2 we simply insert the appropriate information into the Clausius-Clapeyron equation:

$$\ln \frac{0.50 \text{ atm}}{1.00 \text{ atm}} = \frac{40.7 \text{ kJ mol}^{-1}}{8.3145 \times 10^{-3} \text{ kJ K}^{-1} \text{ mol}^{-1}} \left(\frac{1}{373 \text{ K}} - \frac{1}{T_2} \right)$$

$$-1.4\underline{16} \times 10^{-4} \text{ K} = \left(\frac{1}{373 \text{ K}} - \frac{1}{T_2} \right)$$

Solving for T_2 we find a temperature of 354 K or 81°C. Consequently, to achieve an equilibrium between gaseous and liquid water under 0.50 atm pressure, the temperature must be set at 354 K.

36. Figure 13-19 (phase diagram for carbon dioxide) shows that at −60°C and under 1 atm of external pressure, carbon dioxide exists as a gas. In other words, neither solid nor liquid CO_2 can exist at this temperature and pressure. Clearly then, of the three phases, gaseous CO_2 must be the most stable and, hence, have the lowest free energy when T = −60 °C and P_{ext} = 1.00 atm.

Free Energy and Spontaneous Change

37. **(a)** $\Delta H° < 0$ and $\Delta S° < 0$ (since $\Delta n_{gas} < 0$) for this reaction. Thus, this reaction is case 2 of Table 20-1. It is spontaneous at low temperatures and nonspontaneous at high temperatures.

   (b) We are unable to predict the sign of $\Delta S°$ for this reaction, since $\Delta n_{gas} = 0$. Thus, no strong prediction as to the temperature behavior of this reaction can be made. Since $\Delta H° > 0$, we can, however, conclude that the reaction will be non-spontaneous at low temperatures.

   (c) $\Delta H° > 0$ and $\Delta S° > 0$ (since $\Delta n_{gas} > 0$) for this reaction. This is case 3 of Table 20-1. It is nonspontaneous at low temperatures, but spontaneous at high temperatures.

38. **(a)** $\Delta H° > 0$ and $\Delta S° < 0$ (since $\Delta n_{gas} < 0$) for this reaction. This is case 4 of Table 20-1. It is nonspontaneous at all temperatures.

   (b) $\Delta H° < 0$ and $\Delta S° > 0$ (since $\Delta n_{gas} > 0$) for this reaction. This is case 1 of Table 20-1. It is spontaneous at all temperatures.

   (c) $\Delta H° < 0$ and $\Delta S° < 0$ (since $\Delta n_{gas} < 0$) for this reaction. This is case 2 of Table 20-1. It is spontaneous at low temperatures and nonspontaneous at high temperatures.

39. First of all, the process is clearly spontaneous, and therefore $\Delta G < 0$. In addition, the gases are more disordered when they are at a lower pressure and therefore $\Delta S > 0$. We also conclude that $\Delta H = 0$, because the gases are ideal and thus there are no forces of attraction or repulsion between them, producing no energy of interaction.

40. Because an ideal solution forms spontaneously, $\Delta G < 0$. Also, the molecules of solvent and solute are mixed together in the solution, a more disordered state than the separated solvent and solute. Therefore, $\Delta S > 0$. However, in an ideal solution, the attractive forces between solvent and solute molecules equals those forces between solvent molecules and those between solute molecules. Thus, $\Delta H = 0$; there is no energy of interaction.

41. **(a)** An exothermic reaction (one that gives off heat) may not occur spontaneously if, at the same time, the system becomes more ordered, that is, $\Delta S° < 0$. This is particularly true at a high temperature, where the $T\Delta S$ term dominates the ΔG expression. An example of such a process is freezing water (clearly exothermic because the reverse process, melting ice, is endothermic) at temperatures above $0\,°C$.

(b) A reaction in which $\Delta S > 0$ need not be spontaneous if that process also is endothermic. This is particularly true at low temperatures, where the ΔH term dominates the ΔG expression. An example is the vaporization of water (clearly an endothermic process, one that requires heat, and one that produces a gas, so $\Delta S > 0$) at low temperatures, that is, below $100\,°C$.

42. Because this reaction produces more moles of gas than it consumes, $\Delta S > 0$. The reaction also is endothermic, since energy is required to break the A—B bond. Hence $\Delta H > 0$. Therefore, this reaction is of case 3 in Table 20-1. It is non-spontaneous at low temperatures, but eventually becomes spontaneous as the temperature is raised.

Standard Free Energy Change

43. **(a)** $\Delta S° = 2\,S°\left[POCl_3\,(1)\right] - 2\,S°\left[PCl_3\,(g)\right] - S°\left[O_2\,(g)\right]$

$= 2(222.4\text{ J/K}) - 2(311.7\text{ J/K}) - 205.1\text{ J/K} = -383.7\text{ J/K}$

$\Delta G° = \Delta H° - T\Delta S° = -620.2\times10^3\text{ J} - (298\text{ K})(-383.7\text{ J/K}) = -506\times10^3\text{ J} = -506\text{ kJ}$

(b) The reaction proceeds spontaneously in the forward direction when reactants and products are in their standard states, because the value of $\Delta G°$ is less than zero.

44. **(a)** $\Delta S° = S°\left[Br_2\,(1)\right] + 2\,S°\left[HNO_2\,(aq)\right] - 2\,S°\left[H^+\,(aq)\right] - 2\,S°\left[Br^-\,(aq)\right] - 2\,S°\left[NO_2\,(g)\right]$

$= 152.2\text{ J/K} + 2(135.6\text{ J/K}) - 2(0\text{ J/K}) - 2(82.4\text{ J/K}) - 2(240.1\text{ J/K}) = -221.6\text{ J/K}$

$\Delta G° = \Delta H° - T\,\Delta S° = -61.6\times10^3\text{ J} - (298\text{ K})(-221.6\text{ J/K}) = +4.4\times10^3\text{ J} = +4.4\text{ kJ}$

(b) The reaction does not proceed spontaneously in the forward direction when reactants and products are in their standard states, because the value of $\Delta G°$ is greater than zero.

45. We combine the reactions in the same way as for Hess's law calculations.

(a) $N_2O(g) \rightarrow N_2(g) + \tfrac{1}{2}O_2(g)$ $\Delta G° = -\tfrac{1}{2}(+208.4\text{ kJ}) = -104.2\text{ kJ}$

$N_2(g) + 2\,O_2(g) \rightarrow 2\,NO_2(g)$ $\Delta G° = +102.6\text{ kJ}$

Net: $N_2O(g) + \tfrac{3}{2}O_2(g) \rightarrow 2NO_2(g)$ $\Delta G° = -104.2 + 102.6 = -1.6\text{ kJ}$

This reaction reaches an equilibrium condition, a conclusion we reach based on the relatively small absolute value of $\Delta G°$.

(b)

$$2N_2(g) + 6H_2(g) \rightarrow 4NH_3(g) \qquad \Delta G° = 2(-33.0 \text{ kJ}) = -66.0 \text{ kJ}$$

$$4NH_3(g) + 5O_2(g) \rightarrow 4NO(g) + 6H_2O(l) \qquad \Delta G° = -1010.5 \text{ kJ}$$

$$4NO(g) \rightarrow 2N_2(g) + 2O_2(g) \qquad \Delta G° = -2(+173.1 \text{ kJ}) = -346.2 \text{ kJ}$$

Net: $6H_2(g) + 3O_2(g) \rightarrow 6H_2O(l) \quad \Delta G° = -66.0 \text{ kJ} - 1010.5 \text{ kJ} - 346.2 \text{ kJ} = -1422.7 \text{ kJ}$

This reaction is three times the desired reaction, which therefore has

$\Delta G° = -1422.7 \text{ kJ} \div 3 = -474.3 \text{ kJ}$.

The large negative $\Delta G°$ value indicates that this reaction would tend to go to completion at $25°C$.

(c)

$$4NH_3(g) + 5O_2(g) \rightarrow 4NO(g) + 6H_2O(l) \qquad \Delta G° = -1010.5 \text{ kJ}$$

$$4NO(g) \rightarrow 2N_2(g) + 2O_2(g) \qquad \Delta G° = -2(+173.1 \text{ kJ}) = -346.2 \text{ kJ}$$

$$2N_2(g) + O_2(g) \rightarrow 2N_2O(g) \qquad \Delta G° = +208.4 \text{ kJ}$$

$4NH_3(g) + 4O_2(g) \rightarrow 2N_2O(g) + 6H_2O(l) \quad \Delta G° = -1010.5 \text{ kJ} - 346.2 \text{ kJ} + 208.4 \text{ kJ}$

$$= -1148.3 \text{ kJ}$$

This reaction is twice the desired reaction, which, therefore, has

$\Delta G° = -1148.3 \text{ kJ} \div 2 = -574.2 \text{ kJ}$.

The very large negative value of $\Delta G°$ for this reaction indicates that it will go to completion.

46. We combine the reactions in the same way as for Hess's law calculations.

(a)

$$COS(g) + 2CO_2(g) \rightarrow SO_2(g) + 3CO(g) \qquad \Delta G° = -(-246.4 \text{ kJ}) = +246.6 \text{ kJ}$$

$$2CO(g) + 2H_2O(g) \rightarrow 2CO_2(g) + 2H_2(g) \qquad \Delta G° = 2(-28.6 \text{ kJ}) = -57.2 \text{ kJ}$$

$COS(g) + 2H_2O(g) \rightarrow SO_2(g) + CO(g) + 2H_2(g) \quad \Delta G° = +246.6 - 57.2 = +189.4 \text{ kJ}$

This is spontaneous in the reverse direction, because of the large positive value of $\Delta G°$.

(b)

$$COS(g) + 2CO_2(g) \rightarrow SO_2(g) + 3CO(g) \qquad \Delta G° = -(-246.4 \text{ kJ}) = +246.6 \text{ kJ}$$

$$3CO(g) + 3H_2O(g) \rightarrow 3CO_2(g) + 3H_2(g) \qquad \Delta G° = 3(-28.6 \text{ kJ}) = -85.8 \text{ kJ}$$

$COS(g) + 3H_2O(g) \rightarrow CO_2(g) + SO_2(g) + 3H_2(g) \quad \Delta G° = +246.6 - 85.8 = +160.8 \text{ kJ}$

This is spontaneous in the reverse direction, because of the large positive value of $\Delta G°$.

(c)

$$COS(g) + H_2(g) \rightarrow CO(g) + H_2S(g) \qquad \Delta G° = -(+1.4) \text{ kJ}$$
$$\underline{CO(g) + H_2O(g) \rightarrow CO_2(g) + H_2(g) \qquad \Delta G° = -28.6 \text{ kJ} = -28.6 \text{ kJ}}$$
$$COS(g) + H_2O(g) \rightarrow CO_2(g) + H_2S(g) \qquad \Delta G° = -1.4 \text{ kJ} - 28.6 \text{ kJ} = -30.0 \text{ kJ}$$

The negative value of $\Delta G°$ for this reaction indicates that it is spontaneous in the forward direction.

47. The combustion reaction is : $C_6H_6(l) + \frac{15}{2}O_2(g) \rightarrow 6CO_2(g) + 3H_2O(g \text{ or } l)$

(a)
$$\Delta G° = 6\Delta G_f°\left[CO_2(g)\right] + 3\Delta G_f°\left[H_2O(l)\right] - \Delta G_f°\left[C_6H_6(l)\right] - \frac{15}{2}\Delta G_f°\left[O_2(g)\right]$$
$$= 6(-394.4 \text{ kJ}) + 3(-237.1 \text{ kJ}) - (+124.5 \text{ kJ}) - \frac{15}{2}(0.00 \text{ kJ}) = -3202 \text{ kJ}$$

(b)
$$\Delta G° = 6\Delta G_f°\left[CO_2(g)\right] + 3\Delta G_f°\left[H_2O(g)\right] - \Delta G_f°\left[C_6H_6(l)\right] - \frac{15}{2}\Delta G_f°\left[O_2(g)\right]$$
$$= 6(-394.4 \text{ kJ}) + 3(-228.6 \text{ kJ}) - (+124.5 \text{ kJ}) - \frac{15}{2}(0.00 \text{ kJ}) = -3177 \text{ kJ}$$

We could determine the difference between the two values of $\Delta G°$ by noting the difference between the two products: $3H_2O(l) \rightarrow 3H_2O(g)$ and determining the value of $\Delta G°$ for this difference:

$$\Delta G° = 3\Delta G_f°\left[H_2O(g)\right] - 3\Delta G_f°\left[H_2O(l)\right] = 3\left[-228.6 - (-237.1)\right] \text{ kJ} = 25.5 \text{ kJ}$$

48. We wish a value of $\Delta H°$ for the given reaction: $F_2(g) \rightarrow 2F(g)$

$$\Delta S° = 2S°\left[F(g)\right] - S°\left[F_2(g)\right] = 2(158.8 \text{ J K}^{-1}) - (202.8 \text{ J K}^{-1}) = +114.8 \text{ J K}^{-1}$$

$$\Delta H° = \Delta G° + T\Delta S° = 123.9 \times 10^3 \text{ J} + (298 \text{ K} \times 114.8 \text{ J/K}) = 158.1 \text{ kJ/mole of bonds}$$

The value in Table 11.3 is 159 kJ/mol, which is in quite good agreement.

49. **(a)** $\Delta S°_{rxn} = \Sigma S°_{products} - \Sigma S°_{reactants}$
$$= [1 \text{ mol} \times 301.2 \text{ J K}^{-1}\text{mol}^{-1} + 2 \text{ mol} \times 188.8 \text{ J K}^{-1}\text{mol}^{-1}] - [2 \text{ mol} \times 247.4 \text{ J K}^{-1}\text{mol}^{-1}$$
$$+ 1 \text{ mol} \times 238.5 \text{ J K}^{-1}\text{mol}^{-1}] = -54.5 \text{ J K}^{-1}$$

$$\Delta S°_{rxn} = = -0.0545 \text{ kJ K}^{-1}$$

(b) $\Delta H°_{rxn} = \Sigma(\text{bonds broken in reactants (kJ/mol)}) - \Sigma(\text{bonds broken in products(kJ/ mol)})$
$$= [4 \text{ mol} \times (389 \text{ kJ mol}^{-1})_{N-H} + 4 \text{ mol} \times (222 \text{ kJ mol}^{-1})_{O-F}] -$$
$$[4 \text{ mol} \times (301 \text{ kJ mol}^{-1})_{N-F} + 4 \text{ mol} \times (464 \text{ kJ mol}^{-1})_{O-H}]$$

$$\Delta H°_{rxn} = -616 \text{ kJ}$$

(c) $\Delta G°_{rxn} = \Delta H°_{rxn} - T\Delta S°_{rxn} = -616 \text{ kJ} - 298 \text{ K}(-0.0545 \text{ kJ K}^{-1}) = -600 \text{ kJ}$
Since the $\Delta G°_{rxn}$ is <u>negative</u>, the reaction is <u>spontaneous</u>, and hence feasible (at 25 °C). Because both the entropy and enthalpy changes are *negative*, this reaction will be more highly favored at low temperatures (i.e., the reaction is enthalpy driven)

50. The balanced equation for the thermal decomposition of ammonium nitrate is show below:

$$NH_4NO_3(s) \xrightarrow{\Delta} N_2O(g) + 2 H_2O(l)$$

To find $\Delta G°_{rxn}$ we will utilize the $\Delta G°_f$ values provided in Appendix D

$\Delta G°_{298} = [2 \text{ mol}(-237.1 \text{ kJ mol}^{-1}) + 1 \text{ mol}(104.2 \text{ kJ mol}^{-1})] - [1 \text{ mol}(-183.9 \text{ kJ mol}^{-1})]$

$\Delta G°_{298} = -186.1 \text{ kJ}$

The decomposition reaction is <u>spontaneous</u> at 25°C with all substances in their standard states.

$\Delta S°_{rxn} = \Sigma S°_{products} - \Sigma S°_{reactants}$
$= [2 \text{ mol} \times 69.91 \text{ J K}^{-1}\text{mol}^{-1} + 1 \text{ mol} \times 219.9 \text{ J K}^{-1}\text{mol}^{-1}] - [1 \text{ mol} \times 151.1 \text{ J K}^{-1}\text{mol}^{-1}]$

$\Delta S°_{rxn} = 208.6 \text{ J K}^{-1} = -0.2086 \text{ kJ K}^{-1}$

$\Delta H°_{rxn} = \Sigma \Delta H_f°_{products} - \Sigma \Delta H_f°_{reactants}$
$= [2 \text{ mol}(-285.8 \text{ kJ mol}^{-1}) + 1 \text{ mol}(82.05 \text{ kJ mol}^{-1})] - [1 \text{ mol}(-365.6 \text{ kJ mol}^{-1})]$

$\Delta H°_{rxn} = -124.0 \text{ kJ}$

Since $\Delta H°_{rxn}$ is <u>negative</u> and $\Delta S°_{rxn}$ is <u>positive</u>, the decomposition of ammonium nitrate is spontaneous at all temperatures. However, as the temperature increases, the $T\Delta S$ term gets larger and as a result, the decomposition reaction becomes even more spontaneous. Consequently, we can say that the reaction is more highly favored <u>above</u> 298 K (it will also be faster - see Chapter 15)

The Thermodynamic Equilibrium Constant

51. In all three cases, $K_{eq} = K_p$ because only gases, solids, and liquids are present in the chemical equations. There are no factors for solids and liquids in K_{eq} expressions, and gases appear as partial pressures in atmospheres. That makes K_{eq} the same as K_p for these three reactions. We now recall that $K_p = K_c(RT)^{\Delta n}$. Hence, in these three cases we have:

(a) $2SO_2(g) + O_2(g) \rightleftharpoons 2SO_3(g)$; $\Delta n_{gas} = 2 - (2+1) = -1$; $K_{eq} = K_p = K_c(RT)^{-1}$

(b) $HI(g) \rightleftharpoons \frac{1}{2}H_2(g) + \frac{1}{2}I_2(g)$; $\Delta n_{gas} = 1 - (\frac{1}{2} + \frac{1}{2}) = 0$; $K_{eq} = K_p = K_c$

(c) $NH_4HCO_3(s) \rightleftharpoons NH_3(g) + CO_2(g) + H_2O(l)$;

$\Delta n_{gas} = 2 - (0) = +2$ $K_{eq} = K_p = K_c(RT)^2$

52. (a) $K_{eq} = \dfrac{P\{H_2(g)\}^4}{P\{H_2O(g)\}^4}$

(b) Terms for both solids, Fe(s) and $Fe_3O_4(s)$, are properly excluded from the thermodynamic equilibrium constant expression. (Actually, each solid has an activity of 1.00.) Thus, the equilibrium partial pressures of both $H_2(g)$ and $H_2O(g)$ do not depend on the amounts of the two solids present, as long as some of each solid is present. One way to understand this is that any chemical reaction occurs on the surface of the solids, and thus is unaffected by the amount present.

(c) We can produce $H_2(g)$ from $H_2O(g)$ without regard to the proportions of Fe(s) and $Fe_3O_4(s)$ with one qualification, of course. There must always be some Fe(s) present for the production of $H_2(g)$ to continue.

Relationships Involving ΔG, ΔG^0, Q and K

53. First we need to determine the standard free energy change for the reaction:

$$2SO_2(g) + O_2(g) \rightleftharpoons 2SO_3(g)$$

$$\Delta G^\circ = 2\Delta G_f^\circ \left[SO_3(g) \right] - 2\Delta G_f^\circ \left[SO_2(g) \right] - \Delta G_f^\circ \left[O_2(g) \right]$$

$$= 2 \times (-371.1 \text{ kJ/mol}) - 2 \times (-300.2 \text{ kJ/mol}) - 0.0 \text{ kJ/mol} = -141.8 \text{ kJ}$$

Now we can calculate ΔG by employing the equation $\Delta G = \Delta G^\circ + RT \ln Q_p$, where

$$Q_p = \frac{P\{SO_3(g)\}^2}{P\{O_2(g)\} P\{SO_2(g)\}^2} \; ; \; Q_p = \frac{(0.10 \text{ atm})^2}{(0.20 \text{ atm})(1.0 \times 10^{-4} \text{ atm})^2} = 5.0 \times 10^6$$

$$\Delta G = -141.8 \text{ kJ} + (8.3145 \times 10^{-3} \text{ kJ/K·mol})(298 \text{ K})\ln(5.0 \times 10^6)$$

$$\Delta G = -141.8 \text{ kJ} + 38.2 \text{ kJ} = -104 \text{ kJ}.$$

Since ΔG is negative, the reaction is spontaneous in the forward direction.

54. We begin by calculating the standard free energy change for the reaction:

$$H_2(g) + Cl_2(g) \rightleftharpoons 2HCl(g)$$

$$\Delta G^\circ = 2\Delta G_f^\circ \left[HCl(g) \right] - \Delta G_f^\circ \left[Cl_2(g) \right] - \Delta G_f^\circ \left[H_2(g) \right]$$

$$= 2 \times (-95.30 \text{ kJ/mol}) - 0.0 \text{ kJ/mol} - 0.0 \text{ kJ/mol} = -190.6 \text{ kJ}$$

Now we can calculate ΔG by employing the equation $\Delta G = \Delta G^\circ + RT \ln Q_p$, where

$$Q_p = \frac{P\{HCl(g)\}^2}{P\{H_2(g)\} P\{Cl_2(g)\}} \; ; \; Q_p = \frac{(0.5 \text{ atm})^2}{(0.5 \text{ atm})(0.5 \text{ atm})} = 1$$

$$\Delta G = -190.6 \text{ kJ} + (8.3145 \times 10^{-3} \text{ kJ/K·mol})(298 \text{ K})\ln(1)$$

$$\Delta G = -190.6 \text{ kJ} + 0 \text{ kJ} = -190.6 \text{ kJ}.$$

Since ΔG is negative, the reaction is spontaneous in the forward direction.

55. In order to determine in which direction the reaction is spontaneous, we need to calculate the non-standard free energy change for the reaction. To accomplish this, we will employ the equation $\Delta G = \Delta G^{\circ} + RT \ln Q_c$, where

$$Q_c = \frac{[H_3O^+(aq)]\,[CH_3CO_2^-(aq)]}{[CH_3CO_2H(aq)]}\ ;\ Q_c = \frac{\left(1.0\times10^{-3}\ M\right)^2}{(0.10\ M)} = 1.0\times10^{-5}$$

$\Delta G = 27.07\ kJ + (8.3145\times10^{-3}\ kJ/K\cdot mol\ )(298\ K)\ln(1.0\times10^{-5})$

$\Delta G = 27.07\ kJ + (-28.53\ kJ) = -1.46\ kJ.$

Since ΔG is negative, the reaction is spontaneous in the forward direction.

56. As was the case for question 55, we need to calculate the non-standard free energy change for the reaction. Once again, we will employ the equation $\Delta G = \Delta G^{\circ} + RT \ln Q$, but this time

$$Q_c = \frac{[NH_4^+(aq)]\,[OH^-(aq)]}{[NH_3(aq)]}\ ;\ Q_c = \frac{\left(1.0\times10^{-3}\ M\right)^2}{(0.10\ M)} = 1.0\times10^{-5}$$

$\Delta G = 29.05\ kJ + (8.3145\times10^{-3}\ kJ/K\cdot mol)(298\ K)\ln(1.0\times10^{-5})$

$\Delta G = 29.05\ kJ + (-28.53\ kJ) = 0.52\ kJ.$

Since ΔG is positive, the reaction is spontaneous in the reverse direction.

57. The relationship $\Delta S^{\circ} = (\Delta G^{\circ} - \Delta H^{\circ})/T$ (Equation (b)) is incorrect. Rearranging this equation to put ΔG° on one side by itself gives $\Delta G^{\circ} = \Delta H^{\circ} + T\Delta S^{\circ}$. This equation is not valid. The ΔH° term should be subtracted from the $T\Delta S^{\circ}$ term, not added to it.

58. The ΔG° value is a powerful thermodynamic parameter because it can be used to determine the equilibrium constant for the reaction at each and every chemically reasonable temperature via the equation $\Delta G^{\circ} = -RT\ln K$.

59. **(a)** To determine K_p we need the equilibrium partial pressures. In the ideal gas law, each partial pressure is defined by $P = nRT/V$. Because R, T, and V are the same for each gas, and because there are the same number of partial pressure factors in the numerator as in the denominator of the K_p expression, we can use the ratio of amounts to determine K_p.

$$K_p = \frac{P\{CO(g)\}\,P\{H_2O(g)\}}{P\{CO_2(g)\}\,P\{H_2(g)\}} = \frac{n\{CO(g)\}\,n\{H_2O(g)\}}{n\{CO_2(g)\}\,n\{H_2(g)\}}$$

$$= \frac{0.224\ mol\ CO\times0.224\ mol\ H_2O}{0.276\ mol\ CO_2\times0.276\ mol\ H_2} = 0.659$$

(b) $\Delta G^{\circ}_{1000K} = -RT \ln K_p$

$$= -8.3145 \text{ J mol}^{-1}\text{K}^{-1} \times 1000. \text{ K} \times \ln(0.659)$$

$$= 3.467 \times 10^3 \text{ J/mol} = 3.467 \text{ kJ/mol}$$

$$Q_p = \frac{0.0340 \text{ mol CO} \times 0.0650 \text{ mol H}_2\text{O}}{0.0750 \text{ mol CO}_2 \times 0.095 \text{ mol H}_2} = 0.31 < 0.659 = K_p$$

Since Q_p is smaller than K_p, the reaction will proceed to the right, forming products, to attain equilibrium, i.e., $\Delta G = 0$.

60. (a) We know that $K_p = K_c(RT)^{\Delta n}$. For the reaction $2SO_2(g) + O_2(g) \rightleftharpoons 2SO_3(g)$,

$\Delta n_{gas} = 2-(2+1) = -1$, and therefore a value of K_p can be obtained.

$$K_p = K_c(RT)^{-1} = \frac{2.8 \times 10^2}{\dfrac{0.08206 \text{ L atm}}{\text{mol K}} \times 1000 \text{ K}} = 3.4 = K_{eq}$$

We recognize that $K_{eq} = K_p$ since all of the substances involved in the reaction are gases. We can now evaluate ΔG°.

$$\Delta G^{\circ} = -RT \ln K_{eq} = -\frac{8.3145 \text{ J}}{\text{mol K}} \times 1000 \text{ K} \times \ln(3.41) = -1.02 \times 10^4 \text{ J/mol}$$

$$= -10.2 \text{ kJ/mol}$$

(b) We can evaluate Q_c for this situation and compare the value with that of $K_c = 2.8 \times 10^2$.

$$Q_c = \frac{[SO_3]^2}{[SO_2]^2[O_2]} = \frac{\left(\dfrac{0.72 \text{ mol SO}_3}{2.50 \text{ L}}\right)^2}{\left(\dfrac{0.40 \text{ mol SO}_2}{2.50 \text{ L}}\right)^2 \times \left(\dfrac{0.18 \text{ mol O}_2}{2.50 \text{ L}}\right)} = 45 < 2.8 \times 10^2 = K_c$$

Since Q_c is smaller than K_c the reaction will shift right, producing sulfur trioxide and consuming sulfur dioxide and molecular oxygen, until the two values are equal.

61. (a) $K_{eq} = K_c$

$$\Delta G^{\circ} = -RT \ln K_{eq} = -(8.3145 \times 10^{-3} \text{ kJ mol}^{-1} \text{ K}^{-1})(445+273)\text{K} \ln 50.2 = -23.4 \text{ kJ}$$

(b) $K_{eq} = K_p = K_c(RT)^{\Delta n_g} = 1.7 \times 10^{-13}(0.0821 \times 298)^{1/2} = 8.4 \times 10^{-13}$

$$\Delta G^{\circ} = -RT \ln K_p = -(8.3145 \times 10^{-3} \text{ kJ mol}^{-1} \text{ K}^{-1})(298 \text{ K}) \ln(8.4 \times 10^{-13})$$

$$\Delta G^{\circ} = +68.9 \text{ kJ/mol}$$

(c) $\quad K_{eq} = K_p = K_c(RT)^{\Delta n} = 4.61 \times 10^{-3}(0.08206 \times 298)^{+1} = 0.113$

$\quad\quad \Delta G^\circ = -RT \ln K_p = -(8.3145 \times 10^{-3} \text{ kJ mol}^{-1} \text{ K}^{-1})(298\text{K}) \ln(0.113) = +5.40 \text{ kJ / mol}$

(d) $\quad K_{eq} = K_c = 9.14 \times 10^{-6}$

$\quad\quad \Delta G^\circ = -RT \ln K_c = -(8.3145 \times 10^{-3} \text{ kJ mol}^{-1} \text{ K}^{-1})(298 \text{ K}) \ln(9.14 \times 10^{-6})$

$\quad\quad \Delta G^\circ = +28.7 \text{ kJ/mol}$

62. (a) The first equation involves the formation of one mole of $Mg^{2+}(aq)$ from $Mg(OH)_2(s)$ and $2H^+(aq)$, while the second equation involves the formation of only half-a-mole of $Mg^{2+}(aq)$. We would expect a free energy change of half the size if only half the product is formed.

(b) The value of K_{eq} for the first reaction is the square of the value of K_{eq} for the second reaction. This is the same way the equilibrium constant expressions are related.

$$K_1 = \frac{[Mg^{2+}]}{[H^+]^2} = \left(\frac{[Mg^{2+}]^{1/2}}{[H^+]}\right)^2 = (K_2)^2$$

(c) The equilibrium solubilities will be the same regardless of the two expressions are used. The equilibrium conditions (solubilities in this instance) are the same no matter how we choose to express them in an equilibrium constant expression.

63. $\quad \Delta G^\circ = -RT \ln K_p = -(8.3145 \times 10^{-3} \text{ kJ mol}^{-1}\text{K}^{-1})(298\text{K}) \ln(6.5 \times 10^{11}) = -67.4 \text{ kJ/mol}$

$$CO(g) + Cl_2(g) \rightarrow COCl_2(g) \quad\quad\quad\quad\quad \Delta G^\circ = -67.4\,\text{kJ/mol}$$

$$C(\text{graphite}) + \tfrac{1}{2}O_2(g) \rightarrow CO(g) \quad\quad\quad\quad \Delta G_f^\circ = -137.2 \text{ kJ/mol}$$

$$\overline{C(\text{graphite}) + \tfrac{1}{2}O_2(g) + Cl_2(g) \rightarrow COCl_2(g) \quad\quad \Delta G_f^\circ = -204.6 \text{ kJ/mol}}$$

ΔG_f° of $COCl_2(g)$ given in Appendix D is -204.6 kJ/mol, (excellent agreement.).

64. In each case, we determine the value of ΔG° for the solubility reaction. From that, we determine the value of the equilibrium constant, K_{sp}, for that solubility reaction.

(a) $\quad AgBr(s) \rightleftharpoons Ag^+(aq) + Br^-(aq)$

$\quad\quad \Delta G^\circ = \Delta G_f^\circ[Ag^+(aq)] + \Delta G_f^\circ[Br^-(aq)] - \Delta G_f^\circ[AgBr(s)]$

$\quad\quad\quad = 77.11 \text{ kJ/mol} - 104.0 \text{ kJ/mol} - (-96.90 \text{ kJ/mol}) = +70.0 \text{ kJ/mol}$

$$\ln K_{eq} = \frac{-\Delta G^\circ}{RT} = \frac{-70.0 \times 10^3 \text{ J/mol}}{8.3145 \text{ J mol}^{-1}\text{K}^{-1} \times 298.15 \text{ K}} = -28.2; \quad K_{sp} = e^{-28.2} = 6 \times 10^{-13}$$

(b) $CaSO_4(s) \rightleftharpoons Ca^{2+}(aq) + SO_4^{2-}(aq)$

$$\Delta G° = \Delta G_f°\left[Ca^{2+}(aq)\right] + \Delta G_f°\left[SO_4^{2-}(aq)\right] - \Delta G_f°\left[CaSO_4(s)\right]$$

$$= -553.6 \text{ kJ/mol} - 744.5 \text{ kJ/mol} - (-1332 \text{ kJ/mol}) = +34 \text{ kJ/mol}$$

$$\ln K_{eq} = \frac{-\Delta G°}{RT} = \frac{-34 \times 10^3 \text{ J/mol}}{8.3145 \text{ J mol}^{-1}\text{K}^{-1} \times 298.15 \text{ K}} = -14; \qquad K_{sp} = e^{-14} = 8 \times 10^{-7}$$

(c) $Fe(OH)_3(s) \rightleftharpoons Fe^{3+}(aq) + 3OH^-(aq)$

$$\Delta G° = \Delta G_f°\left[Fe^{3+}(aq)\right] + 3\,\Delta G_f°\left[OH^-(aq)\right] - \Delta G_f°\left[Fe(OH)_3(s)\right]$$

$$= -4.7 \text{ kJ / mol} + 3 \times (-157.2 \text{ kJ / mol}) - (-696.5 \text{ kJ / mol}) = +220.2 \text{ kJ / mol}$$

$$\ln K_{eq} = \frac{-\Delta G°}{RT} = \frac{-220.2 \times 10^3 \text{ J/mol}}{8.3145 \text{ J mol}^{-1}\text{K}^{-1} \times 298.15 \text{ K}} = -88.83 \qquad K_{sp} = e^{-88.83} = 2.6 \times 10^{-39}$$

65. (a) We can determine the equilibrium partial pressure from the value of the equilibrium constant.

$$\Delta G° = -RT \ln K_p \qquad \ln K_p = -\frac{\Delta G°}{RT} = -\frac{58.54 \times 10^3 \text{ J/mol}}{8.3145 \text{ J mol}^{-1}\text{K}^{-1} \times 298.15 \text{ K}} = -23.63$$

$$K_p = P\{O_2(g)\}^{1/2} = e^{-23.63} = 5.5 \times 10^{-11} \qquad P\{O_2(g)\} = \left(5.5 \times 10^{-11}\right)^2 = 3.0 \times 10^{-21} \text{ atm}$$

(b) Lavoisier did two things to increase the quantity of oxygen that he obtained. First, he ran the reaction at a high temperature, which favors the products (i.e., the side with molecular oxygen.) Second, the molecular oxygen was removed immediately after it was formed, which causes the equilibrium to shift to the right continuously (the shift towards products as result of the removal of the O_2 is an example of Le Châtelier's principle).

66. (a) We determine the values of $\Delta H°$ and $\Delta S°$ from the data in Appendix D, and then the value of $\Delta G°$ at $25°$ C $= 298$ K.

$$\Delta H° = \Delta H_f°\left[CH_3OH(g)\right] + \Delta H_f°\left[H_2O(g)\right] - \Delta H_f°\left[CO_2(g)\right] - 3\,\Delta H_f°\left[H_2(g)\right]$$

$$= -200.7 \text{ kJ/mol} + (-241.8 \text{ kJ/mol}) - (-393.5 \text{ kJ/mol}) - 3(0.00 \text{ kJ/mol})$$

$$= -49.0 \text{ kJ/mol}$$

$$\Delta S° = S°\left[CH_3OH(g)\right] + S°\left[H_2O(g)\right] - S°\left[CO_2(g)\right] - 3\,S°\left[H_2(g)\right]$$

$$= (239.8 + 188.8 - 213.7 - 3 \times 130.7)\text{ J mol}^{-1}\text{K}^{-1} = -177.2 \text{ J mol}^{-1}\text{K}^{-1}$$

$$\Delta G° = \Delta H° - T\Delta S° = -49.0 \text{ kJ/mol} - 298 \text{ K}\left(-0.1772 \text{ kJ mol}^{-1}\text{K}^{-1}\right) = +3.81 \text{ kJ/mol}$$

Because the value of $\Delta G°$ is positive, this reaction is not favored at $25°$ C.

(b) Because the value of $\Delta H°$ is negative and that of $\Delta S°$ is negative, the reaction is *nonspontaneous* at high temperatures, if reactants and products are *in their standard states*. The reaction will proceed slightly in the forward direction, however, to produce an equilibrium mixture with small quantities of $CH_3OH(g)$ and $H_2O(g)$.

And, because the forward reaction is exothermic, this reaction is favored by lowering the temperature. That is, the value of K_{eq} increases with decreasing temperature.

(c) $\Delta G°_{500K} = \Delta H° - T\,\Delta S° = -49.0 \text{ kJ/mol} - 500.\text{K}\left(-0.1772 \text{ kJ mol}^{-1}\text{K}^{-1}\right) = 39.6 \text{ kJ/mol}$

$= 39.6 \times 10^3 \text{ J/mol} = -RT \ln K_p$

$\ln K_p = \dfrac{-\Delta G°}{RT} = \dfrac{-39.6 \times 10^3 \text{ J/mol}}{8.3145 \text{ J mol}^{-1}\text{K}^{-1} \times 500.\text{ K}} = -9.53; \quad K_p = e^{-9.53} = 7.3 \times 10^{-5}$

(d) Reaction:

	$CO_2(g) +$	$3H_2(g)$	$\rightleftharpoons$	$CH_3OH(g)$	$+H_2O(g)$
Initial:	1.00 atm	1.00 atm		0 atm	0 atm
Changes:	$-x$ atm	$-3x$ atm		$+x$ atm	$+x$ atm
Equil:	$(1.00 - x)$ atm	$(1.00 - 3x)$ atm		x atm	x atm

$K_p = 7.3 \times 10^{-5} = \dfrac{P\{CH_3OH\}\,P\{H_2O\}}{P\{CO_2\}\,P\{H_2\}^3} = \dfrac{x \cdot x}{(1.00-x)(1.00-3x)^3} \approx x^2$

$x = \sqrt{7.3 \times 10^{-5}} = 8.5 \times 10^{-3} \text{ atm} = P\{CH_3OH\}$

Our assumption, that $3x \ll 1.00$ atm, is valid.

$\Delta G°$ and K_{eq} as a Function of Temperature

67. We first determine the value of $\Delta G°$ at $400°C$, from the values of $\Delta H°$ and $\Delta S°$, which are determined from information listed in Appendix D.

$\Delta H° = 2\Delta H_f°\left[NH_3(g)\right] - \Delta H_f°\left[N_2(g)\right] - 3\Delta H_f°\left[H_2(g)\right]$

$= 2(-46.11 \text{kJ/mol}) - (0.00 \text{kJ/mol}) - 3(0.00 \text{ kJ/mol}) = -92.22 \text{kJ/mol N}_2$

$\Delta S° = 2S°\left[NH_3(g)\right] - S°\left[N_2(g)\right] - 3S°\left[H_2(g)\right]$

$= 2(192.5 \text{ J mol}^{-1}\text{K}^{-1}) - (191.6 \text{ J mol}^{-1}\text{K}^{-1}) - 3(130.7 \text{ J mol}^{-1}\text{K}^{-1}) = -198.7 \text{ J mol}^{-1}\text{ K}^{-1}$

$\Delta G° = \Delta H° - T\,\Delta S° = -92.22 \text{ kJ/mol} - 673 \text{ K} \times \left(-0.1987 \text{ kJ mol}^{-1}\text{ K}^{-1}\right)$

$= +41.51 \text{ kJ/mol} = -RT \ln K_p$

$\ln K_p = \dfrac{-\Delta G°}{RT} = \dfrac{-41.51 \times 10^3 \text{ J/mol}}{8.3145 \text{ J mol}^{-1}\text{K}^{-1} \times 673\text{K}} = -7.42; \quad K_p = e^{-7.42} = 6.0 \times 10^{-4}$

68. **(a)** $\Delta H^\circ = \Delta H_f^\circ \left[CO_2(g) \right] + \Delta H_f^\circ \left[H_2(g) \right] - \Delta H_f^\circ \left[CO(g) \right] - \Delta H_f^\circ \left[H_2O(g) \right]$

$\qquad = -393.5 \text{ kJ/mol} - 0.00 \text{ kJ/mol} - \left(-110.5 \text{ kJ/mol} \right) - \left(-241.8 \text{ kJ/mol} \right)$

$\qquad = -41.2 \text{ kJ/mol}$

$\Delta S^\circ = S^\circ \left[CO_2(g) \right] + S^\circ \left[H_2(g) \right] - S^\circ \left[CO(g) \right] - S^\circ \left[H_2O(g) \right]$

$\qquad = 213.7 \text{ J mol}^{-1} \text{ K}^{-1} + 130.7 \text{ J mol}^{-1} \text{ K}^{-1} - 197.7 \text{ J mol}^{-1} \text{ K}^{-1} - 188.8 \text{ J mol}^{-1} \text{ K}^{-1}$

$\qquad = -42.1 \text{ J mol}^{-1} \text{ K}^{-1}$

$\Delta G^\circ = \Delta H^\circ - T\Delta S^\circ = -41.2 \text{kJ/mol} - (298.15\text{K}) - 42.1 \times 10^{-3} \text{kJ mol}^{-1}\text{K}^{-1}$

$\qquad = -41.2 \text{ kJ/mol} + 12.6 \text{ kJ/mol} = -28.6 \text{ kJ/mol}$

(b) $\Delta G^\circ = \Delta H^\circ - T\Delta S^\circ = -41.2 \text{ kJ/mol} - \left(875 \text{ K} \right)\left(-42.1 \times 10^{-3} \text{ kJ mol}^{-1} \text{ K}^{-1} \right)$

$\qquad = -41.2 \text{ kJ/mol} + 36.8 \text{ kJ/mol} = -4.4 \text{ kJ/mol} = -RT \ln K_p$

$\ln K_p = -\dfrac{\Delta G^\circ}{RT} = -\dfrac{-4.4 \times 10^3 \text{ J/mol}}{8.3145 \text{ J mol}^{-1}\text{K}^{-1} \times 875\text{K}} = +0.60 \qquad K_p = e^{+0.60} = 1.8$

69. We assume that both ΔH° and ΔS° are constant with temperature.

$\Delta H^\circ = 2\Delta H_f^\circ \left[SO_3(g) \right] - 2\Delta H_f^\circ \left[SO_2(g) \right] - \Delta H_f^\circ \left[O_2(g) \right]$

$\qquad = 2\left(-395.7 \text{ kJ/mol} \right) - 2\left(-296.8 \text{ kJ/mol} \right) - \left(0.00 \text{ kJ/mol} \right) = -197.8 \text{ kJ/mol}$

$\Delta S^\circ = 2S^\circ \left[SO_3(g) \right] - 2S^\circ \left[SO_2(g) \right] - S^\circ \left[O_2(g) \right]$

$\qquad = 2\left(256.8 \text{ J mol}^{-1} \text{ K}^{-1} \right) - 2\left(248.2 \text{ J mol}^{-1} \text{ K}^{-1} \right) - \left(205.1 \text{ J mol}^{-1} \text{ K}^{-1} \right)$

$\qquad = -187.9 \text{ J mol}^{-1} \text{ K}^{-1}$

$\Delta G^\circ = \Delta H^\circ - T\Delta S^\circ = -RT\ln K_{eq} \qquad \Delta H^\circ = T\Delta S^\circ - RT\ln K_{eq} \qquad T = \dfrac{\Delta H^\circ}{\Delta S^\circ - R\ln K_{eq}}$

$T = \dfrac{-197.8 \times 10^3 \text{ J/mol}}{-187.9 \text{ J mol}^{-1} \text{ K}^{-1} - 8.3145 \text{ J mol}^{-1} \text{ K}^{-1} \ln \left(1.0 \times 10^6 \right)} \approx 650 \text{ K}$

This value compares very favorably with the value of $T = 6.37 \times 10^2$ that was obtained in Example 20-10.

70. We use the van't Hoff equation to determine the value of ΔH°, $448^\circ \text{ C} = 721 \text{ K}$ and $350^\circ \text{ C} = 623 \text{ K}$.

$\ln \dfrac{K_1}{K_2} = \dfrac{\Delta H^\circ}{R}\left(\dfrac{1}{T_2} - \dfrac{1}{T_1} \right) = \ln \dfrac{50.0}{66.9} = -0.291 = \dfrac{\Delta H^\circ}{R}\left(\dfrac{1}{623} - \dfrac{1}{721} \right) = \dfrac{\Delta H^\circ}{R}\left(2.2 \times 10^{-4} \right)$

$\dfrac{\Delta H^\circ}{R} = \dfrac{-0.291}{2.2 \times 10^{-4}} = -1.3 \times 10^3; \Delta H^\circ = -1.3 \times 10^3 \times 8.3145 = -11 \times 10^3 \text{ J/mol} = -11 \text{ kJ/mol}$

71. (a) $\ln\dfrac{K_2}{K_1} = \dfrac{\Delta H^\circ}{R}\left(\dfrac{1}{T_1} - \dfrac{1}{T_2}\right) = \dfrac{57.2\times10^3 \text{ J/mol}}{8.3145 \text{ J mol}^{-1}\text{ K}^{-1}}\left(\dfrac{1}{298 \text{ K}} - \dfrac{1}{273 \text{ K}}\right) = -2.11$

$\dfrac{K_2}{K_1} = e^{-2.11} = 0.121 \qquad K_2 = 0.121\times0.113 = 0.014 \text{ at } 273 \text{ K}$

(b) $\ln\dfrac{K_2}{K_1} = \dfrac{\Delta H^\circ}{R}\left(\dfrac{1}{T_1} - \dfrac{1}{T_2}\right) = \dfrac{57.2\times10^3 \text{ J / mol}}{8.3145 \text{ J mol}^{-1}\text{ K}^{-1}}\left(\dfrac{1}{T_1} - \dfrac{1}{298 \text{ K}}\right) = \ln\dfrac{0.113}{1.00} = -2.180$

$\left(\dfrac{1}{T_1} - \dfrac{1}{298\text{K}}\right) = \dfrac{-2.180\times8.3145}{57.2\times10^3}\text{K}^{-1} = -3.17\times10^{-4}\text{K}^{-1}$

$\dfrac{1}{T_1} = \dfrac{1}{298} - 3.17\times10^{-4} = 3.36\times10^{-3} - 3.17\times10^{-4} = 3.04\times10^{-3}\text{ K}^{-1}; \qquad T_1 = 329 \text{ K}$

72. First we calculate ΔG° at 298 K to obtain a value for K_{eq} at that temperature.

$\Delta G^\circ = 2\Delta G_f^\circ\left[NO_2(g)\right] - 2\Delta G_f^\circ\left[NO(g)\right] - \Delta G_f^\circ\left[O_2(g)\right]$

$\qquad = 2(51.31 \text{ kJ/mol}) - 2(86.55 \text{ kJ/mol}) - 0.00 \text{ kJ/mol} = -70.48 \text{ kJ/mol}$

$\ln K_{eq} = \dfrac{-\Delta G^\circ}{RT} = -\dfrac{-70.48\times10^3 \text{ J/mol K}}{\dfrac{8.3145 \text{ J}}{\text{mol K}}\times298.15 \text{ K}} = 28.43 \qquad K_{eq} = e^{28.43} = 2.2\times10^{12}$

Now we calculate ΔH° for the reaction, which then will be inserted into the van't Hoff equation.

$\Delta H^\circ = 2\Delta H_f^\circ\left[NO_2(g)\right] - 2\Delta H_f^\circ\left[NO(g)\right] - \Delta H_f^\circ\left[O_2(g)\right]$

$\qquad = 2(33.18 \text{ kJ / mol}) - 2(90.25 \text{ kJ / mol}) - 0.00 \text{ kJ / mol} = -114.14 \text{ kJ / mol}$

$\ln\dfrac{K_2}{K_1} = \dfrac{\Delta H^\circ}{R}\left(\dfrac{1}{T_1} - \dfrac{1}{T_2}\right) = \dfrac{-114.14\times10^3 \text{ J/mol}}{8.3145 \text{ J mol}^{-1}\text{ K}^{-1}}\left(\dfrac{1}{298 \text{ K}} - \dfrac{1}{373 \text{ K}}\right) = -9.26$

$\dfrac{K_2}{K_1} = e^{-9.26} = 9.5\times10^{-5}; \quad K_2 = 9.5\times10^{-5}\times2.2\times10^{12} = 2.1\times10^{8}$

Another way, without using the van't Hoff equation is to compute $\Delta H^\circ(-114.14 \text{ kJ / mol})$ from ΔH_f° values and $\Delta S^\circ(-146.5 \text{ J mol}^{-1}\text{ K}^{-1})$ from S° values. Then determine $\Delta G^\circ(-59.5 \text{ kJ/mol})$, and find K_p with the expression $\Delta G^\circ = -RT\ln K_p$. We obtain the same result, $K_p = 2.2\times10^{8}$.

73. First, the van't Hoff equation is used to obtain a value of $\Delta H°$. $200°C = 473K$ and $260°C = 533K$.

$$\ln\frac{K_2}{K_1} = \frac{\Delta H°}{R}\left(\frac{1}{T_1} - \frac{1}{T_2}\right) = \ln\frac{2.15 \times 10^{11}}{4.56 \times 10^8} = 6.156 = \frac{\Delta H°}{8.3145 \text{ J mol}^{-1}\text{ K}^{-1}}\left(\frac{1}{533\text{ K}} - \frac{1}{473\text{ K}}\right)$$

$$6.156 = -2.9 \times 10^{-5}\Delta H° \qquad \Delta H° = \frac{6.156}{-2.9 \times 10^{-5}} = -2.1 \times 10^5 \text{ J / mol} = -2.1 \times 10^2 \text{ kJ / mol}$$

Another route to $\Delta H°$ is the combination of standard enthalpies of formation.

$$CO(g) + 3H_2(g) \rightleftharpoons CH_4(g) + H_2O(g)$$

$$\Delta H° = \Delta H_f°\left[CH_4(g)\right] + \Delta H_f°\left[H_2O(g)\right] - \Delta H_f°\left[CO(g)\right] - 3\Delta H_f°\left[H_2(g)\right]$$

$$= -74.81 \text{ kJ/mol} - 241.8 \text{ kJ/mol} - (-110.5) - 3 \times 0.00 \text{ kJ/mol} = -206.1 \text{ kJ/mol}$$

Within the precision of the data supplied, the results are in agreement.

74. (a)

$t, °C$	T, K	$1/T, K^{-1}$	K_p	$\ln K_p$
30.	303	3.30×10^{-3}	1.66×10^{-5}	-11.006
50.	323	3.10×10^{-3}	3.90×10^{-4}	-7.849
70.	343	2.92×10^{-3}	6.27×10^{-3}	-5.072
100.	373	2.68×10^{-3}	2.31×10^{-1}	-1.465

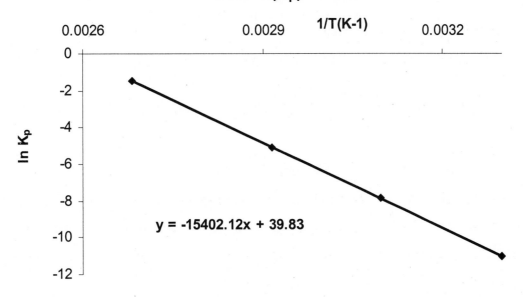

Plot of ln(K_p) versus 1/T

$y = -15402.12x + 39.83$

(b) When the total pressure is 2.00 atm, and both gases have been produced from $NaHCO_3(s)$,

$$P\{H_2O(g)\} = P\{CO_2(g)\} = 1.00 \text{ atm}$$

$$K_p = P\{H_2O(g)\}P\{CO_2(g)\} = (1.00)(1.00) = 1.00$$

Thus, $\ln K_p = \ln(1.00) = 0.000$. The point corresponds to $1/T = 2.59 \times 10^{-3} \text{ K}^{-1}$; $T = 386 \text{ K}$.

We can compute the same temperature from the van't Hoff equation.

$$\ln \frac{K_2}{K_1} = \frac{\Delta H^\circ}{R}\left(\frac{1}{T_1} - \frac{1}{T_2}\right) = \frac{128 \times 10^3 \text{ J/mol}}{8.3145 \text{ J mol}^{-1} \text{ K}^{-1}}\left(\frac{1}{T_1} - \frac{1}{303 \text{ K}}\right) = \ln\frac{1.66 \times 10^{-5}}{1.00} = -11.006$$

$$\left(\frac{1}{T_1} - \frac{1}{303 \text{ K}}\right) = \frac{-11.006 \times 8.3145}{128 \times 10^3} \text{ K}^{-1} = -7.15 \times 10^{-4} \text{ K}^{-1}$$

$$\frac{1}{T_1} = \frac{1}{303} - 7.15 \times 10^{-4} = 3.30 \times 10^{-3} - 7.15 \times 10^{-4} = 2.59 \times 10^{-3} \text{ K}^{-1}; \quad T_1 = 386 \text{ K}$$

This temperature agrees well with the result obtained from the graph.

Coupled Reactions

75. **(a)** We compute ΔG° for the given reaction in the following manner

$$\Delta H^\circ = \Delta H_f^\circ[TiCl_4(l)] + \Delta H_f^\circ[O_2(g)] - \Delta H_f^\circ[TiO_2(s)] - 2\Delta H_f^\circ[Cl_2(g)]$$

$$= -804.2 \text{ kJ/mol} + 0.00 \text{ kJ/mol} - (-944.7 \text{ kJ/mol}) - 2(0.00 \text{ kJ/mol})$$

$$= +140.5 \text{ kJ/mol}$$

$$\Delta S^\circ = S^\circ[TiCl_4(l)] + S^\circ[O_2(g)] - S^\circ[TiO_2(s)] - 2S^\circ[Cl_2(g)]$$

$$= 252.3 \text{ J mol}^{-1} \text{ K}^{-1} + 205.1 \text{ J mol}^{-1} \text{ K}^{-1} - (50.33 \text{ J mol}^{-1} \text{ K}^{-1}) - 2(223.1 \text{ J mol}^{-1} \text{ K}^{-1})$$

$$= -39.1 \text{ J mol}^{-1} \text{ K}^{-1}$$

$$\Delta G^\circ = \Delta H^\circ - T\Delta S^\circ = +140.5 \text{ kJ/mol} - (298 \text{ K})(-39.1 \times 10^{-3} \text{ kJ mol}^{-1} \text{ K}^{-1})$$

$$= +140.5 \text{ kJ/mol} + 11.6 \text{ kJ/mol} = +152.1 \text{ kJ/mol}$$

This reaction is nonspontaneous at 25° C. (We also could have used values of ΔG_f° to calculate ΔG°.)

(b) For the cited reaction, $\Delta G^\circ = 2\Delta G_f^\circ[CO_2(g)] - 2\Delta G_f^\circ[CO(g)] - \Delta G_f^\circ[O_2(g)]$

$\Delta G^\circ = 2(-394.4 \text{ kJ / mol}) - 2(-137.2 \text{ kJ / mol}) - 0.00 \text{ kJ / mol} = -514.4 \text{ kJ / mol}$

Then we couple the two reactions.

$TiO_2(s) + 2Cl_2(g) \longrightarrow TiCl_4(l) + O_2(g)$ $\qquad\qquad$ $\Delta G^\circ = +152.1 \text{ kJ/mol}$

$2CO(g) + O_2(g) \longrightarrow 2CO_2(g)$ $\qquad\qquad$ $\Delta G^\circ = -514.4 \text{ kJ/mol}$

$TiO_2(s) + 2Cl_2(g) + 2CO(g) \longrightarrow TiCl_4(l) + 2CO_2(g); \Delta G^\circ = -362.3 \text{ kJ/mol}$

The coupled reaction has $\Delta G^\circ < 0$, and, therefore, is spontaneous.

76. If $\Delta G^\circ < 0$ for the sum of the coupled reactions, reduction of the oxide with carbon is spontaneous.

(a) $NiO(s) \rightarrow Ni(s) + \frac{1}{2}O_2(g)$ $\qquad\qquad$ $\Delta G^\circ = +115 \text{ kJ}$

$\quad\ C(s) + \frac{1}{2}O_2(g) \rightarrow CO(g)$ $\qquad\qquad$ $\Delta G^\circ = -250 \text{ kJ}$

$\text{Net}: NiO(s) + C(s) \rightarrow Ni(s) + CO(g)$ $\qquad$ $\Delta G^\circ = +115 \text{ kJ} - 250 \text{ kJ} = -135 \text{ kJ}$

Spontaneous

(b) $MnO(s) \rightarrow Mn(s) + \frac{1}{2}O_2(g)$ $\qquad\qquad$ $\Delta G^\circ = +280 \text{ kJ}$

$\quad\ C(s) + \frac{1}{2}O_2(g) \rightarrow CO(g)$ $\qquad\qquad$ $\Delta G^\circ = -250 \text{ kJ}$

$\text{Net}: MnO(s) + C(s) \longrightarrow Mn(s) + CO(g)$ $\qquad$ $\Delta G^\circ = +280 \text{ kJ} - 250 \text{ kJ} = +30 \text{ kJ}$

Nonspontaneous

(c) $TiO_2(s) \longrightarrow Ti(s) + O_2(g)$ $\qquad\qquad$ $\Delta G^\circ = +630 \text{ kJ}$

$\quad\ 2C(s) + O_2(g) \longrightarrow 2CO(g)$ $\qquad\qquad$ $\Delta G^\circ = 2(-250 \text{ kJ}) = -500 \text{ kJ}$

$\text{Net}: TiO_2(s) + 2C(s) \longrightarrow Ti(s) + 2CO(g)$ $\qquad$ $\Delta G^\circ = +630 \text{ kJ} - 500 \text{ kJ} = +130 \text{ kJ}$

Nonspontaneous

FEATURE PROBLEMS

98. **(a)** The first method involves combining the values of ΔG_f°. The second uses

$$\Delta G^\circ = \Delta H^\circ - T\Delta S^\circ$$

$$\Delta G^\circ = \Delta G_f^\circ\left[H_2O(g)\right] - \Delta G_f^\circ\left[H_2O(l)\right]$$
$$= -228.572 \text{ kJ/mol} - \left(-237.129 \text{ kJ/mol}\right) = +8.557 \text{ kJ/mol}$$

$$\Delta H^\circ = \Delta H_f^\circ\left[H_2O(g)\right] - \Delta H_f^\circ\left[H_2O(l)\right]$$
$$= -241.818 \text{ kJ/mol} - \left(-285.830 \text{ kJ/mol}\right) = +44.012 \text{ kJ/mol}$$

$$\Delta S^\circ = S^\circ\left[H_2O(g)\right] - S^\circ\left[H_2O(l)\right]$$
$$= 188.825 \text{ J mol}^{-1} \text{ K}^{-1} - 69.91 \text{ J mol}^{-1} \text{ K}^{-1} = +118.92 \text{ J mol}^{-1} \text{ K}^{-1}$$

$$\Delta G^\circ = \Delta H^\circ - T\Delta S^\circ$$
$$= 44.012 \text{ kJ/mol} - 298.15 \text{ K} \times 118.92 \times 10^{-3} \text{ kJ mol}^{-1} \text{ K}^{-1} = +8.556 \text{ kJ/mol}$$

(b) We use the average value: $\Delta G^\circ = +8.558 \times 10^3 \text{ J / mol} = -RT \ln K_{eq}$

$$\ln K_{eq} = -\frac{8558 \text{ J/mol}}{8.3145 \text{ J mol}^{-1} \text{ K}^{-1} \times 298.15 \text{ K}} = -3.452; \quad K_{eq} = e^{-3.452} = 0.0317 \text{ bar}$$

(c) $P\{H_2O\} = 0.0317 \text{ bar} \times \dfrac{1 \text{ atm}}{1.01325 \text{ bar}} \times \dfrac{760 \text{ mmHg}}{1 \text{ atm}} = 23.8 \text{ mmHg}$

(d) $\ln K_{eq} = -\dfrac{8590 \text{ J/mol}}{8.3145 \text{ J mol}^{-1} \text{ K}^{-1} \times 298.15 \text{ K}} = -3.465$;

$K_{eq} = e^{-3.465} = 0.0312_7 \text{ atm}$;

$P\{H_2O\} = 0.0313 \text{ atm} \times \dfrac{760 \text{ mmHg}}{1 \text{ atm}} = 23.8 \text{ mmHg}$

99. **(a)** When we combine two reactions and obtain the overall value of ΔG°, we subtract the value on the plot of the reaction that becomes a reduction from the value on the plot of the reaction that is an oxidation. Thus, to reduce ZnO with elemental Mg, we subtract the values on the line labeled "$2Zn + O_2 \rightarrow 2ZnO$" from those on the line labeled "$2Mg + O_2 \rightarrow 2MgO$". The result for the overall ΔG° will always be negative because every point on the "zinc" line is above the corresponding point on the "magnesium" line

(b) In contrast, the "carbon" line is only below the "zinc" line at temperatures above about $1000°C$. Only at these elevated temperatures can ZnO be reduced by carbon.

(c) The decomposition of zinc oxide to its elements is the reverse of the plotted reaction, the value of $\Delta G°$ for the decomposition becomes negative, and the reaction becomes spontaneous, where the value of $\Delta G°$ for the plotted reaction becomes positive. This occurs above about $1850°C$.

(d) The "carbon" line has a negative slope, indicating that carbon monoxide becomes more stable as temperature rises. The point where CO(g) would become less stable than 2C(s) and O_2(g) looks to be below $-1000°C$ (by extrapolating the line to lower temperatures). Based on this plot, it is not possible to decompose CO(g) to C(s) and $O_2\left(g\right)$ in a spontaneous reaction.

(e)

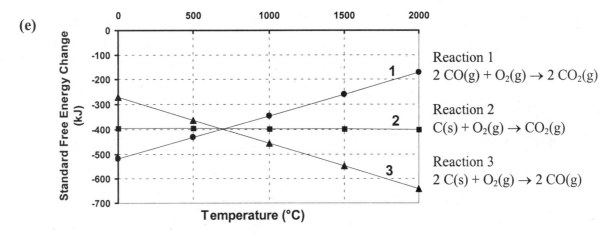

All three lines are straight-line plots of $\Delta G°$ vs. T following the equation $\Delta G° = \Delta H° - T\Delta S°$.

The general equation for a straight line is given below with the slightly modified Gibbs Free-Energy equation as a reference: $\Delta G° = -\Delta S°T + \Delta H°$ (here $\Delta H°$ is assumed to be constant)

$y = mx + b$ ($m = -\Delta S°$ = slope of the line)

Thus, the slope of each line multiplied by minus one is equal to the $\Delta S°$ for the oxide formation reaction. It is hardly surprising, therefore, that the slopes for these lines differ so markedly because these three reactions have quite different $\Delta S°$ values ($\Delta S°$ for Reaction 1 = -173 J K^{-1}, $\Delta S°$ for Reaction 2 = 2.86 J K^{-1}, $\Delta S°$ for Reaction 3 = 178.8 J K^{-1})

(f) Since other metal oxides apparently have positive slopes similar to Mg and Zn, we can conclude that in general, the stability of metal oxides <u>decreases</u> as the temperature increases. Put another way, the decomposition of metal oxides to their elements becomes more spontaneous as the temperature is increased. By contrast, the two reactions involving elemental carbon, namely Reaction 2 and Reaction 3, have negative slopes, indicating that the formation of $CO_2(g)$ and $CO(g)$ from graphite becomes more favorable as the temperature rises, repectively. This means that the $\Delta G°$ for the reduction of metal oxides by carbon becomes more and more negative with increasing temperature. Moreover, there must exist a threshold temperature for each metal oxide above which the reaction with carbon will occur spontaneously. Carbon would appear to be an excellent reducing agent, therefore, because it will reduce virtually any metal oxide to its corresponding metal as long as the temperature chosen for the reaction is higher than the threshold temperature (the threshold temperature is commonly referred to as the transition temperature).

Consider for instance the reaction of MgO(s) with graphite to give $CO_2(g)$ and Mg metal:

$$2\text{ MgO(s)} + \text{C(s)} \rightarrow 2\text{ Mg(s)} + CO_2(g) \quad \Delta S°_{rxn} = 219.4 \text{ J/K and } \Delta H°_{rxn} = 809.9 \text{ kJ}$$

$$T_{transition} = \frac{\Delta H°_{rxn}}{\Delta S°_{rxn}} = \frac{809.9 \text{ kJ}}{0.2194 \text{ kJ K}^{-1}} = 3691 \text{ K} = T_{threshold}$$

Consequently, above 3691 K, carbon will spontaneously reduce MgO to Mg metal.

100. (a) With a 36% efficiency and a condenser temperature (T_l) of 41 °C = 314 K,

$$\text{efficiency} = \frac{T_h - T_l}{T_h} \times 100\% = 36\%$$

$$\frac{T_h - 314}{T_h} = 0.36; \quad T_h = (0.36 \times T_h) + 314 \text{ K}; \quad 0.64 \text{ } T_h = 314 \text{ K}; \quad T_h = 4.9 \times 10^2 \text{ K}$$

(b) The overall efficiency of the power plant is affected by factors in other than the thermodynamic efficiency. For example, a portion of the heat of combustion of the fuel is lost to parts of the surroundings other than the steam boiler; there are frictional losses of energy in moving parts in the engine; and so on. To compensate for these losses, the thermodynamic efficiency must be greater than 36%. To obtain this higher thermodynamic efficiency, T_h must be greater than 4.9×10^2 K.

(c) The steam pressure we are seeking is the vapor pressure of water at 4.9×10^2 K. We also know that the vapor pressure of water at 373 K (100 °C) is 1 atm. The enthalpy of vaporization of water at 298 K is $\Delta H° = \Delta H_f°[H_2O(g) - \Delta H_f°[H_2O(l)] = -241.8$ kJ/mol – (–285.8 kJ/mol) = 44.0 kJ/mol. Although the enthalpy of vaporization is somewhat temperature dependent, we will assume that this value holds from 298 K to 4.9×10^2 K, and make appropriate substitutions into the Clausius-Clapeyron equation.

$$\ln\left(\frac{P_2}{1 \text{ atm}}\right) = \frac{44.0 \text{ kJ mol}^{-1}}{8.3145 \times 10^{-3} \text{ kJ mol}^{-1}}\left(\frac{1}{373 \text{ K}} - \frac{1}{490 \text{ K}}\right) = 5.29 \times 10^{-3}\left(2.68 \times 10^{-3} - 2.04 \times 10^{-3}\right)$$

$$\ln\left(\frac{P_2}{1\ \text{atm}}\right) = 3.39 \,; \quad \left(\frac{P_2}{1\ \text{atm}}\right) = 29._{.7} \,; \quad P_2 \approx 30\ \text{atm}$$

The answer cannot be given with greater certainty because of the weakness of the assumption of a constant $H°_{\text{vapn}}$.

(d)   It is not possible to devise a heat engine with 100% efficiency or greater. For 100% efficiency, $T_1 = 0$ K, which is unattainable. To have an efficiency greater than 100 % would require a *negative T_1*, which is also unattainable.

101. **(a)**   Under biological standard conditions (see page 813)

$$ADP^{3-} + HPO_4^{2-} + H^+ \rightarrow ATP^{4-} + H_2O \quad \Delta G^{o'} = 32.4\ \text{kJ/mol}$$

If all of the energy of combustion of 1 mole of glucose is employed in the conversion of ADP to ATP, then the maximum number of moles ATP produced is

$$\text{Maximum number} = \frac{2870\ \text{kJ mol}^{-1}}{32.4\ \text{kJ mol}^{-1}} = 88.6\ \text{moles ATP}$$

(b)   In an actual cell the number of ATP moles produced is 38, so that the efficiency is:

$$\text{Efficiency} = \frac{\text{number of ATP's actually produced}}{\text{Maximum number of ATP's that can be produced}} = \frac{38}{88.6} = 0.43$$

Thus, the cell's efficiency is about 43%.

(c)   The previously calculated efficiency is based upon the biological standard state. We now calculate the Gibbs energies involved under the actual conditions in the cell. To do this we require the relationship between ΔG and $\Delta G^{o'}$ for the two coupled reactions. For the combustion of glucose we have

$$\Delta G = \Delta G^{o'} + RT \ln\left(\frac{a_{CO_2}^6\, a_{H_2O}^6}{a_{glu}\, a_{O_2}^6}\right)$$

For the conversion of ADP to ATP we have

$$\Delta G = \Delta G^{o'} + RT \ln\left(\frac{a_{ATP}\, a_{H_2O}}{a_{ADP}\, a_{P_i}\left(\left[H^+\right]/10^{-7}\right)}\right)$$

Using the concentrations and pressures provided we can calculate the Gibbs energy for the combustion of glucose under biological conditions. First, we need to replace the activities by the appropriate effective concentrations. That is,

$$\Delta G = \Delta G^{o'} + RT \ln\left(\frac{\left(p/p°\right)_{CO_2}^6\, a_{H_2O}^6}{[glu]/[glu]°\left(p/p°\right)_{O_2}^6}\right)$$

using $a_{H_2O} \approx 1$ for a dilute solution we obtain

$$\Delta G = \Delta G^{\circ'} + RT \ln\left(\frac{(0.050 \text{ bar}/1 \text{ bar})^6 \times 1^6}{[\text{glu}]/1 \times (0.132 \text{ bar}/1 \text{ bar})^6}\right)$$

The concentration of glucose is given in mg/mL and this has to be converted to molarity as follows:

$$[\text{glu}] = \frac{1.0 \text{ mg}}{\text{mL}} \times \frac{\text{g}}{1000 \text{ mg}} \times \frac{1000 \text{ mL}}{\text{L}} \times \frac{1}{180.16 \text{ g mol}^{-1}} = 0.00555 \text{ mol L}^{-1},$$

where the molar mass of glucose is 180.16 g mol^{-1}.
Assuming a temperature of 37 °C for a biological system we have, for one mole of glucose:

$$\Delta G = -2870 \times 10^3 \text{ J} + 8.314 \text{ JK}^{-1} \times 310 \text{K} \times \ln\left(\frac{(0.050\ )^6 \times 1^6}{0.00555/1 \times (0.132)^6}\right)$$

$$\Delta G = -2870 \times 10^3 \text{ J} + 8.314 \text{ JK}^{-1} \times 310 \text{K} \times \ln\left(\frac{(0.3788)^6}{0.00555}\right)$$

$$\Delta G = -2870 \times 10^3 \text{ J} + 8.314 \text{ JK}^{-1} \times 310 \text{K} \times \ln(0.5323)$$

$$\Delta G = -2870 \times 10^3 \text{ J} + 8.314 \text{ JK}^{-1} \times 310 \text{K} \times (-0.6305)$$

$$\Delta G = -2870 \times 10^3 \text{ J} - 1.637 \times 10^3 \text{ J}$$

$$\Delta G = -2872 \times 10^3 \text{ J}$$

In a similar manner we calculate the Gibbs free energy change for the conversion of ADP to ATP:

$$\Delta G = \Delta G^{\circ'} + RT \ln\left(\frac{[ATP]/1 \times 1}{[ADP]/1 \times [P_i]/1 \times ([H^+]/10^{-7})}\right)$$

$$\Delta G = 32.4 \times 10^3 \text{ J} + 8.314 \text{ JK}^{-1} \times 310 \text{K} \times \ln\left(\frac{0.0001}{0.0001 \times 0.0001 \times (10^{-7}/10^{-7})}\right)$$

$$\Delta G = 32.4 \times 10^3 \text{ J} + 8.314 \text{ JK}^{-1} \times 310 \text{K} \times \ln(10^4)$$

$$\Delta G = 32.4 \times 10^3 \text{ J} + 8.314 \text{ JK}^{-1} \times 310 \text{K} \times (9.2103))$$

$$\Delta G = 32.4 \times 10^3 \text{ J} + 23.738 \times 10^3 \text{ J}$$

$$\Delta G = 56.2 \times 10^3 \text{ J}$$

(d) The efficiency under biological conditions is

$$\text{Efficiency} = \frac{\text{number of ATP's actually produced}}{\text{Maximum number of ATP's that can be produced}} = \frac{38}{2872/56.2} = 0.744$$

Thus, the cell's efficiency is about 74%.

The theoretical efficiency of the diesel engine is:

$$\text{Efficiency} = \frac{T_h - T_l}{T_h} \times 100\% = \frac{1923 - 873}{1923} \times 100\% = 55\%$$

Thus, the diesel's actual efficiency is $0.78 \times 55\% = 43\%$.

The cell's efficiency is 74% whereas that of the diesel engine is only 43 %. Why is there such a large discrepancy? The diesel engine supplies heat to the surroundings, which is at a lower temperature than the engine. This dissipation of energy raises the temperature of the surroundings and the entropy of the surroundings. A cell operates under isothermal conditions and the energy not used goes only towards changing the entropy of the cell's surroundings. The cell is more efficient since it does not heat the surroundings.

CHAPTER 21
ELECTROCHEMISTRY
PRACTICE EXAMPLES

1A The conventions state that the anode material is written first, and the cathode material is written last.

Anode, Oxidation: $Sc(s) \rightarrow Sc^{3+}(aq) + 3e^-$

Cathode, Reduction: $\{Ag^+(aq) + e^- \rightarrow Ag(s)\} \times 3$

___

Net Reaction $Sc(s) + 3Ag^+(aq) \rightarrow Sc^{3+}(aq) + 3Ag(s)$

1B Oxidation of Al(s) at the anode: $Al(s) \rightarrow Al^{3+}(aq) + 3e^-$

Reduction of $Ag^+(aq)$ at the cathode: $Ag^+(aq) + e^- \rightarrow Ag(s)$

___

Overall reaction in cell: $Al(s) + 3Ag^+(aq) \rightarrow Al^{3+}(aq) + 3Ag(s)$

Diagram: $Al|Al^{3+}(aq)\|Ag^+(aq)|Ag$

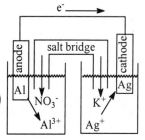

2A We obtain the two balanced half-equations and the half-cell potentials from Table 21-1.

Oxidation: $\{Fe^{2+}(aq) \rightarrow Fe^{3+}(aq) + e^-\} \times 2$ $-E^\circ = -0.771V$

Reduction: $Cl_2(g) + 2e^- \rightarrow 2Cl^-(aq)$ $E^\circ = +1.358V$

___

Net: $2Fe^{2+}(aq) + Cl_2(g) \rightarrow 2Fe^{3+}(aq) + 2Cl^-(aq)$; $E^\circ_{cell} = +1.358V - 0.771V = +0.587V$

2B Since we need to refer to Table 21-1 in any event, it is perhaps a bit easier to locate the two balanced half-equations in the table. There is only one half-equation involving both $Fe^{2+}(aq)$ and $Fe^{3+}(aq)$ ions. It is reversed and written as an oxidation below. The half-equation involving $MnO_4^-(aq)$ is also written below. [Actually, we need to know that in acidic solution $Mn^{2+}(aq)$ is the principal reduction product of $MnO_4^-(aq)$.]

Oxidation: $\{Fe^{2+}(aq) \rightarrow Fe^{3+}(aq) + e^-\} \times 5$ $-E^\circ = -0.771V$

Reduction: $MnO_4^-(aq) + 8H^+(aq) + 5e^- \rightarrow Mn^{2+}(aq) + 4H_2O(l)$ $E^\circ = +1.51V$

___

Net: $MnO_4^-(aq) + 5Fe^{2+}(aq) + 8H^+(aq) \rightarrow Mn^{2+}(aq) + 5Fe^{3+}(aq) + 4H_2O(l)$

$E^\circ_{cell} = +1.51V - 0.771V = +0.74V$

3A We write down the oxidation half-equation with the method of Chapter 5, and obtain the reduction half-equation from Table 21-1, along with the reduction half-cell potential.

Oxidation: $\{H_2C_2O_4(aq) \longrightarrow 2\,CO_2(aq) + 2\,H^+(aq) + 2\,e^-\} \times 3$ $-E°\{CO_2/H_2C_2O_4\}$

Reduction: $Cr_2O_7^{2-}(aq) + 14\,H^+(aq) + 6\,e^- \rightarrow 2\,Cr^{3+}(aq) + 7\,H_2O(l)$ $E° = +1.33\,V$

Net: $Cr_2O_7^{2-}(aq) + 8\,H^+(aq) + 3\,H_2C_2O_4(aq) \rightarrow 2\,Cr^{3+}(aq) + 7\,H_2O(l) + 6\,CO_2(g)$

$E°_{cell} = +1.81\,V = +1.33\,V - E°\{CO_2/H_2C_2O_4\};$

$E°\{CO_2/H_2C_2O_4\} = 1.33\,V - 1.81\,V = -0.48\,V$

3B The 2nd ½-reaction must have $O_2(g)$ as reactant and $H_2O(l)$ as product.

Oxidation: $\{Cr^{2+}+(aq) \longrightarrow Cr^{3+}(aq) + e^-\} \times 4$ $-E°\{Cr^{3+}/Cr^{2+}\}$

Reduction: $O_2(g) + 4\,H^+(aq) + 4\,e^- \rightarrow 2\,H_2O(l)$ $E° = +1.229\,V$

Net: $O_2(g) + 4\,H^+(aq) + 4\,Cr^{2+}(aq) \rightarrow 2\,H_2O(l) + 4\,Cr^{3+}(aq)$

$E_{cell}° = +1.653\,V = +1.229\,V - E°\{Cr^{3+}/Cr^{2+}\};$

$E°\{Cr^{3+}/Cr^{2+}\} = 1.229\,V - 1.653\,V = -0.424\,V$

4A First we write down the two half-equations, obtain the half-cell potential for each, and then calculate $E°_{cell}$. From that value, we determine $\Delta G°$

Oxidation: $\{Al(s) \rightarrow Al^{3+}(aq) + 3e^-\} \times 2$ $-E° = +1.676\,V$

Reduction: $\{Br_2(l) + 2e^- \rightarrow 2\,Br^-(aq)\} \times 3$ $E° = +1.065\,V$

Net: $2\,Al(s) + 3\,Br_2(l) \rightarrow 2\,Al^{3+}(aq) + 6\,Br^-(aq)$ $E°_{cell} = 1.676\,V + 1.065\,V = 2.741\,V$

$\Delta G° = -nFE°_{cell} = -6\,mol\ e^- \times \dfrac{96{,}485\,C}{1\,mol\ e^-} \times 2.741\,V = -1.587 \times 10^6\,J = -1587\,kJ$

4B First we write down the two half-equations, one of which is the reduction equation from the previous example. The other is the oxidation that occurs in the standard hydrogen electrode.

Oxidation: $2H_2(g) \rightarrow 4H^+(aq) + 4e^-$

Reduction: $O_2(g) + 4H^+(aq) + 4e^- \rightarrow 2H_2O(l)$

Net: $2\,H_2(g) + O_2(g) \rightarrow 2\,H_2O(l)$ $n = 4$ in this reaction.

This net reaction is simply twice the formation reaction for $H_2O(l)$ and, therefore,

$\Delta G° = 2\Delta G_f°\left[H_2O(l)\right] = 2 \times (-237.1\,kJ) = -474.2 \times 10^3\,J = -nFE°_{cell}$

$E°_{cell} = \dfrac{-\Delta G°}{nF} = \dfrac{-\left(-474.2 \times 10^3\,J\right)}{4\ mol\ e^- \times \dfrac{96{,}485\,C}{mol\ e^-}} = +1.229\,V = E°$, as we might expect.

5A   Cu(s) will displace metal ions of a metal less active than copper. Silver ion is one example.

Oxidation: $Cu(s) \rightarrow Cu^{2+}(aq) + 2e^-$ $\qquad -E^\circ = -0.340V$ (from Table 21.1)

Reduction: $\{Ag^+(aq) + e^- \rightarrow Ag(s)\} \times 2$ $\qquad E^\circ = +0.800V$

Net: $2Ag^+(aq) + Cu(s) \rightarrow Cu^{2+}(aq) + 2Ag(s)$ $\quad E^\circ_{cell} = -0.340\,V + 0.800\,V = +0.460\,V$

5B   We determine the value for the hypothetical reaction's cell potential.

Oxidation: $\{Na(s) \rightarrow Na^+(aq) + e^-\} \times 2$ $\qquad -E^\circ = +2.713\,V$

Reduction: $Mg^{2+}(aq) + 2e^- \rightarrow Mg(s)$ $\qquad E^\circ = -2.356\,V$

Net: $2\,Na(s) + Mg^{2+}(aq) \rightarrow 2\,Na^+(aq) + Mg(s)$ $\quad E^\circ_{cell} = 2.713\,V - 2.356\,V = +0.357\,V$

The method is not feasible because another reaction occurs that has an even larger cell potential, i.e., Na(s) reacts with water to form $H_2(g)$ and NaOH(aq), which has

$E^\circ_{cell} = 2.713\,V - 0.828\,V = +1.885\,V$.

6A   The oxidation is that of SO_4^{2-} to $S_2O_8^{2-}$, the reduction is that of O_2 to H_2O.

Oxidation: $\{2\,SO_4^{2-}(aq) \rightarrow S_2O_8^{2-}(aq) + 2e^-\} \times 2$ $\qquad -E^\circ = -2.01\,V$

Reduction: $O_2(g) + 4H^+(aq) + 4e^- \rightarrow 2H_2O(l)$ $\qquad E^\circ = +1.229\,V$

Net: $\qquad O_2(g) + 4\,H^+(aq) + 2\,SO_4^{2-}(aq) \rightarrow S_2O_8^{2-}(aq) + 2H_2O(l)$

Because the standard cell potential is negative, we conclude that this cell reaction is nonspontaneous. This would not be a feasible method of producing peroxodisulfate ion.

6B   **(1)**   The oxidation is that of $Sn^{2+}(aq)$ to $Sn^{4+}(aq)$, the reduction is that of O_2 to H_2O.

Oxidation: $\{Sn^{2+}(aq) \rightarrow Sn^{4+}(aq) + 2e^-\} \times 2$ $\qquad -E^\circ = -0.154\ V$

Reduction: $O_2(g) + 4H^+(aq) + 4e^- \rightarrow 2H_2O(l)$ $\quad E^\circ = +1.229\,V$

Net: $\qquad O_2(g) + 4H^+(aq) + 2Sn^{2+}(aq) \rightarrow 2Sn^{4+}(aq) + 2H_2O(l)$

$E^\circ_{cell} = -0.154\,V + 1.229\,V = +1.075\,V$

Since the standard cell potential is positive, this cell reaction is spontaneous.

(2)   The oxidation is that of Sn(s) to $Sn^{2+}(aq)$, the reduction is still that of O_2 to H_2O.

Oxidation: $\{Sn(s) \rightarrow Sn^{2+}(aq) + 2e^-\}$ $\qquad \times 2 \qquad -E^\circ = +0.137\ V$

Reduction: $O_2(g) + 4H^+(aq) + 4e^- \rightarrow 2H_2O(l)$ $\quad E^\circ = +1.229\,V$

Net: $\qquad O_2(g) + 4H^+(aq) + 2Sn(s) \rightarrow 2Sn^{2+}(aq) + 2H_2O(l)$

$E^\circ_{cell} = 0.137\,V + 1.229\,V = +1.366\,V$

The standard cell potential for this reaction is more positive than that for situation (1); reaction (2) should occur preferentially. Also, if $Sn^{4+}(aq)$ is formed, it should react with $Sn(s)$ to form $Sn^{2+}(aq)$.

Oxidation: $Sn(s) \rightarrow Sn^{2+}(aq) + 2e^-$ $\qquad -E^\circ = +0.137\,V$

Reduction: $Sn^{4+}(aq) + 2e^- \rightarrow Sn^{2+}(aq)$ $\qquad E^\circ = +0.154\,V$

Net: $Sn^{4+}(aq) + Sn(s) \rightarrow 2\,Sn^{2+}(aq)$ $\qquad E^\circ_{cell} = +0.137\,V + 0.154\,V = +0.291\,V$

7A For the reaction $2\,Al(s) + 3\,Cu^{2+}(aq) \rightarrow 2\,Al^{3+}(aq) + 3\,Cu(s)$ we know $n = 6$ and $E^\circ_{cell} = +2.013\,V$. We calculate the value of K_{eq}.

$$E^\circ_{cell} = \frac{0.0257}{n}\ln K_{eq}; \quad \ln K_{eq} = \frac{nE^\circ_{cell}}{0.0257} = \frac{6 \times (+2.013)}{0.0257} = 470; \quad K_{eq} = e^{470} = 10^{204}$$

The huge size of the equilibrium constant indicates that this reaction indeed will go to completion.

7B We first determine the value of E°_{cell} from the half-cell potentials.

Oxidation: $Sn(s) \rightarrow Sn^{2+}(aq) + 2e^-$ $\qquad -E^\circ = +0.137\,V$

Reduction: $Pb^{2+}(aq) + 2e^- \rightarrow Pb(s)$ $\qquad E^\circ = -0.125\,V$

Net: $Pb^{2+}(aq) + Sn(s) \rightarrow Pb(s) + Sn^{2+}(aq)$ $\qquad E^\circ_{cell} = +0.137\,V - 0.125\,V = +0.012\,V$

$$E^\circ_{cell} = \frac{0.0257}{n}\ln K_{eq} \quad \ln K_{eq} = \frac{nE^\circ_{cell}}{0.0257} = \frac{2 \times (+0.012)}{0.0257} = 0.93 \quad K_{eq} = e^{0.93} = 2.5$$

The equilibrium constant's small size indicates that this reaction will not go to completion.

8A We first need to determine the standard cell voltage and the cell reaction.

Oxidation: $\{Al(s) \rightarrow Al^{3+}(aq) + 3\,e^-\} \times 2$ $\qquad -E^\circ = +1.676\,V$

Reduction: $\{Sn^{4+}(aq) + 2\,e^- \rightarrow Sn^{2+}(aq)\} \times 3$ $\qquad E^\circ = +0.154\,V$

Net: $2\,Al(s) + 3\,Sn^{4+}(aq) \rightarrow 2\,Al^{3+}(aq) + 3\,Sn^{2+}(aq)$

$E^\circ_{cell} = +1.676\,V + 0.154\,V = +1.830\,V$

Note that $n = 6$. We now set up and substitute into the Nernst equation.

$$E_{cell} = E^\circ_{cell} - \frac{0.0592}{n}\log\frac{[Al^{3+}]^2[Sn^{2+}]^3}{[Sn^{4+}]^3} = +1.830 - \frac{0.0592}{6}\log\frac{(0.36\,M)^2(0.54\,M)^3}{(0.086\,M)^3}$$

$$= +1.830\,V - 0.0149\,V = +1.815\,V$$

8B   We first need to determine the standard cell voltage and the cell reaction.

Oxidation: $2\,Cl^-\,(1.0\ M) \rightarrow Cl_2\,(1\ atm) + 2e^-$ $\qquad\qquad -E^\circ = -1.358\ V$

Reduction: $PbO_2\,(s) + 4\,H^+\,(aq) + 2\,e^- \rightarrow Pb^{2+}\,(aq) + 2\,H_2O(l)$ $\qquad E^\circ = +1.455\ V$

Net: $PbO_2\,(s) + 4\,H^+\,(0.10\ M) + 2\,Cl^-\,(1.0\ M) \rightarrow Cl_2\,(1\ atm) + Pb^{2+}\,(0.050\ M) + 2\,H_2O(l)$

$E^\circ_{cell} = -1.358\ V + 1.455\ V = +0.097\ V$ $\qquad$ Note that $n = 2$. Substitute into Nernst equation.

$$E_{cell} = E^\circ_{cell} - \frac{0.0592}{n}\log\frac{P\{Cl_2\}[Pb^{2+}]}{[H^+]^4[Cl^-]^2} = +0.097 - \frac{0.0592}{2}\log\frac{(1.0\,atm)(0.050\,M)}{(0.10\,M)^4(1.0\,M)^2}$$

$$= +0.097\ V - 0.080\ V = +0.017\ V$$

9A   The cell reaction is $2\,Fe^{3+}\,(0.35\ M) + Cu\,(s) \rightarrow 2\,Fe^{2+}\,(0.25\ M) + Cu^{2+}\,(0.15\ M)$ with

$n = 2$, and $E^\circ_{cell} = -0.337\ V + 0.771\ V = 0.434\ V$ $\qquad$ Next, substitute this voltage into the Nernst equation.

$$E_{cell} = E^\circ_{cell} - \frac{0.0592}{n}\log\frac{[Fe^{2+}]^2[Cu^{2+}]}{[Fe^{3+}]^2} = 0.434 - \frac{0.0592}{2}\log\frac{(0.25)^2(0.15)}{(0.35)^2} = 0.434 + 0.033$$

$E_{cell} = +0.467\ V$ $\qquad$ The reaction is spontaneous as written.

9B   The reaction is not spontaneous in either direction when $E_{cell} = 0.000\ V$. We use the standard cell potential from Example 21-9.

$$E_{cell} = E^\circ_{cell} - \frac{0.0592}{n}\log\frac{[Sn^{2+}]}{[Pb^{2+}]};\qquad 0.000\ V = 0.012\ V - \frac{0.0592}{2}\log\frac{[Sn^{2+}]}{[Pb^{2+}]}$$

$$\log\frac{[Sn^{2+}]}{[Pb^{2+}]} = \frac{0.012 \times 2}{0.0592} = 0.41;\qquad \frac{[Sn^{2+}]}{[Pb^{2+}]} = 10^{0.41} = 2.6$$

10A   In this concentration cell $E^\circ_{cell} = 0.000\ V$ because the same reaction occurs at anode and cathode, only the concentrations of the ions differ. $[Ag^+] = 0.100\ M$ in the cathode compartment. The anode compartment contains a saturated solution of AgCl(aq).

$$K_{sp} = 1.8 \times 10^{-10} = [Ag^+][Cl^-] = s^2;\qquad s = \sqrt{1.8 \times 10^{-10}} = 1.3 \times 10^{-5}\ M$$

Now we apply the Nernst equation. The cell reaction is

$$Ag^+\,(0.100\,M) \rightarrow Ag^+\,(1.3 \times 10^{-5}\ M)$$

$$E_{cell} = 0.000 - \frac{0.0592}{1}\log\frac{1.3 \times 10^{-5}\ M}{0.100\ M} = +0.23\ V$$

10B For this cell because the electrodes are identical, the standard electrode potentials are numerically equal and subtracting one from the other leads to the value $E^{\circ}_{cell} = 0.000\,V$. However, because the ion concentrations differ, there is a potential difference between the two half cells (non-zero nonstandard voltage for the cell). $\left[Pb^{2+}\right] = 0.100\,M$ in the cathode compartment. The anode compartment contains a saturated solution of PbI_2. We use the Nernst equation (with $n = 2$) to determine $\left[Pb^{2+}\right]$ in the saturated solution.

$$E_{cell} = +0.0567\,V = 0.000 - \frac{0.0592}{2}\log\frac{x\,M}{0.100\,M}; \quad \log\frac{x\,M}{0.100\,M} = \frac{2 \times 0.0567}{-0.0592} = -1.92$$

$$\frac{x\,M}{0.100\,M} = 10^{-1.92} = 0.012; \quad \left[Pb^{2+}\right]_{anode} = x\,M = 0.012 \times 0.100\,M = 0.0012\,M;$$

$$\left[I^-\right] = 2 \times 0.0012\,M = 0.0024\,M$$

$$K_{sp} = \left[Pb^{2+}\right]\left[I^-\right]^2 = (0.0012)(0.0024)^2 = 6.9 \times 10^{-9} \text{ compared with } 7.1 \times 10^{-9} \text{ in Appendix D}$$

11A From Table 21-1 we choose one oxidations and one reductions reaction so as to get the least negative cell voltage. This will be the most likely pair of ½ -reactions to occur.

Oxidation: $2I^-(aq) \rightarrow I_2(s) + 2\,e^-$ $-E^{\circ} = -0.535\,V$

      $2\,H_2O(l) \rightarrow O_2(g) + 4\,H^+(aq) + 4\,e^-$ $-E^{\circ} = -1.229\,V$

Reduction: $K^+(aq) + e^- \rightarrow K(s)$ $E^{\circ} = -2.924\,V$

      $2\,H_2O(l) + 2\,e^- \rightarrow H_2(g) + 2\,OH^-(aq)$ $E^{\circ} = -0.828\,V$

The least negative standard cell potential $(-0.535\,V - 0.828\,V = -1.363\,V)$ occurs when $I_2(s)$ is produced by oxidation at the anode, and $H_2(g)$ is produced by reduction at the cathode.

11B We obtain from Table 21-1 all the possible oxidations and reductions and choose one of each to get the least negative cell voltage. That pair is the most likely pair of half-reactions to occur.

Oxidation: $2\,H_2O(l) \rightarrow O_2(g) + 4\,H^+(aq) + 4\,e^-$ $-E^{\circ} = -1.229V$

      $Ag(s) \rightarrow Ag^+(aq) + e^-$ $-E^{\circ} = -0.800V$

      [We cannot further oxidize $NO_3^-(aq)$ or $Ag^+(aq)$.]

Reduction: $Ag^+(aq) + e^- \rightarrow Ag(s)$ $E^{\circ} = +0.800V$

      $2\,H_2O(l) + 2\,e^- \rightarrow H_2(g) + 2\,OH^-(aq)$ $E^{\circ} = -0.828V$

Thus, we expect to form silver metal at the cathode and $Ag^+(aq)$ at the anode.

12A The half-cell equation is $Cu^{2+}(aq) + 2e^- \rightarrow Cu(s)$, indicating that two moles of electrons are required for each mole of copper deposited. Current is measured in amperes, or coulombs per second. We convert the mass of copper to coulombs of electrons needed for the reduction and the time in hours to seconds.

$$\text{Current} = \frac{12.3 \text{ g Cu} \times \dfrac{1 \text{ mol Cu}}{63.55 \text{ g Cu}} \times \dfrac{2 \text{ mol } e^-}{1 \text{ mol Cu}} \times \dfrac{96,485 \text{ C}}{1 \text{ mol } e^-}}{5.50 \text{ h} \times \dfrac{60 \text{ min}}{1 \text{ h}} \times \dfrac{60 \text{ s}}{1 \text{ min}}} = \frac{3.73_5 \times 10^4 \text{ C}}{1.98 \times 10^4 \text{ s}} = 1.89 \text{ amperes}$$

12B We first determine the moles of $O_2(g)$ produced with the ideal gas equation.

$$\text{moles } O_2(g) = \frac{\left(738 \text{ mm Hg} \times \dfrac{1 \text{ atm}}{760 \text{ mm Hg}}\right) \times 2.62 \text{ L}}{\dfrac{0.08206 \text{ L atm}}{\text{mol K}} \times (26.2 + 273.2) \text{ K}} = 0.104 \text{ mol } O_2$$

Then we determine the time needed to produce this amount of O_2.

$$\text{elapsed time} = 0.104 \text{ mol } O_2 \times \frac{4 \text{ mol } e^-}{1 \text{ mol } O_2} \times \frac{96,485 \text{ C}}{1 \text{ mol } e^-} \times \frac{1 \text{ s}}{2.13 \text{ C}} \times \frac{1 \text{ h}}{3600 \text{ s}} = 5.23 \text{ h}$$

REVIEW QUESTIONS

1. **(a)** $E°$ is the symbol for the standard cell potential, the voltage measured when no current is flowing and all cell reagents are in their standard states.

   (b) F is the symbol for Faraday's constant, the charge on one mole of electrons, 96,485 coulombs.

   (c) The anode is the electrode where oxidation occurs and toward which anions move.

   (d) The cathode is the electrode where reduction occurs and toward which cations move.

2. **(a)** A salt bridge is a tube filled with electrolyte that is used to join two half-cells in such a way that the electrochemical circuit is completed but the contents of the half-cells do not mix.

   (b) The standard hydrogen electrode is based on the reduction of hydrogen ion, at 1.000 M concentration, to hydrogen gas at 1.000 atm pressure. It is assigned a half-cell potential of 0.000 V.

   (c) Cathodic protection is achieved by electrically joining a more active metal to a less active one that is to be protected. The more active metal will be oxidized, protecting the less active one from corrosion.

(d)   A fuel cell is a voltaic cell in which a reaction that normally occurs as a combustion reaction serves as the cell reaction. Reactants are continually supplied and products removed from such a cell.

3.   (a)   A half-reaction is either the oxidation reaction or the reduction reaction of a cell. On the other hand, the net reaction is an oxidation-reduction reaction, the combination of two half-reactions.

(b)   In a voltaic or galvanic cell, a chemical change produces electricity. This type of cell has $E > 0$. In an electrolytic cell, the passage of electric current produces a chemical change. This type of cell has $E < 0$.

(c)   A primary battery is nonreversible, like a dry cell. A secondary battery can be recharged and reused.

(d) E°_{cell} is the cell potential when all reactants and all products are in their standard states. E_{cell} is the cell potential when reactants and products are not necessarily in their standard states.

4.   The correct statement is (d). Electrons are produced at the anode and move toward the cathode, regardless of the electrode material. The electrons do not move through the salt bridge; ions do. Electrons do not leave the cell; they provide current within the circuitry. Reduction occurs at the cathode in both galvanic and electrolytic cells—in all types of electrochemical cells, in fact.

5.   (a)   Oxidation: $Fe(s) \longrightarrow Fe^{2+}(aq) + 2e^-$

Reduction: $Cu^{2+}(aq) + 2e^- \rightarrow Cu(s)$

Net: $Fe(s) + Cu^{2+}(aq) \rightarrow Fe^{2+}(aq) + Cu(s)$

(b)   Oxidation: $2Br^-(aq) \rightarrow Br_2(aq) + 2e^-$

Reduction: $Cl_2(aq) + 2e^- \rightarrow 2Cl^-(aq)$

Net: $2Br^-(aq) + Cl_2(aq) \rightarrow Br_2(aq) + 2Cl^-(aq)$

(c)   Oxidation: $Al(s) \rightarrow Al^{3+}(aq) + 3e^-$

Reduction: $\{Fe^{3+}(aq) + e^- \rightarrow Fe^{2+}(aq)\} \quad \times 3$

Net: $Al(s) + 3Fe^{3+}(aq) \rightarrow Al^{3+}(aq) + 3Fe^{2+}(aq)$

(d)   Oxidation: $\{Cl^-(aq) + 3H_2O(l) \rightarrow ClO_3^-(aq) + 6H^+(aq) + 6e^-\} \quad \times 5$

Reduction: $\{MnO_4^-(aq) + 8H^+(aq) + 5e^- \rightarrow Mn^{2+}(aq) + 4H_2O(l)\} \quad \times 6$

Net: $5Cl^-(aq) + 6MnO_4^-(aq) + 18H^+(aq) \rightarrow 5ClO_3^-(aq) + 6Mn^{2+}(aq) + 9H_2O(l)$

(e) Oxidation: $S^{2-}(aq) + 8OH^-(aq) \rightarrow SO_4^{2-}(aq) + 4H_2O(l) + 8e^-$

Reduction: $\{O_2(g) + 2H_2O(l) + 4e^- \rightarrow 4OH^-(aq)\}$ $\times 2$

———————————————————————————————

Net: $S^{2-}(aq) + 2O_2(g) \rightarrow SO_4^{2-}(aq)$

6. Since $Zn^{2+}(aq)$ undergoes reduction, and the other metal (M) undergoes oxidation,

$$E^\circ_{cell} = E^\circ_{Zn} - E^\circ_M \qquad \text{or} \qquad E^\circ_M = E^\circ_{Zn} - E^\circ_{cell}$$

(a) $E^\circ_{Mn} = E^\circ_{Zn} - E^\circ_{cell} = -0.763\ V - 0.417\ V = -1.180\ V$

(b) $E^\circ_{Po} = E^\circ_{Zn} - E^\circ_{cell} = -0.763\ V + 1.13\ V = +0.37\ V$

(c) $E^\circ_{Ti} = E^\circ_{Zn} - E^\circ_{cell} = -0.763\ V - 0.87\ V = -1.63\ V$

(d) $E^\circ_{V} = E^\circ_{Zn} - E^\circ_{cell} = -0.763\ V - 0.37\ V = -1.13\ V$

7. **(a)** Oxidation: $\{Al(s) \rightarrow Al^{3+}(aq) + 3e^-\}\ \times 2$ $-E^\circ = +1.676\ V$

Reduction: $\{Sn^{2+}(aq) + 2e^- \rightarrow Sn(s)\}\ \times 3$ $E^\circ = -0.137\ V$

———————————————————————————————

Net: $2\ Al(s) + 3\ Sn^{2+}(aq) \rightarrow 2\ Al^{3+}(aq) + 3\ Sn(s)$ $E^\circ_{cell} = +1.539\ V$

(b) Oxidation: $Fe^{2+}(aq) \rightarrow Fe^{3+}(aq) + e^-$ $-E^\circ = -0.771\ V$

Reduction: $Ag^+(aq) + e^- \rightarrow Ag(s)$ $E^\circ = +0.800\ V$

———————————————————————————————

Net: $Fe^{2+}(aq) + Ag^+(aq) \rightarrow Fe^{3+}(aq) + Ag(s)$ $E^\circ_{cell} = +0.029\ V$

8. **(a)** Oxidation: $2\ Cl^-(aq) \rightarrow Cl_2(g) + 2\ e^-$ $-E^\circ = -1.358\ V$

Reduction: $PbO_2(s) + 4H^+(aq) + 2e^- \rightarrow Pb^{2+}(aq) + 2H_2O(l)$ $E^\circ = +1.455\ V$

———————————————————————————————

Net: $2\ Cl^-(aq) + PbO_2(s) + 4\ H^+(aq) \rightarrow Cl_2(g) + Pb^{2+}(aq) + 2H_2O(l)$ $E^\circ_{cell} = 0.097\ V$

(b) Oxidation: $\{Mg(s) \rightarrow Mg^{2+}(aq) + 2\ e^-\}\ \times 3$ $-E^\circ = +2.356\ V$

Reduction: $\{Sc^{3+}(aq) + 3\ e^- \rightarrow Sc(s)\}$ $\times 2$ $E^\circ\{Sc^{3+}/Sc\}$

———————————————————————————————

Net: $3\ Mg(s) + 2\ Sc^{3+}(aq) \rightarrow 3\ Mg^{2+}(aq) + 2\ Sc(s)$ $E^\circ_{cell} = +0.33\ V$

$E^\circ\{Sc^{3+}/Sc\} = +0.33\ V - 2.356\ V = -2.03\ V$

(c) Oxidation: $Cu^+(aq) \rightarrow Cu^{2+}(aq) + e^-$ $-E^\circ\{Cu^{2+}/Cu^+\}$

Reduction: $Ag^+(aq) + e^- \rightarrow Ag(s)$ $E^\circ = +0.800\ V$

———————————————————————————————

Net: $Cu^+(aq) + Ag^+(aq) \rightarrow Cu^{2+}(aq) + Ag(s)$ $E^\circ_{cell} = +0.641\ V$

$-E^\circ\{Cu^{2+}/Cu^+\} = +0.641\ V - 0.800\ V = -0.159\ V$ $E^\circ\{Cu^{2+}/Cu^+\} = +0.159\ V$

9. **(a)** Oxidation: $Sn(s) \rightarrow Sn^{2+}(aq) + 2e^-$ $-E° = +0.137$ V

Reduction: $Pb^{2+}(aq) + 2e^- \rightarrow Pb(s)$ $E° = -0.125$ V

Net: $Sn(s) + Pb^{2+}(aq) \rightarrow Sn^{2+}(aq) + Pb(s)$ $E°_{cell} = +0.012$ V   Spontaneous

(b) Oxidation: $2I^-(aq) \rightarrow I_2(s) + 2e^-$ $-E° = -0.535$ V

Reduction: $Cu^{2+}(aq) + 2e^- \rightarrow Cu(s)$ $E° = +0.340$ V

Net: $2I^-(aq) + Cu^{2+}(aq) \rightarrow Cu(s) + I_2(s)$ $E°_{cell} = -0.195$ V   Nonspontaneous

(c) Oxidation: $\{2H_2O(l) \rightarrow O_2(g) + 4H^+(aq) + 4e^-\} \times 3$ $-E° = -1.229$ V

Reduction: $\{NO_3^-(aq) + 4H^+(aq) + 3e^- \rightarrow NO(g) + 2H_2O(l)\} \times 4$ $E° = +0.956$ V

Net: $4NO_3^-(aq) + 4H^+(aq) \rightarrow 3O_2(g) + 4NO(g) + 2H_2O(l)$ $E°_{cell} = -0.273$ V

This is a nonspontaneous cell reaction.

(d) Oxidation: $Cl^-(aq) + 2OH^-(aq) \rightarrow OCl^-(aq) + H_2O(l) + 2e^-$ $-E° = -0.890$ V

Reduction: $O_3(g) + H_2O(l) + 2e^- \rightarrow O_2(g) + 2OH^-(aq)$ $E° = +1.246$ V

Net: $Cl^-(aq) + O_3(g) \rightarrow OCl^-(aq) + O_2(g)$ (basic solution) $E°_{cell} = +0.356$ V

This is a spontaneous cell reaction.

10. It is more difficult to oxidize Hg(l) to $Hg_2^{2+}(-0.797$ V$)$ than it is to reduce H^+ to H_2 (0.000 V); Hg(l) will not dissolve in 1 M HCl. The reduction of nitrate ion to NO(g) in acidic solution is relatively spontaneous in acidic media ($+0.956$ V). This can help overcome the reluctance of Hg to be oxidized. Hg(l) will react with and dissolve in the $HNO_3(aq)$.

11. **(a)** Oxidation: $Mg(s) \rightarrow Mg^{2+}(aq) + 2e^-$ $-E° = +2.356$ V

Reduction: $Pb^{2+}(aq) + 2e^- \rightarrow Pb(s)$ $E° = -0.125$ V

Net: $Mg(s) + Pb^{2+}(aq) \rightarrow Mg^{2+}(aq) + Pb(s)$ $E°_{cell} = +2.231$ V

This reaction occurs to a significant extent.

(b) Oxidation: $Sn(s) \rightarrow Sn^{2+}(aq) + 2e^-$ $-E° = +0.137$ V

Reduction: $2H^+(aq) \rightarrow H_2(g)$ $E° = 0.000$ V

Net: $Sn(s) + 2H^+(aq) \rightarrow Sn^{2+}(aq) + H_2(g)$ $E°_{cell} = +0.137$ V

This reaction will occur to a significant extent.

(c)   Oxidation: $Sn^{2+}(aq) \rightarrow Sn^{4+}(aq) + 2e^-$ $-E^\circ = -0.154\,V$

Reduction: $SO_4^{2-}(aq) + 4H^+(aq) + 2e^- \rightarrow SO_2(g) + 2H_2O(l)$ $E^\circ = +0.17\,V$

Net: $Sn^{2+}(aq) + SO_4^{2-}(aq) + 4H^+(aq) \rightarrow Sn^{4+}(aq) + SO_2(g) + 2H_2O(l)$ $E^\circ_{cell} = +0.02\,V$

This reaction will occur, but not to a large extent.

(d)   Oxidation: $\{H_2O_2(aq) \rightarrow O_2(g) + 2H^+(aq) + 2e^-\} \times 5$ $-E^\circ = -0.695\,V$

Reduction: $\{MnO_4^-(aq) + 8H^+(aq) + 5e^- \rightarrow Mn^{2+}(aq) + 4H_2O(l)\} \times 2$ $E^\circ = +1.51\,V$

Net: $5H_2O_2(aq) + 2MnO_4^-(aq) + 6H^+(aq) \rightarrow 5O_2(g) + 2Mn^{2+}(aq) + 8H_2O(l)$ $E^\circ_{cell} = +0.82\,V$

This reaction will occur to a significant extent.

(e)   Oxidation: $2Br^-(aq) \rightarrow Br_2(aq) + 2e^-$ $-E^\circ = -1.065\,V$

Reduction: $I_2(s) + 2e^- \rightarrow 2I^-(aq)$ $E^\circ = +0.535\,V$

Net: $2Br^-(aq) + I_2(s) \rightarrow Br_2(aq) + 2I^-(aq)$ $E^\circ_{cell} = -0.530\,V$

This reaction will not occur to a significant extent.

12. The relatively small value of E°_{cell} for a reaction indicates that the reaction will proceed in the forward reaction, but will stop short of completion. A much larger value of E°_{cell} would be necessary before we would conclude that the reaction goes to completion. For example, we can compute the value of the equilibrium constant for this reaction. A value of 1000 or more is needed before we can describe the reaction as one that goes to completion.

$$E^\circ_{cell} = \frac{0.0257}{n}\ln K_{eq} \qquad \ln K_{eq} = \frac{n \times E^\circ_{cell}}{0.0257} = \frac{2 \times 0.02}{0.0257} = 2 \qquad K_{eq} = e^2 = 7$$

13. If E°_{cell} is positive, the reaction will occur. For the reduction of $Cr_2O_7^{2-}$ to $Cr^{3+}(aq)$:

$$Cr_2O_7^{2-}(aq) + 14H^+(aq) + 6e^- \rightarrow 2Cr^{3+}(aq) + 7H_2O(l) \qquad E^\circ = +1.33\,V$$

If the oxidation has $-E^\circ$ smaller (more negative) than $-1.33\,V$, the oxidation will not occur.

(a) $Sn^{2+}(aq) \rightarrow Sn^{4+}(aq) + 2e^-$ $-E^\circ = -0.154\,V$

Hence, $Sn^{2+}(aq)$ can be oxidized to $Sn^{4+}(aq)$ by $Cr_2O_7^{2-}(aq)$.

(b) $I_2(s) + 6H_2O(l) \rightarrow 2IO_3^-(aq) + 12H^+(aq) + 10e^-$ $-E^\circ = -1.20\,V$

$I_2(s)$ can be oxidized to $IO_3^-(aq)$ by $Cr_2O_7^{2-}(aq)$.

(c) $Mn^{2+}(aq) + 4H_2O(l) \rightarrow MnO_4^-(aq) + 8H^+(aq) + 5e^-$ $-E^\circ = -1.51\,V$

$Mn^{2+}(aq)$ cannot be oxidized to $MnO_4^-(aq)$ by $Cr_2O_7^{2-}(aq)$.

14. **(a)** Oxidation: $\{Al(s) \rightarrow Al^{3+}(aq) + 3e^-\} \times 2$ $-E^\circ = +1.676\,V$

Reduction: $\{Cu^{2+}(aq) + 2e^- \rightarrow Cu(s)\} \times 3$ $E^\circ = +0.337\,V$

Net: $2\,Al(s) + 3\,Cu^{2+}(aq) \rightarrow 2\,Al^{3+}(aq) + 3\,Cu(s)$ $E^\circ_{cell} = +2.013\,V$

$\Delta G^\circ = -nFE^\circ_{cell} = -(6\ \text{mol e}^-)(96,485\ \text{C/mol e}^-)(2.013\,V)$

$\Delta G^\circ = -1.165 \times 10^6\ J = -1.165 \times 10^3\ kJ$

(b) Oxidation: $\{2I^-(aq) \rightarrow I_2(s) + 2e^-\} \times 2$ $-E^\circ = -0.535\,V$

Reduction: $O_2(g) + 4H^+(aq) + 4e^- \rightarrow 2H_2O(l)$ $E^\circ = +1.229\,V$

Net: $4I^-(aq) + O_2(g) + 4H^+(aq) \rightarrow 2I_2(s) + 2H_2O(l)$ $E^\circ_{cell} = +0.694\,V$

$\Delta G^\circ = -nFE^\circ_{cell} = -(4\ \text{mol e}^-)(96,485\ \text{C/mol e}^-)(0.694\,V) = -2.68 \times 10^5\,J = -268\ kJ$

(c) Oxidation: $\{Ag(s) \rightarrow Ag^+(aq) + e^-\} \times 6$ $-E^\circ = -0.800\,V$

Reduction: $Cr_2O_7^{2-}(aq) + 14\,H^+(aq) + 6\,e^- \rightarrow 2\,Cr^{3+}(aq) + 7\,H_2O(l)$ $E^\circ = +1.33\,V$

Net: $6\,Ag(s) + Cr_2O_7^{2-}(aq) + 14H^+(aq) \rightarrow 6\,Ag^+(aq) + 2\,Cr^{3+}(aq) + 7\,H_2O(l)$

$E^\circ_{cell} = -0.800\,V + 1.33\,V = +0.53\,V$

$\Delta G^\circ = -nFE^\circ_{cell} = -(6\ \text{mol e}^-)(96,485\ \text{C/mol e}^-)(0.53\,V) = -3.1 \times 10^5\,J = -3.1 \times 10^2\ kJ$

15. $\Delta G^\circ = -nFE^\circ_{cell} = -RT\ln K;\ \ \ln K = \dfrac{nFE^\circ_{cell}}{RT};$ This becomes $\ln K = \dfrac{n}{0.0257}E^\circ_{cell}$

(a) Oxidation: $\{Ag(s) \rightarrow Ag^+(aq) + e^-\} \times 2$ $-E^\circ = -0.800\,V$

Reduction: $Sn^{4+}(aq) + 2e^- \rightarrow Sn^{2+}(aq)$ $E^\circ = +0.154\,V$

Net: $2Ag(s) + Sn^{4+}(aq) \rightarrow 2Ag^+(aq) + Sn^{2+}(aq)$ $E^\circ_{cell} = -0.646\,V$

$\ln K_{eq} = \dfrac{n}{0.0257}E^\circ_{cell} = \dfrac{2\ \text{mol e}^- \times (-0.646\,V)}{0.0257} = -50.3$

$K_{eq} = e^{-50.3} = 1 \times 10^{-22} = \dfrac{[Sn^{2+}][Ag^+]^2}{[Sn^{4+}]}$

(b) Oxidation: $2Cl^-(aq) \rightarrow Cl_2(g) + 2\ e^-$ $-E^\circ = -1.358\,V$

Reduction: $MnO_2(s) + 4H^+(aq) + 2e^- \rightarrow Mn^{2+}(aq) + 2H_2O(l)$ $E^\circ = +1.23\,V$

Net: $2\ Cl^-(aq) + MnO_2(s) + 4\ H^+(aq) \rightarrow Mn^{2+}(aq) + Cl_2(g) + 2\ H_2O(l)$ $E^\circ_{cell} = -0.13\,V$

$\ln K_{eq} = \dfrac{2\ \text{mol e}^- \times (-0.13\,V)}{0.0257} = -10._1;\ \ K_{eq} = e^{-10.1} = 4 \times 10^{-5} = \dfrac{[Mn^{2+}]P\{Cl_2(g)\}}{[Cl^-]^2[H^+]^4}$

(c) Oxidation: $4\,OH^-(aq) \rightarrow O_2(g) + 2H_2O(l) + 4\,e^-$ $\qquad -E^\circ = -0.401\ V$

Reduction: $\{OCl^-(aq) + H_2O(l) + 2e^- \rightarrow Cl^-(aq) + 2OH^-\}\ \times 2$ $\qquad E^\circ = +0.890\ V$

Net: $2OCl^-(aq) \longrightarrow 2Cl^-(aq) + O_2(g)$ $\qquad E^\circ_{cell} = +0.489\ V$

$$\ln K_{eq} = \frac{4\ mol\ e^-\,(0.489\ V)}{0.0257} = 76.1 \qquad K_{eq} = e^{76.1} = 1\times 10^{33} = \frac{\left[Cl^-\right]^2 P\{O_2(g)\}}{\left[OCl^-\right]^2}$$

16. (a) Oxidation: $Fe(s) \rightarrow Fe^{2+}(aq) + 2e^-$ $\qquad -E^\circ = +0.440\ V$

Reduction: $Cu^{2+}(aq) + 2e^- \rightarrow Cu(s)$ $\qquad E^\circ = +0.340\ V$

Net: $Fe(s) + Cu^{2+}(aq) \rightarrow Fe^{2+}(aq) + Cu(s)$ $\qquad E_{cell} = +0.780\ V$

Electrons flow from electrode B(Fe) to electrode A(Cu) $E^\circ_{cell} = +0.780\,V$.

(b) Oxidation: $Sn^{2+}(aq) \rightarrow Sn^{4+}(aq) + 2e^-$ $\qquad -E^\circ = -0.154\ V$

Reduction: $\{Ag^+(aq) + e^- \rightarrow Ag(s)\} \times 2$ $\qquad E^\circ = +0.800\ V$

Net: $Sn^{2+}(aq) + 2Ag^+(aq) \rightarrow Sn^{4+}(aq) + 2Ag(s)$ $\qquad E^\circ_{cell} = +0.646\ V$

Electrons flow from electrode A(Pt) to electrode B(Ag) $E_{cell} = +0.646\,V$.

(c) Oxidation: $Zn(s) \rightarrow Zn^{2+}(aq) + 2e^-$ $\qquad -E^\circ = +0.763\ V$

Reduction: $Fe^{2+}(aq) + 2e^- \rightarrow Fe(s)$ $\qquad E^\circ = -0.440\ V$

Net: $Zn(s) + Fe^{2+}(aq) \rightarrow Zn^{2+}(aq) + Fe(s)$ $\qquad E^\circ_{cell} = +0.323\ V$

$$E = E^\circ_{cell} - \frac{0.0592}{n}\log\frac{[Zn^{2+}]}{[Fe^{2+}]} = +0.323\ V - \frac{0.0592}{2}\log\frac{0.10\ M}{1.0\times 10^{-3}\ M}$$

$$= +0.323\ V - 0.0592\ V = +0.264\ V$$

Electrons flow from electrode A(Zn) to electrode B(Fe) $E_{cell} = +0.264\,V$.

17. Oxidation: $Zn(s) \rightarrow Zn^{2+}(aq) + 2e^-$ $\qquad -E^\circ = +0.763\,V$

Reduction: $\{Ag^+(aq) + e^- \rightarrow Ag(s)\} \times 2$ $\qquad E^\circ = +0.800\,V$

Net: $Zn(s) + 2\,Ag^+(aq) \rightarrow Zn^{2+}(aq) + 2Ag(s)$ $\qquad E^\circ_{cell} = +1.563\,V$

$$E = E^\circ_{cell} - \frac{0.0592}{n}\log\frac{[Zn^{2+}]}{\left[Ag^+\right]^2} = +1.563\ V - \frac{0.0592}{2}\log\frac{1.00\ M}{x^2} = +1.250\ V$$

$$\log\frac{1.00\ M}{x^2} = \frac{-2\times(1.250 - 1.563)}{0.0592} = 10.6; \quad x = \sqrt{2.5\times 10^{-11}} = 5\times 10^{-6}\ M$$

18. In each case, we employ the equation $E_{cell} = 0.0592 \, \text{pH}$.

   (a) $\quad E_{cell} = 0.0592 \, \text{pH} = 0.0592 \times 5.25 = 0.311 \, \text{V}$

   (b) $\quad \text{pH} = -\log(0.0103) = 1.987 \qquad E_{cell} = 0.0592 \, \text{pH} = 0.0592 \times 1.987 = 0.118 \, \text{V}$

   (c) $\quad K_a = \dfrac{\left[H^+\right]\left[C_2H_3O_2^{\,-}\right]}{\left[HC_2H_3O_2\right]} = 1.8 \times 10^{-5} = \dfrac{x^2}{0.158 - x} \approx \dfrac{x^2}{0.158}$

$$x = \sqrt{0.158 \times 1.8 \times 10^{-5}} = 1.7 \times 10^{-3} \, \text{M}$$

$$\text{pH} = -\log(1.7 \times 10^{-3}) = 2.77$$

$$E_{cell} = 0.0592 \, \text{pH} = 0.0592 \times 2.77 = 0.164 \, \text{V}$$

19. We predict the possible products at the anode and at the cathode. Then we choose the oxidation and the reduction which, when combined, yield the least negative cell potential.

   (a) Possible products (of oxidation) at the anode are the following.

$$2\,H_2O(l) \rightarrow O_2(g) + 4\,H^+(aq) + 4\,e^- \qquad\qquad -1.229\,V$$

$$2\,Cl^-(aq) \rightarrow Cl_2(g) + 2\,e^- \qquad\qquad -1.358\,V$$

     Possible products (of reduction) at the cathode are the following.

$$2\,H^+(aq) + 2\,e^- \rightarrow H_2(g) \qquad\qquad 0.000\,V$$

$$Cu^{2+}(aq) + 2\,e^- \rightarrow Cu(s) \qquad\qquad +0.337\,V$$

     Because of the high overpotential for the production of $O_2(g)$ (see pg. 852 of text), the products of electrolysis of $CuCl_2(aq)$ will be $Cl_2(g)$ at the anode and $Cu(s)$ at the cathode.

   (b) Possible products (of oxidation) at the anode are the following.

$$2\,H_2O(l) \rightarrow O_2(g) + 4\,H^+(aq) + 4\,e^- \qquad\qquad -1.229\,V$$

$$2\,SO_4^{\,2-}(aq) \rightarrow S_2O_8^{\,2-}(aq) + 2\,e^- \qquad\qquad -2.01\,V$$

     Possible products (of reduction) at the cathode are the following.

$$2\,H^+(aq) + 2\,e^- \rightarrow H_2(g) \qquad\qquad 0.000\,V$$

$$Na^+(aq) + e^- \rightarrow Na(s) \qquad\qquad -2.713\,V$$

     The products of electrolysis of $Na_2SO_4(aq)$ will be $O_2(g)$ at the anode, and $H_2(g)$ and $OH^-(aq)$ at the cathode.

(c)   The only possible product (of oxidation) at the anode is the following.

$$2Cl^-(l) \rightarrow Cl_2(g) + 2e^-$$

The only possible product (of reduction) at the cathode is the following.

$$Ba^{2+}(l) + 2e^- \rightarrow Ba(l)$$

The products of electrolysis of $BaCl_2(l)$ will be $Cl_2(g)$ at the anode and $Ba(l)$ at the cathode.

(d)   The only possible product (of oxidation) at the anode is the following.

$$4OH^- \rightarrow O_2(g) + 2H_2O(l) + 4e^- \qquad\qquad -0.401\,V$$

Possible products (of reduction) at the cathode are the following.

$$2H_2O(l) + 2e^- \rightarrow H_2(g) + 2OH^-(aq) \qquad\qquad -0.828\,V$$

$$K^+(aq) + e^- \rightarrow K(s) \qquad\qquad -2.924\,V$$

The products of electrolysis of KOH(aq) will be $O_2(g)$ at the anode , $H_2(g)$ and $OH^-(aq)$ at the cathode.

20.   When $MgCl_2(l)$ is electrolyzed, Mg(l) is produced at the cathode, with a half-cell voltage of $-2.356V$. On the other hand, when $MgCl_2(aq)$ is electrolyzed, $H_2(g)$ and $OH^-(aq)$ are produced at the cathode, with a standard half-cell voltage of 0.000 V. If we wish to produce elemental magnesium, we must electrolyze molten $MgCl_2$, not its aqueous solution.

21.   We calculate the total amount of charge passed and the number of moles of electrons.

$$mol\ e^- = 75\ min \times \frac{60\ s}{1\ min} \times \frac{2.15\ C}{1\ s} \times \frac{1\ mol\ e^-}{96485\ C} = 0.10\ mol\ e^-$$

(a) $mass\ Zn = 0.10\ mol\ e^- \times \dfrac{1\ mol\ Zn^{2+}}{2\ mol\ e^-} \times \dfrac{1\ mol\ Zn}{1\ mol\ Zn^{2+}} \times \dfrac{65.39\ g\ Zn}{1\ mol\ Zn} = 3.3\ g\ Zn$

(b) $mass\ Al = 0.10\ mol\ e^- \times \dfrac{1\ mol\ Al^{3+}}{3\ mol\ e^-} \times \dfrac{1\ mol\ Al}{1\ mol\ Al^{3+}} \times \dfrac{26.98\ g\ Al}{1\ mol\ Al} = 0.90\ g\ Al$

(c) $mass\ Ag = 0.10\ mol\ e^- \times \dfrac{1\ mol\ Ag^+}{1\ mol\ e^-} \times \dfrac{1\ mol\ Ag}{1\ mol\ Ag^+} \times \dfrac{107.9\ g\ Ag}{1\ mol\ Ag} = 11\ g\ Ag$

(d) $mass\ Ni = 0.10\ mol\ e^- \times \dfrac{1\ mol\ Ni^{2+}}{2\ mol\ e^-} \times \dfrac{1\ mol\ Ni}{1\ mol\ Ni^{2+}} \times \dfrac{58.69\ g\ Ni}{1\ mol\ Ni} = 2.9\ g\ Ni$

22. The two half reactions follow: $Cu^{2+}(aq) + 2\ e^- \rightarrow Cu(s)$ and $2\ H^+(aq) + 2\ e^- \rightarrow H_2(g)$ Thus, two moles of electrons are needed to produce each mole of Cu(s) and two moles of electrons are needed to produce each mole of $H_2(g)$. With this information, we compute the moles of $H_2(g)$ that will be produced.

$$\text{mol } H_2(g) = 3.28 \text{ g Cu} \times \frac{1 \text{ mol Cu}}{63.55 \text{g Cu}} \times \frac{2 \text{ mol } e^-}{1 \text{ mol Cu}} \times \frac{1 \text{ mol } H_2(g)}{2 \text{ mol } e^-} = 0.0516 \text{ mol } H_2(g)$$

Then we use the ideal gas equation to find the volume of $H_2(g)$.

$$\text{Volume of } H_2(g) = \frac{0.0516 \text{ mol } H_2 \times \dfrac{0.08206 \text{ L atm}}{\text{mol K}} \times (273.2 + 28.2) \text{ K}}{763 \text{ mmHg} \times \dfrac{1 \text{ atm}}{760 \text{ mmHg}}} = 1.27 \text{ L}$$

This answer assumes the $H_2(g)$ is not collected over water, and that the $H_2(g)$ formed is the only gas present in the container (i.e. no water vapor present)

EXERCISES

Standard Electrode Potential

23. **(a)** If the metal dissolves in HNO_3, it has a reduction potential that is smaller than $E°\{NO_3^-(aq)/NO(g)\} = 0.956 \text{ V}$. If it also does not dissolve in HCl, it has a reduction potential that is larger than $E°\{H^+(aq)/H_2(g)\} = 0.000 \text{ V}$. If it displaces $Ag^+(aq)$ from solution, then it has a reduction potential that is smaller than $E°\{Ag^+(aq)/Ag(s)\} = 0.800 \text{ V}$. But if it does not displace $Cu^{2+}(aq)$ from solution, then its reduction potential is larger than

$$E°\{Cu^{2+}(aq)/Cu(s)\} = 0.340 \text{ V} \qquad 0.340 \text{ V} < E° < 0.800 \text{ V}$$

(b) If the metal dissolves in HCl, it has a reduction potential that is smaller than $E°\{H^+(aq)/H_2(g)\} = 0.000 \text{ V}$. If it does not displace $Zn^{2+}(aq)$ from solution, its reduction potential is larger than $E°\{Zn^{2+}(aq)/Zn(s)\} = -0.763 \text{ V}$. If it also does not displace $Fe^{2+}(aq)$ from solution, its reduction potential is larger than

$$E°\{Fe^{2+}(aq)/Fe(s)\} = -0.440 \text{ V}. \qquad -0.440 \text{ V} < E° < 0.000 \text{ V}$$

24. We place a strip of solid indium metal into each of the metal ion solutions and see if the dissolved metal plates out on the indium strip. Similarly, strips of all the other metals are immersed in a solution of In^{3+} to see if indium metal plates out. Eventually, we will find one metal whose ions are displaced by indium and another metal that displaces indium from solution, which are adjacent to each other in Table 21-1. The standard electrode potential for the $In/In^{3+}(aq)$ pair will lie between the standard reduction potentials for these two metals. This technique will work only if indium metal does not react with water, that is, if the standard reduction potential of $In^{3+}(aq)/In(s)$ is greater than about -1.8 V. The inaccuracy inherent in this technique is due to overpotentials, which can be as much as 0.200 V. Its imprecision is limited by the closeness of the reduction potentials for the two bracketing metals

25. We separate the given equation into its two half-equations. One of them is the reduction of nitrate ion in acidic solution, whose standard half-cell potential we retrieve from Table 21-1 and use to solve the problem.

Oxidation: $\{Pt(s)+4Cl^-(aq)\longrightarrow[PtCl_4]^{2-}(aq)+2\ e^-\}\times3;\qquad -E°\{[PtCl_4]^{2-}(aq)/Pt(s)\}$

Reduction: $\{NO_3^-(aq)+4H^+(aq)+3e^-\rightarrow NO(g)+2H_2O(l)\}\times2;\ E°=+0.956\,V$

Net: $3Pt(s)+2NO_3^-(aq)+8H^+(aq)+12Cl^-(aq)\rightarrow 3[PtCl_4]^{2-}(aq)+2NO(g)+6H_2O(l)$

$E°_{cell}=0.201\,V=+0.956\ V-E°\{[PtCl_4]^{2-}(aq)/Pt(s)\}$

$E°\{[PtCl_4]^{2-}(aq)/Pt(s)\}=0.956\,V-0.201\,V=+0.755\,V$

26. We separate the given equation into its two half-equations. One of them is the reduction of $Cl_2(g)$ to $Cl^-(aq)$ whose standard half-cell potential we obtain from Table 21-1 and use to solve the problem.

Oxidation: $\{Na(in\,Hg)\rightarrow Na^+(aq)+e^-\}\ \times2$ $\qquad -E°\{Na^+(aq)/Na(in\,Hg)\}$

Reduction: $Cl_2(g)+2e^-\rightarrow2Cl^-(aq)$ $\qquad E°=+1.358\ V$

Net: $2Na(in\,Hg)+Cl_2(g)\rightarrow2Na^+(aq)+2Cl^-(aq)$ $\qquad E°_{cell}=3.20\ V$

$E°_{cell}=3.20\ V=+1.3858\ V-E°\{Na^+(aq)/Na(in\,Hg)\}$

$E°\{Na^+(aq)/Na(in\,Hg)\}=1.358\ V-3.20\ V=-1.84\ V$

27. We divide the net cell equation into two half-equations.

Oxidation: $\{Al(s)+4\ OH^-(aq)\rightarrow[Al(OH)_4]^-(aq)+3\ e^-\}\times4;-E°\{[Al(OH)_4]^-(aq)/Al(s)\}$

Reduction: $\{O_2(g)+2\ H_2O(l)+4e^-\rightarrow4\ OH^-(aq)\}\ \times3;\qquad E°=+0.401V$

Net: $4Al(s)+3O_2(g)+6H_2O(l)+4OH^-(aq)\rightarrow4[Al(OH)_4]^-(aq)\ E_{cell}°=2.71V$

$E°_{cell}=2.71V=+0.401V-E°\{[Al(OH)_4]^-(aq)/Al(s)\}$

$E°\{[Al(OH)_4]^-(aq)/Al(s)\}=0.401V-2.71V=-2.31V$

28. We divide the net cell equation into two half-equations.

Oxidation: $CH_4(g) + 2H_2O(l) \rightarrow CO_2(g) + 8H^+(aq) + 8e^-$ $\qquad -E^\circ\{CO_2(g)/CH_4(g)\}$

Reduction: $\{O_2(g) + 4H^+(aq) + 4e^- \rightarrow 2H_2O(l)\} \times 2$ $\qquad E^\circ = +1.229\,V$

Net: $CH_4(g) + 2O_2(g) \rightarrow CO_2(g) + 2H_2O(l)$ $\qquad E^\circ_{cell} = 1.06\,V$

$E^\circ_{cell} = 1.06\,V = +1.229\,V - E^\circ\{CO_2(g)/CH_4(g)\}$

$E^\circ\{CO_2(g)/CH_4(g)\} = 1.229\,V - 1.06\,V = +0.17\,V$

Predicting Oxidation-Reduction Reactions

29. $Na(s)$ does not displace $Zn^{2+}(aq)$ in an aqueous solution; $Na(s)$ reacts with water instead.

$2Na(s) + Zn^{2+}(aq) \rightarrow 2Na^+(aq) + Zn(s);$ $\qquad E^\circ_{cell} = -(-2.713\,V) + (-0.763\,V) = +1.950\,V$

$2Na(s) + H_2O(l) \rightarrow 2Na^+(aq) + H_2(g) + 2OH^-(aq);$ $E^\circ_{cell} = -(-2.713\,V) + (-0.828\,V) = +1.885\,V$

In contrast, aluminum metal reacts with $Zn^{2+}(aq)$ to displace $Zn(s)$ from solution.

$2Al(s) + 3H_2O(l) \rightarrow 2Al^{3+}(aq) + 3H_2(g) + 6OH^-(aq);$ $E^\circ_{cell} = -(-1.676\,V) + (-0.828\,V) = +0.848\,V$

$2Al(s) + 3Zn^{2+}(aq) \rightarrow 2Al^{3+}(aq) + 3\,Zn(s);$ $\qquad E^\circ_{cell} = -(-1.676\,V) + (-0.763\,V) = +0.913\,V$

The standard cell potentials do not tell the whole story, for we might think that the reaction with the more positive value of E°_{cell} would occur. In truth, $Al(s)$ is coated with a thin layer of tightly adhering $Al_2O_3(s)$, which protects the metal from attack by water. $Na(s)$ has no such protective coating.

30. The reaction that occurs is $Zn(s) \rightarrow Zn^{2+}(aq) + 2e^-$. Some of the electrons produced in this reaction move to the copper surface, where the following reduction occurs. $2H^+(aq) + 2e^- \rightarrow H_2(g)$. The copper does not react with the HCl(aq) even under these circumstances. The copper surface is one on which bubbles of hydrogen form more readily than they do on a zinc metal surface. We say that the copper surface has a lower overpotential. Not as high a voltage is required to produce $H_2(g)$ because of the arrangement of atoms on the copper surface.

31. (a) Oxidation: $\{Ag(s) \rightarrow Ag^+(aq) + e^-\} \times 3$ $\qquad -E^\circ = -0.800\,V$

Reduction: $NO_3^-(aq) + 4H^+(aq) + 3e^- \rightarrow NO(g) + 2H_2O(l)$ $\qquad E^\circ = +0.956\,V$

Net: $3Ag(s) + NO_3^-(aq) + 4H^+(aq) \rightarrow 3Ag^+(aq) + NO(g) + 2H_2O(l)$ $\quad E^\circ_{cell} = +0.156\,V$

$Ag(s)$ reacts with $HNO_3(aq)$ to form a solution of $AgNO_3(aq)$.

(b) Oxidation: $Zn(s) \rightarrow Zn^{2+}(aq) + 2e^-$ $-E° = +0.763\,V$

Reduction: $2H^+(aq) + 2e^- \rightarrow H_2(g)$ $E° = 0.000\,V$

Net: $Zn(s) + 2H^+(aq) \rightarrow Zn^{2+}(aq) + H_2(g)$ $E°_{cell} = +0.763\,V$

$Zn(s)$ reacts with HI(aq) to form a solution of $ZnI_2(aq)$.

(c) Oxidation: $Au(s) \rightarrow Au^{3+}(aq) + 3e^-$ $-E° = -1.52\,V$

Reduction: $NO_3^-(aq) + 4H^+(aq) + 3e^- \rightarrow NO(g) + 2H_2O(l)$ $E° = +0.956\,V$

Net: $Au(s) + NO_3^-(aq) + 4H^+(aq) \rightarrow Au^{3+}(aq) + NO(g) + 2\,H_2O(l)$; $E°_{cell} = -0.56\,V$

$Au(s)$ does not react with 1.00 M $HNO_3(aq)$.

32. In each case, we determine whether $E°_{cell}$ is greater than zero; if so, the reaction will occur.

(a) Oxidation: $Fe(s) \rightarrow Fe^{2+}(aq) + 2e^-$ $-E° = 0.440\,V$

Reduction: $Zn^{2+}(aq) + 2e^- \rightarrow Zn(s)$ $E° = -0.763\,V$

Net: $Fe(s) + Zn^{2+}(aq) \rightarrow Fe^{2+}(aq) + Zn(s)$ $E°_{cell} = -0.323\,V$

The reaction is not spontaneous as written

(b) Oxidation: $\{2Cl^-(aq) \rightarrow Cl_2(g) + 2e^-\} \times 5$ $-E° = -1.358\,V$

Reduction: $\{MnO_4^-(aq) + 8H^+(aq) + 5e^- \rightarrow Mn^{2+}(aq) + 4H_2O(l)\} \times 2$; $E° = 1.51\,V$

Net: $10Cl^-(aq) + 2MnO_4^-(aq) + 16H^+(aq) \rightarrow 5Cl_2(g) + 2Mn^{2+}(aq) + 8H_2O(l)$

$E°_{cell} = +0.15\,V$. The reaction is spontaneous as written.

(c) Oxidation: $\{Ag(s) \rightarrow Ag^+(aq) + e^-\} \times 2$ $-E° = -0.800\,V$

Reduction: $2H^+(aq) + 2e^- \rightarrow H_2(g)$ $E° = +0.000\,V$

Net: $2Ag(s) + 2H^+(aq) \rightarrow 2Ag^+(aq) + H_2(g)$ $E°_{cell} = -0.800\,V$

The reaction is not spontaneous as written.

(d) Oxidation: $\{2Cl^-(aq) \rightarrow Cl_2(g) + 2e^-\} \times 2$ $-E° = -1.358\,V$

Reduction: $O_2(g) + 4H^+(aq) + 4e^- \rightarrow 2H_2O(l)$ $E° = +1.229\,V$

Net: $4Cl^-(aq) + 4H^+(aq) + O_2(g) \rightarrow 2Cl_2(g) + 2H_2O(l)$ $E°_{cell} = -0.129\,V$

The reaction is not spontaneous as written.

Voltaic Cells

33. **(a)** Anode, Oxidation: $Cu(s) \rightarrow Cu^{2+}(aq) + 2e^-$ $-E^\circ = -0.340\,V$

Cathode, Reduction: $\{Fe^{3+}(aq) + e^- \rightarrow Fe^{2+}(aq)\} \times 2$; $E^\circ = +0.771\,V$

Net: $Cu(s) + 2Fe^{3+}(aq) \rightarrow Cu^{2+}(aq) + 2Fe^{2+}(aq)$ $E^\circ_{cell} = +0.431\,V$

(b) Anode, Oxidation: $\{Al(s) \rightarrow Al^{3+}(aq) + 3e^-\} \times 2$ $-E^\circ = +1.676\,V$

Cathode, Reduction: $\{Pb^{2+}(aq) + 2e^- \rightarrow Pb(s)\} \times 3$; $E^\circ = -0.125\,V$

Net: $2Al(s) + 3Pb^{2+}(aq) \rightarrow 2Al^{3+}(aq) + 3Pb(s)$ $E^\circ_{cell} = +1.551\,V$

(c) Anode, Oxidation: $2H_2O(l) \rightarrow O_2(g) + 4H^+(aq) + 4e^-$ $-E^\circ = -1.229\,V$

Cathode, Reduction: $\{Cl_2(g) + 2e^- \rightarrow 2Cl^-(aq)\} \times 2$ $E^\circ = +1.358\,V$

Net: $2H_2O(l) + 2Cl_2(g) \rightarrow O_2(g) + 4H^+(aq) + 4Cl^-(aq)$ $E^\circ_{cell} = +0.129\,V$

(d) Anode, Oxidation: $\{Zn(s) \rightarrow Zn^{2+}(aq) + 2e^-\} \times 3$ $-E^\circ = +0.763\,V$

Cathode, Reduction: $\{NO_3^-(aq) + 4H^+(aq) + 3e^- \rightarrow NO(g) + 2H_2O(l)\} \times 2$; $E^\circ = +0.956\,V$

Net: $3Zn(s) + 2NO_3^-(aq) + 8H^+(aq) \rightarrow 3Zn^{2+}(aq) + 2NO(g) + 4H_2O(l)$ $E^\circ_{cell} = +1.719\,V$

Cell diagram: $Zn(s)|Zn^{2+}(aq)|\,|H^+(aq), NO_3^-(aq)|NO(g)|Pt(s)$

The cells for parts (a) - (d) follow. The anode is on the left in each case.

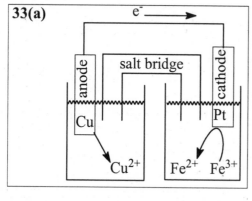

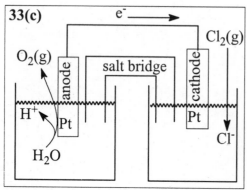

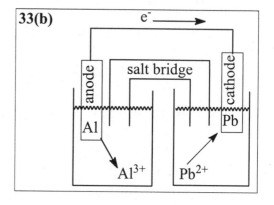

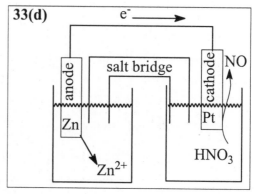

34. **(a)** $Fe(s)|Fe^{2+}(aq)\|\;|Cl^-(aq)Cl_2(g)|Pt(s)$

Oxidation: $Fe(s) \rightarrow Fe^{2+}(aq) + 2e^-$	$-E^\circ = +0.440\,V$
Reduction: $Cl_2(g) + 2e^- \rightarrow 2Cl^-(aq)$	$E^\circ = +1.358\,V$
Net: $Fe(s) + Cl_2(g) \rightarrow Fe^{2+}(aq) + 2Cl^-(aq)$	$E^\circ_{cell} = +1.798\,V$

(b) $Zn(s)|Zn^{2+}(aq)\|\;|Ag^+(aq)|Ag(s)$

Oxidation: $Zn(s) \rightarrow Zn^{2+}(aq) + 2e^-$	$-E^\circ = +0.763\,V$
Reduction: $\{Ag^+(aq) + e^- \rightarrow Ag(s)\} \times 2$	$E^\circ = +0.800\,V$
Net: $Zn(s) + 2Ag^+(aq) \rightarrow Zn^{2+}(aq) + 2Ag(s)$	$E^\circ_{cell} = +1.563\,V$

(c) $Pt(s)|Cu^+(aq), Cu^{2+}(aq)\|\;|Cu^+(aq)|Cu(s)$

Oxidation: $Cu^+(aq) \rightarrow Cu^{2+}(aq) + e^-$	$-E^\circ = -0.159\,V$
Reduction: $Cu^+(aq) + e^- \rightarrow Cu(s)$	$E^\circ = +0.520\,V$
Net: $2Cu^+(aq) \rightarrow Cu^{2+}(aq) + Cu(s)$	$E^\circ_{cell} = +0.361\,V$

(d) $Mg(s)|Mg^{2+}(aq)\|\;|Br^-(aq)Br_2(1)|Pt(s)$

Oxidation: $Mg(s) \rightarrow Mg^{2+}(aq) + 2e^-$	$-E^\circ = +2.356\,V$
Reduction: $Br_2(1) + 2e^- \rightarrow 2Br^-(aq)$	$E^\circ = +1.065\,V$
Net: $Mg(s) + Br_2(1) \rightarrow Mg^{2+}(aq) + 2Br^-(aq)$	$E^\circ_{cell} = +3.421\,V$

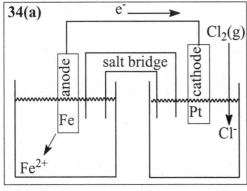

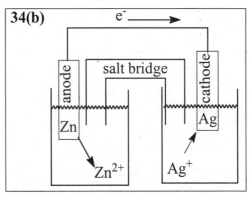

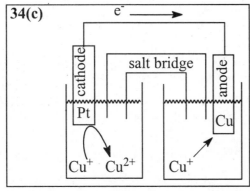

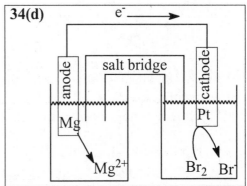

ΔG°, E°_{cell}, and K

35. **(a)** Oxidation: $\{Mn^{2+}(aq)+4H_2O(l)\rightarrow MnO_4^{-}(aq)+8H^{+}(aq)+5e^{-}\}\times 2\,; -E^{\circ}=-1.51\,V$

Reduction: $\{H_2O_2(aq)+2H^{+}(aq)+2e^{-}\rightarrow 2H_2O(l)\}\times 5 \qquad E^{\circ}=+1.763\,V$

Net: $2Mn^{2+}(aq)+5H_2O_2(aq)\rightarrow 2MnO_4^{-}(aq)+6H^{+}(aq)+2H_2O(l)\,; \quad E^{\circ}_{cell}=+0.25\,V$

(b) $\Delta G^{\circ}=-nFE^{\circ}_{cell}=-(10\,mol\ e^{-})(96{,}485\,C/mol\ e^{-})(0.25\ V)=-2.4\times10^{5}\,J=-2.4\times10^{2}\,kJ$

(c) $\ln K_{eq}=-\dfrac{\Delta G^{\circ}}{RT}=-\dfrac{-2.4\times10^{5}\ J}{8.3145\ J\ mol^{-1}\,K^{-1}\times298\ K}=97; \qquad K_{eq}=e^{97}=1\times10^{42}$

(d) Based on the extremely large value of K_{eq}, we conclude that this reaction should go to completion.

36. **(a)** Oxidation: $Fe(s)\rightarrow Fe^{2+}(aq)+2e^{-} \qquad\qquad -E^{\circ}=+0.440\,V$

Reduction: $\{Cr^{3+}(aq)+e^{-}\rightarrow Cr^{2+}(aq)\}\ \times 2 \qquad E^{\circ}=-0.424\,V$

Net: $2Cr^{3+}(aq)+Fe(s)\rightarrow Fe^{2+}(aq)+2Cr^{2+}(aq)\ E^{\circ}_{cell}=+0.016\,V$

(b) $E^{\circ}_{cell}=+0.016\,V$

(c) $\Delta G^{\circ}=-nFE^{\circ}_{cell}=-(2\ mol\ e^{-})(96{,}485\ C/mol\ e^{-})(0.016\ V)=-3.1\times10^{3}\,J=-3.1\,kJ$

(d) $\ln K_{eq}=-\dfrac{\Delta G^{\circ}}{RT}=-\dfrac{-3.1\times10^{3}\ J}{8.3145\ J\ mol^{-1}\,K^{-1}\times298\ K}=1.3; \qquad K_{eq}=e^{1.3}=4$

(e) Based on the very modest value of K_{eq}, we conclude that this reaction will not go to completion.

37. **(a)** A negative value of E°_{cell} $(-0.0050\,V)$ indicates that $\Delta G^{\circ}=-nFE^{\circ}_{cell}$ is positive which in turn indicates that K_{eq} is less than one $(K_{eq}<1.00);\ \Delta G^{\circ}=-RT\ln K_{eq}$.

$$K_{eq}=\dfrac{[Cu^{2+}]^{2}[Sn^{2+}]}{[Cu^{+}]^{2}[Sn^{4+}]}$$

Thus, when all concentrations are the same, the ion product, Q, equals 1.00. From the negative standard cell potential, it is clear that K_{eq} must be (slightly) less than one. Therefore, all the concentrations cannot be 0.500 M at the same time.

(b) In order to establish equilibrium, that is, to have the ion product become less than 1.00, and equal the equilibrium constant, the concentrations of the products must decrease and those of the reactants must increase. A net reaction to the left will occur.

38. (a) We calculate the value of the equilibrium constant from the standard cell potential.

$$E^\circ_{cell} = \frac{0.0257}{n}\ln K_{eq}; \quad \ln K_{eq} = \frac{nE^\circ_{cell}}{0.0257} = \frac{2 \text{ mol e}^- \times 0.0020\,\text{V}}{0.0257} = 0.16; \quad K_{eq} = e^{0.16} = 1.2$$

To determine if the described solution is possible, we compare

K_{eq} with Q. Now $K_{eq} = \dfrac{\left[Ni^{2+}\right]\left[V^{2+}\right]^2}{\left[V^{3+}\right]^2}$. Thus, when $\left[V^{2+}\right] = 0.600$ M and

$\left[V^{3+}\right] = \left[Ni^{2+}\right] = 0.675$M, the ion product, $Q = \dfrac{(0.600)^2 0.675}{(0.675)^2} = 0.533 < 1.2 = K_{eq}$.

Therefore, the described situation cannot occur.

(b) In order to establish equilibrium, that is, to have the ion product (0.533) become equal to 1.2, the equilibrium constant, the concentrations of the products must increase and those of the reactants must decrease. A slight net reaction to the right should occur.

39. Cell reaction: $Zn(s) + Ag_2O(s) \rightarrow ZnO(s) + 2Ag(s)$. We assume that the cell operates at 298 K.

$$\Delta G^\circ = \Delta G^\circ_f\left[ZnO(s)\right] + 2\Delta G^\circ_f\left[Ag(s)\right] - \Delta G^\circ_f\left[Zn(s)\right] - \Delta G^\circ_f\left[Ag_2O(s)\right]$$

$$= -318.3 \text{ kJ/mol} + 2\left(0.00 \text{ kJ/mol}\right) - 0.00 \text{ kJ/mol} - \left(-11.20 \text{ kJ/mol}\right)$$

$$= -307.1 \text{ kJ/mol} = -nFE^\circ_{cell}$$

$$E^\circ_{cell} = -\frac{\Delta G^\circ}{nF} = -\frac{-307.1 \times 10^3 \text{ J/mol}}{2 \text{ mol e}^-/\text{mol rxn} \times 96,485 \text{ C/mol e}^-} = 1.591\,\text{V}$$

40. From equation (21.28) we know $n = 12$ and the cell reaction. First compute value of ΔG°.

$$\Delta G^\circ = -n\,FE^\circ_{cell} = -12 \text{ mol e}^- \times \frac{96485 \text{ C}}{1 \text{ mol e}^-} \times 2.71 \text{ V} = -3.14 \times 10^6 \text{ J} = -3.14 \times 10^3 \text{ kJ}$$

Use this value, the balanced equation and values of ΔG_f° to calculate $\Delta G^\circ_f\left[Al(OH)_4\right]^-$.

$$4\,Al(s) + 3O_2(g) + 6H_2O(l) + 4OH^-(aq) \rightarrow 4\left[Al(OH)_4\right]^-(aq)$$

$$\Delta G^\circ = 4\Delta G^\circ_f\left[Al(OH)_4\right]^- - 4\Delta G^\circ_f[Al(s)] - 3\Delta G^\circ_f[O_2(g)] - 6\Delta G^\circ_f[H_2O(l)] - 4\Delta G^\circ_f[OH^-(aq)]$$

$$-3.14 \times 10^3 \text{ kJ} = 4\Delta G^\circ_f\left[Al(OH)_4\right]^- - 4 \times 0.00 \text{ kJ} - 3 \times 0.00 \text{ kJ} - 6 \times (-237.1 \text{ kJ}) - 4 \times (-157.2)$$

$$= 4\Delta G^\circ_f\left[Al(OH)_4\right]^- + 2051.4 \text{ kJ}$$

$$\Delta G^\circ_f\left[Al(OH)_4\right]^- = \left(-3.14 \times 10^3 \text{ kJ} - 2051.4 \text{ kJ}\right) \div 4 = -1.30 \times 10^3 \text{ kJ}$$

41. From the data provided we can construct the following Latimer diagram.

$$\text{IrO}_2 \xrightarrow{\ 0.223\ \text{V}\ } \text{Ir}^{3+} \xrightarrow{\ 1.156\ \text{V}\ } \text{Ir} \quad \text{(Acidic conditions)}$$
$$\text{(IV)} \qquad\qquad \text{(III)} \qquad\qquad \text{(0)}$$

Latimer diagrams are used to calculate the standard potentials of non-adjacent half-cell couples. Our objective in this question is to calculate the voltage differential between IrO_2 and iridium metal (Ir), which are separated in the diagram by Ir^{3+}. The process basically involves adding two half-reactions to obtain a third half-reaction. The potentials for the two half-reactions cannot, however, simply be added to get the target half-cell voltage because the electrons are not cancelled in the process of adding the two half-reactions. Instead, to find $E^\circ_{1/2\ \text{cell}}$ for the target half-reaction, we must use free energy changes, which are additive. To begin, we will balance the relevant half-reactions in acidic solution:

$$4\ \text{H}^+(aq) + \text{IrO}_2(s) + e^- \ \rightarrow\ \text{Ir}^{3+}(aq) + 2\ \text{H}_2\text{O}(l) \qquad E^\circ_{1/2\text{red(a)}} = 0.223\ \text{V}$$
$$\text{Ir}^{3+}(aq) + 3\ e^- \rightarrow \text{Ir}(s) \qquad\qquad E^\circ_{1/2\text{red(b)}} = 1.156\text{V}$$

$$\overline{4\ \text{H}^+(aq) + \text{IrO}_2(s) + 4e^- \ \rightarrow\ 2\ \text{H}_2\text{O}(l) + \text{Ir}(s) \qquad E^\circ_{1/2\text{red(c)}} = ?}$$

$E^\circ_{1/2\text{red(c)}} \neq E^\circ_{1/2\text{red(a)}} + E^\circ_{1/2\text{red(b)}}$ but $\Delta G^\circ_{(a)} + \Delta G^\circ_{(b)} = \Delta G^\circ_{(c)}$ and $\Delta G^\circ = -nFE^\circ$

$$-4F(E^\circ_{1/2\text{red(c)}}) = -1F(E^\circ_{1/2\text{red(a)}}) + -3F(E^\circ_{1/2\text{red(b)}})$$
$$-4F(E^\circ_{1/2\text{red(c)}}) = -1F(0.223) + -3F(1.156)$$

$$E^\circ_{1/2\text{red(c)}} = \frac{-1F(0.223) + -3F(1.156)}{-4F} = \frac{-1(0.223) + -3(1.156)}{-4} = 0.923\ \text{V}$$

In other words, $E^\circ_{(c)}$ is the weighted average of $E^\circ_{(a)}$ and $E^\circ_{(b)}$

42. This question will be answered in a manner similar to that used to solve question 41. Let's get underway by writing down the appropriate Latimer diagram:

$$\text{H}_2\text{MoO}_4 \xrightarrow{\ 0.646\ \text{V}\ } \text{MoO}_2 \xrightarrow{\ ?\ \text{V}\ } \text{Mo}^{3+} \quad \text{(Acidic conditions)}$$
$$\text{(VI)} \qquad\qquad \text{(IV)} \qquad\qquad \text{(III)}$$

with a $0.428\ \text{V}$ step connecting H_2MoO_4 to Mo^{3+}.

This time we want to calculate the standard voltage change for the 1 e^- reduction of MoO_2 to Mo^{3+}. Once again, we must balance the half-cell reactions in acidic solution:

$$\text{H}_2\text{MoO}_4(aq) + 2\ e^- + 2\ \text{H}^+(aq) \rightarrow \text{MoO}_2(s) + 2\ \text{H}_2\text{O}(l) \qquad E^\circ_{1/2\text{red(a)}} = 0.646\ \text{V}$$
$$\text{MoO}_2(s) + 4\ \text{H}^+(aq) + 1\ e^- \rightarrow \text{Mo}^{3+}(aq) + 2\ \text{H}_2\text{O}(l) \qquad E^\circ_{1/2\text{red(b)}} = ?\ \text{V}$$

$$\overline{\text{H}_2\text{MoO}_4(aq) + 3\ e^- + 6\ \text{H}^+(aq) \rightarrow \text{Mo}^{3+}(aq) + 4\ \text{H}_2\text{O}(l) \qquad E^\circ_{1/2\text{red(c)}} = 0.428\ \text{V}}$$

So, $-3F(E^\circ_{1/2\text{red(c)}}) = -2F(E^\circ_{1/2\text{red(a)}}) + -1F(E^\circ_{1/2\text{red(b)}})$

$$-3F(0.428\ \text{V}) = -2F(0.646) + -1F(E^\circ_{1/2\text{red(b)}})$$
$$-1FE^\circ_{1/2\text{red(b)}} = -3F(0.428\ \text{V}) + 2F(0.646)$$

$$E^\circ_{1/2\text{red(c)}} = \frac{-3F(0.428\ \text{V}) + 2F(0.646)}{-1F} = 1.284\ \text{V} - 1.292\ \text{V} = -0.008\ \text{V}$$

Concentration Dependence of E_{cell}—the Nernst Equation

43. We first calculate E_{cell}° for each reaction and then use the Nernst equation to calculate E_{cell}.

(a) Oxidation: $\{Al(s) \rightarrow Al^{3+}(0.18\ M) + 3\ e^{-}\} \times 2$ $\qquad$ $-E^{\circ} = +1.676\ V$

Reduction: $\{Fe^{2+}(0.85\ M) + 2\ e^{-} \rightarrow Fe(s)\} \times 3$ $\qquad$ $E^{\circ} = -0.440\ V$

Net: $2\ Al(s) + 3\ Fe^{2+}(0.85\ M) \rightarrow 2\ Al^{3+}(0.18\ M) + 3Fe(s)$ $\qquad$ $E_{cell}^{\circ} = +1.236\ V$

$$E_{cell} = E_{cell}^{\circ} - \frac{0.0592}{n}\log\frac{\left[Al^{3+}\right]^{2}}{\left[Fe^{2+}\right]^{3}} = 1.236\ V - \frac{0.0592}{6}\log\frac{(0.18\ M)^{2}}{(0.85\ M)^{3}} = 1.249\ V$$

(b) Oxidation: $\{Ag(s) \rightarrow Ag^{+}(0.34\ M) + e^{-}\} \times 2$ $\qquad$ $-E^{\circ} = -0.800\ V$

Reduction: $Cl_{2}(0.55\ atm) + 2\ e^{-} \rightarrow 2\ Cl^{-}(0.098\ M)$ $\qquad$ $E^{\circ} = +1.358\ V$

Net: $Cl_{2}(0.55\ atm) + 2\ Ag(s) \rightarrow 2\ Cl^{-}(0.098\ M) + 2\ Ag^{+}(0.34\ M)$; $E_{cell}^{\circ} = +0.558\ V$

$$E_{cell} = E_{cell}^{\circ} - \frac{0.0592}{n}\log\frac{\left[Cl^{-}\right]^{2}\left[Ag^{+}\right]^{2}}{P\{Cl_{2}(g)\}} = +0.558 - \frac{0.0592}{2}\log\frac{(0.34)^{2}(0.098)^{2}}{0.55} = +0.638\ V$$

44. (a) Oxidation: $Mn(s) \rightarrow Mn^{2+}(0.40\ M) + 2\ e^{-}$ $\qquad$ $-E^{\circ} = +1.18\ V$

Reduction: $\{Cr^{3+}(0.35\ M) + 1e^{-} \rightarrow Cr^{2+}(0.25\ M)\} \times 2$ $\qquad$ $E^{\circ} = -0.424\ V$

Net: $2\ Cr^{3+}(0.35\ M) + Mn(s) \rightarrow 2\ Cr^{2+}(0.25\ M) + Mn^{2+}(0.40\ M)$ $\quad$ $E_{cell}^{\circ} = +0.76\ V$

$$E_{cell} = E_{cell}^{\circ} - \frac{0.0592}{n}\log\frac{\left[Cr^{2+}\right]^{2}\left[Mn^{2+}\right]}{\left[Cr^{3+}\right]^{2}} = +0.76\ V - \frac{0.0592}{2}\log\frac{(0.25\ M)^{2}(0.40)}{(0.35\ M)^{2}} = +0.78\ V$$

(b) Oxidation: $\{Mg(s) \rightarrow Mg^{2+}(0.016\ M) + 2e^{-}\} \times 3$ $\qquad$ $-E^{\circ} = +2.356\ V$

Reduction: $\{\left[Al(OH)_{4}\right]^{-}(0.25\ M) + 3\ e^{-} \rightarrow 4OH^{-}(0.042\ M) + Al(s)\} \times 2$; $E^{\circ} = -2.310\ V$

Net: $3\ Mg(s) + 2[Al(OH)_{4}]^{-}(0.25\ M) \rightarrow 3Mg^{2+}(0.016\ M) + 8OH^{-}(0.042\ M) + 2\ Al(s)$;

$E_{cell}^{\circ} = +0.046\ V$

$$E_{cell} = E_{cell}^{\circ} - \frac{0.0592}{6}\log\frac{\left[Mg^{2+}\right]^{3}\left[OH^{-}\right]^{8}}{\left[\left[Al(OH)_{4}\right]^{-}\right]^{2}} = +0.046 - \frac{0.0592}{6}\log\frac{(0.016)^{3}(0.042)^{8}}{(0.25)^{2}}$$

$$= 0.046\ V + 0.150\ V = 0.196\ V$$

45. All these observations can be understood in terms of the procedure we use to balance half-equations: the ion—electron method.

(a) The reactions for which E depends on pH are those that contain either $H^+(aq)$ or $OH^-(aq)$ in the balanced half equation. These reactions are those that involve oxoacids and oxoanions in which the central atom changes oxidation state.

(b) $H^+(aq)$ will inevitably be on the left side of the reduction of an oxoanion because reduction is accompanied by not only a decrease in oxidation state, but also by the loss of oxygen atoms, as in $ClO_3^- \rightarrow ClO_2^-$, $SO_4^{2-} \rightarrow SO_2$, and $NO_3^- \rightarrow NO$. These oxygen atoms appear on the right-hand side as H_2O molecules. The hydrogens that are added to the right-hand side with the water molecules are then balanced with $H^+(aq)$ on the left-hand side.

(c) If a half-reaction with $H^+(aq)$ ions present is transferred to basic solution, it may be re-balanced by adding to each side $OH^-(aq)$ ions equal in number to the $H^+(aq)$ originally present. This results in $H_2O(l)$ on the side that had $H^+(aq)$ ions (the left side in this case) and $OH^-(aq)$ ions on the other side (the right side.)

46. Oxidation: $2\,Cl^-(aq) \rightarrow Cl_2(g) + 2\,e^-$ $\qquad\qquad\qquad -E° = -1.358\,V$

Reduction: $PbO_2(s) + 4\,H^+(aq) + 2\,e^- \rightarrow Pb^{2+}(aq) + 2\,H_2O(l)$ $\qquad E° = +1.455\,V$

Net: $PbO_2(s) + 4\,H^+(aq) + 2\,Cl^-(aq) \rightarrow Pb^{2+}(aq) + 2\,H_2O(l) + Cl_2(g)$; $\quad E°_{cell} = +0.097\,V$

We derive an expression for E_{cell} that depends on just the changing $[H^+]$.

$$E_{cell} = E°_{cell} - \frac{0.0592}{2}\log\frac{P\{Cl_2\}[Pb^{2+}]}{[H^+]^4[Cl^-]^2} = +0.097 - 0.0296\log\frac{(1.00\ \text{atm})(1.00\ \text{M})}{[H^+]^4(1.00)^2}$$

$$= +0.097 + 4 \times 0.0296\log[H^+] = +0.097 + 0.118\log[H^+] = +0.097 - 0.118\ \text{pH}$$

(a) $E_{cell} = +0.097 + 0.118\ \log(6.0) = +0.189\,V$ $\qquad$ Forward reaction spontaneous

(b) $E_{cell} = +0.097 + 0.118\ \log(1.2) = +0.106\,V$ $\qquad$ Forward reaction spontaneous

(c) $E_{cell} = +0.097 - 0.118 \times 4.25 = -0.405\,V$ $\quad$ Forward reaction nonspontaneous

Reaction is spontaneous in strongly acidic solutions (very low pH), but is nonspontaneous in basic, neutral, and weakly acidic solutions.

47. Oxidation: $Zn(s) \rightarrow Zn^{2+}(aq) + 2e^-$ $-E^\circ = +0.763\,V$

Reduction: $Cu^{2+}(aq) + 2e^- \rightarrow Cu(s)$ $E^\circ = +0.337\,V$

Net: $Zn(s) + Cu^{2+}(aq) \rightarrow Cu(s) + Zn^{2+}(aq)$ $E^\circ_{cell} = +1.100\,V$

(a) We set $E = 0.000V$, $[Zn^{2+}] = 1.00M$, and solve for $[Cu^{2+}]$ in the Nernst equation.

$$E_{cell} = E^\circ_{cell} - \frac{0.0592}{2} \log \frac{[Zn^{2+}]}{[Cu^{2+}]}; \quad 0.000 = 1.100 - 0.0296 \log \frac{1.0M}{[Cu^{2+}]}$$

$$\log \frac{1.0\,M}{[Cu^{2+}]} = \frac{0.000 - 1.100}{-0.0296} = 37.2; \quad [Cu^{2+}] = 10^{-37.2} = 6 \times 10^{-38}\,M$$

(b) If we work the problem the other way, by assuming initial concentrations of $[Cu^{2+}]_{initial} = 1.0\,M$ and $[Zn^{2+}]_{initial} = 0.0\,M$, we obtain $[Cu^{2+}]_{final} = 6 \times 10^{-38}\,M$ and $[Zn^{2+}]_{final} = 1.0\,M$. We would say that this reaction goes to completion.

48. Oxidation: $Sn(s) \rightarrow Sn^{2+}(aq) + 2e^-$ $-E^\circ = +0.137\ V$

Reduction: $Pb^{2+}(aq) + 2e^- \rightarrow Pb(s)$ $E^\circ = -0.125\ V$

Net: $Sn(s) + Pb^{2+}(aq) \rightarrow Sn^{2+}(aq) + Pb(s)$ $E^\circ_{cell} = +0.012\ V$

Now we wish to find out if $Pb^{2+}(aq)$ will be completely displaced, that is, will $[Pb^{2+}]$ reach 0.0010 M, if $[Sn^{2+}]$ is fixed at 1.00 M? We use the Nernst equation to determine if the cell voltage still is positive under these conditions.

$$E_{cell} = E^\circ_{cell} - \frac{0.0592}{2} \log \frac{[Sn^{2+}]}{[Pb^{2+}]} = +0.012 - \frac{0.0592}{2} \log \frac{1.00}{0.0010} = +0.012 - 0.089 = -0.077\ V$$

No, this reaction will not go to completion under the conditions stated. The reaction stops being spontaneous when $E_{cell} = 0$.

We can work this the another way as well: assume that $[Pb^{2+}] = (1.0 - x)$ M and calculate $[Sn^{2+}] = x$ M at equilibrium, that is, where $E_{cell} = 0$.

$$E_{cell} = 0.00 = E^\circ_{cell} - \frac{0.0592}{2} \log \frac{[Sn^{2+}]}{[Pb^{2+}]} = +0.012 - \frac{0.0592}{2} \log \frac{x}{1.0 - x}$$

$$\log \frac{x}{1.0 - x} = \frac{2 \times 0.012}{0.0592} = 0.41 \quad x = 10^{0.41}(1.0 - x) = 2.6 - 2.6x \quad x = \frac{2.6}{3.6} = 0.72\ M$$

We would expect the final $[Sn^{2+}]$ to equal 1.0 M (or at least 0.999 M) if the reaction went to completion. Instead it equals 0.72 M.

49. (a) The two half-equations and the cell equation are given below. $E_{cell}^{\circ} = 0.000\,\text{V}$

Oxidation: $H_2(g) \rightarrow 2\,H^+(0.65\text{ M KOH}) + 2\,e^-$

Reduction: $2\,H^+(1.0\text{ M}) + 2\,e^- \rightarrow H_2(g)$

Net: $2\,H^+(1.0\text{ M}) \rightarrow 2\,H^+(0.65\text{ M KOH})$

$$\left[H^+\right]_{base} = \frac{K_w}{\left[OH^-\right]} = \frac{1.00 \times 10^{-14}}{0.65} = 1.5 \times 10^{-14}\,\text{M}$$

$$E_{cell} = E_{cell}^{\circ} - \frac{0.0592}{2}\log\frac{\left[H^+\right]_{base}^2}{\left[H^+\right]_{acid}^2} = 0.000 - \frac{0.0592}{2}\log\frac{\left(1.5 \times 10^{-14}\right)^2}{\left(1.0\right)^2} = +0.818\,\text{V}$$

(b) For the reduction of $H_2O(l)$ to $H_2(g)$ in basic solution,

$2\,H_2O(l) + 2\,e^- \rightarrow 2\,H_2(g) + 2\,OH^-(aq)$, $E^{\circ} = -0.828\,\text{V}$. This reduction is the reverse of the reaction that occurs in the anode of the cell described, with one small difference: in the standard half-cell, $[OH^-] = 1.00$ M, while in the anode half-cell in the case at hand, $[OH^-] = 0.65$ M. Or, viewed in another way, in 1.00 M KOH, $[H^+]$ is smaller still than in 0.65 M KOH. The forward reaction (dilution of H^+) should be even more spontaneous with 1.00 M KOH than with 0.65 M KOH. We expect that E_{cell}° (0.828 V) should be a little larger than E_{cell} (0.818 V), which, is in fact, the case.

50. (a) Because $NH_3(aq)$ is a weaker base than KOH(aq), $[OH^-]$ will be smaller than it is in the previous problem. Therefore the $[H^+]$ will be higher. Its logarithm will be less negative, and the cell voltage will be smaller. Or, viewed as in Exercise 49(b), the difference in $[H^+]$ between 1.0 M H^+ and 0.65 M KOH is greater than the difference in $[H^+]$ between 1.0 M H^+ and 0.65 M NH_3. The forward reaction is "less spontaneous" and E_{cell} is smaller.

(b)

Reaction:	$NH_3(aq) + H_2O(l)$	$\rightleftharpoons$	$NH_4^+(aq) +$	$OH^-(aq)$
Initial:	0.65 M		0 M	≈ 0 M
Changes:	$-x$ M		$+x$ M	$+x$ M
Equil:	$(0.65-x)$ M		x M	x M

$$K_b = \frac{\left[NH_4^+\right]\left[OH^-\right]}{\left[NH_3\right]} = 1.8 \times 10^{-5} = \frac{x \cdot x}{0.65 - x} \approx \frac{x^2}{0.65}$$

$$x = \left[OH^-\right] = \sqrt{0.65 \times 1.8 \times 10^{-5}} = 3.4 \times 10^{-3}\,\text{M}; \quad \left[H_3O^+\right] = \frac{1.00 \times 10^{-14}}{3.4 \times 10^{-3}} = 2.9 \times 10^{-12}\,\text{M}$$

$$E_{cell} = E_{cell}^{\circ} - \frac{0.0592}{2}\log\frac{\left[H^+\right]_{base}^2}{\left[H^+\right]_{acid}^2} = 0.000 - \frac{0.0592}{2}\log\frac{\left(2.9 \times 10^{-12}\right)^2}{\left(1.0\right)^2} = +0.683\,\text{V}$$

51. First we need $\left[Ag^+\right]$ in a saturated solution of Ag_2CrO_4.

$$K_{sp} = \left[Ag^+\right]^2\left[CrO_4^{2-}\right] = (2s)^2(s) = 4s^3 = 1.1\times10^{-12} \qquad s = \sqrt[3]{\frac{1.1\times10^{-12}}{4}} = 6.5\times10^{-5}\ M$$

The cell diagrammed is a concentration cell, for which $E^{\circ}_{cell} = 0.000\ V$, $n = 1$,

$$\left[Ag^+\right]_{anode} = 2s = 1.3\times10^{-4}\ M$$

Cell reaction: $Ag(s) + Ag^+(0.125\ M) \rightarrow Ag(s) + Ag^+\left(1.3\times10^{-4}\ M\right)$

$$E_{cell} = E^{\circ}_{cell} - \frac{0.0592}{1}\log\frac{1.3\times10^{-4}\ M}{0.125\ M} = 0.000 + 0.177\ V = 0.177\ V$$

52. We need to determine $\left[Ag^+\right]$ in the saturated solution of Ag_3PO_4.

The cell diagrammed is a concentration cell, for which $E^{\circ}_{cell} = 0.000V$, $n = 1$.

Cell reaction: $Ag(s) + Ag^+(0.140\ M) \rightarrow Ag(s) + Ag^+(x\ M)$

$$E_{cell} = 0.180V = E^{\circ}_{cell} - \frac{0.0592}{1}\log\frac{x\ M}{0.140\ M}; \quad \log\frac{x\ M}{0.140\ M} = \frac{0.180}{-0.0592} = -3.04$$

$$x\ M = 0.140\ M \times 10^{-3.04} = 0.140\ M \times 9.1\times10^{-4} = 1.3\times10^{-4}\ M = \left[Ag^+\right]_{anode}$$

$$K_{sp} = \left[Ag^+\right]^3\left[PO_4^{3-}\right] = (3s)^3(s) = \left(1.3\times10^{-4}\right)^3\left(1.3\times10^{-4}\div3\right) = 9.5\times10^{-17}$$

53. **(a)** Oxidation: $Sn(s) \rightarrow Sn^{2+}(0.075\ M) + 2\ e^-$ $\qquad\qquad -E^{\circ} = +0.137\ V$

Reduction: $Pb^{2+}(0.600\ M) + 2\ e^- \rightarrow Pb(s)$ $\qquad\qquad E^{\circ} = -0.125\ V$

Net: $Sn(s) + Pb^{2+}(0.600\ M) \rightarrow Pb(s) + Sn^{2+}(0.075\ M); E^{\circ}_{cell} = +0.012\ V$

$$E_{cell} = E^{\circ}_{cell} - \frac{0.0592}{2}\log\frac{\left[Sn^{2+}\right]}{\left[Pb^{2+}\right]} = 0.012 - 0.0296\log\frac{0.075}{0.600} = 0.012 + 0.027$$

$$= 0.039\ V$$

(b) E_{cell} decreases with time because the reaction forms $\left[Sn^{2+}\right]$ and uses up $\left[Pb^{2+}\right]$.

(c) When $\left[Pb^{2+}\right] = 0.500\ M = 0.600\ M - 0.100\ M$, $\left[Sn^{2+}\right] = 0.075\ M + 0.100\ M$, because the stoichiometry of the reaction is 1:1 for Sn^{2+} and Pb^{2+}.

$$E_{cell} = E^{\circ}_{cell} - \frac{0.0592}{2}\log\frac{\left[Sn^{2+}\right]}{\left[Pb^{2+}\right]} = 0.012 - 0.0296\log\frac{0.175}{0.500} = 0.012 + 0.013 = 0.025\ V$$

(d) Reaction:

$$\text{Sn(s)} + \text{Pb}^{2+}(\text{aq}) \rightarrow \text{Pb(s)} + \text{Sn}^{2+}(\text{aq})$$

Initial:	0.600 M	0.075 M
Changes:	$-x$ M	$+x$ M
Final:	$(0.600 - x)$ M	$(0.075 + x)$ M

$$E_{cell} = E_{cell}^{\circ} - \frac{0.0592}{2} \log \frac{[\text{Sn}^{2+}]}{[\text{Pb}^{2+}]} = 0.020 = 0.012 - 0.0296 \log \frac{0.075 + x}{0.600 - x}$$

$$\log \frac{0.075 + x}{0.600 - x} = \frac{E_{cell} - 0.012}{-0.0296} = \frac{0.020 - 0.012}{-0.0296} = -0.27; \quad \frac{0.075 + x}{0.600 - x} = 10^{-0.27} = 0.54$$

$$0.075 + x = 0.54(0.600 - x) = 0.324 - 0.54x; \quad x = \frac{0.324 - 0.075}{1.54} = 0.162 \text{ M}$$

$$[\text{Sn}^{2+}] = 0.075 + 0.162 = 0.237 \text{ M}$$

(e) Use the expression developed in part (d).

$$\log \frac{0.075 + x}{0.600 - x} = \frac{E_{cell} - 0.012}{-0.0296} = \frac{0.000 - 0.012}{-0.0296} = +0.41$$

$$\frac{0.075 + x}{0.600 - x} = 10^{+0.41} = 2.6; \quad 0.075 + x = 2.6(0.600 - x) = 1.6 - 2.6x$$

$$x = \frac{1.6 - 0.075}{3.6} = 0.42 \text{ M}$$

$$[\text{Sn}^{2+}] = 0.075 + 0.42 = 0.50 \text{ M}; \qquad [\text{Pb}^{2+}] = 0.600 - 0.42 = 0.18 \text{ M}$$

54. **(a)** Oxidation: $\text{Ag(s)} \rightarrow \text{Ag}^+ (0.015 \text{ M}) + e^- \qquad\qquad -E^{\circ} = -0.800 \text{ V}$

Reduction: $\text{Fe}^{3+}(0.055 \text{ M}) + e^- \rightarrow \text{Fe}^{2+}(0.045 \text{ M}) \qquad\qquad E^{\circ} = +0.771 \text{V}$

Net: $\text{Ag(s)} + \text{Fe}^{3+}(0.055 \text{ M}) \rightarrow \text{Ag}^+ (0.015 \text{ M}) + \text{Fe}^{2+}(0.045 \text{ M}) \quad E_{cell}^{\circ} = -0.029 \text{ V}$

$$E_{cell} = E_{cell}^{\circ} - \frac{0.0592}{1} \log \frac{[\text{Ag}^+][\text{Fe}^{2+}]}{[\text{Fe}^{3+}]} = -0.029 - 0.0592 \log \frac{0.015 \times 0.045}{0.055}$$

$$= -0.029 \text{ V} + 0.113 \text{ V} = +0.084 \text{ V}$$

(b) E_{cell} will decrease with time because, as the reaction proceeds, $[\text{Ag}^+]$ and $[\text{Fe}^{2+}]$ will increase and $[\text{Fe}^{3+}]$ decrease.

(c) When $\left[Ag^+\right] = 0.020\,M = 0.015\,M + 0.005\,M$, $\left[Fe^{2+}\right] = 0.045\,M + 0.005\,M = 0.050\,M$

and $\left[Fe^{3+}\right] = 0.055\,M - 0.005\,M = 0.500\,M$, because, by the stoichiometry of the

reaction, a mole of Fe^{2+} is produced and a mole of Fe^{3+} is consumed for every mole

of Ag^+ produced.

$$E_{cell} = E^\circ_{cell} - \frac{0.0592}{1} \log \frac{\left[Ag^+\right]\left[Fe^{2+}\right]}{\left[Fe^{3+}\right]} = -0.029 - 0.0592\log \frac{0.020 \times 0.050}{0.050}$$

$$= -0.029 \text{ V} + 0.101 \text{ V} = +0.072 \text{ V}$$

(d) Reaction: $Ag(s) +$ $Fe^{3+}\left(0.055\,M\right)$ $\rightleftharpoons$ $Ag^+\left(0.015\text{ M}\right) +$ $Fe^{2+}\left(0.045\text{ M}\right)$

Initial: 0.055 M   0.015 M   0.045 M

Changes: $-x$ M $+x$ M $+x$ M

Final: $\left(0.055 - x\right)$ M $\left(0.015 + x\right)$ M $\left(0.045 + x\right)$ M

$$E_{cell} = E^\circ_{cell} - \frac{0.0592}{1} \log \frac{\left[Ag^+\right]\left[Fe^{2+}\right]}{\left[Fe^{3+}\right]} = -0.029 - 0.0592 \log \frac{\left(0.015 + x\right)\left(0.045 + x\right)}{\left(0.055 - x\right)}$$

$$\log \frac{\left(0.015 + x\right)\left(0.045 + x\right)}{\left(0.055 - x\right)} = \frac{E_{cell} + 0.029}{-0.0592} = \frac{0.010 + 0.029}{-0.0592} = -0.66$$

$$\frac{\left(0.015 + x\right)\left(0.045 + x\right)}{\left(0.055 - x\right)} = 10^{-0.66} = 0.22$$

$$0.00068 + 0.060x + x^2 = 0.22\left(0.055 - x\right) = 0.012 - 0.22x \qquad x^2 + 0.28x - 0.011 = 0$$

$$x = \frac{-b \pm \sqrt{b^2 - 4ac}}{2a} = \frac{-0.28 \pm \sqrt{\left(0.28\right)^2 + 4 \times 0.011}}{2} = 0.035 \text{ M}$$

$$\left[Ag^+\right] = 0.015 \text{ M} + 0.035 \text{ M} = 0.050 \text{ M}$$

$$\left[Fe^{2+}\right] = 0.045 \text{ M} + 0.035 \text{ M} = 0.080 \text{ M}$$

$$\left[Fe^{3+}\right] = 0.055 \text{ M} - 0.035 \text{ M} = 0.020 \text{ M}$$

(e) We use the expression that was developed in part (d).

$$\log \frac{\left(0.015 + x\right)\left(0.045 + x\right)}{\left(0.055 - x\right)} = \frac{E_{cell} + 0.029}{-0.0592} = \frac{0.000 + 0.029}{-0.0592} = -0.49$$

$$\frac{\left(0.015 + x\right)\left(0.045 + x\right)}{\left(0.055 - x\right)} = 10^{-0.49} = 0.32$$

$$0.00068 + 0.060x + x^2 = 0.32(0.055 - x) = 0.018 - 0.32x \qquad x^2 + 0.38x - 0.017 = 0$$

$$x = \frac{-b \pm \sqrt{b^2 - 4ac}}{2a} = \frac{-0.38 \pm \sqrt{(0.38)^2 + 4 \times 0.017}}{2} = 0.040 \, M$$

$$[Ag^+] = 0.015 \, M + 0.040 \, M = 0.055 \, M$$

$$[Fe^{2+}] = 0.045 \, M + 0.040 \, M = 0.085 \, M$$

$$[Fe^{3+}] = 0.055 \, M - 0.040 \, M = 0.015 \, M$$

55. First we will need to come up with a balanced equation for the overall redox reaction. Clearly, the reaction must involve the oxidation of $Cl^-(aq)$ and the reduction of $Cr_2O_7{}^{2-}(aq)$:

$14 \, H^+(aq) + Cr_2O_7{}^{2-}(aq) + 6 \, e^- \rightarrow 2 \, Cr^{3+}(aq) + 7 \, H_2O(l)$	$E^\circ_{1/2red} = 1.33 \, V$
$\{Cl^-(aq) \rightarrow 1/2 \, Cl_2(g) + 1 \, e^-\} \times 6$	$E^\circ_{1/2ox} = -1.358 \, V$

$$14 \, H^+(aq) + Cr_2O_7{}^{2-}(aq) + 6 \, Cl^-(aq) \rightarrow 2 \, Cr^{3+}(aq) + 7 \, H_2O(l) + 3 \, Cl_2(g) \quad E^\circ_{cell} = -0.03 \, V$$

Since the cell voltage is <u>negative</u>, the oxidation of $Cl^-(aq)$ to $Cl_2(g)$ by $Cr_2O_7{}^{2-}(aq)$ at standard conditions will <u>not</u> occur spontaneously. We could obtain some $Cl_2(g)$ from this reaction by driving it to the product side with an external voltage. In other words, the reverse reaction is the spontaneous reaction at standard conditions and if we want to obtain some $Cl_2(g)$ from the system, we must push the non-spontaneous reaction in its forward direction with an external voltage, (i.e., a DC power source). Since E°_{cell} is only slightly negative, we could also drive the reaction by removing products as they are formed and replenishing reactants as they are consumed.

56.

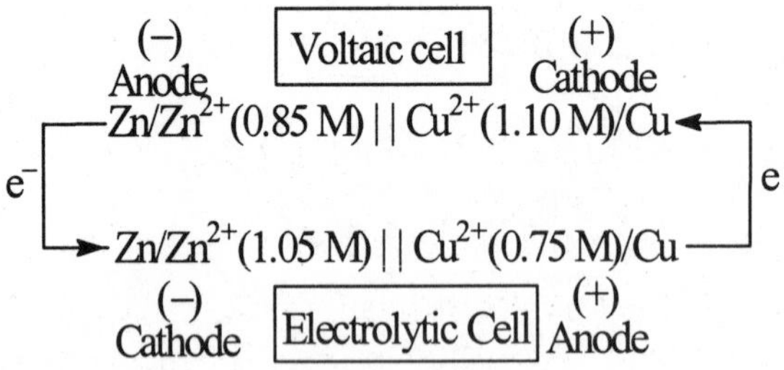

First we must find the voltage for each cell using the Nernst equation:

Cell A(voltaic cell): $Zn(s) + Cu^{2+}(aq) \rightleftharpoons Cu(s) + Zn^{2+}(aq) \qquad E^\circ = 1.100 \, V \, (n = 2 \, e^-)$

$$E_{cell} = E^\circ_{cell} - \frac{0.0592}{n} \log\left(\frac{[Zn^{2+}]}{[Cu^{2+}]}\right) = 1.100 \, V - \frac{0.0592}{2} \log\left(\frac{0.85 \, M}{1.10 \, M}\right) = 1.103 \, V$$

Cell B(electrolytic cell): $Cu(s) + Zn^{2+}(aq) \rightleftharpoons Zn(s) + Cu^{2+}(aq)$; $E° = -1.100$ V ($n = 2$ e$^-$)

$$E_{cell} = E°_{cell} - \frac{0.0592}{n} \log\left(\frac{[Cu^{2+}]}{[Zn^{2+}]}\right) = -1.100 \text{ V} - \frac{0.0592}{2} \log\left(\frac{0.75 \text{ M}}{1.05 \text{ M}}\right) = -1.096 \text{ V}$$

Overall: $E_{overall} = E_{voltaic\ cell} + E_{electrolytic\ cell} = 1.103$ V $+ (-1.096$ V$) = 0.007$ V
(at instant of hook-up)

Batteries and Fuel Cells

57. **(a)** The cell diagram begins with the anode and ends with the cathode.

Cell diagram: $Cr(s)|Cr^{2+}(aq), Cr^{3+}(aq)||Fe^{2+}(aq), Fe^{3+}(aq)|Fe(s)$

(b) Oxidation: $Cr^{2+}(aq) \rightarrow Cr^{3+}(aq) + e^-$ $\qquad\qquad -E° = +0.424$ V

Reduction: $Fe^{3+}(aq) + e^- \rightarrow Fe^{2+}(aq)$ $\qquad\qquad E° = +0.771$ V

Net: $Cr^{2+}(aq) + Fe^{3+}(aq) \rightarrow Cr^{3+}(aq) + Fe^{2+}(aq)$ $\quad E°_{cell} = +1.195$ V

58. **(a)** Oxidation:
$Zn(s) \rightarrow Zn^{2+}(aq) + 2e^-$ $\qquad\qquad\qquad\qquad\qquad -E° = +0.763$ V

Reduction: $2MnO_2(s) + H_2O(l) + 2e^- \rightarrow Mn_2O_3(s) + 2OH^-(aq)$

Acid-base: $\{NH_4^+(aq) + OH^-(aq) \rightarrow NH_3(g) + H_2O(l)\}$ $\qquad \times 2$

Complex: $Zn^{2+}(aq) + 2NH_3(aq) + 2Cl^-(aq) \rightarrow [Zn(NH_3)_2]Cl_2(s)$

Net: $Zn(s) + 2MnO_2(s) + 2NH_4^+(aq) + 2Cl^-(aq) \rightarrow Mn_2O_3(s) + H_2O(l) + [Zn(NH_3)_2]Cl_2(s)$

(b) $\Delta G° = -nFE°_{cell} = -(2 \text{ mol e}^-)(96485 \text{ C/mol e}^-)(1.55 \text{ V}) = -2.99 \times 10^5 \text{ J/mol}$

This is the standard free energy change for the entire reaction, which is composed of the four reactions in part (a). We can determine the values of $\Delta G°$ for the acid-base and complex formation reactions with data from Appendix D and $pK_f = -4.81$ (the negative log of the K_f for $[Zn(NH_3)_2]^{2+}$).

$\Delta G°_{a-b} = -RT\ln K_b^2 = -(8.3145 \text{ J mol}^{-1} \text{ K}^{-1})(298.15 \text{ K})\ln(1.8 \times 10^{-5})^2 = 5.417 \times 10^4 \text{ J/mol}$

$\Delta G°_{cmplx} = -RT\ln K_f = -(8.3145 \text{ J mol}^{-1} \text{ K}^-)(298.15 \text{ K})\ln(10^{4.81}) = -2.746 \times 10^4 \text{ J/mol}$

Then $\Delta G°_{total} = \Delta G°_{redox} + \Delta G°_{a-b} + \Delta G°_{cmplx}$

$$\Delta G_{redox}{}^{o} = \Delta G_{total}{}^{o} - \Delta G_{a-b}{}^{o} - \Delta G_{cmplx}{}^{o}$$

$$= -2.99 \times 10^5 \text{ J/mol} - 5.417 \times 10^4 + 2.746 \times 10^4 \text{ J/mol} = -3.26 \times 10^5 \text{ J/mol}$$

Thus, the voltage of the redox reactions alone is

$$E^{o} = \frac{-3.26 \times 10^5 \text{ J}}{-2\text{mol e}^- \times 96485 \text{ C / mol e}^-} = 1.69 \text{ V}$$

$$1.69\text{V} = +0.763\text{V} + E^{o}\left\{\text{MnO}_2/\text{Mn}_2\text{O}_3\right\}$$

$$E^{o}\left\{\text{MnO}_2/\text{Mn}_2\text{O}_3\right\} = 1.69\text{V} - 0.763\text{V} = +0.93\text{V}$$

The electrode potentials were calculated by using equilibrium constants from Appendix D. These calculations do not take into account the cell's own internal resistance to the flow of electrons, which makes the actual voltage developed by the electrodes less than the theoretical values derived from equilibrium constants. Also because the solid species (other than Zn) do not appear as compact rods, but rather are dispersed in a paste, and since very little water is present in the cell, the activities for the various species involved in the electrochemical reactions will deviate markedly from unity.

As a result, the equilibrium constants for the reactions taking place in the cell will be substantially different from those provided in Appendix D, which apply only to dilute solutions and reactions involving solid reactants and products with small surface areas. The actual electrode voltages, therefore, will end up being different from those calculated here.

59. **(a)** Cell reaction: $2\text{H}_2(\text{g}) + \text{O}_2(\text{g}) \rightarrow 2\text{H}_2\text{O}(\text{l})$

$$\Delta G_{rxn}{}^{o} = 2\Delta G_f^{o}\left[\text{H}_2\text{O}(\text{l})\right] = 2(-237.1 \text{ kJ/mol}) = -474.2 \text{ kJ/mol}$$

$$E_{cell}{}^{o} = -\frac{\Delta G^{o}}{nF} = -\frac{-474.2 \times 10^3 \text{ J/mol}}{4 \text{ mol e}^- \times 96485 \text{ C/mol e}^-} = 1.229 \text{ V}$$

(b) Anode, Oxidation: $\{\text{Zn}(\text{s}) \rightarrow \text{Zn}^{2+}(\text{aq}) + 2\text{e}^-\} \times 2$ $\qquad -E^{o} = +0.763\text{V}$

Cathode, Reduction: $\text{O}_2(\text{g}) + 4\text{ H}^+(\text{aq}) + 4\text{e}^- \rightarrow 2\text{H}_2\text{O}(\text{l})$ $\quad E^{o} = +1.229 \text{ V}$

Net: $2\text{Zn}(\text{s}) + \text{O}_2(\text{g}) + 4\text{H}^+(\text{aq}) \rightarrow 2\text{Zn}^{2+}(\text{aq}) + 2\text{H}_2\text{O}(\text{l})$ $E_{cell}^{o} = +1.992 \text{ V}$

(c) Anode, Oxidation: $\text{Mg}(\text{s}) \rightarrow \text{Mg}^{2+}(\text{aq}) + 2\text{e}^-$ $\qquad -E^{o} = +2.356\text{V}$

Cathode, Reduction: $\text{I}_2(\text{s}) + 2\text{e}^- \rightarrow 2\text{ I}^-(\text{aq})$ $\qquad E^{o} = +0.535\text{V}$

Net: $\text{Mg}(\text{s}) + \text{I}_2(\text{s}) \rightarrow \text{Mg}^{2+}(\text{aq}) + 2\text{ I}^-(\text{aq})$ $\qquad E_{cell}^{o} = +2.891\text{V}$

60. (a) Oxidation: $\quad Zn(s) \rightleftharpoons Zn^{2+}(aq) + 2\ e^-$

Precipitation: $Zn^{2+}(aq) + 2\ OH^-(aq) \rightleftharpoons Zn(OH)_2(s)$

Reduction: $\quad 2\ MnO_2(s) + H_2)(l) + 2\ e^- \rightleftharpoons Mn_2O_3(s) + 2\ OH^-(aq)$

Net : $Zn(s) + 2\ MnO_2(s) + H_2O(l) + 2\ OH^-(aq) \rightleftharpoons Mn_2O_3(s) + Zn(OH)_2(s)$

(b) In question 58, we determined that the standard voltage for the reduction reaction is $+0.93$ V ($n = 2\ e^-$). To convert this voltage to an equilibrium constant (at 25 °C) use:

$$\log K_{red} = \frac{nE^\circ}{0.0592} = \frac{2(0.93)}{0.0592} = 31.\underline{4}; \qquad K_{red} = 10^{31.42} = 3 \times 10^{31}$$

and for $Zn(s) \rightleftharpoons Zn^{2+}(aq) + 2\ e^-$ ($E^\circ = 0.763$ V and $n = 2\ e^-$)

$$\log K_{ox} = \frac{nE^\circ}{0.0592} = \frac{2(0.763)}{0.0592} = 25.\underline{8}; \qquad K_{ox} = 10^{25.78} = 6 \times 10^{25}$$

$$\Delta G^\circ_{total} = \Delta G^\circ_{precipitation} + \Delta G^\circ_{oxidation} + \Delta G^\circ_{reduction}$$

$$\Delta G^\circ_{total} = -RT \ln \frac{1}{K_{sp,\ Zn(OH)_2}} + (-RT \ln K_{ox}) + (-RT \ln K_{red})$$

$$\Delta G^\circ_{total} = -RT\ (\ln \frac{1}{K_{sp,\ Zn(OH)_2}} + \ln K_{ox} + \ln K_{red})$$

$$\Delta G^\circ_{total} = -0.0083145\ \frac{kJ}{K\ mol}\ (298\ K)\ (\ln \frac{1}{1.2 \times 10^{-17}} + \ln(6.0 \times 10^{25}) + \ln(2.6 \times 10^{31}))$$

$$\Delta G^\circ_{total} = -423\ kJ = -nFE^\circ_{total}$$

Hence, $E^\circ_{total} = E^\circ_{cell} = \dfrac{-423 \times 10^3\ J}{-(2\ mol)(96485\ C\ mol^{-1})} = 2.19$ V

61. Aluminum-Air Battery: $2\ Al(s) + 3/2\ O_2(g) \rightarrow Al_2O_3(s)$
Zinc-Air Battery: $\quad Zn(s) + \frac{1}{2}\ O_2(g) \rightarrow ZnO(s)$
Iron-Air Battery $\quad Fe(s) + \frac{1}{2}\ O_2(g) \rightarrow FeO(s)$
Calculate the quantity of charge transferred when 1.00 g of metal is consumed in each cell.

Aluminum-Air Cell:

$$1.00\ g\ Al(s) \times \frac{1\ mol\ Al(s)}{26.98\ g\ Al(s)} \times \frac{3\ mol\ e^-}{1\ mol\ Al(s)} \times \frac{96,485 C}{1\ mol\ e^-} = 1.07 \times 10^4\ C$$

Zinc-Air Cell:

$$1.00 \text{ g Zn(s)} \times \frac{1 \text{ mol Zn(s)}}{65.39 \text{ g Zn(s)}} \times \frac{2 \text{ mol e}^-}{1 \text{ mol Zn(s)}} \times \frac{96,485 \text{ C}}{1 \text{ mol e}^-} = 2.95 \times 10^3 \text{ C}$$

Iron-Air Cell:

$$1.00 \text{ g Fe(s)} \times \frac{1 \text{ mol Fe(s)}}{55.847 \text{ g Fe(s)}} \times \frac{2 \text{ mol e}^-}{1 \text{ mol Fe(s)}} \times \frac{96,485 \text{ C}}{1 \text{ mol e}^-} = 3.46 \times 10^3 \text{ C}$$

As expected, aluminum has the greatest quantity of charge transferred per unit mass (1.00 g) of metal oxidized. This is because aluminum has the smallest molar mass and forms the most highly charged cation (3+ vs 2+ for Zn and Fe).

62. **(a)** A voltaic cell with a voltage of 0.1000 V would be possible by using two half-cells whose standard reduction potentials differ by approximately 0.10 V, such as the following pair.

Oxidation: $2 \text{ Cr}^{3+}(\text{aq}) + 7 \text{ H}_2\text{O(l)} \rightarrow \text{Cr}_2\text{O}_7^{2-}(\text{aq}) + 14 \text{ H}^+(\text{aq}) + 6 \text{ e}^- \quad -E^\circ = -1.33 \text{ V}$

Reduction: $\{\text{PbO}_2(\text{s}) + 4 \text{ H}^+(\text{aq}) + 2\text{e}^- \rightarrow \text{Pb}^{2+}(\text{aq}) + 2 \text{ H}_2\text{O(l)}\} \times 3 \quad E^\circ = +1.455 \text{ V}$

_____

Net: $2\text{Cr}^{3+}(\text{aq}) + 3\text{PbO}_2(\text{s}) + \text{H}_2\text{O(l)} \rightarrow \text{Cr}_2\text{O}_7^{2-}(\text{aq}) + 3\text{Pb}^{2+}(\text{aq}) + 2\text{H}^+(\text{aq}) \quad E^\circ_{\text{cell}} = 0.12\underline{5} \text{ V}$

The voltage can be adjusted to 0.1000 V by suitable alteration of the concentrations. $[\text{Pb}^{2+}]$ or $[\text{H}^+]$ could be increased or $[\text{Cr}^{3+}]$ could be decreased, or any combination of the three of these.

(b) To produce a cell with a voltage of 2.500 V requires that one start with two half-cells whose reduction potentials differ by about that much. An interesting pair follows.

Oxidation: $\text{Al(s)} \rightarrow \text{Al}^{3+}(\text{aq}) + 3\text{e}^- \qquad\qquad -E^\circ = +1.676 \text{ V}$

Reduction: $\{\text{Ag}^+(\text{aq}) + \text{e}^- \rightarrow \text{Ag(s)}\} \times 3 \qquad\qquad E^\circ = +0.800 \text{ V}$

_____

Net: $\qquad \text{Al(s)} + 3\text{Ag}^+(\text{aq}) \rightarrow \text{Al}^{3+}(\text{aq}) + 3 \text{ Ag(s)} \qquad E^\circ_{\text{cell}} = +2.476 \text{ V}$

Again, the desired voltage can be obtained by adjusting the concentrations: in this case increasing $[\text{Ag}^+]$ and/or decreasing $[\text{Al}^{3+}]$.

(c) Since no known pair of half-cells has a potential difference larger than about 6 volts, we conclude that producing a single cell with a potential of 10.00 V is currently impossible. It is possible, however, to join several cells together into a battery that delivers a voltage of 10.00 V. Four of the cells from part (b) would do quite nicely.

Electrochemical Mechanism of Corrosion

63. **(a)** Because copper is a less active metal than is iron, this situation would be similar to that of an iron or steel can plated with tin in which the coating has been scratched. Oxidation of iron metal to Fe^{2+}(aq) should be enhanced in the body of the nail (blue precipitate), and hydroxide ion should be produced in the vicinity of the copper wire (pink color), which serves as the cathode.

(b) Because a scratch stresses the iron and exposes "fresh" metal, it is more susceptible to corrosion. We expect enhanced blue precipitate in the vicinity of the scratch.

(c) Zinc should protect the iron nail from corrosion. There should be almost no blue precipitate; the zinc corrodes instead. The pink color of OH^- continues to form.

64. The anode reaction, that of oxidation, is the formation of Fe^{2+}(aq). This occurs far below the water line. The cathode reaction, that of reduction, is the formation of OH^-(aq) from O_2(g). It is logical that this reaction would occur at or near the water line. This reduction reaction requires O_2(g) from the atmosphere and H_2O(l) from the water. The oxidation reaction, on the other hand simply requires iron from the pipe and also an aqueous solution into which the Fe^{2+}(aq) can disperse and not build up to such a high concentration that corrosion is inhibited.

Anode, Oxidation: $Fe(s) \rightarrow Fe^{2+}(aq) + 2e^-$

Cathode, Reduction: $O_2(g) + 2H_2O(l) + 4e^- \rightarrow 4OH^-(aq)$

65. During the process of corrosion, the metal that corrodes loses electrons. Thus, the metal in these instances behaves as an anode and, hence, carries a negative charge. One way in which we could retard oxidation of the metal would be to convert it into a cathode. Once transformed into a cathode, the metal would develop a positive charge and no longer release electrons (or oxidize). This change in polarity can be accomplished by hooking up the metal to an inert electrode in the ground and then applying a voltage across the two metals in such a way that the inert electrode becomes the anode and the metal that needs protecting becomes the cathode. This way, any oxidation that occurs will take place at the negatively charged inert electrode rather than the positively charged metal electrode.

66. As soon as the iron and copper came into contact, an electrochemical cell was created, in which the more powerfully reducing metal (Fe) was oxidized. In this way, the iron behaved as a sacrificial anode, protecting the copper from corrosion. The two half-reactions and the net cell reaction are shown below:

Anode(Oxidation)	$Fe(s) \rightarrow Fe^{2+}(aq) + 2\ e^-$	$-E° = 0.440$ V
Cathode(Reduction)	$\underline{Cu^{2+}(aq) + 2\ e^- \rightarrow Cu(s)}$	$\underline{E° = 0.337\ \text{V}}$
Net:	$Fe(s) + Cu^{2+}(aq) \rightarrow Fe^{2+}(aq) + Cu(s)$	$E°_{cell} = 0.777$ V

Note that because of the presence of iron and its electrical contact with the copper, any copper that does corrode is reduced back to the metal.

Electrolysis Reactions

67. We determine the standard cell voltage of each chemical reaction. Those voltages that are negative are those of chemical reactions that require electrolysis.

(a) Oxidation: $2\,H_2O(l) \rightarrow 4H^+(aq) + O_2(g) + 4e^-$ $-E° = -1.229\ V$

Reduction: $\{2\,H^+(aq) + 2e^- \rightarrow H_2(g)\} \times 2$ $E° = 0.000\ V$

Net: $2\,H_2O(l) \rightarrow 2\,H_2(g) + O_2(g)$ $E°_{cell} = -1.229\ V$

This reaction requires electrolysis, with an applied voltage greater than $+1.229V$.

(b) Oxidation: $Zn(s) \rightarrow Zn^{2+}(aq) + 2e^-$ $-E° = +0.763\ V$

Reduction: $Fe^{2+}(aq) + 2e^- \rightarrow Fe(s)$ $E° = -0.440\ V$

Net: $Zn(s) + Fe^{2+}(aq) \rightarrow Fe(s) + Zn^{2+}(aq)$ $E°_{cell} = +0.323\ V$

This is a spontaneous reaction.

(c) Oxidation: $\{Fe^{2+}(aq) \rightarrow Fe^{3+}(aq) + e^-\} \times 2$ $-E° = -0.771\ V$

Reduction: $I_2(s) + 2e^- \rightarrow 2I^-(aq)$ $E° = +0.535\ V$

Net: $2\,Fe^{2+}(aq) + I_2(s) \rightarrow 2\,Fe^{3+}(aq) + 2I^-(aq)$ $E°_{cell} = -0.236\ V$

This reaction requires electrolysis, with an applied voltage greater than $+0.236V$.

(d) Oxidation: $Cu(s) \rightarrow Cu^{2+}(aq) + 2e^-$ $-E° = -0.337\ V$

Reduction: $\{Sn^{4+}(aq) + 2e^- \rightarrow Sn^{2+}(aq)\} \times 2$ $E° = +0.154\ V$

Net: $Cu(s) + Sn^{4+}(aq) \rightarrow Cu^{2+}(aq) + Sn^{2+}(aq)$ $E°_{cell} = -0.183\ V$

This reaction requires electrolysis, with an applied voltage greater than $+0.183V$.

68. (a) Because oxidation occurs at the anode, we know that the product cannot be H_2 (since it is produced from H_2O in a reduction reaction), SO_2 (which is a reduction product of SO_4^{2-}), or SO_3 (which is produced from SO_4^{2-} without a change of oxidation state; it is the dehydration product of H_2SO_4). But O_2 indeed is the result of the oxidation of H_2O.

(b) Reduction should occur at the cathode. The possible species that can be reduced are H_2O to $H_2(g)$, $K^+(aq)$ to $K(s)$, and $SO_4^{2-}(aq)$ to perhaps $SO_2(g)$. Because potassium is a highly active metal, it will not be produced in aqueous solution. In order for $SO_4^{2-}(aq)$ to be reduced, a negatively charged ion would have to be attracted to the negatively charged cathode, which should not occur. Thus, $H_2(g)$ is produced at the cathode.

(c) At the anode: $2\,H_2O(l) \rightarrow 4H^+(aq) + O_2(g) + 4\,e^-$ $-E^\circ = -1.229\,V$

At the cathode: $\{2H^+(aq) + 2\,e^- \rightarrow H_2(g)\}\ \times 2$ $E^\circ = 0.000\,V$

Net cell reaction: $2\,H_2O(l) \rightarrow 2\,H_2(g) + O_2(g)$ $E^\circ_{cell} = -1.229\,V$

A voltage greater than 1.229 V is required. Because of the high overpotential required for the formation of gases, we expect that a higher voltage will be necessary.

69. **(a)** The two gases that are produced are $H_2(g)$ and $O_2(g)$.

(b) At the anode: $2\,H_2O(l) \rightarrow 4\,H^+(aq) + O_2(g) + 4\,e^-$ $-E^\circ = -1.229\,V$

At the cathode: $\{2H^+(aq) + 2\,e^- \rightarrow H_2(g)\}\ \times 2$ $E^\circ = 0.000\,V$

Net cell reaction: $2\,H_2O(l) \rightarrow 2\,H_2(g) + O_2(g)$ $E^\circ_{cell} = -1.229\,V$

70. The product of the electrolysis of $Na_2SO_4(aq)$ at the anode is oxygen, $-E^\circ\{O_2(g)/H_2O\} = -1.229$ V. The other possible product is $S_2O_8^{2-}(aq)$, which will not form, since it has a considerably less favorable half-cell potential, $-E^\circ\{S_2O_8^{2-}(aq)/SO_4^{2-}(aq)\} = -2.01$ V. $H_2(g)$ is formed at the cathode.

$$\text{mol } O_2 = 3.75\,h \times \frac{3600\,s}{1\,h} \times \frac{2.83\,C}{1\,s} \times \frac{1\,\text{mol } e^-}{96485\ C} \times \frac{1\,\text{mol } O_2}{4\,\text{mol } e^-} = 0.0990\ \text{mol } O_2$$

The vapor pressure of water at $25^\circ C$, from Table 13-2, is 23.8 mmHg.

$$V = \frac{nRT}{P} = \frac{0.0990\ \text{mol} \times 0.08206\ L\,atm\,mol^{-1}\,K^{-1} \times 298\ K}{(742 - 23.8)\ mmHg \times \dfrac{1\ atm}{760\ mmHg}} = 2.56\ L\ O_2(g)$$

71. **(a)** $Zn^{2+}(aq) + 2\,e^- \rightarrow Zn(s)$

$$\text{mass of } Zn = 42.5\ \text{min} \times \frac{60\,s}{1\,\text{min}} \times \frac{1.87\ C}{1\,s} \times \frac{1\,\text{mol } e^-}{96485\ C} \times \frac{1\,\text{mol } Zn}{2\,\text{mol } e^-} \times \frac{65.39\,g\ Zn}{1\,\text{mol } Zn} = 1.62\ g\ Zn$$

(b) $2I^-(aq) \rightarrow I_2(s) + 2\,e^-$

$$\text{time needed} = 2.79\,g\,I_2 \times \frac{1\,\text{mol}\,I_2}{253.8\,g\,I_2} \times \frac{2\ \text{mol } e^-}{1\,\text{mol}\,I_2} \times \frac{96485C}{1\ \text{mol } e^-} \times \frac{1\,s}{1.75\,C} \times \frac{1\,\text{min}}{60\,s} = 20.2\ \text{min}$$

72. **(a)** $Cu^{2+}(aq) + 2e^- \rightarrow Cu(s)$

$$\text{mmol } Cu^{2+}\text{removed} = 282 \text{ s} \times \frac{2.68 \text{ C}}{1 \text{ s}} \times \frac{1 \text{ mol } e^-}{96485 \text{ C}} \times \frac{1 \text{ mol } Cu^{2+}}{2 \text{ mol } e^-} \times \frac{1000 \text{ mmol}}{1 \text{ mol}}$$

$$= 3.92 \text{ mmol } Cu^{2+}$$

$$\text{decrease in } \left[Cu^{2+}\right] = \frac{3.92 \text{ mmol } Cu^{2+}}{425 \text{ mL}} = 0.00922 \text{ M}$$

$$\text{final } \left[Cu^{2+}\right] = 0.366 \text{ M} - 0.00922 \text{ M} = 0.357 \text{ M}$$

(b) $\text{mmol } Ag^+\text{removed} = 255 \text{ mL}(0.196 \text{ M} - 0.175 \text{ M}) = 5.3_6 \text{ mmol } Ag^+$

$$\text{time needed} = 5.3_6 \text{ mmol } Ag^+ \times \frac{1 \text{ mol } Ag^+}{1000 \text{ mmol } Ag^+} \times \frac{1 \text{ mol } e^-}{1 \text{ mol } Ag^+} \times \frac{96485 \text{ C}}{1 \text{ mol } e^-} \times \frac{1 \text{ s}}{1.84 \text{ C}}$$

$$= 28_1 \text{ s} \approx 2.8 \times 10^2 \text{ s}$$

73. **(a)** $\text{charge} = 1.206 \text{ g } Ag \times \frac{1 \text{ mol } Ag}{107.87 \text{ g } Ag} \times \frac{1 \text{ mol } e^-}{1 \text{ mol } Ag} \times \frac{96,485 \text{ C}}{1 \text{ mol } e^-} = 1079 \text{ C}$

(b) $\text{current} = \dfrac{1079 \text{ C}}{1412 \text{ s}} = 0.7642 \text{ A}$

74. **(a)** Anode, Oxidation: $2H_2O(l) \rightarrow 4H^+(aq) + 4e^- + O_2(g)$ $-E^\circ = -1.229\text{V}$

Cathode, Reduction: $\{Ag^+(aq) + e^- \rightarrow Ag(s)\} \times 4$ $E^\circ = +0.800 \text{ V}$

Net: $2H_2O(l) + 4Ag^+(aq) \rightarrow 4H^+(aq) + O_2(g) + 4Ag(s)$ $E^\circ_{cell} = -0.429 \text{ V}$

(b) $\text{charge} = (25.8639 - 25.0782) \text{ g } Ag \times \frac{1 \text{ mol } Ag}{107.87 \text{ g } Ag} \times \frac{1 \text{ mol } e^-}{1 \text{ mol } Ag} \times \frac{96485 \text{ C}}{1 \text{ mol } e^-} = 702.8 \text{ C}$

$$\text{current} = \frac{702.8 \text{ C}}{2.00 \text{ h}} \times \frac{1 \text{h}}{3600 \text{ s}} = 0.0976 \text{ A}$$

(c) The gas is oxygen.

$$V = \frac{nRT}{P} = \frac{\left(702.8 \text{ C} \times \dfrac{1 \text{ mol } e^-}{96485 \text{ C}} \times \dfrac{1 \text{ mol } O_2}{4 \text{ mol } e^-}\right) 0.08206 \dfrac{\text{L atm}}{\text{mol K}} (23^\circ\text{C} + 273) \text{ K}}{755 \text{ mmHg} \times \dfrac{1 \text{ atm}}{760 \text{ mmHg}}}$$

$$= 0.0445 \text{ L } O_2 \times \frac{1000 \text{ mL}}{1 \text{ L}} = 44.5 \text{ mL of } O_2$$

FEATURE PROBLEMS

98. **(a)** Anode: $H_2(g, 1 \text{ atm}) \rightarrow 2H^+(1 \text{ M}) + 2e^-$ $\qquad\qquad$ $-E° = -0.0000\,\text{V}$

Cathode: $\{Ag^+(x\,\text{M}) + e^- \rightarrow Ag(s)\} \times 2$ $\qquad\qquad$ $E° = 0.800\,\text{V}$

Net: $H_2(g, 1\,\text{atm}) + 2Ag^+(aq) \rightarrow 2H^+(1\,\text{M}) + 2Ag(s)$ $\qquad$ $E°_{cell} = 0.800$

(b) Since the voltage in the anode half-cell remains constant, we use the Nernst equation to calculate the half-cell voltage in the cathode half-cell, with two moles of electrons. This is then added to $-E$ for the anode half-cell. Because $-E° = 0.000$ for the anode half cell, $E_{cell} = E_{cathode}$

$$E = E° - \frac{0.0592}{2}\log\frac{1}{\left[Ag^+\right]^2} = 0.800 + 0.0592\,\log\left[Ag^+\right] = 0.800 + 0.0592\log x$$

(c) **(i)** Initially $\left[Ag^+\right] = 0.0100$; $\quad E = 0.800 + 0.0592\log 0.0100 = 0.682\,\text{V} = E_{cell}$

Note that 50.0 mL of titrant is required for the titration, since both $AgNO_3$ and KI have the same concentrations and they react in equimolar ratios.

(ii) After 20.0 mL of titrant is added, the total volume of solution is 70.0 mL and the unreacted Ag^+ is that in the untitrated 30.0 mL of 0.0100 M $AgNO_3(aq)$.

$$\left[Ag^+\right] = \frac{30.0\,\text{mL} \times 0.0100\,\text{M Ag}^+}{70.0\,\text{mL}} = 0.00429\text{M}$$

$E = 0.800 + 0.0592\log(0.00429) = 0.660\,\text{V} = E_{cell}$

(iii) After 49.0 mL of titrant is added, the total volume of solution is 99.0 mL and the unreacted Ag^+ is that in the untitrated 1.0 mL of 0.0100 M $AgNO_3(aq)$.

$$\left[Ag^+\right] = \frac{1.0\,\text{mL} \times 0.0100\,\text{M Ag}^+}{99.0\,\text{mL}} = 0.00010\,\text{M}$$

$E = 0.800 + 0.0592\log(0.00010) = 0.563\,\text{V} = E_{cell}$

(iv) *At the equivalence point,* we have a saturated solution of AgI, for which

$$\left[Ag^+\right] = \sqrt{K_{sp}(AgI)} = \sqrt{8.5\times10^{-17}} = 9.2\times10^{-9}$$

$E = 0.800 + 0.0592\log(9.2\times10^{-9}) = 0.324\,\text{V} = E_{cell}$.

After the equivalence point, the $\left[Ag^+\right]$ is determined by the $\left[I^-\right]$ resulting from the excess KI(aq).

(v) When 51.0 mL of titrant is added, the total volume of solution is 101.0 mL and the excess I^- is that in 1.0 mL of 0.0100 M KI(aq).

$$\left[I^-\right] = \frac{1.0 \text{ mL} \times 0.0100 \text{ M } I^-}{101.0 \text{ mL}} = 0.000099 \text{ M}$$

$$\left[Ag^+\right] = \frac{K_{sp}}{\left[I^-\right]} = \frac{8.5 \times 10^{-17}}{0.000099} = 8.6 \times 10^{-13} \text{ M}$$

$$E = 0.800 + 0.0592 \log\left(8.6 \times 10^{-13}\right) = 0.086 \text{ V} = E_{cell}$$

(vi) When 60.0 mL of titrant is added, the total volume of solution is 110.0 mL and the excess I^- is that in 10.0 mL of 0.0100 M KI(aq).

$$\left[I^-\right] = \frac{10.0 \text{ mL} \times 0.0100 \text{ M } I^-}{110.0 \text{ mL}} = 0.00091 \text{ M}$$

$$\left[Ag^+\right] = \frac{K_{sp}}{\left[I^-\right]} = \frac{8.5 \times 10^{-17}}{0.00091} = 9.3 \times 10^{-14} \text{ M}$$

$$E = 0.800 + 0.0592 \log\left(9.3 \times 10^{-14}\right) = 0.029 \text{ V} = E_{cell}$$

(d) The titration curve is presented below.

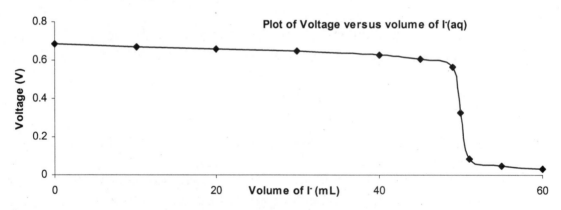

99. **(a)** **(1)** anode: $Na(s) \rightarrow Na^+(\text{in ethylamine}) + e^-$

cathode: $Na^+(\text{in ethylamine}) \rightarrow Na(\text{amalg, 0.206 \%})$

Net: $Na(s) \rightarrow Na(\text{amalg, 0.206\%})$

(2) anode: $2Na(\text{amalg, 0.206\%}) \rightarrow 2Na^+(1\text{ M}) + 2e^-$

cathode: $2H^+(aq, 1M) + 2e^- \rightarrow H_2(g, 1 \text{ atm})$

Net: $2Na\ (\text{amalg, 0.206 \%}) + 2H^+(aq, 1M) \rightarrow 2\ Na^+(1 \text{ M}) + H_2(g, 1 \text{ atm})$

(b) **(1)** $\Delta G = -1 \text{ mol e}^- \times \dfrac{96,485 \text{ C}}{1 \text{mol e}^-} \times 0.8453 \text{ V} = -8.156 \times 10^4 \text{ J or } -81.56 \text{ kJ}$

(2) $\Delta G = -2 \text{ mol e}^- \times \dfrac{96,485 \text{ C}}{1 \text{ mol e}^-} \times 1.8673 \text{ V} = -36.033 \times 10^4 \text{ J or } -360.33 \text{ kJ}$

(c) **(1)** $2\text{Na(s)} \rightarrow 2\text{Na(amalg, 0.206\%)}$ $\qquad\qquad$ $\Delta G_1 = -2 \times 8.156 \times 10^4 \text{ J}$

(2) $2\text{Na (amalg, 0.206\%)} + 2\text{H}^+(\text{aq,1M}) \rightarrow 2\text{Na}^+(1\text{M}) + \text{H}_2(\text{g, 1 atm})$

$\Delta G_2 = -36.033 \times 10^4 \text{ J}$

Overall: $2\text{Na(s)} + 2\text{H}^+(\text{aq}) \rightarrow 2\text{Na}^+(1\text{M}) + \text{H}_2(\text{g, 1 atm})$

$\Delta G = \Delta G_1 + \Delta G_2 = -16.312 \times 10^4 \text{ J} - 36.033 \times 10^4 \text{ J} = -52.345 \times 10^4 \text{ J or } -523.45 \text{ kJ}$

Since standard conditions are implied in the overall reaction, $\Delta G = \Delta G^\circ$.

(d) $E^\circ = -\dfrac{-52.345 \times 10^4 \text{ J}}{2 \text{ mol e}^- \times \dfrac{96,485 \text{ C}}{1 \text{ mol e}^-} \times \dfrac{1 \text{ J}}{1 \text{ V} \cdot \text{C}}} = +2.713 \text{ V}$

$+2.713 \text{ V} = E^\circ \left\{ \text{H}^+ \left(1 \text{ M}\right)/\text{H}_2 \left(1 \text{ atm}\right) \right\} - E^\circ \left\{ \text{Na}^+ \left(1 \text{ M}\right)/\text{Na(s)} \right\}$

$\qquad\qquad\quad = 0.000 \text{ V} - E^\circ \left\{ \text{Na}^+ \left(1 \text{ M}\right)/\text{Na(s)} \right\}$

$E^\circ \left\{ \text{Na}^+ \left(1 \text{ M}\right)/\text{Na(s)} \right\} = -2.713 \text{ V}$. This is precisely the value in Appendix D.

100. The question marks in the original Latimer diagram have been replaced with letters in the diagram below to make the solution easier to follow:

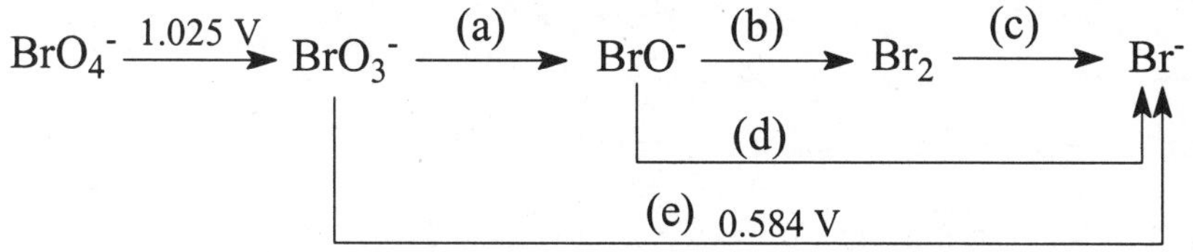

$$\text{BrO}_4^- \xrightarrow{\ 1.025 \text{ V}\ } \text{BrO}_3^- \xrightarrow{\ (a)\ } \text{BrO}^- \xrightarrow{\ (b)\ } \text{Br}_2 \xrightarrow{\ (c)\ } \text{Br}^-$$

$$(d)$$

$$(e) \quad 0.584 \text{ V}$$

By using Appendix D and the methods of page 834 we obtain:

(c) $\text{Br}_2(\text{l}) + 2\text{e}^- \rightarrow 2\text{Br}^-(\text{aq})$ $\qquad\qquad\qquad\qquad$ $E_c^\circ = 1.065 \text{ V}$ (Appendix D)

(d) $\text{BrO}^-(\text{aq}) + \text{H}_2\text{O(l)} + 2\text{e}^- \rightarrow \text{Br}^-(\text{aq}) + 2\text{OH}^-(\text{aq})$ $\qquad$ $E_d^\circ = 0.766 \text{ V}$ (Appendix D)

(b) $2\text{BrO}^-(\text{aq}) + 2\text{H}_2\text{O(l)} + 2\text{e}^- \rightarrow \text{Br}_2(\text{l}) + 4\text{OH}^-(\text{aq})$ $\qquad$ $E_b^\circ = 0.455 \text{ V}$ (Appendix D)

(a) $\{BrO_3^-(aq) + 2H_2O(l) + 4e^- \rightarrow BrO^-(aq) + 4OH^-(aq)\}\ \times 2$ $E_a^{\,\circ} = ?$

(b) $2BrO^-(aq) + 2H_2O(l) + 2e^- \rightarrow Br_2(l) + 4OH^-(aq)$ $E_b^{\,\circ} = 0.455\,V$

(c) $Br_2(l) + 2e^- \rightarrow 2Br^-(aq)$ $E_c^{\,\circ} = 1.065\,V$

(e) $2BrO_3^-(aq) + 6H_2O(l) + 12e^- \rightarrow 12OH^-(aq) + 2Br^-(aq)$ $E_e^{\,\circ} = 0.584\,V$

$$\Delta G_{Total}^{\,\circ} = \Delta G_{(a)}^{\,\circ} + \Delta G_{(b)}^{\,\circ} + \Delta G_{(c)}^{\,\circ}$$

$$-12F\left(E_e^{\,\circ}\right) = -8F\left(E_a^{\,\circ}\right) + -2F\left(E_b^{\,\circ}\right) + -2F\left(E_c^{\,\circ}\right)$$

$$-12F\left(0.584\,V\right) = -8F\left(E_a^{\,\circ}\right) + -2F\left(0.455\,V\right) + -2F\left(1.065\,V\right)$$

$$E_a^{\,\circ} = \frac{-12F\left(0.584\,V\right) + 2F\left(0.455\,V\right) + 2F\left(1.065\,V\right)}{-8F}$$

$$E_a^{\,\circ} = 0.496\,V$$

101. (a)   The capacitance of the cell membrane is given by the following equation,

$$C = \frac{\varepsilon_0 \varepsilon A}{l}$$

where $\varepsilon_0\varepsilon = 3 \times 8.854 \times 10^{-12}\ C^2\,N^{-1}\,m^{-2}$; $A = 1 \times 10^{-6}\ cm^2$; and $l = 1 \times 10^{-6}\ cm$. Together with the factors necessary to convert from cm to m and from cm^2 to m^2, these data yield

$$C = \frac{(3)\left(8.854\times10^{-12}\ \dfrac{C^2}{N^1\,m^2}\right)(1\times10^{-6}\ cm^2)\left(\dfrac{1\ m}{100\ cm}\right)^2}{(1\times10^{-6}cm)\left(\dfrac{1\ m}{100\ cm}\right)}$$

$$C = 2.66 \times 10^{-13}\ \frac{C^2}{N\,m}$$

$$C = \left(2.66\times10^{-13}\ \frac{C^2}{N\,m}\right)\left(\frac{1F}{1\dfrac{C^2}{N\,m}}\right) = 2.66\times10^{-13}\ F$$

(b)   Since the capacitance C is the charge in coulombs per volt, the charge on the membrane, Q, is given by the product of the capacitance and the potential across the cell membrane.

$$Q = 2.66 \times 10^{-13} \frac{C}{V} \times 0.085 \text{ V} = 2.26 \times 10^{-14} \text{ C}$$

(c)   The number of K^+ ions required to produce this charge is

$$\frac{Q}{e} = \frac{2.26 \times 10^{-14} \text{ C}}{1.602 \times 10^{-19} \text{ C/ion}} = 1.41 \times 10^5 \text{ K}^+ \text{ ions}$$

(d)   The number of K^+ ions in a typical cell is

$$\left(6.022 \times 10^{23} \frac{\text{ions}}{\text{mol}}\right)\left(155 \times 10^{-3} \frac{\text{mol}}{\text{L}}\right)\left(\frac{1\text{L}}{1000 \text{ cm}^3}\right)(1 \times 10^{-8} \text{ cm}^3) = 9.3 \times 10^{11} \text{ ions}$$

(e)   The fraction of the ions involved in establishing the charge on the cell membrane is

$$\frac{1.4 \times 10^5 \text{ ions}}{9.3 \times 10^{11} \text{ ions}} = 1.5 \times 10^{-7} \ (\sim 0.000015 \text{ %})$$

Thus, the concentration of K^+ ions in the cell remains constant at 155 mM.

CHAPTER 22
MAIN GROUP ELEMENTS I: METALS
PRACTICE EXAMPLES

1A   From Figure 22-2, the route from sodium chloride to sodium nitrate begins with electrolysis of NaCl(aq) to form NaOH(aq)

$$2\,NaCl(aq) + 2\,H_2O(l) \xrightarrow{\text{electrolysis}} 2\,NaOH(aq) + H_2(g) + Cl_2(g) \text{ followed by addition of}$$

$NO_2(g)$ to NaOH(aq). $2\,NaOH(aq) + 3\,NO_2(g) \rightarrow 2\,NaNO_3(aq) + NO(g) + H_2O(l)$

1B   From Figure 22-2, we see that the route from sodium chloride to sodium thiosulfate

(1) begins with the electrolysis of NaCl(aq) to produce NaOH(aq),

$$2\,NaCl(aq) + 2\,H_2O(l) \xrightarrow{\text{electrolysis}} 2\,NaOH(aq) + H_2(g) + Cl_2(g)$$

(2) continues through the reaction of $SO_2(g)$ with the NaOH(aq) in an acid-base reaction

$[SO_2(g)$ is an acid anhydride] to produce

$Na_2SO_3(aq)$: $2\,NaOH(aq) + SO_2(g) \rightarrow Na_2SO_3(aq) + H_2O(l)$ and (3) concludes with

the addition of S to the boiling solution: $Na_2SO_3(aq) + S(s) \xrightarrow{\text{boil}} Na_2S_2O_3(aq)$

2A   Moles of NaOH $= C \times V = 0.0133\ M \times 0.00759\ L = 1.01 \times 10^{-4}$ mol NaOH

$$[H_3O^+] = \frac{1.01 \times 10^{-4}\ \text{mol}}{0.0250\ \text{L}} = 4.04 \times 10^{-3}\ M \qquad pH = -\log(4.04 \times 10^{-3}); \qquad pH = 2.394$$

2B   185 ppm $Ca^{2+} \longrightarrow \dfrac{185\ mg\ Ca^{2+}}{1\ L} \times \dfrac{1\ mmol\ Ca^{2+}}{40.078\ mg\ Ca^{2+}} \times \dfrac{1\ mol\ Ca^{2+}}{1000\ mmol\ Ca^{2+}} \times \dfrac{2\ mol\ Na^+}{1\ mol\ Ca^{2+}}$

$$= 9.23 \times 10^{-3}\ M\ Na^+$$

REVIEW QUESTIONS

1.   **(a)**   A dimer is a molecule that is formed by the joining together of two identical simpler molecules (called monomers). For instance, N_2O_4 is a dimer of NO_2.

       (b)   An adduct is formed when two simple molecules come together and a covalent bond, often a coordinate covalent bond, forms between them. $NH_3 \cdot BF_3$ is an adduct of NH_3 and BF_3.

       (c)   Calcination is the process of heating calcium carbonate strongly to drive off $CO_2(g)$ and form calcium oxide.

       (d)   An amphoteric oxide is one that reacts with either a strong acid or with a strong base.

2. **(a)** A diagonal relationship refers to the similarity between two elements that are diagonally related to each other in the periodic table, such as Li and Mg, or Be and Al. Even though these elements are in different periodic families, they have some similarities in physical and chemical properties.

(b) Deionized water is prepared by ion exchange by passing it through material in which the ions H^+ or OH^- are present. These ions substitute or exchange for the ions in the water, in a two-step process: first the cations in the water are replaced by H^+, then the anions by OH^-. The result is water virtually free of ionic contaminants.

(c) The thermite reaction refers to the reduction of a metal oxide with another, more active, metal in a highly exothermic reaction, for instance:

$$Fe_2O_3(s) + 2\ Al(s) \rightarrow 2\ Fe(liquid) + Al_2O_3(s)$$

(d) The "inert pair" effect refers to the tendency of heavier representative metals to have oxidation states in which they have lost their np electrons, but not their ns^2 electrons. Thus, they have an oxidation state two units less than their periodic table family number. Examples include Pb^{2+}, Sn^{2+}, Bi^{3+}, Sb^{3+}, Tl^+.

3. **(a)** The peroxide ion is O_2^{2-}; the superoxide ion is O_2^-.

(b) Quicklime is the common name for $CaO(s)$; slaked lime is the common name for $Ca(OH)_2(s)$.

(c) Temporary hard water contains divalent cations, such as Ca^{2+}, Mg^{2+}, and Fe^{2+} and the bicarbonate anion, HCO_3^-. Heating produces, $H_2O(l)$, $CO_2(g)$, and a carbonate precipitate. Permanent hard water does not form a precipitate upon heating; the ions it contains, such as SO_4^{2-} are thermally stable.

(d) A soap is a potassium or sodium salt of a deprotonated carboxylic acid $(—COO^-)$ that has a long hydrocarbon chain. A detergent is a synthetic sodium or potassium salt of a long hydrocarbon chain sulfonic acid $(—OSO_3H)$.

4. **(a)** PbO; lead(II) oxide

(b) SnF_2; tin(II) fluoride

(c) $CaSO_4 \cdot \frac{1}{2}H_2O$; calcium sulfate hemihydrate

(d) Li_3N; lithium nitride

(e) $Ca(OH)_2$; calcium hydroxide

(f) KO_2; potassium superoxide

(g) $Mg(HCO_3)_2$; magnesium hydrogen carbonate

5. **(a)** $Li_2CO_3(s) \xrightarrow{\Delta} Li_2O(s) + CO_2(g)$

(b) $CaCO_3(s) + 2\, HCl(aq) \rightarrow CaCl_2(aq) + CO_2(g) + H_2O(l)$

(c) $2Al(s) + 2Na^+(aq) + 2OH^-(aq) + 6H_2O(l) \rightarrow 2\,Na^+(aq) + 2\left[Al(OH)_4\right]^-(aq) + 3H_2(g)$

(d) $BaO(s) + H_2O(l) \rightarrow Ba(OH)_2$ (s, in limited water)

(e) $2\,Na_2O_2(s) + 2\,CO_2(g) \rightarrow 2\,Na_2CO_3(s) + O_2(g)$

6. **(a)** $MgCO_3(s) + 2\,HCl(aq) \rightarrow MgCl_2(aq) + CO_2(g) + H_2O(l)$

(b) $2\,Na(s) + 2\,H_2O(l) \rightarrow 2\,NaOH(aq) + H_2(g)$

$2\,Al(s) + 2\,NaOH(aq) + 6\,H_2O(l) \rightarrow 2\,Na\left[Al(OH)_4\right](aq) + 3\,H_2(g)$

(c) $2\,NaCl(s) + H_2SO_4(conc., aq) \rightarrow 2\,HCl(g) + Na_2SO_4(s)$

7. **(a)** $K_2CO_3(aq) + Ba(OH)_2(aq) \rightarrow BaCO_3(s) + 2\,KOH(aq)$

(b) $Mg(HCO_3)_2(aq) \xrightarrow{\Delta} MgCO_3(s) + CO_2(g) + H_2O(g)$

(c) $SnO(s) + C(s) \xrightarrow{\Delta} Sn(l) + CO(g)$

(d) $CaF_2(s) + H_2SO_4(conc., aq) \rightarrow 2\,HF(g) + CaSO_4(s)$

(e) $NaHCO_3(s) + HCl(aq) \rightarrow NaCl(aq) + H_2O(l) + CO_2(g)$

(f) $PbO_2(s) + 4\,HBr(aq) \rightarrow PbBr_2(s) + Br_2(l) + 2\,H_2O(l)$

8. Replace the names with chemical formulas and balance the result.

$CaSO_4 \cdot 2H_2O(s) + (NH_4)_2CO_3(aq) \rightarrow (NH_4)_2SO_4(aq) + CaCO_3(s) + 2\,H_2O(l)$

9. Temporary hard water is softened by the addition of an alkaline (basic) material. All the substances listed form alkaline solutions except NH_4Cl. NH_4^+ hydrolyzes to form an acidic solution, and thus, cannot be used to soften temporary hard water.

10. **(a)** $SrCO_3(s) \xrightarrow{\Delta} SrO(s) + CO_2(g)$

(b) $Al_2O_3(s) \xrightarrow{\Delta}$ no reaction

(c) $Li_2CO_3(s) \xrightarrow{\Delta} Li_2O(s) + CO_2(g)$

11. **(a)** $Pb(NO_3)_2(aq) + 2\,NaHCO_3(aq) \rightarrow PbCO_3(s) + H_2O(l) + CO_2(g) + 2\,NaNO_3(aq)$

(b) $Li_2O(s) + (NH_4)_2CO_3(aq) \longrightarrow Li_2CO_3(s) + 2\,NH_3(g) + H_2O(l)$

$NH_4^+(aq)$ is present in very limited amount because the solution is strongly basic; Li_2O is the anhydride of a strong base.

(c) $H_2SO_4(aq) + BaO_2(aq) \longrightarrow H_2O_2(aq) + BaSO_4(s)$

(d) $2PbO(s) + Ca(OCl)_2(aq) \longrightarrow 2PbO_2(s) + CaCl_2(aq)$

12. To answer this one, you really do not need to formally study chemistry. You just have to observe the world around you. Neither aluminum foil nor aluminum cookware reacts with water. But to explain this observation, you have to know that of these four active metals, only aluminum forms an oxide that adheres tightly to its surface, thereby preventing further reaction.

13. The correct answer is (a). Balanced equations for the three pairs follow.

$Ca(s) + 2H_2O(l) \longrightarrow Ca(OH)_2(s) + H_2(g)$ $CaH_2(s) + 2H_2O(l) \longrightarrow Ca(OH)_2(s) + 2H_2(g)$

$2Na(s) + 2H_2O(l) \longrightarrow 2NaOH(aq) + H_2(g)$ $2Na_2O_2(s) + 2H_2O(l) \longrightarrow 4NaOH(aq) + O_2(g)$

$2K(s) + 2H_2O(l) \longrightarrow 2KOH(aq) + H_2(g)$ $4KO_2(s) + 2H_2O(l) \longrightarrow 4KOH(aq) + 3O_2(g)$

14. (a) Stalactites are primarily $CaCO_3(s)$

(b) Gypsum is $CaSO_4 \cdot 2H_2O$

(c) "bathtub ring" is a salt of Ca^{2+} and a long-carbon chain carboxylate anion. An example would be calcium palmitate: $Ca[CH_3(CH_2)_{14}COO]_2(s)$

(d) barium "milkshake" is an aqueous temporary suspension of $BaSO_4(s)$

(e) blue sapphires are Al_2O_3 with Fe^{3+} and Ti^{4+} ions replacing some Al^{3+} ions.

EXERCISES

Group I (Alkali) Metals

15. (a) $2Cs(s) + Cl_2(g) \longrightarrow 2CsCl(s)$

(b) $2Na(s) + O_2(g) \longrightarrow Na_2O_2(s)$

(c) $Li_2CO_3(s) \xrightarrow{\Delta} Li_2O(s) + CO_2(g)$

(d) $Na_2SO_4(s) + 4C(s) \rightarrow Na_2S(s) + 4CO(g)$

(e) $K(s) + O_2(g) \longrightarrow KO_2(s)$

16. (a) $2Rb(s) + 2H_2O(l) \longrightarrow RbOH_2(aq) + H_2(g)$

(b) $2KHCO_3(aq) \xrightarrow{\Delta} K_2CO_3(aq) + H_2O(l) + CO_2(g)$

(c) $2Li(s) + O_2(g) \longrightarrow Li_2O_2(s)$

(d) $2KCl(s) + H_2SO_4(aq) \longrightarrow K_2SO_4(aq) + 2HCl(g)$

(e) $LiH(s) + H_2O(l) \longrightarrow LiOH(aq) + H_2(g)$

17. Both LiCl and KCl are soluble in water, but Li_3PO_4 is not very soluble. Hence the addition of $K_3PO_4(aq)$ to a solution of the white solid will produce a precipitate if the white solid is LiCl, but no precipitate if the white solid is KCl. The best method is a flame test; lithium gives a red color to a flame, while the potassium flame test is violet.

18. When heated, Li_2CO_3 decomposes to $CO_2(g)$ and $Li_2O(l)$. $K_2CO_3(s)$ simply melts when heated. The evolution of $CO_2(g)$ bubbles should be sufficient indication of the difference in behavior. The flame test affords a violet flame if K^+ is present, and a red flame if Li^+ is present.

19. In addition to $OH^-(aq)$, $O_2(g)$ is formed in the case of peroxide and superoxide. The resulting equations are balanced by inspection if one pays attention to charge balance.

Oxide: $O^{2-}(aq) + H_2O(l) \longrightarrow 2\,OH^-(aq)$

Peroxide: $2\,O_2^{2-}(aq) + 2\,H_2O(l) \rightarrow 4\,OH^-(aq) + O_2(g)$

Superoxide: $4\,O_2^-(aq) + 2\,H_2O(l) \longrightarrow 4\,OH^-(aq) + 3\,O_2(g)$

20. We know sodium metal was produced at the cathode from the reduction of sodium ion, Na^+. Thus, hydroxide must have been involved in oxidation at the anode. The hydrogen in hydroxide ion already is in its highest oxidation state and thus cannot be oxidized. This leaves oxidation of the hydroxide ion to elemental oxygen as the remaining reaction.

Cathode, reduction : $\{Na^+ + e^- \longrightarrow Na(l)\} \times 4$

Anode, oxidation : $4\,OH^- \longrightarrow O_2(g) + 2\,H_2O(g) + 4e^-$

$\overline{}$

Net : $4\,Na^+ + 4\,OH^- \longrightarrow 4\,Na(l) + O_2(g) + 2\,H_2O(g)$

21. **(a)** $H_2(g)$ and $Cl_2(g)$ are produced during the electrolysis of NaCl(aq). The electrode reactions are:

Anode, Oxdn: $2\,Cl^-(aq) \longrightarrow Cl_2(g) + 2e^-$

Cathode, Redn: $2\,H_2O(l) + 2e^- \longrightarrow H_2(g) + 2\,OH^-(aq)$

We can compute the amount of OH^- produced at the cathode.

$$\text{mol } OH^- = 2.50\,min \times \frac{60\,s}{1\,min} \times \frac{0.810\,C}{1s} \times \frac{1\,mol\,e^-}{96500\,C} \times \frac{2\,mol\,OH^-}{2\,mol\,e^-} = 1.26 \times 10^{-3}\,mol\,OH^-$$

Then we compute the $[OH^-]$ and, from that, the pH of the solution.

$$[OH^-] = \frac{1.26 \times 10^{-3}\,mol\,OH^-}{0.872\,L\,soln} = 1.45 \times 10^{-3}\,M \quad pOH = -\log(1.45 \times 10^{-3}) = 2.839$$

pH$=14.000 - 2.839 = 11.161$

(b) As long as NaCl is in excess and the volume of the solution is nearly constant, the solution pH only depend on the number of electrons transferred (limiting reagent).

22. (a) total energy $= 3.0\,\text{V} \times 0.50\,\text{A h} \times \dfrac{3600\,\text{s}}{1\,\text{hr}} \times \dfrac{1\,\text{C/s}}{1\,\text{A}} \times \dfrac{1\,\text{J}}{1\,\text{V} \cdot \text{C}} = 5.4 \times 10^3\,\text{J}$

We obtained the first conversion factor for time as follows.

$$5.0\,\mu\text{W} \times \frac{1 \times 10^{-6}\,\text{W}}{1\,\mu\text{W}} \times \frac{1\,\text{J/s}}{1\,\text{W}} = \frac{5.0 \times 10^{-6}\,\text{J}}{1\,\text{s}}$$

$$\text{time} = 5.4 \times 10^3\,\text{J} \times \frac{1\,\text{s}}{5 \times 10^{-6}\,\text{J}} = 1.1 \times 10^9\,\text{s} \times \frac{1\,\text{hr}}{3600\,\text{s}} \times \frac{1\,\text{day}}{24\,\text{hr}} \times \frac{1\,\text{y}}{365\,\text{days}} = 34\,\text{y}$$

(b) The capacity of the battery is determined by the mass of Li present.

$$\text{mass Li} = 0.50\,\text{A h} \times \frac{1\,\text{C/s}}{1\,\text{A}} \times \frac{3600\,\text{s}}{1\,\text{h}} \times \frac{1\,\text{mol e}^-}{96500\,\text{C}} \times \frac{1\,\text{mol Li}}{1\,\text{mol e}^-} \times \frac{6.941\,\text{g Li}}{1\,\text{mol Li}} = 0.13\,\text{g Li}$$

23. (a) We first compute the mass of $NaHCO_3$ that should be produced from 1.00 ton NaCl, assuming that all of the Na in the NaCl ends up in the $NaHCO_3$. We use the unit, ton-mole, to simplify the calculations.

$$\text{mass NaHCO}_3 = 1.00\,\text{ton NaCl} \times \frac{1\,\text{ton-mol NaCl}}{58.4\,\text{ton NaCl}} \times \frac{1\,\text{ton-mol Na}}{1\,\text{ton-mol NaCl}}$$

$$\times \frac{1\,\text{tol-mol NaHCO}_3}{1\,\text{ton-mol Na}} \times \frac{84.0\,\text{ton NaHCO}_3}{1\,\text{ton mol NaHCO}_3} = 1.44\,\text{ton NaHCO}_3$$

$$\% \text{ yield} = \frac{1.03\,\text{ton NaHCO}_3\,\text{produced}}{1.44\,\text{ton NaHCO}_3\,\text{expected}} \times 100\% = 71.5\%\,\text{ yield}$$

(b) NH_3 is used in the principal step of the Solvay proces to produce a solution in which $NaHCO_3$ is formed and from which it will precipitate. The filtrate contains NH_4Cl, from which NH_3 is recovered by treatment with $Ca(OH)_2$. Thus, NH_3 is simply used during the Solvay process to produce the proper conditions for the desired reactions. Any net consumption of NH_3 is the result of unavoidable losses during production.

24. (a) $Ca(OH)_2(s) + SO_4^{2-}(aq) \rightleftharpoons CaSO_4(s) + 2\,OH^-(aq)$

(b) We sum two solubility reactions and combine their values of K_{sp}

$Ca(OH)_2(s) \rightleftharpoons Ca^{2+}(aq) + 2\,OH^-(aq)$ $\qquad\qquad K_{sp} = 5.5 \times 10^{-6}$

$Ca^{2+}(aq) + SO_4^{2-}(aq) \rightleftharpoons CaSO_4(s)$ $\qquad\qquad 1/K_{sp} = 1/9.1 \times 10^{-6}$

$Ca(OH)_2(s) + SO_4^{2-}(aq) \rightleftharpoons CaSO_4(s) + 2\,OH^-(aq)$ $\qquad K_{eq} = \dfrac{5.5 \times 10^{-6}}{9.1 \times 10^{-6}} = 0.60$

Because K_{eq} is not significantly different from 1.00, we conclude that the reaction lies neither very far to the right (it does not go to completion) nor to the left.

(c) Reaction: $Ca(OH)_2(s) + SO_4^{2-}(aq) \rightleftharpoons CaSO_4(s) + 2OH^-(aq)$

Initial:	1.00 M	≈ 0 M
Changes:	$-x$ M	$+2x$ M
Equil:	$(1.00-x)$ M	$2x$ M

$$K = \frac{\left[OH^-\right]^2}{\left[SO_4^{2-}\right]} = 0.60 = \frac{4x^2}{1.00-x} \qquad 4x^2 = 0.60 - 0.60x \qquad 4x^2 + 0.60x - 0.60 = 0$$

$$x = \frac{-b \pm \sqrt{b^2 - 4ac}}{2a} = \frac{-0.60 \pm \sqrt{0.36 + 9.60}}{8} = 0.32 \text{ M}$$

$$\left[SO_4^{2-}\right] = 1.00 - x = 0.68 \text{ M} \qquad \left[OH^-\right] = 2x = 0.64 \text{ M}$$

Group 2 (Alkaline earth) metals

25.

$$CaO \xleftarrow{\Delta} CaCO_3 \xleftarrow{CO_2} Ca(OH)_2 \xrightarrow{HCl} CaCl_2 \xrightarrow{electrolysis} Ca$$

$$CaHPO_4 \xleftarrow{H_3PO_4} \Big| \Big| \xrightarrow{H_2SO_4} CaSO_4$$

The reactions are as follows. $Ca(OH)_2(s) + 2\,HCl(aq) \rightarrow CaCl_2(aq) + 2H_2O(l)$

$CaCl_2(l) \xrightarrow{\Delta,\,electrolysis} Ca(l) + Cl_2(g)$ $Ca(OH)_2(s) + CO_2(g) \rightarrow CaCO_3(s) + H_2O(g)$

$CaCO_3(s) \xrightarrow{\Delta} CaO(s) + CO_2(g)$ $Ca(OH)_2(s) + H_2SO_4(aq) \rightarrow CaSO_4(s) + 2\,H_2O(l)$

$Ca(OH)_2(s) + H_3PO_4(aq) \rightarrow CaHPO_4(aq) + 2H_2O(l)$. Actually $CaO(s)$ is the industrial starting material from which $Ca(OH)_2$ is made. $CaO(s) + H_2O(l) \longrightarrow Ca(OH)_2(s)$

26.

$$MgO \xleftarrow{\Delta} MgCO_3 \xleftarrow{CO_2} Mg(OH)_2 \xrightarrow{HCl} MgCl_2 \xrightarrow{electrolysis} Mg$$

$$MgHPO_4 \xleftarrow{H_3PO_4} \Big| \Big| \xrightarrow{H_2SO_4} MgSO_4$$

Once we return to $Mg(OH)_2$ from $MgSO_4$, other substances can be made by the indicated pathways. The return reaction is:

$MgSO_4(aq) + 2\,NaOH(aq) \rightarrow Mg(OH)_2(s) + Na_2SO_4(aq)$.

Then the other reactions are

$Mg(OH)_2(s) + 2\,HCl(aq) \rightarrow MgCl_2(aq) + 2\,H_2O(l)$

$MgCl_2(l) \xrightarrow{\Delta,\,electrolysis} Mg(l) + Cl_2(g)$; $Mg(OH)_2(s) + CO_2(g) \rightarrow MgCO_3(s) + H_2O(g)$

$MgCO_3(s) \xrightarrow{\Delta} MgO(s) + CO_2(g)$; $Mg(OH)_2(s) + H_2SO_4(aq) \rightarrow MgSO_4(s) + 2H_2O(l)$

$$Mg(OH)_2(s)+H_3PO_4(aq) \rightarrow MgHPO_4(aq)+2H_2O(l)$$

27. The reactions involved:

$$Mg^{2+}(aq) + Ca^{2+}(aq) + 2OH^-(aq) \rightarrow Mg(OH)_2(s) + Ca^{2+}(aq)$$

$$Mg(OH)_2(s) + 2\,H^+(aq) + 2\,Cl^-(aq) \rightarrow Mg^{2+}(aq) + 2\,H_2O(l) + 2\,Cl^-(aq)$$

$$Mg^{2+}(aq) + 2\,Cl^-(aq) \xrightarrow{\Delta} MgCl_2(s)$$

$$MgCl_2(s) \text{ — (electrolysis)} \longrightarrow Mg(l) + Cl_2(g)$$

$$\underline{Mg(l) \rightarrow Mg(s)}$$

$$Mg^{2+} + 2\,Cl^-(aq) \rightarrow Mg(s) + Cl_2(g) \quad \text{(Overall reaction)}$$

As can be seen, the process does not violate the principle of conservation of charge.

28. (a) MgO vs BaO: MgO would have the higher melting point because although Mg^{2+} and Ba^{2+} have the same charge, Mg^{2+} is a smaller ion. Smaller ions have a larger electrostatic attraction to anions (here, in both cases the anion is O^{2-}). This is due to the smaller charge separation (Coulomb's Law).

(b) MgF_2 vs $MgCl_2$ solubility in water. F^- is smaller than Cl^-, hence, electrostatic attraction between Mg^{2+} and the halide is greater in F^- than Cl^-. If we assume that hydration of the ions is similar, we expect that MgF_2 is less soluble than $MgCl_2$. (Note: K_{sp} given for MgF_2 which is sparingly soluble, while no K_{sp} value is given for readily soluble $MgCl_2$).

29. (a) $BeF_2(s) + Mg(s) \xrightarrow{\Delta} Be(s) + MgF_2(s)$

(b) $Ba(s) + Br_2(l) \longrightarrow BaBr_2(s)$

(c) $UO_2(s) + 2\,Ca(s) \longrightarrow U(s) + 2\,CaO(s)$

(d) $MgCO_3 \cdot CaCO_3(s) \xrightarrow{\Delta} MgO(s) + CaO(s) + 2\,CO_2(g)$

(e) $2\,H_3PO_4(aq) + 3\,CaO(s) \longrightarrow Ca_3(PO_4)_2(s) + 3\,H_2O(l)$

30. (a) $Mg(HCO_3)_2(s) \xrightarrow{heat} MgO(s) + 2\,CO_2(g) + H_2O(l)$

(b) $BaCl_2(l) \xrightarrow{electrolysis} Ba(l) + Cl_2(g)$

(c) $Sr(s) + 2\,HBr(aq) \longrightarrow SrBr_2(aq) + H_2(g)$

(d) $H_2SO_4(aq) + Ca(OH)_2(aq) \longrightarrow CaSO_4(s) + 2H_2O(l)$

(e) $CaSO_4 \cdot 2\,H_2O(s) \xrightarrow{\Delta} CaSO_4 \cdot \frac{1}{2}H_2O(s) + \frac{3}{2}\,H_2O(g)$

31. Let us compute the value of the equilibrium constant for each reaction by combining the two solubility product constants. Large values of equilibrium constants indicate that the reaction is displaced far to the right. Values of K that are much smaller than 1 indicate that the reaction is displaced far to the left.

(a) $BaSO_4(s) \rightleftharpoons Ba^{2+}(aq) + SO_4^{2-}(aq)$ $K_{sp} = 1.1 \times 10^{-10}$

   $Ba^{2+}(aq) + CO_3^{2-}(aq) \rightleftharpoons BaCO_3(s)$ $1/K_{sp} = 1/(5.1 \times 10^{-9})$

   $BaSO_4(s) + CO_3^{2-}(aq) \rightleftharpoons BaCO_3(s) + SO_4^{2-}(aq)$ $K = \dfrac{1.1 \times 10^{-10}}{5.1 \times 10^{-9}} = 2.2 \times 10^{-2}$

   Equilibrium lies slightly to the left.

(b) $Mg_3(PO_4)_2(s) \rightleftharpoons 3\, Mg^{2+}(aq) + 2\, PO_4^{3-}(aq)$ $K_{sp} = 2.1 \times 10^{-25}$

   $3\{Mg^{2+}(aq) + CO_3^{2-}(aq) \rightleftharpoons MgCO_3(s)\}$ $1/(K_{sp})^3 = 1/(3.5 \times 10^{-8})^3$

   $Mg_3(PO_4)_2(s) + 3\,CO_3^{2-}(aq) \rightleftharpoons 3\, MgCO_3(s) + 2\, PO_4^{3-}(aq)$

   $K = \dfrac{2.1 \times 10^{-25}}{\left(3.5 \times 10^{-8}\right)^3} = 4.9 \times 10^{-3}$

   Equilibrium lies to the left

(c) $Ca(OH)_2(s) \rightleftharpoons Ca^{2+}(aq) + 2\, OH^-(aq)$ $K_{sp} = 5.5 \times 10^{-6}$

   $Ca^{2+}(aq) + 2\, F^-(aq) \rightleftharpoons CaF_2(s)$ $1/K_{sp} = 1/(5.3 \times 10^{-9})$

   $Ca(OH)_2(s) + 2\,F^-(aq) \rightleftharpoons CaF_2(s) + 2\, OH^-(aq)$ $K = \dfrac{5.5 \times 10^{-6}}{5.3 \times 10^{-9}} = 1.0 \times 10^3$

   Equilibrium lies to the right

32. We expect the reaction to occur to a significant extent if its equilibrium constant is $\gg 1$.

(a) $BaCO_3(s) \rightleftharpoons Ba^{2+}(aq) + CO_3^{2-}(aq)$ $K_{sp} = 5.1 \times 10^{-9}$

   $2\, HC_2H_3O_2(aq) \rightleftharpoons 2H^+(aq) + 2C_2H_3O_2^-(aq)$ $K_a^2 = \left(1.8 \times 10^{-5}\right)^2$

   $H^+(aq) + CO_3^{2-}(aq) \rightleftharpoons HCO_3^-(aq)$ $1/K_{a_2} = 1/(4.7 \times 10^{-11})$

   $H^+(aq) + HCO_3^-(aq) \rightleftharpoons H_2CO_3(aq)$ $1/K_{a_1} = 1/(4.2 \times 10^{-7})$

   $BaCO_3(s) + 2\, HC_2H_3O_2(aq) \rightleftharpoons Ba(C_2H_3O_2)_2(aq) + H_2CO_3(aq)$

   $K = \dfrac{5.1 \times 10^{-9} \times \left(1.8 \times 10^{-5}\right)^2}{4.7 \times 10^{-11} \times 4.2 \times 10^{-7}} = 8.4 \times 10^{-2}$

   This reaction would not occur to any significant extent.

(b) $Ca(OH)_2(s) \rightleftharpoons Ca^{2+}(aq) + 2\,OH^-(aq)$ $K_{sp} = 5.5 \times 10^{-6}$

$2\,NH_4^+(aq) + 2\,OH^-(aq) \rightleftharpoons 2\,NH_3(aq) + 2\,H_2O(l)$ $1/K_b^2 = 1/(1.8 \times 10^{-5})^2$

$$Ca(OH)_2(s) + 2\,NH_4^+(aq) \rightleftharpoons Ca^{2+}(aq) + 2\,NH_3(aq) + 2\,H_2O(l)$$

$$K = \frac{5.5 \times 10^{-6}}{(1.8 \times 10^{-5})^2} = 1.7 \times 10^4$$

This reaction would occur to a significant extent.

(c) $BaF_2(s) \rightleftharpoons Ba^{2+}(aq) + 2\,F^-(aq)$ $K_{sp} = 1.0 \times 10^{-6}$

$2\,H_3O^+(aq) + 2\,F^-(aq) \rightleftharpoons 2\,HF(aq) + 2\,H_2O(l)$ $1/K_a^2 = 1/(6.6 \times 10^{-4})^2$

$$BaF_2(s) + 2H_3O^+(aq) \rightleftharpoons Ba^{2+}(aq) + 2HF(aq) + 2H_2O(l); \quad K = \frac{1.0 \times 10^{-6}}{(6.6 \times 10^{-4})^2} = 2.3$$

This reaction would occur to some extent, certainly not to completion.

Hard Water

33.   Temporary hard water contains HCO_3^-. Quicklime is CaO:

$$CaO(s) + H_2O(l) \longrightarrow Ca(OH)_2(aq)$$

$$2\,HCO_3^-(aq) + Ca(OH)_2(aq) \xrightarrow{\Delta} CaCO_3(s) + H_2O(l) + CO_2(g) + 2\,OH^-(aq)$$

$$2\,HCO_3^-(aq) + CaO(s) \xrightarrow{\Delta} CaCO_3(s) + CO_2(g) + 2\,OH^-(aq)$$

34.   Temporary hard water contains HCO_3^- and divalent cations such as Ca^{2+}

$$2\,H_2O(l) + 2\,NH_3(aq) \longrightarrow 2\,OH^-(aq) + 2\,NH_4^+(aq)$$

$$2\,OH^-(aq) + Ca^{2+}(aq) \longrightarrow Ca(OH)_2(aq)$$

$$2\,HCO_3^-(aq) + Ca(OH)_2(aq) \xrightarrow{\Delta} CaCO_3(s) + H_2O(l) + CO_2(g) + 2\,OH^-(aq)$$

$$H_2O(l) + 2NH_3(aq) + Ca^{2+}(aq) + 2HCO_3^-(aq) \xrightarrow{\Delta} 2NH_4^+(aq) + CaCO_3(s) + CO_2(g) + 2OH^-(aq)$$

35.   **(a)**   There will be two HCO_3^- ions for each Ca^{2+} ion.

$$\frac{g\ Ca^{2+}}{10^6\ g\ soln} = \frac{185.0\ g\ HCO_3^-}{10^6\ g\ soln} \times \frac{1\ mol\ HCO_3^-}{61.02\ g\ HCO_3^-} \times \frac{1\ mol\ Ca^{2+}}{2\ mol\ HCO_3^-} \times \frac{40.08\ g\ Ca^{2+}}{1\ mol\ Ca^{2+}}$$

$$= \frac{60.76\ g\ Ca^{2+}}{10^6\ g\ soln} = 60.76\ ppm\ Ca^{2+}$$

(b)
$$\frac{mg\ Ca^{2+}}{1\ L} = \frac{60.76\ g\ Ca^{2+}}{10^6\ g\ soln} \times \frac{1.00\ g\ soln}{1\ mL} \times \frac{1000\ mL}{1\ L} \times \frac{1000\ mg}{1\ g} = 60.76\ mg\ Ca^{2+}/L$$

(c) First we combine equations representing the neutralization of bicarbonate ion with hydroxide ion and the formation of a generalized carbonate precipitate $[MCO_3(s)]$ to determine the overall stoichiometry of the reaction. We also add to the combination the slaking of lime, CaO(s). Remember that two HCO_3^- ions are associated with each $M^{2+}(aq)$ ion.

$$CaO(s) + H_2O(l) \rightarrow Ca(OH)_2(s)$$

$$Ca(OH)_2(s) \xrightarrow{H_2O} Ca^{2+}(aq) + 2\ OH^-(aq)$$

$$2\{HCO_3^-(aq) + OH^-(aq) \longrightarrow H_2O(l) + CO_3^{2-}(aq)\}$$

$$CO_3^{2-}(aq) + M^{2+}(aq) \longrightarrow MCO_3(s)$$

$$CO_3^{2-}(aq) + Ca^{2+}(aq) \longrightarrow CaCO_3(s)$$

$$\overline{Ca(OH)_2(s) + M^{2+}(aq) + 2\ HCO_3^-(aq) \longrightarrow CaCO_3(s) + MCO_3(s) + 2\ H_2O(l)}$$

Then we determine the mass of $Ca(OH)_2$ needed.

$$mass\ Ca(OH)_2 = 1.00 \times 10^6\ g\ water \times \frac{185.0\ g\ HCO_3^-}{10^6\ g\ water} \times \frac{1\ mol\ HCO_3^-}{61.02\ g\ HCO_3^-}$$

$$\times \frac{1\ mol\ Ca(OH)_2}{2\ mol\ HCO_3^-} \times \frac{74.09\ g\ Ca(OH)_2}{1\ mol\ Ca(OH)_2} = 112.3\ g\ Ca(OH)_2$$

36. (a)
$$mass\ CaCO_3 = 112.3 \times 10^5\ g\ Ca(OH)_2 \times \frac{1\ mol\ Ca(OH)_2}{74.09\ g\ Ca(OH)_2} \times \frac{2\ mol\ CaCO_3}{1\ mol\ Ca(OH)_2}$$

$$\times \frac{100.1\ g\ CaCO_3}{1\ mol\ CaCO_3} = 303.4\ g\ CaCO_3$$

(b) It is clear from the equations that are summed in the answer to Exercise 35, and especially the net equation (that follows), that half the Ca^{2+} is derived from the CaO and half from the hard water itself. This assumes all $M^{2+} = Ca^{2+}$ in question 35 above.

$$Ca(OH)_2(s) + M^{2+}(aq) + 2\ HCO_3^-(aq) \longrightarrow CaCO_3(s) + MCO_3(s) + 2\ H_2O(l)$$

37. pH $= 2.37 = -\log[H^+]$. The $[H^+] = 10^{-2.37} = 4.27 \times 10^{-3}$ M. For every 2 moles of H^+, one mole of Ca^{2+} is absorbed. Hence, the concentration of Ca^{2+} is 2.13×10^{-3} M. Convert this to parts per million (mg Ca^{2+} per 1000 g solution).

$$\text{ppm Ca}^{2+} = \frac{2.13 \times 10^{-3} \text{ mol Ca}^{2+}}{1 \text{ L solution}} \times \frac{1 \text{ L solution}}{1000 \text{ g solution}} \times \frac{40.08 \text{ g Ca}^{2+}}{1 \text{ mol Ca}^{2+}} \times \frac{1000 \text{ mg Ca}^{2+}}{1 \text{ g Ca}^{2+}} = \frac{85 \text{ mg Ca}^{2+}}{1000 \text{ g solution}}$$

The concentration of Ca^{2+} is 85 ppm.

38. Let us determine the $[OH^-]$ in this solution first.

$$2 \text{ OH}^- (aq) + H_2SO_4 (aq) \rightarrow 2 H_2O(l) + SO_4^{2-} (aq)$$

$$[OH^-] = \frac{22.42 \text{ mL acid} \times \dfrac{1.00 \times 10^{-3} \text{ mmol } H_2SO_4}{1 \text{ mL acid}} \times \dfrac{2 \text{ mmol OH}^-}{1 \text{ mmol } H_2SO_4}}{25.00 \text{ mL base}} = 1.79 \times 10^{-3} \text{ M}$$

Now we compute the number of grams of $CaSO_4$ in 10^6 g of hard water. This is the ppm hardness. The anion exchange reaction is

$$R(OH)_2 (s) + SO_4^{2-} (aq) \rightarrow RSO_4 (s) + 2 OH^- (aq)$$

$$\text{mass } CaSO_4 = 10^6 \text{ g water} \times \frac{1 \text{ mL } H_2O}{1.00 \text{ g } H_2O} \times \frac{1 \text{ L } H_2O}{1000 \text{ mL } H_2O} \times \frac{1.79 \times 10^{-3} \text{ mol OH}^-}{1 \text{ L}} \times \frac{1 \text{ mol } SO_4^{2-}}{2 \text{ mol OH}^-}$$

$$\times \frac{1 \text{ mol } CaSO_4}{1 \text{ mol } SO_4^{2-}} \times \frac{136.1 \text{ g } CaSO_4}{1 \text{ mol } CaSO_4} = 122 \text{ g } CaSO_4 \quad (122 \text{ ppm } CaSO_4)$$

39. We first write the balanced equation for the formation of bathtub ring.

$$Ca^{2+} (aq) + 2 CH_3(CH_2)_{16}COO^-K^+ (aq) \rightarrow Ca^{2+}\left[CH_3(CH_2)_{16}COO^-\right]_2 (s, \text{"ring"}) + 2 K^+ (aq)$$

$$\text{"ring" mass} = 32.1 \text{ L} \times \frac{82.6 \text{ g } Ca^{2+}}{1000 \text{ L water}} \times \frac{1 \text{ mol } Ca^{2+}}{40.08 \text{ g } Ca^{2+}} \times \frac{1 \text{ mol "ring"}}{1 \text{ mol } Ca^{2+}} \times \frac{607.0 \text{ g "ring"}}{1 \text{ mol "ring"}}$$

$$= 40.2 \text{ g of bathtub ring}$$

40. We assume that the density of the solution is 1.00 g/mL. The solubility of magnesium palmitate equals the concentration of Mg^{2+}, $[Mg^{2+}]$, because there is one mole Mg^{2+} in each mole of compound.

$$[Mg^{2+}] = \frac{80 \text{ g } Mg^{2+}}{10^6 \text{ g soln}} \times \frac{1 \text{ mol } Mg^{2+}}{24.3 \text{ g } Mg^{2+}} \times \frac{1.00 \text{ g soln}}{1 \text{ mL soln}} \times \frac{10^3 \text{ mL soln}}{1 \text{ L soln}} = 3.3 \times 10^{-3} \text{ M} = s$$

From this, we calculate the value of the solubility product. Stoichiometry gives

$$[Palm^-] = 2[Mg^{2+}];$$

$$K_{sp} = [Mg^{2+}][Palm^-]^2 = (s)(2s)^2 = 4s^3 = 4(3.3 \times 10^{-3})^3 = 1.4 \times 10^{-7}$$

A Group 13 Metal: Aluminum

41. **(a)** $2 \text{ Al(s)} + 6 \text{ HCl(aq)} \rightarrow 2 \text{ AlCl}_3(\text{aq}) + 3 \text{ H}_2(\text{g})$

(b) $2 \text{ NaOH(aq)} + 2 \text{ Al(s)} + 6 \text{ H}_2\text{O(l)} \rightarrow 2 \text{ Na}^+(\text{aq}) + 2\left[\text{Al(OH)}_4\right]^-(\text{aq}) + 3 \text{ H}_2(\text{g})$

(c) Oxidation: $\{\text{Al(s)} \longrightarrow \text{Al}^{3+}(\text{aq}) + 3 \text{ e}^-\}$ $\qquad\qquad\qquad$ $\times 2$
Reduction: $\{\text{SO}_4^{2-}(\text{aq}) + 4 \text{ H}^+(\text{aq}) + 2\text{e}^- \longrightarrow \text{SO}_2(\text{g}) + 2 \text{ H}_2\text{O(l)}\}$ $\qquad$ $\times 3$

$$\text{Net: } 2 \text{ Al(s)} + 3 \text{ SO}_4^{2-}(\text{aq}) + 12 \text{ H}^+(\text{aq}) \longrightarrow 2 \text{ Al}^{3+}(\text{aq}) + 3 \text{ SO}_2(\text{aq}) + 6 \text{ H}_2\text{O(l)}$$

42. **(a)** $2 \text{ Al(s)} + 3 \text{ Br}_2(\text{l}) \longrightarrow 2 \text{ AlBr}_3(\text{s})$

(b) $2 \text{ Al(s)} + \text{Cr}_2\text{O}_3(\text{s}) \xrightarrow{\text{heat}} 2 \text{ Cr(l)} + \text{Al}_2\text{O}_3(\text{s})$

(c) $\text{Fe}_2\text{O}_3(\text{s}) + \text{OH}^-(\text{aq}) \longrightarrow \text{no reaction}$

$\text{Al}_2\text{O}_3(\text{s}) + 2 \text{ OH}^-(\text{aq}) + 3 \text{ H}_2\text{O(l)} \longrightarrow 2\left[\text{Al(OH)}_4\right]^-(\text{aq})$

43. One method of analyzing this reaction is to envision the HCO_3^- ion as a combination of CO_2 and OH^-. Then the OH^- reacts with Al^{3+} and forms Al(OH)_3. [This method of envisioning HCO_3^- is somewhat of a trick, but it does have its basis in reality. After all,

$$\text{H}_2\text{CO}_3\left(= \text{H}_2\text{O} + \text{CO}_2\right) + \text{OH}^- \longrightarrow \text{HCO}_3^- + \text{H}_2\text{O}$$

$$\text{Al}^{3+}(\text{aq}) + 3 \text{ HCO}_3^-(\text{aq}) \longrightarrow \text{Al(OH)}_3(\text{s}) + 3 \text{ CO}_2(\text{g})$$

Another method is to consider the reaction as, first, the hydrolysis of hydrated aluminum ion to produce $\text{Al(OH)}_3(\text{s})$ and an acidic solution, followed by the reaction of the acid with bicarbonate ion.

$$\left[\text{Al(H}_2\text{O)}_6\right]^{3+}(\text{aq}) + 3 \text{ H}_2\text{O(l)} \longrightarrow \text{Al(OH)}_3\left(\text{H}_2\text{O}\right)_3(\text{s}) + 3 \text{ H}_3\text{O}^+(\text{aq})$$

$$3 \text{ H}_3\text{O}^+(\text{aq}) + 3 \text{ HCO}_3^-(\text{aq}) \longrightarrow 6 \text{ H}_2\text{O(l)} + 3 \text{ CO}_2(\text{g})$$

This gives the same net reaction:

$$\left[\text{Al(H}_2\text{O)}_6\right]^{3+}(\text{aq}) + 3 \text{ HCO}_3^-(\text{aq}) \longrightarrow \text{Al(OH)}_3\left(\text{H}_2\text{O}\right)_3(\text{s}) + 3 \text{ CO}_2(\text{g}) + 3 \text{ H}_2\text{O(l)}$$

44. The $\text{Al}^{3+}(\text{aq})$ ion hydrolyzes. $\text{Al}^{3+}(\text{aq}) + 3 \text{ H}_2\text{O(l)} \longrightarrow \text{Al(OH)}_3(\text{s}) + 3 \text{ H}^+(\text{aq})$

And the hydrogen ion that is produced reacts with bicarbonate ion to liberate $\text{CO}_2(\text{g})$.

$$\text{H}^+(\text{aq}) + \text{HCO}_3^-(\text{aq}) \longrightarrow \text{H}_2\text{O(l)} + \text{CO}_2(\text{g})$$

45. Aluminum and its oxide are soluble in both acid and in base.

$$2\ Al(s) + 6\ H^+(aq) \longrightarrow 2\ Al^{3+}(aq) + 3\ H_2(g)$$

$$Al_2O_3(s) + 6\ H^+(aq) \longrightarrow 2\ Al^{3+}(aq) + 3\ H_2O(l)$$

$$2\ Al(s) + 2\ OH^-(aq) + 6\ H_2O(l) \longrightarrow 2\left[Al(OH)_4\right]^-(aq) + 3\ H_2(g)$$

$$Al_2O_3(s) + 2\ OH^-(aq) + 3\ H_2O(l) \longrightarrow 2\left[Al(OH)_4\right]^-(aq)$$

Al(s) is resistant to corrosion only over the pH range 4.5 to 8.5. Thus, aluminum is non-reactive only when the medium to which it is exposed is neither highly acidic nor highly basic.

46. Both Al and Mg are attacked by acid and their ions are both precipitated by hydroxide ion.

$$2\ Al(s) + 6\ H^+(aq) \longrightarrow 2\ Al^{3+}(aq) + 3\ H_2(g) \quad Mg(s) + 2\ H^+(aq) \longrightarrow Mg^{2+}(aq) + H_2(g)$$

$$Al^{3+}(aq) + 3\ OH^-(aq) \longrightarrow Al(OH)_3(s) \quad Mg^{2+}(aq) + 2\ OH^- \longrightarrow Mg(OH)_2(s)$$

But, of these two solid hydroxides, only $Al(OH)_3(s)$ redissolves in excess $OH^-(aq)$.

$$Al(OH)_3(s) + OH^-(aq) \longrightarrow \left[Al(OH)_4\right]^-(aq)$$

Thus, the analytical procedure consists of dissolving the sample in HCl(aq) and then treating the resulting solution with NaOH(aq) until a precipitate forms. If this precipitate dissolves completely when excess NaOH(aq) is added, the sample was aluminum 2S. If at least some of the precipitate does not dissolve, the sample was magnalium.

47. $CO_2(g)$ is, of course, the anhydride of an acid. The reaction here is an acid-base reaction.

$$\left[Al(OH)_4\right]^-(aq) + CO_2(aq) \longrightarrow Al(OH)_3(s) + HCO_3^-(aq)$$

HCl(aq), being a strong acid, can't be used because it will dissolve the Al(OH)₃(s).

48. (a) Oersted: $2\ Al_2O_3(s) + 3\ C(s) + 6\ Cl_2(g) \xrightarrow{\Delta} 4\ AlCl_3(s) + 3\ CO_2(g)$

(b) Wöhler: $AlCl_3(s) + 3\ K(s) \xrightarrow{\Delta} Al(s) + 3\ KCl(s)$

49. 2 KOH(aq) + 2 Al(s) + 6 H₂O(l) → 2 K[Al(OH)₄](aq) + 3 H₂(g)

2 K[Al(OH)₄](aq) + 4 H₂SO₄(aq)→ K₂SO₄(aq) + Al₂(SO₄)₃(aq) — crystallize→ 2 KAl(SO₄)₂

50. HCO_3^- and CO_3^{2-} solutions are basic (producing OH^-). Al^{3+}(aq) in the presence of OH^-(aq) will precipitate as the hydroxide Al(OH)₃(s).

Some Group 14 Metals: Tin and Lead

51. **(a)** $PbO(s) + 2 HNO_3(aq) \longrightarrow Pb(NO_3)_2(s) + H_2O(l)$

(b) $SnCO_3(s) \xrightarrow{\Delta} SnO(s) + CO_2(g)$

(c) $PbO(s) + C(s) \xrightarrow{\Delta} Pb(l) + CO(g)$

(d) $2 Fe^{3+}(aq) + Sn^{2+}(aq) \longrightarrow 2 Fe^{2+}(aq) + Sn^{4+}(aq)$

(e) $2 PbS(s) + 3 O_2(g) \xrightarrow{\Delta} 2 PbO(s) + 2 SO_2(g)$
$2 SO_2(g) + O_2(g) \longrightarrow 2 SO_3(g)$
$SO_3(g) + PbO(s) \longrightarrow PbSO_4(s)$

Or perhaps simply: $PbS(s) + 2 O_2(g) \longrightarrow PbSO_4(s)$

Yet a third possibility:
$PbO(s) + SO_2(s) \longrightarrow PbSO_3(s)$ followed by
$2 PbSO_3(s) + O_2(s) \longrightarrow 2PbSO_4(s)$

52. **(a)** Treat tin(II) oxide with hydrochloric acid.
$SnO(s) + 2 HCl(aq) \longrightarrow SnCl_2(aq) + H_2O(l)$

(b) Attack tin with chlorine. $Sn(s) + 2 Cl_2(g) \longrightarrow SnCl_4(s)$

(c) First we dissolve $PbO_2(s)$ in $HNO_3(aq)$ and then treat the resulting solution with $K_2CrO_4(aq)$ to precipitate $PbCrO_4(s)$
$2 PbO_2(s) + 4 HNO_3(aq) \longrightarrow 2 Pb(NO_3)_2(aq) + O_2(g) + 2 H_2O(l)$
$Pb(NO_3)_2(aq) + K_2CrO_4(aq) \longrightarrow PbCrO_4(s) + 2 KNO_3(aq)$

53. We use the Nernst equation to determine whether the cell voltage still is positive when the reaction has gone to completion.

(a) Oxidation: $Fe^{2+}(aq) \rightarrow Fe^{3+}(aq) + e^-$} $\times 2$ $\qquad -E° = -0.771 V$
Reduction:
$PbO_2(s) + 4 H^+(aq) + 2 e^- \longrightarrow Pb^{2+}(aq) + 2 H_2O(l)$ $\qquad E° = +1.455 V$

Net: $2 Fe^{2+}(aq) + PbO_2(s) + 4 H^+(aq) \longrightarrow 2 Fe^{3+}(aq) + Pb^{2+}(aq) + 2 H_2O(l)$

$E°_{cell} = -0.771 V + 1.455 V = +0.684 V$

In this case, when the reaction has gone to completion,

$\left[Fe^{2+} \right] = 0.001 M, \left[Fe^{3+} \right] = 0.999 M,$ and $\left[Pb^{2+} \right] = 0.500 M.$

$$E_{cell} = E^{\circ}_{cell} - \frac{0.0592}{2} \log \frac{[Fe^{3+}]^2[Pb^{2+}]}{[Fe^{2+}]^2}$$

$$= 0.684\ V - \frac{0.0592}{2} \log \frac{[0.999]^2[0.500]}{[0.001]^2} = 0.684\ V - 0.169\ V = 0.515\ V.$$

Yes, this reaction will go to completion.

(b) Oxidation: $2\ SO_4^{2-}(aq) \longrightarrow S_2O_8^{2-}(aq) + 2\ e^-$ $-E^{\circ} = -2.01V$

Reduction: $PbO_2(s) + 4\ H^+(aq) + 2\ e^- \longrightarrow Pb^{2+}(aq) + 2\ H_2O(l)$ $E^{\circ} = +1.455\ V$

$2\ SO_4^{2-}(aq) + PbO_2(s) + 4\ H^+(aq) \longrightarrow S_2O_8^{2-}(aq) + Pb^{2+}(aq) + 2\ H_2O(l)$

$E^{\circ}_{cell} = -2.01 + 1.455 = -0.56\ V$ This reaction is not even spontaneous initially.

(c) Oxidation: $\{Mn^{2+}(1 \times 10^4\ M) + 4H_2O(l) \longrightarrow MnO_4^-(aq) + 8H^+(aq) + 5\ e^-\} \times 2$;

$-E^{\circ} = -1.51\ V$

Reduction: $\{PbO_2(s) + 4\ H^+(aq) + 2\ e^- \longrightarrow Pb^{2+}(aq) + 2\ H_2O(l)\} \times 5$ $E^{\circ} = +1.455\ V$

Net: $2Mn^{2+}(1 \times 10^4\ M) + 5PbO_2(s) + 4H^+ \longrightarrow 2MnO_4^-(aq) + 5Pb^{2+}(aq) + 2H_2O(l)$

$E^{\circ}_{cell} = -1.51 + 1.455 = -0.06\ V$. The standard cell potential indicates that this reaction is not spontaneous when all concentrations are 1 M. Since the concentration of a reactant (Mn^{2+}) is lower than 1.00 M, this reaction is even less spontaneous than the standard cell potential indicates.

54. A positive value of E°_{cell} indicates that a reaction should occur.

(a) Oxidation : $Sn^{2+}(aq) \longrightarrow Sn^{4+}(aq) + 2\ e^-$ $-E^{\circ} = -0.154\ V$

Reduction : $I_2(s) + 2\ e^- \longrightarrow 2\ I^-(aq)$ $E^{\circ} = +0.535\ V$

Net : $Sn^{2+}(aq) + I_2(s) \longrightarrow Sn^{4+}(aq) + 2\ I^-(aq)$ $E^{\circ}_{cell} = +0.381\ V$

Yes, $Sn^{2+}(aq)$ will reduce I_2 to I^-.

(b) Oxidation : $Sn^{2+}(aq) \longrightarrow Sn^{4+}(aq) + 2\ e^-$ $-E^{\circ} = -0.154\ V$

Reduction : $Fe^{2+}(aq) + 2\ e^- \longrightarrow Fe(s)$ $E^{\circ} = -0.440\ V$

Net : $Sn^{2+}(aq) + Fe^{2+}(aq) \longrightarrow Sn^{4+}(aq) + Fe(s)$ $E^{\circ}_{cell} = -0.594\ V$

No, $Sn^{2+}(aq)$ will not reduce $Fe^{2+}(aq)$ to Fe(s).

(c) Oxidation : $Sn^{2+}(aq) \longrightarrow Sn^{4+}(aq) + 2\ e^-$ $-E^{\circ} = -0.154\ V$

Reduction : $Cu^{2+}(aq) + 2\ e^- \longrightarrow Cu(s)$ $E^{\circ} = +0.337\ V$

Net : $Sn^{2+}(aq) + Cu^{2+}(aq) \longrightarrow Sn^{4+}(aq) + Cu(s)$ $E^{\circ}_{cell} = +0.183\ V$

Yes, $Sn^{2+}(aq)$ will reduce $Cu^{2+}(aq)$ to $Cu(s)$.

(d) Oxidation : $Sn^{2+}(aq) \longrightarrow Sn^{4+}(aq) + 2\ e^-$ $\qquad\qquad -E° = -0.154$ V

Reduction : $\{Fe^{3+}(aq) + e^- \longrightarrow Fe^{2+}(aq)\}\ \times 2$ $\qquad E° = +0.771$ V

Net : $\quad Sn^{2+}(aq) + 2\ Fe^{3+}(aq) \longrightarrow Sn^{4+}(aq) + 2\ Fe^{2+}(aq)$ $E°_{cell} = +0.617$ V

Yes, $Sn^{2+}(aq)$ will reduce $Fe^{3+}(aq)$ to $Fe^{2+}(aq)$.

Feature Problems

71. $Li(s) \rightarrow Li(g)$ $\qquad\qquad$ 159.4 kJ; (Using $\Delta H°_{rxn} = \Sigma\Delta H_f°_{products} - \Sigma\Delta H_f°_{reactants}$ Appendix D)

$Li(g) \rightarrow Li^+(g) + e^-$ $\quad$ 520.2 kJ; (Data given in Table 22.2, Chapter 22)

$\underline{Li^+(g) \rightarrow Li^+(aq)}$ $\quad\underline{-506\ kJ;\ (Given\ in\ question)}$

$Li(s) \rightarrow Li^+(aq)\ \ + e^-$ 174 kJ

$1/2\ H_2(g) \rightarrow H(g)$ $\qquad$ 218.0 kJ $\quad$ (Using $\Delta H°_{rxn} = \Sigma\Delta H_f°_{products} - \Sigma\Delta H_f°_{reactants}$ Appendix D)

$H(g) \rightarrow H^+(g) + e^-$ $\qquad$ 1312 kJ $\quad$ (Use $R_H(N_A)$Bohr theory = 2.179×10^{-18} J(6.022×10^{23}))

$\underline{H^+(g) \rightarrow H^+(aq)}$ $\qquad\underline{-1079\ kJ}$ $\quad$(Given in question)

$1/2\ H_2(g) \rightarrow H^+(aq) + e^-$ $\quad$ 451 kJ

(a) $Li(s) + H^+(aq) \rightarrow Li^+(aq) + 1/2\ H_2(g)$ $\quad \Delta H° = 174$ kJ $- 451$ kJ $= -277$ kJ $\approx \Delta G°$

$$E° = \frac{\Delta G°}{-nF} = \frac{-277 \times 10^3\ J}{-1\ mol\ e^-\left(96{,}485\ \dfrac{C}{mol\ e^-}\right)} = 2.87\ \frac{J}{C} = 2.87\ V\ \ (3.040\ V\ in\ Appendix\ D)$$

(b) $\Delta S = \Sigma S°_{products} - \Sigma S°_{reactants}$

$= [1\ mol\left(13.4\ \dfrac{J}{K\,mol}\right) + 0.5\ mol\left(130.7\ \dfrac{J}{K\,mol}\right)] -$

$[1\ mol\left(29.12\ \dfrac{J}{K\,mol}\right) + 1\ mol\left(0\ \dfrac{J}{K\,mol}\right)]$

$= 49.6\ \dfrac{J}{K}$

$\Delta G° = \Delta H° - T\Delta S° = -277\ kJ - 298.15\ K\left(49.6\ \dfrac{J}{K}\right) \times \dfrac{1\,kJ}{1000\,J} = -292 \times 10^3\ J$

$$E° = \frac{\Delta G°}{-nF} = \frac{-292 \times 10^3\ J}{-1\ mol\ e^-\left(96{,}485\ \dfrac{C}{mol\ e^-}\right)} = 3.03\ \frac{J}{C} = 3.03\ V\ \ (3.040\ V\ in\ Appendix\ D)$$

72.   **(a)**

$$\frac{0.438 \text{ mol NaCl}}{\text{L}} \times \frac{58.443 \text{ g NaCl}}{1 \text{ mol NaCl}} = 26.6 \text{ g NaCl}$$

$$\frac{0.0512 \text{ mol MgCl}_2}{\text{L}} \times \frac{95.211 \text{ g MgCl}_2}{1 \text{ mol MgCl}_2} = 4.87 \text{ g MgCl}_2$$

> 31.5 g of salt per liter of seawater

$$18 \text{ tbs NaHCO}_3 \times \frac{10 \text{ g NaHCO}_3}{1 \text{ tbs NaHCO}_3} = 180 \text{ g NaHCO}_3$$

$$10 \text{ tbs NaCl} \times \frac{10 \text{ g NaCl}}{1 \text{ tbs NaCl}} \doteq 100 \text{ g NaCl}$$

> 307 g of salt per gal. of lake water (3.7854 L/gal)

$$8 \text{ tsp MgSO}_4 \bullet 7 \text{ H}_2\text{O} \times \frac{1 \text{ tbs}}{3 \text{ tsp}} \times \frac{10 \text{ g MgSO}_4 \bullet 7 \text{ H}_2\text{O}}{1 \text{ tbs MgSO}_4 \bullet 7 \text{ H}_2\text{O}} = 27 \text{ g MgSO}_4 \bullet 7 \text{ H}_2\text{O}$$

(b)   The pH of the lake will be determined by the amphiprotic bicarbonate ion, (HCO_3^-), which hydrolyzes in water ($K_{a_1} = 4.4 \times 10^{-7}$ or $pK_{a_1} = 6.36$ and $K_{a_2} = 4.7 \times 10^{-11}$ or $pK_{a_2} = 10.33$). We saw earlier (chapter 18 question 100) that the pH of a solution of alanine, a diprotic species is independent of concentration (as long as it is relatively concentrated). The pH $= \dfrac{(pK_{a_1} + pK_{a_2})}{2} = \dfrac{(6.36 + 10.33)}{2} = 8.35$

This is not as basic as the actual pH of the lake. Addition of Borax (sodium salt of boric acid) would aid in increasing the pH of the solution (since borax is the salt of a weak acid, it is a base). The lake may be more basic due to the presence of other basic anions, namely carbonate ion (CO_3^{2-}).

(c)   Tufa are mostly calcium carbonate ($CaCO_3(s)$). Since they form near underwater springs, one must assume that the springs have a high concentration of calcium ion ($Ca^{2+}(aq)$). We then couple this with the fact that the lake has a high salinity (high carbonate and bicarbonate ion content). We can assume that two major reactions are responsible for the formation of a tufa (see below):

$$Ca^{2+}(aq) + CO_3^{2-}(aq) \rightarrow CaCO_3(s)$$

$$Ca^{2+}(aq) + 2 \text{ HCO}_3^-(aq) \rightarrow Ca(HCO_3)_2(q) \rightarrow CaCO_3(s) + H_2O(l) + CO_2(g)$$

CHAPTER 23
MAIN-GROUP ELEMENTS II: NONMETALS
PRACTICE EXAMPLES

1A This question involves calculating $E°$ for the reduction half-reaction:

$ClO_3^-(aq) + 3 H_2O(l) + 6 e^- \rightarrow Cl^-(aq) + 6 OH^-(aq)$

Here we will consider just one of the several approaches available to solve this problem. The four half-reactions (and their associated $E°$ values) that are used in this method to come up with the "missing $E°$ value" are given below. (Note: the $E°$ for the first reaction was determined in example 23-1)

1) $ClO_3^-(aq) + H_2O(l) + 2 e^- \rightarrow ClO_2^-(aq) + 2 OH^-(aq)$ $E° = 0.295$ V $\Delta G°_1 = -2FE°$

2) $ClO_2^-(aq) + H_2O(l) + 2 e^- \rightarrow OCl^-(aq) + 2 OH^-(aq)$ $E° = 0.681$ V $\Delta G°_2 = -2FE°$

3) $OCl^-(aq) + H_2O(l) + 1 e^- \rightarrow 1/2 Cl_2(aq) + 2 OH^-(aq)$ $E° = 0.421$ V $\Delta G°_3 = -1FE°$

4) $1/2 Cl_2(aq) + 1 e^- \rightarrow Cl^-(aq)$ $E° = 1.358$ V $\Delta G°_4 = -1FE°$

Although the reactions themselves may be added to obtain the desired equation, the $E°$ for this equation is not the sum of the $E°$ values for the above four reactions. The $E°$ values for the desired equation is actually the weighted average of the $E°$ values for reactions 1) to 4). It can be calculated by summing up the free energy changes for the four reactions (the standard voltages for half-reactions of the same type are not additive. The $\Delta G°$ values for these reactions can, however, be summed together).

When 1), 2), 3) and 4) are added together we obtain

$\quad$ 1) $\quad ClO_3^-(aq) + H_2O(l) + 2 e^- \rightarrow ClO_2^-(aq) + 2 OH^-(aq)$ $E°_1 = 0.295$ V

$+$ 2) $\quad ClO_2^-(aq) + H_2O(l) + 2 e^- \rightarrow OCl^-(aq) + 2 OH^-(aq)$ $E°_2 = 0.681$ V

$+$ 3) $\quad OCl^-(aq) + H_2O(l) + 1 e^- \rightarrow 1/2 Cl_2(aq) + 2 OH^-(aq)$ $E°_3 = 0.421$ V

$+$ 4) $\quad 1/2 Cl_2(aq) + 1 e^- \rightarrow Cl^-(aq)$ $E°_4 = 1.358$ V

$\rule{8cm}{0.4pt}$

$\qquad ClO_3^-(aq) + 3 H_2O(l) + 6 e^- \rightarrow Cl^-(aq) + 6 OH^-(aq)$ $E°_5 = ?$

and $\quad \Delta G°_5 = \Delta G°_1 + \Delta G°_2 + \Delta G°_3 + \Delta G°_4$

so, $\quad -6FE°_5 = -2F(0.295 \text{ V}) + -2F(0.681\text{V}) + -1F(0.421 \text{ V}) + -1F(1.358 \text{ V})$

Hence, $E°_5 = \dfrac{-2F(0.295 \text{ V}) + -2F(0.681\text{V}) + -1F(0.421 \text{V}) + -1F(1.358 \text{ V})}{-6F} = 0.622$ V

1B. This question involves calculating $E°$ for the reduction half-reaction:

$$2\,ClO_3^-(aq) + 12H^+(aq) + 10\,e^- \rightarrow Cl_2(aq) + 6\,H_2O(l)$$

The three half-reactions (and their associated $E°$ values) are given below:

(1) $\quad ClO_3^-(aq) + 3H^+(aq) + 2\,e^- \rightarrow HClO_2(aq) + H_2O(l) \qquad E° = 1.181\ V$

(2) $\quad HClO_2(aq) + 2\,H^+(aq) + 2\,e^- \rightarrow HClO(aq) + H_2O(l) \quad E° = 1.645\ V$

(3) $\quad 2\,HClO(aq) + 2\,H^+(aq) + 2\,e^- \rightarrow Cl_2(aq) + 2\,H_2O(l) \quad E° = 1.611\ V$
  (remember that $\Delta G° = -nFE°$)

Although the reactions themselves can be added to obtain the desired equation, the $E°$ for this equation is not the sum of the $E°$ values for the above three reactions. The $E°$ for the desired equation is actually the weighted average of the $E°$ values for reactions (1) to (3). It can be obtained by summing up the free energy changes for the three reactions. (For reactions of the same type, standard voltages are not additive; $\Delta G°$ values are additive, however). When (1) and (2) (each multiplied by two) are added to (3) we obtain:

$2\times(1)\ 2\,ClO_3^-(aq) + 6\,H^+(aq) + 4\,e^- \rightarrow 2\,HClO_2(aq) + 2\,H_2O(l) \qquad E°_1 = 1.181\ V$

$2\times(2)\ 2\,HClO_2(aq) + 4\,H^+(aq) + 4\,e^- \rightarrow 2\,HClO(aq) + 2\,H_2O(l) \qquad E°_2 = 1.645\ V$

$(3)\quad 2\,HClO(aq) + 2\,H^+(aq) + 2\,e^- \rightarrow Cl_2(aq) + 2\,H_2O(l) \qquad\qquad E°_3 = 1.611\ V$

$\overline{\qquad\qquad\qquad\qquad\qquad\qquad\qquad\qquad\qquad\qquad\qquad\qquad\qquad\qquad\qquad}$

$\qquad 2\,ClO_3^-(aq) + 12\,H^+(aq) + 10\,e^- \rightarrow Cl_2(aq) + 6\,H_2O(l) \qquad E°_4 = ?$

and $\qquad \Delta G°_4 = \Delta G°_1 + \Delta G°_2 + \Delta G°_3$

so, $\qquad -10FE°_4 = -4F(1.181\ V) + -4F(1.645\ V) + -2F(1.611\ V)$

Hence, $E°_4 = \dfrac{-4F(1.181\ V) + -4F(1.645\ V) + -2F(1.611\ V)}{-10\,F} = 1.453\ V$

2A The dissociation reaction is the reverse of the formation reaction and thus $\Delta G°$ for the dissociation reaction is the negative of $\Delta G°_f$

$$HF(g) \longrightarrow \tfrac{1}{2}H_2(g) + \tfrac{1}{2}F_2(g) \quad \Delta G° = -\Delta G°_f = -(-273.2\ kJ/mol)$$

We know that

$$\Delta G° = -RT\ln K_P \qquad +273.2\times10^3\ J/mol = -8.3145\ J\ mol^{-1}\ K^{-1} \times 298\ K \times \ln K_P$$

$$\ln K_P = \frac{273.2\times10^3\ J\ mol^{-1}}{-8.3145\ J\ mol^{-1}\ K^{-1} \times 298\ K} = -110 \quad K_P = e^{-110} = 1.7\times10^{-48} \approx 2\times10^{-48}$$

Virtually no dissociation of HF(g) into its elements occurs.

2B   The dissociation reaction with all integer coefficients is twice the reverse of the formation reaction.

$$2HCl(g) \rightleftharpoons H_2(g) + Cl_2(g) \quad \Delta G° = -2 \times (-95.30 \text{ kJ/mol}) = +190.6 \text{ kJ/mol} = -RT \ln K_p$$

$$\ln K_p = \frac{-\Delta G°}{RT} = \frac{-190.6 \times 10^3 \text{ J/mol}}{8.3145 \text{ J mol}^{-1} \text{ K}^{-1} \times 298 \text{ K}} = -76.9 \qquad K_p = e^{-76.9} = 4 \times 10^{-34}$$

We assume an initial HCl(g) pressure of P atm, and calculate the final pressure of $Cl_2(g)$ and $H_2(g)$, x atm.

Reaction:	2 HCl(g)	$\rightleftharpoons$	H₂(g)	+	Cl₂(g)
Initial:	P atm		0 atm		0 atm
Changes:	$-2x$ atm		$+x$ atm		$+x$ atm
Equil:	$(P-2x)$ atm		x atm		x atm

$$K_p = \frac{P\{H_2(g)\}P\{Cl_2(g)\}}{P\{HCl(g)\}^2} = \frac{x \cdot x}{(P-2x)^2} = \left(\frac{x}{P-2x}\right)^2 \qquad \frac{x}{P-2x} = \sqrt{4 \times 10^{-34}} = 2 \times 10^{-17}$$

$$x = 2 \times 10^{-17}(P - 2x) \approx 2 \times 10^{-17} P$$

$$\% \text{ decomposition} = \frac{2x}{P} \times 100\% = 2 \times 2 \times 10^{-17} \times 100\% = 4 \times 10^{-15} \% \text{ decomposed}$$

3A   The first two reactions, are those from Example 23-3, used to produce B_2O_3.

$$Na_2B_4O_7 \cdot 10H_2O(s) + H_2SO_4(l) \longrightarrow 4B(OH)_3(s) + Na_2SO_4(s) + 5H_2O(l)$$

$$2B(OH)_3(s) \xrightarrow{\Delta} B_2O_3(s) + 3H_2O(g)$$

The next reaction is conversion to BCl_3 with heat, carbon, and chlorine.

$$2B_2O_3(s) + 3C(s) + 6Cl_2(g) \xrightarrow{\Delta} 4BCl_3(l) + 3CO_2(g)$$

$LiAlH_4$ is used as a reducing agent to produce diborane.

$$4BCl_3(l) + 3LiAlH_4(s) \rightarrow 2B_2H_6(g) + 3LiCl(s) + 3AlCl_3(s)$$

3B   We first roast ZnS(s) to produce $SO_2(g)$: $\qquad 2ZnS(s) + 3O_2(g) \longrightarrow 2ZnO(s) + 2SO_2(g)$

The $SO_2(g)$ is oxidized to $SO_3(g)$: $\qquad 2SO_2(g) + O_2(g) \xrightarrow[\text{catalyst}]{V_2O_5} 2SO_3(g)$

The $SO_3(g)$ is hydrated in two steps: $\qquad H_2SO_4(l) + SO_3(g) \longrightarrow H_2S_2O_7(l)$

$$H_2S_2O_7(l) + H_2O(l) \longrightarrow 2H_2SO_4(l)$$

REVIEW QUESTIONS

1. **(a)** A polyhalide ion is an anion consisting of two or more halogen atoms, such as I_3^-.

    (b) A polyphosphate is polymeric anion with a chain-like structure composed of $PO_{2.5}$ and PO_3^- units linked to each other and terminal $PO_{3.5}^{2-}$ units by bridging oxygen atoms.

    (c) Allotropy refers to an element existing in several forms in the same state (usually the solid state) of matter. Examples include diamond and graphite, red and white phosphorus, and the metallic and nonmetallic forms of arsenic and antimony.

    (d) Disproportionation refers to a substance being both oxidized and reduced in the same reaction.

2. **(a)** The Frasch process is used to mine underground deposits of sulfur. Superheated water and compressed air are pumped down into the mine, melting the sulfur and forcing it to the surface, where it cools and dries to form solid sulfur.

    (b) The contact process refers to oxidizing $SO_2(g)$ with $O_2(g)$ to $SO_3(g)$ using a $V_2O_5(s)$ catalyst. It is an essential step in the modern industrial production of sulfuric acid.

    (c) Eutrophication is the result of an excess of nutrients in bodies of water, such as ponds, which causes rapid and excessive plant growth in the water. This depletes the oxygen content of the water and marine life dies off. The decaying plant and animal life fill the pond from the bottom, eventually turning it into a marsh.

    (d) A three-center bond is a pair of electrons spanning three atoms, such as occurs in boron hydrides.

3. **(a)** An acid salt's anion is a partly ionized polyprotic acid: HSO_4^- or HPO_4^{2-}. An acid anhydride is a compound that forms an acid on the addition of water: typically nonmetal oxides: SO_3 and CO_2.

    (b) The azide ion is N_3^-, its salts typically are unstable, often explosively so. The nitride ion is N^{3-}; it often forms when active metals, such as magnesium, react with $N_2(g)$.

    (c) A silane is a compound of silicon and hydrogen. A silicone has a —Si—O—Si— backbone terminated by —OH groups, with hydrocarbon groups attached to each of the silicon atoms.

    (d) A colloidal disperson of a solid in a liquid that flows readily is known as a sol. One that has had sufficient liquid removed that it does not flow readily is called a gel.

4. **(a)** $KBrO_3$, potassium bromate **(b)** I_3^-, triiodide ion

    (c) $NaClO$, sodium hypochlorite **(d)** NaH_2PO_4, sodium dihydrogen phosphate

    (e) $Pb(N_3)_2$, lead (II) azide **(f)** BaS_2O_3, barium thiosulfate

5. **(a)** $CaCl_2(s) + H_2SO_4(conc. aq) \xrightarrow{\Delta} CaSO_4(s) + 2HCl(g)$

(b) $I_2(s) + Cl^-(aq) \longrightarrow$ no reaction

(c) $NH_3(aq) + HClO_4(aq) \longrightarrow NH_4ClO_4(aq)$

6. **(a)** $Cl_2(aq) + 2NaOH(aq) \longrightarrow NaCl(aq) + NaOCl(aq) + H_2O(l)$

(b) Oxidation: $2I^-(aq) \longrightarrow I_2(s) + 2e^-$

Reduction: $SO_4^{2-}(aq) + 2e^- + 4H^+(aq) \longrightarrow SO_2(g) + 2H_2O(l)$

Net: $\underline{2I^-(aq) + SO_4^{2-}(aq) + 4H^+(aq) \longrightarrow I_2(s) + SO_2(g) + 2H_2O(l)}$

Hence, overall: $2NaI(aq) + 2H_2SO_4(aq) \longrightarrow I_2(aq) + SO_2(g) + 2H_2O(l) + Na_2SO_4(aq)$

(c) $Cl_2(g) + 2Br^-(aq) \longrightarrow 2Cl^-(aq) + Br_2(l)$

7. **(a)** $KI(s) + H_3PO_4(l) \longrightarrow HI(g) + KH_2PO_4(s)$ Because H_3PO_4 is not a terribly strong acid, this reaction probably does not proceed beyond the first ionization.

(b) $Na_2SiF_6(s) + 4Na(s) \xrightarrow{\Delta} Si(s) + 6NaF(s)$

8. **(a)** $As_4O_6(s) + 6CO(g) \longrightarrow 4As(s) + 6CO_2(g)$

(b) $NH_3(g) + H_3PO_4(aq) \longrightarrow NH_4H_2PO_4(aq)$ ammonium dihydrogen phosphate

$2NH_3(g) + H_3PO_4(aq) \longrightarrow (NH_4)_2HPO_4(aq)$ ammonium hydrogen phosphate

9. First compute the theoretical yield of P from 8.00 tons of phosphate rock, then the percent yield in this case.

$$\text{no. ton P} = 8.00 \text{ ton rock} \times \frac{31 \text{ ton } P_4O_{10}}{100.00 \text{ ton rock}} \times \frac{1 \text{ton-mol } P_4O_{10}}{283.8 \text{ ton } P_4O_{10}} \times \frac{4 \text{ ton-mol P}}{1 \text{ton-mol } P_4O_{10}}$$

$$\times \frac{31.0 \text{ ton P}}{1 \text{ton-mol } P_2O_5} = 1.1 \text{ ton P}$$

The mass of a ton-mole in tons is numerically equal to the mass of a (gram-)mole in grams.

$$\% \text{yield} = \frac{1.00 \text{ ton P produced}}{1.1 \text{ ton P calculated}} \times 100\% = 91\% \text{ yield}$$

10. $3Cl_2(g) + I^-(aq) + 3H_2O(l) \longrightarrow 6Cl^-(aq) + IO_3^-(aq) + 6H^+(aq)$

$Cl_2(g) + 2Br^-(aq) \longrightarrow 2Cl^-(aq) + Br_2(aq)$

11. Number of cubic kilometers of seawater that would have to be processed to yield 45 million tons of H_2SO_4

$$= 45 \times 10^6 \text{ tons } H_2SO_4 \times \frac{2000 \text{ lb}}{1 \text{ ton}} \times \frac{454 \text{ g}}{1 \text{ lb}} \times \frac{1 \text{ mol } H_2SO_4}{98.08 \text{ g } H_2SO_4} \times \frac{1 \text{ mol } SO_4^{2-}}{1 \text{ mol } H_2SO_4} \times$$

$$\frac{96.06 \text{ g } SO_4^{2-}}{1 \text{ mol } SO_4^{2-}} \times \frac{1000 \text{ mg}}{1 \text{ g}} \times \frac{1 \text{ L seawater}}{2650 \text{ mg } SO_4^{2-}} \times \frac{1 \text{ m}^3}{1000 \text{ L}} \times \left(\frac{1 \text{ km}}{1000 \text{ m}}\right)^3$$

$$= 15 \text{ km}^3 \text{ of seawater.}$$

12. **(a)** Reduction: $Cl_2(g) + 2e^- \longrightarrow 2Cl^-(aq) \qquad E^\circ = +1.358 \text{ V}$

Oxidation: $2I^-(aq) \longrightarrow I_2(s) + 2e^- \qquad -E^\circ = -0.535 \text{ V}$

$$2H_2O(l) \longrightarrow O_2(g) + 4H^+(aq) + 4e^- \quad -E^\circ = -1.229 \text{ V}$$

The production of $I_2(s)$ is more likely; it results in the larger standard cell potential.

(b) NH_4^+ has nitrogen in its lowest common oxidation state. NH_4^+ cannot be reduced further. Thus, NH_4^+ is oxidized and H_2O_2 must be reduced to H_2O.

13. **(a)** $2 KI(s) + H_3PO_4(\text{conc, aq}) \xrightarrow{\Delta} K_2HPO_3(aq) + I_2(s) + H_2O(l)$

(b) $K_2O(s) + H_2O(l) \rightarrow 2KOH(aq)$

(c) $I_2(s) + KI(aq) \rightarrow KI_3(aq)$

(d) $5 HSO_3^-(aq) + 2 MnO_4^-(aq) + H^+ \rightarrow 2 Mn^{2+}(aq) + 5 SO_4^{2-}(aq) + 3 H_2O(l)$

14. **(a)** sodium perxenate **(b)** barium peroxide

(c) mercury(II) thiocyanate **(d)** barium nitride

(e) silver (I) thiosulfate

15. **(a)** PCl_2F_3 **(b)** KNCO

(c) $FePO_4$ **(d)** $Ba(N_3)_2$

(e) $Mg_2(P_2O_7)$

16. **(a)** AgAt, silver astatide **(b)** Na_4XeO_6, sodium perxenate

(c) MgPo, magnesium polonide **(d)** H_2TeO_3, tellurous acid

(e) K_2SeSO_3, potassium thioselenate **(f)** $KAtO_4$, potassium perastatate

EXERCISES

Noble Gas Compounds

17. **(a)** $|\overline{\underline{O}}-\overline{Xe}-\overline{\underline{O}}|$

(b) $|\overline{\underline{O}}|$ ... $|\overline{\underline{O}}-Xe-\overline{\underline{O}}|$

(c) $\left[\begin{array}{c} |\overline{F}| \\ \nearrow \\ F\diagdown \\ |Xe-\overline{F}| \\ F\diagup \\ |\underline{F}| \end{array}\right]^{+}$

(a) The Lewis structure has three ligands and one lone pair on Xe. XeO_3 has a trigonal pyramidal shape.

(b) The Lewis structure has four ligands on Xe. XeO_4 has a tetrahedral shape.

(c) There are five ligands and one lone pair on Xe in XeF_5^{+}. Its shape is square pyramidal.

18. We use VSEPR theory to predict the shapes of the species involved.

(a) O_2XeF_2 has a total of $2\times6+8+2\times7=34$ valence electrons $=17$ pairs.

(b) O_3XeF_2 has a total of $3\times6+8+2\times7=40$ valence electrons $=20$ pairs.

(c) $OXeF_4$ has a total of $6+8+4\times7=42$ valence electrons $=21$ pairs.

We draw a plausible Lewis structure for each species below.

(a) $|\overline{O}|$ ... $|\overline{F}-Xe-\overline{F}|$ ... $|\underline{O}|$

(b) $|\overline{F}$ ... $\overline{F}|$ ... $|\overline{O}-Xe-\overline{O}|$ ... $|\underline{O}|$

(c) $|\overline{F}$ ... $\overline{F}|$ ... $|\overline{F}-Xe-\overline{F}|$ ... $|\underline{O}|$

(a) There are four ligands and one lone pair on Xe in O_2XeF_2. Its shape is an irregular tetrahedron.

(b) There are five ligands and no lone pairs on Xe in O_3XeF_2. Its shape is trigonal bipyramidal.

(c) The Lewis structure has five ligands and one lone pair on Xe. $OXeF_4$ has a square pyramidal shape.

19. $3\ XeF_4(aq) + 6\ H_2O(l) \rightarrow 2\ Xe(g) + 3/2\ O_2(g) + 12\ HF(g) + XeO_3(s)$

20. $2\ XeF_6(aq) + 16\ OH^-(aq) \rightarrow Xe(g) + XeO_6^{4-}(aq) + O_2(g) + 8\ H_2O(l) + 12\ F^-(aq)$

The Halogens

21. Iodide ion is slowly oxidized to iodine, which is yellow-brown in aqueous solution, by oxygen in the air.

Oxidation: $\{2\,I^-(aq) \longrightarrow I_2(aq) + 2\,e^-\} \times 2$ $\qquad\qquad -E° = -0.535\,V$

Reduction: $O_2(g) + 4\,H^+(aq) + 4\,e^- \longrightarrow 2\,H_2O(l)$ $\qquad\qquad E° = +1.229\,V$

Net: $\qquad 4\,I^-(aq) + O_2(g) + 4\,H^+(aq) \longrightarrow 2\,I_2(aq) + 2\,H_2O(l)$ $\qquad E°_{cell} = +0.694\,V$

Possibly followed by: $I_2(aq) + \tilde{I}^-(aq) \longrightarrow I_3^-(aq)$

22. $MnF_6^{2-}(s) + 2\,SbF_5(s) \longrightarrow MnF_4(s) + 2\,SbF_6^-(s)$ $\qquad\qquad 2\,MnF_4(s) \longrightarrow 2\,MnF_3(s) + F_2(g)$

23. Displacement reactions involve one element displacing another element from solution The element that dissolves in the solution is more "active" than the element supplanted from solution. Within the halogen group the activity decreases from top to bottom. Thus, each halogen is able to displace the members of the group below it, but not those above it. For instance, molecular bromine can oxidize aqueous iodide ion but molecular iodine is incapable of oxidizing bromide ion:

$Br_2(aq) + 2\,I^-(aq) \rightarrow 2\,Br^-(aq) + I_2(aq)$ $\quad$ however, $\quad I_2(aq) + 2\,Br^-(aq) \rightarrow$ NO RXN

The only halogen with sufficient oxidizing power to displace $O_2(g)$ from water is $F_2(g)$:

$2\,H_2O(l) \rightarrow 4\,H^+ + O_2(g) + 4\,e^-$ $\qquad\qquad E°_{1/2ox} = -1.229\,V$

$\{F_2(g) + 2\,e^- \rightarrow 2\,F^-(aq)\} \times 2$ $\qquad\qquad E°_{1/2red} = 2.866\,V$

$2\,F_2(g) + 2\,H_2O(l) \rightarrow 4\,H^+ + O_2(g) + 4\,F^-(aq)$ $\;E°_{cell} = 1.637\,V$

The large positive standard reduction potential for this reaction indicates that the reaction will occur spontaneously, with products being strongly preferred under standard state conditions.

None of the halogens react with water to form $H_2(g)$. In order to displace molecular hydrogen from water, one must add a strong reducing agent, such as sodium metal.

24. We first list values of halogen properties.

		F	Cl	Br	I
(a)	Covalent radius:	71	99	114	133 pm
(b)	Ionic radius:	133	181	196	220 pm
(c)	First ionization energy: kJ/mol	1681	1251	1140	1008
(d)	electron affinity: kJ/mol	−328.0	−349.0	−324.6	−295.2
(e)	electronegativity:	4.0	3.0	2.8	2.5
(f)	standard reduction potential:	+2.886 V	+1.358 V	+1.065 V	+0.535 V

We can do a reasonably god job of predicting the properties of astatine by simply looking at the difference between Br and I and assuming that the same difference exists between I and At.

(a) Covalent radius: $\approx152\,pm$ **(b)** Ionic radius: $244\,pm$

(c) 1st ionization energy: $876\,kJ/mol$ **(d)** electron affinity: $\approx-265\,kJ/mol$

(e) electronegativity: ≈2.2 **(f)** $E° \approx0.005\,V$ to $0.010V$

Depending on the technique that you use, you may arrive at different answers. For example, -260 kJ/mol is a reasonable estimation of the electron affinity by the following reasoning. The difference between the value for Cl and Br is 24 kJ/mol, the difference between the values for Br and I is 29 kJ/mol, so the difference between I and At should be about 35 kJ/mol.

25. **(a)**
$$mass\ F_2 = 1\ km^3 \times \left(\frac{1000\ m}{1\ km}\times\frac{100\ cm}{1\ m}\right)^3 \times \frac{1.03\ g}{1\ cm^3}\times\frac{1\ lb}{454\ g}\times\frac{1\ ton}{2000\ lb}\times\frac{1\ g\ F^-}{1\ ton}\times\frac{37.996\ g\ F_2}{37.996\ g\ F^-}$$
$$= 1\times10^9\ g\ F_2 = 1\times10^6\ kg\ F_2$$

(b) Bromine is extracted by displacing it from solution with $Cl_2(g)$. Since there is no chemical oxidizing agent that is stronger than $F_2(g)$, this method of displacement would not work for $F_2(g)$ Even if there were a chemical oxidizing agent stronger than $F_2(g)$ it would displace O_2 before it displaced $F_2(g)$. Obtaining $F_2(g)$ would require electrolysis of its molten salts, obtained by evaporating the seawater.

26. $Mass\ F_2 = 1.00\times10^3\ kg\ rock\times\frac{1000\ g}{1\ kg}\times\frac{1\ mol\ 3\ Ca_3(PO_4)_2\cdot CaF_2}{1009\ g}\times\frac{2\ mol\ F}{1\ mol\ 3\ Ca_3(PO_4)_2\cdot CaF_2}$

$\times\frac{19.00\ g\ F}{1\ mol\ F}\times\frac{1\ kg}{1000\ g} = 37.7\ kg\ fluorine$

27. In order for the disproportionation reaction to occur under standard conditions, the $E°$ for the overall reaction must be greater than zero. To answer this question, we must refer to the Latimer diagrams provided in Figure 23-2 and the answer to practice example 23-1B.

(i) Reduction half reaction (acidic solution)

$Cl_2(aq) + 2\ e^- \rightarrow 2\ Cl^-(aq)\quad E°_{1/2\ red} = 1.358\ V$

(ii) Oxidation half reaction (acidic solution)

$Cl_2(aq) + 6\ H_2O(l) \rightarrow 2\ ClO_3^-(aq) + 12\ H^+(aq) + 10\ e^-\quad E°_{1/2}\ ox = -1.453\ V$

Combining (i) × 5 with (ii) × 1, we obtain the desired disproportionation reaction:
$6\ Cl_2(aq) + 6\ H_2O(l) \rightarrow 2\ ClO_3^-(aq) + 12\ H^+(aq) + 10\ Cl^-(aq)\quad E°_{cell} = -0.095\ V$

Since the final cell voltage is negative, the disproportionation reaction will not occur spontaneously under standard conditions. Alternatively, we can calculate K_{eq} by using $\ln K_{eq} = -\Delta G°/RT$ and $\Delta G° = -nFE°$. This method gives a $K_{eq} = 8.6 \times 10^{-17}$. Clearly, the reaction will not go to completion.

28. To find out whether the reaction will go to completion, we must first calculate the standard potential for the disproportionation of hypochlorous acid (HOCl). As was the case for question #27, we will use information contained in Fig. 23.2 to answer this question.

(i) $2\ HOCl(aq) + 2\ H_2O(l) \rightarrow 2\ HClO_2(aq) + 4\ H^+(aq) + 4\ e^-$ $E°_{1/2\ ox} = -1.645$ V

(ii) $2\ HOCl(aq) + 2\ H^+(aq) + 2\ e^- \rightarrow Cl_2(aq) + 2\ H_2O(l)$ $E°_{1/2\ red} = +1.611$ V

(iii) $Cl_2(aq) + 2\ e^- \rightarrow 2\ Cl^-(aq)$ $E°_{1/2\ red} = +1.358$ V

To find the standard voltage difference between HOCl and Cl$^-$ we must sum together the free energy changes for the two reduction half-reactions written above.

Target Equation:

(iv) $2\ HOCl(aq) + 2\ H^+(aq) + 4\ e^- \rightarrow 2\ Cl^-(aq) + 2\ H_2O(l)$ $E°_{1/2\ red(3)}$

          $-4F\ E°_{1/2\ red(3)} = -2F(1.611\ V) + -2F(1.358\ V)$ So, $E°_{1/2\ red(3)} = 1.485$ V

Therefore, by adding equation (i) to equation (iv), we can obtain the standard cell voltage for the disproportionation of HOCl.

(i) $2\ HOCl(aq) + 2\ H_2O(l) \rightarrow 2\ HClO_2(aq) + 4\ H^+(aq) + 4\ e^-$ $E°_{1/2\ ox} = -1.645$ V

(iv) $2\ HOCl(aq) + 2\ H^+(aq) + 4\ e^- \rightarrow 2\ Cl^-(aq) + 2\ H_2O(l)$ $E°_{1/2\ red} = 1.485$ V
_____

          $4\ HOCl(aq) \rightarrow 2\ HClO_2(aq) + 2\ Cl^-(aq) + 2\ H^+(aq)$ $E°_{cell} = -0.160$ V
          (disproportionation reaction of HOCl(aq))

Since the standard cell potential is negative, the equilibrium will favor the reactants and thus the reaction will not go to completion as written (i.e., reactants will predominate at equilibrium and thus the reaction will be far from complete). Alternatively, we can calculate K_{eq} from the reaction by using $\ln K_{eq} = -\Delta G°/RT$ and $\Delta G° = -nFE°$. This gives $K_{eq} = 1.50 \times 10^{-11}$. Clearly, the reaction will not go to completion because K_{eq} is very small (reactants strongly predominate at equilibrium, not products).

29. First we must draw the Lewis structure for all of the species listed. Following this, we will deduce their electron-group geometries and molecular shapes following the VSEPR approach.

(a) BrF_3:   28 valence electrons

          VSEPR class: AX_3E_2

          Thus BrF_3 is T-shaped

(b) IF_5 42 valence electrons

VSEPR class: AX_5E

Thus IF_5 is square pyramidal

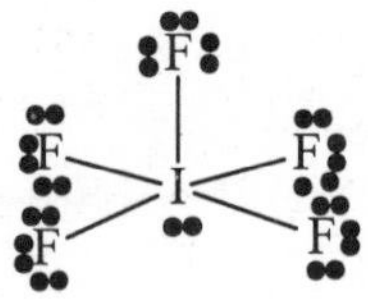

(c) Cl_3IF^- 36 valence electrons

VSEPR class: AX_4E_2

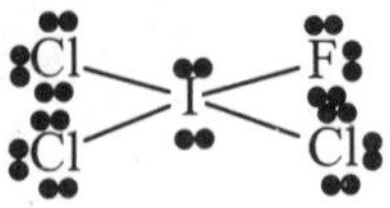

Thus Cl_3IF^- is square planar

30.   To answer this question we need to apply the VSEPR method to each species:

i) ClF_2^+ (total number of valence electrons = 20 e^-)
Molecular shape: AX_2E_2 angular (<109.5° due to
       lone pairs on Cl atom)

ii) $IBrF^-$ (total number of valence electrons = 22 e^-)
Molecular shape: AX_2E_3 linear (~180°)

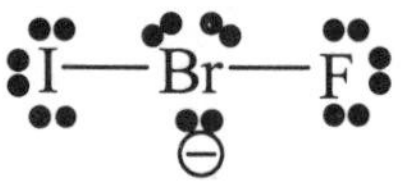

iii) OCl_2 (total number of valence electrons = 20 e^-)
Molecular shape: AX_2E_2 angular (<109.5° due to
       lone pairs on O or Cl atom)
Cl—O—Cl preferred over Cl—Cl—O because
all atoms have a formal charge of zero.

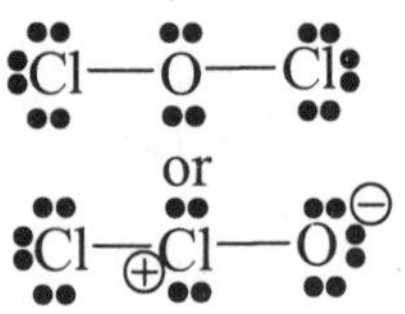

iv) ClF_3 (total number of valence electrons = 28 e^-)
Molecular shape: AX_3E_2 T-shaped molecule

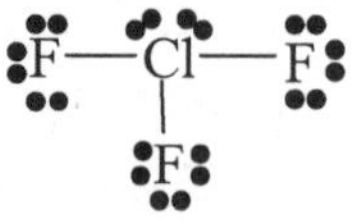

v) SF_4 (total number of valence electrons = 34 e^-)
Molecular shape: AX_4E See-saw molecular
       geometry

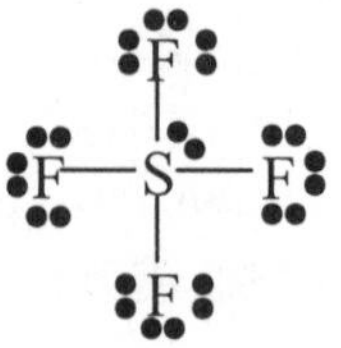

The VSEPR treatment of each species indicates only $IBrF^-$ has a linear structure. Since (i) and (iii) belong to the same VSEPR class (AX_2E_2), it is not unreasonable to assume that they have the same "bent structure".

Oxygen

31.   Since the pK_a for H_2O_2 had been provided to us, we can find the solution pH simply by solving an I.C.E. table for the hydrolysis of a 3.0 % H_2O_2 solution (by mass). Of course, in order to use this method, the mass percent must first be converted to molarity. We must assume that the density of the solution is 1.0 g mL^{-1}.

$$[H_2O_2] = \frac{3.0 \ \text{g} \ H_2O_2}{100 \ \text{g} \ \text{solution}} \times \frac{1\,\text{g solution}}{1\,\text{mL solution}} \times \frac{1 \ \text{mol} \ H_2O_2}{34.015 \ \text{g} \ H_2O_2} \times \frac{1000\,\text{mL}}{1\,\text{L}} = 0.88 \ \text{M}$$

The pK_a for H_2O_2 is 11.75. The K_a for H_2O_2 is therefore $10^{-11.75}$ or 1.8×10^{-12}. By comparison with pure water, which has a K_a of 1.8×10^{-16} at 25 °C, one can see that H_2O_2 is indeed a stronger acid than water but differences in acidity between the two is not that great. Consequently, we cannot ignore the contribution of protons of pure water when we workout the pH of the solution at equilibrium.

Reaction:	$H_2O_2(aq)$	+	$H_2O(l)$	$\xrightleftharpoons{K_a \,=\, 1.8 \,\times\, 10^{-12}}$	$H_3O^+(aq)$	+	$HO_2^-(aq)$
Initial	0.88 M		—		$1.0 \times 10^{-7}\,M$		0 M
Change	$-x$ M				$+x$ M		$+x$ M
Equilibrium	0.88$-x$ M($\sim$0.88 M)				$(1.0 \times 10^{-7} + x)$M		x M

So, $1.8 \times 10^{-12} = \dfrac{x(x \ + \ 1.0 \times 10^{-7})}{\sim 0.88}$

$x^2 + 1.0 \times 10^{-7}x - 1.58 \times 10^{-12} = 0$

$$x = \frac{- \ 1.0 \times 10^{-7} \pm \ \sqrt{1.0 \times 10^{-14} \ + \ 4(1.58 \times 10^{-12})}}{2}$$

The root that makes sense in this context is $x = 1.2 \times 10^{-6}$ M.

Thus, the $[H_3O^+] = 1.2 \times 10^{-6}$ M $+ 1.0 \times 10^{-7}$ M $= \ 1.3 \times 10^{-6}$ M

Consequently, the pH for the 3.0 % H_2O_2 solution (by mass) should be 5.89 (i.e., the solution is weakly acidic)

32.   Oxide ion reacts with water to form hydroxide ion.

$Li_2O(s) + H_2O(l) \longrightarrow 2\,Li^+(aq) + 2\,OH^-(aq)$

We first calculate $\left[OH^- \right]$ and then the solution's pH.

$$\left[OH^- \right] = \frac{0.050\ \text{g}\ Li_2O}{750.0\ \text{mL}} \times \frac{1000\ \text{mL}}{1\ \text{L}} \times \frac{1\ \text{mol}\ Li_2O}{29.88\ \text{g}\ Li_2O} \times \frac{2\ \text{mol}\ OH^-}{1\ \text{mol}\ Li_2O} = 0.00446\ \text{M}$$

$pOH = -\log(0.00446) = 2.350 \qquad pH = 14.000 - pOH = 14.000 - 2.350 = 11.650$

33. $3\ \overline{\text{O}}\!=\!\overline{\text{O}} \longrightarrow 2\ \overline{\text{O}} \bullet\!\bullet\!\bullet \overline{\text{O}} \bullet\!\bullet\!\bullet \overline{\text{O}}$

Bonds broken: $3 \times (\text{O}\!=\!\text{O}) = 3 \times 498\ \text{kJ/mol} = 1494\ \text{kJ/mol}$

Bonds formed: $4 \times (\text{O} \bullet\!\bullet\!\bullet \text{O})$

$\Delta H^\circ = +285\ \text{kJ/mol} = \text{bonds broken} - \text{bonds formed} = 1494\ \text{kJ/mol} - 4 \times (\text{O} \bullet\!\bullet\!\bullet \text{O})$

$4 \times (\text{O} \bullet\!\bullet\!\bullet \text{O}) = 1494\ \text{kJ/mol} - 285\ \text{kJ/mol} = 1209\ \text{kJ/mol}$

$\text{O} \bullet\!\bullet\!\bullet \text{O} = 1209\ \text{kJ/mol} + 4 = 302\ \text{kJ/mol}$

34. $3\ \overset{\bullet\bullet}{\underset{\bullet\bullet}{\text{O}}}\!=\!\overset{\bullet\bullet}{\underset{\bullet\bullet}{\text{O}}} \longrightarrow 2\ :\!\overset{\bullet\bullet}{\text{O}}\!-\!\overset{\bullet\bullet}{\text{O}}\!=\!\overset{\bullet\bullet}{\text{O}}\!:$

$$\text{Average bond energy} = \frac{\text{O}\!=\!\text{O} + \text{O}\!-\!\text{O}}{2} = \frac{498 + 142}{2} = 320\ \text{kJ/mol}$$

Since bond energies actually are average values taken from many sources, the result of 320 kJ/mol is in reasonably good agreement with the specific value of 302 kJ/mol obtained in Exercise 33.

35. **(a)** H_2S, while polar, forms only weak hydrogen bonds. H_2O forms much stronger hydrogen bonds, leading to a higher boiling point.

(b) All electrons are paired in O_3, producing a diamagnetic molecule. $\mathsf{I}\overline{\underline{\text{O}}}\!-\!\overline{\underline{\text{O}}}\!=\!\overline{\underline{\text{O}}}$

36. **(a)** $\overline{\underline{\text{O}}}\!=\!\overline{\underline{\text{O}}} \qquad \mathsf{I}\overline{\underline{\text{O}}}\!-\!\overline{\underline{\text{O}}}\!=\!\overline{\underline{\text{O}}} \qquad \text{H}\!-\!\overline{\underline{\text{O}}}\!-\!\overline{\underline{\text{O}}}\!-\!\text{H}$

The O—O bond in O_2 is a double bond, which should be short (121 pm). That in O_3 is a "one-and-a-half" bond, of intermediate length (128 pm). And that in H_2O_2 is a longer (148 pm) single bond.

(b) The $O_2{}^+$ ion has one less electron than does the O_2 molecule. Based on Lewis structures, $\qquad \overline{\underline{\text{O}}}\!=\!\overline{\underline{\text{O}}} \quad \overline{\underline{\text{O}}}\!\overset{\bullet}{-}\!\overline{\underline{\text{O}}}$

one would predict that $O_2{}^+$ would have a bond order of 1.5, thereby making its O—O bond weaker than the double bond of O_2, and therefore longer, in contradiction to the experimental evidence. The molecular orbital picture of the two species suggests the opposite. According to this model, O_2 has a bond order of 2.0 while that for $O_2{}^+$ is 2.5. Thus MO theory predicts a stronger bond for $O_2{}^+$.

	σ_{1s}	σ^*_{1s}	σ_{2s}	σ^*_{2s}	σ_{2p}	π_{2p}	π^*_{2p}	σ^*_{2p}
O_2	↑↓	↑↓	↑↓	↑↓	↑↓	↑↓	↑ ↑	☐

$\text{bond order} = (\text{no. bonding electrons} - \text{no. antibonding electrons}) \div 2$

$\text{bond order} = (10 - 6) \div 2 = 2.0$

$$\sigma_{1s} \quad \sigma^*_{1s} \quad \sigma_{2s} \quad \sigma^*_{2s} \quad \sigma_{2p} \quad \pi_{2p} \quad \pi^*_{2p} \quad \sigma^*_{2p}$$

O_2^+

bond order $= \left(\text{no. bonding electrons} - \text{no. antibonding electrons}\right) \div 2$

bond order $= (10 - 5) \div 2 = 2.5$

37. Reactions that have K_{eq} values greater than 1000 are considered to be essentially quantitative (i.e. they go virtually to completion). So to answer this question we need only calculate the equilibrium constant for each reaction via. the equation $E°_{cell} = (0.0257/n)\ln K_{eq}$

(a) $\quad H_2O_2(aq) + 2\,H^+(aq) + 2\,e^- \rightleftharpoons 2\,H_2O(l) \qquad\qquad E°_{1/2red} = +1.763\ V$

$\qquad\qquad\qquad 2\,I^-(aq) \rightleftharpoons I_2(s) + 2\,e^- \qquad\qquad\qquad E°_{1/2ox} = -0.535\ V$

$\overline{H_2O_2(aq) + 2\,H^+(aq) + 2\,I^-(aq) \rightleftharpoons I_2(s) + 2\,H_2O(l) \quad E°_{1/2cell} = +1.228\ V (n = 2\ e^-)}$

$\ln K_{eq} = \dfrac{1.228\ V \times 2}{0.0257\ V} = 95.56 \qquad K_{eq} = 3.2 \times 10^{41}$

Therefore the reaction goes to completion (or very nearly so).

(b) $\quad O_2(g) + 2\,H_2O(l) + 4\,e^- \rightleftharpoons 4\,OH^-(aq) \qquad\qquad E°_{1/2red} = +0.401\ V$

$4\,Cl^-(aq) \rightleftharpoons 2\,Cl_2(g) + 4\,e^- \qquad\qquad\qquad\qquad E°_{1/2ox} = -1.358\ V$

$\overline{O_2(g) + 2\,H_2O(l) + 4\,Cl^-(aq) \rightleftharpoons 2\,Cl_2(g) + 4\,OH^-(aq)\ E°_{1/2cell} = -0.957\ V(n = 4\ e^-)}$

$\ln K_{eq} = \dfrac{-0.957\ V \times 4}{0.0257\ V} = -148.95 \qquad K_{eq} = 2.1 \times 10^{-65}$

The extremely small value of K_{eq} indicates that reactants are strongly preferred and thus, the reaction does not even come close to going to completion.

(c) $\quad O_3(g) + 2\,H^+(aq) + 2\,e^- \rightleftharpoons O_2(g) + H_2O(l) \qquad\qquad E°_{1/2red} = +2.075\ V$

$Pb^{2+}(aq) + 2\,H_2O(l) \rightleftharpoons PbO_2(s) + 4H^+(aq) + 2\,e^- \qquad\quad E°_{1/2ox} = -1.455\ V$

$\overline{O_3(g) + Pb^{2+}(aq) + H_2O(l) \rightleftharpoons PbO_2(s) + 2\,H^+(aq) + O_2(g)\ E°_{cell} = 0.620\ V}$

$\qquad\qquad\qquad\qquad\qquad\qquad\qquad\qquad\qquad\qquad\qquad\qquad (n = 2\ e^-)$

$\ln K_{eq} = \dfrac{+0.62\ V \times 2}{0.0257\ V} = 48.25 \qquad K_{eq} = 9.0 \times 10^{20}$

Therefore the reaction goes to completion (or very nearly so).

(d) $HO_2^-(aq) + H_2O(l) + 2\ e^- \rightleftharpoons 3\ OH^-(aq)$ $E°_{1/2red} = +0.878\ V$

                         $2\ Br^-(aq) \rightleftharpoons Br_2(l) + 2\ e^-$ $E°_{1/2ox} = -1.065\ V$

$HO_2^-(aq) + H_2O(l) + 2\ Br^-(aq) \rightleftharpoons Br_2(s) + 3\ OH^-(aq)$ $E°_{1/2cell} = -0.187\ V$

                                                                            $(n = 2\ e^-)$

$$\ln K_{eq} = \frac{-0.187\ V \times 2}{0.0257\ V} = -14.55 \qquad K_{eq} = 4.8 \times 10^{-7}$$

The extremely small value of K_{eq} indicates the reaction heavily favors reactants at equilibrium and thus, the reaction does not even come close to going to completion.

38.   **(a)** $HgO(s) \xrightarrow{\Delta} Hg(l) + 1/2\ O_2(g)$

        (b) $2KClO_4(s) \xrightarrow{\Delta} 2\ KCl(s) + 4\ O_2(g)$

        (c) $Hg(NO_3)_2(s) \xrightarrow{\Delta} Hg(l) + 2\ NO_2(g) + O_2(g)$
                                or   }   Depending on the temperature

             $Hg(NO_3)_2(s) \xrightarrow{\Delta} Hg(NO_2)_2(g) + O_2(g)$

        (d) $H_2O_2(l) \xrightarrow{\Delta} H_2O(l) + 1/2\ O_2(g)$

Sulfur

39.   **(a)**   ZnS, zinc sulfide

        (b) $KHSO_3$, potassium hydrogen sulfite

        (c) $K_2S_2O_3$, potassium thiosulfate

        (d) SF_4, sulfur tetrafluoride

40.   **(a)** $CaSO_4 \cdot 2H_2O$, calcium sulfate dihydrate

        (b) $H_2S(aq)$, hydrosulfuric acid

        (c) $NaHSO_4$, sodium hydrogen sulfate

        (d) $H_2S_2O_7(aq)$, disulfuric acid

41.   **(a)** $FeS(s) + 2\ HCl(aq) \longrightarrow FeCl_2(aq) + H_2S(aq)$
               MnS(s), ZnS(s), etc. also are possible.

        (b) $CaSO_3(s) + 2\ HCl(aq) \longrightarrow CaCl_2(aq) + H_2O(l) + SO_2(g)$

        (c)   Oxidation: $SO_2(aq) + 2\ H_2O(l) \longrightarrow SO_4^{2-}(aq) + 4H^+(aq) + 2\ e^-$

                Reduction: $MnO_2(s) + 4H^+(aq) + 2\ e^- \longrightarrow Mn^{2+}(aq) + 2\ H_2O(l)$

                Net: $SO_2(aq) + MnO_2(s) \longrightarrow Mn^{2+}(aq) + SO_4^{2-}(aq)$

(d) Oxidation: $S_2O_3^{2-}(aq) + H_2O(l) \longrightarrow 2SO_2(g) + 2H^+(aq) + 4e^-$

Reduction: $S_2O_3^{2-}(aq) + 6H^+(aq) + 4e^- \longrightarrow 2S(s) + 3H_2O(l)$

Net: $S_2O_3^{2-}(aq) + 2H^+(aq) \longrightarrow S(s) + SO_2(g) + H_2O(l)$

42. (a) $S(s) + O_2(g) \longrightarrow SO_2(g)$

$2Na(s) + 2H_2O(l) \longrightarrow 2NaOH(aq) + H_2(g)$

$2NaOH(aq) + SO_2(g) \longrightarrow Na_2SO_3(aq) + H_2O(l)$

(b) $S(s) + O_2(g) \longrightarrow SO_2(g)$

$SO_2(g) + 2H_2O(l) + Cl_2(g) \longrightarrow SO_4^{2-}(aq) + 2Cl^-(aq) + 4H^+(aq)$ $\left(E^\circ_{cell} = +1.19V\right)$

$2Na(s) + 2H_2O(l) \longrightarrow 2NaOH(aq) + H_2(g)$

$2NaOH(aq) + 2H^+(aq) + SO_4^{2-}(aq) \longrightarrow Na_2SO_4(aq) + 2H_2O(l)$

(c) $S(s) + O_2(g) \longrightarrow SO_2(g)$

$2Na(s) + 2H_2O(l) \longrightarrow 2NaOH(aq) + H_2(g)$

$2NaOH(aq) + SO_2(g) \longrightarrow Na_2SO_3(aq) + H_2O(l)$

$Na_2SO_3(aq) + S(s) \longrightarrow NaS_2O_3(aq)$ (Boil the reactants in an alkaline solution.)

43. The decomposition of thiosulfate ion is promoted in an acidic solution. If the white solid is Na_2SO_4, there will be no reaction with strong acids such as HCl. By contrast, if the white solid is $Na_2S_2O_3$, $SO_2(g)$ will be liberated and a pale yellow precipitate of S(s,rhombic) will form upon addition of HCl(aq).

$$S_2O_3^{2-}(aq) + 2H^+(aq) \longrightarrow S(s) + SO_2(g) + H_2O(l)$$

Consequently, the solid can be identified by adding a strong mineral acid such as HCl(aq).

44. Sulfites are easily oxidized to sulfates by, for example, O_2 in the atmosphere. On the other hand, there are no oxidizing agents naturally available in reasonable concentrations that can oxidize SO_4^{2-} to a higher oxidation state, such as in $S_2O_8^{2-}$. Similarly, the atmosphere of Earth is an oxidizing one, reducing agents are not present to reduce SO_4^{2-} to a species with a lower oxidation state, except in localized areas.

45. $Na^+(aq)$ will not hydrolyze, being the cation of a strong base. But $HSO_4^-(aq)$ will ionize further, $K_2 = 1.1 \times 10^{-2}$ for $HSO_4^-(aq)$. We set up the situation, and solve the quadratic equation to obtain $\left[H_3O^+\right]$.

$$\left[HSO_4^-\right] = \frac{12.5 \text{ g NaHSO}_4}{250.0 \text{ mL soln}} \times \frac{1000 \text{ mL}}{1 \text{ L soln}} \times \frac{1 \text{ mol NaHSO}_4}{120.1 \text{ g NaHSO}_4} \times \frac{1 \text{ mol HSO}_4^-}{1 \text{ mol NaHSO}_4} = 0.416 \text{ M}$$

Reaction: $HSO_4^-(aq) + H_2O(l) \rightleftharpoons SO_4^{2-}(aq) + H_3O^+(aq)$

Initial:	0.416 M	0 M	≈ 0 M
Changes:	$-x$ M	$+x$ M	$+x$ M
Equil:	$(0.416 - x)$ M	x M	x M

$$K_2 = \frac{\left[H_3O^+\right]\left[SO_4^{2-}\right]}{\left[HSO_4^-\right]} = 0.011 = \frac{x^2}{0.416 - x}$$

$x^2 = 0.0046 - 0.011x$

$0 = x^2 + 0.011x - 0.0046$

$$x = \frac{-b \pm \sqrt{b^2 - 4ac}}{2a} = \frac{-0.011 \pm \sqrt{1.2 \times 10^{-4} + 1.8 \times 10^{-2}}}{2} = 0.062 = \left[H_3O^+\right]$$

$pH = -\log(0.062) = 1.21$

46. Oxidation: $\{SO_3^{2-}(aq) + H_2O(l) \longrightarrow SO_4^{2-}(aq) + 2H^+(aq) + 2e^-\}$ $\times 5$

Reduction: $\{MnO_4^-(aq) + 8H^+(aq) + 5e^- \longrightarrow Mn^{2+}(aq) + 4H_2O(l)\}$ $\times 2$

Net: $5 SO_3^{2-}(aq) + 2 MnO_4^-(aq) + 6H^+(aq) \longrightarrow 5 SO_4^{2-}(aq) + 2 Mn^{2+}(aq) + 3 H_2O(l)$

$$\text{mass Na}_2SO_3 = 26.50 \text{ mL} \times \frac{1 \text{ L}}{1000 \text{ mL}} \times \frac{0.0510 \text{ mol MnO}_4^-}{1 \text{ L soln}} \times \frac{5 \text{ mol SO}_3^{2-}}{2 \text{ mol MnO}_4^-} \times \frac{1 \text{ mol Na}_2SO_3}{1 \text{ mol SO}_3^{2-}}$$

$$\times \frac{126.0 \text{ g Na}_2SO_3}{1 \text{ mol Na}_2SO_3} = 0.426 \text{ g Na}_2SO_3$$

47. The question is concerned with assaying for the mass percent of copper in an ore. The assay in this instance involves the quantitative determination of the amount of metal in an ore by chemical analysis. The titration for copper in the sample does not occur directly, but rather indirectly via the number of moles of $I_3^-(aq)$ produced from the reaction of Cu^{2+} with I^-:

$2 Cu^{2+} + 5 I^-(aq) \rightarrow 2 CuI(s) + I_3^-(aq)$

The number of moles of $I_3^-(aq)$ produced is determined by titrating the iodide-treated sample with sodium thiosulfate. The balanced oxidation reaction that forms the basis for the titration is:

$I_3^-(aq) + 2\,S_2O_3^{2-}(aq) \rightarrow 3\,I^-(aq) + S_4O_6^{2-}(aq)$

The stoichiometric ratio is one $I_3^-(aq)$ reacting with two $S_2O_3^{2-}(aq)$ in this titration.

The number of moles of I_3^- formed

$= 0.01212\ L\ S_2O_3^{2-}(aq) \times \dfrac{0.1000\ \text{moles } S_2O_3^{2-}}{1\ L\ S_2O_3^{2-}} \times \dfrac{1\ \text{mol } I_3^-}{2\ \text{mol } S_2O_3^{2-}} = 6.060 \times 10^{-4}\ \text{moles } I_3^-$

Therefore, the number of moles of Cu^{2+} released when the sample is dissolved is

$= 6.060 \times 10^{-4}\ \text{moles } I_3^- \times \dfrac{2\ \text{mol } Cu^{2+}}{1\ \text{mol } I_3^-} = 1.212 \times 10^{-3}\ \text{moles of } Cu^{2+}$

Consequently, the mass percent for copper in the ore is

$1.212 \times 10^{-3}\ \text{moles of } Cu^{2+} \times \dfrac{63.546\ \text{g } Cu^{2+}}{1\ \text{mol } Cu^{2+}} \times \dfrac{1}{1.100\ \text{g of Cu ore}} \times 100\% = 7.002\%$

48. This question is similar to Question 47 in that it is also a titration question. In this case, however, the titration involves a metathesis rather than a redox reaction. What we need to do here is to use the mass of solid lead(II) sulfide formed in reaction #1 shown below to determine the mass of sulfur (present as $H_2S(g)$) that can be recovered per cubic meter as a result of lead sulfide precipitation:

$Pb^{2+}(aq) + H_2S(aq) \rightarrow 2\,H^+(aq) + PbS(s)\downarrow$

Let's begin by finding out how many grams of sulfur (as $H_2S(g)$) are present in 25.0 L of natural gas at 25 °C and a pressure of 740 torr:

$= 0.535\ \text{g PbS} \times \dfrac{1\ \text{mol PbS}}{239.3\ \text{g PbS}} \times \dfrac{1\ \text{mol } H_2S}{1\ \text{mol PbS}} \times \dfrac{1\ \text{mol S}}{1\ \text{mol } H_2S} \times \dfrac{32.066\ \text{g S}}{1\ \text{mol S}} = 0.0717\ \text{g "sulfur"}$

So, the number of grams that can be recovered per cubic meter of natural gas is:

$= \dfrac{0.0717\ \text{g S}}{25.0\ \text{L natural gas}} \times \dfrac{1\ L}{1000\ cm^3} \times \dfrac{(100\ cm)^3}{1\ m^3} = \dfrac{2.87\ \text{g S (as } S^{2-})}{1\ m^3\ \text{natural gas}}$

Nitrogen

49. **(a)** $2\,NO_2(g) \rightleftharpoons N_2O_4(g)$

    (b)   i) $HNO_2(aq) + N_2H_5^+(aq) \rightarrow HN_3(aq) + 2H_2O(l) + H^+(aq)$

           ii) $HN_3(aq) + HNO_2(aq) \rightarrow N_2(g) + H_2O(l) + N_2O(g)$

    (c) $H_3PO_4(aq) + 2\,NH_3(aq) \rightarrow (NH_4)_2HPO_4(aq)$

50. (a) $3\ Ag(s) + 4\ H^+(aq) + 4\ NO_3^-(aq) \longrightarrow 3\ AgNO_3(aq) + NO(g) + 2\ H_2O(l)$

(b) $(CH_3)_2\ NNH_2(l) + 4\ O_2(g) \longrightarrow 2\ CO_2(g) + 4\ H_2O(l) + N_2(g)$

(c) $NaH_2PO_4(s) + 2Na_2HPO_4(s) \xrightarrow{\Delta} 2H_2O(l) + Na_5P_3O_{10}(s)$

51.

(a)	$(CH_3)_2NNH_2$: Each molecule has a total of 26 valence electrons.	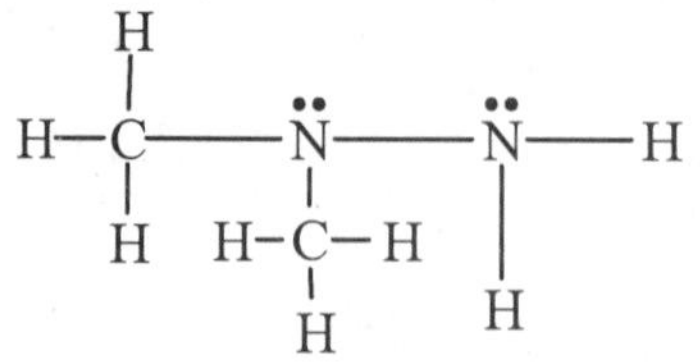
(b)	$ClNO_2$: Each molecule has a total of 24 valence electrons.	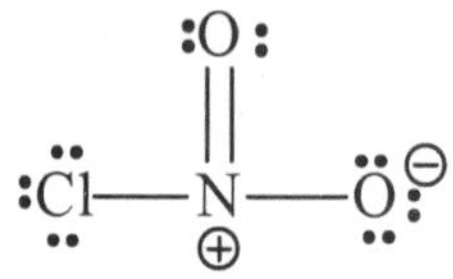
(c)	H_3PO_3 Each molecule has a total of 26 valence electrons.	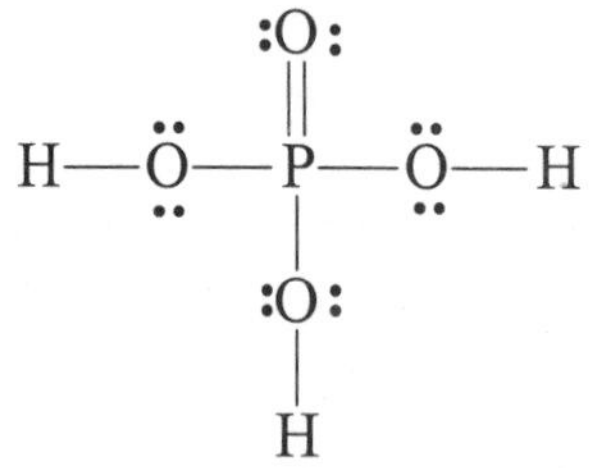

52. Hyponitrous acid is a weak diprotic acid while nitramide has an amide group, —NH$_2$

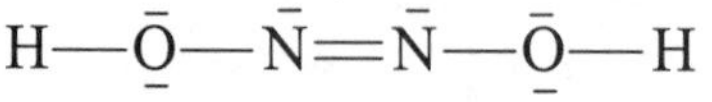

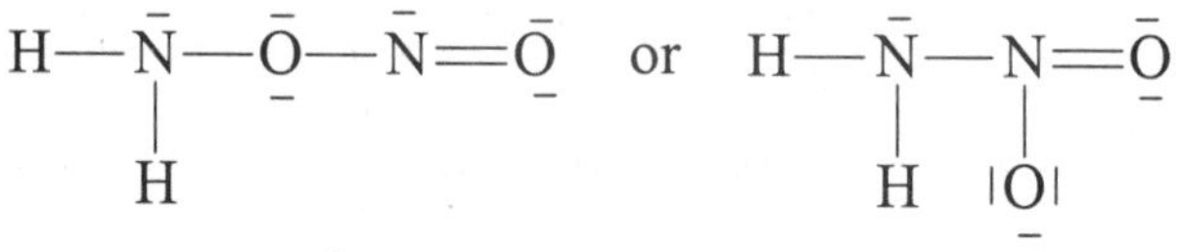

Both Lewis structures for nitramide are plausible based on the information supplied. To choose between them would require further information, such as whether the molecule contains a nitrogen-nitrogen bond. Experimental evidence indicates the structure to be the one with the nitrogen-nitrogen bond.

53. (a) HPO_4^{2-}, hydrogen phosphate ion

(b) $Ca_2P_2O_7$, calcium pyrophosphate or calcium diphosphate

(c) $H_6P_4O_{13}$, tetrapolyphosphoric acid

54. (a) $HONH_2$, hydroxylamine

(b) $CaHPO_4$, calcium hydrogen phosphate

(c) Li_3N, lithium nitride

55. i) $2 H^+(aq) + N_2O_4(aq) + 2 e^- \rightarrow 2 HNO_2(aq)$ $E^\circ_1 = 1.065$ V

     ii) $2 HNO_2(aq) + 2 H^+(aq) + 2 e^- \rightarrow 2 NO(aq) + 2 H_2O(l)$ $E^\circ_2 = 0.996$ V

     iii) $N_2O_4(aq) + 4 H^+(aq) + 4 e^- \rightarrow 2 NO(aq) + 2 H_2O(l)$ $E^\circ_3 = ?$ V

Recall that $\Delta G^\circ = -nFE^\circ$ and that ΔG° values, not standard voltages are additive for reactions in which the number of electrons do not cancel out.

So, $-4FE^\circ_3 = -2F(1.065$ V$) + -2F(0.996$ V$)$ $E^\circ_3 = 1.031$ V (4 sig figs)

56. i) $2 NO_3^-(aq) + 2 H_2O(l) + 2 e^- \rightarrow N_2O_4(aq) + 4 OH^-(aq)$ $E^\circ_1 = -0.86$ V

     ii) $N_2O_4(aq) + 2 e^- \rightarrow 2 NO_2^-(aq)$ $E^\circ_2 = +0.87$ V

     iii) $2 NO_3^-(aq) + 2 H_2O(l) + 4 e^- \rightarrow 2 NO_2^-(aq) + 4 OH^-(aq)$ $E^\circ_3 = ?$ V

Recall that $\Delta G^\circ = -nFE^\circ$ and that ΔG° values, not standard voltages are additive for reactions in which the number of electrons do not cancel out.

So, $-4FE^\circ_3 = -2F(-0.86$ V$) + -2F(0.87$ V$)$

         $-4FE^\circ_3 = 1.72F - 1.74 F = 0.02F$ (1 sig fig)

         $E^\circ_3 = 0.005$ V (1 sig fig)

Carbon and Silicon

57. In the sense that diamonds react imperceptibly slowly at room temperature (either with oxygen to form carbon dioxide, or in its transformation to the more stable graphite), it is true that "diamonds last forever." However, at elevated temperatures, diamond will burn to form $CO_2(g)$ and thus the statement is false. Also, the transformation

$C(\text{diamond}) \longrightarrow C(\text{graphite})$ might occur more rapidly under other conditions.

Eventually, of course, the conversion to graphite occurs.

58. The graphite in the pencil "lead" is a good dry lubricant that will lubricate the stickiness (or reluctance) in the lock and enable it to work smoothly. The key carries the graphite to the site that needs to be lubricated within the lock mechanism.

59. **(a)** $3 SiO_2(s) + 4 Al(s) \xrightarrow{\Delta} 2 Al_2O_3(s) + 3 Si(s)$

     (b) $K_2CO_3(s) + SiO_2(s) \xrightarrow{\Delta} CO_2(g) + K_2SiO_3(s)$

     (c) $Al_4C_3(s) + 12 H_2O(l) \longrightarrow 3 CH_4(g) + 4Al(OH)_3(s)$

60. **(a)** $2 KCN(aq) + AgNO_3(aq) \longrightarrow KAg(CN)_2(aq) + KNO_3(aq)$

     (b) $Si_3H_8(l) + 5 O_2(g) \longrightarrow 3 SiO_2(s) + 4 H_2O(l)$

     (c) $N_2(g) + CaC_2(s) \xrightarrow{\Delta} CaNCN(s) + C(s)$

61. A silane is a silicon-hydrogen compound, with the general formula Si_nH_{2n+2}. A silanol is a compound in which one or more of the hydrogens of silane is replaced by an —OH group. Then, the general formula becomes $Si_nH_{2n+1}(OH)$. In both of these classes of compounds the number of silicon atoms, n, ranges from 1 to 6. Silicones are produced when silanols condense into chains, with the elimination of a water molecule between every two silanol molecules.

$$HO{-}Si_nH_{2n}{-}OH + HO{-}Si_nH_{2n}{-}OH \longrightarrow HO{-}Si_nH_{2n}{-}O{-}Si_nH_{2n}{-}OH + H_2O$$

62. Both the alkali metal carbonates and the alkali metal silicates are soluble in water. Hence, they also are soluble in acids. However, the carbonates will produce gaseous carbon dioxide— CO_2 —on reaction with acid

$$\left(CO_3^{2-}(aq) + 2\ H^+(aq) \longrightarrow \text{``}H_2CO_3\text{''} \longrightarrow H_2O(l) + CO_2(g)\right),$$ while the silicates will

produce silica— SiO_2 — in an analogous reaction

$$\left(SiO_4^{4-}(aq) + 4H^+(aq) \longrightarrow \text{``}H_4SiO_4\text{''} \longrightarrow 2\ H_2O(l) + SiO_2(s)\right).$$ The silica is produced in

many forms depending on reaction conditions: a colloidal dispersion, a gelatinous precipitate, or a semisolid gel. Silicates of cations other than those of alkali metals are insoluble in water, as are the analogous carbonates. However, the carbonates will dissolve in acids (witness the reaction of acid rain on limesone carvings and marble statues, both forms of $CaCO_3$), while the silicates will not. (Silicate rocks are not significantly affected by acid rain.)

63. **(1)** $2\ CH_4(g) + S_8(g) \longrightarrow 2\ CS_2(g) + 4\ H_2S(g)$

(2) $CS_2(g) + 3\ Cl_2(g) \longrightarrow CCl_4(l) + S_2Cl_2(l)$

(3) $4\ CS_2(g) + 8\ S_2Cl_2(g) \longrightarrow 4\ CCl_4(l) + 3\ S_8(s)$

64. **(a)** $(CH_3)_3SiCl + H_2O \longrightarrow (CH_3)_3Si{-}OH + HCl$

$2(CH_3)_3Si{-}OH \longrightarrow (CH_3)_3Si{-}O{-}Si(CH_3)_3 + H_2O$

(b) A silicone polymer does not form from $(CH_3)_3Si{-}Cl$. Only a dimer is produced.

(c) The product that results from the treatment of CH_3SiCl_3 is two long Si—O—Si chains, with CH_3 groups on the outside, and joined by Si—O—Si bridges. Part of one of these chains and the beginnings of the bridges are shown below.

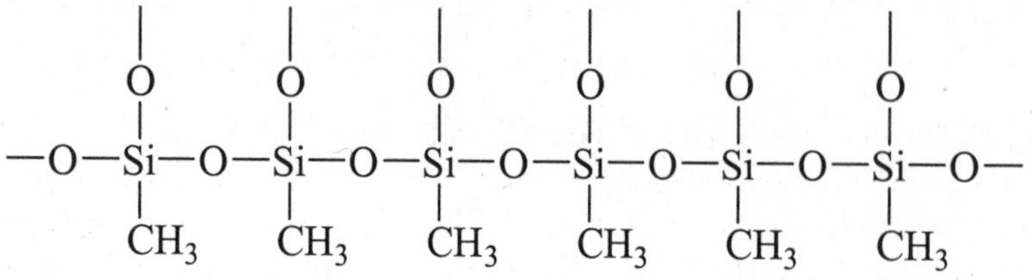

65. Muscovite or white mica has the formula $KAl_2(OH)_2(AlSi_3O_{10})$. Since they are not segregated into O_2 units in the formula, all of the oxygen atoms in the mineral must be in the -2 oxidation state. Potassium is obviously in the +1 oxidation state as are the hydrogen atoms in the hydroxyl groups. Up to this point we have -24 from the twelve oxygen atoms and +3 from the potassium and hydrogen atoms for a net number of -21 for the oxidation state. We still have three aluminum atoms and three silicon atoms to account for. In oxygen-rich salts such as mica, we would expect that the silicon and the aluminum atoms would be in their highest possible oxidation states, namely +4 and +3, respectively. Since the salt is neutral, the oxidation numbers for the silicon and aluminum atoms must add up to +21. This is precisely that total that is obtained if the silicon and aluminum atoms are in their highest possible oxidation states: $(3 \times (+3) + 3 \times (+4) = +21)$. Consequently, the empirical formula for white Muscovite is consistent with the expected oxidation state for each element present.

66. Chrysotile asbestos has the formula $[Mg_3Si_2O_5(OH)_4]$. Since they are not segregated to O_2 units in the formula, all of the oxygen atoms in the mineral must be in the -2 oxidation state. The three magnesium atoms are obviously in the +2 oxidation state, while hydrogen atoms in the hydroxyl groups have an oxidation state of +1. Up to this point then, the sum of the oxidation numbers equals -8 (-18 from O-atoms, +6 from Mg-atoms and + 4 from H-atoms). Since the mineral is neutral, the two silicon atoms must have oxidation states that sum to +8 ($-8 + 8 =$ neutral mineral), therefore, each silicon would need to be in the +4 oxidation state. In oxygen–rich salts such as asbestos, we would expect that the silicon atoms would be in their highest possible oxidation state, namely +4. Thus, the oxidation states for the element in this mineral are precisely consistent with expectations.

Boron

67. **(a)** B_4H_{10} contains a total of $4 \times 3 + 10 \times 1 = 22$ valence electrons or 11 pairs. Ten of these pairs could be allocated to form 10 B—H bonds, leaving but one pair to bond the four B atoms together, clearly an electron deficient situation.

(b) In our analysis in the first part, we noted that the four B atoms had but one electron pair to bond them together. To bond these four atoms into a chain requires three electron pairs. Since each electron pair in a bridging bond replaces two "normal" bonds, there must be at least two bridging bonds in the B_4H_{10} molecules. By analogy with B_2H_6, we might write the structure below left. But this structure uses only a total of 20 electrons. (The bridge bonds are shown as dots, normal bonds—electron pairs—as dashes.) In the structure at right below, we have retained some of the form of B_2H_6, and produced a compound with the formula B_4H_{10} and 11 electron pairs. (The experimentally determined structure of B_4H_{10} consists of a four-membered ring of alternating B and H atoms, held together by bridging bonds. Two of the B atoms have two H atoms bonds to each of them by normal covalent bonds. The other two B atoms have one H atom covalently bonded to each. One final B—B bond joins these last two B atoms, across the diameter of the ring.). See the diagram that follows:

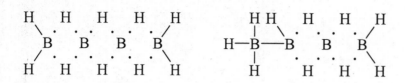

(c) C_4H_{10} contains a total of $4\times4+10\times1=26$ valence electrons or 13 pairs. A plausible Lewis structure follows.

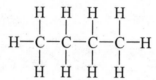

68. **(a)**

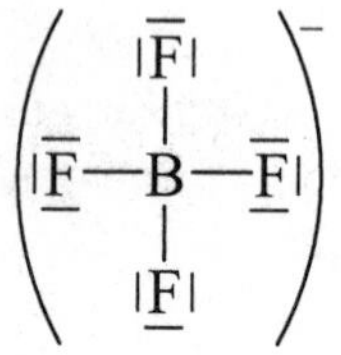

$$\left(\begin{array}{c} |\overline{F}| \\ | \\ |\overline{F}-B-\overline{F}| \\ | \\ |\overline{F}| \end{array}\right)^{-}$$

(b)

```
      :F:  H  H  H
      -|  +|  |  |
:F—B—N—C—C—H
      :F:  H  H  H
```

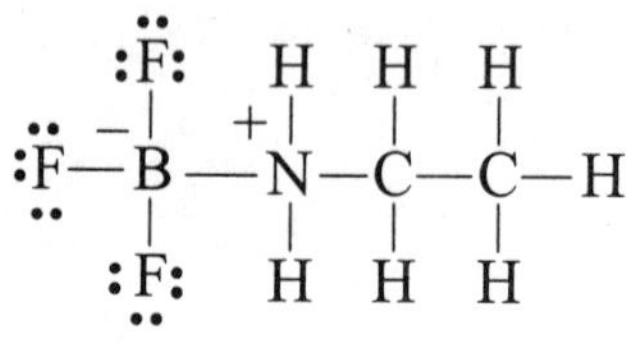

69.   **(a)** $2BBr_3(l) + 3H_2(g) \longrightarrow 2B(s) + 6HBr(g)$

(b)   i) $B_2O_3(s) + 3\,C(s) \xrightarrow{\Delta} 3\,CO(g) + 2\,B(s)$

ii) $2\,B(s) + 3\,F_2(g) \xrightarrow{\Delta} 2\,BF_3(g)$

(c) $2\,B(s) + 3\,N_2O(g) \xrightarrow{\Delta} 3\,N_2(g) + B_2O_3(s)$

70.   Each boron atom has an oxidation number of +3. The hydroxyl oxygens are each –2 while the bridging oxygens are each –1. Finally, the hydroxyl H atoms are all in the +1 oxidation state. The oxidation numbers for the all the constituent atoms add up to the charge on the perborate ion, viz 2–.

FEATURE PROBLEMS

89. We begin by calculating the standard voltages for the two steps in the decomposition mechanism. Step 1 involves the reduction of Fe^{3+} and the oxidation of H_2O_2. The two half-reactions that constitute this step are:

(i) $Fe^{3+}(aq) + e^- \rightleftharpoons Fe^{2+}(aq)$ $\qquad\qquad$ $E^{\circ}_{1,red} = 0.771$ V

(ii) $H_2O_2(aq) \rightleftharpoons O_2(g) + 2\,H^+(aq) + 2\,e^-$ $\qquad$ $E^{\circ}_{2,ox} = -0.695$ V

The balanced reaction is obtained by combining reaction (i), multiplied by two, with reaction (ii).

2(i) $2(Fe^{3+}(aq) + e^- \rightleftharpoons Fe^{2+}(aq))$

(ii) $H_2O_2(aq) \rightleftharpoons O_2(g) + 2\,H^+(aq) + 2\,e^-$

$\overline{}$

(iii) $2\,Fe^{3+}(aq) + H_2O_2(aq) \rightleftharpoons 2\,Fe^{2+}(aq) + O_2(g) + 2\,H^+(aq),$

for which $E^{\circ}_{cell} = E^{\circ}_{i(red)} + E^{\circ}_{ii(ox)} = 0.771$ V $+ (-0.695$ V$) = 0.076$ V

Since the overall cell potential is positive, this step is spontaneous. The next step involves oxidation of $Fe^{2+}(aq)$ back to $Fe^{3+}(aq)$, (i.e. the reverse of reaction (i) and the reduction of $H_2O_2(aq)$ to $H_2O(l)$ in acidic solution, for which the half-reaction is

(iv) $H_2O_2(aq) + 2\,H^+(aq) + 2\,e^- \rightleftharpoons 2\,H_2O(l)$ $\qquad$ $E^{\circ}_{iv(red)} = 1.763$ V

Combining (iv) with two times the reverse of (i) gives the overall reaction for the second step:

(i) $\{Fe^{2+}(aq) \rightleftharpoons Fe^{3+}(aq) + e^-\} \times 2$

iv) $H_2O_2(aq) + 2\,H^+(aq) + 2\,e^- \rightleftharpoons 2\,H_2O(l)$

$\overline{}$

(v) $2\,Fe^{2+}(aq) + H_2O_2(aq) + 2\,H^+(aq) \rightleftharpoons 2\,Fe^{3+}(aq) + 2\,H_2O(l)$

Thus the overall cell potential for the second step in the mechanism, via equation (v) is

$E^{\circ}_{cell} = -E^{\circ}_{i(red)} + E^{\circ}_{iv(red)} = -0.771$ V $+ (1.763$ V$) = 0.992$ V.

Since the overall standard cell potential is positive, like step 1, this reaction is spontaneous.

The overall reaction arising from the combination of these two steps is:

Step 1 $\quad$ $2\,Fe^{3+}(aq) + H_2O_2(aq) \rightleftharpoons 2\,Fe^{2+} + (aq) + O_2(g) + 2\,H^+(aq)$

Step 2 $\quad$ $2\,Fe^{2+}(aq) + H_2O_2(aq) + 2\,H^+(aq) \rightleftharpoons 2\,Fe^{3+}(aq) + 2\,H_2O(l)$

$\overline{}$

Overall $\quad$ $2\,H_2O_2(aq) \rightleftharpoons O_2(g) + 2\,H_2O(l)$

The overall potential, $E°_{overall} = E°_{step1} + E°_{step2} = 0.076$ V $+ 0.992$ V $= 1.068$ V
Therefore, the reaction is spontaneous at standard conditions.

To determine the minimum and maximum $E°$ values necessary for the catalyst, we need to consider each step separately.

In step 1, if $E°_{(1)}$ is less than 0.695 V, the overall voltage for the first step will be negative and hence non-spontaneous. In step 2, if the oxidation half-reaction has a potential that is more negative than -1.763 V, the overall potential for this step will be negative, and hence non-spontaneous. Consequently, $E°_{(1)}$ must fall between 0.695 V and 1.763 V in order for both steps to be spontaneous. On this basis we find that

(a) $Cu^{2+}(aq) + 2 e^- \rightleftharpoons Cu(s)$ $E°_{1/2red} = 0.337$ V cannot catalyze the reaction.

(b) $Br_2(l) + 2 e^- \rightleftharpoons 2 Br^-(aq)$ $E°_{1/2red} = 1.065$ V may catalyze the reaction.

(c) $Al^{3+}(aq) + 3 e^- \rightleftharpoons Al(s)$ $E°_{1/2red} = -1.676$ V cannot catalyze the reaction.

(d) $Au^{3+}(aq) + 2 e^- \rightleftharpoons Au^+(s)$ $E°_{1/2red} = 1.36$ V may catalyze the reaction.

In the reaction of hydrogen peroxide with iodic acid in acidic solution, the relevant half-reactions are:

$2 IO_3^-(aq) + 12 H^+(aq) + 10 e^- \rightleftharpoons I_2(s) + 6 H_2O(l)$ $E°_{1/2red} = 1.20$ V

$H_2O_2(aq) \rightleftharpoons O_2(g) + 2 H^+(aq) + 2e^-$ $E°_{1/2ox} = -0.695$ V

Thus, both steps in the decomposition of H_2O_2, as described above, are spontaneous if IO_3^- is used as the catalyst. As the iodate gets reduced to I_2, the I_2 forms a highly colored complex with the starch, resulting in the appearance of a deep blue color solution. Some of the H_2O_2 will be simultaneously oxidized to $O_2(g)$ by the iodic acid When sufficient $I_2(s)$ accumulates, the reduction of H_2O_2 by I_2 begins to take place and the deep blue color fades as I_2 is consumed in the reaction. Additional iodate is formed concurrently, and this goes on to oxidize the $H_2O_2(aq)$, thereby causing the cycle to repeat itself. Each cycle results in some H_2O_2 being depleted. Thus, the oscillations of color can continue until the H_2O_2 has been largely consumed.

90.

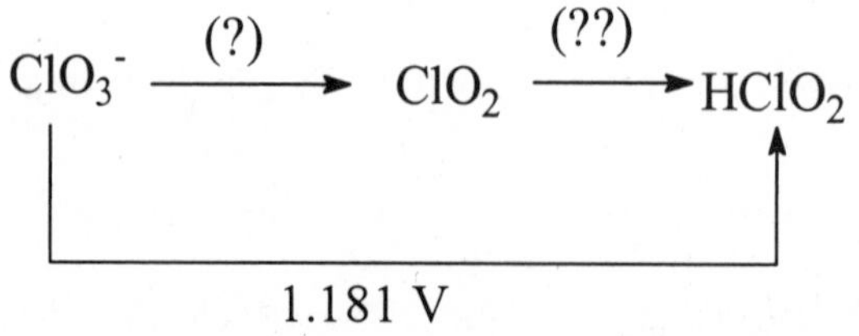

$$ClO_3^- \xrightarrow{(?)} ClO_2 \xrightarrow{(??)} HClO_2$$

1.181 V

In order to add ClO_2 to the Latimer diagram drawn above, we must calculate the voltages denoted by (?) and (??) . The equation associated with the reduction potential (?) is

(i) $2 H^+(aq) + ClO_3^-(aq) + 1 e^- \rightarrow ClO_2(g) + H_2O(l)$

The standard voltage for this half-reaction is given in Appendix D:

To finish up this problem, we just need to calculate the standard voltage (??) for the half-reaction (ii):

(ii) $H^+(aq) + ClO_2(aq) + 1 e^- \rightarrow HClO_2(g)$ $E^o = (??)$

To obtain the voltage for reaction (ii), we need to subtract reaction (i) from reaction (iii) below, which has been taken from Figure 23-2:

(iii) $3 H^+(aq) + ClO_3^-(aq) + 2 e^- \rightleftharpoons HClO_2(g) + H_2O(l)$ $E^o = 1.181$ V

Thus,

(iii) $3 H^+(aq) + ClO_3^-(aq) + 2 e^- \rightleftharpoons HClO_2(g) + H_2O(l)$ $E^o(iii) = 1.181$ V

$-1\times$ (i) $2 H^+(aq) + ClO_3^-(aq) + 1 e^- \rightarrow ClO_2(g) + H_2O(l)$ $E^o(i) = 1.175$ V

Net (ii) $H^+(aq) + ClO_2(g) + 1 e^- \rightarrow HClO_2(g)$ $E^o(ii) = (??)$

Since reactions (i) and (iii) are both reduction half reactions, we cannot simply subtract the potential for (i) from the potential for (iii). Instead, we are forced to obtain the voltage for (ii) via the free energy changes for the three half reactions. Thus,

$\Delta G(ii) = \Delta G(iii) - \Delta G(i) = -1FE^o(ii) = -2F(1.181 \text{ V}) + 1F(1.175 \text{ V})$

Dividing both sides by $-F$ gives

$E^o(ii) = 2(1.181 \text{ V}) - 1.175$ V

So, $E^o(ii) = 1.187$ V.

91.   **(a)**   As before, we can organize a solution around the balanced chemical equation.

Equation: $I_2(aq) \rightleftharpoons \qquad I_2(CCl_4)$

Initial: $1.33\times10^{-3} M$ 0 M

Initial: 0.0133 mmol

Changes: $-x$ mmol $+x$ mmol

Equil: $(0.0133-x)$ mmol x mmol

$$K_c = 85.5 = \frac{[I_2(CCl_4)]}{[I_2(aq)]} = \frac{\dfrac{x}{10.0 \text{ mL}}}{\dfrac{0.0133-x}{10.0 \text{ mL}}}$$

$$x = 1.13_7 - 85.5\,x$$

$$86.5x = 1.13_7$$

$$x = \frac{1.13_7}{86.5} = 0.01315 \text{ mmol } I_2 \text{ in } CCl_4$$

$$I_2\left(aq\right) = \left(0.0133 - 0.01315\right) \text{ mmol} = 0.00015 \text{ mmol}$$

$$\text{mass } I_2 = 0.00015 \text{ mmol} \times \frac{253.8 \text{ mg } I_2}{1 \text{ mmol } I_2} = 3.8 \times 10^{-2} \text{ mg } I_2 = 0.038 \text{ mg } I_2$$

(b) We have the same set-up, except that the initial amount $I_2\left(aq\right) = 0.00015\,\text{mmol}$.

$$K_c = 85.5 = \frac{[I_2(CCl_4)]}{[I_2(aq)]} = \frac{x}{0.00015 - x}$$

$$x = 0.0128 - 85.5x$$

$$86.5x = 0.0128$$

$$x = \frac{0.0128}{86.5} = 0.0001480 \text{ mmol} = I_2\left(CCl_4\right),$$

$$I_2\left(aq\right) = 0.00015 - 0.0001480 = 2.0 \times 10^{-6} \text{ mmol } I_2$$

$$\text{mass } I_2 = 2.0 \times 10^{-6} \text{ mmol in } H_2O \times \frac{253.8 \text{ mg } I_2}{1 \text{ mol } I_2} = 5.1 \times 10^{-4} \text{ mg } I_2 = 0.00051 \text{ mg } I_2$$

(c) If twice the volume of CCl_4 were used for the initial extraction, the equilibrium concentrations would have been different from part 1.

Equation: $\quad I_2\left(aq\right) \quad \rightleftharpoons \quad I_2\left(CCl_4\right)$

Initial: $\quad 1.33 \times 10^{-3}\,M \qquad 0\,M$

Initial: $\quad 0.0133 \text{ mmol}$

Changes: $\quad -x \text{ mmol} \qquad\qquad +x \text{ mmol}$

Equil: $\quad \left(0.0133 - x\right) \text{ mmol} \qquad x \text{ mmol}$

$$K_c = 85.5 = \frac{[I_2(CCl_4)]}{[I_2(aq)]} = \frac{\dfrac{x}{20.0 \text{ mL}}}{\dfrac{0.0133 - x}{10.0 \text{ mL}}}$$

$$x \div 2 = 1.13_7 - 85.5x$$

$$86.0x = 1.13_7$$

$$x = \frac{1.13_7}{86.0} = 1.32_2 \times 10^{-2} \text{ mmol} = \text{amount } I_2\left(CCl_4\right)$$

$$\text{total mass I}_2 = \frac{1.33 \times 10^{-3} \text{ mmol I}_2}{1 \text{ mL}} \times 10.0 \text{ mL soln} \times \frac{253.8 \text{ mg I}_2}{1 \text{ mol I}_2} = 3.376 \text{ mg I}_2$$

$$\text{mass I}_2 \text{ in CCl}_4 = 1.32_2 \times 10^{-2} \text{mmol} \times \frac{253.8 \text{ mg I}_2}{1 \text{ mol I}_2} = 3.356 \text{ mg I}_2$$

$$\text{mass I}_2 \text{ remaining in water} = 3.376 \text{ mg} - 3.356 \text{ mg} = 0.020 \text{ mg}$$

Thus, two smaller volume extractions are much more efficient than one large volume extraction.

92. **(a)** The pyroanions, a series of structurally analogous anions with the general formula $X_2O_7^{n-}$, are known for Si, P and S. The Lewis structures for these three anions are drawn below: (Note: Every member of the series has 56 valence e^-)

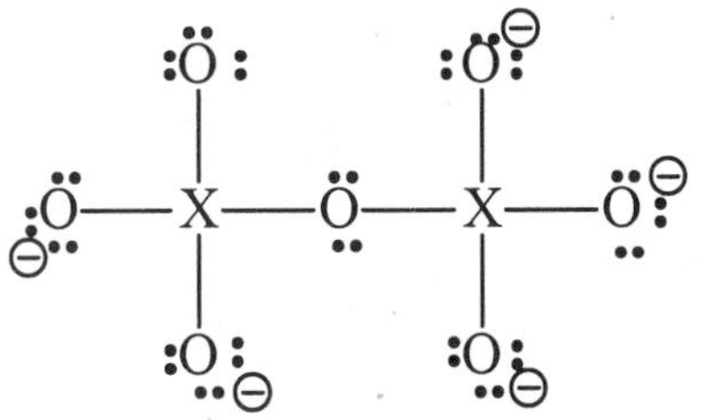

$$X = Si : Si_2O_7^{6-}$$
$$X = P_\oplus : P_2O_7^{4-}$$
$$X = S_\oplus : S_2O_7^{2-}$$

Based upon a VSEPR approach, we would predict tetrahedral geometry for each X atom (i.e. Si, P^+, S^{2+}) and a bent geometry for each bridging oxygen atom Therefore, a maximum of five atoms in each pyroanion can lie in a plane:

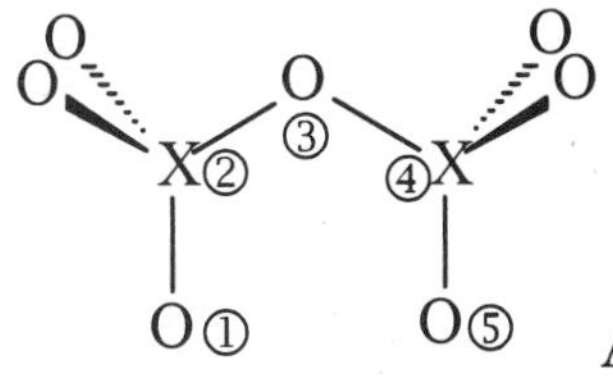

Atoms labeled 1-5 are all in the same plane.

(b) The related "mononuclear" acids, which contain just one third-row element are H_4SiO_4, H_3PO_3 and H_2SO_4. A series of pyroacids can be produced by strongly heating the "mononuclear" acids in the absence of air. In each case, the reaction proceeds via loss of water:

$$2 H_4SiO_4 \xrightarrow{\Delta} H_6Si_2O_7(l) + H_2O(l)$$

$$2 H_3PO_3 \xrightarrow{\Delta} H_4P_2O_7(l) + H_2O(l)$$

$$2 H_2SO_4 \xrightarrow{\Delta} H_2S_2O_7(l) + H_2O(l)$$

(c) The highest oxidation state that Cl can achieve is VII (+7); consequently, the chlorine containing compound that is analogous to the mononuclear acids mentioned earlier is $HClO_4$. The strong heating of perchloric acid in the absence of air, should, in principle, afford Cl_2O_7, which is the neutral chlorine analogue of the pyroanions. Thus, Cl_2O_7 is the anhydride of perchloric acid.

$$2 HClO_4 \xrightarrow{\Delta} Cl_2O_7(l) + H_2O(l)$$

93. **(a)** The bonding in the XeF_2 molecule can be explained quite simply in terms of a 3–center, 4 electron bond that spans all three atoms in the molecule. The bonding in this molecular orbital description involves the filled $5p_z$ orbital of Xe and the half-filled $2p_z$ orbitals of the two F-atoms. The linear combination of these three atomic orbitals affords one bonding, one non-bonding and one anti-bonding orbital, as depicted below:

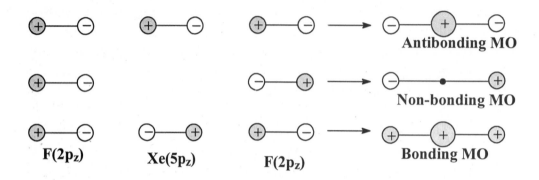

(b) If one assumes that the order of energies for molecular orbitals is:

bonding MO < non-bonding MO < antibonding MO,

the following molecular orbital representation of the bonding can be sketched.

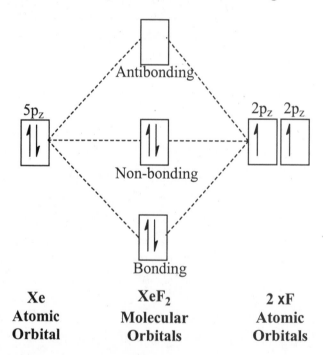

Thus, a single bonding pair of electrons is responsible for holding all three atoms together. The non-bonding pair of electrons is localized primarily on the two F–atoms. This suggests that the bond possesses substantial ionic character. Bond order is defined as one-half the difference between the number of bonding electrons and the number of antibonding electrons. In the case of XeF_2, the overall bond order is

therefore: $\dfrac{2\,e^- - 0\,e^-}{2\,e^-} = 1.0$. (i.e., each Xe—F bond has a bond order of 0.5)

(c) By invoking a molecular orbital description based upon three-center bonding for the three F—Xe—F units in XeF_6, one obtains an octahedral structure in which there are six identical Xe—F bonds. The extra nonbonding pair of electrons is delocalized over the entire structure in this scheme. In spite of its manifest appeal, this explanation of the bonding in XeF_6 is incorrect, or at best a stretch, because the actual structure for this molecule is a distorted octahedron. A more accurate description of the stereochemistry adopted by XeF_6 is provided by VSEPR theory. In this approach, the shape of the molecule is determined by repulsions between bonding and non-bonding electrons in the valence shell of the central Xe atom.

Accordingly, in XeF_6, six bonding pairs and one lone pair of electrons surround the Xe. Having the repulsions of 7 pairs of electrons to cope with, the XeF_6 molecule adopts a distorted structure that approaches either a monocapped octahedron or a pentagonal bipyramidal arrangement of electron pairs, depending on how one chooses to view the structure. In either case, these two shapes are much closer to the true shape for XeF_6 than that predicted by the molecular orbital treatment involving three, three-center bonds. By contrast, the molecular orbital description involving three-center bonds gives far better results when applied to XeF_4. In this instance, with two three-center F—Xe—F bonds in the structure, molecular orbital theory predicts that XeF_4 should adopt a square planar structure.

The result here is quite satisfactory because XeF_4 does in fact exhibit square planar geometry. It is worth noting, however, that a square planar shape for XeF_4 is also predicted by VSEPR theory. Despite the fact that the molecular orbital method has made some inroads as of late, VSEPR is still the best approach available for rationalizing the molecular geometries of noble gas compounds.

94. (a) In the phase diagram sketched in the problem, extend the liquid-vapor equilibrium curve (the vapor pressure curve) to lower temperatures (supercooled liquid region). Extend the S_α-vapor equilibrium curve (sublimation curve) to the temperature at which it intersects the extended vapor pressure curve. This should come at 113 °C. Draw a line from this point of intersection to the "peak" of the S_β phase region. Erase the three lines that bound the S_β region in the original sketch. The remaining three lines would represent the equilibria, S_α-V, S_α-L, and L-V, producing the phase diagram if S_α (rhombic) were the only solid form of sulfur.

(b) If rhombic sulfur is heated rapidly, the transition to monoclinic sulfur at 95.3 °C might not occur. (Solid-state transitions are often very slow.) In this case, rhombic sulfur would melt at 113 °C, as described in the modified phase diagram in part (a). If the molten rhombic sulfur is further heated and then cooled, the liquid very likely will freeze at the equilibrium temperature of 119 °C, producing *monoclinic*, not rhombic, sulfur.

CHAPTER 24
THE TRANSITION ELEMENTS
PRACTICE EXAMPLES

1A **(a)** Cu_2O should form. $2\ Cu_2S(s) + 3\ O_2(g) \rightarrow 2\ Cu_2O(s) + 2\ SO_2(g)$

      (b) $W(s)$ is the reduction product. $WO_3(s) + 3\ H_2(g) \rightarrow W(s) + 3\ H_2O(g)$

      (c) $Hg(l)$ forms. $2\ HgO(s) \xrightarrow{\ \Delta\ } 2\ Hg(l) + O_2(g)$

1B **(a)** $SiO_2(s)$ is the oxidation product of Si $3\ Si(s) + 2\ Cr_2O_3(s) \xrightarrow{\ \Delta\ } 3\ SiO_2(s) + 4\ Cr(l)$

      (b)   Roasting is simply heating in air. $2\ Co(OH)_3(s) \xrightarrow{\ \Delta,\,Air\ } Co_2O_3(s) + 3H_2O(g)$

      (c) $MnO_2(s)$ forms; (acidic solution). $Mn^{2+}(aq) + 2H_2O(l) \rightarrow MnO_2(s) + 4H^+(aq) + 2e^-$

2A   We write and combine the half-equations for oxidation and reduction. If $E° > 0$, the reaction is spontaneous.

Oxidation: $\{V^{3+}(aq) + H_2O(l) \rightarrow VO^{2+}(aq) + 2H^+(aq) + e^-\} \times 3$ $E° = -0.337V$

Reduction: $NO_3^-(aq) + 4H^+(aq) + 3e^- \rightarrow NO(g) + 2H_2O$ $E° = +0.956V$

Net: $NO_3^-(aq) + 3V^{3+}(aq) + H_2O(l) \rightarrow NO(g) + 3VO^{2+}(aq) + 2H^+(aq)$ $E_{cell}° = +0.619V$

Nitric acid can be used to oxidize $V^{3+}(aq)$ to $VO^{2+}(aq)$.

2B   The reducing couple that we seek must have a half-cell potential of such a size and sign that a positive sum results when this half-cell potential is combined with $E°\{VO^{2+}(aq)|V^{2+}(aq)\} = 0.041\,V$ (this is the weighted average of the $VO^{2+}|V^{3+}$ and $V^{3+}|V^{2+}$ reduction potentials) *and* a negative sum must be produced when this half-cell potential is combined with $E°\{V^{2+}(aq)|V(s)\} = -1.13\,V$.

So, $-E°$ for the couple must be> -0.041 V and <+ 1.13 V (i.e. it cannot be more positive than 1.13V, nor more negative than -0.041 V)

Some possible reducing couples from Table 21-1 are:
$-E°\{Cr^{3+}(aq)|Cr^{2+}(aq)\} = +0.42\ V$; $-E°\{Fe^{2+}(aq)|Fe(s)\} = +0.440\ V$
$-E°\{Zn^{2+}(aq)|Zn(s)\} = +0.763\ V$; thus $Fe(s)$, $Cr^{2+}(aq)$, $Zn(s)$ will do the job.

REVIEW QUESTIONS

1.   **(a)**   A domain is a region within a metal in which the magnetic moments of the metal atoms align. In some ways, it behaves as a small magnet.

(b)   Flotation is a process of separating metal ore from the extra rock, called *gangue*. It involves agitating the crushed rock and ore with water and a detergent. The ore particles stick to the bubbles and are skimmed off.

(c)   Leaching refers to treating ore-bearing material with a chemical solution to dissolve the desired metallic elements. The solution generally is trickled through the material, and the resulting metal-bearing solution is collected at the bottom.

(d)   An amalgam is an alloy (or solution) of another metal in mercury.

2.   **(a)**   The lanthanide contraction refers to the steady decrease in the size of atoms and (particularly) +3 ions as one proceeds across the lanthanide series from La to Lu.

(b)   Zone refining uses a ring oven to melt a portion of a cylindrical metal bar. Impurities are more soluble in the molten metal and tend to remain in the molten portion. The molten region then is swept down the bar to the end, carrying the impurities with it. The process can be repeated several times, concentrating impurities at one end of the bar, which is cut off and returned to be refined.

(c)   The basic oxygen process is a relatively recently (1950s) developed technique of steel making in which molten pig iron is treated with oxygen gas (at about 10 atm pressure) and powdered limestone to reduce the carbon content and remove the undesirable impurities.

(d)   Slag formation occurs when a flux, such as limestone, is added to an ore being reduced or a metal being refined. $CaO(s)$ forms from thermal decomposition and this unites with SiO_2 to form $CaSiO_3$. In this molten slag, impurities dissolve. The slag also floats on top of the reduced metal, protecting it from oxidation. Other nonmetal oxides, such as P_4O_{10}, can take the place of SiO_2 giving slags of other formulas: $Ca_3(PO_4)_2$. In this way low-level impurities are removed from the metal.

3.   **(a)**   Paramagnetism indicates that atoms, molecules, or ions have one or more unpaired electrons. In ferromagnetism, the unpaired electrons on several adjacent atoms or ions align with each other, producing a stronger magnetic field.

(b)   Roasting refers to heating a concentrated ore, usually to convert hydroxides, carbonates, or sulfides to oxides. Refining is the process of purifying the crude metal that results from the reduction of the ore.

(c)   In hydrometallurgy, reactions in aqueous solution are used in concentration, reduction, and/or refining. In pyrometallurgy, the process of reduction and/or refining are carried out at high temperatures.

(d)   Chromate ion is CrO_4^{2-}, a good precipitating agent; dichromate ion is $Cr_2O_7^{2-}$, a good oxidizing agent.

<u>**4.**</u>   **(a)** $Sc(OH)_3$, scandium hydroxide

(b) Cu_2O,   copper(I) oxide

(c) $TiCl_4$,   titanium(IV) chloride or titanium tetrachloride

    (d) V_2O_5,   vanadium(V) oxide

    (e) K_2CrO_4, potassium chromate

    (f) K_2MnO_4, potassium manganate

5. **(a)** CrO_3,   chromium(VI) oxide   **(b)** $FeSiO_3$,   iron(II) silicate

    (c) $BaCr_2O_7$,   barium dichromate   **(d)** $CuCN$,   copper(I) cyanide

    (e) $CoCl_2 \cdot 6H_2O$, cobalt(II) chloride hexahydrate

6. **(a)** Pig iron is the iron obtained from the blast furnace. Once poured into molds it is called cast iron. "It contains about 95% Fe, 3% to 4% C, and various other impurities."

    (b) Ferromanganese is an alloy of iron and manganese, produced by reducing the mixed oxides of the two metals with carbon.

    (c) Chromite ore is the principal ore of chromium, $Fe(CrO_2)_2$

    (d) Brass is a mixture of copper and zinc, with small quantities of Sn, Pb, and Fe.

    (e) Aqua regia is a mixture of three parts HCl(aq) and one part $HNO_3(aq)$. It is both strongly oxidizing acid can form complex ions with cations in solution, making it the only acid mixture that can dissolve sparingly soluble substances, such as Au and HgS.

    (f) Blister copper is the impure copper that results from the reduction process. The blisters are frozen bubbles of the $SO_2(g)$ that form during that process.

    (g) Stainless steel contains iron, and a significant proportion (about 10% each) of chromium and nickel. It is not ferromagnetic.

7. **(a)** $TiCl_4(g) + 4Na(l) \xrightarrow{\Delta} Ti(s) + 4NaCl(l)$

    (b) $Cr_2O_3(s) + 2Al(s) \xrightarrow{\Delta} 2\,Cr(l) + Al_2O_3(s)$

    (c) $Ag(s) + HCl(aq) \rightarrow$ no reaction

    (d) $K_2Cr_2O_7(aq) + 2KOH(aq) \rightarrow 2\,K_2CrO_4(aq) + H_2O(l)$

    (e) $MnO_2(s) + 2\,C(s) \xrightarrow{\Delta} Mn(l) + 2CO(g)$

8. **(a)** Oxidation: $\{Fe_2S_3(s)+6\ OH^-(aq)\rightarrow 2Fe(OH)_3(s)+3S(s)+6e^-\}$ $\times 2$

Reduction: $\{O_2(g)+2\ H_2O(l)+4e^-\rightarrow 4\ OH^-(aq)\}$ $\times 3$

Net: $2\ Fe_2S_3(s)+3\ O_2(g)+6\ H_2O(l)\rightarrow 4\ Fe(OH)_3(s)+6S(s)$

(b) Oxidation: $\{Mn^{2+}(aq)+4\ H_2O(l)\rightarrow MnO_4^-(aq)+8\ H^+(aq)+5e^-\}$ $\times 2$

Reduction: $\{S_2O_8^{2-}(aq)+2e^-\rightarrow 2\ SO_4^{2-}(aq)\}$ $\times 5$

Net: $2\ Mn^{2+}(aq)+8\ H_2O(l)+5\ S_2O_8^{2-}(aq)\rightarrow 2\ MnO_4^-(aq)+16H^+(aq)+10\ SO_4^{2-}(aq)$

(c) Oxidation: $\{Ag(s)+2\ CN^-(aq)\rightarrow\left[Ag(CN)_2\right]^-(aq)+e^-\}$ $\times 4$

Reduction: $O_2(g)+2\ H_2O(l)+4e^-\rightarrow 4\ OH^-(aq)$

Net: $4\ Ag(s)+8\ CN^-(aq)+O_2(g)+2\ H_2O(l)\rightarrow 4\left[Ag(CN)_2\right]^-(aq)+4OH^-(aq)$

9. **(a)** $Cr(s)+2HCl(aq)\rightarrow CrCl_2(aq)+H_2(g)$ Virtually any first period transition metal except Cu can be substituted for Cr, with a different metallic oxidation state, as well.

(b) $Cr_2O_3(s)+2\ OH^-(aq)+3\ H_2O(l)\rightarrow 2\ Cr(OH)_4^-(aq)$

The oxide must be amphoteric. Thus, Sc_2O_3, TiO_2, ZrO_2, and ZnO could be substituted for Cr_2O_3.

(c) $2\ La(s)+6\ HCl(aq)\rightarrow 2\ LaCl_3(aq)+3\ H_2(g)$

Any lanthanide or actinide element can be substituted for lanthanum.

10. **(a)** Ti   [Ar] $3d$ $4s$ **(b)** V^{3+} [Ar] $3d$ $4s$

(c) Cr^{2+} [Ar] $3d$ $4s$ **(d)** Mn^{4+} [Ar] $3d$ $4s$

(e) Mn^{2+} [Ar] $3d$ $4s$ **(f)** Fe^{3+} [Ar] $3d$ $4s$

11. We first give the orbital diagram for each of the species, and then count the number of unpaired electrons.

Fe   [Ar] $3d$ $4s$ 4 unpaired electrons

Sc^{3+} [Ar] $3d$ $4s$ 0 unpaired electrons

Ti^{2+} [Ar] $3d$ $4s$ 2 unpaired electrons

Mn^{4+} [Ar] $3d$ $4s$ 3 unpaired electrons

Cr   [Ar] $3d$ ↑↑↑↑↑ $4s$ ↑   6 unpaired electrons

Cu^{2+} [Ar] $3d$ ↑↓↑↓↑↓↑↓↑ $4s$ 1 unpaired electron

Finally, arrange in order of decreasing number of unpaired electrons:

$Cr > Fe > Mn^{4+} > Ti^{2+} > Cu^{2+} > Sc^{3+}$

12.   Transition metals tend to have <u>higher melting points</u> than representative metals. Because they are metals, transition elements have relatively <u>low ionization energies</u>. Ions of transition metals often <u>are colored</u> in aqueous solution. Because they are metals and thus readily form cations, they have <u>negative standard reduction potentials</u>. Their compounds often have unpaired electrons because of the diversity of d-electron configurations, and thus, they often are <u>paramagnetic</u>. Consequently, the correct answers are (c) and (e).

13. Sc^{3+} has the electron configuration of the preceding noble gas (Ar). This is obviously an electron configuration in which all electrons are paired, a diamagnetic species. All of the other species have at least one unpaired electron, as orbital diagrams show.

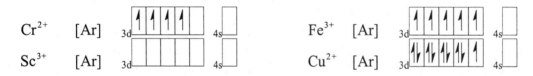

Cr^{2+} [Ar] $3d$ ↑↑↑↑ $4s$ Fe^{3+} [Ar] $3d$ ↑↑↑↑↑ $4s$

Sc^{3+} [Ar] $3d$ $4s$ Cu^{2+} [Ar] $3d$ ↑↓↑↓↑↓↑↓↑ $4s$

14.   Ti should not display a +6 oxidation state, because in order for Ti to display a +6 oxidation state, two electrons would have to be removed from the noble gas electron configuration of Ar. This is quite unlikely.

15.   The ion that is the best oxidizing agent in aqueous solution is the ion that can most readily be reduced. And we can assess that property either by standard reduction potentials

—$E°\{Ag^+(aq)|Ag(s)\} = +0.800$ V,

$E°\{Cu^{2+}(aq)|Cu(s)\} = +0.337$ V, $E°\{Zn^{2+}(aq)|Zn(s)\} = -0.763$ V,

$E°\{Na^+(aq)|Na(s)\} = -2.714$ V, with the most positive value indicating the best oxidizing agent, or from our own experience with the reverse reaction. We know that silver is the least likely of all the metals listed to be oxidized, and thus its ion is the most readily reduced.

16.   The +3 state probably is the most stable oxidation state of iron because in this oxidation state iron has a d^5 electron configuration. As we have seen before, an electron configuration containing a half-filled subshell is unexpectedly stable energetically, compared to other configurations with partially filled subshells. In the cases of cobalt and nickel, the ions with d^5 configurations are Co^{4+} and Ni^{5+}. The slight advantage of a d^5 configuration is not sufficient to stabilize ions with such high charges.

EXERCISES

Properties of the Transition Elements

17. A given main group metal typically displays one oxidation state, usually equal to its family number in the periodic table. Exceptions are elements such as Tl (+1 and +3), Pb (+2 and +4), and Sn (+2 and +4) in which the lower oxidation state represents a pair of s electrons not being ionized (a so-called "inert pair").

Main group metals do not form a wide variety of complex ions, with Al^{3+}, Sn^{2+}, Sn^{4+}, and Pb^{2+} being major exceptions. On the other hand, most transition metal ions form an extensive variety of complex ions. Most compounds of main group metals are colorless; exceptions occur when the anion is colored. On the other hand, many of the compounds of transition metal cations are colored, due to d-d electron transitions. Virtually every main group metal cation has no unpaired electrons and hence is diamagnetic. On the other hand, many transition metals cations have one or more unpaired electrons and therefore are paramagnetic.

18. As we proceed from Sc to Cr the valence electron configuration has an increasing number of unpaired electrons, which are capable of forming bonds to adjacent atoms. As we continue beyond Cr, however, these electrons become paired, and the resulting atoms are less able to form bonds with their neighbors. Also, smaller atoms tend to bond more tightly, making their metallic agglomerations more difficult to melt. (i.e., the smaller the metal atom, the higher the melting point). This trend of increasing melting point with decreasing size is clearly visible for the first transition series from Sc to Zn.

19. When an electron is added to a main group element to create the element of next highest atomic number, this electron is added to the outer shell of the atom, far from the nucleus. Thus, it has a major influence on the size of the atom. However, when an electron is added to a transition metal atom to create the atom of next highest atomic number, it is added to the electronic shell inside the outermost. The electron thus has been added to a position close to the nucleus to which it is attracted quite strongly and thus it has small effect on the size of the atom.

20. The reason why the radii of Pd (138 pm) and Pt (139 pm) are so similar, and so different from the radius of Ni (125 pm) is because the lanthanide series intrudes between Pd and Pt. Because of the lanthanide contraction, elements in the second transition row are almost identical in size to their congeners (family members) in the third transition row.

21. Of the first transition series, manganese exhibits the greatest number of different oxidation states in its compounds, every state from +1 to +7 . One possible explanation might be its $3d^5 4s^2$ electron configuration. Removing one electron produces an electron configuration ($3d^5 4s^1$) with two half-filled subshells, removing two produces one with a half-filled and an empty subshell. Then there is no point of semistability until the remaining five d electrons are removed. These higher oxidation states all are stabilized by being present in oxides (MnO_2) or oxoanions (e.g. MnO_4^-).

22.   At the beginning of the series, there are few electrons beyond the last noble gas than can be ionized and thus the maximum oxidation state is limited. Toward the end of the series, many of the electrons are paired up or the d subshell is filled, and thus a somewhat stable situation would be disrupted by ionization.

23.   The greater ease of forming lanthanide cations compared to forming transition metal cations, is due to the larger size of lanthanide atoms. The valence (outer shell) electrons of these larger atoms are much further from the nucleus, much less strongly attracted to the positive charge of the nucleus, and thus are removed much more readily.

24.   As we proceed across the transition series, electrons are being added to a shell next to the valence shell. Thus, the electronic character of the outside of each atom is changing somewhat, leading to changes in their chemical properties. (Recall that this valence shell consists of only two $4s$ electrons.) As we proceed across the lanthanides, on the other hand, the electrons are being added to a shell two removed from the outermost shell; there is less of an effect on the electronic character of the valence shell. (Recall also that there are 10 electrons in subshells outside the one being filled: $5s^2$, $5p^6$ and $6s^2$.)

Reactions of Transition Metals and Their Compounds

25.   (a) $Sc(OH)_3(s) + 3H^+(aq) \rightarrow Sc^{3+}(aq) + 3 H_2O(l)$

      (b) $3 Fe^{2+}(aq) + MnO_4^-(aq) + 2 H_2O(aq) \rightarrow 3 Fe^{3+}(aq) + MnO_2(s) + 4 OH^-(aq)$

      (c) $2 KOH(l) + TiO_2(s) \xrightarrow{\Delta} K_2TiO_3(s) + H_2O(g)$

      (d) $Cu(s) + 2 H_2SO_4(conc, aq) \rightarrow CuSO_4(aq) + SO_2(g) + 2 H_2O(l)$

26.   (a) $2 Sc_2O_3(l, in\ Na_3ScF_6) + 3 C(s) \xrightarrow{electrolysis} 4 Sc(l) + 3 CO_2(g)$
            [By analogy with the equation for the electrolytic production of Al]

      (b) $Cr(s) + 2 HCl(aq) \rightarrow Cr^{2+}(aq) + 2 Cl^-(aq) + H_2(g)$

      (c) $4 Cr^{2+}(aq) + O_2(g) + 4 H^+(aq) \rightarrow 4 Cr^{3+}(aq) + 2 H_2O(l)$

      (d) $Ag(s) + 2 HNO_3(aq) \rightarrow AgNO_3(aq) + NO_2(g) + H_2O(l)$

27.   We write some of the following reactions as total equations rather than as net ionic equations so that the reagents used are indicated.

      (a) $FeS(s) + 2 HCl(aq) \rightarrow FeCl_2(aq) + H_2S(g)$

            $4 Fe^{2+}(aq) + O_2(g) + 4H^+(aq) \rightarrow 4Fe^{3+}(aq) + 2 H_2O(l)$

            $Fe^{3+}(aq) + 3 OH^-(aq) \rightarrow Fe(OH)_3(s)$

(b) $BaCO_3(s) + 2\ HCl(aq) \rightarrow BaCl_2(aq) + H_2O(l) + CO_2(g)$

$2\ BaCl_2(aq) + K_2Cr_2O_7(aq) + 2\ NaOH(aq)$
$$\rightarrow 2\ BaCrO_4(s) + 2\ KCl(aq) + 2\ NaCl(aq) + H_2O(l)$$

28. **(a)** i) $CuO(s) + 2H^+(aq) \longrightarrow Cu^{2+}(aq) + H_2O(l)$

ii) $Cu^{2+}(aq) + 2OH^-(aq) \longrightarrow Cu(OH)_2(s)$

(b) $(NH_4)_2 Cr_2O_7(s) \xrightarrow{\Delta} N_2(g) + 4\ H_2O(g) + Cr_2O_3(s)$

$Cr_2O_3(s) + 6\ HCl(aq) \rightarrow CrCl_3(aq) + 3\ H_2O(l)$

29. $HgS(s) + O_2(g) \xrightarrow{\Delta} Hg(l) + SO_2(g)$

$4\ HgS(s) + 4\ CaO(s) \xrightarrow{\Delta} 4\ Hg(l) + 3\ CaS(s) + CaSO_4(s)$

30. **(a)** $C(s) + O_2(g) \rightarrow CO_2(g)$ $\Delta S° = 213.7\ J\ K^{-1} - (5.74\ J\ K^{-1} + 205.1\ J\ K^{-1})$
$$\Delta S° = 2.86\ J\ K^{-1}$$

Since $\Delta S° \approx 0$, $\Delta G°$ will be relatively constant

(b) $2\ CO(g) + O_2(g) \rightarrow 2\ CO_2(g)$ $\Delta S° = (2(213.7\ J\ K^{-1}) - (205.1\ J\ K^{-1} + 2(197.7\ J\ K^{-1}))$
$$\Delta S° = -173.1\ J\ K^{-1}$$

Since ΔS is less than zero, $\Delta G°$ will increase (less negative or more positive) as temperature increases.

31. The plot of $\Delta G°$ versus T will consist of three lines of increasing positive slope. The first line is joined to the second line at the melting point for Ca(s), while the second line is joined to the third at the boiling point for Ca(l).
$2\ Ca(s) + O_2(g) \rightarrow 2\ CaO(s)$ $\Delta H°_f = -1270.2\ kJ$
$\Delta S° = 2(39.75\ J\ K^{-1}) - 2(41.42\ J\ K^{-1} + 205.1\ J\ K^{-1}) = -208.4\ J\ K^{-1}$

The graph should be similar to that for $2\ Mg(s) + O_2(g) \rightarrow 2\ MgO(s)$. We expect a positive slope with slight changes in the slope after the melting point(839 °C) and boiling points(1484 °C) ,owing mainly to changes in entropy. The plot will always below the $\Delta G°$ line for $2\ Mg(s) + O_2(g) \rightarrow 2\ MgO(s)$ at all temperatures.

32. $2\ Na_2CrO_4(s) + 3\ C(s) + 4HCl(aq) \rightarrow Cr_2O_3(s) + 2H_2O(l) + 3\ CO(g) + 4NaCl(aq)$
$2\ Cr_2O_3(s) + 3\ Si(s) \rightarrow 4\ Cr(s) + 3\ SiO_2(s)$

Oxidation-reduction

33. **(a)** Reduction: $\quad VO^{2+}(aq) + 2H^+(aq) + e^- \rightarrow V^{3+}(aq) + H_2O\,(l)$

    (b) Oxidation: $\quad Cr^{2+}(aq) \rightarrow Cr^{3+}(aq) + e^-$

34. **(a)** Oxidation: $\quad Fe(OH)_3\,(s) + 5\ OH^-(aq) \rightarrow FeO_4{}^{2-}(aq) + 4\ H_2O(l) + 3e^-$

    (b) Reduction: $\quad \left[Ag(CN)_2\right]^-(aq) + e^- \rightarrow Ag(s) + 2\ CN^-(aq)$

35. **(a)** First we need the reduction potential for the couple $VO_2{}^+(aq)/V^{2+}(aq)$. We will use a method learned in Chapter 21 (see page 834).

$$VO_2{}^+(aq) + 2\ H^+(aq) + e^- \rightarrow VO^{2+}(aq) + H_2O(l) \quad \Delta G^\circ = -1\ F(+1.000\ V)$$

$$VO^{2+}(aq) + 2\ H^+(aq) + e^- \rightarrow V^{3+}(aq) + H_2O(l) \quad \Delta G^\circ = -1\ F(+0.337\ V)$$

$$V^{3+}(aq) + e^- \rightarrow V^{2+}(aq) \quad \Delta G^\circ = -1\ F(-0.255\ V)$$

$$VO_2{}^+(aq) + 4\ H^+(aq) + 3\ e^- \rightarrow V^{2+}(aq) + 2\ H_2O \quad \Delta G^\circ = -3\ FE^\circ$$

$$E^\circ = \frac{1.000\ V + 0.337\ V - 0.255\ V}{3} = +0.361\ V$$

We next analyze the oxidation-reduction reaction.

Oxidation:
$$\{2\ Br^-(aq) \rightarrow Br_2\,(l) + 2e^-\} \qquad\qquad \times 3 \qquad\qquad -E^\circ = -1.065\ V$$
Reduction: $\{VO_2{}^+(aq) + 4H^+(aq) + 3e^- \rightarrow V^{2+}(aq) + 2H_2O(l)\} \times 2\ \ E^\circ = +0.361V$

Net: $6\ Br^-(aq) + 2\ VO_2{}^+(aq) + 8\ H^+(aq) \rightarrow 3\ Br_2\,(l) + 2\ V^{2+}(aq) + 4\ H_2O$

$E^\circ_{cell} = -0.704V \qquad$ This reaction does not occur to a significant extent as written.

    (b) Oxidation: $Fe^{2+}(aq) \rightarrow Fe^{3+}(aq) + e^- \qquad\qquad -E^\circ = -0.771V$

          Reduction: $VO_2{}^+(aq) + 2\ H^+(aq) + e^- \rightarrow VO^{2+}(aq) + H_2O(l) \quad E^\circ = +1.000V$

Net: $\qquad Fe^{2+}(aq) + VO_2{}^+(aq) + 2\ H^+(aq) \rightarrow Fe^{3+}(aq) + VO^{2+}(aq) + H_2O(l)$

$E^\circ_{cell} = +0.229V$

This reaction does occur to a significant extent under standard conditions.

(c) Oxidation: $H_2O_2 \rightarrow 2\,H^+(aq) + 2\,e^- + O_2(g)$ $-E^\circ = -0.695$ V

Reduction: $MnO_2(s) + 4\,H^+(aq) + 2\,e^- \rightarrow Mn^{2+}(aq) + 2\,H_2O(l)$ $E^\circ = +1.23$V

Net: $H_2O_2 + MnO_2(s) + 2\,H^+(aq) \rightarrow O_2(g) + Mn^{2+}(aq) + 2\,H_2O(l)$ $E_{cell}{}^\circ = +0.54$V

This reaction does occur to a significant extent under standard conditions.

36. When a species acts as a reducing agent, it is oxidized. From Appendix D we obtain the potentials for each of the following couples.

$-E^\circ\{Zn^{2+}(aq)/Zn(s)\} = +0.763\,V$; $-E^\circ\{Sn^{4+}(aq)/Sn^{2+}(aq)\} = -0.154V$

$-E^\circ\{I_2(s)/I^-(aq)\} = -0.535$ V

Each of these potentials is combined with the reduction potential cited in each part. If the resulting value of E°_{cell} is positive, then the reducing agent will be effective in accomplishing the desired reduction, which we indicate with "yes"; if not, we write "no".

(a) $E^\circ\{Cr_2O_7{}^{2-}(aq)/Cr^{3+}(aq)\} = +1.33$ V

$E^\circ_{cell} = E^\circ\{Cr_2O_7{}^{2-}(aq)/Cr^{3+}(aq)\} - E^\circ\{Zn^{2+}(aq)/Zn(s)\}$

$E^\circ_{cell} = +1.33\;V + 0.763\;V = +2.09\;V \Rightarrow$ Yes

$E^\circ_{cell} = E^\circ\{Cr_2O_7{}^{2-}(aq)/Cr^{3+}(aq)\} - E^\circ\{Sn^{4+}(aq)/Sn^{2+}(aq)\}$

$E^\circ_{cell} = +1.33\;V - 0.154\;V = +1.18\;V \Rightarrow$ Yes

$E^\circ_{cell} = E^\circ\{Cr_2O_7{}^{2-}(aq)/Cr^{3+}(aq)\} - E^\circ\{I_2(s)/I^-(aq)\}$

$E^\circ_{cell} = +1.33\;V - 0.535\;V = +0.80\;V \Rightarrow$ Yes

(b) $E^\circ\{Cr^{3+}(aq)/Cr^{2+}(aq)\} = -0.424$ V

$E^\circ_{cell} = E^\circ\{Cr^{3+}(aq)/Cr^{2+}(aq)\} - E^\circ\{Zn^{2+}(aq)/Zn(s)\}$

$E^\circ_{cell} = -0.424\;V + 0.763\;V = +0.339\;V \Rightarrow$ Yes

$E^\circ_{cell} = E^\circ\{Cr^{3+}(aq)/Cr^{2+}(aq)\} - E^\circ\{Sn^{4+}(aq)/Sn^{2+}(aq)\}$

$E^\circ_{cell} = -0.424\;V - 0.154\;V = -0.578\;V \Rightarrow$ No

$E^\circ_{cell} = E^\circ\{Cr^{3+}(aq)/Cr^{2+}(aq)\} - E^\circ\{I_2(s)/I^-(aq)\}$

$E^\circ_{cell} = -0.424\;V - 0.535\;V = -0.959\;V \Rightarrow$ No

(c) $E^\circ\{SO_4{}^{2-}(aq)/SO_2(g)\} = +0.17$ V

$E^\circ_{cell} = E^\circ\{SO_4{}^{2-}(aq)/SO_2(g)\} - E^\circ\{Zn^{2+}(aq)/Zn(s)\}$

$E^\circ_{cell} = = +0.17\;V + 0.763\;V = +0.93\;V \Rightarrow$ Yes

$$E_{cell}^{o} = E^{o}\{SO_4^{\,2-}(aq)/SO_2(g)\} - E^{o}\{Sn^{4+}(aq)/Sn^{2+}(aq)\}$$

$$E_{cell}^{o} = +0.17\ V - 0.154\ V = +0.02\ V \quad \Rightarrow \quad Yes$$

$$E_{cell}^{o} = E^{o}\{SO_4^{\,2-}(aq)/SO_2(g)\} - E^{o}\{I_2(s)/I^-(aq)\}$$

$$E_{cell}^{o} = +0.17\ V - 0.535\ V = -0.37\ V \quad \Rightarrow \quad No$$

37. The reducing couple that we seek must have a half-cell potential of such a size and sign that a positive sum results when this half-cell potential is combined with $E^{o}\{VO^{2+}(aq)|V^{3+}(aq)\} = 0.337\ V$ *and* a negative sum must be produced when this half-cell potential is combined with $E^{o}\{V^{3+}(aq)|V^{2+}(aq)\} = -0.255V$.

So, $-E^{o}$ for the couple must be $> -0.337\ V$ and $< +0.255\ V$ (i.e. it cannot be more positive than 0.255V, nor more negative than $-0.337\ V$)

Some possible reducing couples from Table 21-1 are:
$-E^{o}\{Sn^{2+}(aq)|Sn(s)\} = +0.137\ V$; $-E^{o}\{H^+(aq)|H_2(g)\} = 0.000\ V$
$-E^{o}\{Pb^{2+}(aq)|Pb(s)\} = +0.125\ V$; thus Pb(s), Sn(s), $H_2(g)$, to name but a few, will do the job.

38. There are two methods that may be used to determine the MnO_4^-/Mn^{2+} reduction potential.
Method 1: $MnO_4^- - 1.70\ V \rightarrow MnO_2 - 1.23\ V \rightarrow Mn^{2+}$

$$E^{o}_{MnO_4^{--}/Mn^{2+}} = \frac{3\,(1.70\ V) + 2\,(1.23\ V)}{5} = 1.51\underline{2}\ V \sim 1.51\ V$$

Method 2: $MnO_4^- \rightarrow MnO_4^{2-} \rightarrow MnO_2 \rightarrow Mn^{3+} \rightarrow Mn^{2+}$

$$E^{o}_{MnO_4^{--}/Mn^{2+}} = \frac{0.56\ V + 2\,(2.27\ V) + 0.95\ V + 1.49\ V}{5} = 1.50\underline{8}\ V \sim 1.51\ V$$

Compares favorably to Table 21.1, where $E^{o}_{MnO_4^{--}/Mn^{2+}} = 1.51\ V$

39. Table D-4 contains the following data: Cr^{3+}/Cr^{2+} reduction potential = -0.424 V, $Cr_2O_7^{2-}/Cr^{3+}$ reduction potential = 1.33 V and Cr^{2+}/Cr reduction potential = -0.90 V. By using the additive nature of free energies and the fact that $\Delta G^{o} = -nFE^{o}$, we can determine the two unknown potentials and complete the diagram.

(i) $Cr_2O_7^{2-}/Cr^{2+}$: $E^{o} = \dfrac{3(1.33\,V) - 0.424\,V}{4} = 0.892\ V$

(ii) Cr^{3+}/Cr: $E^{o} = \dfrac{-0.424\,V + 2(-0.90)\,V}{3} = -0.74\ V$

40. From Table 24.4 we are given: VO_2^+/VO^{2+} reduction potential = 1.000 V, VO^{2+}/V^{3+} reduction potential = 0.337 V, V^{3+}/V^{2+} reduction potential = -0.255 V and V^{2+}/V reduction potential = -1.13 V. By taking advantage the additive nature of free energies and the fact that $\Delta G° = -nFE°$, we can determine the three unknown potentials and complete the diagram.

(i) VO_2^+/V^{3+}: $E° = \dfrac{1.000\,V + 0.337\,V}{2} = 0.669\,V$

(ii) VO_2^+/V^{2+}: $E° = \dfrac{1.000\,V + 0.337\,V + (-0.255\,V)}{3} = 0.361\,V$

(iii) VO_2^+/V: $E° = \dfrac{1.000\,V + 0.337\,V + (-0.255\,V) + 2(-1.13\,V)}{5} = -0.24\,V$

Chromium and Chromium Compounds

41. Orange dichromate ion is in equilibrium with yellow chromate ion in aqueous solution.
$Cr_2O_7^{2-}(aq) + H_2O(l) \rightleftharpoons 2CrO_4^{2-}(aq) + 2H^+(aq)$

The chromate ion in solution then reacts with lead(II) ion to form a precipitate of yellow lead(II) dichromate. $Pb^{2+}(aq) + CrO_4^{2-}(aq) \rightleftharpoons PbCrO_4(s)$

$PbCrO_4(s)$ will form until $[H^+]$ from the first equilibrium increases to a significant value and both equilibria are simultaneously satisfied.

42. The initial dissolving reaction forms orange dichromate ion.

$2\,BaCrO_4(s) + 2\,HCl(aq) \rightleftharpoons 2\,Ba^{2+}(aq) + Cr_2O_7^{2-}(aq) + 2\,Cl^-(aq) + H_2O(l)$

Dichromate ion is a good oxidizing agent, sufficiently strong to oxidize $Cl^-(aq)$ to $Cl_2(g)$ if the concentrations of reactants are high, the solution is acidic, and the product $Cl_2(g)$ is allowed to escape.

$6Cl^-(aq) + Cr_2O_7^{2-}(aq) + 14H^+(aq) \rightleftharpoons 3Cl_2(g) + 2Cr^{3+}(aq) + 7H_2O(l)$

Chromium(III) can hydrolyze in solution to produce green $[Cr(OH)_4]^-(aq)$, but the solution needs to be alkaline (pH >7) for this to occur. A more likely source of the green color is a complex ion such as $[Cr(H_2O)_4Cl_2]^+$.

43. Oxidation: $\{Zn(s) \rightarrow Zn^{2+}(aq) + 2e^-$ $\} \times 3$
Reduction: $Cr_2O_7^{2-}(aq, orange) + 14H^+(aq) + 6e^- \rightarrow 2\,Cr^{3+}(aq, green) + 7\,H_2O(l)$
_____
Net: $3\,Zn(s) + Cr_2O_7^{2-}(aq) + 14\,H^+(aq) \rightarrow 3\,Zn^{2+}(aq) + 2\,Cr^{3+}(aq) + 7\,H_2O(l)$

Oxidation: $Zn(s) \rightarrow Zn^{2+}(aq) + 2e^-$

Reduction: $\{Cr^{3+}(aq, green) + e^- \rightarrow Cr^{2+}(aq, blue)\}$ ×2

_____

Net: $Zn(s) + 2\ Cr^{3+}(aq) \rightarrow Zn^{2+}(aq) + 2\ Cr^{2+}(aq)$

The green color is most likely due to a chloro complex of Cr^{3+}, such as $\left[Cr(H_2O)_4 Cl_2\right]^+$.

Oxidation: $\{Cr^{2+}(aq, blue) \rightarrow Cr^{3+}(aq, green) + e^-\}$ ×4

Reduction: $O_2(g) + 4\ H^+(aq) + 4e^- \rightarrow 2\ H_2O(l)$

_____

Net: $4\ Cr^{2+}(aq) + O_2(g) + 4\ H^+(aq) \rightarrow 4\ Cr^{3+}(aq) + 2\ H_2O(l)$

44. $CO_2(g)$, as the oxide of a nonmetal, is an acid anhydride. Its function is to make the solution acidic. A reasonable guess of the reaction that occurs follows.

$2\ CrO_4^{2-}(aq) + 2\ H^+(aq) \rightleftharpoons Cr_2O_7^{2-}(aq) + H_2O(l)$

$2\ H_2O(l) + 2CO_2(aq) \rightleftharpoons 2H^+(aq) + 2\ HCO_3^-(aq)$

_____

$2\ CrO_4^{2-}(aq) + 2\ CO_2(aq) + H_2O(l) \rightleftharpoons Cr_2O_7^{2-}(aq) + 2\ HCO_3^-(aq)$

<u>45.</u> Simple substitution into expression (24.19) yields $\left[Cr_2O_7^{2-}\right]$ in each case. In fact, the expression is readily solved for the desired concentration:

$\left[Cr_2O_7^{2-}\right] = 3.2 \times 10^{14}\left[H^+\right]^2\left[CrO_4^{2-}\right]^2$. In each case, we use the value of pH to determine $\left[H^+\right] = 10^{-pH}$.

(a) $\left[Cr_2O_7^{2-}\right] = 3.2 \times 10^{14}(10^{-6.62})^2(0.20)^2 = 0.74\ M$

(b) $\left[Cr_2O_7^{2-}\right] = 3.2 \times 10^{14}\left(10^{-8.85}\right)^2(0.20)^2 = 2.6 \times 10^{-5}\ M$

46. We use expression (24.19).

$$\frac{\left[Cr_2O_7^{2-}\right]}{\left[CrO_4^{2-}\right]^2} = 3.2 \times 10^{14}\left[H^+\right]^2 = 3.2 \times 10^{14}\left(10^{-7.55}\right)^2 = 3.2 \times 10^{14}\left(2.8 \times 10^{-8}\right)^2 = 0.254$$

$$\left[CrO_4^{2-}\right]_{initial} = \frac{1.505\ g\ Na_2CrO_4}{0.345\ L\ soln} \times \frac{1\ mol\ Na_2CrO_4}{161.97\ g\ Na_2CrO_4} \times \frac{1\ mol\ CrO_4^{2-}}{1\ mol\ Na_2CrO_4} = 0.0269\ M$$

Reaction: $2CrO_4^{2-}(aq) + 2H^+(aq) \rightleftharpoons Cr_2O_7^{2-}(aq) + H_2O(l)$

Initial: 0.0269 M   0 M

Changes: $-2x$ M $+x$ M

Equil: $(0.0269 - 2x)$ M x M

$$\frac{\left[Cr_2O_7^{2-}\right]}{\left[CrO_4^{2-}\right]^2} = 0.254 = \frac{x}{(0.0269 - 2x)^2} \qquad x = 0.254\left(0.000724 - 0.108x + 4x^2\right)$$

$$x = 0.000184 - 0.0274\,x + 1.016\,x^2 \qquad 1.016\,x^2 - 1.0274\,x + 0.000184 = 0$$

$$x = \frac{-b \pm \sqrt{b^2 - 4ac}}{2a} = \frac{1.0274 \pm \sqrt{1.0556 - 0.000736}}{2.032} = 1.0111\ \text{M},\ \ 0.00016\ \text{M}$$

We choose the second root because the first root gives a negative $\left[CrO_4^{2-}\right]$.

$$\left[Cr_2O_7^{2-}\right] = x = 0.00016\ \text{M} \quad \left[CrO_4^{2-}\right] = 0.0269 - 2x = 0.0266\ \text{M}$$

We carry extra significant figures to avoid a significant rounding error in this problem.

47. Each mole of chromium metal plated out from a chrome plating bath (i.e., CrO_3 and H_2SO_4) requires six moles of electrons.

$$\text{mass Cr} = 1.00\ \text{h} \times \frac{3600\ \text{s}}{1\ \text{hr}} \times \frac{3.4\ \text{C}}{1\ \text{s}} \times \frac{1\ \text{mol e}^-}{96485\ \text{C}} \times \frac{1\ \text{mol Cr}}{6\ \text{mol e}^-} \times \frac{52.00\ \text{g Cr}}{1\ \text{mol Cr}} = 1.10\ \text{g Cr}$$

48. First we compute the amount of Cr(s) deposited.

$$\text{amount Cr} = \left(0.0010\ \text{mm} \times \frac{1\ \text{cm}}{10\ \text{mm}} \times 0.375\ \text{m}^2 \times \frac{10^4\ \text{cm}^2}{1\ \text{m}^2}\right) \times \frac{7.14\ \text{g Cr}}{1\ \text{cm}^3\ \text{Cr}} \times \frac{1\ \text{mol Cr}}{52.0\ \text{g Cr}}$$

$$= 5.1 \times 10^{-2}\ \text{mol Cr}$$

Recall that the deposition of each mole of chromium requires six moles of electrons. We now compute the time required to deposit the $5.1 \times 10^{-2}\,\text{mol Cr}$.

$$\text{time} = 5.1 \times 10^{-2}\ \text{mol Cr} \times \frac{6\ \text{mol e}^-}{1\ \text{mol Cr}} \times \frac{96485\ \text{C}}{1\ \text{mol e}^-} \times \frac{1\ \text{s}}{3.5\ \text{C}} \times \frac{1\ \text{h}}{3600\ \text{s}} = 2.3\ \text{h}$$

49. Dichromate ion is the prevalent species in acidic solution. Oxoanions are better oxidizing agents in acidic solution because increasing the concentration of hydrogen ion favors formation of product. The half-equation is: $Cr_2O_7^{2-}(aq) + 14H^+(aq) + 6e^- \longrightarrow 2\,Cr^{3+}(aq) + 7H_2O(l)$ Note that precipitation occurs most effectively in alkaline solution. In fact, adding an acid to a compound is often an effective way of dissolving a water-insoluble compound. Thus, we expect to see the form that predominates in alkaline solution be the most effective precipitating agent. Notice also that CrO_4^{2-} is smaller than is $Cr_2O_7^{2-}$, giving it a higher lattice energy in its compounds and making those compounds harder to dissolve.

50. Both metal ions precipitate as hydroxides when $\left[OH^-\right]$ is moderate.

$$Mg^{2+}(aq) + 2\ OH^-(aq) \rightarrow Mg(OH)_2(s) \qquad\qquad Cr^{3+}(aq) + 3\ OH^-(aq) \rightarrow Cr(OH)_3(s)$$

Because chromium(III) oxides and hydroxides are amphoteric (and those of magnesium ion are not), $Cr(OH)_3(s)$ will dissolve in excess base.

$$Cr(OH)_3(s) + NaOH(aq) \rightarrow Na^+(aq) + \left[Cr(OH)_4\right]^-(aq)$$

51. $4\ Fe^{2+}(aq) + O_2(g) + 4\ H^+ \rightarrow 4\ Fe^{3+}(aq) + 2\ H_2O(l) \qquad E° = 0.44\ V$

$[Fe^{2+}] = [Fe^{3+}];\ pH = 3.25\ or\ [H^+] = 5.6 \times 10^{-4}$ and $P_{O_2} = 0.20$ atm

$$E = E° - \frac{0.0592}{n}\log\left(\frac{[Fe^{3+}]^4}{[Fe^{2+}]^4[H^+]^4 P_{O_2}}\right) = 0.44\ V - \frac{0.0592}{4}\log\left(\frac{\cancel{[Fe^{3+}]^4}}{\cancel{[Fe^{2+}]^4}[10^{-3.25}]^4 0.20}\right)$$

$E = 0.24\ V$ (spontaneous under these conditions)

52. $\{NiO(OH)(s) + e^- + H^+(aq) \rightarrow Ni(OH)_2(s) \quad\}\times 2 \qquad E°_{red}$

$\{Cd(s) + 2\ OH^-(aq) \rightarrow Cd(OH)_2(s) + 2\ e^- \quad\}\times 1 \qquad E°_{ox}$

$2\ NiO(OH)(s) + Cd(s) + 2\ H_2O(l) \rightarrow Cd(OH)_2(s) + 2\ Ni(OH)_2(s)\ E°_{cell} \approx 1.50\ V$
(Used $2\ H^+(aq) + 2\ OH^-(aq) = 2\ H_2O(l)$)

$E°_{ox}$ may be obtained by combining the K_{sp} value for $Cd(OH)_2(s)$
$(K_{sp\,Cd(OH)_2} = 2.5 \times 10^{-14})$ and $E°_{red} = -0.403\ V$ for $Cd^{2+}(aq) + 2\ e^- \rightarrow Cd(s)$. Hence,

$Cd(s) \rightarrow Cd^{2+}(aq) + 2\ e^- \qquad\qquad E°_{ox} = 0.403\ V$

$Cd^{2+}(aq) + 2\ OH^-(aq) \rightarrow Cd(OH)_2(s) \quad E°_{cell} = \dfrac{0.0592}{n}\log\left(\dfrac{1}{K_{sp}}\right) = \dfrac{0.0592}{2}\log\left(\dfrac{1}{2.5 \times 10^{-14}}\right) = 0.40\underline{3}\ V$

$Cd(s) + 2\ OH^-(aq) \rightarrow Cd(OH)_2(s) + 2\ e^-\ E°_{ox} = 0.80\underline{6}\ V$
Now plug this result back into first set of redox reactions:

$\{NiO(OH)(s) + e^- + H^+(aq) \rightarrow Ni(OH)_2(s) \quad\}\times 2 \qquad E°_{red}$

$\{Cd(s) + 2\ OH^-(aq) \rightarrow Cd(OH)_2(s) + 2\ e^- \quad\}\times 1 \qquad E°_{ox} = 0.80\underline{6}\ V$

$2\ NiO(OH)(s) + Cd(s) + 2\ H_2O(l) \rightarrow Cd(OH)_2(s) + 2\ Ni(OH)_2(s)\ E°_{cell} \approx 1.50\ V$

$E°_{red} = 1.50 \text{ V} - 0.80\underline{6} \text{ V} = 0.69 \text{ V}$

53. $Fe^{3+}(aq) + K_4[\overset{[II]}{Fe}(CN)_6](aq) \rightarrow K\overset{[III]}{Fe}[\overset{[II]}{Fe}(CN)_6](s) + 3 K^+(aq)$

Alternate formulation: $4Fe^{3+} + 3[\overset{[II]}{Fe}(CN)_6]^{4-}(aq) + Fe_4[\overset{[II]}{Fe}(CN)_6]_3$

54. $K^+(aq) + Fe^{2+}(aq) + K_3[\overset{[III]}{Fe}(CN)_6](aq) \rightarrow Fe^{3+}(aq) + K_4[\overset{[II]}{Fe}(CN)_6](aq)$

$Fe^{3+}(aq) + K_4[\overset{[II]}{Fe}(CN)_6](aq) \rightarrow K\overset{[III]}{Fe}[\overset{[II]}{Fe}(CN)_6](s) + 3 K^+(aq)$

Group 11 Metals

55. **(a)** $Cu^{2+}(aq) + H_2(g) \rightarrow Cu(s) + 2H^+(aq)$

(b) $Au^+(aq) + Fe^{2+}(aq) \rightarrow Au(s) + Fe^{3+}(aq)$

(c) $2Cu^{2+}(aq) + SO_2(g) + 2 H_2O(l) \rightarrow 2 Cu^+(aq) + SO_4^{2-}(aq) + 4 H^+(aq)$

56. In either case, the acid acts as an oxidizing agent and dissolves silver; the gold is unaffected.

Oxidation: $\{Ag(s) \rightarrow Ag^+(aq) + e^-\}$ ×3

Reduction: $NO_3^-(aq) + 4H^+(aq) + 3e^- \rightarrow NO(g) + 2H_2O(l)$

Net: $3Ag(s) + NO_3^-(aq) + 4H^+(aq) \rightarrow 3Ag^+(aq) + NO(g) + 2H_2O(l)$

Oxidation: $\{Ag(s) \rightarrow Ag^+(aq) + e^-\}$ ×2

Reduction: $SO_4^{2-}(aq) + 4H^+(aq) + 2e^- \rightarrow SO_2(g) + 2H_2O(l)$

Net: $2 Ag(s) + SO_4^{2-}(aq) + 4 H^+(aq) \rightarrow 2 Ag^+(aq) + SO_2(g) + 2 H_2O(l)$

57. In the Integrative Example we determined that $K_c = 1.2 \times 10^6 = \dfrac{[Cu^{2+}]}{[Cu^+]^2}$ or

$$[Cu^{2+}] = 1.2 \times 10^6 [Cu^+]^2$$

(a) When $[Cu^+] = 0.20$ M, $[Cu^{2+}] = 1.2 \times 10^6 (0.20)^2 = 4.8 \times 10^4$ M. This is an impossibly high concentration. Thus $[Cu^+] = 0.20$ M can never be achieved.

(b) When $\left[Cu^+\right] = 1.0 \times 10^{-10}$ M, $\left[Cu^{2+}\right] = 1.2 \times 10^6 \left(1.0 \times 10^{-10}\right)^2 = 1.2 \times 10^{-14}$ M. This is an entirely reasonable (even though small) concentration; $\left[Cu^+\right] = 1.0 \times 10^{-10}$ M can be maintained in solution.

58. Oxidation: $\{Cu(s) \rightarrow Cu^{2+} + 2\,e^-$ $\}\times 2$

Reduction: $\{O_2(g) + 2\,H_2O(l) + 4\,e^- \rightarrow 4\,OH^-(aq)\,\}\times 1$

Overall: $2\,Cu(s) + O_2(g) + 2\,H_2O(l) \rightarrow 2\,Cu^{2+}(aq) + 4\,OH^-\,(aq)$

Note: This produces an alkaline solution. CO_2 dissolves in this alkaline solution to produce carbonate ion.

Step(a): $CO_2(g) + OH^-\,(aq) \rightarrow HCO_3^-\,(aq)$

Step(b): $HCO_3^-\,(aq) + OH^-\,(aq) \rightarrow CO_3^{2-}\,(aq) + H_2O(l)$

Total: $CO_2(g) + 2\,OH^-\,(aq) \rightarrow CO_3^{2-}\,(aq) + H_2O(l)$

Combination of the reactions labeled "overall" and "total" gives the following result.

$2\,Cu(s) + O_2(g) + H_2O(l) + CO_2(g) \rightarrow 2\,Cu^{2+}(aq) + 2\,OH^-\,(aq) + CO_3^{2-}\,(aq)$

The ionic product of the resultant reaction combine in a precipitation reaction to form

$2\,Cu^{2+}(aq) + 2\,OH^-\,(aq) + CO_3^{2-}\,(aq) \rightarrow Cu_2(OH)_2CO_3\,(s)$

Group 12 Metals

59. Given: Hg^{2+}/Hg reduction potential = 0.854 V and Hg_2^{2+}/Hg reduction potential = 0.796 V Using the additive nature of free energies and the fact that $\Delta G^\circ = -nFE^\circ$, we may determine the

Hg^{2+}/Hg_2^{2+} potential as $\dfrac{2(0.854V) - 0.796V}{1} = 0.912$ V

60. $\Delta G^\circ = -25,000$ J mol^{-1} = $-RT\ln K_{eq}$ = $-(8.3145$ J K^{-1} mol$^{-1})(673$ K$)(\ln K_{eq})$

$K_{eq} = K_p = 87$ atm$^{-1} = \dfrac{1}{P_{O_2}}$ Therefore, $P_{O_2} = \dfrac{1}{K_p} = \dfrac{1}{87} = 0.011$ atm

61. **(a)** Estimate K_p for $ZnO(s) + C(s) \rightarrow Zn(l) + CO(g)$ at 800 °C (Note: $Zn(l)$ boils at 907 °C)

$$\{2\ C(s) + O_2(g) \rightarrow 2\ CO(g)\} \times 1/2 \qquad \Delta G° = (-415\ kJ)(1/2)$$

$$\underline{\{2\ ZnO(s) \rightarrow 2\ Zn(l) + O_2(g)\} \times 1/2 \qquad \Delta G° = (+485\ kJ(1/2)}$$

$$ZnO(s) + C(s) \rightarrow Zn(l) + CO(g) \qquad \Delta G° = 35\ kJ$$

Use $\Delta G° = -RTlnK_{eq}$ where $T = 800\ °C$, $K_{eq} = K_p$

$35\ kJ\ mol^{-1} = -(8.3145 \times 10^{-3}\ kJ\ K^{-1}\ mol^{-1})((273.15 + 800)\ K)(ln\ K_p)$

$lnK_p = -3.9$ or $K_p = 0.02\ ^{\cdot}$

(b) $K_p = P_{CO} = 0.02$ Hence, $P_{CO} = 0.02$ atm

62. $T = 298\ K$. Hence, $\log (P_{mmHg}) = \dfrac{-0.05223\ (61,960)}{298} + 8.118 = -2.742$

$P_{mmHg} = 0.001811$ mmHg

$$PV = nRT \rightarrow \frac{n}{V} = \frac{P}{RT} = [Hg] = \frac{(0.001811\,mmHg)\left(\dfrac{1\,atm}{760\,mmHg}\right)}{\left(0.08206\dfrac{L\,atm}{K\,mol}\right)(298K)} = 9.744 \times 10^{-8}\ mol\ L^{-1}$$

Next, convert to mg Hg m^{-3}:

$$[Hg] = \left(9.874\ x\ 10^{-8}\ \frac{mol}{L}\right)\left(\frac{1000\,L}{m^3}\right)\left(\frac{200.59\,g\,Hg}{1\,mol\,Hg}\right)\left(\frac{1000\,mg}{1\,g\,Hg}\right) = 19.5\ mg\ Hg\ m^{-3}$$

(This is approximately 400 times greater than the permissible level of 0.05 mg Hg m^{-3})

63. We calculate the wavelength of light absorbed in order to promote an electron across each band gap.

First a few relationships. $E_{mole} = N_A E_{photon}$ $\qquad E_{photon} = h\nu$ $\qquad c = \nu\lambda$ or $\nu = c/\lambda$

Then, some algebra. $E_{mole} = N_A E_{photon} = N_A h\nu = N_A hc/\lambda$ $\qquad$ or $\qquad \lambda = N_A hc/E_{mole}$

For ZnO, $\lambda = \dfrac{6.022 \times 10^{23}\ mol^{-1} \times 6.626 \times 10^{-34}\ J\ s \times 2.998 \times 10^8\ m\ s^{-1}}{290 \times 10^3\ J\ mol^{-1}} \times \dfrac{10^9\ nm}{1\,m}$

$\qquad = \dfrac{1.196 \times 10^8\ J\ mol^{-1}\ nm}{290 \times 10^3\ J\ mol^{-1}} = 413\ nm$ $\quad$ violet light

For CdS, $\lambda = \dfrac{1.196 \times 10^8\ J\ mol^{-1}\ nm}{250 \times 10^3\ J\ mol^{-1}} = 479\ nm$ blue light

The blue light absorbed by CdS is subtracted from the white light incident on the surface of the solid. The remaining reflected light is colored, yellow in this case. When the violet light is subtracted from the white light incident on the ZnO surface, the reflected light appears white.

64. The color that we see is the complementary color to the color absorbed. The color absorbed, in turn, is determined by the energy separation of the band gap. If the wavelength for the absorbed color is short, the energy separation for the band gap is large.

The yellow color of CdS means that the complementary color, violet, is absorbed. Since violet light has a quite short wavelength (about 410 nm), CdS must have a very large band gap energetically.

HgS is red, meaning that the color absorbed is green, which has a moderate wavelength (about 520 nm). HgS must have a band gap of intermediate energy.

CdSe is black, meaning that visible light of all wavelengths and thus all energies in the visible region are absorbed. This would occur if CdSe has a very small band gap, smaller than that of least energetic red light (about 650 nm).

Feature Problems

82. **(a)** If $\Delta n_{gas} = 0$, then $\Delta S° \sim 0$ and $\Delta G°$ is essentially independent of temperature

$$(C(s) + O_2(g) \rightarrow CO_2(g))$$

If $\Delta n_{gas} > 0$, then $\Delta S° > 0$ and $\Delta G°$ will become more negative with increasing temperature, hence the graph has a negative slope $(2\ C(s) + O_2(g) \rightarrow 2\ CO\ (g))$

If $\Delta n_{gas} < 0$, then $\Delta S° < 0$ and $\Delta G°$ will become more positive with increasing temperature, hence the graph has a positive slope $(2\ CO(g) + O_2(g) \rightarrow 2\ CO_2\ (g))$

(b) The additional blast furnace reaction, $C(s) + CO_2(g) \rightarrow 2\ CO(g))$, has

$\Delta H° = [2 \times -110.5\ kJ] - [-393.5\ kJ + 0\ kJ] = 172.5\ kJ$ and

$\Delta S° = [2 \times 197.7\ J\ K^{-1}] - [1 \times 5.74\ J\ K^{-1} + 213.7\ J\ K^{-1}] = 176.0\ J\ K^{-1}$

It can be obtained by adding reaction (b) to the reverse of reaction (c) (both appear in the provided figure)

$$C(s) + O_2(g) \rightarrow CO_2(g) \qquad \text{Reaction (b)}$$
$$+ \quad 2CO_2(s) \rightarrow 2CO(g) + O_2(g) \quad \text{Reverse of Reaction (c)}$$

Net: $C(s) + CO_2(g) \rightarrow 2\ CO(g)$ (Additional Blast Furnace Reaction)

Consequently, the plot of $\Delta G°$ for the *net* reaction as a function of temperature will be a straight line with a slope of $-[\{\Delta S°\ (Rxn\ b)\} - \{\Delta S°\ (Rxn\ c)\}]$ (in kJ/K) and a y-intercept of $[\{\Delta H°\ (Rxn\ b)\} - \{\Delta H°\ (Rxn\ c)\}]$ (in kJ).

Since ΔH° (Rxn b) = -393.5 kJ, ΔH° (Rxn c) = -566 kJ, ΔS° (Rxn b) = 2.9 J/K and ΔS° (Rxn c) = -173.1 J/K, the plot of ΔG° vs. T for the reaction C(s) + CO_2(g) $\rightarrow$ 2 CO(g) will follow the equation $y = -0.176x + 172.5$

From the graph, we can see that the difference in ΔG° (line b – line c at 1000 °C) is ~ -40 kJ/mol. K_p is readily calculated using this value of ΔG°.

ΔG° = $-RT\ln K_p$ = -40 kJ = $-(8.3145 \times 10^{-3}$ kJ K^{-1} $mol^{-1})($ 1273 K)(ln K_p)
$\ln K_p$ = 3.$\underline{8}$ Hence, K_p = 4$\underline{4}$

The equilibrium partial pressure for CO(g) is then determined by using the K_p expression:

$$K_p = \frac{\left(P_{CO}\right)^2}{\left(P_{CO_2}\right)} = 4\underline{4} = \frac{\left(P_{CO}\right)^2}{(0.25 \text{ atm})}$$ Hence, $(P_{CO})^2 = 1\underline{1}$ and $P_{CO} = 3.\underline{3}$ atm or 3 atm.

Alternatively, we can determine the partial pressure for CO_2 at 1000 °C via the calculated ΔH° and ΔS° values for the reaction C(s) + CO_2(g) $\rightarrow$ 2 CO(g) to find ΔG° at 1000 °C , and ultimately K_p with the relationship ΔG° = $-RT\ln K_p$ (here we are making the assumption that ΔH° and ΔS° are relatively constant over the temperature range 298 K to 1273 K). The calculated values of H° and ΔS° (using Appendix D) are given below:

ΔH° = $[2 \times -110.5$ kJ$]$ $-[-393.5$ kJ $+ 0$ kJ$]$ = 172.5 kJ

ΔS° = $[2 \times 197.7$ J $K^{-1}]$ $- [1 \times 5.74$ J K^{-1} $+ 213.7$ J $K^{-1}]$ = 176.0 J K^{-1}

To find ΔG° at 1000°C, we simply plug $x = 1273$ K into the straight-line equation we developed above and solve for y (ΔG°).

So, $y = -0.176(1273 \text{ K}) + 172.5$; $y = -51.5 \text{ kJ}$
Next we need to calculate the K_p for the reaction at 1000°C.

ΔG° = $-RT\ln K_p$ = -51.5 kJ = $-(8.3145 \times 10^{-3}$ kJ K^{-1} $mol^{-1})($ 1273 K)(ln K_p)

$\ln K_p$ = 4.87 Hence, K_p = 1.3×10^2

The equilibrium partial pressure for CO(g) is then determined by using the K_p expression:

$$K_p = \frac{\left(P_{CO}\right)^2}{\left(P_{CO_2}\right)} = 130 = \frac{\left(P_{CO}\right)^2}{(0.25 \text{ atm})}$$ Hence, $(P_{CO})^2 = 32.\underline{5}$ and $P_{CO} = 5.7$ atm

83. **(a)** The amphoteric cations are Al^{3+}, Cr^{3+} and Zn^{2+}.

$Cr^{3+}(aq) + 4OH^-(aq) \rightarrow [Cr(OH)_4]^-(aq)$

$Al^{3+}(aq) + 4OH^-(aq) \rightarrow [Al(OH)_4]^-(aq)$

$Zn^{2+}(aq) + 4OH^-(aq) \rightarrow [Zn(OH)_4]^{2-}(aq)$

$Fe^{3+}(aq) + 3OH^-(aq) \rightarrow Fe(OH)_3(s)$

$Ni^{2+}(aq) + 2OH^-(aq) \rightarrow Ni(OH)_2(s)$

$Co^{2+}(aq) + 2OH^-(aq) \rightarrow Co(OH)_2(s)$

(b) Of the three hydroxide precipitates, only Co^{2+} is easily oxidized: (refer to the Standard Reduction Potential table in Appendix D)

$2Co(OH)_2(s) + H_2O_2(aq) \longrightarrow 2Co(OH)_3(s)$

(c) We know that the chromate ion is yellow. Thus,

$[Cr(OH)_4]^-(aq) + 3/2\ H_2O_2(aq) + OH^-(aq) \longrightarrow CrO_4^{2-}(aq,\ YELLOW) + 4\ H_2O(l)$

(d) Co^{3+} is reduced to Co^{2+} by the H_2O_2 in solution. The other hydroxides simply dissolve in strong acid:

i) $Co(OH)_3(s) + 3H^+(aq) \longrightarrow Co^{3+}(aq) + 3H_2O(l)$

ii) $2\ Co^{3+}(aq) + 2\ Cl^-(aq) \longrightarrow 2\ Co^{2+}(aq) + Cl_2(g)$

$Fe(OH)3(s) + 3HCl(aq) \longrightarrow FeCl_3(aq) + 3H_2O(l)$

$Ni(OH)_2(s) + 2H^+(aq) \longrightarrow Ni^{2+}(aq) + 2H_2O(l)$

(e) The $Fe^{3+}(aq)$ ions in the presence of concentrated ammonia(6 M) form an insoluble precipitate with the hydroxide ions generated by the ammonia hydrolysis reaction:

$NH_3(aq) + H_2O(l) \rightleftharpoons NH_4^+(aq) + OH^-(aq)$

$Fe^{3+}(aq) + 3\ OH^-(aq) \rightleftharpoons Fe(OH)_3(s);$ $\qquad K_{sp}\ Fe(OH)_3 = 4 \times 10^{-38}$

Both Co^{3+} and Ni^{2+} form soluble complex ions with ammonia ligands rather than hydroxide precipitates.

CHAPTER 25
COMPLEX IONS AND COORDINATION COMPOUNDS
PRACTICE EXAMPLES

1A   There are two different kinds of ligands in this complex ion, I^- and CN^-. Both are monodentate ligands, that is, they each form only one bond to the central atom. Since there are five ligands in total for the complex ion, the coordination number is 5: C.N.$= 5$. Each CN^- ligand has a charge of -1, as does the I^- ligand. Thus, the O.S. must be such that:

$$O.S. + \left[(4+1) \times (-1) \right] = -3 = O.S. - 5. \text{ Therefore, } O.S. = +2.$$

1B   The ligands are CN^-. Fe^{3+} is the central metal ion. The complex ion is $\left[Fe(CN)_6 \right]^{3-}$.

2A   There are six "Cl^-" ligands (chloride), each with a charge of $1-$. Platinum is the metal ion with an oxidation state of $+4$. Thus, the complex ion is $\left[PtCl_6 \right]^{2-}$, and we need two K^+ to balance charge: $K_2 \left[PtCl_6 \right]$

2B   The "SCN^-" ligand is the thiocyanato group, with a charge of $1-$, bonding to the central metal ion through the sulfur atom. The "NH_3" ligand is ammonia, a neutral ligand. There are five (penta) ammine ligands bonded to the metal. The oxidation state of cobalt is $+3$. The complex ion is not negatively charged, so its name does not end with "ate". The name of the compound is pentaamminethiocyanato-*S*-cobalt(III) chloride

3A   The oxalato ligand must occupy two *cis*- positions. Either the two NH_3 or the two Cl^- ligands can be coplanar with the oxalate ligand, leaving the other two ligands axial. The other isomer has one NH_3 and one Cl^- ligand coplanar with the oxalate ligand. The structures are sketched below.

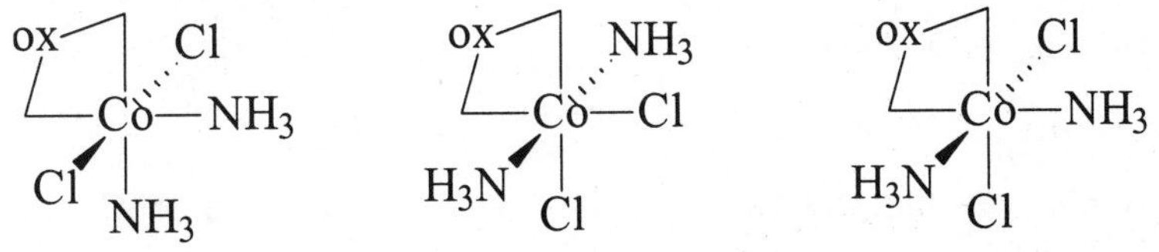

3B   We start out with the two pyridines, C_5H_5N, located *cis* to each other. With that imposed, we can have the two Cl^- ligands *trans* and the two CO ligands *cis*, the two CO ligands *trans* and the two Cl^- ligands *cis*, or both the Cl^- ligands and the two CO ligands *cis*. If we now consider the two pyridines *trans*, we can either have both other pairs *trans*, or both other pairs *cis*. There are five geometric isomers. They follow, in the order described.

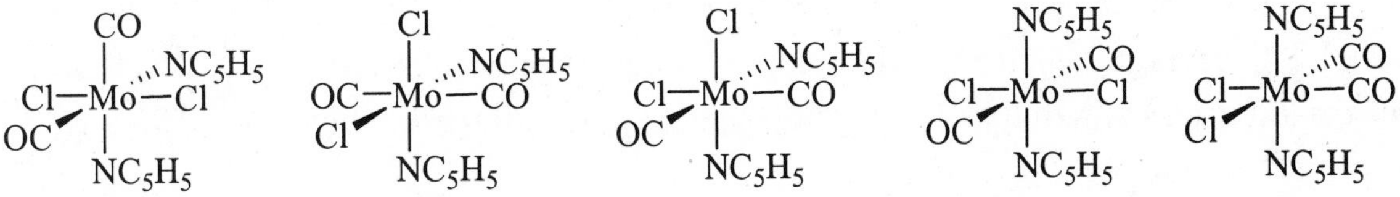

4A The F^- ligand is a weak field ligand. $[MnF_6]^{2-}$ is an octahedral complex. Mn^{4+} has three $3d$ electrons. The ligand field splitting diagram for $[MnF_6]^{2-}$ is sketched below. There are three unpaired electrons.

e_g ☐☐

t_{2g} ⇡⇡⇡   ↑E

4B Co^{2+} has seven $3d$ electrons. Cl^- is a weak field ligand. H_2O is a moderate field ligand. There are three unpaired electrons in each case. The number of unpaired electrons has no dependence on geometry for either metal ion complex.

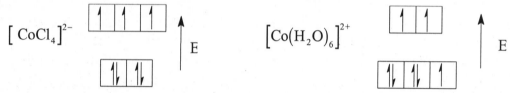

$[CoCl_4]^{2-}$ $[Co(H_2O)_6]^{2+}$

5A CN^- is a strong field ligand. Co^{2+} has seven $3d$ electrons. In the absence of a crystal field, all five d orbitals of the same energy, seven d electrons will result in three unpaired electrons. Thus, we need an orbital splitting diagram in which there are three orbitals of the same energy at the highest energy. This is the case with a tetrahedral orbital diagram.

⇅ ⇅ ↑ ↑ ↑ t_{2g} ↑ ↑ ↑ e_g ↑ —

in absence of a

crystal field e_g ⇅ ⇅ t_{2g} ⇅ ⇅ ⇅

   Tetrahedral Geometry   Octahedral Geometry

Thus, $[Co(CN)_4]^{2-}$ must be tetrahedral (3 unpaired electrons) and not octahedral (1 unpaired electron) because the magnetic property would agree with the experimental observations (3 unpaired electrons).

5B NH_3 is a strong field ligand. Cu^{2+} has nine $3d$ electrons. There is only one way to arrange nine electrons in five d-orbitals, that is, to have four fully occupied orbitals (to electrons in each orbital), and one half-filled orbital. Thus, the complex ion must be paramagnetic to the extent of one unpaired electron, regardless of the geometry the ligands adopt around the central metal ion.

6A We are certain that $[Co(H_2O)_6]^{2+}$ is octahedral with a moderate field ligand. Tetrahedral $[CoCl_4]^{2-}$ has a weak field ligand. The relative values of ligand field splitting for the same ligand are $\Delta_t = 0.44\,\Delta_o$. Thus, $[Co(H_2O)_6]^{2+}$ absorbs light of higher energy, blue or green light, leaving a light pink as the complementary color we observe. $[CoCl_4]^{2-}$ absorbs lower energy red light, leaving blue light to pass through and be seen.

6B   In order, the two complex ions are $\left[\text{Fe}(\text{H}_2\text{O})_6\right]^{2+}$ and $\left[\text{Fe}(\text{CN})_6\right]^{4-}$. We know that CN^- is a strong field ligand; it should give rise to a large value of Δ_o and result in the absorption of light of the shorter wavelength. We would expect the cyano complex to absorb blue or violet light and thus $\text{K}_4\left[\text{Fe}(\text{CN})_6\right]\cdot 3\,\text{H}_2\text{O}$ should appear yellow. The compound $\left[\text{Fe}(\text{H}_2\text{O})_6\right](\text{NO}_3)_2$, containing the weak field complex thus should be green, because the weak field would result in the absorption of light of long wavelength, namely, red light.

REVIEW QUESTIONS

1.   **(a)**   The coordination number of a complex refers to the number of atoms in a ligands that bind (bond) to the central metal ion. If all ligands are monodentate, the coordination number equals the number of ligands attached to the central metal ion.

   (b) Δ is the crystal field splitting energy: the energy difference between the lower- and higher-energy d orbitals in the crystal field splitting diagram.

   (c)   An ammine complex is one with ammonia molecule(s) as the ligand(s).

   (d)   An enantiomer is one of two optical isomers: molecules of identical formula and bonding which are nonsuperimposable mirror images of each other.

2.   **(a)**   The spectrochemical series is a listing of common ligands, ordered in terms of their ability to produce a splitting in the d-orbitals, (Δ), around a central metal ion.

   (b)   Crystal field theory describes bonding in complexes in terms of electrostatic attractions between ligands and the central metal (both of which are assumed to be point charges). Particular attention is focused on the energies of the d-orbitals on the central metal.

   (c)   Optical isomerism refers to two compounds that differ in physical and chemical properties only in the direction in which each rotates the plane of polarized light.

   (d)   Structural isomerism refers to differences in the ligands that are bonded to the central atom, and in the ligand atoms that are directly attached to the central atom.

3.   **(a)**   The coordination number of a central metal atom is the total number of ligand atoms that are directly bonded to that central metal atom. It also is known as the secondary valence. The oxidation number, or primary valence, is equal to the charge of the isolated central metal ion.

   (b)   A monodentate ligand is one that binds to the central metal atom using only one donor atom (lone pair). A monodentate ligand only occupies one position in the coordination sphere. A polydentate ligand is one that binds to the central metal atom using two or more donor atoms and occupies two or more positions in the coordination sphere of the complex.

(c) In a *cis* isomer, two ligands have a small angular distance between their bonds to the central atom (~90°). In a *trans* isomer, two ligands are on opposite sides of the central atom (~180°).

(d) Dextrorotatory and levorotatory isomers differ in the direction in which they rotate the plane of polarized light. A dextrorotatory isomer rotates this plane to the right (clockwise), while a levorotatory isomer rotates it to the left (counterclockwise).

(e) In a high-spin complex, a weak crystal field splitting results in the *d* orbital electrons of the central atom to arrange themselves so as to maximize the number of unpaired electrons(pairing energy is comparable to crystal field splitting energy). Conversely, in a low-spin complex, a strong ligand field results in the *d* orbital electrons of the central atom to arrange themselves so as to minimize the number of unpaired electrons(pairing energy is small compared to crystal field splitting energy).

4. **(a)** $\left[CrCl_4(NH_3)_2\right]^-$ diamminetetrachlorochromate(III) ion

(b) $\left[Fe(CN)_6\right]^{3-}$ hexacyanoferrate(III) ion

(c) $\left[Cr(en)_3\right]_2\left[Ni(CN)_4\right]_3$ tris(ethylenediamine)chromium(III) tetracyanonickelate(II) ion

5. **(a)** $\left[Co(NH_3)_6\right]^{2+}$ The coordination number of Co is 6; there are six monodentate NH_3 ligands attached to Co. Since the NH_3 ligand is neutral, the oxidation state of cobalt is $+2$, the same as that of the complex ion; hexaamminecobalt(III) ion

(b) $\left[AlF_6\right]^{3-}$ The coordination number of Al is six; F^- is monodentate. Each F^- has a $1-$ charge; thus the oxidation state of Al is $+3$; hexafluoroaluminate(III) ion

(c) $\left[Cu(CN)_4\right]^{2-}$ The coordination number of Cu is 4; CN^- is monodendate. CN^- has a $1-$ charge; thus the oxidation state of Cu is $+2$; tetracyanocuprate(II) ion

(d) $\left[CrBr_2(NH_3)_4\right]^+$ The coordination number of Cr is 6; NH_3 and Br^- are monodentate. NH_3 has no charge; Br^- has a $1-$ charge. The oxidation state of chromium is $+3$; tetraamminedibromochromium(III) ion

(e) $\left[Co(ox)_3\right]^{4-}$ The coordination number of Co is 6; oxalate is bidentate. $C_2O_4^{2-}(ox)$ has a $2-$ charge; thus the oxidation state of cobalt is $+2$; trioxalatocobaltate(II) ion.

(f) $\left[Ag(S_2O_3)_2\right]^{3-}$ The coordination number of Ag is 2; $S_2O_3^{2-}$ is monodentate. $S_2O_3^{2-}$ has a $2-$ charge; thus the oxidation state of silver is $+1$; dithiosulfatoargentate(I) ion. (Although +1 is by far the most common oxidation state for silver in its compounds, stable silver(III) complexes are known. Thus, strictly speaking, silver is not a non-variable metal, and hence when naming silver compounds, the oxidation state(s) for the silver atom(s) should be specified).

6. **(a)** $\left[AgI_2\right]^-$ diiodoargentate(I) ion

   (b) $\left[Al(OH)(H_2O)_5\right]^{2+}$ pentaaquahydroxoaluminum(III) ion

   (c) $\left[Zn(CN)_4\right]^{2-}$ tetracyanozincate(II) ion

   (d) $\left[Pt(en)_2\right]^{2+}$ bis(ethylenediamine)platinum(II) ion

   (e) $[CoCl(NO_2)(NH_3)_4]^+$ tetraamminechloronitrito-N-cobaltate(III) ion

7. **(a)** $\left[CoBr(NH_3)_5\right]SO_4$ pentaamminebromocobalt(III) sulfate

   (b) $\left[CoSO_4(NH_3)_5\right]Br$ pentaamminesulfatocobalt(III) bromide

   (c) $\left[Cr(NH_3)_6\right]\left[Co(CN)_6\right]$ hexaamminechromium(III) hexacyanocobaltate(III)

   (d) $Na_3\left[Co(NO_2)_6\right]$ sodium hexanitrocobaltate(III)

   (e) $\left[Co(en)_3\right]Cl_3$ tris(ethylenediamine)cobalt(III) chloride

8. **(a)** $\left[Ag(CN)_2\right]^-$ dicyanoargentate(I) ion

   (b) $\left[Pt(NO_2)(NH_3)_3\right]^+$ triamminenitroplatinum(II) ion

   (c) $\left[CoCl(en)_2(H_2O)\right]^{2+}$ aquachlorobis(ethylenediamine)cobalt(III) ion

   (d) $K_4\left[Cr(CN)_6\right]$ potassium hexacyanochromate(II)

9. The Lewis structures are grouped together at the end.

   (a) H_2O has $2\times1+6=8$ valence electrons, or 4 pairs.

   (b) CH_3NH_2 has $4+3\times1+5+2\times1=14$ valence electrons, or 7 pairs.

   (c) ONO^- has $2\times6+5+1=18$ valence electrons, or 9 pairs. The structure has a 1−
        formal charge on the oxygen that is singly bonded to N.

   (d) SCN^- has $6+4+5+1=16$ valence electrons, or 8 pairs. This structure,
        appropriately, gives a 1− formal charge to N.

   (a) H—$\bar{\text{O}}$—H   **(b)**   H
                          |
                      H—C—$\bar{\text{N}}$—H
                          | |
                          H   H

   (c) $|\bar{\text{O}}-\bar{\text{N}}=\bar{\text{O}}|^-$ **(d)** $|\bar{\text{S}}=\text{C}=\bar{\text{N}}|^-$

10. **(a)** manganese(II) sulfate hexahydrate, $MnSO_4 \cdot 6\ H_2O$ or $\left[Mn(H_2O)_6\right]SO_4$

(b) potassium hexacyanochromate(II) trihydrate, $K_4[Cr(CN)_6]\cdot 3\ H_2O$

11. **(a)** $\left[PtCl_4\right]^{2-}$ \qquad tetrachloroplatinate(IV) ion

(b) $\left[FeCl_4(en)\right]^-$ \qquad tetrachloro(ethylenediamine)ferrate(III) ion

(c) $cis-\left[FeCl_2(ox)(en)\right]^-$ \qquad cis-dichloro(ethylenediamine)(oxalato)ferrate(III) ion.

(d) $trans-\left[CrCl(OH)(NH_3)_4\right]^+$ \qquad trans-tetraamminechlorohydroxochromium(III) ion

(a)

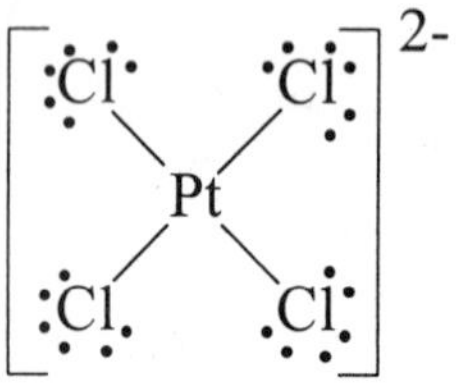

(b)

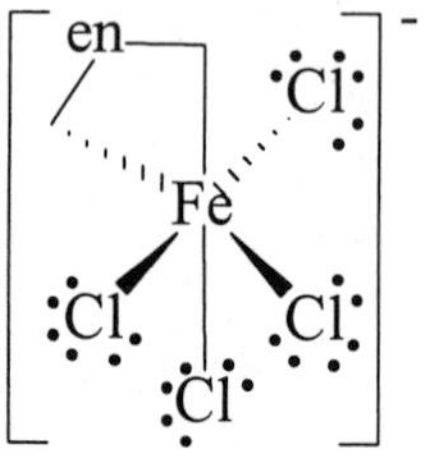

(c)

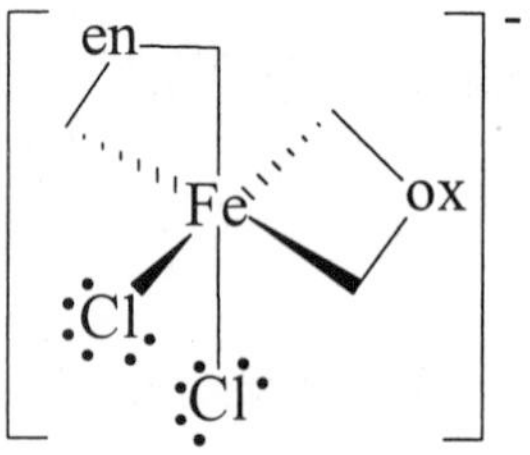

(d)

12. We assume that all of these complexes are octahedral in shape.

(a) $\left[Co(H_2O)(NH_3)_5\right]^{3+}$ has just one isomer.

(b) $\left[Co(H_2O)_2(NH_3)_4\right]^{3+}$ has two isomers, a *cis*-isomer (drawn on the left following) and a *trans*-isomer (drawn on the right). The two H_2O ligands are $90°$ from each other (*cis*-) or $180°$ from each other (*trans*-).

(a)

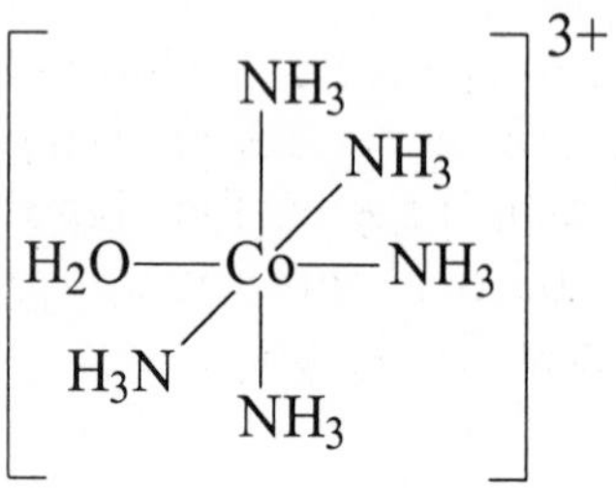

(b)

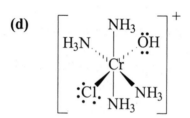

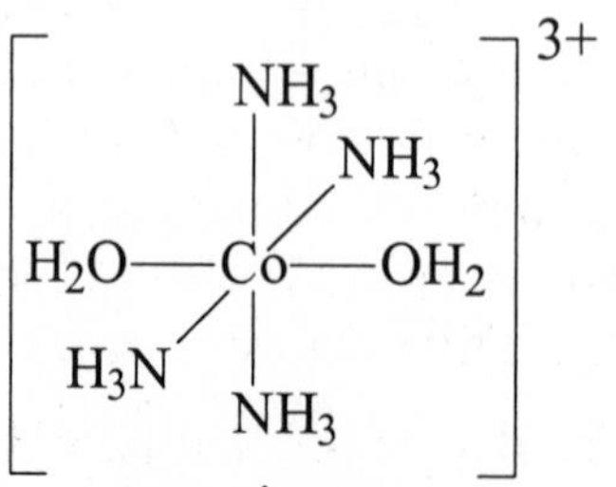

trans-isomer 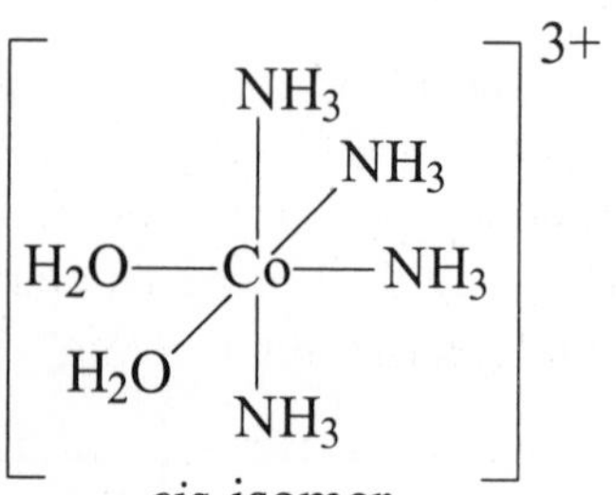 \qquad *cis*-isomer

(c) $\left[Co(H_2O)_3(NH_3)_3\right]^{3+}$ has two isomers, a *fac*-isomer (in which the three NH_3 ligands are $90°$ to each other, also called the *cis*-isomer) and a *mer*-isomer (in which two of the NH_3 ligands are $180°$ from each other, on opposite sides of the central Co atom, also referred to as the *trans*-isomer).

(d) $\left[Co(H_2O)_4(NH_3)_2\right]^{3+}$ has two isomers, a *cis*-isomer and a *trans*-isomer. In this case it is the two NH_3 groups that are $90°$ from each other (*cis*-) or $180°$ from each other (*trans*-).

(c)

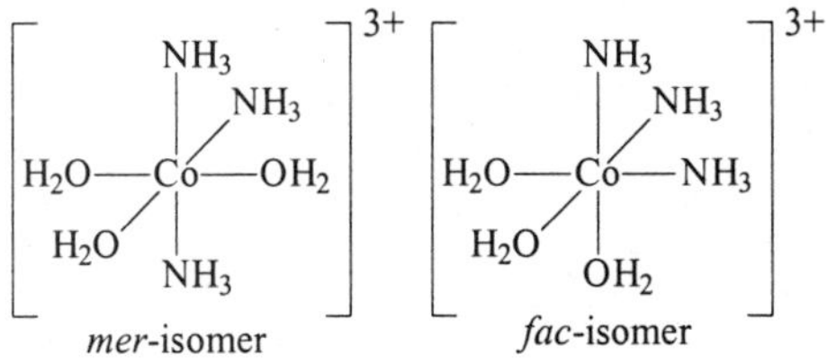

mer-isomer   fac-isomer

(d)

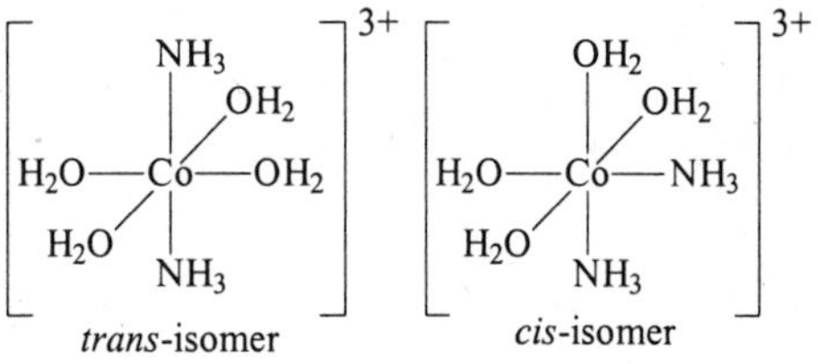

trans-isomer   cis-isomer

13. **(a)** $\left[Zn(NH_3)_4\right]\left[CuCl_4\right]$ can display coordination isomerism. Another isomer is $\left[Cu(NH_3)_4\right]\left[ZnCl_4\right]$.

(b) $\left[Fe(CN)_5SCN\right]^{4-}$ displays linkage isomerism. The other isomer is $\left[Fe(CN)_5NCS\right]^{4-}$

(c) $\left[NiCl(NH_3)_5\right]^{+}$ does not display isomerism.

(d) $\left[PtBrCl_2(py)\right]^{-}$ displays geometric isomerism, because the complex is square planar.

(e) $\left[Cr(OH)_3(NH_3)_3\right]^{-}$ displays geometric isomerism. There is a *fac*-isomer and a mer-isomer. These could also be labeled *cis*- and *trans*-, respectively.

(d)

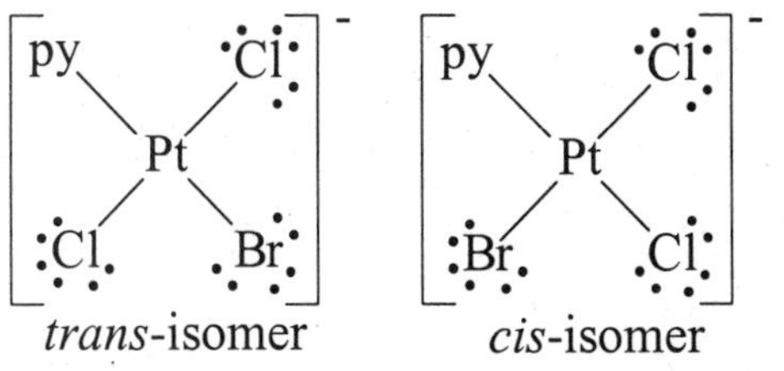

trans-isomer   cis-isomer

(e)

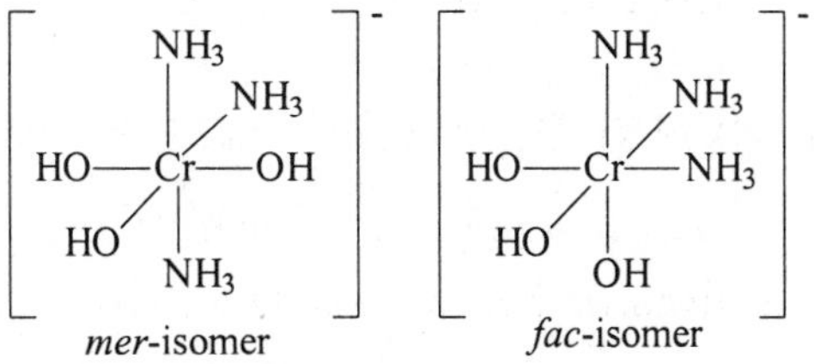

mer-isomer   fac-isomer

14. A yellow color indicates that the complex absorbs blue light, while a blue color indicates that the complex absorbs red and yellow light. Blue light has more energy per photon than do red and yellow light. The complex with the larger value of Δ will absorb the higher energy light. The ethylenediamine ligand produces a larger crystal field splitting energy than does the aqua ligand, according to the spectrochemical series. Thus, the yellow complex is $\left[Co(en)_3\right]^{3+}$ and the blue complex is $\left[Co(H_2O)_6\right]^{3+}$

EXERCISES

Nomenclature

15. **(a)** $\left[Co(OH)(H_2O)_4(NH_3)\right]^{2+}$ amminetetraaquahydroxocobalt(III) ion

    (b) $\left[Co(ONO)_3(NH_3)_3\right]$ triamminetrinitrito-O-cobalt(III)

    (c) $\left[Pt(H_2O)_4\right]\left[PtCl_6\right]$ tetraaquaplatinum(II) hexachloroplatinate(IV)

    (d) $\left[Fe(ox)_2(H_2O)_2\right]^-$ diaquadioxalatoferrate(III) ion

    (e) $Ag_2\left[HgI_4\right]$ silver(I) tetraiodomercurate(II)

16. **(a)** $K_3\left[Fe(CN)_6\right]$ potassium hexacyanoferrate(III)

    (b) $\left[Cu(en)_2\right]^{2+}$ bis(ethylenediamine)copper(II) ion

    (c) $\left[Al(OH)(H_2O)_5\right]Cl_2$ pentaaquahydroxoaluminum(III) chloride

    (d) $\left[CrCl(en)_2NH_3\right]SO_4$ amminechlorobis(ethylenediammine)chromium(III) sulfate

    (e) $\left[Fe(en)_3\right]_4\left[Fe(CN)_6\right]_3$ tris(ethylenediamine)iron(III) hexacyanoferrate(II)

Bonding and Structure in Complex Ions

17. We assume that $\left[PtCl_4\right]^{2-}$ is square planar by analogy with $\left[PtCl_2(NH_3)_2\right]$ in Figure 25-5. The other two complex ions are octahedral.

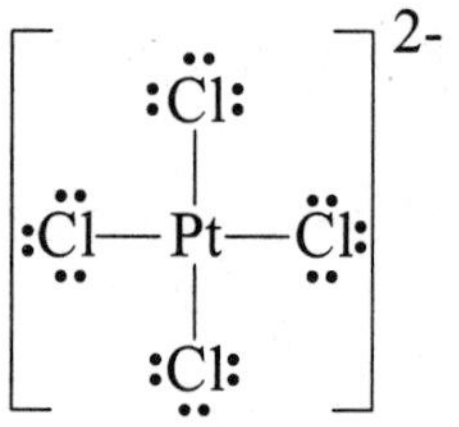

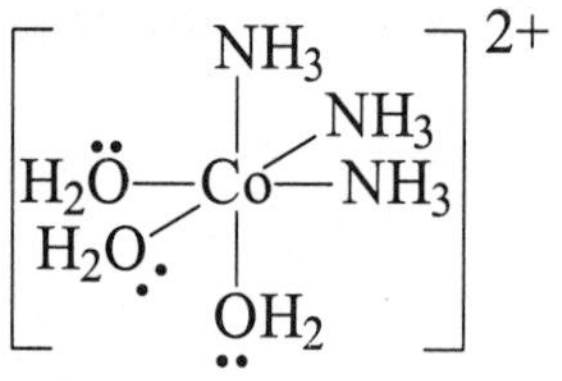

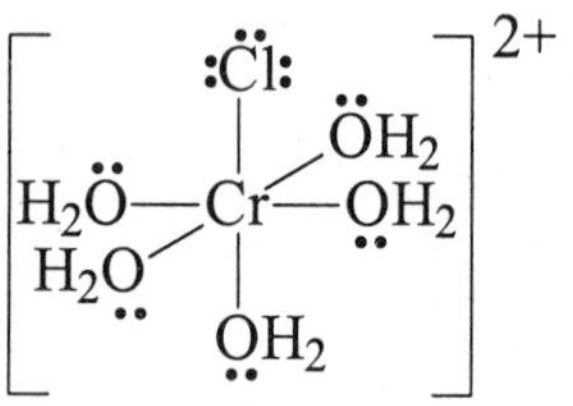

18. Structures of $\left[Pt(ox)_2\right]^{2-}$ and $\left[Cr(ox)_3\right]^{3-}$ are drawn below. The structure of $\left[Fe(EDTA)\right]^{2-}$ is the same as the structure of $\left[M(EDTA)\right]^{2-}$ drawn in Figure 25-23, with $M^{n+} = Fe^{2+}$.

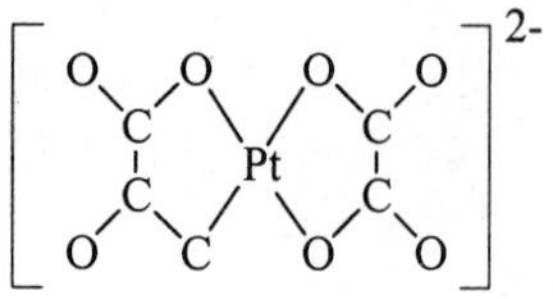

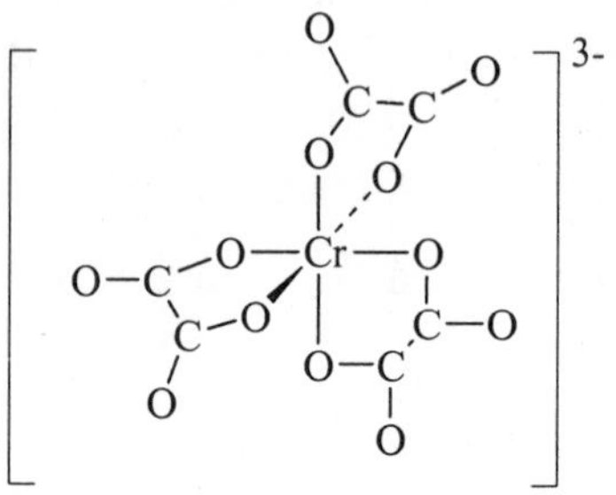

    (c) See figure 25-23

19. **(a)**

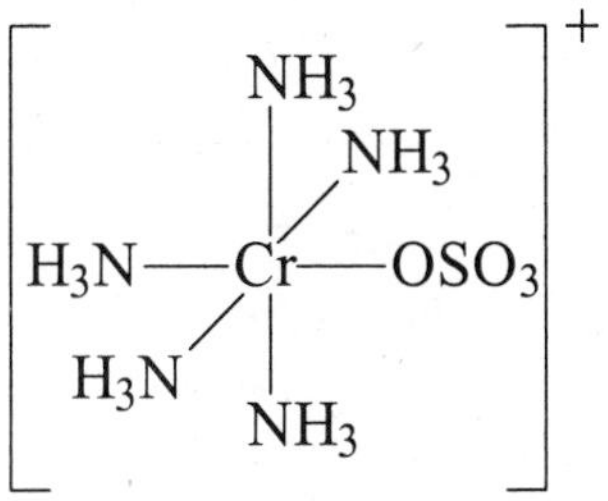

$$\left[\begin{array}{c} \text{NH}_3 \\ \text{NH}_3 \\ \text{H}_3\text{N}-\text{Cr}-\text{OSO}_3 \\ \text{H}_3\text{N} \\ \text{NH}_3 \end{array} \right]^{+}$$

pentaamminesulfateochromium(III)

(b)

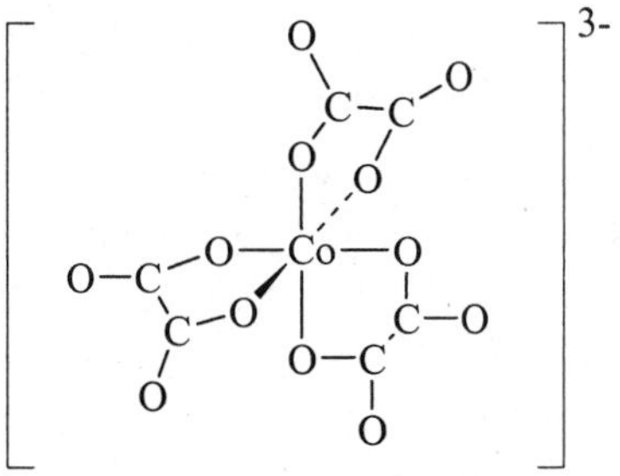

$$\left[\begin{array}{c} \text{O} \\ \text{C}-\text{C} \\ \text{O} \\ \text{O}-\text{C}-\text{O}-\text{Co}-\text{O} \\ \text{C}-\text{O} \\ \text{O}-\text{C} \\ \text{O} \end{array} \right]^{3-}$$

trioxalatocobaltate(III)

(c)

fac-triamminedichloronitro-O-cobalt(III)

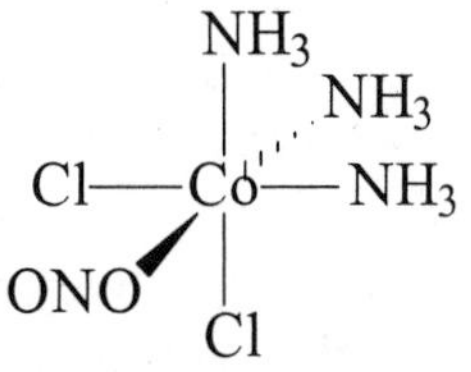

mer-triammine-*cis*-dichloronitro-O-cobalt(III)

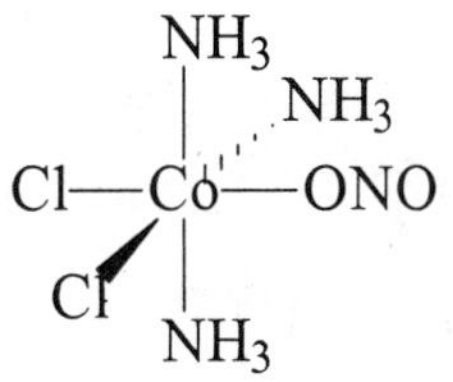

mer-triammine-*trans*-dichloronitro-O-cobalt(III)

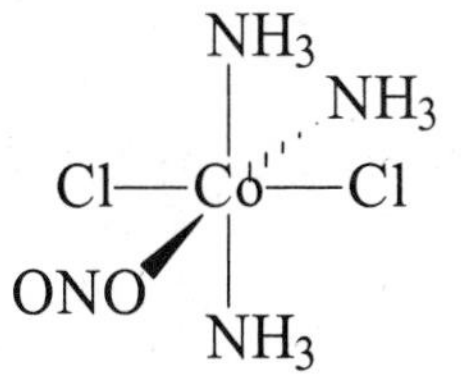

20. **(a)** pentaamminenitroto-N-cobalt(III) ion

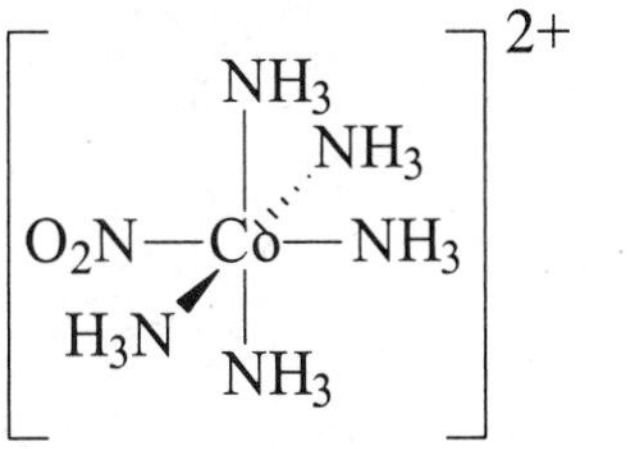

(b) ethylenediaminedithiocyanato-S-copper(II)

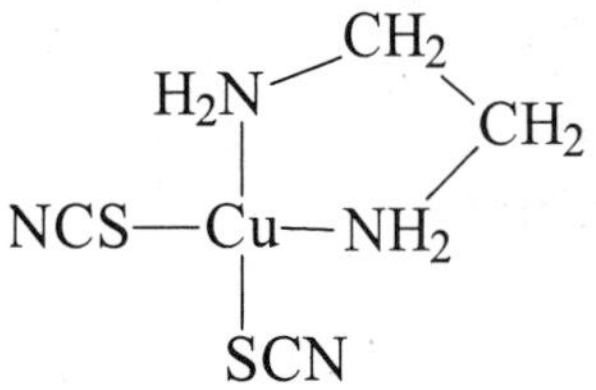

(c) hexaaquanickel(II) ion

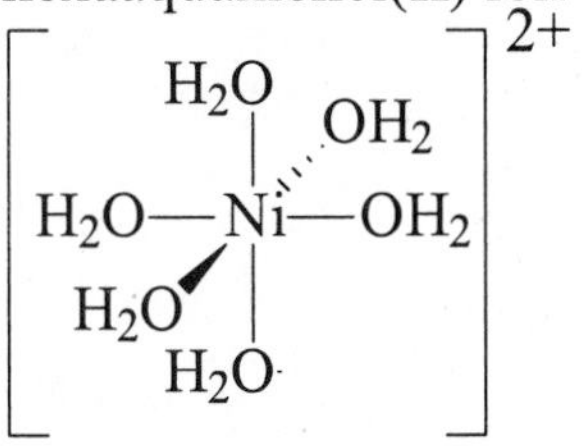

Isomerism

21. **(a)** *cis-trans* isomerism cannot occur with tetrahedral structures because all of the ligands are separated by the same angular distance from each other. One ligand cannot be on the other side of the central atom from another.

(b) Square planar structures can show *cis-trans* isomerism. Examples are drawn following, with the *cis*-isomer drawn on the left, and the *trans*-isomer drawn on the right.

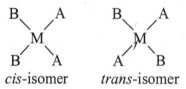

cis-isomer   *trans*-isomer

(c) Linear structures do not display *cis-trans* isomerism; there is only one way to bond the two ligands to the central atom.

22. All of the isomers are drawn together after the answer to this question.

(a) $\left[CrOH(NH_3)_5\right]^{2+}$ has one isomer.

(b) $\left[CrCl_2(H_2O)(NH_3)_3\right]^{+}$ has three isomers.

(c) $\left[CrCl_2(en)_2\right]^{+}$ has two geometric isomers, *cis*- and *trans*-.

(d) $\left[CrCl_4(en)\right]^{-}$ has only one isomer since the ethylenediamine (en) ligand cannot bond *trans* to the central metal atom.

(e) $\left[Cr(en)_3\right]^{3+}$ has only one geometric isomer; it has two optical isomers.

(a)

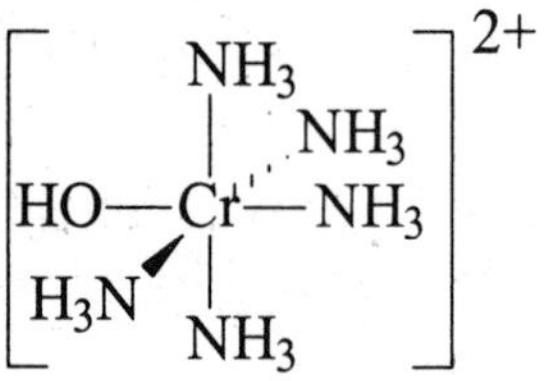

$$\left[\begin{array}{c} NH_3 \\ | \quad NH_3 \\ HO-Cr-NH_3 \\ H_3N \quad | \\ NH_3 \end{array} \right]^{2+}$$

(b)

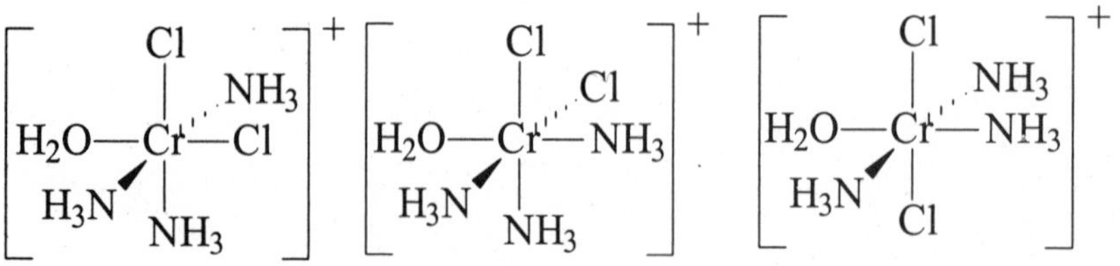

$$\left[\begin{array}{c} Cl \\ | \quad NH_3 \\ H_2O-Cr-Cl \\ H_3N \quad | \\ NH_3 \end{array} \right]^{+} \left[\begin{array}{c} Cl \\ | \quad Cl \\ H_2O-Cr-NH_3 \\ H_3N \quad | \\ NH_3 \end{array} \right]^{+} \left[\begin{array}{c} Cl \\ | \quad NH_3 \\ H_2O-Cr-NH_3 \\ H_3N \quad | \\ Cl \end{array} \right]^{+}$$

(c)

$$\left[\begin{array}{c} Cl \\ | \quad Cl \\ N-Cr-N \\ \end{array} \right]^{+} \left[\begin{array}{c} Cl \\ | \quad Cl \\ N-Cr-N \\ \end{array} \right]^{+}$$

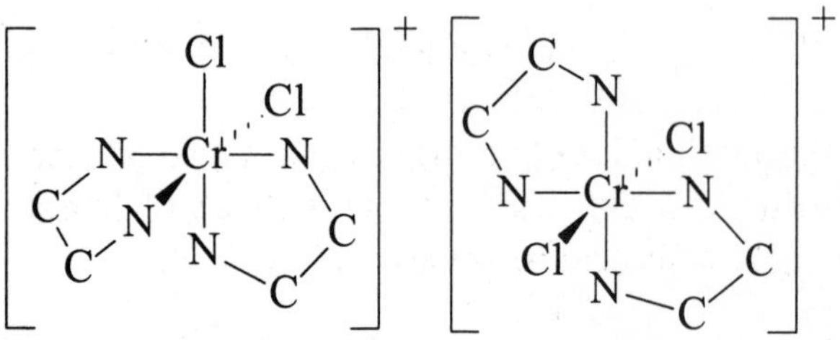

(d) 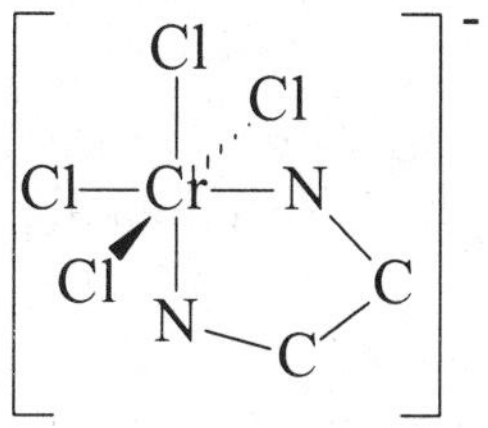 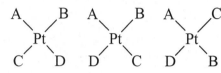   **(e)**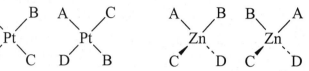

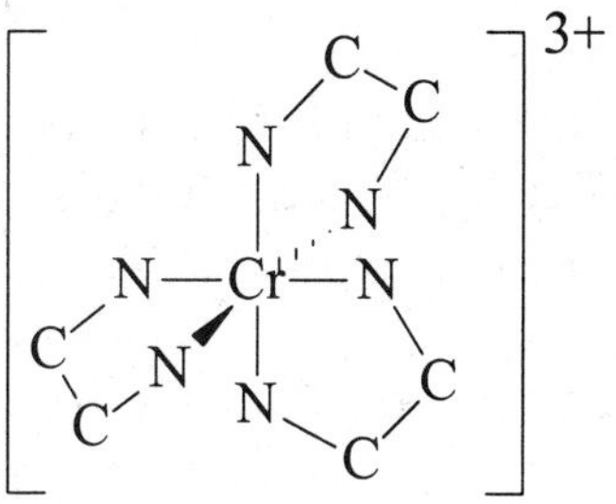

23. **(a)** There are three different square planar isomers, with D, C, and B, respectively, trans to the A ligand. They are drawn below.

(b) Tetrahedral $[ZnABCD]^{2+}$ does display optical isomerism. The two optical isomers are drawn above.

24. There are a total of four coordination isomers. They are listed below. We assume that the oxidation state of each central metal ion is $+3$.

$[Co(en)_3][Cr(ox)_3]$ tris(ethylenediamine)cobalt(III) trioxalatochromate(III)

$[Co(ox)(en)_2][Cr(ox)_2(en)]$ bis(ethylenediamine)oxalatocobalt(III)
(ethylenediamine)dioxalatochromate(III)

$[Cr(ox)(en)_2][Co(ox)_2(en)]$ bis(ethylenediamine)oxalatochromium(III)
(ethylenediamine)dioxalatocobaltate(III)

$[Cr(en)_3][Co(ox)_3]$ tris(ethylenediamine)chromium(III) trioxalatocobaltate(III)

25. The *cis*-dichlorobis(ethylenediamine)cobalt(III) ion is optically active. The two optical isomers are drawn below. The *trans*-isomer is not optically active: the ion and its mirror image are superimposable.

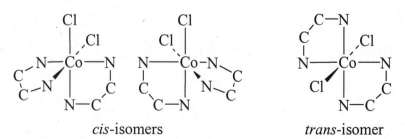

cis-isomers   *trans*-isomer

26. Complex ions (a) and (b) are identical; complex ions (a) and (d) are geometric isomers; complex ions (b) and (d) are geometric isomers; complex ion (c) is distinctly different from the other three complex ions (it has a different chemical formula).

Crystal Field Theory

27. In crystal field theory, the five *d* orbitals of a central transition metal ion are split into two (or more) groups of different energies. The energy spacing between these groups often corresponds to the energy of a photon of visible light. Thus, the complex ion will absorb light with energy corresponding to this spacing. If white light is incident on the complex ion, the light remaining after absorption will be missing some of its components. Thus, light of certain wavelengths (corresponding to the energies absorbed) will no longer be present in the formerly white light. The resulting light is colored. For example, if blue light is absorbed from white light, the remaining light will be yellow in color.

28. The difference in color is due to the difference in the value of Δ, the ligand field splitting energy. When the value of Δ is large, short wavelength light, which has a blue color, is absorbed, and the substance or its solution appears yellow. On the other hand, when the value of Δ is small, light of long wavelength, which has a red or yellow color, is absorbed, and the substance or its solution appears blue. The cyano ligand is a strong field ligand, producing a large value of Δ, and thus yellow complexes. On the other hand, the aqua ligands are weak field ligands, which produce a small value of Δ, and hence blue complexes.

29. We begin with the 7 electron *d*-orbital diagram for Co^{2+} [Ar] ↑↓ ↑↓ ↑ ↑ ↑

The strong field and weak field diagrams for octahedral complexes follow, along with the number of unpaired electrons in each case.

Strong Field ↑ __

(1 unpaired electron) ↑↓ ↑↓ ↑↓

Weak Field ↑ ↑

(3 unpaired electrons) ↑↓ ↑↓ ↑

30. We begin with the 8 electron *d*-orbital diagram for Ni^{2+} [Ar] ↑↓ ↑↓ ↑↓ ↑ ↑

The strong field and weak field diagrams for octahedral complexes follow, along with the number of unpaired electrons in each case.

Strong Field ↑ ↑

(2 unpaired electrons) ↑↓ ↑↓ ↑↓

Weak Field ↑ ↑

(2 unpaired electrons) ↑↓ ↑↓ ↑↓

The number of unpaired electrons is the same in both cases.

31. **(a)** Both of the central atoms have the same oxidation state, +3. We give the electron configuration of the central atom to the left, then the completed crystal field diagram in the center, and finally the number of unpaired electrons. The chloro ligand is a weak field ligand in the spectrochemical series.

Mo^{3+} $[Kr]\,4d^3$ *weak field*

$[Kr]_{4d}$ ⊞⊞⊞⊞⊞

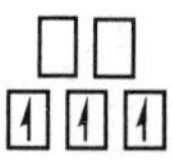

3 unpaired electrons; paramagnetic

The ethylenediamine ligand is a strong field ligand in the spectrochemical series.

Co^{3+} $[Ar]\,3d^6$ *strong field*

$[Ar]$ ⊞⊞⊞⊞⊞

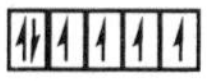

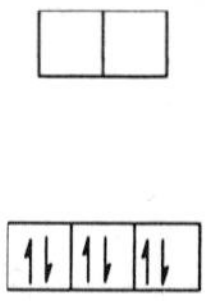

no unpaired electrons; diamagnetic

(b) In $[CoCl_4]^{2-}$ the oxidation state of cobalt is 2+. Chloro is a weak field ligand. The electron configuration of Co^{2+} is $[Ar]\,3d^7$ or $[Ar]$ ⊞⊞⊞⊞⊞

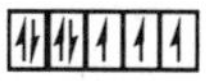

The tetrahedral ligand field diagram is shown on the right.   *weak field*

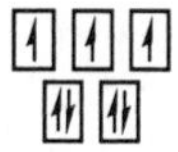

3 unpaired electrons

32. **(a)** In $\left[Cu(py)_4\right]^{2+}$ the oxidation state of copper is +2. Pyridine is a strong field ligand. The electron configuration of Cu^{2+} is $[Ar]\,3d^9$ or $[Ar]$ ⊞⊞⊞⊞⊞

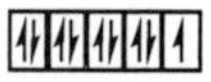

There is no possible way that an odd number of electrons can be paired up, without at least one electron being unpaired. $\left[Cu(py)_4\right]^{2+}$ is paramagnetic.

(b) In $\left[Mn(CN)_6\right]^{3-}$ the oxidation state of manganese is +3. Cyano is a strong field ligand. The electron configuration of Mn^{3+} is $[Ar]\,3d^4$ or $[Ar]$ ⊞⊞⊞⊞⊞

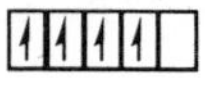

The ligand field diagram follows, at left. In $[FeCl_4]^-$ the oxidation state of iron is +3. Chloro is a weak field ligand. The electron configuration of Fe^{3+} is $[Ar]\,3d^5$ or $[Ar]$ ⊞⊞⊞⊞⊞

The ligand field diagram follows, below.

strong field   ▢▢

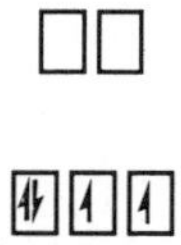

⊞⊞⊞

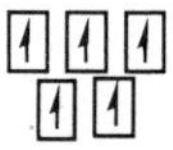

2 unpaired electrons

weak field ⊞⊞⊞ ⊞⊞

5 unpaired electrons

There are more unpaired electrons in $[FeCl_4]^-$ than in $\left[Mn(CN)_6\right]^{3-}$.

33. The electron configuration of Ni^{2+} is $[Ar]\,3d^8$ or $[Ar]$ ▨▨▨▨▨

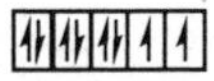

Ammine is a strong field ligand. The ligand field diagrams follow, octahedral at left, tetrahedral in the center and square planar at right.

Octahedral	Tetrahedral	Square Planar

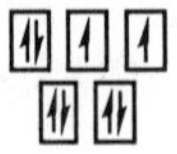

Since the octahedral and tetrahedral configurations have the same number of unpaired electrons (that is, 2 unpaired electrons), we cannot use magnetic properties to determine whether the ammine complex of nickel(II) is octahedral or tetrahedral. But we can determine if the complex is square planar, since the square planar complex is diamagnetic with zero unpaired electrons.

34. The difference is due to the fact that $\left[Fe(CN_6)\right]^{4-}$ is a strong field complex ion, while $\left[Fe(H_2O)_6\right]^{2+}$ is a weak field complex ion. The electron configurations for an iron atom and an iron(II) ion, and the ligand field diagrams for the two complex ions follow.

$$\left[Fe(CN)_6\right]^{4-} \qquad \left[Fe(H_2O)_6\right]^{2+}$$

iron atom $\qquad [Ar]\,_{3d}$ ▨▨▨▨▨ $_{4s}$ ▨

iron(II) ion $\qquad [Ar]\,_{3d}$ 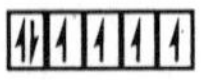 ▨▨▨▨▨ $_{4s}$ □

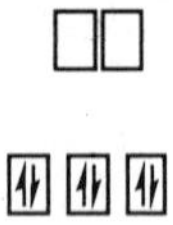

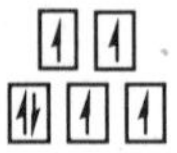

Complex Ion Equilibria

35. **(a)** $Zn(OH)_2(s) + 4\,NH_3(aq) \rightleftharpoons \left[Zn(NH_3)_4\right]^{2+}(aq) + 2\,OH^-(aq)$

(b) $Cu^{2+}(aq) + 2\,OH^-(aq) \rightleftharpoons Cu(OH)_2(s)$

The blue color is most likely due to some unreacted $\left[Cu(H_2O)_4\right]^{2+}$ (aq, pale blue)

$Cu(OH)_2(s) + 4\,NH_3(aq) \rightleftharpoons \left[Cu(NH_3)_4\right]^{2+}$ (aq, dark blue) $+ 2\,OH^-(aq)$

$\left[Cu(NH_3)_4\right]^{2+}(aq) + 4\,H_3O^+(aq) \rightleftharpoons \left[Cu(H_2O)_4\right]^{2+}(aq) + 4\,NH_4^+(aq)$

36. **(a)** $CuCl_2(s) + 2 Cl^-(aq) \rightleftharpoons [CuCl_4]^{2-}(aq, \text{ yellow})$

$2[CuCl_4]^{2-}(aq) + 4H_2O(l)$

$\updownarrow$

$[CuCl_4]^{2-}(aq, \text{ yellow}) + [Cu(H_2O)_4]^{2+}(aq, \text{ pale blue}) + 4 Cl^-(aq)$

or

$2[Cu(H_2O)_2 Cl_2](aq, \text{ green}) + 4 Cl^-(aq)$

Either $\quad [CuCl_4]^{2-}(aq) + 4 H_2O(l) \rightleftharpoons [Cu(H_2O)_4]^{2+}(aq) + 4 Cl^-(aq)$

Or $\quad [Cu(H_2O)_2 Cl_2](aq) + 2H_2O(l) \rightleftharpoons [Cu(H_2O)_4]^{2+}(aq) + 2 Cl^-(aq)$

(b) The blue solution is that of $[Cr(H_2O)_6]^{2+}$. This is quickly oxidized to Cr^{3+} by $O_2(g)$ from the atmosphere. The green color is due to $[CrCl_2(H_2O)_4]^{3+}$.

$4 [Cr(H_2O)_6]^{2+}(aq, \text{ blue}) + 4 H^+(aq) + 8 Cl^-(aq) + O_2(g)$

$\downarrow$

$4 [CrCl_2(H_2O)_4]^+(aq, \text{ green}) + 10 H_2O(l)$

Over a period of time, we might expect volatile HCl(g) to escape, leading to the formation of complex ions with more H_2O and less Cl^-.

$H^+(aq) + Cl^-(aq) \rightleftharpoons HCl(g)$

$[CrCl_2(H_2O)_4]^+(aq, \text{ green}) + H_2O(l) \rightleftharpoons [CrCl(H_2O)_5]^{2+}(aq, \text{ blue-green}) + Cl^-(aq)$

$[CrCl(H_2O)_5]^{2+}(aq, \text{ blue-green}) + H_2O(l) \rightleftharpoons [Cr(H_2O)_6]^{3+}(aq, \text{ blue}) + Cl^-(aq)$

Actually, to ensure that these final two reactions actually do occur in a timely fashion, it would be helpful to dilute the solution with water after the chromium metal has dissolved.

37. $[Co(en)_3]^{3+}$ should have the largest overall K_f value. We expect a complex ion with polydentate ligands to have a larger value for its formation constant than complexes that contain only monodentate ligands. This is an expression of the chelate effect. Once one end of a polydentate ligand becomes attached to the central metal, the attachment of the remaining electron pairs is relatively easy because they already are close to the central metal (and do not have to migrate in from a distant point in the solution).

38. **(a)** $\left[\text{Zn}(\text{NH}_3)_4\right]^{2+}$

$\beta_4 = K_1 \times K_2 \times K_3 \times K_4 = 3.9 \times 10^2 \times 2.1 \times 10^2 \times 1.0 \times 10^2 \times 50. = 4.1 \times 10^8$.

(b) $\left[\text{Ni}(\text{H}_2\text{O})_2(\text{NH}_3)_4\right]^{2+}$

$\beta_4 = K_1 \times K_2 \times K_3 \times K_4 = 6.3 \times 10^2 \times 1.7 \times 10^2 \times 54 \times 15 = 8.7 \times 10^7$

39. First: $\left[\text{Fe}(\text{H}_2\text{O})_6\right]^{3+}(\text{aq}) + \text{en}(\text{aq}) \rightleftharpoons \left[\text{Fe}(\text{H}_2\text{O})_4(\text{en})\right]^{3+}(\text{aq}) + 2\text{H}_2\text{O}(\text{l})$ $K_1 = 10^{4.34}$

Second: $\left[\text{Fe}(\text{H}_2\text{O})_4(\text{en})\right]^{3+}(\text{aq}) + \text{en}(\text{aq}) \rightleftharpoons \left[\text{Fe}(\text{H}_2\text{O})_2(\text{en})_2\right]^{3+}(\text{aq}) + 2\text{H}_2\text{O}(\text{l})$ $K_2 = 10^{3.31}$

Third: $\left[\text{Fe}(\text{H}_2\text{O})_2(\text{en})_2\right]^{3+}(\text{aq}) + \text{en}(\text{aq}) \rightleftharpoons \left[\text{Fe}(\text{en})_3\right]^{3+}(\text{aq}) + 2\text{H}_2\text{O}(\text{l})$ $K_3 = 10^{2.05}$

Net: $\left[\text{Fe}(\text{H}_2\text{O})_6\right]^{3+}(\text{aq}) + 3\,\text{en}(\text{aq}) \rightleftharpoons \left[\text{Fe}(\text{en})_3\right]^{3+}(\text{aq}) + 6\text{H}_2\text{O}(\text{l})$

$K_f = K_1 \times K_2 \times K_3$

$\log K_f = 4.34 + 3.31 + 2.05 = 9.70$ $K_f = 10^{9.70} = 5.0 \times 10^9 = \beta_3$

40. Since the overall formation constant is the product of the individual stepwise formation constants, the logarithm of the overall formation constant is the sum of the logarithms of the stepwise formation constants.

$\log K_f = \log K_1 + \log K_2 + \log K_3 + \log K_4 = 2.80 + 1.60 + 0.49 + 0.73 = 5.62$

$K_f = 10^{5.62} = 4.2 \times 10^5$

41. **(a)** Aluminum(III) forms a stable (and soluble) hydroxo complex but not a stable ammine complex.

$$\left[\text{Al}(\text{H}_2\text{O})_3(\text{OH})_3\right](\text{s}) + \text{OH}^-(\text{aq}) \rightleftharpoons \left[\text{Al}(\text{H}_2\text{O})_2(\text{OH})_4\right]^-(\text{aq}) + \text{H}_2\text{O}(\text{l})$$

(b) Although zinc(II) forms a soluble stable ammine complex ion, its formation constant is not sufficiently large to dissolve highly insoluble ZnS. However, it is sufficiently large to dissolve the moderately insoluble ZnCO_3. Said another way, ZnS does not produce sufficient $\left[\text{Zn}^{2+}\right]$ to permit the complex ion to form.

$$\text{ZnCO}_3(\text{s}) + 4\,\text{NH}_3(\text{aq}) \rightleftharpoons \left[\text{Zn}(\text{NH}_3)_4\right]^{2+}(\text{aq}) + \text{CO}_3{}^{2-}(\text{aq})$$

(c) Chloride ion forms a stable complex ion with silver(I) ion, that dissolves the AgCl(s) that formed when $\left[\text{Cl}^-\right]$ is low.

$$\text{AgCl}(\text{s}) \rightleftharpoons \text{Ag}^+(\text{aq}) + \text{Cl}^-(\text{aq}) \quad \text{and} \quad \text{Ag}^+(\text{aq}) + 2\,\text{Cl}^-(\text{aq}) \rightleftharpoons \left[\text{AgCl}_2\right]^-(\text{aq})$$

overall: $\text{AgCl}(\text{s}) + \text{Cl}^-(\text{aq}) \rightleftharpoons \left[\text{AgCl}_2\right]^-(\text{aq})$

42. (a) Because of the large value of the formation constant for the complex ion, $\left[Co(NH_3)_6\right]^{3+}(aq)$, the concentration of free $Co^{3+}(aq)$ is too small to enable it to oxidize water to $O_2(g)$. Since there is not a complex ion present—except, of course, $\left[Co(H_2O)_6\right]^{3+}(aq)$—when $CoCl_3$ is dissolved in water, the $\left[Co^{3+}\right]$ is sufficiently high for the oxidation-reduction reaction to be spontaneous.

$$4\ Co^{3+}(aq) + 2\ H_2O(l) \rightarrow 4\ Co^{2+}(aq) + 4\ H^+(aq) + O_2(g)$$

(b) Although AgI(s) is insoluble, there is a small concentration of $Ag^+(aq)$ present because of the solubility equilibrium:

$$2\ AgI(s) \ \rightleftharpoons \ Ag^+(aq) + I^-(aq)$$

These silver ions react with thiosulfate ion to form the stable dithiosulfatoargentate(I) complex ion:

$$Ag^+(aq) + 2S_2O_3^{2-}(aq) \rightleftharpoons \left[Ag(S_2O_3)_2\right]^-(aq)$$

Acid-Base Properties

43. $\left[Al(H_2O)_6\right]^{3+}(aq)$ is capable of releasing H^+:

$$\left[Al(H_2O)_6\right]^{3+}(aq) + H_2O(l) \rightleftharpoons \left[AlOH(H_2O)_5\right]^{2+}(aq) + H_3O^+(aq)$$

The value of its ionization constant ($pK_a = 5.01$) approximates that of acetic acid.

44. (a) $\left[CrOH(H_2O)_5\right]^{2+}(aq) + OH^-(aq) \rightarrow \left[Cr(OH)_2(H_2O)_4\right]^+(aq) + H_2O(l)$

(b) $\left[CrOH(H_2O)_5\right]^{2+}(aq) + H_3O^+(aq) \rightarrow \left[Cr(H_2O)_6\right]^{3+}(aq) + H_2O(l)$

Applications

45. (a) Solubility: $\qquad\qquad AgBr(s) \xrightleftharpoons[\ ]{K_{sp}=5.0\times10^{-13}} Ag^+(aq) + Br^-(aq)$

Cplx. Ion Formation: $\quad Ag^+(aq) + 2S_2O_3^{2-}(aq) \xrightleftharpoons[\ ]{K_f=1.7\times10^{13}} \left[Ag(S_2O_3)_2\right]^{3-}(aq)$

Net: $AgBr(s) + 2S_2O_3^{2-}(aq) \rightleftharpoons \left[Ag(S_2O_3)_2\right]^{3-}(aq) \qquad K_{overall} = K_{sp} \times K_f$

$$K_{overall} = 5.0\times10^{-13} \times 1.7\times10^{13} = 8.5$$

With a reasonably high $\left[S_2O_3^{2-}\right]$, this reaction will go essentially to completion.

(b) $NH_3(aq)$ cannot be used in the fixing of photographic film because of the relatively small value of K_f for $\left[Ag(NH_3)_2\right]^+ (aq)$, $K_f = 1.6 \times 10^7$. This would produce a value of $K = 8.0 \times 10^{-6}$ in the expression above, too small to indicate a reaction that goes to completion.

46.   Oxidation: $\left\{\left[Co(NH_3)_6\right]^{2+}(aq) \rightarrow \left[Co(NH_3)_6\right]^{3+}(aq) + e^-\right\} \times 2$ $-E^\circ = -0.10 V$.

Reduction: $\left\{H_2O_2(aq) + 2e^- \rightarrow 2OH^-(aq)\right\}$ $E^\circ = 0.88 V$

Net: $H_2O_2(aq) + 2\left[Co(NH_3)_6\right]^{2+}(aq) \rightarrow 2\left[Co(NH_3)_6\right]^{3+}(aq) + 2OH^-(aq)$

$E^\circ_{cell} = +0.88\ V + -0.10\ V = +0.78\ V.$

The positive value of the standard cell potential indicates that this is a spontaneous reaction.

Feature Problems

71. **(a)**   A trigonal prismatic structure predicts three geometric isomers for $[CoCl_2(NH_3)_4]^+$, which is one more than the actual number of geometric isomers found for this complex ion. All three geometric isomers arising from a trigonal prism are shown below.

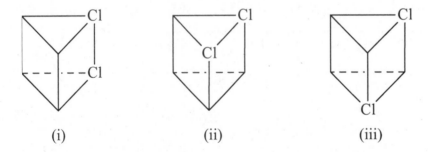

(i)   (ii)   (iii)

The fact that the trigonal prismatic structure does not afford the correct number of isomers is a clear indication that the ion actually adopts some other structural form (i.e., the theoretical model is contradicted by the experimental result). We know now of course, that this ion has an octahedral structure and as a result, it can exist only in *cis* and *trans* configurations.

(b) All attempts to produce optical isomers of $[Co(en)_3]^{3+}$ based upon a trigonal prismatic structure are shown below. The ethylenediamine ligand appears as an arc in each diagram below:

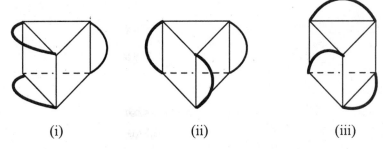

(i)   (ii)   (iii)

Only structure (iii), which has an ethylenediamine ligand connecting the diagonal corners of a face can give rise to optical isomers. Structure (iii) is highly unlikely, however, because the ethylenediamine ligand is simply too short to effectively span the diagonal distance across the face of the prism. Thus, barring any unusual stretching of the ethylenediamine ligand, a trigonal prismatic structure cannot account for the optical isomerism that was observed for $[Co(en)_3]^{3+}$.

72. Assuming that each hydroxide ligand bears it normal 1– charge, and that each ammonia ligand is neutral, the total contribution of negative charge from the ligands is 6 –. Since the net charge on the complex ions is 6+, the average oxidation state for each Co atom must be +3 (i.e., each Co in the complex can be viewed as a Co^{3+} $3d^6$ ion surrounded by six ligands.) The five $3d$ orbitals on each Co are split by the octahedrally arranged ligands into three lower energy orbitals, called t_{2g} orbitals, and two higher energy orbitals, called e_g orbitals. We are told in the question that the complex is *low spin*. This is simply another way of saying that all six $3d$ electrons on each Co are paired up in the t_{2g} set as a result of the e_g and t_{2g} orbitals being separated by a relatively large energy gap (see below). Hence, there should be no unpaired electrons in the hexacation (i.e., the cation is expected to be diamagnetic).

Low Spin Co^{3+}
(Octahedral Environment):

e_g ☐☐

Relatively Large Crystal Field Splitting

t_{2g} [↑↓|↑↓|↑↓]

The Lewis structures for the two optical isomers are depicted below:

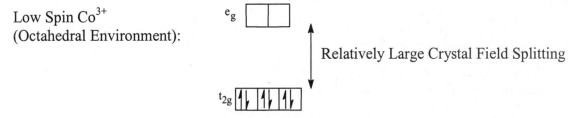

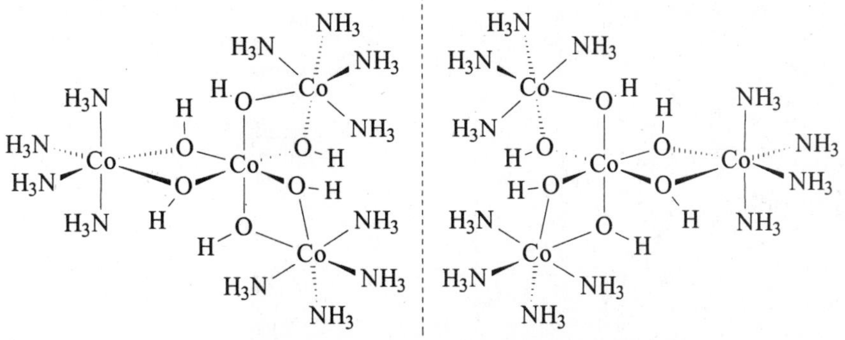

Non-superimposable mirror images of one another

73.   The data used to construct a plot of hydration energy as a function of metal ion atomic number is collected in the table below. The graph of hydration energy (kJ/mol) versus metal ion atomic number is located beneath the table.

Metal Ion	Atomic number	Hydration Energy
Ca^{2+}	20	−2468 kJ/mol
Sc^{2+}	21	−2673 kJ/mol
Ti^{2+}	22	−2750 kJ/mol
V^{2+}	23	−2814 kJ/mol
Cr^{2+}	24	−.2799 kJ/mol
Mn^{2+}	25	−2743 kJ/mol
Fe^{2+}	26	−2843 kJ/mol
Co^{2+}	27	−2904 kJ/mol
Ni^{2+}	28	−2986 kJ/mol
Cu^{2+}	29	−2989 kJ/mol
Zn^{2+}	30	−2936 kJ/mol

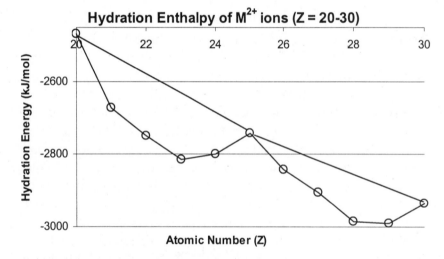

Hydration Enthalpy of M^{2+} ions (Z = 20-30)

(b)   When a metal ion is placed in an octahedral field of ligands, the five d-orbitals are split into e_g and t_{2g} subsets, as shown in the diagram below:

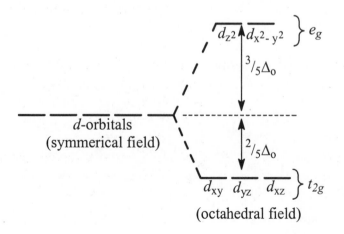

Since water is a weak field ligand, the magnitude of the splitting is relatively small. As a consequence, high-spin configurations result for all of the hexaaqua complexes. The electron configurations for the metal ions in the high-spin hexaaqua complexes and their associated crystal field stabilization energies (CFSE) are provided in the table below

Metal Ion	Configuration	t_{2g}	e_g	# of unpaired e^-	CFSE(Δ_o)
Ca^{2+}	$3d^0$	0	0	0	0
Sc^{2+}	$3d^1$	1	0	1	$-^2/_5$
Ti^{2+}	$3d^2$	2	0	2	$-^4/_5$
V^{2+}	$3d^3$	3	0	3	$-^6/_5$
Cr^{2+}	$3d^4$	3	1	4	$-^3/_5$
Mn^{2+}	$3d^5$	3	2	5	0
Fe^{2+}	$3d^6$	4	2	4	$-^2/_5$
Co^{2+}	$3d^7$	5	2	3	$-^4/_5$
Ni^{2+}	$3d^8$	6	2	2	$-^6/_5$
Cu^{2+}	$3d^9$	6	3	1	$-^3/_5$
Zn^{2+}	$3d^{10}$	6	4	0	0

Thus, the crystal field stabilization energy is zero for Ca^{2+}. Mn^{2+} and Zn^{2+}.

(c)   The lines drawn between those ions that have a CFSE = 0 show the trend for the enthalpy of hydration after the contribution from the crystal field stabilization energy has been subtracted from the experimental values. The Ca to Mn and Mn to Zn lines both have a negative slope. This trend shows that as one proceeds from left to right across the periodic table, the energy of hydration for dications becomes increasingly more negative. The hexaaqua complexes become progressively more stable because the Z_{eff} experienced by the bonding electrons in the valence shell of the metal ion steadily increases as we move further and further to the right. Put another way, the Z_{eff} climbs steadily as we move from left to right and this leads to the positive charge density on the metal becoming larger and larger, which results in the water ligands steadily being pulled closer and closer to the nucleus. Of course, the closer the approach of the water ligands to the metal, the greater is the energy released upon successful coordination of the ligand.

(d)   Those ions that exhibit crystal field stabilization energies greater than zero have heats of hydration that are more negative (i.e. more energy released) than the hypothetical heat of hydration for the ion with CFSE subtracted out. The heat of hydration without CFSE for a given ion falls on the line drawn between the two flanking ions with CFSE = 0 at a position directly above the point for the experimental hydration energy. The energy difference between the observed heat of hydration for the ion and the heat of hydration without CFSE is, of course, approximately equal to the CFSE for the ion.

(e)   As was mentioned in the answer to part (c), the straight line drawn between manganese and zinc (both ions with CFSE = 0) on the previous plot, describes the enthalpy trend after the ligand field stabilization energy has been subtracted from the experimental values for the hydration enthalpy. Thus, Δ_o for Fe^{2+} in $[Fe(H_2O)_6]^{2+}$ is approximately equal to $\frac{5}{2}$ of the energy difference (in kJ/mol) between the observed hydration energy for $Fe^{2+}(g)$ and the point for Fe^{2+} on the line connecting Mn^{2+} and Zn^{2+}, which is the expected enthalpy of hydration after the CFSE has been subtracted out. Remember that the crystal field stabilization energy for Fe^{2+} that is obtained from the graph is not Δ_o, but rather $\frac{2}{5}\Delta_o$, since the CFSE for a $3d^6$ ion in an octahedral field is just $\frac{2}{5}$ of Δ_o. Consequently, to obtain Δ_o, we must multiply the enthalpy difference by $\frac{5}{2}$. According to the graph, the high-spin CFSE for Fe^{2+} is -2843 kJ/mol $- (-2782$ kJ/mol$)$ or -61 kJ/mol. Consequently, $\Delta_o = \frac{5}{2}(-61$ kJ/mol$)$ $= -153$ kJ/mol, or 1.5×10^2 kJ/mol is the energy difference between the e_g and t_{2g} orbital sets.

(f)   The color of an octahedral complex is caused by the promotion of an electron on the metal from a t_{2g} orbital $\rightarrow e_g$ orbital. The energy difference between the e_g and t_{2g} orbital sets is Δ_o. As the metal-ligand bonding becomes stronger, the separation between the t_{2g} and e_g orbitals becomes larger. If the e_g set is not full, then the metal complex will exhibit an absorption band corresponding to a $t_{2g} \rightarrow e_g$ transition. Thus, the $[Fe(H_2O)_6]^{2+}$ complex ion should absorb electromagnetic radiation that has $E_{photon} = \Delta_o$. Since $\Delta_o \approx 150$ kJ/mol (calculated in part (e) of this question) for a mole of $[Fe(H_2O)_6]^{2+}(aq)$,

$$E_{photon} = \frac{1.5 \times 10^2 \, kJ}{1 \, mol[Fe(H_2O)_6]^{2+}} \times \frac{1 \, mol \, [Fe(H_2O)_6]^{2+}}{6.022 \times 10^{23}[Fe(H_2O)_6]^{2+}} \times \frac{1000 \, J}{1 \, kJ}$$

$$= 2.5 \times 10^{-19} \, J \text{ per ion}$$

$$v = \frac{E}{h} = \frac{2.5 \times 10^{-19} \, J}{6.626 \times 10^{-34} \, J \, s} = 3.8 \times 10^{14} \, s^{-1}$$

$$\lambda = \frac{c}{h} = \frac{2.998 \times 10^8 \, m \, s^{-1}}{3.8 \times 10^{14} \, s^{-1}} = 7.8 \times 10^{-7} \, m \, (780 \, nm)$$

So, the $[Fe(H_2O)_6]^{2+}$ ion will absorb radiation with a wavelength of 780 nm, which is red light in the visible part of the electromagnetic spectrum.

CHAPTER 26
NUCLEAR CHEMISTRY
PRACTICE EXAMPLES

1A   A β^- has a mass number of zero and an "atomic number" of -1. Emission of this electron has the effect of transforming a neutron into a proton. $^{241}_{94}Pu \rightarrow ^{241}_{95}Am + ^{0}_{-1}\beta$

1B ^{58}Ni has a mass number of 58 and an atomic number of 28. A positron has a mass number of 0 and an effective atomic number of $+1$. Emission of a positron has the seeming effect of transforming a proton into a neutron. The parent nuclide must be copper-58.

$^{58}_{29}Cu \rightarrow ^{58}_{28}Ni + ^{0}_{+1}\beta$

2A   The sum of the mass numbers $(139 + 12 = ? + 147)$ tells us that the other product species has $A = 4$. The atomic number of La is 57, that of C is 6, and that of Eu is 63. The atomic number sum $(57 + 6 = ? + 63)$ indicates that the atomic number of this product species is zero. Therefore, four neutrons must have been emitted. $^{139}_{57}La + ^{12}_{6}C \rightarrow ^{147}_{63}Eu + 4^{0}_{1}n$

2B   An alpha particle is $^{4}_{2}He$ and a positron is $^{0}_{+1}\beta$. We note that the total mass number in the first equation is 125; the mass number of the additional product is 1.
The total atomic number is 53; the atomic number of the additional product is 0; it is a neutron.

$^{121}_{51}Sb + ^{4}_{2}He \rightarrow ^{124}_{53}I + ^{1}_{0}n$

In the second equation, the positron has a mass number of 0, meaning that the mass number of the product is 124. Because the atomic number of the positron is $+1$, that of the product is 52; it is $^{124}_{52}Te$.

$^{124}_{53}I \rightarrow ^{0}_{+1}\beta + ^{124}_{52}Te$

3A   **(a)**   The decay constant is found from the 8.040-day half-life.

$$\lambda = \frac{0.693}{8.040 \text{ d}} = 0.0862 \text{ d}^{-1} \times \frac{1 \text{ d}}{24 \text{ h}} \times \frac{1 \text{ h}}{60 \text{ min}} \times \frac{1 \text{ min}}{60 \text{ s}} = 9.98 \times 10^{-7} \text{ s}^{-1}$$

(b)   The number of ^{131}I atoms is used to find the activity.

$$\text{no.}^{131}I \text{ atoms} = 2.05 \text{ mg} \times \frac{1 \text{ g}}{1000 \text{ mg}} \times \frac{1 \text{ mol } ^{131}I}{131 \text{ g } ^{131}I} \times \frac{6.022 \times 10^{23} \text{ atoms}}{1 \text{ mol } ^{131}I}$$

$$= 9.42 \times 10^{18} \text{ atoms } ^{131}I$$

$$\text{Activity} = \lambda N = 9.98 \times 10^{-7} \text{ s}^{-1} \times 9.42 \times 10^{18} \text{ atoms} = 9.40 \times 10^{12} \text{ disintegrations/second}$$

(c) We now determine the number of atoms remaining after 16 days. Because two half-lives elapse in 16 days, the number of atoms has halved twice, to one-forth (25%) the original number of atoms.

$$N_t = 0.25 \times N_0 = 0.25 \times 9.42 \times 10^{18} \text{ atoms} = 2.36 \times 10^{18} \text{ atoms}$$

(d) The rate after 14 days is determined by the number of atoms present.

$$\text{rate} = \lambda N_t = 9.98 \times 10^{-7} \text{ s}^{-1} \times 2.36 \times 10^{18} \text{ atoms} = 2.36 \times 10^{12} \text{ dis/s}$$

3B First we determine the value of λ: $\quad \lambda = \dfrac{0.693}{t_{1/2}} = \dfrac{0.693}{11.4 \text{ d}} = 0.0608 \text{ d}^{-1}$

Then we allow $N_t = 1\% N_0 = 0.010 N_0$ in equation (26.12).

$$\ln \frac{N_t}{N_0} = -\lambda t = \ln \frac{0.010 N_0}{N_0} = \ln (0.010) = -4.61 = -(0.0608 \text{ d}^{-1})t$$

$$t = \frac{-4.61}{-0.0608 \text{ d}^{-1}} = 75.8 \text{ d}$$

4A The half-life of ^{14}C is 5730 y and $\lambda = 1.21 \times 10^{-4}$ y^{-1}. The activity of ^{14}C when the object supposedly stopped growing was 15 dis/min per g C. We use equation (26.12) with activities (λN) in place of numbers of atoms (N).

$$\ln \frac{A_t}{A_0} = -\lambda t = \ln \frac{8.5 \text{ dis/min}}{15 \text{ dis/min}} = -(1.21 \times 10^{-4} \text{ y}^{-1})t = -0.56_8 \; ; t = \frac{0.57}{1.21 \times 10^{-4} \text{ y}^{-1}} = 4.7 \times 10^3 \text{ y}$$

4B The half-life of ^{14}C is 5730 y and $\lambda = 1.21 \times 10^{-4}$ y^{-1}. The activity of ^{14}C when the object supposedly stopped growing was 15 dis/min per g C. We use equation (26.12) with activities (λN) in place of numbers of atoms (N).

$$\ln \frac{A_t}{A_0} = -\lambda t = \ln \frac{A_t}{15 \text{ dis/min}} = -(1.21 \times 10^{-4} \text{ y}^{-1})(1100 \text{ y}) = -0.13$$

$$\frac{A_t}{15 \text{ dis/min}} = e^{-0.13} = 0.88 \; , A_t = 0.88 \times 15 \text{ dis/min} = 13 \text{ dis/min (per gram of C)}$$

5A Mass defect. $= 145.913053 \text{ u} (^{146}\text{Sm}) - 141.907719 \text{ u} (^{142}\text{Nd}) - 4.002603 \text{ u} (^{4}\text{He}) = 0.002731 \text{ u}$

Then, from the text, we have $931.5 \text{ MeV} = 1 \text{ u}$ $\quad E = 0.002731 \text{ u} \times \dfrac{931.5 \text{ MeV}}{1 \text{ u}} = 2.544 \text{ MeV}$

5B Unfortunately, we cannot use the result of Example 26–5 ($0.0045 \text{ u} = 4.2 \text{ MeV}$) because it is expressed to only two significant figures, and we begin with four significant figures. But, we essentially work backwards through that calculation. The last conversion factor is from Table 2-1.

$$E = 5.590 \text{ MeV} \times \frac{1.602 \times 10^{-13} \text{ J}}{1 \text{ MeV}} = 8.955 \times 10^{-13} \text{ J} = mc^2 = m\left(2.9979 \times 10^8 \text{ m/s}\right)^2$$

$$m = \frac{8.955 \times 10^{-13} \text{ J}}{\left(2.9979 \times 10^8 \text{ m/s}\right)^2} \times \frac{1000 \text{ g}}{1 \text{ kg}} \times \frac{1.0073 \text{ u}}{1.673 \times 10^{-24} \text{ g}} = 0.005999 \text{ u}$$

Or we could use $m = 5.590 \text{ MeV} \times \dfrac{1 \text{ u}}{931.5 \text{ MeV}} = 0.006001 \text{ u}$

6A **(a)** ^{88}Sr has an even atomic number (38) and an even neutron number (50); its mass number (88) is not too far from the average mass (87.6) of Sr. It should be stable.

(b) ^{118}Cs has an odd atomic number (55) and a mass number (118) that is pretty far from the average mass of Cs (132.9). It should be radioactive.

(c) ^{30}S has an even atomic number (16) and an even neutron number (14); but its mass number (30) is too far from the average mass of S (32.1). It should be radioactive.

6B We know that ^{19}F is stable, with approximately the same number of neutrons and protons: 9 protons, and 10 neutrons. Thus, nuclides of light elements with approximately the same number of neutrons and protons will be stable. In Practice Example 26–1 we saw that positron emission has the effect of transforming a proton into a neutron. β^- emission has the opposite effect: transforming a neutron into a proton. The mass number does not change in either case. Now let us analyze our two nuclides.
^{17}F has 9 protons and 8 neutrons. Replacing a proton with a neutron would produce a more stable nuclide. Thus, we predict positron emission by ^{17}F to produce ^{17}O.
^{22}F has 9 protons and 14 neutrons. Replacing a neutron with a proton would produce a more stable nuclide. Thus, we predict β^- emission by ^{22}F to produce ^{22}Ne.

REVIEW QUESTIONS

1. **(a)** α refers to an alpha particle, a helium-4 nucleus: $^4_2\text{He}^{2+}$.

(b) β^- refers to a beta particle, that is, an electron.

(c) β^+ refers to a positron, identical in all respects to an electron but with opposite (positive) charge.

(d) γ refers to gamma radiation, electromagnetic radiation with energy measured in MeV per photon. In contrast, visible light has energies measured in eV per photon.

(e) $t_{1/2}$ indicates half-life, the time needed for half of a sample to decay radioactively.

2.   **(a)**   A radioactive decay series is the series of successive products resulting from one
             nuclide, which decays to a second nuclide, which in turn decays to yet a third, and so
             forth, until a stable nuclide is produced.

     (b)   A charged-particle accelerator is a device that produces particles of very high speeds.
             It imparts energy and thus speed to these particles by using the influence of either an
             electric field or a magnetic field on the charge of the particle.

     (c)   The neutron-to-proton ratio is the quotient of the number of neutrons divided by the
             number of protons. It is a general indication of the stability of a nuclide (when
             combined with atomic number). Certainly, for light elements (up to about $Z = 20$), a
             neutron-to-proton ration of 1-to-1 is that of a stable nuclide.

     (d)   The mass–energy relationship was first promulgated by Einstein: $E = mc^2$.

     (e)   Background radiation is that radiation consistently present on earth from natural
             sources: cosmic rays, radioactive elements in the soil and the air, etc.

3.   **(a)**   Both electrons and positrons have the same mass, about 0.00055 u. However, the
             electron is negatively charged, while the positron has the same magnitude of charge
             but a positive one.

     (b)   The half-life, $t_{1/2}$, of a radioactive nuclide is the time necessary for half of a sample
             to decay. The decay constant is the rate constant for that first-order decay process:
             $\lambda = 0.693 / t_{1/2}$.

     (c)   The mass defect is the difference between the mass sum of the protons, neutrons, and
             electrons that constitute a nuclide and the nuclidic mass. The binding energy is that
             defect expressed as an energy.

     (d)   Nuclear fission refers to the process in which the nucleus of an atom is split into
             smaller fragments, usually with the release of energy. Nuclear fusion refers to the
             process in which two lighter nuclei are combined to give a heavier nucleus

     (e)   Primary ionization refers to those ions created by collision of the radiation with the
             atoms of the sample. However, when these atoms ionize, they give off electrons,
             which often are energetic enough to produce further ionization, called secondary
             ionization, in the sample.

4.   **(a)** γ rays penetrate through matter the greatest distance, largely because they are
             uncharged and thus do not interact with matter extensively.

     (b) α particles have the greatest ionizing power, principally because they have the
             largest charge and mass.

     (c) β particles are deflected the most in a magnetic field, because of their small mass
             and relatively large charge, that is, because they have the largest charge-to-mass
             ratio.

5. **(a)** $^{160}_{74}W \rightarrow ^{156}_{72}Hf + ^{4}_{2}He$

(b) $^{38}_{17}Cl \rightarrow ^{38}_{18}Ar + ^{0}_{-1}\beta$

(c) $^{214}_{83}Bi \rightarrow ^{214}_{84}Po + ^{0}_{-1}\beta$

(d) $^{32}_{17}Cl \rightarrow ^{32}_{16}S + ^{0}_{+1}\beta$

6. **(a)** $^{23}_{11}Na + ^{2}_{1}H \rightarrow ^{24}_{11}Na + ^{1}_{1}H$

(b) $^{59}_{27}Co + ^{1}_{0}n \rightarrow ^{56}_{25}Mn + ^{4}_{2}He$

(c) $^{238}_{92}U + ^{2}_{1}H \rightarrow ^{240}_{94}Pu + ^{0}_{-1}\beta$

(d) $^{246}_{96}Cm + ^{13}_{6}C \rightarrow ^{254}_{102}No + 5^{1}_{0}n$

(e) $^{238}_{92}U + ^{14}_{7}N \rightarrow ^{246}_{99}Es + 6^{1}_{0}n$

7. **(a)** $^{214}_{88}Ra \rightarrow ^{210}_{86}Rn + ^{4}_{2}He$

(b) $^{205}_{85}At \rightarrow ^{205}_{84}Po + ^{0}_{+1}\beta^{+}$

(c) $^{212}_{87}Fr + ^{0}_{-1}e \rightarrow ^{212}_{86}Rn$

(d) $^{2}_{1}H + ^{2}_{1}H \rightarrow ^{3}_{2}He + ^{1}_{0}n$

(e) $^{241}_{95}Am + ^{4}_{2}He \rightarrow ^{243}_{97}Bk + 2^{1}_{0}n$

8. **(a)** Since the decay constant is inversely related to the half-life, the nuclide with the smallest half-life also has the largest value of its decay constant. This is the nuclide $^{214}_{84}Po$ with a half-life of 1.64×10^{-4} s.

(b) The nuclide that displays a 75% reduction in its radioactivity has passed through two half-lives in a period of one month. Thus, this is the nuclide with a half-life of approximately two weeks. This is the nuclide $^{32}_{15}P$, with a half-life of 14.3 days.

(c) If more than 99% of the radioactivity is lost, less than 1% remains. Thus $(\frac{1}{2})^{n} < 0.010$. Now, when $n = 7$, $(\frac{1}{2})^{n} = 0.0078$. Thus, seven half-lives have elapsed in one month, and each half-life approximates 4.3 days. The longest lived nuclide that fits this description is $^{222}_{86}Rn$, which has a half-life of 3.823 days. Of course, all other nuclides with shorter half-lives also meet this criterion, specifically the following nuclides: $^{13}_{8}O(8.7 \times 10^{-3}$ s$)$, $^{28}_{12}Mg(21\,h)$, $^{80}_{35}Br(17.6\,min)$, and $^{214}_{84}Po(1.64 \times 10^{-4}$ s$)$.

9. Since $16 = 2^{4}$, four half-lives have elapsed in 18.0 h, and each half-life equals 4.50 h. The half-life of isotope B thus is $2.5 \times 4.50\,h = 11.25\,h$. Now, since $32 = 2^{5}$, five half-lives must elapse before the decay rate of isotope B falls to $\frac{1}{32}$ of its original value. Thus, the time elapsed for this amount of decay is:

$$\text{time elapsed} = 5\,\text{half lives} \times \frac{11.25}{1\,\text{half life}} = 56.3\,h$$

10. We use expression (26.12), substituting the disintegration rate for the number of atoms, since we recognize that in this first-order reaction the rate is directly proportional to the amount of reactant, that is, the number of atoms. (All radioactive decay processes follow first-order kinetics.) We also use equation (26.13), rearranged to $\lambda = \dfrac{0.693}{t_{1/2}}$. Thus

$$\ln\frac{N_t}{N_0} = -\frac{0.693t}{t_{1/2}}.$$

(a) $\ln\dfrac{253 \text{ dis/min}}{1.00\times10^3 \text{ dis/min}} = -\dfrac{0.693\,t}{87.9 \text{ d}} = -1.374; \quad t = \dfrac{1.374\times87.9 \text{ d}}{0.693} = 174 \text{ d}$

(b) $\ln\dfrac{104 \text{ dis/min}}{1.00\times10^3 \text{ dis/min}} = -\dfrac{0.693\,t}{87.9 \text{ d}} = -2.263; \quad t = \dfrac{2.263\times87.9 \text{ d}}{0.693} = 287 \text{ d}$

(c) $\ln\dfrac{52 \text{ dis/min}}{1.00\times10^3 \text{ dis/min}} = -\dfrac{0.693\,t}{87.9 \text{ d}} = -2.96; \quad t = \dfrac{2.96\times87.9 \text{ d}}{0.693} = 375 \text{ d}$

11. The principal equation that we shall employ is $E = mc^2$, along with conversion factors.

(a) $E = 6.02\times10^{-23} \text{ g}\times\dfrac{1 \text{ kg}}{1000 \text{ g}}\times\left(3.00\times10^8 \text{ m/s}\right)^2 = 5.42\times10^{-9} \text{ kg m}^2 \text{ s}^{-2} = 5.42\times10^{-9} \text{ J}$

(b) $E = 4.0015 \text{ u}\times\dfrac{931.5 \text{ MeV}}{1 \text{ u}} = 3727 \text{ MeV}$

12. Mass of individual particles $= \left(47 \text{ p}\times\dfrac{1.0073 \text{ u}}{1 \text{ p}}\right)\times\left(60 \text{ n}\times\dfrac{1.0087 \text{ u}}{1 \text{ n}}\right)$

$$= 47.3431 \text{ u} + 60.5220 \text{ u} = 107.8651 \text{ u}$$

$$\frac{\text{Binding energy}}{\text{nucleon}} = \frac{107.8651 \text{ u} - 106.879289 \text{ u}}{107 \text{ nucleons}}\times\frac{931.5 \text{ MeV}}{1 \text{ u}} = 8.58 \text{ MeV/nucleon}$$

13. **(c)** ^{80}Br does not occur naturally; it has an odd number of protons (35) and an odd number of neutrons (45).

(d) ^{132}Cs also does not occur naturally, for the same reason.

14. **(a)** A radioisotope with a long half-life gives off few disintegrations per second, and thus is not exceedingly hazardous unless a large quantity of it is present. Radioisotopes with short half-lives, however, give off large quantities of radiation in a short time, but they also "burn themselves out" quickly and thus are hazardous for only a short time. On the other hand, radioisotopes with an intermediate half-life are both giving off reasonably large quantities of radiation and are in existence for relatively long times.

(b) The radioisotopes that are hazardous from a distance are those that give off γ or high energy β radiation, radiation with a large or moderate penetrating power. On the other hand, α particles are not hazardous at long distances because of their short penetrating power, but they are highly ionizing, and thus do significant damage when they are encountered at short distances.

(c) Potassium-40 is a radioisotope that decays by electron capture to argon-40. This accounts for a large majority of the argon in the atmosphere. Helium-4 is also produced by radioactive decay, but it is so light that it escapes from the atmosphere at a much faster rate than does argon-40.

(d) Francium is produced by radioactive decay. It is quite short-lived and thus there is not much present on the Earth at any one time.

(e) Fusion involves joining positively charged nuclei together. Such a process requires quite energetic nuclei. This, in turn, means that the nuclei must be at very high temperatures.

EXERCISES

Radioactive Processes

15. **(a)** $^{234}_{94}\text{Pu} \rightarrow ^{230}_{92}\text{U} + ^{4}_{2}\text{He}$

(b) $^{248}_{97}\text{Bk} \rightarrow ^{248}_{98}\text{Cf} + ^{0}_{-1}\text{e}$

(c) $^{196}_{82}\text{Pb} + ^{0}_{-1}\text{e} \rightarrow ^{196}_{81}\text{Tl}$; $\quad ^{196}_{81}\text{Tl} + ^{0}_{-1}\text{e} \rightarrow ^{196}_{80}\text{Hg}$

16. **(a)** $^{214}_{82}\text{Pb} \rightarrow ^{214}_{83}\text{Bi} + ^{0}_{-1}\text{e}$; $\quad ^{214}_{83}\text{Bi} \rightarrow ^{214}_{84}\text{Po} + ^{0}_{-1}\text{e}$

(b) $^{226}_{88}\text{Ra} \rightarrow ^{222}_{86}\text{Rn} + ^{4}_{2}\text{He}$; $\quad ^{222}_{86}\text{Rn} \rightarrow ^{218}_{84}\text{Po} + ^{4}_{2}\text{He}$; $\quad ^{218}_{84}\text{Po} \rightarrow ^{214}_{82}\text{Pb} + ^{4}_{2}\text{He}$

(c) $^{69}_{33}\text{As} \rightarrow ^{69}_{32}\text{Ge} + ^{0}_{+1}\text{e}$

17. We would expect a neutron:proton ratio that is closer to 1:1 than that of ^{14}C. This would be achieved if the product were ^{14}N, which will be the result of β^- decay:

$^{14}_{6}\text{C} \rightarrow ^{14}_{7}\text{N} + ^{0}_{-1}\text{e}$.

18. A nuclide with a closer to 1:1 neutron:proton ratio (than that of tritium) is helium-3, arrived at by beta emission: $^{3}_{1}\text{H} \rightarrow ^{3}_{2}\text{He} + ^{0}_{-1}\text{e}$. Another possible product is deuterium, which is arrived at by neutron emission: $^{3}_{1}\text{H} \rightarrow ^{2}_{1}\text{H} + ^{1}_{0}\text{n}$

Radioactive Decay Series

19. We first write conventional nuclear reactions for each step in the decay series.

$$^{232}_{90}\text{Th} \rightarrow\ ^{228}_{88}\text{Ra} +\ ^{4}_{2}\text{He} \qquad\qquad ^{228}_{88}\text{Ra} \rightarrow\ ^{228}_{89}\text{Ac} +\ ^{0}_{-1}\text{e} \qquad\qquad ^{228}_{89}\text{Ac} \rightarrow\ ^{228}_{90}\text{Th} +\ ^{0}_{-1}\text{e}$$

$$^{228}_{90}\text{Th} \rightarrow\ ^{224}_{88}\text{Ra} +\ ^{4}_{2}\text{He} \qquad\qquad ^{224}_{88}\text{Ra} \rightarrow\ ^{220}_{86}\text{Rn} +\ ^{4}_{2}\text{He} \qquad\qquad ^{220}_{86}\text{Rn} \rightarrow\ ^{216}_{84}\text{Po} +\ ^{4}_{2}\text{He}$$

Now for a branch in the series

these two $\qquad ^{216}_{84}\text{Po} \rightarrow\ ^{212}_{82}\text{Pb} +\ ^{4}_{2}\text{He} \qquad\qquad ^{212}_{82}\text{Pb} \rightarrow\ ^{212}_{83}\text{Bi} +\ ^{0}_{-1}\text{e}$

or these two $\qquad ^{216}_{84}\text{Po} \rightarrow\ ^{216}_{85}\text{At} +\ ^{0}_{-1}\text{e} \qquad\qquad ^{216}_{85}\text{At} \rightarrow\ ^{212}_{83}\text{Bi} +\ ^{4}_{2}\text{He}$

And now a second branch

these two $\qquad ^{212}_{83}\text{Bi} \rightarrow\ ^{208}_{81}\text{Tl} +\ ^{4}_{2}\text{He} \qquad\qquad ^{208}_{81}\text{Tl} \rightarrow\ ^{208}_{82}\text{Pb} +\ ^{0}_{-1}\text{e}$

or these two $\qquad ^{212}_{83}\text{Bi} \rightarrow\ ^{212}_{84}\text{Po} +\ ^{0}_{-1}\text{e} \qquad\qquad ^{212}_{84}\text{Po} \rightarrow\ ^{208}_{82}\text{Pb} +\ ^{4}_{2}\text{He}$

Both branches end with the isotope $^{208}_{82}\text{Pb}$. The graph, similar to Figure 26-2, is drawn below.

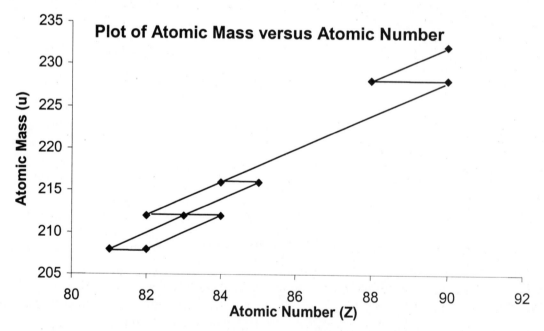

20.   The series begins with uranium-235, and ends with lead-207.

$$^{235}_{92}\text{U} \rightarrow {}^{231}_{90}\text{Th} + {}^{4}_{2}\text{He} \qquad\qquad {}^{231}_{90}\text{Th} \rightarrow {}^{231}_{91}\text{Pa} + {}^{0}_{-1}\text{e} \qquad\qquad {}^{231}_{91}\text{Pa} \rightarrow {}^{227}_{89}\text{Ac} + {}^{4}_{2}\text{He}$$

Now the series branches

these two $\qquad {}^{227}_{89}\text{Ac} \rightarrow {}^{223}_{87}\text{Fr} + {}^{4}_{2}\text{He} \qquad\qquad {}^{223}_{87}\text{Fr} \rightarrow {}^{223}_{88}\text{Ra} + {}^{0}_{-1}\text{e}$

or these two $\qquad {}^{227}_{89}\text{Ac} \rightarrow {}^{227}_{90}\text{Th} + {}^{0}_{-1}\text{e} \qquad\qquad {}^{227}_{90}\text{Th} \rightarrow {}^{223}_{88}\text{Ra} + {}^{4}_{2}\text{He}$

then $\;{}^{223}_{88}\text{Ra} \rightarrow {}^{219}_{86}\text{Rn} + {}^{4}_{2}\text{He} \qquad\qquad {}^{219}_{86}\text{Rn} \rightarrow {}^{215}_{84}\text{Po} + {}^{4}_{2}\text{He} \qquad\qquad {}^{215}_{84}\text{Po} \rightarrow {}^{211}_{82}\text{Pb} + {}^{4}_{2}\text{He}$

$$^{211}_{82}\text{Pb} \rightarrow {}^{211}_{83}\text{Bi} + {}^{0}_{-1}\text{e}$$

The series branches again

these two $\qquad {}^{211}_{83}\text{Bi} \rightarrow {}^{207}_{81}\text{Tl} + {}^{4}_{2}\text{He} \qquad\qquad {}^{207}_{81}\text{Tl} \rightarrow {}^{207}_{82}\text{Pb} + {}^{0}_{-1}\text{e}$

or these two $\qquad {}^{211}_{83}\text{Bi} \rightarrow {}^{211}_{84}\text{Po} + {}^{0}_{-1}\text{e} \qquad\qquad {}^{211}_{84}\text{Po} \rightarrow {}^{207}_{82}\text{Pb} + {}^{4}_{2}\text{He}$

The plot of atomic mass versus atomic number for these decay series is shown below.

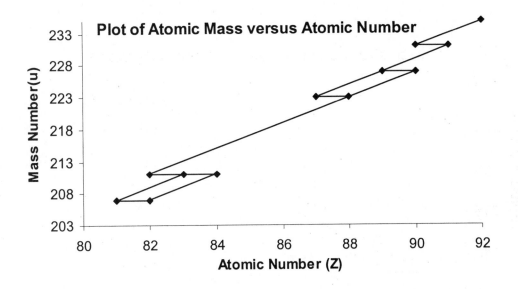

21.   In Figure 26–2, only the following mass numbers are represented: 206, 210, 214, 218, 222, 226, 230, 234, and 238. We see that these mass numbers are separated from each other by 4 units. The first of them, 206, equals $(4 \times 51) + 2$, that is $4n + 2$, where $n = 51$.

22. The series to which each nuclide belongs is determined by dividing its mass number by 4 and obtaining the remainder.

   (a) The mass number of $^{214}_{83}\text{Bi}$ is 214, and the remainder following its division by 4 is 2. This nuclide is a member of the $4n+2$ series.

   (b) The mass number of $^{216}_{84}\text{Po}$ is 216, and the remainder following its division by 4 is 0. This nuclide is a member of the $4n$ series.

   (c) The mass number of $^{215}_{85}\text{At}$ is 215, and the remainder following its division by 4 is 3. This nuclide is a member of the $4n+3$ series.

   (d) The mass number of $^{235}_{92}\text{U}$ is 235, and the remainder following its division by 4 is 3. This nuclide is a member of the $4n+3$ series.

Nuclear Reactions

23. **(a)** $^{7}_{3}\text{Li} + ^{1}_{1}\text{H} \rightarrow ^{8}_{4}\text{Be} + \gamma$ **(b)** $^{9}_{4}\text{Be} + ^{2}_{1}\text{H} \rightarrow ^{10}_{5}\text{B} + ^{1}_{0}\text{n}$ **(c)** $^{14}_{7}\text{N} + ^{1}_{0}\text{n} \rightarrow ^{14}_{6}\text{C} + ^{1}_{1}\text{H}$

24. **(a)** $^{238}_{92}\text{U} + ^{4}_{2}\text{He} \rightarrow ^{239}_{94}\text{Pu} + 3^{1}_{0}\text{n}$ **(b)** $^{3}_{1}\text{H} + ^{2}_{1}\text{H} \rightarrow ^{4}_{2}\text{He} + ^{1}_{0}\text{n}$ **(c)** $^{33}_{16}\text{S} + ^{1}_{0}\text{n} \rightarrow ^{33}_{15}\text{P} + ^{1}_{1}\text{H}$

25. $^{209}_{83}\text{Bi} + ^{64}_{28}\text{Ni} \rightarrow ^{272}_{111}\text{E} + ^{1}_{0}\text{n}$; $^{272}_{111}\text{E} \rightarrow 5^{4}_{2}\text{He} + ^{252}_{101}\text{Md}$

26. $^{208}_{82}\text{Pb} + ^{86}_{36}\text{Kr} \rightarrow ^{293}_{118}\text{E} + ^{1}_{0}\text{n}$; $^{293}_{118}\text{E} \rightarrow 6^{4}_{2}\text{He} + ^{269}_{106}\text{Sg}$

Rate of Radioactive Decay

27. We use expression (26.13) to determine λ and then expression (26.11) to determine the number of atoms.

$$\lambda = \frac{0.693}{5.2 \text{ y}} \times \frac{1 \text{ y}}{365.25 \text{ d}} \times \frac{1 \text{ d}}{24 \text{ h}} = 1.5 \times 10^{-5} \text{ h}^{-1}$$

$$N = \frac{\text{rate of decay}}{\lambda} = \frac{6740 \text{ atoms/h}}{1.5 \times 10^{-5} \text{ h}^{-1}} = 4.4 \times 10^{8} \ ^{60}_{27}\text{Co atoms}$$

28. This follows first-order kinetics (as do all radioactive decay processes) with a rate of decay directly proportional to the number of atoms. We therefore use expression (26.12), with rates substituted for numbers of atoms.

$$6740 \frac{\text{dis}}{\text{h}} \times \frac{1 \text{ h}}{60 \text{ min}} = 112 \frac{\text{dis}}{\text{min}}$$

$$\ln \frac{R_t}{R_o} = -\lambda t = \ln \frac{101 \text{ dis/min}}{112 \text{ dis/min}} = -1.5 \times 10^{-5} \text{ h}^{-1} t = -0.103$$

$$t = \frac{0.103}{1.5 \times 10^{-5} \text{ h}^{-1}} = 6.9 \times 10^{3} \text{ h} \times \frac{1 \text{ d}}{24 \text{ h}} \times \frac{1 \text{ y}}{365.25 \text{ d}} = 0.78 \text{ y}$$

29. Let us use the first and the last values to determine the decay constant.

$$\ln\frac{R_t}{R_o} = -\lambda t = \ln\frac{138 \text{ cpm}}{1000 \text{ cpm}} = -\lambda 250 \text{ h} = -1.981 \qquad \lambda = \frac{1.981}{250 \text{ h}} = 0.00792 \text{ h}^{-1}$$

$$t_{1/2} = \frac{0.693}{\lambda} = \frac{0.693}{0.00792 \text{ h}^{-1}} = 87.5 \text{ h}$$

A slightly different value of $t_{1/2}$ may result from other combinations of R_o and R_t.

30. First calculate the decay constant.

$$\lambda = \frac{0.693}{1.7\times10^7 \text{ y}} \times \frac{1 \text{ y}}{365.25 \text{ d}} \times \frac{1 \text{ d}}{24 \text{ h}} \times \frac{1 \text{ h}}{3600 \text{ s}} = 1.3\times10^{-15} \text{ s}^{-1}$$

$$N = 1.00 \text{ mg } ^{129}\text{I} \times \frac{1 \text{ g}}{1000 \text{ mg}} \times \frac{1 \text{ mol } ^{129}\text{I}}{129 \text{ g}} \times \frac{6.022\times10^{23} \text{ atoms}}{1 \text{ mol } ^{129}\text{I}} = 4.67\times10^{18} \ ^{129}\text{I atoms}$$

decay rate $= \lambda N = 1.3\times10^{-15} \text{ s}^{-1} \times 4.67\times10^{18} \text{ atoms} = 6.1\times10^3 \text{ dis/s}$

31. $^{32}_{15}\text{P}$ half-life = 14.3 d. We need to determine the time necessary to get to the detectable

limit, $\dfrac{1}{1000}$ of the initial value. Use $\lambda = \dfrac{0.693}{t_{1/2}} = \dfrac{0.693}{14.3 \text{ d}} = 0.0485 \text{ d}^{-1}$

$$\ln\left(\frac{1}{1000}\right) = -0.0485 \text{ d}^{-1}(t) \qquad t = 142 \text{ days}$$

32. $1.00 \text{ mCi} = 1.00 \times 10^{-3} (3.70 \times 10^{10} \text{ dis s}^{-1}) = 3.70 \times 10^7 \text{ dis s}^{-1}$

$$\lambda = \frac{0.693}{t_{1/2}} = \frac{0.693}{5730 \text{ y}} = 1.21 \times 10^{-4} \text{ y}^{-1} \quad (1 \text{ y} = 365.25 \text{ d} = 3.15\underline{6} \times 10^7 \text{ s})$$

$$\lambda = \frac{1.21\times10^{-4}}{\text{y}} \times \frac{1 \text{ y}}{3.15_6 \times 10^7 \text{ s}} = 3.83 \times 10^{-12} \text{ s}^{-1}$$

$1.00 \text{ mCi} = 3.70 \times 10^7 \text{ dis s}^{-1} = \lambda N = 3.83 \times 10^{-12} \text{ s}^{-1}(N)$

$N = 9.66 \times 10^{18}$ atoms of ^{14}C or $1.60\underline{4} \times 10^{-5}$ mol ^{14}C

mass of $^{14}\text{C} = 1.60\underline{4} \times 10^{-5}$ mol $^{14}\text{C} \times \dfrac{14.00 \text{ g } ^{14}\text{C}}{1 \text{ mol } ^{14}\text{C}} = 2.25 \times 10^{-4}$ g ^{14}C

Radiocarbon Dating

33. Again we use expression (26.12) and (26.13) to determine the time elapsed. The initial rate of decay is about 15 dis/min. First we compute the decay constant.

$$\lambda = \frac{0.693}{5730 \text{ y}} = 1.21 \times 10^{-4} \text{ y}^{-1}$$

$$\ln \frac{10 \text{ dis/min}}{15 \text{ dis/min}} = -0.40_5 = -\lambda t; \qquad t = \frac{0.40_5}{1.21 \times 10^{-4} \text{ y}^{-1}} = 3.35 \times 10^3 \text{ y}$$

The object is a bit more than 3000 years old, probably not dating from the pyramid era, about 3000 B.C.

34. We use the value of λ from the previous exercise.

$$\ln \frac{R_t}{R_o} = -\lambda t = -\left(1.21 \times 10^{-4} \text{ y}^{-1}\right) t = \ln \frac{0.03 \text{ dis min}^{-1} \text{ g}^{-1}}{15 \text{ dis min}^{-1} \text{ g}^{-1}} = -6.2$$

$$t = \frac{6.2}{1.21 \times 10^{-4} \text{ y}^{-1}} = 5.1 \times 10^4 \text{ y}$$

35. First we determine the decay constant. $\lambda = \dfrac{0.693}{1.39 \times 10^{10} \text{ y}} = 4.99 \times 10^{-11} \text{ y}^{-1}$

Then we can determine the ratio of (N_t), the number of thorium atoms after 2.7×10^9 y, to (N_0), the initial number of thorium atoms:

$$\ln \frac{N_t}{N_0} = -kt = -\left(4.99 \times 10^{-11} \text{ y}^{-1}\right)\left(2.7 \times 10^9 \text{ y}\right) = -0.13 \qquad \frac{N_t}{N_0} = 0.88$$

Thus, for every mole of ^{232}Th present initially, after 2.7×10^9 y there are

0.88 mol ^{232}Th and 0.12 mol ^{208}Pb. From this information, we can compute the mass ratio.

$$\frac{0.12 \text{ mol } ^{208}\text{Pb}}{0.88 \text{ mol } ^{232}\text{Th}} \times \frac{1 \text{ mol } ^{232}\text{Th}}{232 \text{ g } ^{232}\text{Th}} \times \frac{208 \text{ g } ^{208}\text{Pb}}{1 \text{ mol } ^{208}\text{Pb}} = \frac{0.12 \text{ g } ^{208}\text{Pb}}{1 \text{ g } ^{232}\text{Th}}$$

36. First we determine the decay constant. $\lambda = \dfrac{0.693}{1.39 \times 10^{10} \text{ y}} = 4.99 \times 10^{-11} \text{ y}^{-1}$

The rock currently contains 1.00 g ^{232}Th and 0.25 g ^{208}Pb. We can calculate the mass of ^{232}Th that must have been present to produce this 0.25 g ^{208}Pb, and from that find the original mass of ^{232}Th.

$$\text{original mass } ^{232}\text{Th} = 1.00 \text{ g } ^{232}\text{Th now} + \left(0.25 \text{ g } ^{208}\text{Pb} \times \frac{232 \text{ g } ^{232}\text{Th}}{208 \text{ g } ^{208}\text{Pb}}\right)$$

$$= (1.00 + 0.28) \text{ g} = 1.28 \text{ g}$$

$$\ln\frac{N_t}{N_0} = -\lambda t = \ln\frac{1.00 \text{ g } ^{232}\text{Th now}}{1.28 \text{ g originally}} = -0.247 = -4.99\times10^{-11} \text{ y}^{-1}t;$$

$$t = \frac{0.247}{4.99\times10^{-11} \text{ y}^{-1}} = 4.95\times10^9 \text{ y}$$

Energetics of Nuclear Reactions

37. The mass defect is the difference between the mass of the nuclide and the sum of the masses of its constituent particles. The binding energy is this mass defect expressed as an energy.

$$\begin{aligned}
\text{Particle mass} &= 9p + 10n + 9e = 9(p+n+e) + n \\
&= 9(1.0073 + 1.0087 + 0.0005486) \text{ u} + 1.0087 \text{ u} = 19.1576 \text{ u}
\end{aligned}$$

$$\text{mass defect} = 19.1576 \text{ u} - 18.998403 \text{ u} = 0.1592 \text{ u}$$

$$\text{binding energy per nucleon} = \frac{0.1592\,\text{u}\times\dfrac{931.5\,\text{MeV}}{1\text{u}}}{19\,\text{nucleons}} = 7.805 \text{ MeV/nucleon}$$

38. The mass defect is the difference between the mass of the nuclide and the sum of the masses of its constituent particles. The binding energy is this mass defect expressed as an energy.

$$\begin{aligned}
\text{Particle mass} &= 26p + 30n + 26e = 26(p+n+e) + 4n \\
&= 26(1.0073 + 1.0087 + 0.0005486) \text{ u} + 4\times1.0087 \text{ u} = 56.4651 \text{ u}
\end{aligned}$$

$$\text{mass defect} = 56.4651 \text{ u} - 55.934939 \text{ u} = 0.5302 \text{ u}$$

$$\text{binding energy per nucleon} = \frac{0.5302\,\text{u}\times\dfrac{931.5\,\text{MeV}}{1\text{u}}}{56\,\text{nucleons}} = 8.819 \text{ MeV/nucleon}$$

39. $\text{mass defect} = (10.01294 \text{ u} + 4.00260 \text{ u}) - (13.00335 \text{ u} + 1.00783 \text{ u}) = 0.00436 \text{ u}$

$$\text{energy} = 0.00436 \text{ u}\times\frac{931.5 \text{ MeV}}{1 \text{ u}} = 4.06 \text{ MeV}$$

40. $\text{mass defect} = (6.01513 \text{ u} + 1.008665 \text{ u}) - (4.00260 \text{ u} + 3.01604 \text{ u}) = 0.00516 \text{ u}$

$$\text{energy} = 0.00516 \text{ u}\times\frac{931.5 \text{ MeV}}{1 \text{ u}} = 4.81 \text{ MeV}$$

41.   1 neutron $\approx$ 1 amu $= 1.66 \times 10^{-27}$ kg

$E = mc^2 = 1.66 \times 10^{-27}$ kg$(2.998 \times 10^8$ m s$^{-1})^2 = 1.49 \times 10^{-10}$ J (1 neutron)

1 eV $= 1.602 \times 10^{-19}$ J,

Hence, 1 neutron $= 1.49 \times 10^{-10}$ J $\times \dfrac{1 \text{ eV}}{1.602 \times 10^{-19} \text{ J}} = 9.30 \times 10^8$ eV or 930. MeV

6.75×10^6 MeV $\times \dfrac{1 \text{ neutron}}{930 \text{ MeV}} = 7.26 \times 10^3$ neutrons

42. $\beta^+ + \beta^-$ collide $\rightarrow$ produce two γ-rays.

Basically the mass of $\beta^+ = \beta^-$ = mass of an electron $(9.11 \times 10^{-31}$ kg$)$
Each γ-ray has the same energy as the complete conversion of one electron into pure energy.
$E = mc^2 = (9.11 \times 10^{-31}$ kg$)$ $(2.998 \times 10^8$ m s$^{-1})^2 = 8.19 \times 10^{-14}$ J

In electron volts: 8.19×10^{-14} J $\times \dfrac{1 \text{ eV}}{1.602 \times 10^{-19} \text{ J}} = 5.11 \times 10^5$ eV or 0.511 MeV

Each γ-ray has an energy of 0.511 MeV

Nuclear Stability

43.   **(a)**   We expect ^{20}Ne to be more stable than ^{22}Ne. A neutron-to-proton ratio of 1-to-1 is associated with stability for elements of low atomic number (with $Z \leq 20$).

   (b)   We expect ^{18}O to be more stable than ^{17}O. An even number of protons and an even number of neutrons are associated with a stable isotope.

   (c)   We expect ^{7}Li to be more stable than ^{6}Li. Both isotopes have an odd number of protons, but only ^{7}Li has an even number of neutrons.

44.   **(a)**   We expect ^{40}Ca to be more stable than ^{42}Ca. A neutron-to-proton ratio of 1-to-1 is associated with stability for elements of low atomic number (with $Z \leq 20$).

   (b)   We expect ^{31}P to be more stable than ^{32}P. Both isotopes have an odd number of protons, but only ^{31}P has an even number of neutrons.

   (c)   We expect ^{64}Zn to be more stable than ^{63}Zn. An even number of protons and an even number of neutrons are associated with a stable isotope.

45. β^- emission has the effect of "converting" a neutron to a proton. β^+ emission, on the other hand, has the effect of "converting" a proton to a neutron.

   (a)   The most stable isotope of phosphorus is ^{31}P, with a neutron-to-proton ratio of close to 1-to-1 and an even number of neutrons. Thus, ^{29}P has "too few" neutrons, or too many protons. It should decay by β^+ emission. In contrast, ^{33}P has "too many" neutrons, or "too few" protons. Therefore, ^{33}P should decay by β^- emission.

(b) Based on the atomic mass of I (126.90447), we expect the isotopes of iodine to have mass numbers close to 127. This means that ^{120}I has "too few" neutrons and therefore should decay by β^+ emission, whereas ^{134}I has "too many" neutrons (or "too few" protons) and therefore should decay by β^- emission.

46. β^- emission has the effect of converting a neutron to a proton, while β^+ emission has the effect of converting a proton to a neutron.

(a) Based on the fact that elements of low atomic number have about the same number of protons as neutrons, $^{28}_{15}$P—with 15 protons and 13 neutrons—has too few neutrons. Therefore, it should decay by β^+ emission.

(b) Once again, elements of low atomic number have about the same number of protons as neutrons. $^{45}_{19}$K with 19 protons and 26 neutrons—has too many neutrons. Therefore, it should decay by β^- emission.

(c) Based on the atomic mass of zinc (65.39) we expect most of its isotopes to have about 36 neutrons. There are 42 neutrons in $^{72}_{30}$Zn, more than we expect. Thus we expect this nuclide to decay by β^- emission.

47. A "doubly magic" nuclide is one in which the atomic number is a magic number (2, 8, 20, 28, 50, 82, 114) and the number of neutrons also is a magic number (2, 8, 20, 28, 50, 82, 126, 184). Nuclides that fit this description are given below.

nuclide	^{4}He	^{16}O	^{40}Ca	^{56}Ni	^{208}Pb
no. of protons	2	8	20	28	82
no. of neutrons	2	8	20	28	126

48. In isotopes of high atomic number, stable nuclides are characterized by a neutron-to-proton ratio greater than 1, which increases with increasing atomic number. Naturally occurring isotopes of high atomic number decrease their atomic number by losing an alpha particle, which has a neutron-to-proton ratio of 1. This leaves the neutron-to-proton ratio of the daughter higher than that of the parent, when it should be slightly lower. In order to redress this, the number of neutrons needs to be decreased and the number of protons increased. Beta emission accomplishes this. In contrast, artificially produced isotopes have no definite neutron-to-proton ratio. Thus, sometimes, the number of neutrons needs to be decreased, which is accomplished by beta emission, while at other times the number of protons needs to be decreased, which is accomplished by positron emission.

Fission and Fusion

49. We use the conversion factor between number of curies and mass of ^{131}I which was developed in the Summarizing Example.

$$\text{no. g I-131} = 170 \text{ curies} \times \frac{18.8 \text{ g I-131}}{2.33 \times 10^6 \text{ curie}} = 1.37 \times 10^{-3} \text{ g} = 1.37 \text{ mg}$$

50. Nuclear fission is the process by which a heavy nucleus disintegrates into neutrons and stable nuclei with smaller mass numbers. For instance, uranium-238 undergoes fission according to the equation

$$^{238}_{92}U \rightarrow ^{234}_{90}Th + ^{4}_{2}He$$

The nuclear binding energy for uranium-238 is less than the sum of the binding energies for thorium-234 and helium-4. Consequently, when a uranium-238 nucleus splits apart, energy is released. Nuclear fusion, by contrast, involves the amalgamation of light nuclei into heavier, more stable nuclei. For instance, part of the energy released by our Sun is believed to come from the fusion of hydrogen to form deuterium:

$$^{1}_{1}H + ^{1}_{1}H \rightarrow ^{2}_{1}H + ^{0}_{-1}e$$

Although both fusion and fission release vast amounts of energy, fusion releases far more energy on a per nucleon basis. To understand why this is so, we need to refer to Figure 26-6, which is a plot of Average Binding Energy per Nucleon as a Function of Atomic Number. The graph clearly shows that the increase in binding energy observed for the formation of the lightest nuclides (e.g. deuterium, tritium, helium-3) is much more dramatic than the decrease in binding energy that is seen for the fragmentation of heavier nuclei such as uranium-235. Thus, the plot indicates that more energy should be released by the combination of light nuclei(nuclear fusion) than by the disintegration of heavy nuclei(nuclear fission).

Effect of Radiation on Matter

51. The term "rem" is an acronym for "radiation equivalent-man," and takes into account the quantity of biological damage done by a given dosage of radiation. On the other hand, the rad is the dosage which places 0.010 J of energy into each kg of irradiated matter. Thus, for living tissue, the rem provides a good idea of how much tissue damage a certain kind and quantity of radiation damage will do. But for nonliving materials, the rad is often preferred, and indeed is often the only unit of utility.

52. Low-level radiation is very close in its dosage to background radiation and one problem is to separate out the effects of the two sources (low-level and background). The other problem is that low-level radiation does not produce severe damage in a short period of time. Thus the effects of low-level radiation will only be observed over a long time period. Of course other effects, such as chemical and biological toxins, will also be observed over these time periods, and we have to try to separate these two types of effects. (There also is the heritage of the organism to consider, of course.)

53. One reason why ^{90}Sr is hazardous is because strontium is in the same family of the periodic table as calcium, and hence often reacts in a similar fashion to calcium. One site where calcium is incorporated into the body is in bones, where it resides for a long time. Strontium is expected to behave in a similar fashion. Thus, it will be retained in the body for a long time. Bone is an especially dangerous place for a radioisotope to be present—even if it has low penetrating power, as do β^- rays—because blood cells are produced in bone marrow.

54. It is not particularly hazardous to be near a flask of ^{222}Rn, because it is unlikely that the alpha particles can get through the walls of the flask. (Note that since radon is a gas, the flask must be sealed.) The decay products of ^{222}Rn may produce other forms of radiation that are more penetrating, such as β^- particles and γ rays, so being near the flask may still pose a risk. ^{222}Rn can be potentially hazardous if one breathes the gas.

Application of Radioisotopes

55. Mix a small amount of tritium with the $H_2(g)$ and detect where the radioactivity appears.

56. In neutron activation analysis, the sample is bombarded with neutrons. Radioisotopes are produced by this process. These radioisotopes can be easily detected even in very small quantities, much smaller than the quantities that can be detected by conventional means of quantitative analysis. These radioisotopes are produced in quantities that are proportional to the quantity of each element originally present in the sample. And each radioisotope is characteristic of the element from which it was produced by neutron bombardment. Even microscopic samples can be analyzed by this technique. Finally, neutron activation analysis is a nondestructive technique, while the conventional techniques of precipitation or titration require that the sample, or at least part of it, be destroyed.

57. The recovered sample will be radioactive. When NaCl(s) and $NaNO_3(s)$ are dissolved in solution, the ions (Na^+, Cl^-, and NO_3^-) are free to move throughout the solution. A given anion does not remain associated with a particular cation. Thus, all the anions and cations are shuffled and some of the radioactive ^{24}Na will end up in the crystallized $NaNO_3$.

58. We would expect the tritium label to appear in both the $NH_3(g)$ and the $H_2O(l)$. When $NH_4^+(aq)$ is formed, one of the four chemically and spatially equivalent H atoms is occasionally a tritium atom. In the subsequent reaction between the occasionally marked NH_4Cl and NaOH to form $NH_3(g)$ and $H_2O(l)$, there are three chances in four that a tritium atom will remain attached to N in NH_3 and one chance in four that a tritium ion will react with a hydroxide ion to form $H_2O(l)$.

Feature Problems

75. First tabulate the isotopes symbols, the mass of isotope and its associated packing fraction.

Isotope Symbol	Mass of Isotope (u)	Packing Fraction
^{1}H	1.007825	0.007825
^{4}He	4.002603	0.000651
^{9}Be	9.012186	0.001354
^{12}C	12	0
^{16}O	15.994915	−0.000318
^{20}Ne	19.992440	−0.000378
^{24}Mg	23.985042	−0.000623
^{32}S	31.972074	−0.000873
^{40}Ar	39.962384	−0.000940
^{40}Ca	39.962589	−0.000935
^{48}Ti	47.947960	−0.001084
^{52}Cr	51.940513	−0.001144
^{56}Fe	55.934936	−0.001162
^{58}Ni	57.935342	−0.001115
^{64}Zn	63.929146	−0.001107
^{80}Se	79.916527	−0.001043
^{84}Kr	83.911503	−0.001054
^{90}Zr	89.904700	−0.001059
^{102}Ru	101.904348	−0.000938
^{114}Cd	113.903360	−0.000848
^{130}Te	129.906238	−0.000721
^{138}Ba	137.905000	−0.000688
^{142}Nd	141.907663	−0.000650
^{158}Gd	157.924178	−0.000480
^{166}Er	165.932060	−0.000409

Plot of Packing Fraction versus Mass Number

Data plotted above

This graph and Fig. 26-6 are almost the inverse of one another, with the maxima of one being the minima of the other. Actual nuclidic mass is often a number slightly less than the number of nucleons (mass number). This difference divided by the number of nucleons (packing fraction) is proportional to the negative of the mass defect per nucleon.

76. (a) The rate of decay depends on both the half-life and on the number of radioactive atoms present. In the early stages of the decay chain, the larger number of radium-226 atoms multiplied by the very small decay constant is still larger than the product of the very small number of radon-222 atoms and its much larger decay constant. Only after some time has elapsed, does the rate of decay of radon-222 approach the rate at which it is formed from radium-226 and the amount of radon-222 reaches a maximum. Beyond this point, the rate of decay of radon-222 exceeds its rate of formation.

(b) $\dfrac{dD}{dt} = \lambda_p P - \lambda_d D = \lambda_p P_o e^{-\lambda_p t} - \lambda_d D$

(c) The number of radon-222 atoms at various times are 2.90×10^{15} atoms after 1 day; 1.26×10^{16} after 1 week; 1.75×10^{16} after 1 year; 1.68×10^{16} after one century; and 1.13×10^{16} after 1 millennium. The actual maximum comes after about 2 months, but the amount after 1 year is only slightly smaller.

77. (a)

$\text{Zr(s)} + 6 H_2O(l) \rightarrow ZrO_2(s) + 4 H_3O^+(aq) + 4 e^- \qquad 1.43 \text{ V}$

$4 H_2O(l) + 4 e^- \rightarrow 2 H_2(g) + 4 OH^-(aq) \qquad -0.828 \text{ V}$

$\overline{\qquad\qquad\qquad\qquad\qquad\qquad\qquad\qquad\qquad\qquad\qquad\qquad}$

$\text{Zr(s)} + 2 H_2O(l) \rightarrow ZrO_2(s) + 2 H_2(g) \qquad 0.602 \text{ V (spont)}$

Yes, Zr can reduce water under standard conditions.

(b) $E° = \dfrac{0.0592}{n} \log K_{eq} \qquad 0.602 \text{ V} = \dfrac{0.0592}{4} \log K_{eq} \qquad K_{eq} = 5 \times 10^{40}$

(c) pH = 7 Therefore, $[OH^-] = [H_3O^+] = 1.0 \times 10^{-7}$

$E_{ox} = E°_{ox} - \dfrac{0.0592}{n} \log Q = 1.43 \text{ V} - \dfrac{0.0592}{4} \log(1.0 \times 10^{-7})^4 = 1.84 \text{ V}$

$E_{red} = E°_{red} - \dfrac{0.0592}{n} \log Q = -0.828 \text{ V} - \dfrac{0.0592}{4} \log(1 \times 10^{-7})^4 = -0.414 \text{ V}$

$E_{cell} = E_{ox} + E_{red} = 1.84 + (-0.414) = 1.43 \text{ V (spontaneous)}$

(d) Zr may be the culprit responsible for the $H_2(g)$ formation. In the Chernobyl accident the reaction of carbon with superheated steam played a major role.
Reaction: $H_2O(g) + C(s) \rightarrow CO_{(g)} + H_2(g)$

78. (a) Average atomic mass of Sr in the rock

$\dfrac{^{87}Sr}{^{86}Sr} = 2.25 \qquad \dfrac{^{86}Sr}{^{88}Sr} = 0.119 \qquad \dfrac{^{84}Sr}{^{88}Sr} = 0.007 \qquad$ Given: 15.5 ppm Sr

Let $x = {}^{86}Sr$, $y = {}^{88}Sr$, $z = {}^{87}Sr$, $w = {}^{84}Sr \qquad x + y + z + w = 15.5$ ppm

$\dfrac{z}{x} = 2.25$, $\dfrac{x}{y} = 0.119$, $\dfrac{w}{y} = 0.007$

Hence, $z = 2.25x$; $y = \dfrac{x}{0.119}$; $w = 0.007; y = \dfrac{x}{0.119} 0.007$

$x + y + z + w = 15.5 \text{ ppm} = x + \dfrac{x}{0.119} + 2.25x + 0.007\left(\dfrac{x}{0.119}\right)$

$11.712x = 15.5$ ppm

$x = 1.32$ ppm $= {}^{86}S$. Plugging into the ratios above, we find:

$y = 11.1$ ppm $= {}^{88}Sr$; $z = 2.97$ ppm $= {}^{87}Sr$; $w = 0.08$ ppm $= {}^{84}Sr$;
% abundance in sample and isotopic mass are tabulated below:

1.32 ppm of ${}^{86}Sr \rightarrow 8.52$ % 85.909 u

2.97 ppm of ${}^{87}Sr \rightarrow 19.2$ % 86.909 u

11.1 ppm of ${}^{88}Sr \rightarrow 71.6$% 87.906 u

0.08 ppm of ${}^{84}Sr \rightarrow 0.5$% 83.913 u

Atomic mass
$= 8.52$ % (85.909 u)+ 19.2 % (86.909 u) + 71.6% (87.906 u) + 0.5%(83.913 u)

Average Atomic mass $= 87.4$ u

(b)   Original Rb in rock?

Currently 265.4 ppm for Rb, however, $\dfrac{{}^{87}Rb}{{}^{85}Rb} = 0.330$

Let $x = {}^{87}Rb$, therefore, $\dfrac{x}{265.4 - x} = 0.330$ (Solve for x)

$x = 87.582 - 0.330x$ or $1.330x = 87.582$

$x = 65.85$ ppm $= {}^{87}Rb$ and thus, ${}^{85}Rb = 199.6$ ppm

Original $\dfrac{{}^{87}Sr}{{}^{86}Sr} = 0.700$ (Since the half-life for strontium-86 is very long, we can

assume that the concentration of strontium-86 has remained essentially constant)

${}^{86}Sr = 1.32$ ppm, hence, ${}^{87}Sr = 0.700(1.32$ ppm$) = 0.924$ ppm

The original Rb concentration is 265.4 ppm + 0.924 ppm = 266.3 ppm

(c)   % ${}^{87}Rb$ decayed $= \left(\dfrac{0.924 \text{ ppm}}{0.924 \text{ ppm} + 65.85 \text{ ppm}} \right) 100\% = 1.38\%$

(d) $\ln(0.9862) = -\lambda t$ $\left(\lambda = \dfrac{0.693}{t_{1/2}} = \dfrac{0.693}{4.8 \times 10^{10} \text{ y}} = 1.44 \times 10^{-11} \text{ y}^{-1} \right)$

$\ln(0.9862) = -1.44 \times 10^{-11} \text{ y}^{-1} t$; $t = 9.65 \times 10^{8}$ years

CHAPTER 27
ORGANIC CHEMISTRY
PRACTICE EXAMPLES

1A We show only the C atoms and the bonds between them. Remember that there are four bonds to each C atom; the remaining bonds not shown are to H atoms. First we realize there is only one isomer with all six C atoms in one line. Then we draw the isomers with one 1-C branch. These are the center two structures below. The isomers with two 1-C branches can have them both on the same atom or on different atoms, as in the rightmost two structures. This accounts for all five isomers.

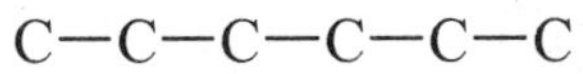

C—C—C—C—C—C

$CH_3CH_2CH_2CH_2CH_2CH_3$

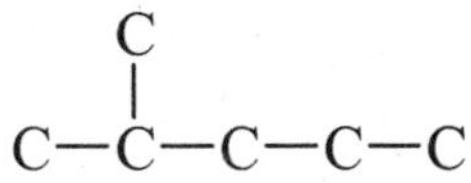

C—C—C—C—C (with C branch on 2nd C)

$(CH_3)_2CHCH_2CH_2CH_3$

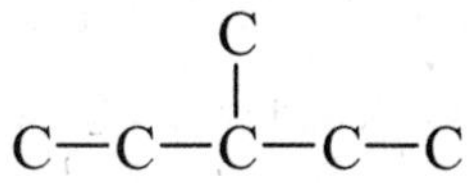

C—C—C—C—C (with C branch on 3rd C)

$CH_3CH_2CH(CH_3)CH_2CH_3$

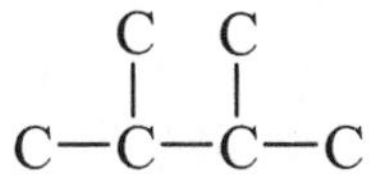

C—C—C—C (with C branches on 2nd and 3rd C)

$(CH_3)_2CHCH(CH_3)_2$

C—C—C—C (with two C branches)

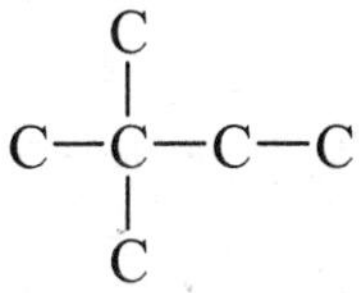

$(CH_3)_3CCH_2CH_3$

1B We show only the C atoms and the bonds between them. Remember that there are four bonds to each C atom; the remaining bonds not shown are to H atoms.

C—C—C—C—C—C—C

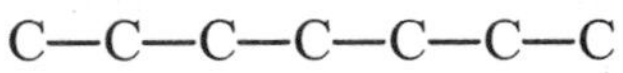

$CH_3CH_2CH_2CH_2CH_2CH_2CH_3$

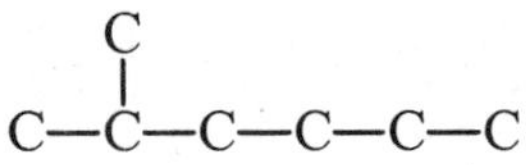

C—C—C—C—C—C (with C branch)

$(CH_3)_2CHCH_2CH_2CH_2CH_3$

C—C—C—C—C—C (with C branch)

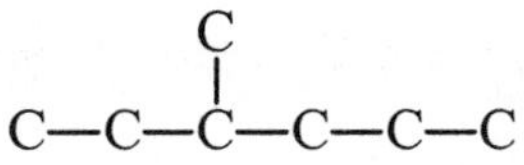

$CH_3CH_2CH(CH_3)CH_2CH_2CH_3$

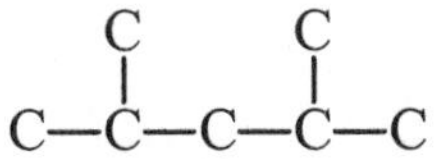

C—C—C—C—C (with two C branches)

$(CH_3)_2CHCH_2CHC(CH_3)_2$

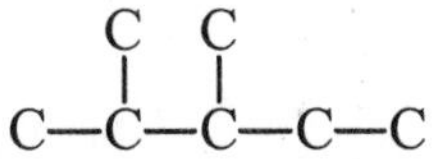

C—C—C—C—C (with two C branches)

$(CH_3)_2CHCH(CH_3)CH_2CH_3$

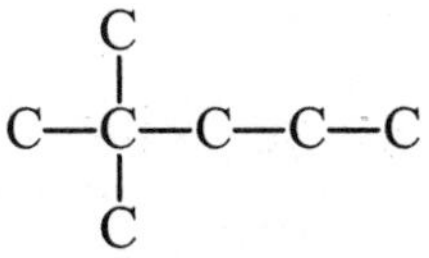

C—C—C—C (with branches)

$(CH_3)_3CCH_2CH_2CH_3$

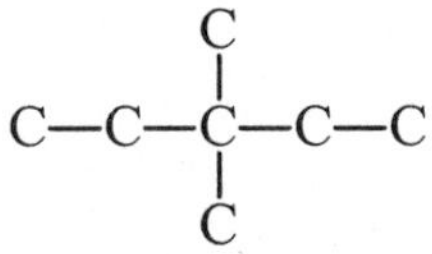

C—C—C—C—C (with branches)

$CH_3CH_2C(CH_3)_2CH_2CH_3$

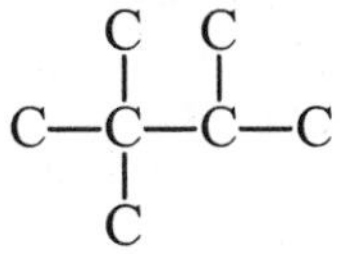

C—C—C—C (with branches)

$(CH_3)_3CCH(CH_3)_2$

C—C—C—C (with branch)

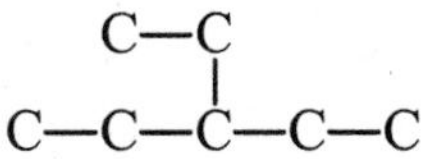

$CH(CH_2CH_3)_3$

2A

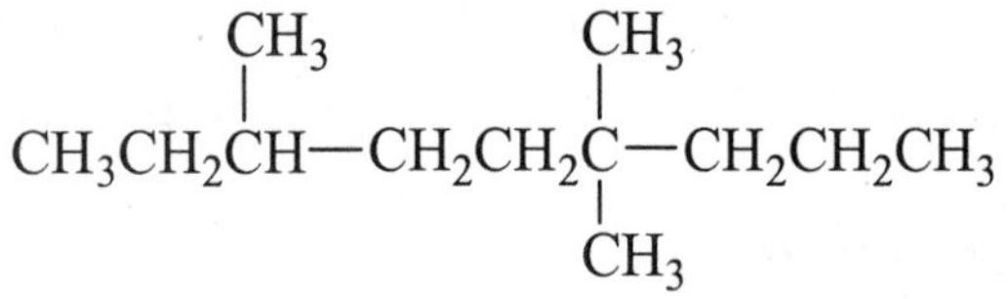

$$CH_3CH_2\overset{\overset{\displaystyle CH_3}{|}}{CH}-CH_2CH_2\overset{\overset{\displaystyle CH_3}{|}}{\underset{\underset{\displaystyle CH_3}{|}}{C}}-CH_2CH_2CH_3$$

Numbering starts from the left and goes right so that the substituents appear with the lowest numbers possible. This is 3,6,6-trimethylnonane.

2B   In the structural formula we show only the C atoms and the bonds between them. Remember that there are four bonds to each C atom; the remaining bonds not shown are to H atoms. The longest chain has 8 C atoms, making this an octane.

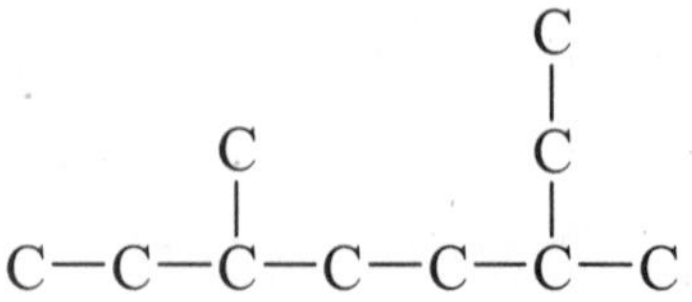

$$C-C-\overset{\overset{\displaystyle C}{|}}{C}-C-C-\overset{\overset{\overset{\displaystyle C}{|}}{\overset{\displaystyle C}{|}}}{C}-C$$

There are two methyl groups on the chain. The IUPAC name is 3,6-dimethyloctane.

3A   We write the structural formula as before, showing only the C skeleton. The main chain is 7 C's long; we number it from left to right. Methyl groups are 1 carbon chains, an ethyl group is on carbon 2.

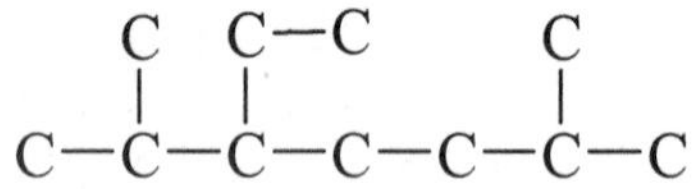

$$C-\overset{\overset{\displaystyle C}{|}}{C}-\overset{\overset{\displaystyle C-C}{|}}{C}-C-C-\overset{\overset{\displaystyle C}{|}}{C}-C$$

condensed structural formula: $CH_3CH(CH_3)CH(CH_2CH_3)CH_2CH_2CH(CH_3)_2$

3B   Pentane is a five-carbon chain. There is a — CH_3 group on carbon 2, a sec-butyl group on carbon 3.

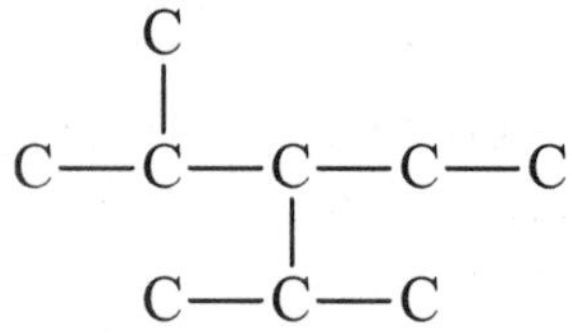

$$C-\overset{\overset{\displaystyle C}{|}}{C}-\underset{\underset{\underset{\displaystyle C-C-C}{|}}{}}{C}-C-C$$

Condensed structural formula: $CH_3CH(CH_3)CH(CH(CH_3)_2)CH_2CH_3$

4A   The aldehyde group is a meta director. Thus, the product of the mononitration of benzaldehyde should be the compound whose structure is drawn below:

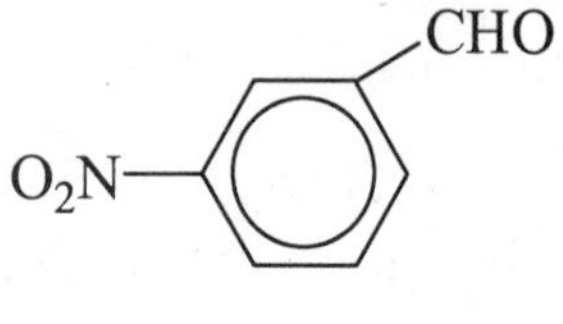

m-nitrobenzaldehyde or 3-nitrobenzaldehyde.

4B — Cl is an ortho, para director. The possible products are 1,3-dichloro-2-nitrobenzene and 2,4-dichloro-1-nitrobenzene.

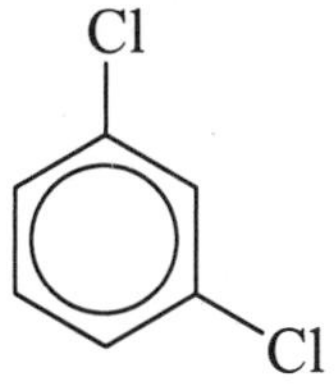

1,3-dichlorobenzene 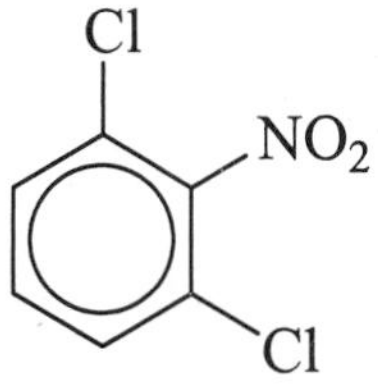   1,3-dichloro-2-nitrobenzene   2,4-dichloro-1-nitrobenzene

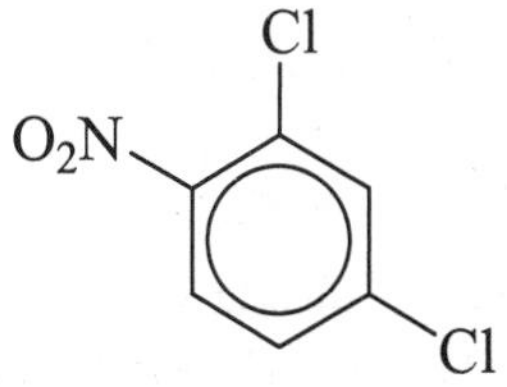

5A **(a)**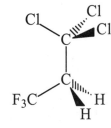

All three carbon atoms in this molecule are attached to at least two groups of the same type; thus, the molecule is achiral

(b)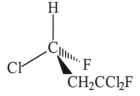

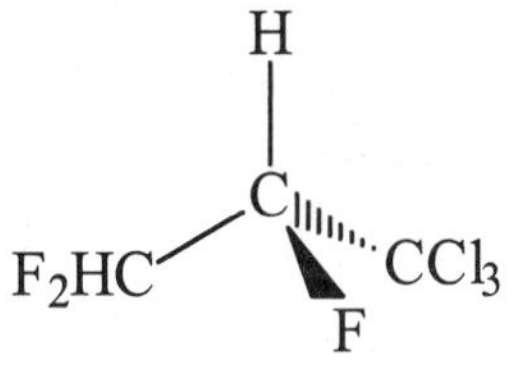

This molecule contains a carbon atom that is bonded to four different groups; consequently, the molecule is chiral.

(c)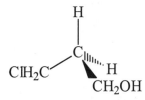

This molecule contains a carbon atom that is attached to four different groups; consequently, the molecule is chiral.

5B **(a)**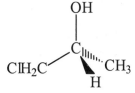

None of the three carbon atoms in this alcohol are bonded to four different group; consequently the molecule is achiral

(b)

This molecule contains a carbon atom that is bonded to four different groups; consequently, the molecule is chiral.

(c)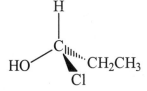

This molecule contains a carbon atom that is bonded to four different groups; consequently, the molecule is chiral.

6A (a)

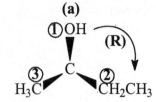

①OH
(R)
C
③ H3C
② CH2CH3

(b)

②Cl
(R)
C
①Br
③ CH2CH3

(c)

①OH
(R)
C
③ H3C
② COOH

6B (a)

③CH3
C
① HO
② CH2CH3
(S)

①OH
(R)
C
③ H3C
② CH2CH3

Thus, the structures are enantiomers.

(b)

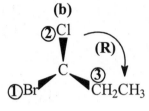

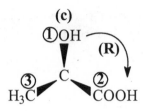

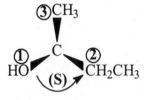

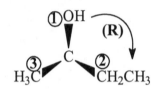

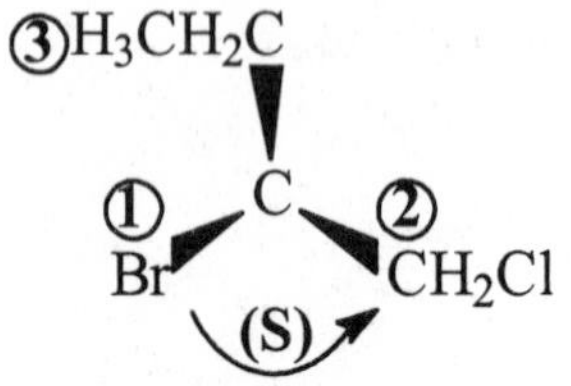

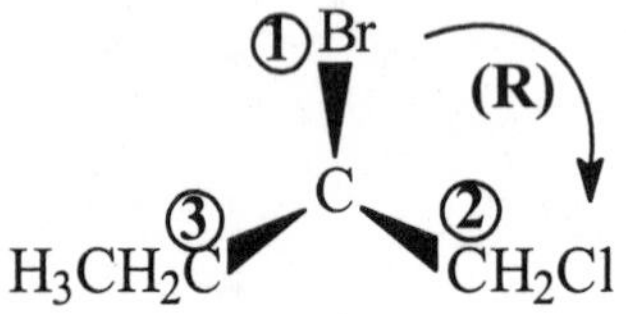

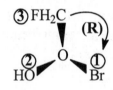

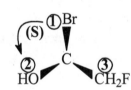

③H3CH2C
C
① Br
② CH2Cl
(S)

①Br
(R)
C
③ H3CH2C
② CH2Cl

Thus, the structures are enantiomers.

(c)

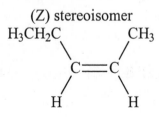

③ FH2C
(R)
O
② HO
① Br

①Br
(S)
C
② HO
③ CH2F

Thus, the structures are enantiomers.

7A (a) This is the (E) stereoisomer

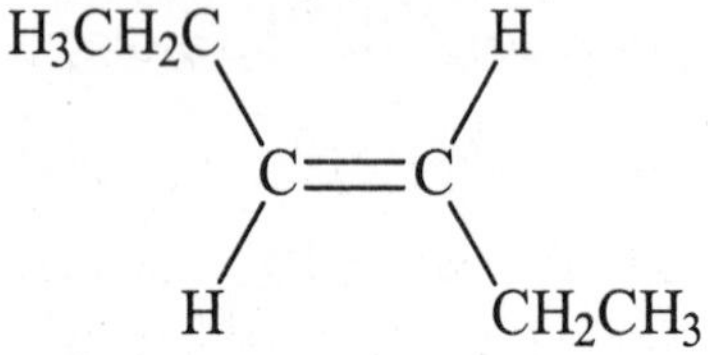

H3CH2C
H
C=C
H
CH2CH3

(b) This is the (Z) stereoisomer

Br
C≡CH
C=C
Cl
CH3

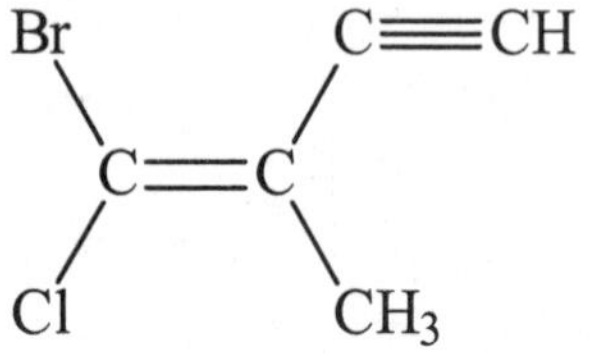

(c) This is the (Z) stereoisomer

ClH2CH2C
F
C=C
H3CH2C
H

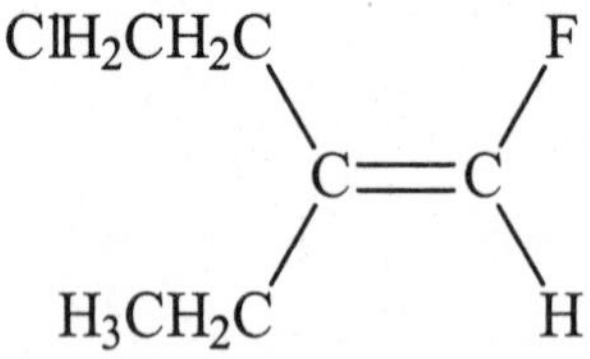

(d) This is the (Z) stereoisomer

H3C
CH2Br
C=C
H
CH2CH3

7B (a) (Z) stereoisomer

H3CH2C
CH3
C=C
H
H

(E) stereoisomer

H3CH2C
H
C=C
H
CH3

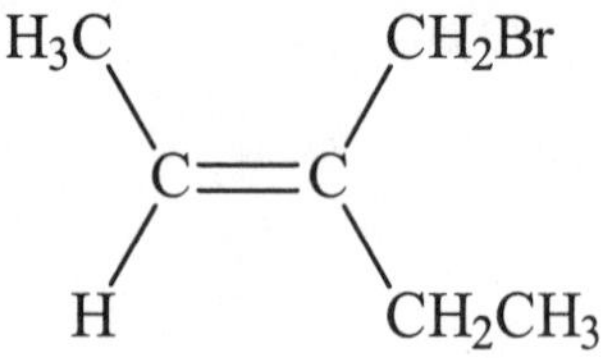

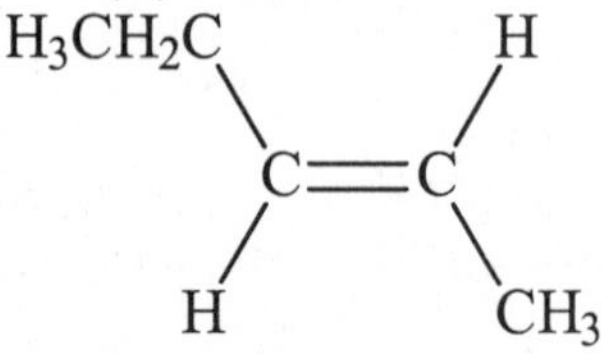

(b)

(Z) stereoisomer

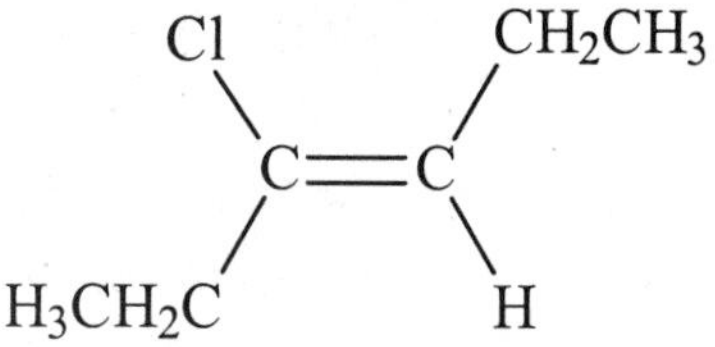

(E) stereoisomer

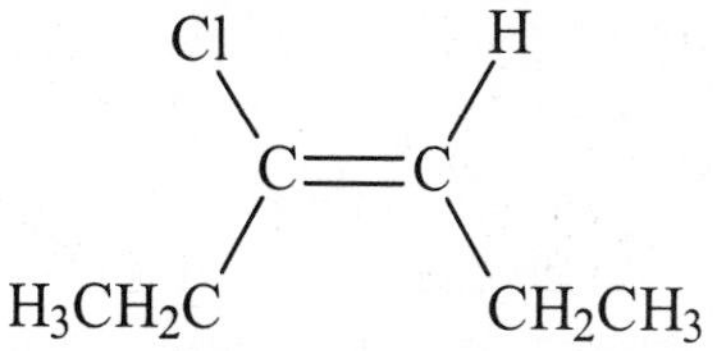

(c)

(Z) stereoisomer

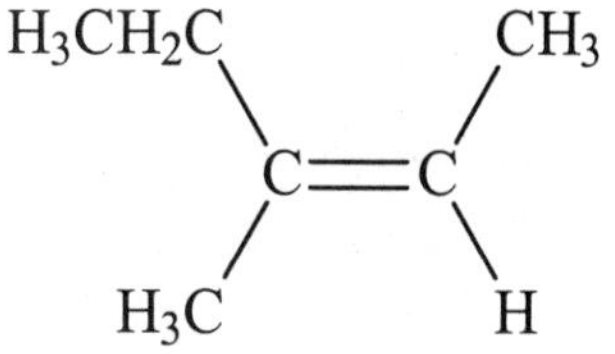

(E) stereoisomer

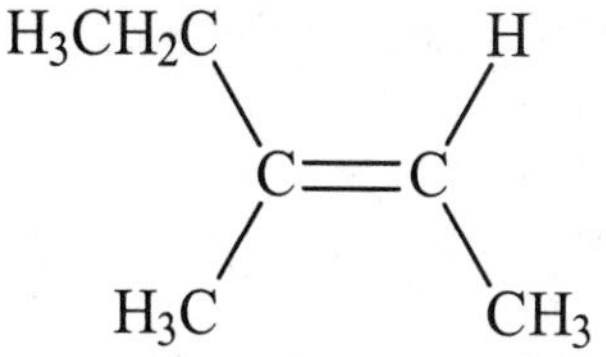

(d)

(Z) stereoisomer

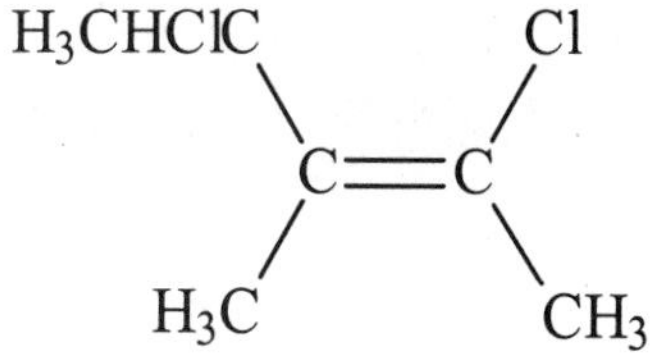

(E) stereoisomer

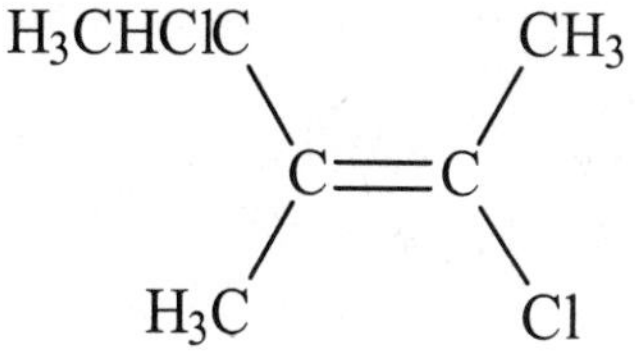

8A   The hydride ion is a nucleophile because it has a lone pair of electrons that can be shared with another atom.

The central Al atom is electron deficient. Thus, $AlCl_3$ is an electrophile.

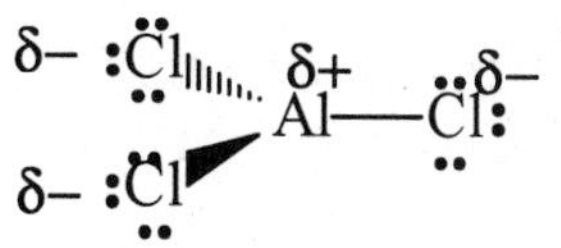

Ethylene can be viewed as a nucleophile because it has a pair of pi-electrons that can be shared with another atom

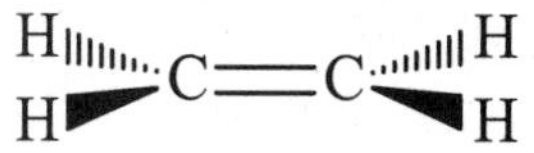

The negatively charged S atom possesses three lone pairs of electrons that can potentially be shared with other atoms. Thus methylthiolate anion is a nucleophile.

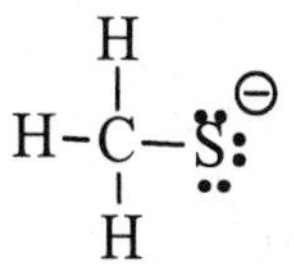

8B   **(a)** CH_3I + NH_3 → $H_3CNH_3^+I^-$
                (electrophile)   (nucleophile)

  (b) CH_3CH_2Cl + CH_3S^- → $H_3CCH_2SCH_3 + Cl^-$
            (electrophile)   (nucleophile)

  (c) $CH_3CH_2CHClCH_3$ + CN^- → $CH_3CH_2CHCNCH_3 + Cl^-$
            (electrophile)   (nucleophile)

9A **(a)** $CH_3C\equiv C^- + CH_3Br \xrightarrow{\;S_N2\;} CH_3C\equiv CCH_3 + Br^-$

(b) $Cl^- + CH_3CH_2CN \rightarrow$ NO REACTION

The nucleophile in this reaction is Cl^- while the leaving group is the CN^-. The CN^- ion is a much stronger nucleophile than Cl^-, so the equilibrium will strongly favor the reactants. In other words, no reaction is expected.

(c) $CH_3NH_2 + (CH_3)_3CCl \xrightarrow{\;S_N1\;} CH_3\overset{+}{N}H_2(C(CH_3)_3) + Cl^-$

9B **(a)**

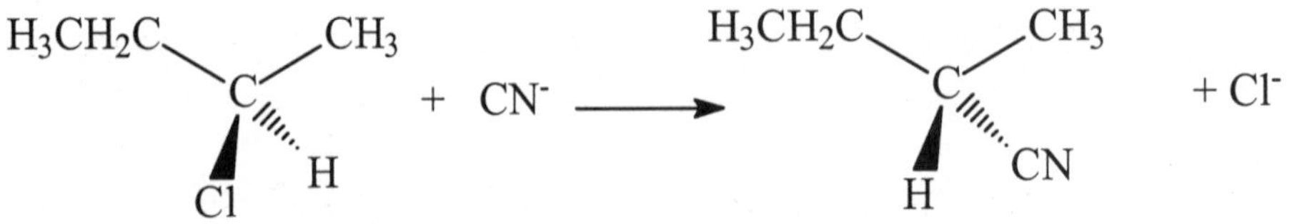

(R)-2-chlorobutane   (S)-2-cyanobutane

Because the configuration at the stereogenic carbon has undergone an inversion, we can conclude that the reaction has occurred via an S_N2 mechanism.

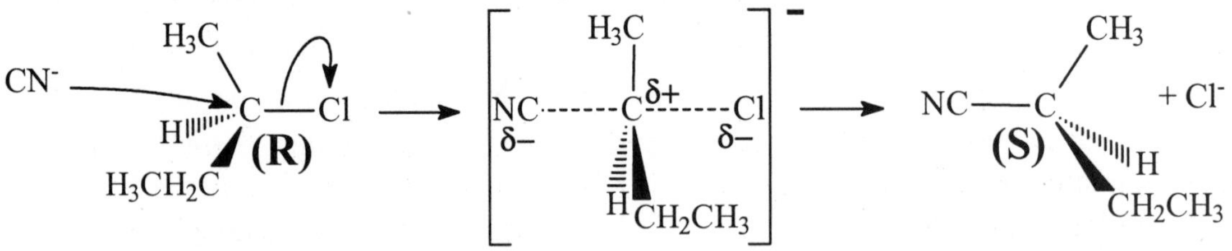

(b)

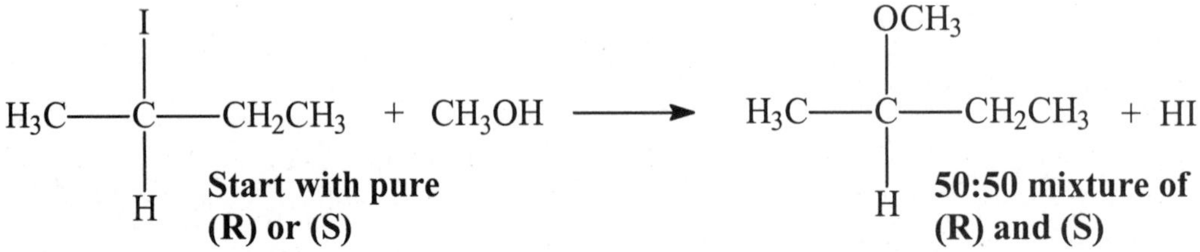

Clearly, since a racemic mixture forms, the reaction must occur via an S_N1 mechanism.

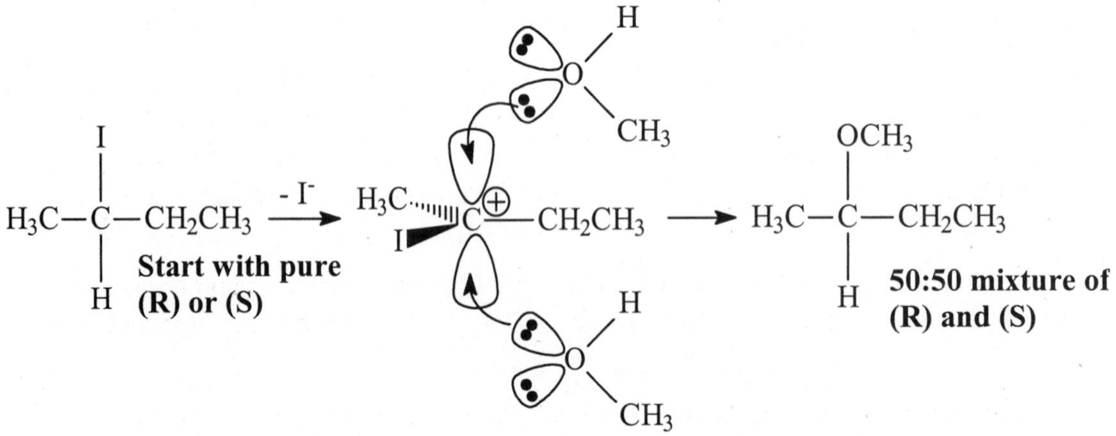

REVIEW QUESTIONS

1. **(a)** *t* is the symbol for tertiary; it refers to a carbon atom to which three other carbon atoms are singly bonded.

   (b) R— is the symbol for an alkyl group, a carbon-hydrogen chain with no multiple bonds.

   (c) A hexagon with an inscribed circle is the symbol for a benzene ring, the molecule with formula C_6H_6.

   (d) A carbonyl group is the C=O moiety that is a characteristic feature in aldehydes, ketones, esters, amides and carboxylic acids.

   (e) A primary amine is a compound that contains the $-\overset{..}{N}H_2$ functional group.

2. **(a)** In a substitution reaction, one atom or group of atoms replaces another atom or group of atoms.

   (b) The octane rating of gasoline is a measure of its quality. Gasoline's octane number is equal to the percent of isooctane in an isooctane-*n*-heptane mixture that has the same combustion characteristics as the gasoline.

   (c) Stereoisomerism refers to isomers (same chemical formula) that have exactly the same bonding arrangements, only differing in the three dimensional spatial orientation of the atoms. Geometric and optical isomerism are two examples of stereoisomerism.

   (d) An ortho, para director is a substituent on an aromatic ring that guides substitution reactions to occur preferentially at positions ortho- and para- to it.

   (e) Step reaction (condensation) polymerization occurs when the monomers join the polymer chain by the reaction of two functional groups and the elimination of a small molecule such as water.

3. **(a)** An alkane is a hydrocarbon in which there are no multiple bonds; in an alkene there is at least one double bond in the hydrocarbon.

   (b) An aliphatic compound is an alkane, a compound of carbon and hydrogen in which all bonds are single bonds and in which the carbon skeleton consists of straight or branched chains. In an aromatic compound there is at least one planar, cyclic arrangement of carbon atoms held together by both σ-bonds and delocalized π-bonds.

   (c) An alcohol is has an — OH (hydroxyl) functional group attached to an aliphatic (sp^3 hybridized) carbon atom. In phenols, the — OH group is attached to a benzene ring.

   (d) An ether has a C— O— C linkage. An ester has a $C-O-\overset{\overset{\textstyle O}{\|}}{C}-$ linkage.

   (e) An amine contains N—C bond; ammonia is NH_3.

4. (a) Structural (or skeletal) isomers differ in how the atoms are joined together.

    (b) Positional isomers differ in the location of a functional group along the carbon chain of the molecule.

    (c) *cis*-, *trans*- isomers differ in the arrangement of atoms around a double bond.

    (d) ortho-, meta-, and para- isomers differ in the location of substituent groups on an aromatic ring.

5. Just in terms of straight and branched chains, C_4H_8 has more isomers than C_4H_{10}, as shown by the structures below, in which the C_4H_8 structures are at right. In addition, only C_4H_8 can form rings: one four-membered ring isomer and one three-membered ring isomer.

C_4H_{10} C_4H_8

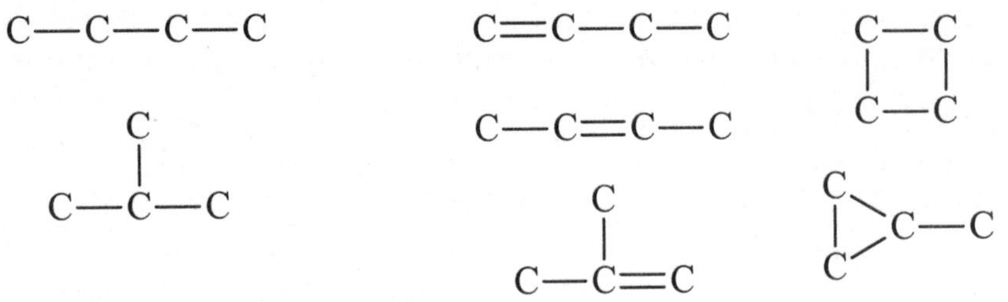

$$C-C-C-C$$

$$C=C-C-C$$

$$\begin{array}{c} C-C \\ |\quad| \\ C-C \end{array}$$

$$\underset{\displaystyle C-\underset{\displaystyle C}{\overset{\displaystyle |}{C}}-C}{}$$

$$C-C=C-C$$

$$\underset{\displaystyle C-\underset{\displaystyle C}{\overset{\displaystyle |}{C}}=C}{}$$

$$\begin{array}{c} C \\ | \diagdown \\ | \diagup C-C \\ C \end{array}$$

6. Cyclobutane has the formula C_4H_8; there is one carbon for every two hydrogens. $CH_3CH=CHCH_3$ has the formula C_4H_8, and $CH_3C\equiv CCH_3$ has the formula C_4H_6. Only in $CH_3CH=CHCH_3$ is there the same carbon-to-hydrogen ratio as in cyclobutane.

7. (a) 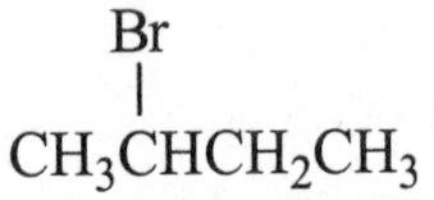 $\underset{\displaystyle CH_3CHCH_2CH_3}{\overset{\displaystyle Br}{\overset{\displaystyle |}{}}}$ alkyl halide (bromide)

         2-Bromobutane

    (b) 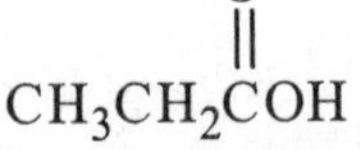 $\underset{\displaystyle CH_3CH_2COH}{\overset{\displaystyle O}{\overset{\displaystyle \|}{}}}$ carboxylic acid

         propanoic acid

    (c) 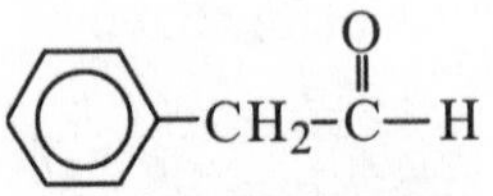 $\langle\bigcirc\rangle-CH_2-\overset{\displaystyle O}{\overset{\displaystyle \|}{C}}-H$ aldehyde

         benzylaldehyde

    (d) $(CH_3)_2CHCH_2OCH_3$ ether

         sec-butylmethyl ether

(e)

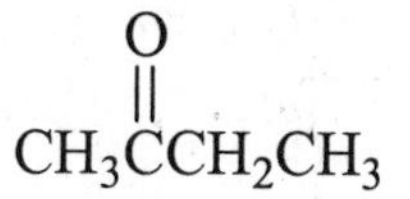

$$O$$
$$\|$$
$$CH_3CCH_2CH_3$$

methyl ethyl ketone

ketone

(f) $CH_3CH(NH_2)CH_2CH_3$

2-aminobutane

amine

(g) $HO-C_6H_4-OH$

2,3 or 4 hydroxyphenol

phenol, hydroxyl group, phenyl group

(h)

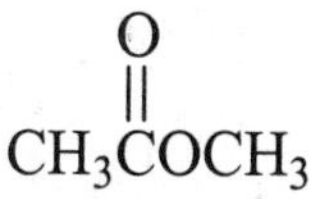

$$O$$
$$\|$$
$$CH_3COCH_3$$

methyl acetate

ester

8. (a)

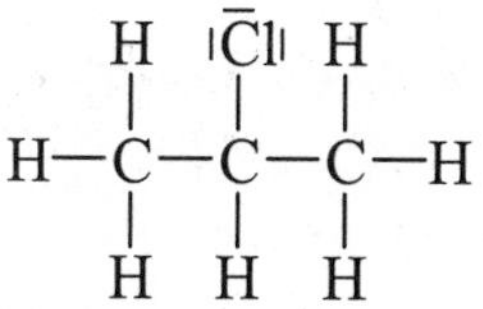

(b)

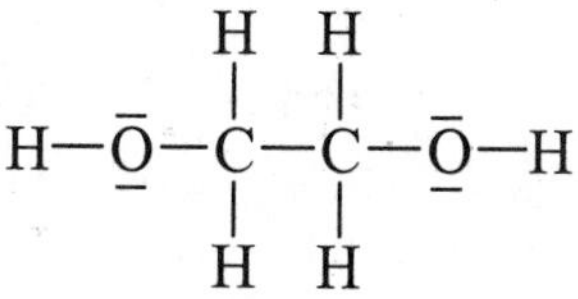

(c)

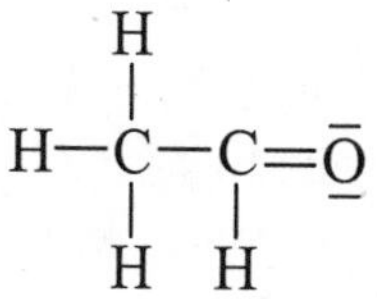

9. We draw only the carbon and chlorine atoms in each structure. Remember that there are four bonds to each carbon atom; the bonds not shown in these structures are C—H bonds.

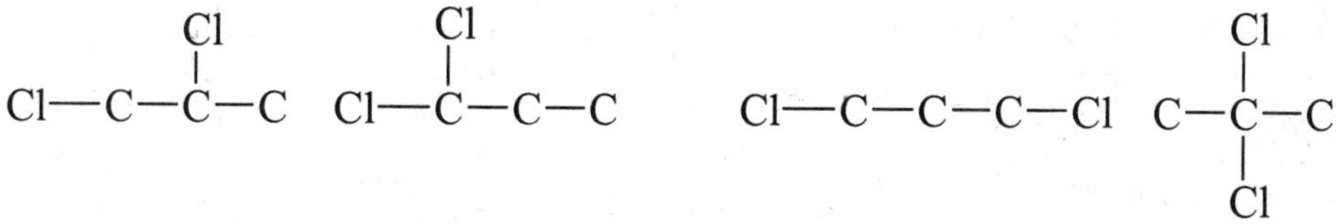

10. (a) These are not isomers, because they have different formulas: C_4H_{10} and C_4H_8.

(b) These two compounds are structural isomers. (Only the carbon skeleton is shown below).

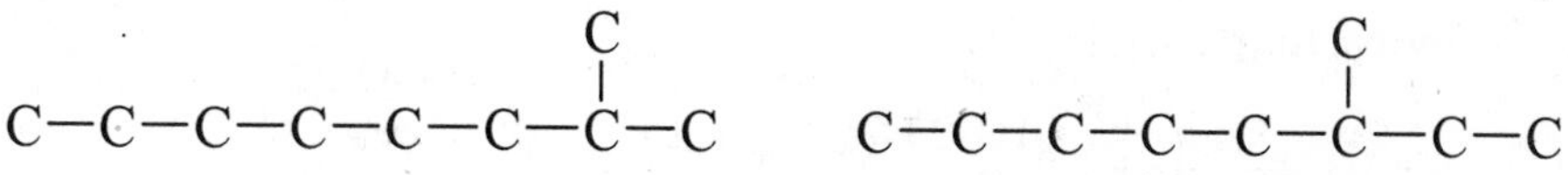

(c) The compounds are not isomers; they have different formulas: C_4H_9Cl and $C_5H_{11}Cl$.

(d) These two compounds are identical; simply rotated.

(e) These two compounds are identical; simply rotated.

(f) These are two ortho-para isomers, ortho-nitrophenol on the left, and para-nitrophenol on the right.

11. (a) The longest chain is eight carbons long, the two substituent groups are methyl groups, and they are attached to the number 3 and number 5 carbon atom. This is 3,5-dimethyloctane.

     (b) The longest carbon chain is three carbons long, the two substituent groups are methyl groups, and they are both attached to the number 2 carbon atom. This is 2,2-dimethylpropane.

     (c) The longest carbon chain is 7 carbon atoms long, there are two chloro groups attached to carbon atom 3, and an ethyl group attached to carbon 5. This is 3,3-dichloro-5-ethylheptane.

12. (a) There are 2 chloro groups at the 1 and 3 positions on a benzene ring. This is 1,3-dichlorobenzene or, more appropriately, meta-dichlorobenzene.

     (b) There is a methyl group at position 1 on a benzene ring, and a nitro group at position 3. This is 3-nitrotoluene or meta-nitrotoluene.

     (c) There is a — COOH group at position 1 on the benzene ring, and a — NH_2 group at position 4, or para to the — COOH group. This is 4-aminobenzoic acid or p-aminobenzoic acid.

13. In the structural formulas drawn below, we omit the hydrogen atoms. Remember that there are four bonds to each C atom. The bonds that are not shown are C— H bonds.
     (a) 3-isopropyloctane   (b) 2-chloro-3-methylpentane

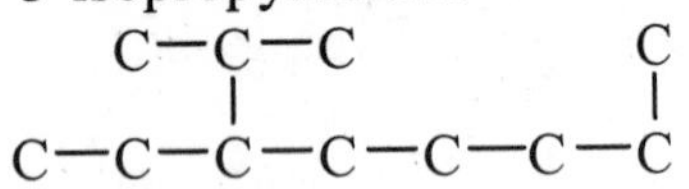

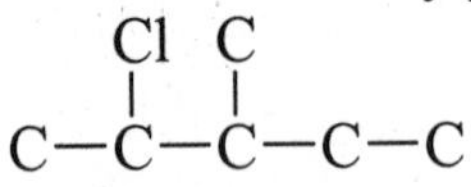

          C—C—C   C   Cl C
              | | | |
          C—C—C—C—C—C—C   C—C—C—C—C

     (c) 2-pentene   (d) dipropyl ether

          C–C=C–C–C   C–C–C–O–C–C–C

     (e) p-bromophenol

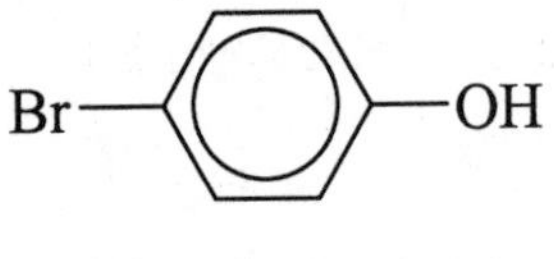

          Br—⬡—OH

14. (a) isopropyl alcohol $CH_3CH(OH)CH_3$

     (b) 1,1,1-chlorodifluoroethane $CClF_2CH_3$

     (c) 2-methyl-1,3-butadiene $CH_2 = C(CH_3)CH = CH_2$

15. (a)   (b) 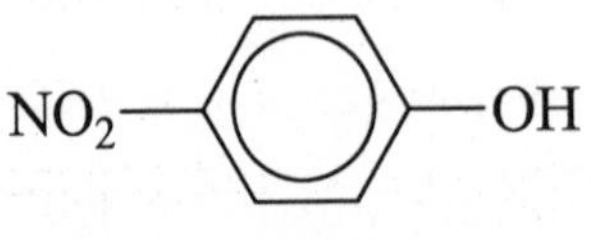   (c)
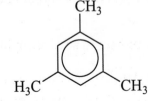

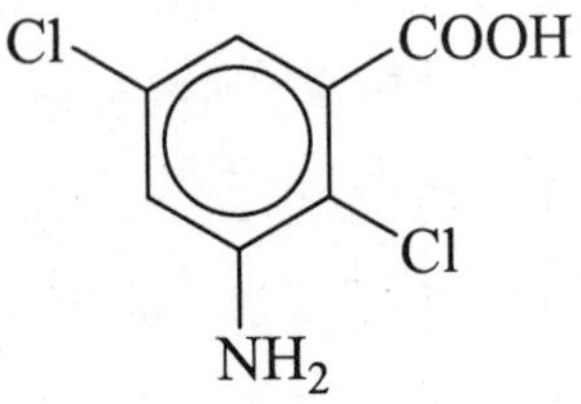

16. Methyl ethyl ketone has the abbreviated structure shown on the right. Thus, the alcohol from which it is produced must have an — OH group on carbon 2 and it must be a four-carbon-atom straight-chain alcohol. Of the alcohols listed, 2-butanol satisfies these criteria.

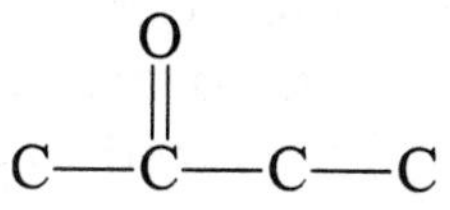

$$C-\overset{\overset{\textstyle O}{\|}}{C}-C-C$$

17. (a) $CH_3CH_2CH{=}CH_2 \xrightarrow{H_2O, H_2SO_4} CH_3CH_2CH(OH)CH_3$

(b) $CH_3CH_2CH_3 + Cl_2 \xrightarrow{hv} CH_3CH_2CH_2Cl + CH_3CHClCH_3$
$\cdot$ + other polychlorinated propanes

(c) $\langle\!\langle\bigcirc\rangle\!\rangle{-}COOH + (CH_3)_2CHOH \xrightarrow{\Delta} \langle\!\langle\bigcirc\rangle\!\rangle{-}COOCH(CH_3)_2 + H_2O$

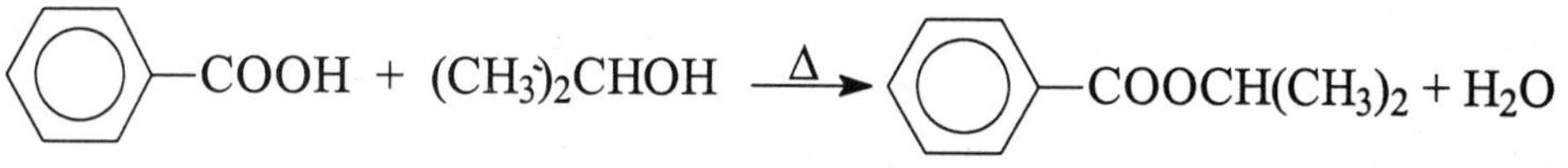

(d) $CH_3CH(OH)CH_2CH_3 \xrightarrow{Cr_2O_7^{2-}, H^+} H_3C-\overset{\overset{\textstyle O}{\|}}{C}-CH_2-CH_3$

18. (a) C_6H_6 would have a higher boiling point that C_6H_{12} because its London forces are stronger; C_6H_6 has many π electrons which interact with each other. There are no dipole-dipole forces or hydrogen bonds between these molecules.

(b)   Both C_3H_7OH and $C_7H_{15}OH$ are alcohols that can hydrogen bond, but $C_7H_{15}OH$ has a much longer nonpolar hydrocarbon chain, interfering with its aqueous solubility. C_3H_7OH is more water soluble.

(c)   Functional groups determine acidity of these two otherwise alike molecules. The aldehyde hydrogen is not notably acidic while the carboxylic hydrogen is. C_6H_5COOH is more acidic in aqueous solution.

19. (a) $:NH_3$ (nucleophile, posses a chemically active lone pair)

(b) CH_3Cl (electrophile, the polar C-Cl bond makes the molecule an electrophile)

(c) Br^- (nucleophile, posses chemically active lone pairs)

(d) CH_3OH (nucleophile: lone pairs on oxygen can be used to form dative bonds)

(e) $CH_3NH_3^+$ (neither: though positively charged, this cation will not accept any additional electron pairs).

20. (a) N_3^- (nucleophile, posses lone pairs of electrons that can be used to form dative bonds)

(b) NH_4^+ (neither: though positively charged, this cation will not accept any additional electron pairs)

(c) $CH_3CHClCH_3$ (electrophile: the polar C-Cl bond makes the molecule an electrophile)

(d) OH^- (nucleophile: lone pairs on oxygen can be used to form dative bonds)

(e) $H_2C{=}O$ (both: as a nucleophile via the oxygen, donating a pair of electrons or as an electrophile via the carbon accepting a pair of electrons from a nucleophile).

21. **(a)**

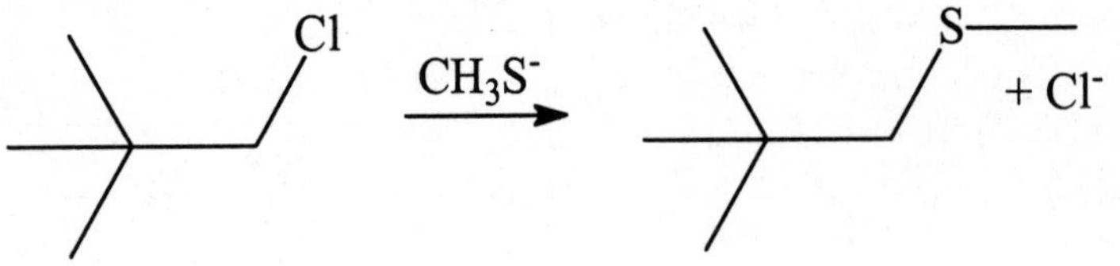

$\xrightarrow{CH_3S^-}$

S—— + Cl⁻   Bimolecular
Nucleophilic   (S$_N$2)
Substitution

(b)

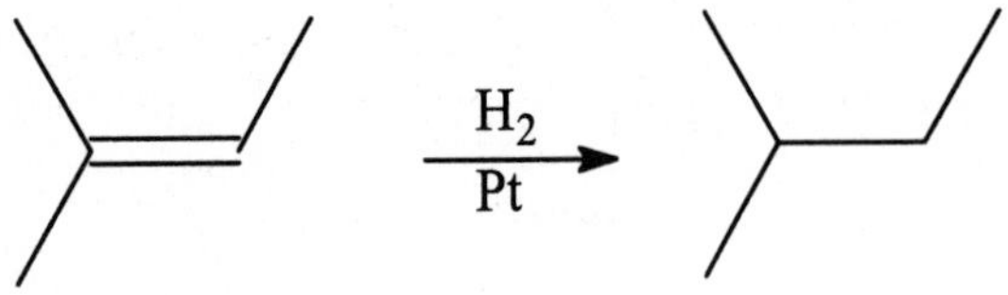

$\xrightarrow[Pt]{H_2}$ Hydrogenation

(c)

+

$\xrightarrow{H^+}$

+ H$_2$O

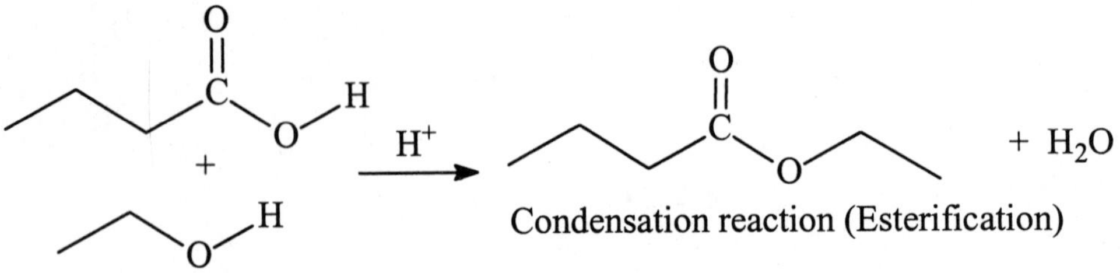

Condensation reaction (Esterification)

22. **(a)**

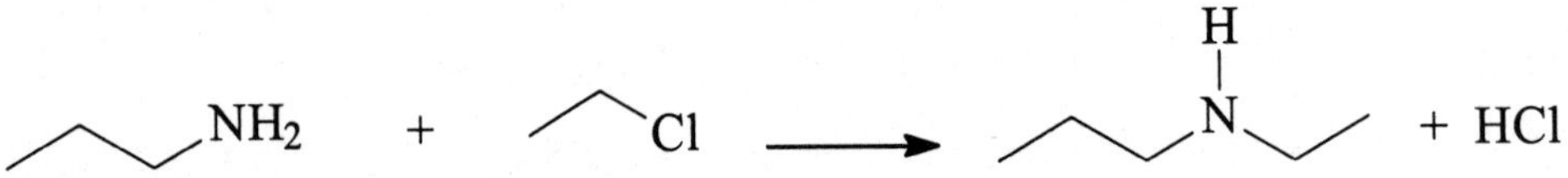

NH$_2$ + Cl ⟶ N + HCl

Substitution Reaction (formation of a secondary amine)

(b) OH + Cl ⟶ O + HCl

Substitution Reaction (formation of an ether)

(c) OH [O]⟶

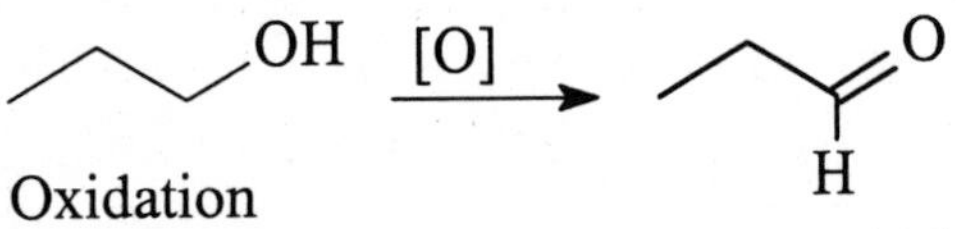

Oxidation
(conversion of an alcohol into an aldehyde)

EXERCISES

Organic Structures

23. In the following structural formulas, the hydrogen atoms are omitted for simplicity. Remember that there are four bonds to each carbon atom. The missing bonds are C—H bonds.

(a) $CH_3CH_2CHBrCHBrCH_3$

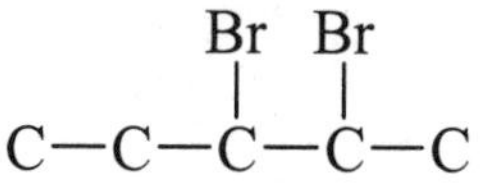

$$\begin{array}{ccccc} & & Br & Br & \\ & & | & | & \\ C{-}C{-}C&{-}&C&{-}&C \end{array}$$

(b) $(CH_3)_3CCH_2C(CH_3)_2CH_2CH_2CH_3$

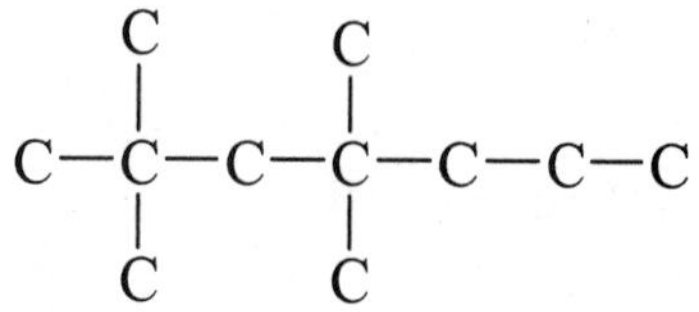

$$\begin{array}{ccccccc} & C & & C & & & \\ & | & & | & & & \\ C{-}C{-}C{-}C{-}C{-}C{-}C \\ & | & & | & & & \\ & C & & C & & & \end{array}$$

(c) $(C_2H_5)_2CHCH{=}CHCH_2CH_3$

$$\begin{array}{ccccccc} & & C & & & & \\ & & | & & & & \\ & & C & & & & \\ & & | & & & & \\ C{-}C{-}C{-}C{=}C{-}C{-}C \end{array}$$

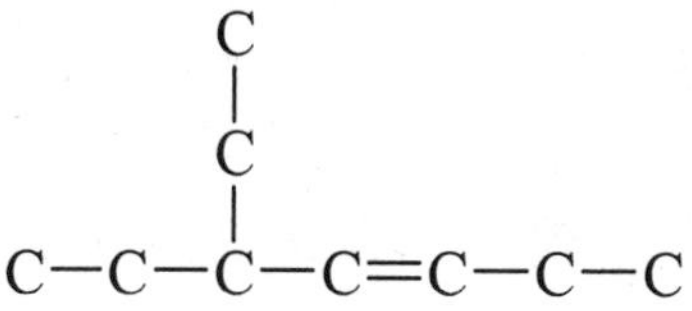

24. In the following structural formulas, the hydrogen atoms are omitted for simplicity. Remember that there are four bonds to each carbon atom. The missing bonds are C—H bonds.

(a) $(CH_3)_3CCH_2CH(CH_3)CH_2CH_2CH_3$

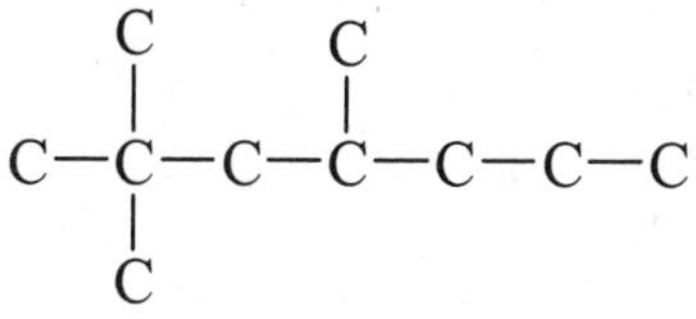

$$\begin{array}{ccccccc} & C & & C & & & \\ & | & & | & & & \\ C{-}C{-}C{-}C{-}C{-}C{-}C \\ & | & & & & & \\ & C & & & & & \end{array}$$

(b) $(CH_3)_2CHCH_2C(CH_3)_2CH_2Br$

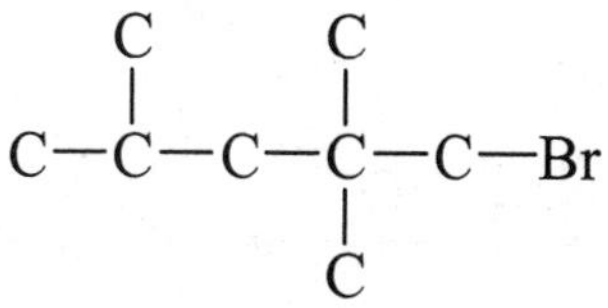

$$\begin{array}{ccccccc} & C & & C & & & \\ & | & & | & & & \\ C{-}C{-}C{-}C{-}C{-}Br \\ & & & | & & & \\ & & & C & & & \end{array}$$

(c) $Cl_3CCH_2CH(CH_3)CH_2Cl$

$$\begin{array}{ccccc} & Cl & & CH_2Cl & \\ & | & & | & \\ Cl{-}C{-}C{-}C{-}C \\ & | & & & \\ & Cl & & & \end{array}$$

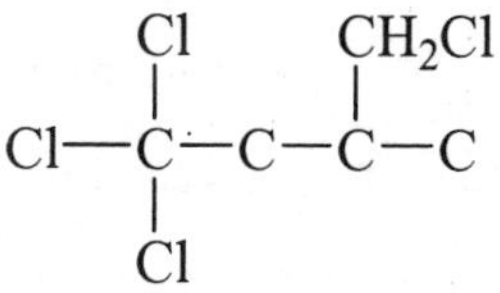

25. (a) Each carbon atom is sp^3 hybridized. All of the C–H bonds in the structure (drawn below) are sigma bonds, between the $1s$ orbital of H and the sp^3 orbital of C. The C–C bond is between sp^3 orbitals on each C atom.

(b)   Both carbon atoms are sp^2 hybridized. All of the C–H bonds in the structure (drawn below) are sigma bonds between the $1s$ orbital of H and the sp^2 orbital of C. The C–Cl bond is between the sp^2 orbital on C and the $3p$ orbital on Cl. The C=C double bond is composed of a sigma bond between the sp^2 orbitals on each C atom and a pi bond between the $2p_z$ orbitals on the two C atoms.

(c)   The left-most C atom (in the structure drawn below) is sp^3 hybridized, and the C–H bonds to that C atom are between the sp^3 orbitals on C and the $1s$ orbital on H. The other two C atoms are sp hybridized. The right-hand C–H bond is between the sp orbital on C and the $1s$ orbital on H. The C≡C triple bond is composed of one sigma bond formed by overlap of sp orbitals, one from each C atom, and two pi bonds, each formed by the overlap of two $2p$ orbitals, one from each C atom (that is a $2p_y$—$2p_y$ overlap and a $2p_z$—$2p_z$ overlap).

(a)

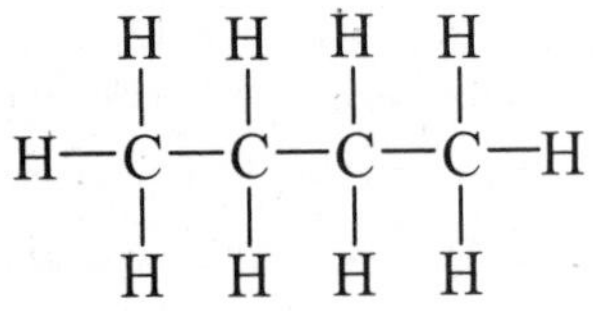

$$
\begin{array}{c}
\ \ \ \ \text{H}\ \ \text{H}\ \ \text{H}\ \ \text{H} \\
\ \ \ \ |\ \ \ \ |\ \ \ \ |\ \ \ \ | \\
\text{H}-\text{C}-\text{C}-\text{C}-\text{C}-\text{H} \\
\ \ \ \ |\ \ \ \ |\ \ \ \ |\ \ \ \ | \\
\ \ \ \ \text{H}\ \ \text{H}\ \ \text{H}\ \ \text{H}
\end{array}
$$

(b)

$$
\begin{array}{c}
\text{H}-\text{C}=\text{C}-\text{Cl} \\
\ \ \ \ \ \ |\ \ \ \ | \\
\ \ \ \ \ \ \text{H}\ \ \text{H}
\end{array}
$$

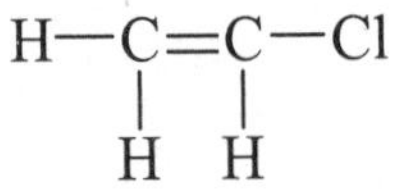

(c)

$$
\begin{array}{c}
\ \ \ \ \ \ \text{H} \\
\ \ \ \ \ \ | \\
\text{H}-\text{C}-\text{C}\equiv\text{C}-\text{H} \\
\ \ \ \ \ \ | \\
\ \ \ \ \ \ \text{H}
\end{array}
$$

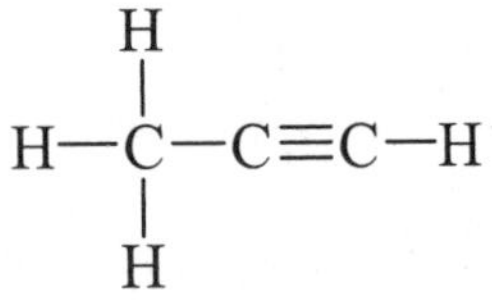

26.   **(a)**   The left- and right-most C atoms in the structure (drawn below) are sp^3 hybridized. All C–H bonds are sigma bonds formed by the overlap of an sp^3 orbital on C with a 1s orbital on H. The central C atom is sp^2 hybridized; both C–C bonds are sigma bonds, formed by the overlap between the sp^3 orbital on the terminal C atom and an sp^2 orbital on the central C atom. The C=O double bond is composed of a σ bond between the sp^2 orbital on the central C atom and a $2p_y$ orbital on the O atom, and a π bond between the $2p_z$ orbital on the central C atom and the $2p_z$ orbital on the O atom.

(b)   The left C atom in the structure (drawn below) is sp^3 hybridized. All C–H bonds are sigma bonds formed by the overlap of an sp^3 orbital on C with a 1s orbital on H. The central C atom is sp^2 hybridized; the C–C bond is a sigma bond, formed by the overlap between the sp^3 orbital on the terminal C atom and an sp^2 orbital on the central C atom. The C=O double bond is composed of a σ bond between the sp^2 orbital on the central C atom and a $2p_y$ orbital on the O atom, and a π bond between the $2p_z$ orbital on the central C atom and the $2p_z$ orbital on the O atom. The right O atom is sp^3 hybridized; the C–O sigma bond forms by the overlap of $\text{C}\left(sp^2\right)$ with $\text{O}\left(sp^3\right)$ and the O–H bond forms by overlap of the $\text{O}\left(sp^3\right)$ with the $\text{H}(1s)$.

(c)   The two end C's are sp^2 hybridized; all C−H bonds form by the overlap of C(sp^2) with H(1s). The central C is sp hybridized. Both C=C bonds consist of a σ bond formed by C$_{central}(sp)$ with C$_{end}(sp^2)$ overlap and a π bond formed by C$_{central}(2p)$ with C$_{end}(2p)$ overlap. Note: The 2p orbitals that make up each pi bond are mutually perpendicular relative the left or right side of the molecule.

(a)   (b)   (c)

Isomers

27.   Structural (skeletal) isomers differ from each other in the length of their carbon atom chains and in the length of the side chains. The carbon skeleton differs between these isomers. Positional isomers differ in the location or position where functional groups are attached to the carbon skeleton. Geometric isomers differ in whether two substituents are on the same side of the molecule or on opposite sides of the molecule from each other; usually they are on opposite sides or the same side of a double bond.

(a)   The chloro group is attached to the terminal carbon in $CH_3CH_2CH_2Cl$ and to the central carbon atom in $CH_3CHClCH_3$. These are positional isomers.

(b) $CH_3CH(CH_3)CH_2CH_3$, methylbutane, and $CH_3(CH_2)_3CH_3$, pentane, are structural isomers.

(c)   Ortho-aminotoluene and meta-aminotoluene differ in the position of the —NH_2 group on the benzene ring. They are positional isomers.

28.   (a)   The two chloro groups in CHCl=CHCl are on different carbon atoms. In $CH_2=CCl_2$ they are on the same atoms. These are positional isomers.

(b)   1-butene, $CH_2=CHCH_2CH_3$, and $CH_2=C(CH_3)_2$, methylpropene, are structural isomers. They might be termed positional isomers, differing as they do in the position of a CH_3 group. But since that group's location changes the structure of the molecule, they are structural isomers.

(c)   These two compounds differ in the location of the two—COOH groups; the left-hand one has the two groups on the same side of the double bond (the cis isomer) and the right-hand one has the two groups on opposite sides of the double bond (the trans isomer). These are geometric isomers.

29. We show only the carbon atom skeleton in each case. Remember that there are four bonds to carbon. The bonds that are not indicated in these structures are C–H bonds.

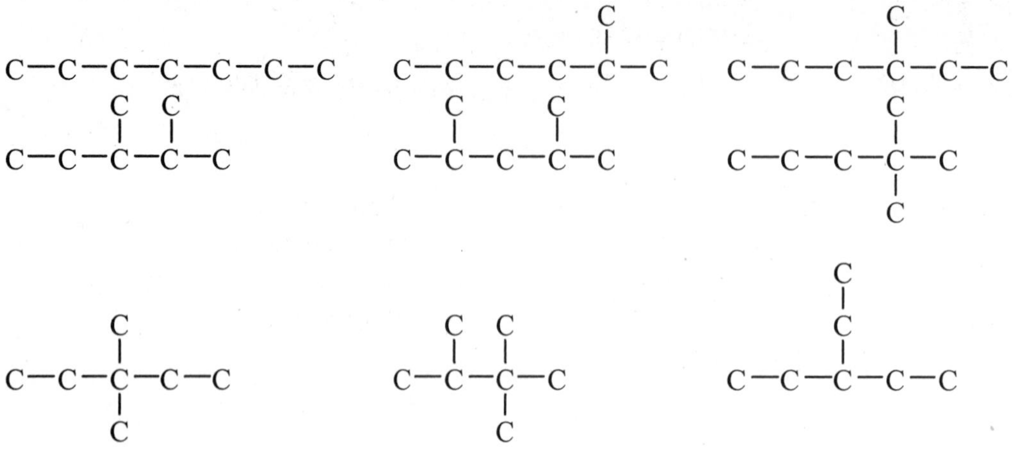

30. In each case, we draw only the carbon skeleton. It is understood that there are four bonds to each carbon atom. The bonds that are not shown are C–H bonds.

(a)

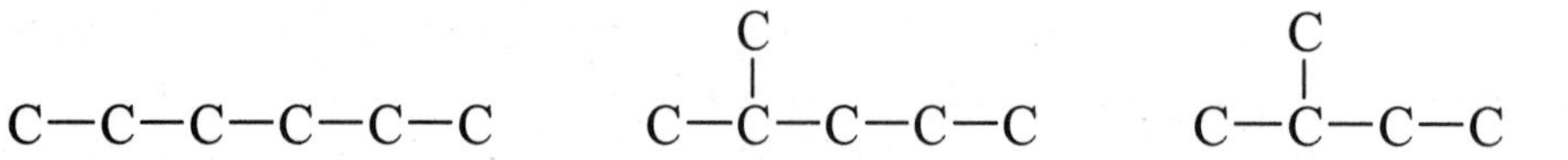

n-hexane   2-methylpentane   3-methylpentane

2,3-dimethylbutane   2,2-dimethylbutane

(b)

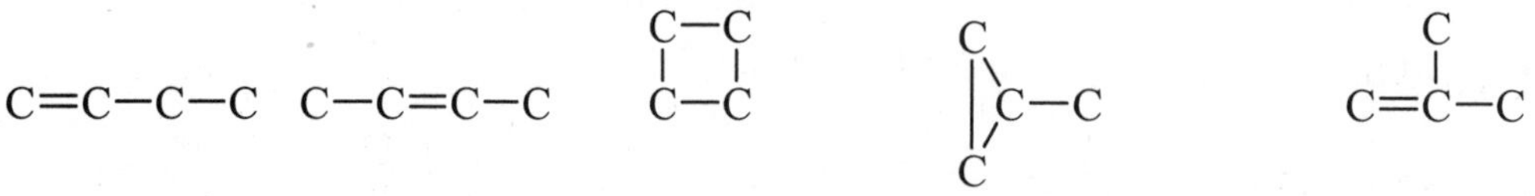

1-butene   2-butene   cyclobutane   methylcyclopropane   2-methyl-1-propene

(c)

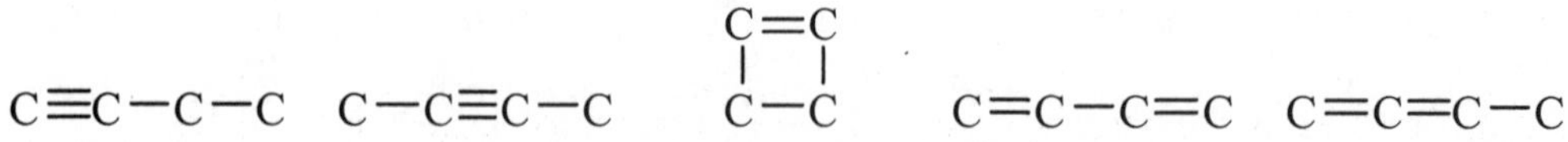

1-butyne   2-butyne   cyclobutene   1,3-butadiene   1,2-butadiene

31. **(a)**

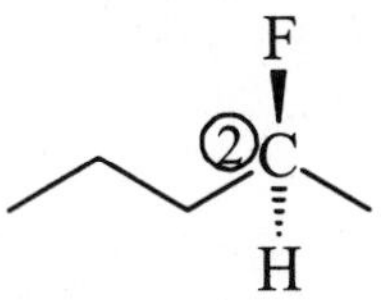

F
|
②C
/   \
H

Carbon② is chiral

(b) There are no chiral carbon atoms in compound (b)

(c)

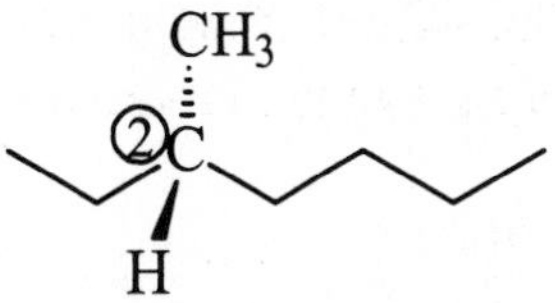

CH₃
⋮
②C
/   \
H

Carbon② is chiral

32. **(a)** There are no chiral carbon atoms in compound (a)

(b)

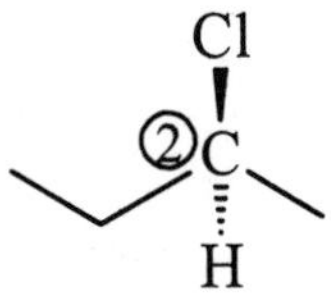

Cl
|
②C
/   \
H

Carbon② is chiral

(c) There are no chiral carbon atoms in compound (c)

33. **(a)**

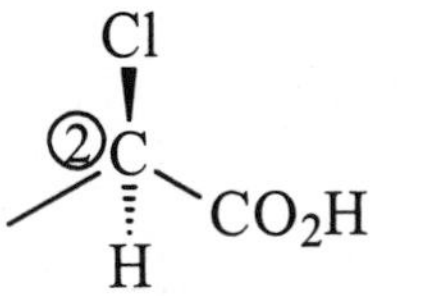

Cl
|
②C—CO₂H
|
H

Carbon② is chiral

(b)

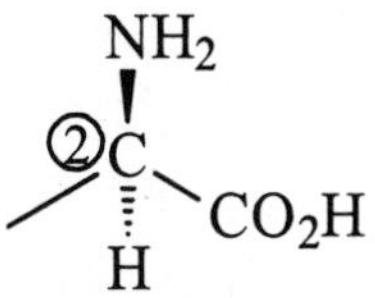

NH₂
|
②C—CO₂H
|
H

Carbon② is chiral

(c) There are no chiral carbon atoms in compound (c)

34. **(a)**

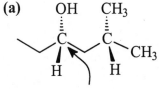

chiral carbon

(b)

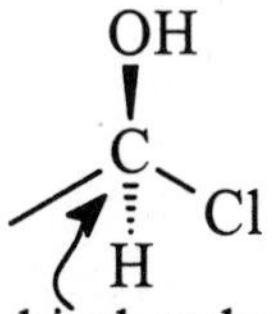

OH
|
C—Cl
/ |
H

chiral carbon

(c) There are no chiral carbon atoms in compound (c)

Functional Groups

35. **(a)** A carbonyl group is
$$\underset{\ \ \ }{-\overset{\displaystyle O}{\overset{\|}{C}}-}$$, while a carboxyl group is $-\overset{\displaystyle O}{\overset{\|}{C}}-OH$

The essential difference between them is the hydroxyl group, — OH.

(b) An aldehyde has a carbonyl group at the end of a molecule, $R-\overset{\displaystyle O}{\overset{\|}{C}}-H$

In a ketone, $R-\overset{\displaystyle O}{\overset{\|}{C}}-R'$ the carbonyl group is in the center of the molecule.

(c) Acetic acid is $H_3C-\overset{\displaystyle O}{\overset{\|}{C}}-OH$, while an acetyl group is $H_3C-\overset{\displaystyle O}{\overset{\|}{C}}-$

The essential difference is the presence of the — OH group in acetic acid.

36. **(a)** aromatic nitro compound, $C_6H_5NO_2$, nitrobenzene

(b) aliphatic amine, $C_2H_5NH_2$, ethylamine

(c) chlorophenol, p-ClC_6H_4OH, para-chlorophenol (drawn below)

(d) aliphatic diol, $HOCH_2CH_2OH$, 1,2-ethanediol

(e) unsaturated aliphatic alcohol, $CH_2{=}CHCH_2CH_2OH$, 3-butene-1-ol

(f) alicyclic ketone, $C_6H_{10}O$, cyclohexanone (drawn below)

(g) halogenated alkane, $CH_3CHICH_2CH_3$, 2-iodobutane

(h) aromatic dicarboxylic acid, $o-$ $C_6H_4(COOH)_2$, ortho-phthalic acid (drawn below)

(c)
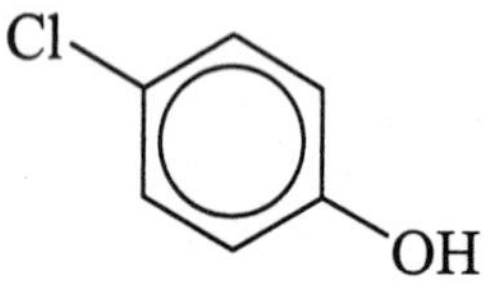
Cl
OH

(f)
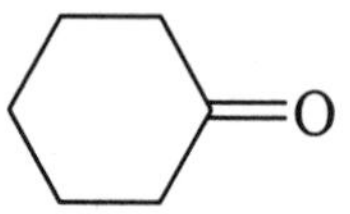
O

(h)
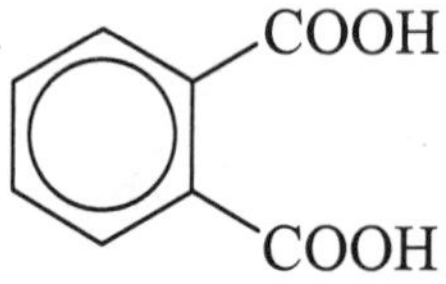
COOH
COOH

37. **(a)**
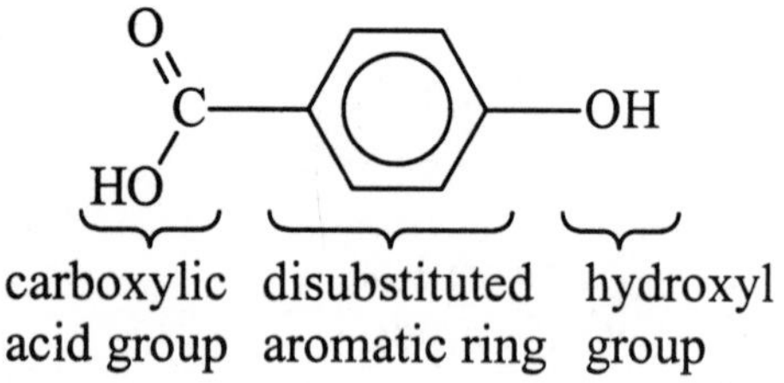
O
C
HO
OH
carboxylic acid group
disubstituted aromatic ring
hydroxyl group

(b)
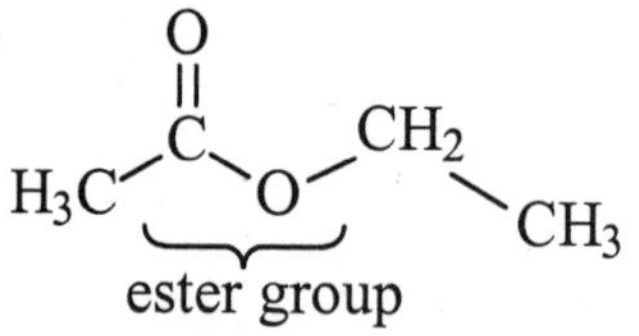
O
C
H3C
O
CH2
CH3
ester group

(c)
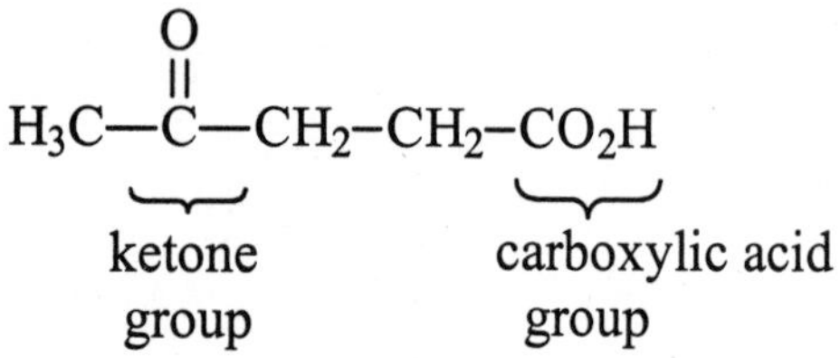
O
H3C—C—CH2–CH2–CO2H
ketone group
carboxylic acid group

(d)
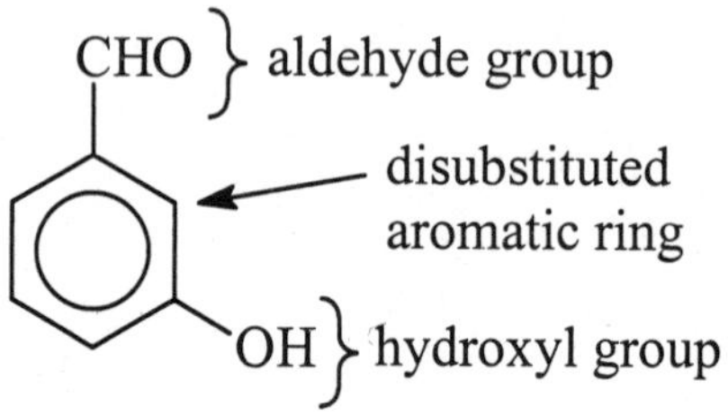
CHO
aldehyde group
disubstituted aromatic ring
OH
hydroxyl group

38. **(a)**
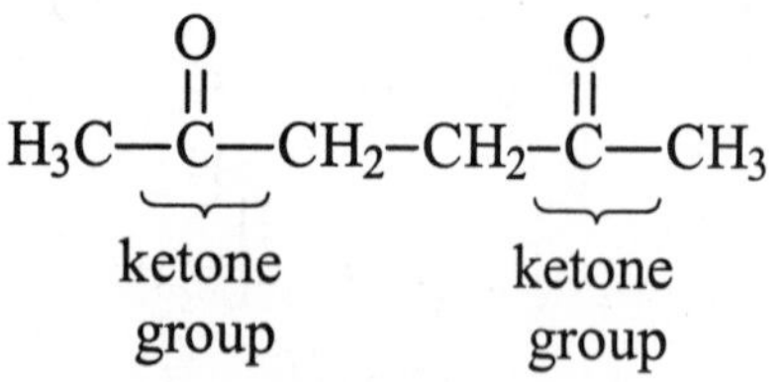
O
O
H3C—C—CH2–CH2–C—CH3
ketone group
ketone group

(b)
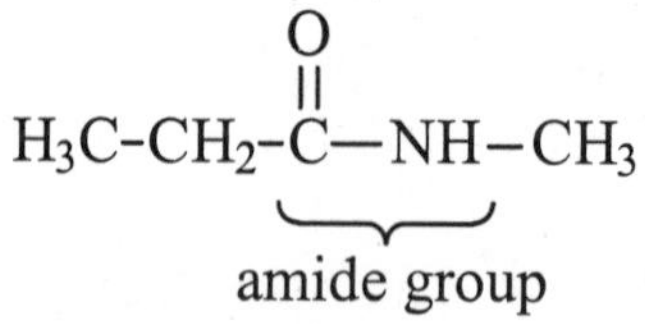
O
H3C-CH2-C—NH–CH3
amide group

(c)
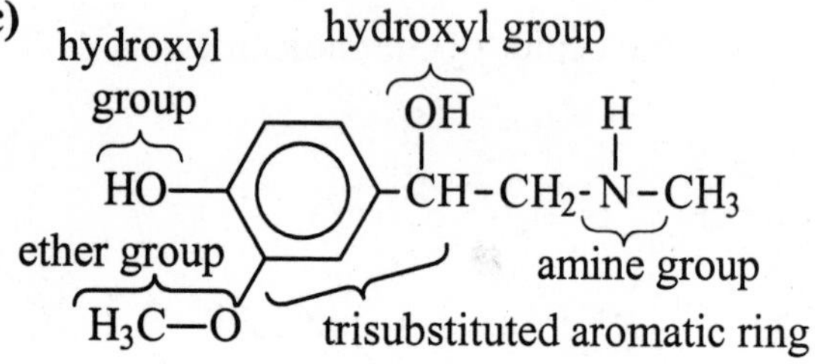
hydroxyl group
hydroxyl group
OH
H
HO
CH-CH2-N-CH3
ether group
amine group
H3C—O
trisubstituted aromatic ring

(d)
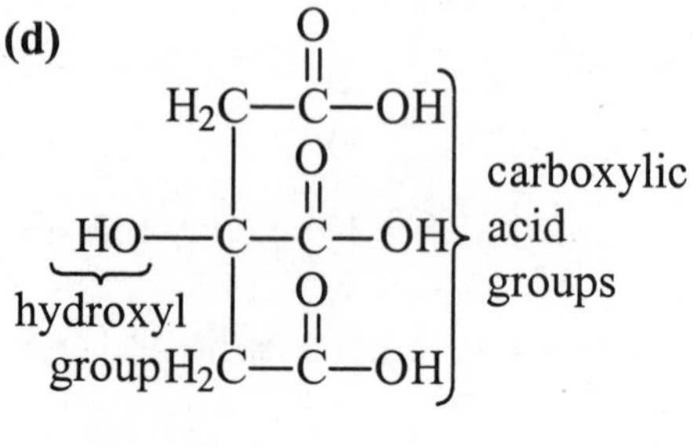
O
H2C—C—OH
O
HO—C—C—OH
O
hydroxyl group
H2C—C—OH
carboxylic acid groups

39.

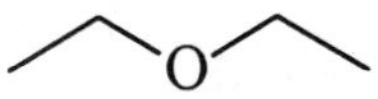

diethyl ether    methyl propyl ether 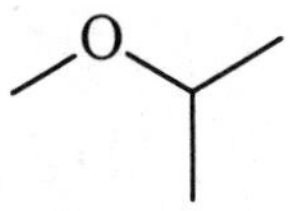   methyl isopropyl ether

40.

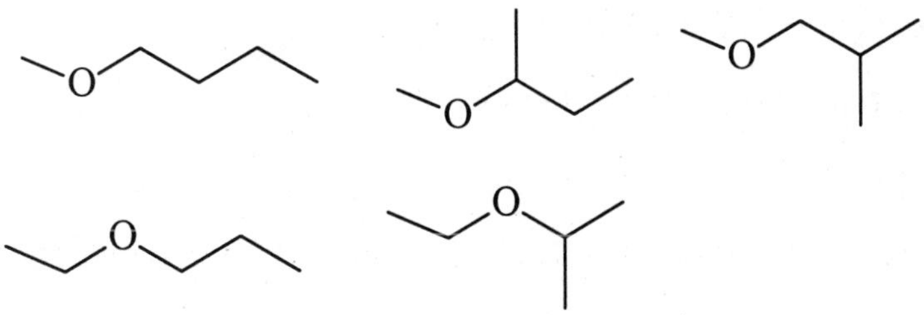

41.

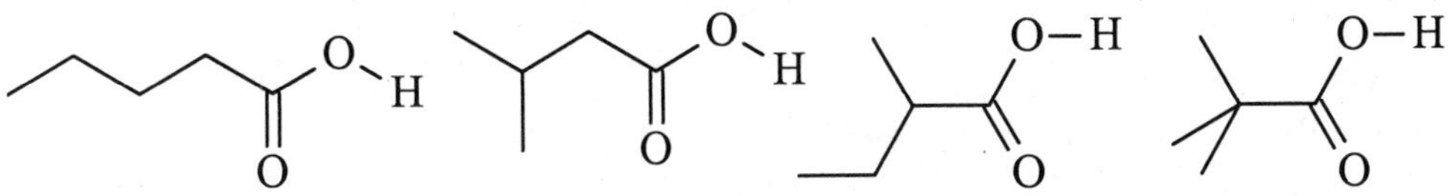

42.

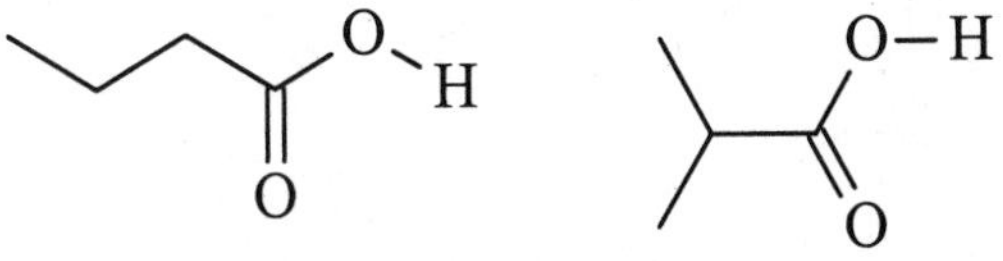

43.

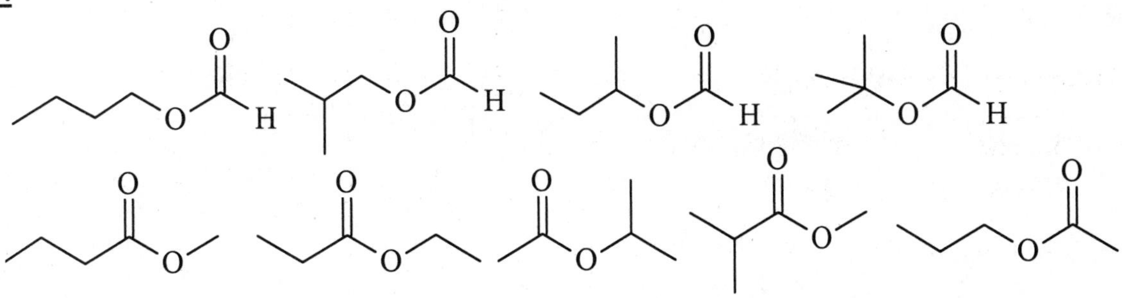

44.

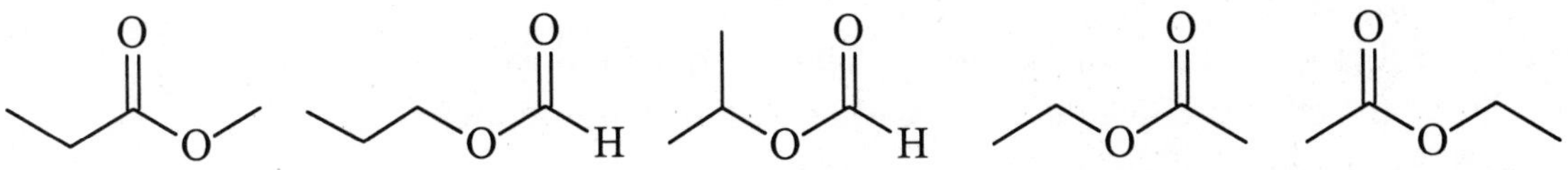

45.

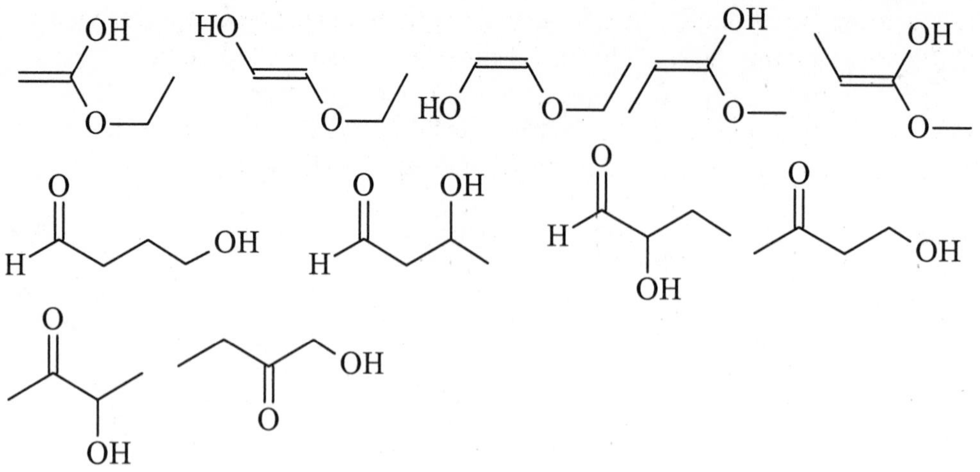

46. There are many isomers; here are a few of them

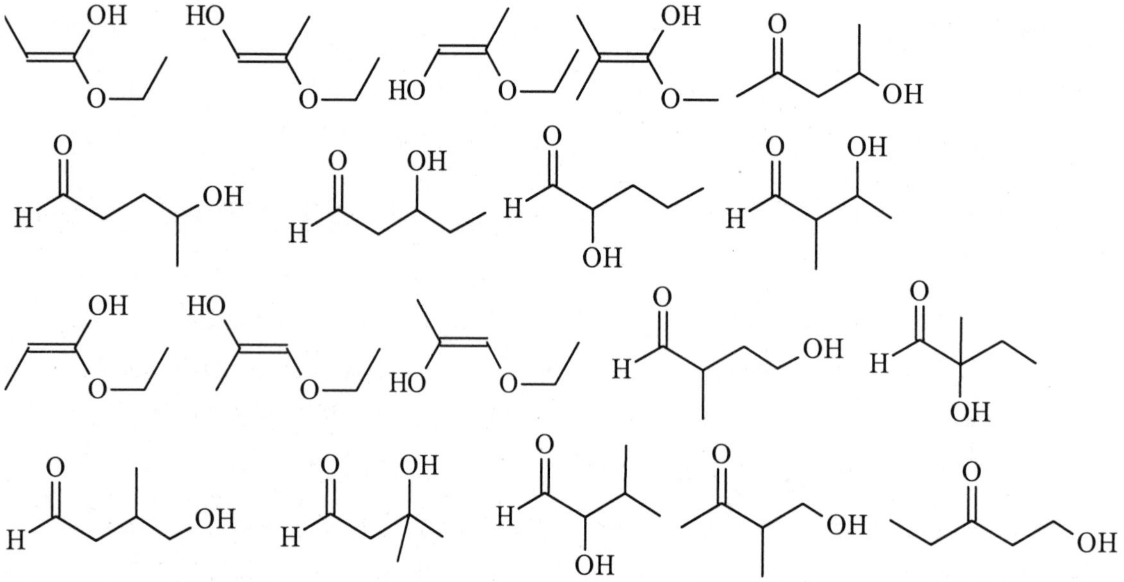

Nomenclature and Formulas

47. (a) The longest carbon chain has four carbon atoms and there are 2 methyl groups attached to carbon 2. This is 2,2-dimethylbutane.

(b) The longest chain is three carbons long, there is a double bond between carbons 1 and 2, and a methyl group attached to carbon 2. This is 2-methylpropene.

(c) Two methyl groups are attached to a three-carbon ring. This is 1,2-dimethylcyclopropane.

(d) The longest chain is 5 carbons long, there is a triple bond between carbons 2 and 3 and a methyl group attached to carbon 4. This is 4-methyl-2-pentyne.

(e) The longest chain is 5 carbons long. There is an ethyl group on carbon 2 and a methyl group on carbon 3. This compound is 2-ethyl-3-methylpentane.

(f)   The longest carbon chain containing the double bond is 5 carbons long. The double bond is between carbons 1 and 2. There is a propyl group on carbon 2, and carbons 3 and 4 each have one methyl group. This is 3,4-dimethyl-2-propyl-1-pentene.

48.   Again we do not show the hydrogen atoms in the structures below. But we realize that there are four bonds to each C atom and the missing bonds are C–H bonds.

(a)   isopentane

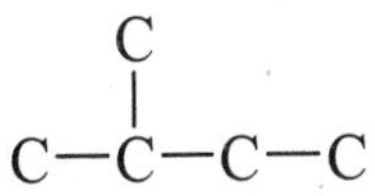

$$
\begin{array}{c}
\quad\quad C \\
\quad\quad | \\
C - C - C - C
\end{array}
$$

(b)   cyclohexene

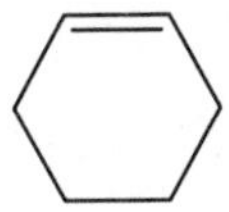

(c)   2-methyl-3-hexyne

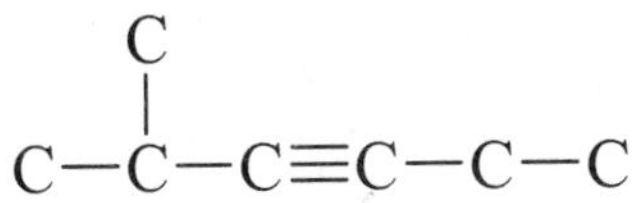

$$
\begin{array}{c}
\quad\quad C \\
\quad\quad | \\
C - C - C \equiv C - C - C
\end{array}
$$

(d)   2-butanol

$$
\begin{array}{c}
\quad\quad\quad OH \\
\quad\quad\quad | \\
C - C - C - C - C
\end{array}
$$

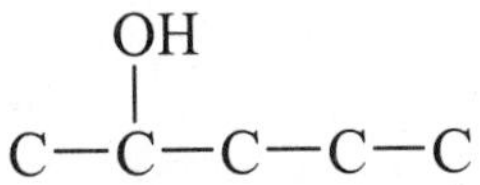

(e)   ethyl isopropyl ether

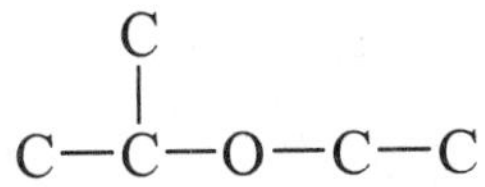

$$
\begin{array}{c}
\quad\quad C \\
\quad\quad | \\
C - C - O - C - C
\end{array}
$$

(f)   Propionaldehyde

$$
\begin{array}{c}
\quad\quad\quad O \\
\quad\quad\quad || \\
C - C - C - H
\end{array}
$$

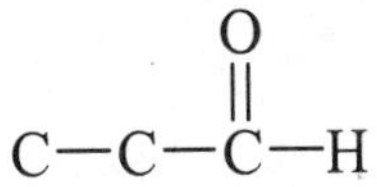

49.   (a)   The name pentene is insufficient. 1-pentene is $CH_2=CHCH_2CH_2CH_3$ and 2-pentene is $CH_3CH=CHCH_2CH_3$.

(b)   The name butanone is sufficient. There is only one four-carbon ketone.

(c)   The name butyl alcohol is insufficient. There are numerous butanols. 1-butanol is $HOCH_2CH_2CH_2CH_3$, 2-butanol is $CH_3CH(OH)CH_2CH_3$, isobutanol is $HOCH_2CH(CH_3)_2$, and t-butanol is $(CH_3)_3COH$.

(d)   The name methylaniline is insufficient. It specifies $CH_3-C_6H_4-NH_2$, with no relative locations for the $-NH_2$ and $-CH_3$ substituents.

(e)   The name methylcyclopentane is sufficiently precise. It does not matter where on a five-carbon ring with only C–C single bonds a methyl group is placed.

(f)   Dibromobenzene is insufficient. It specifies $Br-C_6H_4-Br$, with no indication of the relative locations of the two bromo groups on the ring.

50.   (a)   3-pentene specifies $CH_3CH_2CH=CHCH_3$. The proper name for this compound is 2-pentene. The number for the functional group should be as small as possible.

(b)   Pentadiene is insufficient. Possible pentadienes are 1,2 pentadiene, $CH_2=C=CHCH_2CH_3$, 1,3-pentadiene, $CH_2=CHCH=CHCH_3$, and 2,3 pentadiene, $CH_3CH=C=CHCH_3$.

(c) 1-propanone is incorrect. There cannot be a ketone on the first carbon of a chain. The compound specified is either propionaldehyde, CH_3CH_2CHO, or propanone (2-propanone).

(d) Bromopropane is insufficient. It could be either 1-bromopropane, $BrCH_2CH_2CH_3$, or 2-bromopropane, $CH_3CHBrCH_3$.

(e) Although 2,6-dichlorobenzene conveys enough information, the proper name for the compound is meta-dichlorobenzene, or 1,3-dichlorobenzene. Substituent numbers should be as small as possible.

(f) 2-methyl-3-pentyne is $(CH_3)_2CHC \equiv CCH_3$. The proper name for this compound is 4-methyl-2-pentyne. The number of the triple bond should be as small as possible.

51. **(a)** 2,4,6-trinitrotoluene: **(b)** methylsalicylate:

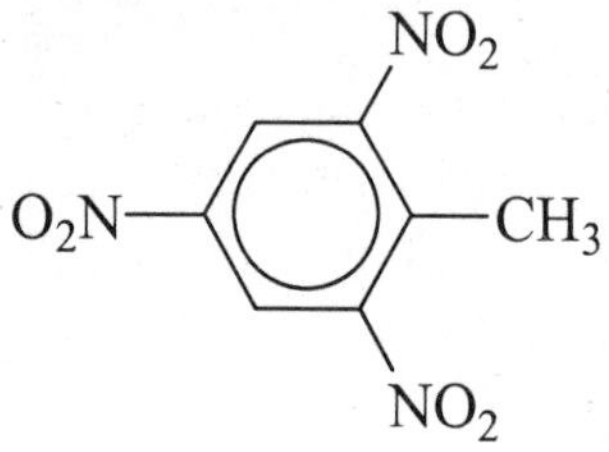

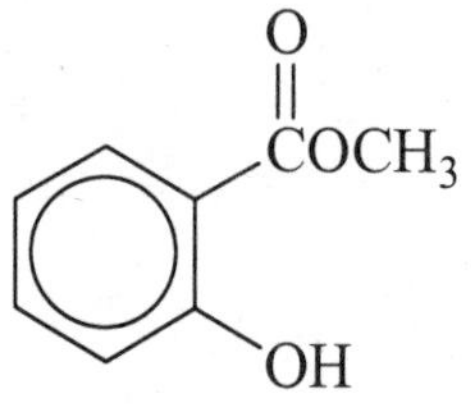

(c) 2-hydroxy-1,2,3-propanetricarboxylic acid: $HOOCCH_2C(OH)(COOH)CH_2COOH$

52. **(a)** o-t-butylphenol   **(b)** 1-phenyl-2-aminopropane   **(c)** 2-methylheptadecane

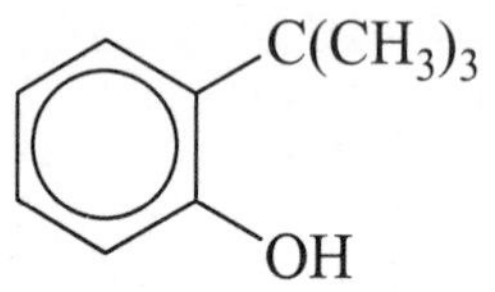

$C_6H_5CH_2CH(NH_2)CH_3$ $(CH_3)_2CH(CH_2)_{14}CH_3$

53. **(a)** $HN(CH_2CH_3)_2$
dimethylamine

(b) 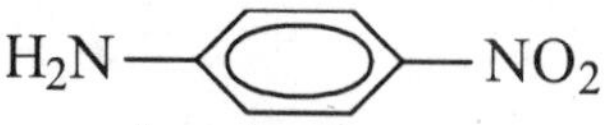
p-aminonitrobenzene

(c) 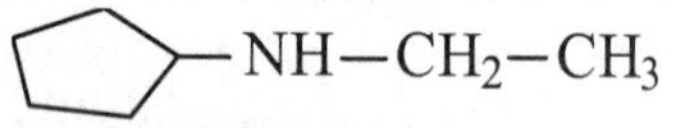
cyclopentylethylamine

(d) $CH_3CH_2N(CH_3)CH_2CH_3$
diethylmethylamine

54. **(a)** ethylamine:
H_3C-CH_2-NH_2

(b) 3-chloroaniline:

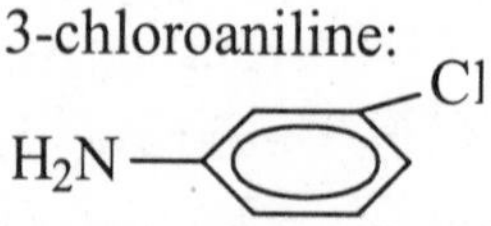

(c) Dicyclopropylamine

(d) 2-chloroethylamine

Alkanes

55. The general formula of an alkane is C_nH_{2n+2}. Thus an alkane with a molecular mass of 72 u has the molecular formula C_5H_{12}.

    (a) If the compound forms four different monochlorination products, there must be four different kinds of carbons in the molecule, each with H atoms attached. This compound is 2-methylbutane, on the left.

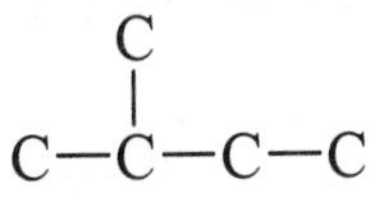

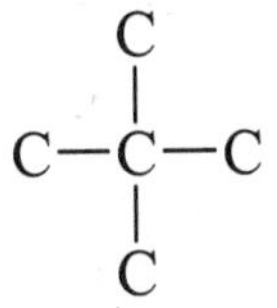

    (b) If the compound forms but one monochlorination product, every carbon atom in the molecule with H atoms attached must be the same. This compound is 2,2-dimethylpropane, the compound above and to the right.

56. **(a)** An alkane with a molecular mass of 44 u must be a propane, since 3 carbons have a molecular mass of 36 u. There is only one propane, n-propane, and it has the condensed formula $CH_3CH_2CH_3$. Its two monochlorination products are $CH_3CH_2CH_2Cl$ and $CH_3CHClCH_3$

    (b) An alkane with a molecular mass of 58 u cannot be a pentane, since five carbons have a mass of 60 u, so it must be a butane. Both normal butane and isobutane have only two monobromination products.

        From *n*-butane,

        $CH_3CH_2CH_2CH_3$ $CH_3CH_2CH_2CH_2Br$ and $CH_3CH_2CHBrCH_3$

        From iso-butane,

        $CH_3CH(CH_3)_2$ $BrCH_2CH(CH_3)_2$ and $CH_3CBr(CH_3)_2$

Alkenes

57. In the case of ethene there are only two carbon atoms between which there can be a double bond. Thus, specifying the compound as 1-ethene is unnecessary. In the case of propene, there can be a double bond only between the central carbon atom and a terminal carbon atom. Thus here also, specifying the compound as 1-propene is unnecessary. The case of butene is different, however, since 1-butene, $CH_2 = CHCH_2CH_3$, is distinct from 2-butene, $CH_3CH = CHCH_3$.

58. In an alkene there is a $C = C$ double bond. On the other hand, in an cyclic alkane there are no double bonds, but rather a chain of carbon atoms joined at the ends into a ring.

59. These reactions either saturate or create double bonds.

(a) $CH_2 = CHCH_3 + H_2 \xrightarrow{Pt, \text{ heat}} CH_3CH_2CH_3$

(b) $CH_3CHOHCH_2CH_3 \xrightarrow{H_2SO_4, \text{ heat}} CH_3CH=CHCH_3 + CH_2=CHCH_2CH_3$

60. (a) $CH_3CC(Cl)=CH_2 + HCl \rightarrow CH_3CCl_2CH_3$

(b) $CH_3C \equiv CH + HCN \rightarrow CH_3C(CN)=CH_2$

(c) $CH_3CH = C(CH_3)_2 + HCl \rightarrow CH_3CH_2CCl(CH_3)_2$

Aromatic Compounds

61. (a) phenylacetylene   (b) meta-dichlorobenzene   (c) 3,5-dihydroxyphenol

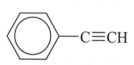

62. (a) *p*-phenylphenol   (b) 2-hydroxy-4-isopropyltoluene

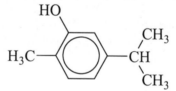

(c) meta-dinitrobenzene

Organic Reactions

63. (a) In an aliphatic substitution reaction, an atom (usually H) of an alkane molecule is replaced by another atom or group of atoms. An example is equation (27.4).

(b) In an aromatic substitution reaction, an atom (usually H) of a phenyl group is replaced by another atom or group of atoms. Examples are in equations (27.12) and (27.13).

(c) In an addition reaction, a small molecule breaks apart into two fragments each of which bond to one of the adjacent atoms in a molecule that are engaged in multiple bonding (i.e. C=C, C=O, C≡C to name a few). An example is equation (27.10).

(d) In an elimination reaction, two groups from within the same molecule join to form a small molecule— such as H_2 or H_2O — and are eliminated from the larger molecule. Often a multiple bond results as well. An example is equation (27.6).

64. **(a)** $HC \equiv CH \xrightarrow{Br_2} CHBr = CHBr \xrightarrow{Br_2} CHBr_2CHBr_2$

(b) $HC \equiv CH \xrightarrow{Pt, H_2} CH_2 = CH_2 \xrightarrow{H_2O, H_2SO_4} CH_3CH_2OH \xrightarrow{Cr_2O_7^{2-}, H^+} CH_3CHO$

65. **(a)** $CH_3CH_2CH_2CH_2CH_2OH \xrightarrow{Cr_2O_7^{2-}, H^+} CH_3CH_2CH_2CH_2COOH$

(b) $CH_3CH_2CH_2COOH + HOCH_2CH_3 \xrightarrow{H^+} CH_3CH_2CH_2 \overset{\overset{\displaystyle O}{\|}}{C} OCH_2CH_3 + H_2O$

(c)

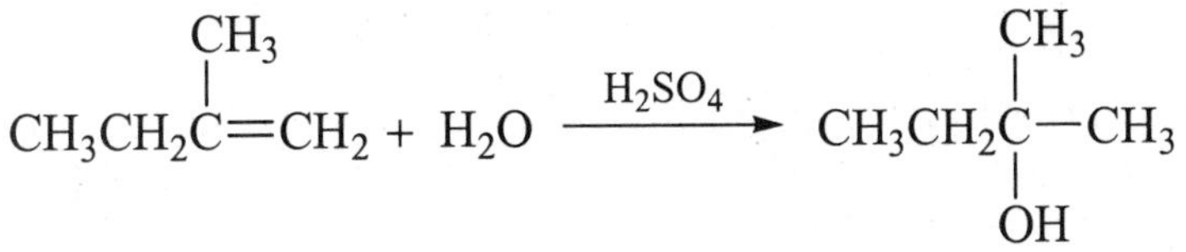

$$CH_3CH_2\underset{\underset{\displaystyle}{|}}{\overset{\overset{\displaystyle CH_3}{|}}{C}}=CH_2 + H_2O \xrightarrow{H_2SO_4} CH_3CH_2\underset{\underset{\displaystyle OH}{|}}{\overset{\overset{\displaystyle CH_3}{|}}{C}}-CH_3$$

66. **(a)** Nitro groups are meta directors. Thus the only product formed is 5-bromo-1, 3-dinitrobenzene, drawn below.

(b) Amine is an ortho, para director. There are two products formed: ortho-bromoaniline and para-bromoaniline, both of which are drawn below.

(a)

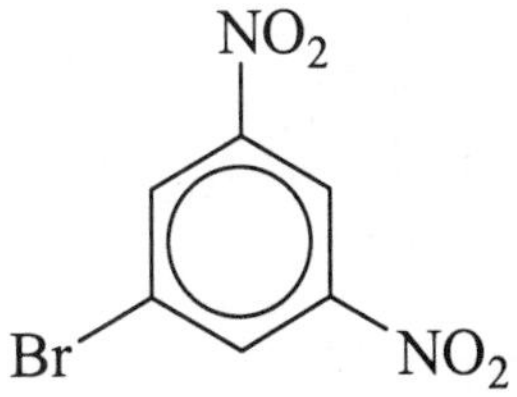

(b)

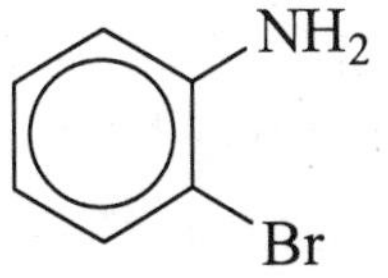

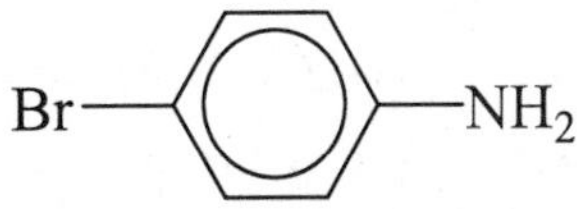

(c) Both an ether and a bromo group are ortho, para directors, but an ether is a stronger ortho, para director. The principal product we expect is 2,4-dibromoanisole, drawn below.

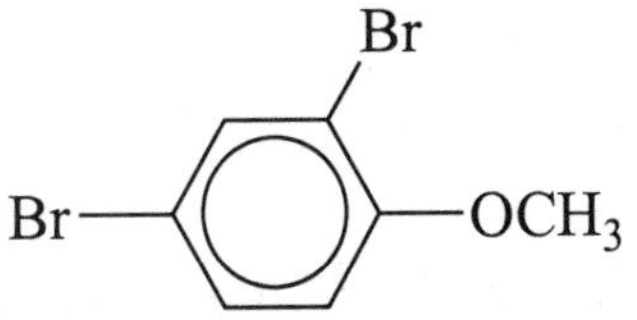

67. **(a)** $H_3C\text{-}CH=CH_2 + H_2 \xrightarrow{Ether} H_3C\text{-}CH_2\text{-}CH_3$

(b) $H_3C\text{-}CH=CH_2 + Cl_2 \xrightarrow{Ether} H_3C\text{-}CHCl\text{-}CH_2Cl$

(c) $H_3C\text{-}CH=CH_2 + HCN \longrightarrow H_3C\text{-}CH(CN)\text{-}CH_3$

(d) $H_3C\text{-}CH=CH_2 + HCl \longrightarrow H_3C\text{-}CHCl\text{-}CH_3$

(e) $H_3C\text{-}CH=CH_2(g) + H_2O(l) \xrightarrow{H_3O^+} H_3C\text{-}CH(OH)\text{-}CH_3(aq)$

68. **(a)** $H_3C\text{-}CH=CH\text{-}CH_3 + H_2 \xrightarrow{\text{Ether}} H_3C\text{-}CH_2\text{-}CH_2\text{-}CH_3$

(b) $H_3C\text{-}CH=CH\text{-}CH_3 + Cl_2 \xrightarrow{\text{Ether}} H_3C\text{-}CHCl\text{-}CHCl\text{-}CH_3$

(c) $H_3C\text{-}CH=CH\text{-}CH_3 + HCN \longrightarrow H_3C\text{-}CH(CN)\text{-}CH_2\text{-}CH_3$

(d) $H_3C\text{-}CH=CH\text{-}CH_3 + HCl \longrightarrow H_3C\text{-}CHCl\text{-}CH_2\text{-}CH_3$

(e) $H_3C\text{-}CH=CH\text{-}CH_3 + H_2O \xrightarrow{H_3O^+} H_3C\text{-}CH(OH)\text{-}CH_2\text{-}CH_3$

69. **(a)** $H_3CCH_2CO_2^-(aq) + HCl(aq) \rightarrow H_3CCH_2CO_2H(aq) + Cl^-(aq)$

(b) $H_3CCH_2CH_2CO_2CH_3(aq) + H_2O(l) \rightarrow H_3CCH_2CH_2CO_2H(aq) + HOCH_3(aq)$

(c) $H_3CCH_2CO_2CH_2CH_3(aq) + NaOH(aq) \rightarrow H_3CCH_2CH_2CO_2Na(aq) + HOCH_2CH_3(aq)$

70. **(a)** $H_3CCH_2CO_2CH_3(sol) + LiAlH_4(s) \xrightarrow{\text{workup } H^+} H_3CCH_2COH(aq) + HOCH_3(aq)$

(b) $H_3CCH_2CO_2CH_3(sol) + NaBH_4(s) \longrightarrow \text{No reaction}$

(c) $H_3CCH_2CO_2H(aq) + H_3O^+(aq) \longrightarrow \text{No reaction}$

71. **(a)** $H_3CCO_2H + NH_3 \longrightarrow H_3CC(O)NH_2 + H_2O$

(b) $H_3CCH_2CONH_2 \xrightarrow{\Delta} H_3CCH_2C\equiv N + H_2O$

(c) $H_3CCH_2C(O)NH_2(aq) + NaOH(aq) \rightarrow H_3CCH_2C(O)ONa(aq) + NH_3(aq)$

72. **(a)** $H_3CCH_2CO_2H + H_3CNH_2 \xrightarrow{\Delta} H_3CCH_2C(O)NHCH_3 + H_2O$

(b) $H_3CCH_2C(O)NHCH_3 \xrightarrow[\text{2. } H_2O(l)]{\text{1. LiAlH}_4} H_3CCH_2CH_2NHCH_3$

(c) $H_3CCH_2C(O)NHCH_3 + HCl(aq) \xrightarrow{H_2O} H_3CCH_2CO_2H(aq) + H_3CNH_3^+Cl^-(aq)$

73. **(a)** $CH_3C(O)O^-(aq) + HCl(aq) \rightarrow CH_3C(O)OH(aq) + Cl^-(aq)$

(b) $H_3CCH_2C(O)OCH_3(aq) + H_3O^+(aq) \rightarrow CH_3OH(aq) + H_3CCH_2C(O)OH(aq)$

(c) $H_3CCH_2COOH(aq) + NaOH(aq) \rightarrow H_3CCH_2COO^-Na^+(aq) + H_2O(l)$

74. **(a)** **(2)** $H_3CCH_2NH_2(aq) + HCl(aq) \rightarrow H_3CCH_2NH_3^+Cl^-(aq)$

(b) **(1)** $H_3CCH_2(O)NH_2(aq) + H_2O(l) \rightarrow H_3CCH_2COOH + NH_3(aq)$

(c) **(3)** $H_3CCH_2CH_2NH_3^+Cl^-(aq) + NaOH(aq) \rightarrow H_3CCH_2CH_2NH_2(aq) + NaCl(aq) + H_2O(l)$

75. $H_3CCH_2CH_2\,CH_2\,NH_2(aq) + H_2O(l) \rightleftharpoons H_3CCH_2CH_2\,CH_2\,NH_3^+(aq) + OH^-(aq)$

$$K_b = \frac{[H_3CCH_2CH_2CH_2\overset{\oplus}{N}H_3][OH^-]}{[H_3CCH_2CH_2CH_2NH_2]}$$

76. $H_3CCH_2CH_2\,CH_2\,NHCH_3(aq) + H_2O(l) \rightleftharpoons H_3CCH_2CH_2\,CH_2\,N^+H_2CH_3(aq) + OH^-(aq)$

$$K_b = \frac{[H_3CCH_2CH_2CH_2\,\overset{\oplus}{N}H_2CH_3][OH^-]}{[H_3CCH_2CH_2CH_2NHCH_3]}$$

77. **(a)** $H_3CCH_2NH_2(aq) + HCl(aq) \rightarrow H_3CCH_2NH_3^+Cl^-(aq)$

  (b) $(CH_3)_3N(aq) + HBr(aq) \rightarrow (CH_3)_3NH^+Br^-(aq)$

  (c) $H_3CCH_2NH_3^+(aq) + H_3O^+(aq) \rightarrow$ No Reaction

  (d) $H_3CCH_2NH_3^+(aq) + OH^-(aq) \rightarrow H_3CCH_2NH_2(aq) + H_2O(l)$

78. **(a)**

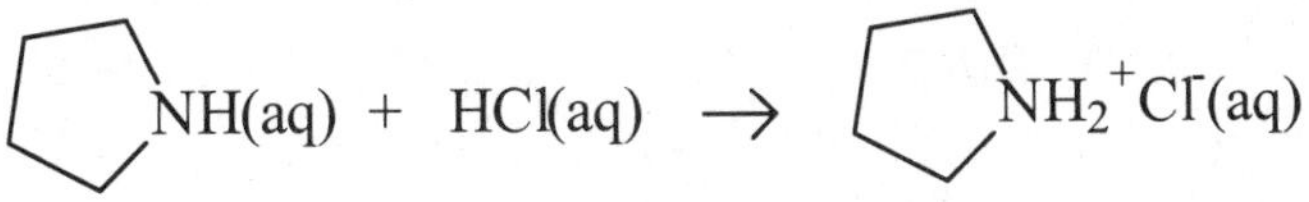

  (b) $(CH_3)_4N^+(aq) + HCl(aq) \rightarrow$ No Reaction

  (c) $H_3CCH_2NH_2(aq) + OH^-(aq) \rightarrow$ No Reaction

  (d) $(CH_3)_3NH^+(aq) + OH^-(aq) \rightarrow (CH_3)_3N(aq) + H_2O(l)$

79. **(a)** (1)

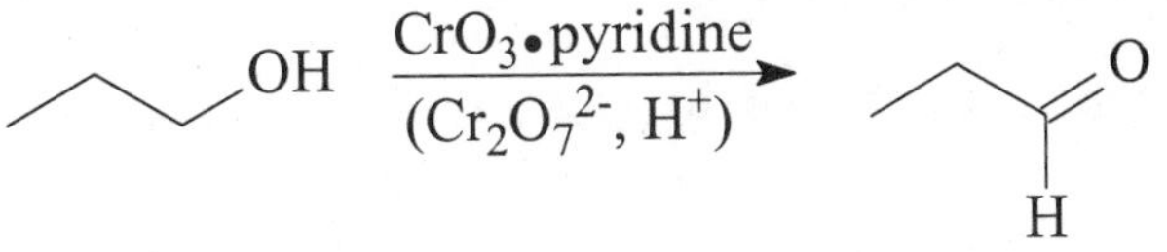

  (b) (2)

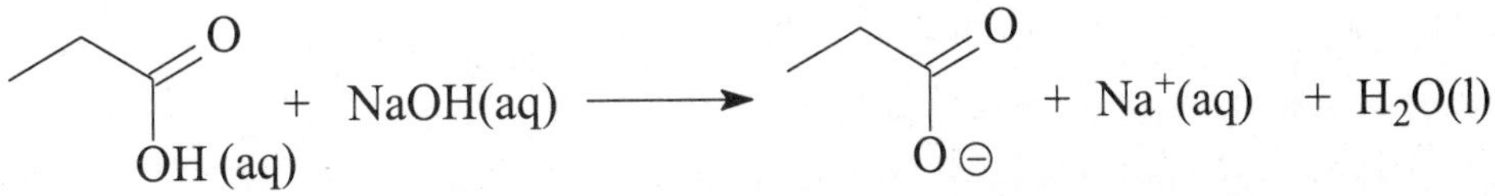

  (c) (2)

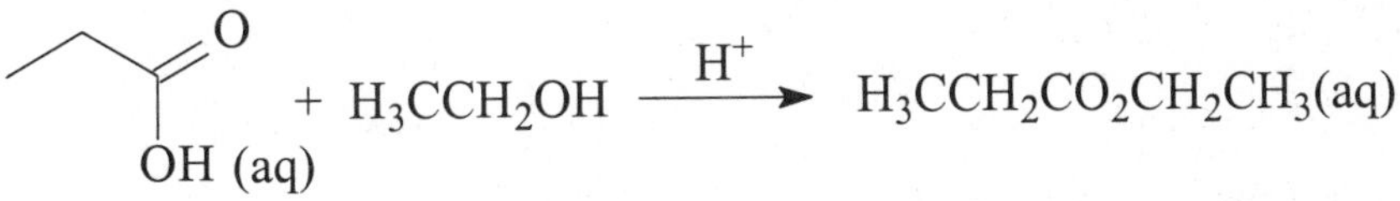

  (d) (1)

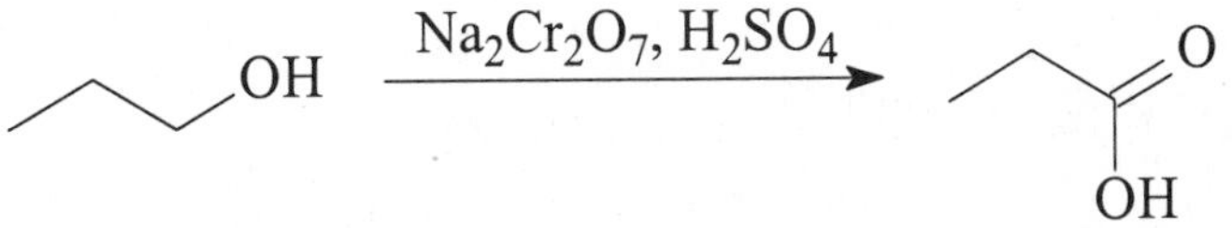

  (d) (3)

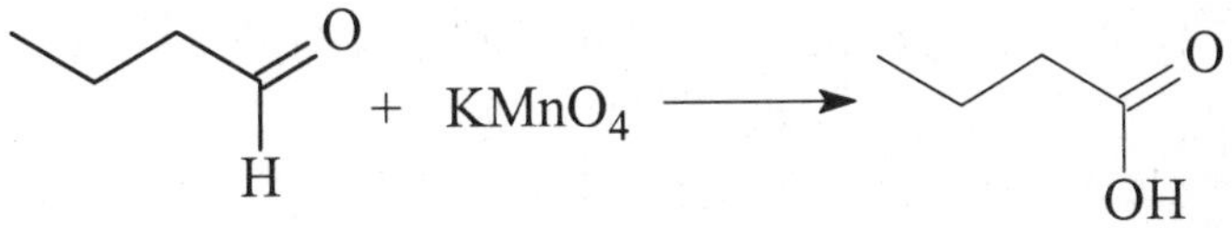

  (e) None of the compounds (1) - (3) can be dehydrated to an alcohol

80. **(a)** (1) 
$$\text{~~NH}_2\text{(aq)} + \text{HCl(aq)} \longrightarrow \text{~~NH}_3{}^+\text{Cl}^-\text{(aq)}$$

(b) (2) 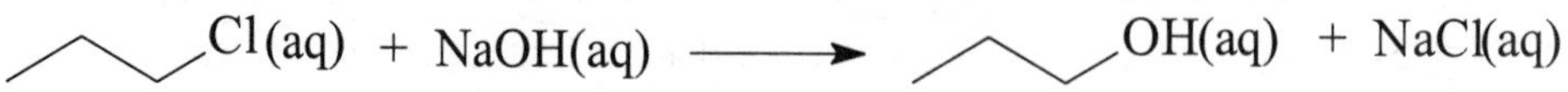
$$\text{~~Cl(aq)} + \text{NaOH(aq)} \longrightarrow \text{~~OH(aq)} + \text{NaCl(aq)}$$

(c) (1) 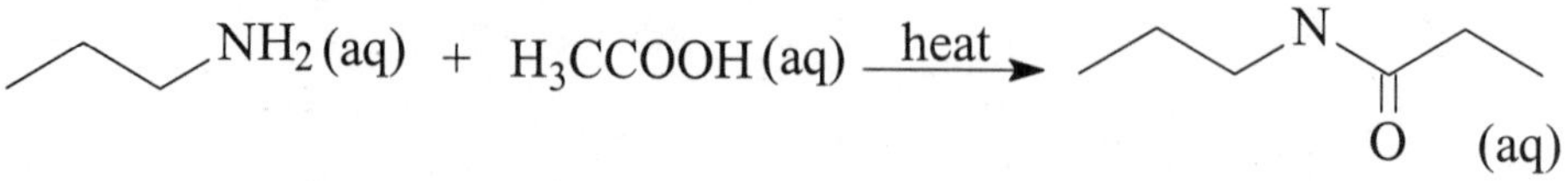
$$\text{~~NH}_2\text{(aq)} + \text{H}_3\text{CCOOH(aq)} \xrightarrow{\text{heat}} \text{~~N~~} \underset{\text{O}}{\|} \text{~~} \quad \text{(aq)}$$

(d) (2) 
$$\text{~~Cl (aq)} + \text{NH}_3\text{(aq)} \longrightarrow \text{~~NH}_3{}^+\text{Cl}^-\text{(aq)}$$

(e) (2) 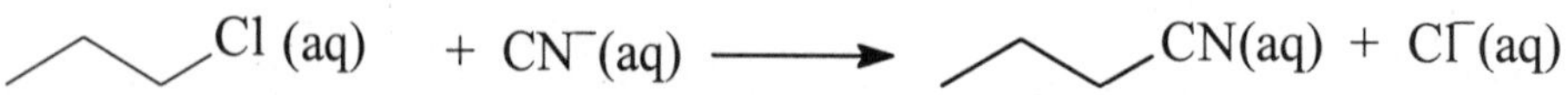
$$\text{~~Cl (aq)} + \text{CN}^-\text{(aq)} \longrightarrow \text{~~CN(aq)} + \text{Cl}^-\text{(aq)}$$

81. **(a)** Identical molecules (Both are achiral)
(b) Identical molecules (Both are the *R* enantiomers)
(c) Structural isomers (one is optically active, the other is not)
(d) Enantiomers (*R* and S optical isomers)
(e) Identical molecules (Both are the *R* enantiomers)
(f) Identical molecules (Both are the *R* enantiomers)

82. **(a)** Enantiomers (*R* and S optical isomers)
(b) Identical molecules (Both are the *R* enantiomers)
(c) Enantiomers (*R* and S optical isomers)
(d) Identical molecules (Both are the *R* enantiomers)
(e) Enantiomers (*R* and S optical isomers)
(f) Structural Isomers(both are optically active)

83. **(a)** (*S*)-3-bromo-2-methylpentane
(b) (*S*)-1,2-dibromopentane
(c) (*R*)-3-bromomethyl-5-chloro-*n*-pentan-3-ol
(d) (*S*)-1-bromo-*n*-propan-2-ol

84. **(a)** (*R*)-pentan-2-ol
(b) (*S*)-3-methyl-2-butanamine
(c) (*S*)-1-bromo-2-chlorobutane
(d) (*S*)-2-hydroxypentan-1-ol

85. **(a)** H₃CH₂C, CH₃

C=C

H, H

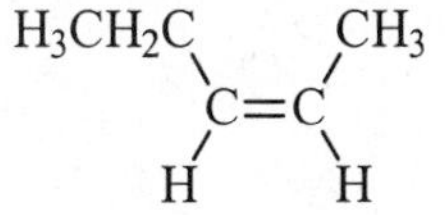

(Z)-2-pentene

(b) Cl, CH₃

C=C

H, CH₂CH₃

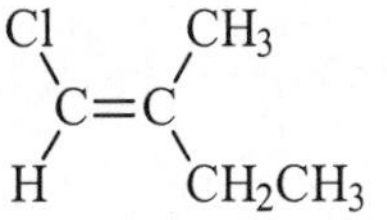

(E)-1-chloro-2-methyl-1-butene

(c) H₃C, CH₂Cl

C=C

H₃CH₂C, CH₂CH₂CH(CH₃)₂

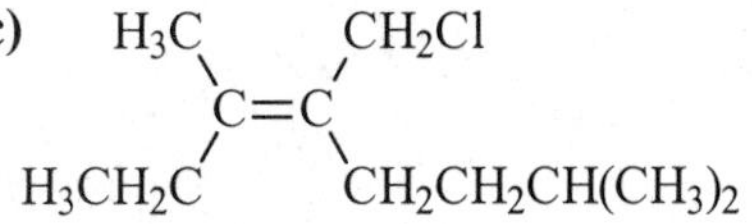

(E)-3-7-dimethyl-4-chloromethyl-3-octene

(d) O, ‖, HC, CH₂Br

C=C

H₃C, CH₂CH₂Br

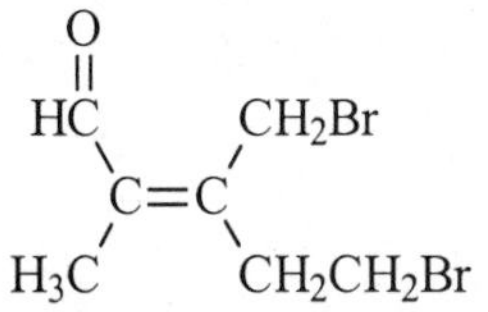

(Z)-5-bromo-3-bromomethyl-2-methylpent-2-enal

86. **(a)** H₃C, CH₂Cl

C=C

ClH₂C, CH₃

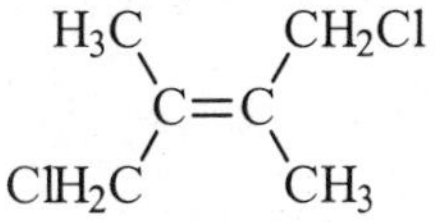

(E)-1,4-dichlor-2,3-dimethyl-2-butene

(b) H₃C, CH₂Cl

C=C

H₃CH₂C, CH₂OH

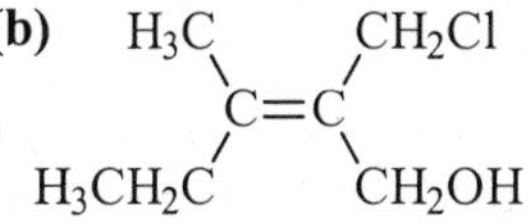

(E)-2-chloromethyl-3-methyl-pent-2-en-1-ol

(c) O, ‖, H₃CC, CH₂OH

C=C

HC≡C, CH₂CHClCH₃

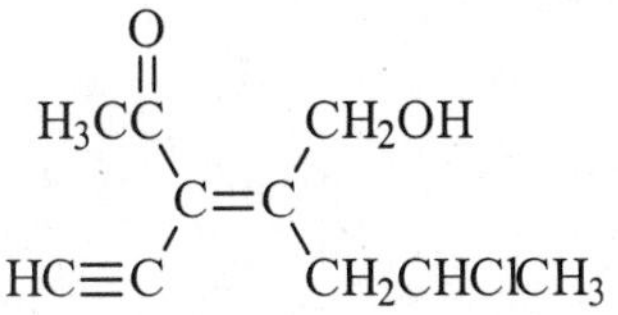

(Z)-3-ethynyl-6-chloro-4-hydroxymethyl-hept-3-en-2-one

(d) H₃CCH₂, CH₃

C=C

HOH₂C, CH₃

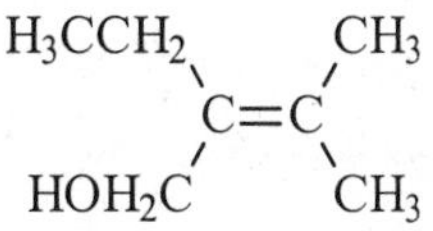

(3-hydroxymethyl)-2-methyl-2-pentene

87. **(a)** BrH₂C, Br

C=C

H, CH₂CH₂Br

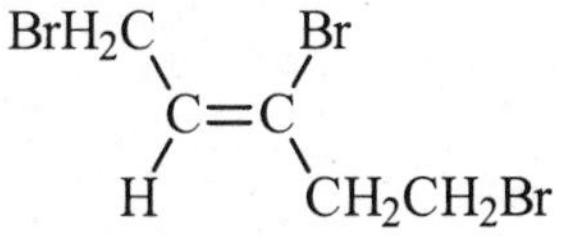

(Z)-1,3,5-tribromo-2-pentene

(b) BrH₂C, CH₂CH₂CH₃

C=C

Br, CH₃

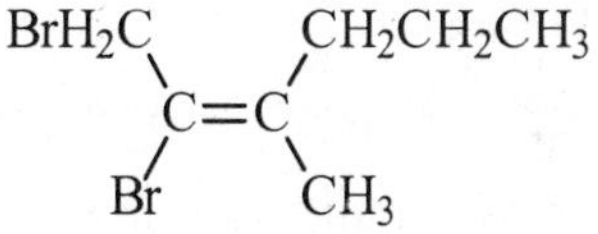

(E)-1,2-dibromo-3-methyl-2-hexene

(c) Cl

H₃CH₂CH₂C—C⋯H

Br

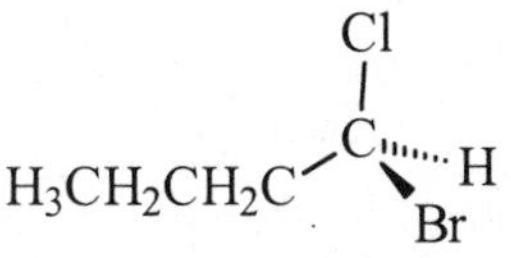

(S)-1-bromo-1-chlorobutane

(d) Br

H₃CCH₂H₂C—C⋯H

CH₂CH₂Br

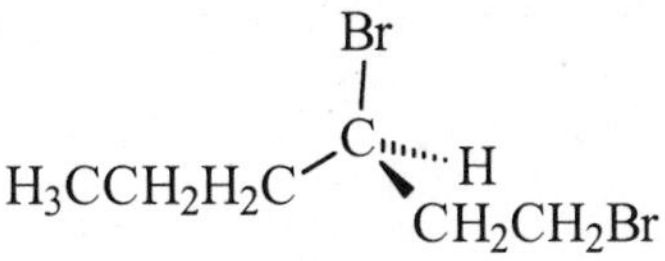

(R)-1,3-dibromohexane

(e) OH

ClH₂C—C⋯H

CH₃

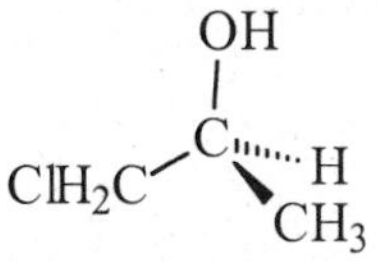

(S)-1-chloro-2-propanol

88. **(a)** Br

H₃C—C⋯H

Cl

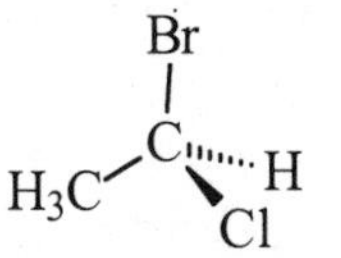

(R)-1-bromo-1-chloroethane

(b) H₃C, CH₂CH₃

C=C

Br, H

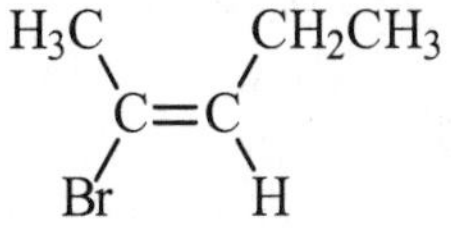

(E)-2-bromo-2-pentene

(c) ClH₂CH₂C, CH₂CH₂CH₃

C=C

H₃CH₂C, H

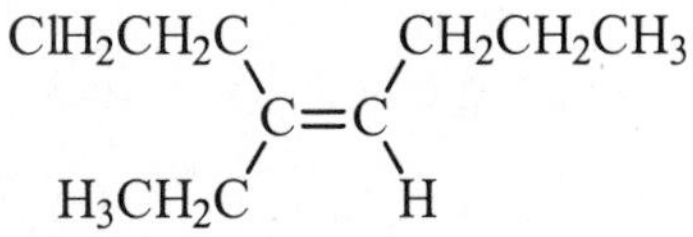

(Z)-1-chloro-3-ethyl-3-heptene

(d) CO₂H

OH—C⋯H

CH₃

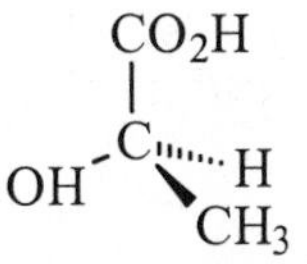

(R)-2-hydroxypropanoic acid

(e) CH₃

H₂N—C⋯H

COO⁻

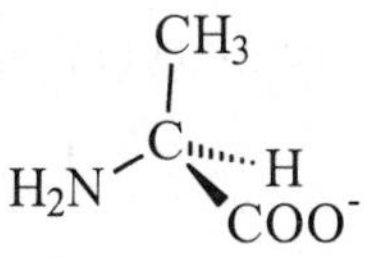

(S)-2-aminopropanoate anion

89. **(a)** Reaction rate = $k_{obs}[OH^-][BrCH_2CH_2CH_2CH_3]$ (S_N2 mechanism)

(b) See diagram below:

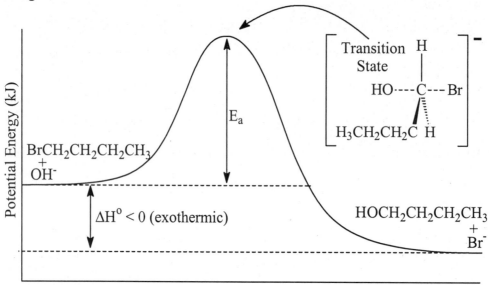

(c) If the concentration of n-butyl bromide were doubled, the reaction rate would increase two fold

(d) If the concentration of hydroxide ion were halved, the reaction rate would decrease by a factor of two.

90. **(a)** Reaction rate = $k_{obs}[BrC(CH_3)_2CH_2CH_2CH_3]$ (S_N1 mechanism)

(b) See diagram below:

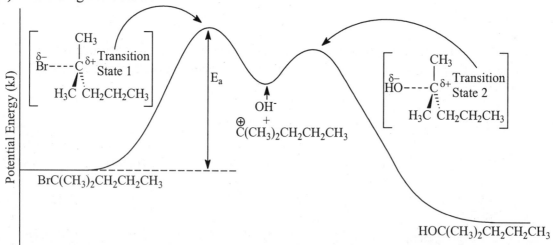

(c) If the concentration of 1-bromo-1-methylpentane were doubled, the reaction rate would increase by a factor of two.

(d) Adding more ethanol would cause the concentration of 1-bromo-1-methylpentane to fall, which would lead to a decrease in the rate of reaction.

91. **(a)** $BrCH_2CH_2CH_2CH_3(aq) + NaOH(aq) \rightarrow NaBr(aq) + HOCH_2CH_2CH_2CH_3(aq)$

  (b) $BrCH_2CH_2CH_2CH_3(aq) + NaC\equiv N(aq) \rightarrow NaBr(aq) + N\equiv C-CH_2CH_2CH_2CH_3(aq)$

  (c) $BrCH_2CH_2CH_2CH_3(aq) + NH_3(aq) \rightarrow Br^-(aq) + {}^+H_3NCH_2CH_2CH_2CH_3(aq)$

  (d) $BrCH_2CH_2CH_2CH_3 + Na^{+-}O\text{-}CH_2CH_3 \rightarrow NaBr + H_3CH_2C\text{-}O\text{-}CH_2CH_2CH_2CH_3$

92. **(a)** $BrCH_2CH_2CH_2CH_2CH_3 + NaN_3 \rightarrow NaBr + N_3CH_2CH_2CH_2CH_2CH_3$

  (b) $Br(CH_2)_4CH_3(aq) + N(CH_3)_3(aq) \rightarrow Br^-(aq) + (CH_3)_3N^+(CH_2)_4CH_3(aq)$

  (c) $BrCH_2CH_2CH_2CH_2CH_3 + Na^{+-}C\equiv CCH_2CH_3 \rightarrow NaBr + N\equiv C-CH_2CH_2CH_2CH_2CH_3$

  (d) $BrCH_2CH_2CH_2CH_2CH_3 + Na^{+-}SCH_2CH_3 \rightarrow NaBr + CH_3CH_2SCH_2CH_2CH_2CH_2CH_3$

93. **(a)**

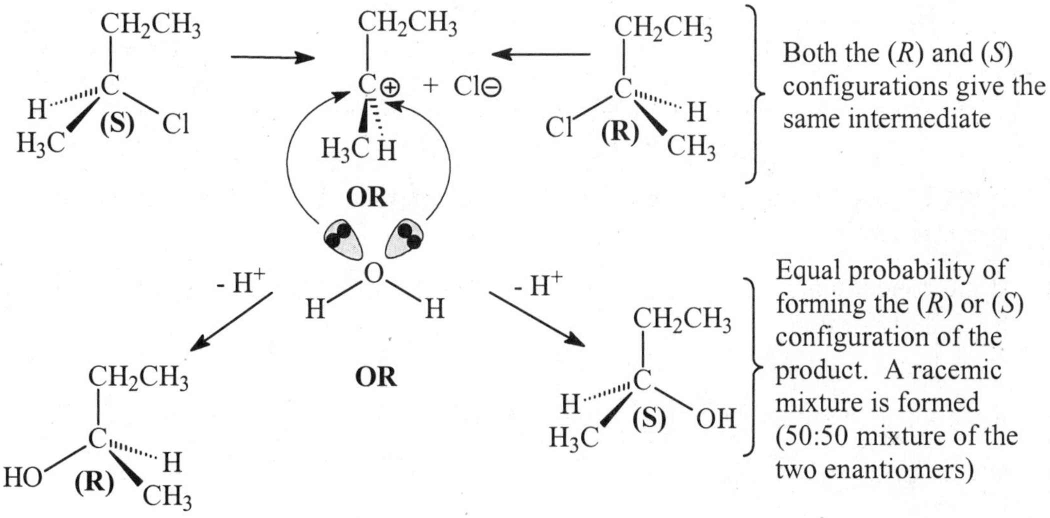

  (b)   Since the resulting product solution is optically inactive, we can conclude that a racemic mixture has been formed and hence, the reaction has occurred via an S_N1 mechanism.

94. **(a)**

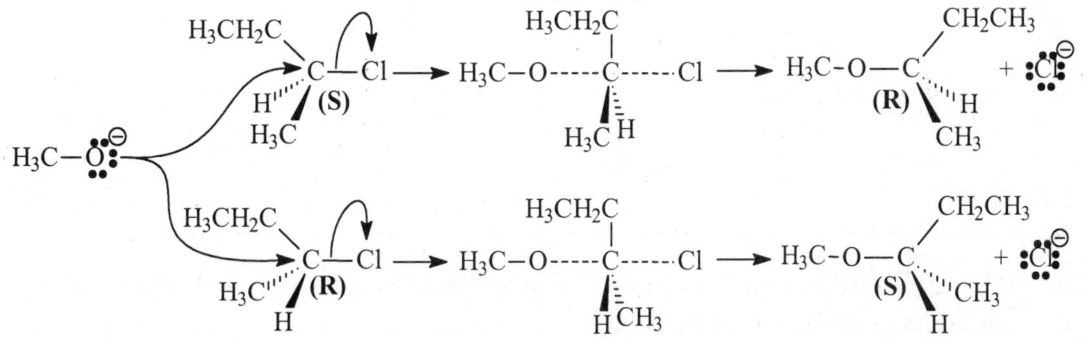

Optically pure starting alkyl chloride (R or S) $\longrightarrow$ Optically pure ether
(inversion of configuration)

(b) Since the product is optically active (i.e., since just one enantiomer is produced), we can conclude that the reaction must have occurred in a concerted fashion via an S_N2 mechanism.

95. **(a)**

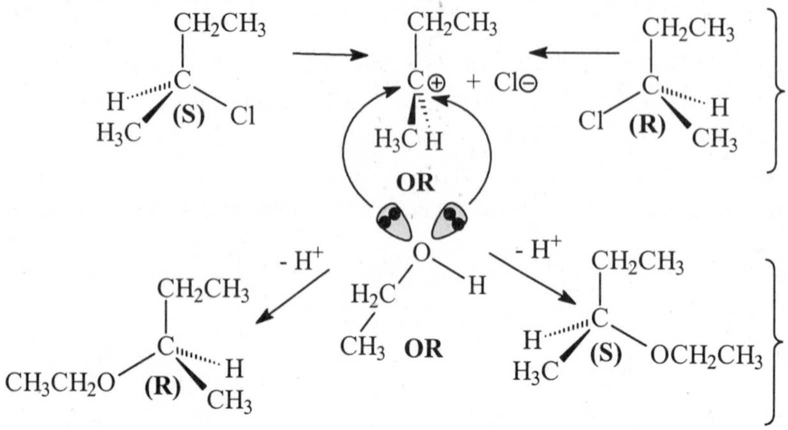

Both the (R) and (S) configurations give the same intermediate

Equal probability of forming the (R) or (S) configuration of the product. A racemic mixture is formed (50:50 mixture of the two enantiomers)

(b) Since the resulting product solution is optically inactive. We can conclude that a racemic mixture has been formed and hence, the reaction has occurred via an S_N1 mechanism.

96. **(a)**

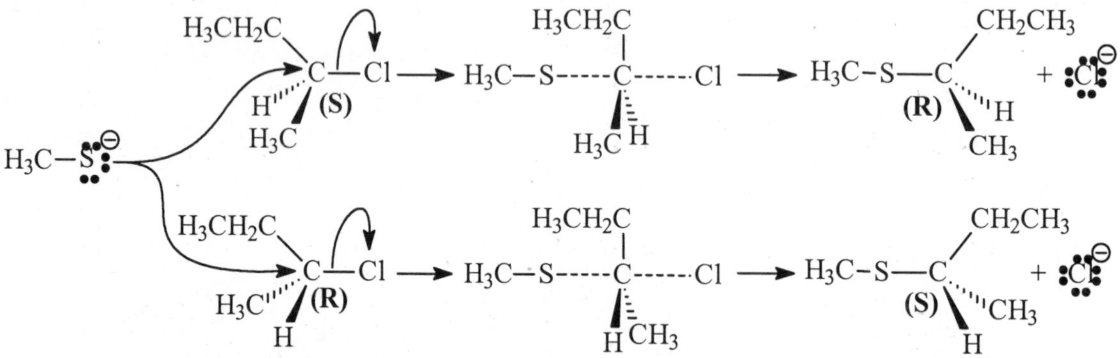

Optically pure starting alkyl chloride (*R* or *S*) $\longrightarrow$ Optically pure thioether (inversion of configuration)

(b) Since the product is optically active (i.e., since just one enantiomer is produced), we can conclude that the reaction must have occurred in a concerted fashion via an S_N2 mechanism.

Polymerization Reactions

97. In a simple molecular substance like benzene, all molecules are identical (C_6H_6). No matter how many of these molecules are present in one sample, any other sample with the same number of molecules has the same mass. The mass in grams of one mole of molecules— the molar mass—is a unique quantity. In a polymer the situation is quite widely variable. Thus individual polymer molecules differ very much in mass. Thus, a mass of a mole of their molecules also is quite variable. However, when we take a sample for analysis, we obtain many molecules of each chain length or molecular mass. The resulting determination of molar mass obtains the average mass of all of these different sized polymer molecules.

98. An ester linkage is formed by the condensation of a carboxylic acid and an alcohol. Accompanying this condensation is the elimination of a water molecule between the ester and the carboxylic acid. Dacron is formed by the condensation of a dicarboxylic acid with a diol. Thus, Dacron contains ester linkages. Because there are a large number of these ester linkages in Dacron— joining many subunits together— it is appropriate to call the polymer a polyester.

   To determine the percent of oxygen in Dacron, we refer to the basic unit of the polymer, which is shown on page 1106. This unit has the formula $C_{10}H_8O_4$.

$$\% \; O \frac{(4 \times 16.00) \; g \; O}{192.2 \; g \; polymer} \times 100\% = 33.30\% \; O$$

99. The polymerization of 1,6-hexanediamine with sebacyl chloride proceeds in the following manner. The italicized H and Cl atoms are removed in this condensation reaction.

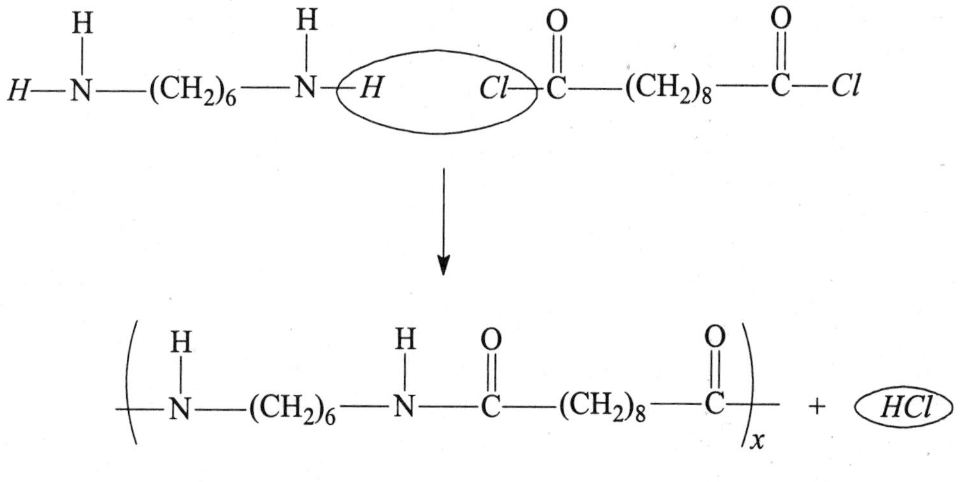

100. In order to form long-chain molecules, every monomer must have at least two functional groups, one on each end of the molecule. Ethyl alcohol has only one functional group, a hydroxyl group (—OH). It cannot participate in a polymerization reaction with dimethyl terephthalate. But glycerol, with three hydroxyl functional groups, can participate in this polymerization reaction.

FEATURE PROBLEMS

129. **(a)** $HC(O)H + CH_3CH_2CH_2CH_2\text{-Mg-Br} \rightarrow CH_3CH_2CH_2CH_2CH_2OH$

(non-aqueous reaction of Grignard, followed by acid/water work-up of final product)

(b) $CH_3MgBr + CH_3CH_2CH_2CH_2C(O)H \rightarrow CH_3CH_2CH_2CH_2CH(OH)CH_3$ or

$CH_3C(O)H + CH_3CH_2CH_2CH_2\text{-Mg-Br} \rightarrow CH_3CH_2CH_2CH_2CH(OH)CH_3$ (non-

aqueous reaction of Grignard, followed by acid/water work-up of final product)

(c) $CH_3\text{-Mg-Br} + CH_3CH_2CH_2CH_2C(O)CH_3 \rightarrow CH_3CH_2CH_2CH_2C(OH)(CH_3)_2$ or

$CH_3C(O)CH_3 + CH_3CH_2CH_2CH_2\text{-Mg-Br} \rightarrow CH_3CH_2CH_2CH_2C(OH)(CH_3)_2$

(non-aqueous reaction of Grignard, followed by acid/water work-up of final product)

(d) $C_6H_5\text{-Mg-Cl} + CH_3C(O)CH_3 \rightarrow C_6H_5C(OH)(CH_3)_2$

Alternatively, this product can be prepared as follows:

$C_6H_5C(O)CH_3 + CH_3\text{-Mg-Cl} \rightarrow C_6H_5C(OH)(CH_3)_2$

(non-aqueous reaction of Grignard, followed by acid/water work-up of final product)

(e) $CH_3CH_2\text{-Mg-Br} + HC\equiv CCH_2CH_2CH_2CH_3 \rightarrow Br\text{-Mg-C}\equiv CCH_2CH_2CH_2CH_3 + CH_3CH_3$

(f) $CH_3CH_2\text{-Mg-Br} + HC\equiv CCH_2CH_2CH_2CH_3 \rightarrow Br\text{-Mg-C}\equiv CCH_2CH_2CH_2CH_3 + CH_3CH_3$

then, $Br\text{-Mg-C}\equiv CCH_2CH_2CH_2CH_3 + HC(O)H \rightarrow HOH_2CC\equiv CCH_2CH_2CH_2CH_3$

(non-aqueous reaction of Grignard, followed by acid/water work-up of final product)

CHAPTER 28
CHEMISTRY OF THE LIVING STATE
PRACTICE EXAMPLES

1A

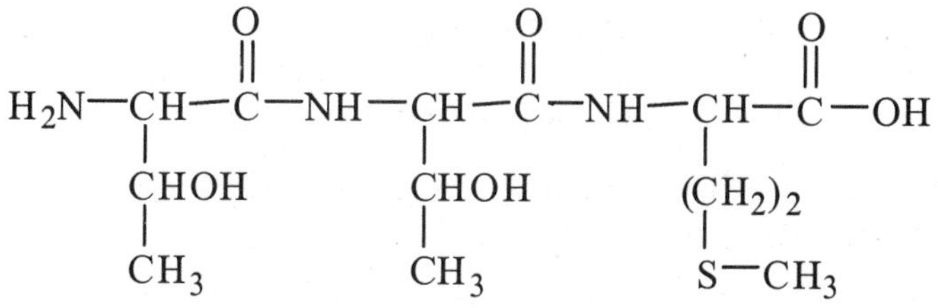

$$H_2N-CH-\overset{\overset{\displaystyle O}{\|}}{C}-NH-CH-\overset{\overset{\displaystyle O}{\|}}{C}-NH-CH-\overset{\overset{\displaystyle O}{\|}}{C}-OH$$

with substituents: CHOH / CH$_3$; CHOH / CH$_3$; (CH$_2$)$_2$ / S–CH$_3$

The amino acids are threonine, threonine, and methionine. This tripeptide is dithreonylmethionine.

1B The amino acids are: serine, glycine, and valine. The N terminus is first.

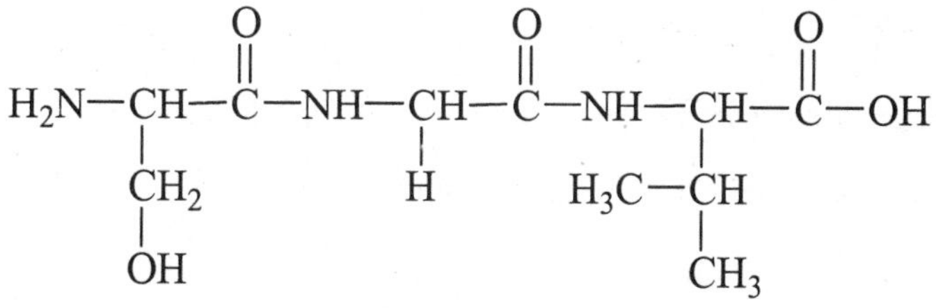

$$H_2N-CH-\overset{\overset{\displaystyle O}{\|}}{C}-NH-CH-\overset{\overset{\displaystyle O}{\|}}{C}-NH-CH-\overset{\overset{\displaystyle O}{\|}}{C}-OH$$

with substituents: CH$_2$ / OH ; H ; H$_3$C–CH / CH$_3$

2A Because it is a pentapeptide and five amino acids have been identified, no amino acid is repeated. The sequences fall into place, as follows.

Gly	Cys				second fragment
	Cys	Val	Phe		third fragment
		Val	Phe		first fragment
			Phe	Tyr	fourth fragment
pentapeptide sequence Gly	Cys	Val	Phe	Tyr	

2B Because it is a hexapeptide and there are five distinct amino acids, one amino acid must appear twice. The fragmentation pattern indicates that the doubled amino acid is glycine. The sequences fall into place if we begin with the N-terminal end.

Ser	Gly	Gly			third fragment
	Gly	Gly	Ala		second fragment
		Ala	Val	Trp	fourth fragment
			Val	Trp	first fragment
hexapeptide sequence Ser	Gly	Gly	Ala	Val	Trp

REVIEW QUESTIONS

1. (a) (+) is another way of designating a dextrorotatory compound.

    (b) L indicates that, in the Fisher projection of the compound, the — OH group on the penultimate carbon atom is to the left and the — H group is to the right.

    (c) A sugar is a carbohydrate that is either a monosaccharide or an oligosaccharide.

    (d) An α-amino acid is a carboxylic acid that has an amine group on the carbon next to the — COOH group, that is, on the α carbon atom. Glycine (H_2NCH_2COOH) is the simplest α-amino acid.

    (e) The pH at which the zwitterion form of an amino acid predominates in solution is known as the isoelectric point. The isoelectric point of glycine is $pI = 6.03$.

2. (a) Saponification is the hydrolysis of a triglyceride in alkaline solution to produce glycerol and soaps: salts of fatty acids.

    (b) A chiral carbon atom is one that exhibits optical isomerism, one to which four different groups are attached.

    (c) A racemic mixture is one composed of equal amounts of an optically active compound and its enantiomer. Since these two compounds rotate polarized light by the same amount but in opposite directions, such a mixture does not rotate the plane of polarized light.

    (d) The denaturation of a protein is the process in which the protein is treated with somewhat harsh conditions and loses at least some of its secondary or tertiary structure, temporarily or permanently, leading to a loss of biological activity.

3. (a) A fat is a triglyceride in which the fatty acid chains are predominantly saturated hydrocarbon chains. Fats are solids at room temperature. An oil is a triglyceride in which the fatty acid chains are to predominantly unsaturated. Oils are liquids at room temperature.

    (b) Enantiomers are optically active isomers of a compound that are mirror images of each other. Diastereomers are optically active isomers that are not mirror images.

    (c) The primary structure of a protein refers to the sequence of amino acids. The secondary structure describes the shape of that polypeptide chain.

    (d) DNA is deoxyribonucleic acid, RNA is ribonucleic acid. They differ by an O atom. DNA is found in the cell nucleus, RNA in the cytoplasm.

    (e) ADP is adenosine diphosphate, ATP is adenosine triphosphate. They differ by a phosphate group.

4.   **(a)** $C_{15}H_{31}COOH$ is palmitic acid. $C_{17}H_{29}COOH$ is linolenic acid or eleosteric acid. $C_{11}H_{23}COOH$ is lauric acid. Thus, the given compound is glyceryl palmitolinolenolaurate or glyceryl palmitoeleosterolaurate.

   (b) $C_{17}H_{33}COOH$ is oleic acid. Thus, the compound is glyceryl trioleate or triolein.

   (c) $C_{13}H_{27}COOH$ is myristic acid. Thus, the compound is sodium myristate.

5.   **(a)**   glyceryl palmitolauroeleosterate

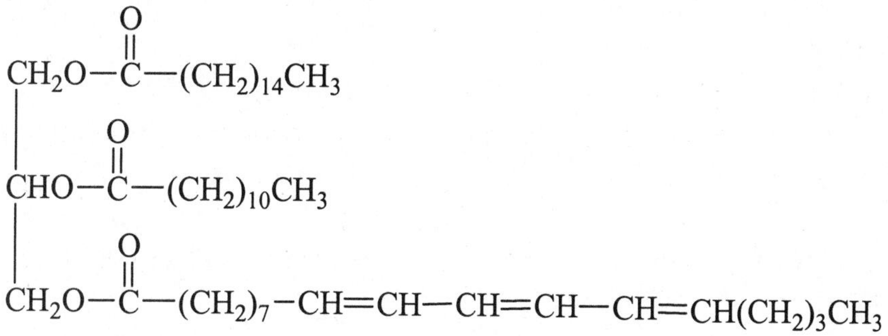

$$
\begin{array}{l}
\overset{\displaystyle O}{\overset{\|}{CH_2O-C-(CH_2)_{14}CH_3}} \\
\quad\quad\overset{\displaystyle O}{\overset{\|}{CHO-C-(CH_2)_{10}CH_3}} \\
\quad\quad\overset{\displaystyle O}{\overset{\|}{CH_2O-C-(CH_2)_7-CH=CH-CH=CH-CH=CH(CH_2)_3CH_3}}
\end{array}
$$

   (b)   tripalmitin

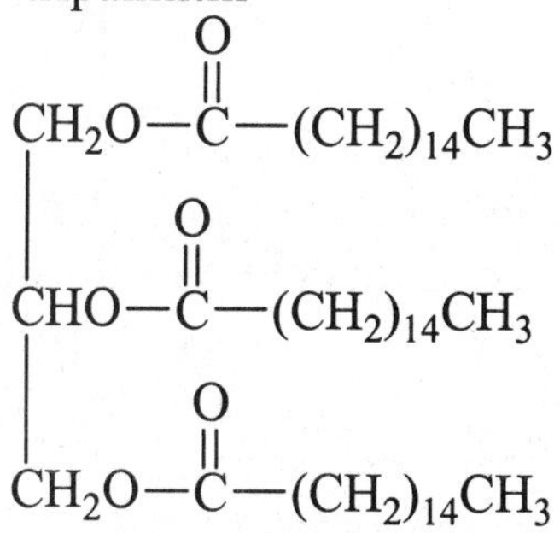

$$
\begin{array}{l}
\overset{\displaystyle O}{\overset{\|}{CH_2O-C-(CH_2)_{14}CH_3}} \\
\quad\quad\overset{\displaystyle O}{\overset{\|}{CHO-C-(CH_2)_{14}CH_3}} \\
\quad\quad\overset{\displaystyle O}{\overset{\|}{CH_2O-C-(CH_2)_{14}CH_3}}
\end{array}
$$

   (c)   potassium myristate

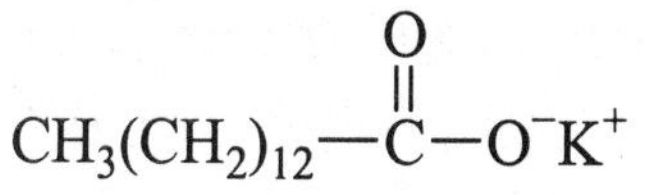

$$
CH_3(CH_2)_{12}-\overset{\displaystyle O}{\overset{\|}{C}}-O^-K^+
$$

   (d)   butyl oleate

$$
CH_3(CH_2)_3-O-\overset{\displaystyle O}{\overset{\|}{C}}-(CH_2)_7-CH=CH-(CH_2)_7-CH_3
$$

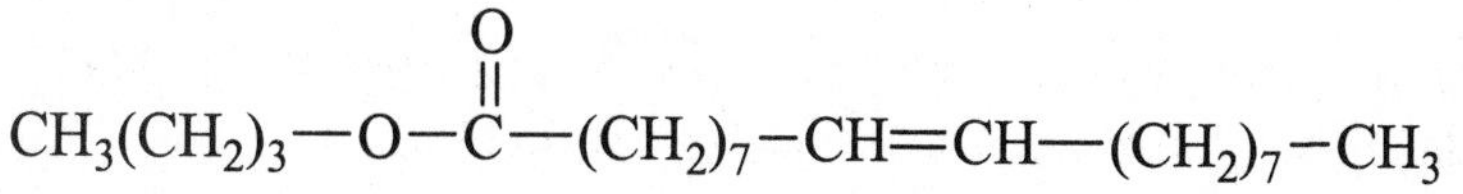

6.   A DL mixture contains equal amounts of both enantiomers. This is also known as a racemic mixture. The optical activities of the two enantiomers cancel each other out, and the mixture rotates the plane of polarized light neither to the right nor to the left. Thus, statement (4) is the most appropriate.

7.   **(a)**   D-(−)-arabinose is the optical isomer of L-(+)-arabinose. Its structure is shown below.

(b)   A diastereomer of L-(+)-arabinose is a molecule that is its optical isomer, but not its mirror image. There are several such diastereomers, some of which follows, at right.

(a)   **(b)**

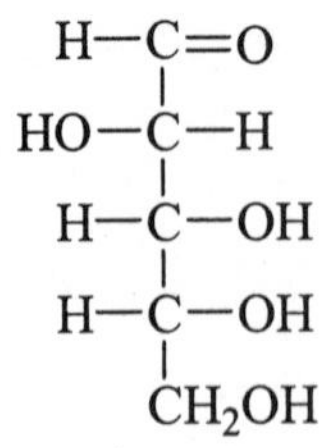

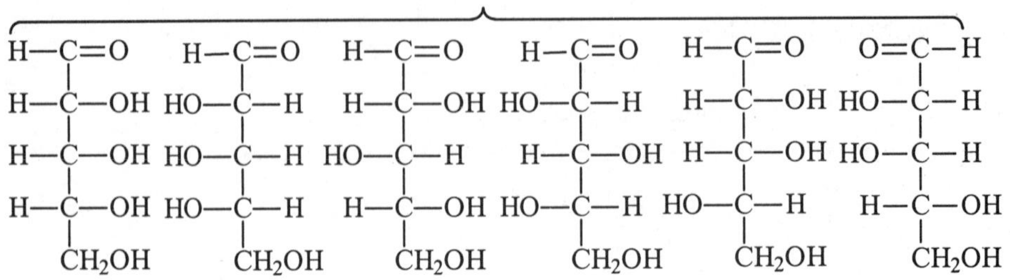

```
  H—C=O        H—C=O    H—C=O    H—C=O    H—C=O    H—C=O    O=C—H
 HO—C—H        H—C—OH  HO—C—H    H—C—OH  HO—C—H    H—C—OH  HO—C—H
  H—C—OH        H—C—OH  HO—C—H  HO—C—H    H—C—OH    H—C—OH  HO—C—H
  H—C—OH        H—C—OH  HO—C—H    H—C—OH  HO—C—H  HO—C—H    H—C—OH
   CH₂OH         CH₂OH   CH₂OH    CH₂OH    CH₂OH    CH₂OH    CH₂OH
```

8.

(a) β has OH equitorial at C₁

(b)

```
   H    O
    \\ //
     C
HO—C—H
 H—C—OH
 H—C—OH
   CH₂OH
```

(c)

```
   H    O
    \\ //
     C
 H—C—OH
HO—C—H
 H—C—OH
 H—C—OH
   CH₂OH
```

All three molecules are optically active

cyclized/open-chain forms

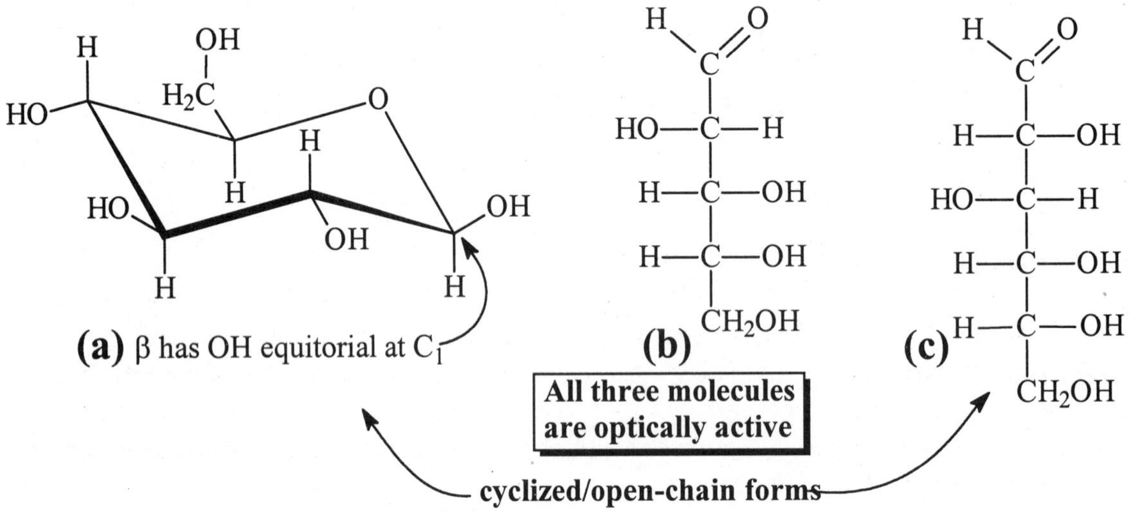

9.   The pI of phenylalanine is 5.74. Thus, phenylalanine is in the form of a cation in 1.0 M HCl (pH = 0.0), in the form of an anion in 1.0 M NaOH (pH = 14.0), and in the form of a zwitterion at pH = 5.7 . These three structures follow.

(a)

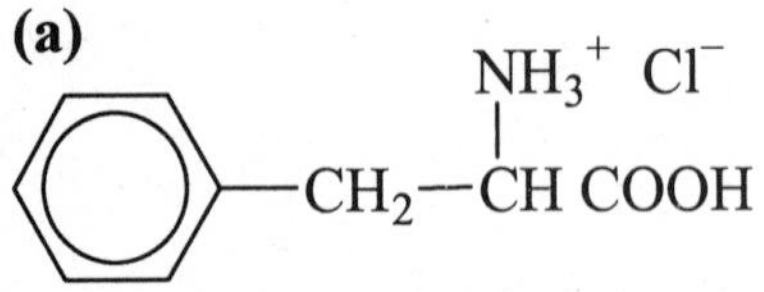

NH_3^+ Cl^-
—CH₂—CH COOH

(b)

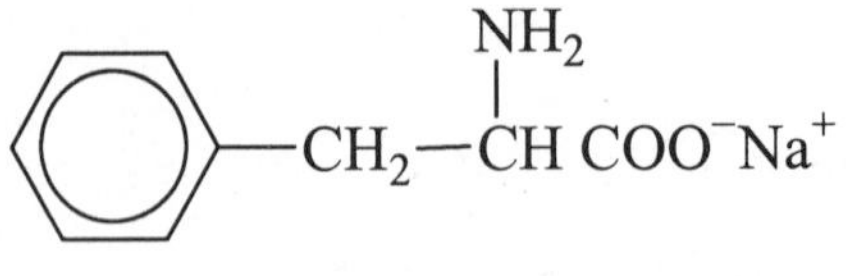

NH_2
—CH₂—CH COO⁻ Na⁺

(c)

NH_3^+
—CH₂—CH COO⁻

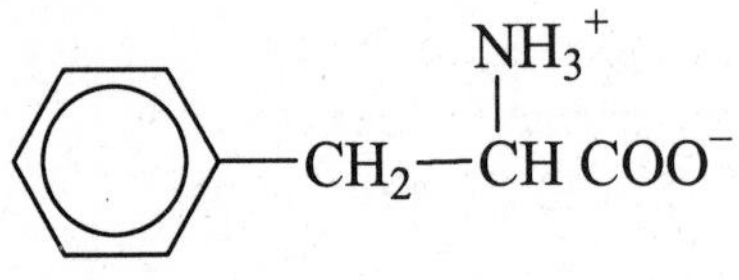

10. **(a)** alanylcysteine

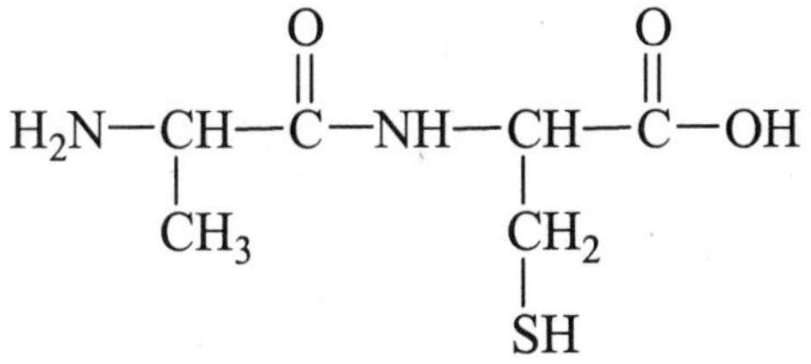

$$H_2N-\underset{\underset{CH_3}{|}}{CH}-\overset{\overset{O}{\parallel}}{C}-NH-\underset{\underset{\underset{SH}{|}}{CH_2}}{CH}-\overset{\overset{O}{\parallel}}{C}-OH$$

(b) threonylvalylglycine

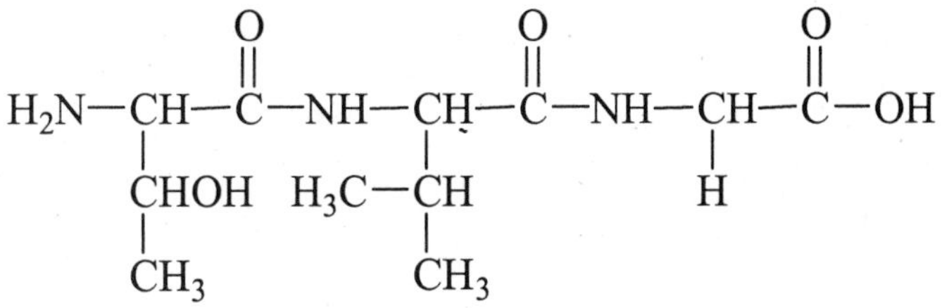

$$H_2N-\underset{\underset{\underset{CH_3}{|}}{CHOH}}{CH}-\overset{\overset{O}{\parallel}}{C}-NH-\underset{\underset{\underset{CH_3}{|}}{H_3C-CH}}{CH}-\overset{\overset{O}{\parallel}}{C}-NH-\underset{\underset{H}{|}}{CH}-\overset{\overset{O}{\parallel}}{C}-OH$$

11. **(a)**

$$H_2N-\underset{\underset{\underset{S-CH_3}{|}}{(CH_2)_2}}{CH}-\overset{\overset{O}{\parallel}}{C}-NH-\underset{\underset{\underset{CH_3}{|}}{H_3C-CH}}{CH}-\overset{\overset{O}{\parallel}}{C}-NH-\underset{\underset{\underset{CH_3}{|}}{CHOH}}{CH}-\overset{\overset{O}{\parallel}}{C}-NH-\underset{\underset{\underset{SH}{|}}{CH_2}}{CH}-\overset{\overset{O}{\parallel}}{C}-OH$$

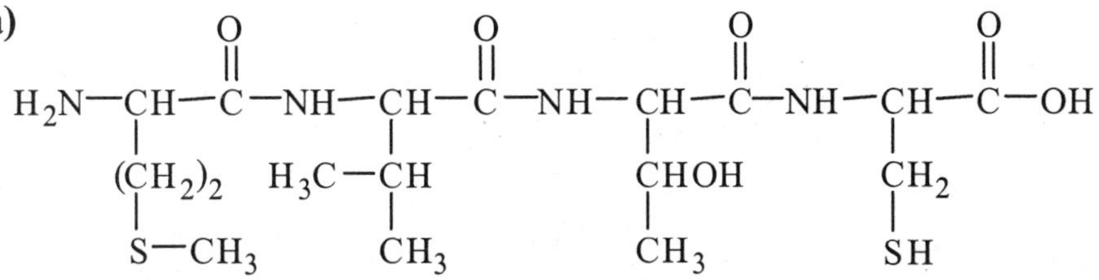

(b) methionylvalylthreonylcysteine

12. The bases are on the right side of the Figure. From the top down, they are: adenine (purine), uracil (pyrimidine), guanine (purine), and cytosine (pyrimidine). The pentose sugars have five-membered rings as their carbon skeletons; they are ribose sugars, and hence this is a chain of RNA. The phosphate groups are five-membered groups centered on P atoms and surrounded by O atoms.

EXERCISES

Structure and Composition of the Cell

13. The volume of a cylinder is given by $V = \pi r^2 h = \pi d^2 h / 4$.

$$V = \left[3.14159\left(1\times10^{-6}\ m\right)^2 \left(2\times10^{-6}\ m\right) \div 4\right] \times \frac{1000\ L}{1\ m^3} = 1.6\times10^{-15}\ L$$

The volume of the solution in the cell is $V_{soln} = 0.80 \times 1.6 \times 10^{-15}\ L = 1.3 \times 10^{-15}\ L$.

(a) $\left[H^+\right] = 10^{-6.4} = 4 \times 10^{-7}\ M$

$$\text{no. } H_3O^+ \text{ ions} = 1.3\times10^{-15}\ L \times \frac{4\times10^{-7}\ \text{mol } H^+ \text{ ions}}{1\ L\ \text{soln}} \times \frac{6.022\times10^{23}\ \text{ions}}{1\ \text{mol ions}}$$

$$= 3\times10^2\ H_3O^+ \text{ ions}$$

(b)

$$\text{no. K}^+ \text{ ions} = 1.3 \times 10^{-15} \text{ L} \times \frac{1.5 \times 10^{-4} \text{ mol K}^+ \text{ ions}}{1 \text{ L soln}} \times \frac{6.022 \times 10^{23} \text{ ions}}{1 \text{ mol ions}}$$

$$\text{no. K}^+ \text{ ions} = 1.2 \times 10^5 \text{ K}^+ \text{ ions}$$

14. Mass of all lipid molecules $= 0.02 \times 2 \times 10^{-12} \text{ g} = 4 \times 10^{-14} \text{ g}.$

$$\text{lipid molecules} = 4 \times 10^{-14} \text{ g} \times \frac{1 \text{ u}}{1.66 \times 10^{-24} \text{ g}} \times \frac{1 \text{ lipid molecule}}{700 \text{ u}} = 3 \times 10^7 \text{ lipid molecules}$$

15. mass of protein in cytoplasm $= 0.15 \times 0.90 \times 2 \times 10^{-12} \text{ g} = 2._7 \times 10^{-13} \text{ g}.$

$$\text{no. of protein molecules} = 2._7 \times 10^{-13} \text{ g} \times \frac{1 \text{ mol protein}}{3 \times 10^4 \text{ g}} \times \frac{6.022 \times 10^{23} \text{ molecules}}{1 \text{ mol protein}}$$

$$= 5 \times 10^6 \text{ protein molecules}$$

16. DNA length $= 4.5 \times 10^6 \text{ mononucleotides} \times \dfrac{450 \text{ pm}}{1 \text{ mononucleotide}} \times \dfrac{10^{-12} \text{ m}}{1 \text{ pm}} = 2 \times 10^{-3} \text{ m} = 2 \text{ mm}$

$2 \text{ mm} = 2 \times 10^3 \, \mu\text{m}$, the length of the stretched out DNA is one thousand times $2 \, \mu\text{m}$, the length of the cell. Thus, the DNA must be wrapped up, or coiled, within the cell.

Lipids

17. **(a)**

Trilaurin	Trilinolein
$\begin{array}{c} \quad\quad\;\; O \\ \quad\quad\;\; \| \\ CH_2\text{-O-C-C}_{11}H_{23} \\ \mid \quad\quad O \\ \mid \quad\quad \| \\ CH\text{-O-C-C}_{11}H_{23} \\ \mid \quad\quad O \\ \mid \quad\quad \| \\ CH_2\text{-O-C-C}_{11}H_{23} \end{array}$	$\begin{array}{c} \quad\quad\;\; O \\ \quad\quad\;\; \| \\ CH_2\text{-O-C-C}_{17}H_{31} \\ \mid \quad\quad O \\ \mid \quad\quad \| \\ CH\text{-O-C-C}_{17}H_{31} \\ \mid \quad\quad O \\ \mid \quad\quad \| \\ CH_2\text{-O-C-C}_{17}H_{31} \end{array}$
A triglyceride or glycerol ester	A triglyceride or glycerol ester
Saturated triglyceride -made using saturated acid	Unsaturated triglyceride -made using unsaturated acid
A fat (usually solid at room temperature)	An oil (usually liquid at room temperature)

(b)   Soaps: salt of fatty acids (from saponification of triglycerides)

Phospholipids: derived from glycerols, fatty acids, phosphoric acid and a nitrogen
containing base   (Both have hydrophilic heads and hydrophobic tails)

18.

Monoglyceride   Diglyceride

19.   Polyunsaturated fatty acids are characterized by a large number of $C = C$ double bonds in
their hydrocarbon chain. Stearic acid has no $C = C$ double bonds and therefore is not
unsaturated, let alone polyunsaturated. But eleostearic acid has three $C = C$ double bonds
and thus is polyunsaturated. Polyunsaturated fatty acids are recommended in dietary
programs since saturated fats are linked to a high incidence of heart disease. Of the lipids
listed in Table 28-2, safflower oil has the highest percentage of unsaturated fatty acids,
predominately linoleic acid, an unsaturated fatty acid with two C=C bonds.

20.   Safflower oil contains a larger percentage of the unsaturated fatty acid, linoleic acid (two
C=C bonds) (75 – 80%) than does corn oil (34 – 62%). It also contains a smaller
percentage of saturated fatty acids, particularly palmitic acid (6 – 7%) than does corn oil
(8 – 12%). And the two oils contain about the same proportion of theunsaturated fatty acid,
oleic acid (one C=C bond). Consequently, safflower oil should require the greater amount
of $H_2(g)$ for its complete hydrogenation to a solid fat.

21. tripalmitin

$$\begin{array}{l} CH_2OOC(CH_2)_{14}CH_3 \\ CHOOC(CH_2)_{14}CH_3 \\ CH_2OOC(CH_2)_{14}CH_3 \end{array}$$

saponification products of tripalmitin: sodium palmitate and glycerol

$CH_2OHCHOHCH_2OH$

$NaOOC(CH_2)_{14}CH_3$

22. First we write the equation for the saponification reaction.

$$\begin{array}{l} CH_2OOCC_{13}H_{27} \\ CHOOCC_{13}H_{27} + 3\ NaOH \rightarrow CH_2OHCHOHCH_2OH + 3\ NaOOCC_{13}H_{27} \\ CH_2OOCC_{13}H_{27} \end{array}$$

$$\text{mass of soap} = 105\text{ g triglyceride} \times \frac{1\text{ mol triglyceride}}{723.2\text{ g triglyceride}} \times \frac{3\text{ mol soap}}{1\text{ mol triglyceride}} \times \frac{250.4\text{ g soap}}{1\text{ mol soap}}$$

$$= 109\text{ g soap}$$

Carbohydrates

23.

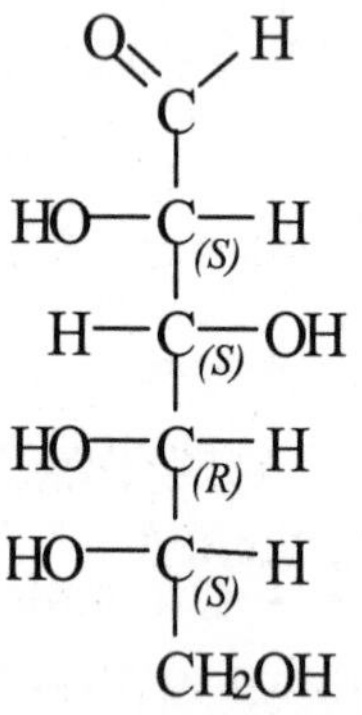

D-Mannose
aldohexose

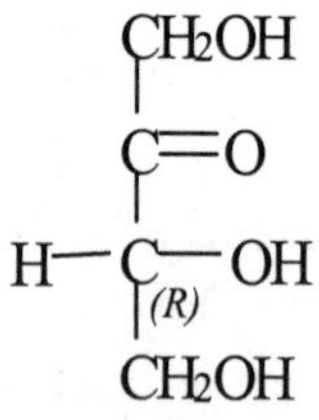

D-erythrulose
ketotetrose

24.

$$H-C=O \qquad O=C-H$$

D-Glucose	L-Glucose
H—C—OH	HO—C—H
HO—C—H	H—C—OH
H—C—OH	HO—C—H
H—C—OH	HO—C—H
CH₂OH	CH₂OH

The designation "L" in L-Glucose simply arises from nomenclature and does not convey the dextrorotatory or levorotatory designations. The structure of a "chiral" molecule will determine whether it is dextrorotatory or levorotatory. Put another way, the D- and L- designation arises from the rules of nomenclature, whereas, the dextrorotatory or levorotatory designation must be experimentally determined

◄——— These molecules are enantiomers

25. **(a)** A dextrorotatory compound rotates the plane of polarized light to the right: clockwise.

(b) A levorotatory compound is one that rotates the plane of polarized light to the left: counterclockwise.

(c) A racemic mixture has equal amounts of an optically active compound and its enantiomer. Since these two compounds rotate polarized light by the same amount but in opposite directions, such a mixture does not rotate the plane of polarized light.

(d) (R) In organic nomenclature, this designation is given to a chiral carbon atom. First, we must assign priorities to the four substituents on the chiral carbon atom. With the lowest priority group pointing directly away from the viewer, we say that the stereogenic center has an R-configuration if a curved arrow from the group of highest priority through to the one of lowest priority is drawn in a clockwise direction.

26. **(a)** Two compounds that are optical isomers of each other— they have different locations of the substituent groups around their chiral carbons— but which are not mirror images of each other are diastereomers.

(b) Two isomers that are nonsuperimposable mirror images of one other are enantiomers.

(c) (–) is another way of designating a levorotatory compound.

(d) D indicates that, in the Fisher projection of the compound, the — OH group on the penultimate carbon atom is to the right and the — H group is to the left.

27. A reducing sugar has a sufficient amount of the straight-chain form present in equilibrium with its cyclic form such that the sugar will reduce $Cu^{2+}(aq)$ to insoluble, red $Cu_2O(s)$. Only free aldehyde groups are able to reduce the copper(II) ion down to copper (I).

Next, we need to calculate the mass of Cu_2O expected when 0.500 g of glucose is oxidized in the reducing sugar test:

$$\text{Mass } Cu_2O \text{ (g)} = \frac{1 \text{ mol glucose}}{180.2 \text{ g glucose}} \times \frac{2 \text{ mol } Cu^{2+}}{1 \text{ mol glucose}} \times \frac{1 \text{ mol } Cu_2O}{2 \text{ mol } Cu^{2+}} \times \frac{143.1 \text{ g } Cu_2O}{1 \text{ mol } Cu_2O} = 0.397 \text{g } Cu_2O$$

28. The eight aldopentoses are drawn below as Fischer structures.
Note: There are 4 pairs of enantiomers and each structure has 6 diasteromers.

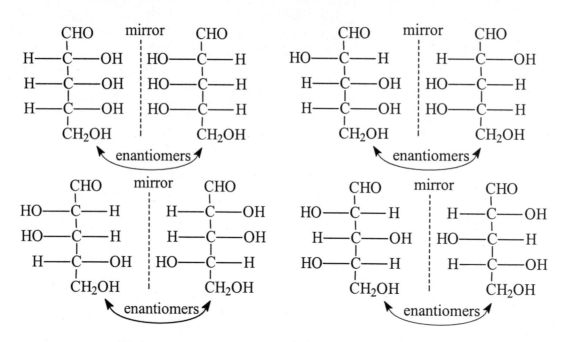

29. Enantiomers are alike in all respects, including in the degree to which they rotate polarized
light. They differ only in the direction in which this rotation occurs. Since α-glucose and
β-glucose rotate the plane of polarized light by different degrees, and in the same
direction, they are not enantiomers, but rather diastereomers.

30. We let x represent the fraction of α-D-glucose. Then $1-x$ is the fraction of β-D-glucose.
$$+52.7° = x(112°)+(1-x)(18.7°)=112°x+18.7°-18.7°x=93°x+18.7°$$

$$x=\frac{52.7°-18.7°}{93°}=0.37 \quad \text{The solution is 37% } \alpha\text{-D-glucose; and thus 63% } \beta\text{-D-glucose.}$$

Fischer Projections and *R,S* Nomenclature

31. **a)**   Enantiomers: *S*-config. (leftmost structure), *R*-config. (rightmost structure)
     b)   Same molecule: both *R*-configuration
     c)   Diasteriomers: *S,R* config.(leftmost structure-top to bottom),
              S,S-config. (rightmost structure)
     d)   Diasteriomers: *R,R*-config (leftmost structure-top to bottom),
              R, S-config. (rightmost structure-top to bottom)

32. **a)**   Same molecule: both *R*-configuration
     b)   Same molecule: both *R*-configuration
     c)   Enantiomers: *S,S*-config. (leftmost structure), *R,R*-config. (rightmost structure)
     d)   Diasteriomers: *S,R*-config (leftmost structure-top to bottom),
              R, R-config. (rightmost structure)

33.

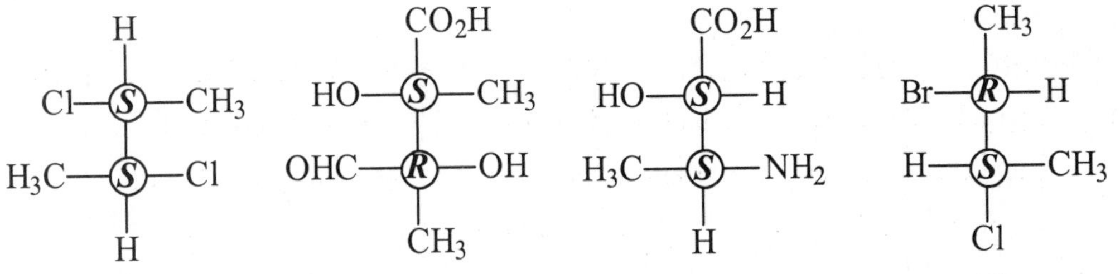

34.

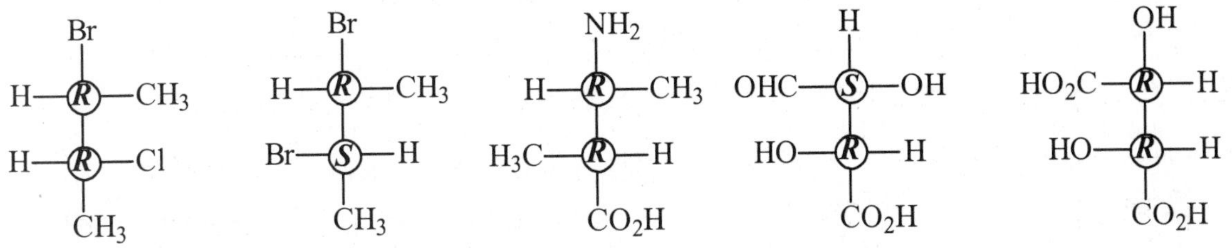

Amino acids, Polypeptides, and Proteins

35. **(a)** An α-amino acid has an amine group (—NH$_2$) bonded to the same carbon as the carboxyl group (—COOH). For example: glycine (H_2NCH_2COOH) is the simplest α-amino acid.

(b) A zwitterion is a form of an amino acid where the amine group is protonated $(—NH_3^+)$ and the carboxyl group is deprotonated $(—COO^-)$. The zwitterion form of glycine is $^+H_3NCH_2COO^-$.

(c) The pH at which the zwitterion form of an amino acid predominates in solution is known as the isoelectric point. The isoelectric point of glycine is $pI = 6.03$.

(d) The peptide bond is the bond that forms between the carbonyl group of one amino acid and the amine group of another, with the elimination of a water molecule between them. The peptide bond between two glycine molecules is shown as an outlined bold dash (━━) in the structure below.

$$H_2N—CH_2—\overset{\overset{\displaystyle O}{\|}}{C}—O━━NH—CH_2—\overset{\overset{\displaystyle O}{\|}}{C}—OH$$

(e) Tertiary structure describes how a coiled protein chain further interacts with itself to wrap into a cluster through a combination of salt linkages, hydrogen bonding and disulfide linkages, to name a few.

36. **(a)** A polypeptide is a long chain of amino acids, joined together by peptide bonds.

(b) A protein is another name for a polypeptide, but there is a distinction that often is drawn. Proteins are longer chains than are polypeptides, and proteins are biologically active.

(c) The N-terminal amino acid in a polypeptide is the one at the end of a molecule where there is a free NH_2 group at the end of the polypeptide chain. The N-terminal amino acid is at the left end of the structure of diglycine in the answer to the previous exercise.

(d) An α helix is a natural secondary structure adopted by many proteins. It is rather like a spiral rising upward to the right (that is, clockwise as viewed from the bottom). This is a right-handed screw. The alpha helix is shown in Figure 28-12.

(e) Denaturation is that process in which at least some of the structure of a protein is disrupted, either thermally (with heat), mechanically, or by changing the pH or changing the ionic strength of the medium in which the protein is enveloped. Denaturation is accompanied by a decrease in the biological activity. Denaturation can be temporary or permanent.

37. pH = 6.3 is near the isoelectric point of proline (6.21). Thus proline will not migrate very effectively under these conditions. But pH = 6.3 is considerably more acidic than the isoelectric point of lysine ($pI = 9.74$). Thus, lysine is positively charged in this solution and will migrate toward the negatively charged cathode. And pH = 6.3 is much less acidic than the isoelectric point of aspartic acid ($pI = 2.96$). Aspartic acid, therefore, is negatively charged in this solution and will migrate toward the positively charged anode.

38. $pI = 5.74$ is the isoelectric point of phenylalanine. Thus phenylalanine will not migrate. But pH = 5.7 is more acidic than the isoelectric point of histidine ($pI = 7.58$). Thus, histidine is positively charged in this solution and will migrate toward the negatively charged cathode. And pH = 5.7 is less acidic than the isoelectric point of glutamic acid ($pI = 3.22$). Glutamic acid, is negatively charged in this solution; it will migrate to the positively charged anode.

39. **(a)** in strongly acidic solution
$^+H_3NCH(CHOHCH_3)COOH$

$$^+H_3N-\overset{\displaystyle |}{\underset{\displaystyle \underset{\displaystyle CH_3}{|}}{\underset{\displaystyle |}{CHOH}}}CH-\overset{\displaystyle O}{\overset{\displaystyle ||}{C}}-OH$$

(b) at the isoelectric point
$^+H_3NCH(CHOHCH_3)COO^-$

$$^+H_3N-\overset{\displaystyle |}{\underset{\displaystyle \underset{\displaystyle CH_3}{|}}{\underset{\displaystyle |}{CHOH}}}CH-\overset{\displaystyle O}{\overset{\displaystyle ||}{C}}-O^-$$

(c) in strongly basic solution $H_2NCH(CHOHCH_3)COO^-$

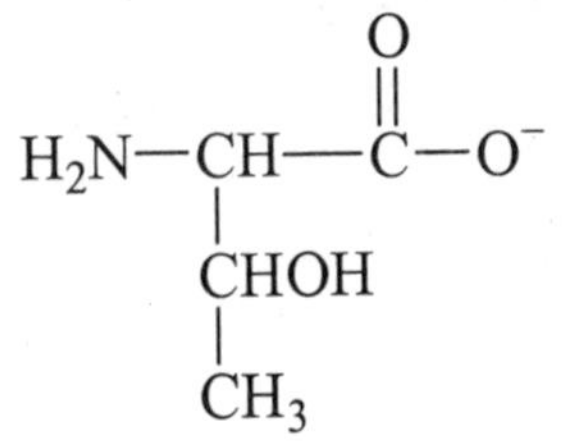

$$H_2N-CH-\overset{\overset{\displaystyle O}{\|}}{C}-O^-$$
$$|$$
$$CHOH$$
$$|$$
$$CH_3$$

40. (a) aspartic acid **(b)** lysine **(c)** alanine

$$H_2N-CH-\overset{\overset{\displaystyle O}{\|}}{C}-O^-$$
$$|$$
$$CH_2$$
$$|$$
$$COOH$$

$$^+H_3N-CH-\overset{\overset{\displaystyle O}{\|}}{C}-OH$$
$$|$$
$$(CH_2)_4$$
$$|$$
$$NH_2$$

$$^+H_3N-CH-\overset{\overset{\displaystyle O}{\|}}{C}-O^-$$
$$|$$
$$CH_3$$

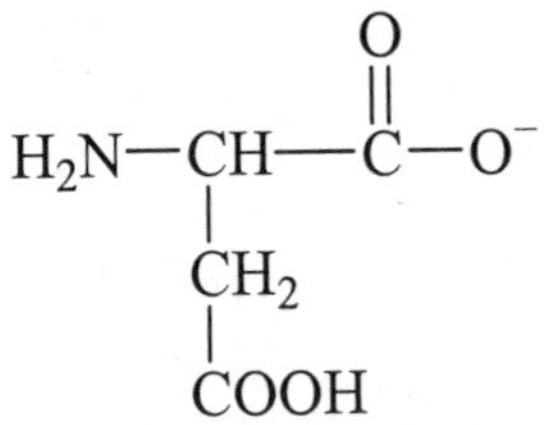

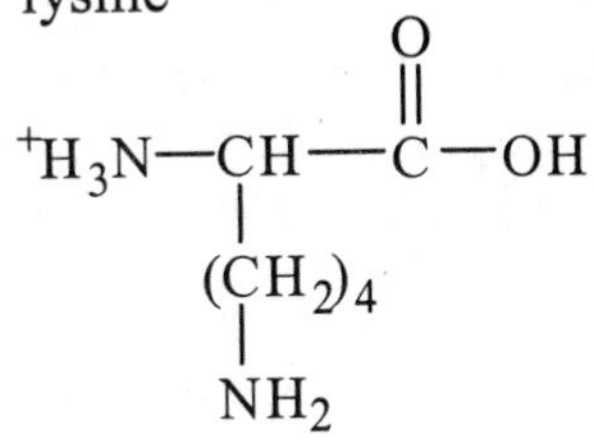

$H_2NCH(CH_2COOH)COO^-$ $^+H_3NCH((CH_2)_4NH_2)COOH$ $^+H_3NCH(CH_3)COO^-$

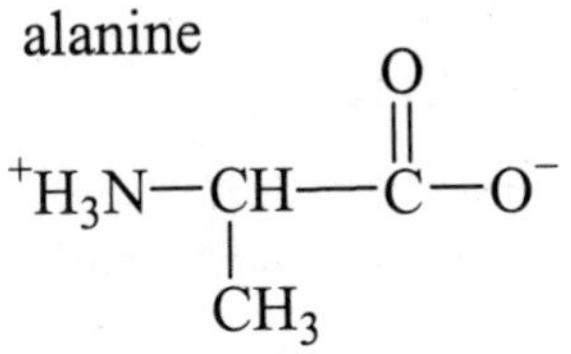

41. (a) The structures of the six tripeptides that contain one each of the amino acids alanine, serine, and lysine are drawn below (in no particular order).

Lys-Ser-Ala (1 of 6)

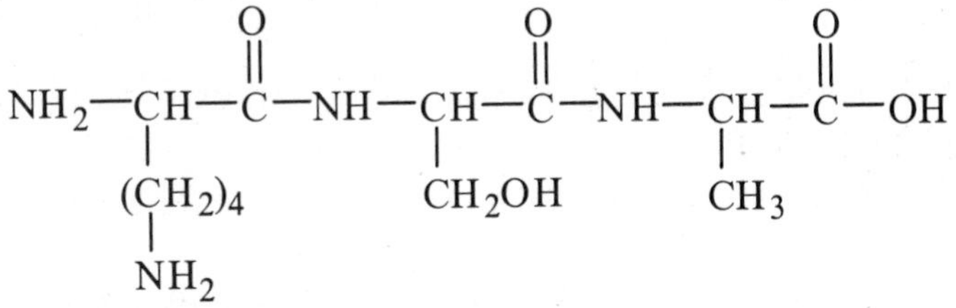

$$NH_2-CH-\overset{\overset{\displaystyle O}{\|}}{C}-NH-CH-\overset{\overset{\displaystyle O}{\|}}{C}-NH-CH-\overset{\overset{\displaystyle O}{\|}}{C}-OH$$
$$|\qquad\qquad\quad |\qquad\qquad\quad |$$
$$(CH_2)_4\qquad CH_2OH\qquad CH_3$$
$$|$$
$$NH_2$$

Lys-Ala-Ser (2 of 6)

$$NH_2-CH-\overset{\overset{\displaystyle O}{\|}}{C}-NH-CH-\overset{\overset{\displaystyle O}{\|}}{C}-NH-CH-\overset{\overset{\displaystyle O}{\|}}{C}-OH$$
$$|\qquad\qquad\quad |\qquad\qquad\quad |$$
$$(CH_2)_4\qquad CH_3\qquad CH_2OH$$
$$|$$
$$NH_2$$

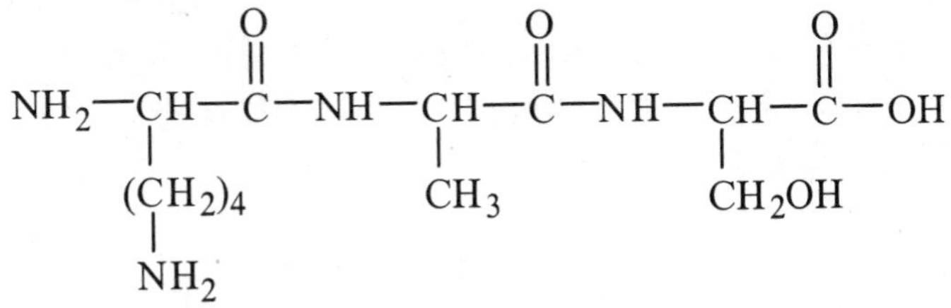

Ser-Lys-Ala (3 of 6)

$$NH_2-CH-\overset{\overset{\displaystyle O}{\|}}{C}-NH-CH-\overset{\overset{\displaystyle O}{\|}}{C}-NH-CH-\overset{\overset{\displaystyle O}{\|}}{C}-OH$$
$$|\qquad\qquad\quad |\qquad\qquad\quad |$$
$$CH_2OH\qquad (CH_2)_4\qquad CH_3$$
$$|$$
$$NH_2$$

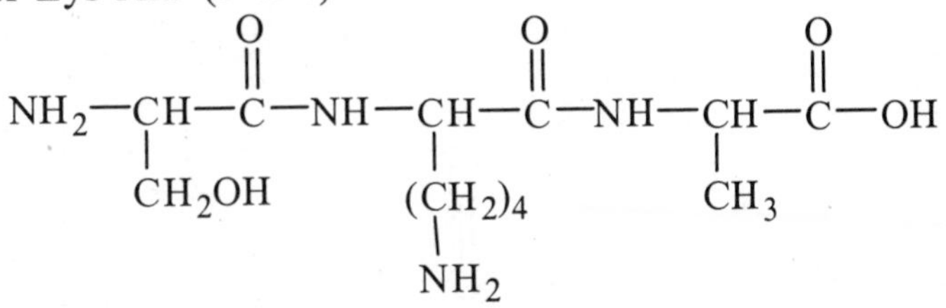

Ser-Ala-Lys (4 of 6)

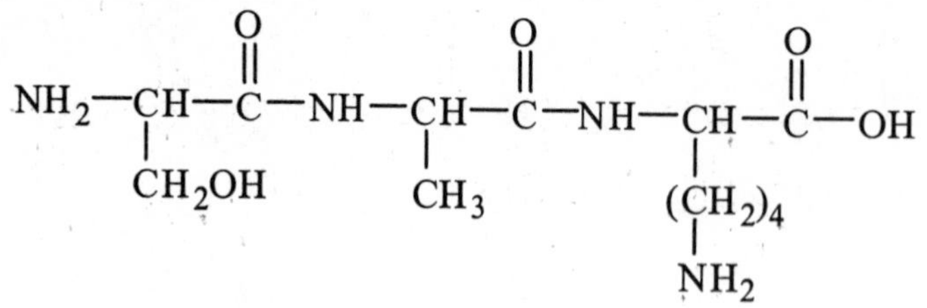

$$NH_2-\underset{\underset{CH_2OH}{|}}{CH}-\overset{\overset{O}{\|}}{C}-NH-\underset{\underset{CH_3}{|}}{CH}-\overset{\overset{O}{\|}}{C}-NH-\underset{\underset{\underset{NH_2}{|}}{(CH_2)_4}}{CH}-\overset{\overset{O}{\|}}{C}-OH$$

Ala-Ser-Lys (5 of 6)

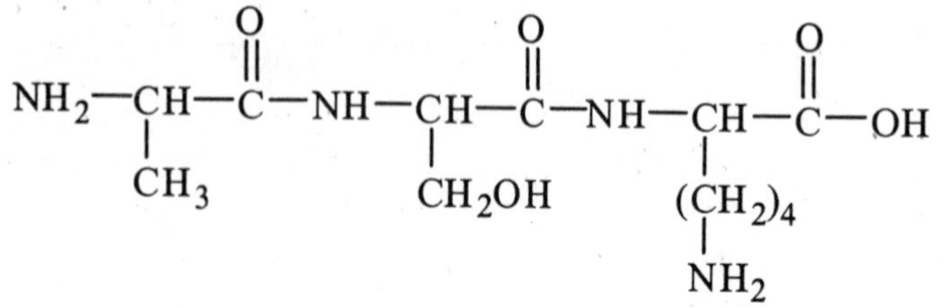

$$NH_2-\underset{\underset{CH_3}{|}}{CH}-\overset{\overset{O}{\|}}{C}-NH-\underset{\underset{CH_2OH}{|}}{CH}-\overset{\overset{O}{\|}}{C}-NH-\underset{\underset{\underset{NH_2}{|}}{(CH_2)_4}}{CH}-\overset{\overset{O}{\|}}{C}-OH$$

Ala-Lys-Ser (6 of 6)

$$NH_2-\underset{\underset{CH_3}{|}}{CH}-\overset{\overset{O}{\|}}{C}-NH-\underset{\underset{\underset{NH_2}{|}}{(CH_2)_4}}{CH}-\overset{\overset{O}{\|}}{C}-NH-\underset{\underset{CH_2OH}{|}}{CH}-\overset{\overset{O}{\|}}{C}-OH$$

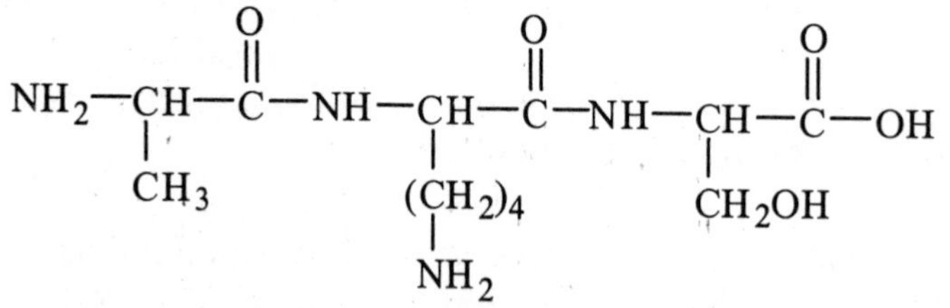

(b) The structures of the six tetrapeptides that contain two serine and two alanine amino acids each follow (in no particular order).

Ala-Ser-Ala-Ser (1 of 6)

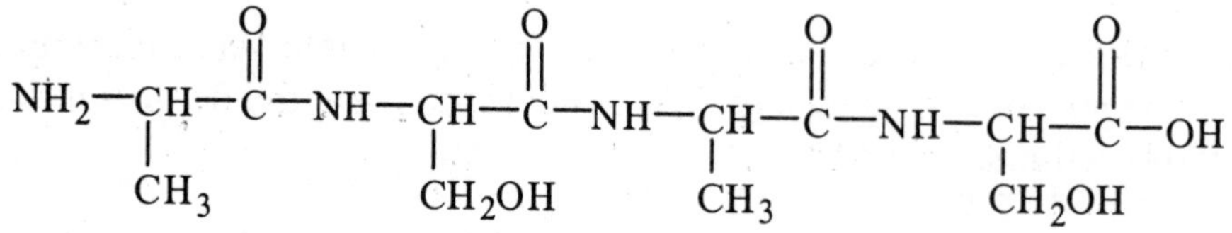

$$NH_2-\underset{\underset{CH_3}{|}}{CH}-\overset{\overset{O}{\|}}{C}-NH-\underset{\underset{CH_2OH}{|}}{CH}-\overset{\overset{O}{\|}}{C}-NH-\underset{\underset{CH_3}{|}}{CH}-\overset{\overset{O}{\|}}{C}-NH-\underset{\underset{CH_2OH}{|}}{CH}-\overset{\overset{O}{\|}}{C}-OH$$

Ala-Ala-Ser-Ser (2 of 6)

$$NH_2-\underset{\underset{CH_3}{|}}{CH}-\overset{\overset{O}{\|}}{C}-NH-\underset{\underset{CH_3}{|}}{CH}-\overset{\overset{O}{\|}}{C}-NH-\underset{\underset{CH_2OH}{|}}{CH}-\overset{\overset{O}{\|}}{C}-NH-\underset{\underset{CH_2OH}{|}}{CH}-\overset{\overset{O}{\|}}{C}-OH$$

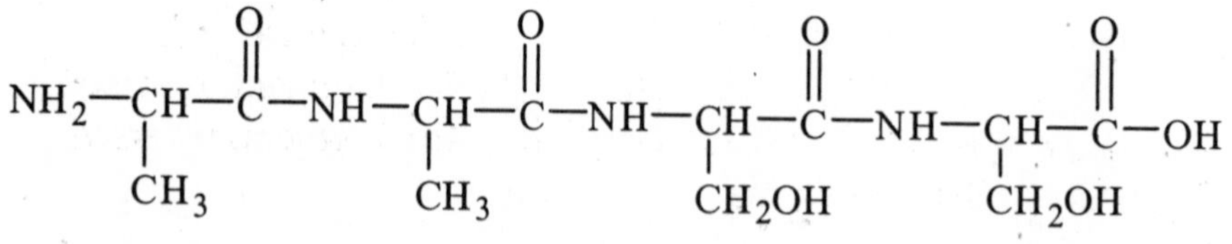

Ala-Ser-Ser-Ala (3 of 6)

$$NH_2-\underset{\underset{CH_3}{|}}{CH}-\overset{\overset{O}{\|}}{C}-NH-\underset{\underset{CH_2OH}{|}}{CH}-\overset{\overset{O}{\|}}{C}-NH-\underset{\underset{CH_2OH}{|}}{CH}-\overset{\overset{O}{\|}}{C}-NH-\underset{\underset{CH_3}{|}}{CH}-\overset{\overset{O}{\|}}{C}-OH$$

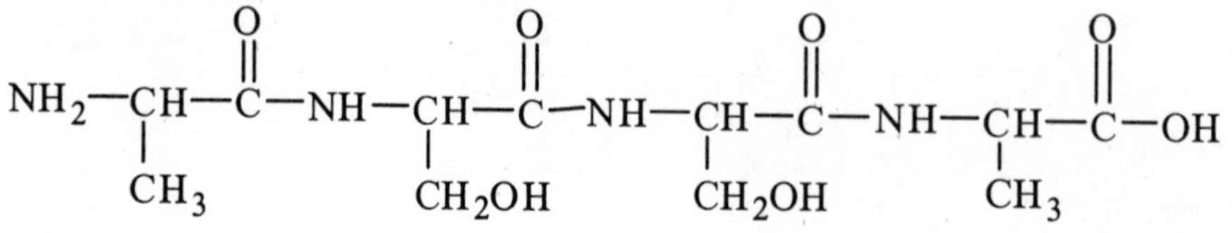

Ser-Ser-Ala-Ala (4 of 6)

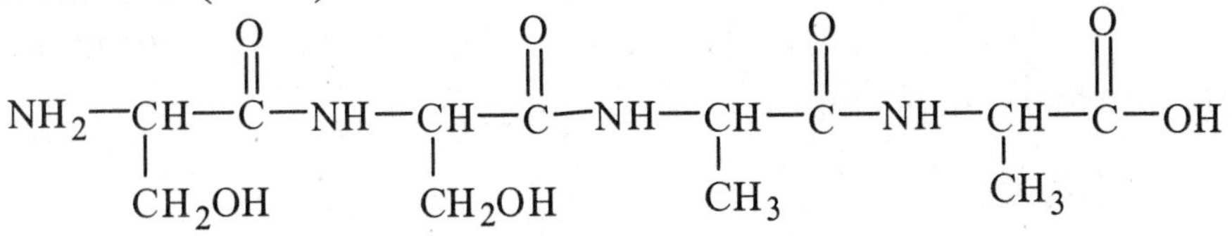

$$\underset{\underset{CH_2OH}{|}}{NH_2-CH}-\overset{\overset{O}{||}}{C}-NH-\underset{\underset{CH_2OH}{|}}{CH}-\overset{\overset{O}{||}}{C}-NH-\underset{\underset{CH_3}{|}}{CH}-\overset{\overset{O}{||}}{C}-NH-\underset{\underset{CH_3}{|}}{CH}-\overset{\overset{O}{||}}{C}-OH$$

Ser-Ala-Ser-Ala (5 of 6)

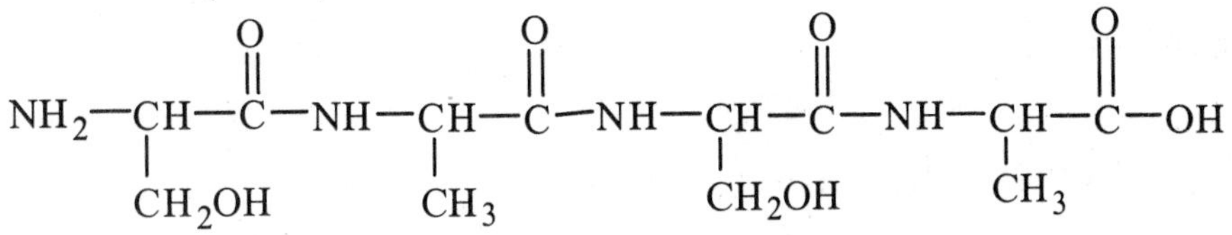

$$\underset{\underset{CH_2OH}{|}}{NH_2-CH}-\overset{\overset{O}{||}}{C}-NH-\underset{\underset{CH_3}{|}}{CH}-\overset{\overset{O}{||}}{C}-NH-\underset{\underset{CH_2OH}{|}}{CH}-\overset{\overset{O}{||}}{C}-NH-\underset{\underset{CH_3}{|}}{CH}-\overset{\overset{O}{||}}{C}-OH$$

Ser-Ala-Ala-Ser (6 of 6)

$$\underset{\underset{CH_2OH}{|}}{NH_2-CH}-\overset{\overset{O}{||}}{C}-NH-\underset{\underset{CH_3}{|}}{CH}-\overset{\overset{O}{||}}{C}-NH-\underset{\underset{CH_3}{|}}{CH}-\overset{\overset{O}{||}}{C}-NH-\underset{\underset{CH_2OH}{|}}{CH}-\overset{\overset{O}{||}}{C}-OH$$

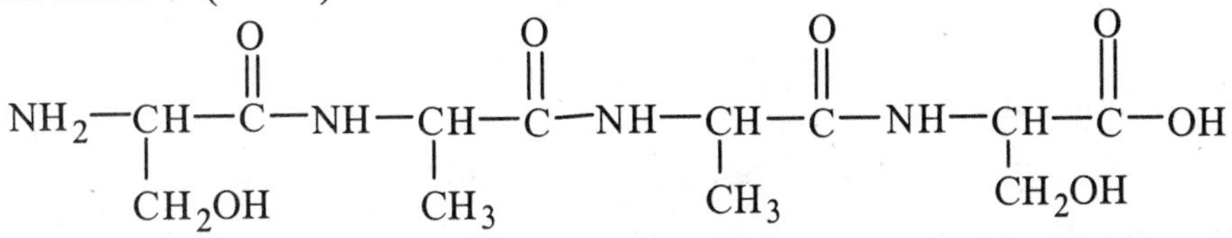

42. There are twenty four possibilities. They are listed below.

Ala-Lys-Ser-Phe	Lys-Ala-Ser-Phe	Ser-Ala-Phe-Lys	Phe-Ala-Ser-Lys
Ala-Lys-Phe-Ser	Lys-Ala-Phe-Ser	Ser-Ala-Lys-Phe	Phe-Ala-Lys-Ser
Ala-Ser-Lys-Phe	Lys-Ser-Phe-Ala	Ser-Lys-Phe-Ala	Phe-Lys-Ser-Ala
Ala-Ser-Phe-Lys	Lys-Ser-Ala-Phe	Ser-Lys-Ala-Phe	Phe-Lys-Ala-Ser
Ala-Phe-Ser-Lys	Lys-Phe-Ser-Ala	Ser-Phe-Ala-Lys	Phe-Ser-Ala-Lys
Ala-Phe-Lys-Ser	Lys-Phe-Ala-Ser	Ser-Phe-Lys-Ala	Phe-Ser-Lys-Ala

43. **(a)** We put the fragments together as follows, starting from the Ala end, simply placing them down in a matching pattern. We do not assume that the fragments are given with the N-terminal end first.

Fragments:	Ala	Ser					3rd fragment
		Ser	Gly	Val			1st fragment
			Gly	Val	Thr		5th fragment
				Val	Thr		2nd fragment, reversed
				Val	Thr	Leu	4th fragment, reversed
Result:	Ala	Ser	Gly	Val	Thr	Leu	

(b) alanyl-seryl-glycyl-valyl-threonyl-leucine, or alanylserylglycylvalylthreonylleucine

44. **(a)** We put the fragments together as follows, starting from the Ala end, simply placing them down in a matching pattern. We do not assume that the fragments are given with the N-terminal end first.

Fragments:	Ala	Lys	Ser				
		Lys	Ser	Gly			5th fragment
			Ser	Gly			3rd fragment
				Gly	Phe		4th fragment
				Gly	Phe	Gly	2nd fragment
Result:	Ala	Lys	Ser	Gly	Phe	Gly	

Fragments: Ala   Lys   Ser   1st fragment

(b) alanyl-lysyl-seryl-glycyl-phenylalanyl-glycine, or
alanyllysylserylglycylphenylalanylglycine

45. The *primary* structure of an amino acid is the sequence of amino acids in the chain of the polypeptide. The secondary structure describes how the protein chain is folded, coiled, or convoluted. Possible structures include α-helices and β-pleated sheets. This secondary structure is held together principally by hydrogen bonds. The *tertiary* structure of a protein refers to how different parts of the molecules, often quite distant from each other, are held together into crudely spherical shapes. Although hydrogen bonding is involved here as well, disulfide linkages, hydrophobic interactions, and hydrophilic interactions (salt linkages) are responsible as well for tertiary structure. Finally, quaternary structure refers to how two or more protein molecules pack together into a larger protein complex. Not all proteins have a quaternary structure since many proteins have but one polypeptide chain.

46. The sole difference between sickle cell hemoglobin and normal hemoglobin is due to the substitution of one amino acid for another (valine for glutamic acid) at one position in the entire protein. This changes the quaternary structure of the hemoglobin. We can think of the fault as being due to the mistaken incorporation of one molecule for another during protein synthesis. This is why the name "molecular disease" is apt.

47.

$$H_2N-\overset{\displaystyle CH_3}{\underset{\displaystyle CO_2H}{(R)}}-H \qquad R\text{-alanine}$$

48.

$$H_2N-\overset{\displaystyle CH_2OH}{\underset{\displaystyle CO_2H}{(R)}}-H \qquad R\text{-serine}$$

49.

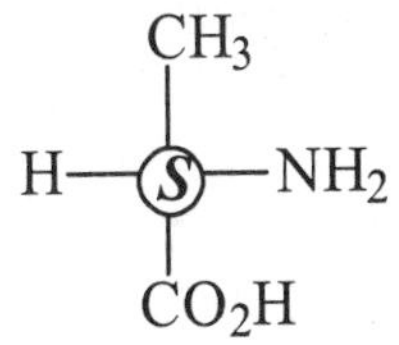

S-alanine 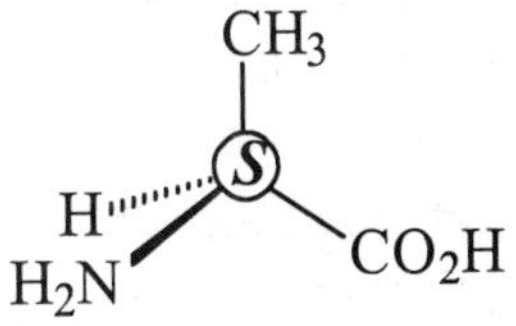 S-alanine 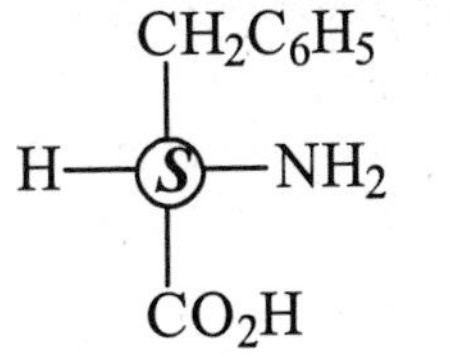 S-phenylalanine 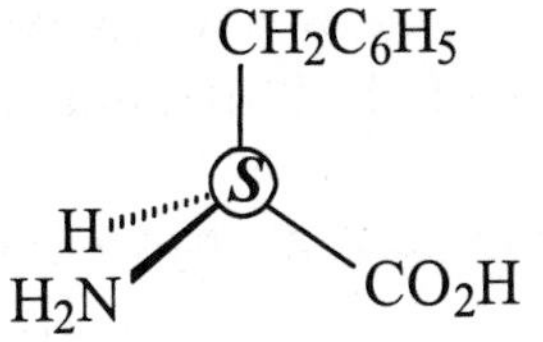 S-phenylalanine

50.

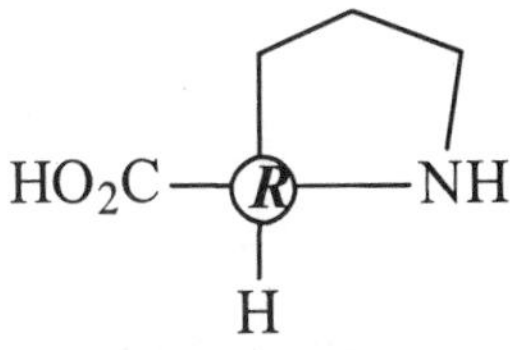

R-proline 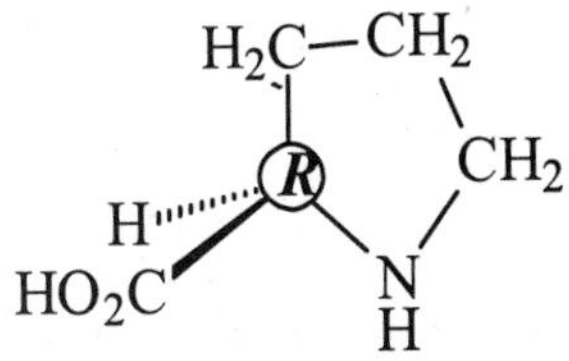 R-proline 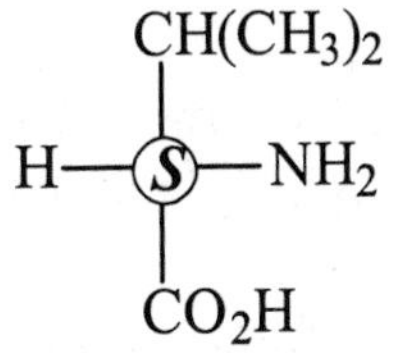 R-valine 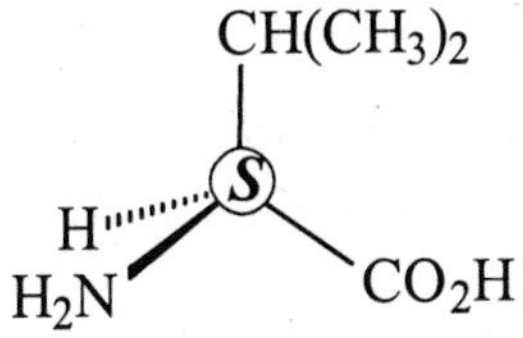 R-valine

Nucleic Acids

51.   The two major types of nucleic acids are DNA, deoxyribonucleic acid, and RNA, ribonucleic acid. Both of them contain phosphate groups. These phosphate groups alternate with sugars to form the backbone of the molecule. These sugars are deoxyribose in the case of DNA and ribose in the case of RNA. Attached to each sugar is a purine or a pyrimidine base. The purine bases are adenine and guanine. One pyrimidine base is cytosine. In the case of RNA the other pyrimidine base is uracil, while in the case of DNA the other pyrimidine base is thymine.

52.   The "thread of life" is an apt name for DNA, being both a literal and a figurative description. It is literal in that it is thread-like, long and narrow, in shape and is an essential molecule for life. It is figurative in that DNA is essential for the continuance of life and runs like a thread through all stages in the life of the organism, from origin through growth and reproduction to final death.

53. The complementary sequence to AGC is TCG. The hydrogen bonding is shown in the drawing below. One polynucleotide chain is completely shown, as is the hydrogen bonding to the bases in the other polynucleotide chain. Because of the distortions that result from depicting a 3-D structure in two dimensions, the H- bonds themselves are distorted (they are all of approx. equal length) and the second sugar phosphate chain has been omitted.

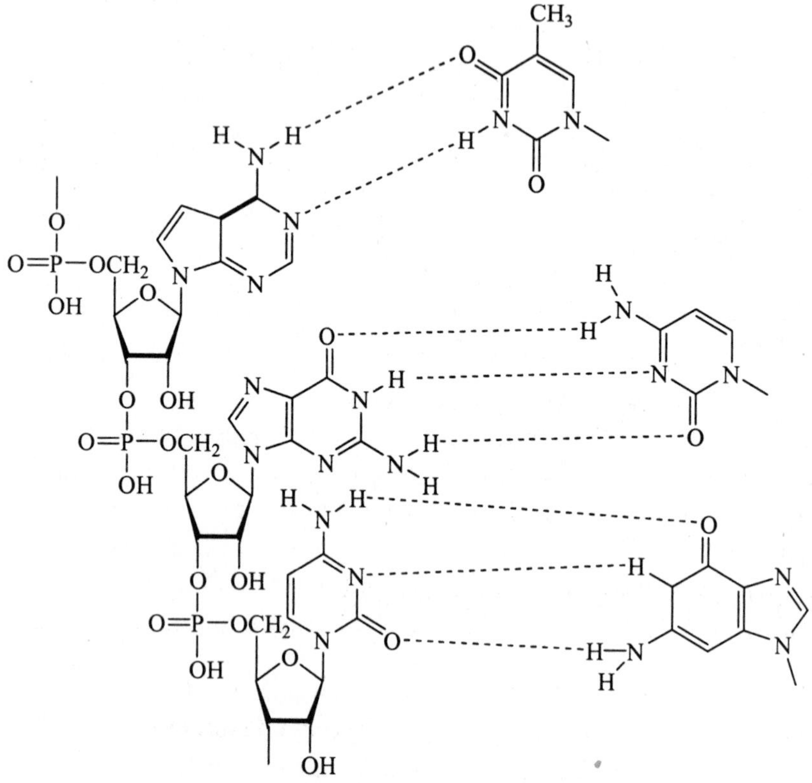

54. The complementary sequence to TCT is AGA. The hydrogen bonding is shown in the drawing on the following page. One polynucleotide chain is completely shown, as is the hydrogen bonding to the bases in the other polynucleotide chain. Because of the distortions that result from depicting a three-dimensional structure in two dimensions, the hydrogen bonds themselves are distorted (they are all of approximately equal length) in the diagram.

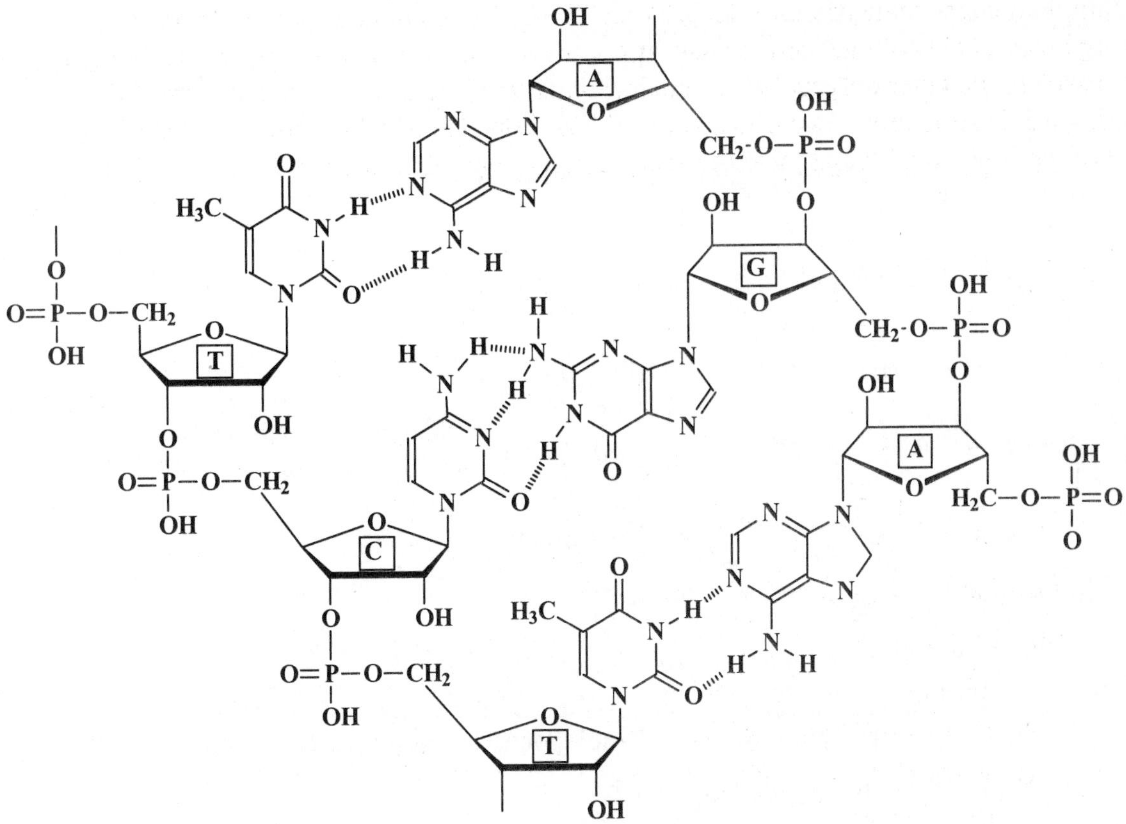

FEATURE PROBLEMS

<u>**70.**</u>

Glyceryl Tristearate

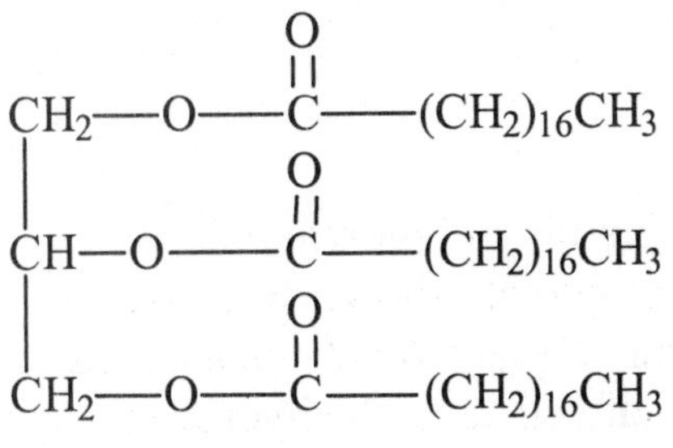

Molar Mass = 891.5 g mol^{-1}

Glyceryl Trioleate

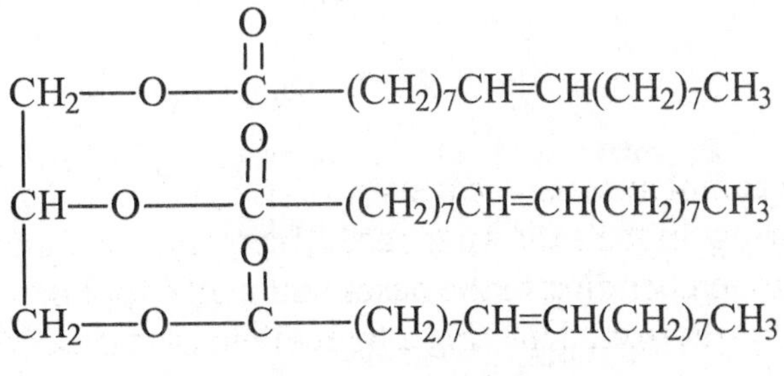

Molar Mass = 885.45 g mol^{-1}

(a) Saponification value of glyceryl tristearate: mass of KOH

$$= 1.00 \text{ g glyceryl tristearate} \times \frac{1 \text{ mol glyceryl tristearate}}{891.5 \text{ g glyceryl tristearate}} \times \frac{3 \text{ mol KOH}}{1 \text{ mol glyceryl tristearate}}$$

$$\times \frac{56.1056 \text{ g KOH}}{1 \text{ mol KOH}} = 0.189 \text{ g KOH or } 189 \text{ mg KOH}$$

Saponification value = 189

Iodine number for glyceryl trioleate: mass of I_2

$= 100$ g glyceryl trioleate $\times \dfrac{1 \text{ mol glyceryl trioleate}}{885.45 \text{ g glyceryl trioleate}} \times \dfrac{3 \text{ mol } I_2}{1 \text{ mol glyceryl trioleate}}$

$\times \dfrac{253.81 \text{ g } I_2}{1 \text{ mol } I_2} = 86.0$ g I_2 Iodine number $= 86.0$

(b)   If we assume that castor oil is all glyceryl tririciniolate

$CH_2-O-\overset{\overset{\displaystyle O}{\|}}{C}-(CH_2)_7CH{=}CHCH_2CHOH(CH_2)_5CH_3$

$CH-O-\overset{\overset{\displaystyle O}{\|}}{C}-(CH_2)_7CH{=}CHCH_2CHOH(CH_2)_5CH_3$

$CH_2-O-\overset{\overset{\displaystyle O}{\|}}{C}-(CH_2)_7CH{=}CHCH_2CHOH(CH_2)_5CH_3$

Glyceryl Tririciniolate
Molar Mass $= 933.4$ g mol^{-1}

Iodine number for glyceryl tririciniolate: mass of I_2

$= 100$ g glyceryl tririciniolate

$\times \dfrac{1 \text{ mol glyceryl tririciniolate}}{933.4 \text{ g glyceryl tririciniolate}} \times \dfrac{3 \text{ mol } I_2}{1 \text{ mol glyceryl tririciniolate}} \times \dfrac{253.81 \text{ g } I_2}{1 \text{ mol } I_2}$

$= 81.6$ g I_2 Iodine number $= 81.6$

Saponification value of glyceryl tririciniolate: mass of KOH

$= 1.00$ g glyceryl tririciniolate

$\times \dfrac{1 \text{ mol glyceryl tririciniolate}}{933.4 \text{ g glyceryl tririciniolate}} \times \dfrac{3 \text{ mol KOH}}{1 \text{ mol glyceryl tririciniolate}} \times \dfrac{56.1056 \text{ g KOH}}{1 \text{ mol KOH}}$

$= 0.180$ g KOH or 180. mg KOH   Saponification value $= 180.$

c)   Safflower oil: Consider each component

Palmitic acid: Molar mass: $C_3H_5(C_{15}H_{31}CO_2)_3 = 807.34$ g mol^{-1}

Iodine number $= 0$ (saturated)

Saponification value of palmitic acid: mass of KOH

$= 1.00$ g palmitic acid

$\times \dfrac{1 \text{ mol palmitic acid}}{807.33 \text{ g palmitic acid}} \times \dfrac{3 \text{ mol KOH}}{1 \text{ mol palmitic acid}} \times \dfrac{56.1056 \text{ g KOH}}{1 \text{ mol KOH}}$

$= 0.208$ g KOH or 208 mg KOH   Saponification value $= 208$

Stearic acid: Molar mass: $C_3H_5(C_{17}H_{35}CO_2)_3 = 891.49$ g mol^{-1}

Iodine number $= 0$ (saturated)

Saponification value of stearic acid: mass of KOH

$$= 1.00 \text{ g stearic acid} \times \frac{1 \text{ mol stearic acid}}{891.49 \text{ g stearic acid}} \times \frac{3 \text{ mol KOH}}{1 \text{ mol stearic acid}} \times \frac{56.1056 \text{ g KOH}}{1 \text{ mol KOH}}$$

$= 0.189$ g KOH or 189 mg KOH   Saponification value $= 189$

Oleic acid: Molar mass: $C_3H_5(C_{17}H_{33}CO_2)_3 = 885.45$ g mol^{-1}

Iodine number for oleic acid: mass of I_2

$$= 100 \text{ g oleic acid} \times \frac{1 \text{ mol oleic acid}}{885.45 \text{ g oleic acid}} \times \frac{3 \text{ mol } I_2}{1 \text{ mol oleic acid}} \times \frac{253.81 \text{ g } I_2}{1 \text{ mol } I_2} = 86.0 \text{ g } I_2$$

Iodine number $= 86.0$

Saponification value of oleic acid: mass of KOH

$$= 1.00 \text{ g oleic acid} \times \frac{1 \text{ mol oleic acid}}{885.45 \text{ g oleic acid}} \times \frac{3 \text{ mol KOH}}{1 \text{ mol oleic acid}} \times \frac{56.1056 \text{ g KOH}}{1 \text{ mol KOH}}$$

$= 0.190$ g KOH or 190. mg KOH   Saponification value $= 190.$

Linoleic acid: Molar mass: $C_3H_5(C_{17}H_{31}CO_2)_3 = 879.402$ g mol^{-1}

Iodine number for linoleic acid: mass of I_2

$$= 100 \text{ g linoleic acid} \times \frac{1 \text{ mol linoleic acid}}{879.402 \text{ g linoleic acid}} \times \frac{6 \text{ mol } I_2}{1 \text{ mol linoleic acid}} \times \frac{253.81 \text{ g } I_2}{1 \text{ mol } I_2}$$

$= 173$ g I_2 Iodine number $= 173$

Saponification value of linoleic acid: mass of KOH

$$= 1.00 \text{ g linoleic acid} \times \frac{1 \text{ mol linoleic acid}}{879.402 \text{ g linoleic acid}} \times \frac{3 \text{ mol KOH}}{1 \text{ mol linoleic acid}} \times \frac{56.1056 \text{ g KOH}}{1 \text{ mol KOH}}$$

$= 0.191$ g KOH or 191 mg KOH   Saponification value $= 191$

Linolenic acid: Molar mass: $C_3H_5(C_{17}H_{29}CO_2)_3 = 873.348$ g mol^{-1}

Iodine number for linolenic acid: mass of I_2

$$= 100 \text{ g linolenic acid} \times \frac{1 \text{ mol linolenic acid}}{873.354 \text{ g linolenic acid}} \times \frac{9 \text{ mol } I_2}{1 \text{ mol linolenic acid}} \times \frac{253.81 \text{ g } I_2}{1 \text{ mol } I_2}$$

$= 261.\underline{6}$ g I_2 Iodine number $= 261.\underline{6} \approx 262$

Saponification value of linolenic acid: mass of KOH

$$= 1.00 \text{ g linolenic acid} \times \frac{1 \text{ mol linolenic acid}}{873.354 \text{ g linolenic acid}} \times \frac{3 \text{ mol KOH}}{1 \text{ mol linolenic acid}} \times \frac{56.1056 \text{ g KOH}}{1 \text{ mol KOH}}$$

$= 0.193$ g KOH or 193 mg KOH   Saponification value = 193

Summary:

Acid	I_2#	Sap#	%	High I_2 # %	Low I_2 # %	High Sap# %	Low Sap# %
Palmitic	0	208	6-7	6	7	7	6
Stearic	0	189	2-3	2	3	2	3
Oleic	86	190	12-14	12	14	12	14
Linoleic	173	191	75-80	78.5	75.5	87.5	76.5
Linolenic	262	193	0.5-1.5	1.5	0.5	1.5	0.5

Hence: The iodine number for safflower oil may range between 150 – 144
The saponification value for safflower oil may range between 211 – 179
(Note:the high iodine number contribution for each acid in the mixture is
calculated by multiplying its percentage by its iodine number. The sum of all of
the high iodine number contributions for all of the components in the mixture
equals the high iodine number for safflower oil. Similar calculations were used to
obtain the Low iodine number, along with the High/Low saponification numbers.

71.   (a)

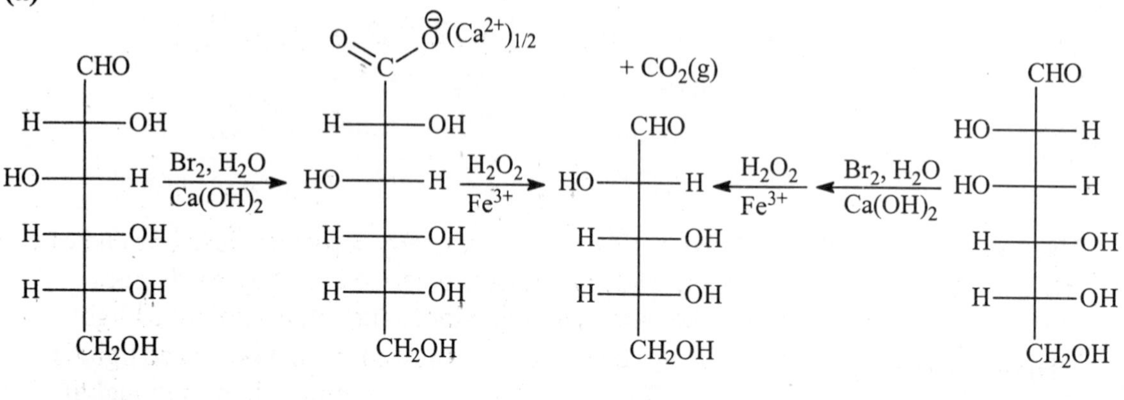

D-Glucose   D-Arabinose   D-Manose

(b)

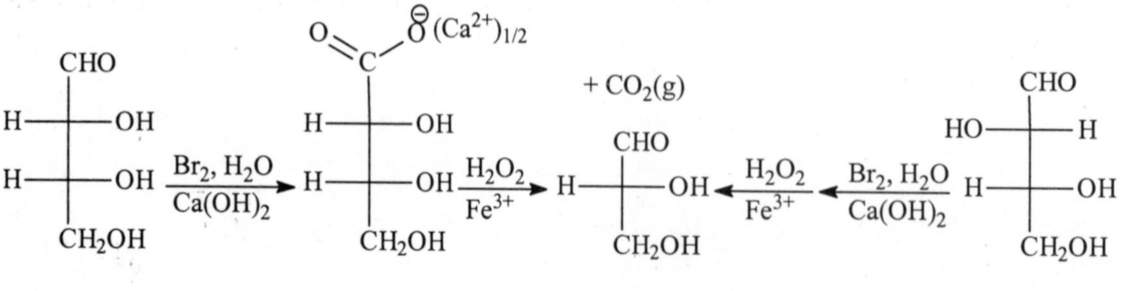

D-Erythrose   D-Glyceraldehyde   D-Threose

72. (a-c)

(d)   Meso form of Tartaric acid

Meso Form is not optically active because one end of the molecule rotates polarized light by +x degrees, the other end of the molecule rotates polarized light by −x degrees. The net result is that polarized light is unaffected by this type of compound, even though it has two chiral carbons. Meso forms must have a mirror image running through them, for every chiral carbon, there must exist its mirror image on the other side of the molecule.